Introductory Statistics

A Problem-Solving Approach

Stephen Kokoska
Bloomsburg University

■■ W. H. Freeman and Company • New York

Senior Publisher: Craig Bleyer
Publisher: Ruth Baruth
Senior Acquisitions Editor: Roland Cheyney
Senior Development Editor: Shona Burke
Freelance Development Editor: Leslie Lahr
Executive Marketing Manager: Jennifer Somerville
Market Development Manager: Kirsten Watrud
Marketing Assistant: Betsy Breen
Media Editor: Laura Capuano
Associate Editor: Katrina Wilhelm
Assistant Media Editor: Catriona Kaplan
Editorial Assistant: Lauren Kimmich
Photo Editor: Christine Buese
Photo Researcher: Elyse Rieder
Art Director: Diana Blume
Text Designer: Jerry Wilke Design
Project Editor: Aptara®, Inc.
Illustrations: Aptara, Inc.
Production Manager: Julia DeRosa
Composition: Aptara, Inc.
Printing and Binding: World Color (USA) Corp

Library of Congress Control Number: 2009935446

Student Edition ISBN-13: 978-1-4292-3196-1
ISBN-10: 1-4292-3196-3
ISBN-13: 978-1-4292-3976-9
ISBN-10: 1-4292-3976-X (packaged with CD)

Instructor Comp Book ISBN-13: 978-1-4292-3968-4
ISBN-10: 1-4292-3968-9
ISBN-13: 978-1-4292-3975-2
ISBN-10: 1-4292-3975-1 (packaged with CD)

Printed in the United States of America

Third printing

W.H. Freeman and Company
41 Madison Avenue
New York, NY 10010
Houndmills, Basingstoke RG21 6Xs, England
www. whfreeman.com

Table I Binomial Distribution Cumulative Probabilities (*Continued*)

n = 20

x	0.01	0.05	0.10	0.20	0.25	0.30	0.40	0.50	0.60	0.70	0.75	0.80	0.90	0.95	0.99
0	0.8179	0.3585	0.1216	0.0115	0.0032	0.0008	0.0000								
1	0.9831	0.7358	0.3917	0.0692	0.0243	0.0076	0.0005	0.0000							
2	0.9990	0.9245	0.6769	0.2061	0.0913	0.0355	0.0036	0.0002							
3	1.0000	0.9841	0.8670	0.4114	0.2252	0.1071	0.0160	0.0013	0.0000						
4		0.9974	0.9568	0.6296	0.4148	0.2375	0.0510	0.0059	0.0003						
5		0.9997	0.9887	0.8042	0.6172	0.4164	0.1256	0.0207	0.0016	0.0000					
6		1.0000	0.9976	0.9133	0.7858	0.6080	0.2500	0.0577	0.0065	0.0003	0.0000				
7			0.9996	0.9679	0.8982	0.7723	0.4159	0.1316	0.0210	0.0013	0.0002	0.0000			
8			0.9999	0.9900	0.9591	0.8867	0.5956	0.2517	0.0565	0.0051	0.0009	0.0001			
9			1.0000	0.9974	0.9861	0.9520	0.7553	0.4119	0.1275	0.0171	0.0039	0.0006			
10				0.9994	0.9961	0.9829	0.8725	0.5881	0.2447	0.0480	0.0139	0.0026	0.0000		
11				0.9999	0.9991	0.9949	0.9435	0.7483	0.4044	0.1133	0.0409	0.0100	0.0001		
12				1.0000	0.9998	0.9987	0.9790	0.8684	0.5841	0.2277	0.1018	0.0321	0.0004		
13					1.0000	0.9997	0.9935	0.9423	0.7500	0.3920	0.2142	0.0867	0.0024	0.0000	
14						1.0000	0.9984	0.9793	0.8744	0.5836	0.3828	0.1958	0.0113	0.0003	
15							0.9997	0.9941	0.9490	0.7625	0.5852	0.3704	0.0432	0.0026	
16							1.0000	0.9987	0.9840	0.8929	0.7748	0.5886	0.1330	0.0159	0.0000
17								0.9998	0.9964	0.9645	0.9087	0.7939	0.3231	0.0755	0.0010
18								1.0000	0.9995	0.9924	0.9757	0.9308	0.6083	0.2642	0.0169
19									1.0000	0.9992	0.9968	0.9885	0.8784	0.6415	0.1821

n = 25

x	0.01	0.05	0.10	0.20	0.25	0.30	0.40	0.50	0.60	0.70	0.75	0.80	0.90	0.95	0.99
0	0.7778	0.2774	0.0718	0.0038	0.0008	0.0001	0.0000								
1	0.9742	0.6424	0.2712	0.0274	0.0070	0.0016	0.0001								
2	0.9980	0.8729	0.5371	0.0982	0.0321	0.0090	0.0004	0.0000							
3	0.9999	0.9659	0.7636	0.2340	0.0962	0.0332	0.0024	0.0001							
4	1.0000	0.9928	0.9020	0.4207	0.2137	0.0905	0.0095	0.0005	0.0000						
5		0.9988	0.9666	0.6167	0.3783	0.1935	0.0294	0.0020	0.0001						
6		0.9998	0.9905	0.7800	0.5611	0.3407	0.0736	0.0073	0.0003						
7		1.0000	0.9977	0.8909	0.7265	0.5118	0.1536	0.0216	0.0012	0.0000					
8			0.9995	0.9532	0.8506	0.6769	0.2735	0.0539	0.0043	0.0001					
9			0.9999	0.9827	0.9287	0.8106	0.4246	0.1148	0.0132	0.0005	0.0000				
10			1.0000	0.9944	0.9703	0.9022	0.5858	0.2122	0.0344	0.0018	0.0002	0.0000			
11				0.9985	0.9893	0.9558	0.7323	0.3450	0.0778	0.0060	0.0009	0.0001			
12				0.9996	0.9966	0.9825	0.8462	0.5000	0.1538	0.0175	0.0034	0.0004			
13				0.9999	0.9991	0.9940	0.9222	0.6550	0.2677	0.0442	0.0107	0.0015			
14				1.0000	0.9998	0.9982	0.9656	0.7878	0.4142	0.0978	0.0297	0.0056	0.0000		
15					1.0000	0.9995	0.9868	0.8852	0.5754	0.1894	0.0713	0.0173	0.0001		
16						0.9999	0.9957	0.9461	0.7265	0.3231	0.1494	0.0468	0.0005		
17						1.0000	0.9988	0.9784	0.8464	0.4882	0.2735	0.1091	0.0023	0.0000	
18							0.9997	0.9927	0.9264	0.6593	0.4389	0.2200	0.0095	0.0002	
19							0.9999	0.9980	0.9706	0.8065	0.6217	0.3833	0.0334	0.0012	
20							1.0000	0.9995	0.9905	0.9095	0.7863	0.5793	0.0980	0.0072	0.0000
21								0.9999	0.9976	0.9668	0.9038	0.7660	0.2364	0.0341	0.0001
22								1.0000	0.9996	0.9910	0.9679	0.9018	0.4629	0.1271	0.0020
23									0.9999	0.9984	0.9930	0.9726	0.7288	0.3576	0.0258
24									1.0000	0.9999	0.9992	0.9962	0.9282	0.7226	0.2222

Brief Contents

Optional Sections

(available on the *Introductory Statistics: A Problem-Solving Approach* CD and web site: **www.whfreeman.com/introstats**):

Contents

Preface

Students frequently ask me why they need to take an introductory statistics course. My answer is simple. In almost every occupation and in ordinary daily life, you will have to make data-driven decisions and inferences, as well as assess risk. In addition, you must be able to translate complex problems into manageable pieces, recognize patterns, and, most importantly, solve problems. This text helps students develop the fundamental lifelong skill of solving problems and interpreting solutions in real-world terms.

One of my goals was to make this problem-solving approach accessible and easy to apply in many situations. I certainly want students to appreciate the beauty of statistics and the connections to so many other disciplines. However, it is even more important for students to be able to apply problem-solving skills to a wide range of academic and career pursuits, including business, science and technology, and education.

Introductory Statistics: A Problem-Solving Approach presents long-term, universal skills for students taking a one- or two-semester introductory-level statistics course. Examples include guided, explanatory Solution Trails that emphasize problem-solving techniques. Example solutions are presented in a numbered, step-by-step format. An abundance and variety of exercises provide ample opportunity for practice and review. Concepts, examples, and exercises are presented from a practical, realistic perspective. Data sets have been thoroughly researched to ensure that they are the most current and relevant available. The text uses mathematically correct notation and symbols and precise definitions to clearly illustrate statistical processes.

My text aims to help students fully understand the steps in basic statistical arguments, emphasizing the importance of assumptions in order to follow valid arguments or identify inaccurate conclusions. Most importantly, students will understand the process of statistical inference. A four-step process (Claim, Experiment, Likelihood, Conclusion) is used throughout the text to present the smaller pieces of introductory statistics upon which the larger, essential statistical inference puzzle is built.

FEATURES

Focus on Statistical Inference. The main theme of this text is statistical inference and decision making through the interpretation of numerical results. The process of statistical inference is introduced in a variety of applications and statistical settings all using a similar, carefully delineated, four-step approach: Claim, Experiment, Likelihood, and Conclusion.

SOLUTION

a. Let X be the number of people (out of the 30 selected) who experience tachycardia after taking the medication. X is a binomial random variable with $n = 30$ and $p = 0.10$: $X \sim B(30, 0.10)$.

Translate the words into a mathematical probability statement, convert to cumulative probability if necessary, and use Table I in the Appendix.

The probability that at most 1 person will experience tachycardia

$$= P(X \le 1) \qquad \text{Translate the words into mathematics.}$$
$$= 0.1837 \qquad \text{Already cumulative probability; use Table I in the Appendix.}$$

The probability of at most 1 person experiencing tachycardia is 0.1837.

b. Warrick Pharmaceuticals, Ltd. claims $p = 0.10$. This implies the random variable X has a binomial distribution with $n = 30$ and $p = 0.10$.

Claim: $p = 0.10 \rightarrow X \sim B(30, 0.10)$.

The experimental outcome is: seven people experience tachycardia.

Experimental outcome: $x = 7$.

It seems reasonable to consider $P(X = 7)$ and draw a conclusion based on this probability. However, in order to be conservative (to give the person making the claim the benefit of the doubt), we always consider a *tail probability*. We accumulate the probability in a tail of the distribution, and if it is small, then there is evidence to suggest the claim is false.

So, which tail? It depends on the mean of the distribution (and later on, the alternative hypothesis). Formulas for the mean, variance, and standard deviation of a binomial random variable are given below. Intuitively, however, the mean of a binomial random variable is $\mu = np$. If $n = 30$ and $p = 0.10$, we expect to see $\mu = (30)(0.10) = 3$ people experience bad side effects. Since $x = 7$ is to the right of the mean, we'll consider a right-tail probability. See Figure 5.9.

Likelihood:
$$P(X \ge 7) = 1 - P(X < 7) \qquad \text{The Complement Rule.}$$
$$= 1 - P(X \le 6) \qquad \text{The first value } X \text{ takes on } \textit{less than } 7 \text{ is } 6.$$
$$= 1 - 0.9742 \qquad \text{Cumulative probability; use Table I in the Appendix.}$$
$$= 0.0258$$

Conclusion: Since this tail probability is so small (less than 0.05), it is very unusual to observe 7 or more people with tachycardia. But it happened! This is either an incredibly lucky occurrence, or someone is lying. We usually discount the lucky possibility, and conclude that there is evidence to suggest Warrick's claim is false.

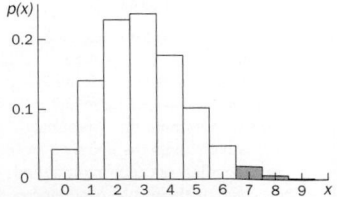

Figure 5.9 A portion of the probability histogram for the random variable X. The right-tail probability $P(X \ge 7)$ is the sum of the heights of the rectangles above 7, 8, 9, . . . , 30.

Chapter 12 Challenge

Are windmills too noisy?

Modern windmills are quickly becoming an efficient, clean alternative to fossil fuels for producing energy in the United States and other countries. On a typical three-blade windmill, the length of a blade is approximately 50–60 feet with diameter 12 feet, and each blade is angled. While a cost-effective windmill can utilize low wind velocity, the ideal locations for wind farms are along ocean coastlines or in mountainous areas, where there is a consistent wind speed of at least 15 mph.

Windmills cannot be constructed too near a town due to noise-level regulations. A typical windmill produces approximately 56 dB 200 feet away. This is softer than the sound of human speech (which is about 70 dB). Suppose a small coastal community is considering constructing a windmill in order to generate electricity for the town hall. An experiment is conducted in order to measure the windmill's noise level (in dB) at various distances (in feet) from the proposed site. The data are summarized in the following table.

(Beverett/Dreamstime.com)

Distance	10	50	75	120	150	160	200	250	400	500
Noise level	75	110	73	52	58	77	56	57	28	4

(*Source*: American Wind Energy Association.)

The techniques presented in this chapter can be used to determine whether there is a significant linear relationship between distance and noise level. The Chapter Challenge Wrap-Up (page 623) shows how regression analysis can be used to predict a value of the noise level for a given distance from the windmill.

Chapter 12 Challenge Wrap-Up

In the windmill experiment, residents would like to be able to predict the noise level at a certain distance from a windmill. The distance (x, in feet) is the predictor variable, and the noise level (y, in dB) is the response variable. A scatter plot of the data (Figure 12.73) suggests a linear relationship. Find the estimated regression line, and complete the ANOVA table.

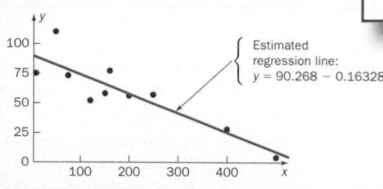

Estimated regression line: $y = 90.268 - 0.16328x$

Figure 12.73 Scatter plot of the windmill data and the estimated regression line.

Chapter Challenge. Each chapter opens with a unique Chapter Challenge case study, providing interesting applications to begin discussion. The case study is revisited and the solution explained at the end of each chapter in the **Chapter Challenge Wrap-Up.**

Review and Preview. At the beginning of each chapter, the "Review" box includes reminders of specific concepts from earlier chapters that will be used to develop new skills. The "Preview" box lists the learning objectives for the chapter.

Review
- Recall the point estimates for the parameters μ, p, and σ^2.
- Remember how to construct and interpret confidence intervals.
- Think about the concept of a sampling distribution for a statistic and the process of standardization.

Preview
- Use the available information in a sample to make a specific decision about a population parameter.
- Understand the formal decision process and learn the four-part hypothesis test procedure.
- Conduct formal hypothesis tests concerning the population parameters μ, p, and σ^2.

Solution Trail

KEYWORDS
- Is there any evidence?
- Greater than the long term mean.
- Standard deviation 0.70.

TRANSLATION
- Conduct a one-sided, right-tailed test about μ.
- $\mu_0 = 2.4$.
- $\sigma = 0.70$.

CONCEPTS
Hypothesis test concerning a population mean when σ is known.

VISION
Use the template for a one-sided, right-tailed test about μ. The underlying population distribution is unknown, but n is large and σ is known. Determine the appropriate alternative hypothesis and the corresponding rejection region, find the value of the test statistic, and draw a conclusion.

Solution Trail. The Solution Trail, a structured technique and visual aid for solving problems, appears in the text margins alongside selected examples. This feature serves as a guide for approaching and solving the problems before moving to the solution steps within the example. The four steps of the Solution Trail are as follows:

1. Find the *keywords*.
2. Correctly *translate* these words into statistics.
3. Determine the applicable *concepts*.
4. Develop a *vision* for the solution.

The *keywords* lead to a *translation* into statistics. Then, the statistics question is solved with the use of specific *concepts*. Finally, the keywords, translation, and concepts are all used to develop a *vision* for the solution. This method encourages students to think conceptually before making calculations. Although the technique is not applicable or necessary in every example, it is most appropriate for probability through hypothesis testing, the foundation of most introductory statistics courses.

Step-by-Step Solutions. The solutions to selected examples are presented in logical, systematic steps. This approach helps guide the reader through necessary calculations in order to find a solution and interpret results.

SOLUTION

STEP 1 The assumed mean is $\mu = 55 \ (= \mu_0)$; $n = 12$, $s = 4.8$, and $\alpha = 0.05$.

We are looking for evidence to suggest that the VO$_2$ max is *greater* in this group of men. The relevant alternative hypothesis is one-sided and right-tailed.

The underlying population is assumed to be normal, but σ is unknown. A t test is appropriate.

STEP 2 The four parts of the hypothesis test are

H_0: $\mu = 55$

H_a: $\mu > 55$

TS: $T = \dfrac{\bar{X} - \mu_0}{S/\sqrt{n}}$

RR: $T \geq t_\alpha = t_{0.05} = 1.7959$ df = 12 − 1 = 11

STEP 3 The value of the test statistic is

$t = \dfrac{\bar{x} - \mu_0}{s/\sqrt{n}} = \dfrac{58.6 - 55}{4.8/\sqrt{12}} = 2.5981 \geq 1.7959$

STEP 4 Since 2.5981 lies in the rejection region, we reject the null hypothesis at the $\alpha = 0.05$ level. There is evidence to suggest that the mean VO$_2$ max is greater than 55 in this group of men.

Calculation Explanations. Each line in a calculation is explained so that the reader can easily follow each step in a solution.

SOLUTION

P(at least one herbicide works)

$= 1 - $ P(no herbicide works)   Complement Rule.

$= 1 - $ P($M' \cap N' \cap Y'$)   All three do *not* work.

$= 1 - $ P(M') $\cdot$ P(N') $\cdot$ P(Y')   Independent events.

$= 1 - [1 - $P($M$)$] \cdot [1 - $P($N$)$] \cdot [1 - $P($Y$)]$ Complement Rule.

$= 1 - (1 - 0.75)(1 - 0.55)(1 - 0.45)$ Use given probabilities.

$= 1 - (0.25)(0.45)(0.55)$

$= 1 - 0.0619 = 0.9381$

STEP 3 The value of the test statistic is

$z = \dfrac{\bar{x} - \mu_0}{\sigma/\sqrt{n}} = \dfrac{2.75 - 2.4}{0.70/\sqrt{35}} = 2.958 \geq 1.96$

STEP 4 Since 2.958 lies in the rejection region, we reject the null hypothesis at the $\alpha = 0.025$ level. There is evidence to suggest that the current mean inventory-to-sales ratio is greater than 2.4 months.

Figures 9.7 through 9.9 together show a technology solution.

Technology Solutions. Wherever possible, a technology solution using the TI-84 Plus, Minitab, or Excel is presented at the end of each text example. Screen illustrations are color coded to allow for easy association with hand calculations shown in examples.

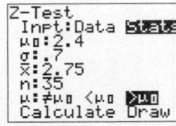

Figure 9.7 TI-84 Plus Z-Test input screen.

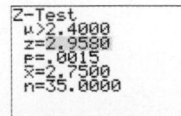

Figure 9.8 TI-84 Plus Z-Test Calculate results.

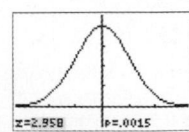

Figure 9.9 TI-84 Plus Z-Test Draw results.

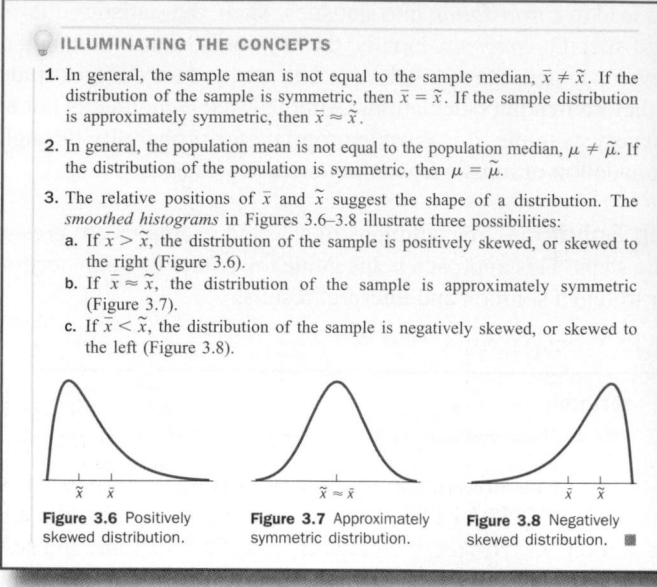

Illuminating the Concepts. These numbered lists offer straightforward explanations of various definitions and concepts. The itemized specifics, including hints, tips, and reminders, make it easier for the reader to comprehend and learn important statistical ideas.

Theory Symbols. More advanced material is offset with blue, half-moon symbols. This material can be skipped by the typical reader, but provides more complete explanations of various topics.

◀ As in Section 8.2, start with an appropriate probability statement involving the random variable Z in Equation 8.11. Find a symmetric interval about 0 such that the probability that Z lies in this interval is $1 - \alpha$.

$$P(-z_{\alpha/2} < Z < z_{\alpha/2}) = 1 - \alpha$$

$$P\left(-z_{\alpha/2} < \frac{\hat{P} - p}{\sqrt{\dfrac{p(1 - p)}{n}}} < z_{\alpha/2}\right) = 1 - \alpha \qquad (8.12)$$

Trying to sandwich p in Equation 8.12 is a little tricky because p appears in both the numerator and the denominator (inside a square root). Instead, since n is large, we usually use the sample proportion, $\hat{p}$, as a good estimate of p *in the denominator*. Manipulating the inequality (inside the probability statement) leads to the following general result. ▶

How To Boxes. This feature provides clear steps for constructing basic graphs or performing essential calculations. "How To" boxes are color coded and easy to locate within each chapter.

HOW TO COMPUTE PERCENTILES

Suppose $x_1, x_2, \ldots, x_n$ is a set of n observations.

1. Arrange the observations in ascending order, from smallest to largest.

2. To find p_r, compute $d_r = \dfrac{n \cdot r}{100}$.

   a. If d_r is a whole number, then the depth of p_r (position in the ordered list) is $d_r + 0.5$. p_r is the mean of the observations in positions d_r and $d_r + 1$ in the ordered list.

   b. If d_r is not a whole number, round up to the next whole number for the depth of p_r.

DEFINITION

A **confidence interval** (CI) for a population parameter is an interval of values constructed so that, with a specified degree of confidence, the value of the population parameter lies in this interval.

The **confidence coefficient** is the probability that the CI encloses the population parameter in repeated samplings.

The **confidence level** is the confidence coefficient expressed as a percentage.

Definition/Formula Boxes. Definitions and formulas are highlighted for emphasis and easy reference.

What happens if $k = 1$?

CHEBYSHEV'S RULE

Let $k > 1$. For *any* set of observations, the proportion of observations within k standard deviations of the mean [lying in the interval $(\bar{x} - k\,s, \bar{x} + k\,s)$, where s is the standard deviation] is at least $1 - \dfrac{1}{k^2}$.

Technology Corner. This feature, at the end of most sections, presents step-by-step instructions for using the TI-84 Plus, Minitab, and Excel to solve the examples presented in that section. Keystrokes, menu items, specific functions, and screen illustrations are presented.

TECHNOLOGY CORNER

Procedure: Hypothesis tests and confidence intervals concerning two population proportions.
Reconsider: Example 10.11, page 485, solution, and interpretation.

TI-84 Plus

Use 2-PropZTest to conduct a hypothesis test portions, $\Delta_0 = 0$, and 2-PropZInt to construct a ence in population proportions. There is no built-in test if $\Delta_0 \neq 0$.
1. Select STAT; TESTS; 2-PropZTest.
2. Enter the number of successes and the number o n_2. Highlight the appropriate alternative hypothes
3. The Calculate and Draw results are show page 486.
4. To construct a confidence interval, select STA
5. Enter the number of successes and the number x_2, n_2, and the confidence level. See Figure 10.
6. Highlight Calculate and press ENTER. Th on the Home Screen. See Figure 10.41.

```
2-PropZInt
x1:92
n1:250
x2:114
n2:300
C-Level:.95
Calculate
```

Figure 10.40
2-PropZInt input
screen.

Minitab

Use the function 2 Proportions to conduct a hypothesis test and to construct a confidence interval. Input is samples in one column, samples in different columns, or summarized data.
2. Choose Summarized data and enter the number of successes (Events) and number of trials for each sample.
3. Choose the Options button. Enter a Confidence level and the (hypothesized) Test difference (Δ_0), and select the appropriate Alternative hypothesis. If $\Delta_0 = 0$, check the box for Use pooled estimate of p for test.
4. The hypothesis test results and confidence interval are displayed in a session window. See Figure 10.42.

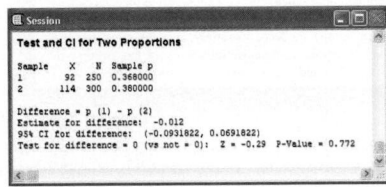

Test and CI for Two Proportions

```
Sample   X    N   Sample p
1       92   250  0.368000
2      114   300  0.360000


Difference = p (1) - p (2)
Estimate for difference:  -0.012
95% CI for difference:  (-0.0931822, 0.0691822)
Test for difference = 0 (vs not = 0):  Z = -0.29  P-Value = 0.772
```

Figure 10.42 Hypothesis test results and confidence interval.

Excel

There are no built-in functions to conduct a hypothesis test concerning two population proportions or to construct a confidence interval for the difference in population proportions. However, functions associated with the standard normal distribution may be used to find critical values and p values. Use ordinary spreadsheet calculations where necessary.

esses and number of trials for each sample.

test statistic, z, and use the function NORMSDIST to

ce interval, compute the difference, $\hat{p}_1 - \hat{p}_2$, use the to find the critical value, and compute

d the right endpoint of the confidence interval. See

	B	C	D
_1		114	= x_2
_1		300	= n_2
A1/A2 = phat_1		0.3800	= C1/C2 = phat_2
(A1+C1)/(A2+c2) = phat_c			
(A3-C3)/(SQRT(A4*(1-A4)*(1/A2+1/C2))) = z			
2*NORMSDIST(A5) = p value			
A3 - C3 = phat_1 - phat_2			
NORMSINV(0.025) = critical value			
SQRT(A3*(1-A3)/A2+C3*(1-C3)/C2)			
A8-A9*A10 = left endpoint			
A8+A8*A10 = right endpoint			

Figure 10.43 Hypothesis test results and confidence interval.

Figure
Resulti
interval.

Data Sets. All data sets presented in the examples and exercises are included on the *Introductory Statistics: A Problem-Solving Approach* CD and companion web site (www.whfreeman.com/introstats). The data are stored in an Excel workbook. Each worksheet is labeled and corresponds to the appropriate chapter. The first row of each worksheet names the exercise number and specific item or variable.

	A	B	C	D	E	F	G	H
1	**8.6**	**8.8**	**8.9**	**8.11**	**8.19**	**8.24**	**8.30 US**	**8.30 SA**
2	15.8	7.3	102	0.04	1.4	0.2	6.1	10.0
3	18.7	7.4	103	1.48	1.9	0.8	7.3	10.5
4	10.4	7.4	96	0.07	1.6	0.8	6.0	10.0
5	13.7	7.2	100	0.21	1.9	0.7	6.5	9.8
6	16.5	7.2	101	0.57	1.7	1.3	5.2	10.4
7	15.6	7.8	98	0.23	1.6	0.2	6.4	10.4
8	11.0	7.2	97	1.10	1.7	0.5	7.8	9.8
9	12.2	7.5	95	0.05	1.4	0.5	7.4	10.6
10	11.7	7.7	100	0.25	1.9	1.3	7.7	10.3
11	17.1	7.9	100	0.81	1.6	0.1	7.5	10.9
12	17.9	7.7	104	0.78	1.9	0.6	6.4	10.8
13	12.8	7.4	104	0.20	1.7	1.5	7.0	11.3
14	19.3	7.3	95	1.21	1.6	0.3	8.3	10.2
15	16.8	7.1	97	1.26	1.8	0.3	8.7	10.6
16	14.6	7.5	103	1.26	1.0	0.5	7.0	9.6
17	16.3	7.3	97	0.46	1.8	1.8	6.2	10.9
18	14.4	7.0	102	0.66	1.1	0.5	6.2	10.7
19	12.8	7.7	96	0.76	1.3	1.2	6.9	11.3
20	13.8	7.0	98	1.03	1.2	1.3	6.8	10.4
21	18.7	7.2	100	1.26	1.7	0.5	7.0	8.8

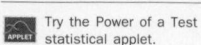

Try the Power of a Test statistical applet.

$$\beta(\mu_a) = P(\overline{X} > \overline{X}_c)$$
$$\beta(20.5) = P(\overline{X} > 21.02)$$

Definition of a type II error.
Use values for μ_a and $\overline{X}_c$.

$$= P\left(\frac{\overline{X} - 20.5}{3.0/\sqrt{36}} > \frac{21.02 - 20.5}{3.0/\sqrt{36}}\right)$$

Standardize using $\mu_a = 20.5$.
Assume $\sigma = 3.0$ is unchanged even if the true mean is 20.5.
Equation 6.8.

$$= P(Z > 1.04)$$
$$= 1 - P(Z \le 1.04)$$
$$= 1 - 0.8508 = 0.1492$$

Use cumulative probability.
Use Table III in the Appendix.

If the true mean weight is really 20.5 pounds, the probability of a type II error is 0.1492, at the $\alpha = 0.025$ significance level. ●

3.39 Fuel Consumption and Cars The tune-up information for a 2006 Subaru Forester L.L. Bean edition with automatic transmission indicates that the idle speed should be 700 rpm. Sixteen of these automobiles in for service at Nate Wade Subaru in Salt Lake City, Utah, were selected at random. The idle speed for each Forester is given in the following table.

| 745 | 650 | 670 | 730 | 700 | 690 | 670 | 665 |
| 640 | 700 | 700 | 690 | 690 | 700 | 660 | 705 |

(*Source:* cars101.com.)

a. Find the sample variance and the sample standard deviation.
b. Find the first and third quartiles.
c. Find the interquartile range and the *quartile deviation* (another measure of variability), QD = $(Q_3 - Q_1)/2$.

Applet Icons. These references in examples, exercises, and marginal notes indicate statistical applets that are available on the *Introductory Statistics: A Problem-Solving Approach* CD and companion web site (www.whfreeman.com/introstats).

Grouped Exercises. There is a wide variety of interesting, engaging exercises on relevant topics, based on current data, at the end of each section. These problems provide plenty of opportunity for practice, review, and application of concepts. Exercises are grouped according to:

PRACTICE Basic, introductory problems to familiarize students with the concepts and solution methods.

SECTION 9.3 EXERCISES

Practice

9.45 Consider a hypothesis test concerning a population mean with H_0: $\mu = 170$, H_a: $\mu < 170$, $n = 38$, and $\sigma = 15$.
a. Write the appropriate test statistic.
b. Write the rejection region corresponding to each value of α.
    (i) $\alpha = 0.01$ (ii) $\alpha = 0.025$ (iii) $\alpha = 0.05$
    (iv) $\alpha = 0.10$ (v) $\alpha = 0.001$ (vi) $\alpha = 0.0001$

Applications

9.55 Business and Management The mean income per year (in dollars) of employees who produce internal and external newsletters and magazines for corporations was reported to be 51,500. These editors and designers work on corporate publications but not on marketing materials. Due to poor economic conditions and oversupply, these corporate communications workers may be experiencing a decrease in salary. A random sample of 38 corporate communications workers revealed their mean salary was $\overline{x} = 49,762$. Conduct a hypothesis test to determine whether there is any evidence to suggest that the mean income per year of corporate communications workers has decreased. Assume $\sigma = 3750$ and use $\alpha = 0.01$.

APPLICATIONS Realistic, appealing exercises to build confidence and promote routine understanding.

Extended Applications

9.71 The four parts of a hypothesis test concerning a population mean from a normal population are shown below.

$$H_0: \mu = 50$$
$$H_a: \mu > 50$$
$$\text{TS: } Z = \frac{\overline{X} - \mu_0}{\sigma/\sqrt{n}}$$
$$\text{RR: } Z \ge z_a$$

Assume that the sample size is $n = 25$; $\sigma = 7.5$ and $\alpha = 0.01$.
a. Find the probability of a type II error for the alternative mean $\mu_a = 54$; that is, find $\beta(54)$.
b. Find $\beta(55)$ and $\beta(56)$.
c. Repeat parts (a) and (b) for $\alpha = 0.025$.

EXTENDED APPLICATIONS Applied problems that require extra care and thought.

Challenge

9.80 The Power of a Test The probability of a type II error, β, represents the likelihood of accepting the null hypothesis when the alternative is true. The *power* of a statistical test is $\pi = 1 - \beta$. This is the probability of (correctly) rejecting the null hypothesis, of detecting a difference in the hypothesized value of the population parameter. Try the Power statistical applet on the text web site.

A hot torsion test is used to determine the workability of a metal. Suppose a carbon steel rod is designed to fail (break) with mean axial load of 800 N (under certain temperature and speed conditions). A random sample of 25 rods is obtained, and the axial load failure is measured for each. Suppose the underlying distribution of axial load is normal and $\sigma = 50$.
a. Consider a hypothesis test of H_0: $\mu = 800$ versus H_a: $\mu < 800$ with $\alpha = 0.025$. Find the probability of a type II error and the power of this test for $\mu = 775$; that is, find $\beta(775)$ and $\pi(775)$.
b. Use a calculator or computer to find the power, $\pi(\mu_a)$, for $\mu_a = 730, 735, 740, \ldots, 800$.
c. Use the values from part (b) to carefully sketch a plot of $\pi(\mu_a)$ versus μ_a. The resulting plot is called a *power curve*.

CHALLENGE Additional exercises and technology projects that allow students to discover more advanced concepts and connections.

At the end of each chapter, **Chapter Exercises** help to test students' overall understanding of that chapter's concepts and to practice for exams. Answers to odd-numbered section and chapter exercises are given at the back of the book.

CHAPTER 5 SUMMARY

Concept	Page	Notation / Formula / Description
Random variable	184	A function that assigns a unique numerical value to each outcome in a sample space.
Discrete random variable	186	The set of all possible values is finite, or countably infinite.
Continuous random variable	186	The set of all possible values is an interval of numbers.
Probability distribution for a discrete random variable	188, 194	A method for conveying all the possible values and the probability associated with each value of the random variable
Mean, or expected value, of a discrete random variable X	198	$\mu = E(X) = \sum_{\text{all } x} [x \cdot p(x)]$.
Variance of a discrete random variable X	200	$\sigma^2 = \text{Var}(X) = \sum_{\text{all } x} [(x - \mu)^2 \cdot p(x)]$. $= E[(X - \mu)^2] = E(X^2) - E(X)^2$ $= E(X^2) - \mu^2$

(continued)

Chapter Summary. A table at the end of each chapter provides a short summary and page reference for each key concept, notation, and formula.

CAREFUL, PRECISE DEVELOPMENT

W. H. Freeman is committed to publishing high-quality, precise textbooks and supplements. From this project's inception and throughout, its development, production, quality, and precision have been given significant priority. We have in place unparalleled procedures to ensure the accuracy of the text.

In addition to careful editorial development, we incorporated suggestions from students and instructors in order to reflect the market's most pressing needs in introductory statistics.

These are the steps we took to ensure an accurate first edition for you:

CLASS TESTING: Fourteen colleges have used the preliminary edition in its entirety and provided us with valuable information on what was working and what needed further development.

FOCUS GROUPS: We conducted focus groups that included over 20 instructors to examine features and give us feedback on the style and readability of the book. These instructors also provided us with feedback on which mathematical notation was appropriate for an introductory statistics course. We also held a focus group at Bloomsburg University, the author's school.

REVIEWS: We had over 70 instructors examine chapter contents for accuracy, clarity, and coverage of the topics presented.

STUDENT INPUT: We sought student opinions of the pedagogical features, particularly the Solution Trail. With their help, we tailored the features to the needs of future students using the text.

ACCURACY REVIEWS: The book was thoroughly reviewed for accuracy prior to publication of both the preliminary and first edition.

Together, these procedures far exceed previous industry standards that safeguard the quality and precision of an introductory statistics textbook.

SUPPLEMENTS

The following electronic and print supplements are available with *Introductory Statistics: A Problem-Solving Approach:*

FOR STUDENTS:

STATS P⊿RTAL

courses.bfwpub.com/introstats (**Access code or online purchase required.**)

StatsPortal® is the digital gateway to *Introductory Statistics,* designed to enrich the course and enhance students' study skills through a collection of Web-based tools. *StatsPortal* integrates a rich suite of diagnostic, assessment, tutorial, and enrichment features, enabling students to master statistics at their own pace. It is organized around three main teaching and learning components:

1. **Interactive eBook** integrates a complete and customizable online version of the text with all of its media resources. Students can quickly search the text and can personalize the eBook just as they would the print version, with highlighting, bookmarking, and note-taking features. Instructors can add, hide, and reorder content, integrate their own material, and highlight key text.

2. **Resources** organize all the resources for *Introductory Statistics* into one location for ease of use. These resources include the following:

   - **Stats@Work Simulations** put students in the role of consultants, helping them better understand statistics within the context of real-life scenarios.
   - **Statistical Applets** are 16 interactive applets to help students master key statistical concepts and work exercises from the text.
   - **EESEE Case Studies** developed by The Ohio State University Statistics Department teach students to apply their statistical skills by exploring actual case studies, using real data, and answering questions about the study.
   - **CrunchIt! 2.0®** New CrunchIt! 2.0® Statistical Software, powered by R, is an easy-to-use program for students and offers capabilities well beyond those needed for a first course. Access codes are available in each new copy of *Introductory Statistics* or can be purchased online.
   - **Data sets** are available in ASCII, Excel, TI-83/84, Minitab, SPSS, JMP, and S-Plus formats.
   - **Statistical Software Manuals** are available for TI-83/84, Excel, Minitab, and SPSS.
   - **Student Solutions Manual** contains detailed solutions to odd-numbered text exercises.
   - **Interactive Table Reader** allows students to quickly find values in any of the statistical tables used in the course.
   - **Tables and Formulas** provide each table and formula from each chapter.
   - **Additional Material** Section 12.6, "The Polynomial and Qualitative Predictor Models," and Section 12.7, "Model Selection Procedures," allow further study of multiple linear regression.

Resources for Instructors only:

   - **Instructor's Solutions Manual** contains detailed solutions to all text exercises.
   - **Test Bank** offers hundreds of multiple-choice questions.
   - **Lecture PowerPoint Slides** offer a detailed lecture presentation for each chapter of *Introductory Statistics.*
   - **SolutionMaster** is a web-based version of the solutions in the *Instructor's Solutions Manual.* This easy-to-use tool allows instructors to generate a solution file for any set of homework exercises. Solutions can be downloaded in PDF

format for convenient printing and posting. For more information or a demonstration, contact your W.H. Freeman sales representative.

3. **Assignments** organizes assignments and guides instructors through an easy-to-create assignment process with access to questions from the Test Bank, Web Quizzes, and Exercises from the text, including many algorithmic problems.

Online Study Center 2.0 www.whfreeman.com/osc/introstats (Access code required. Available for purchase online.) The Online Study Center offers all the resources available in StatsPortal, except the eBook and Assignment Center.

Companion Web site www.whfreeman.com/introstats The open-access web site includes statistical applets, data sets, self-quizzes, tables and formulas, and optional Sections 12.6 and 12.7.

Interactive Student CD-ROM Included with every new copy of *Introductory Statistics*, the CD contains access to data sets, applets, tables and formulas, and optional Sections 12.6 and 12.7. (These items are also available on the companion Web site.)

Printed Student Solutions Manual The manual offers detailed solutions to all odd-numbered exercises. ISBN: 1-4292-4281-7

Statistical Software Manuals Statistical software manuals covering Minitab, Excel, SPSS, and TI-83/84 are offered within StatsPortal and the Online Study Center. These manuals are available in printed versions through custom publishing. They serve as basic introductions to popular statistical software packages as well as providing specific guidance to using these packages with *Introductory Statistics*.

FOR INSTRUCTORS ONLY

The Instructor's Web site www.whfreeman.com/introstats Password protected, the Instructor's Web site features access to all student Web materials on the companion Web site, plus

- Instructor version of EESEE (Electronic Encyclopedia of Statistical Examples and Exercises), with solutions to the exercises in the student version.
- All textbook figures and tables provided in JPEG format.
- Lecture PowerPoint slides offering a detailed lecture presentation of statistical concepts covered in each chapter of *Introductory Statistics*.

Instructor's Solutions Manual The manual contains full solutions to all exercises from *Introductory Statistics*. ISBN: 1-4292-4280-9

Test Bank The test bank contains multiple-choice questions for each chapter of the text. The test bank is also available electronically on CD-ROM (for Windows and Mac), allowing questions to be downloaded, edited, and resequenced. Print version, ISBN: 1-4292-4283-3 CD-ROM version, ISBN: 1-4292-4284-1

Instructor's Resource CD-ROM Allows instructors to search and export (by key term or chapter):

- All text images and tables
- Statistical applets and data sets
- Instructor's Solutions Manual and Test Bank files
- PowerPoint lecture slides

ISBN: 1-4292-4282-5

COURSE MANAGEMENT SYSTEMS

W.H. Freeman and Company provides courses for Blackboard, WebCT (Campus Edition and Vista), and Angel course management systems. They are completely integrated courses that you can easily customize and adapt to meet your teaching goals and course objectives. Contact your sales representative for more information.

xvi Acknowledgments

ACKNOWLEDGMENTS

I would like to thank the following colleagues who offered specific comments and suggestions on both the preliminary and first edition manuscripts throughout various stages of development:

ALABAMA Edward R. Mansfield, *University of Alabama* **ALASKA** Chris Hay-Jahans, *University of Alaska Southeast*; Kanapathi Thiru, *University of Alaska Anchorage* **ARKANSAS** George N. Bratton, *University of Central Arkansas*; Hassan Elsalloukh, *University of Arkansas* **ARIZONA** Tracey Ann Gust, *Arizona State University*; Jacob J. Oleson, *Arizona State University*; Kathryn Prewitt, *Arizona State University* **CALIFORNIA** Maryanne Anthony, *Santa Ana College*; Kelly D. Brooks, *Pierce College*; Kevin A. Fox, *Shasta College*; Maurice A. Geraghty, *California State University, East Bay*; Susan Herring, *Sonoma State University*; Mark Littrell, *Rio Hondo College*; Simon Tong, *Antelope Valley College*; James Wan, *Long Beach City College*; Steve Waters, *Pacific Union College* **CANADA** Pierce Bailey, *College of the North Atlantic*; Jean-Marc Perreault, *Yukon College* **CONNECTICUT** Chun Jin, *Central Connecticut State University* **FLORIDA** Janette H. Campbell, *Palm Beach Community College*; Ali A. Choudhry, *Florida International University*; Mary Beth Headlee, *Manatee Community College*; Nancy R. Johnson, *Manatee Community College*; B. M. Golam Kibria, *Florida International University*; Susan Schott, *University of Central Florida*; Nizam Uddin, *University of Central Florida* **GEORGIA** Ken Frerichs, *Columbus State University*; Shahryar Heydari, *Piedmont College*; Samuel Kwakye, *Chattahoochee Technical College*; Chandler Pike, *University of Georgia*; Gina Reed, *Gainesville State College*; Betty C. Rogers, *Piedmont College*; Martha Tapia, *Berry College* **ILLINOIS** Richard Diefenbach, *Shawnee Community College* **IOWA** Michael Collyer, *Iowa State University*; Sergio Loch, *Grand View College*; Timothy McDaniel, *Buena Vista University* **KANSAS** Satya Mandal, *University of Kansas*; John Vogt, *Newman University* **KENTUCKY** Lloyd R. Jaisingh, *Morehead State University* **LOUISIANA** Arun K. Agarwal, *Grambling State University*; Barbara Gonzalez-Arevalo, *University of Louisiana at Lafayette* **MAINE** Seth J. Ramus, *Bowdoin College* **MARYLAND** Casey T. Cremins, *University of Maryland*; Matthew Konicki, *University of Maryland*; Annette Noble, *University of Maryland Eastern Shore*; Stephen Prehoda, *Frederick Community College* **MASSACHUSETTS** Stanley Bezuszka, S.J., *Boston College*; Claire McAndrew, *Fitchburg State College*; Daniel C. Weiner, *Boston University*; Bonnie-Lou Wicklund, *Mount Wachusett Community College* **MICHIGAN** Jerrold W. Grossman, *Oakland University*; Elizabeth J. Reed, *Michigan Technological University*; Daniela Szatmari-Voicu, *Central Michigan University* **MINNESOTA** Edith Bogue. O.S.B., *The College of St. Scholastica*; Christopher J. Malone, *Winona State University*; Mike Marzinske, *Inver Hills Community College* **MISSOURI** Siva Balakumar, *Lincoln University of Missouri*; Tim Chappell, *Metropolitan Community College*; Christina Morian, *Lincoln University of Missouri*; Lawrence D. Ries, *University of Missouri–Columbia*; Paul Speckman, *University of Missouri–Columbia*; Suzanne Tourville, *Columbia College of Missouri*; Shingmin Wang, *Truman State University* **MONTANA** Debra J. Wiens, *Rocky Mountain College* **NEW HAMPSHIRE** George Fair banks, *University of New Hampshire*; Robert Rock, *Daniel Webster College* **NEW JERSEY** Boualem Bendjilali, *Raritan Valley Community College*; David Holmes, *The College of New Jersey*; Kelly Jackson, *Camden County College*; Chris Lacke, *Rowan University*; Carla A. Monticelli, *Camden County College*; Katarzyna Potocka, *Ramapo College of New Jersey*; Ricardo Sanchez, *Kean University* **NEW MEXICO** Mir Mortazavi, *Eastern New Mexico University* **NEW YORK** Kathleen A. Cantone, *Onondaga Community College*; Samuel Kohn, *Touro College*; Kenneth O'Brien. *Farmingdale State College*; Nicholas A. Zaino, Jr., *University of Rochester* **OHIO** Aaron Bogan, *Miami University*; G. Andy Chang, *Youngstowm State University*; Michael Hughes, *Miami University*; Arjun Gupta, *Bowling Green State University*; Rasul A. Khan, *Cleveland State University*; Xiaoping Annie Shen, *Ohio University*; Mahbobeh Vezvaci, *Kent State University* **OKLAHOMA** John Nichols, *Oklahoma Baptist University*; Mike Turegun, *Oklahoma City Community College* **OREGON** Jong Sung Kim, *Portland State University* **PENNSYLVANIA** Paul Ache, *Kutztown University*; Andrea Boito, *Penn State Altoona*; Harold S. Hayford, *Penn State Altoona*; Wei-Min Huang, *Lehigh University*; Ryan Savitz, *Neumann College*; Carol Hecht Serotta, *Cabrini College*; Steven J. Tedford, *Franklin & Marshall College*; Janet M. Winter, *The Pennsylvania State University* **RHODE ISLAND** Mary Sullivan, *Rhode Island College* **SOUTH CAROLINA** Georgie Baker, *University of South Carolina*; Aileen Solomon, *Trident Technical College* **TENNESSEE** Aniekan Ebiefung, *The University of Tennessee at Chattanooga*; J. Michael Jackson, *The University of Tennessee at Martin*; Xinxin Jiang, *Rhodes College*; Maggie Williams Flint, *Northeast State Technical Community College* **TEXAS** Ananda Bandulasiri, *Sam Houston State University*; David D. Marshall, *Texas Woman's University*; Melissa Reeves, *East Texas Baptist University*; James Stamey, *Baylor University* **UTAH** Cheryl Whitelaw, *Southern Utah University* **VIRGINIA** Brian Bradie, *Christopher Newport University*; Claude S. Moore, *Danville Community College*; Anthony J. Vavra, *West Virginia Northern Community College* **WASHINGTON** John Kellermeier, *Tacoma Community College*; Scott MacDonald, *Tacoma Community College* **WISCONSIN** William Applebaugh, *University of Wisconsin–Eau Claire*; Barb Barnet, *University of Wisconsin–Platteville*; John E. Baum, *Concordia University Wisconsin*; Lirong Ding, *University of Wisconsin–Fox Valley*; Todd Hoff, *Wisconsin Indianhead Technical College*

In particular, I would like to thank the instructors at Baylor University, Borough of Manhattan Community College (CUNY), Florida International University, Oklahoma Baptist University, Penn State Altoona, The University of Akron, The University of Texas at El Paso, and University of Central Arkansas, who class-tested the manuscript. Additionally, I thank the students at Hunter College for their helpful feedback on the Solution Trail feature.

A special thanks to Ruth Baruth, Craig Bleyer, Diana Blume, Betsy Breen, Shona Burke, Laura Capuano, Roland Cheyney, Julia DeRosa, Patrick Farace, Laura Hanrahan, Catriona Kaplan, Lauren Kimmich, Jennifer Somerville, Kirsten Watrud, and Katrina Wilhelm of W.H. Freeman and Company. I appreciate your support, confidence, encouragement, and enthusiasm. Thanks to James Lapp and Karen Muller for help with the Preliminary Edition. I am very grateful to Jackie Miller for offering insights and suggestions throughout the entire manuscript, as well as accuracy in reviewing all answers and solutions, and to Aaron Bogan for his detailed reviews, very specific comments, and enthusiasm. Thanks to Dennis Free from Aptara for his patience and typesetting expertise. And I certainly could not have completed this project without Leslie Lahr, her superb editing skills, gentle suggestions, and unwavering support.

I am very grateful to the entire Antoniewicz family for providing the foundation for a wide variety of problems, including those that involve nephelometric turbidity units, floor slip testers, and the weight of squab eggs.

I have learned a great deal by writing this text; I believe it has made me a better teacher, and I've had a lot of fun.

To Joan, Mark, and Jen—thank you for your patience, understanding, and inspiration for so many exercises.

ABOUT THE AUTHOR

Steve received his undergraduate degree from Boston College and his M.S. and Ph.D. from the University of New Hampshire. His initial research interests included the statistical analysis of cancer chemoprevention experiments. He has published a number of research papers in mathematics journals, including *Biometrics*, *Anticancer Research*, and *Computer Methods and Programs in Biomedicine*; presented results at national conferences; and written several books. He has been awarded grants from the National Science Foundation, the Center for Rural Pennsylvania, and the Ben Franklin Program.

Steve is a longtime consultant for the College Board, is an Exam Leader for the grading of the Advanced Placement Calculus test, and has been involved with calculus reform and the use of technology in the classroom. He has been teaching at Bloomsburg University for 20 years, regularly uses *Mathematica* and LaTeX, and recently has become involved with cell phone and GPS forensics.

Steve has been teaching introductory statistics classes throughout his academic career, and there is no doubt that this is his favorite course. This class (and text) provides students with necessary, basic quantitative skills that they will use in almost any job and teaches them how to think and reason logically. Steve believes very strongly in data-driven decisions.

Steve's uncle, Fr. Stanley Bezuszka, a Jesuit and professor at Boston College, was one of the creators of the so-called new math in the 1950s and 1960s. He had a huge influence on Steve's career. Steve helped Fr. B. with text accuracy checks, as a teaching assistant, and even writing projects through high school and college. Steve learned about the precision, order, and elegance of mathematics and developed an unbounded enthusiasm to teach.

An Introduction to Statistics and Statistical Inference

1

Chapter 1 Challenge

Is your bottled water really pure?

Bottled water has become a very popular consumer product. This is due to health reasons, fears related to terrorism and possible disruption of supply, or simply taste. There are many different types of bottled water: spring water, purified water, filtered water, distilled water, ozonated water, and even glacier water. Some of these waters come from a municipal source, while others originate from springs, wells, or glacial runoff. These products all contain various amounts of dissolved solids such as minerals, metals, and salts.

The primary method for determining water purity is to measure total dissolved solids (TDS). For example, deionized water used exclusively in industrial applications usually has a TDS measure of 12 parts per million (ppm), while municipal water may have TDS levels between 150 and 300 ppm.[1]

(Edyta Pawowska/iStockphoto)

The manufacturer of a new bottled water claims the TDS level is at most 10 ppm. A consumer group would like to check this claim. A random sample of 100 bottles of this water is obtained and each is measured for TDS. The methods presented in this chapter will enable us to identify the population of interest and the sample, to understand the definition and importance of a *random* sample, and most importantly, to identify using statistics for an extraordinary event that simply cannot be attributed to luck. These concepts are discussed in relation to this problem in the Chapter Challenge Wrap-Up (page 15).

Preview

- Recognize that data and statistics are pervasive; see that statistics are used to describe typical values and variability and to make decisions that affect everyone.
- Understand the relationship between a population and a sample and their connection to probability and statistics.
- Learn the basic steps in a statistical inference procedure.

1.1 STATISTICS TODAY

Statistics data are everywhere: newspapers, magazines, the Internet, the evening weather forecast, medical studies, and even sports reports. They are used to describe typical values and variability and to make decisions that affect every one of us. It is important to be able to read and understand statistical summaries and arguments with a critical eye. This chapter presents the basic elements of every statistics problem—a population and a sample and their connection to probability and statistics. Two common methods for data collection, observational sampling and experimentation, are also introduced.

Statistics data are used by professionals in many different disciplines. Actuaries probably are the biggest users of statistics. They conduct statistical analyses, assess risk, and estimate financial outcomes. An actuary helped compute your last automobile insurance bill.

Statistical analyses are used in a variety of settings. The National Agricultural Statistics Service publishes statistics on food production and supply, prices, farm labor, and even the price of land. Pollsters use statistical methods to predict a candidate's chances of winning an election. Using complex statistical analyses, companies make decisions about new products.

Traditional statistical techniques and new sophisticated methods are used every day in order to make decisions that directly affect our lives. Pharmaceutical companies use a battery of standard statistical tests in order to determine a new drug's efficacy and possible side effects. Data mining, a combination of computer science and statistics, is a new technique used for constructing theoretical models and detecting patterns. This technique is used by many companies to understand customers better and to respond quickly to their needs. Predictive microbiology is used to ensure that our food is not contaminated and is safe to consume. Given certain food properties and environmental parameters, a mathematical model is used to predict safety and shelf life.

Statistics is the science of collecting and interpreting data and drawing logical conclusions from available information in order to solve real-world problems. This text presents several numerical and graphical procedures for organizing and summarizing data. The constant theme throughout the course, however, is statistical inference using a four-step approach: claim, experiment, likelihood, and conclusion.

Here are some examples of statistics in the news.

1. **Statistical Inference:** In an article in *The Lancet*,[2] researchers concluded that a daily 300-mg aspirin reduces the chance of developing colon cancer. This study involved over 7500 people in Great Britain who were part of large-scale clinical trials started in the 1970s and 1980s. The researchers concluded that a daily dose of aspirin reduces the incidence rate of colon cancer by 63–74%.

2. **Summary Statistics:** During July 2007, the National Association of Colleges and Employers reported an increase in many *average* starting salaries of trades related to engineering. Offers to students awarded a bachelor's degree in chemical engineering rose 5.4% to $59,361, and starting salaries for students awarded a bachelor's degree in computer engineering rose 4.8% to an average of $56,201.[3] The average starting salaries in this report are summary statistics that suggest the *middle*, or *central tendency*, of a data set.

3. **Probability and Odds:** In an article in *Transportation Research*, the authors investigated the consequences of allowing long combination vehicles (LCVs) to use two-lane highways. They considered the risk of passing longer vehicles, taking into account automobile performance, the aggressiveness of the driver, the volume of oncoming traffic, and the length of the LCV. The authors concluded that the odds of failing to pass a 120-foot LCV (in moderate traffic) versus a 65-foot standard truck are two to six times greater.[4] A solid background in probability is necessary to understand statistical inference.

4. **Likelihood and Inference:** In 2004, the San Francisco police began using DNA evidence in unsolved homicides. The crime lab found evidence that

suggested John Puckett was a DNA *match*, based on $5\frac{1}{2}$ genetic locations, for a crime committed three decades earlier. Usually 13 genetic markers are used to distinguish between two people. Jurors were told that the chance of randomly finding the defendant's DNA profile at the crime scene was 1 in 1.1 million. A "combined probability of inclusion" suggested the chance of a random match was 1 in 152 billion. Since the chance of this happening is so small, jurors should infer that the two DNA samples came from the same person. The jurors found the DNA evidence compelling and convicted Puckett of first-degree murder.[5]

5. **Relative Frequency and Probability:** Each year in the United States, approximately 80 people are killed by lightning.[6] If all 307 million Americans are equally likely to be hit, then the probability that a randomly selected individual will be killed by lightning during a particular year is 0.0000002606. The relative frequency of occurrence is a good intuitive estimate of probability and is often used to develop statistical models and make predictions.

There has been an explosion of numerical information in stories like those above, in business, in consumer reports, and even in casual conversation. Interpretation of graphs and evaluation of statistical arguments are not reserved for academics and researchers any longer. It is essential for all of us to be able to understand arguments that are based on acquired data. This numerical, or quantitative, literacy is a vital lifelong tool.

No matter how you are employed or where you live, you will have to make decisions based on available information or data. Here are some questions you may have to consider.

1. Do you have enough information (data) to make a confident decision? How were the data obtained? If more information is necessary, how will these data be gathered?

2. How are the data summarized? Are the graphical and/or numerical techniques appropriate? Does the summary accurately represent the data?

3. What is the appropriate statistical technique for analyzing the data? Are the conclusions reasonable and reliable?

1.2 POPULATIONS, SAMPLES, PROBABILITY, AND STATISTICS

There are two very general applications of statistics: **descriptive statistics** and **inferential statistics**. Descriptive statistics involves summarizing and organizing the given information graphically and/or numerically. The focus of this text is statistical inference. The procedures of inferential statistics allow us to use the given data to draw conclusions and assess risk.

Here's a dictionary definition for inference: a deduction or logical conclusion.

DEFINITION

Descriptive Statistics: Graphical and numerical methods used to describe, organize, and summarize data.

Inferential Statistics: Techniques and methods used to analyze a *small*, specific set of data in order to draw a conclusion about a large, more general collection of data.

Example 1.1 Population Summary Statistics The United States Census Bureau maintains a huge database of people and households, business and industry, geography, and other special topics. This information may be neatly organized using tables, bar charts, pie charts, histograms, or stem-and-leaf plots, or summarized numerically using the mean, median, quartiles, percentiles, variance, or standard deviation. These simple *descriptive* statistics reveal characteristics of the entire data set. Part of a table is shown in Table 1.1. ●

Table 1.1 A portion of a table summarizing education attainment of the population 18 years and over, by age, 2007. All numbers are in hundred thousands.

			Educational Attainment					
Age	High school graduate	Some college, no degree	Associate's degree, occupational	Associate's degree, academic	Bachelor's degree	Master's degree	Professional degree	Doctoral degree
18 to 24 years	8,618	9,876	689	778	2,266	149	23	9
25 to 29 years	6,056	3,950	911	942	4,821	1,017	203	82
30 to 34 years	5,352	3,284	973	901	4,255	1,520	299	178
35 to 39 years	6,009	3,323	1,064	889	4,776	1,705	364	289
40 to 44 years	6,697	3,670	1,163	981	4,710	1,506	315	271
45 to 49 years	7,404	3,824	1,128	1,069	4,411	1,588	355	259
50 to 54 years	6,566	3,593	1,112	961	3,964	1,667	404	286
55 to 59 years	5,402	3,342	870	748	3,480	1,627	318	325
60 to 64 years	4,533	2,403	585	518	2,246	1,090	287	312
65 to 69 years	3,895	1,656	362	324	1,311	735	167	209
70 to 74 years	3,091	1,185	255	220	1,019	413	114	109
75 years and over	6,487	2,243	485	310	1,666	739	265	168

(*Source:* U.S. Census Bureau).

(*Hkratky/Dreamstime.com*)

Example 1.2 Airline Accidents The International Air Transport Association publishes industry information concerning passenger traffic, fuel, airfares, cargo traffic, and delays, and even maintains a world jet inventory. Much of this information can be summarized or organized—in tables or charts, with a variety of graphs, and numerically—in order to describe typical values and variability. These summary *descriptive* procedures might be used to indicate preference for a certain airline or to promote safety records. Figure 1.1 shows a bar graph of total accidents by year. ●

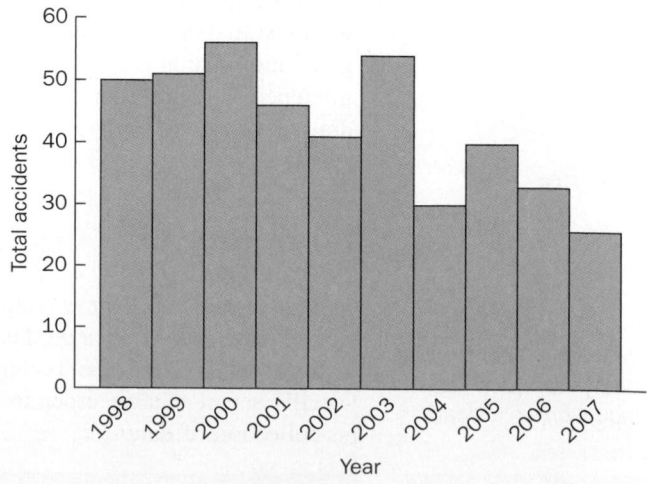

Figure 1.1 The total accidents per year (*Source:* National Transportation Safety Board).

The Solution Trail is a technique and visual aid for problem solving (illustrated in the margin of this page). It is a guide to help us plan how to solve a problem. Look at the Solution Trail before you read the steps of the solution. Start this hike by identifying keywords and phrases. The four steps to solving each problem are

1. Find the *keywords*.

2. Correctly *translate* these words into statistics.

3. Determine the applicable *concepts*.

4. Develop a *vision* for the solution.

The keywords lead to a translation into statistics. The statistics question is solved using specific concepts. The keywords, translation, and concepts are all used to develop a vision for the solution. This technique is not applicable or necessary in every problem. It is most appropriate for probability through hypothesis testing, the foundation of most introductory statistics courses.

Solution Trail

KEYWORDS

- Potassium level.
- Typical banana.
- One hundred bananas.

↓

TRANSLATION

- Characteristic of each banana.
- All bananas.
- Subset of all bananas.

↓

CONCEPTS

- Variable.
- Population.
- Sample.

↓

VISION

Determine the set of all objects of interest, the subset, and the attribute to be measured.

Example 1.3 Faster Drying Spackling Paste The manager of an Ace Hardware Store claims DAP spackling paste dries faster than any other brand. In order to check this claim, DAP spackling paste is applied in a variety of situations. In each test case, the drying time is carefully measured and recorded. These times are used to determine whether there is evidence to suggest that DAP spackling paste really does dry faster. The collected data are used in *inferential* statistics to draw a conclusion regarding the claim. ●

Example 1.4 Benefits of Engine Oil Additives Many automobile parts stores sell engine-oil additives that are designed to improve efficiency (increase gas mileage, reduce wear, etc.). An independent agency tested cars driven with and without a specific additive and recorded the efficiency (miles per gallon) of each automobile. Each car was classified by transmission type, engine size, and make. The data collected are used to determine whether there is a difference in efficiency. If there is a difference in miles per gallon, further *inferential* statistical techniques will be used to isolate this difference. It may be due to the additive, the type of transmission, the engine size, or the automobile make. ●

Whether we are summarizing data or making an inference, every statistics problem involves a population and a sample. Consider the definitions below.

DEFINITION

A **population** is the entire collection of individuals or objects to be considered or studied.
A **sample** is a subset of the entire population, a small selection of individuals or objects taken from the entire collection.
A **variable** is a characteristic of an individual or object in a population of interest.

💡 **ILLUMINATING THE CONCEPTS**

1. A **population** consists of all objects of a particular type. There are usually infinitely many objects in a population, or at least so many that we cannot look at all of them.

2. A **sample** is simply a (usually) small part of a population.

3. A **variable** may be a *qualitative* (categorical) or a *quantitative* (numerical) attribute of each individual in a population. ■

The following examples illustrate the relationships among populations, samples, and variables.

Example 1.5 Potassium Level in Bananas Various research studies suggest that bananas may be a natural cure for several different ailments. Bananas are high in iron, which can help people suffering from anemia, and there is a natural antacid in bananas that may help those who suffer from heartburn.[7] A new study is concerned with the potassium level in a typical banana. One hundred bananas are selected at random from around the world, and the potassium level in each is carefully measured and recorded. Describe the population, sample, and variable in this problem.

Solution Trail

KEYWORDS

- Electronically filed federal tax returns.
- Six.
- Time required to process.

↓

TRANSLATION

- All electronically filed federal tax returns.
- Subset of all electronically filed federal tax returns.
- Characteristic of electronically filed federal tax returns.

↓

CONCEPTS

- Population.
- Sample.
- Variable.

↓

VISION

Determine the set of all objects of interest, the subset, and the attribute to be measured for each object.

SOLUTION

STEP 1 The population consists of all of the bananas in the entire world. Although this population is not infinite, we certainly could not examine every single banana.

STEP 2 The sample is the 100 bananas selected at random. This is a subset or selection (or, OK, a bunch) of the population.

STEP 3 The variable in this problem is the potassium level. This characteristic will be carefully measured for each banana, and the data will be summarized or used to draw a conclusion. ●

Example 1.6 Faster Refunds from the IRS The Internal Revenue Service now allows taxpayers to complete and file their tax returns electronically. For those fortunate enough to receive a refund, the IRS claims electronic filing decreases the time to receive the refund. In order to check this claim, six electronically filed federal returns in which each taxpayer indicated a refund were selected. The time (in days) required to process the return and send a refund was recorded for each return. Describe the population, sample, and variable in this problem.

SOLUTION

STEP 1 The population consists of all of the electronically filed federal returns that indicated a refund. This population is not infinite, but it is so large that it would be costly and difficult to consider every one of these returns.

STEP 2 The sample consists of the six returns selected.

STEP 3 The variable in this problem is the number of days required to process the return and send a refund. ●

💡 **ILLUMINATING THE CONCEPTS**

This example raises some important issues regarding the sample of six federal returns.

1. How large a sample is necessary in order for us to be confident in our conclusion? Six returns do not seem like enough. But how many do we need? 60? 600? We will consider the problem of sample size in Chapter 7 and beyond.

2. This problem does not say how the sample was obtained. Perhaps the first six returns that indicated a refund were selected on a specific day. Or maybe only those from one part of the country were included. We need to be certain the sample is *representative* of the entire population in order to draw a valid conclusion. The formal definition of a representative sample is presented in Section 1.3. ■

Statistical inference is based on, and follows from, basic probability concepts. **Probability** and inferential **statistics** are both related to a population and a sample, but have different perspectives. For the rest of this chapter, *statistics* really means *inferential statistics*.

DEFINITION

In order to solve a **probability problem**, certain characteristics of a population are assumed to be known. We then answer questions concerning a sample from that population.

In a **statistics problem**, we assume very little about a population. We use the information about a sample to answer questions concerning the population.

Figure 1.2 illustrates this definition. Picture an entire population of individuals or objects. Suppose we know everything about the population and we select a sample from this population. A probability problem would involve answering a question concerning the sample. In a typical statistics problem, we assume very little about the population. We select a sample and analyze it completely. We use this information to draw a conclusion about the population.

Because Figure 1.2 is a circular diagram, it may seem like we can start our study anywhere. However, we need to understand probability before we can learn statistics. A solid background in probability is necessary before we can actually do statistical inference.

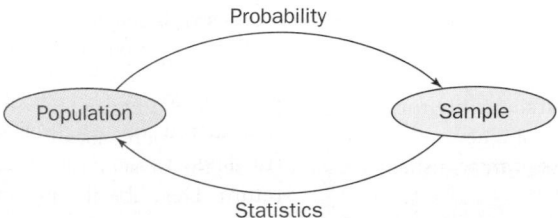

Figure 1.2 Relationships among population, sample, probability, and statistics.

(20th Century Fox/The Kobal Collection)

Example 1.7 Top Television Moments   According to Nielsen ratings, the final episode of *M*A*S*H* is one of the top 10 television moments. Other top shows include the "Who Shot J.R.?" episode of *Dallas* and the Beatles' live performance on *The Ed Sullivan Show*.[8] Consider the population consisting of all television viewers and a sample of 20 from this population.

Population: All television viewers at the time *M*A*S*H* was aired.

Sample: The 20 television viewers from this population.

Here is a probability question. The final episode of *M*A*S*H* was watched by 47% of all television viewers. What is the probability that 10 or more (of the 20) viewers watched this episode? We know something about the population, and try to answer a question about the sample.

Here is a statistics question. Suppose we interview the 20 viewers in the sample and find that 9 of the 20 watched the final episode of *M*A*S*H*. What can we conclude about the percentage of *all* television viewers who watched the final episode of *M*A*S*H*? We know about the sample and try to answer a question about the (whole or general) population.   ●

Example 1.8 Benefits of a Fishy Diet   Some research suggests that the fatty acids in fish oils are good for the heart and may also contribute to strong nerve cell membranes. Some people try to eat more fish for just these reasons. For others, fish is simply a major component of their diet. Consider the population consisting of all the people in the world and a sample of 100 people from this population.

A probability question: Suppose 25% of all people in the world eat fish at least once per week. What is the probability that at most 20 (of the 100) people in the sample eat fish at least once per week?

A statistics question: All 100 people selected are asked to complete an extensive health and diet questionnaire. The information indicates that 35 of the 100 eat fish at least once per week. What does this suggest about the proportion of the world's population that eat fish once per week?   ●

SECTION 1.2 EXERCISES

Applications

1.1 Descriptive or Inferential Statistics Determine whether each of the following is a descriptive or inferential statistics problem.

a. The Nebraska Department of Transportation maintains records concerning all trucks stopped for inspection. A report of these inspections lists the proportion of all trucks stopped, by cargo.

b. Eric Knudsen, a researcher at Stanford University Medical Center, obtains a random sample of wild owls and measures how far each can turn its neck. The data are used to conclude that an owl can turn its neck more than 120 degrees from the forward position.

c. A Navy research facility runs several tests to check the structural integrity of a new submarine. A laboratory report states that the vessel can withstand pressure at depths of at most 800 feet.

d. A safety inspector in Atlanta selects a sample of apartment buildings and checks the fire ladders on each. The proportion of broken ladders in the sample is used to estimate the proportion of broken fire ladders in the entire city.

e. Each 42-gallon barrel of oil produces a wealth of energy products. Besides gas and heating oil, a barrel of oil also contains key components used to make a variety of other products, such as aspirin, deodorants, DVDs, golf balls, pacemakers, contact lenses, and toothpaste. The pie chart in Figure 1.3 shows the percentage of each main product derived from a barrel of oil.[9]

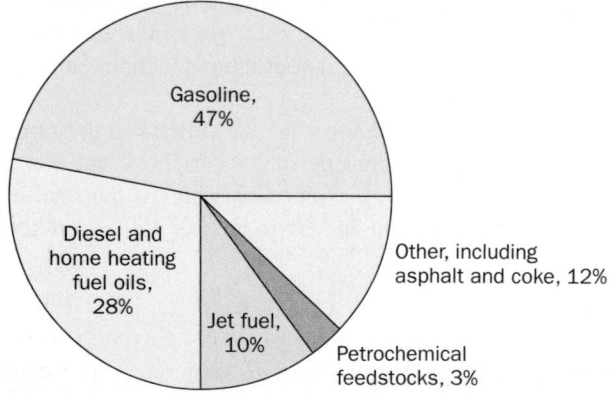

Products produced from oil

Figure 1.3 Pie chart illustrating the percentage of main products derived from a barrel of oil.

f. A report from the Bibb Medical Center in Centreville, Alabama, includes a table of the total number of West Nile virus cases treated, and the outcome of each case.

1.2 Descriptive or Inferential Statistics Determine whether each of the following is a descriptive or inferential statistics problem.

a. A pie chart published by the International Institute for Democracy and Electoral Assistance shows the number and percentage of countries using each electoral system (as in Figure 1.4).

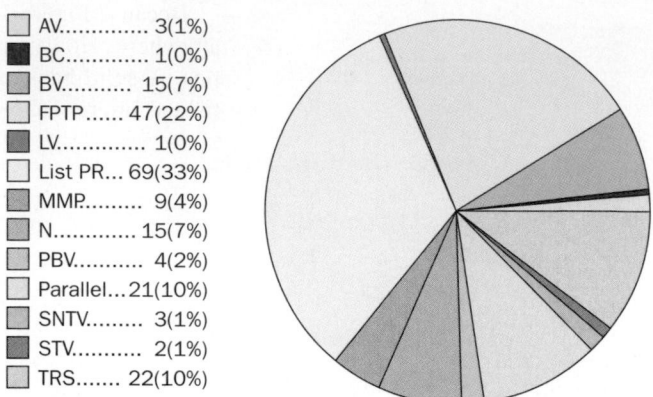

AV.............. 3(1%)
BC............. 1(0%)
BV.......... 15(7%)
FPTP...... 47(22%)
LV.............. 1(0%)
List PR... 69(33%)
MMP......... 9(4%)
N............. 15(7%)
PBV.......... 4(2%)
Parallel...21(10%)
SNTV........ 3(1%)
STV.......... 2(1%)
TRS....... 22(10%)

Figure 1.4 Pie chart illustrating the number and percentage of countries using each electoral system.

b. The resting heart rate was measured for adult males from two separate groups: those who exercise at least three days per week, and those who do not exercise regularly. The resulting data are used to suggest that regular exercise decreases the resting heart rate in adult males.

c. Interior Exterior Remodeling in Northridge, California, maintains a comprehensive list of each home constructed by type, size, exterior color, and so on.

d. Researchers at the Center for Food Safety selected a sample of frozen toaster apple strudel sold in grocery stores. Measurements indicated that the producer was baking each piece of strudel with less apple than advertised on the box.

e. A report issued by the athletic department at Brigham Young University listed each item in a trainer's bag and the number of times each was used.

1.3 Medicine and Clinical Studies Managers at Cedarcrest Hospital in Newington, Connecticut, are interested in the length of stay (in days) of patients admitted for open-heart surgery. Hospital managers have decided to limit their investigation to open-heart patients who were operated on within the last year. Thirty open-heart surgery patients admitted to the hospital within the last year are selected.

a. What is the population of interest?
b. What is the sample?
c. Describe the variable of interest.

1.4 Marketing and Consumer Behavior Since T-shirt labels irritate many people's skin, *ChoiceShirts* would like to produce shirts without a label. The company wants to know whether there is an advantage to this type of T-shirt. Fifty people are surveyed about whether they cut the tags off their T-shirts.

a. What is the population of interest?
b. What is the sample?
c. Describe the variable of interest.

1.5 Psychology and Human Behavior Managers at Citigroup, Inc., in New York are concerned about the number of employees who eat and/or drink at their desks while working. Some managers believe this is an unnecessary distraction, and spills can cause computer failures and ruin documents. Thirty-five employees are selected and each is questioned about eating/drinking while working. Describe the population and the sample in this problem.

1.6 Public Policy and Political Science Senator Kay Hutchison is unsure of her vote on an emotional and controversial issue. Before she votes, she would like to know what her constituents think about the proposed bill. An aide for the senator selects 500 people from Texas and asks each whether they believe the bill should become law. Describe the population and the sample in this problem.

1.7 Economics and Finance A recent flood along the Susquehanna River affected 10,000 families. Every family filed some sort of insurance claim for damage to their home and/or car. An insurance company serving the area is interested in the typical amount of a claim as a result of this flood. Seventy-five affected families are selected and their total claims are recorded. Describe the population and the sample in this problem.

1.8 Probability or Statistics In each problem below, identify the population and the sample and determine whether the question involves probability or statistics.

a. Seventy-five percent of all people who buy a dining room table purchase matching chairs. Five people who purchased a dining room table within the last month are selected at random. What is the probability that all five purchased matching chairs?

b. Twenty-five people entering a rest area and food court on Highway 59 near Houston are selected at random. Of these 25, 20 purchased food from at least one of the eateries. Estimate the true proportion of people stopping at this rest area who purchase food.

c. Historical records indicate that 1 out of every 500 people using a particularly steep water slide suffer some kind of injury. Fifty people using the slide are selected at random. How many do you expect to be injured?

d. A building inspector in Henderson, Nevada, is checking public buildings with doors that open automatically. One hundred doors are randomly selected. Careful inspection reveals that 12 doors are broken. Use this information to estimate the percentage of automatic doors in Henderson that are broken.

e. One thousand people entering the Los Angeles Airport (LAX) are selected at random. Each person is asked to complete a short survey regarding travel. The survey results show that 637 carry a frequent-flier card. Is there evidence to suggest that the true proportion of travelers entering LAX who carry frequent-flier cards is greater than 0.60?

f. The Risdall Advertising Agency reports that 65% of all women have purchased perfume within the last three months. Thirty-four women are selected at random. Is it likely that more than 20 of these women purchased perfume within the last three months?

g. Representatives from the Occupational Safety and Health Administration inspected several for-profit and Medicare nursing homes for violations. The resulting data will be used to determine whether there is any evidence to suggest that the quality of treatment is different in the two types of nursing homes.

1.9 Psychology and Human Behavior During each summer, many families spend part of their vacation time at a beach along the East or West Coast. Due to the popularity of movies like *Jaws*, and recent shark attacks at U.S. beaches, Americans have become increasingly concerned about water activities. Research suggests that 70% of all Americans believe that sharks are dangerous.[10] One thousand Americans were selected, and each was asked whether they believe sharks are dangerous.
a. What is the population of interest?
b. What is the sample?
c. Describe the variable of interest.

1.10 Manufacturing and Product Development Certain kinds of shampoo may contain sulfur compounds that can affect the fibers in clothes. Shampoo containing sulfur can ruin clothes by causing fiber to turn brown, permanently.[11] Twenty shampoos are selected and the amount of sulfur in each is measured. Describe the population, sample, and variable in this problem.

1.11 Medicine and Clinical Studies There is some evidence to suggest people with chronic hepatitis C have a liver enzyme level that fluctuates between normal and abnormal.[12] Fifty patients diagnosed with hepatitis C were selected and their liver enzyme levels were recorded each day for one month. Describe the population, sample, and variable in this problem.

1.12 Manufacturing and Product Development Paper towel manufacturers constantly advertise their products' strength, amount of stretch, and softness. A consumer group is interested in testing the absorption of Bounty™ paper towels. Thirty-five rolls are selected, and the amount of absorption for a single paper towel from each roll is recorded.
a. What is the population of interest?
b. What is the sample?
c. Describe the variable of interest.

Extended Applications

1.13 Manufacturing and Product Development While much of the cheddar cheese consumed around the world is processed, some is still produced in the traditional manner: made in small batches, wrapped in cloth to breathe, and allowed to age. Most traditional cheddar is aged one to two years; like fine wines, older cheddars assume their own character and flavor. Suppose 75% of all cheddars are aged less than two years, and a sample of 20 cheddar cheeses from around the world is obtained.
a. Describe the population and the sample in this problem.
b. Write a probability question and a statistics question involving this population and sample.

1.14 Marketing and Consumer Behavior DVD players have become increasingly popular. According to the Digital Entertainment Group, more than 80% of TV-owning households in the United States have at least one DVD player. The price of DVD players has decreased dramatically, the cost of renting a DVD movie has declined, and the quality of a DVD movie is superior to VHS.

Suppose a sample of 500 TV-owning households in the United States is obtained.

a. Describe the population and the sample in this problem.

b. Write a probability question and a statistics question involving this population and sample.

1.15 Business and Management Many American companies have been eliminating IT jobs locally and moving them to locations overseas. Studies by Forrester Research suggest that 40% of all U.S. firms have IT workers overseas.[13] Seventy-five manufacturers are selected. Employee records are inspected to determine whether each company has overseas IT workers.

a. What is the population of interest?

b. What is the sample?

c. Describe the variable of interest.

d. Write a probability question and a statistics question involving this population and sample.

1.3 EXPERIMENTS AND RANDOM SAMPLES

Statisticians perform two types of experiments: **observational studies** and **experimental studies**. The definitions are given below.

DEFINITION

In an **observational study,** we observe the response for a specific variable for each individual or object.

In an **experimental study,** we investigate the effects of certain conditions on individuals or objects in the sample.

The data collected in an observational study may be summarized in a variety of ways or used to draw a conclusion about the entire population. The following is an example of an observational study.

(Eastwestimaging/Dreamstime.com)

Example 1.9 Is There Time for Breakfast?   Lorraine GaNun, a guidance counselor at Rice School in Marlton, New Jersey, is interested in the amount of time each student spends in the morning eating breakfast. Some students wake up an hour before the bus arrives, have a leisurely breakfast, read the newspaper comics, and complete last-minute homework. Others roll out of bed and onto the school bus. Mrs. GaNun decides to measure the amount of time from wake-up to school bus arrival. A random sample of students is selected, and each is asked for the amount of school-day preparation time. The data are summarized graphically and numerically in this observational study.   ●

In almost all statistical applications it is important for the data to be *representative* of the relevant population. A representative sample, also known as a **random sample**, has characteristics similar to those of the entire population and therefore can be used to draw a conclusion about the (general) population. The following definition describes a method for obtaining data in an *observational study* to ensure the resulting sample is representative of the corresponding population.

DEFINITION

A **(simple) random sample (SRS) of size** n is a sample selected in such a way that every possible sample of size n has the same chance of being selected.

How can we be absolutely certain every possible sample of size *n* is equally likely?

💡 **ILLUMINATING THE CONCEPTS**

1. In practice, a random sample may be very difficult to achieve. Statisticians employ various techniques, including random number tables and random number generators, to select a random sample.

2. If a sample is not random, then it may be *biased*. There are many different kinds of bias and factors that contribute to a biased sample.

3. *Nonresponse bias* is very common when data are collected using surveys. The majority of people who receive a survey in the mail simply discard it. The original collection of people receiving the survey may be random, but the final sample of completed surveys is not. Since the sample is biased, it is impossible to draw a valid conclusion.

4. *Self-selection bias* occurs when the individuals (or objects) choose to be included in the sample, as opposed to being selected. For example, a television news program may ask viewers to respond to a yes/no question by dialing one of two phone numbers to cast their vote. Viewers *choose* to participate, and usually those with strong opinions (either way) vote. There are many more who did not choose to respond—every single sample is *not* equally likely. Certainly this sample is biased and hence no valid conclusion is possible.

5. If the population is infinite, then the number of simple random samples is also infinite. For finite populations, the formula for the number of possible random samples is presented in Chapter 7.

6. A simple random sample is vital for sound statistical practice. You should always ask how the data were obtained prior to performing any analysis. If there is any evidence of a pattern in selection, if the observations are associated or linked in some way, or if there is some connection among the observations, then the sample is not random. There is simply no way to transform *bad* data into good statistics. ■

Example 1.10 Summer Jobs Joseph Augustin, a reporter for the Iowa State daily newspaper, is conducting a study regarding past summer jobs. He plans to choose 100 students from the total student body of 25,741 and ask each selected person to complete a short questionnaire. The results will be summarized and presented in the next edition of the paper.

The reporter would like a simple random sample of size 100, a representative sample of the entire student body. There is one basic selection procedure: Write each student's name on a piece of paper and place all of them in a hat. Then thoroughly mix the papers and select 100 names.

While this procedure is clear-cut and uncomplicated, it can be very tedious if the number of individuals or objects in the population is large. In addition, it is hard to guarantee a complete mixing of the slips of paper.

More practical methods for selecting a simple random sample include the use of a random number table or a random number generator (available in most statistical software packages). In this example, we might assign each student a number, from 1 to 25741, and use a random number generator to produce a list of 100 numbers in this range. The students associated with these 100 numbers would comprise the random sample. ●

Researchers often investigate the effects of certain conditions on individuals or objects. The data obtained are from an *experimental study*. Individuals are randomly assigned to specific groups, and certain factors are systematically controlled, or imposed, in order to investigate and isolate specific effects. The following example is of an experimental study.

Example 1.11 To Fertilize or Not to Fertilize   The manager of Gardener's Supply Company claims that a new organic fertilizer, in comparison with the leading brand, increases the yield and size of tomatoes. In order to test this claim, tomato plants are randomly assigned to one of two groups. One group is grown with the leading fertilizer and the other is cultivated with the new product. At harvest time the size and weight of each tomato is recorded, along with the total yield per plant. The data collected from this experiment are used to compare the two fertilizers. ●

In an experimental study, researchers must be careful to ensure that significant effects are indeed due to an imposed *treatment*, or controlled factor. *Confounding* occurs when several factors together contribute to an effect, but no single cause can be isolated. Suppose the tomato plants in one of the groups above are watered more and/or exposed to more sunlight and warmer temperatures. If the tomato plants that received the new fertilizer were subject to these different (favorable) growing conditions, a difference in yield cannot be attributed to the new product.

The focus of this text is statistical inference, and many formal procedures will be presented beginning in Chapter 9. Most statistical inference is based on determining the likelihood of an observed experimental outcome. This strategy will be used informally in early chapters. We will follow the four-step process presented below.

STATISTICAL INFERENCE PROCEDURE

The process of checking a claim can be divided into four parts.

Claim   This is a statement of what we assume to be true.

Experiment   In order to check the claim, we conduct a relevant experiment.

Likelihood   Consider how likely it is that the observed experimental outcome occurs assuming the claim is true. We will use many techniques to determine whether the experimental outcome is a reasonable observation (subject to reasonable variability) or whether it is a rare occurrence.

Conclusion   There are only two possible conclusions: (1) If the outcome is reasonable, then we cannot doubt the claim. We usually write, "There is no evidence to suggest the claim is false." (2) If the outcome is rare, we disregard the lucky alternative and question the claim. A rare outcome is a contradiction. It shouldn't happen (often) if the claim is true. In this case we write, "There is evidence to suggest the claim is false."

Example 1.12 Perfect Pens   Precise Write ships a box containing 1000 fine-point pens and claims that 999 are in perfect condition and only 1 is defective. Upon receipt of the shipment, a quality control inspector reaches into the box, mixes the pens around a bit, and selects one at random, and it's defective!

Claim: There were 999 good pens and 1 defective pen in the box.

Experiment: The quality-control inspector selected one pen from the box, tested it, and found it to be defective.

Likelihood: One of two things has happened.

1. The quality-control inspector could be incredibly lucky. Intuitively, the chance of selecting the 1 defective pen from among the 1000 total pens is very small. It is *possible* to select the one defective pen, but this is very *unlikely*.

We have found evidence that the claim is false by showing the observed experimental outcome is unreasonable, an outcome that is so rare, it should almost never happen if the claim is really true.

2. The claim (999 perfect pens, 1 defective) is false. Since the chance of selecting the single defective pen is so small, it is more likely the manufacturer (Precise Write) lied about the number of defective pens in the shipment. (Perhaps there are really 999 *defective* pens and only 1 good pen in the box.)

Conclusion: Typically, statistical inference discounts the *lucky* alternative. Selecting the single defective pen is an extremely rare occurrence. Therefore, there is evidence to suggest the manufacturer's claim is false, since this outcome is very rare. ●

We will use this four-step process for checking a claim in many different contexts. The method for determining likelihood is the key to this valuable tool for logical reasoning.

SECTION 1.3 EXERCISES

Applications

1.16 Fuel Consumption and Cars The administration at the University of Nebraska in Lincoln is interested in student reaction to a planned parking garage on campus. A dormitory near the proposed site is selected and several Student Senate members volunteer to solicit responses. One Thursday evening, the volunteers each take a specific dorm wing, knock on doors, and record student answers to several prepared questions.
a. Is this an observational or experimental study?
b. Describe the sample in this problem.
c. Is this a random sample? Justify your answer.

1.17 Demographics and Population Statistics State Farm Insurance Company would like to estimate the proportion of volunteer firemen across the country who are full-time teachers. The 25 largest volunteer fire companies in the United States are identified. Each is contacted and asked to complete a short survey regarding the number of volunteers and the occupation of each volunteer.
a. Is this an observational or experimental study?
b. Describe the sample in this problem.
c. Is this a random sample? Justify your answer.

1.18 Manufacturing and Product Development The Visniak Bottling Plant in Cheektowaga, New York, has been accused of systematically underfilling 12-ounce bottles of soda. An inspection team enters the plant one afternoon and selects bottled soda ready for shipment from various locations within the plant. The contents of each selected bottle are carefully measured.
a. Describe the population and the sample in this problem.
b. Is this a random sample? Justify your answer.

1.19 Biology and Environmental Science Oregon Scientific has come under suspicion for purposely shipping defective wireless weather stations. The Attorney General's office in Delaware would like to estimate the proportion of defective products being shipped by this company. Describe a method for obtaining a simple random sample of shipped wireless weather stations.

1.20 Fuel Consumption and Cars The Massachusetts State Police union is interested in the number of miles driven by each officer during an 8-hour shift. Twelve officers are selected from the 11:00 P.M. to 7:00 A.M. shift, and the number of miles traveled by each officer is recorded.
a. Is this an observational or experimental study?
b. Describe the population and the sample in this problem.
c. Is this a random sample? Justify your answer.

1.21 Manufacturing and Product Development Gillette claims that a new disposable razor provides a *closer* shave than any other brand currently on the market. One hundred men who are observed buying a disposable razor are selected and asked to participate in a shaving study.
a. Describe the population and the sample in this problem.
b. Is this a random sample? Justify your answer.

1.22 Manufacturing and Product Development Midwest Pet Supplies claims its K9 Chain Link Dog Kennel can be set up in less than 30 minutes. An investigative reporter would like to check this claim. Describe a method for obtaining a simple random sample of customers who set up this kennel.

1.23 Sports and Leisure A National Football League (NFL), coach is permitted to initiate two *challenges* to referee calls per game (outside of the final 2 minutes in each half). During a challenge, the referee reviews the play in question on a replay monitor on the field, and the play either stands or the challenge is upheld. The NFL reports the time required for a coach's challenge is less than 5 minutes. A sports statistician would like to check this claim. Describe a method for obtaining a simple random sample of challenges during NFL games.

1.24 Travel and Transportation The Department of Public Works in Bismarck, North Dakota, would like to estimate the number of potholes per mile (after a long, snowy winter). Each selected mile-long stretch will be thoroughly examined for potholes, and the number in each section will be recorded.
a. Describe a method for obtaining a simple random sample of mile-long road segments.
b. Is this an observational or experimental study?

1.25 Biology and Environmental Science The Faber Floral Company in Kankakee, Illinois, claims to have developed a special spray for roses that causes the blossom to last longer than an untreated flower. Fifty long-stemmed roses are obtained and randomly assigned to one of two groups: treated versus untreated. The treated roses are sprayed, and the lifetime of each blossom is carefully recorded.

a. Is this an observational or experimental study?

b. What is the variable of interest?

c. Describe a technique to randomly assign each rose to a group.

1.26 Fuel Consumption and Cars Hybrid cars run on gas and electric power. They are designed for drivers who want to help the environment and save money on gas. In addition, the federal government may allow a tax deduction for the purchase of a hybrid car.[14] While there are certainly some benefits to owning a hybrid car, many people complain about the slow acceleration, repair expense, and overall comfort. Thirty-five passengers are randomly selected. Each is blindfolded and taken for a ride in a traditional combustion engine automobile and in a comparably sized hybrid car (over the same route). The passenger is then asked to select the car with the most comfortable ride.

a. Is this an observational or experimental study?

b. What is the variable of interest?

c. Describe possible sources of bias in these results.

1.27 Manufacturing and Product Development The ceramic tile used to construct the floors in a mall must be sturdy, easy to clean, and long lasting. Before installing a specific tile, a construction firm orders a box of 25 tiles and uses a standard strength test on each. The results are used to determine whether the tiles will be used throughout the new mall.

a. Describe the population and the sample in this problem.

b. Is this a random sample? If so, justify your answer. If not, describe a technique for obtaining a random sample.

1.28 Manufacturing and Product Development Many comforters contain both white feathers and down in order to provide a warm, soft cover. A bed-and-bath company would like to expand its line of products and sell comforters for queen- and king-size beds. Before manufacturing begins, a random sample of comforters is obtained from other companies and the proportions of white feathers, down, and other components are measured and recorded. These data will be used to determine the exact mixture of feathers and down for the new line of comforters.

a. Is this an observational or experimental study?

b. What are the variables of interest?

c. Describe a method for obtaining a random sample of comforters from current manufacturers.

Chapter 1 Challenge Wrap-Up

A manufacturer claims that the total dissolved solids (TDS) of a new bottled water is at most 10 ppm. A random sample of 100 bottles of this water is obtained and each is measured for TDS.

The population of interest is the entire supply of bottled water produced by the manufacturer. The sample consists of the 100 bottles selected at random. The variable of interest is the TDS in each bottle.

This is an observational study. There are no special conditions imposed or effects tested. We are simply selecting a subset of the relevant population.

The claim (made by the manufacturer) is that the TDS is at most 10 ppm in each bottle of water. The experiment consists of selecting 100 bottles at random and measuring the TDS in each bottle. One of several methods may be used to compute the likelihood of observing this particular sample of 100 TDS measurements (assuming the claim is true). Based on this calculation (likelihood), we can draw a conclusion. If the observed sample (or experimental outcome) is likely, then there *is no* evidence to suggest the claim is false. If the observed sample is very unlikely, there *is* evidence to suggest the claim is false.

(Edyta Pawowska/iStockphoto)

CHAPTER 1 SUMMARY

Concept	Page	Notation/Formula/Description
Descriptive statistics	3	Graphical and numerical methods used to describe, organize, and summarize data.
Inferential statistics	3	Techniques and methods used to draw a conclusion or make an inference.
Population	5	The entire collection of individuals or objects to be considered or studied.
Sample	5	A subset of the entire population.
Variable	5	A characteristic of an individual or object in a population of interest.
Probability problem	6	Certain properties of a population are assumed known. Questions involve a sample taken from this population.
Statistics problem	6	Information about a sample is used to answer questions concerning a population.
Observational study	10	We observe the response for a specific variable for each individual or object in the sample.
Experimental study	10	We investigate the effects of certain conditions on individuals or objects in the sample.
Simple random sample of size n	10	A sample selected in such a way that every possible sample of size n has the same chance of being selected.
Statistical inference procedure	12	Four-step process: Claim, Experiment, Likelihood, and Conclusion.

Solution Trail

KEYWORDS

- At most 10 ppm.
- Random sample of 100 bottles.
- Each is measured for TDS.

TRANSLATION

- TDS in each bottle of water is ≤ 10.
- Subset of the entire supply of bottled water.
- Characteristic of each object.

CONCEPTS

- Claim.
- Sample.
- Variable.

VISION

Identify the population of interest and use the measurements of the sample to test the manufacturer's claim. The statistical inference procedure is used to determine whether there is any evidence to suggest the claim is false.

CHAPTER 1 EXERCISES

Applications

1.29 Descriptive or Inferential Statistics Determine whether each of the following is a descriptive or inferential statistics problem.

a. The Society of Government Economists conducted a salary and working condition survey of top bank executives in the United States. A report issued by this group included a table that listed the number of bank executives in each state with salaries above one million dollars.

b. The Flowers Canada Growers obtained a sample of people who sent roses for Valentine's Day and recorded the color of the roses purchased. This information was used to construct a table listing the proportion of each color of rose purchased on Valentine's Day.

c. The Intergovernmental Panel on Climate Change collected data associated with global warming and predicted the extinction of up to 30% of plant and animal species in the world.

d. American Express conducted a survey of travelers in the Los Angeles International Airport. The information was used to estimate the proportion of all travelers who make a purchase in an airport Duty Free shop.

1.30 Descriptive or Inferential Statistics Determine whether each of the following is a descriptive or inferential statistics problem.

a. A report by NASA listed each weather satellite orbiting the earth and the number of years each has been in service.

b. The Agricultural Research Service obtained samples of natural cocoa from a variety of sources and measured the total antioxidant capacity in each sample. The resulting data was used to suggest that eating a moderate amount of chocolate may help prevent cancer, heart disease, and stroke.

c. The U.S. government patent office issued a report listing every company that was granted a patent in 2009 and the number of patents awarded to each company.

d. A researcher at Emory University used brain scans to conclude that Zen meditation may help treat disorders characterized by distracting thoughts.

1.31 Public Health and Nutrition A parents association would like to determine the proportion of teenagers who have the ability to prepare an entire meal. A sample of teenagers was obtained and each was asked if he or she could cook. Describe the population of interest, the sample, and the variable of interest in this problem.

1.32 Marketing and Consumer Behavior Hallmark is interested in the proportion of adults who sent a greeting card on Mother-in-Law Day. A sample of 400 adults was obtained and each was asked whether they sent a greeting card on this holiday. Describe the population and the sample in this problem.

1.33 Medicine and Clinical Studies A study was conducted to determine whether exhaustion is related to cortisol levels.[15] Seventy-eight individuals were selected and their cortisol levels were measured 30 minutes after awakening. Describe the population, sample, and variable in this problem.

1.34 Public Policy and Political Service The Office of the Privacy Commissioner (OPC) of Canada's Contributions Program is interested in reaction to a proposal to allow police to obtain cell phone records without a subpoena. One thousand people in British Columbia were called and each was asked to respond to several questions.

a. Is this an observational or experimental study?

b. Describe the sample in this problem.

c. Is this a random sample? Justify your answer.

1.35 Travel and Transportation Amtrak would like to estimate the proportion of travelers on the Sunset Limited, from New Orleans to Los Angeles, who utilize the Sightseer Lounge on route. At the end of the trip on March 15, an Amtrak representative stopped every third person getting off the train and asked each if they used the Sightseer Lounge to buy food, a drink, or souvenirs.

a. Is this an observational or experimental study?

b. Describe the sample in this problem.

c. Is this a random sample? If not, suggest a method for obtaining a random sample.

1.36 Physical Sciences The Air Liquide Company has developed a new deicing chemical for airplanes. The chemical consists of glycol and several proprietary additives. The new chemical was designed to keep aircraft wings ice-free for a longer period of time. Ten typical Dehaviland commuter airplanes were obtained and randomly assigned to one of two groups: new chemical versus old chemical. Each plane was subject to constant icing conditions in a controlled environment and treated with one of the chemicals. The length of time until ice formed on the wings was recorded for each plane.

a. Is this an observational or experimental study?

b. What is the variable of interest?

c. Describe a technique to randomly assign each plane to a chemical group.

Extended Applications

1.37 Technology and Internet NationMaster.com reported the 2007 software piracy rate in the United States was 20%. They define the piracy rate as the number of units of pirated software deployed divided by the total number of units of software installed. One thousand installed software titles are selected. Each is carefully examined to determine whether the software was pirated.

a. What is the population of interest?

b. What is the sample?

c. Describe the variable of interest.

d. Write a probability question and a statistics question involving this population and sample.

1.38 Travel and Transportation The Channel Tunnel, or Chunnel, is a 31.4-mile railroad tunnel beneath the English Channel between Folkstone, Kent, in England and Coquelles in France. In order to ensure passenger safety, engineers selected the 35 deepest areas in the tunnel and measured the pressure on each section.

a. Is this an observational or experimental study?

b. What is the variable of interest?

c. Is this sample random? If so, justify your answer. If not, describe a technique to obtain a random sample.

Tables and Graphs for Summarizing Data

2

Chapter 2 Challenge

How far does a typical employee commute to work?

With close to 4 million miles of roads, the United States has the largest highway system in the world. Utilizing this transportation network, people often commute great distances to work each day. It is not uncommon for commuters to spend two hours per day driving to and from work.

In April 2008, NPR reported that homes farther from the center of cities that require longer commutes decreased significantly in value during the economic slowdown.

In order to characterize the one-way commute distance, a large corporation obtained a random sample of its employees and recorded the distances they travel to work (in miles). The data are given in the following table.

(Patrick Herrera/iStockphoto)

5.3	3.4	2.7	22.9	19.9	3.3	4.0	2.9	13.6	1.4
40.8	18.0	7.1	10.4	3.7	14.9	2.1	1.5	22.8	9.0
11.1	4.6	1.8	8.2	15.3	1.7	12.0	15.5	16.5	1.4
14.7	7.8	32.6	7.1	1.9	4.6	24.5	9.4	12.6	9.3
5.3	6.4	8.9	1.6	4.3	1.0	6.9	1.3	4.1	4.3

The tabular and graphical techniques presented in this chapter will be used to describe the shape, center, and spread of this distribution of one-way commuting distances and to identify any outlying values. The results are given in the Chapter Challenge Wrap-up (page 59).

Review
- Realize the difference between a sample and the population.
- Recognize the importance of a simple random sample in the statistical inference procedure.
- Understand the difference between descriptive and inferential statistics.

Preview
- Be able to classify a data set as categorical or numerical, discrete or continuous.
- Learn several graphical summary techniques.
- Construct bar charts, pie charts, stem-and-leaf plots, and histograms.

CONTENTS

17

2.1 TYPES OF DATA

As members of an information society, we have access to all kinds of descriptive statistics in newspapers, in research journals, on CD-ROM, and even via the Internet. Whether the information is obtained from a carefully designed experiment or an observational study, the first step in analyzing it is to organize and summarize the data. For example, Figure 2.1 shows a stacked bar chart of the number of Political Action Committees by year. Tables, charts, and graphs may reveal characteristics about the shape, center, and variability of a data set, or distribution. The shape of a distribution may be symmetric or skewed. The center of a distribution refers to the position of the

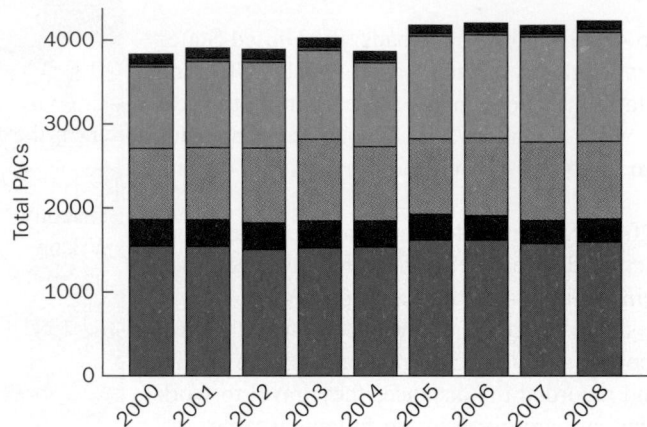

Figure 2.1 The growth of Political Action Committees (*Source:* Federal Election Commission).

majority of the data, and measures of variability indicate the *spread* of the data. The variability (or dispersion) of a distribution describes how much the measurements vary, and how compact or how spread out the data are. While not suitable for inference, the tabular and graphical techniques introduced in this chapter help to initially describe the distribution of data and identify unusual characteristics.

The summary table or graph to be used, and later the statistical analysis to be performed, depends on the type of data. Consider golfers arriving at a public country club on a Saturday morning. Here are several characteristics we could record: brand of golf clubs, handicap, whether the patron wears a golf hat, even the number of days since the golfer last played at this course.

DEFINITION

A data set consisting of observations of only a single characteristic, or attribute, is a **univariate** data set.
If we measure, or record, two observations of each individual or object, the data set is **bivariate**.
If there are more than two observations of the same individual or object, the data set is **multivariate**.

We'll do more with bivariate data in Chapter 12.

Suppose we record only the make of car driven by each person arriving at the country club—a **univariate** data set. The observations, for example, Ford, Honda, or Lexus, are **categorical**. There is no natural ordering of the data and each observation falls into only one category or class. We might instead ask each person who arrives how long it took to reach the country club. This time the responses, for example, 10, 15, or 45 minutes, are **numerical**.

> **DEFINITION**
>
> A **categorical**, or **qualitative**, univariate data set consists of nonnumerical observations that may be placed in categories.
> A **numerical**, or **quantitative**, univariate data set consists of observations that are numbers.

The following examples illustrate the two basic types of data sets.

Example 2.1 Sneaker Preference   A random sample of 12 people at a health club is selected and each person is asked to name his or her favorite brand of sneaker. The responses are given in the following table.

| Adidas | Saucony | Nike | New Balance | Adidas | Reebok |
| Avia | Wilson | Nike | Nike | Reebok | AND1 |

Each response is nonnumerical, since there is no natural ordering. This is a (univariate) categorical data set.   ●

Example 2.2 Egg Weights   To improve egg production, producers often test alternative feeds and enhanced nutrients. Suppose a new enzyme is being tested and 20 eggs are randomly selected and weighed. The resulting weights (in grams) are given in the following table.

| 58.8 | 59.9 | 60.7 | 59.0 | 55.9 | 56.5 | 54.4 | 60.6 | 62.1 | 59.9 |
| 57.1 | 58.9 | 58.1 | 56.6 | 55.2 | 57.7 | 55.8 | 56.2 | 57.9 | 56.5 |

Since each observation is numerical, this is a (univariate) numerical data set.   ●

We can classify numerical data even further. Consider the following examples.

On a hot summer day in the Southeast, suppose we record the number of lightning strikes within a specified county during the next 24 hours. The possible values are 0, 1, 2, 3, up to, say 10. There are only a finite number of possible numerical values, and these values are **discrete**, isolated points on a number line (Figure 2.2). Instead, suppose we record the barometric pressure, in millibars, at 4:00 P.M. The possible values are not discrete and isolated. Rather, the barometric pressure can (theoretically) be *any* number in the **continuous** interval 960 to 1070 mmHg (Figure 2.3).

The number of lightning strikes is discrete. For instance, the number of possible lightning strikes can be 1, 2, 3, etc., but not, for example, 2.5. However, if we had an instrument that could measure barometric pressure accurately enough, any number between 960 and 1070 mmHg is possible—for example, 995.466347789 mmHg.

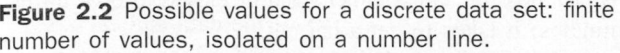

Figure 2.2 Possible values for a discrete data set: finite number of values, isolated on a number line.

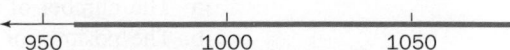

Figure 2.3 Possible values for a continuous data set: numerical values in some interval.

The two types of numerical data are defined on the next page.

> **DEFINITION**
>
> A numerical data set is **discrete** if the set of all possible values is finite, or countably infinite. Discrete data sets are usually associated with *counting*. A numerical data set is **continuous** if the set of all possible values is an interval of numbers. Continuous data sets are usually associated with *measuring*.

ILLUMINATING THE CONCEPTS

1. In order to decide whether a data set is discrete or continuous, consider all the possible values. Finite or countably infinite means discrete. An interval of possible values means continuous.

2. Countably infinite means there are infinitely many possible values, but they are countable. We may not ever be able to finish counting all of the possible values, but there exists a method for actually counting them.

3. The interval for a continuous data set can be any interval, of any length, open or closed. We may not know the exact interval, only that there is some interval of possible values.

4. In practice, we have no measurement device that is precise enough to return any number in some interval. We may only be able to achieve up to 10 digits of accuracy. So a continuous data set may contain any number in some interval in theory, but not in reality. ∎

Mathematically, a set is countably infinite if it can be put into one-to-one correspondence with the counting numbers (1, 2, 3, 4, . . .). The three dots, . . . , mean the list continues in the same manner.

The classifications of univariate data are shown in Figure 2.4. Here is an example to illustrate these classifications.

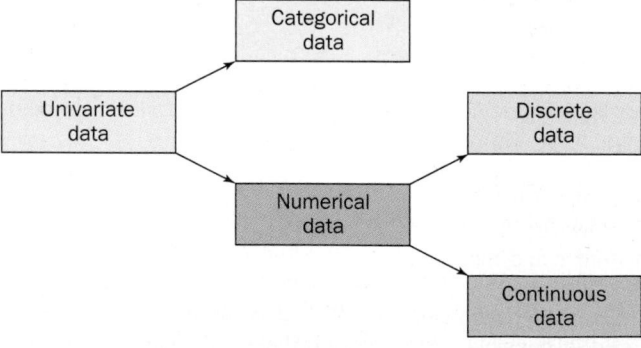

Figure 2.4 Classifications of univariate data.

Example 2.3 Univariate Data Classifications   A researcher obtained the following observations. Classify each resulting data set as categorical or numerical. If the data set is numerical, determine whether it is discrete or continuous.

a. The number of books read by middle-school students during an academic year.
b. The position of the drawbridge in Belmar, New Jersey at noon on days in July. Assume the drawbridge is not moving, and is either open or closed to boat traffic.
c. The length of time (in minutes) it takes to get a haircut.
d. The number of security guards on duty at office buildings.
e. The types of candy received at houses on Halloween.
f. The air pressure in footballs at the beginning of college games.

SOLUTION

a. The observations are numbers, so the data set is numerical. The set of possible values is finite. We don't know the maximum number of books read, but the possible numbers in the data set represent counts. The data set is discrete.

b. The observations are categorical: Up (open) or down (closed). There is no natural ordering; the possible responses fall into groups or classes. This data set is categorical.

c. The observations are numbers and the set of possible values is some interval, perhaps 5 to 45 minutes. This is a numerical continuous data set.

d. The observations are numbers and the set of possible values is finite. We can count the number of security guards. The minimum number of guards on duty may be 1 and the maximum may be 10. This is a numerical discrete data set.

e. The observations may be Milky Way, Snickers, Crunch Bar, etc. Although there may be some personal preference and an individual ranking, this is a categorical data set.

f. The observations are numbers and the set of possible values is some interval. This is a numerical continuous data set.

Methods for summarizing and displaying categorical data are discussed in Section 2.2, while tables and graphs for numerical data are presented in Sections 2.3 and 2.4.

SECTION 2.1 EXERCISES

Practice

2.1 Univariate Data Classifications A set of observations is obtained as indicated below. In each case, classify the resulting data set as categorical or numerical. If the data set is numerical, determine whether it is discrete or continuous.
a. The weights of several reams of paper.
b. The number of cars towed from the Pennsylvania Turnpike during given 24-hour periods.
c. The first ingredient in the product listing of boxes of cereal.
d. The number of games the Red Sox win during several seasons.
e. The amount of sand used on roads during winters in a small town.
f. The diagnoses of patients in an emergency ward.

2.2 Univariate Data Classifications A set of observations is obtained as indicated below. In each case, classify the resulting data set as categorical or numerical. If the data set is numerical, determine whether it is discrete or continuous.
a. The lengths of the spans of bridges in New York State.
b. The number of people hired by a company during certain weeks.
c. The cloud ceiling at airports around the country.
d. The temperature of the coffee purchased at several fast-food restaurants.
e. The type of notebook used by students in a statistics class.
f. The classifications of Forward Operating Air Force bases (Main Air Base, Air Facility, Air Site, or Air Point).

2.3 Univariate Data Classifications A set of observations is obtained as indicated below. In each case, classify the resulting data set as categorical or numerical. If the data set is numerical, determine whether it is discrete or continuous.
a. The number of steps on apartment fire escapes.
b. The number of leaves on maple trees.
c. The reason several automobiles fail inspection.
d. The weight of fully loaded tractor trailers.
e. The area of several Nebraska farms.
f. The cellular calling plan selected by customers.

2.4 Univariate Data Classifications A set of observations is obtained as indicated below. In each case, classify the resulting data set as categorical or numerical. If the data set is numerical, determine whether it is discrete or continuous.
a. The number of engine revolutions per minute in automobiles.
b. The thickness of the polar ice cap in several locations.
c. The state in which families vacationed last summer.
d. The type of Internet connection in county households.
e. The make of watch worn by people entering a certain department store.
f. The number of raisins in 24-ounce boxes.

2.5 Numerical Observations A set of numerical observations is obtained as described below. Classify each resulting data set as discrete or continuous.
a. The widths of posters at an art gallery.
b. The time it takes to compile computer programs.
c. The number of radioactive particles that escape from special containers during a 1-hour period.
d. The time it takes to bake batches of banana muffins.
e. The concentration of carbon monoxide in homes during the winter.
f. The number of pages in best-selling murder-mystery novels.

2.6 Numerical Observations A set of numerical observations is obtained as described below. Classify each resulting data set as discrete or continuous.
a. The weight of baseball bats.
b. The area of selected dorm rooms.
c. The number of bees in hives.

d. The height of storm surges during hurricanes.

e. The amount of ink used in office printers during a week.

f. The number of fish in office aquariums.

2.7 Numerical Observations A set of numerical observations is obtained as described below. Classify each resulting data set as discrete or continuous.

a. The time it takes giant slalom skiers to cover a race course.

b. The number of magazines available for sale at newsstands.

c. The number of black squares in crossword puzzles.

d. The length of time spent waiting in line at grocery-store check-out lanes.

e. The number of french fries in a small order from fast-food restaurants.

f. The length in words of email messages received.

2.8 Univariate Data Classifications Classify each data set as categorical, discrete, or continuous.

a. A random sample of mature Eastern tent caterpillars is obtained from a tree branch in a neighborhood yard. The length of each caterpillar is recorded.

b. Randomly selected prime-time television shows are selected and the number of violent acts is recorded for each show.

c. A representative sample of employees from a large company is obtained, and the overtime hours for the past month are recorded for each employee.

d. An HMO selects a random sample of subscribers and records the number of office visits over the past year for each patient.

e. Thirty-six apples are randomly selected from an orchard. Each is graded for quality of appearance: excellent, good, fair, or poor.

f. A random sample of mattresses is obtained and the firmness (medium, medium firm, firm, or extra firm) of each is recorded.

2.9 Univariate Data Classifications Classify each data set as categorical, discrete, or continuous.

a. A random sample of cheeses is obtained and the number of months each is allowed to age before sale is recorded.

b. Sixteen universities are selected and each computer network system is carefully analyzed. The computer virus threat is assessed for each campus: low, medium, or high.

c. A random sample of Waterford Normandy dinner plates is selected and the weight of each plate is recorded.

d. Thirty-five new customers at a health club are selected and the body fat percentage of each member is computed and recorded.

e. A random sample of CDs is obtained from a local music store. The company that produced each CD is noted.

f. A collection of pens is obtained from employees at a large company. For each pen, the outside diameter of the barrel at its widest point is measured and recorded.

2.2 BAR CHARTS AND PIE CHARTS

The natural summary measures for a categorical data set are the number of times each category occurred and the proportion of times each category occurred. These values are usually displayed in a table as in Table 2.1.

Table 2.1 A frequency distribution summarizing the results of a survey on eyeglasses purchased at a mall vision store.

Class	Frequency	Relative frequency
Myopic glasses	230	0.46
Sunglasses	75	0.15
Contact lenses	45	0.09
Anastigmatic glasses	55	0.11
Hypermetropic glasses	40	0.08
Presbyopic glasses	30	0.06
Special glasses	25	0.05
Total	**500**	**1.00**

DEFINITION

A **frequency distribution** for categorical data is a summary table that presents categories, counts, and proportions.

1. Each unique value in a categorical data set is a label, or **class**. In Table 2.1, the classes are myopic glasses, sunglasses, contact lenses, etc.

2. The **frequency** is the count for each class. In Table 2.1, the frequency for the anastigmatic glasses class is 55 (i.e., 55 of the people surveyed purchased anastigmatic glasses).

3. The **relative frequency**, or sample proportion, for each class is the frequency of the class divided by the total number of observations. In Table 2.1, the relative frequency for the special-glasses class is 25/500 = 0.05.

A **frequency distribution** for a categorical data set is illustrated in the next example.

Example 2.4 Cruise Ship Destinations
A random sample of cruise ships leaving from the Port of New York showed the following destinations.

Bermuda	Southampton	Mediterranean	Southampton	Caribbean
Southampton	Bermuda	Southampton	Caribbean	Caribbean
Caribbean	Bermuda	Mediterranean	Caribbean	Southampton
Caribbean	Southampton	Mediterranean	Southampton	Southampton
Bahamas	Bermuda	Bahamas	Southampton	Southampton

Construct a frequency distribution to describe these data. What proportion of cruise ships did not go to Southampton?

SOLUTION

STEP 1 Each unique destination is a label, or **class**. This is a categorical data set. There are 5 unique classes and 25 observations in total.

STEP 2 Draw a table and list each unique class in the left-hand column. Find the **frequency** and **relative frequency** for each class. For example, since Bermuda appears four times in the sample, the frequency for this class is 4. The relative frequency for Bermuda is 4/25 = 0.16.

Class	Tally	Frequency	Relative frequency
Bahamas	||	2	0.08 (= 2/25)
Bermuda	||||	4	0.16 (= 4/25)
Caribbean	⁚⁚⁚⁚ |	6	0.24 (= 6/25)
Mediterranean	|||	3	0.12 (= 3/25)
Southampton	⁚⁚⁚⁚ ⁚⁚⁚⁚	10	0.40 (= 10/25)
Total		**25**	**1.00**

The proportion of cruise ships that did go to Southampton is 10/25 = 0.40. The total proportion is always 1.00. Therefore, the proportion of cruise ships that did not go to Southampton is 1.00 − 0.40 = 0.60. ●

(Jimmy Lopes/FeaturePics)

Solution Trail

KEYWORDS

- Construct a frequency distribution.
- What proportion of cruise ships did not go to Southampton?

↓

TRANSLATION

- Summary table of categories, counts, and proportions.
- Proportion of cruise ships in certain categories.

↓

CONCEPTS

Use a frequency distribution to describe a data set.

↓

VISION

Construct a frequency distribution using each destination as a class. Count the number of observations in each class, and compute the frequency of occurrence for each class. Find the proportion of cruise ships that went to Southampton, U.K., and subtract from the total proportion, 1.00.

A tally mark is a short line drawn for each count up to four. On number five, draw a diagonal line across the other four. Count in sets of five.

💡 ILLUMINATING THE CONCEPTS

1. If you have to construct a frequency distribution by hand, an additional tally column is helpful. Insert this after the class column, and use a tally mark or tick mark to count observations as you read them from the table.

2. The last (total) row is optional, but is a good check of your calculations. The frequencies should sum to the total number of observations, and the relative frequencies should sum to 1.00 (subject to round-off error).

3. There is no rule for ordering the classes. In Example 2.4, the classes happen to be presented in alphabetical order. ■

A **bar chart** is a graphical representation of a frequency distribution for categorical data. An example of a bar chart is shown in Figure 2.5.

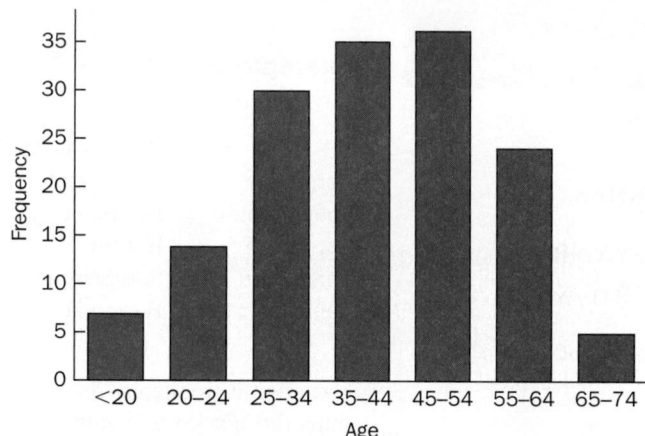

Figure 2.5 Bar chart showing the number of fatal occupational injuries in New York State in 2006 by age category (*Source:* New York State Department of Health).

HOW TO CONSTRUCT A BAR CHART

1. Draw a horizontal axis with equally spaced tick marks, one for each class.
2. Draw a vertical axis for the frequency (or relative frequency) and use appropriate tick marks. Label each axis.
3. Draw a rectangle centered at each tick mark (class) with height equal to, or proportional to, the frequency of each class (also called the class frequency). The bars should be of equal width, but do not necessarily have to abut one another; there can be spaces between them.

Example 2.5 Cruise Ship Destinations: Continued   Construct a bar chart for the cruise ship data in Example 2.4.

SOLUTION

STEP 1   Use the frequency distribution for the cruise ship data. There are five classes and the frequencies range from 2 to 10.

STEP 2   Draw a horizontal and a vertical axis. On the horizontal axis, draw five ticks for the five classes and label them with the class names. Since the greatest frequency is 10, draw and label tick marks from 0 to 10 on the vertical axis.

STEP 3   The height of each vertical bar is determined by the frequency of the class. For example, since the frequency of trips to Bermuda is 4, the height of the bar representing Bermuda is 4. The resulting bar chart is shown in Figure 2.6.

Either frequency or relative frequency may be used on the vertical axis. Both are acceptable because the resulting graphical representations of the distribution are identical. The only difference between the two graphs is the labels on the vertical axis. Unless it is stated otherwise, frequencies are used.

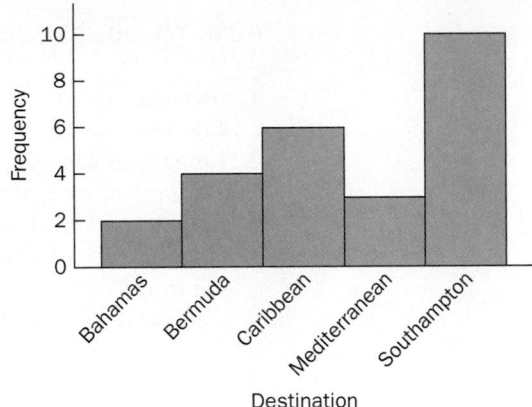

Figure 2.6 Bar chart for the cruise ship data.

Figure 2.7 shows a technology solution.

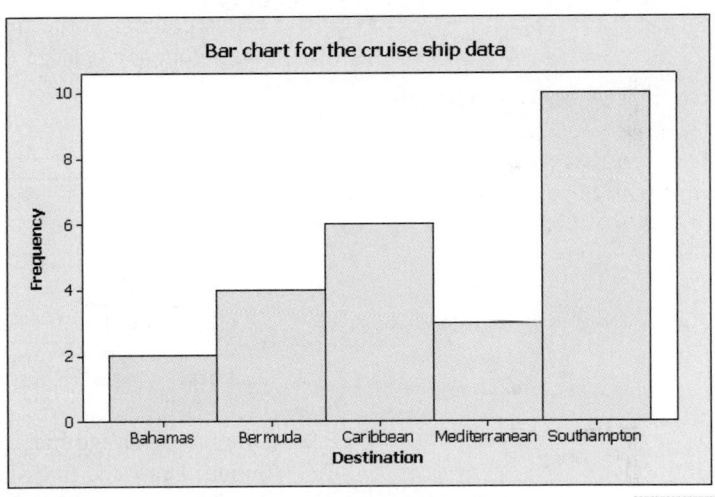

Figure 2.7 Minitab bar chart.

A **pie chart** is another graphical representation of a frequency distribution for categorical data. An example of a pie chart is shown in Figure 2.8.

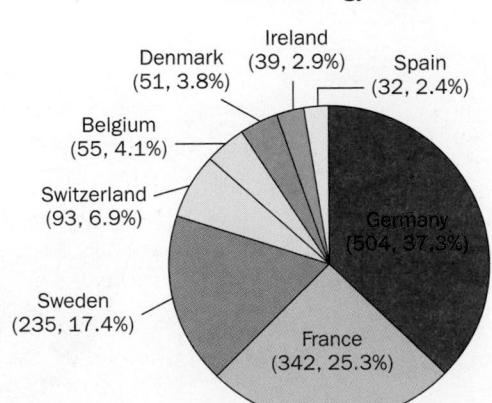

Figure 2.8 Pie chart showing number and percentage of biotechnology firms in selected countries (*Source:* NationMaster.com, 2008).

HOW TO CONSTRUCT A PIE CHART

1. Divide a circle (or pie) into slices or wedges so that each slice corresponds to a class.
2. The size of each slice is measured by the angle of the slice. To compute the angle of each slice, multiply the relative frequency by 360° (the number of degrees in a whole or complete circle).
3. The first slice of a pie chart is usually drawn with an edge horizontal and to the right (0°). The angle is measured counterclockwise. Each successive slice is added counterclockwise with the appropriate angle.

Example 2.6 Cruise Ship Destinations: Another Stop Construct a pie chart for the cruise ship data in Example 2.4.

SOLUTION

STEP 1 Add a column to the frequency distribution for slice angle. Use the relative frequency of each class to find the slice angle.

Class	Relative frequency	Angle
Bahamas	0.08	$0.08 \times 360° = 28.8°$
Bermuda	0.16	$0.16 \times 360° = 57.6°$
Caribbean	0.24	$0.24 \times 360° = 86.4°$
Mediterranean	0.12	$0.12 \times 360° = 43.2°$
Southampton	0.40	$0.40 \times 360° = 144.0°$
Total	**1.00**	**360.0°**

STEP 2 Draw a circle and mark slices using the angles in the frequency distribution. Draw the first slice with an edge extending from the center of the circle to the right. The remaining slices are drawn moving around the pie counterclockwise. It may be helpful to use a protractor and compass to draw the circle and measure the angles.

Pie chart for the cruise ship data

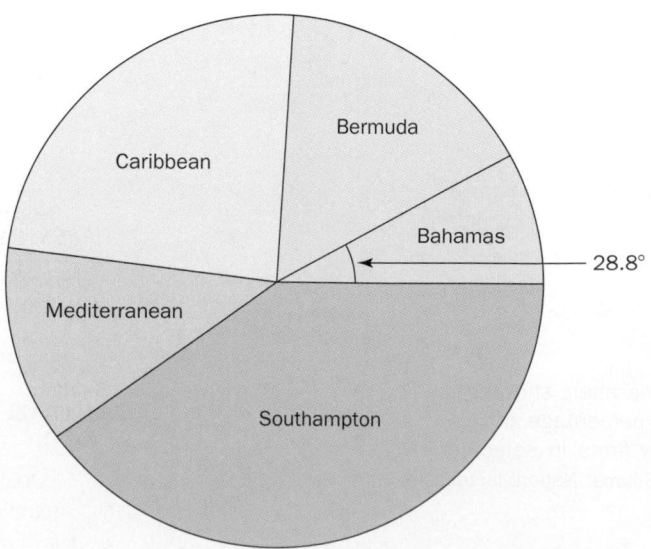

Figure 2.9 shows a technology solution.

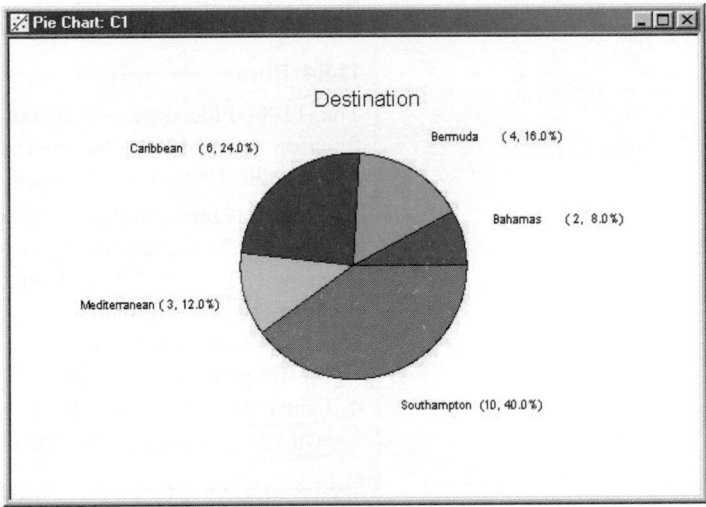

Note: Southampton corresponds to the biggest slice of the pie (chart). This class has the greatest frequency (and relative frequency); it is the destination that occurred most often in the sample.

Figure 2.9 Minitab pie chart.

💡 ILLUMINATING THE CONCEPTS

1. A pie chart is hard to draw accurately by hand, even with a protractor and compass. A graphing calculator or computer is quicker and more efficient for constructing this graph.

2. There are lots of pie chart variations, for example, exploding pie charts and 3D pie charts (an example of a 3D pie chart is shown in Figure 2.10). Each is simply a visual representation of a frequency distribution for categorical data. ■

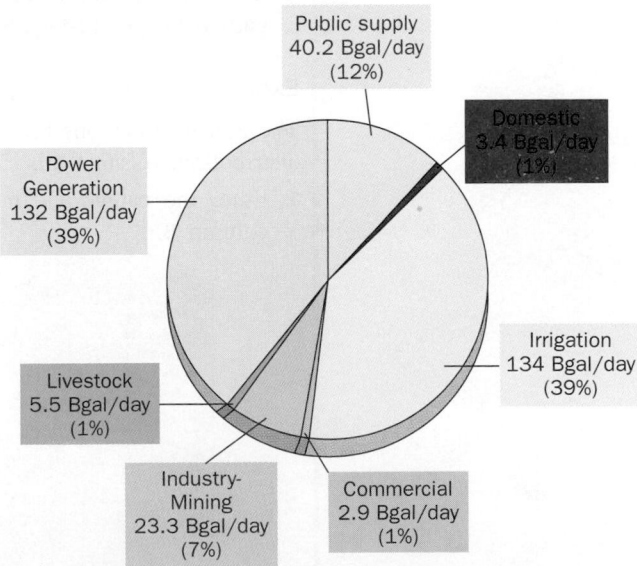

Figure 2.10 Daily freshwater use in the United States by category (Bgal/day = billions of gallons of water used per day) (*Source:* U.S. Geological Survey, Water Resources Division).

TECHNOLOGY CORNER

Procedure: Construct a bar chart.
Reconsider: Example 2.5, page 24, solution, and interpretations.

TI-84 Plus

The TI-84 Plus does not accept categorical data. Therefore, there is no built-in function to construct a bar chart. However, you may assign a number to each class and use the Histogram statistical plot to construct a bar chart.

1. Enter integers corresponding to each class in list L1 and the frequency for each class in the corresponding row in list L2. See Figure 2.11.
2. Press `STAT PLOT` and select `Plot1` from the STAT PLOTS menu.
3. Turn the plot `On` and select `Type` histogram. For `Xlist`, enter the name of the list containing the categories. For `Freq`, enter the name of the list containing the frequencies. See Figure 2.12.
4. Consider each *class* to have width 1. Enter appropriate WINDOW settings. Press `GRAPH` to display the bar chart. See Figure 2.13.

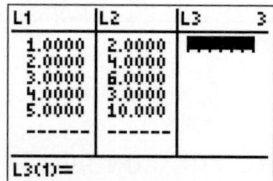

Figure 2.11 The *categories* and frequencies.

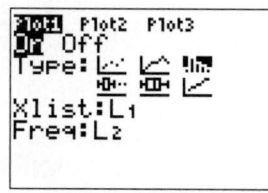

Figure 2.12 The `Plot1` setup screen.

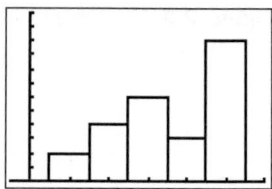

Figure 2.13 TI-84 Plus bar chart.

Minitab

The input can be either the entire data set in a single column or a summary table of categories and frequencies in two columns.

1. Enter the data into column C1.
2. Select Graph; Bar chart. Choose Counts of unique values and Simple.
3. Enter C1 under Categorical variables.
4. Edit graph attributes as necessary, for example, the axes labels, plot title, and gaps between clusters. See Figure 2.7 (page 25).

Excel

The built-in functions Frequency or Sum-If may be used to construct a frequency distribution. Assume this summary information is available.

1. Enter the categories into column A and the corresponding frequencies into column B.

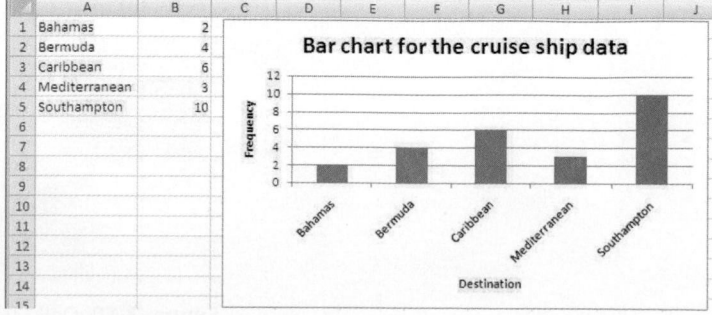

Figure 2.14 Excel bar chart.

2. Select the range of cells A1:B5. Under the Insert tab, select Column; 2-D Column; Clustered Column.
3. Use Chart Tools to format the bar chart as necessary. See Figure 2.14, page 28.

Procedure: Construct a pie chart.
Reconsider: Example 2.6, page 26, solution, and interpretations.

TI-84 Plus

The TI-84 Plus does not have a built-in function to construct a pie chart. However, there are several freely available calculator programs that draw a pie chart based on class frequencies or class relative frequencies.

1. Enter the class frequencies (or relative frequencies) into list L1.
2. Execute the program CIRCLE.[1]
3. Enter the name of the list containing the class frequencies and select DATA. See Figures 2.15–2.17.

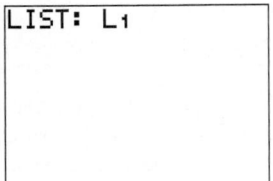

Figure 2.15 The list containing the class frequencies.

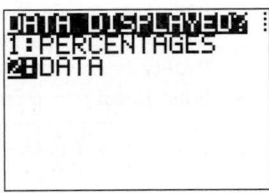

Figure 2.16 The DATA option corresponds to class frequencies.

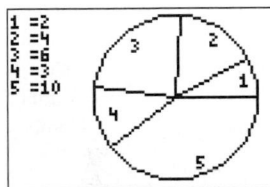

Figure 2.17 A TI-84 Plus pie chart.

Minitab

The input can be either the entire data set in a single column or a summary table of categories and frequencies in two columns.

1. Enter the data into column C1.
2. Select Graph; Pie chart. Choose Chart counts of unique values.
3. Enter C1 under Categorical variables.
4. Edit graph attributes as necessary, for example, the Slice Legend, Labels, and Title. See Figure 2.9, page 27.

Excel

The built-in functions Frequency or Sum-If may be used to construct a frequency distribution. Assume this summary information is available.

1. Enter the categories into column A and the corresponding frequencies into column B.
2. Select the range of cells A1:B5. Under the Insert tab, select Pie; 2-D Pie; Pie.
3. Use Chart Tools to format the bar chart as necessary. See Figure 2.18.

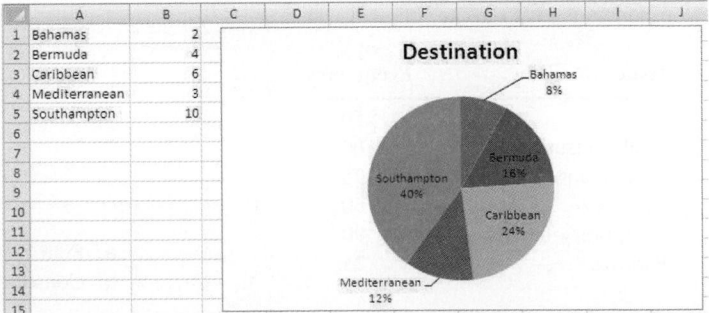

Figure 2.18 Excel pie chart.

SECTION 2.2 EXERCISES

Applications

2.10 Psychology and Human Behavior A random sample of TV viewers was obtained and each person was asked to select the entertainment category of his or her favorite show. The results are given in the following table.

Comedy	Comedy	Drama	Soap
Reality	Sports	Reality	Sports
Sports	Comedy	Drama	Reality
Drama	Soap	Soap	Soap
Educational	Reality	Drama	Drama
Sports	Educational	Drama	Comedy
Comedy	Soap	Reality	Soap
Sports	Drama	Drama	Soap
Soap	Reality	Reality	Soap
Soap	Drama	Educational	Drama
Comedy	Comedy		

Construct a frequency distribution for these data.

2.11 Psychology and Human Behavior A random sample of patrons visiting the Rena Branston Gallery in San Francisco was obtained. The patrons were asked the type of art they most enjoy viewing. The results are given in the following table.

Abstract	Abstract	Surrealist	Expressionist
Realist	Realist	Realist	Realist
Surrealist	Abstract	Abstract	Abstract
Realist	Realist	Abstract	Abstract
Surrealist	Surrealist	Abstract	Realist
Expressionist	Expressionist	Abstract	Surrealist
Surrealist	Surrealist	Abstract	Abstract
Surrealist	Surrealist	Abstract	Realist
Expressionist	Realist	Realist	Expressionist
Abstract	Abstract	Abstract	Expressionist
Realist	Realist		

Construct a frequency distribution for these data.

2.12 Education and Child Development In order to prepare for negotiations, the Faculty Association at Youngstown State University asked members to name the most important contract issue. A summary of their responses is given in the following table.

Issue	Frequency
Salary	50
Health insurance	100
Retirement benefits	75
Class size	60
Temporary faculty	90
Parking	25

a. Find the relative frequency for each issue.
b. Construct a bar chart for these data using frequency on the vertical axis.

2.13 Biology and Environmental Science The following table lists the number of farms for various counties in Missouri.

County	Frequency
Adair	915
Carroll	1081
Chariton	1095
Grundy	735
Linn	969
Livingston	903
Macon	1351
Mercer	569
Putnam	723
Schuyler	480
Sullivan	850

(*Source*: Missouri Economic Research and Information Center.)

a. Find the relative frequency for each county.
b. Construct a bar chart for these data using relative frequency on the vertical axis.

2.14 Fuel Consumption and Cars The business manager of a Chrysler automobile dealership sent a survey to randomly selected owners in order to gauge customer satisfaction. One question was, "How likely are you to buy another car of the same make and model?" Survey participants could answer Very Likely (VL), Likely (L), Neutral (N), Unlikely (U), or Very Unlikely (VU). The results are given on the data CD and book's web site.
a. Construct a frequency distribution for these data.
b. Use the table in part (a) to construct a pie chart for these data.

2.15 Public Policy and Political Science A random sample of voters was obtained in an election to replace a state senator who decided to retire due to health reasons. The political affiliation of each voter is given on the data CD and web site: Democrat (D), Republican (R), or Independent (I).
a. Construct a frequency distribution for these data.
b. Use the table in part (a) to construct a pie chart for these data.

2.16 Demographics and Population Statistics The following table lists the number of Nobel Prize winners by country.

Country	Frequency
United States	270
United Kingdom	101
Germany	76
France	49
Sweden	30
Switzerland	22
All others	130

(*Source*: nobelprize.org.)

a. Find the relative frequency for each country.
b. Construct a pie chart for these data.

2.17 Education and Child Development The grade distribution for a large psychology class at Louisiana State University is given in the following frequency distribution.

Grade	Frequency
A	10
B	43
C	54
D	26
F	15

a. Find the relative frequency for each grade.
b. Construct a bar chart using frequency on the vertical axis and a pie chart from the frequency distribution.
c. How many students were in this psychology class? What proportion of students passed (i.e., received a D or better)?

2.18 Public Health and Nutrition A random survey of 200 customers who purchased ice cream at Brigham's showed the following proportions.

Ice cream	Relative frequency
The Big Dig	0.100
Cashew Turtle	0.185
Chocolate Chip	0.260
Pistachio	0.150
Strawberry	0.080
Vanilla with Oreos	0.225

a. Find the frequency of each ice cream (class).
b. Construct a bar chart using frequency on the vertical axis and a pie chart for these ice cream data.

2.19 Sports and Leisure A random sample of long-time subscribers to *Popular Woodworking* was obtained, and each person was asked to name the brand of table saw he or she uses. The results are given in the following table.

DeWalt	DeWalt	Craftsman
DeWalt	Craftsman	Delta
Craftsman	Craftsman	Delta
DeWalt	Black & Decker	Makita
Black & Decker	DeWalt	Delta
Makita	Delta	Makita
Makita	DeWalt	Black & Decker
Delta	Delta	Makita
Black & Decker	Makita	Craftsman
DeWalt		

a. Construct a frequency distribution for these data.
b. Carefully sketch a bar chart using frequency on the vertical axis and a pie chart for these data.
c. What proportion of people in this sample use a Craftsman or Black & Decker table saw?
d. What proportion of people in this sample do not use a Delta table saw?

2.20 Marketing and Consumer Behavior Suppose there were 253 exhibitors at the NFPA World Fire Safety Conference in Las Vegas, Nevada in June 2005. Each exhibitor was classified according to the type of product or service offered for sale. The proportions are given in the following table.

Product	Proportion
Alarms	0.2964
Training	0.0632
Extinguishers	0.0514
Pumps	0.0237
Sprinklers	0.0632
Building materials	0.0751
Electrical equipment	0.1265
Hazmat storage	0.0870
Security products	0.1621
Signaling systems	0.0514

a. Find the number of exhibitors in each classification.
b. Carefully sketch a bar chart and a pie chart using the proportions for each class.

2.21 Psychology and Human Behavior Using the Library of Congress classification scheme, the Brookings Public Library in South Dakota recorded the type of book borrowed by 30 randomly selected patrons. The data are given in the following table.

Medicine	Science	Medicine	Medicine
Science	Education	Medicine	Science
Education	Law	Medicine	Technology
Education	Technology	Literature	Education
Technology	Medicine	Technology	Science
Science	Literature	Medicine	Literature
Law	Literature	Law	Technology
Technology	Education		

a. Construct a frequency distribution for these data.
b. Carefully sketch a bar chart using relative frequency on the vertical axis and a pie chart for these data.
c. Do you think the public library should try to purchase more books in one particular subject area? Why or why not?

2.22 Marketing and Consumer Behavior Cardinal Glass Industries produces several products for residential buildings, vehicles, and ordinary consumer use. The proportion of each type of manufactured product is given in the following table.

Building window	Vehicle window	Container	Tableware	Lamp
0.35	0.15	0.10	0.25	0.15

Construct a bar chart and a pie chart for these data using the proportions in the table.

2.23 Psychology and Human Behavior According to Divorce Magazine.com, Infoplease, and Americans for Divorce Reform, approximately 49% of all marriages in the United States end in divorce. Belarus has the highest divorce rate (68%) while Sri Lanka has the lowest (0.15%). Many factors contribute to divorce, for example, religion, occupation, and expectations. In a recent survey, American adults were asked to name the most important issue to discuss before marriage. The proportion of adults who indicated each issue in given in the following table.

Children	Finances	Religion	Work–life issues	Others
0.36	0.32	0.19	0.07	0.06

(*Source*: Findlaw.com as reported by D. Stuckey and G. Kereselidze for *USA Today*, December 12, 2005.)

a. Construct a bar chart and a pie chart for these data using the proportions in the table.

b. Suppose 1002 adults participated in this survey. Find the frequency for each issue, that is, the number of adults who indicated each key issue.

2.24 Marketing and Consumer Behavior A survey of new homes built in the Sleepy Creek Mountains of West Virginia produced the following results for the type of siding.

Siding	Frequency
Aluminum	20
Brick	15
Stucco	12
Vinyl	45
Wood	24

a. Find the relative frequency for each siding classification.

b. Construct a bar chart using frequency on the vertical axis and a pie chart for these data.

2.25 Public Policy and Political Science There are many *think tanks* in the United States, or groups of independent scholars with academic, government, and/or private experience. These think-tank scholars publish articles in appropriate journals and offer advice on politics, economics, and federal policy matters. Media coverage of think tanks is an important indicator of their influence and ability to obtain grants. The table below lists the frequency of citation in 2007 of several think tanks.

Think tank	Frequency
Brookings Institution	2380
Council on Foreign Relations	1191
Heritage Foundation	1168
RAND Corporation	740
Cato Institute	640
Urban Institute	558
Carter Center	341
Carnegie Endowment	287
Aspen Institute	209

(*Source*: Dolny, M., March/April 2008, The Incredible Shrinking Think Tank, *FAIR*.)

a. Find the relative frequency of citation for each think tank.

b. Construct a bar chart using frequency on the vertical axis and a pie chart for these data.

2.26 Marketing and Consumer Behavior In August 2005, Hurricane Katrina had a devastating effect on New Orleans and many other cities and towns along the Gulf Coast. A few months later, a survey was conducted regarding the relief and rebuilding efforts. Respondents were asked how long companies should continue to support Katrina relief efforts. The results are given in the following table.

Length of time	Relative frequency
One month	0.02
Six months	0.11
One year	0.13
Two years	0.09
More than two years	0.07
Until affected areas and people are thriving	0.53
Don't know	0.05

(*Source*: Cone survey as reported in *USA Today Snapshots*. *Note*: The proportions reported did not sum to 1 due to rounding. The *Don't know* category was adjusted here.)

a. Construct a pie chart for these data.

b. The survey reported that there were 1044 respondents. Using this information, find the frequency for each length of time category.

Extended Applications

2.27 Sports and Leisure Complete the following frequency distribution from a random sample of people visiting Atlantic City casinos.

Class	Frequency	Relative frequency
Bally's	40	
Caesars	25	0.125
Harrah's	32	
Resorts		0.110
Sands	25	
Trump Plaza		0.280
Total		**1.000**

a. What is the size of the random sample?

b. Which casino is most preferred by people in this survey? Justify your answer.

2.28 Travel and Transportation Families traveling to Disney World in Florida often rent a car rather than using airport and hotel shuttle buses. A recent survey asked families to indicate the rental car agency they used. The results are presented in the following bar chart.

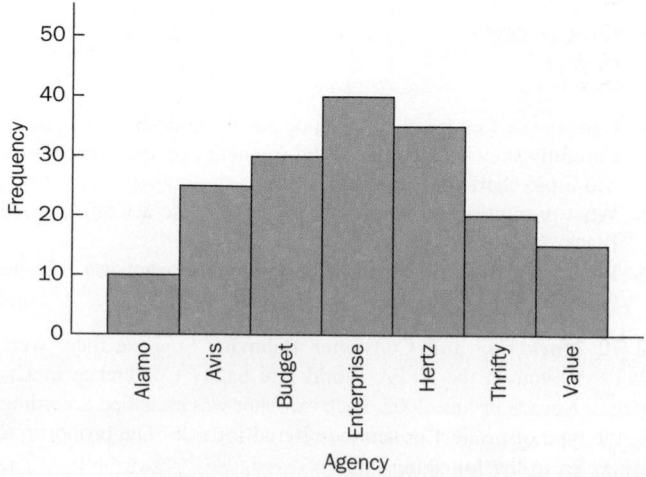

a. Construct a frequency distribution for these survey results.
b. How many observations were in this data set?
c. What proportion of people did not use Hertz or Enterprise?
d. Construct a pie chart for these data.

2.29 Marketing and Consumer Behavior One thousand customers entering the Mall of America in Bloomington, Minnesota, were randomly selected and asked to rank the variety of stores. The results are given in the following table.

Response	Frequency	Relative frequency
Excellent	50	
Very Good	152	
Good	255	
Fair		0.4250
Poor		0.1180

a. Complete the frequency distribution.
b. Construct a bar chart using frequency on the vertical axis and a pie chart for these data.
c. What proportion of customers did not rank the store variety as very good or excellent?

2.30 Travel and Transportation An automobile seatbelt is designed to prevent an occupant from coming into contact with the interior of the car and to stop the occupant from being ejected during an accident. Front seatbelts became standard equipment in 1964, and rear seatbelts became standard in 1968. The National Highway Traffic Safety Administration studies several characteristics related to passenger vehicle fatalities. The following table shows the passenger vehicle occupant fatalities age 16 and older in motor vehicle traffic crashes in 2006 by time of day and restraint (seatbelt) use.

Time of day	Restrained	Unrestrained
6 a.m. to 9 a.m.	1498	1611
9 a.m. to noon	1635	1261
Noon to 3 p.m.	2205	1647
3 p.m to 6 p.m.	2306	2112
6 p.m. to 9 p.m.	1644	2109
9 p.m. to midnight	1375	2461
Midnight to 3 a.m.	1179	2937
3 a.m. to 6 a.m.	938	1747
Unknown	71	175

(*Source*: NHTSA Traffic Safety Facts, May 2008.)

a. Find the relative frequency for *restrained* for each time of day. Construct a bar chart using relative frequency on the vertical axis (and time of day on the horizontal axis).
b. Find the relative frequency for *unrestrained* for each time of day. Construct a bar chart using relative frequency on the vertical axis (and time of day on the horizontal axis).
c. How do these two bar charts compare? Describe any similarities or differences.

2.31 Biology and Environmental Science An avalanche is a major danger facing skiers, snowboarders, and snowmobilers. The number of avalanche fatalities per year in the United States varies from approximately 5 to 35, but in general has increased steadily since 1956. The following table shows the number of avalanche fatalities from 1997–2007 in the United States by activity.

Activity	Frequency
Misc. recreation	4
In bounds skier	4
Snowshoer	20
Snowboarder	34
Out of bounds skier	30
Backcountry skier	49
Snowmobiler	126
Climber/hiker	32
Highway personnel	1
Others at work	2
Resident	6

(*Source*: Colorado Avalanche Information Center.)

a. Find the relative frequency of fatalities for each activity.
b. Construct a bar chart using frequency on the vertical axis, and a bar chart using relative frequency on the vertical axis. Which of these two graphs do you think is a better graphical description of avalanche fatalities by activity? Why?

2.32 Sports and Leisure In early February 2006, Vice President Dick Cheney accidentally shot Austin lawyer Harry Whittington while hunting quail. This event gave political writers and comedians a lot of material, and many articles regarding safe hunting appeared in newspapers and magazines. The following tables show the weapons and intended game involved in accidental hunting injuries in the United States in 2007.

Weapon	Frequency	Game	Frequency
Shotgun	157	Deer	85
Rifle	56	Pheasant	18
Handgun	10	Turkey	28
Unknown	3	Squirrel	17
Other	10	Cottontail	8
Crossbow	1	Dove/pigeon	17
Bow	1	Duck/goose	18
N/A	1	Other	48

(*Source*: Hunter Incident Clearinghouse, 2007.)

a. Construct a pie chart for the weapon data.
b. Construct a pie chart for the game data.
c. Using these two charts, is it reasonable to conclude that most hunting injuries occur with a shotgun while hunting deer? Why or why not?

2.33 Public Policy and Political Science A side-by-side or a stacked bar chart may be used to compare categorical data obtained from two (or more) different sources or groups. Figures 2.19 and 2.20 show an example of each—a comparison of test grades in two different sections of an introductory statistics course. The dark rectangles represent students from Section 01; the light rectangles represent students from Section 02.

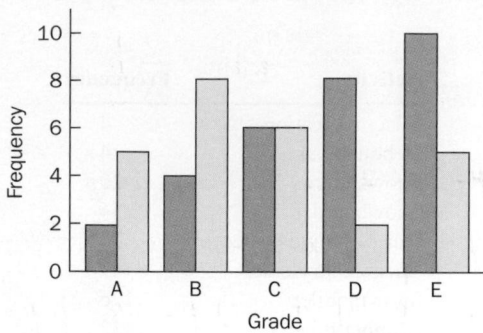

Figure 2.19 Side-by-side bar chart. Bars corresponding to the same category are placed side-by-side for easy comparison.

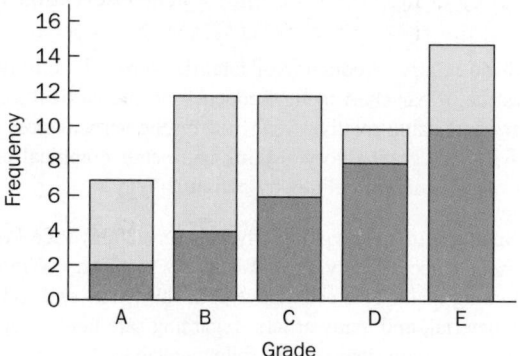

Figure 2.20 Stacked bar chart. Within each category, bars are stacked for comparison.

A recent poll conducted by *The Detroit Free Press* recorded the job rating of a state senator. The frequency of occurrence of each response, grouped by gender, is given in the following table.

Rating	Men frequency	Women frequency
Excellent	368	350
Very good	550	375
Good	426	165
Fair	450	360
Poor	206	250

a. Compute the relative frequency for each rating, for both groups.
b. Construct a side-by-side bar chart using the relative frequency of each class.
c. Why should relative frequency be used for comparison in the side-by-side bar chart rather than frequency?

2.34 Demographics and Population Statistics The number of women in the workforce has increased steadily since 1970. In 2007, there were approximately 71 million working women age 16 or older representing 59% of the total United States labor force.[2] The following table presents the relative frequency of men and women working, by age, in 2007.

Age group	Relative frequency	
	Men	Women
16 to 19	0.411	0.415
20 to 24	0.787	0.701
25 to 34	0.922	0.745
35 to 44	0.923	0.755
45 to 54	0.882	0.760
55 to 64	0.696	0.583
65 and older	0.205	0.126

(*Source*: U.S. Department of Labor, Bureau of Labor Statistics.)

a. Construct a side-by-side bar chart using the relative frequency of each class.
b. Using these data, is there any way to determine the relative frequency of *all* women (or men) age 16 and older working in 2007? Why or why not?

2.3 STEM-AND-LEAF PLOTS

We will eventually need a quantitative measure of *very far away.*

This section introduces the **stem-and-leaf plot,** a graphical technique for describing numerical data. In Section 2.4, you will learn about some other tables and graphs for summarizing numerical data. The goal of all of these techniques is the same: to get a quick idea of the distribution of the data in terms of *shape, center,* and *variability.* In addition, we are always watching for *outliers,* values that are very far away from the rest.

A stem-and-leaf plot is a relatively new graphical procedure used to describe numerical data. It is fairly easy to construct, even by hand, and most statistical software packages have options for drawing this graph. A stem-and-leaf plot is a combination of sorting and graphing. One advantage of this plot is that the actual data are used to create the graph; we do not lose the original data values as we do when we use tally marks to count them.

A stem-and-leaf plot can be used to describe the *shape, center,* and *variability* of the distribution. In Section 2.4, some specific terms and expressions used to describe shape are defined and illustrated. To estimate the center of a distribution, or to find a typical value, first arrange the observations in increasing order. Simply approximate a middle value, or range of values, in this list. More precise definitions and computations are presented in Chapter 3. The variability refers to the spread or compactness of the data. In addition, we always check for outliers.

The center of a distribution, or typical value, often occurs where the data are clustered.

HOW TO CREATE A STEM-AND-LEAF PLOT

To create a stem-and-leaf plot, each observation in the data set must have at least two digits. Think of each observation as consisting of two pieces (a stem and a leaf). For example, suppose we consider the number of people watching a movie, and in one theater there are 372 people. The number 372 could be split into the pieces 37 (the first two digits) and 2 (the last digit).

There are, of course, exceptions to this *two-digit rule.*

1. Split each observation into a
   Stem: one or more of the leading, or left-hand, digits, and a
   Leaf: the trailing, or remaining, digit(s) to the right. Each observation in the data set must be split at the same place, for example, between the tens place and the ones place.
2. Write a sequence of stems in a column, from the smallest occurring stem to the largest. Include all stems between the smallest and largest, even if there are no corresponding leaves.
3. List all the digits of each leaf next to its corresponding stem. It is not necessary to put the leaves in increasing order, but make sure the leaves line up vertically.
4. Indicate the units for the stems and leaves.

Example 2.7 Math SAT Scores Many colleges require applicants to take the Scholastic Aptitude Test (SAT). The table below shows the mean mathematics SAT score in 2008 for all 50 states.

The *mean* is an average, a measure of how the entire state performed on this exam.

(*Digital Vision/Alamy*)

AK	520	AL	557	AR	567	AZ	522	CA	515
CO	570	CT	513	DE	498	FL	497	GA	493
HI	502	IA	612	ID	540	IL	601	IN	508
KS	589	KY	570	LA	564	MA	525	MD	502
ME	466	MI	598	MN	609	MO	597	MS	556
MT	548	NC	511	ND	604	NE	585	NH	523
NJ	513	NM	548	NV	506	NY	504	OH	544
OK	572	OR	527	PA	501	RI	498	SC	497
SD	596	TN	570	TX	505	UT	557	VA	512
VT	523	WA	533	WI	604	WV	501	WY	574

(*Source*: The College Board.)

Construct a stem-and-leaf plot for these data.

SOLUTION

STEP 1 There are only two options for splitting each observation:
   a. split between the hundreds place and the tens place (e.g., split 520 as 5 and 20), or
   b. split between the tens place and the ones place (e.g., split 520 as 52 and 0).

If we split between the hundreds and tens place, there will be only three stems, because the only numbers in the hundreds place are 4, 5, and 6. The resulting plot will not reveal much about the distribution of the data. The better split is between the tens place and the ones place.

STEP 2 Scan the data to find the smallest and largest stems, and list all of the stems in a vertical column. Write each leaf next to its corresponding stem. For example,

$$520 \Rightarrow \quad 52 \mid 0 \qquad \text{A 0 is placed in the 52 stem row.}$$
$$\qquad\qquad\quad \uparrow \quad \uparrow$$
$$\qquad\qquad \text{stem} \quad \text{leaf}$$

For 557 (AL), a 7 is placed in the 55 stem row.
For 567 (AR), a 7 is placed in the 56 stem row.
For 522 (AZ), a 2 is placed in the 52 stem row.

STEP 3 Continue in this manner to produce the following stem-and-leaf plot.

A technology solution:

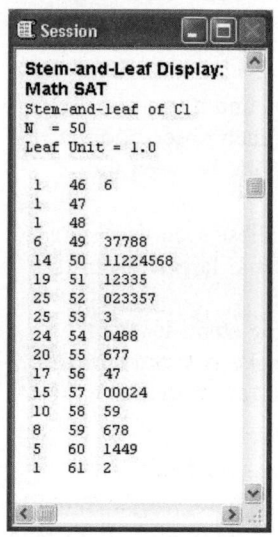

Figure 2.21 Minitab stem-and-leaf plot.

Stem	Leaf
46	6
47	
48	
49	8 7 3 8 7
50	2 8 2 6 4 1 5 1
51	5 3 1 3 2
52	0 2 5 3 7 3
53	3
54	0 8 8 4
55	7 6 7
56	7 4
57	0 0 2 0 4
58	9 5
59	8 7 6
60	1 9 4 4
61	2

Stem = hundreds and tens
Leaf = ones

Reading from the graph, the smallest SAT score is 466 and the largest is 612. The data seem to tail off slightly toward the larger values. The *center* of a data set is a typical value or values near the middle of the observations when they are arranged in (increasing) order. For these data, the center appears to be in the 52 or 53 stem row. There is one possible outlying value, 466.

ILLUMINATING THE CONCEPTS

1. As a general rule of thumb, try to construct the plot with 5 to 20 stems. With fewer than 5, the graph is too compact; with more than 20, the observations are too spread out. Neither extreme reveals much about the distribution.

2. Sometimes, to help us find the center of the data, we put the leaves in increasing order, to make an ordered stem-and-leaf plot.

3. Some advantages of a stem-and-leaf plot are that each observation is a visible part of the graph and (in an ordered stem-and-leaf plot, as when using a computer) the data are sorted. However, a stem-and-leaf plot can get very big, very fast. ■

If a stem-and-leaf plot is made for a very large data set, the stems may be divided, usually in half or fifths. Consider the following example.

Example 2.8 Hotel Room Rates   Hotels.com offers last-minute deals on hotel rooms around the country. A random sample of room rates (in dollars) associated with *Back to School Deals* was obtained, and the data are given in the following table.

75	84	78	79	72	73	50	90	85	69
76	77	77	78	61	58	80	81	75	75
89	89	74	73	79	86	85	94	64	72
77	83	78	81	91	70	93	75	78	54
55	60	63	69	65	73	93	81	79	79
77	75	61	71	68	72	77			

(*Source*: Hotels.com.)

Construct an ordered stem-and-leaf plot for these data.

SOLUTION

STEP 1   If we split each observation between the tens place and the ones place, there will be five stems. However, the leaves will extend far to the right and the shape, center, and spread of the distribution will be unclear.

STEP 2   Divide each stem in half. The first 5 stem row holds numbers 50–54, the second 5 stem row holds numbers 55–59, the first 6 stem row holds numbers 60–64, etc.

STEP 3   The resulting stem-and-leaf plot, with divided stems, offers a better graphical description of the distribution. Note that the leaves have been ordered.

Stem-and-leaf plot for the hotel rate data

```
5 | 0 4
5 | 5 8
6 | 0 1 1 3 4
6 | 5 8 9 9
7 | 0 1 2 2 2 3 3 3 4
7 | 5 5 5 5 5 6 7 7 7 7 7 8 8 8 8 9 9 9
8 | 0 1 1 1 3 4
8 | 5 5 6 9 9                          Stem = tens
9 | 3 3 4                              Leaf = ones
9 | 0 1
```

STEP 4   Notice, we can draw a straight line across the stem-and-leaf plot near the 75–79 row and the graph is almost a mirror image, or reflection, on each side of this line. Therefore, the distribution of the data is approximately *symmetric,* centered near 75–79. In addition, the distribution is compact and there are no outlying values.   ●

Two sets of data can be compared graphically using a back-to-back stem-and-leaf plot. Two plots are constructed using the same stem column. List the leaves for one data set to the left and those for the other to the right.

Example 2.9 Cholesterol Levels   Your total cholesterol level is the sum of low-density lipoproteins (LDLs) and high-density lipoproteins (HDLs). A total cholesterol level of less than 200 mg/dL (milligrams per deciliter) is desirable while 240 mg/dL or higher is considered high risk.[3] According to the National Health and Nutrition Survey, in 2006 the mean total cholesterol level for men was 195 mg/dL and for women 201 mg/dL. Suppose a random sample of total cholesterol levels was obtained for men and women. The data are given in the following table.

Men					Women				
110	124	132	147	157	183	190	201	211	154
164	172	180	193	201	186	212	213	169	173
210	224	112	158	165	177	195	203	203	189
173	181	193	205	216	207	158	218	213	205
194	194	179	185	185	205	204	189	179	177

Construct a back-to-back ordered stem-and-leaf plot for these data.

SOLUTION

STEP 1   The following graph is a back-to-back stem-and-leaf plot for these data.

Men		Women
2 0	11	
4	12	
2	13	
7	14	
8 7	15	4 8
5 4	16	9
9 3 2	17	3 7 7 9
5 5 1 0	18	3 6 9 9
4 4 3 3	19	0 5
5 1	20	1 3 3 4 5 5 7
6 0	21	1 2 3 3 8
4	22	

Stem = hundreds and tens
Leaf = ones

STEP 2   The center column of numbers (11, 12, 13, . . .) represents the stems for both groups. The stem-and-leaf plot for the men's data is constructed to the left, while the plot for the women's data is constructed to the right. Note that the leaves have been placed in increasing order, starting from the stem and proceeding outward.

STEP 3   The distribution for the men seems more spread out, or has more variability, while the distribution of women's cholesterol levels is more compact and has a slightly higher center.

In mathematics, *truncate* means discard the digits to the right of a specific place. In order to round a number to a certain position, consider the digit to the right of the rounding position. If this digit is a five or greater, then round up. Otherwise, leave the rounding digit unchanged (and replace all digits to the right with zero).

When constructing a stem-and-leaf plot, if there are two or more digits in each leaf, the trailing digits may be *truncated* or the entire leaf may be *rounded*. Suppose a data set includes the total yardage for randomly selected golf courses. Consider the three observations: 6518, 6523, and 6576, and suppose each observation is split between the hundreds place and the tens place.

The following diagram shows the 65 stem row for three stem-and-leaf plots. The first is constructed with two-digit leaves. The second is constructed by simply truncating the ones, or last, digit. The third plot is constructed by rounding each leaf to the nearest ten.

Two-digit leaf		Truncate each leaf		Round each leaf	
65 | 18	leaf = 18	65 | 18	leaf = 1	65 | 18	rounds to 20, leaf = 2
65 | 23	leaf = 23	65 | 23	leaf = 2	65 | 23	rounds to 20, leaf = 2
65 | 76	leaf = 76	65 | 76	leaf = 7	65 | 76	rounds to 80, leaf = 8

Stem	Leaves		Stem	Leaves		Stem	Leaves
⋮	⋮		⋮	⋮		⋮	⋮
65	18 23 76		65	1 2 7		65	2 2 8
⋮	⋮		⋮	⋮		⋮	⋮

Stem = thousands and hundreds
Leaf = tens and ones

Stem = thousands and hundreds
Leaf = tens

Stem = thousands and hundreds
Leaf = tens

Here is one more example of a stem-and-leaf plot.

Example 2.10 Fill It Up   Four Eagles Service Station selected a random sample of customers. The amount of gasoline (in gallons) purchased by each customer is given in the following table.

7.3	9.7	8.9	9.0	9.2	8.0	14.0	7.9	9.1	12.4
9.5	7.7	10.1	10.4	8.3	6.0	9.1	8.2	11.5	6.1
9.2	10.0	7.4	8.6	8.6	9.4	8.2	7.3	9.7	10.5

Construct a stem-and-leaf plot for these data.

SOLUTION

STEP 1   Split each observation between the ones place and the tenths place. List the stems in one column and each leaf next to its corresponding stem. Here is an example of one split.

$$7.3 \;\Rightarrow\; 7 \mid 3$$

$$\qquad\quad \uparrow \;\; \uparrow$$

$$\qquad \text{stem} \;\; \text{leaf}$$

STEP 2   Here is the resulting stem-and-leaf plot.

6	0 1
7	3 9 7 4 3
8	9 0 3 2 6 6 2
9	7 0 2 1 5 1 2 4 7
10	1 4 0 5
11	5
12	4
13	
14	0

Stem = tens and ones
Leaf = tenths

Even though there are no leaves in the 13 stem row, it must be included in the graph.

STEP 3   The graph is mound-shaped, tailing off slower at the bottom of the graph, or in the high end of the distribution. The data set seems compact (with little variability), and is centered near 8–9 gallons. The observation 14.0 is a possible outlier, since it is quite far away from the rest of the data. ●

TECHNOLOGY CORNER

Procedure: Construct a stem-and-leaf plot.
Reconsider: Example 2.7, page 35, solution, and interpretations.

There is no built-in command on the TI-84 Plus or in Excel to construct a stem-and-leaf plot. There are some calculator programs available and there are several add-ins for Excel for drawing a stem-and-leaf plot.

Minitab

1. Enter the data into column C1.
2. Select Graph; Stem-and-Leaf.
3. Enter C1 under Graph variables. Click OK.
4. The increment (distance between stems) is automatically selected and the leaves are placed in order. The numbers in the left column represent the cumulative counts from each end. The stem row containing the middle value is marked by only a count, in parentheses, of the number of observations in that row. N is the total number of observations. See Figure 2.21 (page 36).

SPSS

1. Enter the data into variable VAR00001.
2. Select Analyze; Descriptive Statistics; Explore.
3. Enter VAR00001 in the Dependent List. Select the Plots option in the Display box. Check Stem-and-leaf.

```
VAR00001 Stem-and-Leaf Plot

 Frequency     Stem &  Leaf

    6.00        4 .  699999
   24.00        5 .  000000001111122222234444
   15.00        5 .  555667777788999
    5.00        6 .  00001

Stem width:      100.00
Each leaf:         1 case(s)
```

Figure 2.22 SPSS stem-and-leaf plot.

SECTION 2.3 EXERCISES

Practice

2.35 Construct a stem-and-leaf plot for the following data.

4.7	5.1	6.6	3.9	5.0	2.9	3.6	5.5	4.2	5.1
4.9	5.4	6.1	4.1	3.6	6.4	4.7	4.1	5.7	3.6
6.8	3.5	6.4	6.4	7.1	2.7	5.8	5.2	5.9	5.7

Determine a range of numbers to indicate the center of the data. Within this range, select one number that is a *typical* value for this data set.

2.36 Construct a stem-and-leaf plot for the data given on the data CD and book's web site.

2.37 Construct a stem-and-leaf plot for the data given on the data CD and book's web site. Split each observation between the tens place and the ones place, and divide each stem in half. Determine a range of numbers to indicate the center of the data. Within this range, select one number that is a *typical* value for this data set.

2.38 Construct a stem-and-leaf plot for the data given on the data CD and book's web site. Use the stem-and-leaf plot to identify any outliers in this distribution.

2.39 Consider the following stem-and-leaf plot.

50	3
51	5
52	3 7
53	4 6
54	3 3 9
55	0 0 3 3 7
56	1 1 1 3 3 4 6 7 7
57	0 0 1 1 3 4 4 4 4 5 7 8 8
58	0 1 2 2 3 4 4 6 6 6 7 7
59	3 3 3 5 5 6 9
60	1 1 2

Stem = hundreds and tens
Leaf = ones

a. List the actual observations in the 54 stem row.
b. What is a typical value for this data set?
c. Do the data seem to be evenly distributed, or does one end tail off slower than the other?
d. Does the stem-and-leaf plot suggest there are any outliers in this data set? If so, what are they?

2.40 Consider the data given in the table below.

1717	1719	1645	3739	3024	3664	3830
2991	2430	2730	3469	5086	2119	3021
3292	2844	3426	2067	3215	2767	3124
2573	2840	2449	2584	1505	1390	1645
2497	3466	3228	3192			

a. Construct a stem-and-leaf plot by splitting each observation between the thousands place and the hundreds place.
b. Construct a stem-and-leaf plot by splitting each observation between the hundreds place and the tens place (using two-digit leaves).
c. Which plot presents a better picture of the distribution? Why?

Applications

2.41 Psychology and Human Behavior A random sample of patients involved in a psychology experiment was selected and the reaction time (in seconds) for each was recorded. The data are given on the data CD and book's web site.
a. Construct a stem-and-leaf plot by splitting each observation between the ones place and the tenths place. Truncate the hundredths digit so that each leaf has a single digit.
b. Construct a stem-and-leaf plot by splitting each observation between the ones place and the tenths place. Round each leaf to the nearest tenth so that each leaf has a single digit.
c. Describe any differences between the two plots. What is a typical value?

2.42 Public Health and Nutrition The owner of Copperfield Racquet and Health Club randomly selected 50 people and recorded the number of calories they burned after 20 minutes on a treadmill. The data are given on the data CD and book's web site.

a. Construct a stem-and-leaf plot by splitting each observation between the tens place and the ones place.
b. Construct a stem-and-leaf plot by splitting each observation between the tens place and the ones place, and by dividing each stem in half.
c. Which stem-and-leaf plot is *better*? Why?

2.43 Physical Sciences A random sample of hot water temperatures (°F) on lower floors and upper floors in the Renaissance Dallas Hotel was obtained. The data are given on the data CD and book's web site.
a. Construct a back-to-back stem-and-leaf plot to compare these two distributions.
b. Using the plot in part (a), describe any similarities and/or differences between the distributions.

2.44 Physical Sciences The intensity of light is measured in foot-candles or in lux. Television studios have a light intensity of approximately 1000 lux, and moonlight has an intensity of about 1 lux. The recommended level of light in offices is 300–500 lux.[4] A random sample of 50 businesses was obtained and the lux measurement at a typical work area was recorded for each. The data are given in the following table.

381	402	406	374	372	422	394	370	410	384
422	436	392	403	402	378	395	412	426	408
401	390	413	413	405	442	397	395	386	432
383	395	409	412	385	402	363	397	397	415
414	414	382	378	385	335	385	427	400	372

a. Construct a stem-and-leaf plot for these light-intensity data.
b. What is a typical light intensity? Are there any outliers? If so, what are they?

2.45 Public Health and Nutrition Every patient that visits a hospital emergency room is classified by the immediacy with which the patient should be seen. Approximately 37% of all patients who went to an emergency room in the United States in 2006 were classified as urgent.[5] Suppose a random sample of hospital emergency rooms was obtained and yearly records were examined. The percentage of patient visits classified as urgent is given on the data CD and book's web site.
a. Construct a stem-and-leaf plot for these data.
b. What is a typical value? Are there any outliers? If so, what are they?

2.46 Business and Management The port of Tacoma (Washington) handles approximately 14 million short tons (ST) of cargo each year. A random sample of outbound international containers was obtained and the weight of each (in ST) was recorded. The data are given on the data CD and book's web site.
a. Construct a stem-and-leaf plot for these data. Split each observation between the tenths place and the hundredths place.
b. Describe the distribution of container weight in terms of shape, center, and spread. Are there any outliers? If so, what are they?

2.47 Business and Management When Texaco merged with two other energy companies, all Texaco gas stations were converted to

either Chevron or Shell stations. A random sample of stations set for conversion was obtained. The number of years each station has been in operation is given on the data CD and book's web site.
a. Construct a stem-and-leaf plot for these data.
b. What is a typical number of years a Texaco station has been in operation? Are there any outliers? If so, what are they?

2.48 Manufacturing and Product Development *House Beautiful* magazine conducted a survey on the lifetime of dishwashers. Forty random users were contacted and asked to report the number of years their dishwasher lasted before needing replacement. The data are given on the data CD and book's web site.
a. Construct a stem-and-leaf plot for these data. Split each observation between the ones place and the tenths place.
b. Describe the distribution of dishwasher lifetimes in terms of shape, center, and spread.
c. What is a typical lifetime? Are there any outliers? If so, what are they?

2.49 Biology and Environmental Science One of the most popular pumpkin varieties is the Autumn King. These orange, oblong pumpkins with medium ribbing are commonly used as Halloween decorations. A random sample of Autumn King pumpkins was obtained, and the weight (in pounds) of each is given in the following table.

17	18	9	13	17	16	23	33	23	15
20	14	10	17	8	17	13	25	13	15
21	9	22	20	24					

(*Source*: the University of Tennessee Experiment Station Vegetable Initiative).

a. Construct a stem-and-leaf plot for these data.
b. What is a typical weight for this pumpkin variety? Are there any outliers in this data set? If so, what are they?

Extended Applications

2.50 Sports and Leisure A greyhound race handicapper uses several factors to predict the winner, such as past performance, track condition, early speed, form, and competition. Races on a $\frac{5}{16}$-mile track were randomly selected at the Jacksonville Kennel Club, and the winning time (in seconds) was recorded for each. The data are given in the following table.

31.12	31.46	31.93	31.51	31.52	31.52
31.52	31.86	31.29	31.27	31.29	31.57
31.08	31.41	31.17	31.31	31.47	31.32
31.09	31.23	31.32	31.52	31.67	30.80
31.54	31.61	31.59	31.47	31.77	31.48

(*Source*: Jacksonville Greyhound Racing, Inc.)

a. Construct a stem-and-leaf plot for these data. Split each observation between the tenths place and the hundredths place.
b. What is a typical winning time? If a dog has never run better than 31.70 seconds in a $\frac{5}{16}$-mile race, do you think it has a chance of winning? Justify your answer.
c. Could a stem-and-leaf plot be constructed with the split between the ones place and the tenths place? How about between the tens place and the ones place? Explain.

2.51 Biology and Environmental Science Many piano sellers recommend using a special humidifier, especially for more expensive pianos. This device is installed inside the piano and works to keep the instrument in tune by maintaining a stable humidity. In order to test whether a humidifier really helps, several pianos with and without humidifiers were tuned and checked 6 months later. Middle C was used as a measure of how well each piano stayed in tune. In a perfectly tuned piano, middle C has a frequency of 256 cycles per second. The frequency (in cycles per second) of middle C for each group, after 6 months, is given on the data CD and book's web site.
a. Construct a back-to-back stem-and-leaf plot for this data.
b. Use the plot in part (a) to describe any differences between the groups. Do you think a humidifier helps a piano stay in tune? Justify your answer.

2.52 Travel and Transportation There are national standards for every road sign, pavement marking, and traffic signal. While state laws take precedence, there are no formal state policies regarding the duration of an amber light. Suppose the federally recommended duration is 3 to 6 seconds, with longer times on roads with higher speed limits. A random sample of traffic signals in Norman, Oklahoma was selected, and the duration of the amber light was recorded for each. The data are given on the data CD and book's web site.
a. Construct a stem-and-leaf plot for these data. Divide each stem into five parts.
b. Based on the plot in part (a), do you believe this city has set the amber light duration to meet federal recommendations? Justify your answer.

2.4 FREQUENCY DISTRIBUTIONS AND HISTOGRAMS

Stem-and-leaf plots can be used to describe the shape, center, and variability of a numerical data set, but they can become huge and complex if the number of observations is large. A summary table like a **frequency distribution** for categorical data would be helpful. However, when the data set is numerical, there are no natural categories as there are for qualitative data. The solution is to use intervals as categories, or classes. We can then construct a frequency distribution for continuous data (similar to the categorical case), and a histogram (analogous to a bar chart for categorical data). For a small section of the sky, Figure 2.23 shows a histogram of the number of galaxies at each distance (measured by red-shift) from the earth. Redshifts of 2.6–3.4 correspond to galaxies approximately 12.5 billion light years away. The data for this graph were obtained from the Palomar Hale Telescope.

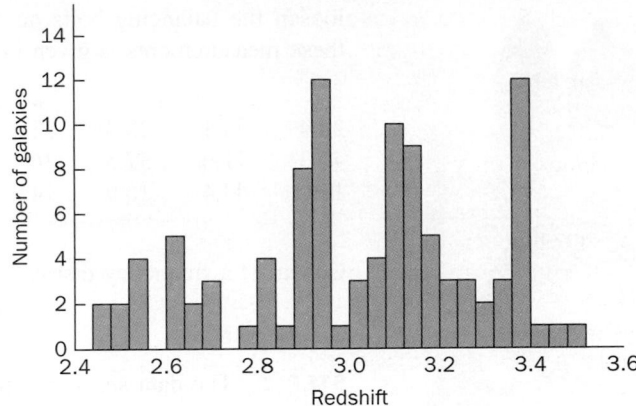

Figure 2.23 An example of a histogram.

Here is a procedure for constructing a frequency distribution, along with the necessary definitions.

HOW TO CONSTRUCT A FREQUENCY DISTRIBUTION FOR NUMERICAL DATA

In other words: *partition* the measurement axis into 5–20 subintervals.

1. Choose a range of values that captures all of the data. Divide it into non-overlapping (usually equal) intervals. Each interval is called a **class** or **class interval**. The endpoints of each class are the *class boundaries*.
2. We use the left-endpoint convention. An observation equal to an endpoint is allocated to the class with that value as its lower endpoint. Hence, the lower class boundary is always included in the interval, and the upper class boundary is never included. This ensures that each observation falls into exactly one interval.
3. As a rule of thumb, there should be 5–20 intervals. Use *friendly* numbers; for example, 10–20, 20–30, etc., not 15.376–18.457, 18.457–21.538, etc.
4. Count the number of observations in each class interval. This count is the **class frequency** or simply the **frequency**.
5. Compute the proportion of observations in each class. This ratio, the class frequency divided by the total number of observations, is the **relative frequency**.
6. Find the **cumulative relative frequency** (CRF) for each class: the sum of all the relative frequencies of classes up to and including that class. This column is a *running total* or *accumulation* of relative frequency, by row.

Example 2.11 Nuts and Bolts Torque is a measure of the force needed to cause an object to rotate. It is usually measured in foot–pounds (ft–lbs). As part of a quality-control program, Whirlpool inspectors measure the initial torque needed to

loosen the balancing bolts on each leg of a clothes washer. A random sample of these measurements is given in the following table.

20.4	24.1	28.4	53.4	62.1	31.7	57.2	45.7	38.1	51.1
41.3	11.0	37.5	36.4	25.6	43.5	23.1	24.2	35.5	26.4
13.0	44.4	16.9	14.9	63.7					

Construct a frequency distribution for these data.

SOLUTION

STEP 1 The data set is numerical (continuous). The observations are measurements, and each can be any number in some interval. Scan the data to find the smallest and largest observations (11.0 and 63.7). Choose between 5 and 20 reasonable (equal) intervals that capture all of the data.

STEP 2 The range of values 10–70 captures all of the data. Divide this range using the friendly numbers 10, 20, 30, . . . into the class intervals 10–20, 20–30, 30–40, etc.

STEP 3 Count the number of observations in each interval. For example, in the interval 10–20, there are four observations (16.9, 14.9, 11.0, and 13.0), so the frequency is 4.

STEP 4 Compute the proportion of observations in each class. For example, in the interval 10–20, the relative frequency is 4 (observations) divided by 25 (total number of observations).

STEP 5 Find the CRF for each class. For example, for the class 30–40, the cumulative relative frequency is the sum of the relative frequencies of this class and of all those listed above it: 0.16 + 0.28 + 0.20 = 0.64.

Class	Frequency	Relative frequency	Cumulative relative frequency
10–20	4	4/25 = 0.16	0.16 = 0.16
20–30	7	7/25 = 0.28	0.16 + 0.28 = 0.44
30–40	5	5/25 = 0.20	0.44 + 0.20 = 0.64
40–50	4	4/25 = 0.16	0.64 + 0.16 = 0.80
50–60	3	3/25 = 0.12	0.80 + 0.12 = 0.92
60–70	2	2/25 = 0.08	0.92 + 0.08 = 1.00
Total	**25**	**1.00**	

STEP 6 As for categorical data, if you must construct a frequency distribution by hand, an additional *tally column* is helpful (as introduced in Section 2.2). Insert this after the class column, and use a *tally mark* or *tick mark* to count observations as you read them from the table. ●

The last (*total*) row in a frequency distribution is optional, but is a good check of your calculations. The frequencies should sum to the total number of observations (25 in the example above), and the relative frequencies should sum to 1.00 (subject to round-off error).

The CRF of the first class row is equal to the relative frequency of the first class. There are no other observations before the first class. The CRF of the last class should be 1.00 (subject to round-off error). You must accumulate all of the data by the last class.

CRF gives the proportion of observations in that class and all previous classes. In Example 2.11, the CRF of the class 40–50 is 0.80. Interpretation: the proportion of observations less than 50 is 0.80.

ILLUMINATING THE CONCEPTS

This idea of working *backward* from cumulative relative frequency to obtain relative frequency is a handy technique for answering many probability questions.

1. Suppose you were given just the CRF for each class. To find the *relative frequency* for a class, take the class CRF and subtract the previous class CRF. In the example above, to find the relative frequency for the class 50–60: 0.92 (CRF for the class 50–60) − 0.80 (CRF for the previous class 40–50) = 0.12.

2. If the data set is numerical and discrete, use the same procedure outlined above for constructing a frequency distribution. If the number of discrete observations is small, then each value may be a class, or category. In addition, certain liberties are sometimes acceptable in listing the classes. For example, suppose a discrete data set consists of integers from 1 to 30. One might use the classes 1–5, 6–10, 11–15, 16–20, 21–25, and 26–30. This is not a strict partition of the interval 1–30 even though these classes are disjoint, or do not overlap. They do not allow for all numbers between 1 and 30. For example, the value 5.5 is between 1 and 30, but does not fall into any of these classes.

However, these classes work fine in this case since each observation is an integer. The resulting frequency distribution is perfectly valid. ■

A **histogram** is a graphical representation of a frequency distribution, a plot of frequency versus class interval. Given a frequency distribution, here is a procedure for constructing a histogram.

HOW TO CONSTRUCT A HISTOGRAM

1. Draw a horizontal (measurement) axis and place tick marks corresponding to the class boundaries.
2. Draw a vertical axis and place tick marks corresponding to frequency. Label each axis.
3. Draw a rectangle above each class with height equal to frequency.

Example 2.12 Nuts and Bolts (Continued)   Construct a frequency histogram for the torque data presented in the previous example.

For reference, here is the frequency distribution from the example above.

Class	Frequency	Relative frequency	Cumulative relative frequency
10–20	4	0.16	0.16
20–30	7	0.28	0.44
30–40	5	0.20	0.64
40–50	4	0.16	0.80
50–60	3	0.12	0.92
60–70	2	0.08	1.00
Total	**25**	**1.00**	

SOLUTION

STEP 1 Draw a horizontal axis and place tick marks corresponding to the class boundaries, or endpoints: 10 through 70 by tens.

STEP 2 Draw a vertical axis for frequency and place appropriate tick marks by checking the frequency distribution. Since the frequencies range from 0 to 7, draw tick marks at 0 to 7 on the vertical axis.

STEP 3 Draw a rectangle above each class with height equal to frequency. The resulting histogram is shown in Figure 2.24.

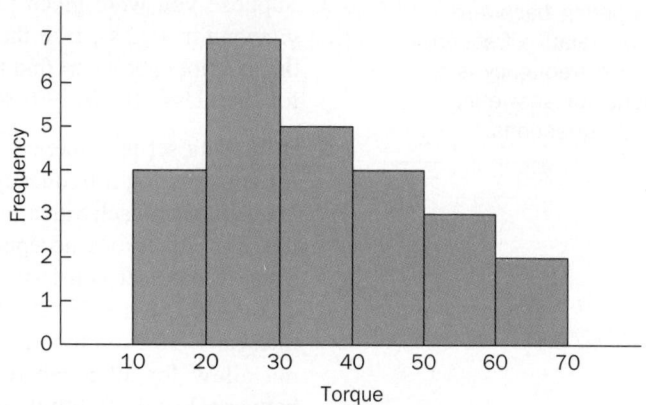

Figure 2.24 Frequency histogram for torque data.

Figures 2.25, 2.26, and 2.27 together show a technology solution.

Figure 2.25 The TI-84 Plus Plot1 setup screen for a histogram.

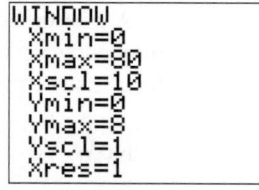

Figure 2.26 The WINDOW settings.

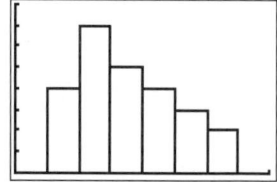

Figure 2.27 TI-84 Plus histogram.

💡 **ILLUMINATING THE CONCEPTS**

1. A histogram tells us about the shape, center, and variability of the distribution. In addition, it allows us to quickly identify any outliers.

2. If you must draw a histogram by hand, then you need to construct the frequency distribution first. However, calculators and computers construct histograms directly from the data. The frequency distribution is in the background and is usually not displayed.

3. To construct a *relative* frequency histogram, plot relative frequency versus class interval. The only difference between a frequency histogram and a relative frequency histogram is the scale on the vertical axis. The two graphs are identical in appearance. In Example 2.13 both a frequency histogram and a relative frequency histogram are shown.

4. Histograms should not be used for inference. They provide a quick look at the distribution of data and only *suggest* certain characteristics. ∎

Histogram usually means *frequency histogram*.

(*Roman Milert/iStockphoto*)

Example 2.13 Highway Tunnels   A random sample of highway tunnel lengths (in feet) was obtained, and the resulting frequency distribution is shown in the following table.

Class	Frequency	Relative frequency	Cumulative relative frequency
0–500	16	0.08	0.08
500–1000	28	0.14	0.22
1000–1500	54	0.27	0.49
1500–2000	48	0.24	0.73
2000–2500	36	0.18	0.91
2500–3000	18	0.09	1.00
Total	**200**	**1.00**	

Use this table to construct a frequency histogram and a relative frequency histogram for these data.

SOLUTION

STEP 1   For each graph, draw a horizontal axis and place tick marks at the class boundaries: 0, 500, 1000, . . . , 3000.

STEP 2   For the frequency histogram:
    a. Draw a vertical axis for frequency. Since the largest frequency is 54, use the tick marks at 0, 10, 20, . . . , 60.
    b. Draw a rectangle above each class with height equal to frequency. The resulting frequency histogram is shown in Figure 2.28.

STEP 3   For the relative frequency histogram:
    a. Draw a vertical axis for relative frequency. Since the largest relative frequency is 0.27, use the tick marks at 0, 0.05, 0.10, . . . , 0.30.
    b. Draw a rectangle above each class with height equal to relative frequency. The resulting relative frequency histogram is shown in Figure 2.29.

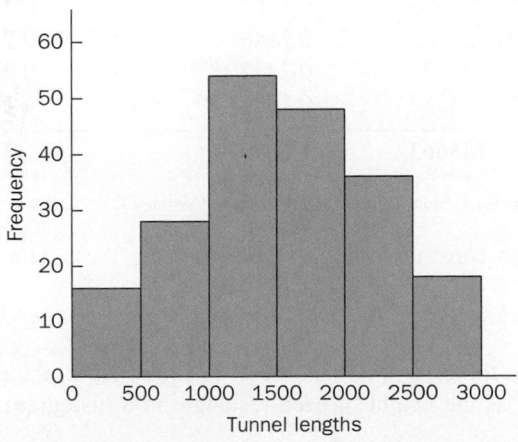

Figure 2.28 Frequency histogram for the tunnel-length data.

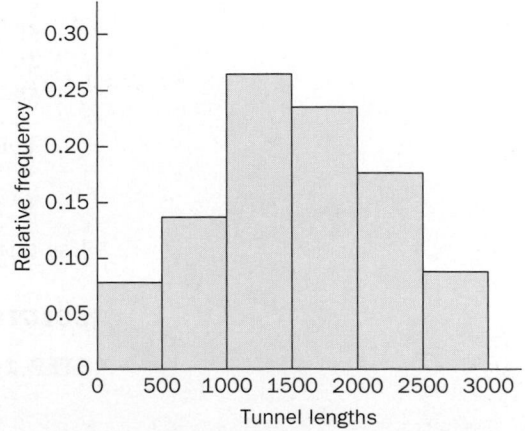

Figure 2.29 Relative frequency histogram for the tunnel-length data.

If the class widths are *unequal* in a frequency distribution, then neither the frequency nor the relative frequency should be used on the vertical axis of the corresponding histogram. To account for the unequal class widths, set the area of each rectangle equal to the relative frequency. In this case, the height of each rectangle is called the **density** and is equal to the relative frequency divided by the class width.

HOW TO FIND THE DENSITY

To find the density for each class,

1. Set the *area* of each rectangle equal to relative frequency. The *area* of each rectangle is *height* times class *width*.
Area of rectangle = Relative frequency
$$= \text{(Height)} \times \text{(Class width)}$$
2. Solve for the height.
Density = Height = (Relative frequency) / (Class width)

If two classes have the same frequency, but one class has double the width, then the corresponding rectangle in a traditional histogram would have double the area. This misrepresents the distribution.

The following example shows an extended frequency distribution with the density of each class included, and the corresponding **density histogram**.

Example 2.14 Accident Demographics The New York State Department of Motor Vehicles records the age and gender of each driver involved in an automobile accident. A subset of these data for female drivers in 2006 is summarized in the table below. The width of each class and the density calculations are also shown.

Class	Frequency	Relative frequency	Width	Density
16–18	5901	0.0405	2	0.0405/2 = 0.0203
18–21	14640	0.1005	3	0.1005/3 = 0.0335
21–25	17603	0.1208	4	0.1208/4 = 0.0302
25–30	17585	0.1207	5	0.1207/5 = 0.0241
30–40	31261	0.2146	10	0.2146/10 = 0.0215
40–50	34121	0.2342	10	0.2342/10 = 0.0234
50–60	24552	0.1686	10	0.1686/10 = 0.0169
Total	**145663**	**1.0000**		

(*Source*: New York State Department of Motor Vehicles.)

Use this table to construct a density histogram for these data.

SOLUTION

STEP 1   Since the class intervals are of unequal width, the class density must be used as the height of each rectangle in a histogram.

STEP 2   Draw a horizontal axis corresponding to age. Since the classes range from 16 to 60, use tick marks at 15, 20, 25, . . . , 60, or tick marks corresponding to the endpoints of each class.

STEP 3   Add a vertical axis for density. Since the largest density is 0.0335, use the tick marks 0, 0.005, 0.010, . . . , 0.035.

STEP 4   Draw a rectangle above each class with height equal to density. The resulting density histogram is shown in Figure 2.30.

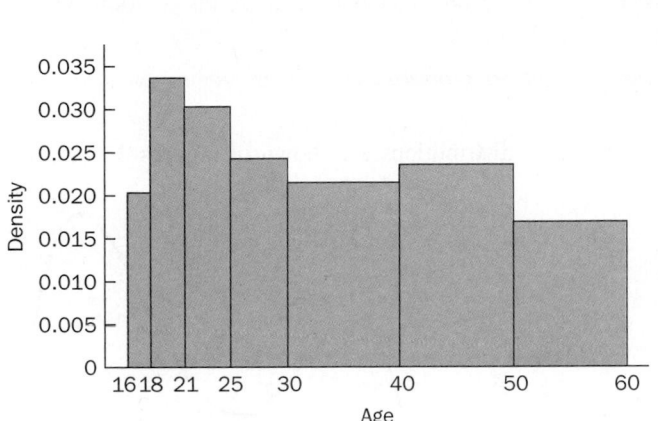

Figure 2.30 Histogram for unequal class widths: density histogram for the age data.

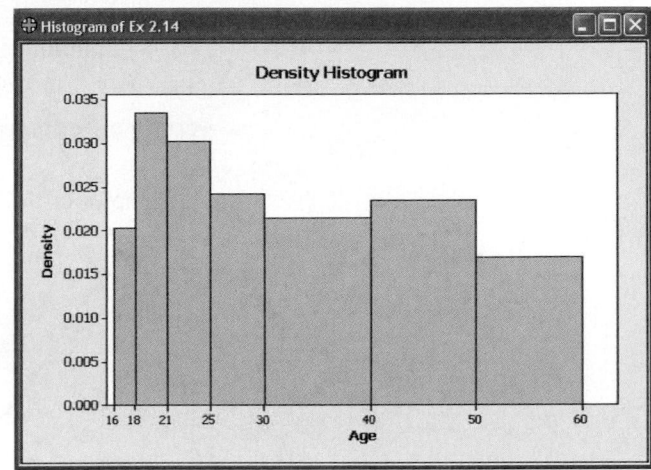

Figure 2.31 A technology solution: Minitab density histogram.

Shape of a Distribution

Since the relative frequency is equal to the area of each rectangle in a *density* histogram, the sum of the *areas* of all the rectangles is 1. This is an important concept as we begin to associate area with probability.

The shape of a distribution, represented in a histogram, is an important characteristic. To help describe the various shapes, we draw a smooth curve along the tops of the rectangles that captures the general nature of the distribution (as shown in Figure 2.32). To help identify and describe distributions quickly, a *smoothed histogram* is often drawn on a graph without a vertical axis, without any tick marks on the measurement axis, and without any rectangles (as shown in Figure 2.33).

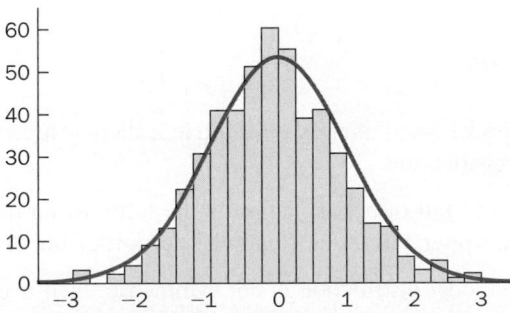

Figure 2.32 Smooth curve that captures the general shape of the distribution.

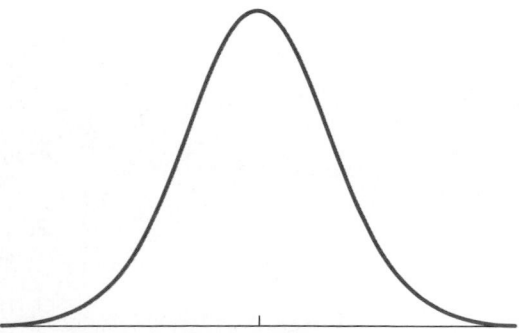

Figure 2.33 Typical smoothed histogram.

The first important characteristic of a distribution is the number of peaks.

DEFINITION

1. A **unimodal** distribution has one peak. This is very common; almost all distributions have a single peak.

2. A **bimodal** distribution has two peaks. This shape is not very common and may occur if data from two different populations are accidentally mixed.

3. A **multimodal** distribution has more than one peak. A distribution with more than two distinct peaks is very rare.

Examples of these three types of distributions are shown in Figures 2.34–2.36.

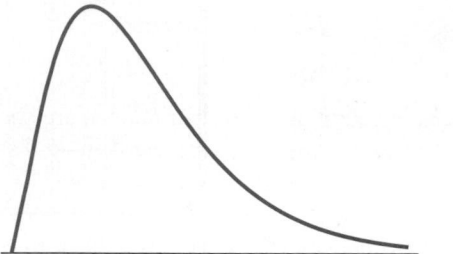

Figure 2.34 Unimodal distribution.

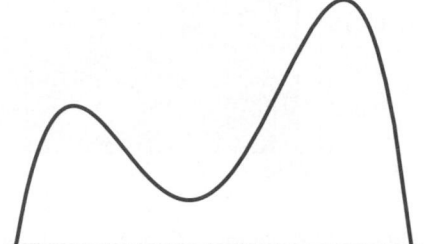

Figure 2.35 Bimodal distribution.

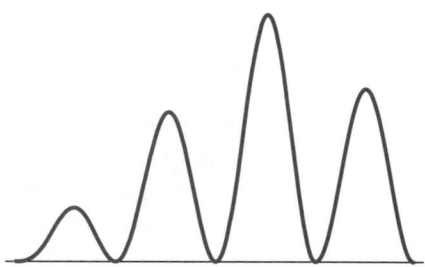

Figure 2.36 Multimodal distribution with four peaks.

The following characteristics are used to further classify and identify **unimodal** distributions.

DEFINITION

1. A unimodal distribution is **symmetric** if there is a vertical line of symmetry in the distribution.

2. The **lower tail** of a distribution is the leftmost portion of the distribution, and the **upper tail** is the rightmost portion of the distribution.

3. If a unimodal distribution is not symmetric, then it is **skewed**.
   a. In a **positively skewed** distribution, or a distribution that is **skewed to the right**, the upper tail extends farther than the lower tail.
   b. In a **negatively skewed** distribution, or a distribution that is **skewed to the left**, the lower tail extends farther than the upper tail.

Figures 2.37 and 2.38 show examples of **symmetric**, unimodal distributions. Each shows the (dashed) line of symmetry. The left half of the distribution is a mirror image of the right half. A **bimodal** or **multimodal** distribution may also be symmetric, and many distributions are *approximately* symmetric.

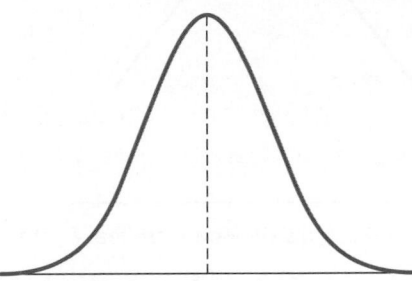

Figure 2.37 Symmetric distribution.

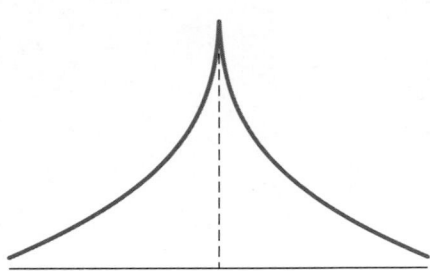

Figure 2.38 Symmetric distribution.

Examples of **skewed** distributions are shown in Figures 2.39 and 2.40. Positively skewed distributions are more common. The distribution of the lifetime of an electronics part might be positively skewed.

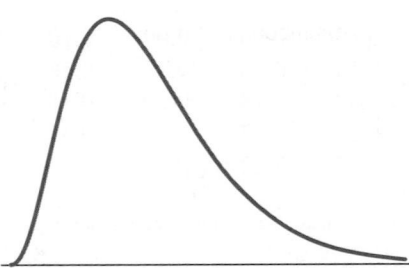

Figure 2.39 Positively skewed distribution.

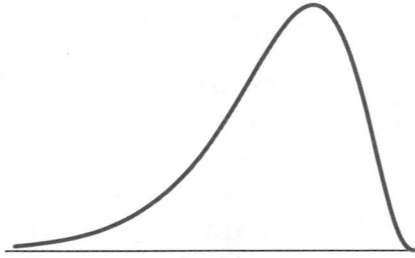

Figure 2.40 Negatively skewed distribution.

We will learn much more about the normal curve in Chapter 6.

The vertical cross section of a bell is a normal curve.

The most common unimodal distribution shape is a **normal curve** (as shown in Figure 2.41). This curve is symmetric and *bell-shaped*, and can be used to model, or approximate, many populations.

A curve with heavy tails has more observations in the tails of the distribution than a comparable normal curve. The tails do not drop down to the measurement axis as fast as a normal curve. A curve with light tails has fewer observations in the tails of the distribution than a comparable normal curve. The tails drop to the measurement

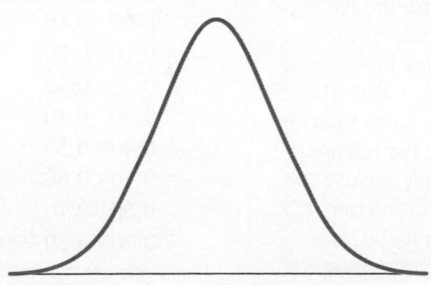

Figure 2.41 A normal curve.

axis quickly. Examples of curves with heavy and light tails are shown in Figures 2.42 and 2.43. Both of these characteristics are subtle and tricky to spot.

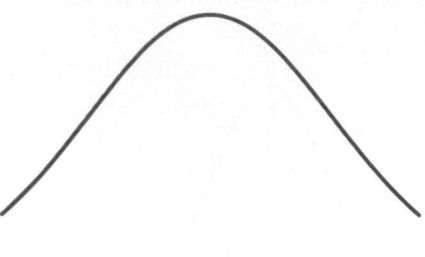

Figure 2.42 A distribution with heavy tails.

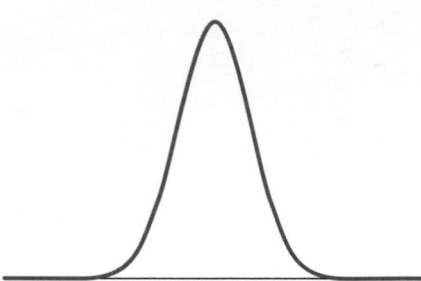

Figure 2.43 A distribution with light tails.

Example 2.15 Radiation Exposure   The U.S. Regulatory Commission monitors the radiation exposure at nuclear power reactors and other facilities. The individual radiation dose per year is measured in rems (roentgen equivalent man). A sample of 50 individual radiation measurements was obtained from employees and the data are given in the following table.

0.62	0.29	0.06	0.09	0.10	0.24	0.06	0.38	0.32	0.46
0.71	0.53	0.28	0.19	0.16	0.40	0.08	0.24	0.57	0.11
0.30	0.32	0.18	0.29	0.15	0.13	0.42	0.18	0.28	0.39
0.14	0.18	0.27	0.20	0.37	0.22	0.26	0.31	0.11	0.29
0.19	0.12	0.22	0.21	0.12	0.05	0.22	0.26	0.49	0.43

(*Source*: U.S. Nuclear Regulatory Commission.)

a. Construct a frequency distribution and a histogram for these data using the class intervals 0–0.10, 0.10–0.20, etc.
b. Describe the shape, center, and spread of the distribution.
c. What proportion of observations are less than 0.40 rem?
d. What proportion of observations are at least 0.50 rem?

SOLUTION

a. The class intervals are given. Create a frequency distribution and compute the frequency, relative frequency, and cumulative relative frequency for each class.

Class	Frequency	Relative frequency	Cumulative relative frequency
0.00–0.10	5	5/50 = 0.10	0.10 = 0.10
0.10–0.20	14	14/50 = 0.28	0.10 + 0.28 = 0.38
0.20–0.30	15	15/50 = 0.30	0.38 + 0.30 = 0.68
0.30–0.40	7	7/50 = 0.14	0.68 + 0.14 = 0.82
0.40–0.50	5	5/50 = 0.10	0.82 + 0.10 = 0.92
0.50–0.60	2	2/50 = 0.04	0.92 + 0.04 = 0.96
0.60–0.70	1	1/50 = 0.02	0.96 + 0.02 = 0.98
0.70–0.80	1	1/50 = 0.02	0.98 + 0.02 = 1.00
Total	**50**	**1.00**	

Use the frequency distribution to sketch the histogram.

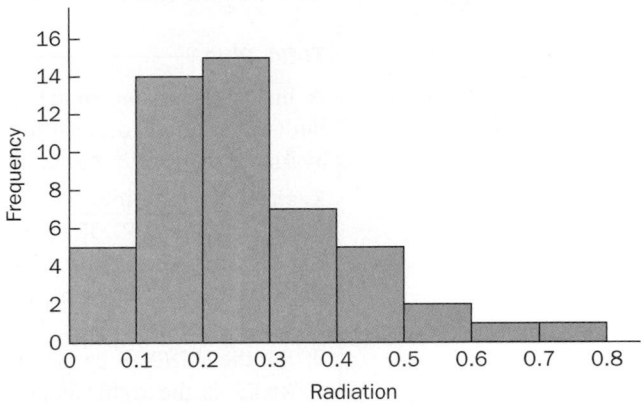

b. The distribution is positively skewed. There are more observations in the lower tail than in the upper tail. The upper tail extends farther than the lower tail. To estimate the center of the distribution, use the histogram to identify a value such that approximately half of the observations are below that number and half are above that number. A number between 0.2 and 0.3 appears to divide the ordered data in half. Typical values for this data set are in this range and an estimate of the center is 0.25.

The variability is typically described as either compact (data that are compressed or squeezed together) or spread out (observations that extend over a wide range). While somewhat subjective for now, this data set is fairly compact. All of the observations lie between 0.5 and 0.71 (even though the smallest class boundary is 0.00 and the largest class boundary is 0.80).

c. Using the cumulative relative frequency column of the frequency distribution, the proportion of observations less than 0.40 is 0.82.

d. There are two ways to find the proportion of observations that are at least 0.50.

i. Add the relative frequencies that correspond to the classes that are at least 0.50.

$$\underset{0.50\text{–}0.60}{0.04} + \underset{0.60\text{–}0.70}{0.02} + \underset{0.70\text{–}0.80}{0.02} = 0.08$$

ii. Find the cumulative relative frequency up to 0.50 and subtract this value from 1.

Proportion of observations ≥ 0.50

$$= 1 - (\text{proportion of observations} < 0.50)$$
$$= 1 - 0.92 = 0.08$$

> Other estimates of the center are also valid here. 0.24 is reasonable. So is 0.27. More precise measurements of the center of a data set are presented in Chapter 3.

Figures 2.44, 2.45, and 2.46 together show a technology solution.

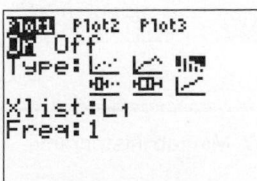

Figure 2.44 TI-84 Plus Plot1 setup screen.

Figure 2.45 The WINDOW settings.

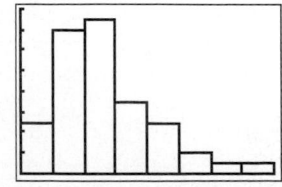

Figure 2.46 TI-84 Plus histogram. ●

TECHNOLOGY CORNER

Procedure: Construct a histogram.
Reconsider: Example 2.15, page 52, solution, and interpretations.

TI-84 Plus

A histogram is one of the six types of TI-84 Plus statistical plots. There is no built-in function to construct a density histogram. There are calculator programs available that will produce this graph.

1. Enter the data into list L1.
2. Select STAT PLOT ; Plot1 to define, or set up, the histogram. Turn the plot On, select Type histogram, set Xlist to the name of the list containing the data, and set Freq (frequency of occurrence of each observation) to 1. See Figure 2.44, page 53.
3. Set the WINDOW parameters so that Xmin is the left endpoint on the first class, Xmax is the right endpoint on the last class, and Xscl is the width of each class. Ymin should be 0 (the smallest frequency) and Ymax should be at least the largest frequency. Set Yscl to a reasonable distance between tick marks. See Figure 2.45, page 53.
4. Press GRAPH to display the histogram. See Figure 2.46, page 53. Note: The TRACE key is used to move on the graph between rectangles (classes). The corresponding class boundaries and frequency are displayed.

Minitab

The input may be either a single column containing the data or summary information in two columns: observations and frequencies.

1. Enter the data into column C1.
2. Select Graph; Histogram and highlight Simple histogram. Enter C1 in the Graph variables window. Click Ok to view the histogram.
3. Edit the horizontal axis scale to use the correct class intervals. Under the Binning tab, select Interval type: Cutpoint and enter the Midpoint/Cutpoint positions (class boundaries).

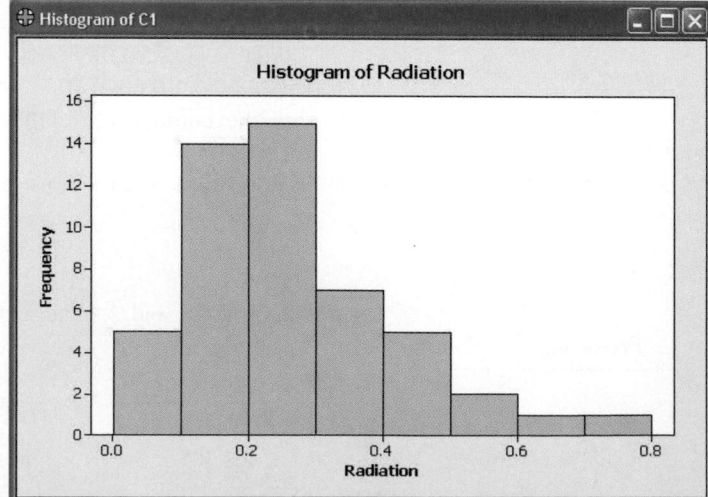

Figure 2.47 Minitab histogram.

Excel

The input may be either a single column containing the data or summary information in two columns: right endpoint of each class (bin limits) and corresponding frequencies. There are several methods to construct a frequency distribution in Excel using FREQUENCY, SUM-IF, or COUNTIFS, for example.

1. Enter the data into column A and the right endpoint of each class into column B.
2. Under the Data tab, select Data Analysis and choose Histogram. Enter the Input Range, the Bin Range, and the Output Range, and select Chart Output.
3. Each class is labeled with its right endpoint. In addition, Excel places observations on a boundary in the smaller class. See Figure 2.48.

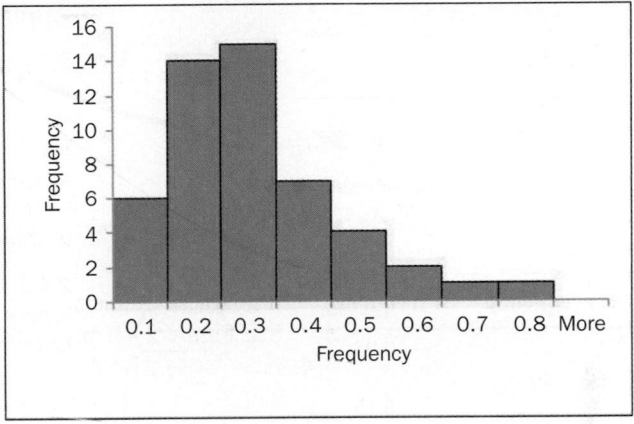

Figure 2.48 Excel histogram.

SECTION 2.4 EXERCISES

Practice

2.53 Consider the data given in the following table.

87	81	86	90	88	85	79	91	87	82
91	86	86	87	88	85	92	85	87	86
91	81	89	89	83	90	83	80	90	80
89	85	86	90	90	89	78	91	83	92

Construct a frequency distribution to summarize these data using the class intervals 78–80, 80–82, 82–84, ...

2.54 Consider the data given on the data CD and book's web site. Construct a frequency distribution to summarize these data.

2.55 Consider the following frequency distribution.

Class	Frequency	Relative frequency	Cumulative relative frequency
400–410	5	0.0758	0.0758
410–420	8	0.1212	0.1970
420–430	10	0.1515	0.3485
430–440	12	0.1818	0.5303
440–450	9	0.1364	0.6667
450–460	8	0.1212	0.7879
460–470	5	0.0758	0.8637
470–480	4	0.0606	0.9243
480–490	3	0.0455	0.9698
490–500	2	0.0303	1.0001

Draw the corresponding *frequency* histogram. (Notice the last entry in the Cumulative Relative Frequency column is not exactly 1. This is due to round-off error.)

2.56 Consider the following frequency distribution.

Class	Frequency	Relative frequency	Cumulative relative frequency
0.5–1.0	6	0.03	0.03
1.0–1.5	8	0.04	0.07
1.5–2.0	10	0.05	0.12
2.0–2.5	16	0.08	0.20
2.5–3.0	24	0.12	0.32
3.0–3.5	34	0.17	0.49
3.5–4.0	36	0.18	0.67
4.0–4.5	22	0.11	0.78
4.5–5.0	18	0.09	0.87
5.0–5.5	12	0.06	0.93
5.5–6.0	8	0.04	0.97
6.0–6.5	4	0.02	0.99
6.5–7.0	2	0.01	1.00

Draw the corresponding *relative frequency* histogram.

2.57 Complete the following frequency distribution.

Class	Frequency	Relative frequency	Cumulative relative frequency
100–150	155		
150–200	120		
200–250	130		
250–300	145		
300–350	150		
350–400	100		
Total			

2.58 Complete the following frequency distribution.

Class	Frequency	Relative frequency	Cumulative relative frequency
1.0–1.1		0.05	
1.1–1.2	20		
1.2–1.3		0.15	
1.3–1.4	65		
1.4–1.5		0.25	
1.5–1.6	35		
1.6–1.7	25		
1.7–1.8			
Total	**300**		

2.59 Complete the following frequency distribution and draw the corresponding histogram.

Class	Frequency	Relative frequency	Cumulative relative frequency
0–25			0.150
25–50			0.350
50–75			0.525
75–100			0.675
100–125			0.800
125–150			0.900
150–175			0.975
175–200			1.000
Total	**1000**		

2.60 Consider the data given on the data CD and book's web site.
a. Construct a frequency distribution to summarize these data using the class intervals 0–1, 1–2, 2–3, etc., and draw the corresponding histogram.
b. Use the histogram to describe the shape of the distribution. Are there any outliers?

2.61 Consider the data given on the data CD and book's web site.
a. Construct a frequency distribution to summarize these data and draw the corresponding histogram.
b. Use the histogram to describe the shape of the distribution.

c. Use the frequency distribution to estimate the *middle* of the data: a number M such that 50% of the observations are below M and 50% are above M.
d. Use the frequency distribution to estimate a number Q_1 such that 25% of the observations are below Q_1 and 75% are above Q_1.
e. Use the frequency distribution to estimate a number Q_3 such that 75% of the observations are below Q_3 and 25% are above Q_3.

Applications

2.62 Biology and Environmental Science A weather station located along the Maine coast recorded the maximum wind speeds (in km/h) for 50 days selected at random. The data are given on the data CD and book's web site.[6]
a. Construct a frequency distribution to summarize these data, and draw the corresponding histogram.
b. Describe the shape of the distribution. Are there any outliers?

2.63 Fuel Consumption and Cars The quality of an automobile battery is often measured by the Cold Cranking Amps (CCA), a measure of the current supplied at 0°F. Thirty automobile batteries were randomly selected and subjected to subfreezing temperatures. The resulting CCA data are given in the following table.

63	87	302	4	259	106	198	55	99	134
122	514	91	117	325	39	30	164	75	16
340	199	77	217	64	320	145	84	47	232

a. Construct a frequency distribution to summarize these data, and draw the corresponding histogram.
b. Describe the shape of the distribution.
c. Estimate the middle of the distribution, a number M such that 50% of the data are below M and 50% are above M.

2.64 Marketing and Consumer Behavior The weights of diamonds and other precious stones are usually measured in carats. One carat is traditionally equal to 200 milligrams. A random sample of the weights (in carats) of diamond engagement rings is given in the following table.

0.7	2.1	0.4	0.1	0.4	1.1	1.3	0.1	0.6	0.5
2.7	1.1	0.3	0.7	0.1	0.8	1.3	0.4	0.1	1.2
1.5	1.1	0.6	0.4	1.2	1.2	1.1	0.5	1.8	0.9

(*Source*: www.diamond.com.)

a. Construct a frequency distribution and a histogram for these data.
b. Multiply each observation in the table by 200 to convert the weights into milligrams. Construct a frequency distribution and a histogram for these new, transformed data.
c. Compare the two histograms. Are the shapes similar? Describe any differences.

2.65 Public Health and Nutrition Vitamin B3 (niacin) helps to detoxify the body, aids digestion, can ease the pain of migraine headaches, and helps promote healthy skin. A random sample of adults in the United States and in Europe was obtained and the daily in take of niacin (in milligrams) was recorded. The data are summarized in the following table.

Class	United States frequency	Europe frequency
0–3	15	4
3–6	23	6
6–9	21	12
9–12	14	17
12–15	12	32
15–18	9	25
18–21	3	20
21–24	2	10

a. Construct two *relative frequency* histograms, one for the United States and one for Europe.

b. Describe the shape of each histogram. Does a comparison of the two histograms suggest any differences in niacin intake between the two samples? Explain.

2.66 Manufacturing and Product Development Yarn is often sold in hanks. One hank of woolen yarn is approximately 1463 meters. A quality control inspector uses a special machine to quickly measure each hank. A random sample was obtained during the manufacturing process, and the length (in meters) of each hank is given on the data CD and book's web site.

a. Construct a histogram for these data.

b. Describe the distribution in terms of shape, center, and variability.

2.67 Sports and Leisure The National Hockey League is concerned about the number of penalty minutes assessed to each player. While some people in attendance hope to see a lot of fighting (and penalty minutes), the League Office believes that most fans are interested in good, clean hockey. A sample of penalty minutes per player for the 2007–2008 postseason was obtained, and the data are given in the following table.

141	70	66	103	96	85	84	85	60	182
36	121	81	81	74	73	119	59	55	55
71	116	67	53	116	96	73	57	53	38
59	66	64	70	100	68	100	79	79	58

(*Source*: nhl.com.)

a. Construct a histogram for these data.

b. Describe the distribution in terms of shape, center, and variability.

c. Find a value m for the number of minutes such that 90% of all games have fewer than m penalty minutes.

Extended Applications

2.68 Biology and Environmental Science Fruits such as cherries and grapes are harvested and placed in a shallow box or crate called a lug. Lugs vary in size, but typically hold between 16 and 28 pounds. A random sample of the weight (in pounds) of full lugs holding peaches was obtained, and the data are summarized in the following table.

Class	Frequency	Class	Frequency
20.0–20.5	6	22.5–23.0	25
20.5–21.0	12	23.0–23.5	19
21.0–21.5	17	23.5–24.0	15
21.5–22.0	21	24.0–24.5	11
22.0–22.5	28	24.5–25.0	10

a. Complete the frequency distribution.

b. Construct a histogram corresponding to this frequency distribution.

c. Estimate the weight w such that 90% of all full peach lugs weigh more than w.

2.69 Travel and Transportation Maglev trains operate in Germany and Japan at speeds of up to 300 miles per hour. Magnets create a frictionless system in which the train operates at a distance of 100–150 millimeters from the rail. The size of this air gap is monitored constantly to ensure a safe ride. A random sample of the size of air gaps (in mm) at one specific location in the track was obtained. The frequency distribution for these data is shown in the following table.

Class	Frequency	Relative frequency	Cumulative relative frequency
100–105			0.050
105–110			0.425
110–115			0.625
115–120			0.750
120–125			0.850
125–130			0.925
130–135			0.975
135–140			1.000
Total	**200**		

a. Complete the frequency distribution.

b. Draw a histogram corresponding to this frequency distribution.

c. What proportion of air gaps were between 110 mm and 125 mm?

2.70 Biology and Environmental Science NOAA trawlers recently analyzed a portion of the ocean floor near a productive commercial fishing area. Random measurements of depth (in meters) were recorded and are summarized in the following table.

Class	Frequency	Relative frequency	Width	Density
0–30	12			
30–50	68			
50–60	72			
60–70	80			
70–80	55			
80–90	43			
90–100	24			
100–150	18			
150–200	14			
Total	**386**			

(*Source*: National Geophysical Data Center.)

a. Complete the frequency distribution.
b. A traditional frequency histogram or relative frequency histogram is not appropriate in this case. Why not?
c. Construct a density histogram corresponding to this frequency distribution.

2.71 Biology and Environmental Science Mount Kilimanjaro is 19,340 feet high and is located in Tanzania, close to the border of Kenya. The total ice area on Mount Kilimanjaro has decreased steadily from approximately 12 km^2 in 1912 to 2.6 km^2 in 2000. It has been suggested that if current climate trends continue, the ice on Kilimanjaro will disappear between 2015 and 2020.[7] A random sample of total ice area mea-surements (in km^2) on Mount Kilimanjaro from 1950 to 2000 was obtained, and the data are given in the following table.

4.8	5.2	3.3	5.6	2.9	2.8	2.9	5.6	2.9	4.3
5.3	2.8	4.4	3.3	3.1	5.7	4.6	3.5	4.4	5.4
4.3	5.7	3.6	5.6	3.8	5.7	3.6	3.4	4.2	5.9
4.6	4.0	4.3	4.6	3.8	5.4	5.0	4.4	3.3	5.0

a. Construct a histogram for these data.
b. Describe the distribution in terms of shape, center, and variability.
c. Find a number Q_1 such that 25% of the ice area measurements are less than Q_1. Find a number Q_3 such that 25% of the ice area measurements are greater than Q_3.
d. How many values should be between Q_1 and Q_3? Find the actual number of values between Q_1 and Q_3. Explain any difference between these two values.

Chapter 2 Challenge Wrap-Up

The data set for the one-way commuting distances of the corporation's employees is continuous. The data are summarized in a stem-and-leaf plot (with stems divided in half and leaves truncated); in a frequency distribution; and in Figure 2.49, a histogram.

(Patrick Herrera/iStockphoto)

```
0    1 1 1 1 1 1 1 1 1 2 2 2 3 3 3 4 4 4 4 4 4
0    5 5 6 6 7 7 7 8 8 9 9 9
1    0 1 2 2 3 4 4
1    5 5 6 8 9
2    2 2 4
2
3    2
3                                    Stem = tens
4    0                               Leaf = ones
```

Stem-and-leaf plot for the commuting data.

Class	Frequency	Relative frequency	Cumulative relative frequency
0–5	21	0.42	0.42
5–10	12	0.24	0.66
10–15	7	0.14	0.80
15–20	5	0.10	0.90
20–25	3	0.06	0.96
25–30	0	0.00	0.96
30–35	1	0.02	0.98
35–40	0	0.00	0.98
40–45	1	0.02	1.00
Total	**50**	**1.00**	

Frequency distribution for the commuting data.

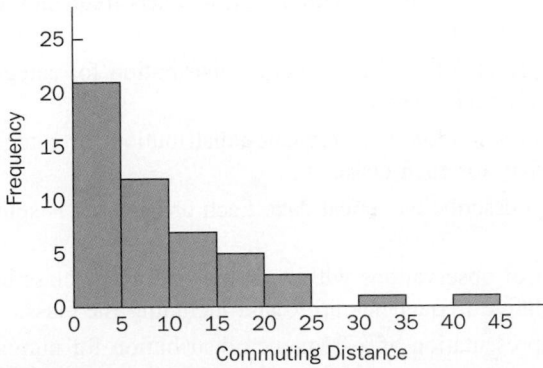

Figure 2.49 Histogram for the commuting data.

The stem-and-leaf plot and the histogram suggest that the data are positively skewed. Most of the observations are in the lower tail of the distribution; the upper tail has few observations. The center of the distribution (the middle number) is approximately 7.0. Most of the observations are between 0 and 25, making the

KEYWORDS

- Tabular and graphical techniques.
- Shape, center, and spread.

TRANSLATION

- Stem-and-leaf plot, frequency distribution, histogram.
- Describe the distribution.

CONCEPTS

- How to construct a stem-and-leaf plot, frequency distribution, and histogram.
- Interpretation of graphs for summarizing data.

VISION

Determine where to split each observation and construct a stem-and-leaf plot. Select appropriate classes; construct a frequency distribution and histogram. Describe the resulting summary graphs in terms of shape, center, and spread.

distribution fairly compact. The proportion of observations less than 25 is 0.96. The observations 32.6 and 40.8 are possible outliers; these observations are far away from the rest.

Figure 2.50 shows a technology solution.

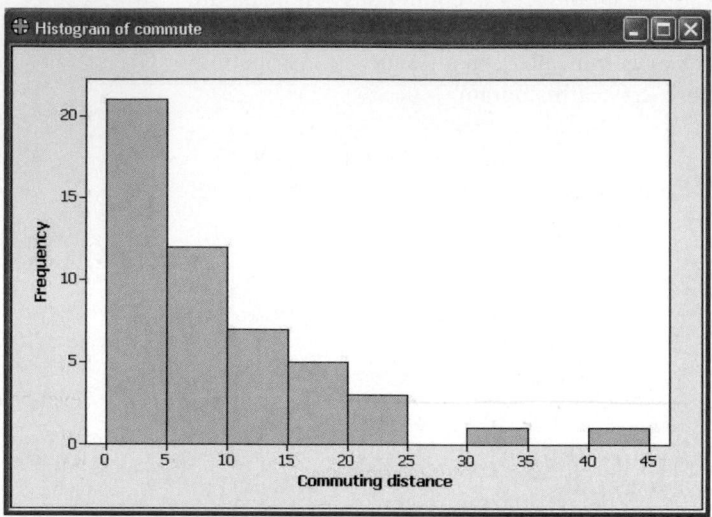

Figure 2.50 Minitab histogram of commuting distances.

CHAPTER 2 SUMMARY

Concept	Page	Notation/Formula/Description
Categorical data set	19	Consists of observations that may be placed into categories.
Numerical data set	19	Consists of observations that are numbers.
Discrete data set	20	The set of all possible values is finite, or countably infinite.
Continuous data set	20	The set of all possible values is an interval of numbers.
Frequency distribution	22, 43	A table used to describe a data set. It includes the class, frequency, and relative frequency (and cumulative relative frequency, if the data set is numerical).
Class frequency	23, 43	The number of observations within a class.
Class relative frequency	23, 43	The proportion of observations within a class: class frequency divided by total number of observations.
Bar chart	24	A graphical representation of a frequency distribution for categorical data with a vertical bar for each class.
Pie chart	25	A graphical representation of a frequency distribution for categorical data with a slice, or wedge, for each class.
Stem-and-leaf plot	34	A graph used to describe numerical data. Each observation is split into a stem and a leaf.
Class cumulative relative frequency	43	The proportion of observations within a class and every class before it: the sum of all the relative frequencies up to and including the class.
Histogram	45	A graphical representation of a frequency distribution for numerical data.
Density histogram	48	A graphical representation of a frequency distribution for numerical data containing class intervals of unequal width.
Unimodal distribution	50	A distribution with one peak.
Bimodal distribution	50	A distribution with two peaks.
Multimodal distribution	50	A distribution with more than one peak.

Symmetric distribution	50	A distribution with a vertical line of symmetry.
Positively skewed distribution	50	A distribution in which the upper tail extends farther than the lower tail.
Negatively skewed distribution	50	A distribution in which the lower tail extends farther than the upper tail.
Normal curve	51	The most common distribution, a bell-shaped curve.

CHAPTER 2 EXERCISES

Applications

2.72 Business and Management A laborshed is a region from which an employment center draws its workforce. In order to understand the potential work-force in a laborshed, the Walker County Development Authority in Alabama sampled residents and reported the data in the following table.

Employment status	Frequency
Employed (white collar)	125
Employed (blue collar)	200
Unemployed	30
Homemaker	50
Retired	95

a. Add a relative frequency column to the table.
b. Construct a bar chart and a pie chart for these data.

2.73 Fuel Consumption and Cars The coefficient of drag (C_d) is a measure of a car's aerodynamics. This unitless number is directly related to the speed of the car, overall performance, and miles per gallon. A low coefficient of drag indicates good performance. A random sample of new automobiles was examined, and the coefficient of drag was computed. The results are given on the data CD and book's web site.
a. Construct a stem-and-leaf plot for these data.
b. Use the plot in part (a) to describe the distribution in terms of shape, center, and variability.

2.74 Public Policy and Political Science A consulting company conducted a survey of U.S. residents for the U.S. Department of Commerce Bureau of Economic Analysis. Each resident was asked to indicate what he or she believes to be the most important social issue. The results are given in the following table.

Social issue	Frequency
Housing	245
Transportation	112
Health Care	153
Education	71
Food	133
Other	306

(*Source*: Bureau of Economic Analysis.)

a. Find the relative frequency for each category.
b. Construct a pie chart for these data.
c. What proportion of people believe housing or transportation is the most important social issue?

d. What proportion of people believe education is not the most important social issue?

2.75 Physical Sciences Construction equipment used to build homes, businesses, and roads (for example, cranes, backhoes, and front loaders) can be exceptionally loud. The peak noise level (in dBA) measured 50 feet away from several construction related machines are given on the data CD and book's web site.[8]
a. Construct a frequency distribution for these data.
b. Draw the corresponding histogram.
c. What proportion of construction equipment had a peak noise level below 80 dBA?
d. What proportion of construction equipment have peak noise levels of at least 90 dBA?

2.76 Technology and Internet Many computer sellers and most software vendors maintain help lines for customers. A random sample of the duration (in minutes) of technical support calls to Hewlett-Packard was obtained, and the resulting stem-and-leaf plot is given below.

0	11223344555566678888999	
1	00012222223335668999	
2	012334556678	
3	000123478	
4	334468	
5	125	
6	15	Stem = tens
7	7	Leaf = ones

a. Describe the shape of this distribution of the duration of technical support calls.
b. Use the plot to construct a frequency distribution using the class intervals 0–5, 5–10, 10–15, etc.
c. What proportion of support calls last less than 15 minutes?
d. If a call lasts at least 25 minutes, a supervisor monitors the conversation. What proportion of calls were monitored?

2.77 Technology and Internet Many police departments have been experimenting with and implementing state-of-the-art 9-1-1 equipment. This equipment is designed to allow a faster response time without voice contact. Caller information is displayed on a monitor, printed, and then processed. In order to compare the two procedures, a random sample of police response times (in minutes) was obtained. The data are given on the data CD and book's web site.
a. Construct a back-to-back stem-and-leaf plot for these data.
b. Use the plot in part (a) to describe any similarities and/or differences between the distributions.
c. Based on the plot in part (a), which procedure is better? Justify your answer.

2.78 Manufacturing and Product Development Microwave ovens are often rated by their output power, for example, 900 watts. However, the actual output of a microwave oven tends to decrease with age. If the *actual* output is more than 400 watts below the *rated* output, then service is recommended. A random sample of 5-year-old, 1000-watt-rated microwave ovens was obtained and tested for output. The data are given on the data CD and book's web site.

a. Construct a frequency distribution for these data and draw the corresponding histogram.

b. Based on this random sample, what proportion of 5-year-old, 1000-watt microwave ovens need service?

c. Suppose the performance of these microwave ovens is *graded* by actual output power, according to the following chart.

Power	Grade	Power	Grade
900–1000	Excellent	600–700	Fair
800–900	Very good	500–600	Poor
700–800	Good	0–500	Not serviceable

Classify each power output, construct a frequency distribution by grade, and draw the resulting pie chart.

Extended Applications

2.79 Economics and Finance In 2008, new home sales in the United States were greatly affected by the depressed economy and tight credit. In September, the government announced a huge federal bailout of the banking system in an attempt to unfreeze the credit markets. The partial frequency distribution below shows cumulative number of homes sold in the last 12 months ending September 2008, by price (hundreds of thousands of dollars), in Huntsville, Alabama.

Class	Frequency	Relative frequency	Cumulative relative frequency
000–100			0.3820
100–200			0.7999
200–300			0.9291
300–400			0.9725
400–500			0.9873
500–600			0.9944
600–700			0.9978
700–800			0.9990
800–900			1.0000
Total	**8646**		

(*Source*: Rudolph/Brander Monthly Residential Real Estate Report, July 1, 2008.)

a. Complete the frequency distribution and draw the corresponding histogram.

b. Describe the distribution in terms of shape, center, and variability.

c. What is a typical selling price? Are there any outliers?

d. What proportion of homes sold for at least $300,000?

2.80 Fuel Consumption and Cars Remanufactured parts are common in the automotive industry. In order to ensure quality, Hite Parts Exchange routinely checks the maximum output of rebuilt alternators. Each day a random sample is obtained and the output delivered (in amps) at 2500 rpm is recorded. The results from a recent day are presented in the following table.

Class	Frequency
30.0–32.0	8
32.0–33.0	7
33.0–34.0	10
34.0–34.5	25
34.5–35.0	30
35.0–35.5	40
35.5–36.0	45
36.0–50.0	5
Total	**170**

a. Find the width and the density for each class.

b. Construct a density histogram for these data.

2.81 Medicine and Clinical Studies A common cold usually lasts from 3 to 14 days. Some studies suggest that echinacea, zinc, or vitamin C can prevent colds and/or shorten their duration. In a new study of the effect of vitamin C, patients with colds were randomly assigned to a placebo group or a vitamin C group. The duration of each cold (in days) was recorded, and the data are summarized in the following table.

Duration	Placebo frequency	Vitamin C frequency
3	0	3
4	0	6
5	8	7
6	7	10
7	21	18
8	10	15
9	26	17
10	15	10
11	8	9
12	3	2
13	1	3
14	1	0

a. Use appropriate graphical procedures to compare the placebo and vitamin C data sets.

b. Do the graphs suggest any differences in shape, center, or variability?

c. Is there any graphical evidence to suggest that vitamin C reduced the duration of a cold?

2.82 Fuel Consumption The performance of a gas furnace can be measured by the annual fuel utilization efficiency (AFUE). This number depends on many furnace properties, and is an indication of the proportion of fuel energy delivered as heat energy during an entire heating season. The U.S. Department of Energy (DOE) requires all new gas furnaces to operate at an AFUE of at least 78%.[9] A gas company selected a random sample of customers, carefully tested each furnace, and recorded the AFUE number. The data are given on the data CD and book's web site.

a. Construct a stem-and-leaf plot for these data.

b. Construct a frequency distribution for these data and draw the corresponding histogram.

c. Describe the distribution in terms of shape, center, and variability. Are there any outliers? If so, what are they?

d. Using the frequency distribution in part (a), approximately what proportion of furnaces do not meet the DOE's minimum AFUE requirement?

e. The gas company classifies each AFUE reading according to the following scheme: 90 or above, excellent; at least 80 but below 90, good; at least 70 but below 80, fair; and less than 70, poor. Classify each reading in the table above, and construct a bar chart for these classification data.

Challenge

2.83 Sports and Leisure An *ogive,* or *cumulative relative frequency polygon*, is another type of visual representation of a frequency distribution. To construct an ogive:

• Plot each point (upper endpoint of class interval, cumulative relative frequency) and (lower endpoint of first class, the o).

• Connect the points with line segments.

The following show a frequency distribution and the corresponding ogive. The values to be used in the plot are shown in bold in the table.

Class	Frequency	Relative frequency	Cumulative relative frequency
12–16	8	0.08	**0.08**
16–20	10	0.10	**0.18**
20–24	20	0.20	**0.38**
24–28	30	0.30	**0.68**
28–32	15	0.15	**0.83**
32–36	10	0.10	**0.93**
36–40	8	0.07	**1.00**
Total	**100**	**1.00**	

Frequency distribution.

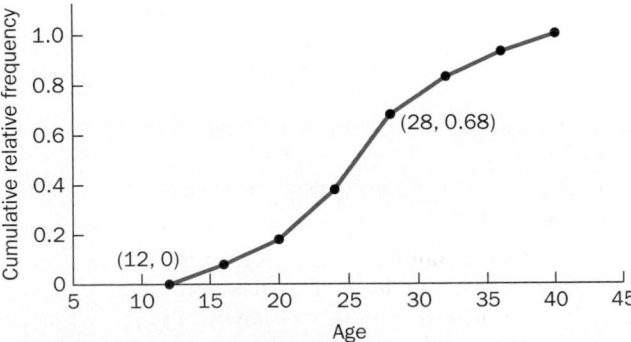

Figure 2.51 Resulting ogive.

A random sample of game scores from a Thursday evening bowling league was obtained, and the data are given on the data CD and book's web site.

a. Construct a frequency distribution for these data.

b. Draw the resulting ogive for these data.

2.84 Public Health and Nutrition A *doughnut graph* is another graphical representation of a frequency distribution for categorical data.

To construct a doughnut graph:

a. Divide a (flat) doughnut (or washer) into pieces, so that each piece (bite of the doughnut) corresponds to a class.

b. The size of each piece is measured by the angle made at the center of the doughnut. To compute the angle of each piece, multiply the relative frequency times 360° (the number of degrees in a whole, or complete, circle).

The manager at a Whole Foods Market obtained a random sample of customers who purchased at least one popular herb (for cooking or medicinal purposes). The following show a frequency distribution and the corresponding doughnut graph.

Herb	Frequency	Relative frequency
Echinacea (1)	25	0.125
Ephedra (2)	15	0.075
Feverfew (3)	10	0.050
Garlic (4)	45	0.225
Ginkgo (5)	40	0.200
Kava (6)	30	0.150
Saw Palmetto (7)	20	0.100
St. John's Wort (8)	15	0.075
Total	**200**	**1.000**

Frequency distribution.

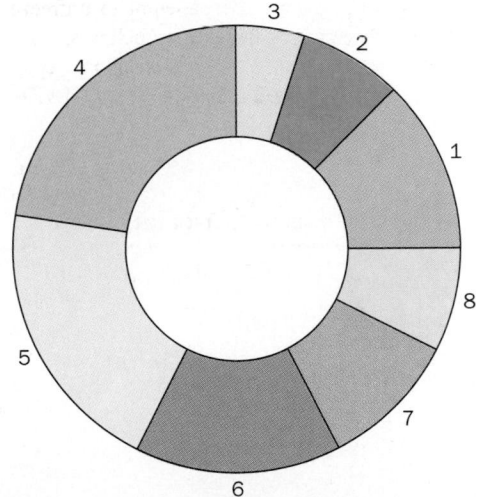

Figure 2.52 Resulting doughnut graph.

A random sample of house fires in Bismarck, North Dakota, was selected and the cause of each was recorded. The resulting data are shown in the following table.

Class	Frequency
Smoking or smoking materials	70
Heating equipment	85
Cooking and cooking equipment	205
Children playing with matches	105
Arson/suspicious	35

a. Find the relative frequency for each class.

b. Draw a doughnut graph for these data.

Numerical Summary Measures

Chapter 3 Challenge

How high are the most thrilling roller coasters?

Roller-coaster enthusiasts all over the world belong to clubs, maintain museums, and travel to conferences at amusement parks. There are approximately 1430 coasters worldwide.

Factors that contribute to an exciting ride include track height, track length, number of hills, maximum velocity, acceleration, centripetal force, wood versus steel, number of loops, the designer, and the builder. The *Pittsburgh Tribune-Review* reported that the tallest roller coaster in the world is at Six Flags Great Adventure in Jackson, New Jersey, and is 151 feet taller than the Statue of Liberty. The *Los Angeles Times* reported (August 15, 2007) that these thrilling rides can raise a person's heart rate to 155 and even change heart rhythm. The table below shows the height (in feet) of 30 randomly selected operating roller coasters from around the world.

66	80	186	131	171	112	178	196	125	189
38	150	60	115	160	185	49	135	125	99
245	170	230	65	115	43	107	61	50	82

(*Source*: Roller-Coaster Database.)

The procedures presented in this chapter will be used to describe the *center* and *variability* of these data, and to search for any unusual observations. The results are presented in the Chapter Challenge Wrap-Up (page 112).

(*Cedar Point Amusement Park/Resort, Sandusky, Ohio*)

Review
- Be familiar with several common tabular and graphical summary procedures.
- Be able to construct a bar chart, pie chart, frequency distribution, stem-and-leaf plot, and histogram.

Preview
- Learn how to compute and interpret common numerical summary measures that describe central tendency, variability, or relative standing.
- Learn how to measure *distance* in statistics.
- Find a five-number summary and construct box plots.

CONTENTS

3.1   MEASURES OF CENTRAL TENDENCY

As we learned in Chapter 2, tabular and graphical procedures provide some very useful summaries of data. However, these techniques are not sufficient for statistical inference. For example, since there are no definite rules for constructing a histogram, two people may construct very different-looking displays for the same data, which could lead to different conclusions. The numerical summary measures presented in this chapter are more precise, combine information in the data into a single number, and allow us to draw a conclusion about an entire population. The two most common types of numerical summary measures describe the *center* and the *variability* of the data.

A numerical summary measure is a single number computed from a sample that conveys a specific characteristic of the entire sample. Measures of *central tendency* indicate where the majority of the data is centered, bunched, or clustered. There are many different measures of central tendency. They all combine information in a sample into a single number, and each has advantages and disadvantages.

In order to properly define and understand numerical summary measures, the following notation will be used.

Note: A capital, or uppercase, X has a very different meaning (introduced in Chapter 5).

x : This stands for a specific, fixed observation on a variable. In general, lowercase letters are used to represent observations on a variable; y and z are also commonly used.

n : This is usually used to denote the number of observations in a data set, or the sample size. If there are two relevant data sets, then m and n may be used to denote their sample sizes. If there are two or more relevant data sets, then n_1, n_2, n_3, . . . may be used to denote their sample sizes.

The three dots, . . . , mean the list continues in the same manner.

$x_1, x_2, x_3, \ldots, x_n$: A set of fixed observations on a variable. The subscripts indicate the order in which the observations were selected, not magnitude. For example, x_5 is the fifth observation drawn from a population, not the fifth largest.

$\sum_{i=1}^{n} x_i = x_1 + x_2 + \cdots + x_n$: This is an example of summation notation, often used to write long mathematical expressions more concisely. Here, the sum of n observations can be written more compactly by using the notation on the left side. Σ is the Greek capital letter sigma; i is the *index of summation*; 1 is the *lower bound*; and n is the *upper bound*. In order to make the notation more compact and less threatening, we will usually omit the subscript $i = 1$ and superscript n. Unless specifically indicated, each summation applies to all values of the variable. For example, the following notation is used to represent the sum of each squared observation: $\sum x_i^2 = x_1^2 + x_2^2 + \cdots + x_n^2$

The following example illustrates the use of this notation and some of the computations used throughout this text.

Example 3.1 Sum Practice   Suppose $x_1 = 5$, $x_2 = 9$, $x_3 = 12$, $x_4 = -6$, $x_5 = 17$, and $x_6 = -2$. Compute the following sums.

a. $\left(\sum x_i \right)^2$ **b.** $\sum x_i^2$ **c.** $\sum (x_i - 7)^2$

SOLUTION

In each case, i is the index of summation, 1 is the lower bound, and 6 is the upper bound. Apply the definition of summation notation to each expression.

a. In words, expression (a) says add all of the observations, and square the result.

$$\left(\sum x_i \right)^2 = (x_1 + x_2 + x_3 + x_4 + x_5 + x_6)^2 \qquad \text{Expand summation notation.}$$

$$= [5 + 9 + 12 + (-6) + 17 + (-2)]^2 \qquad \text{Use given data.}$$

$$= (35)^2 = 1225 \qquad \text{Add, and square the sum.}$$

b. In words, expression (b) says square each observation, and add the resulting values.

$$\sum x_i^2 = x_1^2 + x_2^2 + x_3^2 + x_4^2 + x_5^2 + x_6^2 \qquad \text{Expand summation notation.}$$
$$= (5)^2 + (9)^2 + (12)^2 + (-6)^2 + (17)^2 + (-2)^2 \qquad \text{Use given data.}$$
$$= 25 + 81 + 144 + 36 + 289 + 4 \qquad \text{Square each observation.}$$
$$= 579 \qquad \text{Add.}$$

c. In words, expression (c) says subtract 7 from each observation, square each difference, and add the resulting values.

$$\sum (x_i - 7)^2$$
$$= (x_1 - 7)^2 + (x_2 - 7)^2 + (x_3 - 7)^2 + (x_4 - 7)^2 + (x_5 - 7)^2 + (x_6 - 7)^2$$
$$\text{Expand summation notation.}$$
$$= (5 - 7)^2 + (9 - 7)^2 + (12 - 7)^2 + (-6 - 7)^2 + (17 - 7)^2 + (-2 - 7)^2$$
$$\text{Use given data.}$$
$$= (-2)^2 + (2)^2 + (5)^2 + (-13)^2 + (10)^2 + (-9)^2 \qquad \text{Compute each difference.}$$
$$= 4 + 4 + 25 + 169 + 100 + 81 \qquad \text{Square each difference.}$$
$$= 383 \qquad \text{Add.}$$

The most common measure of central tendency is the **sample**, or **arithmetic**, **mean**.

DEFINITION

The **sample (arithmetic) mean**, denoted $\bar{x}$, of the n observations $x_1, x_2, \ldots, x_n$ is the sum of the observations divided by n. Written mathematically, the sample mean is

$$\bar{x} = \frac{1}{n} \sum x_i = \frac{x_1 + x_2 + \cdots + x_n}{n} \qquad (3.1)$$

💡 **ILLUMINATING THE CONCEPTS**

1. The notation $\bar{x}$ is used to represent the sample mean for a set of observations denoted by $x_1, x_2, \ldots, x_n$. Similarly, $\bar{y}$ would represent the sample mean for a set of observations denoted by $y_1, y_2, \ldots, y_n$.

2. The **population mean** is denoted by μ, the Greek letter mu. ■

Example 3.2 AP® Courses   Advanced Placement (AP) courses offer high school students the opportunity to experience college-level material and to prepare for the AP exams, usually given in May. Depending on the exam score, a student may earn college credit. An article in *The Boston Globe* (July, 2005) listed the number of AP courses offered at each high school in Massachusetts. The values for 12 schools are given in the table below.

5	7	8	12	3	15	7	6	4	5	5	1

Find the sample mean number of AP courses per school.

SOLUTION

Use Equation 3.1 to find the sample mean.

Add all the numbers and divide by $n = 12$.

$$\bar{x} = \frac{1}{12}\sum x_i = \frac{1}{12}(x_1 + x_2 + \cdots + x_{12})$$

$$= \frac{1}{12}(5 + 7 + 8 + 12 + 3 + 15 + 7 + 6 + 4 + 5 + 5 + 1)$$

$$= \frac{1}{12}(78) = \boxed{6.5}$$

Figures 3.1 and 3.2 show two ways to find the sample mean using the TI-84.

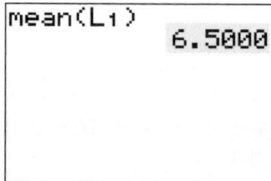

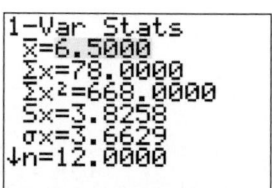

Figure 3.1 The sample mean may be found using the [LIST]; MATH function mean.

Figure 3.2 The sample mean is also part of the output from the 1-Var Stats function.

The sample mean is a *balance point*, or *center of mass*. Take a handful of marbles, all weighing the same. For each observation, place a marble at the corresponding point on a number line. If a fulcrum is placed at the sample mean on the number line, then the line will balance. See Figure 3.3.

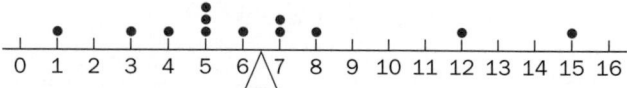

Figure 3.3 This figure illustrates the sample mean as a balance point for the data in Example 3.2.

💡 **ILLUMINATING THE CONCEPTS**

1. $\bar{x}$ is a *sample characteristic*. It describes the center of a *specific* collection of data. There is no set rule to determine the number of included decimal places. Often, at least one extra decimal place to the right is used to write the result; then the sample mean has one more decimal place than the original data values.

2. The sample mean is *an average*. There are many other averages, for example, the geometric mean, the harmonic mean, a weighted mean, the median, and the mode. People usually associate *the* average with the sample mean.

3. μ is a *population characteristic*. It describes the center of an entire population. If the population happens to be of finite size N, then μ is the sum of all the values divided by N. Most populations of interest are infinite, or at least very large and therefore μ is an unknown constant that cannot be measured. It seems reasonable to use $\bar{x}$ to estimate and draw conclusions about μ.

4. The population mean μ is a fixed constant. $\bar{x}$ varies from sample to sample. It is reasonable to think that two sample means computed using samples from the same population should be close, but different. ■

If a data set contains outliers—observations *very far away from the rest*—then the sample mean may not be a very good measure of central tendency. An outlier has lots of influence on the sample mean and tends to pull the mean in its direction. Example 3.3 shows how an outlier can affect the sample mean.

Example 3.3 AP® Courses (Modified) Modify the data in Example 3.2: suppose one high school in Massachusetts offers 27 AP courses, not 12. So the data set is now

| 5 | 7 | 8 | 27 | 3 | 15 | 7 | 6 | 4 | 5 | 5 | 1 |

The observation 27 is an obvious outlier. The new sample mean is

$$\bar{y} = \frac{1}{2}(5 + 7 + 8 + 27 + 3 + 15 + 7 + 6 + 4 + 5 + 5 + 1)$$

$$= \frac{1}{12}(93) = 7.75$$

Since $\bar{x} = 6.5$, $\bar{y} > \bar{x}$. The sample mean is pulled in the direction of the outlier and therefore is not an adequate measure of central tendency. Only three observations are greater than the mean.

The **sample median** is another measure of central tendency that is not as sensitive to outlying values. ●

$\tilde{x}$ is read as "x tilde."

DEFINITION

The **sample median**, denoted $\tilde{x}$, of the n observations $x_1, x_2, \ldots, x_n$ is the *middle number* when the observations are arranged in order from smallest to largest.

1. If n is odd, the sample median is the single middle value.

2. If n is even, the sample median is the mean of the two middle values.

💡 **ILLUMINATING THE CONCEPTS**

1. The median divides the data set into two parts, so that half of the observations lie below and half lie above the median.

2. There is only one calculation necessary in order to find the median (no calculations are needed if n is odd). Put the observations in ascending order of magnitude (*not* the order in which the observations were selected), and find the middle value.

3. Similarly, $\tilde{y}$ would represent the sample median for a set of observations denoted by $y_1, y_2, \ldots, y_n$.

4. The **population median** is denoted by $\tilde{\mu}$. ∎

Example 3.4 Median Calculations   The following three examples show how to find the median under various circumstances and show the effect of an outlying value. The observations are already arranged in order from smallest to largest.

Observations	Median
a. 10 11 14 16 17	There are $n = 5$ observations. The middle number is in the third position. $\tilde{x} = 14$.
b. 10 11 14 16 57	There are still $n = 5$ observations. The middle number is in the third position, and $\tilde{x} = 14$. The outlier 57 does not affect the median.
c. 10 11 14 16 17 20	There are $n = 6$ observations. There is no single middle value. The median is the mean of the observations in the third and fourth positions. $\tilde{x} = \frac{1}{2}(14 + 16) = 15$. ●

Example 3.5 Real Estate Prices   A Phoenix, Arizona real estate office sold 12 homes during the past week. The sale prices are given in the table below. Find the median sale price.

137,000	145,500	117,900	275,500	125,000	97,900
151,000	95,400	131,400	147,000	110,200	105,700

(*Source*: Arizona Real Estate Center.)

SOLUTION

STEP 1   Arrange the observations in order.

95,400   97,900   105,700   110,200   117,900   125,000   131,400   137,000   145,500   147,000   151,000   275,500

STEP 2   There are $n = 12$ observations. The median is the mean of the two middle values (in the sixth and seventh positions).

$$\tilde{x} = \frac{1}{2}(125,000 + 131,400) = \boxed{128,200}$$

Figures 3.4 and 3.5 show technology solutions:

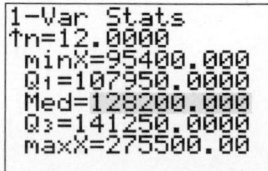

```
median(L1)
         128200.0000
```

```
1-Var Stats
↑n=12.0000
 minX=95400.000
 Q1=107950.0000
 Med=128200.000
 Q3=141250.0000
 maxX=275500.00
```

Figure 3.4 The sample median may be found using the [LIST]; MATH function median

Figure 3.5 The sample mean is also part of the output from the 1-Var Stats function (second screen, symbol Med). ●

 Try the Mean and Median statistical applet on the book companion web site www.whfreeman.com/ introductorystat.

Recall: A histogram consists of rectangles drawn above each class with height proportional to frequency or relative frequency. We draw a curve along the tops of the rectangles to *smooth out* the histogram and display an enhanced graphical representation of the distribution.

💡 ILLUMINATING THE CONCEPTS

1. In general, the sample mean is not equal to the sample median, $\bar{x} \neq \tilde{x}$. If the distribution of the sample is symmetric, then $\bar{x} = \tilde{x}$. If the sample distribution is approximately symmetric, then $\bar{x} \approx \tilde{x}$.

2. In general, the population mean is not equal to the population median, $\mu \neq \tilde{\mu}$. If the distribution of the population is symmetric, then $\mu = \tilde{\mu}$.

3. The relative positions of $\bar{x}$ and $\tilde{x}$ suggest the shape of a distribution. The *smoothed histograms* in Figures 3.6–3.8 illustrate three possibilities:
   a. If $\bar{x} > \tilde{x}$, the distribution of the sample is positively skewed, or skewed to the right (Figure 3.6).
   b. If $\bar{x} \approx \tilde{x}$, the distribution of the sample is approximately symmetric (Figure 3.7).
   c. If $\bar{x} < \tilde{x}$, the distribution of the sample is negatively skewed, or skewed to the left (Figure 3.8).

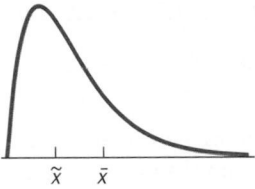

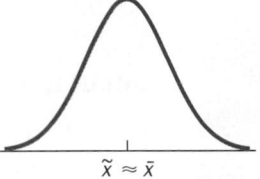

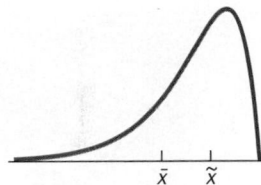

Figure 3.6 Positively skewed distribution.

Figure 3.7 Approximately symmetric distribution.

Figure 3.8 Negatively skewed distribution. ■

Since the sample mean is extremely sensitive to outliers, and the sample median is very insensitive to outliers, it seems reasonable to search for a compromise measure of central tendency. A **trimmed mean** is moderately sensitive to outliers.

DEFINITION

A **$100p\%$ trimmed mean**, denoted $\bar{x}_{\text{tr}(p)}$, of the n observations $x_1, x_2, \ldots, x_n$ is the sample mean of the *trimmed* data set.

1. Order the observations from smallest to largest.

2. Delete, or trim, the smallest $100p\%$ and the largest $100p\%$ of the observations from the data set.

3. Compute the sample mean for the remaining data.

$100p$ is the **trimming percentage**, the percentage of observations deleted from *each end* of the ordered list.

 ILLUMINATING THE CONCEPTS

1. We compute a trimmed mean by deleting the smallest and largest values, which are possible outliers. Some statisticians believe that deleting any data is a bad idea because every observation contributes to the big picture.

2. A **$100p\%$ trimmed mean** is computed by deleting the smallest $100p\%$ and the largest $100p\%$ of the observations. Therefore, $2(100p)\%$ of the observations are removed.

3. There is no set rule for determining the value of p. It seems reasonable to delete only a few observations, and to select p so that np (the number of observations deleted from each end of the ordered data) is an integer.

4. Here is a specific example using the notation: $\bar{x}_{tr(0.05)}$ is a $(100)(0.05) = 5\%$ trimmed mean. 10% of the observations are thrown out. ■

(David Woolley/Getty Images)

Example 3.6 Overtime and Stress According to an article in *The New Yorker*, Americans spend more time at their jobs than workers in Europe.[1] Dr. Paul Landsbergis, an epidemiologist at Mt. Sinai Medical Center, studies job stress, and he warns that too many overtime hours may increase the chance of heart disease. Suppose the following December overtime hours for tellers at the Kaw Valley State Bank and Trust Company in Topeka, Kansas were obtained. Find a 10% trimmed mean.

0.2 0.8 1.5 1.5 1.6 1.7 1.7 1.8 2.0 2.0 2.2 2.5 2.7 2.7 3.0 3.0 3.2 3.5 4.0 5.0

SOLUTION

STEP 1 The trimming percentage is 10%. $p = 10/100 = 0.10$. Find the number of observations to delete from each end of the ordered list.

There are $n = 20$ observations.

$np = (20)(0.10) = 2$ Trim two observations from each end.

Note, np may not be an integer. Computer software packages have algorithms for dealing with this problem.

STEP 2 The resulting data set is

~~0.2~~ ~~0.8~~ 1.5 1.5 1.6 1.7 1.7 1.8 2.0 2.0 2.2 2.5 2.7 2.7 3.0 3.0 3.2 3.5 ~~4.0~~ ~~5.0~~

A technology solution:

B	C
2.2875	=TRIMMEAN(A1:A20,0.2)

Figure 3.9 Calculation of a trimmed mean using Excel.

STEP 3 Find the sample mean for the remaining data. Since there are 16 observations remaining, use $n = 16$.

$$\bar{x}_{tr(0.10)} = \frac{1}{16}(1.5 + 1.5 + 1.6 + \cdots + 3.0 + 3.2 + 3.5) = \frac{1}{16}(36.6) = 2.29$$

2.29 is the 10% trimmed mean.   ●

Another commonly used measure of central tendency is the **mode**.

DEFINITION

The **mode**, denoted M, of the n observations $x_1, x_2, \ldots, x_n$ is the value that occurs most often, or with the greatest frequency.
If all the observations occur with the same frequency, then the mode does not exist.
If two or more observations occur with the same greatest frequency, then the mode is not unique. If there are two modes, the distribution is bimodal; if there are three modes, it is trimodal, etc.

The mode is easy to compute and, intuitively, it does return a reasonable measure of central tendency. For example, consider a bell-shaped distribution. A random sample from this distribution should contain lots of (identical) values near the center. Therefore, the mode should suggest the middle of the distribution (Figure 3.10). For symmetric distributions, the mean, the median, and the mode will be about the same.

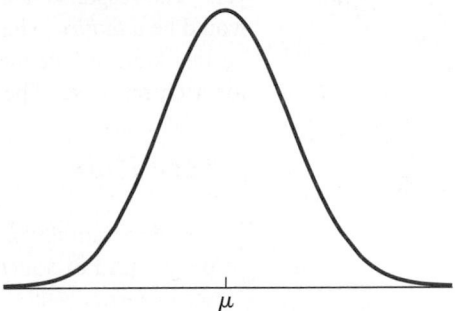

Figure 3.10 We expect the mode M of a sample from this distribution to be near the population mean μ.

💡 ILLUMINATING THE CONCEPTS

1. Sometimes, only a data summary table, or *grouped data*, is available. Let $x_1, x_2, \ldots, x_k$ be a set of (representative) observations with corresponding frequencies $f_1, f_2, \ldots, f_k$. For example, x_7 occurs f_7 times. The total number of observations is $n = \sum f_i$. If the data are grouped, there are corresponding formulas for the (approximate) measures of central tendency defined above.

2. Remember, there are many other *averages*, for example, a weighted mean, the geometric mean, and the harmonic mean. ■

The remaining part of this section describes summary measures for qualitative data.

The natural summary measures for observations on a qualitative variable are simply the frequency and relative frequency of occurrence for each category. We have already done this! Recall Example 2.4 in which 25 cruise ships were randomly selected, and the destination of each recorded. Each response is categorical (destination), and the data were summarized in a table listing only category, frequency of occurrence for each category, and relative frequency of occurrence for each category.

Suppose the commuter students at a small college were asked to complete a survey to identify the make of car used to drive to school. Numerical summary measures for this categorical variable should include frequencies and relative frequencies, or proportions, as shown in the following table. (The cumulative relative frequency is only used for numerical data sets, and doesn't really make sense here since there is no natural ordering.)

Category	Frequency	Relative frequency
Buick	137	0.0938
Chevrolet	288	0.1971
Ford	202	0.1383
Honda	336	0.2300
Mazda	175	0.1198
Saturn	323	0.2211
Total	**1461**	**1.0001**

A dichotomous or Bernoulli variable is a special categorical variable that has only two possible responses. One response is often associated with, or called, a *success*, denoted S, and the other response is called a *failure*, denoted F. The two possible actual responses are ignored. For example, suppose a medical researcher selects children at random and asks each whether they have had an ear infection within the past year. The response had an ear infection might be a *success*, and had no ear infection would be a *failure*. The same numerical measures are used to summarize observations on this kind of categorical variable: frequency and relative frequency of occurrence for each response. The relative frequency of successes has a special name.

DEFINITION

For observations on a categorical variable with only two responses, the **sample proportion of successes**, denoted $\hat{p}$, is the relative frequency of occurrence of successes:

$$\hat{p} = \frac{\text{Number of } S\text{'s in the sample}}{\text{Total number of responses}} = \frac{\text{N}(S)}{n}. \tag{3.2}$$

$\hat{p}$ is read as "*p* hat."

The symbol *p* is used in notation to represent several quantities. It represents the population proportion of successes and appears in the sample proportion of successes, and in the definition of the trimmed mean. The context in which the notation is used implies the appropriate concept.

 ILLUMINATING THE CONCEPTS

1. The **population proportion of successes** is denoted by p.

2. The success response is *not* necessarily associated with a good thing. For example, a researcher may be interested in the proportion of laboratory animals that die when exposed to a certain toxic chemical. A *success* may be associated with the death of an animal.

3. The **sample proportion of successes** $\hat{p}$ can be thought of as a sample mean in disguise. Suppose every S is changed to a 1, and every F to a 0. The sample mean for this new numerical data is

$$\bar{x} = \frac{1}{n}(\text{a sum of 0's and 1's}) = \frac{\text{N}(S)}{n} = \hat{p}. \blacksquare$$

Example 3.7 Seatbelt Checkpoint   In many states it is against the law to drive without a seatbelt fastened. The state police recently established a checkpoint along a heavily traveled road. A success was recorded for a driver wearing a seatbelt, and a failure recorded otherwise. The observations from this checkpoint are given in the table below.

S	S	F	F	S	S	F	S	F	S	F	S	S	S
S	F	S	S	S	S	F	S	S	F	S	F	S	F

The sample contains 28 observations and 18 successes. The sample proportion of successes is

$$\hat{p} = \frac{\text{N}(S)}{n} = \frac{18}{28} = 0.6429.$$

Approximately 64% of the drivers stopped at the checkpoint were wearing their seatbelts. It is reasonable to assume the value of $\hat{p}$ is *close* to the population proportion of successes: in this example, the true proportion of drivers who wear a seatbelt.   ●

TECHNOLOGY CORNER

Procedure: Compute the sample mean, sample median, a trimmed mean, and the mode.

Reconsider: Example 3.2, page 67, solution, and interpretations.

TI-84 Plus

There are several ways to find the sample mean and the sample median using the graphing calculator. There is no built-in function to compute a trimmed mean nor a sample mode.

1. Enter the data into list L1.
2. Use the command LIST; MATH; mean to compute the sample mean (Figure 3.1, page 68). Use the command LIST; MATH; median to compute the sample median (Figure 3.4, page 70).
3. The function STAT; CALC; 1-Var Stats returns several summary statistics. The sample mean is on the first output screen and the sample median is on the second, denoted by Med. See Figures 3.2 (page 68) and 3.5 (page 70).

Minitab

There are several ways to find the summary statistics using Minitab. In addition to the general Describe command, there are Calc; Column statistics functions, Calc; Calculator functions, and various macros.

1. Enter the data into column C1.
2. Select Stat; Basic Statistics; Display Descriptive Statistics. Enter C1 in the Variables window.
3. Choose the Statistics option button and check the summary statistics Mean, Median, Mode, and Trimmed mean. Note: Minitab computes only a 5% trimmed mean. Other macros allow any percentage.

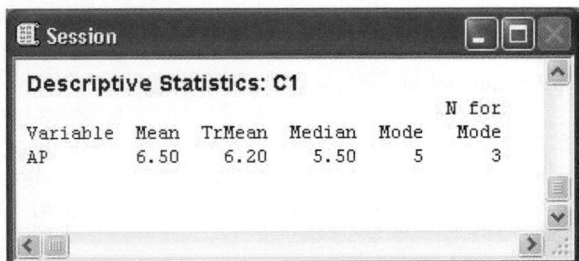

Figure 3.11 Minitab descriptive statistics.

Excel

Excel has built-in functions for these four descriptive statistics. Under the Data tab, choose Data Analysis and then Descriptive Statistics to compute several summary statistics simultaneously.

1. Enter the data into column A.
2. Use the appropriate Excel function to compute the sample mean, sample median, trimmed mean, and mode. Note: the second argument in TRIMMEAN is the total proportion of data trimmed. Excel rounds the number of trimmed observations down to the nearest multiple of two.

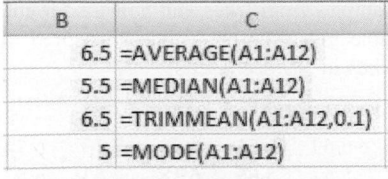

Figure 3.12 Excel descriptive statistics.

SECTION 3.1 EXERCISES

Practice

3.1 Compute each summation using the following random sample.

$x_1 = -15$, $x_2 = 6$, $x_3 = 40$, $x_4 = 13$, $x_5 = 38$.

a. $\sum x_i$ b. $\sum x_i^2$ c. $\sum (x_i - 10)$

d. $\sum (x_i - 5)^2$ e. $\sum (2x_i)$ f. $2\sum x_i$

3.2 Suppose the following random sample is obtained.

43.3	52.7	67.7	52.1	54.7
54.1	46.7	47.2	48.5	45.8

Compute the following sums.

a. $\sum x_i^2$ b. $\sum x_i^3$

c. $\sum (x_i - 50)^2$ d. $\left(\sum x_i\right)^2$

e. $\sum (x_i - 51.28)$ f. $\sum \dfrac{x_i}{7}$

3.3 Compute the mean for each sample with known sum.

a. $\sum x_i = 1057, n = 10$

b. $\sum x_i = 356, n = 27$

c. $\sum x_i = 250.5, n = 36$

d. $\sum x_i = 1.355, n = 11$

e. $\sum x_i = -37.4, n = 15$

f. $\sum x_i = 496.81, n = 28$

3.4 Find the position, or location, of the sample median in an ordered data set of size n.

a. $n = 22$

b. $n = 37$

c. $n = 117$

d. $n = 64$

3.5 Find the sample mean and the sample median for each data set.

a. 5, 3, 7, 9, 11, 5, 6, 7, 7

b. −7, 10, 25, 22, 36, −24, 0, 1, 12, 9, −11

c. 5.4, 3.3, 6.0, 10.1, 13.6, 7.7, 16.6, 28.9, 4.6

d. −103.7, −110.35, −109.1, −99.7, −115.6

3.6 Consider the modified data set in Example 3.3, including the school with 27 AP courses.

5	7	8	27	3	15	7	6	4	5	5	1

Find the sample median. Note that this summary statistic is a better measure of central tendency than the sample mean for this data set. Why?

3.7 Use the values of the sample mean and the sample median to determine whether the distribution is symmetric, skewed to the left, or skewed to the right.

a. $\bar{x} = 37$, $\tilde{x} = 49$

b. $\bar{x} = 63.5$, $\tilde{x} = 62.75$

c. $\bar{x} = -37$, $\tilde{x} = -16$

d. $\bar{x} = -12.56$, $\tilde{x} = 12.56$

3.8 Compute the indicated trimmed mean for each data set.

a. {24, 36, 26, 30, 28, 35, 33, 33, 34, 27}, $\bar{x}_{tr(0.10)}$.

b. {72, 76, 76, 77, 85, 76, 80, 86, 62, 70}, $\bar{x}_{tr(0.20)}$.

c. {182, 169, 180, 166, 173, 101, 188, 124, 182, 137, 100, 137, 118, 111, 137, 181, 189, 130, 168, 133}, $\bar{x}_{tr(0.20)}$.

d. {5.5, 7.5, 7.3, 6.4, 5.3, 9.5, 7.2, 5.8, 7.0, 6.7, 9.0, 8.1, 8.4, 5.8, 5.4, 7.2, 7.4, 7.5, 5.9, 7.5}, $\bar{x}_{tr(0.15)}$.

3.9 Find the mode for each data set, if it exists.

a. 3, 5, 6, 7, 3, 4, 6, 6, 8, 11, 13, 2, 1

b. −17, −10, 0, 3, −5, 4.3, 12, 0, 5, −2.1, 1.7, −7

c. 6.6, 7.3, 5.2, 6.2, 8.3, 9.8, 4.1, 3.7

3.10 Find the sample proportion of successes for each data set.

a. S, F, S, F, F, F, F, F, S, S, S, F, F, S

b. F, S, S, F, S, F, F, S, S, S, S, S, S, S, S, S, F, S, S, S, S, S

c. S, F, S, F, F, F, F, F, S, S, S, S, F, F, S, F, F, S, F, S, S, S, S, F, F, F, F, F, F, F, S, S, F, S, F, F

Applications

3.11 **Travel and Transportation** Tractor trailers tend to exceed the speed limit (65 mph) on one downhill stretch of Route 80 in Pennsylvania. Using a radar gun, the following tractor trailer speeds (in mph) were observed.

81	66	67	69	79	62	70	73	67	60	61
67	74	65	77	74	64	71	64	67	61	

a. Find the sample mean, $\bar{x}$.

b. Find the sample median, $\tilde{x}$.

c. What do your answers in parts (a) and (b) suggest about the shape of the distribution of speeds?

3.12 **Biology and Environmental Science** The data CD and book's web site contain a table that lists the 2007 wheat yield (in Hg/ha, hectograms per hectare) for several countries.[2]

a. Find the sample mean and the sample median for these data.

b. What do the summary statistics in part (a) suggest about the shape of the distribution of wheat yield?

3.13 **Biology and Environmental Science** The Massachusetts Fly Fishing Association is concerned that acid rain from Midwest factories is adversely affecting lakes. The pH scale is used to measure acids and bases. It ranges from 0 to 14. A pH of 7 is neutral, while values less than 7 indicate greater acidity. Citizen volunteers collect and analyze water samples from lakes and streams around the state. The following table gives the pH values measured in 2008 for selected water bodies in Massachusetts.

6.50	6.22	7.94	6.16	6.69	6.53	4.48	7.13	6.54
6.39	7.33	7.37	6.26	5.67	5.66	6.18	7.53	6.57

(*Source*: Acid Rain Monitoring Project, University of Massachusetts, Amherst.)

a. Find the sample mean and the sample median.

b. Suppose the last observation was 10.57 instead of 6.57. Find the sample mean and sample median for this revised data set. Explain how this change in the data has affected the mean and median found in part (a).

3.14 Fuel Consumption and Cars The data CD and book's web site contain a table that lists the carbon footprint (tons of carbon dioxide emitted annually) for 30 cars of the 2009 model year.[3]

a. Find the sample mean and sample median for this data.

b. A certain consumer group considers any automobile that has CO_2 emissions less than 6.5 a success (not harmful to the environment). Find the sample proportion of successes.

3.15 Education and Child Development The Math SAT scores for all students in an introductory statistics class at Bemidji State University are given on the data CD and book's web site.

a. Find the sample mean and the sample median.

b. Find a 5% trimmed mean.

c. Using these three numerical summary measures, describe the shape of the distribution.

3.16 Manufacturing and Product Development A random sample of 12-ounce cans of Dr Pepper was obtained from E.M. Heaths supermarket. The exact amount of soda (in ounces) in each can was measured, and the data are given on the data CD and book's web site.

a. Find the sample mean and the sample median.

b. What do the summary statistics in part (a) suggest about the shape of the distribution of the amount of soda in each can?

c. Suppose any amount of 12 ounces or greater is considered a *success*. Find the sample proportion of successes.

3.17 Biology and Environmental Science The 2006 crustaceans and molluscs catch (in tons) for various regions around the world is given on the data CD and book's web site.[4]

a. Find the sample mean and the sample median for this data set.

b. Which statistic is a better measure of central tendency for these data? Justify your answer.

3.18 Sports and Leisure Some critics of Major League Baseball believe the ball is *juiced* (livelier) because it is manufactured to give hitters an advantage. To investigate this claim, a sample of the earned run average (ERA) for American League starting pitchers was obtained. The data are given on the data CD and book's web site.[5]

a. Find the sample mean and the sample median.

b. Suppose the pitcher with the highest ERA plays in Colorado, where the air is thin, and the home runs are many. To eliminate such outliers, find a 2.5% trimmed mean.

c. Find the mode for this data set, if it exists.

3.19 Biology and Environmental Science The water temperature (in degrees Fahrenheit) during the summer of 2008 at several North Atlantic Coast locations is given in the following table.

54	46	54	54	61	61	61	62	56	46
57	56	63	65	63	66	58	50	60	60
66	70	68	68	60	51	61	60	67	72

(*Source*: National Oceanographic Data Center.)

a. Find the sample mean and the sample median.

b. Find a 10% trimmed mean for these data.

c. Find the mode for these data, if it exists.

3.20 Education and Child Development An educational study was designed to compare cooperative learning versus traditional lecture style. Two sections of an introductory statistics class were used. Seven students were randomly selected from each section. The scores on the second test (a 29-item exam) are given in the following table.

| Traditional | 21 | 28 | 25 | 25 | 21 | 19 | 23 |
| Cooperative | 25 | 30 | 28 | 25 | 24 | 24 | 29 |

Which group of students did better *on average*? Justify your answer.[6]

Extended Applications

3.21 Sports and Leisure Competitors in the 2008 Kaiser Permanente National Diving Championship, senior division, performed five dives each. Each dive was scored by seven judges. The data CD and book's web site contain the scores for dives with the same degree of difficulty for various participants.[7]

a. Find the sample mean and the sample median for this data set.

b. Find the mode, if it exists.

c. Multiply each score by the degree of difficulty, 2.7. Find the sample mean for this new data set. How does this sample mean compare with the sample mean found in part (a)?

3.22 Manufacturing and Product Development Snowmobile noise has become a problem in several national parks. Starting Line Products tested the noise level (in decibels, dB) using the SAE J1192 test on several snowmobiles. The data are given in the following table.

| 81.6 | 82.4 | 78.3 | 81.1 | 78.2 | 79.3 | 78.5 | 84.5 | 76.1 |

(*Source:* Mathews, J., Nov. 1, 2005, *What Is Noise?*)

a. Find the sample mean for these data.

b. Some researchers believe that the wind velocity gradient can add as much as 4 dB to each reading. Add 4 dB to each observation in the data set. Compute the new sample mean. How does this compare with the sample mean found in part (a)?

3.23 Travel and Transportation The following table contains the estimated unlinked transit passenger trips (in thousands) for various transit agencies during June 2008.

4059.2	1420.4	3459.7	4241.4	963.3	1602.2
32.4	413.7	7096.6	644.5	875.7	421.4
1793.3	608.2	319.8	3246.4	2554.3	634.3
104.4	1687.8	2.8	979.6	1207.3	39.1
69.8	9.4				

(*Source*: American Public Transportation Association.)

a. Find the sample mean.

b. Use each June observation to estimate the yearly number of trips. That is, multiply each observation by 12. Find the sample mean for this new data set. How does this sample mean compare with the sample mean found in part (a)?

3.24 Manufacturing and Product Development A new quality control program recently started at a Hyundai manufacturing facility. Several times each day, randomly selected panels from a stamping press are inspected for defects. A nondefective panel is a *success* (S). A defective panel is a failure (F) and must be restamped at an additional cost. During a recent inspection, the following 32 observations were recorded.

S	S	S	S	F	S	F	F	S	S	S	S	S	S
S	S	F	S	S	S	S	S	S	S	S	F	S	S
S	S	S	S										

a. Find the sample proportion of successes.
b. Change each S to a 1, and each F to a 0. Find the sample mean for these new data. How does the mean compare with the sample proportion of successes found in part (a)?
c. Suppose 8 additional panels were selected and inspected (for a total of 40 panels). Is it possible for the sample proportion of successes to be 0.9? Why or why not?

3.25 Sports and Leisure The sixth man on a basketball team is usually the first substitution. Often this player receives more playing time than some starters. A random sample of playing times (in minutes) for sixth men from National Basketball Association teams was obtained. The data are given in the following table.

8.9	19.1	23.3	8.1	15.1	21.7	23.2	25.4	26.3
13.6	20.1	8.2	19.2	13.4	10.2	13.3	12.4	

(*Source*: National Basketball Association.)

a. Find the sample mean and the sample median.
b. What do the summary statistics in part (a) suggest about the shape of the distribution of playing time for sixth men?
c. Can you change the maximum observation (26.3) so that the sample mean is equal to the sample median? Why or why not?

3.26 Medicine and Clinical Studies In a random sample of 13 patients with calcaneus bone fractures, the sample mean number of days until fracture healing was $\bar{x} = 37.85$ and the sample median was $\tilde{x} = 40$. Suppose an additional patient is added to the sample so that $x_{14} = 44.5$.
a. Find the sample mean for all 14 patients.
b. Is there any way to determine the sample median for all 14 patients? Explain.

3.27 Fuel Consumption and Cars The estimated oil reserves (in millions of barrels) of four wells are given by

$$x_1 = 1078 \quad x_2 = 5833 \quad x_3 = 10772 \quad x_4 = 7320.$$

a. Find x_5 so that the mean for all five observations is 6883.4.
b. Find x_5 so that the sample mean is equal to the sample median.

3.28 Manufacturing and Product Development A consumer group has tested the drying time for 15 samples of exterior latex paint. The sample mean drying time is 83.8 minutes. What must the 16th drying time be if the 16th observation decreases the mean drying time by 30 seconds? By 1 minute?

3.29 Biology and Environmental Science The beaches along the coast of New Hampshire are famous for chilly waters, even during the hottest summer days. A recent sample of the water temperature on 24 randomly selected summer days was obtained. The following temperatures are in degrees Fahrenheit (°F).

58	58	53	53	59	57	54	61	56	60
57	61	56	55	59	60	55	53	55	58
59	53	59	63						

a. Find the sample mean and the sample median.
b. Convert each temperature to degrees Celsius (°C). Use the formula $C = (F - 32)/1.8$. Find the mean for all the water temperatures in degrees Celsius.
c. What is the relationship between the sample means in parts (a) and (b)?

3.30 Technology and Internet A recent survey of students at Minneapolis North High School included a question about the number of computers at home. The (grouped) data are summarized below.

Number of computers	Frequency of occurrence
0	3
1	27
2	23
3	7
4	3
5	1

Find the sample mean and the sample median number of computers at home.

3.2 MEASURES OF VARIABILITY

Measures of central tendency are only one characteristic of a data set. These numerical summary measures alone are not sufficient to completely describe a sample. It is possible to have two very *different* data sets with (approximately) the same mean (and median). Figures 3.13 and 3.14 show two smoothed histograms to illustrate the problem.

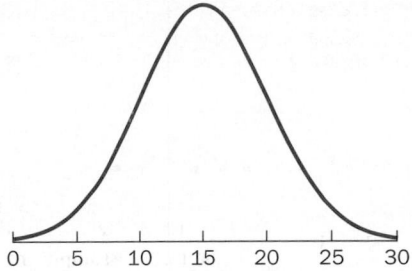

Figure 3.13 Sample 1: $x_1, x_2, \ldots, x_n$. The smoothed histogram suggests a compact distribution.

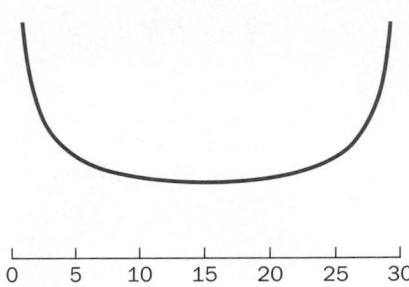

Figure 3.14 Sample 2: $y_1, y_2, \ldots, y_m$. The smoothed histogram suggests that the data are more dispersed, or spread out.

The measures of central tendency (sample mean and sample median) are approximately the same ($\bar{x} \approx \bar{y} \approx 15$ and $\tilde{x} \approx \tilde{y} \approx 15$), but the data in Sample 1 are more compact because more of the data are clustered about the mean $\bar{x} = 15$. In order to describe the difference between the data sets, we need to consider variability.

DEFINITION

The (**sample**) **range**, denoted R, of the n observations $x_1, x_2, \ldots, x_n$ is the largest observation minus the smallest observation. Written mathematically, the range is

$$R = x_{\max} - x_{\min}, \qquad (3.3)$$

where $x_{\max}$ denotes the maximum, or largest, observation, and $x_{\min}$ stands for the minimum, or smallest, observation.

💡 **ILLUMINATING THE CONCEPTS**

1. In theory, the **sample range** does measure, or describe, variability. A data set with a small range has little variability and is compact. A data set with a large range has lots of variability and is spread out.

2. The sample range is used in many quality control applications. For example, a production supervisor may want to maintain small variability in a manufacturing process. The sample range may be used to determine whether the process is still well controlled, or whether there is abnormal variation. ∎

Despite being very easy to compute and a logical measure, the sample range is not adequate for describing variability. The sample range may not accurately represent the variability of a distribution if the maximum and minimum values are outliers.

The sample range for each data in figures 3.13 and 3.14 set is approximately the same: $R \approx 30 - 0 = 30$. In fact, the two data sets have approximately the same mean. Therefore, it is necessary to use a better, more sensitive measure of variability. To derive a more precise measure of variability, consider how far each observation lies from the mean.

A graph may be used to visualize the spread of data and to suggest another measurement. A **dot plot** is a graph that simply displays a dot corresponding to each observation along a number line. The *stacked* dot plot in Figure 3.15 may be used to compare the variability in Sample 1 (x's) versus Sample 2 (y's).

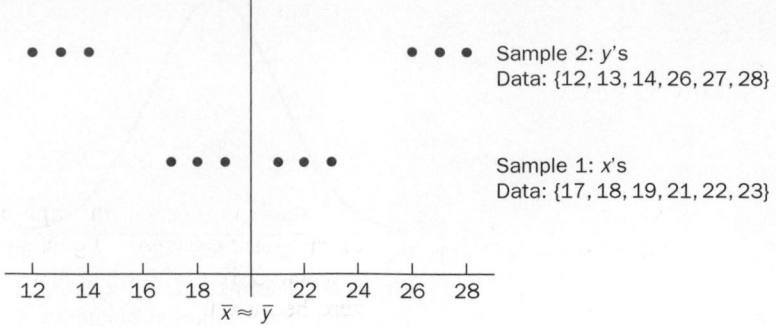

Figure 3.15 Stacked dot plot.

In Sample 1, the data set is compact; each observation is *very close to the mean.* In Sample 2, the data set is more spread out; each observation is *far away from the mean.* This analysis of Figure 3.15 suggests a better measure of variability might include the distances from the mean.

DEFINITION

Given a set of n observations $x_1, x_2, \ldots, x_n$, the **ith deviation about the mean** is $x_i - \bar{x}$.

ILLUMINATING THE CONCEPTS

1. Given a data set, in order to calculate the **ith deviation about the mean**: find $\bar{x}$, then compute the difference $x_i - \bar{x}$. For example, the seventh deviation about the mean is the value $x_7 - \bar{x}$.

2. We usually do not need any one deviation about the mean; all of the deviations about the mean together will be used to find a suitable measure of variability.

3. If the ith deviation about the mean is positive, then the observation is to the right of the mean: if $x_i - \bar{x} > 0$ then $x_i > \bar{x}$.

   If the ith deviation about the mean is negative, then the observation is to the left of the mean: if $x_i - \bar{x} < 0$ then $x_i < \bar{x}$. ■

◖ Here are some ideas for using the deviations about the mean to measure variability.

1. $\sum (x_i - \bar{x})$ (the sum of all the ith deviations about the mean)

   A data set with little variability should have small deviations about the mean, and a data set with lots of variability should have large deviations about the mean. So, it seems reasonable to simply add up all the deviations about the mean. The sum should be small for a data set with little variability and large for a data set with a lot of variability.

   Consider the following data: $\{4, 9, 7, 18, 4\}$.

   The sample mean is $\bar{x} = \dfrac{1}{5}(4 + 9 + 7 + 18 + 4) = 8.4$.

The sum of the ith deviations about the mean is

$$\sum (x_i - \bar{x}) = (4 - 8.4) + (9 - 8.4) + (7 - 8.4) + (18 - 8.4) + (4 - 8.4)$$

Use data and $\bar{x}$.

$$= (-4.4) + (0.6) + (-1.4) + (9.6) + (-4.4) = 0$$

Compute differences and add.

Alas, even though this approach is intuitive, the sum is always zero! For *every* possible data set, the sum of all the deviations about the mean is zero. Problem 3.51 in the Exercises section asks for a proof of this. The sum is zero because there are always some positive deviations and some negative deviations, and they tend to cancel each other out when added together.

So, if we still want to use the deviations about the mean to measure variability, we need to address the problem of positive deviations and negative deviations *canceling out*. In order to make all deviations about the mean positive, we could use absolute value.

2. $\sum |x_i - \bar{x}|$ (the sum of the absolute value of all the ith deviations)

This is actually a pretty good idea. A data set with little variability has small deviations about the mean and small absolute deviations about the mean. The sum should be small positive. Similarly, a data set with lots of variability has large deviations about the mean and large absolute deviations about the mean. The sum should be large positive.

Although this is a reasonable measure of variability, the absolute value function presents some difficult mathematical theory. The absolute value function has many applications in mathematics and statistics, but there is another, more convenient, way to *transform* all of the deviations about the mean to positive values and preserve the information about variability. ▶

DEFINITION

The **sample variance**, denoted s^2, of the n observations $x_1, x_2, \ldots, x_n$ is the sum of the squared deviations about the mean divided by $n - 1$. Written mathematically, the sample variance is

$$s^2 = \frac{1}{n-1} \sum (x_i - \bar{x})^2 \tag{3.4}$$

$$= \frac{1}{n-1}[(x_1 - \bar{x})^2 + (x_2 - \bar{x})^2 + \cdots + (x_n - \bar{x})^2]$$

The **sample standard deviation**, denoted s, is the positive square root of the sample variance. Written mathematically, it is

$$s = \sqrt{s^2} \tag{3.5}$$

ILLUMINATING THE CONCEPTS

1. The **population variance**, a measure of variability for an entire population, is denoted by σ^2, and the **population standard deviation** is denoted by σ, the Greek letter sigma.

2. Just knowing s^2 doesn't seem to say much about variability. If $s^2 = 6$, for example, it is hard to infer anything about variability. However, the **sample**

The sample variance s^2 is often called an average of the squared deviations about the mean, yet we divide the sum of the squared deviations by $n - 1$. While this does not seem correct, dividing by $n - 1$ makes s^2 an unbiased estimator of σ^2. We will see later in the text that an unbiased statistic is, in some sense, a good thing. There are $n - 1$ *degrees of freedom*, a kind of dimension of variability, associated with the sample variance s^2.

variance s^2 *is* a measure of variability and is useful in comparisons. For example, if Sample 1 and Sample 2 have similar units, $s_1^2 = 14$, and $s_2^2 = 10$, then the data in Sample 2 are more compact.

3. The **sample standard deviation** s is used (rather than s^2) in many statistical inference problems. So, if we need to find s (by hand), we need to compute s^2 first, then take the positive square root to find s.

4. The units for the sample standard deviation are the same as those for the original data. And a value of $s = 0$ means there is no variability in the data set.

5. The notation s_x^2 is used to represent the sample variance for a set of observations denoted by $x_1, x_2, \ldots, x_n$. Similarly, s_y^2 would represent the sample variance for a set of observations $y_1, y_2, \ldots, y_n$. ■

Example 3.8 Furniture Delays   Plaza Gallery in Chicago sells home furniture online and is concerned about some of their shipments to customers arriving late. Five days were randomly selected and the number of late shipment complaints were recorded. The observations were 3, 5, 1, 9, 6. Find the sample variance and the sample standard deviation for these data.

SOLUTION

STEP 1   Find the sample mean: $\bar{x} = \dfrac{1}{5}(3 + 5 + 1 + 9 + 6) = \dfrac{1}{5}(24) = 4.8$

STEP 2   Use Equation 3.4 to find the sample variance.

A technology solution:

```
variance(L₁
            9.2000
√(Ans)
            3.0332
```

Figure 3.16 Sample variance and sample standard deviation.

$$s^2 = \frac{1}{4}\left[(3 - 4.8)^2 + (5 - 4.8)^2 + (1 - 4.8)^2 + (9 - 4.8)^2 + (6 - 4.8)^2\right]$$

Use data and $\bar{x}$.

$$= \frac{1}{4}\left[(-1.8)^2 + (0.2)^2 + (-3.8)^2 + (4.2)^2 + (1.2)^2\right]$$

Compute differences.

$$= \frac{1}{4}\left[3.24 + 0.04 + 14.44 + 17.64 + 1.44\right]$$

Square each difference.

$$= \frac{1}{4}(36.8) = \boxed{9.2}$$

Add, divide by 4.

STEP 3   Take the positive square root of the variance to find the standard deviation.

$$s = \sqrt{9.2} \approx \boxed{3.0332}$$

●

Equation 3.4 is the definition of the sample variance and may be used to find s^2, but there is actually a more efficient technique for computing s^2.

DEFINITION

The computational formula for the sample variance is

$$s^2 = \frac{1}{n - 1}\left[\sum x_i^2 - \frac{1}{n}\left(\sum x_i\right)^2\right] \tag{3.6}$$

This is a convenient shortcut method for calculating s^2 without having to find all the deviations about the mean. Suppose $x_1, x_2, \ldots, x_n$ is a set of observations. In order to find s^2, Equation 3.6 says:

1. Find the sum of the *squared* observations: $\sum x_i^2$.

2. Find the sum of the observations: $\sum x_i$.

3. Square the sum of the observations: $(\sum x_i)^2$.

4. Multiply the square of the sum of the observations by $1/n$: $\dfrac{1}{n}(\sum x_i)^2$.

5. Subtract the two quantities, and multiply the difference by $1/(n-1)$:

$$s^2 = \frac{1}{n-1}\left[\sum x_i^2 - \frac{1}{n}(\sum x_i)^2\right].$$

Example 3.9 Furniture Delays (Continued)   Use the computational formula for s^2 to find the sample variance for the data in Example 3.8. The data on late-shipment complaints are 3, 5, 1, 9, 6.

SOLUTION

STEP 1   Find the sum of the squared observations:

$$\sum x_i^2 = 3^2 + 5^2 + 1^2 + 9^2 + 6^2 = 9 + 25 + 1 + 81 + 36 = 152$$

STEP 2   Find the sum of the observations:

$$\sum x_i = 3 + 5 + 1 + 9 + 6 = 24$$

STEP 3   Square this sum and multiply by $1/n$:

$$\frac{1}{5}(\sum x_i)^2 = \frac{1}{5}(24)^2 = 115.2$$

STEP 4   Subtract the two quantities, and multiply by $1/(n-1)$:

$$s^2 = \frac{1}{4}(152 - 115.2) = \frac{1}{4}(36.8) = 9.2 \qquad \text{(The same answer as above.)} \qquad \bullet$$

It can be shown that Equation 3.4 and Equation 3.6 are equivalent. Problem 3.53 in the Exercises section asks for a proof. If you must find a sample variance by hand, then use the computational formula. It has fewer calculations (is more efficient) and is usually more accurate (has less round-off error). In fact, most calculator and computer programs that find the sample variance use the computational formula.

The sample variance is always greater than or equal to zero: $s^2 \geq 0$. This is easy to see by looking at the definition in Equation 3.4. We sum *squared* deviations about the mean (always greater than or equal to zero) and divide by a positive number $(n-1)$. There are two *special cases*.

1. $s^2 = 0$: This occurs if all the observations are the same. If all the observations are equal to c, the mean is c, and all the deviations about the mean are zero. Hence, $s^2 = 0$. This makes sense intuitively also: if all the observations are the same, there is no variability.

2. $n = 1$: This is a strange case, but it can occur. If $n = 1$, the variance is undefined; another way to think of this is that we cannot measure variability. The denominator in Equation 3.4 is zero, and anything divided by zero is undefined.

The sample variance (and the sample standard deviation) can be greatly influenced by outliers. An observation very far away from the rest has a large deviation about the mean and a large squared deviation about the mean, and therefore it contributes a lot to the sum (in the definition of the sample variance). The **interquartile range** is another measure of variability and is resistant to outliers.

Note that the definition for Q_1 and Q_3 involves the median, not the mean.

In smoothed histograms, the area under the curve and above the horizontal axis between two values corresponds to the proportion of observations between those values. Interpreting Figure 3.17: 25% of the observations are between Q_1 and $\tilde{x}$.

DEFINITION

Let $x_1, x_2, \ldots, x_n$ be a set of observations. The **quartiles** divide the data into four parts.

1. The **first (lower) quartile**, denoted Q_1 (Q_L), is the median of the lower half of the observations when arranged in ascending order.

2. The **second quartile** is the median $\tilde{x} = Q_2$.

3. The **third (upper) quartile**, denoted Q_3 (Q_U), is the median of the upper half of the observations when they are arranged in ascending order.

4. The **interquartile range**, denoted IQR, is the difference $IQR = Q_3 - Q_1$.

The quartiles are illustrated in Figure 3.17.

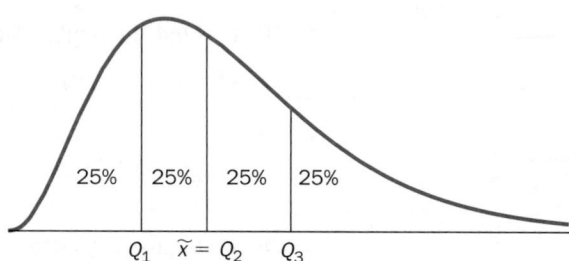

Figure 3.17 Smoothed histogram and quartiles.

ILLUMINATING THE CONCEPTS

There is a very intuitive method for finding the quartiles. Arrange the data in order from smallest to largest. The median, $\tilde{x} = Q_2$, is the middle value. The **first quartile**, Q_1, is the median of the lower half, and the **third quartile**, Q_3, is the median of the upper half. In practice, a more general method is used for locating the *position*, or *depth*, of the first and third quartile (in the ordered data set). ∎

HOW TO COMPUTE QUARTILES

Suppose $x_1, x_2, \ldots, x_n$ is a set of n observations.

1. Arrange the observations in ascending order, from smallest to largest.
2. To find Q_1, compute $d_1 = n/4$.
   a. If d_1 is a whole number, then the depth of Q_1 (position in the ordered list) is $d_1 + 0.5$. Q_1 is the mean of the observations in positions d_1 and $d_1 + 1$ in the ordered list.
   b. If d_1 is not a whole number, round up to the next whole number for the depth of Q_1.
3. To find Q_3, compute $d_3 = 3n/4$.
   a. If d_3 is a whole number, then the depth of Q_3 is $d_3 + 0.5$. Q_3 is the mean of the observations in positions d_3 and $d_3 + 1$ in the ordered list.
   b. If d_3 is not a whole number, round up to the next whole number for the depth of Q_3.

Example 3.10 Pulse Rates   The following 10 observations represent the resting pulse rate for patients involved in an exercise study.

| 68 | 71 | 64 | 58 | 61 | 76 | 73 | 62 | 72 | 66 |

a. Find the first quartile, the third quartile, and the interquartile range.
b. Suppose there were 12 patients in the study, with $x_{11} = 78$ and $x_{12} = 81$. Find the first quartile, the third quartile, and the interquartile range for this modified data set.

SOLUTION

STEP 1   Arrange the observations in order from smallest to largest.

Observation	58	61	62	64	66	68	71	72	73	76
Position	1	2	3	4	5	6	7	8	9	10

STEP 2   Find the depth of the first quartile.

$$d_1 = \frac{n}{4} = \frac{10}{4} = 2.5$$ Since d_1 is not a whole number, round up. The depth of the first quartile is 3.

Q_1 is in the third position in the ordered list.
Using the table above, $Q_1 = \boxed{62}$.

STEP 3   Find the depth of the third quartile.

$$d_3 = \frac{3n}{4} = \frac{(3)(10)}{4} = 7.5$$ Since d_3 is not a whole number, round up. The depth of the third quartile is 8.

Q_3 is in the eighth position in the ordered list.
Using the table above, $Q_3 = \boxed{72}$.

STEP 4   Find the interquartile range $IQR = Q_3 - Q_1$.

$$IQR = 72 - 62 = \boxed{10}$$

STEP 5   Arrange the observations in order from smallest to largest in the modified data set.

Observation	58	61	62	64	66	68	71	72	73	76	78	81
Position	1	2	3	4	5	6	7	8	9	10	11	12

STEP 6   Find the depth of the first quartile.

$$d_1 = \frac{n}{4} = \frac{12}{4} = 3$$ Since d_1 is a whole number, add 0.5. The depth of the first quartile is 3.5.

Q_1 is the mean of the observations in the third and fourth positions in the ordered list.

$$Q_1 = \frac{1}{2}(62 + 64) = \boxed{63}$$

STEP 7   Find the depth of the third quartile.

$$d_3 = \frac{3n}{4} = \frac{(3)(12)}{4} = 9$$ Since d_3 is a whole number, add 0.5. The depth of the third quartile is 9.5.

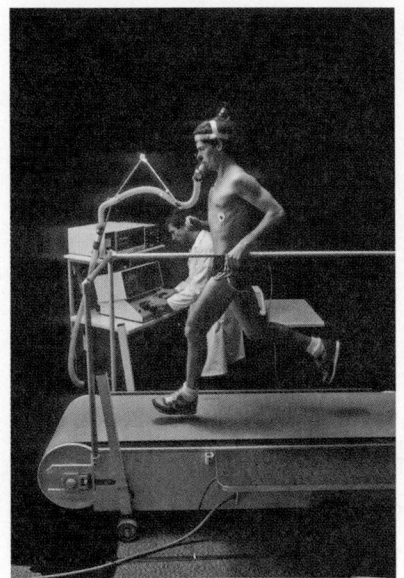

(Tom Tracy Photography/Alamy)

A technology solution:

```
1-Var Stats
↑n=10.0000
 minX=58.0000
 Q₁=62.0000
 Med=67.0000
 Q₃=72.0000
 maxX=76.0000
```

Figure 3.18 1-Var Stats is used to compute the quartiles.

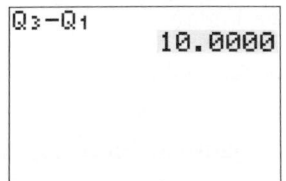

```
Q₃-Q₁
            10.0000
```

Figure 3.19 Compute *IQR* on the Home Screen.

Q_3 is the mean of the observations in the ninth and tenth positions in the ordered list.

$$Q_3 = \frac{1}{2}(73 + 76) = \boxed{74.5}$$

STEP 8 Find the interquartile range.

$$IQR = 74.5 - 63 = \boxed{11.5}$$

A technology solution:

```
1-Var Stats
↑n=12.0000
 minX=58.0000
 Q₁=63.0000
 Med=69.5000
 Q₃=74.5000
 maxX=81.0000
```

Figure 3.20 1-Var Stats is used to compute the quartiles.

 **APPLET** Try the One Variable Statistical Calculator statistical applet on the book web site.

```
variance(L₁)
          51.9697
√(Ans)
           7.2090
```

Figure 3.21 Sample variance and sample standard deviation.

A technology solution:

```
Q₃−Q₁
          11.5000
```

Figure 3.22 Compute *IQR* on the Home Screen.

ILLUMINATING THE CONCEPTS

1. The interquartile range is the length of an interval that includes the middle half (middle 50%) of the data.

2. The interquartile range is not sensitive to outlying values. The lower and/or upper 25% of the distribution can be extreme without affecting Q_1 and/or Q_3. ■

TECHNOLOGY CORNER

Procedure: Compute the sample variance, sample standard deviation, first quartile, third quartile, and interquartile range.

Reconsider: Example 3.10(b), page 85, solutions, and interpretations.

TI-84 Plus

1. Enter the data into list L1.
2. Select $\boxed{\text{LIST}}$; MATH; variance. Take the square root of the variance to find the standard deviation. See Figure 3.21.
3. Select $\boxed{\text{STAT}}$; CALC; 1-Var Stats.
4. The quartiles are displayed on the second output screen. See Figure 3.20. Note: the sample standard deviation is displayed on the first output screen and the value is stored in the statistics variable Sx.
5. Compute the interquartile range on the Home Screen. Use the TI-84 Plus statistics variables that represent the quartiles. See Figure 3.22.

Minitab

1. Enter the data into column C1.
2. Select Stat; Basic Statistics; Display Descriptive Statistics. Enter C1 in the Variables window.
3. Choose the Statistics option button and check the summary statistics Standard deviation, Variance, First quartile, Third quartile, and Interquartile range. Note: Minitab computes quartiles using a slightly different algorithm.

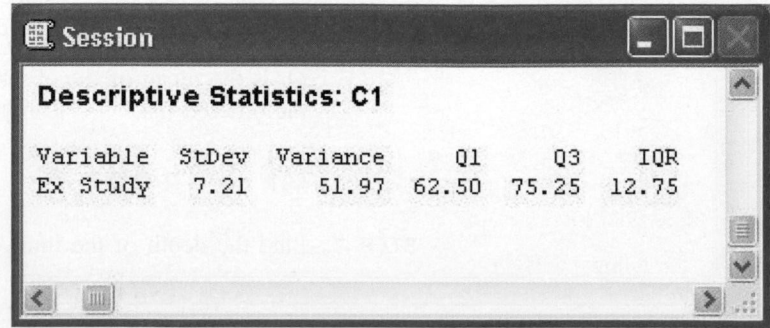

Figure 3.23 Measures of variability computed using Minitab.

Excel ────────────────────────────

1. Enter the data into column A.
2. Use the function STDEV to compute the sample standard deviation, VAR to compute the sample variance, and QUARTILE to compute the first and third quartile. Compute the interquartile range using the results. Note: Excel computes quartiles using a different algorithm.

B	C
7.2090	=STDEV(A1:A12)
51.9697	=VAR(A1:A12)
63.5000	=QUARTILE(A1:A12,1)
73.7500	=QUARTILE(A1:A12,3)
10.2500	=B4-B3

Figure 3.24 Measures of variability computed using Excel.

SECTION 3.2 EXERCISES

Practice

3.31 Find the sample range, sample variance, and sample standard deviation for each data set.
a. {2.7, 6.0, 5.7, 5.4, 4.0, 3.1, 6.6, 5.7, 6.1, 3.0}
b. {18.5, 23.5, 15.7, 15.7, 36.3, 20.8, 21.1, 20.2, 26.8, 19.9, 17.6, 17.5, 21.5, 22.4, 25.7}
c. {23.94, −31.04, 37.09, 22.64, −61.23, 1.59, 23.09, 1.14}
d. {0.13, 0.96, −0.50, 0.10, −1.65, −0.14, 1.43, −2.57, −1.28, −0.24, −0.90, −1.27, 1.53, 3.00, −1.28, 1.04, −0.90, 2.44, 1.70, 3.13}

3.32 Compute the sample variance and the sample standard deviation for each sample with known sums.
a. $\sum x_i = 1219.29$, $\sum x_i^2 = 58945.1$, $n = 30$
b. $\sum x_i = 35.2918$, $\sum x_i^2 = 7748.98$, $n = 17$
c. $\sum x_i = 218.291$, $\sum x_i^2 = 3615.96$, $n = 15$
d. $\sum (x_i - \bar{x})^2 = 49.784$, $n = 21$

3.33 Find the depth of the first quartile and the third quartile in an ordered data set of size n.
a. $n = 60$ b. $n = 37$ c. $n = 100$ d. $n = 48$

3.34 Find the first quartile, the third quartile, and the interquartile range for each data set.
a. {20, 17, 37, 33, 29, 50, 20, 33}
b. {13.1, 7.8, 11.9, 2.3, 6.7, 2.3, 7.4, 2.7, 8.9, 6.6, 6.8, 5.1, 2.2, 5.6, 5.5, 2.1, 7.7, 13.9, 1.6, 1.7}
c. {−15, −13, −7, −15, −22, −12, −21, −21, −26, −17}
d. {43.6, 44.1, 59.5, 52.3, 50.9, 39.7, 42.4, 58.5, 40.9, 38.5, 44.2, 60.3, 72.2, 34.8, 46.0, 54.7, 51.0, 54.3, 49.7, 62.9, 44.6, 61.3, 52.4, 43.9, 68.8, 59.2, 57.1, 70.5, 52.3, 49.5}

3.35 Consider the following data set.

21	28	38	12	33	47	51	11	81	36

a. Find the sample variance and the sample standard deviation.
b. If 20 is subtracted from each observation in part (a), a new data set is formed.

1	8	18	−8	13	27	31	−9	61	16

Find the sample variance and the sample standard deviation for this new data set. How are these values related to the sample variance and the sample standard deviation found in part (a)?
c. If each observation in part (a) is multiplied by 20, the following data set is formed.

420	560	760	240	660
940	1020	220	1620	720

Find the sample variance and the sample standard deviation for this new data set. How are these values related to the sample variance and the sample standard deviation found in part (a)?

3.36 How does an outlier affect each of the following?
a. The sample variance
b. The sample standard deviation
c. The first quartile and the third quartile
d. The interquartile range

Applications

3.37 Sports and Leisure The following times (in seconds) are for the first 50 meters from the women's 200 meter backstroke at the Beijing Olympics in 2008.

29.62	29.83	30.14	30.63	31.16
30.06	30.62	30.80	30.83	29.98

(*Source*: MSNBC.)

a. Find the sample range, R.
b. Find the sample variance, s^2, and the sample standard deviation, s.

c. Find the first and third quartiles, Q_1 and Q_3, and the interquartile range, *IQR*.

3.38 Physical Sciences The following table includes some of the data from an experiment performed by H. S. Lew for the Center for Building Technology at the National Institute of Standards and Technology (NIST). These data are used to certify computational results and evaluate statistical software. Each observation represents the deflection of a steel–concrete beam while it is subjected to periodic pressure.

−213	−564	−35	−15	141	115	−420
−360	203	−338	−431	194	−220	−513
154	−125	−559	92	−21	−579	

(*Source*: NIST.)

a. Find the sample standard deviation.
b. Find the interquartile range.
c. Which statistic, *s* or *IQR*, is a better measure of variability for this data set? Why?

 3.39 Fuel Consumption and Cars The tune-up information for a 2006 Subaru Forester L.L. Bean edition with automatic transmission indicates that the idle speed should be 700 rpm. Sixteen of these automobiles in for service at Nate Wade Subaru in Salt Lake City, Utah, were selected at random. The idle speed for each Forester is given in the following table.

745	650	670	730	700	690	670	665
640	700	700	690	690	700	660	705

(*Source*: cars101.com.)

a. Find the sample variance and the sample standard deviation.
b. Find the first and third quartiles.
c. Find the interquartile range and the *quartile deviation* (another measure of variability), QD = $(Q_3 - Q_1)/2$.

 3.40 Education and Child Development Many educators believe that success in school is directly related to the amount of time students spend completing homework assignments. A research study compared the academic ability of 17-year-olds who spend less than 1 hour on homework every day and those who spend more than 2 hours on homework every day. The National Assessment of Educational Progress (NAEP) scores for each student in each group are given in the following table.

Less than 1 hour

290	289	291	289	289	294	288	291
293	290	290	291	290	290	296	292

More than 2 hours

303	305	302	297	294	303	299	297	303	299
300	295	297	297	297	293	296	297	302	294

(*Source*: National Center for Education Statistics.)

a. Find the sample variance, sample standard deviation, and interquartile range of the progress scores for students who spend less than 1 hour on homework.

b. Find the sample variance, sample standard deviation, and interquartile range of the progress scores for students who spend more than 2 hours on homework.
c. Use your answers in parts (a) and (b) to determine which data set has more variability.

3.41 Travel and Transportation The number of passengers travelling on a regularly scheduled flight from Allentown, Pennsylvania to Chicago, Illinois was recorded on seven randomly selected days. The data are given in the following table.

51	76	47	61	53	68	79

a. Compute s^2 using the definition in Equation 3.4.
b. Compute s^2 using the computational formula in Equation 3.6.
c. How do your answers in parts (a) and (b) compare?

3.42 Public Policy and Political Science The president of the United States has the authority to grant clemencies, pardons, and commutations of sentences to convicted criminals. A sample of U.S. presidents was obtained, and the number of presidential clemency actions for each was recorded. The data are given in Table 3.1.

Table 3.1

President	Clemency actions
Calvin Coolidge	1545
Jimmy Carter	566
Woodrow Wilson	2480
John F. Kennedy	575
Thomas Jefferson	119
Millard Fillmore	170
Rutherford B. Hayes	893
Richard Nixon	926
Andrew Jackson	386
James Madison	196
Zachary Taylor	38
Martin Van Buren	168
Ulysses S. Grant	1332
Lyndon B. Johnson	1187

(*Source*: *Jurist*, University of Pittsburgh School of Law.)

a. Find Q_1, Q_3, and *IQR* for the clemency actions data.
b. Find s^2 and *s*.
c. Franklin D. Roosevelt has the highest number of clemency actions of any president, 3687. Add this value to the data set. Find *IQR* and s^2 for this expanded data set.
d. How do *IQR* and s^2 compare in these two data sets? Explain why these values are the same/different.

3.43 Physical Sciences The following operating temperatures (°F) for a certain steam turbine were measured on 10 randomly selected days.

298	313	305	292	283	348	291	286	346	304

a. Find Q_1, Q_3, and *IQR*.
b. Find s^2 and *s*.

c. Suppose the smallest observation (283) is changed to 226. Find IQR and s^2 for this modified data set.

d. How do IQR and s^2 compare in these two data sets? Which measurement is more sensitive to outliers?

3.44 Marketing and Consumer Behavior Two measures designed to give a relative measure of variability are the **coefficient of variation**, denoted CV, and the **coefficient of quartile variation**, denoted CQV. These measures are defined by

$$CV = 100 \cdot \frac{s}{x}, \qquad CQV = 100 \cdot \frac{Q_3 - Q_1}{Q_3 + Q_1}.$$

The areas (in square feet) for homes constructed in two new residential developments in San Antonio (one in North Central and one on the city's West Side) were recorded and are given in the following table.

North Central development

2038	1939	2024	1990	2109	2102	1918	2022

West Side development

2061	2383	2638	2142	2382	1489	2070	2340
1725	2368	1674	1877				

a. Compute CV and CQV for each development.

b. Compare the coefficient of variation and the coefficient of quartile variation for each development. Which data set has more variability?

3.45 Physical Sciences The international space station, Mir, was constructed from 1986 to 1996. This scientific laboratory in space had a combined mass of 250 tons. Several U.S.–Russian missions to Mir included extravehicular activities (EVAs), or spacewalks. The following table lists times (in minutes) of several EVAs associated with missions to Mir.

19	298	352	304	166	363	359	347	265	321
411	386	235	14	340	270	321	328	270	

(*Source*: Sandcastle VI.)

a. Find the sample variance and the sample standard deviation.

b. Find the Q_1, Q_3, and IQR for these data.

c. Remove the two smallest EVA times from the data set. Answer parts (a) and (b) for this reduced data set. Compare the sample standard deviation and IQR in these two data sets and explain how these values have changed.

3.46 Public Health and Nutrition The Center for Science in the Public Interest (CSPI), a consumer group concerned about nutrition labeling, has defined a new measure of breakfast cereal called the nutritional index (NI) that is based on calories, vitamins, minerals, and sugar content per serving. A larger NI indicates greater nutritional value. The NI was measured for randomly selected cereals sold by Kellogg's and General Mills. The results are given in the following table.

Kellogg's

86	70	77	79	71	80	88	62	81	82
75	83	70	67	72	68	74	80	62	74

General Mills

54	49	50	31	46	29	81	63	41	60
66	68	39	59	47	80	41	91	41	33

a. Find s^2, s, and IQR for Kellogg's.

b. Find s^2, s, and IQR for General Mills.

c. Use the results in parts (a) and (b) to compare the variability in NI for the two companies.

3.47 Biology and Environmental Science The stream velocity (in ft/s) at various times in October and November 2008 3.5 miles south of the Mid-Hudson bridge is given on the data CD and book's web site. (*Source*: United States Geological Survey.)

a. Find Q_1, Q_3, and IQR for these data.

b. How large could the minimum stream velocity be without changing the IQR?

c. Find the coefficient of quartile variation, CQV (defined in Exercise 3.44).

Extended Applications

3.48 Biology and Environmental Science The number of wildland fires in 2007 in selected states are listed in the following table.

State	AK	CA	CO	CT	DE	FL
Fires	538	5825	1962	74	13	2468

State	IN	KY	LA	MA	MD	ME
Fires	413	904	969	2776	651	528

State	NC	NH	NJ	OH
Fires	5299	372	1617	150

(*Source*: National Interagency Fire Center.)

a. Find the sample variance and the sample standard deviation.

b. Find Q_1, Q_3, and IQR for these data.

c. Verify that the sum of the deviations about the mean is 0 (subject to round-off error).

3.49 Marketing and Consumer Behavior An Internet search for the best deal on a certain digital camera revealed the following prices (in dollars).

316.00	894.95	300.00	344.00	253.00
401.00	279.00	399.99	405.00	352.95

(*Source*: Consumer Electronics HQ, Inc.)

a. Find Q_1, Q_3, and IQR for these price data.

b. Suppose the highest price (894.95) is changed to 652.50. Find Q_1, Q_3, and IQR for this modified data set.

c. How large could the maximum price be without changing *IQR*?

d. How much could the minimum price be raised before Q_1 changes?

3.50 Travel and Transportation The Bureau of Transportation maintains detailed records associated with air travel including on-time data and frequently delayed flights. Table 3.2 lists the number of available seats (in millions) for selected carriers during July 2008.

Table 3.2

Carrier	Available seats
JetBlue Airways	2208
Air Wisconsin Airlines	682
American Airlines	8006
Comair	960
Continental Airlines	3842
Delta Airlines	6398
Frontier Airlines	1232
Northwest Airlines	4502
Southwest Airlines	14049
US Airways	5516
United Airlines	5937

(*Source*: Bureau of Transportation Statistics.)

a. Find the sample variance and the sample standard deviation of the number of available seats.

b. Suppose each airline cuts the number of available seats by 10% in August. Find the sample variance and the sample standard deviation for August.

c. How do your answers in parts (a) and (b) compare?

3.51 Travel and Transportation In its annual study, the International Telework Association & Council asks each survey participant how many miles they drive to work each day. Six study participants were selected at random and their mileage was recorded.

25	39	16	35	18	45

a. Find each deviation about the mean.

b. Verify that the sum of the deviations about the mean is 0 (subject to round-off error).

c. Prove that in general $\sum (x_i - \bar{x}) = 0$.
(*Hint*: Write as two separate sums, and use the definition of the sample mean.)

3.52 Biology and Environmental Science The Virginia Estuarine & Coastal Observing System monitors the Chesapeake Bay and records values of several variables including salinity, temperature, and turbidity. The measurements of maximum wind speed (in mph) at a selected location for various time periods during May 2008 are given on the data CD and book's web site.

a. Find the sample variance and the sample standard deviation for these wind speed data.

b. Convert each observation to m/s (multiply each observation by 0.44704). Find the sample variance and the sample standard deviation for the wind speed data in m/s.

c. How do your answers in parts (a) and (b) compare?

3.53 Proof Prove that Equation 3.4 (the definition of the sample variance) can be written as Equation 3.6. That is, show that

$$\frac{1}{n-1}\sum (x_i - \bar{x})^2 = \frac{1}{n-1}\left[\sum x_i^2 - \frac{1}{n}\left(\sum x_i\right)^2\right].$$

3.54 Public Health and Nutrition A nutritional study recently found the following number of calories in one slice of plain pizza at 10 different national chains.

228	281	274	408	364	259	317	299	302	231

(*Source*: Food Science and Human Nutrition, Colorado State University.)

a. Find the sample variance and the sample standard deviation.

b. Add 15 (calories) to each observation. Find the sample variance and the sample standard deviation for this modified data set.

c. How do your answers in parts (a) and (b) compare?

d. Suppose a data set (x's) has variance s_x^2 and standard deviation s_x. A new (transformed) data set is created using the equation $y_i = x_i + b$ where b is a constant. How are the variance and standard deviation of the new data set (s_y^2 and s_y) related to s_x^2 and s_x?

3.55 Technology and Internet A benchmark computer program was executed on eight different machines and the following times (in seconds) to completion were recorded.

12.592	14.152	12.396	6.801
13.646	12.075	15.377	7.602

a. Find the sample variance and the sample standard deviation.

b. Multiply each observation by 7. Find the sample variance and the sample standard deviation for this modified data set.

c. How do your answers in parts (a) and (b) compare?

d. Suppose a data set (x's) has variance s_x^2 and standard deviation s_x. A new (transformed) data set is created using the equation $y_i = a * x_i$ where a is a constant. How are the variance and standard deviation of the new data set (s_y^2 and s_y) related to s_x^2 and s_x?

3.56 Transformed Data Combine the results obtained in the previous two problems. Suppose a data set (x's) has variance s_x^2 and standard deviation s_x. A new (transformed) data set is created using the equation $y_i = a * x_i + b$ where a and b are constants. How are the variance and standard deviation of the new data set (s_y^2 and s_y) related to s_x^2 and s_x?

3.57 Is This Possible? Consider the following set of observations:

$$\{5, 7, 3, 2, 4, 6, 9, 11, 13\}$$

Can you find a subset of size $n = 7$ with $\bar{x} = 5$ and $s^2 = 6$? If not, why not?

Challenge

3.58 Biology and Environmental Science A whale-watching tour off the coast of Maine is considered a success if at least one whale is sighted. Thirty-two randomly selected summer tours are classified in the following table.

```
S   S   S   S   F   S   S   S   S   S   S   S
S   S   S   S   S   S   S   F   S   S   S
S   F   S   S   S   S   S   S
```

a. Find the sample proportion of successes.
b. Change each S to a 1 and each F to a 0. Find the sample variance for these new data. Write the sample variance in terms of the sample proportion.
c. If a population happens to be of finite size N, then the population mean and population variance are defined by

$$\mu = \frac{1}{N}\sum x_i, \qquad \sigma^2 = \frac{1}{N}\sum (x_i - \mu)^2.$$

Suppose the table represents an entire population. Find the population variance for the data (consisting of 0's and 1's). Write the population variance in terms of the sample proportion.

3.59 Other Summary Statistics There are many other summary statistics that describe various characteristics of a numerical data set. Suppose $x_1, x_2, \ldots, x_n$ is a set of observations.

For $r = 1, 2, 3, \ldots$, the **rth moment about the mean** $\bar{x}$ is defined as

$$m_r = \frac{1}{n}\sum (x_i - \bar{x})^r.$$

For example, the second moment about the mean is

$$m_2 = \frac{1}{n}\sum (x_i - \bar{x})^2.$$

Certain moments about the mean are used to define the **coefficient of skewness** (g_1) and the **coefficient of kurtosis** (g_2):

$$g_1 = \frac{m_3}{m_2^{3/2}}, \qquad g_2 = \frac{m_4}{m_2^2}.$$

The statistic g_1 is a measure of the lack of symmetry, and g_2 is a measure of the extent of the peak in a distribution.

Use technology to compute the values g_1 and g_2 for various distributions: skewed, symmetric, unimodal, uniform. Use your results to determine the values of g_1 that suggest more skewness in the distribution and the values of g_2 that indicate a flatter, more uniform distribution.

3.3 THE EMPIRICAL RULE AND MEASURES OF RELATIVE STANDING

Measures of central tendency and measures of variability are used to describe the general nature of a data set. These two types of measures may be combined in order to describe the distribution of a data set more precisely. In addition, these measures may be used to define measures of relative standing, quantities used to compare observations from different data sets (with different units), or even to draw a conclusion or make an inference.

The first result combines the mean and the standard deviation to describe a distribution.

What happens if $k = 1$?

CHEBYSHEV'S RULE

Let $k > 1$. For *any* set of observations, the proportion of observations within k standard deviations of the mean [lying in the interval $(\bar{x} - ks, \bar{x} + ks)$, where s is the standard deviation] is at least $1 - \frac{1}{k^2}$.

Recall interval notation: (a, b) denotes an open interval, with the endpoints not included, from a to b. Therefore, $(\bar{x} - ks, \bar{x} + ks)$ means the set of all x's such that $\bar{x} - ks < x < \bar{x} + ks$.

The diagram in Figure 3.25, and the accompanying table, illustrate this idea. For any set of observations, the smoothed histogram shows that the proportion of observations captured in the interval $(\bar{x} - ks, \bar{x} + ks)$ is at least $1 - (1/k^2)$. For example, the proportion of observations within 1.5 standard deviations of the mean is at least 0.56 (or 56%). The proportion of observations within three standard deviations of the mean is at least 0.89 (or 89%).

Recall: In smoothed histograms, the area under the curve between two values a and b corresponds to the proportion of observations between a and b.

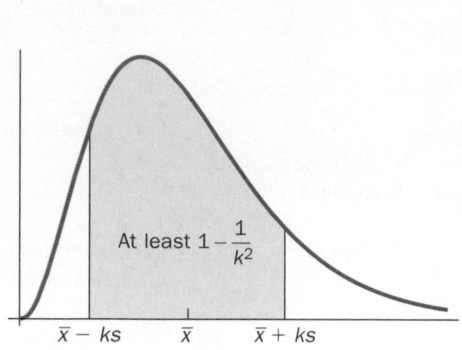

k	$1 - \dfrac{1}{k^2}$
1.5	$1 - \dfrac{1}{1.5^2} \approx 0.56$
2.0	$1 - \dfrac{1}{2.0^2} \approx 0.75$
2.3	$1 - \dfrac{1}{2.3^2} \approx 0.81$
3.0	$1 - \dfrac{1}{3.0^2} \approx 0.89$

Figure 3.25 Illustration of Chebyshev's Rule.

A symmetric interval about the mean is centered at the mean and has endpoints that are the same distance from the mean.

ILLUMINATING THE CONCEPTS

1. **Chebyshev's Rule** simply helps to describe a set of observations using symmetric intervals about the mean. If we move k standard deviations from the mean in both directions, then the proportion of observations captured is at least $1 - (1/k^2)$.

2. The total area under the curve (the sum of all the proportions) is 1. Hence, Chebyshev's Rule also implies that the proportion of observations in the tails of the distribution, outside the interval $(\bar{x} - ks, \bar{x} + ks)$, is at most $1/k^2$.

3. As indicated in the statement of Chebyshev's Rule and as suggested in the table, you may use any value of k greater than 1, including decimals. The two most common values for k are $k = 2$ and $k = 3$. The actual proportions of observations within two and within three standard deviations can be compared to the values predicted by Chebyshev's Rule and the Empirical Rule (on page 94). This comparison may be used to determine whether a distribution is approximately normal (or not normal). In addition, $k = 2$ and $k = 3$ provide the fundamental background to statistical inference.

4. Chebyshev's Rule is very conservative because it applies to any set of observations. Usually, the proportion of observations within k standard deviations of the mean is bigger than $1 - (1/k^2)$.

5. Chebyshev's Rule may also be used to describe a population. If the mean and standard deviation are known, then μ and σ may be used in place of $\bar{x}$ and s. For any population, the proportion of observations that lie in the interval $(\mu - k\sigma, \mu + k\sigma)$ is at least $1 - (1/k^2)$. ∎

Solution Trail

KEYWORDS

Approximate proportion of observations between.

↓

TRANSLATION

What proportion of observations is captured by the interval?

↓

CONCEPTS

Chebyshev's Rule.

↓

VISION

We don't know anything about the shape of the distribution of the length of songs. However, Chebyshev's Rule applies to any distribution, tells us about the proportion of observations captured by certain intervals, and may be used here if the questions involve symmetric intervals about the mean.

Example 3.11 Automobile Battery Lifetime In a random sample of the lifetime (in months) of an Odyssey automobile battery, $\bar{x} = 54$ and $s = 5.3$. Use Chebyshev's Rule with $k = 2$ and $k = 3$ to describe this distribution of battery lifetimes.

SOLUTION

STEP 1 For $k = 2$: $1 - \dfrac{1}{k^2} = 1 - \dfrac{1}{2^2} = 1 - \dfrac{1}{4} = \dfrac{3}{4} = 0.75$

At least $\frac{3}{4}$ (or 75%) of the observations lie in the interval
$(\bar{x} - 2s, \bar{x} + 2s) = (54 - 2(5.3), 54 + 2(5.3)) = (43.4, 64.6)$

STEP 2 For $k = 3$: $1 - \dfrac{1}{k^2} = 1 - \dfrac{1}{3^2} = 1 - \dfrac{1}{9} = \dfrac{8}{9} \approx 0.89$

At least $\frac{8}{9}$ (or 89%) of the observations lie in the interval $(\bar{x} - 3s, \bar{x} + 3s) = (54 - 3(5.3), 54 + 3(5.3)) = (38.1, 69.9)$

STEP 3 Note also:

At most $\frac{1}{4}$ (or 25%) of the observations lie *outside* the interval (43.4, 64.6).

At most $\frac{1}{9}$ (or 11%) of the observations lie *outside* the interval (38.1, 69.9).

Example 3.12 How Unusual Was "In-A-Gadda-Da-Vida"? In 1968, the group Iron Butterfly recorded the 17-minute song In-A-Gadda-Da-Vida. Most popular songs are much shorter; for example, Viva la Vida by Coldplay is approximately 4 minutes long. Suppose in a random sample of the length (in minutes) of songs on CDs produced by hard rock bands, $\bar{x} = 3.35$ and $s = 0.5$.
a. Find the approximate proportion of observations between 2.35 and 4.35.
b. Find the approximate proportion of observations less than 1.85 or greater than 4.85.
c. Approximately what proportion of songs last more than 5 minutes?

SOLUTION

Since no values of k are specified, try $k = 2$ and $k = 3$.
a. $(\bar{x} - 2s, \bar{x} + 2s) = (3.35 - 2(0.5), 3.35 + 2(0.5)) = (2.35, 4.35)$ $k = 2$.
At least $1 - \left(\frac{1}{4}\right) = \frac{3}{4}$ (or 75%) of the observations lie between 2.35 and 4.35.

b. $(\bar{x} - 3s, \bar{x} + 3s) = (3.35 - 3(0.5), 3.35 + 3(0.5)) = (1.85, 4.85)$ $k = 3$.
At least $1 - \left(\frac{1}{9}\right) = \frac{8}{9}$ (or 89%) of the observations lie between 1.85 and 4.85.

At most $\frac{1}{9}$ (or 11%) of the observations are less than 1.85 or greater than 4.85.

c. Since Chebyshev's Rule measures intervals in terms of the number of standard deviations from $\bar{x}$, find out how far 5 is from $\bar{x}$ in standard deviations.

$\bar{x} + ks = 3.35 + k(0.5) = 5 \Rightarrow k = 3.3$

We cannot assume anything about the shape of the distribution.

$$1 - \frac{1}{k^2} = 1 - \frac{1}{3.3^2} \approx 0.91$$

At least 0.91 (or 91%) of the observations lie in the interval

$(\bar{x} - 3.3s, \bar{x} + 3.3s) = (3.35 - 3.3(0.5), 3.35 + 3.3(0.5)) = (1.7, 5.0)$

Therefore, at most $1 - 0.91 = 0.09$ (or 9%) of the observations are outside this interval, either less than 1.7 or greater than 5.0. Since we cannot assume the distribution is symmetric, we do not know what part of the 9% is less than 1.7 and what part is more than 5.0. To be conservative, the best we can say is at most 9% of the observations are more than 5 minutes long.

A normal curve is bell-shaped and symmetric, centered at the mean.

If a set of observations can be reasonably modeled by a normal curve, then we can describe this distribution more precisely. The Empirical Rule involves the mean and standard deviation also, and the results apply to three specific symmetric intervals about the mean.

THE EMPIRICAL RULE

If the shape of the distribution of a set of observations is approximately normal, then

1. The proportion of observations within **one standard deviation** of the mean is approximately 0.68.
2. The proportion of observations within **two standard deviations** of the mean is approximately 0.95.
3. The proportion of observations within **three standard deviations** of the mean is approximately 0.997.

Figure 3.26 illustrates **the Empirical Rule**, the symmetric intervals about the mean and the proportions. The Empirical Rule conclusions are more accurate than Chebyshev's Rule because we know (assume) more about about the shape of the distribution (normality).

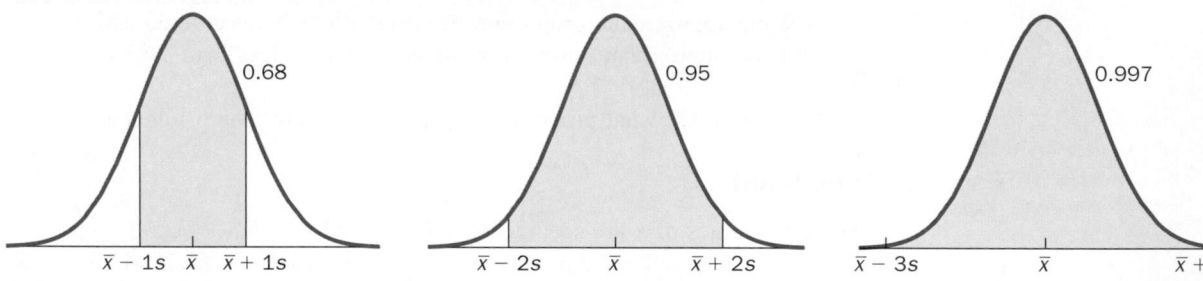

Figure 3.26 Symmetric intervals and proportions associated with the Empirical Rule.

💡 ILLUMINATING THE CONCEPTS

For now, the reasons for the proportions 0.68, 0.95, and 0.997 remain a mystery. We will discover where these numbers come from in Chapter 6.

1. Given a set of observations, the Empirical Rule may be used to *check* normality. To numerically test for normality, find the mean, standard deviation, and the three symmetric intervals about the mean $(\bar{x} - ks, \bar{x} + ks)$, $k = 1, 2, 3$. Compute the *actual* proportion of observations in each interval. If the actual proportions are close to 0.68, 0.95, and 0.997, then normality seems reasonable. Otherwise, there is evidence to suggest that the shape of the distribution is not normal. This process is sort of a *backward* Empirical Rule.

2. The Empirical Rule may also be used to describe a population. If the distribution of the population is approximately normal, and the mean and standard deviation are known, then μ and σ may be used in place of $\bar{x}$ and s.

3. The proportion of observations beyond three standard deviations from the mean is $1 - 0.997 = 0.003$ (pretty small). Therefore, if the shape of a (population) distribution is approximately normal, it would be unusual to have an observation more than three standard deviations from the mean. What if there is one? (See Example 3.14.) ■

Example 3.13 Distribution of Check Amounts The manager of a Bi-Lo grocery store is interested in characterizing the distribution of purchases made by personal check. In a random sample of check amounts (in dollars), the shape of the distribution is approximately normal with $\bar{x} = 37.5$ and $s = 5.2$.

Solution Trail

KEYWORDS

- Approximately normal.
- Approximate proportion of observations between.

↓

TRANSLATION

What proportion of observations is captured by the interval?

↓

CONCEPTS

Empirical Rule.

↓

VISION

Since the shape of the distribution is approximately normal, the Empirical Rule may be used to determine the proportion of observations captured by certain intervals, related in some way to three special symmetric intervals about the mean.

a. Approximately what proportion of observations is between 27.1 and 47.9?
b. Approximately what proportion of observations is greater than 53.1 or less than 21.9?
c. Approximately what proportion of observations is greater than 53.1?
d. Approximately what proportion of observations is between 32.3 and 47.9?

SOLUTION

a. Find the values one, two, and three standard deviations about the mean in each direction.
   Notice, $(37.5 - 2(5.2), 37.5 + 2(5.2)) = (27.1, 47.9)$
   So, 27.1 to 47.9 is a symmetric interval about the mean, two standard deviations in each direction. The Empirical Rule states that approximately 0.95 (or 95%) of the observations lie in this interval. See Figure 3.27.
b. Notice, $(37.5 - 3(5.2), 37.5 + 3(5.2)) = (21.9, 53.1)$
   So, 21.9 to 53.1 is a symmetric interval about the mean, three standard deviations in each direction. The Empirical Rule states that approximately 0.997 (or 99.7%) of the observations lie in this interval. The remaining proportion, $1 - 0.997 = 0.003$ (or 0.3%) of observations, lie outside this interval and are greater than 53.1 or less than 21.9. See Figure 3.28.

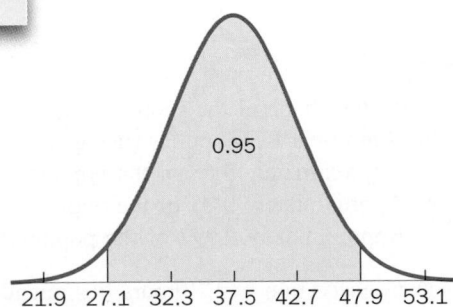

Figure 3.27 Approximately 0.95 (or 95%) of the observations lie within two standard deviations of the mean, in the interval (27.1, 47.9).

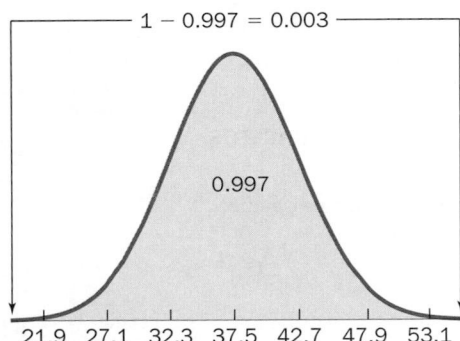

Figure 3.28 Approximately $1 - 0.997 = 0.003$ (or 0.3%) of the observations lie outside the interval (21.9, 53.1).

c. Since a normal distribution is symmetric about the mean, the remaining proportion outside three standard deviations from the mean $(1 - 0.997 = 0.003)$ is divided evenly between the two tails. Therefore, approximately $0.003/2 = 0.0015$ (or 0.15%) of the observations are greater than 53.1. See Figure 3.29.
d. (32.3, 47.9) is not a symmetric interval about the mean. However, approximately 0.68 of the observations lie in the interval (32.3, 42.7) (one standard deviation from the mean). Approximately 0.95 of the observations lie in the interval (27.1, 47.9) (two standard deviations from the mean). This means that $0.95 - 0.68 = 0.27$ of the observations lie in the intervals (27.1, 32.3) and (42.7, 47.9). Since a normal distribution is symmetric, $0.27/2 = 0.135$ of the observations lie between 42.7 and 47.9. Therefore, a total of approximately $0.68 + 0.135 = 0.815$ (or 81.5%) of the observations lie between 32.3 and 47.9. See Figure 3.30.

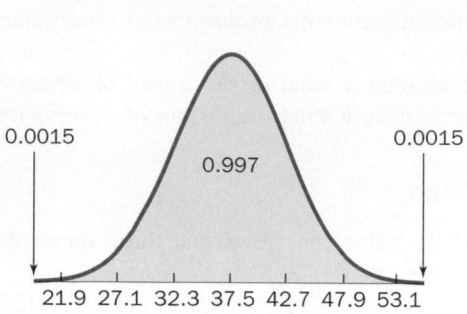

Figure 3.29 Approximately 0.0015 (or 0.15%) of the observations are greater than 53.1.

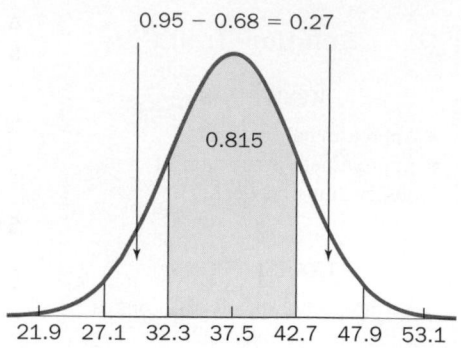

Figure 3.30 Approximately 0.27 (or 27%) of the observations lie in the intervals (27.1, 32.3) and (42.7, 47.9). ●

Example 3.14 For How Long Will the Medicine Work?   First Horizon Pharmaceutical has just developed a new medicine for treatment of routine aches and pains. The company claims the distribution of pain-relief times (in hours) is approximately normal with mean $\mu = 8$ and standard deviation $\sigma = 0.2$. A patient with a typical muscle ache is randomly selected and the medicine is administered. The patient reports pain relief for only 7 hours. Is there any evidence to refute the manufacturer's claim?

SOLUTION

STEP 1   Since the shape of the (population) distribution is approximately normal, the Empirical Rule applies (using $\mu = 8$ and $\sigma = 0.2$).
   a. Approximately 0.68 of the population lies in the interval (7.8, 8.2).
   b. Approximately 0.95 of the population lies in the interval (7.6, 8.4).
   c. Approximately 0.997 of the population lies in the interval (7.4, 8.6).

STEP 2   The observation $x = 7$ hours lies outside the largest interval (7.4, 8.6). Only $1 - 0.997 = 0.003$ of the population lies outside this interval. More precisely (because of symmetry), only $0.003/2 = 0.0015$ of the population lies below 7.4. Seven hours is a very rare observation. Two things may have occurred.
   a. Seven hours is an incredibly lucky (in the sense that it is unusual) observation. Even though the proportion of observations below 7.4 is small, it is still possible for the manufacturer's claim to be true and for the pain reliever to have lasted only 7 hours in this patient.
   b. The manufacturer's claim is false. Since an observation of 7 hours is so rare, it is more likely that one of the assumptions is wrong. The shape of the distribution may not be normal, the mean may be different from 8, and/or the standard deviation might be different from 0.2.

STEP 3   Typically, statistical inference discounts the *lucky* alternative. Therefore, since 7 hours is such an unlikely observation, there is evidence to suggest the manufacturer's claim is false. Something is awry. We should not see pain relief of only 7 hours if the claim is true.

Note: We may be too quick to make an inference based on only a single observation. We will learn how to use more observations (information) to reach a more confident conclusion. ●

One method for comparing observations from different samples (with different units) is to use a *standardized score*. For a given observation, this relative measure is used to determine the distance from the mean in standard deviations.

DEFINITION

Suppose $x_1, x_2, \ldots, x_n$ is a set of n observations with mean $\bar{x}$ and standard deviation s. The **z-score** corresponding to the ith observation, x_i, is given by

$$z_i = \frac{x_i - \bar{x}}{s}. \tag{3.7}$$

z_i is a measure associated with x_i that indicates the distance from $\bar{x}$ in standard deviations.

ILLUMINATING THE CONCEPTS

1. z_i may be positive or negative (or zero). A positive **z-score** indicates that the observation is to the right of the mean. A negative z-score indicates that the observation is to the left of the mean.

2. A z-score is a measure of relative standing; it indicates where an observation lies in relation to the rest of the values in the data set. There are other methods of standardization, but this is the most common.

3. Given a set of n observations, the sum of all the z-scores is 0; $\sum z_i = 0$. Can you prove this? ∎

Example 3.15 Starting Salary Most college career counselors agree that starting salary is associated with academic major. Even if a person's first job is not directly related to his or her course of study, the salary may still be related to the person's academic major. A recent survey of academic major and starting salary of graduates showed the following information.

Major	Mean	Standard deviation
English	$32,300	$1,175
Computer Science	$42,500	$2,375

(*Source*: National Association of Colleges and Employers.)

One Computer Science major who responded to the survey received a starting salary of $45,000, and an English major received an offer of $35,000. Which salary is *better*, in terms of statistics?

SOLUTION

STEP 1 The higher starting salary is probably better (subject to working conditions, benefits, location, etc.), but to answer this question in terms of statistics, consider the z-scores.

STEP 2 Computer Science major: $z = \dfrac{45{,}000 - 42{,}500}{2{,}375} \approx 1.05$

Statisticians tend to measure *distances* in standard deviations, not miles, feet, inches, or meters. We often ask, "How many standard deviations from the mean is a given observation?"

Solution Trail

KEYWORDS

Which salary is better?

↓

TRANSLATION

Which salary is farther away from the mean (to the right) in standard deviations?

↓

CONCEPTS

z-score.

↓

VISION

Compute and compare the z-scores for each salary. This will allow us to determine how many *statistical steps* each observation is from the mean. The higher the z-score, the *better* the salary.

$45,000 is approximately 1.05 standard deviations to the right of the mean.

$$\text{English major: } z = \frac{35,000 - 32,300}{1,175} \approx 2.30$$

$35,000 is approximately 2.30 standard deviations to the right of the mean.

STEP 3   The English major's starting salary is actually better, because the salary is much higher than those of most English majors.   ●

Example 3.16 Pet Return Policy   The owner of the Jungle Pet Store is trying to establish a policy for the return of animals. In a random sample of the lifetime (in months) of pet guinea pigs, $\bar{x} = 60$ and $s = 15.5$. One of the guinea pigs in this sample lived 48 months. Is this a reasonable lifetime, or should the store provide some sort of refund (or a new guinea pig)? (*Source*: Animal Adventure, Inc.)

SOLUTION

STEP 1   To determine whether 48 months is a *reasonable* lifetime, consider the z-score corresponding to this observation.

STEP 2 $z = \dfrac{48 - 60}{15.5} = -0.80$

The observation (48 months) is only 0.8 standard deviation to the left of the mean. Since 48 months is within 1 standard deviation of the mean (regardless of the shape of the distribution), this is a very conservative, reasonable observation.

STEP 3   The guinea pig lived a very *normal* life. No refund is necessary.   ●

Another indication of relative standing is a **percentile**. Do you remember all of those standardized tests in grade school? The results were usually reported in terms of percentiles. The 90th percentile was a good score and the 25th percentile meant more homework in your future.

> ### Solution Trail
>
> #### KEYWORDS
> Is this a reasonable lifetime?
>
> ↓
>
> #### TRANSLATION
> Draw a conclusion. Do you think the guinea pig should have lived longer?
>
> ↓
>
> #### CONCEPTS
> - Inference procedure.
> - z-score.
>
> ↓
>
> #### VISION
> For any distribution, most observations are within three standard deviations of the mean, or have a z-score between -3 and $+3$. Compute the z-score for this guinea pig's lifetime, the number of statistical steps from the mean.

> **DEFINITION**
>
> Let $x_1, x_2, \ldots, x_n$ be a set of observations. The **percentiles** divide the data set into 100 parts. For any integer r $(0 < r < 100)$ the **rth percentile**, denoted p_r, is a value such that r percent of the observations lie at or below p_r (and $100 - r$ percent lie above p_r).

The **rth percentile** has the same units as the observations, not a percent. Figure 3.31 shows a smoothed histogram and illustrates the location of the 75th percentile on the measurement axis.

ILLUMINATING THE CONCEPTS

1. The 50th percentile is the median, $p_{50} = \tilde{x}$.

2. The 25th percentile is the first quartile and the 75th percentile is the third quartile: $p_{25} = Q_1, p_{75} = Q_3$. ■

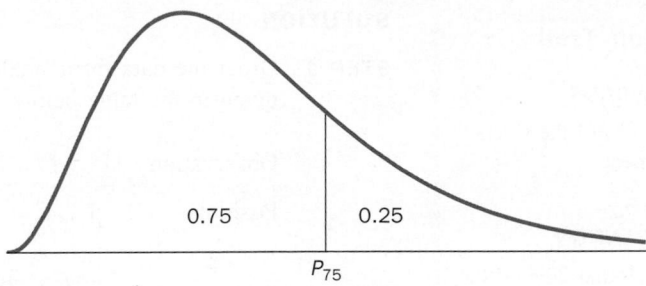

Remember, the area under the curve between *a* and *b* corresponds to the proportion of observations between *a* and *b*. So the total area under the curve is 1.

Figure 3.31 The 75th percentile is illustrated using a smoothed histogram.

HOW TO COMPUTE PERCENTILES

Suppose $x_1, x_2, \ldots, x_n$ is a set of n observations.

1. Arrange the observations in ascending order, from smallest to largest.

2. To find p_r, compute $d_r = \dfrac{n \cdot r}{100}$.

    a. If d_r is a whole number, then the depth of p_r (position in the ordered list) is $d_r + 0.5$. p_r is the mean of the observations in positions d_r and $d_r + 1$ in the ordered list.

    b. If d_r is not a whole number, round up to the next whole number for the depth of p_r.

Example 3.17 Technical Support Hotline   The calls to a computer software technical support line are carefully monitored and counted. On a recent randomly selected day, there were 1189 calls, a number that lies at the 73rd percentile. Interpret these results.

SOLUTION

1189 is a single observation from the population of calls per day to technical support, and percentiles divide these observations into 100 parts. Since 1189 lies at the 73rd percentile, 73% of the days had 1189 calls or fewer, and $100 - 73 = 27\%$ of the days had more than 1189 calls.

Note: We do not know anything about the shape of the distribution, nor do we know the mean or standard deviation. There is no way of telling how far 1189 is from the mean in standard deviations. ●

Example 3.18 Mountain Bike Maintenance   A Kettler Dynamite mountain bike has a special gear-shifting mechanism that requires adjustment after a break-in period. In a recent customer survey, 30 riders were asked the number of hours they rode before they adjusted their shifting mechanism. The results are given in the following table.

44	25	33	19	21	50	40	28	29	26
17	29	39	27	15	18	36	41	30	20
48	39	43	48	43	32	50	30	32	24

How long was it before 20% of the riders in this sample made the adjustment?

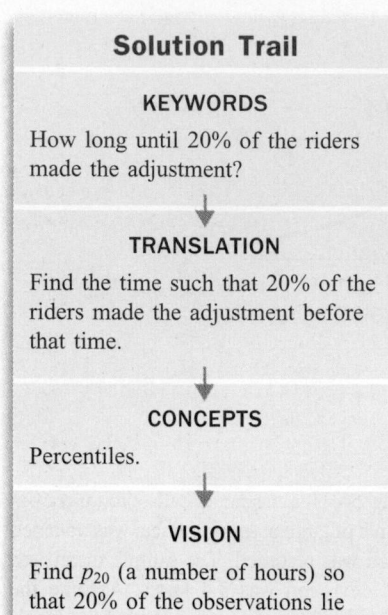

Solution Trail

KEYWORDS

How long until 20% of the riders made the adjustment?

↓

TRANSLATION

Find the time such that 20% of the riders made the adjustment before that time.

↓

CONCEPTS

Percentiles.

↓

VISION

Find p_{20} (a number of hours) so that 20% of the observations lie below and 80% lie above. Follow the steps for computing percentiles.

SOLUTION

STEP 1 Order the data from smallest to largest. A portion of this ordered list is given in the table below.

Observation	15	17	18	19	20	21	24	25	26	27
Position	1	2	3	4	5	6	7	8	9	10

STEP 2 Compute $d_{20} = \dfrac{n \cdot r}{100} = \dfrac{30 \cdot 20}{100} = 6$

STEP 3 Since d_{20} is a whole number, add 0.5. The depth of p_{20} is $d_{20} + 0.5 = 6 + 0.5 = 6.5$.

STEP 4 The 20th percentile, p_{20}, is the mean of the sixth and seventh observations.

$$p_{20} = \frac{1}{2}(21 + 24) = 22.5$$

STEP 5 Twenty percent of the riders made the adjustment before 22.5 hours, and 80% made the adjustment after 22.5 hours. ●

TECHNOLOGY CORNER

Procedure: Compute the rth percentile.
Reconsider: Example 3.18, page 99, solution, and interpretation.
Neither the TI-84 Plus nor Minitab has a built-in function to compute the rth percentile. There are functions in Excel and SPSS to compute percentiles. However both use slightly different algorithms.

Excel

1. Enter the data into column A.
2. Use the function PERCENTILE to compute the 20th percentile. See Figure 3.32.

A technology solution:

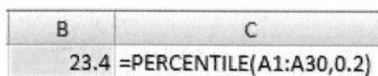

B	C
23.4	=PERCENTILE(A1:A30,0.2)

Figure 3.32 The Excel function PERCENTILE.

SPSS

1. Enter the data into variable VAR00001.
2. Select Analyze; Descriptive Statistics; Explore.
3. Enter VAR00001 in the Dependent List. Select the Statistics option button and check Percentiles. The output contains the 5th, 10th, 25th, 50th, 75th, 90th, and 95th percentile by default.
4. In order to compute a different percentile, select File; New; Syntax. Use the SPSS command language to find the 20th percentile. See figure 3.33.

Percentiles

		Percentiles
		20
Weighted Average (Definition 1)	VAR00001	21.6

Figure 3.33 SPSS output: the 20th percentile.

SECTION 3.3 EXERCISES

Practice

3.60 For each data set with $\bar{x}$ and s given, find a symmetric interval k standard deviations about the mean, and use Chebyshev's Rule to compute the approximate proportion of observations within this interval.
a. $\bar{x} = 50$, $s = 5$, $k = 2$
b. $\bar{x} = 352$, $s = 10.5$, $k = 3$
c. $\bar{x} = 17$, $s = 3.5$, $k = 1.6$
d. $\bar{x} = 36.5$, $s = 10.45$, $k = 1.75$
e. $\bar{x} = 158$, $s = 25$, $k = 2.5$
f. $\bar{x} = -55$, $s = 0.125$, $k = 2.8$
g. $\bar{x} = 1.7$, $s = 25.8$, $k = 2.25$

3.61 Assume the distribution of each data set is approximately normal, with $\bar{x}$ and s given. Find the intervals (referred to by the Empirical Rule) that are one, two, and three standard deviations about the mean. Carefully sketch the corresponding normal curve for each data set, indicating the endpoints of each interval.
a. $\bar{x} = 20$, $s = 5$
b. $\bar{x} = 37$, $s = 0.2$
c. $\bar{x} = 675$, $s = 250$
d. $\bar{x} = -5.5$, $s = 12$
e. $\bar{x} = 98.6$, $s = 1.7$
f. $\bar{x} = 5280$, $s = 150$

3.62 For each data set with $\bar{x}$ and s given, find the z-score corresponding to the given observation x.
a. $\bar{x} = 8$, $s = 3$, $x = 17$
b. $\bar{x} = 100$, $s = 16$, $x = 80$
c. $\bar{x} = 15$, $s = 3$, $x = 17.5$
d. $\bar{x} = 27$, $s = 4.5$, $x = 22$
e. $\bar{x} = 212$, $s = 32$, $x = 175$
f. $\bar{x} = -105$, $s = 33$, $x = -90$
g. $\bar{x} = 6.55$, $s = 0.25$, $x = 6$
h. $\bar{x} = 64$, $s = 8.75$, $x = 100$
i. $\bar{x} = 0.025$, $s = 0.0018$, $x = 0.027$
j. $\bar{x} = 407$, $s = 16$, $x = 500$

3.63 For each data set with $\bar{x}$ and s given, find an observation corresponding to the z-score given.
a. $\bar{x} = 25$, $s = 5$, $z = 2.3$
b. $\bar{x} = 9.8$, $s = 1.2$, $z = -0.7$
c. $\bar{x} = -456$, $s = 37$, $z = 1.25$
d. $\bar{x} = 37.6$, $s = 5.9$, $z = -1.96$
e. $\bar{x} = 55$, $s = 0.05$, $z = 3.5$
f. $\bar{x} = 3.14$, $s = 0.5$, $z = 1.28$
g. $\bar{x} = 2.35$, $s = 0.94$, $z = -2.5$
h. $\bar{x} = 0.529$, $s = 1.9$, $z = 0.55$

3.64 Find the position, or depth, of the indicated percentile in an ordered data set of size n.
a. $n = 150$, p_{80}
b. $n = 257$, p_{35}
c. $n = 36$, p_{60}
d. $n = 75$, p_{40}
e. $n = 100$, p_{20}
f. $n = 5035$, p_{70}

Applications

3.65 Demographics and Population Statistics The FBI uses public assistance in tracking criminals by maintaining the "Ten Most Wanted Fugitives" list. Fugitives are removed from this list if they are captured, the charges are dropped, or they no longer fit a certain profile. In a random sample of fugitives, the mean time on the list was 26.5 months, with a standard deviation of 4.3 months.
a. What values are one standard deviation away from the mean? What values are two standard deviations away from the mean?
b. Without assuming anything about the shape of the distribution of times, approximately what proportion of times is between 17.9 months and 35.1 months?

3.66 Travel and Transportation Cunard recently launched the largest passenger ship ever built—the Queen Mary 2. This ship has 15 restaurants, five swimming pools, a casino, a ballroom, and even a planetarium. A random sample of large passenger liners was obtained and the cruising speed of each was recorded. The sample mean was 25.6 knots and the standard deviation was 3.4 knots. Assume the shape of the speed distribution is approximately normal.
a. What values are two standard deviations away from the mean? What values are three standard deviations away from the mean?
b. Approximately what proportion of speeds is between 22.2 and 29.0 knots?

3.67 Sports and Leisure During the Big Rock Blue Marlin fishing tournament, participants must bring their catch to a central location and have each fish weighed. At a recent event, Josh Ruskey caught a fish with a weight in the 85th percentile of all fish caught. Interpret this value.

3.68 Biology and Environmental Science Many people in North America try to attract Purple Martins since they eat mosquitoes and other insects, up to 2000 per bird per day. In a random sample of assembly times for the Health Aluminum Deluxe Martin House, the mean time to assemble this product was 150 minutes, with a standard deviation of 12 minutes. (*Source*: Birdhouse.com.)
a. What values are one standard deviation away from the mean? What values are two standard deviations away from the mean?
b. Without assuming anything about the shape of the distribution of times, approximately what proportion of times is between 126 and 174 minutes?
c. Without assuming anything about the shape of the distribution of times, approximately what proportion of times is either less than 114 or greater than 186 minutes?
d. Assuming the distribution of times is normal, what proportion of times is between 126 and 174? What proportion is either less than 114 or greater than 186?

3.69 Biology and Environmental Science The Commonwealth of Pennsylvania is concerned about the dwindling number of family-owned farms and the number of smaller, less efficient farms. In a random sample the total acreage of each farm was recorded. The mean was 1125 acres, with a standard deviation of 250. The shape of the distribution of areas is not normal.
a. Approximately what proportion of areas is between 625 and 1625?
b. Approximately what proportion of areas is between 375 and 1875?

c. Approximately what proportion of areas is less than 375?

d. Approximately what proportion of areas is between 750 and 1500?

3.70 Physical Sciences During the spring, many rivers are monitored very carefully in an attempt to warn residents of an impending flood. The depth (in feet) of the Susquehanna River at the Bloomsburg bridge is measured and reported daily. In a random sample of depths, $\bar{x} = 16.7$, $s = 2.1$, and the shape of the distribution is approximately normal.

a. Approximately what proportion of depths is between 14.6 and 18.8?

b. Approximately what proportion of depths is less than 14.6?

c. Approximately what proportion of depths is between 14.6 and 23?

3.71 Biology and Environmental Science Many farmers use the height of their corn on July 4th as an indication of the entire crop. In a random sample of corn-stalk heights on July 4th in Columbia County, $\bar{x} = 25.6$, $s = 0.9$ inch, and a histogram of the observations was bell-shaped.

a. Approximately what proportion of observations is between 23.8 and 27.4 inches?

b. Approximately what proportion of observations is between 22.9 and 26.5 inches?

c. Approximately what proportion of observations is less than 27.4 inches?

3.72 Education and Child Development The Iowa Test of Basic Skills (ITBS) is a multiple-choice exam given to students in various grades in each state each year. The purpose is to test fundamental skills in reading, mathematics, language, social studies, and science. Scores are reported as state and/or national percentile points. Results from the 2004–2005 academic year indicate that third graders in Washington state scored at the 58th and 66th (national) percentiles in reading and mathematics, respectively.[8]

a. Interpret these values.

b. The 50th percentile (in any subject area) is the national average. Explain the meaning of *average* in this context.

c. Suppose a third grader scored at the 99th percentile (nationally) in mathematics. Interpret this result.

3.73 Travel and Transportation An architectural firm would like to evaluate two bicycle delivery services in New York City. The first service has a mean and standard deviation for delivery (in minutes) of 37 and 5. The second service has a mean of 42 with a standard deviation of 7. The company sent two test packages to the same location, one with each delivery service. The times to delivery were 33 and 35 minutes, respectively. Use *z*-scores to determine which service performed better.

3.74 Business and Management The Green Mill Restaurant and Bar in Wausau, Wisconsin, is advertising quick lunches with a mean waiting time of 11 minutes and a standard deviation of 2.5 minutes. The general manager (Rob Meyer, a former statistician) also claims the distribution of waiting times is approximately normal.

a. Suppose your waiting time is 13 minutes. Is there any reason to believe the general manager's claim is false? (Use a *z*-score.)

b. Suppose your waiting time is 20 minutes. *Now* is there any evidence to refute the general manager's claim?

3.75 Marketing and Consumer Behavior The time spent in a grocery store is an important issue for shoppers and for companies trying to market new products. A random sample of shoppers at a local grocery store was obtained and the shopping time (in minutes) for each was recorded. The data are given on the data CD and web site.

a. Construct a histogram for these data.

b. Use your histogram in part (a) to approximate the following percentiles: (i) 45th, (ii) 80th, (iii) 10th.

c. Compute the exact percentiles in part (b) and compare your results.

Extended Applications

3.76 Manufacturing and Product Development The lifetime of a motorcycle tire is affected by several factors, including tire pressure, riding conditions, and brand. The stock rear tire on a BMW R1200RT motorcycle, a Bridgestone 020, is very sensitive to road heat, and is expected to last between 3500 and 16,500 miles. A random sample of lifetimes of these tires was obtained and the data are given on the data CD and book's web site.[9]

a. Find the mean and the standard deviation of these tire lifetimes.

b. Find the actual proportion of observations within one standard deviation of the mean, within two standard deviations of the mean, and within three standard deviations of the mean.

c. Using the results in part (b), do you think the shape of the distribution of tire lifetimes is normal? Why or why not?

3.77 Sports and Leisure In 1974 Ernö Rubik created an imaginative and best-selling puzzle—the Rubik's cube. Many countries hold competitions in which participants try to solve this puzzle as fast as possible. The data CD and book's web site contain a sample from the top 100 United States times (in seconds), as of November 2008, to solve a 3×3×3 Rubik's cube.[10]

a. Find the actual proportion of observations within one standard deviation of the mean, within two standard deviations of the mean, and within three standard deviations of the mean.

b. Using the results in part (a), do you think the shape of the distribution of national record times is normal? Why or why not?

c. Construct a histogram for these data. Describe the shape of the distribution.

3.78 Manufacturing and Product Development Paint viscosity is a measure of thickness that determines whether the paint will cover in a single coat. A random sample of latex paint viscosities (in KU, or Krebs Units) was obtained, and the data are given in the following table.

113	124	141	115	115	129	113	129	112	112

a. Find the mean and the standard deviation for these data.

b. Find the *z*-score for each observation.

c. Find the mean and the standard deviation for all of the *z*-scores.

d. For *any* set of observations, can you predict the mean and standard deviation of the corresponding *z*-scores? Try to prove this result.

Challenge

3.79 Travel and Transportation According to the Massachusetts Bay Transportation Authority (MBTA), the ride from Chestnut Hill to Boston's Logan Airport on the MBTA takes less than 45 minutes. A random sample of travel times (in minutes) was obtained, and the results are given in the following table.

| 46.5 | 38.3 | 39.1 | 41.1 | 42.0 | 37.6 | 41.6 | 45.5 |
| 39.0 | 34.8 | 36.5 | 38.6 | 38.4 | 44.4 | 42.4 | |

a. Find the mean ($\bar{x}$) and the standard deviation (s) for these data set.

b. Compute each z-score and find $\sum z_i^2$.

c. Find a general formula for $\sum z_i^2$ for any data set.

3.80 Reconsider Example 3.16. Find a good *minimum* guaranteed life. That is, if a guinea pig fails to reach such an age, then the store would provide a refund.

3.4 FIVE-NUMBER SUMMARY AND BOX PLOTS

A **box plot**, or box-and-whisker plot, is a compact graphical summary procedure that conveys information about central tendency, symmetry, skewness, variability, and outliers. A standard box plot is constructed using the minimum and maximum values in the data set, the first and third quartiles, and the median. This collection of values is called the **five-number summary**.

DEFINITION

The **five-number summary** for a set of n observations $x_1, x_2, \ldots, x_n$ consists of the minimum value, the maximum value, the first and third quartiles, and the median.

> Recall that the range of a data set is the largest observation (maximum value) minus the smallest observation (minimum value). This descriptive statistic was our first attempt at measuring variability in a data set.

These five numbers do provide a glimpse of symmetry, central tendency, and variability in a data set. For example, minimum and maximum values that are very far apart suggest lots of variability. If the median is approximately halfway between the minimum and maximum values, that suggests the distribution is symmetric. A box plot is constructed as described below.

HOW TO CONSTRUCT A STANDARD BOX PLOT

Given a set of n observations $x_1, x_2, \ldots, x_n$:

1. Find the five-number summary $x_{min}, Q_1, \tilde{x}, Q_3, x_{max}$.
2. Draw a (horizontal) measurement axis. Carefully sketch a box with edges at the quartiles: left edge at Q_1, right edge at Q_3. (The height of the box is irrelevant.)
3. Draw a vertical line in the box at the median.
4. Draw a horizontal line (whisker) from the left edge of the box to the minimum value (from Q_1 to x_{min}). Draw a horizontal line (whisker) from the right edge of the box to the maximum value (from Q_3 to x_{max}).

> Recall that x_{min} denotes the minimum value and x_{max} denotes the maximum value.

Figure 3.34 illustrates this step-by-step procedure and shows a standard box plot with the five numbers indicated on a measurement axis. Note that the length of the box is the interquartile range. The box contains the middle half of the values.

The position of the vertical line in the box (median) and the lengths of the horizontal lines (whiskers) indicate symmetry or skewness and variability. Figure 3.35 shows a standard box plot for a distribution of data that is skewed to the right. The

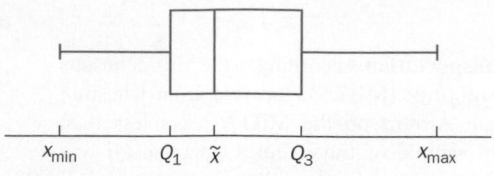

Figure 3.34 Standard box plot.

lower half of the data is in the interval from 3 to 4.5, while the upper half of the data is much more spread out, from 4.5 to 11. Figure 3.36 shows a standard box plot for a fairly symmetric distribution with lots of variability. The lower and upper half of the data are evenly distributed, but the whiskers extend far from each edge of the box. That is, 25% of the data are between 0 and 4, and 25% are between 7 and 11.

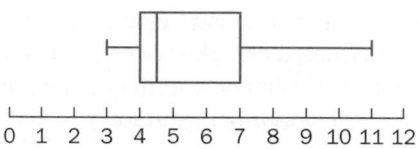

Figure 3.35 Standard box plot for data skewed to the right.

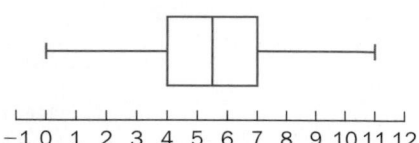

Figure 3.36 Standard box plot for a symmetric distribution.

Example 3.19 Blood Pressure The systolic blood pressures for 30 randomly selected subjects involved in a genetics research case study are given in the following table. Construct a standard box plot for these data.

177	122	128	191	180	142	197	196	67	160
167	138	107	188	102	116	138	114	188	176
148	175	169	203	135	142	168	181	168	150

(*Source*: Blood Pressure Case Study, Raymond Lam, M.D., GlaxoSmithKline, Canada.)

SOLUTION

STEP 1 Find the five-number summary:
$x_{min} = 67$, $Q_1 = 135$, $\tilde{x} = 163.5$, $Q_3 = 180$, $x_{max} = 203$.

STEP 2 Draw a measurement axis and sketch a box with edges at $Q_1 = 135$ and $Q_3 = 180$.

STEP 3 Draw a vertical line at the median $\tilde{x} = 163.5$.

STEP 4 Draw a horizontal line from $Q_1 = 135$ to $x_{min} = 67$ and another horizontal line from $Q_3 = 180$ to $x_{max} = 203$. The resulting box plot is shown in Figure 3.37.

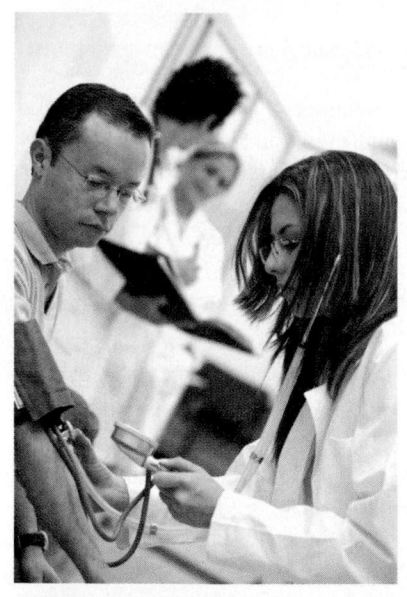

(*Andresr/Dreamstime.com*)

Tick marks for Q_1, $\tilde{x}$, and Q_3 are added to this graph for clarity.

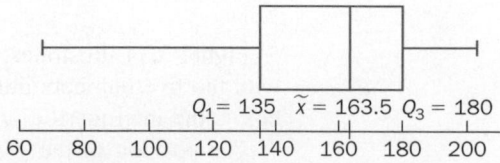

Figure 3.37 Standard box plot for the systolic blood pressure data.

A technology solution:

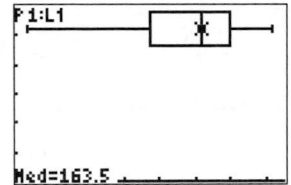

Figure 3.38 TI-84 Plus standard box plot.

STEP 5 The box plot suggests the data are negatively skewed, or skewed to the left. The lower half of the data is much more spread out than the upper half. ●

ILLUMINATING THE CONCEPTS

1. A box plot has only one measurement axis, and it may be horizontal or vertical. Many software packages, like Minitab, draw box plots with a vertical measurement axis by default. The construction and interpretations are the same. The `Transpose` option produces a box plot with a horizontal measurement axis.

2. The software does *not* usually include tick marks on the measurement axis for the five-number summary. The tick marks and scale are selected simply for convenience. ■

There are some disadvantages to using a standard box plot based on the five-number summary to describe a data set. By examining the graph, there is no way of knowing how many observations are between each quartile and the extreme. Each whisker is drawn from the quartile to the extreme, regardless of the number of observations in between. In addition, there are no provisions for identifying outliers. A standard (graphical) technique for distinguishing outliers is important since these values play an important role in statistical inference. Therefore, many statisticians prefer to use a **modified box plot** to graphically describe a data set. This type of graph still conveys information about center, variability, symmetry, and skewness, but it is also more precise and plots outliers.

HOW TO CONSTRUCT A MODIFIED BOX PLOT

Given a set of n observations $x_1, x_2, \ldots, x_n$:

1. Find the quartiles, the median, and the interquartile range:
   $Q_1, \tilde{x}, Q_3, IQR = Q_3 - Q_1$.

2. Compute the two inner *fences* (low and high) and two outer fences (low and high) using the following formulas:
   $$IF_L = Q_1 - 1.5(IQR) \quad IF_H = Q_3 + 1.5(IQR)$$
   $$OF_L = Q_1 - 3(IQR) \quad OF_H = Q_3 + 3(IQR)$$

   Think of the interquartile range as a *step*. The inner fences are 1.5 steps away from the quartiles, and the outer fences are 3 steps away from the quartiles.

3. Draw a (horizontal) measurement axis. Carefully sketch a box with edges at the quartiles: left edge at Q_1, right edge at Q_3. Draw a vertical line in the box at the median.

4. Draw a horizontal line (whisker) from the left edge of the box to the most extreme observation within the low inner fence. This line will extend from Q_1 to at most IF_L.
   Draw a horizontal line (whisker) from the right edge of the box to the most extreme observation within the high inner fence. This line will extend from Q_3 to at most IF_H.

5. Any observations between the inner and outer fences (between IF_L and OF_L, or between IF_H and OF_H) are classified as *mild outliers* and are plotted separately with shaded circles.
   Any observations outside the outer fences (less than OF_L, or greater than OF_H) are classified as *extreme outliers* and are plotted separately with open circles.

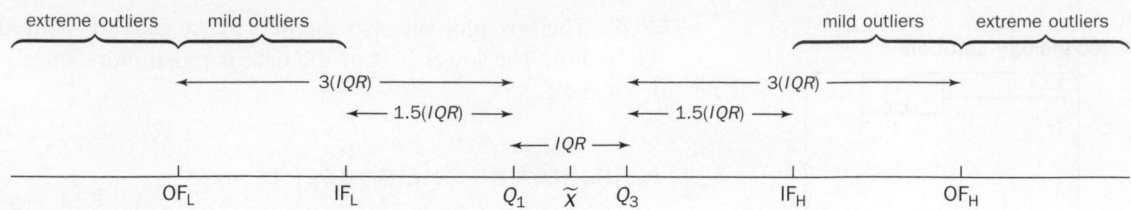

Figure 3.39 Construction points for a modified box plot.

Figure 3.39 shows the relationship between construction points for a modified box plot and the location of any outliers.

Example 3.20 Well Pump Lifetime Rural homes often have separate wells for their water supply. In a recent study on the extra costs of rural living, 30 homes were randomly selected and the lifetime (in years) of each well pump was recorded. The data are given in the following table. Construct a modified box plot for these data.

3.3	2.0	15.1	1.7	10.5	0.5	13.6	7.9	3.1	5.1
0.4	21.6	0.5	1.3	3.1	2.5	2.6	2.0	1.7	2.9
3.3	0.3	5.8	5.3	0.9	10.5	0.5	11.1	1.3	3.5

SOLUTION

STEP 1 Find the quartiles, the median, and the interquartile range.

$$Q_1 = 1.3, \tilde{x} = 3.0, Q_3 = 5.8, IQR = 5.8 - 1.3 = 4.5.$$

STEP 2 Find the inner and outer fences.

$$IF_L = 1.3 - (1.5)(4.5) = 1.3 - 6.75 = -5.45$$
$$IF_H = 5.8 + (1.5)(4.5) = 5.8 + 6.75 = 12.55$$
$$OF_L = 1.3 - (3)(4.5) = 1.3 - 13.5 = -12.2$$
$$OF_H = 5.8 + (3)(4.5) = 5.8 + 13.5 = 19.3$$

STEP 3 Draw a (horizontal) measurement axis. Carefully sketch a box with edges at the quartiles: left edge at Q_1, right edge at Q_3. Draw a vertical line in the box at the median.

STEP 4 Draw a horizontal line (whisker) from the left edge of the box to the most extreme observation within the inner fence IF_L (0.3).

Draw a horizontal line (whisker) from the right edge of the box to the most extreme observation within the inner fence IF_H (11.1).

STEP 5 Plot any mild outliers, observations between -12.2 and -5.45, or between 12.55 and 19.3. There are two mild outliers, 13.6 and 15.1.

Plot any extreme outliers, observations less than -12.2 or greater than 19.3. There is one extreme outlier, 21.6.

STEP 6 The resulting modified box plot is shown in Figure 3.41. The box plot suggests the data are positively skewed, or skewed to the right. The upper half of the data is much more spread out than the lower half. There are two mild outliers and one extreme outlier.

A technology solution:

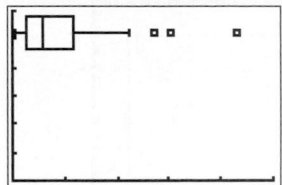

Figure 3.40 TI-84 Plus modified box plot.

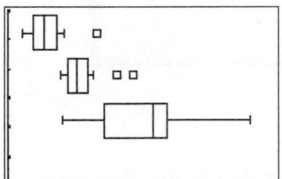

Figure 3.42 TI-84 Plus box plots for gasoline data.

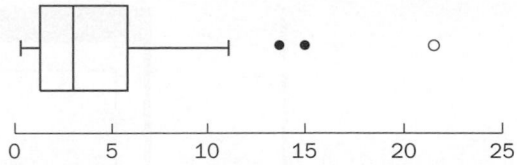

Figure 3.41 Modified box plot for the well pump data.

Note: IF$_L$ and OF$_L$ are negative even though an observed pump lifetime cannot be less than 0 years. That's OK. This is a correct statistical calculation, not a contradiction, even though it seems odd. ●

💡 ILLUMINATING THE CONCEPTS

When we compare two (or more) data sets graphically, the corresponding box plots may be placed on the same measurement axis (one above the other using a horizontal axis, or side-by-side with a vertical axis). Figure 3.42 shows three box plots on the same measurement axis, representing the number of gallons of gasoline pumped in randomly selected vehicles at three different stations. Exercises 3.94 and 3.95 ask for this kind of comparative plot. ■

TECHNOLOGY CORNER

Procedure: Construct a box plot.
Reconsider: Example 3.20, page 106, solution, and interpretation.

TI-84 Plus

The TI-84 Plus has two built-in statistical plots, a standard and a modified box plot. The modified box plot does not distinguish between mild and extreme outliers.

1. Enter the data into list L1.
2. Press ⟨STATPLOT⟩ and select Plot1 from the STATPLOT menu.
3. Turn the plot On and select Type box plot (modified or standard). Set Xlist to the name of the list containing the data and Freq to 1. Choose a Mark for outliers (if constructing a modified box plot).
4. Enter appropriate WINDOW settings. Press ⟨GRAPH⟩ to display the box plot. A standard box plot is shown in Figure 3.38, page 105, and a modified box plot is shown in Figure 3.40.

Minitab

The Minitab modified box plot does not distinguish between mild and extreme outliers.

1. Enter the data into column C1.
2. Select Graph; Boxplot and choose a One Y; Simple box plot.
3. Enter C1 in the Graph variables window. Select the Scale options button and check Transpose value and category scales to construct a box plot with a horizontal measurement axis.
4. Select the Data view options button and check Interquartile range box, and Outlier symbols for a modified box plot. Note there are only two outliers in this box plot due to the numerical method Minitab uses to compute quartiles.

Excel

The following steps may be used to construct a standard box plot. Additional calculations and options are necessary to construct a modified box plot.

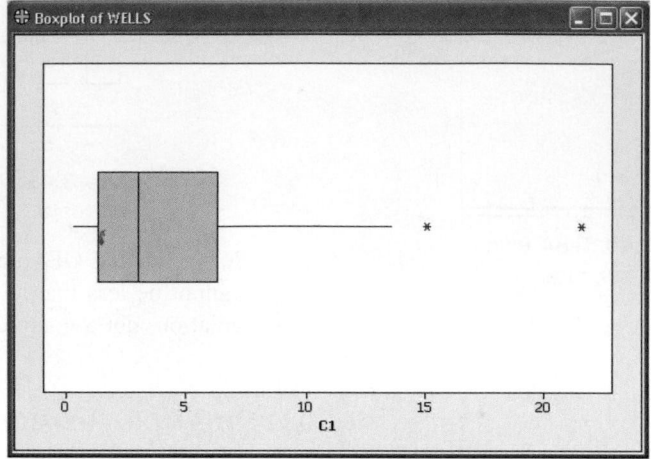

Figure 3.43 Minitab modified box plot.

1. Enter the data into column A.
2. Find the five-number summary in the order shown. Highlight these five cells.
3. Under the Insert tab, select a Line chart with Markers.
4. Under Chart tools; Design, select Switch Row/Column.
5. Right-click on the point representing the minimum value. Select Format Data Series; Line Color and choose No line. Repeat this process for each point.
6. Select any point on the graph. Under the Chart Tools; Layout tab, select Line; High-Low Lines. In addition, select Up/Down Bars; Up/Down Bars.
7. Format other graph items as appropriate.

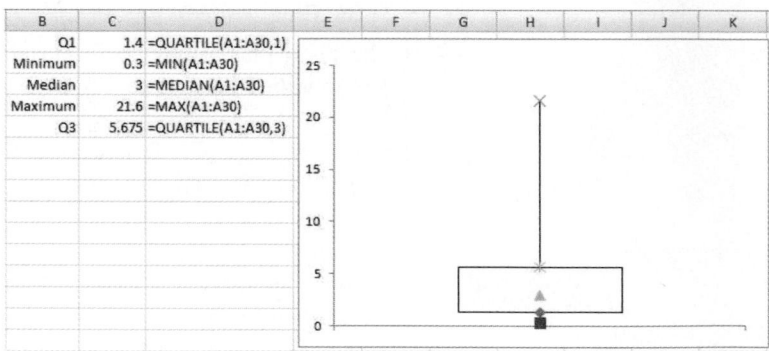

Figure 3.44 Excel standard box plot.

SECTION 3.4 EXERCISES

Practice

3.81 Find the five-number summary for each data set.
a. {34, 40, 34, 32, 32, 40, 35, 35, 28, 35}
b. {57, 65, 70, 71, 67, 56, 52, 66, 74, 57, 67, 78}
c. {94, 80, 91, 94, 83, 92, 83, 93, 96, 80, 87, 98, 81, 93}
d. {2.3, 1.8, 2.1, 1.0, 2.4, 2.3, 0.4, 9.8, 0.6, 1.4, 3.1, 10.9, 3.8, 0.5, 0.9, 2.2, 1.3, 1.3}
e. {166.8, 103.1, 119.9, 141.9, 110.6, 189.8, 121.6, 141.6, 133.6, 178.2, 158.9, 145.9, 139.1, 148.6, 135.0, 174.0, 152.4, 119.7, 196.9, 118.7, 159.7, 150.3, 113.8, 108.9, 163.2}
f. {−33.8, −9.8, −18.5, −11.5, −36.3, −33.1, −21.1, −26.2, −25.4, −32.1, −35.9, −28.0, −38.2, −12.0, −29.2, −40.1, −13.1}

3.82 Construct a standard box plot for each five-number summary.
a. $x_{min} = 15.3$, $Q_1 = 21.8$, $\tilde{x} = 25.3$, $Q_3 = 28.2$, $x_{max} = 34.2$
b. $x_{min} = 70.9$, $Q_1 = 167.8$, $\tilde{x} = 187.1$, $Q_3 = 225.3$, $x_{max} = 329.3$
c. $x_{min} = 0.06$, $Q_1 = 5.3$, $\tilde{x} = 13.7$, $Q_3 = 30.8$, $x_{max} = 122.3$
d. $x_{min} = 10.1$, $Q_1 = 10.7$, $\tilde{x} = 11.3$, $Q_3 = 12.5$, $x_{max} = 26.7$

3.83 For each data set with Q_1 and Q_3 given, find the interquartile range and the inner and outer fences.
a. $Q_1 = 22$, $Q_3 = 46$
b. $Q_1 = 1255$, $Q_3 = 1306$
c. $Q_1 = 65.75$, $Q_3 = 75.21$
d. $Q_1 = 914.9$, $Q_3 = 1140.5$

e. $Q_1 = 1.275$, $Q_3 = 4.07$
f. $Q_1 = 0.265$, $Q_3 = 2.51$
g. $Q_1 = -33.67$, $Q_3 = -23.90$
h. $Q_1 = 98.43$, $Q_3 = 98.81$

3.84 For each data set with Q_1 and Q_3 given, determine whether the observation x is a mild outlier, an extreme outlier, or neither.
a. $Q_1 = 20$, $Q_3 = 29$, $x = 35$
b. $Q_1 = 486.1$, $Q_3 = 510.9$, $x = 440$
c. $Q_1 = 5.18$, $Q_3 = 6.32$, $x = 4.2$
d. $Q_1 = 96.3$, $Q_3 = 101.1$, $x = 116.5$
e. $Q_1 = 68.92$, $Q_3 = 69.07$, $x = 68.4$
f. $Q_1 = 101.26$, $Q_3 = 144.59$, $x = 132.6$

3.85 For each box plot, find the five-number summary. (Estimate these numbers the best you can by using the tick marks on each graph.)

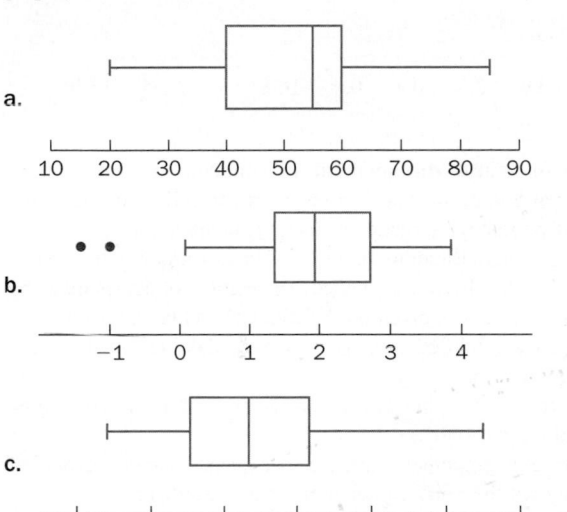

a.

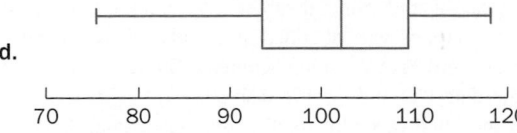

b.

c.

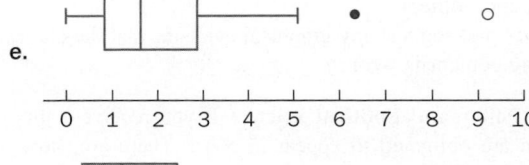

d.

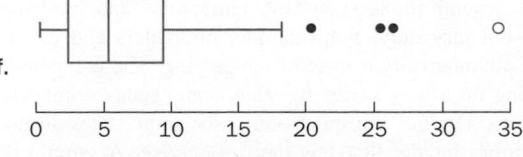

e.

f.

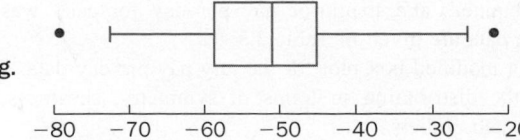

g.

Applications

3.86 Business and Management A Roth's supermarket in Salem, Oregon, is using a new statistical tool to help in ordering bottles of raspberry iced tea. A random sample of the number of bottles sold per day is given in the following table. Construct a modified box plot for these data. Describe the distribution of the number of bottles sold.

48	52	46	58	50	46	59	51	46	48
45	47	50	48	49	49	49	48	48	45

3.87 Travel and Transportation Major airlines compete for customers by advertising on-time arrival. A recent survey from seven major airports and randomly selected airlines compared the actual arrival time with the scheduled arrival time. The differences (in minutes) are given in the following table (negative numbers indicate the flight arrived before the scheduled arrival time). Construct a modified box plot for these data. Describe the distribution in terms of symmetry, skewness, and variability. Are there any outliers? If so, are they mild or extreme?

22	19	28	11	50	16	23	26	22
-11	17	45	7	6	-18	16	33	20
25	25	19	24	51	-2	48		

(*Source*: Bureau of Transportation Statistics, U.S. Department of Transportation.)

3.88 Public Policy and Political Science A recent study reported the school tax bills (in dollars) for randomly selected families in Hillsboro, New Hampshire, a small rural community. Use the following modified box plot to describe the distribution of the data.

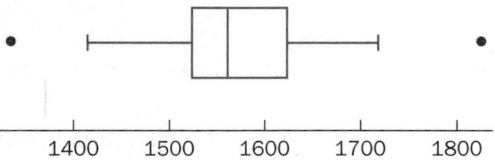

3.89 Education and Child Development Dr. Jan Remer, a psychologist in Hermitage, Tennessee, randomly selected 6-year-olds and recorded the time (in minutes) each child needed to complete a 20-piece picture puzzle. The data were used to predict *readiness* for first grade. The following standard box plot for the data was drawn. Describe the distribution of completion times.

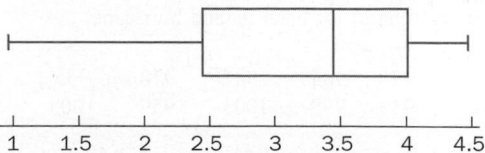

3.90 Public Health and Nutrition As part of a new physical fitness program, the Crooked Oak Middle School in Oklahoma City, Oklahoma, records the number of sit-ups each sixth-grade student can do in one minute. A random sample for males and females was obtained and the following modified box plots were drawn. Describe the male and female data separately. What similarities and/or differences do the box plots suggest?

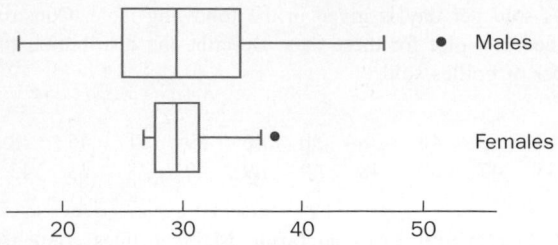

3.91 Economics and Finance Many government program budgets are determined by the annual inflation rate. The inflation rate (as a percent change in the consumer price index since 1913) for the United States for the years 1965–2007 are given on the data CD and book's web site.[11] Construct a modified box plot for these data. Describe the distribution of inflation rate over the past 40 years in terms of symmetry, skewness, variability, and outliers.

3.92 Public Policy and Political Science According to US Vending Company, the national average commission per machine is approximately 35 dollars (*Source*: www.usvending.com, customer care frequently asked questions). Suppose the Culver City, California, City Council recently entered into an agreement with Coca-Cola Bottling Company. As part of this contract, two US Vending snack machines were installed, one at Veteran's Park and the other at the Teen Center. The 2008 monthly commission checks (in dollars) associated with these machines are given in the following table.

51.24	27.60	37.80	44.76	24.36	14.04
119.52	123.48	38.48	35.28	16.92	47.25
25.65	31.95	40.50	22.50	33.75	59.85
36.90	18.45	31.95	18.45	19.50	22.65

a. Construct a modified box plot for these data. Describe the distribution in terms of symmetry, skewness, variability, and outliers.
b. Construct a standard box plot for these data. Which plot do you think is more descriptive? Why?

3.93 Biology and Environmental Science Several weather centers across the country carefully track and record data for tropical disturbances during the hurricane season. The following table from the 2008 Atlantic hurricane season contains the minimum central pressure (in millibars) for each named hurricane.

1005	948	964	941	978	935	994
984	993	998	1005	959	1003	943

(*Source*: Unisys Corporation.)

Construct a modified box plot for the hurricane data. Describe the distribution in terms of symmetry, skewness, variability, and outliers. How would the graph change if a standard box plot were used?

Extended Applications

3.94 Public Policy and Political Science The Beverly Hills Police Department has decided to use cameras to catch motorists who run red lights. Information at www.highwayrobbery.net suggests the city has shortened the amber-light time at intersections with cameras. A random sample of amber-light times (in seconds) at intersections with and without cameras is given. Construct a modified box plot for each set of data on the same measurement axis. Describe and compare the distributions. Do the box plots support the web site's claim of shorter amber-light times?

No camera

7.0	8.3	7.7	4.5	9.7	7.2	5.5	11.5	7.4
8.1	8.2	8.3	10.3	3.0	5.0	8.4		

Camera

4.9	3.2	2.3	0.8	0.4	16.8	3.2	1.4	15.0

3.95 Business and Management In an attempt to control costs, a manager at San José State University is carefully looking at the number of photocopies made by faculty members at the campus duplicating center. Random samples of liberal arts faculty and of natural science faculty were obtained. The number of copies made by each faculty member is given on the data CD and book's web site.
a. Construct a modified box plot for each data set on the same measurement axis.
b. Compare the distributions in terms of symmetry, skewness, variability, and outliers.
c. Is there any graphical evidence to suggest that one group of faculty uses the copy center more than the other?

3.96 Public Health and Nutrition The Food and Drug Administration has become concerned about claims made by companies selling vitamin and mineral tablets. A random sample of 400-mg vitamin C tablets was obtained and analyzed by an independent laboratory for exact vitamin C content. The results (in mg) are given on the data CD and book's web site.
a. Construct a modified box plot for the vitamin C data.
b. Describe the distribution in terms of symmetry, skewness, variability, and outliers.
c. Does the box plot suggest any graphical evidence that the claim of a 400-mg content is wrong?

3.97 Public Policy and Political Science If you receive a jury summons, you are obligated to appear in court. There are, however, several general (honest) instant, temporary, and hardship excuses to avoid jury duty. For example, firefighters and physicians may be automatically excused from serving. The compensation for serving on a jury varies by state; some states reimburse childcare expenses and/or transportation costs; some states do not compensate jurors for the first few days of service. A sample of states was obtained, and the juror pay per day for each was recorded. The data are given in Table 3.3.
a. Construct a modified box plot for the jury pay per day data.
b. Describe the distribution in terms of symmetry, skewness, variability, and outliers.

Table 3.3

State	Pay	State	Pay
Alabama	10.00	Alaska	25.00
Arkansas	20.00	California	15.00
Delaware	20.00	Georgia	5.00
Idaho	10.00	Illinois	4.00
Kentucky	12.50	Louisiana	25.00
Maryland	15.00	Massachusetts	50.00
Missouri	6.00	Nebraska	35.00
New Mexico	5.15	North Carolina	12.00
Pennsylvania	9.00	Texas	6.00
Virginia	30.00	Wyoming	30.00

(*Source*: National Center for State Courts.)

c. In January 2006, the $6.00 per day rate in Texas was increased to $40.00. Change the value for Texas to $40.00 and construct a new modified box plot. How does this new box plot compare with the one in part (a)? Describe any similarities and/or differences.

3.98 Economics and Finance In January 2006, Wisconsin joined at least 18 other states by posting on web sites the names of people and businesses owing back taxes. The law in Wisconsin requires the Department of Revenue to list those who owe at least $25,000. These "web sites of shame" are designed to help states collect additional tax money during tight budget times. A random sample of names from the Wisconsin list was obtained, and the taxes owed by each individual or business are given on the data CD and book's web site. (*Source*: Wisconsin Department of Revenue.) Construct a modified box plot for these data and describe the distribution in terms of symmetry, skewness, variability, and outliers. Find a symmetric interval about the mean in which 50% of the values in this data set lie.

Empirical Rule	94	If a distribution is approximately normal, the proportion of observations within one, two, and three standard deviations about the mean is approximately 0.68, 0.95, and 0.997, respectively.
z-score	97	$z_i = \dfrac{x_i - \bar{x}}{s}$, how far an observation is from the mean in standard deviations.
Percentiles	98	The percentiles divide a data set into 100 parts.
Five-number summary	103	$x_{min}, Q_1, \tilde{x}, Q_3, x_{max}$
Box plot	103	A graphical description of a data set, constructed using the five-number summary. The graph conveys information about central tendency, symmetry, skewness, and variability.
Modified box plot	105	A graphical description of a data set, constructed using $\tilde{x}, Q_1, Q_3,$ IQR, and the inner and outer fences. This box plot also indicates any outliers.

CHAPTER 3 EXERCISES

Applications

3.99 Public Health and Nutrition Most multivitamins contain calcium to maintain strong bones and to lower the risk of heart disease. A random sample of multivitamins was obtained, and the calcium content for each is given in the following table.

156	151	173	201	182	166	173	180	174	185
160	178	173	169	203	190	187	202	173	171

a. Find the mean, the variance, and the standard deviation.
b. Find the proportion of observations within one, two, and three standard deviations about the mean.
c. Using the proportions obtained in part (b), do you think the distribution of observations is normal? Why or why not?

3.100 Manufacturing and Product Development Many box cake mixes include special high-altitude baking instructions. To determine any difference between baking times at low and high altitudes, the consumer group Public Citizen made several similar cakes in 9-inch round pans in Miami and Denver, and carefully recorded the time it took the cakes to bake (in minutes). The data are given in the following table.

Low-altitude times (Miami)

25.1	25.6	24.9	23.7	25.5	22.4	24.7	24.2	25.6
24.8	23.9	24.4	24.7	24.4	26.4	24.7	24.7	26.8
24.9	24.3							

High-altitude times (Denver)

22.8	30.0	27.3	30.3	28.3	31.1	27.0	26.8	26.3
29.1	23.5	26.2	29.2	23.0			

a. Construct a modified box plot for each data set on the same measurement axis.
b. Describe each box plot in terms of center, shape, and spread.
c. Describe the similarities and differences between the two distributions.

3.101 Sports and Leisure The longest running U.S.-produced, fictional content television show by number of episodes is *Gunsmoke*. Other long-running shows of this type include *Lassie*, *Ozzie and Harriett*, and *Bonanza*. *Law and Order* is currently number 10 with approximately 350 shows. The number of episodes for some of the top 115 shows are given in the following table.

633	588	435	430	381	369	361	344	336	296
203	203	202	201	201	200	199	197	196	194
284	265	251	236	193	174				

(*Source*: www.angelfire.com/trek/proutsy)

a. Find the median, the first and third quartiles, and the interquartile range.
b. Find the 30th and the 95th percentiles.
c. Suppose the *Simpsons* currently have 438 episodes (as of November 2008). Using the data in the table, in what percentile does this episode count lie?

3.102 Manufacturing and Product Development A typical wind turbine in the United Kingdom has a rated mean electrical generating capacity of 0.66 MW, with a standard deviation of 0.07 MW. A quality-control engineer is trying to develop a plan for routine maintenance based on z-scores.

a. Suppose a randomly selected wind turbine is inspected and found to have a generating capacity of 0.74 MW. Is there any reason to believe this generating capacity is unusual? Why or why not?
b. Suppose another randomly inspected wind turbine has a generating capacity of 0.44 MW. Is there any reason to believe this generating capacity is unusual? Why or why not?

3.103 Sports and Leisure String tension in tennis rackets is usually measured in pounds. Recommended string tensions are usually in the mid 60s (pounds) for oversize rackets, and high 50s to low 60s for mid-overs. Higher tensions tend to decrease the size of the sweet spot and reduce power, but increase control. The data CD and book's web site present the results from a random sample of string tension from tennis rackets of players on the professional tour.

a. Find the range, sample variance, interquartile range, coefficient of variation, and coefficient of quartile variation for each type of racket.
b. Using the results from part (a), compare the variability in string tension for the two types of rackets.
c. Construct a modified box plot for each type of racket on the same measurement axis. Does this graphical comparison support your numerical comparison in part (b)?

3.104 Manufacturing and Product Development Many homes that use forced hot air for heat have air ducts installed in every room. A system using galvanized pipe is constructed to distribute heat throughout the house. A random sample of 6-inch-diameter, 5-foot-long, 28-gauge galvanized pipe was obtained from various manufacturers and the weights (in pounds) are given on the data CD and book's web site.

a. Find the sample mean and the sample median.
b. Use your results in part (a) to describe the symmetry of the distribution.
c. Find a 10% trimmed mean. Is the use of a trimmed mean to measure central tendency justified (or necessary) in this case? Why or why not?

3.105 Medicine and Clinical Studies Some health reports suggest that 300 mg of caffeine (the amount in about three cups of coffee) is a moderate intake (*Source*: National Institute of Environmental Health Sciences). The amount of caffeine in a cup of coffee varies according to coffee bean, brewing technique, filter, etc. A random sample of 8-ounce cups of coffee was obtained and the caffeine content (in mg) was measured. The data are given in the following table.

89	75	90	115	88	96	107	106	93
97	95	101	115	112	100	71	109	89

a. Find the mean, median, variance, and standard deviation.
b. Construct a modified box plot for these data.
c. Use your results in parts a and b to describe the data.
d. Based on your results in parts (a) and (b), do you believe a person who drinks three cups of coffee ingests a moderate amount of caffeine?

3.106 Sports and Leisure A video game is often judged by the amount of time (in hours) it takes to *beat* the game. A new video game called "Blue Dragon" is being test marketed. The game is given to randomly selected players and the time it takes for each player to beat the game is recorded. The times (in hours) are given on the data CD and book's web site.

a. Find the mean, variance, and standard deviation.
b. Find the proportion of observations within one, two, and three standard deviations about the mean. Use these proportions to determine whether the distribution of times is approximately normal.
c. The video game manufacturer, Namco, would like to reward the best (fastest) players with gift certificates. Using the data above, find a time that will give only 10% of all players a gift certificate.

3.107 Education and Child Development The time (in minutes) it takes to read a certain passage is part of an elementary school assessment test. Two different groups of children were given the same passage to read. One group received a standard reading curriculum, and the other was given reading instruction that was based on whole language. The results are given on the data CD and book's web site.

a. Find the mean, variance, and standard deviation for each group.
b. Construct a modified box plot for each group and display the graphs on the same measurement axis.
c. Based on your results in parts (a) and (b), how, if at all, has whole-language instruction changed reading speed?

3.108 Physical Sciences A standard often used for measuring brightness is lux. For example, bright moonlight has 0.1 lux and bright sunshine has 100,000 lux. The light required for general office work is approximately 400 lux. A random sample of the brightness in typical office cubicles was obtained and the data are given on the data CD and book's web site.

a. Find the mean, variance, and standard deviation.
b. Construct a modified box plot for these data. Classify any outliers as mild or extreme.
c. Using the data in the table, in what percentile does 400 lux lie?
d. Use Chebyshev's Rule to describe this data set ($k = 2, 3$).

3.109 Manufacturing and Product Development The density of tires is an important selling point for serious mountain-bike riders. The tire industry uses a type A durometer to measure the indentation hardness for mountain bike tires. Suppose the distribution of tire hardness is approximately normal with mean 45 and standard deviation 7.

a. Carefully sketch the normal curve for tire hardness.
b. Is a tire hardness of 30 unusually soft? Justify your answer.
c. A certain bicycle shop claims the hardness of all its tires is in the 84th percentile. If this is true, what is the minimum hardness of any tire in the store?

3.110 Physical Sciences In 2008, there were approximately 24,675 earthquakes around the world as of November 11, 2008. A random sample of the magnitudes (on the Richter scale) of these earthquakes during the first 11 days of November is given in the following table.

1.3	1.4	1.4	2.0	4.5	1.6	2.5	1.3	1.0	1.7
2.9	2.5	2.0	2.6	1.4	1.8	1.0	1.6	2.5	3.3
1.6	4.9	3.0	1.7	1.7	1.9	1.6	1.9		

(*Source*: National Earthquake Information Center, U.S. Geological Survey.)

a. Find the mean, median, variance, and standard deviation of the magnitudes.
b. Find the 40th and the 80th percentiles.
c. How likely is a magnitude of 5.7? Justify your answer.

Extended Applications

3.111 **Physical Sciences** In a random sample of depths (in feet) of U.S. drilling rigs actively developing oil in 2001, $\bar{x} = 5033$ and $s = 1250$. Assume the distribution of depths is approximately normal.

a. What proportion of observations is between 3783 and 6283 feet?

b. What proportion of observations is less than 1283 feet?

c. What proportion of observations is between 5033 and 8783 feet?

d. Suppose a new oil well was drilled in 2005 to a depth of 6515 feet. Is there any evidence to suggest that the average depth of wells has changed?

3.112 **Physical Sciences** A building code officer inspected random home fire extinguishers for pressure (in psi) and the data are given on the data CD and book's web site.

a. Construct a modified box plot for these data.

b. Use the Empirical Rule to decide whether this distribution of pressures is approximately normal.

c. Create a new set of observations, $y_i = \ln(x_i)$, where ln is the natural logarithm function. Construct a modified box plot for this new set of data. Use this graph and the Empirical Rule to decide whether the distribution of the transformed data is approximately normal.

3.113 **Manufacturing and Product Development** Jelly Belly Candy Company claims its jelly beans are produced with mean volume 3.53 cm^3 and standard deviation 0.1 cm^3.

a. Without assuming anything about the shape of the distribution of jelly bean volumes, what proportion of jelly beans have volumes between 3.33 and 3.78 cm^3? Less than 3.23 cm^3?

b. Suppose a random jelly bean has volume 3.1 cm^3. Do you believe the manufacturer's claim about the mean volume? Justify your answer.

3.114 **Manufacturing and Product Development** The actual width of a 2 × 4 piece of lumber is approximately 1.75 inches, but can vary considerably. The Lumber Yard in Martinsburg, West Virginia advertises consistent dimensions for better building, and claims that all 2 × 4s sold have a mean width of 1.75 inches with a standard deviation of 0.02 inch.

a. Assume the distribution of widths is approximately normal. Find a symmetric interval about the mean that contains almost all of the 2 × 4 widths.

b. Suppose a random 2 × 4 has width 1.79 inches. Is there any evidence to suggest that The Lumber Yard's claim is wrong? Justify your answer.

c. Suppose a random 2 × 4 has width 1.68 inches. Is there any evidence to suggest that The Lumber Yard's claim is wrong? Justify your answer.

3.115 **Biology and Environmental Science** Some fish have been found to have mercury levels greater than 1 ppm (parts per million), a level considered safe by the Food and Drug Administration. Suppose the mean mercury level for smallmouth bass in the Susquehanna River is 0.7 ppm with standard deviation 0.1 ppm, and the distribution of mercury level is approximately normal.

a. Is it likely a fisherman will catch a smallmouth bass with mercury level greater than 1 ppm? Justify your answer.

b. Suppose the standard deviation is 0.05 ppm. Now, is it likely a fisherman will catch a smallmouth bass with mercury level greater than 1 ppm? Justify your answer.

c. Carefully sketch the normal curves in parts (a) and (b) on the same measurement axis.

3.116 **Sports and Leisure** In January 2006 *Phantom of the Opera* became the longest running Broadway show with over 7486 performances, surpassing *Cats*. A sample of Broadway shows was obtained, and the number of performances of each was recorded. The data are given in Table 3.4.

a. Find the sample mean and the sample median number of performances. What do these values suggest about the shape of the distribution?

b. Find the sample variance and the sample standard deviation. Find the proportion of observations within one, two, and three standard deviations of the mean. What do these proportions suggest about the shape of the distribution?

c. Find the first quartile, the third quartile, and the interquartile range. Construct a modified box plot for the performance data. Use this graph to describe the distribution in terms of symmetry, skewness, variability, and outliers. Does your description based on the box plot agree with your answers in parts (a) and (b)? Why or why not?

d. Find out how many performances there have been to date for *Phantom of the Opera* and add this value to the data set. How will this value affect the sample mean, sample median, sample variance, and quartiles? Find these values and verify your predictions.

Table 3.4

Show	Performances	Show	Performances	Show	Performances
Cats	7485	*Les Misérables*	6680	*A Chorus Line*	6137
Miss Saigon	4097	*42nd Street*	3486	*Grease*	3388
Fiddler on the Roof	3242	*Life With Father*	3224	*Tobacco Road*	3182
My Fair Lady	2717	*Cabaret* (Revival)	2378	*Annie*	2377
Man of La Mancha	2328	*Oklahoma!*	2212	*Pippin*	1944
South Pacific	1925	*The Magic Show*	1920	*Aida*	1852
Deathtrap	1793	*Harvey*	1775	*La Cage Aux Folles*	1761
Hair	1750	*The Wiz*	1672	*Crazy for You*	1622
Ain't Misbehavin'	1604	*Evita*	1567	*Jekyll & Hyde*	1543
Barefoot in the Park	1530	*Brighton Beach Memoirs*	1530	*Dreamgirls*	1522
Mame	1508	*Same Time, Next Year*	1453	*Arsenic and Old Lace*	1444
The Sound of Music	1443	*Me and My Girl*	1420	*The Music Man*	1375
Funny Girl	1348	*Movin' Out*	1303	*Angel Street*	1295
Lightnin'	1291	*Promises, Promises*	1281	*The King and I*	1246
Sleuth	1222	*1776*	1217	*Equus*	1209
Guys and Dolls	1200	*Amadeus*	1181	*Cabaret*	1165
Mister Roberts	1157	*Annie Get Your Gun*	1147	*Guys and Dolls* (Revival)	1144
Butterflies Are Free	1128	*Pins and Needles*	1108	*Plaza Suite*	1097
They're Playing Our Song	1082	*Grand Hotel*	1077	*Kiss Me, Kate*	1070
The Pajama Game	1063	*Shenandoah*	1050	*Damn Yankees*	1019
Contact	1010	*Never Too Late*	1007	*Any Wednesday*	982
Sunset Boulevard	977	*The Odd Couple*	964	*Anna Lucasta*	957
Kiss and Tell	956	*Show Boat* (Revival)	951	*Dracula* (Revival)	925
The Moon Is Blue	924	*Beatlemania*	920	*Proof*	917
The Elephant Man	916	*Kiss of the Spider Woman*	906	*The Who's Tommy*	900
Chicago	898	*Applause*	896	*Carousel*	890
Hats Off to Ice	889	*Children of a Lesser God*	887	*Follow the Girls*	882
City of Angels	878	*Camelot*	873	*The Bat*	867
My Sister Eileen	864	*Ragtime*	861	*Song of Norway*	860
A Streetcar Named Desire	855	*Barnum*	854	*Raisin*	847
Blood Brothers	839	*You Can't Take It With You*	837	*The Subject Was Roses*	832
Black and Blue	824	*Inherit the Wind*	806	*Titanic*	804

(*Source*: Hernandez, Long Runs on Broadway, *Playbill*, January 4, 2006.)

Probability

4

Chapter 4 Challenge

"I tested positive, but I didn't use anything illegal!" Should you believe this athlete?

The International Olympic Committee (IOC) has become increasingly concerned about athletes using banned substances, especially the endurance-boosting hormone erythropoietin (EPO). This substance is one of the most widely abused drugs in sports. It enhances endurance by stimulating the production of red blood cells. Athletes involved in cycling, long-distance running, and many winter sports have been known to use this drug. Endurance athletes are subject to blood tests at least one day before competitions. The testing procedure can detect the use of EPO up to five days earlier. Athletes may also be subject to blood tests on the morning of competition.

(View Stock/AgeFotostock)

ESPN reported that over 5,000 doping tests were conducted at the 2008 Beijing Olympics, an increase of 1,400 from the Athens Games in 2004. Approximately 3,000 doping tests were done at the 2000 Sydney Olympics. A record 24 doping cases were reported by the IOC during the Athens Games and The Associated Press (September 19, 2008) reported that six athletes tested positive at the Beijing Olympics.

Suppose an athlete tests positive for the presence of EPO. Is he or she necessarily guilty of using this banned substance? Should the athlete be barred from competition? Let's not be too quick to judge. No drug test is perfect, and a positive test result may be due to the use of a simple over-the-counter supplement or some other cause. The techniques presented in this chapter will be used in the Chapter Challenge Wrap-Up (page 178) to determine the probability that an athlete with a positive test result really is using a banned substance.

Review

Understand the relationships among a population, a sample, probability, and statistics.

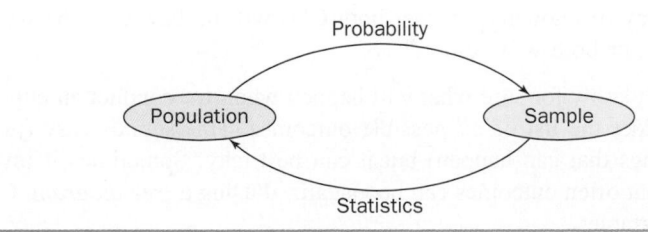

CONTENTS

Preview

- Learn the definition of probability and useful probability concepts.
- Compute the probability of various events involving counting techniques, independence, or conditional probability.

4.1 EXPERIMENTS, SAMPLE SPACES, AND EVENTS

In order to understand probability concepts, we need to think closely about **experiments**. Consider the activity, or act, of tossing a coin, selecting a card from a standard poker deck, measuring the number of contaminants in 1 cm^3 of drinking water, or even testing a computer chip for defects. In every one of these activities, the outcome is uncertain. For example, when we test a computer chip, we do not know (for sure) whether it will be defect free. This idea of uncertainty leads to the definition of an experiment.

> **DEFINITION**
>
> An **experiment** is an activity in which there are at least two possible outcomes and the result of the activity cannot be predicted with absolute certainty.

Here are some examples of experiments.

1. Roll a six-sided die and record the number that lands face-up.

   Since we cannot say with certainty whether the number face-up will be a 1, or a 2, etc., this activity is an experiment.

2. Using a radar gun, record the speed of a pitch at a Red Sox baseball game.

   We're not sure whether the pitch will be a fastball, curveball, slider, etc. And even if we steal the signal from the catcher, we cannot predict the speed of the pitch with certainty.

3. Count the number of patients who arrive at the emergency room of a city hospital during a 24-hour period.

   Although past records might help us form an estimate, there is no way of predicting the exact number of emergency room patients.

4. Select two manufactured CDs at random and inspect them for flaws (scratches, warping, etc.).

   There is no way of knowing whether both CDs will be flawless, one will contain a flaw, or both will be defective.

Since we don't know for sure what will happen when we conduct an experiment, we need to consider the list of *all* possible outcomes. This sounds easy (just think about all the things that can happen) but it can be tricky. Sometimes it involves a lot of counting, but often outcomes can be visualized using a *tree diagram*. Consider the following examples.

This is an experiment, since we cannot predict the last digit with certainty.

Example 4.1 Social Security Numbers   Suppose a U.S. citizen is selected and the last digit of her social security number is recorded. How many possible outcomes are there, and what are they?

SOLUTION

STEP 1   The last digit of a person's social security number could be any integer from 0 to 9.

STEP 2   There are 10 possible outcomes.

The outcomes are 0, 1, 2, 3, 4, 5, 6, 7, 8, 9.   ●

Example 4.2 Buckle Up   Two drivers on the Pennsylvania Turnpike are selected at random and checked for compliance with the seatbelt law. How many possible outcomes are there, and what are they?

SOLUTION

STEP 1   If a driver is wearing a seatbelt, denote this observation by R (for restrained), and if the driver is not wearing a seatbelt, use U (for unrestrained).

There are lots of other ways to denote these four outcomes. There is no single *correct* notation. Write the outcomes so that others can understand and interpret your list.

STEP 2   Each outcome is a pair of observations, one on each driver. There are four possible outcomes: RR, RU, UR, UU.

The first letter indicates the observation on the first driver, and the second letter indicates the observation on the second driver.

RU is a different outcome from UR. RU means the first driver was wearing a seatbelt and the second driver was not. UR means the first driver was not wearing a seatbelt and the second driver was.   ●

Tree diagrams will also be extremely useful for determining probabilities in problems involving Bayes' Rule. Problems of this type are presented in Section 4.5.

All of the outcomes from the experiment in Example 4.2 can be determined by constructing a **tree diagram**, a visual road map of possible outcomes. Figure 4.1 is a tree diagram associated with this experiment.

The *first generation branches* indicate the possible choices associated with the first driver and the *second generation branches* represent the choices for the second driver. A path from left to right represents a possible experimental outcome.

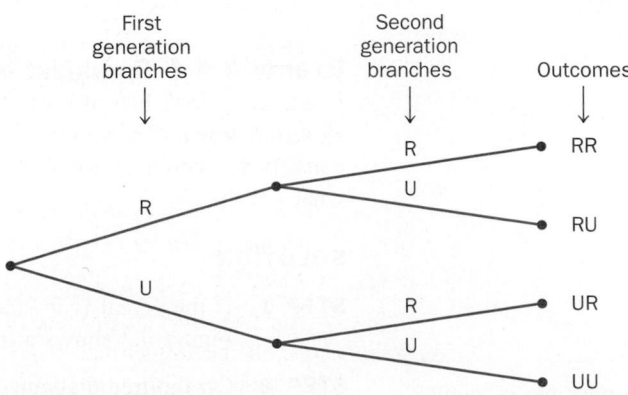

Figure 4.1 Tree diagram.

Example 4.3 Buckle Up (Continued) Extend the previous example. How many outcomes are there if we stop *three* drivers and record their seatbelt usage?

SOLUTION

Now there are eight possible outcomes: RRR, RRU, RUR, RUU, URR, URU, UUR, UUU.

Figure 4.2 is a tree diagram for this extended experiment. Again, every path from left to right represents a possible outcome.

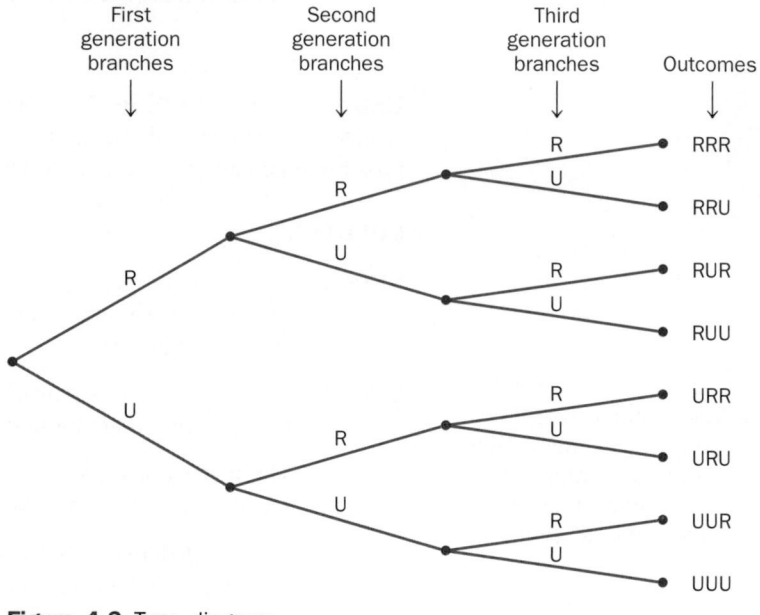

Figure 4.2 Tree diagram.

Tree diagrams are also used to prove the *multiplication rule* (Section 4.3), an arithmetic technique used to count the number of possible outcomes in certain experiments.

💡 **ILLUMINATING THE CONCEPTS**

1. Tree diagrams are a fine technique for finding all the possible outcomes for an experiment. However, they can get very big, very fast.

2. A tree diagram does not have to be symmetric, as they are in Figures 4.1 and 4.2. The branches and paths depend on the experiment. Consider the next example. ■

Example 4.4 Breakfast of Champions A consumer in Clarkdale, Arizona is searching for a box of his favorite breakfast cereal. He will check all three grocery stores in town if necessary, but will stop if he finds the cereal. The experiment consists of searching for the cereal. How many possible outcomes are there, and what are they?

SOLUTION

STEP 1 If the cereal is in stock, use the letter I; if it is out of stock, use O. Figure 4.3 shows a tree diagram for this experiment.

Why isn't IO a possible outcome?

STEP 2 On the tree diagram, there are four possible paths from left to right. The outcomes are

Outcome	Experiment result
I	The cereal is in stock in store 1.
OI	The cereal is not in stock in store 1, but it is in stock in store 2.
OOI	The cereal is not is stock in stores 1 and 2, but it is in stock in store 3.
OOO	The cereal is not in stock in any store.

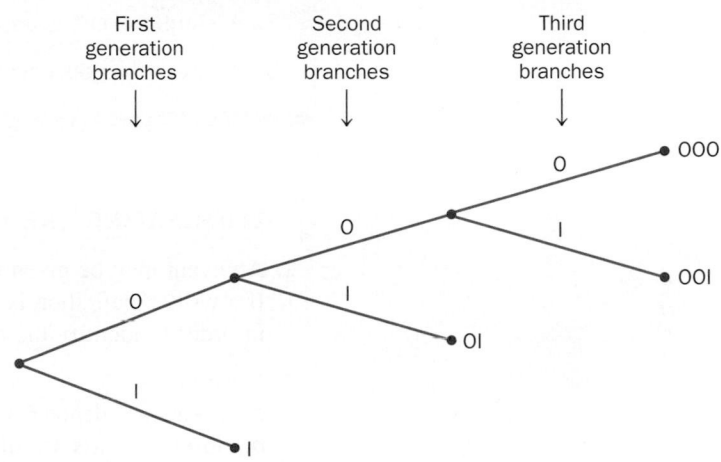

Figure 4.3 Tree diagram.

This tree diagram is not symmetric, but all possible outcomes are represented by left-to-right paths.

The paths representing the outcomes have different lengths. Some of the outcomes are shorter because the experiment ends early if the consumer finds the cereal in the first or second store. ●

DEFINITION

The **sample space** associated with an experiment is a listing of all the possible outcomes *using set notation*. It is the collection of all outcomes written mathematically, with curly braces, and denoted by S.

Example 4.5 Sample Spaces   Find the **sample space** for each of the four experiments above.

SOLUTION

We determined the outcomes for each experiment. Write the sample space using set notation.

STEP 1   Last digit of social security number: $S = \{0, 1, 2, 3, 4, 5, 6, 7, 8, 9\}$.

STEP 2   Seatbelt experiment: $S = \{RR, RU, UR, UU\}$.

STEP 3   Extended seatbelt experiment: $S = \{RRR, RRU, RUR, RUU, URR, URU, UUR, UUU\}$.

STEP 4   Cereal experiment: $S = \{I, OI, OOI, OOO\}$. ●

Given an experiment and the sample space, we usually study and find the probability of specific collections of outcomes, called **events**.

DEFINITION

1. An **event** is any collection (or set) of outcomes from an experiment (any subset of the sample space).

2. A **simple event** is an event consisting of exactly one outcome.

3. An event has **occurred** if the resulting outcome is contained in the event.

💡 **ILLUMINATING THE CONCEPTS**

1. An event may be given in standard set notation, or it may be defined in words. If a written definition is given, we need to translate the words into mathematics in order to identify the event outcomes.

2. Notation:
   a. Events are denoted with capital letters, for example, $A, B, C, \ldots$
   b. **Simple events** are often denoted by $E_1, E_2, E_3, \ldots$

3. It is possible for an event to be empty. An event containing no outcomes is denoted by $\{\}$ or ϕ (the empty set). ■

Translate at most *and* at least *carefully. These expressions appear frequently in probability and statistics questions.*

Example 4.6 College Dining Two resident students at Lafayette College are selected and asked whether they purchased a meal plan (M) or cook for themselves (C). The experiment consists of recording the response from both students.

There are four possible outcomes. A tree diagram is shown in the margin. The sample space is $S = \{MM, MC, CM, CC\}$.

There are four relevant simple events:
$E_1 = \{MM\}, E_2 = \{MC\}, E_3 = \{CM\}, E_4 = \{CC\}$.

Here are some other events, in words and in set notation.

Let A be the event that both students made the same choice.
$A = \{MM, CC\}$.

Let B be the event that at most one student purchased a meal plan.
$B = \{CC, MC, CM\}$ contains observations with at most one M.

Let D be the event that at least one student cooks.
$D = \{CM, MC, CC\}$ contains observations with one or more Cs.   ●

Tree diagram:

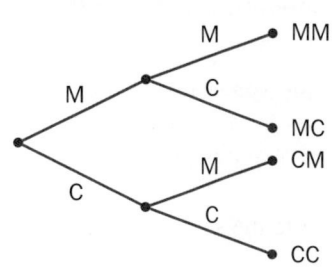

Example 4.7 On-Time Delivery A UPS driver may deliver to floors 2 through 6 in an office building and may use one of three elevators (labeled A, B, and C). The experiment consists of recording the floor and elevator used.

There are 15 possible outcomes because there are three elevators for each of the five floors. A tree diagram works again. The sample space is
$S = \{2A, 3A, 4A, 5A, 6A, 2B, 3B, 4B, 5B, 6B, 2C, 3C, 4C, 5C, 6C\}$

(Courtesy UPS)

The number in each outcome represents the floor, and the letter represents the elevator.

Let E be the event that the delivery is made on an odd floor using elevator B.
$E = \{3B, 5B\}$.

Let F be the event that the delivery is made on an even floor.

This definition says nothing about the elevator used. There are no restrictions on the elevator in this event.
$F = \{2A, 4A, 6A, 2B, 4B, 6B, 2C, 4C, 6C\}$.

Let G be the event that the delivery is made using elevator C.
$G = \{2C, 3C, 4C, 5C, 6C\}$.   ●

When an experiment is conducted, only one outcome can occur. For example, if the UPS driver used elevator B to deliver to the third floor, the experimental outcome is 3B. The observed outcome may be contained in several relevant events. In the delivery example above, if the outcome 2C is observed, then the events F and G have occurred. The event E did not occur.

Given an experiment, the sample space, and some relevant events, we often combine events in various ways to create and study new events. Since events are really sets, the methods of combining events are set operations.

A' is read as "A prime" or "A complement."

DEFINITION

Let A and B denote two events associated with a sample space S.

1. The event A **complement**, denoted A', consists of all outcomes in the sample space S *not* in A.

2. The event A **union** B, denoted $A \cup B$, consists of all outcomes in A or B or both.

3. The event A **intersection** B, denoted $A \cap B$, consists of all outcomes in both A and B.

4. If A and B have no elements in common, they are **disjoint** or **mutually exclusive**, written $A \cap B = \{\}$.

💡 **ILLUMINATING THE CONCEPTS**

1. The event A' is also called **not** A. The word *not* in the text of a probability question usually means you need to find the **complement** of an event.

2. *Or* usually means **union**; A or B means $A \cup B$.

3. *And* usually means **intersection**; A and B means $A \cap B$.

4. Any outcome in *both* A and B is included only once in the event $A \cup B$.

5. A', $A \cup B$ and $A \cap B$ are traditional mathematical symbols to denote complement, union, and intersection. But the three events defined above could be denoted using new symbols.

6. It is possible for one of these new events to contain all the outcomes in the sample space. ■

Example 4.8 One Coat Coverage A hardware store sells Benjamin Moore interior paint in one of three finishes: flat (F), satin (T), or gloss (G). The manager is interested in customer preferences and conducts an experiment by recording the interior paint finish for the next two customers who buy paint.

The sample space for this experiment has nine outcomes. See Figure 4.4.

$S = \{FF, FT, FG, TF, TT, TG, GF, GT, GG\}$.

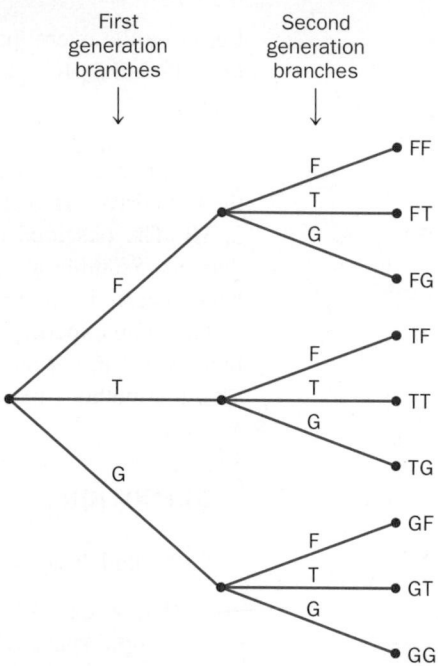

First
generation
branches

Second
generation
branches

Figure 4.4 Tree diagram.

Consider the following events.

$A = \{FF, TT, GG\}$	Both buy the same finish.
$B = \{FF, FT, TF, TT\}$	Neither buys gloss.
$C = \{FF, FT, FG, TF, GF\}$	At least one buys flat.
$D = \{FT, TF, TG, GT\}$	Exactly one buys satin.

Here are some new events related to the four given events.

$D' = \{FF, FG, TT, GF, GG\}$
    $= D$ complement, neither or both buy satin.
    $=$ All outcomes in S *not* in D.
$A \cup C = \{FF, TT, GG, FT, FG, TF, GF\}$
    $=$ Both buy the same finish or at least one buys flat.
    $=$ All outcomes in A *or* C (or both).
$A \cap D = \{\}$
    $=$ Both buy the same finish and exactly one buys satin.
    $=$ All outcomes in A *and* D. A and D are **disjoint**.
$(A \cup C)' = \{TG, GT\}$
    $= A$ union C, complement.
    $=$ All outcomes in S *not* in $A \cup C$.
$(A \cap D)' = S$
    $= A$ intersection D, complement.
    $=$ All outcomes in S *not* in $A \cap D$.

A **Venn diagram** may be used to visualize a sample space and events, to determine outcomes in combinations of events, and to answer probability questions in later sections. To construct this diagram, draw a rectangle to represent the sample space. Various figures (often circles) are drawn inside the rectangle to represent events. The Venn diagrams in Figure 4.5 illustrate various combinations of events.

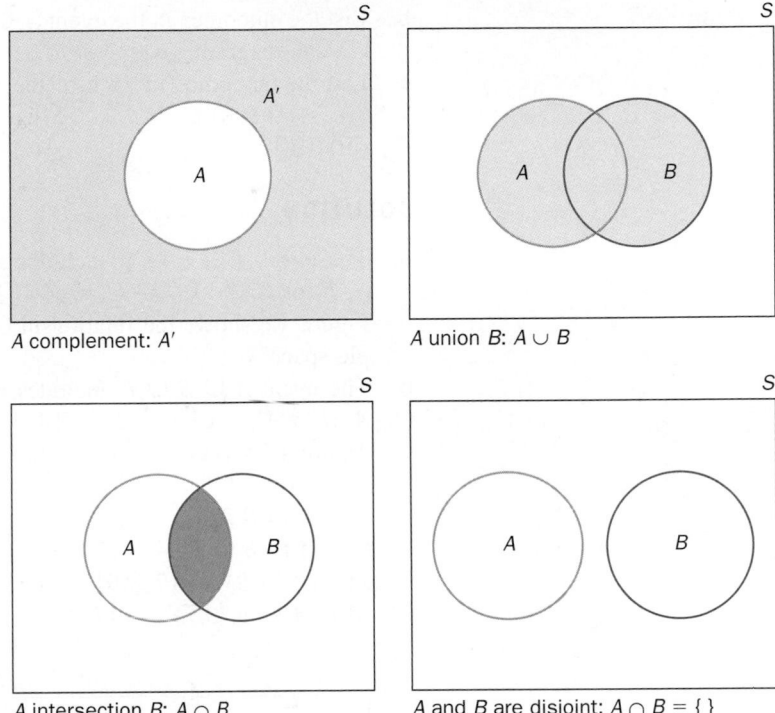

A complement: *A*′   *A* union *B*: $A \cup B$

A intersection *B*: $A \cap B$ *A* and *B* are disjoint: $A \cap B = \{\,\}$

Figure 4.5 Venn diagrams.

In a Venn diagram, plane regions represent events. We often add labeled points to denote outcomes. Later, probabilities assigned to events will be added to the graphs.

The definitions for union, intersection, and disjoint events can be extended to a collection consisting of more than two events.

DEFINITION

Let $A_1, A_2, A_3, \ldots, A_k$ be a collection of k events.

1. The event $A_1 \cup A_2 \cup \cdots \cup A_k$ is a **generalized union** and consists of all outcomes in at least one of the events $A_1, A_2, A_3, \ldots, A_k$.

2. The event $A_1 \cap A_2 \cap \cdots \cap A_k$ is a **generalized intersection** and consists of all outcomes in every one of the events $A_1, A_2, A_3, \ldots, A_k$.

3. The k events $A_1, A_2, A_3, \ldots, A_k$ are **disjoint** if no two have any element in common.

Example 4.9 Priority Request A university computer technician attaches a priority code to each help request. The range is 0 to 9, with 0 as the lowest priority and 9 as the highest priority. Consider an experiment in which a random request is selected and the priority is recorded. The sample space is
$S = \{0, 1, 2, 3, 4, 5, 6, 7, 8, 9\}$

and consider the events

$A = \{0, 1, 2, 3, 4\}$ $B = \{3, 4, 5, 6\}$
$C = \{7, 8\}$ $D = \{2, 4, 6, 9\}$

a. List the outcomes in the event $A \cup B \cup C$ and illustrate these three events using a Venn diagram.

b. List the outcomes in the event $A \cup B \cup D$ and illustrate these three events using a Venn diagram.

c. List the outcomes in each of the following events.
   (i) $A \cap B \cap C$ (ii) $A \cap B \cap D$
   (iii) $(A \cup B)'$ (iv) $(A \cup B \cup D)'$

SOLUTION

a. The event $A \cup B \cup C$ includes all the outcomes in at least one of the events A, B, or C. $A \cup B \cup C = \{0, 1, 2, 3, 4, 5, 6, 7, 8\}$.
   Figure 4.6 shows the relationships among the events A, B, and C, and the sample space S.

b. The event $A \cup B \cup D$ includes all the outcomes in at least one of the events A, B, or D. $A \cup B \cup D = \{0, 1, 2, 3, 4, 5, 6, 9\}$.
   Figure 4.7 shows the relationship among the events A, B, and D, and the sample space S.

c. (i) $A \cap B \cap C = \{\ \}$ There are no outcomes in all three events.
   (ii) $A \cap B \cap D = \{4\}$ 4 is the only outcome in all three events.
   (iii) $(A \cup B)' = \{7, 8, 9\}$ All outcomes in S *not* in $A \cup B$.
   (iv) $(A \cup B \cup D)' = \{7, 8\}$ All outcomes in S *not* in $A \cup B \cup D$.

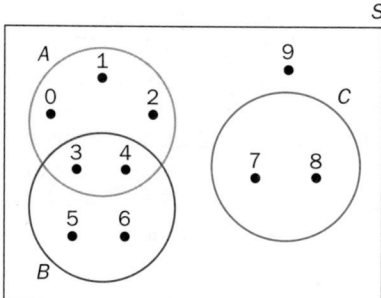

Figure 4.6 The events A, B, and C.

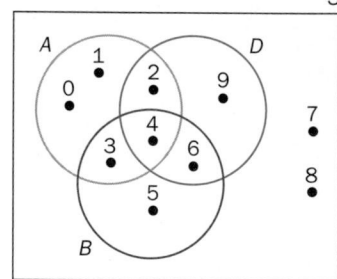

Figure 4.7 The events A, B, and D. ●

SECTION 4.1 EXERCISES

Practice

4.1 An experiment consists of rolling a six-sided die, recording the number that lands face-up, and then tossing a coin and recording heads or tails. Carefully sketch a tree diagram and find the sample space for this experiment.

4.2 A basketball player is going to select a sneaker with red, blue, green, or black stripes, and in either low- or high-top style. An experiment consists of recording the color and style he chooses. Carefully sketch a tree diagram and find the sample space for this experiment.

4.3 An experiment consists of selecting one letter from B, I, N, G, O, and one of five rows. How many possible outcomes are there in this experiment? Carefully sketch the corresponding tree diagram.

4.4 One playing card is selected from a regular 52-card deck. An experiment consists of recording the denomination (ace, 2, 3, 4, 5, 6, 7, 8, 9, 10, jack, queen, king) and suit (club, diamond, heart, or spade). How many possible outcomes are there in this experiment?

4.5 Consider an experiment with sample space
$S = \{0, 1, 2, 3, 4, 5, 6, 7, 8, 9\}$
and the events
$A = \{0, 2, 4, 6, 8\}$ $B = \{1, 3, 5, 7, 9\}$
$C = \{0, 1, 2, 3, 4\}$ $D = \{5, 6, 7, 8, 9\}$

Find the outcomes in each of the following events.

a. A' b. C' c. D'
d. $A \cup B$ e. $A \cup C$ f. $A \cup D$

4.6 Use the sample space and the events in Exercise 4.5 to find the outcomes in each of the following events.

a. $B \cap C$ b. $B \cap D$ c. $A \cap B$
d. $A \cap C$ e. $(B \cap C)'$ f. $B' \cup C'$

4.7 Consider an experiment with sample space
$S = \{a, b, c, d, e, f, g, h, i, j, k\}$
and the events
$A = \{a, c, e, g\}$ $B = \{b, c, f, j, k\}$
$C = \{c, f, g, h, i\}$ $D = \{a, b, d, e, g, h, j, k\}$
Find the outcomes in each of the following events.

a. A' b. C' c. D'
d. $A \cap B$ e. $A \cap C$ f. $C \cap D$

4.8 Use the sample space and the events in Exercise 4.7 to find the outcomes in each of the following events.

a. $A \cup B \cup D$ b. $B \cup C \cup D$
c. $B \cap C \cap D$ d. $A \cap B \cap C$

4.9 Use the sample space and the events in Exercise 4.7 to find the outcomes in each of the following events.

a. $(A \cap B \cap C)'$ b. $A \cup B \cup C \cup D$
c. $(B \cup C \cup D)'$ d. $B' \cap C' \cap D'$

4.10 The Venn diagram given below shows the relationship between two events.

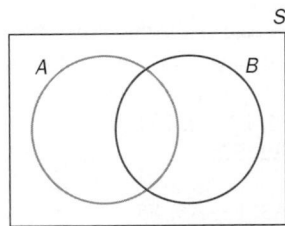

Redraw the Venn diagram for each part of this problem and carefully shade in the region corresponding to each new event.

a. $(A \cup B)'$ b. $(A \cap B)'$ c. $A' \cap B$
d. $A \cap B'$ e. $A' \cap B'$ f. $A' \cup B'$

4.11 The Venn diagram given below shows the relationships among three events.

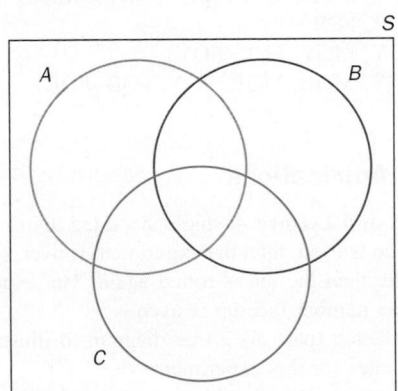

Redraw the Venn diagram for each part of this problem and carefully shade in the region corresponding to each new event.

a. $A \cup B \cup C$ b. $A \cap B \cap C$
c. $A \cup C$ d. $B \cap C$
e. $B \cap C'$ f. $(A \cup B)' \cap C$
g. $(A \cup B \cup C)'$ h. $A' \cap B' \cap C'$
i. $B \cap C \cap A'$

4.12 Consider an experiment with sample space
$S = \{$YYY, YYN, YNY, YNN, NYY, NYN, NNY, NNN$\}$
a. Find the outcomes in each of the following events.

  $A = $ Exactly one Y.
  $B = $ Exactly two Ns.
  $C = $ At least one Y.
  $D = $ At most one N.

Find the outcomes in each of the following events and write each as a combination of the events A, B, C, and D.

b. Exactly one Y or at most one N.
c. Two or more Ns.
d. Exactly two Ns and at least one Y.
e. Two or more Ys.

4.13 Consider an experiment with sample space
$S = \{0, 1, 2, 3, 4, 5, 6, 7, 8, 9\}$
and the events

  $A = \{0, 1, 2, 7, 8, 9\}$
  $B = \{0, 1, 2, 4, 8\}$
  $C = \{0, 1, 3, 9\}$
  $D = \{1, 4, 9\}$

Draw a separate Venn diagram to illustrate the relationships among each collection of events and the sample space S.

a. B and C.   b. A and D.
c. A, B, and C.   d. A, C, and D.

Applications

4.14 **Physical Sciences** An experiment consists of recording the time zone (E, C, M, P) and strength (L, M, H) for the next earthquake in the 48 contiguous states.

a. Carefully sketch a tree diagram to illustrate the possible outcomes for this experiment.
b. Find the sample space S for this experiment.

4.15 **Economics and Finance** Three taxpayers are selected at random and asked whether they itemized their tax deductions last year or used the standard deduction. An experiment consists of recording each response. Construct a tree diagram to represent this experiment and find the outcomes in the sample space.

4.16 **Travel and Transportation** Two people who work in New York City are selected at random and asked how they get to work: drive, take a train, or take a bus. An experiment consists of recording each response. Construct a tree diagram to represent this experiment and find the outcomes in the sample space.

4.17 **Business and Management** In a survey conducted by the Society for Human Resource Management, 66% of U.S. companies have a one-day-a-week casual dress policy.[1] Suppose the Human Resource department at Time Warner is concerned about employee attire on dress-down Fridays. An experiment consists of selecting a random employee on a Friday and recording

a. whether the employee is wearing sneakers or some other kind of shoe.

b. whether the employee is wearing jeans or some other kind of pants.

c. whether the employee is wearing a T-shirt, sport shirt, or button-down collar shirt.

Construct a tree diagram to represent this experiment and find the outcomes in the sample space.

4.18 Physical Sciences A construction crew excavating a site for a building foundation must remove the rock and prepare a trench for concrete footers. An experiment consists of recording the type of rock present (I, igneous; S, sedimentary; M, metamorphic) and the number of days needed to prepare the site (1 to 5).

a. Carefully sketch a tree diagram to illustrate the possible outcomes for this experiment.

b. Find the sample space S for this experiment.

4.19 Manufacturing and Product Development One of four calculator batteries is bad. An experiment consists of testing each one until the dead battery is found.

a. How many possible outcomes are there for this experiment?

b. Is the outcome GBGG (Good, Bad, Good, Good) possible? Why or why not?

4.20 Sports and Leisure An experiment consists of recording the number of pins knocked down on each roll during a frame of a bowling game. A bowler may take a maximum of two rolls per frame. How many outcomes are in the sample space for this experiment? Hint: If the first roll is a 10 (a strike), the experiment is over.

4.21 Sports and Leisure A sports statistician must carefully chart opposition football plays in preparation for the next game. An experiment consists of recording the type of play (pass or rush) and the yards gained ($-99, -98, -97, \ldots, -2, -1, 0, 1, 2, \ldots, 97, 98, 99$) on a randomly selected first down. How many outcomes are in the sample space for this experiment?

4.22 Medicine and Clinical Studies The emergency room in a rural hospital is staffed in four 6-hour shifts (1, 2, 3, 4). During any shift an emergency room patient is attended to by either a general physician (G), a surgeon (R), or an intern (I). An experiment consists of coding the next emergency room patient by shift and attending doctor. Consider the following events.

A = The attending doctor is the general physician.
B = The patient is admitted during the second shift.
C = The patient is admitted during shift 3 or is seen by the intern.
D = The patient is admitted during shift 4 and is seen by the general physician.

a. Find the sample space S for this experiment.

b. List the outcomes in each of the events A, B, C, and D.

c. List the outcomes in the events $A \cup B$ and $A \cap B$.

4.23 Psychology and Human Behavior Drivers entering a local mall parking lot at the main entrance may turn left, turn right, or

go straight. An experiment consists of recording the direction of the next car entering the mall and the vehicle style (sedan, SUV, van, or pickup). Consider the following events.

A = The next vehicle is a van.
B = The next vehicle is a sedan or pickup.
C = The next vehicle turns left.
D = The next vehicle goes straight or turns right.

a. Find the sample space S for this experiment.

b. List the outcomes in each of the events A, B, C, and D.

c. List the outcomes in the events $C \cup D$ and $C \cap D$.

4.24 Public Health and Nutrition Each patient with a regular appointment at a local dentist's office is classified by the number of cavities found (assume four is the maximum) and by whether the patient arrived late (L) or on time (T) for the appointment.

a. Find the sample space S for this experiment.

b. Describe the following events in words.

$A = \{0L, 1L, 2L, 3L, 4L\}$
$B = \{3L, 4L, 3T, 4T\}$
$C = \{1L, 3L, 1T, 3T\}$
$D = \{0L, 0T\}$
$E = \{0L, 0T, 1L, 2L, 3L, 4L\}$
$F = \{4T\}$

4.25 Travel and Transportation Every passenger arriving at the Las Vegas McCarran International Airport is classified as American (A) or Foreign (F), and by the number of checked bags (assume five is the maximum).

a. Find the sample space S for this experiment.

b. Describe the following events in words.

$A = \{A0, F0\}$
$B = \{F0, F1, F2, F3, F4, F5\}$
$C = \{A1, F1, A2, F2\}$
$D = \{F0, F5\}$
$E = \{A1, F1, A3, F3, A5, F5\}$

4.26 Public Health and Nutrition A researcher working for a Hardee's fast-food restaurant in Cleveland selects random customers and classifies each according to gender [male (M) or female (F)], order type [combo (C) or other (O)], and age group [young (Y), middle-aged (D), or senior (R)].

a. Find the sample space S for this experiment.

b. Describe the following events in words.

$A = \{MCY, MCD, MCR, MOY, MOD, MOR\}$
$B = \{MCR, FCR\}$
$C = \{MCY, MOY, FCY, FOY\}$
$D = \{MOY, MOD, MOR, FOY, FOD, FOR\}$

Extended Applications

4.27 Sports and Leisure A single six-sided die is rolled. If the number face-up is even, then the experiment is over. If the number face-up is odd, then the die is rolled again. The experiment continues until the number face-up is even.

a. Carefully sketch (part of) a tree diagram to illustrate the possible outcomes for this experiment.

b. Find the sample space for this experiment.

4.28 Economics and Finance A taxpayer in need of advice will call the IRS repeatedly until she can get through (no busy signal). If she receives a busy signal, she will hang up and try again later, and will stop calling as soon as she reaches an agent. An experiment consists of recording the calling pattern. A possible outcome is BBH: a busy signal (B) on the first two calls, and (finally) help (H) on the third call.
a. How many possible outcomes are there in this experiment?
b. List some of the outcomes for this experiment.

4.29 Marketing and Consumer Behavior Music-notes.com sells sheet music in the following genres: rock, jazz, new age, and country. An experiment consists of recording the preferred genre for the next customer and the number of songs purchased (assume five is the maximum). Consider the following events.

A = The next customer prefers rock.
B = The next customer prefers jazz and buys at least three songs.
C = The next customer buys at most two songs.
D = The next customer prefers country and buys one song.

a. Find the sample space S for this experiment.
b. Find the outcomes in each of the following events.
(i) A'
(ii) $A \cup C$
(iii) $A \cap D$
(iv) $C \cap D$
(v) $A \cap C \cap D$
(vi) $(A \cap B)$

4.30 Marketing and Consumer Behavior A phone company offers six wireless plans, each with a different number of Anytime minutes. An experiment consists of selecting a customer and recording the plan number and whether the customer was under or over the number of Anytime minutes last month. Consider the following events.

A = The customer was over the number of Anytime minutes.
B = The customer uses plan 1, 2, or 3.
C = The customer uses plan 5 or 6 and was over the number of Anytime minutes.
D = The customer uses an even-numbered plan.

a. Find the sample space S for this experiment.
b. Find the outcomes in each of the following events.
(i) B'
(ii) $A \cup B$
(iii) $A \cap B$
(iv) $C \cap D$
(v) $A \cap B \cap D$
(vi) $(A \cap D)'$

4.31 Travel and Transportation An experiment consists of selecting a random passenger on a train and recording the purpose of travel (business or pleasure) and the number of pieces of luggage (zero to four). Consider the following events.

A = The passenger is traveling on business.
B = The passenger has no luggage.
C = The passenger has at most one piece of luggage.
D = The passenger has three pieces of luggage or is traveling for pleasure.

a. Find the sample space S for this experiment.
b. Find the outcomes in each of the following events.
(i) $A \cup B$
(ii) $A \cap B$
(iii) $B \cup C$
(iv) $B \cap C$
(v) $A \cap D$
(vi) $A \cap B \cap C \cap D$

4.32 Marketing and Consumer Behavior A coffee shop offers chai tea in the following variations: hot or cold; with whipped cream or without; and in small, medium, or large size. An experiment consists of recording these three options for the next customer.

a. Carefully sketch a tree diagram to illustrate the possible outcomes for this experiment.
b. Find the sample space S for this experiment.
c. Consider the following events.

A = The next customer order is small.
B = The next customer order is cold.
C = The next customer order is small or hot.

Find the outcomes in each of the following events.
(i) $A \cup B$
(ii) $B \cup C$
(iii) $B \cap C$
(iv) C'

4.2 AN INTRODUCTION TO PROBABILITY

P(A) works like a function. The inputs are events; the outputs are probabilities.

In any given experiment, some events are more likely to occur than others. For any event A, we need to assign a number to A that corresponds to this intuitive *likelihood of occurrence*. The likelihood that A will occur *is* simply the probability of the event A. For example, the probability that an asteroid 100 meters in diameter will strike the Earth in any given year is 0.001 (a pretty unlikely event). The probability of wind gusts over 40 miles per hour at the Mount Washington Observatory on any given winter day is 0.07 (a more likely event). The notation P(A) is used to denote this likelihood, the probability of an event A. To begin our discussion of probability, consider the following working definition.

DEFINITION

The **probability of an event** A is a number between 0 and 1 (including those end points) that conveys the likelihood that A will occur.

1. If the probability of an event is close to 1, then the event is likely to occur.

2. If the probability of an event is close to 0, then the event is not likely to occur.

Would you enroll in a class where the probability of receiving an *A* is 1?

If the **probability of an event** A is 1, then the event is a certainty; it will occur. If the probability of an event A is 0, then A is definitely not going to occur. What about events with probabilities in between? How do we decide to assign a probability of 0.3, for example? We need a reasonable, all-purpose rule for linking an event to its likelihood of occurrence. The natural (theoretical) definition for assigning a probability to an event is very intuitive.

DEFINITION

The **relative frequency of occurrence of an event** is the number of times the event occurs divided by the total number of times the experiment is conducted.

This is illustrated in the following example.

It seems like the answer should be 1/4. Why?

Example 4.10 Pick a Card, Any Card In a regular 52-card deck there are 13 clubs, 13 diamonds, 13 hearts, and 13 spades. Suppose an experiment consists of selecting one card from the deck and recording the suit. What is the probability of selecting a club?

SOLUTION

STEP 1 Let C be the event that a club is selected. We want the probability of the event C, which is denoted by P(C).

STEP 2 In order to estimate the probability of C, it seems reasonable to conduct the experiment several times and see how often a club is selected. If C occurs often (we get a club a lot of the time), then the likelihood (probability) should be high. If C rarely happens, then the probability should be close to 0.

Relative frequency was defined in Chapter 2 in the context of frequency distributions.

STEP 3 To estimate the likelihood of selecting a club, we use the **relative frequency of occurrence** of a club, which is the frequency divided by total trials, or

$$\text{Relative frequency} = \frac{\text{Number of times a club is selected}}{\text{Total number of selections}}$$

After every selection, the observed card is placed back in the deck. The deck is shuffled, and another selection is made.

STEP 4 Suppose after 10 tries, a club was selected only twice. The relative frequency is 2/10 = 0.2. This is an estimate of P(C). It's quick and easy, but doesn't seem too accurate.

STEP 5 Suppose we try the experiment a few more times. With more observations we should be able to make a better guess at P(C). The table below shows values for N, the number of trials, and $\hat{p}$, the relative frequency of occurrence of a club.

N	10	50	100	200	300	400	500	600	700
$\hat{p}$	0.2	0.3	0.29	0.23	0.223	0.205	0.228	0.252	0.267

N	800	900	1000	1100	1200	1300	1400	1500
$\hat{p}$	0.245	0.243	0.227	0.254	0.260	0.256	0.261	0.249

For example, after 300 draws, the relative frequency of occurrence of the event C was 0.223.

STEP 6 Figure 4.8 shows a scatter plot of relative frequency versus number of trials. The graph shows a remarkable pattern. As N increases, the points are noticeably closer to the dashed line. The relative frequencies seem to be homing in on one number (around 0.25); this relative frequency, whatever it is, should be the probability of the event C.

Try the Probability statistical applet on the book companion web site.

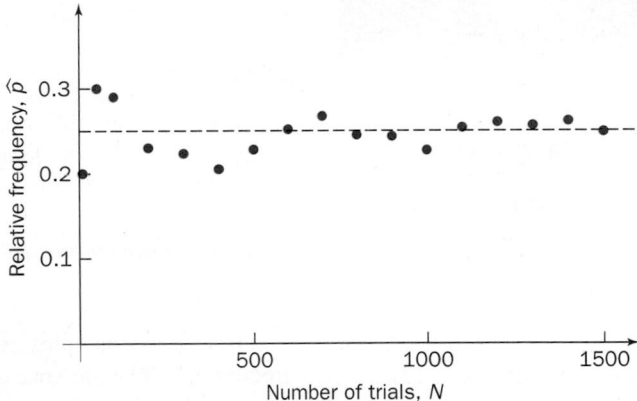

Figure 4.8 Scatterplot of relative frequency versus number of trials.

STEP 7 In the long run, the relative frequencies tend to stabilize, or even out, and become almost constant. They close in on one number, the *limiting relative frequency*. The probability of the event C is the limiting relative frequency. ●

If an experiment is conducted N times and an event occurs n times, then the probability of the event is *approximately n/N* (the relative frequency of occurrence). The **probability of an event A**, P(A), is the *limiting* relative frequency, the proportion of time the event A will occur in the long run. This is a basic and sensible definition, a rule for assigning probability to an event. Given an event, all we need to do is find the limiting relative frequency.

While this definition makes sense, and Example 4.10 and Figure 4.8 support and illustrate our intuition, there is a real practical problem. We cannot conduct experiments over and over, compute relative frequencies, and only then estimate the true probability. How will we ever know the true limiting relative frequency? How large should N be? When are we close enough? Will we ever hit the limiting relative frequency exactly? The definition is nice, but there seems little hope of ever finding the true probability of an event. Fortunately there is another way to determine the exact probability in some cases. Consider the next two examples.

If we were to conduct this experiment over and over, the relative frequency of occurrence of H would close in on 1/2.

(SuperStock)

The relative frequency of occurrence of a 1 would get closer and closer to 1/6 as the number of rolls gets larger and larger.

The word *chance* is also used to express likelihood. A 10% chance means that the probability is 0.10.

Example 4.11 Call It in the Air   Suppose an experiment consists of tossing a fair coin and recording the side that lands face-up. The event H is the coin landing with heads face-up. Find P(H).

SOLUTION

There are only two possible outcomes on each flip of the coin, and they are both *equally likely* to occur. In the long run, we expect heads to occur half of the time.

Therefore, P(H) = 1/2.

Without flipping the coin thousands of times, making estimates, or guessing at the limiting relative frequency, we are certain the probability is 1/2.   ●

Example 4.12 Roll the Die   An experiment consists of tossing a fair six-sided die and recording the number that lands face-up. Consider the event $E = \{1\}$, rolling a 1. Find P(E).

SOLUTION

There are six possible outcomes on each roll of the die, and they are all equally likely to occur. In the long run, we expect 1 to occur one-sixth of the time.

Therefore, P(E) = 1/6.

We *can* identify the exact limiting relative frequency.   ●

These two examples suggest it is indeed possible to find the limiting relative frequency! They are special cases, however, because in each individual experiment, all of the outcomes are *equally likely*.

PROPERTIES OF PROBABILITY

1. For any event A, $0 \leq P(A) \leq 1$.
   The probability of any event is a *limiting relative frequency*, and a relative frequency is a number between 0 and 1. An event with probability close to 0 is very unlikely to occur, and an event with probability close to 1 is very likely to occur.
2. For any event A, P(A) is the sum of the probabilities of all of the outcomes in A.
   To compute P(A), just add up the probability of each outcome or simple event in A.
3. The sum of the probabilities of all possible outcomes in a sample space is 1: P(S) = 1.
   The sample space S is an event. If an experiment is conducted, S is guaranteed to occur.
4. The probability of the empty set is 0: P($\{\}$) = P(ϕ) = 0. This event contains no outcomes.

In the next example, the probability (limiting relative frequency) of each simple event is assumed to be known. We will use the properties above and some earlier definitions to develop some common tools and strategies for solving similar probability questions.

Example 4.13 Try the Easy Button   There are five sales associates (indicated by their initials) on duty in a Staples office supply store: three women (MK, JB, and KY) and two men (DN and PD). An experiment consists of classifying the next customer's action. The customer either will make a purchase from one of the sales associates (indicated by their initials) or will buy nothing (NO). The probability of each simple event is given in the table below.

Simple event	DN	PD	MK	JB	KY	NO
Probability	0.08	0.12	0.10	0.25	0.15	0.30

Consider the following events.
$A = \{DN, PD\}$
   = The next customer buys something from a male sales associate.
$B = \{MK, JB, KY\}$
   = The next customer buys something from a female sales associate.
$C = \{NO\}$
   = The next customer buys nothing.
$D = \{DN, JB\}$
   = The next customer buys from one of these two sales associates.
Find $P(A)$, $P(C)$, $P(B \cup D)$, $P(A \cap D)$, and $P(A \cap B)$.

SOLUTION

STEP 1 $P(A) = P(DN) + P(PD)$ Add the probabilities of each
          $= 0.08 + 0.12 = 0.20$ outcome in A.

STEP 2 $P(C) = P(NO) = 0.30$ There is only one outcome in C.

STEP 3 $P(B \cup D) = P(DN, MK, JB, KY)$ Find the outcomes in the event $B \cup D$.
          $= P(DN) + P(MK) + P(JB) + P(KY)$ Add up the probabilities of
          $= 0.08 + 0.10 + 0.25 + 0.15 = 0.58$ each outcome.

STEP 4 $P(A \cap D) = P(DN) = 0.08$ The intersection is one outcome. Check
          the probability in the table above.

STEP 5 $P(A \cap B) = P(\{\}) = 0$ The intersection is empty, so the
          probability is 0. ●

In order to find probabilities in the previous example, we looked at each event piece by piece. We broke down each event into simple events. Let's apply the same properties in an **equally likely outcome experiment**.

Suppose an experiment has n equally likely outcomes, $S = \{e_1, e_2, e_3, \ldots, e_n\}$. Since each simple event has the same chance of occurring, the probability of each is $1/n$; $P(e_i) = 1/n$. The limiting relative frequency of e_i is $1/n$. This is exactly what we found in Examples 4.11 and 4.12. Consider an event $A = \{e_1, e_2, e_3, e_4, e_5\}$. To find $P(A)$, add up the probabilities of each simple event in A.

$$P(A) = P(e_1) + P(e_2) + P(e_3) + P(e_4) + P(e_5)$$

$$= \frac{1}{n} + \frac{1}{n} + \frac{1}{n} + \frac{1}{n} + \frac{1}{n} = \frac{5}{n}$$

$$= \frac{\text{Number of outcomes in } A}{\text{Number of outcomes in the sample space } S} = \frac{N(A)}{N(S)}$$

Think about tossing a fair coin, or rolling a fair die, or blindly selecting a student in a class to answer a question.

You will not always see the phrase *equally likely outcomes* in these probability questions. We will identify some keywords and work with familiar experiments that imply equally likely outcomes.

FINDING PROBABILITIES IN AN EQUALLY LIKELY OUTCOME EXPERIMENT

In an **equally likely outcome experiment**, the probability of *any* event A is the number of outcomes in A divided by the total number of outcomes in the sample space S. Finding the probability of any event, in this case, means counting the number of outcomes in A, counting the number of outcomes in the sample space S, and dividing.

$$P(A) = \frac{N(A)}{N(S)}$$

Section 4.3 presents some special counting rules to help compute probabilities associated with common experiments and events. However, we can solve some of these problems already, and even use our results to make a statistical inference.

Example 4.14 Bank Teller Jobs   The Legend Bank in Bowie, Texas, has five tellers: 1 and 2 are trainees; 3, 4, and 5 are veterans. Tellers 2, 3, and 4 are female, and tellers 1 and 5 are male. At the end of the day, two tellers will be randomly selected and all of their transactions for the day will be audited.
a. What is the probability that both trainees will be selected for the audit?
b. What is the probability that one male and one female will be selected for the audit?
c. What is the probability that two females will be selected for the audit?

SOLUTION

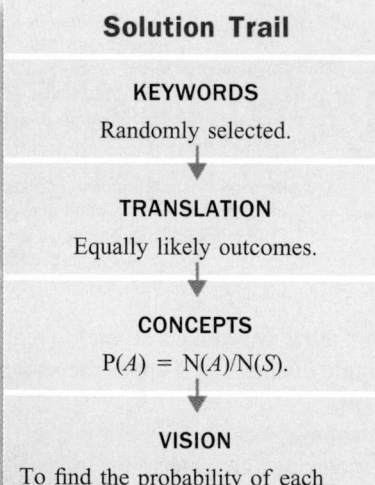

Solution Trail

KEYWORDS
Randomly selected.
↓
TRANSLATION
Equally likely outcomes.
↓
CONCEPTS
$P(A) = N(A)/N(S)$.
↓
VISION
To find the probability of each event, count the number of outcomes in that event and divide by the total number of outcomes in the sample space.

The experiment consists of selecting two tellers at random. The outcomes consist of two tellers that can be represented by their numbers. Therefore, 12 represents the outcome that tellers 1 and 2 were selected. The order of selection does not matter. For example, 12 and 21 both represent the event in which tellers 1 and 2 were selected. We can (a) list all possible outcomes systematically, (b) sketch a tree diagram, or (c) use combinations (to be presented in Section 4.3). There are 10 outcomes in the sample space.

$$S = \{12, 13, 14, 15, 23, 24, 25, 34, 35, 45\}.$$

a. Let A = both trainees are selected for the audit. Since the trainees are tellers 1 and 2, there is only one outcome in the event A: $A = \{12\}$.

$$P(A) = \frac{\text{Number of outcomes in } A}{\text{Number of outcomes in } S} = \frac{N(A)}{N(S)} = \frac{1}{10} = 0.10$$

b. Let B = one male and one female selected. Tellers 2, 3, and 4 are female, and tellers 1 and 5 are male. Check the sample space carefully to list the outcomes in B.

$$B = \{12, 13, 14, 25, 35, 45\} \Rightarrow P(B) = \frac{N(B)}{N(S)} = \frac{6}{10} = 0.60$$

c. Let C = two females are selected. Tellers 2, 3, and 4 are female. Check the sample space again, and pick out the matching outcomes.

$$C = \{23, 24, 34\} \Rightarrow P(C) = \frac{N(C)}{N(S)} = \frac{3}{10} = 0.30$$

●

a. Solution Trail

KEYWORDS

- Demand for each kind is the same.
- Selected at random.

↓

TRANSLATION

Equally likely outcomes.

↓

CONCEPTS

$P(A) = N(A)/N(S)$.

↓

VISION

To find the probability of each event, count the number of outcomes in that event and divide by the total number of outcomes in the sample space.

b. Solution Trail

KEYWORDS

- All five customers purchase a plain bagel.
- Is there any evidence?

↓

TRANSLATION

- Experimental outcome.
- Draw a conclusion.

↓

CONCEPTS

Inference procedure.

↓

VISION

Find the probability of the experimental outcome that all five customers buy a plain bagel. By computing how likely this outcome is, we can draw a conclusion about the claim. If this probability is large, then the outcome is reasonable (given the claim). If this probability is small, then there is evidence to suggest that the claim is wrong.

The next example involves an equally likely outcome experiment and an inference question. We'll need to compute the likelihood of the *observed* event to help us draw a conclusion.

Example 4.15 Buttered Bagels The Pleasantville Bagel Shop sells only two different varieties: plain (P) and cinnamon raisin (C). The owner believes that the demand for each kind is the same and that the shop should continue to bake these varieties in equal numbers. Five customers are selected at random. Each customer buys only one bagel and the bagel purchase is noted.

a. Find the probability that exactly one person buys a plain bagel.

b. Suppose all five customers purchase a plain bagel. Is there any evidence to suggest that demand is weighted more toward one variety?

SOLUTION

The experiment consists of selecting five customers at random and recording their bagel purchase. Each outcome is a sequence of five letters: Cs and/or Ps. For example, the outcome CCPCP stands for: the first customer buys a cinnamon raisin bagel, the second customer buys a cinnamon raisin bagel, the third customer buys a plain bagel, the fourth buys a cinnamon raisin bagel, and the fifth buys a plain bagel. There are 32 possible outcomes: a systematic listing helps, and a tree diagram works (but is big). (The multiplication rule also works here. This very useful counting technique is presented in Section 4.3.) Here is the sample space:

S = {PPPPP, PPPPC, PPPCP, PPPCC, PPCPP, PPCPC, PPCCP, PPCCC, PCPPP, PCPPC, PCPCP, PCPCC, PCCPP, PCCPC, PCCCP, PCCCC, CPPPP, CPPPC, CPPCP, CPPCC, CPCPP, CPCPC, CPCCP, CPCCC, CCPPP, CCPPC, CCPCP, CCPCC, CCCPP, CCCPC, CCCCP, CCCCC}

a. Let A = exactly one person buys a plain bagel. Check the sample space and carefully list all the outcomes in A.

A = {PCCCC, CPCCC, CCPCC, CCCPC, CCCCP}

$$P(A) = \frac{N(A)}{N(S)} = \frac{5}{32} = 0.15625$$ Equally likely outcomes.

b. The claim is that the demand for each type of bagel is equal. If this is true, then all of the outcomes in the sample space S are equally likely. The experiment consists of observing the bagel purchase for the next five customers. Let B = the observed outcome, everyone buys a plain bagel.

Find the likelihood of the event B occurring. There is only one outcome in B, so the probability of the event B is

$$P(B) = \frac{N(B)}{N(S)} = \frac{1}{32} = 0.03125$$ Count and divide.

The conclusion: Since this probability is so small, all five people buying a plain bagel is a rare event. But it happened! This suggests that the assumption is wrong—there is evidence to suggest that the demand for each type of bagel is *not* equal.

Note: There is really evidence to suggest that *some* assumption is wrong. It could be, for example, that the five customers were not selected at random. To draw a conclusion about the demand for these two types of bagels, we must accept all other assumptions as true. ●

Consider an experiment, two events A and B, and known probabilities $P(A)$ and $P(B)$. Suppose we use A and B to create a new event using complement, union, or intersection. Sometimes we can use the *known probabilities* $P(A)$ and $P(B)$ to quickly calculate the probability of the *new event*. We may not have to break down the new event into simple events, or even count all the outcomes in the new event (if it is an equally likely outcome experiment). The **Complement Rule** and the **Addition Rule for Two Events** are two rules that help with probability calculations.

THE COMPLEMENT RULE

For any event A, $P(A) = 1 - P(A')$.

💡 **ILLUMINATING THE CONCEPTS**

1. The Complement Rule is easy to visualize and justify by looking at a Venn diagram. Figure 4.9 shows an event A and its complement A'.

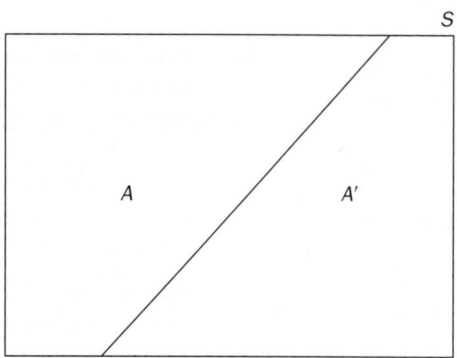

Figure 4.9 Venn diagram for visualizing the Complement Rule.

$(A')'$ is read as "A complement, complement." What is $(A')'$? All outcomes not in A', which is A!

Remember, the area of a region represents probability. $P(A) + P(A') = P(S) = 1$, which can be written as $P(A) = 1 - P(A')$ or $P(A') = 1 - P(A)$.

2. The Complement Rule is incredibly handy; it is used in various contexts throughout probability and statistics. The problem is, how do you know when to use it? Look for keywords like *not*, *at least*, and *at most*. A rule of thumb: If you are faced with a very long probability calculation involving many simple events, or one that may require lots of counting, try looking at the complement. ■

Example 4.16 Law and Order   Three public defenders are assigned to cases randomly. An experiment consists of recording the lawyer (by number) assigned to the next three cases. The outcome 132 means lawyer 1 was assigned case 1, lawyer 3 was assigned case 2, and lawyer 2 was assigned case 3.
a. Find the probability that all three cases are assigned to different lawyers.
b. Find the probability that lawyer 2 is not assigned to any of the three cases.
c. Find the probability that lawyer 2 is assigned to at least one case.

SOLUTION

There are 27 possible outcomes; a tree diagram works.
Note: Each case can be assigned to one of three lawyers:

Number of possible assignments for each case.

$$\underset{\text{Case 1}}{\underline{3}} \times \underset{\text{Case 2}}{\underline{3}} \times \underset{\text{Case 3}}{\underline{3}} = 27$$

Here's the sample space:

$$S = \{111, 112, 113, 121, 122, 123, 131, 132, 133,$$
$$211, 212, 213, 221, 222, 223, 231, 232, 233,$$
$$311, 312, 313, 321, 322, 323, 331, 332, 333\}$$

a. Let A = all three cases are assigned to different lawyers. Find all the outcomes in S with a 1, a 2, and a 3.

$$A = \{123, 132, 213, 231, 312, 321\}$$

$$P(A) = \frac{N(A)}{N(S)} = \frac{6}{27} = 0.2222$$
Equally likely outcomes.

b. Let B = lawyer 2 is not assigned to any of the three cases. Find all the outcomes without a 2.

$$B = \{111, 113, 131, 133, 311, 313, 331, 333\}$$

$$P(B) = \frac{N(B)}{N(S)} = \frac{8}{27} = 0.2963$$

8/27 is really *approximately* equal to 0.2963. Many answers in this text are rounded (here, to four decimal places) and an equal sign is used for simplicity and convenience.

c. Let C = lawyer 2 is assigned to at least one case. The outcomes in C include those with one 2, two 2s, and three 2s. That seems like a lot of counting. This is a good opportunity to use the Complement Rule.

$$P(C) = 1 - P(C')$$
Complement Rule.

$$= 1 - P(\text{lawyer 2 is assigned 0 cases})$$
Interpretation of C'.

$$= 1 - P(B)$$
$C' = B$ in this example.

$$= 1 - \frac{8}{27} = 1 - 0.2963 = 0.7037$$ ●

c. Solution Trail

KEYWORDS

At least one case.

↓

TRANSLATION

Let C be the event that lawyer 2 has at least one case.

↓

CONCEPTS

Complement Rule.

↓

VISION

Consider the complement, C', the event that lawyer 2 has no cases. Count the outcomes that have no 2s and use the Complement Rule: $P(C) = 1 - P(C')$.

THE ADDITION RULE FOR TWO EVENTS

1. For any two events A and B: $P(A \cup B) = P(A) + P(B) - P(A \cap B)$.
2. For any two *disjoint* events A and B: $P(A \cup B) = P(A) + P(B)$.

ILLUMINATING THE CONCEPTS

1. Figure 4.10 helps to illustrate and justify this rule.

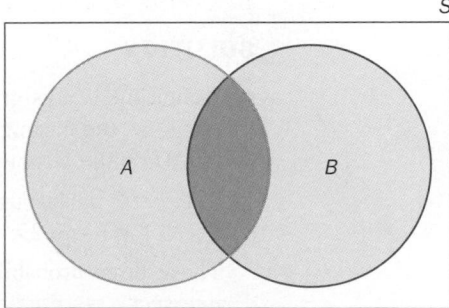

Figure 4.10 Venn diagram illustrating the Addition Rule.

To find the probability of the union, start by adding $P(A) + P(B)$. This sum includes the region of intersection, $P(A \cap B)$, twice. Adjust this total by subtracting the intersection area once.

2. $A \cup B = B \cup A$; order doesn't matter here. So, $P(A \cup B) = P(B \cup A)$.

3. The first, more general, formula *always* works. If A and B are disjoint, then $P(A \cap B) = 0$.

4. ◀ The Addition Rule can be extended.
For any three events A, B, and C:

$$P(A \cup B \cup C) = P(A) + P(B) + P(C)$$
$$-P(A \cap B) - P(A \cap C) - P(B \cap C)$$
$$+ P(A \cap B \cap C)$$

You can also visualize and derive this by using a Venn diagram. In this case, the sum $P(A) + P(B) + P(C)$ includes the double intersections twice and the triple intersection three times. We therefore need to adjust the total accordingly. ▶

5. Let $A_1, A_2, A_3, \ldots, A_k$ be a collection of k *disjoint* events.

$$P(A_1 \cup A_2 \cup \cdots \cup A_k) = P(A_1) + P(A_2) + \cdots + P(A_k).$$

If the events are disjoint, to find the probability of a union, just add up the corresponding probabilities. This is especially useful in questions that ask about the number of individuals or objects with a specific attribute. For example, suppose 10 people are asked whether they received a flu shot this winter. The probability of at least 3 is the probability of 0, plus the probability of 1, plus the probability of 2, plus the probability of 3.

6. Complement, union, and intersection are operations applied to *events*. It doesn't make sense to take the union of probabilities (which are numbers). Similarly, addition and subtraction are operations on real numbers. You shouldn't try to add or subtract events. ■

Beware of these common errors:
$P(A + B)$: can't add two events.
$P(A) \cup P(B)$: can't union two numbers.

Probabilities are given as percentages in this example. Divide each by 100 to convert to a probability.

Example 4.17 Milk or Sugar?   Marketing research by The Coffee Beanery in Detroit, Michigan, indicates that 70% of all customers put sugar in their coffee, 35% add milk, and 25% use both. Suppose a Coffee Beanery customer is selected at random.

a. Draw a Venn diagram to illustrate the events in this problem.
b. What is the probability that the customer uses at least one of these two items?
c. What is the probability that the customer uses neither?
d. What is the probability that the customer uses just sugar?
e. What is the probability that the customer uses just one of these two items?

SOLUTION

a. Define the events given in the problem.
Let G = the customer adds sugar; $P(G) = 0.70$.
Let M = the customer adds milk; $P(M) = 0.35$.

Use both means uses sugar *and* milk, which means intersection. Therefore, $P(G \cap M) = 0.25$.

These three probabilities add up to more than 1. That's OK because the events intersect.

Remember, area of a region corresponds to probability. To complete the picture, start at the inside and work your way out.

(i) The shaded area represents the probability that the customer uses both sugar *and* milk, that is, $P(G \cap M)$. We know that $P(G \cap M) = 0.25$. Since $P(G) = 0.70$, the remaining area representing G corresponds to $0.70 - 0.25 = 0.45$.

(ii) Similarly, since P(G ∩ M) = 0.25 and P(M) = 0.35, the remaining area representing M corresponds to 0.35 − 0.25 = 0.10.

(iii) The total probability in the entire sample space must sum to 1; the remaining probability is 1 − (0.45 + 0.25 + 0.10) = 0.20.

Figure 4.11 is the Venn diagram that corresponds to this problem.

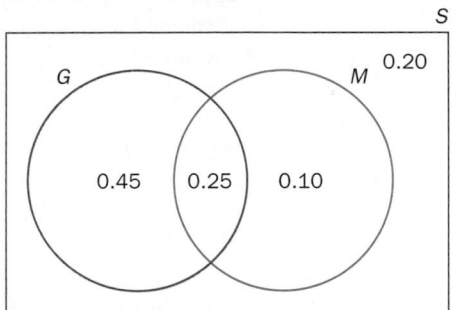

Figure 4.11 Venn diagram.

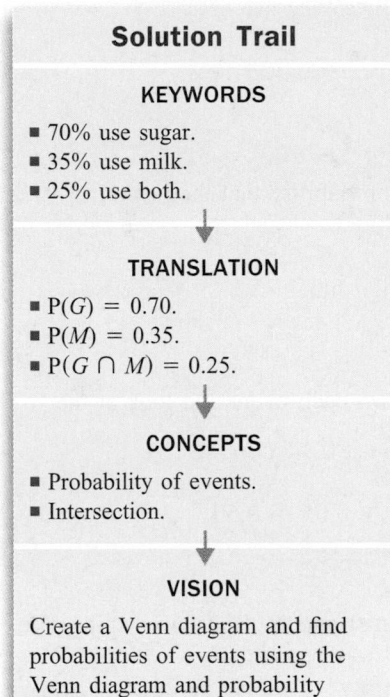

b. The probability of using *at least one* item means using sugar, or milk, or both. That's a union of two events.

$$P(G \cup M) = P(G) + P(M) − P(G \cap M) \qquad \text{Addition Rule.}$$
$$= 0.35 + 0.70 − 0.25 = 0.80 \qquad \text{Use the known probabilities.}$$

The Venn diagram supports this answer. Look at the region that represents P(G ∪ M), and add up the corresponding probabilities.

c. *Uses neither* means does *not* use sugar or milk. Since G ∪ M means sugar or milk, *neither* suggests the complement of G ∪ M.

$$P[(G \cup M)'] = 1 − P(G \cup M) \qquad \text{Complement Rule applied to the event } G \cup M.$$
$$= 1 − 0.80 = 0.20 \qquad \text{Use the previous answer.}$$

In the Venn diagram, this is the region outside of G ∪ M.

d. *Uses just* sugar means uses sugar *but not both* sugar and milk. This is not simply P(G), because this probability includes more than just sugar. Start with the probability of using sugar, and subtract the probability of using both.

$$P(\text{just sugar}) = P(G) − P(G \cap M) \qquad \text{Use the Venn diagram.}$$
$$= 0.70 − 0.25 = 0.45 \qquad \text{Use the known probabilities.}$$

e. *Uses just one* of these items means uses sugar or milk, but not both. Start with the union, and subtract off the intersection.

$$P(\text{exactly one}) = P(G \cup M) − P(G \cap M) \qquad \text{Use the Venn diagram.}$$
$$= 0.80 − 0.25 = 0.55 \qquad \text{Use the known probabilities.} \; \bullet$$

Example 4.18 Movie Receipts The Cheswick Theatre in Pittsburgh, Pennsylvania, has six screens, each showing a different movie. Receipts from a recent weekend were used to compile the following table, showing the probability of watching each movie.

Movie	M_1	M_2	M_3	M_4	M_5	M_6
Probability	0.10	0.25	0.20	0.30	0.10	0.05

Consider the following events.

$A = \{M_1, M_2\}$ (movies rated PG)
$B = \{M_2, M_3, M_6\}$ (action adventures)
$C = \{M_4, M_5\}$ (dramas)
$D = \{M_6\}$ (foreign films)

Suppose a patron is randomly selected. Find the probability that the person
a. watched a movie rated PG or an action adventure.
b. watched a movie rated PG or a drama.
c. watched a movie rated PG, a drama, or a foreign film.

SOLUTION

Find the probability of A, B, C, and D. Break down each event and look at the individual outcomes.

$P(A) = P(M_1) + P(M_2) = 0.10 + 0.25 = 0.35$
$P(B) = P(M_2) + P(M_3) + P(M_6) = 0.25 + 0.20 + 0.05 = 0.50$
$P(C) = P(M_4) + P(M_5) = 0.30 + 0.10 = 0.40$
$P(D) = P(M_6) = 0.05$

a. *Or* means union. Find the corresponding events, and translate everything into mathematics.

$$P(A \cup B) = P(A) + P(B) - P(A \cap B)$$ (General) Addition Rule.
$$= P(A) + P(B) - P(M_2)$$ Find the events in $A \cap B$.
$$= 0.35 + 0.50 - 0.25 = 0.60$$ Use known probabilities.

b. Part (b) is the same kind of question; *or* means union.

$$P(A \cup C) = P(A) + P(C)$$ A and C are disjoint.
$$= 0.35 + 0.40 = 0.75$$ Use known probabilities.

c. *Or* means union again in part (c), but with three events.

$$P(A \cup C \cup D) = P(A) + P(C) + P(D)$$ Three *disjoint* events.
$$= 0.35 + 0.40 + 0.05 = 0.80$$ Use known probabilities. ●

SECTION 4.2 EXERCISES

Practice

4.33 Consider an experiment with the probability of each simple event given in the table below.

Simple event	e_1	e_2	e_3	e_4
Probability	0.07	0.09	0.13	0.18

Simple event	e_5	e_6	e_7
Probability	0.22	0.15	0.16

The events A, B, C, and D are defined by
$A = \{e_1, e_2, e_3\}$ $B = \{e_2, e_4, e_6, e_7\}$
$C = \{e_1, e_5, e_7\}$ $D = \{e_3, e_4, e_5, e_6, e_7\}$
Find the following probabilities.
a. $P(A)$ **b.** $P(C)$ **c.** $P(D)$
d. $P(A \cup B)$ **e.** $P(A \cap C)$ **f.** $P(B \cap D)$
g. $P(A')$ **h.** $P(A \cap C')$ **i.** $P(A' \cap D)$
j. $P(C')$ **k.** $P(B \cap C \cap D)$ **l.** $P[(B \cup C)']$
How do you know there is no other possible simple event in this experiment?

4.34 An experiment consists of rolling a special 18-sided die. All of the numbers, 1 through 18, are equally likely. Find the probability of each event.
a. $A = $ rolling an even number.
b. $B = $ rolling a number divisible by 3.
c. $C = $ rolling a number less than 7.
d. $D = $ rolling at least a 10.

4.35 An experiment consists of rolling a special 22-sided die. All of the numbers, 1 through 22, are equally likely. Find the probability of each event.
a. $A = $ rolling a number greater than 10 and even.
b. $B = $ rolling a prime number or a number divisible by 5.
c. $C = $ rolling at most an 11.
d. $D = $ rolling a number divisible by 2 and 3.

4.36 Consider an experiment, the events A and B, and probabilities $P(A) = 0.55$, $P(B) = 0.45$, and $P(A \cap B) = 0.15$. Find the probability of
a. A or B occurring.
b. A and B occurring.
c. Just A occurring.
d. Just A or just B occurring.

4.37 Consider an experiment, the events A and B, and probabilities $P(A) = 0.26$, $P(B) = 0.68$, and $P(A \cup B) = 0.80$. Find each probability.
a. $P(A \cap B)$
b. $P(A')$
c. $P[(A \cap B)']$
d. $P[(A \cup B)']$

4.38 Consider an experiment, the events A and B, and probabilities $P(A) = 0.355$, $P(B) = 0.406$, and $P(A \cap B) = 0.229$. Find each probability.
a. $P(A \cup B)$
b. $P[(A \cup B)']$
c. $P(B')$
d. $P[(A \cap B)']$

4.39 Carefully sketch a Venn diagram showing the relationship between two events. Add probabilities to the appropriate regions so that the following statements are true: $P(A \cap B) = 0.31$, $P(A) = 0.57$, and $P(B) = 0.48$.

4.40 Carefully sketch a Venn diagram showing the relationships among three events. Add probabilities to the appropriate regions so that the following statements are true:

$P(A) = 0.46$	$P(B) = 0.35$	$P(C) = 0.44$
$P(A \cap B) = 0.05$	$P(A \cap C) = 0.18$	
$P(B \cap C) = 0.14$	$P(A \cap B \cap C) = 0.03$	

Applications

4.41 Manufacturing and Product Development Valassis, a marketing services company, offers a cafeteria-style benefit program; an employee may select three benefits from five. The five possible benefits are health insurance, life insurance, a prescription plan, dental insurance, and vision insurance.
a. How many different benefit packages can an employee select? List them.
b. If all benefit packages are equally likely, what is the probability that an employee selects a package that includes health insurance?
c. If all benefit packages are equally likely, what is the probability that an employee selects a package that includes life insurance and a prescription plan?

4.42 Travel and Transportation Suppose a bridge has nine toll booths in the east-bound lane: three are only for E-Z Pass holders, two are only for exact change, one takes only tokens, and the remainder are staffed by toll collectors who accept only cash. During heavy traffic hours it is difficult to see the signs indicating the type of toll booth. Suppose a driver randomly selects a toll booth.
a. What is the probability that an exact-change toll booth is selected?
b. What is the probability that a manual-collection toll booth or the token toll booth is selected?
c. What is the probability that an E-Z Pass toll booth is not selected?
d. Suppose the driver only has tokens. What is the probability that he or she selected the appropriate toll booth?

4.43 Demographics and Population Statistics As of July 2005, the Federal Bureau of Prisons reported the following proportions concerning the sentence imposed on inmates.

Sentence imposed (in years)	Inmate proportion
0–1	0.022
1–3	0.141
3–5	0.161
5–10	0.289
10–15	0.179
15–20	0.084
20 or more	0.091
Life	0.032
Death	0.001

(*Source*: Federal Bureau of Prisons, Quick Facts).

Suppose an inmate is selected at random.
a. What is the probability that the sentence is between 1 and 10 years (including 1 but not 10)?
b. What is the probability that the sentence is for life or death?
c. What is the probability that the sentence is not the death penalty?

4.44 Marketing and Consumer Behavior A local Pizza Palace offers five different toppings on its pizza: pepperoni, sausage, olives, mushrooms, and anchovies. A large pizza comes with any two different toppings.
a. How many different two-topping pizzas are possible?
b. Suppose all of the pizzas are equally likely. What is the probability that the next pizza ordered has at least one meat topping?
c. What is the probability that the next pizza ordered does not have anchovies?
d. Suppose one more large pizza choice is added: plain cheese with no toppings. Answer parts (b) and (c) with this added assumption.

4.45 Demographics and Population Statistics The Brady Act allows people to transfer firearms if they pass a criminal history background check. The following table shows the reasons for rejection of firearm transfer applications by state agencies and the proportion of applications that are rejected in each category.

Reason for rejection	Proportion
Felony indictment or conviction	0.386
Misdemeanor conviction	0.067
Restraining order	0.036
State law prohibition	0.089
Fugitive	0.060
Illegal alien	0.002
Mental illness or disability	0.030
Drug addiction	0.012
Local law prohibition	0.001
Other	0.317

(*Source*: Sourcebook of Criminal Justice Statistics Online, 2007. Note: 0.001 added to Local law prohibition.)

Suppose a rejected application is selected at random.
a. What is the probability that the rejection is for state or local law prohibition?
b. What is the probability that the rejection is for a reason different from domestic violence (domestic violence includes misdemeanor convictions and restraining orders)?
c. What is the probability that the rejection is for neither a felony nor other?

4.46 Public Policy and Political Science The following table lists the number of states by state tree.

State tree	Pine	Oak	Maple	Spruce	Other
Frequency	10	7	5	4	24

(*Source*: *USA Today Snapshot* and The National Arbor Day Foundation.)

Suppose a state is randomly selected.
a. Find the probability that the state has some *other* state tree.
b. Find the probability that the state has a pine or spruce as a state tree.
c. Find the probability that the state has an oak or some other state tree.

4.47 Marketing and Consumer Behavior A marketing firm can place an advertisement using several media. The table below shows the probability that a randomly selected person in a targeted region will see the advertisement in the given medium.

Medium	Newspaper	Radio	Magazine	TV
Probability	0.15	0.10	0.08	0.30

Medium	Internet	Billboard	Not seen
Probability	0.12	0.05	0.20

Consider the following events.
A = {Magazine, Newspaper}
B = {TV, Radio, Internet}
C = {Magazine, Newspaper, Internet, Billboard}

Find the following probabilities.
a. $P(A), P(B), P(C)$
b. $P(A \cup B), P(A \cap B), P(B \cap C)$
c. $P(A'), P(A' \cap C), P(A \cap B \cap C)$
d. $P(B' \cap C'), P[(B \cup C)']$

4.48 Sports and Leisure The Pennsylvania State Lottery Daily Number consists of three digits, each 0–9.
a. How many possible Daily Numbers are there?
b. If all of the Daily Numbers are equally likely, find the probability that all three digits are the same.
c. If all of the Daily Numbers are equally likely, find the probability that all three digits are 8s or 9s.

4.49 Home Heating The following table shows the type of fuel used to heat homes in the Boston-Cambridge-Quincy area in Massachusetts and the corresponding probabilities.

Heating fuel	Probability
Utility gas	0.479
Bottled, tank, or LP gas	0.030
Electricity	0.133
Fuel oil, kerosene, etc.	0.345
Coal or coke	0.001
Wood	0.006
Other fuel	0.004
No fuel used	0.002

(*Source*: U.S. Census Bureau, American Community Survey, 2007.)

Consider the events:
A = {heats with utility gas or bottled, tank, or LP gas}
B = {does not heat with fuel oil}
C = {heats with electricity}

Find the following probabilities.
a. $P(A), P(B), P(C)$.
b. $P(A \cap B), P(A \cup C), P(A \cap C)$.
c. $P(A' \cup C), P(A \cup B \cup C')$.

4.50 Technology and Internet A research study concerning teenagers and video games reported that of those teenagers who play video games, 74% play racing games and 49% play strategy games.[2] Suppose 35% play both genres and suppose a teenager who plays video games is selected at random.
a. What is the probability that the teenager plays racing games or strategy games?
b. What is the probability that the teenager does not play racing games or strategy games?
c. What is the probability that the teenager plays only racing games? Only strategy games?

Extended Applications

4.51 Sports and Leisure According to a Yankelovich Partners Survey, 50% of all bicycle riders regularly wear a helmet while riding a bike and 61% regularly ride on neighborhood streets with little traffic.[3] Suppose 30% of all riders regularly wear a helmet and ride on neighborhood streets with little traffic, and a bicycle rider is selected at random.
a. Carefully sketch a Venn diagram with probabilities to illustrate the relationship between the two events A = regularly wears a helmet, and B = regularly rides on neighborhood streets with little traffic.
b. What is the probability that the rider regularly wears a helmet or rides on neighborhood streets with little traffic?
c. What is the probability that the rider neither wears a helmet nor rides on neighborhood streets with little traffic?
d. What is the probability that the rider only regularly wears a helmet?

4.52 Marketing and Consumer Behavior Of all those people who enter Uncle's Stereo, a discount electronics store in New York City, 28% purchase a DVD player, 5% buy a large-screen TV, and 4% buy both. Suppose a customer is selected at random.
a. What is the probability that the customer buys a large-screen TV or a DVD player?
b. What is the probability that the customer buys either a large-screen TV or a DVD player, but not both?
c. What is the probability that the customer buys only a DVD player?
d. What is the probability that the customer does not buy a large-screen TV?

4.53 Demographics and Population Statistics The following table shows the ABO and Rh blood type probabilities for people in the United States. (This table is called a *joint probability table*.) Each number in the table can be thought of as the probability of an intersection; for example, the probability of blood type A *and* negative Rh is 0.063.

		ABO type			
		O	A	B	AB
Rh type	Positive	0.374	0.357	0.085	0.034
	Negative	0.066	0.063	0.015	0.006

(*Source*: Department of Transfusion Medicine, National Institutes of Health.)

Suppose a U.S. resident is selected at random. Find the following probabilities.
a. The person has Rh positive blood.
b. The person has type B blood.
c. The person does not have type O blood.
d. The person has type AB or Rh negative blood.

4.54 Manufacturing and Product Development A tire manufacturer has started a program to monitor production. In every batch of eight tires, two will be randomly selected and electronically tested for defects. An experiment consists of recording the condition of these two tires: defect-free (G) or reject (B). Suppose two of the eight tires in a batch actually contain defects.
a. List the outcomes in this experiment.
b. What is the probability that both tires selected will be defect-free?

c. What is the probability that at least one of the tires selected will contain a defect?
d. What is the probability that both tires selected will contain a defect?

4.55 Travel and Transportation Speeding and alcohol are the most common factors contributing to traffic accidents. For all male drivers under 21 involved in a fatal crash in 2003, 34% were speeding, 25% were intoxicated, and 23% were both speeding and intoxicated.[4] Suppose a fatal crash involving a male under 21 is selected at random.
a. Carefully sketch a Venn diagram showing the relationship between the events speeding and intoxicated, and add probabilities to the appropriate regions.
b. What is the probability that the driver was speeding or intoxicated?
c. What is the probability that the crash was due to some other cause?
d. What is the probability that the driver was either speeding or intoxicated, but not both?

Challenge

4.56 Reconsider Example 4.15 (page 137). Suppose the owner records the type of bagel purchased for the next 10 customers. Find the probability that everyone buys a plain bagel. What do you think about the assumption of equal demand now?

4.3   COUNTING TECHNIQUES

In an equally likely outcome experiment, computing probabilities means counting. To find the probability of an event A, count the number of outcomes in the event A and divide by the number of outcomes in the entire sample space S: $P(A) = N(A)/N(S)$. If $N(S)$ is large, drawing a tree diagram or listing all of the possible outcomes is impractical. For certain experiments, the following rules may be used instead to count outcomes in an event and/or a sample space.

THE MULTIPLICATION RULE

Suppose an outcome in an experiment consists of an ordered list of k items selected using the following procedure:

1. There are n_1 choices for the first item.
2. There are n_2 choices for the second item, no matter which first item was selected.
3. The process continues until there are n_k choices for the kth item, regardless of the previous items selected.
   There are $N(S) = n_1 \cdot n_2 \cdot n_3 \cdots n_k$ outcomes in the sample space S.

ILLUMINATING THE CONCEPTS

1. You can picture (and even prove) this rule by drawing a tree diagram and counting the number of paths from left to right.

2. To use this rule, think of each choice as a slot, or a position, to fill.

$$\underbrace{n_1}_{\text{Item 1}} \times \underbrace{n_2}_{\text{Item 2}} \times \cdots \times \underbrace{n_k}_{\text{Item } k} = n_1 \cdot n_2 \cdots n_k$$

Number of choices for each slot.

3. This counting technique can also be used for events, not just for sample spaces. ∎

Example 4.19 Surround Sound   A home theater system consists of a receiver, speakers, and a DVD player. Vann's Inc. sells seven different receivers, 12 types of speakers, and nine different DVD players. How many possible systems can be constructed?

SOLUTION

STEP 1   This is a counting problem, and there are three slots to fill: receiver, speakers, and DVD player. We'll assume all components are compatible and that the choice of any one item does not depend on any other item.

STEP 2   Here's how to apply the **Multiplication Rule**.

$$\underbrace{7}_{\text{Receiver}} \times \underbrace{12}_{\text{Speakers}} \times \underbrace{9}_{\text{DVD Player}} = 756$$

There are 756 possible systems.   ●

Example 4.20 Vanity Plates   A Connecticut license plate consists of three letters followed by three numbers.
a. How many different license plates are possible?
b. How many license plates end in 555?

SOLUTION

a. This is a counting problem. There are six slots to fill: 3 letters followed by 3 numbers. There are 26 possible letters for each of the first three positions, and 10 possible numbers for each of the last three positions. Use the Multiplication Rule.

$$\underbrace{26}_{\text{Letter}} \times \underbrace{26}_{\text{Letter}} \times \underbrace{26}_{\text{Letter}} \times \underbrace{10}_{\text{Number}} \times \underbrace{10}_{\text{Number}} \times \underbrace{10}_{\text{Number}} = 17{,}576{,}000$$

There are 17,576,000 possible different license plates.

b. If the license plate ends in 555, then each of the number positions is fixed; there is only one choice. We are still free to choose any letter in each of the first three positions. The Multiplication Rule still works.

$$\underbrace{26}_{\text{Letter}} \times \underbrace{26}_{\text{Letter}} \times \underbrace{26}_{\text{Letter}} \times \underbrace{1}_{\text{Number}} \times \underbrace{1}_{\text{Number}} \times \underbrace{1}_{\text{Number}} = 17{,}576$$

There are 17,576 license plates that end in 555.   ●

The actual number of possible license plates is smaller, since some three-letter words aren't allowed.

(Keeweeboy/Dreamstime.com)

Example 4.21 Five-of-a-Kind   In the game of Yahtzee, five fair dice are rolled and the numbers that land face-up are recorded.
a. How many different rolls are possible?
b. What is the probability of rolling a Yahtzee (all five dice with the same number)?

SOLUTION

a. There are five slots to fill, one for each die. Use the Multiplication Rule.

$$\underset{\text{Die 1}}{6} \times \underset{\text{Die 2}}{6} \times \underset{\text{Die 3}}{6} \times \underset{\text{Die 4}}{6} \times \underset{\text{Die 5}}{6} = 7776$$

There are 7776 possible rolls, or outcomes, in the sample space.

b. There are only six possible Yahtzees: 11111, 22222, 33333, 44444, 55555, and 66666. Since all the outcomes are equally likely (fair dice), the probability of rolling a Yahtzee is

$$P(\text{Yahtzee}) = \frac{\text{Number of Yahtzees}}{\text{Number of different rolls}} = \frac{6}{7776} = 0.0007716$$ ●

Example 4.22 Win, Place, or Show   Suppose there are 12 entries in the Preakness Stakes horse race. An experiment consists of recording the *finish*: the first-, second-, and third-place horse. For example, the outcome (7, 9, 2) means horse 7 came in first, horse 9 came in second, and horse 2 came in third.
a. How many different finishes are possible?
b. What is the probability of a finish with horse 4 or 5 in first place?
c. What is the probability that horse 7 will not finish first, second, or third?

SOLUTION

a. There are three positions to fill, but the number of choices in the second slot depends on the first choice and the number of choices in the third slot depends on the first two choices. Even though we are drawing from the same collection, we can still use the Multiplication Rule.

There are 12 horses that could finish first. Once a first-place horse is selected, there are only 11 left that could come in second. After a first and second-place horse are selected, there are only 10 possible for third place. The Multiplication Rule is used here to count the number of *permutations*.

$$\underset{\text{First}}{12} \times \underset{\text{Second}}{11} \times \underset{\text{Third}}{10} = 1320$$

There are 1320 possible different finishes. N(S) = 1320.

b. Let *A* be the event that horse 4 or 5 wins the race. We'll assume all of the outcomes are equally likely so that P(*A*) = N(*A*)/N(*S*) = N(*A*)/1320.

There are 2 choices for first place (horse 4 or 5). There are now 11 choices for second place (the horse not selected for first, plus the remaining 10), and 10 choices for third place.

The Multiplication Rule is used to find the number of outcomes in *A*:

$$\underset{\text{First}}{2} \times \underset{\text{Second}}{11} \times \underset{\text{Third}}{10} = 220$$

Finally, P(*A*) = 220/1320 = 0.1667.

c. The word *not* suggests the use of a complement, but a direct approach may be easier. Let the event B = horse 7 does not finish first, second, or third. $P(B) = N(B)/N(S) = N(B)/1320$.

Use the Multiplication Rule again to count the number of outcomes in the event B. We do not want horse 7 in the top three. That leaves 11 possible horses for first place, 10 for second, and 9 for third.

$$\underset{\text{First}}{11} \times \underset{\text{Second}}{10} \times \underset{\text{Third}}{9} = 990$$

$P(B) = 990/1320 = 0.75$.   ●

The following notation is often used in order to write large numbers associated with counting problems more concisely.

DEFINITION

For any positive whole number n, the symbol $n!$ (read "***n* factorial**") is defined by

$$n! = n(n-1)(n-2)\cdots(3)(2)(1).$$

In addition, $0! = 1$ (0 factorial is 1).

💡 ILLUMINATING THE CONCEPTS

1. To find $n!$, just start with n, multiply by $(n-1)$, then multiply by $(n-2)$, . . . , down to 1. For example,

$7! = (7)(6)(5)(4)(3)(2)(1) = 5040$

$10! = (10)(9)(8)(7)(6)(5)(4)(3)(2)(1) = 3{,}628{,}800$

2. Factorials get really big, really fast. Try finding $50!$. If you absolutely have to find a large factorial, then you should probably use a good calculator or computer. ■

Consider a generalization of the horse-racing problem. Suppose there are n items to choose from and r positions to fill, and the order of selection matters. There are n choices for the first position, $n - 1$ choices for the second position, and $n - 2$ choices for the third position. This process continues until there are $n - (r - 1)$ choices for the rth position. The product of these numbers is the total number of **permutations**.

DEFINITION

Given a collection of n different items, an ordered arrangement, or subset, of these items is called a **permutation**. The number of permutations n items, taken r at a time, is given by

$$_nP_r = n(n-1)(n-2)\cdots[n-(r-1)]$$

Using the definition of factorial,

$$_nP_r = \frac{n!}{(n-r)!}$$ In the denominator, do the subtraction first, then do the factorial.

$_nP_r$ is also referred to as n items permuted r at a time.

💡 **ILLUMINATING THE CONCEPTS**

1. All n items must be *different* in order for this formula to be used.

2. A distinguishing characteristic of a permutation is that order matters. For example, if the outcome AB is *different* from the outcome BA, that suggests a permutation. Suppose an experiment consists of selecting 2 students from a class of 35. The first one selected will be the president and the second will be the vice president. Order certainly matters here; we will be counting permutations. If the 2 students selected will form a committee, however, then the order of selection does not matter. Counting in this case involves a *combination*, which will be introduced below.

3. Here is an example to justify this formula.

$${}_{12}P_3 = (12)(11)(10)$$ Definition of ${}_{12}P_3$, $n = 12$, $r = 3$.

$$= (12)(11)(10) \times \frac{(9)(8)(7)(6)(5)(4)(3)(2)(1)}{(9)(8)(7)(6)(5)(4)(3)(2)(1)}$$ Multiply by **1** in a useful form.

$$= \frac{(12)(11)(10)(9)(8)(7)(6)(5)(4)(3)(2)(1)}{(9)(8)(7)(6)(5)(4)(3)(2)(1)}$$ Rewrite as one fraction.

$$= \frac{12!}{9!} = \frac{12!}{(12-3)!} = \frac{n!}{(n-r)!}$$ Definition of factorial. ■

Example 4.23 Vending Machine Selection A vending machine has room for 6 types of soda. The soda can be arranged in any order to correspond with the selection buttons on the front of the machine. If the operator has 10 different types of soda to choose from, how many machine selection arrangements are possible?

SOLUTION

If you compute ${}_nP_r$ by hand, there is always a lot of canceling. ${}_nP_r$ is a count, so the answer has to be an integer.

A technology solution:

```
10 nPr 6
            151200
```

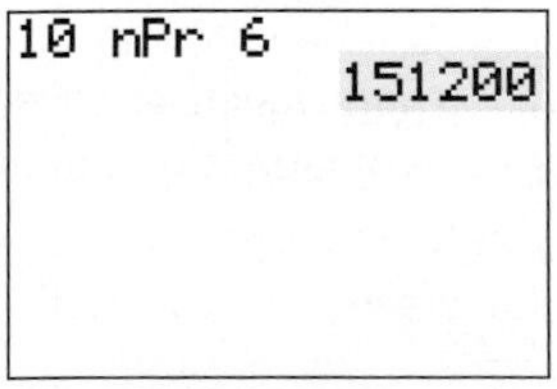

Figure 4.12 TI-84 Plus permutation function.

STEP 1 There are $n = 10$ items, we need to choose $r = 6$, and the order in which the soda is arranged matters. For example, if capital letters represent soda types, then the arrangement ABCDEF is *different* from ABCEDF. We must count the number of permutations of 10 items, taken 6 at a time.

STEP 2 $${}_{10}P_6 = \frac{10!}{(10-6)!} = \frac{10!}{4!}$$ Definition of ${}_nP_r$, using factorials.

$$= \frac{(10)(9)(8)(7)(6)(5)(4)(3)(2)(1)}{(4)(3)(2)(1)}$$ Definition of factorial.

$$= (10)(9)(8)(7)(6)(5) = 151{,}200$$ Cancel; multiply.

There are 151,200 ordered arrangements of soda types in the vending machine. ●

Example 4.24 Sipowicz and the 15th Squad An NYPD Blue fan has recorded nine episodes (on nine different DVDs) from the last season of this show. However, he only has time to watch four episodes. Suppose he selects four DVDs at random.

a. How many different ordered arrangements of episodes are possible?

b. If the season finale is on DVD, what is the probability that he will select and watch this episode last?

SOLUTION

a. There are $n = 9$ episodes to choose from. We need to count the number of ordered arrangements of $r = 4$ DVDs.

$$_9P_4 = \frac{9!}{(9-4)!} = \frac{9!}{5!}$$ Definition of $_nP_r$, using factorials.

$$= \frac{(9)(8)(7)(6)(5)(4)(3)(2)(1)}{(5)(4)(3)(2)(1)}$$ Definition of factorial.

$$= (9)(8)(7)(6) = 3024$$ Cancel; multiply.

There are 3024 different ordered arrangements of four episodes.

b. Let A = the last DVD selected is the season finale. There are four positions to fill, but the last slot is fixed (with the season finale). The first three positions can be filled by any of the remaining eight DVDs, in any order.

$$\overbrace{\underset{\text{Tape 1}}{8} \times \underset{\text{Tape 2}}{7} \times \underset{\text{Tape 3}}{6}}^{_8P_3} \times \underset{\text{Tape 4}}{1} = 336$$

$$P(A) = \frac{336}{3024} = 0.1111$$ ●

In many experiments, the order in which the items are selected does *not* matter. For example, selecting 5 manufactured items from a batch of 50 for inspection, choosing 9 people from 35 for a search committee, or picking 3 tax returns from 100 for a federal audit. In each case, the order of selection is not important; the collection, or group selected, is a single outcome. These *unordered* arrangements are called **combinations**.

DEFINITION

Given a collection of n different items, an unordered arrangement, or subset, of these items is called a **combination**. The number of combinations of n items, taken r at a time, is given by

$$_nC_r = \binom{n}{r} = \frac{n!}{r!(n-r)!} = \frac{_nP_r}{r!}.$$

💡 **ILLUMINATING THE CONCEPTS**

1. $\binom{n}{r}$ is read as "n choose r."

2. To find $_nC_r$ from $_nP_r$ we need to *collapse* all ordered arrangements of the same r items into one possible outcome. Dividing by $r!$ does this because every unordered set of r distinct items can be arranged in $r!$ ways.

3. If you have to calculate $_nC_r$ by hand, there is always a lot of cancelation. The final answer must be an integer because it is a count. ∎

A technology solution:

```
(8 nPr 3)/3024
           .1111
```

Figure 4.13 Find the number of permutations and divide by the total number of outcomes.

A technology solution:

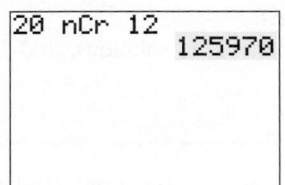

```
20 nCr 12
             125970
```

Figure 4.14 The TI-84 Plus combination function.

Solution Trail

KEYWORDS

Randomly selects.

↓

TRANSLATION

Equally likely outcomes.

↓

CONCEPTS

▪ Probability of an event means counting.

▪ Combinations.

↓

VISION

The order in which the cartons are selected does not matter. The installer has the same collection of five cartons regardless of the order selected. To find the number of outcomes in the sample space, we'll need to count combinations. To count the number of outcomes in each event, we will use the Multiplication Rule and the formula for $_nC_r$.

A technology solution:

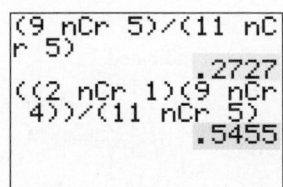

```
(9 nCr 5)/(11 nC
r 5)
             .2727
((2 nCr 1)(9 nCr
 4))/(11 nCr 5)
             .5455
```

Figure 4.15 Probability calculations.

Example 4.25 Jury Duty   How many different ways are there to select a jury of 12 people from a pool of 20?

SOLUTION

STEP 1   There are $n = 20$ prospective jurors, and we need to choose $r = 12$, without regard to order. A jury is an *unordered* arrangement of 12 people. We need to count the number of combinations of 20 items, taken 12 at a time.

STEP 2 $_{20}C_{12} = \begin{pmatrix} 20 \\ 12 \end{pmatrix} = \dfrac{20!}{12!(20-12)!} = \dfrac{20!}{12!8!}$ Definition of $\begin{pmatrix} n \\ r \end{pmatrix}$.

$$= \frac{(20)(19)(18)(17)(16)(15)(14)(13)}{8!} = 125{,}970$$

Cancelation; computation.

There are 125,970 ways to select a jury of 12 from a pool of 20 candidates. ●

Example 4.26 Hardwood Floors   Lumber Liquidators ships $2\text{-}\frac{1}{4}$-inch solid oak wood flooring in cartons containing 20 square feet. Suppose there are 2 cartons in a shipment of 11 that contain defective pieces. An installer randomly selects 5 cartons.
a. What is the probability that there are no defective pieces in all 5 cartons?
b. What is the probability the installer picks exactly 1 carton with defective pieces?

SOLUTION

There are $n = 11$ cartons, and we need to choose $r = 5$, without regard to order.

$$_{11}C_5 = \begin{pmatrix} 11 \\ 5 \end{pmatrix} = \frac{11!}{5!(11-5)!} = \frac{11!}{5!6!} = \frac{(11)(10)(9)(8)(7)}{5!} = 462$$

There are 462 outcomes in the sample space, all equally likely.

a. Let $A =$ select no cartons with defects. Count the number of ways to select no cartons with defects. Choosing no cartons with defects is the same as choosing 5 good cartons. Since there are 9 good cartons, we count the number of ways to select 5 cartons from the 9 *good* cartons, without regard to order.

$$N(A) = \begin{pmatrix} 9 \\ 5 \end{pmatrix} = \frac{9!}{5!(9-5)!} = \frac{9!}{5!4!} = \frac{(9)(8)(7)(6)}{4!} = 126$$

Since this is an equally likely outcome experiment,

$$P(A) = \frac{N(A)}{N(S)} = \frac{126}{462} = 0.2727$$

b. Let $B =$ select 1 carton with defects (and, therefore, 4 good cartons). To find the number of outcomes in B, there are two *cases* to consider: the number of ways to select 1 bad carton, and the number of ways to select 4 good cartons.

$$\begin{pmatrix} 2 \\ 1 \end{pmatrix} \times \begin{pmatrix} 9 \\ 4 \end{pmatrix} = 2 \times 126 = 252$$

↑ The number of ways to select 1 defective carton from 2, without regard to order.

↑ The number of ways to select 4 good cartons from 9, without regard to order.

$$P(B) = \frac{N(B)}{N(S)} = \frac{252}{462} = 0.5455$$ ●

TECHNOLOGY CORNER

Procedure: Compute permutations and combinations.
Reconsider: Examples 4.23, page 149 and 4.25, page 151, solutions, and interpretations.

TI-84 Plus

There are built-in functions to compute permutations, combinations, and even factorials.

1. On the Home Screen, enter the value of n.
2. Select MATH ; PRB; nPr or MATH ; PRB; nCr.
3. Enter the value of r. See Figures 4.12 (page 149) and 4.14 (page 151).

Excel

Use the function PERMUT to compute permutations and the function COMBIN to compute combinations. Use intermediate results to compute probabilities.

A	B
151200	=PERMUT(10,6)
125970	=COMBIN(20,12)

Figure 4.16 Excel functions for permutations and combinations.

Minitab

Use the functions Permutations and Combinations in either a Session window or in Calc; Calculator. See Figure 4.17.

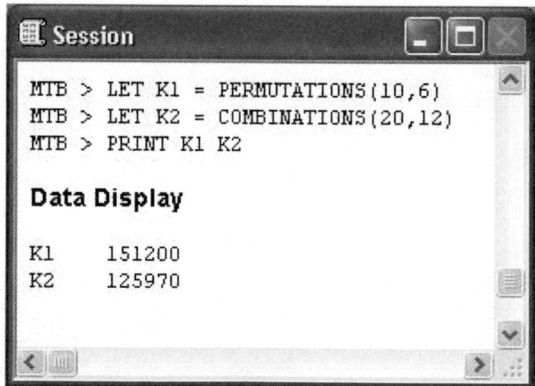

```
MTB > LET K1 = PERMUTATIONS(10,6)
MTB > LET K2 = COMBINATIONS(20,12)
MTB > PRINT K1 K2

Data Display

K1    151200
K2    125970
```

Figure 4.17 Minitab functions for permutations and combinations.

SECTION 4.3 EXERCISES

Practice

4.57 Find the number of permutations indicated.
a. $_8P_4$ **b.** $_{11}P_7$ **c.** $_{12}P_4$
d. $_{10}P_{10}$ **e.** $_{10}P_1$ **f.** $_{10}P_0$
g. $_9P_2$ **h.** $_{20}P_2$ **i.** $_{100}P_2$

4.58 Find the number of combinations indicated.
a. $\binom{9}{5}$ **b.** $\binom{9}{4}$ **c.** $\binom{14}{7}$ **d.** $\binom{10}{10}$ **e.** $\binom{10}{1}$
f. $\binom{10}{0}$ **g.** $\binom{12}{3}$ **h.** $\binom{16}{7}$ **i.** $\binom{20}{18}$

4.59 How many permutations of the letters in the word HISTOGRAM are possible?

4.60 A businessman's outfit consists of a pair of pants, a shirt, and a tie. Suppose he can choose from among five pairs of pants, eight shirts, and 15 ties.
a. How many different outfits are possible?
b. Suppose a winter outfit includes a sweater and he can select one of 7 sweaters. Now how many different winter outfits are possible?

4.61 A disc jockey has 20 songs to choose from and can only play 7 in the next half hour. How many different playlists are possible?

4.62 A grocery store has six cashiers on duty, 10 baggers, and four people who will help customers load their groceries into a car. How many different checkout crews are possible?

4.63 A television station is developing a new identifying 3-note theme. How many different 3-note themes are possible if there are 20 notes to choose from and no note can be repeated?

4.64 A small basket contains 17 good apples and 3 rotten apples.
a. How many different handfuls of 6 apples are possible?
b. How many different handfuls of 5 good apples and 1 rotten apple are possible?
c. How many different handfuls of 3 good apples and 3 rotten apples are possible?

Applications

4.65 Manufacturing and Product Development Suppose Walmart sells a *combination* lock that is really a *permutation* lock with 40 numbers, 0 to 39. The combination for each lock is set at the factory and consists of 3 numbers.
a. How many lock combinations are possible if numbers can be repeated?
b. If all lock combinations are equally likely, what is the probability of selecting a lock with only single-digit numbers in the combination?
c. Answer parts (a) and (b) if the lock combination must have 3 different numbers.

4.66 Public Policy and Political Science Suppose 14 carpenters report to the union hall hoping for a chance to work. Three of the 14 do not have their union cards, and 6 carpenters will be selected at random for construction jobs.
a. What is the probability that all 6 carpenters selected will have their union cards?
b. What is the probability that exactly 1 carpenter selected will not have a union card?
c. What is the probability that at least 1 carpenter selected will not have a union card?

4.67 Fuel Consumption and Cars Suppose Allstate offers automobile insurance with specific levels of coverage according to the table given below.

Coverage: Levels
Medical: $10,000; $20,000; $50,000; $100,000
Bodily injury liability: $50,000; $100,000
Property damage liability: $25,000; $50,000; $100,000
Uninsured motorists: $50,000; $100,000; $200,000
Comprehensive: $250,000; $500,000; $1,000,000

Suppose an automobile policy must have all five coverages.
a. How many different automobile policies are possible?
b. How many policies have comprehensive coverage of at least $500,000?
c. How many policies have bodily injury liability and property damage liability of $100,000?

4.68 Manufacturing and Product Development A mail-order company offers backpacks in five styles, three sizes, and 10 colors.
a. How many different backpacks does the company offer?
b. Midnight blue and dark green are the two most popular colors. How many different backpacks in these colors does the company offer?
c. The urban style backpack is the least popular. If the company eliminates this style, how many different backpacks will it offer?

4.69 Fuel Consumption and Cars A small tool and die shop manufactures kneuter valves. A shipment of 15 valves to a Swedish automobile assembly plant contains 3 defective valves. Suppose the assembly plant randomly selects 4 valves from the shipment.
a. What is the probability that all 4 valves will be defect-free?
b. What is the probability that the plant will select all 3 defectives?
c. What is the probability that the plant will select at least 1 defective?

4.70 Medicine and Clinical Studies A physician routinely visits a local nursing home on Thursday mornings to examine patients. Suppose the facility has 20 residents, but the physician only has time to check 8. The supervisor places 8 random patients on an ordered list and presents the schedule to the physician.
a. How many different schedules are possible?
b. If there are 15 women and 5 men in the facility, what is the probability that all the appointments will be with women?

4.71 Psychology and Human Behavior A telemarketer has 12 people on his contact list. Suppose he will randomly select 8 people to call during the next shift.
a. How many different calling schedules are possible?
b. Suppose only 2 of the 12 will definitely purchase the product when contacted. What is the probability that these 2 people will be the first ones called?
c. Suppose another 2 of the 12 will ask to be placed on the do-not-call list when contacted. What is the probability these people will not be called?

4.72 Sports and Leisure In preparation for the coming season, a bass fisherman decides to buy 5 random lures out of the 10 new ones in the local tackle shop.
a. How many different collections of 5 new lures are possible?
b. Suppose that 1 of the 10 lures is a Crazy Crawler. What is the probability that the fisherman will not select this lure?
c. Suppose that 3 of the 10 are Excalibur lures. What is the probability that at least 1 of the 5 selected will be an Excalibur lure?

4.73 Technology and Internet A student has just finished downloading 15 new songs to her computer. Because her MP3 player is almost full, she will only be able to select and transfer 3 of the songs.
a. How many different ways are there to select 3 new songs for her MP3 player?

b. Suppose the songs are selected at random and 5 are by female vocalists. What is the probability that all 3 selected will be by female vocalists?

c. What is the probability that none of the 3 selected will be by female vocalists?

d. Suppose that 1 of the downloaded songs contains explicit lyrics. What is the probability that this song is not selected?

4.74 Business and Management An art gallery has 20 stored paintings, but has just made room to display several of them. Seven paintings will be randomly selected and offered to the public for sale.

a. How many different collections of 7 paintings are possible?

b. Suppose 10 of the 20 stored works are by the same local artist. What is the probability that all 7 of the selected paintings will be by this artist?

c. The featured room in the gallery receives the most attention, and the order in which the paintings are displayed in this room is related to buyer interest. Suppose the 7 selected paintings will be placed in this featured room. How many different arrangements are possible?

4.75 Economics and Finance The purchasing agent for a state office building placed a call for bids on replacing the entry doors. Suppose eight sealed bids are received by the deadline. The bids will be opened in random order.

a. In how many different ways can the bids be opened?

b. What is the probability that the lowest bid will be opened last?

4.76 Marketing and Consumer Behavior While remodeling a kitchen, a builder decides to place a splash-guard behind the sink consisting of eight 6-inch-square ceramic tiles decorated with different botanical herbs. The tiles will be installed in a custom-made wooden panel. The tile supplier has 12 different herb designs to choose from, and the builder selects 8 of these 12 at random. Suppose the order in which the tiles are arranged on the splash-guard does not matter.

a. Two of the 12 herb tiles contain a blue tint that matches the kitchen color scheme. What is the probability that these 2 tiles will be included in the splashguard?

b. The family actually grows 5 of the 12 herbs in a backyard garden. What is the probability that all 5 of these will be included on the splashguard?

4.77 Travel and Transportation An airline flight crew consists of a pilot, copilot, and navigator. Suppose a supervisor is preparing the schedule for a flight for which 10 pilots, 15 copilots, and 17 navigators are available.

a. How many different flight crews are possible?

b. Suppose flight crews are selected at random, and there is 1 pilot who has a personality conflict with 1 copilot. What is the probability that neither of these individuals will be on the flight crew?

c. Eight of the navigators have already passed the test to become a copilot. What is the probability that the flight crew will include one of these copilot-qualified navigators?

4.78 Education and Child Development A university library is preparing a display case of books written by faculty members. There are 25 new faculty books, but there is room for only 10 in the display case. Suppose 10 books are selected at random.

a. How many different faculty book collections can be displayed?

b. If 15 of the new books are written by faculty members from the College of Science and Technology, what is the probability that all 10 displayed books are written by faculty members from this college?

c. If none of the 10 displayed books is written by faculty members from the College of Science and Technology, is there any evidence to suggest that the selection process was not random? Justify your answer.

4.79 Psychology and Human Behavior In a family with five children, two of the five are selected at random each evening to do the dishes. The first one selected washes, and the second one dries.

a. How many different wash-dry crews are possible?

b. Suppose there are two girls and three boys in the family. If the two girls are selected to wash and dry, is there any evidence to suggest that the selection process was not random? Justify your answer.

Extended Applications

4.80 Manufacturing and Product Development A remote-control garage door opener has a series of 10 two-position (0 or 1) switches used to set the access code. The code is initially set at the factory, and the switch sequence on the remote control and the opener must match in order to use the system.

a. How many different access codes are possible?

b. If all access codes are equally likely, what is the probability that a randomly selected system will have a code with exactly one 0?

c. To increase security and ensure that customers will have different access codes, new systems have 10 three-position switches (0, 1, or 2). Answer parts (a) and (b) for the new system.

4.81 Psychology and Human Behavior An annual family picture following Thanksgiving dinner is arranged with all 10 family members in a row in front of a fireplace.

a. How many different arrangements of family members are possible?

b. Suppose the family includes one set of twins, and all arrangements are equally likely. What is the probability that the twins will be in the middle two places (positions 5 and 6)?

c. What is the probability that the twins will be side-by-side in the picture?

d. Suppose the family includes five males and five females. What is the probability that the picture arrangement will alternate male, female, male, female, etc., or female, male, female, male, etc.?

4.82 Public Policy and Political Science A special committee on community development has 4 members from the town council. The full town council has 14 members: six Democrats and eight Republicans.

a. How many different committees on community development are possible?

b. Suppose the committee members are selected at random. What is the probability of a committee consisting of all Republicans?

c. Suppose every member of the committee selected is a Democrat. Do you believe the selection process was random? Justify your answer.

4.83 Sports and Leisure Texas Hold 'Em poker has become very popular in Las Vegas casinos and is seen on ESPN and the Travel Channel. (As of 11/13/08, Daniel Negreanu had won over 10.4 million dollars playing poker.) The game is played with a standard 52-card deck and starts with each player being dealt 2 (random) cards face down (hole cards). There is a round of betting, the dealer then flips 3 cards face up (the flop), there is more betting, 1 card is flipped (the turn), there is more betting, a fifth card is flipped (the river), and there is more betting. Let's focus on the 2 hole cards, called a (pre-flop) *hand*, in this problem.
a. How many (2-card, pre-flop) hands are possible in Texas Hold 'Em?
b. What is the probability that a pre-flop hand consists of two aces?
c. What is the probability that a pre-flop hand consists of a pair, that is, 2 cards of the same rank?
d. What is the probability that a pre-flop hand consists of 2 cards of the same suit?

Challenge

4.84 The Complement Rule Reconsider Example 4.22. Verify the probability in part (c) using the Complement Rule.

4.85 Combination Patterns Find the following sums.

a. $\binom{2}{0} + \binom{2}{1} + \binom{2}{2}$

b. $\binom{3}{0} + \binom{3}{1} + \binom{3}{2} + \binom{3}{3}$

c. $\binom{4}{0} + \binom{4}{1} + \binom{4}{2} + \binom{4}{3} + \binom{4}{4}$

d. $\binom{n}{0} + \binom{n}{1} + \binom{n}{2} + \cdots + \binom{n}{n}$

4.86 Sports and Leisure Consider a regular deck of 52 playing cards. For a 5-card poker hand, find the probability of
a. One pair.
b. Two pairs.
c. Three of a kind: 3 cards of the same rank and 2 others of different ranks, for example JJJ74.
d. A straight: 5 cards in sequence; the ace can be either high or low.
e. A flush: 5 cards of the same suit.

4.87 Psychology and Human Behavior How many different ways are there to arrange n people at a round table? (*Hint*: A simple rotation of a seating plan, shifting each person around the table but keeping the order the same, is not a different arrangement.)

4.88 Travel and Transportation Suppose there are n items of which n_1 are of one type, n_2 are of a second type, . . ., and n_k are of the kth type, and $n_1 + n_2 + \cdots + n_k = n$. The number of unordered arrangements of the n items is a *generalized combination* given by

$$\binom{n}{n_1 \, n_2 \cdots n_k} = \frac{n!}{n_1! \, n_2! \cdots n_k!}$$

(Think about arranging a string of colored Christmas-tree lights.) The Amtrak Auto Train from Washington, D.C., to Florida has 10 sleeper cars, two diner cars, and 14 car carriers.[5] Discounting the engine and caboose, how many *different* arrangements of cars in the train are there?

4.89 Public Policy and Political Science The U.S. Senate Committee on Commerce, Science, and Transportation has 22 members. The full Senate has 45 Democrats and 55 Republicans.
a. How many different 22-member Senate committees are possible?
b. If the committee members are selected at random, what is the probability of a committee consisting of all Democrats?
c. What is the probability that the committee has more Democrats than Republicans?

4.4 CONDITIONAL PROBABILITY

The probability questions we have considered so far have all been examples of *unconditional* probability. There were no special conditions imposed, nor was any extra information given. However, sometimes two events are related so that the probability of one depends on whether the other has occurred. In this case, knowing something extra may affect the probability assignment. This type of situation usually involves two events. The extra information may be expressed as an event separate from the event whose probability is desired.

Example 4.27 Morning Commute Consider a banker who commutes 30 miles to work every day. Due to several factors (weather, road construction, family obligations, etc.) the probability that she makes it to work on time on any random day is 0.5. If the event T is

T = the banker makes it to work on time,

then P(T) = 0.5. This is an *unconditional* probability statement; no extra information related to the event T is known or given.

Suppose a random day is selected, and the road conditions are terrible due to a snowstorm. The probability that the banker arrives at work on time is surely lower, perhaps around 0.1. Knowing the extra information (a snowstorm) changes the probability assignment for T.

The statement, "What is the probability that the banker arrives at work on time if it is snowing?" is a *conditional* probability question. The extra information is that it's snowing outside. If the event F is defined as

F = a snowstorm,

The vertical bar, |, in the probability statement is read as "given."

then this conditional probability is written as P($T|F$) = 0.1; the probability that the banker arrives at work on time given it is snowing is 0.1.

Suppose another random day is selected, but this time the banker wakes up before the alarm goes off and leaves the house early. The probability that she makes it to work on time is certainly higher, say close to 0.95. Once again, knowing some extra information changes the probability assignment for T. If the event E is

E = the banker leaves her house early,
then P($T|E$) = 0.95.

Knowing something extra *may* change the probability assignment. How do we use any added information to compute the (possibly) new probability? Consider the next example.

Example 4.28 Roll the Die Consider an experiment in which a fair, six-sided die is rolled and the number that lands face-up is recorded. The sample space is S = {1, 2, 3, 4, 5, 6}. Consider the following events:

A = {1} = roll a 1, and B = {1, 3, 5} = roll an odd number.

Finding P(A) is an unconditional probability question because no extra information is known. Since all of the outcomes in the experiment are equally likely, there is one outcome in A, and there are six outcomes in the sample space, P(A) = 1/6.

Suppose someone rolls the die, covers it with her hands, peeks at the number, and reports, "I rolled an odd number." With this added information, the probability of a 1 is now 1/3; P($A|B$) = 1/3. This conditional probability is reasonable because now we only have to consider three possibilities, that is, we have reduced the sample space from six outcomes to three, and the number of outcomes in A is 1.

The idea of reducing, or shrinking, the sample space is key to calculating conditional probabilities. The definition of **conditional probability**, and some justification for it, are given below.

DEFINITION

Suppose A and B are events with P(B) > 0. The **conditional probability of the event A given the event B has occurred**, P($A|B$), is

$$P(A|B) = \frac{P(A \cap B)}{P(B)}.$$

What goes wrong with this definition if P(B) = 0?

1. The unconditional probability of an event A can be written as

$$P(A) = \frac{P(A)}{1} = \frac{P(A)}{P(S)} = \frac{\text{Probability of the event } A}{\text{Probability of the relevant sample space}}$$

We use this same reasoning to find $P(A|B)$.

2. Given B has occurred, the relevant sample space has changed. It is *reduced* from S to B. (See Figure 4.18.)

3. Given B has occurred, the only way A can occur is if $A \cap B$ has occurred, because the sample space has been reduced to B.

4. $P(A|B)$ is the probability A has occurred, $P(A \cap B)$, divided by the probability of the relevant sample space $P(B)$.

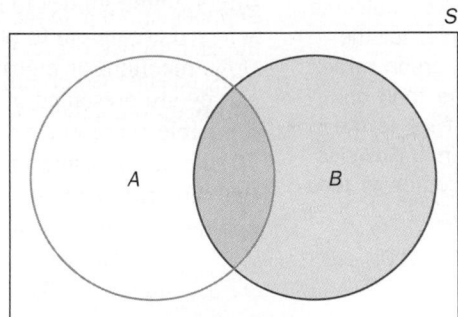

Figure 4.18 An illustration for calculating conditional probability. ∎

In the following example, the formula for conditional probability is used to confirm our intuitive answer to the die problem above.

Example 4.29 Roll the Die (Continued)   The experiment consists of rolling a fair, six-sided die and recording the number that lands face-up. $S = \{1, 2, 3, 4, 5, 6\}$. Consider the following events:

$A = \{1\}$ = roll a 1,   and $B = \{1, 3, 5\}$ = roll an odd number.

Find $P(A|B)$, the probability of rolling a 1 given an odd number was rolled.

SOLUTION

STEP 1   We will need the following probabilities:

$$P(B) = 3/6, \quad \text{and} \quad P(A \cap B) = P(1) = 1/6.$$

STEP 2 $P(A|B) = \dfrac{P(A \cap B)}{P(B)} = \dfrac{1/6}{3/6} = \dfrac{1}{6} \cdot \dfrac{6}{3} = \dfrac{1}{3}$

This answer agrees with the intuitive answer above (thank goodness). ●

Here are some facts about union, intersection, and conditional probability to help translate and solve many of the problems that follow.

1. $P(A \cup B) = P(B \cup A)$

This is always true, because $A \cup B = B \cup A$ (all the outcomes in A or B or both).

2. $P(A \cap B) = P(B \cap A)$

This is also always true, because $A \cap B = B \cap A$.

3. $P(A|B) \neq P(B|A)$.

These two probabilities *could* be equal, but in general they are different. It's all right to switch A and B with union and intersection, but not with conditional probability.

4. The keywords *given* and *suppose* often signal partial information and, therefore, indicate a conditional probability question. ∎

When are these two conditional probabilities equal?

You can also think of this as representing all of the sample events in an equally likely outcome experiment. For example, let the outcome *HA* mean a person rated the prices high and the food one star. The probability of *HA* is the number of outcomes in *HA* divided by the number of outcomes in the sample space:

N(*HA*)/N(*S*) = 25/510.

Example 4.30 Do You Have a Reservation? The Zagat Survey, started in 1979 by two Yale-educated lawyers, invites diners to rate and review restaurants. The first survey included only New York City restaurants, but the company now offers dining guides to more than 70 major cities.[6] Suppose Zagat asked 510 people selected at random to rate Charlie Trotter's Restaurant in Chicago according to price (low, medium, or high) and food (one, two, three, or four stars). The results of this survey are presented in the **two-way**, or **contingency**, **table** below. The numbers in this table represent frequencies. For example, in the third row and fourth column, 30 people rated the prices high and the food four stars. The last column contains the sum for each row, and similarly, the bottom row contains the sum for each column. These sums are often called *marginal totals*.

	Food rating				
	1 Star (*A*)	2 Stars (*B*)	3 Stars (*C*)	4 Stars (*D*)	
Low (*L*)	20	35	90	15	160
Price Medium (*M*)	50	80	95	25	250
High (*H*)	25	5	40	30	100
	95	120	225	70	510

Assume these results are representative of the entire city of Chicago, so that the relative frequency of occurrence is the true probability of the event. A person from Chicago is randomly selected.

a. Find the probability that the person rates the prices medium.

b. Find the probability that the person rates the food two stars.

c. Suppose the person selected rates the prices high. What is the probability that he rates the restaurant one star?

d. Suppose the person selected does not rate the food four stars. What is the probability that she rates the prices high?

SOLUTION

a. This is an unconditional probability question, asking only about the event *M*. Compute the relative frequency of occurrence of *M*, that is, the proportion of responses that rated the restaurant medium priced.

$$P(M) = \frac{50 + 80 + 95 + 25}{510} = \frac{250}{510} = 0.4902$$

b. This is just another unconditional probability question. Find the relative frequency of occurrence of two stars.

$$P(B) = \frac{35 + 80 + 5}{510} = \frac{120}{510} = 0.2353$$

c. This is a conditional probability question; the keyword is *suppose*. The *given* information is *rates prices high*. We need the probability of the event A given the event H has occurred. Using the formula for conditional probability:

$$P(A|H) = \frac{P(A \cap H)}{P(H)} = \frac{25/510}{100/510} = \frac{25}{510} \cdot \frac{510}{100} = \frac{25}{100} = 0.2500$$

This probability can also be obtained directly by *reducing* the sample space in the two-way table. The shaded row is the reduced sample space.

	1 Star (A)	2 Stars (B)	3 Stars (C)	4 Stars (D)	
Low (L)	20	35	90	15	160
Medium (M)	50	80	95	25	250
High (H)	25	5	40	30	100
	95	120	225	70	510

In the reduced sample space, 25 outcomes are in the event A. Therefore,

$$P(A|H) = \frac{\text{Number of outcomes in } A \text{ and in the reduced sample space}}{\text{Number of outcomes in the reduced sample space}}$$

$$= \frac{25}{100} = 0.2500$$

d. Solve this conditional probability question by reducing the sample space via the two-way table.

	1 Star (A)	2 Stars (B)	3 Stars (C)	4 Stars (D)	
Low (L)	20	35	90	15	160
Medium (M)	50	80	95	25	250
High (H)	25	5	40	30	100
	95	120	225	70	510

There are 440 ($= 95 + 120 + 225$) outcomes in the reduced sample space, and 70 ($= 25 + 5 + 40$) people rated the prices high.

$$P(H|D') = \frac{70}{440} = 0.1591$$

Solution Trail

KEYWORDS

- *If* the worker has a defined-contribution plan.
- Probability the worker also has a defined-benefits plan.

↓

TRANSLATION

Given the event *the worker has a defined-contribution plan*, find the probability of the event *the worker has a defined-benefits plan*.

↓

CONCEPTS

Conditional probability.

↓

VISION

Use the formula for conditional probability to find $P(B|C)$.

Example 4.31 Fringe Benefits Over the past 15 years, there has been a decline in traditional benefits packages offered to workers (for example, lump sum payments upon retirement) and an increase in the number of defined-contribution plans (for example 401(k) and IRA accounts). According to the Department of Labor, 43% of all workers in the private sector have a defined-contribution plan and 14% have both a defined-contribution plan and a defined-benefits package.[7] Suppose a worker in the private sector is selected at random. If the worker has a defined-contribution plan, what is the probability the worker also has a defined-benefits plan?

SOLUTION

STEP 1 Consider the events:
$C =$ the worker has a defined-contribution plan;
$B =$ the worker has a defined-benefits plan.

The statement of the problem includes two probabilities involving these two events.

$$P(C) = 0.43 \qquad \text{Percentage converted to unconditional probability.}$$

$$P(B \cap C) = 0.14 \qquad \text{The word } both \text{ means intersection.}$$

STEP 2 $\quad P(B \mid C) = \dfrac{P(B \cap C)}{P(C)} \qquad$ Translated conditional probability; definition.

$$= \frac{0.14}{0.43} = 0.3256 \qquad \text{Use known probabilities.}$$

If a worker has a defined-contribution plan, the probability the worker has a defined-benefits plan is 0.3256. ●

The solution here requires careful translation of the words into mathematics.

Steps for Calculating a Conditional Probability

To find the conditional probability of the event *A* given the event *B* has occurred:
(a) Calculate P(*B*) and P(*A* ∩ *B*).
(b) Find

$$P(A \mid B) = \frac{P(A \cap B)}{P(B)}.$$

Example 4.32 Gender, Marital Status, and the Census The United States Constitution directs the government to conduct a census of the population every 10 years. Population totals are used to allocate congressional seats, electoral votes, and funding for many government programs. The Census Bureau also compiles information related to income and poverty, living arrangements for children, and marital status. The following **joint probability table** lists the probabilities corresponding to marital status and gender of persons 15 years and over.

Gender	Marital Status				
	Married (*R*)	Never married (*N*)	Widowed (*W*)	Divorced or separated (*D*)	
Male (*M*)	0.266	0.159	0.011	0.049	0.485
Female (*F*)	0.263	0.135	0.048	0.069	0.515
	0.529	0.294	0.059	0.118	1.000

(*Source*: U.S. Census Bureau, 2007.)

Suppose a U.S. resident 15 years or older is selected at random.
a. Find the probability that the person is female and widowed.
b. Suppose the person is male. What is the probability he was never married?
c. Suppose the person is married. What is the probability the person is female?

SOLUTION

a. The keyword is *and*, which means intersection. The probability of female (*F*) and widowed (*W*) is found by reading the appropriate cell.

The body of the table contains *intersection* probabilities: the probability of a row event *and* a column event. For example, the probability that a person is male and divorced is 0.040, the intersection of the first row and the first column. The probabilities obtained by summing across rows or down columns are called *marginal* probabilities. The total probability in the table is 1.000.

	Married (*R*)	Never married (*N*)	Widowed (*W*)	Divorced (*D*)	
Male (*M*)	0.266	0.159	0.011	0.049	0.485
Female (*F*)	0.263	0.135	0.048	0.069	0.515
	0.529	0.294	0.059	0.118	1.000

$$P(F \cap W) = 0.048$$

b. The keyword is *suppose*. That suggests conditional probability. The extra information is *male*.

$$P(N \mid M) = \frac{P(N \cap M)}{P(M)} \qquad \text{Translated conditional probability; definition.}$$

$$= \frac{0.159}{0.485} = 0.328 \qquad \text{Use known probabilities.}$$

c. This is another conditional probability. This time, the event *R* is given.

$$P(F \mid R) = \frac{P(F \cap R)}{P(R)} = \frac{0.263}{0.529} = 0.497 \qquad ●$$

ILLUMINATING THE CONCEPTS

1. ◀ In Example 4.32:

$$P(R) = P(R \cap M) + P(R \cap F)$$
$$= P(R \cap M) + P(R \cap M')$$

In general, for any two events A and B:

$$P(A) = P(A \cap B) + P(A \cap B')$$

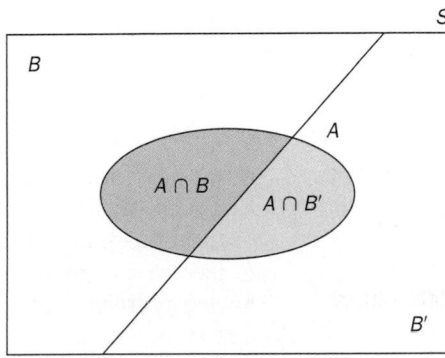

Figure 4.19 Venn diagram showing decomposition of the event A.

This *decomposition* technique is often needed in order to find $P(A)$. The Venn diagram in Figure 4.19 illustrates this equation.

The events B and B' make up the entire sample space: $S = B \cup B'$.

Try to draw the Venn diagram to illustrate this equality.

---

2. Suppose B_1, B_2, and B_3 are mutually exclusive and *exhaustive*: $B_1 \cup B_2 \cup B_3 = S$. For any other event A:

$$P(A) = P(A \cap B_1) + P(A \cap B_2) + P(A \cap B_3)$$ ▶ ■

SECTION 4.4 EXERCISES

Practice

4.90 Identify each of the following statements as conditional or unconditional probability questions.
a. The probability that a randomly selected car will start in the morning.
b. The probability that a person will remember to bring home a loaf of bread after work if he leaves a Post-It note reminder on the steering wheel.
c. The probability that the next player will get a hit in a baseball game.
d. The probability that a randomly selected heart transplant operation will be successful.
e. Of all one-way streets in a large city, the probability that the street has more than two lanes.

4.91 Identify each of the following statements as a conditional or unconditional probability.

a. The probability that a randomly selected circuit board will be defective given it was manufactured during the third shift.
b. The probability that a waitress receives a tip over 15% of the cost of the meal.

c. The probability that the next customer in a bookstore will buy a magazine.
d. The probability that a company's sales will increase given more money is spent on advertising.
e. The probability that a bowler will make three strikes in a row.

4.92 Consider the following joint probability table describing the events A, B, C, D, E, F, and G.

	F	G
A	0.12	0.05
B	0.15	0.07
C	0.17	0.04
D	0.19	0.02
E	0.11	0.08

a. Verify this is a valid joint probability table; that is, each probability must be greater than or equal to 0 and the sum of all probabilities must equal 1.
b. Compute the marginal probabilities.
c. Find $P(A \cap F)$, $P(B \cap G)$, and $P(D \cap G)$.
c. Find $P(A|G)$, $P(F|D)$, and $P(E|C)$.
d. Verify $P(C) = P(C \cap F) + P(C \cap G)$.

4.93 Consider the following joint probability table.

	B_1	B_2	B_3
A_1	0.095	0.016	0.007
A_2	0.205	0.188	0.003
A_3	0.155	0.238	0.093

a. Find $P(A_1)$, $P(A_2)$, and $P(A_3)$.
b. Find $P(B_1)$, $P(B_2)$, and $P(B_3)$.
c. Find $P(A_1 \cap B_1)$, $P(A_2 \cap B_2)$ and $P(A_3 \cap B_3)$.
d. Find $P(A_1 | B_1)$, $P(B_1 | A_1)$, and $P(A_1' \cap B_1')$.
e. Find $P(B_2 | A_2)$, and $P(B_3 | A_3)$.

4.94 Consider the following joint probability table.

	C_1	C_2	C_3	
A	0.135	0.125	0.206	0.466
B	0.145	0.174	0.215	0.534
	0.280	0.299	0.421	1.000

a. Find $P(A)$ and $P(C_2)$.
b. Find $P(A \cap C_1)$ and $P(B \cap C_3)$.
c. Find $P(C_2 | B)$, $P(A | C_3)$, and $P(A | C'_3)$.
d. Verify $P(B) = P(B \cap C_1) + P(B \cap C_2) + P(B \cap C_3)$.
   Carefully sketch a Venn diagram to illustrate this equality.

4.95 A recent survey classified each person according to the following two-way table.

	B_1	B_2	B_3	
A_1	178			815
A_2		150	244	
A_3	165	202		
	466	583	985	

a. Complete the two-way table.
b. How many people participated in this survey?
Assume the results from this survey are representative of the entire population, and one person from this population is randomly selected.
c. Find $P(A_1)$, $P(A_2)$, and $P(A_3)$.
d. Find $P(B_1 \cap A_1)$, $P(B_2 \cap A_2)$ and $P(B_3 \cap A_3)$.
e. Find $P(A_3 | B_1)$, $P(B_2 | A_2)$, and $P(A_3 \cap B_3')$.

4.96 Consider an experiment and three events A, B, and C defined in the Venn diagram below.

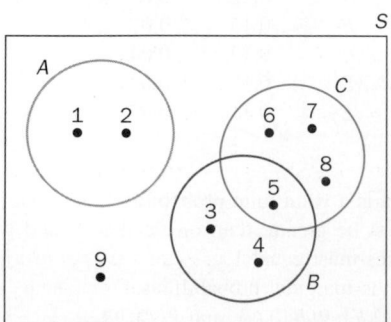

The following table gives the probability of each outcome.

Outcome	1	2	3	4	5
Probability	0.01	0.12	0.11	0.10	0.15
Outcome	6	7	8	9	
Probability	0.25	0.14	0.08	0.04	

Find the following probabilities.
a. $P(A)$, $P(B)$, and $P(C)$.
   Why don't these three probabilities sum to 1?
b. $P(A \cap B)$ and $P(B \cap C)$.
c. $P(B | C)$ and $P(C | B)$.
d. $P(A | B')$, $P(C | A')$ and $P[1 | (A \cup B)']$.
e. $P(3 | B)$, $P(4 | B)$, and $P(5 | B)$.
   Why do these three probabilities sum to 1?

4.97 Consider an experiment and three events A, B, and C defined in the Venn diagram below.

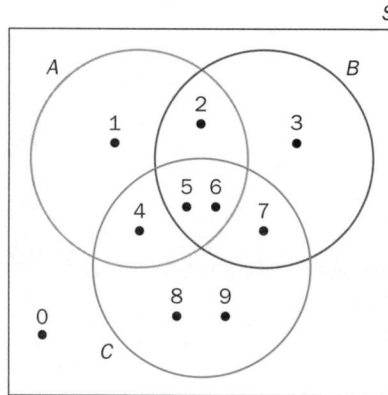

The following table gives the probability of each outcome.

Outcome	0	1	2	3	4
Probability	0.135	0.130	0.142	0.128	0.147
Outcome	5	6	7	8	9
Probability	0.083	0.072	0.063	0.055	0.045

Find the following probabilities.
a. $P(A)$, $P(B)$, and $P(C)$.
b. $P(A \cap B)$ and $P(B \cap C)$.
c. $P(A | B)$, $P(B | C)$, and $P[(A \cap B) | C]$.
d. $P(0 | C')$, $P(7 | C)$, and $P[(A \cup B) | C']$.
e. $P(2 | B)$, $P(3 | B)$, and $P(7 | B)$.

Applications

4.98 Sports and Leisure Consider a regular 52-card deck of playing cards. Suppose 2 cards are drawn at random from the deck without replacement.
a. What is the probability that the second card is an ace, given the first card is a king?

b. What is the probability that the second card is an ace, given the first card is an ace?

c. What is the probability that the second card is a heart, given the first card is a heart?

d. Suppose 2 cards are drawn at random from the deck with replacement. What is the probability the second card is a heart, given the first card is a heart?

4.99 Economics and Finance There is some evidence to suggest that people who participate in an office football pool are more likely to cheat on their income taxes. Suppose 25% of all people participate in an office football pool. The IRS estimates that 15% of all people participate in an office football pool and cheat on their income tax returns. Suppose a person is randomly selected. If the person is known to participate in an office football pool, what is the probability that the person cheats on his or her income tax return?

4.100 Marketing and Consumer Behavior According to the Energy Information Administration, 65.1% of all housing units have a ceiling fan, 55% of all households have ceiling fans and are owned, and 42.6% of all housing units have a ceiling fan and are fully air conditioned.[8] Suppose a household is randomly selected.

a. Given the household has a ceiling fan, what is the probability that it is owned?

b. If the household has a ceiling fan, what is the probability that it is fully air conditioned?

c. If the household has a ceiling fan, what is the probability that it is not fully air conditioned?

4.101 Travel and Transportation In a rural area in upstate New York, 80% of all people use chains on their car tires (for winter driving), 60% carry snow shovels in their cars and use chains, and 15% carry a shovel but do not use chains. Suppose a person from this area is selected at random.

a. If the person uses chains, what is the probability that he or she carries a shovel?

b. Given the person does not use chains, what is the probability that he or she carries a shovel?

4.102 Fuel Consumption and Cars As the price of gasoline continues to rise, many drivers are avoiding high-octane fuel and filling up with regular instead. The following table lists the probabilities of purchase corresponding to automobile engine type and octane rating.

		Octane rating		
		87	90	92
Engine type	four-cylinder	0.50	0.03	0.03
	six-cylinder	0.34	0.02	0.08

(*Source:* Energy Information Administration/Petroleum Marketing Monthly, August 2005.)

Suppose a consumer purchasing gasoline is selected at random.

a. Suppose the consumer buys gas with an octane rating of 90. What is the probability that the car is six-cylinder?

b. Suppose the consumer's car is four-cylinder. What is the probability that the gas purchased has an octane rating of 92?

c. Suppose the consumer buys gas with an octane rating of 90 or above. What is the probability that the car is not four-cylinder?

4.103 Sports and Leisure An assistant football coach at a Division II school helps his team prepare for the next opponent by charting plays. He looks at game films and records the down distance (first, second, or third down, and a categorical measure of the number of yards needed for a first down) and type of play (rush or pass), to look for tendencies. The two-way table below shows the number of plays that fall into each category. (Fourth-down plays are not charted since they usually involve a punt or field-goal attempt.)

		Down/Distance				
		1st	2nd short	2nd long	3rd short	3rd long
Play	Rush	126	35	46	65	12
	Pass	87	16	67	23	59

Suppose this table represents the true tendencies of the next opponent.

a. What is the probability that the opponent rushes the ball?

b. Suppose the opponent has a first down. What is the probability of a pass?

c. Suppose it is a first or second down. What is the probability that the opponent rushes the ball?

d. Suppose the opponent passes the ball. What is the probability that it is a third down?

4.104 Public Policy and Political Science Due to decreasing revenue and economic conditions, many states have tried to legalize gambling or, in states where gambling is already legal, expand casino operations. For example, New Hampshire legislators recently added games of chance and raised the limit on a single wager. They also passed legislation to more closely regulate gambling by charitable organizations.[9] To measure public opinion in Kansas, a random sample of residents was selected and each response was categorized according to revenue preference and age. The results are given in the following two-way table.

		Revenue source		
		Gambling	Liquor stores	Other
Age	18–21	33	68	12
	21–30	55	121	50
	30–45	117	109	132
	≥ 45	158	110	90

Assume this table is representative of the entire state's population and suppose a resident is randomly selected.

a. What is the probability that the resident is in favor of legalized gambling?

b. Suppose the person is in favor of state-owned liquor stores. What is the probability that the person is 30–45 years old?

c. Suppose the person selected is under 21. What is the probability that the person is in favor of some other option?

d. Suppose the resident is not in favor of legalized gambling. What is the probability that the respondent is 21–30 years old?

e. If the person selected is under 21 or at least 45, what is the probability that this resident is in favor of state-owned liquor stores?

4.105 Technology and Internet Internet service providers in Portugal offer four types of connections for customers: dial-up, ADSL, cable modem, and dedicated access. Assume a customer may only use one type of connection. During the first quarter of 2005, 81.3% used dial-up, 9.6% used ADSL, and 0.1% used dedicated access.[10] Suppose cable modem connections have three levels of service, determined by speed. Records indicate that the probability of a customer using a cable modem and silver service (slowest speed) is 0.045, using a cable modem and gold service (medium speed) is 0.036, and using a cable modem and premium service (fastest speed) is 0.009. Suppose a customer is randomly selected.

a. What is the probability that the customer uses a cable modem?

b. Suppose the customer uses a cable modem. What is the probability that he or she has the premium service?

c. Suppose the customer does not use a cable modem. What is the probability that he or she has dial-up service?

4.106 Marketing and Consumer Behavior The manager at Of The Land Gallery in Red Lake Falls, Minnesota, is looking at many ways to increase pottery sales for next year. The probability that she will advertise more is 0.65, and the probability of advertising more and increasing revenue is 0.35.

a. Suppose the store manager decides to advertise more. What is the probability that revenue will increase?

b. If the store manager does advertise more, what is the probability that revenue will not increase?

c. What is the probability of not advertising more and revenue increasing?

4.107 Medicine and Clinical Studies A random sample of seniors 65 and older was obtained, 1000 men and 900 women. A survey showed that 220 men and 252 women were regularly taking five or more drugs for chronic health problems.[11] Suppose a person included in this survey is randomly selected.

a. What is the probability that the person selected is a woman and takes five or more drugs per day?

b. Suppose the person selected is a man. What is the probability that he takes five or more drugs per day?

c. Suppose the person selected takes five or more drugs per day. What is the probability that the person is a woman?

Extended Applications

4.108 Demographics and Population Statistics According to the Census Bureau, 86.5% of the adults in the United States aged 65 or older are white and 8.6% are African American. In this age group, 1.33% are white and have no health insurance, and 0.25% are African American and have no health insurance. (*Source:* U.S. Census Bureau, Current Population Survey, 2007.) Suppose an adult 65 or older is selected at random.

a. Suppose the adult is white. What is the probability that the adult had no health insurance?

b. Suppose the adult is African American. What is the probability that the adult had no health insurance?

c. Suppose the adult is African American. What is the probability that the adult had health insurance?

4.109 Biology and Environmental Science Homeowners who cultivate small backyard gardens are often worried about pests (for example, rabbits and groundhogs) ruining plants. Some gardeners protect their gardens with a fence, others spread chemicals around the perimeter of the garden to keep animals away, and some do nothing. The joint probability table below shows the relationship between these garden protection methods and success.

		Garden defense		
		Fence	Chemicals	Nothing
Result	Pests	0.05	0.08	0.34
	No pests	0.30	0.20	0.03

Suppose a backyard gardener is selected at random.

a. Suppose the garden had pests. What is the probability that the gardener used nothing?

b. Suppose the gardener used chemicals. What is the probability that there were pests?

c. Given the garden had no pests, which method of defense did the gardener most likely use? Justify your answer.

4.110 Biology and Environmental Science The Environmental Protection Agency is concerned with issues involving human health and, of course, the environment. One specific issue is the biological quality of the wadeable streams in the United States. The results from a January 2008 report are summarized in the following joint probability table, per stream mile.

		Biological quality			
		Good	Fair	Poor	Not assessed
Region	West	0.102	0.059	0.062	0.004
	Plains	0.105	0.105	0.144	0.007
	East	0.075	0.084	0.213	0.040

(*Source:* Environmental Protection Agency, National Aquatic Resource Surveys: A Progress Report, January 2008.)

Suppose this table is representative of all wadeable streams in the United States and a wadeable stream mile is selected at random.

a. What is the probability that the stream mile is in the plains and has poor quality?

b. If the stream mile is in the west, what is the probability that it is good quality?

c. Suppose the stream mile was not assessed. What is the probability that it is in the East?

d. If the water quality is good or fair, which region is it most likely from? Justify your answer.

4.111 Psychology and Human Behavior Table 4.1, a partial two-way table, lists the number of adult criminal cases in Canada by case type and sentence.

Table 4.1

Criminal code	Type of sentence			
	Prison	Conditional sentence	Probation	
Crimes of violence	15302	2791		54559
Property crimes	24443		33193	61255
Administration of justice	21412	1026	13635	36073
Other criminal code offenses		671	9940	17482
Criminal code offenses (traffic)	7327	833	6659	14819
Other federal statute	7292	2214	8584	18090
	82647		108477	

(The left-side vertical label reads "Criminal code")

(*Source*: Statistics Canada, 2005–2006.)

a. Complete the table.
b. Suppose a criminal case is selected at random. What is the probability that it is a property crime and a conditional sentence?
c. Suppose the case results in probation. What is the probability that it was a conviction for a criminal code offense (traffic)?
d. Suppose the case was a crime of violence conviction. What is the probability that it resulted in a prison sentence?

4.112 Public Health and Nutrition The McPherson Middle School in Clyde, Ohio, is examining its school lunch program. A survey asked 2200 students about their lunch types and how they got to school in the morning. The following (partial) two-way table is assumed to represent the entire student body.

		Transportation		
		Bus	Car	Walk
Lunch	Carries		466	142
	Buys	345		500	967
		970		

a. Complete the two-way table.
b. Suppose a student at the school is randomly selected. What is the probability that the student carries a lunch and gets to school by car?
c. Suppose the student takes the bus to school. What is the probability that the student buys lunch?
d. Suppose the student does not walk to school. What is the probability that the student carries a lunch?
e. If the student buys lunch, how did he or she most likely get to school?

4.113 Medicine and Clinical Studies In the movie *A Christmas Story*, Ralph "Ralphie" Parker wanted an official Red Ryder carbine action 200-shot range model BB gun (with a compass in the stock). Everyone in the movie (including Santa) tells Ralphie he will shoot his eye out with this present. According to a survey of emergency room patients, approximately 6% of BB gun-related injuries are to the eye.[12] In a survey of BB- and pellet gun-related injuries treated at hospitals, each injury was classified by primary body part injured and victim-shooter relationship. The results are given in Table 4.2.

Table 4.2

Body part injured	Victim shooter relationship					
	Self	Friend/ acquaintance	Relative	Stranger	Other/ shooter not seen	Not stated
Extremity	200	128	89	17	25	189
Trunk	57	40	21	5	10	58
Face	51	35	20	4	14	49
Head/neck	40	34	19	6	12	44
Eye	23	20	9	3	7	26
Other	2	3	3	2	1	5

Suppose this table is representative of the entire population of BB- and pellet gun-related injuries, and a person in an emergency room suffering from this type of injury is selected at random.

a. What is the probability that the injury is to the eye and the shooter is a relative?
b. Suppose the injury was caused by a stranger. What is the probability that the body part injured is an extremity?
c. Suppose the injury is to the head/neck. What is the probability that the shooter is a friend/acquaintance?
d. Suppose the injury is not to the eye. What is the probability that the shooter is a relative?

Challenge

4.114 Public Policy and Political Science A survey asked whether voters in a certain district favored a return to stronger isolationism. The following *three-way* table classifies each response by gender, political party, and response.

	Male			Female		
	Dem	Rep	Ind	Dem	Rep	Ind
Yes	202	126	105	234	101	95
No	124	288	85	312	66	150

Suppose a random voter is selected from this district.
a. What is the probability that the voter is in favor of isolationism, a female, and a Republican?
b. What is the probability that the voter is not in favor of isolationism?
c. Suppose the voter is female. What is the probability that she is a Democrat?
d. Suppose the voter is not in favor of isolationism. What is the probability that the voter is a Republican and male?
e. Suppose the voter is not an Independent. What is the probability that he/she is in favor of isolationism?

4.5   INDEPENDENCE

In the last section we learned about conditional probability, that is, how knowing extra information *may* change a probability assignment. Often, however, additional information has no effect on the probability assignment. Consider the following examples.

Example 4.33 The Common Cold   Hundreds of viruses can cause the common cold. Many people are able to develop a resistance to some of these viruses, but may still contract a cold from a different virus. Catching a cold is *not* related to cold temperatures or bad weather, exercise, diet, enlarged tonsils, or adenoids.[13]

Let the event C = catching a cold. Suppose the (unconditional) probability that a certain person contracts a cold this winter is 0.45; $P(C) = 0.45$.

If this person decides to exercise more this winter, the *cold* facts above mean this extra exercise has no effect on contracting a cold. What is the probability that this person contracts a cold this winter given the event E = they exercise more?

$$P(C \mid E) = P(C) = 0.45.$$

Knowing extra information here does not change the conditional probability assignment. Intuitively, the events C (contracting a cold) and E (exercising more) are unrelated, or *independent*.   ●

Example 4.34 No Purchase Necessary   There are lots of sweepstakes in which a consumer is automatically entered by making a purchase. However, almost all sweepstakes entry rules explain that "no purchase is necessary to enter." A person may make a purchase to enter the sweepstakes or instead may enter by mailing a postcard or completing an online form. If this statement in the rules is true, then making a purchase *cannot* change the probability of winning the sweepstakes.

Suppose the event A = winning the sweepstakes, and the event B = making a purchase.

$$P(A \mid B) = P(A) \qquad \text{and} \qquad P(A \mid B') = P(A).$$

Whether or not you make a purchase has no effect on the chance of your entry being the winner. The events winning the sweepstakes and making a purchase are *independent*.   ●

In these two examples, the occurrence or nonoccurrence of one event has no effect on the occurrence of the other. In this case, the two events are **independent**.

DEFINITION

Two events A and B are **independent** if and only if

$$P(A \mid B) = P(A).$$

If A and B are *not* independent, they are said to be **dependent** events.

One way to verify independent events: is $P(A \mid B) = P(A)$? If so, then A and B are independent; if not, they are dependent.

ILLUMINATING THE CONCEPTS

1. If we know the events A and B are independent, then

$$P(A \mid B) = P(A) \quad and \quad P(B \mid A) = P(B).$$

Similarly, if either one of these equations is true, then the other is also true, and the events are independent.

2. If A and B are independent events, then so are all combinations of these two events and their complements.

Mathematical translation: If $P(A \mid B) = P(A)$ then $P(A \mid B') = P(A)$, $P(A' \mid B) = P(A')$, and $P(A' \mid B') = P(A')$.

3. Unfortunately, independent events *cannot* be shown on a Venn diagram. In problems that involve independent events, we'll have to translate the words into a probability question and then use an appropriate formula.

4. It is reasonable to think of independent events as unrelated. One might conclude that they are therefore disjoint. This is not true!

Suppose A and B are mutually exclusive and $P(A) \neq 0$ (there is some positive probability associated with the event A). Then

$$P(A \mid B) = 0 \neq P(A).$$

The probability of A given B has to be 0 since A and B are disjoint. Once B occurs, A cannot occur. Hence, disjoint events are **dependent**. ■

In Section 4.4, we learned the formula for finding conditional probability:

$$P(A \mid B) = \frac{P(A \cap B)}{P(B)} \quad or \quad P(B \mid A) = \frac{P(A \cap B)}{P(A)}.$$

We can solve both of these equations for $P(A \cap B)$ to obtain the following **Probability Multiplication Rule**

THE PROBABILITY MULTIPLICATION RULE

For any two events A and B:

$$\left. \begin{aligned} P(A \cap B) &= P(B) \cdot P(A \mid B) \\ &= P(A) \cdot P(B \mid A) \end{aligned} \right\} \text{Always true.}$$

$$ = P(A) \cdot P(B) \qquad \textbf{True only if } \textit{A} \textbf{ and } \textit{B} \textbf{ are independent.}$$

 ILLUMINATING THE CONCEPTS

1. The real skill in applying this rule is knowing which equality to use. The first two equalities are *always* true. Use one of these only if A and B are dependent and you need to find $P(A \cap B)$. Read the problem carefully to determine which conditional and unconditional probabilities are given.

If A and B are independent, use the third equality to compute the probability of the intersection. The word *independent* will not always appear in the problem. It may be implied or can be inferred from the type of experiment described.

2. If events are dependent, a *modified tree diagram* can be used to apply the Probability Multiplication Rule. In Figure 4.20 the probability of *traveling* along any branch is written along the appropriate leg. Second-generation branch probabilities are conditional. For example, P(C|A) (the probability of C given A) is the probability of taking path C, given path A.

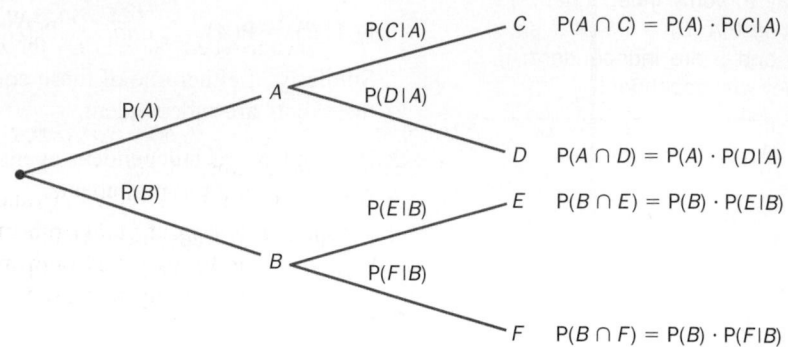

Figure 4.20 The Probability Multiplication Rule on a tree diagram.

On this road map, to determine a final probability, we *multiply* probabilities along the way.

All probabilities coming from a single node must sum to 1. To find the probability of traveling along a complete path from left to right (equivalent to the probability of an intersection), we *multiply* probabilities along the path.

A modified tree diagram is useful here also. Try drawing one to illustrate this extended rule.

3. The Probability Multiplication Rule can be extended. For any three events A, B, and C: $P(A \cap B \cap C) = P(A) \cdot P(B|A) \cdot P(C|A \cap B)$.

4. If the events $A_1, A_2, \ldots, A_k$ are mutually independent, then
$$P(A_1 \cap A_2 \cap \cdots \cap A_k) = P(A_1) \cdot P(A_2) \cdots P(A_k).$$

In words: if the events are mutually independent, the probability of an intersection is the product of the corresponding probabilities. ∎

There are lots of probability rules, formulas, and diagrams in this section. Here are some examples (along with Solution Trails to help you translate the words into mathematics) to illustrate these concepts.

Example 4.35 Emergency Road Service   The probability that a randomly selected family belongs to the AAA automobile club is 0.25 (*Source:* American Automobile Association). If a family belongs to AAA, the probability that they have more than one car is 0.45. Suppose a family is randomly selected. What is the probability that they have more than one car and belong to AAA?

SOLUTION

STEP 1   Define the following events:

A = belong to AAA;
M = have more than one car.

We are given that the probability the family belongs to AAA is 0.25, that is, $P(A) = 0.25$. In addition, we are told that the probability they own more than one car if they belong to AAA is 0.45. This is a conditional probability statement that can be written $P(M|A) = 0.45$.

STEP 2 $P(M \cap A) = P(A) \cdot P(M|A)$ Probability Multiplication Rule.
$= (0.25)(0.45) = 0.1125$ Use the given probabilities.

Note: Using the Probability Multiplication Rule, we can also write:

$P(M \cap A) = P(M) \cdot P(A|M)$.

Solution Trail

KEYWORDS

More than one car *and* belong to the AAA automobile club.

↓

TRANSLATION

What is the probability of the intersection of the event *have more than one car* and the event *belong to AAA*?

↓

CONCEPTS

Probability Multiplication Rule.

↓

VISION

Determine whether the events are independent, determine which probabilities are given, and use the appropriate form of the Probability Multiplication Rule.

This is a correct application of the rule, but it doesn't help in this problem, because the probabilities on the right-hand side are not given. An accurate but inappropriate use of the Probability Multiplication Rule is evident because the probabilities given in the problem and in the equality are mismatched. Simply try the other equality.   ●

Example 4.36 It's Made for Sleep   Better Bedding in East Hartford, Connecticut, claims that 99.4% of all mattress deliveries are on time. Suppose two mattress deliveries are selected at random.
a. What is the probability that both mattresses will be delivered on time?
b. What is the probability that both mattresses will be delivered late?
c. What is the probability that exactly one mattress will be delivered on time?

SOLUTION

a. Let M_i = mattress i is delivered on time; $P(M_i) = 0.994$ (given).

Both mattresses delivered on time means mattress 1 is on time *and* mattress 2 is on time.

$$
\begin{aligned}
P(M_1 \cap M_2) &= P(M_1) \cdot P(M_2) && \text{Both mattresses on time; independent events.}\\
&= (0.994)(0.994) && \text{Probability of each mattress delivered on time.}\\
&= 0.988036
\end{aligned}
$$

The probability that both mattresses will be delivered on time is 0.9880.

b. Both mattresses delivered late means mattress 1 delivered late *and* mattress 2 delivered late. Delivered late is the complement of delivered on time.

$$
\begin{aligned}
P(M_1' \cap M_2') &= P(M_1') \cdot P(M_2') && \text{Independent events.}\\
&= [1 - P(M_1)] \cdot [1 - P(M_2)] && \text{Complement Rule.}\\
&= (0.006)(0.006)\\
&= 0.000036
\end{aligned}
$$

The probability that both mattresses will be delivered late is 0.000036.

c. Exactly one mattress on time means
(i) Mattress 1 is on time *and* mattress 2 is late, *or*
(ii) Mattress 1 is late *and* mattress 2 is on time.

$P(\text{Exactly one mattress is on time}) = P[(M_1 \cap M_2') \cup (M_1' \cap M_2)]$

We don't usually see this step, but there really is a union of two events in the background. (Notice the *or* separating (i) and (ii) above.) These two events are disjoint, and the probability of the union of disjoint events is the sum of the corresponding probabilities.

$$
\begin{aligned}
&= P(M_1 \cap M_2') + P(M_1' \cap M_2)\\
&= P(M_1) \cdot P(M_2') + P(M_1') \cdot P(M_2) && \text{Independent events.}\\
&= (0.994)(0.006) + (0.006)(0.994) && \text{Use known probabilities.}\\
&= 0.011928
\end{aligned}
$$

The probability that exactly one mattress will be on time is 0.0119.

Note: There is another way to solve part (c). With two mattresses to deliver, one of three things must happen: 0 are on time, 1 is on time, or 2 are on time; and the probabilities of these three events must sum to 1. (Why?) From part (a), P(2 on time) = 0.9880; from part (b), P(0 on time) = 0.000036.

$$
\begin{aligned}
P(1 \text{ on time}) &= 1 - [P(0 \text{ on time}) + P(2 \text{ on time})] && \text{Complement Rule.}\\
&= 1 - (0.000036 + 0.988036) && \text{Use known probabilities.}\\
&= 1 - 0.988072\\
&= 0.011928
\end{aligned}
$$

   ●

Solution Trail

KEYWORDS

Both mattresses will be delivered on time.

↓

TRANSLATION

What is the probability of the intersection of the event *mattress 1 is delivered on time* and the event *mattress 2 is delivered on time?*

↓

CONCEPTS

Probability Multiplication Rule.

↓

VISION

Determine whether the events are independent, determine which probabilities are given, and use the appropriate form of the Probability Multiplication Rule.

In Chapter 5, we will convert all (symbolic) outcomes into real numbers and use the probabilities of experimental outcomes to find the probabilities associated with real numbers.

(Blend Image/Punchstock)

Example 4.37 Immunizations   Federal health officials recently reported that a record number of toddlers (19- to 35-month olds) have received a full series of inoculations against nine diseases, including diphtheria, tetanus, measles, and mumps.[14] The probability that a randomly selected toddler in Alabama has received a full set of inoculations is 0.823, for a toddler in Georgia, 0.847, and for a toddler in Utah, 0.713.[15] Suppose a toddler from each of these three states is randomly selected.
a. Find the probability that all three toddlers have received these inoculations.
b. Find the probability that none of the three has received these inoculations.
c. Find the probability that exactly one of the three has received these inoculations.

SOLUTION

a. Define the following three events:
   A = toddler A from Alabama has received these inoculations,
   G = toddler G from Georgia has received these inoculations, and
   U = toddler U from Utah has received these inoculations.

   Assume these three events are independent.

   $P(A \cap G \cap U)$ All three means intersection.
   $= P(A) \cdot P(G) \cdot P(U)$ Independent events.
   $= (0.823)(0.847)(0.713) = 0.4970$ Use given probabilities.

   The probability that all three toddlers have received these inoculations is 0.4970.

b. *None of the three has received inoculations* means toddler A has *not* received the inoculations *and* toddler G has *not* received the inoculations *and* toddler U has *not* received the inoculations. Translate this sentence into mathematics using intersection and complement.

   $P(A' \cap G' \cap U')$ Math translation: intersection.
   $= P(A') \cdot P(G') \cdot P(U')$ Independent events.
   $= [1 - P(A)] \cdot [1 - P(G)] \cdot [1 - P(U)]$ Complement Rule.
   $= (1 - 0.823)(1 - 0.847)(1 - 0.713)$ Use given probabilities.
   $= (0.177)(0.153)(0.287)$ Simplify.
   $= 0.0078$

   The probability that none of the three has received the inoculations is 0.0078.

c. To write a probability statement for *exactly one has received the inoculations* ask "How can that happen?" Toddler A has received the inoculations and toddlers G and U have not, or toddler G has received the inoculations and toddlers A and U have not, or toddler U has received the inoculations and toddlers A and G have not. Translate this sentence into probability using intersection and complement.

   $P(A \cap G' \cap U') + P(A' \cap G \cap U') + P(A' \cap G' \cap U)$
   Three ways exactly one toddler has received the inoculations.
   $= P(A) \cdot P(G') \cdot P(U') + P(A') \cdot P(G) \cdot P(U') + P(A') \cdot P(G') \cdot P(U)$
   Independent events.
   $= (0.823)(0.153)(0.287) + (0.177)(0.847)(0.287) + (0.177)(0.153)(0.713)$
   Known probabilities; Complement Rule.
   $= 0.0361 + 0.0430 + 0.0193$
   $= 0.0984$

   The probability that exactly one toddler of the three has received the inoculations is 0.984.

Solution Trail

KEYWORDS

At least one herbicide works.

↓

TRANSLATION

The words *at least one* mean the event *one, two, or three herbicides work.* This is the same as the complement of the event *none of the herbicides work.*

↓

CONCEPTS

Complement Rule.

↓

VISION

Define the event *none of the herbicides work* and use the Complement Rule to find the probability that at least one works.

Example 4.38 Grapes versus Weeds   Horseweed has become resistant to many herbicides and has become a severe problem on grape farms. The weed spreads quickly, grows taller than grape vines, and can clog irrigation ditches.[16] New herbicides are being tested on farms in various grape farming districts in California. Suppose the probability that the herbicide works on a farm in Mendocino County is 0.75 (event *M*), on a farm in Napa County 0.55 (event *N*), and on a farm in Yolo County 0.45 (event *Y*). Find the probability that at least one herbicide works.

SOLUTION

P(at least one herbicide works)

$= 1 - $ P(no herbicide works)	Complement Rule.
$= 1 - $ P$(M' \cap N' \cap Y')$	All three do *not* work.
$= 1 - $ P$(M') \cdot $ P$(N') \cdot $ P(Y')	Independent events.
$= 1 - [1 - $ P$(M)] \cdot [1 - $ P$(N)] \cdot [1 - $ P$(Y)]$	Complement Rule.
$= 1 - (1 - 0.75)(1 - 0.55)(1 - 0.45)$	Use given probabilities.
$= 1 - (0.25)(0.45)(0.55)$	
$= 1 - 0.0619 = 0.9381$	

The probability that at least one herbicide works is 0.9381.

Challenge: Find this probability using a direct approach, without using the Complement Rule.   ●

Example 4.39 A Traveling Salesperson   During frequent trips to a certain city a traveling salesperson stays at hotel A 50% of the time, at hotel B 30% of the time, and at hotel C 20% of the time. When the salesperson checks in, there is some problem with the reservation 3% of the time at hotel A, 6% of the time at hotel B, and 10% of the time at hotel C. Suppose the salesperson travels to this city.

a. Find the probability that the salesperson stays at hotel A and has a problem with the reservation.

b. Find the probability that the salesperson has a problem with the reservation.

c. Suppose the salesperson has a problem with the reservation. What is the probability that the salesperson is staying at hotel *A*?

SOLUTION

Define the following events:

A = stays at hotel A; B = stays at hotel B;
C = stays at hotel C; and R = has a problem with the reservation.

Convert all the given percentages into probabilities.
The phrase *of the time* indicates conditional probability.

P$(A) = 0.50$	P$(B) = 0.30$	P$(C) = 0.20$
P$(R \vert A) = 0.03$	P$(R \vert B) = 0.06$	P$(R \vert C) = 0.10$

This experiment can be represented with a modified tree diagram (Figure 4.21). Remember, the probabilities along all paths coming from a node must sum to 1, and second-generation branch probabilities are conditional.

a. The events A and R are dependent. The likelihood of having a problem with a reservation depends on the hotel.

P$(A \cap R) = $ P$(A) \cdot $ P$(R \vert A)$	Probability Multiplication Rule.
$\qquad = (0.50)(0.03)$	Use known probabilities.
$\qquad = 0.0150$	

The probability of staying at hotel A and having a problem with the reservation is 0.0150.

a. Solution Trail

KEYWORDS

Staying at hotel A, problem with the reservation.

↓

TRANSLATION

What is the probability of the intersection of the event A and the event R?

↓

CONCEPTS

Probability Multiplication Rule.

↓

VISION

Determine whether the events are independent, determine which probabilities are given, and use the appropriate form of the Probability Multiplication Rule.

To find P(R'|A), apply the Complement Rule to a conditional probability statement: P(R'|A) = 1 − P(R|A).

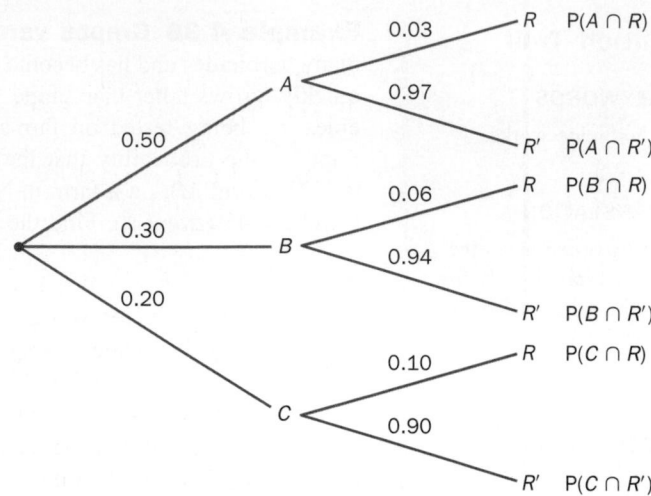

Figure 4.21 Tree diagram.

b. Solution Trail

KEYWORDS

Problem with the reservation.

↓

TRANSLATION

The event R.

↓

CONCEPTS

Unconditional probability.

↓

VISION

To find P(R) ask, "How can that happen?" Which paths from left to right involve the event R? The tree diagram suggests there are three compound events (paths) that involve a problem with the reservation. Figure 4.22 shows this decomposition of R using a Venn diagram.

b. $P(R) = P(A \cap R) + P(B \cap R) + P(C \cap R)$ Decomposition of R.

$= P(A) \cdot P(R|A) + P(B) \cdot P(R|B) + P(C) \cdot P(R|C)$ Probability Multiplication Rule.

$= (0.50)(0.03) + (0.30)(0.06) + (0.20)(0.10)$ Use known probabilities.

$= 0.0150 + 0.0180 + 0.0200 = 0.0530$

The probability of having a problem with the reservation (regardless of the hotel) is 0.0530.

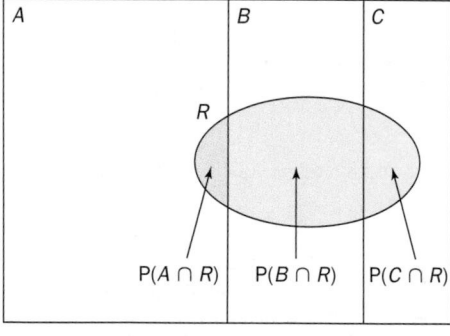

Figure 4.22 Venn diagram showing decomposition of R.

c. Solution Trail

KEYWORDS

Problem with the reservation, staying at hotel A.

↓

TRANSLATION

Given the event R, find the probability of the event A.

↓

CONCEPTS

Conditional probability with the event R given.

↓

VISION

Find P(A|R).

c. $P(A|R) = \dfrac{P(A \cap R)}{P(R)}$ Formula for conditional probability.

$= \dfrac{0.0150}{0.0530}$ Use answers to (a) and (b).

$= 0.2830$

The probability that the salesperson stayed at hotel A, given there was a problem with the reservation, is 0.2830.

ILLUMINATING THE CONCEPTS

1. Part (c) of the hotel example illustrates **Bayes' Rule**. This theorem loosely states that given certain conditional probabilities (and other unconditional probabilities), we are able to solve for a new conditional probability where the events are inverted, or swapped.

   In the hotel example, we were given the conditional probabilities $P(A|R)$, $P(R|B)$, and $P(R|C)$. Using these probabilities and the unconditional probabilities $P(A)$, $P(B)$, and $P(C)$, we were able to find $P(A|R)$, a conditional probability with the events A and R switched.

2. Suppose $P(A)$, $P(B)$, and $P(A \cap B)$ are known. To decide whether A and B are independent, check the equation $P(A \cap B) \stackrel{?}{=} P(A) \cdot P(B)$. If the probability of the intersection is equal to the product of the probabilities, then the events are independent. If not, they are dependent.

3. There are many applications in probability and statistics that involve repeated sampling from a population *with replacement*. In this case each draw is independent of any other draw. ∎

Other applications involve sampling *without replacement*, for example exit polls and telephone surveys. Consider each individual response as an event. These events are definitely dependent. However, if the population is *large enough* and the sample is small relative to the size of the population, then the events are *almost independent*. Calculating probabilities assuming independence results in little loss of accuracy. Exercise 4.134 on page 175 illustrates this idea.

SECTION 4.5 EXERCISES

Practice

4.115 Decide whether each pair of events is independent or dependent.
a. A = make an error on your income tax return, and B = file Form 1040.
b. C = correctly put together a swing set, and D = read the directions.
c. E = run out of milk, and F = the refrigerator breaks.
d. G = break your pencil lead while writing, and H = feel overly stressed.

4.116 Decide whether each pair of events is independent or dependent.
a. A = a randomly selected CD has a scratch, and B = a random email message is spam.
b. C = one paper towel is enough to completely clean a spill, and D = you use a generic paper towel.
c. E = no accidents are reported in 24 hours in a county, and F = there are no storms in the area.
d. G = your automobile insurance bill increases, and H = you had one speeding ticket within the last year.

4.117 Suppose the following probabilities are known: $P(A)$ = 0.25, $P(B|A)$ = 0.34, and $P(C|A \cap B)$ = 0.62.
a. Find $P(A \cap B)$, $P(B'|A)$, and $P(A \cap B')$.
b. Find $P(A \cap B \cap C)$, $P(C'|A \cap B)$, and $P(A \cap B \cap C')$.
c. Are the events A and B independent? Justify your answer.

4.118 Suppose the events A, B, and C are independent and $P(A)$ = 0.55, $P(B)$ = 0.45, and $P(C)$ = 0.35. Find the following probabilities.
a. $P(A \cap B)$, $P(A \cap C)$, and $P(B \cap C)$.
b. $P(A \cap B \cap C)$, and $P(A' \cap B' \cap C')$.
c. $P(A \cap B' \cap C')$ and $P(A' \cap B \cap C)$.

4.119 Suppose the following probabilities are known: $P(A)$ = 0.40, $P(B|A)$ = 0.25, $P(C|A)$ = 0.45, and $P(D|A)$ = 0.30.
a. Find $P(A \cap B)$, $P(A \cap C)$, and $P(A \cap D)$.
b. Are the events A and B independent? Justify your answer.
c. Find $P(B'|A)$. If the event A occurs, are there any other events in addition to B, C, and D that can occur? Justify your answer.

4.120 Suppose the probability that an individual has blue eyes is 0.41. Four people are randomly selected.
a. Find the probability that all four have blue eyes.
b. Find the probability that none has blue eyes.
c. Find the probability that exactly two have blue eyes.

4.121 Consider the modified tree diagram below.

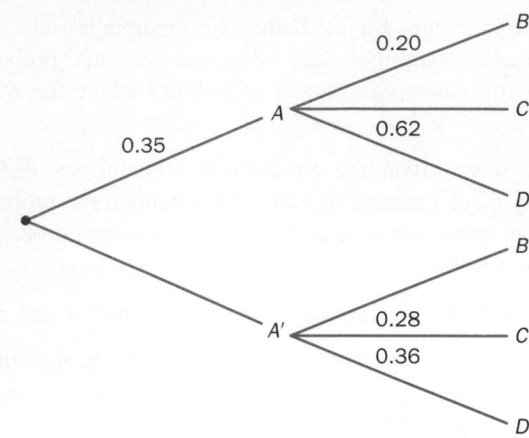

a. Identify and determine each missing path probability.
b. Find P($A \cap C$) and P($A' \cap B$).
c. Find P(D).

4.122 Consider Figure 4.23.

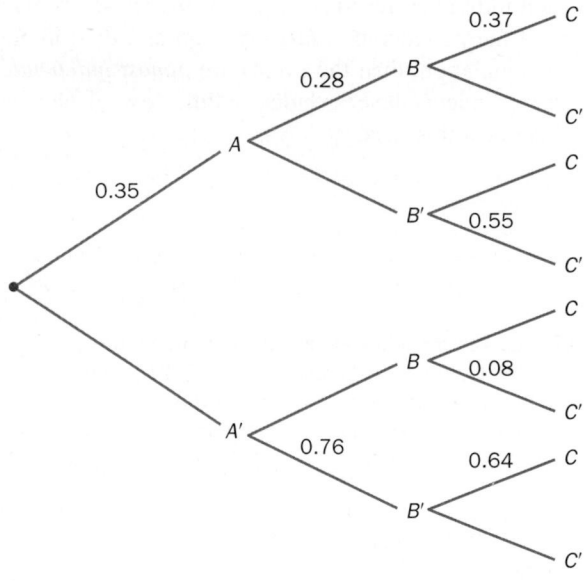

Figure 4.23 Tree diagram.

a. Identify and determine each missing path probability.
b. Find P($A \cap B \cap C$) and P($A' \cap B \cap C'$).
c. Find P(C). Are the events B and C independent? Justify your answer.

Applications

4.123 Demographics and Population Statistics Since 1970, the production by coal miners has nearly tripled while related injuries and fatalities have declined dramatically. According to the Bureau of Labor Statistics, the 2007 fatality rate for coal miners was 0.0248%. Suppose two coal miners are selected at random.

a. What is the probability that both miners will be fatally injured during a given year?
b. What is the probability that neither will be fatally injured during a given year?
c. What is the probability that exactly one will be fatally injured during a given year?

4.124 Public Policy and Political Science The port of South Louisiana is the largest tonnage port in the United States. Inspectors randomly select ships at one of the facilities and check for safety violations. Past records indicate that 90% of all ships inspected have no safety violations. Suppose two ships are selected at random.

a. What is the probability that both ships have safety violations?
b. What is the probability that neither ship has a safety violation?
c. What is the probability that exactly one ship has a safety violation?

4.125 Biology and Environmental Science A krait is a venomous snake found in Southeast Asia. Its bite causes rapid sleepiness and is fatal 50% of the time even if an antivenom is used.[17] Suppose two people bitten by a krait are selected at random.

a. What is the probability that both people will die?
b. What is the probability that exactly one person will die?
c. What is the probability that at most one person will die?

4.126 Demographics and Population Statistics The infant mortality rate is the number of deaths to individuals under the age of 1 year per 1000 births. In 2007, the infant mortality rate in Canada was 4.6 and in the United States was 6.4.[18] Suppose three Canadian children are selected at random.

a. What is the probability that all three will live past the age of 1 year?
b. What is the probability that at least one will die within 1 year?
c. Answer parts (a) and (b) for three randomly selected American children.

4.127 Biology and Environmental Science The International Volcano Research Centre reported the following 2008 eruption probabilities for certain volcanoes.

| Sakura-Jima | 0.9063 | Dukono | 0.7216 |
| Fuego | 0.7418 | Karymsky | 0.5254 |

(*Source:* Southwest Volcano Research Centre.)

Suppose the volcanoes erupt independently of one another.

a. What was the probability that all four would erupt before the end of the year?
b. What was the probability that none of the four would erupt before the end of the year?
c. What was the probability that at least one would erupt before the end of the year?
d. Find out which volcanoes actually erupted before the end of the year. Using the probabilities given above, find the probability of that outcome.

4.128 Economics and Finance Research indicates that 45% of all customers are delinquent on payments (and therefore have a *bad account*) after three months, and 15% are still delinquent after six months. According to the Commercial Collection Agency, after three months the chance of recovering a bad account is 73% and after six months the chance is 50%.[19]

a. What is the probability that a randomly selected account is bad after three months and will be recovered?

b. What is the probability that a randomly selected account is bad after six months and will be recovered?

4.129 Biology and Environmental Science Only about 20% of exploratory wells drilled in geologically favorable locations yield commercially viable quantities of oil.[20] Suppose four wells are independently drilled in favorable locations.

a. What is the probability that all four wells will yield oil?

b. What is the probability that exactly one of the four wells will yield oil?

c. What is the probability that at least two wells will yield oil?

4.130 Psychology and Human Behavior According to a survey by Frank N. Magid Associates,[21] 42% of Americans never watch programs recorded from TV. Suppose four Americans are selected at random.

a. What is the probability that all four never watch programs recorded from TV?

b. What is the probability that all four do watch programs recorded from TV?

c. What is the probability that exactly two of the four watch programs recorded from TV?

4.131 Economics and Finance Detailed analysis of two technology stocks indicates that over the next 6 months the probability that the price of stock 1 will rise is 0.42 and for stock 2 the probability is 0.63. Suppose the stock prices react independently.

a. What is the probability that both stock prices will rise over the next 6 months?

b. What is the probability that stock 1 will rise and stock 2 will sink?

c. Suppose both stocks are in the technology sector, and stock 2 tends to follow stock 1. If stock 1 rises over the next 6 months, the chance of stock 2 rising is 81%. Now what is the probability of both stock prices rising over the next 6 months?

4.132 Sports and Leisure The PGA Tour maintains statistical reports on variables such as money leaders, driving distance, and driving accuracy. The table below lists the probability that selected players were able to hit the green in regulation (GIR). According to the PGA Tour, a green is considered hit in regulation if any portion of the ball is touching the putting surface after the GIR stroke has been taken.

Golfer	Probability
Vijay Singh	0.6845
Bubba Watson	0.6760
Bob Tway	0.6736

(*Source:* PGA Tour, as of 11/11/2008.)

Suppose these three golfers are playing a round at Sawgrass and they tee up on number 11, one of the most difficult holes to play on the professional tour.

a. What is the probability that all three players will hit the green in regulation?

b. What is the probability that none of the three players will hit the green in regulation?

c. What is the probability that exactly one of the players will hit the green in regulation?

d. What is the probability that all three players will hit the green in regulation on all four rounds?

4.133 Sports and Leisure As of November 2008, Larry Bird had the second-highest career free-throw percentage in Celtics history, 88.6%. Ray Allen was number one.[22] Suppose Larry steps up to the free-throw line for two shots. It is unlikely the two shots are independent. If he misses the first shot, the probability that he makes the second is 0.95, and if he makes the first shot, the probability that he makes the second is 0.85.

a. What is the probability that he makes both shots?

b. What is the probability that he misses both shots?

c. What is the probability that he makes only one shot?

4.134 Economics and Finance As Baby Boomers reach retirement age, many are beginning to carefully examine their savings plans and government benefits. This generation is worried about remaining financially independent as many companies cut or even eliminate pension plans. The Coughlin Courier reported (April 2008) 64% of Canadian baby boomers expect to maintain their current lifestyles after retirement. Suppose three Canadian baby boomers are selected at random.

a. What is the probability that exactly two of the three expect to maintain their current lifestyles after retirement?

b. What is the probability that all three expect to maintain their current lifestyles after retirement?

c. Suppose another random sample of five Canadian baby boomers is obtained. What is the probability that exactly one adult from each sample expects to maintain his or her current lifestyle?

4.135 Psychology and Human Behavior In September 2008, the United States Congress passed a financial bailout bill to help the economy. A Pew Research Center survey conducted September 19–22, 2008 asked Americans how this bailout bill made them feel. The results indicated that 56% of Republicans, 61% of Democrats, and 66% of Independents were angry. Suppose one adult is selected at random from each of these three political parties.

a. What is the probability that all three are angry?

b. What is the probability that all three are not angry?

c. What is the probability that exactly one of the three is angry?

Extended Applications

4.136 Medicine and Clinical Studies When a person has a certain type of leukemia, a physician may perform a bone marrow transplant in order to restore a healthy blood supply. Among the general population, the chances of an acceptable bone marrow match are 1 in 20,000.[23] Suppose a person needs a bone marrow transplant and four people from the general population are selected at random.

a. What is the probability that none of the four will match?

b. What is the probability that at least one will match?

c. How many people would have to be tested in order for the probability of at least one match to be 0.50?

4.137 Travel and Transportation A family trying to arrange a vacation is using the Internet to name their own price for a rental car. The software reports that 50% of all people name a price of $30 per day, 40% bid $25 per day, and 10% bid $20 per day. The Internet company also reports that 90% of all $30 bids, 60% of all $25 bids, and 5% of all $20 bids are accepted.

a. What is the probability that the family will submit a bid of $25 and have it accepted?

b. What is the probability that their bid will be accepted?

c. Suppose their bid is accepted. What is the probability that it is for $20?

4.138 Biology and Environmental Science Opponents of the U.S. Navy SURTASS LFA Sonar System argue that it constitutes a substantial risk to marine life, causing extraordinary numbers of stranded, or beached, whales. Consider the following statements concerning the use of this system near a remote island in the South Pacific.

- On any given day, the probability of a mass stranding of whales in this area is 0.01.
- The probability of a military exercise on any given day is 0.001.
- If there is a military exercise, the probability of a mass stranding is 0.17.

a. Define events and write a probability statement for each fact above.

b. On a randomly selected day, what is the probability of a mass stranding of whales and a military exercise?

c. Are the events mass stranding and military exercise independent? Justify your answer.

4.139 Medicine and Clinical Studies A tine test is a common method used to determine whether a person has been exposed to tuberculosis. Approximately 5% of people in the United States have been exposed to tuberculosis.[24] Using the tine test, 95% of all people who have been exposed test positive, and 98% of those not exposed test negative. Suppose a person is randomly selected and given the tine test.

a. What is the probability that the person tests positive and has been exposed to tuberculosis?

b. What is the probability that the person tests positive?

c. Suppose the test is positive. What is the probability that the person actually has been exposed?

4.140 Travel and Transportation There are three major air carriers with flights from Boston to Los Angeles: 55% of all passengers take American Airlines, 16% take American West, and 29% take United Airlines. Data from 2005 indicate 23% of all American Airlines flights from Boston to Los Angeles are late, 27% of America West flights from Boston to Los Angeles are late, and 24% of United Airlines flights from Boston to Los Angeles are late.[25] Suppose a passenger taking a flight from Boston to Los Angeles is randomly selected.

a. What is the probability that the passenger takes American Airlines and is late?

b. What is the probability that the passenger is late? On time?

c. Suppose the passenger arrives late. Which airline did the passenger most likely fly?

4.141 Manufacturing and Product Development The Italian Aspide missile, a licensed version of the American Sparrow, has a sophisticated homing guidance system and single-shot hit probability of 0.80.[26] Suppose an enemy plane is within range of three missile firing stations, all three fire an Aspide surface-to-air missle, and the missiles operate independently.

a. What is the probability that the plane is hit?

b. What is the probability that all three missiles miss?

c. How many missiles would have to be fired at the plane in order to be 99.99% sure it would be hit?

4.142 Medicine and Clinical Studies A new vaccine for children combines diphtheria, tetanus, and acellular pertussis into one injection. The NPI Reference Guide on Vaccines and Vaccine Safety indicates approximately 5% of all children who receive this vaccine will experience tenderness in the area of the injection.[27] Suppose five children who received the vaccine are selected at random.

a. What is the probability that exactly one of the five experiences tenderness after the shot?

b. What is the probability that only the first child and the fifth child experience tenderness after the shot?

c. Suppose all five children experience tenderness. Do you believe the Reference Guide's claim concerning this adverse reaction? Justify your answer.

4.143 Fuel Consumption and Cars After a small collision, a driver must take his car to one of two body shops in the area. Consider the following events.

D = driver takes his car to shop D.

L = driver takes his car to shop L.

T = the work is completed on time.

B = the cost is less than or equal to the estimate (under budget).

Figure 4.24 describes the relationships among these events.

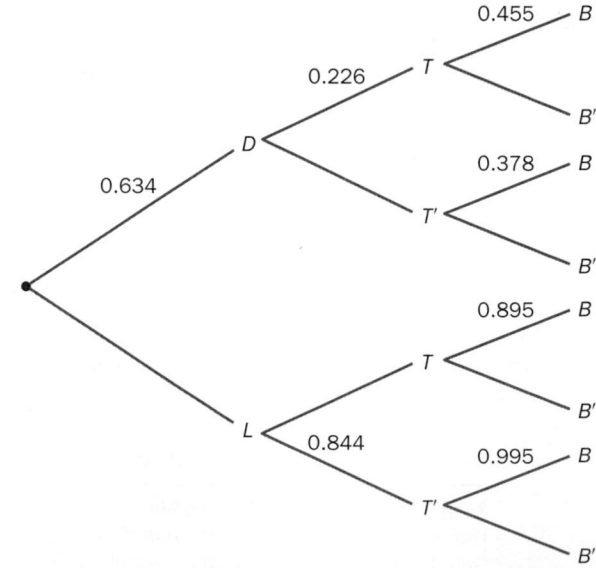

Figure 4.24 Tree diagram.

a. Complete the tree diagram by filling in the missing path probabilities.

b. What is the probability that the car is repaired under budget, on time, and with company D?

c. What is the probability that the cost of the repair is over the estimate?

d. What is the probability that the car is repaired under budget given it is ready on time?

4.144 Sports and Leisure Suppose 2 cards are drawn without replacement from a regular deck of 52 playing cards. Consider the events

A_1 = an ace is selected on the first draw;

A_2 = an ace is selected on the second draw.

a. Find $P(A_2|A_1)$ and $P(A_2)$. Are the events A_1 and A_2 independent? Justify your answer.

b. Suppose the 2 cards are drawn without replacement from *six* regular 52-card decks shuffled together. Find $P(A_2|A_1)$ and $P(A_2)$ for this experiment. Are the events A_1 and A_2 independent? Justify your answer.

c. In part (b), the events are *almost independent*. For six decks, find $P(A_1 \cap A_2)$ exactly, and then find the same probability assuming the two events are independent (with the probability of an ace on any draw being 24/312).

Challenge

4.145 The Traveling Salesperson Reconsider Example 4.39 (page 171). Suppose the salesperson has a problem with the reservation. In which hotel did the salesperson most likely stay?

(View Stock/AgeFotostock)

Chapter 4 Challenge Wrap-Up

Suppose 3% of all athletes are actually using the endurance-boosting hormone EPO. The test for the presence of this substance is very reliable; the probability of a positive test result given the presence of EPO is 0.99, and the probability of a negative test result when EPO is not present is 0.90.

Using the events

E = presence of EPO in the blood,
Pos = positive test result, and Neg = negative test result, here is the modified tree diagram.

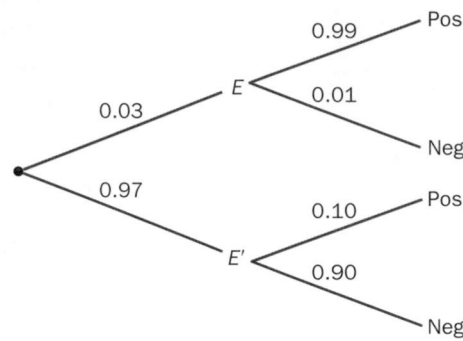

The probability that a randomly selected athlete is using EPO and tests positive is given by

$$P(E \cap \text{Pos}) = P(E) \cdot P(\text{Pos}|E)$$

Probability Multiplication Rule.

$$= (0.03)(0.99)$$

Use known probabilities.

$$= 0.0297$$

The probability that a randomly selected athlete tests positive for the presence of EPO is

$$P(\text{Pos}) = P(E \cap \text{Pos}) + P(E' \cap \text{Pos})$$

Decomposition of Pos.

$$= P(E) \cdot P(\text{Pos}|E) + P(E') \cdot P(\text{Pos}|E')$$

Probability Multiplication Rule.

$$= (0.03)(0.99) + (0.97)(0.10)$$

Use known probabilities.

$$= 0.1267$$

Suppose an athlete tests positive for the presence of EPO. The probability of the athlete actually using the banned substance is computed using the formula for conditional probability.

$$P(E \mid \text{Pos}) = \frac{P(E \cap \text{Pos})}{P(\text{Pos})}$$

Definition of conditional probability.

$$= \frac{0.0297}{0.1267} = 0.2344$$

Use probabilities from above.

This analysis suggests the majority of positive test results are *false positives*. Only 23% of athletes who test positive for the presence of EPO are actually using the banned substance! Since no test is 100% reliable, it might be more reasonable (but too expensive and time consuming) to test for the presence of banned substances using several techniques simultaneously.

CHAPTER 4 SUMMARY

Concept	Page	Notation/Formula/Description
Experiment	120	An activity in which there are at least two possible outcomes and the result cannot be predicted with certainty.
Tree diagram	121	A visual road map of possible outcomes in an experiment.
Sample space	123	S: a listing of all possible outcomes, using set notation.
Event	124	Any collection of outcomes from an experiment.
Simple event	125	An event consisting of exactly one outcome.
Complement	125	A': all outcomes in the sample space S not in A.
Union	125	$A \cup B$: all outcomes in A or B or both.
Intersection	125	$A \cap B$: all outcomes in both A and B.
Disjoint events	125	Two events are disjoint if their intersection is empty: $A \cap B = \{ \ \}$.
Venn diagram	127	Geometric representation of a sample space and events.
Relative frequency of occurrence of an event	132	The number of times the event occures divided by the total number of times the experiment is conducted.
Probability of an event	133	The limiting relative frequency of occurrence of the event.
Equally likely outcome experiment	136	All outcomes in the experiment have the same chance of occurring.
Probabilities in an equally likely outcome experiment	136	$P(A) = N(A)/N(S)$.
Complement Rule	138	$P(A) = 1 - P(A')$.
Addition Rule for two events	139	$P(A \cup B) = P(A) + P(B) - P(A \cap B)$ $P(A \cup B) = P(A) + P(B)$ if A and B are disjoint.
Multiplication Rule	145	$N(S) = n_1 \cdot n_2 \cdot n_3 \cdots n_k$.
n factorial	148	$n! = n(n-1)(n-2) \cdots (3)(2)(1). \qquad 0! = 1$
Permutation	148	An ordered arrangement. $\ _nP_r = n(n-1)(n-2) \cdots [n-(r-1)] = \dfrac{n!}{(n-r)!}$
Combination	150	An unordered arrangement. $\ _nC_r = \dbinom{n}{r} = \dfrac{n!}{r!(n-r)!}$
Conditional probability	156	The conditional probability of the event A given the event B has occurred is $P(A \mid B) = P(A \cap B)/P(B)$.
Two-way (contingency) table	158	A two-way table with observed frequencies corresponding to classifications of two variables.
Joint probability table	160	A two-way table with probabilities corresponding to the intersection of two events.
Independent events	166	Two events A and B are independent if $P(A \mid B) = P(A)$; if the occurrence of the event B does not affect the occurrence or nonoccurrence of the event A.
Dependent events	166	If two events A and B are not independent, then they are dependent.
Probability Multiplication Rule	167	$P(A \cap B) = P(B) \cdot P(A \mid B) = P(A) \cdot P(B \mid A).$ $P(A \cap B) = P(A) \cdot P(B)$ if A and B are independent.

CHAPTER 4 EXERCISES

Applications

4.146 Marketing and Consumer Behavior A home decorating store received a shipment of 20 different Tiffany-style lamps and the store manager selects three lamps at random for display.

a. How many different displays are possible?

b. Suppose 3 of the lamps were damaged during shipping. What is the probability that 2 of the lamps selected will be broken?

c. What is the probability that at least 1 of the lamps selected will be broken?

4.147 Physical Sciences At a state middle-school science fair, students launch bottle rockets designed and built from plastic 2-liter beverage containers. An experiment consists of recording the general appearance of a rocket [bad (B), good (G), or excellent (E)] and the maximum altitude [low (L), medium (M), or high (H)]. Consider the following events.

A = The rocket is rated as excellent.

B = The rocket flies to a high altitude.

C = The rocket is rated as bad or flies low.

D = The rocket is good and flies to a medium altitude.

a. Find the sample space S for this experiment.

b. List the outcomes in each of the events A, B, C, and D.

c. List the outcomes in $A \cup B$, $B \cup C$, and D'.

d. List the outcomes in $A \cap B$, $C \cap D$, and $(B \cup D)'$.

4.148 Sports and Leisure The Boston Bruins play in the Northeast Division of the Eastern Conference in the National Hockey League. For each game played against another team in the Eastern Conference, the division [Atlantic (A), Northeast (N), or Southeast (S)] and the outcome [win (W), loss (L), tie (T), or overtime loss (O)] are recorded. Consider the following events.

E = The opponent is in the Southeast Division.

F = The Bruins win the game.

G = The opponent is in the Northeast or the game is an overtime loss.

H = The Bruins lose and the opponent is from the Atlantic Division.

a. Carefully sketch a tree diagram to illustrate the possible outcomes for this experiment.

b. Find the sample space for this experiment.

c. Find the outcomes in each of the events E, F, G, and H.

d. List the outcomes in $E \cup F$, $F \cup G$, and H'.

e. List the outcomes in $E \cup H'$, $E \cup F \cup G'$, and $F \cup G'$.

4.149 Demographics and Population Statistics In a recent population survey, the U.S. Census Bureau reported the following classifications and corresponding probabilities.

Educational attainment	Probability
No degree	0.193
High school graduate	0.289
Some college, no degree	0.191
Associate degree	0.080
Bachelor's degree	0.167
Master's degree	0.057
Professional degree	0.013
Doctorate degree	0.010

(*Source:* Educational Attainment in the United States, U.S. Census Bureau.)

Consider the events:

A = Has a bachelor's, master's or doctorate degree

B = Does not have an associate degree

C = Does not have a degree

Find the following probabilities.

a. $P(A)$, $P(B)$, and $P(C)$.

b. $P(A \cap B)$, $P(A \cup C)$, and $P(B \cup C)$.

c. $P(C')$, $P(A' \cup B)$, and $P(B' \cap C')$.

4.150 Biology and Environmental Science The germination rate for pumpkin seeds is directly related to the prevailing weather conditions. The Autumn Gold Pumpkin is medium-sized and ripens to a deep orange. If conditions are seasonable, the probability of germination is 0.85.[28] If it is dry, suppose the probability that a random seed will germinate is 0.75. Recent weather history suggests there is a 40% chance of a dry start to the growing season. Suppose an Autumn Gold pumpkin seed is randomly selected.

a. What is the probability that the growing season will be dry and the seed will germinate?

b. What is the probability that the seed will germinate?

c. Suppose the seed did not germinate. What is the probability that the growing season had a dry start?

4.151 Economics and Finance Many Americans use savings bonds to supplement retirement funds or to pay for qualified higher education expenses. The U.S. Treasury even sells savings bonds online. Approximately one in every five Americans owns savings bonds.[29] Suppose four Americans are randomly selected.

a. What is the probability that all own bonds?

b. What is the probability that none of the four owns bonds?

c. What is the probability that exactly two of the four own bonds?

4.152 Marketing and Consumer Behavior At Elmo's, an old-fashioned barber shop in Melbourne, Florida, 70% of all customers get a haircut, 40% get a shave, and 15% get both.

a. What is the probability that a randomly selected customer gets a shave or a haircut?

b. What is the probability that a randomly selected customer gets neither?

c. What is the probability that a randomly selected customer gets only a shave?

d. What is the probability a randomly selected customer gets a shave, given they get a haircut?

e. Suppose two customers are selected at random. What is the probability that both get only a haircut?

4.153 Medicine and Clinical Studies More and more people are trying herbal remedies, including gooseberry juice, eucalyptus oil, and crushed ajwain, for relief from the common cold. The following joint probability table shows the relationship between having tried an herbal remedy and highest educational degree earned.

	Highest degree earned			
	Vocational	High school	College degree	Graduate degree
Tried	0.23	0.17	0.06	0.05
Not tried	0.04	0.12	0.15	0.18

Suppose one person is randomly selected.

a. What is the probability the person has tried an herbal remedy, given the highest degree earned is from college?

b. If the person has not tried an herbal remedy, what is the probability that the highest degree earned is from high school?

c. Suppose the person has not earned a graduate degree. What is the probability that the person has tried an herbal remedy?

d. Suppose two people are selected at random. What is the probability that exactly one has tried an herbal remedy?

4.154 Psychology and Human Behavior Do you read your horoscope every day? According to a Harris Poll,[30] 29% of all Americans believe in astrology. Suppose that of those Americans who believe in astrology, 10% have called a psychic hot line. Suppose an American is randomly selected.

a. If the person believes in astrology, what is the probability that he/she has never called a psychic hot line?

b. What is the probability that the person believes in astrology and has called a psychic hot line?

c. What is the probability that the person believes in astrology and has not called a psychic hot line?

4.155 Travel and Transportation As the price of gasoline increases, more Americans are using public transportation in order to commute to work. It has been reported that 54.6% of commuters in New York City, 37.7% in Washington, and 32.7% in San Francisco use public transportation to get to work.[31] Suppose three commuters are selected at random, one from each city.

a. What is the probability that all three use public transportation to commute to work?

b. What is the probability that none of the three uses public transportation to commute to work?

c. What is the probability that only the commuter from Washington uses public transportation to commute to work?

d. What is the probability that exactly two commuters use public transportation to commute to work?

4.156 Manufacturing and Product Development At a glass manufacturing facility, crystal stemware is carefully inspected for correct dimensions, quality, and production trends. After lengthy studies, the factory is known to produce 15% defectives. Most of these pieces are discovered through inspection and are reworked or discarded. Suppose two pieces are randomly selected for inspection.

a. What is the probability that both pieces are defect free?

b. What is the probability that neither piece is defect free?

c. Suppose at least one of the pieces has a flaw. What is the probability that both are defective?

4.157 Public Health and Nutrition In a study conducted by Southern Methodist University, University Case Western Reserve, and UT Southwestern Medical Center, it was reported that one in four donated hearts used for a transplant patient came from a person without any health insurance.[32] These results raise some serious ethical and moral questions. Suppose four heart transplant patients are selected at random.

a. What is the probability that all four received hearts from people without health insurance?

b. What is the probability that none of the four received a heart from someone without health insurance?

c. What is the probability that exactly two of the four transplant patients received hearts from people without health insurance?

4.158 Psychology and Human Behavior According to a survey compiled by TV-Free America,[33] 59% of Americans can name The Three Stooges but only 17% can name at least three justices of the U.S. Supreme Court. Suppose two Americans are selected at random.

a. What is the probability that both can name The Three Stooges?

b. What is the probability that both can name at least three justices of the U.S. Supreme Court?

c. What is the probability that at least one of them can name at least three justices of the U.S. Supreme Court?

d. Do you believe that being able to name The Three Stooges and being able to name at least three justices of the U.S. Supreme Court are independent events? Why or why not?

Extended Applications

4.159 Medicine and Clinical Studies Family history plays a large role in whether a child will develop asthma. If both parents suffer from asthma, there is a 90% chance that their child will develop asthma. If only one parent suffers from asthma, then the chance drops to 40%, and if neither parent has asthma, then only 20% of children develop asthma.[34] A recent research study indicates that in 15% of families, one parent has asthma, and in 5% of families, both parents have asthma. Suppose a child is randomly selected.

a. What is the probability that the child will have asthma and have parents who do not have asthma?

b. What is the probability that the child will develop asthma?

c. What is the probability that one parent has asthma given the child develops asthma?

4.160 Travel and Transportation The probability of a jet engine failure on a three-hour commercial airline flight is approximately 0.0001.[35] Suppose engines operate independently and a plane must have at least half of its engines working in order to continue flying.

a. On a randomly selected three-hour flight, what is the probability that a plane with two jet engines will reach its destination? How about a plane with three jet engines?

b. Some larger commercial airliners have four jet engines. A new airline design includes six jet engines. Which plane has a better chance of reaching its destination during a 3-hour flight?

4.161 Economics and Finance Customers at a Safeway grocery store in Scottsdale, Arizona can pay for purchases with cash, an ATM card, or a credit card. Fifty-five percent of all customers use cash and 38% use an ATM card. Careful research has shown that of those paying with cash, 75% use coupons; of those using an ATM card, 35% use coupons; and of those using a credit card, only 10% use coupons. Suppose a customer is randomly selected.

a. What is the probability that the customer pays with a credit card and does not use coupons?

b. What is the probability that the customer does not use coupons?

c. If the customer does not use coupons, what is the probability that he or she paid with an ATM card?

4.162 Demographics and Population Statistics A recent study showed that 86% of Americans have access to curbside or drop-off recycling programs.[36] Suppose four U.S. residents are selected at random.

a. What is the probability that all four have access to curb side or drop-off recycling programs?

b. What is the probability that at least one of the four does not have access to curbside or drop-off recycling programs?

c. Suppose none of the four has access to curbside or drop-off recycling programs. Is there any evidence to suggest the study's claim is not correct? Justify your answer.

4.163 Fuel Consumption and Cars AutoZone offers a wide variety of parts and accessories for cars. Consider the following events:

A = a randomly selected customer purchases a manual.
B = a randomly selected customer purchases trim accessories.
C = a randomly selected customer purchases a car care product.
Suppose the following probabilities are known.

$P(A) = 0.44$, $P(B) = 0.52$, $P(C) = 0.39$
$P(A \cap B) = 0.19$, $P(A \cap C) = 0.10$, $P(B \cap C) = 0.23$,
$P(A \cap B \cap C) = 0.08$.

a. Carefully sketch a Venn diagram illustrating the relationships among these three events and label each region with the corresponding probability.
b. Find the probability of just event A occurring.
c. Find the probability of none of the events (A, B, or C) occurring.
d. Find $P(A \mid C)$, $P(B \mid A \cap C)$, and $P(A \cap B \cap C \mid A)$.

4.164 Travel and Transportation Pasco County in Florida has special evacuation plans in the event of a hurricane. Suppose residents can take one of five different major highways out of the county. Department of Transportation officials have produced the following table indicating the probability that a resident will use a selected road.

Road	A	B	C	D	E
Probability	0.20	0.18	0.26	0.32	0.04

Suppose three Pasco County residents are selected at random and a hurricane strikes.

a. What is the probability that all three will take the same escape route?
b. What is the probability that exactly one will take escape route E?
c. What is the probability that two will take escape route C?
d. Suppose all three Pasco County residents hear a traffic report indicating route A is flooded and impassable. What is the probability that all three take route B?

4.165 Travel and Transportation Since 1999 state legislatures have struggled with laws designed to eliminate racial profiling, especially during traffic stops. Table 4.3 presents the number of drivers stopped by primary initial reason and race/ethnicity, in Maryland. Suppose this table is representative of all Maryland drivers and one such driver stopped by police is selected at random.

a. What is the probability that the driver stopped is Hispanic?
b. Suppose the driver is stopped for a Title 21-3 violation. What is the probability that the driver is white?
c. Suppose the driver is African American. What is the probability that the driver was stopped for an unknown violation?
d. Suppose the driver is not white. What is the probability that the driver was stopped for a Title 22 violation?
e. Suppose the driver was not stopped for a Title 24 violation. What is the probability that the driver is not Hispanic?
f. Are the events White and All other TR independent? Justify your answer.

Table 4.3

		Race/Ethnicity				
		Asian	African American	Hispanic	White	Other
Primary initial reason	Title 13	82	394	157	703	30
	Title 21-2	986	17393	2170	19340	2234
	Title 21-3	480	8660	1232	14486	1073
	Title 21-4	55	1173	155	2017	177
	Title 21-6	37	990	88	1155	158
	Title 21-7	631	9070	1029	12138	1351
	Title 21-8	2357	37769	3865	59011	3476
	Title 21-9	107	2491	637	5977	188
	Title 21-11	99	2428	490	5017	163
	Title 21-13	6	58	8	170	0
	Title 21-14	31	704	63	1337	53
	Title 22	1494	38926	4202	70983	6054
	Title 24	8	179	28	371	9
	All other TR	2108	64794	8530	62348	7373
	All SERO Codes	788	19612	2206	31068	1260
	Unknown	2378	38765	7624	36653	4094

(*Source:* Maryland Statistical Analysis Center, Governor's Office on Crime Control and Prevention, September 2007.)

g. Suppose two drivers are stopped by police. What is the probability that both were stopped for a Title 21 violation and are white?

Challenge

4.166 Free Nights During the month of August, one guest at the Ritz-Carlton will be selected at random to participate in a contest to win free lodging. A fair quarter will be tossed until the first head is recorded. If the first head occurs on toss x, the contestant will win x free nights at the Ritz-Carlton.

So, if a head is obtained on the first coin toss, the contest is over, and the guest wins one free night. If the first head appears on the fourteenth toss (13 tails and then a head), the guest wins 14 free nights. Theoretically, a guest could win any number of free nights: 1,2,3,4, . . ., although it seems unlikely that someone could win, for example, 100 free nights.

a. Use technology to model this contest. Try your simulation 10 times and record the number of free nights awarded each time. Did anyone win 5 or more free nights?
b. Consider the event A = the guest wins 5 or more free nights at the Ritz-Carlton. Simulate the contest $n = 50$ times and compute the relative frequency of occurrence of the event A. Repeat this process for $n = 100, 150, 200, \ldots, 2000$.
c. Construct a scatterplot of the relative frequency versus the number of simulations. Describe any patterns.
d. Use your results in (b) and (c) to estimate the probability of winning 5 or more free nights at the Ritz-Carlton.
e. Find the exact probability of winning 5 or more free nights at the Ritz-Carlton. *Hint:* Consider the complement of the event A.

Random Variables and Discrete Probability Distributions

5

Chapter 5 Challenge

Is a flu shot really effective?

Each year, the Centers for Disease Control and Prevention recommend a flu shot for certain groups of people who are classified as *at-risk* for serious complications from the most common strains of influenza virus. Adults aged 50 or older, residents of nursing homes, people with chronic heart or lung conditions, and even people who simply hate the flu are all advised to get a shot when the shots become available, usually during the fall.

Advocates claim this vaccine is 75% effective in preventing the flu.[1] Therefore, a vaccinated individual exposed to the virus has only a 1 in 4 chance of contracting the flu. There were between 143 and 146 million doses of flu vaccine available during the 2008–2009 flu season and health officials believed they had accurately predicted the strains of flu that would be circulating that year.

(Zuma/Newscom)

In order to check this claim (75% effective), a random sample of 50 at-risk people was selected and each was given a flu shot. During the flu season, all 50 people were exposed to the flu and 19 actually contracted the disease. Is there any evidence to suggest the claim about the vaccine is false, that the chance of contracting the flu is greater than 25% for a vaccinated person?

The techniques presented in this chapter will allow us to compute the likelihood of at least 19 people (out of 50) contracting the flu. This probability and conclusion about the claimed efficacy of a flu shot is presented in the Chapter Challenge Wrap-Up (page 231).

Review

- Recall the definition of an experiment, how to find a sample space, and how to perform operations on events.
- Remember the properties and rules used to compute the probability of various events.

Preview

- Learn the concept of a random variable—a bridge between the experimenter's world and the statistician's world—and learn how information is transferred between worlds.
- Understand the connection between an experimental *outcome* and the *number* associated with that outcome.
- Understand probability distributions for discrete random variables and work with several special discrete distributions.

CONTENTS

5.1   RANDOM VARIABLES

The idea of *assignment* suggests the need for a *function*.

A *function f* is a rule that takes an input value and computes an output value (according to the rule). Suppose the function *f* is defined by $f(x) = x^2 + 4$. This rule indicates that *f* takes an input *x* and *assigns*, or *maps*, *x* to the value $x^2 + 4$. For example, the function *f* assigns the input 1 to the output 5 because $f(1) = 1^2 + 4 = 5$. A **random variable** is just a special kind of function.

> **DEFINITION**
>
> A **random variable** is a function that assigns a unique numerical value to each outcome in a sample space.

 ILLUMINATING THE CONCEPTS

1. Such functions are called random variables because their values cannot be predicted with certainty before the experiment is performed.

2. Capital letters, like X and Y, are used to represent random variables.

$X: S \rightarrow \mathbf{R}$

A random variable maps elements of a sample space to the real numbers.

---

3. A random variable is a rule for assigning each outcome in a sample space to a unique real number. If *e* is an experimental outcome and *x* is a real number, here is a formal way to picture this assignment: $X(e) = x$. The random variable X takes an outcome *e* and maps, or assigns, it to the number *x*. The number *x* is associated with the outcome *e*, and is a value that the random variable can take on, or assume.

4. Figures 5.1 and 5.2 help us understand how a random variable works. These figures illustrate the random variable X as the link between experimental outcomes and numerical values. The rule for a random variable may be given by a formula, as a table, or even in words. Note that several outcomes may be assigned to the same number, but each outcome is assigned to only one number. ■

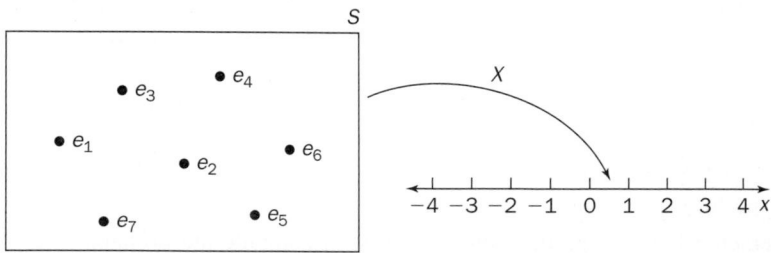

Figure 5.1 A random variable assigns a numerical value to each outcome.

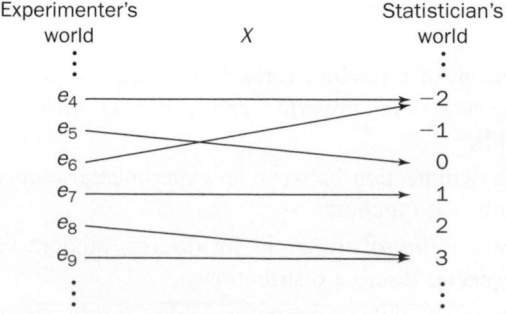

Figure 5.2 Another visualization of the definition of a random variable.

The next example shows how a specific random variable maps outcomes to numbers. The notation will get shorter and more concise as the concept of assignment becomes clearer.

(NASA/Kennedy Space Center)

Example 5.1 Space Shuttle Tiles Each space shuttle is covered with more than 23,000 special thermal tiles to protect the spacecraft from intense heat when it reenters Earth's atmosphere.[2] Following a flight, each tile is carefully inspected for damage and repaired if necessary. Suppose 3 tiles are selected at random and each is labeled as A (acceptable) or D (damaged). Let the random variable X be defined to be the number of damaged tiles.

SOLUTION

STEP 1 The experiment consists of recording the condition of 3 tiles. Each outcome consists of a sequence of three letters; each letter is an A or a D. There are eight possible outcomes (from the Multiplication Rule). Here is the sample space.

$$S = \{AAA, AAD, ADA, ADD, DAA, DAD, DDA, DDD\}$$

STEP 2 The random variable X takes each outcome and returns the number of damaged tiles (Ds). Here is a table that illustrates this mapping and the values the random variable X can assume.

Outcome	Value of X
AAA	0
AAD	1
ADA	1
ADD	2
DAA	1
DAD	2
DDA	2
DDD	3

More formally, one could write:

$X(AAA) = 0$, $X(AAD) = 1$,
$X(ADA) = 1$, . . .

A number is assigned to each outcome. Note that the outcomes AAD, ADA, and DAA are mapped to the same number (1), and the outcomes ADD, DAD, and DDA are mapped to 2.

STEP 3 Here's the key: We are no longer interested in the sequence of letters, or outcomes, but rather in the *numbers* associated with the outcomes. We need to consider the number of possible values X can assume and the probability X assumes each value.

The statement $X = 1$ is an event defined in terms of a random variable.

STEP 4 In order to find, say, the probability that X takes on the value 1, think about which outcomes are assigned to 1 and sum the probabilities of those outcomes. The probability that the random variable X equals 1 is

$$P(X = 1) = P(AAD) + P(ADA) + P(DAA)$$

because these three outcomes are mapped to 1. As shown in Figure 5.3, the random variable X links these three outcomes and their associated probabilities to the number 1.

x is a possible value of the random variable X.

Outcome	X	x
AAA		
AAD		
ADA		0
ADD		1
DAA		2
DAD		3
DDA		
DDD		

Figure 5.3 The random variable X maps three outcomes to the number 1.

There are two types of random variables. The type depends on the number of possible values the random variable can assume.

DEFINITION

A random variable is **discrete** if the set of all possible values is finite, or countably infinite.

A random variable is **continuous** if the set of all possible values is an interval of numbers.

These definitions are analogous to those for discrete and continuous data sets. The following remarks are also similar.

 ILLUMINATING THE CONCEPTS

1. **Discrete** random variables are usually associated with *counting,* and **continuous** random variables are usually associated with *measuring.*

2. In order to decide whether a random variable is discrete or continuous, consider *all* the *possible* values the random variable could assume. Finite or countably infinite means discrete. An interval of possible values means continuous.

3. Recall: Countably infinite means there are infinitely many possible values, but they are countable. You may not be able to ever finish counting all of the possible values, but there exists a method for actually counting them.

4. The interval of possible values for a continuous random variable can be *any* interval, of any length, open or closed. We may not know the exact interval, only that there is *some* interval of possible values.

5. In practice, we have no measurement device that is precise enough to return *any* number in some interval. In theory a continuous random variable may assume any value in some interval (but not in reality). ■

Remember, an experiment may result in a numerical value right away, not a symbol or a token. In this case we do not need any extra link or connection to the real numbers. The description of the experiment is the same as the definition of the random variable. The values the random variable can assume are the possible distinct experimental outcomes. In the following example, several experiments are described and each associated random variable is identified.

Example 5.2 Discrete or Continuous   Consider each experiment below and determine whether the associated random variable is discrete or continuous.

a. A Kohl's department store has 65 cash registers. At the end of the day, the receipts are carefully audited to determine whether each cash register balances. Let the random variable X be the number of cash registers that balance on a randomly selected day.

b. Patients undergoing a tonsillectomy are administered a general anesthetic. Let the random variable Y be the length of time from injection of the anesthetic until a patient is rendered unconscious.

c. Jones Soda Company is sponsoring a contest in which winning game pieces are printed on the inside of the bottle cap. An experiment consists of purchasing bottles until a winning game piece is found. Let W be a winning game piece and

L stand for a losing bottle. The sample space is $S = \{$W, LW, LLW, LLLW, LLLLW, . . .$\}$. Let the random variable X be the number of bottles purchased until a winning game piece is found.

d. Let the random variable Y be the length of the largest fish caught on the next party boat arriving back to the dock in Rockport, Massachusetts.

SOLUTION

a. There is no need to use a collection of symbols to represent experimental outcomes for the cash registers. The possible values for X (and the distinct experimental outcomes) are finite: 0, 1, 2, 3, . . . , 65. These values are distinct, disconnected points on a number line. The random variable X is discrete.

b. Y is a measurement, the time elapsed until a patient is unconscious. The possible values for Y are all the numbers in some *interval*, say, 0 to 60 minutes. The random variable Y is continuous.

c. The values X can assume are 1, 2, 3, 4, ... The number of possible values is countably infinite; the values are disconnected on a number line. The random variable X is discrete.

d. Y is a measurement, and can (theoretically) take on any value in some interval. The random variable Y is continuous. ●

SECTION 5.1 EXERCISES

Practice

5.1 Classify each random variable as discrete or continuous.
a. The number of boll weevils in one acre of a Louisiana cotton farm.
b. The volume of ice cream in one scoop.
c. The area of a randomly selected baseball field including foul territory.
d. The number of late deliveries in one month by a package delivery service.
e. The number of girls born in a rural hospital during the next year.
f. The interest rate on a savings account at a randomly selected bank in Philadelphia.
g. The number of tickets sold in the next Powerball lottery.
h. The number of oil tankers registered to a certain country at a given time.

5.2 Classify each random variable as discrete or continuous.
a. The number of visitors to the Museum of Science in Boston on a randomly selected day.
b. The camber-angle adjustment necessary for a front-end alignment.
c. The total number of pixels in a photograph produced by a digital camera.
d. The number of days until a rose begins to wilt after it is purchased from a flower shop.
e. The running time for the latest James Bond movie.
f. The blood alcohol level of the next person arrested for DUI in a particular county.

5.3 Classify each random variable as discrete or continuous.
a. The number of people requesting vegetarian meals on a flight from New York to Las Vegas.
b. The exact thickness (in millimeters) of a paper towel.

c. The time it takes a driver to react after the car in front stops suddenly.
d. The number of escapees in the next prison breakout.
e. The length of time a deep space probe remains in contact with Earth.
f. The number of points on the antlers of a randomly selected buck. The definition of a point is an antler projection at least 1 inch in length from the base to tip. The brow tine and main beam tip shall be counted as points regardless of length.

Applications

5.4 Marketing and Consumer Behavior T.J. Maxx sells home fashions and men's, women's, boys', and girls' apparel. An experiment consists of classifying the next two items purchased, each as men's, women's, boys', or girls' apparel. Let the random variable X be the number of sales of women's or girls' apparel.
a. List the outcomes in the sample space.
b. What are the possible values for X? Is X discrete or continuous? Justify your answer.

5.5 Education and Child Development An experiment consists of showing a four-year-old child an interactive instructional video and then asking the child to tie his or her shoelaces. The random variable Y is the length of time the child takes to tie the first shoelace. Is Y discrete or continuous? Justify your answer.

5.6 Biology and Environmental Science The Waynesburg Lions Club receives a shipment of 300 Christmas trees from Wending Creek Farms in Coudersport, Pennsylvania, to sell as a fundraiser. Classify each of the following random variables as discrete or continuous.
a. The number of trees over 6 feet tall.
b. The moisture content (expressed as a percentage) of a randomly selected tree.

c. The number of Douglas fir trees in the shipment.

d. The diameter of the trunk at the bottom of a randomly selected tree.

5.7 Biology and Environmental Science In order to map the current of bottom water in a certain part of the Atlantic Ocean, a dye is released and used to trace the water flow. Let the random variable X be the maximum distance (in meters) from release at which the dye is detected after 1 day. Is X discrete or continuous? Justify your answer.

5.8 Psychology and Human Behavior An experiment consists of recording the behavior of a randomly selected Duluth cab driver as a traffic signal changes from red to green. Let the random variable X be the acceleration (in ft/s^2) of the cab 1 second after the light changes. Is X discrete or continuous? Justify your answer.

5.2 PROBABILITY DISTRIBUTIONS FOR DISCRETE RANDOM VARIABLES

A random variable is a rule that assigns each experimental outcome to a real number. In order to complete the description of a discrete random variable so that we can understand and answer questions involving the random variable, we need to know all the possible values the random variable can assume and all the associated probabilities. This collection of values and probabilities is called a *probability distribution*. Since random variables are used to model populations, a probability distribution is a theoretical description of a population.

A random variable provides the link between experimental outcomes and real numbers. An experimental outcome and the probability assigned to that outcome are both associated with exactly the same value of the random variable. This connection determines probability assignments for a random variable.

DEFINITION

The **probability distribution for a discrete random variable** X is a method for specifying *all* of the possible values of X and the probability associated with each value.

 ILLUMINATING THE CONCEPTS

1. A **probability distribution for a discrete random variable** may be presented in the form of an itemized listing, a table, a graph, or a function.

2. A probability mass function (pmf) is denoted with a small p and is the probability that a discrete random variable is equal to some specific value. In symbols, it is defined by

$$p(x) = \underbrace{P(X = x)}_{\text{Rule}}.$$

In words: The *rule* for the function p evaluated at an input x is the probability of an event, the probability that the random variable X takes on the specific value x. The function p and its probability rule are used interchangeably.

Suppose X is a discrete random variable. $p(7)$ means find the probability that the random variable X equals 7, or $P(X = 7)$. ∎

A probability distribution is constructed using the definition of a random variable and the links between experimental outcomes and real numbers. The next example illustrates this concept.

Example 5.3 Construct a Probability Distribution Suppose an experiment has eight possible outcomes, each denoted by a sequence of three letters, each an N or a D. The probability of each outcome is given in the following table.

Outcome	NNN	NND	NDN	DNN	NDD	DND	DDN	DDD
Probability	0.336	0.224	0.144	0.084	0.096	0.056	0.036	0.024

The random variable X is defined to be the number of Ds in an outcome. Find the probability distribution for X.

SOLUTION

STEP 1 The probability distribution for X consists of all the possible values X can assume along with the associated probabilities. The table below shows the random variable assignment and a technique for calculating the probability of each value.

Experiment			Probability distribution	
Probability	**Outcome**	**X**	**Value, x**	**Probability**
0.336	NNN	0	$P(X = 0) = 0.336$	
0.224	NND			
0.144	NDN	1	$P(X = 1) = 0.224 + 0.144 + 0.084 = 0.452$	
0.084	DNN			
0.096	NDD			
0.056	DND	2	$P(X = 2) = 0.096 + 0.056 + 0.036 = 0.188$	
0.036	DDN			
0.024	DDD	3	$P(X = 3) = 0.024$	

To find the probability that X takes on a specific value x, find all the outcomes that are mapped to x, and add the probabilities of these outcomes.

STEP 2 The random variable X takes on the values 0, 1, 2, and 3. The probability distribution can be presented in a table as follows.

x	0	1	2	3
$p(x)$	0.336	0.452	0.188	0.024

Looking at just this probability distribution table, how do you know X cannot assume any other value?

💡 **ILLUMINATING THE CONCEPTS**

1. Think about this process of constructing a probability distribution. To find the probability that X takes on the value x, look back at the experiment, and find all the outcomes that are mapped to x. Drag along these probabilities and sum them.

2. The probability distribution for a random variable X is a reference for use in answering probability questions about the random variable. For example, we'll need to answer probability questions such as "find $P(X = 3)$." Think of $X = 3$ as an *event* stated in terms of a random variable. The details needed for answering this question are in the probability distribution. ∎

The next example illustrates various methods for presenting a probability distribution.

Don't worry about where these actual probabilities came from. In this example, focus only on the *methods* for conveying all the values and probabilities.

Example 5.4 Cash Dispenser   Suppose the random variable Y represents the number of people who use a certain ATM machine within the next 10 minutes. The probabilities of Y taking on various values are as follows: 5/15 for no people; 4/15 for one person; 3/15 for two people; 2/15 for three people; and 1/15 for four people. Here are several ways to represent the probability distribution for Y.

SOLUTION

STEP 1   A complete listing of all possible values and associated probabilities (use either the probability mass function, p, or the assignment rule):

$P(Y = 0) = 5/15$
$P(Y = 1) = 4/15$
$P(Y = 2) = 3/15$
$P(Y = 3) = 2/15$
$P(Y = 4) = 1/15$

The random variable Y can take on the values 0, 1, 2, 3, or 4, and the probability of each value is given. There can be no other value of Y since the probabilities sum to 1.

STEP 2   A table of values and probabilities:

y	0	1	2	3	4
$p(y)$	5/15	4/15	3/15	2/15	1/15

This kind of table is a common way to present a probability distribution for a discrete random variable. It concisely lists all the values that Y can assume and the associated probabilities.

STEP 3   A probability histogram:

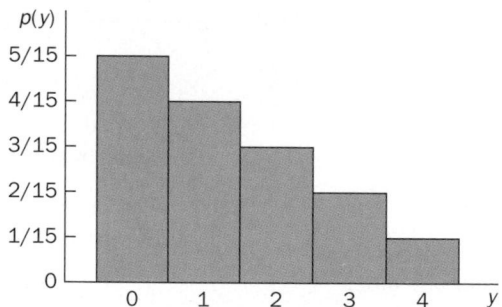

The distribution of Y is represented graphically. A rectangle is drawn for each value y, centered at y, with height equal to $p(y)$.

STEP 4   Point representation:

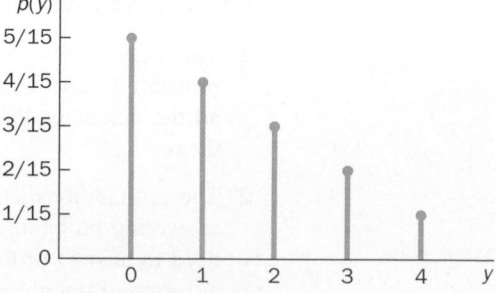

Plot the points $(y, p(y))$ and draw a line from $(y, 0)$ to $(y, p(y))$.

STEP 5   Formula:

$$p(y) = \frac{5-y}{15}, \; y = 0, 1, 2, 3, 4.$$

This shows the rule for the probability mass *function*. For example, to find $p(2)$, the probability $Y = 2$, let $y = 2$ in the formula:

$$p(2) = \frac{5-2}{15}$$

$$= \frac{3}{15}.$$

All of the techniques presented in the previous example are valid methods for presenting a probability distribution. Use the style that is most convenient or appropriate or that is called for in the question. Often, a graphical representation of the distribution is helpful. Sometimes, having a formula for the probability distribution is more useful. In the next example we'll construct another probability distribution and consider some probability questions involving a random variable.

Example 5.5 Who Wants Coffee?   The Hard Rock Café in Dallas carefully monitors customer orders and has found that 70% of all customers ask for some kind of coffee (C), while the remainder order a specialized tea (T). Suppose four customers are selected at random. Let the random variable X be the number of customers who order coffee.
a. Find the probability distribution for X.
b. Find the probability that more than two customers order coffee.
c. Suppose at least two customers order coffee. What is the probability that all four customers order coffee?

a. Solution Trail

KEYWORDS

Probability distribution.

↓

TRANSLATION

Find all the values X can assume and all the associated probabilities.

↓

CONCEPTS

Connection between experimental outcomes and real numbers.

↓

VISION

Think about the experiment first. Use the definition of the random variable to link experimental outcomes with values of the random variable, and drag along all of the probabilities. Construct a table listing all the values of X and the associated probabilities.

SOLUTION

The experiment consists of observing four customer choices. Each outcome consists of a sequence of four letters, each a C or a T. From the multiplication rule, there are 16 possible outcomes: CCCC, CCCT, CCTC, etc.

Since the customers are selected at random, each choice is independent, and the probability of each outcome is obtained by multiplying the corresponding probabilities. For example,

$$P(CTCT) = P(C \cap T \cap C \cap T)$$ First customer buys coffee *and* second customer buys tea *and* . . .

$$= P(C) \cdot P(T) \cdot P(C) \cdot P(T)$$ Events are independent. Multiply corresponding probabilities.

$$= (0.70)(0.30)(0.70)(0.30)$$ $P(T) = 1 - P(C)$

$$= 0.0441$$

The following table lists all the possible experimental outcomes, the probability of each outcome (computed as above), and the value of the random variable assigned to each outcome.

Outcome	Probability	x	Outcome	Probability	x
TTTT	0.0081	0	CTTC	0.0441	2
TTTC	0.0189	1	CTCT	0.0441	2
TTCT	0.0189	1	CCTT	0.0441	2
TCTT	0.0189	1	TCCC	0.1029	3
CTTT	0.0189	1	CTCC	0.1029	3
TTCC	0.0441	2	CCTC	0.1029	3
TCTC	0.0441	2	CCCT	0.1029	3
TCCT	0.0441	2	CCCC	0.2401	4

This table shows that the values of X are 0, 1, 2, 3, and 4.

a. Use the links in this table to construct the probability distribution for X.

$$p(0) = P(X = 0) = P(TTTT) = 0.0081$$

There is only one outcome assigned to a 0, and the probability of that outcome is 0.0081.

$$
\begin{aligned}
p(1) &= P(X = 1) && \text{Definition of a probability mass function.}\\
&= P(TTTC \text{ or } TTCT \text{ or } TCTT \text{ or } CTTT) && \text{These outcomes are mapped to 1.}\\
&= P(TTTC) + P(TTCT) + P(TCTT) + P(CTTT) && \\
&&& \text{\textit{Or} means union; the outcomes are disjoint.}\\
&= 0.0189 + 0.0189 + 0.0189 + 0.0189 = 0.0756
\end{aligned}
$$

Continue in this manner to obtain the probability distribution for X.

x	0	1	2	3	4
$p(x)$	0.0081	0.0756	0.2646	0.4116	0.2401

b. $P(X > 2) = P(X = 3) + P(X = 4)$ Only values greater than 2.
$= 0.4116 + 0.2401 = 0.6517$ Use the probability distribution table.

The probability of more than two customers ordering coffee is 0.6517.

c. Given X is at least 2, find the probability that X is exactly 4.

$$
\begin{aligned}
P(X = 4 \mid X \geq 2) &= \frac{P(X = 4 \cap X \geq 2)}{P(X \geq 2)} && \text{Definition of conditional probability.}\\
&= \frac{P(X = 4)}{P(X \geq 2)} && \text{Intersection of } X = 4 \text{ and } X \geq 2.\\
&= \frac{0.2401}{0.2646 + 0.4116 + 0.2401} && \text{Use the probability distribution.}\\
&= \frac{0.2401}{0.9163} = 0.2620
\end{aligned}
$$

Given at least two people order coffee, the probability that exactly four order coffee is 0.2620.

Here is a way to picture this conditional probability using the probability distribution table.

x	0	1	2	3	4
$p(x)$	0.0081	0.0756	0.2646	0.4116	0.2401

Given that X is either 2, 3, or 4, the *reduced*, or relevant, probability is 0.9163. The proportion of time that X is equal to 4, given X is 2, 3, or 4, is 0.2401/0.9163. ●

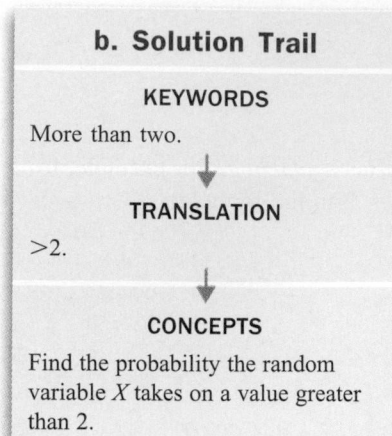

b. Solution Trail

KEYWORDS

More than two.

↓

TRANSLATION

>2.

↓

CONCEPTS

Find the probability the random variable X takes on a value greater than 2.

↓

VISION

We have the probability distribution for X. Use the table to determine which values are greater than 2, and add the associated probabilities.

How would this probability change if the question asked for the probability that *two or more* customers order coffee?

---

c. Solution Trail

KEYWORDS

- Suppose at least two.
- All four.

↓

TRANSLATION

- *Given* at least two.
- Find the probability $X = 4$.

↓

CONCEPTS

Conditional probability.

↓

VISION

Given at least two customers order coffee, the number of values X can assume is reduced. Use the definition of conditional probability with events involving the random variable.

The probability distribution of a random variable reveals which values of the random variable are most likely to occur. This information is extremely helpful in making a statistical inference. Consider the following example.

Example 5.6 In Case There Is a Power Outage The Carson City Hospital has three emergency generators for use in case of a power failure. Each generator operates independently and the manufacturer claims that the probability each generator will function properly during a power failure is 0.95. Suppose a power failure occurs and all three generators fail. Do you have reason to doubt the manufacturer's claim? Justify your answer.

SOLUTION

STEP 1 In the event of a power failure, let F stand for a generator that fails, and let S represent a generator that functions properly (i.e., starts). There are eight possible experimental outcomes. Let X be the number of failures.

STEP 2 The table below lists each outcome, the probability of each outcome, and the value of the random variable associated with each outcome.

Since each generator operates independently, the probability of each outcome is the product of the corresponding probabilities. For example,

$$P(SFS) = P(S \cap F \cap S)$$
$$= P(S) \cdot P(F) \cdot P(S)$$
$$= (0.95)(0.05)(0.95) = 0.0451$$

Note: The probabilities are rounded to four places to the right of the decimal.

Outcome	Probability	x
SSS	0.8574	0
SSF	0.0451	1
SFS	0.0451	1
SFF	0.0024	2
FSS	0.0451	1
FSF	0.0024	2
FFS	0.0024	2
FFF	0.0001	3

STEP 3 Use the links in this table to construct the probability distribution for X.

x	0	1	2	3
$p(x)$	0.8574	0.1353	0.0072	0.0001

STEP 4 Use the four-step inference procedure.

Claim: The probability that each generator will function properly during a power failure is 0.95.

Experimental outcome: The value of the random variable observed is $x = 3$.

Likelihood: The likelihood of the observed outcome is $P(X = 3) = 0.0001$.

Conclusion: Since this probability is so small, the outcome of observing three failures is very rare. But it happened! This small probability suggests that the assumption is wrong. There is evidence to suggest that the claim of 0.95 (start probability) is wrong. ●

Solution Trail

KEYWORDS

Reason to doubt the manufacturer's claim?

↓

TRANSLATION

Use the available evidence to draw a reasonable conclusion.

↓

CONCEPTS

Inference procedure.

↓

VISION

In order to decide whether three generator failures are reasonable, we need to follow the four-step inference procedure. Consider the claim or assumption, the experiment, and the likelihood of the experimental outcome. Then draw your conclusion. If the probability is large, then the outcome is reasonable (given the assumption). If the probability is small, then there is evidence to suggest the assumption is wrong. Consider the experiment, define an appropriate random variable, and use the probability distribution to determine the probability that all three generators fail.

As the previous examples suggest, the following properties must be true for every probability distribution for a discrete random variable X.

PROPERTIES OF A VALID PROBABILITY DISTRIBUTION FOR A DISCRETE RANDOM VARIABLE

1. $0 \leq p(x) \leq 1$

    The probability that X takes on any value, $p(x) = P(X = x)$, must be between 0 and 1.

2. $\sum_{\text{all } x} p(x) = 1$

    The sum of all the probabilities in a probability distribution for a discrete random variable must equal 1.

The following example involves a probability distribution for a discrete random variable and illustrates these two properties.

(Photodisc/Alamy)

Example 5.7 Video Game Points   In the video game *The Kings of Madra*, users solve logic puzzles in order to become more powerful and influential. The number of experience points earned at each level is a random variable, Y, and is the square of the number of puzzles solved. Suppose Y has the following probability distribution:

y	0	1	4	9	16	25
$p(y)$	0.01	0.05	?	0.25	0.35	0.30

a. Find $p(4)$.
b. Find $P(1 \leq Y \leq 16)$ and $P(1 < Y < 16)$.
c. Construct the corresponding probability histogram.

SOLUTION

a. The sum of all the probabilities must equal 1.

$$p(4) = 1 - [p(0) + p(1) + p(9) + p(16) + p(25)]$$
$$= 1 - (0.01 + 0.05 + 0.25 + 0.35 + 0.30)$$
$$= 1 - 0.96 = 0.04$$

b. The values Y takes on between 1 and 16 inclusive are 1, 4, 9, and 16.

$$P(1 \leq Y \leq 16) = p(1) + p(4) + p(9) + p(16)$$
$$= 0.05 + 0.04 + 0.25 + 0.35 = 0.69$$

The values Y takes on *strictly* between 1 and 16 are 4 and 9.

$$P(1 < Y < 16) = p(4) + p(9)$$
$$= 0.04 + 0.25 = 0.29$$

In this example, including (or excluding) an endpoint (a single value) changes the probability assignment. It is important to remember that a single value *may* make a difference in a probability assignment for a discrete random variable.

c. To construct the probability histogram, draw a rectangle for each value y, centered at y, with height equal to $p(y)$.

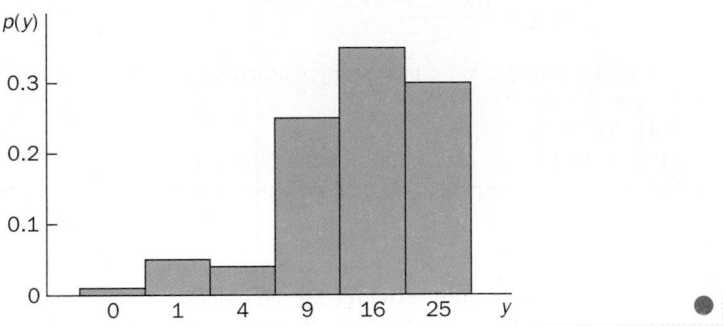

SECTION 5.2 EXERCISES

Practice

5.9 The probability distribution for the random variable X is given in the following table.

x	1	2	3	4	5	6	7
$p(x)$	0.35	0.20	0.15	0.12	?	0.08	0.03

a. Find $p(5)$.
b. Find $P(2 \le X \le 6)$ and $P(2 < X \le 6)$.
c. Find $P(X < 4)$.
d. Find the probability that X takes on the value 1 or 7.

5.10 The probability distribution for the random variable Y is given in the following table.

y	10	20	25	30	45	50
$p(y)$	0.155	0.237	0.184	0.122	?	0.258

a. Find $p(45)$.
b. Find $P(Y \ge 25)$ and $P(Y > 25)$.
c. Find the probability that Y is divisible by 10.
d. Construct the corresponding probability histogram.

5.11 The probability distribution for the random variable X is given in the following table.

x	−3	−2	−1	0	1	2	3
$p(x)$	0.20	0.10	0.05	0.30	0.05	0.10	0.20

a. Find $P(X \ge 0)$ and $P(X > 0)$.
b. Find $P(X^2 > 1)$.
c. Find $P(X \ge 2 \mid X \ge 0)$.
d. Construct the corresponding probability histogram.

5.12 Determine whether each probability distribution below is valid. Justify your answers.

a.

x	2	4	6	8	10	12
$p(x)$	0.15	0.16	0.17	0.18	0.19	0.20

b.

x	2	4	6	8	10	12
$p(x)$	0.25	0.25	0.25	−0.25	0.25	0.25

c.

x	2	4	6	8	10	12
$p(x)$	0.05	0.20	0.25	0.25	0.20	0.05

5.13 The table below lists all of the possible outcomes for an experiment, the probability of each outcome, and the value of a random variable assigned to each outcome. Use this table to construct the probability distribution for X. Construct the corresponding probability histogram.

Outcome	Probability	x	Outcome	Probability	x
AA	0.01	1	CA	0.03	3
AB	0.02	2	CB	0.06	3
AC	0.03	3	CC	0.09	3
AD	0.04	4	CD	0.12	4
BA	0.02	2	DA	0.04	4
BB	0.04	2	DB	0.08	4
BC	0.06	3	DC	0.12	4
BD	0.08	4	DD	0.16	4

5.14 The table on the data CD and book's web site lists all of the possible outcomes for an experiment, the probability of each outcome, and the value of a random variable assigned to each outcome. Use this table to construct the probability distribution for Y. Construct the corresponding probability histogram.

5.15 The probability distribution for a discrete random variable X is given by the formula

$$p(x) = \frac{x(x+1)}{112}, \quad x = 1, 2, \ldots, 6.$$

a. Verify that this is a valid probability distribution.
b. Find $P(X = 4)$.
c. Find $P(X > 2)$.
d. Find the probability that X takes on the value 3 or 4.
e. Construct the corresponding probability histogram.

Applications

5.16 Manufacturing and Product Development A wooden kitchen cabinet is carefully inspected at the manufacturing facility before it is sent to a retailer. The random variable X is the number of defects found in a randomly selected cabinet. The probability distribution for X is given in the table below.

x	0	1	2	3	4	5
$p(x)$	0.900	0.050	0.025	0.020	0.004	0.001

Suppose a cabinet is selected at random.
a. What is the probability that the cabinet is defect free?
b. What is the probability that the cabinet has at most two defects?
c. What is the probability that two randomly selected cabinets both have at least three defects?
d. Find $P(2 \leq X \leq 4)$ and $P(2 < X < 4)$.

5.17 Fuel Consumption and Cars An automobile insurance policy depends on several factors: age, gender, marital status, where you live, commuting miles, make and model of car, driving record, and even credit history.[3] Suppose for some driver category, the probability distribution for the random variable Y, the amount (in dollars) paid to policy holders due to claims in one year, is given in the table below.

y	0	500	1000	5000	10000
$p(y)$	0.65	0.20	0.10	0.04	0.01

a. Find $P(Y > 0)$.
b. Find $P(Y \leq 1000)$.
c. What is the probability that a randomly selected driver is paid $5000?
d. Suppose two drivers are selected at random. What is the probability that both are paid $1000?
e. Suppose two drivers are selected at random. What is the probability that at least one is paid $500 or more?

5.18 Public Policy and Political Science The police department in Detroit is attempting to clean up the rough Pendleton section and help transform the area into an arts district. The number of times a police cruiser drives through the neighborhood during a 1-hour period is a random variable X, with probability distribution given in the table below.

x	0	1	2	3
$p(x)$	0.3679	0.3679	0.1839	0.0613

x	4	5	6	7
$p(x)$	0.0153	0.0031	0.0005	0.0001

Suppose a one-hour period is randomly selected.
a. What is the probability that no police cruiser will drive through the neighborhood?
b. What is the probability that at least one police cruiser will drive through the neighborhood?
c. What is the probability that at most two police cruisers will drive through the neighborhood?
d. What is the probability that more than seven police cruisers will drive through the neighborhood?
e. Suppose at least two police cruisers were sighted in the neighborhood during the 1-hour period. What is the probability that there were at least four during this time?

5.19 Marketing and Consumer Behavior In a paper by Simosohn and Ariely,[4] the authors studied eBay auctions for DVD movies. For a randomly selected patron who places a $10 bid on a DVD movie with a starting price of $6, the probability of winning the auction is 0.30. Suppose three patrons (in different auctions) who bid $10 on a DVD movie with starting price $6 are selected at random. Let the random variable X be the number of patrons who win their auctions.
a. Construct the probability distribution for X. Construct the corresponding probability histogram.
b. What is the probability that all three patrons win their auctions?
c. What is the probability that at least one of the three will win an auction?

5.20 Travel and Transportation During September 2008, the probability that a randomly selected flight arrived on time at Logan Airport in Boston was 0.7999; at New York La Guardia, 0.7317; at O'Hare Airport in Chicago, 0.7621; and at Los Angeles airport, 0.8797.[5] Suppose one flight arriving at each airport is randomly selected and let random variable Y be the total number of flights arriving on time.
a. Construct the probability distribution for Y.
b. What is the probability that at least one flight arrives on time?
c. Suppose at least two flights arrive on time. What is the probability that all four flights arrive on time?

5.21 Manufacturing and Product Development The Bellevue Community College bookstore has six special drafting pencils for sale, two of which are defective. A student buys two of these drafting pencils, selected at random. Let the random variable X be the number of defective pencils purchased. Construct the probability distribution for X.

5.22 Business and Management Two packages are independently shipped from Fort Collins, Colorado, to the same address in Seattle, Washington, and each is guaranteed to arrive within 4 days. The probability that a package arrives within 1 day is 0.10, within 2 days is 0.15, within 3 days is 0.25, and on the fourth day is 0.50. Let the random variable X be the total number of days it takes for both packages to arrive. Construct the probability distribution for X.

Extended Applications

5.23 Biology and Environmental Science Agway, a farm and garden supply store, sells winter fertilizer in 50-pound bags. For customers purchasing this product, the probability distribution for the random variable X, the number of bags sold, is given in the table that follows.

x	1	2	3	4	5
$p(x)$	0.55	0.35	0.07	0.02	0.01

Suppose a person buying winter fertilizer is randomly selected.
a. What is the probability that the customer buys more than two bags?
b. What is the probability that the customer does not buy two bags $[P(X \neq 2)]$?
c. Find the probability that two randomly selected customers each buy one bag.
d. Suppose two customers are randomly selected. What is the probability that the total number of bags purchased will be at least eight?
e. Let the random variable Y be the number of *pounds* sold to a randomly selected customer buying winter fertilizer. Find the probability distribution for Y.

5.24 Sports and Leisure Suppose the probability that a person *says* he or she was at the Woodstock Festival and Concert is 0.20. An experiment consists of randomly selecting people in Green Bay and asking them whether they were at Woodstock. The experiment stops as soon as one person says he or she was there. The random variable X is the number of people stopped and questioned (until one person says he or she was there). Let Y and N stand for a Yes and No response, respectively.
a. List the first several outcomes in the sample space.
b. Find the probability of each outcome in part (a).
c. Find the value of the random variable associated with each outcome in (a).
d. Find a formula for the probability distribution of X.

5.25 Sports and Leisure A game show contestant on Let's Make a Deal selects two envelopes with prize money enclosed. Two of the envelopes contain $100, one envelope contains $250, two envelopes contain $500, and the last envelope contains $1000. Let the random variable M be the maximum of the two prizes.
a. Find the probability distribution for M.
b. Suppose two contestants independently select prize envelopes on two different days. What is the probability that both win the top prize?

5.26 Economics and Finance A Chick-Fil-A fast-food restaurant recently installed a new computer system that allows customers to pay for their meals with a bank debit card. The manager of the restaurant claims this new system will decrease the waiting time and that the probability of getting a meal in under 2 minutes (with this system in place) is 0.75. Suppose four customers are selected at random. Let the random variable X be the number of customers who receive their meal in under 2 minutes.
a. Construct the probability distribution for X.
b. Suppose none of the four customers receives the meal in under 2 minutes. Is there any evidence to suggest that the manager's claim is false? Justify your answer.

5.3 MEAN, VARIANCE, AND STANDARD DEVIATION FOR A DISCRETE RANDOM VARIABLE

(David Cannon/Getty Images)

Just as there are descriptive measures of a sample (for example, $\bar{x}$, s^2, and s), there are corresponding descriptive measures of a population (μ, σ^2, and σ). As we said in Chapter 3, these population *parameters* describe the center and variability of the *entire* population. They are usually unknown values that we would like to estimate.

However, since a random variable may be used to model a population, these (population) descriptive measures are inherent in and determined by the probability distribution. This section presents the methods used to compute the mean, variance, and standard deviation of a random variable (or population). The next example suggests a definition of **expected value**.

Example 5.8 The Caddy Pool A teenager is a member of the caddy program at the Montebello Country Club. Each morning he arrives at the golf course and enters his name into the caddy pool. The probability of being selected on any day is 4/5, and if selected he will earn $50. On days when he is not selected, he earns nothing. How much money does this caddy earn per day on *average*? Or, in the long run, how much does the caddy earn each day?

SOLUTION

STEP 1   This question is concerned with the amount earned each day *on average*, not on any one particular day. Consider the probabilities given and consider 5 typical days.

On four of five days, the caddy earns $50. On the fifth day, he earns $0. The total earned for the 5 *typical* days is $200. To find the *average* amount earned each day, divide by five: $200/5 = $40.

STEP 2   Consider a random variable X that takes on only two values, 0 and 50, with probabilities 0.20 and 0.80, respectively. Another way to compute the *average* earned each day is to use this probability distribution.

$$40 = \underset{\text{value}}{0} \times \underset{\text{probability}}{0.20} + \underset{\text{value}}{50} \times \underset{\text{probability}}{0.80}$$

The long-run average earnings per day can be found by using a probability distribution. Multiply each value by its corresponding probability, and sum these products. ●

DEFINITION

Let X be a discrete random variable with probability mass function $p(x)$. The **mean**, or **expected value**, of X is

$$\underbrace{E(X) = \mu = \mu_X}_{\text{Notation}} = \underbrace{\sum_{\text{all } x}[x \cdot p(x)]}_{\text{Calculation}} \tag{5.1}$$

💡 **ILLUMINATING THE CONCEPTS**

1. The capital E stands for *expected value* and is a function. The function E takes a random variable as an input and returns the expected value. More generally, E accepts as an input any *function* of a random variable. For example, suppose $f(X)$ is a function of a discrete random variable X. The expected value of $f(X)$ is

$$E[f(X)] = \sum_{\text{all } x}[f(x) \cdot p(x)] \tag{5.2}$$

2. μ is the **mean**, or expected value, of a random variable (that may model a population). If necessary, the associated random variable is used as a subscript for identification, for example μ_X or μ_Y.

3. The mean is very easy to compute. Multiply each value of the random variable by its corresponding probability, and add the products.

4. The mean of a random variable is a *weighted average* and is only what happens on average. The mean may not be any of the possible values of the random variable. ∎

Example 5.9 Road Construction Next 30 Miles Thirty miles of interstate highway I-16 near Macon, Georgia, were recently repaved. At the end of each work day, the project supervisor estimated the number of hours that the crew was behind or ahead of schedule. Suppose this estimate is a discrete random variable, X, with probability distribution given in the following table.

x	-20	-10	30	60
$p(x)$	0.2	0.3	0.4	0.1

Find the mean of X.

SOLUTION

STEP 1 X is a discrete random variable. To find the mean, use the definition given above.

STEP 2 $\mu = \sum_{\text{all } x}[x \cdot p(x)]$ Equation 5.1.

$= (-20)(0.2) + (-10)(0.3) + (30)(0.4) + (60)(0.1)$

Multiply each value by its probability, and sum.

$= (-4) + (-3) + (12) + (6) = 11$

As the sample size increases, the sample mean, $\bar{x}$, would tend to, or approach, the population mean $\mu = 11$.

STEP 3 The mean, or long-run value, of X is 11. In this example the mean is *not* a possible value of X. ●

Example 5.10 Extended Missions A space-shuttle mission is sometimes extended due to bad weather conditions at the landing site, for additional payload activities, or to continue scientific research. Suppose X is a random variable that represents the number of extended mission days. The 123 missions flown from April 1981 to October 2008 were used to construct the probability distribution for X in the table below.

x	0	1	2	3	4	15
$p(x)$	0.7643	0.1626	0.0488	0.0081	0.0081	0.0081

(*Source:* NASA John F. Kennedy Space Center)

Find the expected number of extended mission days.

SOLUTION

STEP 1 X is a discrete random variable (it takes on only a finite number of values). Find the mean using Equation 5.1.

STEP 2 $\mu = \sum_{\text{all } x}[x \cdot p(x)]$ Equation 5.1.

$= (0)(0.7643) + (1)(0.1626) + (2)(0.0488) + (3)(0.0081) + (4)(0.0081)$

$+ (15)(0.0081)$ Sum of each value times probability.

$= 0 + 0.1626 + 0.0976 + 0.0243 + 0.0324 + 0.1215 = 0.4384$

STEP 3 The mean number of extended mission days is 0.4384. ●

The **variance** and **standard deviation** of a random variable measure the spread of the distribution. The variance is computed using the expected value function, and the standard deviation together with the mean can be used to determine the most likely values of the random variable.

DEFINITION

Let X be a discrete random variable with probability mass function $p(x)$. The **variance** of X is

$$\underbrace{\text{Var}(X) = \sigma^2 = \sigma_X^2}_{\text{Notation}} = \underbrace{\sum_{\text{all } x}[(x - \mu)^2 \cdot p(x)]}_{\text{Calculation}} = \underbrace{E[(X - \mu)^2]}_{\substack{\text{Definition in terms of} \\ \text{expected value}}} \quad (5.3)$$

The **standard deviation** of X is the positive square root of the variance:

$$\underbrace{\sigma = \sigma_X}_{\text{Notation}} = \underbrace{\sqrt{\sigma^2}}_{\text{Calculation}} \quad (5.4)$$

ILLUMINATING THE CONCEPTS

1. In words, the variance is the expected value of the *squared deviations about the mean.*

2. The symbol Var stands for variance and is a function. The function Var takes a random variable as an input and returns the variance.

3. To compute the variance using Equation 5.3,
    a. Find the mean, μ, of X using Equation 5.1.
    b. Find each difference: $(x - \mu)$.
    c. Square each difference: $(x - \mu)^2$.
    d. Multiply each squared difference by the associated probability.
    e. Sum the products.

4. There is a *computational formula* for the variance of a random variable. ■

COMPUTATIONAL FORMULA FOR σ^2

$$\sigma^2 = E(X^2) - E(X)^2 = E(X^2) - \mu^2 \quad (5.5)$$

In words, the variance is the expected value of X squared minus the square of the expected value of X.

Example 5.11 Children in Day Care Suppose the discrete random variable X, the age of a randomly selected child at the Unikids Academy Day Care Center in New Orleans, has the probability distribution given in the following table.

x	1	2	3	4	5	6	7
$p(x)$	0.05	0.10	0.15	0.25	0.20	0.15	0.10

a. Find the expected value, variance, and standard deviation of X.
b. Find the probability that the random variable X takes on a value within one standard deviation of the mean.

SOLUTION

a. *Expected value:* Find the expected value of X using Equation 5.1.

$$E(X) = \sum_{\text{all } x} [x \cdot p(x)] \qquad \text{Equation 5.1}$$

$$= (1)(0.05) + (2)(0.10) + (3)(0.15) + (4)(0.25) + (5)(0.20)$$
$$+ (6)(0.15) + (7)(0.10)$$
$$= 0.05 + 0.20 + 0.45 + 1.00 + 1.00 + 0.90 + 0.70$$
$$= 4.30 = \mu$$

Variance: Find the variance of X using Equation 5.3.

$$\text{Var}(X) = \sum_{\text{all } x} (x - \mu)^2 \cdot p(x) \qquad \text{Equation 5.3}$$

$$= (1 - 4.30)^2(0.05) + (2 - 4.30)^2(0.10) + (3 - 4.30)^2(0.15)$$
$$+ (4 - 4.30)^2(0.25) + (5 - 4.30)^2(0.20) + (6 - 4.30)^2(0.15)$$
$$+ (7 - 4.30)^2(0.10) \qquad \text{Sum over all values of } x.$$

$$= (10.89)(0.05) + (5.29)(0.10) + (1.69)(0.15)$$
$$+ (0.09)(0.25) + (0.49)(0.20) + (2.89)(0.15)$$
$$+ (7.29)(0.10) \qquad \text{Square each difference.}$$

$$= 0.5445 + 0.5290 + 0.2535 + 0.0225$$
$$+ 0.0980 + 0.4335 + 0.7290 \qquad \text{Compute each product.}$$
$$= 2.6100 = \sigma^2 \qquad \text{Sum the products.}$$

Here is a tabular method for computing the variance using the definition. Sum the last column to obtain σ^2.

x	$x - \mu$	$(x - \mu)^2$	$p(x)$	$(x - \mu)^2 \cdot p(x)$
1	-3.30	10.89	0.05	0.5445
2	-2.30	5.29	0.10	0.5290
3	-1.30	1.69	0.15	0.2535
4	-0.30	0.09	0.25	0.0225	Sum this column.
5	0.70	0.49	0.20	0.0980
6	1.70	2.89	0.15	0.4335
7	2.70	7.29	0.10	0.7290

$$2.6100 \quad \leftarrow \sigma^2$$

Variance: Using the *computational formula.*

Find $E(X^2)$, the expected value of X^2.

$$E(X^2) = \sum_{\text{all } x} [x^2 \cdot p(x)] \qquad \text{Equation 5.2.}$$

$$= 1^2(0.05) + 2^2(0.10) + 3^2(0.15) + 4^2(0.25) + 5^2(0.20)$$
$$+ 6^2(0.15) + 7^2(0.10) \qquad \text{Sum over all values of } x.$$

$$= 1(0.05) + 4(0.10) + 9(0.15) + 16(0.25) + 25(0.20)$$
$$+ 36(0.15) + 49(0.10) \qquad \text{Square each } x.$$

$$= 0.05 + 0.40 + 1.35 + 4.00 + 5.00 + 5.40 + 4.90 \qquad \text{Compute each product.}$$
$$= 21.10 \qquad \text{Sum the products.}$$

Use this result to find the variance.

$$\sigma^2 = E(X^2) - \mu^2 \qquad \text{Equation 5.5.}$$
$$= 21.10 - (4.30)^2 \qquad \text{Use previous results.}$$
$$= 21.10 - 18.49 \qquad \text{Find } \mu^2.$$
$$= 2.61 \qquad \text{Find the difference.}$$

Variance: Using a tabular method.

Start at the middle of each row, and work toward the ends. Sum the outer columns to obtain μ and $E(X^2)$.

$x^2 \cdot p(x)$	x^2	$p(x)$	x	$x \cdot p(x)$
0.05	1	0.05	1	0.05
0.40	4	0.10	2	0.20
1.35	9	0.15	3	0.45
4.00	16	0.25	4	1.00
5.00	25	0.20	5	1.00
5.40	36	0.15	6	0.90
4.90	49	0.10	7	0.70

Sum this column for $E(X^2)$.

Sum this column for μ.

$$E(X^2) \rightarrow \quad 21.10 \qquad\qquad\qquad 4.30 \quad \leftarrow \mu$$

$$\sigma^2 = E(X^2) - \mu^2 = 21.10 - (4.30)^2 = 2.61$$

Standard deviation: The positive square root of the variance.

$$\sigma = \sqrt{\sigma^2} = \sqrt{2.61} \approx 1.6155$$

b. $P(\mu - \sigma \le X \le \mu + \sigma)$ <small>Translation to a probability statement.</small>

$\quad = P(4.30 - 1.6155 \le X \le 4.30 + 1.6155)$ <small>Use values for μ and σ.</small>

$\quad = P(2.6845 \le X \le 5.9155)$ <small>Compute the difference and sum.</small>

$\quad = P(X = 3) + P(X = 4) + P(X = 5)$ <small>Find values of X in the interval.</small>

$\quad = 0.15 + 0.25 + 0.20$ <small>Use corresponding probabilities.</small>

$\quad = 0.60$ <small>Compute the sum.</small>

The probability that X takes on a value within one standard deviation of the mean is 0.60.   ●

---

b. Solution Trail

KEYWORDS

Within one standard deviation of the mean.

↓

TRANSLATION

In the interval $(\mu - \sigma, \mu + \sigma)$, within one *step* in each direction from the mean.

↓

CONCEPTS

- Probability distribution.
- Probability statement.

↓

VISION

In order to solve probability questions involving random variables, we need two important facts: (1) the probability distribution, and (2) the probability statement. The probability distribution is given in this problem. Use the *Translation* to write a mathematical probability statement. Find the value(s) of X that lie in the interval, and add the corresponding probabilities.

💡 **ILLUMINATING THE CONCEPTS**

1. The computational formula for the variance is quicker and produces less round-off error. Use this formula (Equation 5.5) to find the variance of a discrete random variable.

2. In the example above, the random variable X is not (approximately) normal; the Empirical Rule does *not* apply. In addition, even though Chebyshev's Rule applies to any distribution, it should not be used if the probability distribution is known. Chebyshev's Rule provides only an *estimate*, a lower bound for the probability that X is within k standard deviations of the mean. The exact probability can be determined using the known probability distribution. (Actually, Chebyshev's Rule can't help at all here because k must be greater than 1.)

3. Neither the TI-84 nor Minitab has a built-in menu function to find the mean, variance, and standard deviation of a discrete random variable. However, it's easy to perform list operations on the calculator and column operations in Minitab or Excel in order to produce these summary statistics. See the technology manuals for details. ■

(iStockphoto)

Example 5.12 One-Armed Bandit A typical three-reel mechanical slot machine has different payoffs determined by the number (and sometimes the position) of various pictures. Suppose the payoff (in dollars) is a discrete random variable with probability distribution given in the following table.

x	2400	1199	200	150	80	50	40
$p(x)$	0.000004	0.000011	0.000160	0.000160	0.004574	0.000687	0.000801

x	25	20	10	5	2	1	0
$p(x)$	0.001442	0.000431	0.001648	0.030430	0.008114	0.125000	0.826538

(*Source:* The Wizard of Odds, 2007).

a. Find the expected payoff and the variance and standard deviation of the payoff.
b. Interpret the values found in part (a). In particular, if it costs $1.00 to play, what happens in the long run?

SOLUTION

a. Use the tabular method and the computational formula for the variance.

$x^2 \cdot p(x)$	x^2	$p(x)$	x	$x \cdot p(x)$
23.0700	5760000	0.000004	2400	0.0096
15.8136	1437601	0.000011	1199	0.0132
6.4000	40000	0.000160	200	0.0320
3.6000	22500	0.000160	150	0.0240
29.2786	6400	0.004574	80	0.3659
1.7175	2500	0.000687	50	0.0344
1.2816	1600	0.000801	40	0.0320
0.9013	625	0.001442	25	0.0361
0.1724	400	0.000431	20	0.0086
0.1648	100	0.001648	10	0.0165
0.7608	25	0.030430	5	0.1522
0.0325	4	0.008114	2	0.0162
0.1250	1	0.125000	1	0.1250
0.0000	0	0.826538	0	0.0000

$E(X^2) \rightarrow$ 82.2830   0.8657 $\leftarrow \mu$

$$\sigma^2 = E(X^2) - \mu^2 = 82.2830 - (0.8657)^2 = 82.53$$

$$\sigma = \sqrt{\sigma^2} = \sqrt{82.53} \approx 9.08$$

b. The mean payoff from this slot machine is approximately 87 cents, with variance 82.53 dollars^2 and standard deviation $9.08. If it costs $1.00 to play this machine, in the long run the player loses $1.00 - 0.87 = 0.13$, or 13 cents, on average. This doesn't seem like much. However, from the casino's point of view, every time someone pulls the lever on a slot machine, the casino makes 13 cents (on average). ●

SECTION 5.3 EXERCISES

Practice

5.27 Suppose X is a discrete random variable. Complete the table below and find the mean, variance, and standard deviation of X.

$x^2 \cdot p(x)$	x^2	$p(x)$	x	$x \cdot p(x)$
		0.10	2	
		0.16	4	
		0.20	6	
		0.24	8	
		0.18	10	
		0.12	12	

5.28 The probability distribution for a random variable X is given in the table below.

x	5	10	15	20
$p(x)$	0.10	0.15	0.70	0.05

a. Find the mean, variance, and standard deviation of X.
b. Find the probability that X takes on a value smaller than the mean.
c. Using the probability distribution, explain why the value of the mean of X makes sense.

5.29 Suppose Y is a discrete random variable with probability distribution given in the table below.

y	-20	-10	0	10	20
$p(y)$	0.30	0.15	0.10	0.15	0.30

a. Find μ, σ^2, and σ.
b. Find $P(\mu - 2\sigma \leq Y \leq \mu + 2\sigma)$.
c. Find $P(Y \geq \mu)$ and $P(Y > \mu)$.

5.30 Suppose the random variable X has the probability distribution given in the table below.

x	2	3	5	7	11	13
$p(x)$	0.15	0.25	0.15	0.10	0.30	0.05

a. Find the mean, variance, and standard deviation of X.
b. Suppose the random variable Y is defined by $Y = 2X + 1$. Find the mean, variance, and standard deviation of Y.
c. Suppose the random variable W is defined by $W = X^2 + 1$. Find the mean, variance, and standard deviation of W.

5.31 Suppose X is a discrete random variable with probability distribution given in the table below.

x	1	2	3	5	8	13	21
$p(x)$	0.05	0.10	0.15	0.20	0.25	0.20	0.05

a. Find the mean, variance, and standard deviation of X.
b. Find the probability that X is more than one standard deviation from the mean.
c. Find $P(X \leq \mu + 2\sigma)$.

Applications

5.32 Manufacturing and Product Development In a quarter-pound bag of red pistachio nuts, some shells are too difficult to pry open by hand. Suppose the random variable X, the number of pistachios in a randomly selected bag that cannot be opened by hand, has the probability distribution given in the table below.

x	0	1	2	3	4	5
$p(x)$	0.500	0.250	0.100	0.050	0.075	0.025

a. Is this a valid probability distribution? Justify your answer.
b. Find the mean, variance, and standard deviation of X.
c. Find the probability that X takes on a value less than the mean.
d. Suppose two bags of pistachios are selected at random. What is the probability that both bags have four or more pistachios that are too difficult to open by hand?

5.33 Marketing and Consumer Behavior Suppose the weight (in carats) of a diamond engagement ring purchased at a certain jewelry store is a random variable with probability distribution given in the table below.

x	0.25	0.50	1.00	1.25	1.50	2.00	2.50
$p(x)$	0.04	0.10	0.50	0.20	0.10	0.05	0.01

(*Source:* www.diamonds.com.)

a. Find the mean weight of a diamond engagement ring.
b. Find the variance and standard deviation of the weight of diamond engagement rings.
c. Find the probability that the next diamond engagement ring purchased has a weight within one standard deviation of the mean.
d. Find the probability that a randomly selected diamond engagement ring purchased has a weight less than one standard deviation above the mean.

5.34 Public Policy and Political Science Approximately 100 children's products are recalled every year. Children's clothing is recalled for a variety of reasons; for example, drawstrings that are too long and pose a hazard, small buttons that may break off and cause choking, and material that fails to meet federal flammability standards. Suppose the number of recalls of children's clothing during a given month is a random variable with probability distribution given in the table below.

x	0	1	2	3	4	5	6
$p(x)$	0.005	0.185	0.275	0.305	0.200	0.020	0.010

(*Source:* Kids in Danger.)

a. Find the mean, variance, and standard deviation of the number of recalls of children's clothing during a given month.
b. Suppose the number of recalls in a given month is at least three. What is the probability that the number of recalls that month will be at least five?
c. If the number of recalls in a given month is above one standard deviation from the mean, the federal government issues a special warning directed toward parents. What is the probability that a special warning will be issued during a given month?

5.35 Public Health and Nutrition The exposure time for a tanning session depends on each individual model of sunlamp and the specific equipment, as well as skin type.[6] Most sessions range from 10 to 30 minutes. Suppose the duration of a tanning session (in minutes) at the Paradise Island Day Spa in Augusta, Georgia, is a random variable with probability distribution given in the table that follows.

x	10	12	15	20	25	30
$p(x)$	0.30	0.25	0.15	0.12	0.10	0.08

a. Find the mean, variance, and standard deviation of the duration of a tanning session time.
b. Find the probability that a randomly selected session has a duration within one standard deviation of the mean.
c. Find the probability that a randomly selected session has duration within two standard deviations of the mean.
d. Suppose a sunlamp lasts for 100 hours. After approximately how many tanning sessions will the sunlamp have to be replaced?

5.36 Sports and Leisure The best-selling basketball sneaker of all time is the Converse 8220 Chuck Taylor 8221 All Star.[7] Suppose Chuck Taylor 8221s are sold in Holabird Sports and are available in sizes from 7 to 13. The size purchased is a random variable with probability distribution given in the table below.

x	7	8	9	10	11	12	13
$p(x)$	0.05	0.08	0.10	0.14	0.28	0.21	0.14

a. Find the mean, variance, and standard deviation of the sneaker size purchased.
b. Find the probability that a randomly selected customer buys a pair of these sneakers with size greater than one standard deviation to the right of the mean.
c. Suppose three randomly selected customers each purchase a pair of these sneakers. What is the probability that exactly two of the three buy size 11 sneakers?

5.37 Psychology and Human Behavior A certain elevator in Tampa's tallest office building, 100 North Tampa, is used heavily between 8:00 A.M. and 9:00 A.M. as employees arrive for work. Suppose the number of people who board the elevator on the ground floor going up is a random variable with probability distribution given in the table below.

x	1	2	3	4	5	6
$p(x)$	0.002	0.010	0.050	0.060	0.080	0.090

x	7	8	9	10	11	12
$p(x)$	0.100	0.120	0.140	0.150	0.150	0.048

a. Find μ, σ^2, and σ.
b. For a randomly selected elevator ride from the ground floor going up, what is the probability that the number of riders is within one standard deviation of the mean?
c. For two randomly selected elevator rides from the ground floor going up, what is the probability that both trips have a number of riders more than two standard deviations from the mean?

Extended Applications

5.38 Manufacturing and Product Development A cordless drill has several torque settings for driving different screws into different materials. A manufacturer models the torque setting required for a randomly selected task with a probability distribution given by

$$p(x) = \frac{(x - 12)^2}{247}, \quad x = 1, 5, 10, 15, 20$$

a. Verify this is a valid probability distribution.
b. Find the mean, variance, and standard deviation of X.
c. A torque setting is classified as *rare* if it is more than one standard deviation from the mean. Find the probability of a task requiring a rare torque setting.

5.39 Business and Management The number of songs on a commercial compact disc recorded by an artist is a random variable with probability distribution given in the following table.

x	10	11	12	13	14	15	16
$p(x)$	0.075	0.115	0.275	0.225	0.205	0.065	0.040

(*Source:* Digital Intelligence Center.)

a. Find the mean, variance, and standard deviation of the number of songs on a randomly selected CD.
b. Find $P(X \geq \mu - \sigma)$.
c. Suppose the cost of a CD, Y, is a random variable related to the number of songs on the CD by the formula $Y = 2X - 9$. Find the expected cost, variance, and standard deviation of the cost of a randomly selected CD.

5.40 Marketing and Consumer Behavior While the temperature range of most household ovens is approximately 200–600 °F, most consumers only use four or five common settings. Suppose the probability distribution for the oven temperature setting for a randomly selected use is given in the table below.

x	300	325	350	375	400	500
$p(x)$	0.040	0.205	0.400	0.075	0.200	0.080

a. Find the mean, variance, and standard deviation of the oven temperature settings.
b. Suppose three different uses are randomly selected. Find the probability that the temperature settings for all three are at least 400 °F.
c. Suppose three different uses are randomly selected. Find the probability that exactly one use is for 350 °F.

5.41 Education and Child Development An elementary school class rarely remains the same size from the beginning of the school year until the end. Families move in and out of the district, some students are reassigned, and scheduling conflicts necessitate changes. Suppose the change in the number of students in a class is a random variable with probability distribution given by

$$p(x) = \frac{|x| + 1}{19}, \quad x = -3, -2, -1, 0, 1, 2, 3.$$

a. Verify this is a valid probability distribution.
b. Find the mean, variance, and standard deviation of the change in class size.
c. Suppose two classes are selected at random. Find the probability that both classes remain the same size for the entire year.

Challenge

5.42 Dichotomous Random Variable Suppose the random variable X takes on only two values, according to the probability distribution given below.

x	0	1
$p(x)$	0.4	0.6

a. Find the mean, variance, and standard deviation of X.
b. Suppose $P(X = 1) = 0.7$, and therefore $P(X = 0) = 1 - 0.7 = 0.3$. Find the mean, variance, and standard deviation of X.
c. Suppose $P(X = 1) = 0.8$. Find the mean, variance, and standard deviation of X.

d. Suppose $P(X = 1) = p$ and $P(X = 0) = 1 - p = q$. Find the mean, variance, and standard deviation of X in terms of p and q.
e. For what values of p (and q) is the variance of X greatest?

5.43 Linear Function Suppose X is a discrete random variable with mean μ_X and variance σ_X^2. Let Y be a linear function of X, such that $Y = aX + b$, where a and b are constants. Find the mean and variance of Y in terms of μ_X and σ_X^2.

5.44 Variance Computation Formula Suppose X is a discrete random variable that takes on a finite number of values. Prove the variance *computation formula*. That is, show

$$E[(X - \mu)^2] = E(X^2) - \mu^2.$$

Hint: Write $E[(X - \mu)^2]$ as a sum using the probability mass function *p(x)*. Expand and simplify.

5.45 Standardization Suppose X is a discrete random variable with mean μ_X and variance σ_X^2. Let Y be defined in terms of X by

$$Y = \frac{X - \mu_X}{\sigma_X}.$$

Find the mean, variance, and standard deviation of Y.

5.4 THE BINOMIAL DISTRIBUTION

In the previous sections, the general definition and probability distribution for a discrete random variable were introduced. This section presents a specific discrete random variable that is common and very important. The **binomial random variable** can be used to model many real-world populations and to do more formal inference.

As with any random variable, there is a related experiment in the background. Consider the following experiments (and look for similarities):

1. Toss a coin 50 times and record the sequence of heads and tails.

2. In a random sample of 100 voters, ask each one whether he or she is going to vote for a particular candidate. Record the sequence of yes and no responses.

3. Select a random sample of 25 customers at a fast-food restaurant and record whether or not each pays with exact change.

4. Drill a series of randomly selected test oil wells. Each well will either yield oil worth drilling or be classified as dry. Record the result for each well.

There are four common properties in all of these experiments. These properties are used to describe a **binomial experiment** and are necessary in order to define a **binomial random variable**.

PROPERTIES OF A BINOMIAL EXPERIMENT

1. The experiment consists of n identical trials.
2. Each trial can result in only one of two possible (mutually exclusive) outcomes. One outcome is usually designated a success (S) and the other a failure (F).
3. The outcomes of the trials are independent.
4. The probability of a success, p, is constant from trial to trial.

💡 **ILLUMINATING THE CONCEPTS**

1. A trial is a small part of the larger experiment. A trial results in a single occurrence of either a success or a failure. For example, flipping a coin once, or drilling one test oil well, is a single trial. A typical binomial experiment might consist of $n = 50$ trials.

2. A *success* does not have to be a *good* thing. For example, the experiment may consist of injecting animals with a potential carcinogen and checking for the development of tumors. A *success* might be an animal that develops at least one tumor. *Success* and *failure* could stand for *heads* and *tails*, *acceptable* and *non-acceptable*, or even *dead* and *alive*.

3. Trials are *independent* if whatever happens during one trial has no effect on any other trial. For example, any one voter response has no effect on any other voter response.

4. The probability of a success on every trial is exactly the same. For example, the probability of the tossed (fair) coin landing with head face-up is always 1/2. ■

In a binomial experiment, outcomes consist of sequences of Ss and Fs. For example, SSFSFSFS is a possible outcome in a binomial experiment with $n = 8$ trials.

THE BINOMIAL RANDOM VARIABLE

The binomial random variable maps each outcome in a binomial experiment to a real number and is defined to be the *number of successes* in n trials.

Notation:

Why is P(F) = 1 − *p*?

1. The probability of a success is denoted by p. Therefore, $P(S) = p$ and $P(F) = 1 - p = q$.

2. A binomial random variable X is completely determined by the number of trials n and the probability of a success p. If we know those two values, then we will be able to answer any probability question involving X.

   The shorthand notation $X \sim B(n, p)$ means X is (distributed as) a binomial random variable with n trials and probability of a success p.

   For example, $X \sim B(25, 0.4)$ means X is a binomial random variable with 25 trials and probability of success 0.4.

   Our goal now is to find the probability distribution for a binomial random variable. Given n and p, we want to find the probability of obtaining x successes in n trials, $P(X = x) = p(x)$. We will solve this problem by first considering a simple case with $n = 5$.

For example, suppose five people are selected at random. Let *p* be the probability that a randomly selected person snores.

◀ **Example 5.13 Binomial Experiment with $n = 5$** Consider a binomial experiment with $n = 5$ trials and probability of success p.

a. A typical outcome with two successes is SFFSF. Find the probability of this outcome.
b. Another possible outcome with two successes is FSFSF. Find the probability of this outcome.
c. Compare your results from (a) and (b).
d. Find the probability that $X = 2$ successes.

SOLUTION

a. The probability of this outcome is

$$P(SFFSF) = P(S \cap F \cap F \cap S \cap F)$$

Probability of a success on the first trial *and* a failure on the second trial *and...*

$$= P(S) \cdot P(F) \cdot P(F) \cdot P(S) \cdot P(F)$$

Trials are independent (property of a binomial experiment).

$$= p \cdot (1 - p) \cdot (1 - p) \cdot p \cdot (1 - p)$$ $P(S) = p$, $P(F) = 1 - p$.

$$= p^2(1 - p)^3$$ Multiplication is commutative.

b. The probability of this outcome is

$$P(FSFSF) = P(F) \cdot P(S) \cdot P(F) \cdot P(S) \cdot P(F)$$ Trials are independent.

$$= (1 - p) \cdot p \cdot (1 - p) \cdot p \cdot (1 - p)$$ $P(S) = p$, $P(F) = 1 - p$.

$$= p^2(1 - p)^3$$ Multiplication is commutative.

c. The results are identical. Every other outcome with two successes, and therefore three failures, has exactly the sample probability, $p^2(1 - p)^3$. Therefore, the probability of an outcome depends on the *number of successes* (and failures), **not** on the order in which they appear.

d. To compute the probability that $X = 2$ successes, find all the outcomes mapped to a 2, and add the corresponding probabilities. But every outcome that is mapped to a 2 has the *same* probability. So, all we need to know is *how many* outcomes are mapped to a 2.

The number of successes and the number of failures must sum to n, the total number of trials.

$$P(X = 2) = (\text{Number of outcomes with 2 successes})p^2(1 - p)^3$$

◀ Generalizing, suppose $X \sim B(n, p)$. We want to find the probability of obtaining x successes in n trials. The probability of any *single outcome* with x successes, and therefore $n - x$ failures, is $p^x(1 - p)^{n-x}$. The probability of obtaining x successes is

$$P(X = x) = (\text{Number of outcomes with } x \text{ successes})p^x(1 - p)^{n-x}.$$

We need a method for quickly counting the number of outcomes with x successes. Recall: For any positive whole number n, the symbol $n!$ (read "n factorial") is defined by

$$n! = n(n - 1)(n - 2) \cdots (3)(2)(1).$$

In addition, $0! = 1$ (0 factorial is 1). Given a collection of n items, the number of combinations of size x is given by

$$_nC_x = \binom{n}{x} = \frac{n!}{x!(n - x)!}.$$

The number of outcomes with x successes is determined using combinations. Suppose

Can you figure out why this is true?

$X \sim B(n, p)$. The number of outcomes with x successes is $\binom{n}{x}$. We can now write an expression for the probability of obtaining x successes in n trials. ▶

THE BINOMIAL PROBABILITY DISTRIBUTION

Suppose X is a binomial random variable with n trials and probability of a success p: $X \sim B(n, p)$. Then

$$p(x) = P(X = x) = \underbrace{\binom{n}{x}}_{\substack{\text{Number of outcomes} \\ \text{with } x \text{ successes}}} \underbrace{p^x(1 - p)^{n-x}}_{\substack{\text{Probability of } x \text{ successes and } n - x \\ \text{failures in any single outcome}}}, \quad x = 0, 1, 2, 3, \ldots, n. \qquad (5.6)$$

Example 5.14 Digital versus Film Cameras Seventy-five percent of all cameras sold at New York Camera and Video are digital. The remaining 25% are traditional film cameras. Suppose 10 customers buying cameras at this store are selected at random.

a. Find the probability that exactly 6 customers will buy digital cameras.

b. Find the probability that at least 7 customers will buy digital cameras.

SOLUTION

Let X be the number of customers (out of the 10 selected) who purchase digital cameras. Since the experiment exhibits the properties of a binomial experiment, $X \sim B(10, 0.75)$ ($n = 10$, $p = 0.75$).

a. Translate the words into a probability statement involving the random variable X. *Exactly* 6 means $X = 6$.

Use Equation 5.6 to find the relevant probability.

$$P(X = 6) = \binom{10}{6}(0.75)^6\,(1 - 0.75)^{10-6}$$ *Equation 5.6.*

$$= (210)(0.75)^6(0.25)^4$$ *Compute $_{10}C_6$.*

$$= 0.1460$$

The probability that exactly 6 of the 10 randomly selected customers will purchase digital cameras is 0.1460.

b. The probability that X is greater than or equal to 7 means the probability that X is 7 or 8 or 9 or 10.

$$P(X \geq 7) = P(X = 7 \text{ or } X = 8 \text{ or } X = 9 \text{ or } X = 10)$$
$$= P(X = 7) + P(X = 8) + P(X = 9) + P(X = 10)$$

Or means union; the outcomes are disjoint.

$$= \binom{10}{7}(0.75)^7(0.25)^3 + \binom{10}{8}(0.75)^8(0.25)^2$$
$$+ \binom{10}{9}(0.75)^9(0.25)^1 + \binom{10}{10}(0.75)^{10}(0.25)^0$$

Use Equation 5.6 four times.

$$= 0.2503 + 0.2816 + 0.1877 + 0.0563$$

Compute combinations and powers, and multiply.

$$= 0.7759$$

The probability that at least 7 customers will buy digital cameras is 0.7759. ●

Note:

a. The two most important elements for solving this problem are (1) the probability distribution and (2) the probability statement.

b. Often, the properties of a binomial experiment will not be explicitly stated in the problem. Usually we must read into the problem to *see* the n trials, to *identify* a success, to *recognize* independence, and to *presume* that the probability of a success remains constant from trial to trial.

💡 **ILLUMINATING THE CONCEPTS**

1. Even for small values of n, many of the probabilities associated with a binomial random variable are a little tedious to calculate and are subject to lots of round-off error. Technology helps, and Table I in the Appendix presents **cumulative probabilities** for a binomial random variable, for various values of n and p.

2. Cumulative probability is an important concept. If $X \sim \mathrm{B}(n, p)$, the probability that X takes on a value *less than or equal to x* is cumulative probability. Accumulate all the probability associated with values up to and including x. Symbolically, cumulative probability is

$$P(X \leq x) = \sum_{k=0}^{x} P(X = k) \tag{5.7}$$
$$= P(X = 0) + P(X = 1) + P(X = 2) + \cdots + P(X = x)$$

3. Graphically, cumulative probability is like standing on a special staircase, looking down (or back), and measuring the height. The steps are labeled $0, 1, 2, \ldots, n$, the height of step x is $P(X = x)$, and the total height of the staircase is 1. Figure 5.4 illustrates $P(X \leq 3)$.

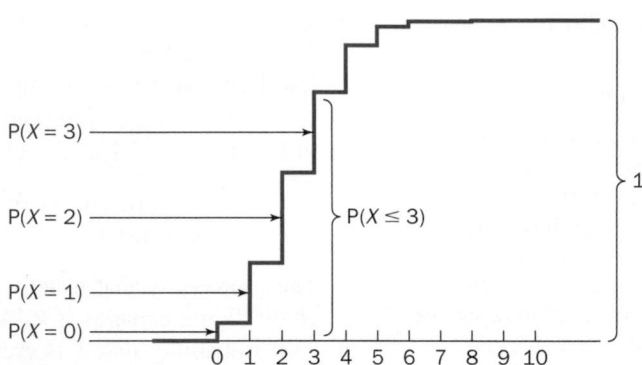

Figure 5.4 Staircase analogy to cumulative probability.

The number of steps is $n + 1$, and the height of each step depends on n and p. In this example, $n = 10$ and $p = 0.25$. The largest steps (the highest probabilities) are associated with $X = 1, 2,$ and 3. Steps 7, 8, 9, and 10 are hard to see since $P(X = 7)$, $P(X = 8)$, $P(X = 9)$, and $P(X = 10)$ are so small. ■

Every probability question about a binomial random variable can be answered using *cumulative* probability. There may also be other, faster methods, but cumulative probability always works. The following example illustrates some of the techniques for converting to and using cumulative probability.

Solution Trail

KEYWORDS

■ 48%.
■ 20 Americans.
■ Randomly selected.

TRANSLATION

■ $p = 0.48$.
■ $n = 20$.

CONCEPTS

Probabilities associated with a binomial distribution.

VISION

Although it is not explicitly stated, there is a binomial experiment in the background. n and p are given, and the four characteristics of a binomial experiment are assumed to apply. Identify the probability distribution and write each probability question in terms of the random variable. Use appropriate probability rules.

Example 5.15 Numismatics Approximately 48% of all Americans are saving state quarters as they are minted and distributed.[8] Suppose 20 Americans are randomly selected.
a. Find the probability that at most 8 collect state quarters.
b. Find the probability that exactly 10 collect state quarters.
c. Find the probability that at least 7 collect state quarters.
d. Find the probability that between 5 and 11 (inclusive) collect state quarters.

SOLUTION

Let X be the number of Americans (out of the 20 selected) who collect state quarters. X is a binomial random variable with $n = 20$ and $p = 0.48$: $X \sim \mathrm{B}(20, 0.48)$.
a. The probability that at most 8 collect state quarters

$= P(X \leq 8)$ Translate the words into mathematics.
$= 0.3127$ Cumulative probability; use Table I in the Appendix.

b. The probability that exactly 10 collect state quarters

$$
\begin{aligned}
&= P(X = 10) &&\text{Translate the words into mathematics.}\\
&= P(X \le 10) - P(X \le 9) &&\text{Convert to cumulative probability.}\\
&= 0.6568 - 0.4834 &&\text{Use Table I in the Appendix.}\\
&= 0.1734
\end{aligned}
$$

This solution may also be found by using the probability mass function for a binomial random variable (Equation 5.6). In addition, most statistical software can compute binomial probabilities for single values.

c. The probability that at least 7 collect state quarters

$$
\begin{aligned}
&= P(X \ge 7) &&\text{Translate the words into mathematics.}\\
&= 1 - P(X < 7) &&\text{The Complement Rule.}\\
&= 1 - P(X \le 6) &&\text{The first value } X \text{ takes on \textit{less than} 7 is 6.}\\
&= 1 - 0.0814 &&\text{Use Table I in the Appendix.}\\
&= 0.9186
\end{aligned}
$$

d. The probability that between 5 and 11 (inclusive) collect state quarters

$$
\begin{aligned}
&= P(5 \le X \le 11) &&\text{Translate the words into mathematics.}\\
&= P(X \le 11) - P(X \le 4) &&\text{Convert to cumulative probability.}\\
&= 0.8023 - 0.0096 &&\text{Use Table I in the Appendix}\\
&= 0.7927
\end{aligned}
$$

Figure 5.5 through 5.8 show technology solutions:

```
binomcdf(20,.48,
8)
            .3127
```

```
binompdf(20,.48,
10)
            .1734
```

Figure 5.5 $P(X \le 8)$; cumulative probability.

Figure 5.6 $P(X = 10)$; using the probability mass function.

```
binomcdf(20,.48,
6)
            .0814
1-Ans
            .9186
```

```
binomcdf(20,.48,
11)-binomcdf(20,
.48,4)
            .7927
```

Figure 5.7 $P(X \ge 7)$; using cumulative probability.

Figure 5.8 $P(5 \le X \le 11)$; using cumulative probability.

The next example shows how the binomial distribution can be used to make an inference.

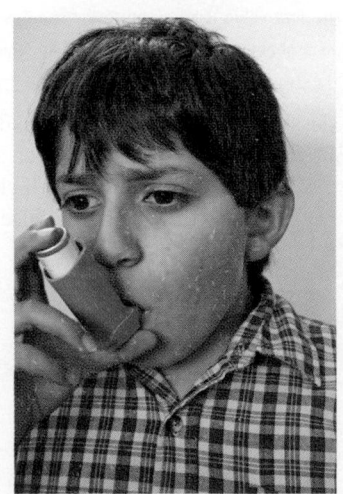

(Pelvidge/Dreamstime.com)

Example 5.16 Adverse Reactions Albuterol aerosol is manufactured by Warrick Pharmaceuticals, Ltd., and is used to treat and prevent wheezing, shortness of breath, and conditions caused by asthma. Based on a clinical trial,[9] Warrick claims that only 10% of patients at least 12 years of age using this medication will experience tachycardia, or a rapid heart rate, which is considered an adverse reaction. Suppose 30 people (at least 12 years old) who need this medication are selected at random. Each is given albuterol aerosol, and the number of people who experience tachycardia is recorded.

a. Find the probability that at most 1 person will experience tachycardia.

b. Suppose 7 people experience tachycardia. Is there any evidence to suggest that the company's claim is wrong? Justify your answer.

SOLUTION

a. Let X be the number of people (out of the 30 selected) who experience tachycardia after taking the medication. X is a binomial random variable with $n = 30$ and $p = 0.10$: $X \sim B(30, 0.10)$.

Translate the words into a mathematical probability statement, convert to cumulative probability if necessary, and use Table I in the Appendix.

The probability that at most 1 person will experience tachycardia

$$= P(X \le 1)$$ Translate the words into mathematics.
$$= 0.1837$$ Already cumulative probability; use Table I in the Appendix.

The probability of at most 1 person experiencing tachycardia is 0.1837.

b. Warrick Pharmaceuticals, Ltd. claims $p = 0.10$. This implies the random variable X has a binomial distribution with $n = 30$ and $p = 0.10$.

Claim: $p = 0.10 \rightarrow X \sim B(30, 0.10)$.

The experimental outcome is: seven people experience tachycardia.

Experimental outcome: $x = 7$.

It seems reasonable to consider $P(X = 7)$ and draw a conclusion based on this probability. However, in order to be conservative (to give the person making the claim the benefit of the doubt), we always consider a *tail probability*. We accumulate the probability in a tail of the distribution, and if it is small, then there is evidence to suggest the claim is false.

So, which tail? It depends on the mean of the distribution (and later on, the alternative hypothesis). Formulas for the mean, variance, and standard deviation of a binomial random variable are given below. Intuitively, however, the mean of a binomial random variable is $\mu = np$. If $n = 30$ and $p = 0.10$, we expect to see $\mu = (30)(0.10) = 3$ people experience bad side effects. Since $x = 7$ is to the right of the mean, we'll consider a right-tail probability. See Figure 5.9.

Likelihood:
$$P(X \ge 7) = 1 - P(X < 7)$$ The Complement Rule.
$$= 1 - P(X \le 6)$$ The first value X takes on *less than 7* is 6.
$$= 1 - 0.9742$$ Cumulative probability; use Table I in the Appendix.
$$= 0.0258$$

Conclusion: Since this tail probability is so small (less than 0.05), it is very unusual to observe 7 or more people with tachycardia. But it happened! This is either an incredibly lucky occurrence, or someone is lying. We usually discount the lucky possibility, and conclude that there is evidence to suggest Warrick's claim is false.

Figure 5.9 A portion of the probability histogram for the random variable X. The right-tail probability $P(X \ge 7)$ is the sum of the heights of the rectangles above 7, 8, 9, . . . , 30.

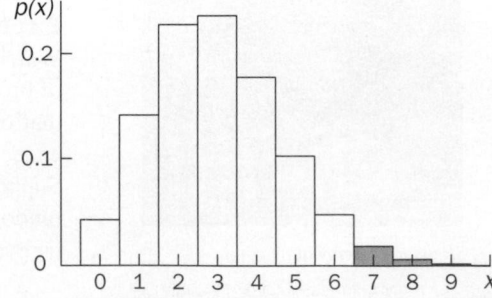

---

a. Solution Trail

KEYWORDS

- 10%.
- 30 people.
- At most one.

↓

TRANSLATION

- $p = 0.10$.
- $n = 30$.
- $X \le 1$.

↓

CONCEPTS

Binomial random variable.

↓

VISION

There is a binomial experiment in the background. There are $n = 30$ trials with two outcomes (tachycardia or no tachycardia), the trials are independent (random sample), and the probability of a success (experience tachycardia) is constant from trial to trial. Define a random variable, identify its probability distribution, and write an appropriate probability statement.

---

b. Solution Trail

KEYWORDS

Is there any evidence?

↓

TRANSLATION

Use the experimental outcome to draw a conclusion concerning the Warrick's claim.

↓

CONCEPTS

Inference procedure.

↓

VISION

In order to decide whether seven people experiencing tachycardia is reasonable, we need to follow the four-step inference procedure. This process now involves a random variable. Consider the assumption (claim), the experiment, and the likelihood of the experimental outcome.

Technology solutions:

Figure 5.10 $P(X \le 1)$; cumulative probability.

Figure 5.11 $P(X \ge 7)$; using cumulative probability.

A random variable is often described, or characterized, by its mean and variance (or standard deviation): μ and σ^2 (or σ). If we know μ and σ, we can use Chebyshev's Rule to determine the most likely values of the random variable. For *any* population (random variable), most (at least 89%) of the values are within three standard deviations of the mean. This fact provides another approach to statistical inference, for determining the likelihood of an experimental outcome.

To find the mean and variance of a binomial random variable, we could use the mathematical definitions (Equations 5.1 and 5.3). These formulas are used to produce the general results below. However, the mean is intuitive. Consider a binomial random variable $X \sim B(10, 0.5)$. We *expect* to see $(10)(0.5) = 5 = np$ successes in 10 trials. (Think about tossing a fair coin 10 times.) Similarly, if $X \sim B(100, 0.75)$, we expect to see $100(0.75) = 75 = np$ successes. The mean of a binomial random variable with n trials and probability of a success p is $\mu = np$.

MEAN, VARIANCE, AND STANDARD DEVIATION OF A BINOMIAL RANDOM VARIABLE

If X is a binomial random variable with n trials and probability of a success p, $X \sim B(n, p)$, then
$$\mu = np, \quad \sigma^2 = np(1 - p), \quad \text{and} \quad \sigma = \sqrt{np(1 - p)}. \quad (5.8)$$

Given a binomial random variable, n, and p, we know the mean, variance, and standard deviation immediately. There is no need to create a table of values and probabilities, nor to use the formulas to find μ and σ^2. Here is an example to illustrate the use of this concept.

Example 5.17 Impulse Buying Research concerning purchasing habits at the grocery store suggests that impulse buys make up approximately 60% of all purchases.[10] Suppose 100 customers are selected at random.
a. Find the mean, variance, and standard deviation of the number of customers who make an impulse purchase.
b. Suppose 55 of the 100 customers make an impulse purchase. Is there any evidence to suggest that the study's claim is false? Justify your answer.

SOLUTION

a. Let X be the number of people (out of the 100 selected) who make an impulse purchase. X is a binomial random variable with $n = 100$ and $p = 0.60$: $X \sim B(100, 0.60)$.

---

a. Solution Trail

KEYWORDS

- 60%.
- 100 customers.

↓

TRANSLATION

- $p = 0.60$.
- $n = 100$.

↓

CONCEPTS

Binomial random variable, mean, variance, standard deviation.

↓

VISION

There is a binomial experiment in the background. There are $n = 100$ trials with two outcomes (impulse purchase or no impulse purchase), the trials are independent (random sample), and the probability of a success (impulse purchase) is constant from trial to trial. Define a random variable, identify its probability distribution, and use the formulas to find the mean, variance, and standard deviation.

b. Solution Trail

KEYWORDS
Is there any evidence?

↓

TRANSLATION
Use the experimental outcome to draw a conclusion concerning the grocery store's claim.

↓

CONCEPTS
Inference procedure; most likely values of a binomial random variable.

↓

VISION
Follow the four-step inference procedure. This time, use the mean and standard deviation to determine the most likely values of the random variable. If the number of customers who make an impulse buy is reasonable, then there is no reason to doubt the claim. If the number of customers who make an impulse buy is very unlikely (very far away from the mean, in standard deviations), then there is a reason to doubt the claim.

Use Equation 5.8 to find the mean, variance, and standard deviation.

$$\mu = np = (100)(0.60) = 60$$
$$\sigma^2 = np(1 - p) = (100)(0.60)(1 - 0.60) = (100)(0.60)(0.40) = 24$$
$$\sigma = \sqrt{\sigma^2} = \sqrt{24} \approx 4.9$$

The expected number of impulse buyers is 60, with a variance of 24 and a standard deviation of approximately 4.9.

b. The grocery store claims $p = 0.60$. This implies that the random variable X has a binomial distribution with $n = 100$ and $p = 0.60$.

Claim: $p = 0.60 \rightarrow X \sim B(100, 0.60)$.

The experimental outcome is: 55 people make impulse buys.

Experimental outcome: $x = 55$.

Likelihood:
From part (a), $\mu = 60$ and $\sigma = 4.9$. Most observations are within three standard deviations of the mean. Therefore, most values of X are in the interval

$$(\mu - 3\sigma, \mu + 3\sigma) = (60 - 3(4.9), 60 + 3(4.9))$$
$$= (60 - 14.7, 60 + 14.7)$$
$$= (45.3, 74.7)$$

Conclusion: Since 55 lies in this interval, it is a reasonable observation. There is no evidence to lead us to doubt the claim of $p = 0.60$.   ●

💡 ILLUMINATING THE CONCEPTS

1. Recall: In statistics, we usually measure *distance* in standard deviations, not miles, feet, inches, meters, or other units. We often want to know how many standard deviations from the mean is a given observation.

2. The inference problem in part (b) of the example above could also have been answered by using the *tail probability* approach. (Try it!) This method leads to the same conclusion and is generally more precise than constructing an interval about the mean. A more formal process for checking claims (hypothesis tests) is introduced in Chapter 9.

3. Whenever we test a claim, there are only two possible conclusions:
   a. There *is* evidence to suggest the claim is false.
   b. There *is no* evidence to suggest the claim is false.

   Note, in either case, we never state with absolute certainty that the claim is true or the claim is false. This is because we never look at the *entire* population, only at a sample. With a large, random (representative) sample, we can be pretty confident in our conclusion, but never absolutely sure. ■

TECHNOLOGY CORNER

Procedure: Compute probabilities associated with a binomial random variable.
Reconsider: Example 5.15 (page 210), solution, and interpretations.

TI-84 Plus

Suppose $X \sim B(n, p)$. There are built-in functions to compute a value of the probability mass function (the probability that X takes on a single value) and cumulative probability.

1. For cumulative probability, use $\boxed{\text{DISTR}}$; DISTR; `binomcdf`. $P(X \le x) = $ `binomcdf(n,p,x)`. See Figure 5.5 (page 211).
2. To find the probability that X takes on a single value, use $\boxed{\text{DISTR}}$; DISTR; `binompdf`. $P(X = x) = $ `binompdf(n,p,x)`. See Figure 5.6 (page 211).

Minitab

There are built-in functions accessed via input windows or the command language to compute a value of the probability mass function (the probability that X takes on a single value) and cumulative probability. The command language may be necessary to perform additional calculations involving probabilities.

1. Select Calc; Probability Distributions; Binomial. Choose Cumulative probability. Enter the Number of trials (n), the Event probability (p); and the Input constant (x). See Figure 5.12.
2. To find the probability that X takes on a single value, Select Calc; Probability Distributions; Binomial. Choose Probability. Enter the Number of trials (n), the Event probability (p); and the Input constant (x). See Figure 5.13.

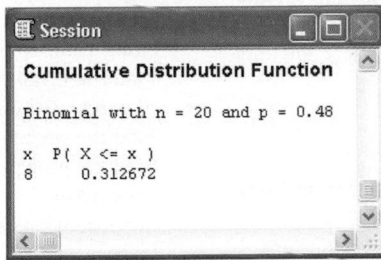

Figure 5.12 Cumulative probability.

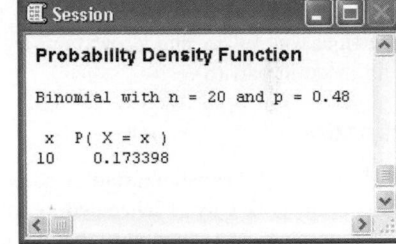

Figure 5.13 The probability density that $X = 10$.

Excel

There is a single built-in function to find the probability that X takes on a single value or cumulative probability. The last argument of the function is either `True` for cumulative probability or `False` for the probability mass function. Additional spreadsheet calculations may be necessary to find the final answer.

1. To find cumulative probability, use the function `BINOMDIST`. Enter x, n, p, and `True`. See Figure 5.14.
2. To find the probability that X takes on a single value, use `BINOMDIST`. Enter x, n, p, and `False`. See Figure 5.14.

	A	B
1	0.3127	=BINOMDIST(8,20,0.48,TRUE)
2	0.1734	=BINOMDIST(10,20,0.48,FALSE)

Figure 5.14 Excel function for finding cumulative probability and for evaluating the probability mass function.

SECTION 5.4 EXERCISES

Practice

5.46 Suppose $X \sim B(15, 0.25)$. Find the following probabilities.
a. $P(X \le 2)$ **b.** $P(X < 2)$
c. $P(X = 7)$ **d.** $P(X > 6)$
e. $P(3 \le X \le 10)$

5.47 Suppose $X \sim B(20, 0.40)$. Find the following probabilities.
a. $P(X \ge 12)$ **b.** $P(X \ne 10)$
c. $P(X \le 15)$ **d.** $P(2 < X \le 8)$

5.48 Suppose $X \sim B(25, 0.70)$. Find the following probabilities.
a. $P(X \ge 1)$ **b.** $P(X \ge 10)$
c. $P(X \ge 17.5)$ **d.** $P(10.1 \le X \le 19)$

5.49 Suppose X is a binomial random variable with $n = 25$ and $p = 0.80$.

a. Find the mean, variance, and standard deviation of X.

b. Find the probability that X is within one standard deviation of the mean.

c. Find the probability that X is more than two standard deviations from the mean.

5.50 Suppose X is a binomial random variable with $n = 30$ and $p = 0.40$.

a. Find the mean, variance, and standard deviation of X.

b. Find the intervals $\mu \pm \sigma$, $\mu \pm 2\sigma$, and $\mu \pm 3\sigma$.

c. Find $P(X > \mu + 3\sigma)$.

d. Find $P(X \leq \mu - 2\sigma)$.

5.51 Suppose X is a binomial random variable with $n = 10$ and $p = 0.50$.

a. Create a table of values of X and associated probabilities. (*Hint:* This is very quick and easy using technology.)

b. Use the table in part (a) and the definitions of expected value and variance (Equations 5.1 and 5.3) to find μ, σ^2, and σ.

c. Use Equation 5.8 to find μ, σ^2, and σ. Check these answers with those in part (b).

Applications

5.52 Travel and Transportation A commuter train from Long Island to New York City is scheduled to arrive at 8:07 A.M. every weekday morning. The probability that the train arrives on time is 0.90.[11] Suppose 20 weekdays are randomly selected.

a. Find the probability that the train will arrive on time on at most 15 days.

b. Find the probability that the train will arrive on time on at least 12 days.

c. Find the expected number of days that the train will arrive on time.

d. Suppose the train arrives on time on at least 15 days. What is the probability that it arrives on time on all 20 days?

5.53 Fuel Consumption and Cars The battery manufacturer Varta sells a car battery with 800 cold-cranking amps and advertises great performance even in biting cold weather. Varta claims that after sitting on a frozen Minnesota lake for 10 days at temperatures below 32 °F, this battery will still have enough power to start a car. Suppose the actual probability of starting a car following this experiment is 0.75, and 15 randomly selected cars (equipped with this battery) are subjected to these grueling conditions.

a. Find the probability that fewer than 10 cars will start.

b. Find the probability that more than 12 cars will start.

c. Suppose 9 cars actually start. Is there any evidence to suggest that the probability of starting a car is different from 0.75? Justify your answer.

5.54 Marketing and Consumer Behavior Levain Bakery, on West 74th Street in New York City, is trying to determine the number of loaves of raisin bread to make each day. Over the past few months the store has baked 50 loaves each day and has sold out with probability 0.80. Suppose the owner continues this practice and 30 days are selected at random.

a. What is the expected number of days on which all 50 loaves will be sold?

b. Find the probability of selling all 50 loaves on at least 20 days.

c. Find the probability of selling all 50 loaves on at most 18 days.

5.55 Sports and Leisure A Six Flags Great Adventure Theme Park now offers a wild safari drive-thru with more than 1000 animals roaming freely on over 400 acres. The park claims that the probability of some car damage due to an animal during a safari drive-thru is 0.60. Suppose 20 cars are selected at random.

a. Find the probability that exactly 10 cars will be damaged.

b. Find the probability that at least 15 cars will be damaged.

c. Find the probability that no more than 12 cars will be damaged.

d. Suppose 19 cars are damaged. Is there any evidence to suggest that the claim of 0.60 is false? Justify your answer.

5.56 Public Health and Nutrition Parents tend to be very good at diagnosing their children's routine medical problems, such as ear infections, sinus infections, or strep throat. If an ailment is correctly identified, a trip to the doctor's office may be avoided. A physician may confer with a parent by telephone and simply call a pharmacy with a prescription for an antibiotic. Suppose parents are correct 90% of the time and 50 families with a child suffering from some minor illness are selected at random.

a. Find the mean, variance, and standard deviation of the number of parents who correctly identify their child's illness.

b. Find the probability that at least 42 parents are correct.

c. Find the probability that between 42 and 47 (inclusive) parents are correct.

d. Suppose 41 parents are actually correct. Is there any evidence to suggest that the claim of 90% of parents being correct is false? Justify your answer.

5.57 Public Policy and Political Science A building inspector enforces building, electrical, mechanical, plumbing, and energy code requirements for the safety and health of people in a certain city, county, or state. In Santa Cruz County, the probability that a building inspector will find at least one code violation at a commercial building is 0.25. Suppose 30 commercial buildings are selected at random.

a. Find the mean, variance, and standard deviation of the number of commercial buildings with at least one violation.

b. Find the probability that the number of commercial buildings with at least one violation will be within one standard deviation of the mean.

c. Find the probability that the number of commercial buildings with at least one violation will be more than two standard deviations from the mean.

d. Suppose the actual number of commercial buildings with at least one violation is 10. Is there any evidence to suggest that code violations are found in more than 25% of commercial buildings? Justify your answer.

5.58 Public Policy and Political Science During a major disaster, for example, a hurricane, earthquake, or blizzard, it is very important to have certain recommended household items on hand. In a recent survey, only 47% of Americans had a three-day supply of bottled water in case of emergency.[12] Suppose a sample of 35 Americans is selected at random.

a. What is the probability that exactly 12 Americans have a three-day supply of bottled water on hand?

b. What is the probability that at least 18 Americans have a three-day supply of bottled water on hand?

c. Suppose 10 Americans have a three-day supply of bottled water on hand. Is there any evidence to suggest that the proportion of Americans who have a three-day supply of water on hand is different from 0.47? Justify your answer.

5.59 Manufacturing and Product Development A manufacturer of rewritable compact discs claims that only 1 out of every 50 is defective. A large retailer receives a shipment of 100,000 CDs from this manufacturer. In order to check the manufacturer's claim, the retailer selects a random sample of 100 CDs and thoroughly inspects each one.
a. What is the probability of finding no defective CDs?
b. What is the expected number of defective CDs?
c. What is the probability of finding more than the expected number of defective CDs?
d. Suppose 6 defective CDs are actually found. Is there any evidence to suggest that the manufacturer's claim is false? Justify your answer.

5.60 Physical Sciences As showcased on many television dramas, new technology allows crime investigators to use DNA evidence in order to convict a suspect. However, even if a lab is able to match crime-scene DNA with a existing genetic profile, a conviction is not guaranteed. According to the Virginia Department of Forensic Science, as of November 2005, only 597 of 3000 matches led to convictions. Suppose 40 new DNA matches are selected at random and that the probability of a conviction in each case is 597/3000.
a. What is the probability of at most five convictions?
b. What is the probability of between four and eight convictions?
c. Suppose 14 of the 40 matches lead to convictions. Is there any evidence to suggest that the probability of a conviction has increased? Justify your answer.

5.61 Public Health and Nutrition According to Health Grades, Inc. (November 27, 2008), the probability that a person catches a cold during a calendar year is 0.2279. A report in *Medical News Today* (June 26, 2007) suggests that using the herbal product echinacea reduces the chance of catching a cold by 58%. Thirty people are randomly selected and each regularly drinks echinacea tea.
a. If echinacea has no effect, find the mean, variance, and standard deviation for the number of people who will get colds.
b. If echinacea has no effect, find the probability that more than 10 people get colds.
c. If echinacea has no effect, find the probability that exactly 14 people get colds.
d. If only 5 people get colds, is there any evidence to suggest that using echinacea reduces the chance of a cold? Justify your answer.

5.62 Business and Management As of November 2008, 40% of all privately held business in the United States were owned by women.[13] Twenty-five nonfarm, private businesses are randomly selected and the owner of each is noted.
a. Find the mean, variance, and standard deviation of the number of businesses (out of the 25 selected) owned by women.
b. Construct intervals one, two, and three standard deviations from the mean.
c. Suppose 7 businesses (out of the 25 selected) are owned by women. How many standard deviations from the mean is this observation? What does this *distance* measure indicate about the likelihood of observing 7 businesses owned by women?

5.63 Economics and Finance The 2007 PNC Wealth Management Survey indicated that of those Americans with an income of at least $500,000, 44% were pessimistic about the economy in 2008.[14] In order to check this claim, 50 Americans with an income of at least $500,000 were randomly selected and asked whether they feel pessimistic about the economy in 2008.
a. If the claim is true, find the probability that exactly 30 Americans are pessimistic about the economy.
b. If the claim is true, find the probability that more than 23 Americans are pessimistic about the economy.
c. Suppose 17 Americans are pessimistic about the economy. Is there any evidence to suggest that the claim is false? Justify your answer.

Extended Applications

5.64 Business and Management In the movie *Lethal Weapon*, the character played by Joe Pesci was concerned about drive-thru windows at fast-food restaurants. Suppose the probability that an order from a drive-thru window at a fast-food restaurant will be filled correctly is 0.75. Twenty orders are selected at random.
a. What is the probability that exactly 15 orders will be filled correctly?
b. What is the probability that at most 12 orders will be filled correctly?
c. What is the probability that between 10 and 14 (inclusive) orders will be filled correctly?
d. Suppose *two* groups of 20 random orders are independently selected. What is the probability that at least 16 orders will be filled correctly in both groups?

5.65 Marketing and Consumer Behavior According to an article in *redOrbit* (November 15, 2008), high definition Blu-ray players account for approximately 10% of total home video sales. Twenty home video sales are selected at random.
a. Find the probability that at least 5 sales are Blu-ray players.
b. Find the expected number, μ, of Blu-ray sales. Find the probability of exactly μ Blu-ray sales.
c. Suppose at most 4 are Blu-ray sales. What is the probability that none are Blu-ray sales?

5.66 Marketing and Consumer Behavior A photo developing center offers standard three-day film development and premium one-hour developing. The probability that a randomly selected customer will ask for one-hour service is 0.25. Suppose 10 customers are selected at random. Let the random variable X be the number of customers who ask for one-hour service.
a. Construct a probability histogram for the random variable X.
b. Find the mean, variance, and standard deviation of X. Indicate the mean on the graph from part (a).
c. Find $P(\mu - \sigma \leq X \leq \mu + \sigma)$ and indicate this probability on the graph from part (a).

5.67 Manufacturing and Product Development A company has developed a very inexpensive explosive-detection machine for use at airports. If an explosive is in a suitcase, the probability of it being detected by this machine is 0.60. Several of these machines will be used simultaneously to independently screen each piece of luggage. Suppose a piece of luggage actually contains an explosive.

a. If three machines screen this luggage, what is the probability that exactly one detects the explosive? What is the probability that none of the three detects the explosive?

b. If four machines screen this luggage, what is the probability that at least one device detects the explosive?

c. If five machines screen this luggage, what is the probability that at least one device detects the explosive?

d. How many machines are necessary for screening in order to be certain that at least one device detects the explosive with probability 0.999 or greater?

5.68 Marketing and Consumer Behavior Nine West sells women's flip flops in eight different colors. Despite the available selection, 50% of all flip flops purchased are white. Suppose 30 buyers are selected at random.

a. Find the mean, variance, and standard deviation of the number of buyers who purchase white flip flops.

b. Find the probability that the number of white flip flops purchased will be within two standard deviations of the mean. Compare this with the predicted result from Chebyshev's Rule.

c. Suppose two groups of 30 customers are independently selected. What is the probability of at least one group having exactly 15 people who buy white flip flops?

5.69 Manufacturing and Product Development US Relays & Technology, Inc., produces reed switches used in a variety of devices and claims the proportion of components that are defective is 0.02. Ark Logix, a manufacturer of security systems, receives a shipment of 100,000 reed switches for use in wireless burglar alarms. Before accepting the entire lot, Ark Logix selects a random sample of 25 components and thoroughly tests each. If 4 or more components are found to be defective, the entire shipment will be sent back. Otherwise the shipment will be accepted.

a. Suppose the claim is true; the actual proportion of defectives is $p = 0.02$. What is the probability that the shipment will be rejected? (This is one type of *error probability*. The company would be making a mistake if this event occurred. They would be rejecting the shipment when the proportion of defectives is as claimed.)

b. Suppose the actual proportion of defectives is $p = 0.05$. What is the probability that the shipment will be accepted? (This is another type of error probability. In this case, the company would also be making a mistake. They would be accepting the shipment when the proportion of defectives is too high.)

c. Suppose the actual proportion of defectives is $p = 0.07$. What is the probability that the shipment will be accepted?

5.70 Psychology and Human Behavior Due to stricter training requirements, fewer big fires, higher paying jobs in cities, and changes in society, the number of volunteer firefighters is declining. According to *USA Today* (November 7, 2005), approximately three-fourths of all firefighters in the United States are volunteers and the total number of volunteer firefighters has decreased by 10% over the past two decades. Suppose 30 U.S. firefighters are selected at random.

a. Find the probability that exactly 22 of the firefighters are volunteers.

b. Find the probability that more than 25 of the firefighters are volunteers.

c. Suppose 17 of the firefighters are volunteers. Is there any evidence to suggest that the proportion of volunteer firefighters has decreased? Justify your answer.

d. Suppose 50 firefighters are selected at random from the West and 50 firefighters are selected from the Northeast. What is the probability that at least 40 of the firefighters in each group will be volunteers?

5.5 OTHER DISCRETE DISTRIBUTIONS

There are many other common discrete probability distributions. This section presents three of these distributions along with brief background, properties, and examples. Remember that many of the problems involving these distributions are solved using the same general technique:

1. Define a random variable and identify its probability distribution (distribution statement).

2. Translate the words into a probability question where the event is stated in terms of the random variable (probability statement).

3. If necessary, try to convert the probability statement into an equivalent expression involving *cumulative* probability. Use tables and technology wherever possible.

The **geometric distribution** is closely related to the binomial distribution. In a binomial experiment, *n* (the number of trials) is fixed and the number of successes varies. The binomial random variable is the number of successes in *n* trials. In a geometric experiment, *the number of successes is fixed* at 1, and the number of trials varies.

PROPERTIES OF A GEOMETRIC EXPERIMENT

1. The experiment consists of identical trials.
2. Each trial can result in only one of two possible outcomes: a success (S) or a failure (F).
3. The trials are independent.
4. The probability of a success, p, is constant from trial to trial.

The experiment ends when the first success is obtained.

THE GEOMETRIC RANDOM VARIABLE

The **geometric random variable** is the number of trials necessary to realize the first success.

Think of an experiment in which you continue to phone a friend until you get through. The number of calls necessary until the first success (reaching your friend) is the value of a **geometric random variable**.

The derivation of the probability distribution involves the properties given above. Let X be a geometric random variable, the number of trials until the first success (including the trial on which the success is obtained). Given p, the probability of a success, find the probability of needing x trials, $P(X = x) = p(x)$.

$$P(X = 1) = P(S) = p.$$

$X = 1$ means the first trial results in a success, and the experiment is over. The probability of a success is simply p.

$$P(X = 2) = P(F \cap S) = P(F) \cdot P(S) = (1 - p)p.$$

$X = 2$ means the first trial is a failure *and* the second trial is a success. Since trials are independent, we multiply the corresponding probabilities.

$$P(X = 3) = P(F \cap F \cap S) = P(F) \cdot P(F) \cdot P(S)$$
$$= (1 - p)(1 - p)p = (1 - p)^2 p$$

$X = 3$ means the first two trials are failures and the third trial is a success. We use independence again, and multiply the corresponding probabilities.

Why isn't FSF a possible outcome in a geometric experiment?

$$P(X = 4) = P(F \cap F \cap F \cap S) = P(F) \cdot P(F) \cdot P(F) \cdot P(S)$$
$$= (1 - p)(1 - p)(1 - p)p = (1 - p)^3 p$$

$X = 4$ means the first three trials are failures and the fourth trial is a success. We use independence again, and multiply the corresponding probabilities.

In general,

$$P(X = x) = \underbrace{P(F) \cdot P(F) \cdots P(F)}_{x - 1 \text{ failures}} \cdot P(S) = \underbrace{(1 - p)(1 - p) \cdots (1 - p)}_{x - 1 \text{ terms}} \cdot p$$
$$= (1 - p)^{x-1} p$$

$X = x$ means the first $x - 1$ trials are failures and the xth trial is the first success. This generalization is the formula for the probability distribution.

THE GEOMETRIC PROBABILITY DISTRIBUTION

Suppose X is a geometric random variable with probability of a success p. Then

$$p(x) = P(X = x) = (1 - p)^{x-1}p, \quad x = 1, 2, 3, \ldots \tag{5.9}$$

$$\mu = \frac{1}{p} \quad \text{and} \quad \sigma^2 = \frac{1 - p}{p^2}. \tag{5.10}$$

ILLUMINATING THE CONCEPTS

1. The geometric random variable is discrete. The number of possible values is countably infinite: 1, 2, 3, ...

2. The geometric distribution is completely characterized, or defined, by one parameter, p.

3. We do not need a table to find cumulative probabilities associated with a geometric random variable because there is an easy formula for computing these values. If X is a geometric random variable with probability of success p, then

$$P(X \le x) = 1 - (1 - p)^x. \tag{5.11}$$

4. Equation 5.9 is a valid probability distribution. ◀ Each probability is between 0 and 1, and the sum of all the probabilities is an *infinite series*. The sum

$$\sum_{x=1}^{\infty} P(X = x) = \sum_{x=1}^{\infty} (1 - p)^{x-1}p$$

is called a *geometric series* and it does sum to 1! ▶ ■

There is a formula for the sum of a geometric series. Can you use it to show that this sum is 1?

Solution Trail

KEYWORDS

First uninsured motorist.

↓

TRANSLATION

First success.

↓

CONCEPTS

Geometric probability distribution.

↓

VISION

This police procedure has the characteristics of a geometric experiment. We assume each trial is stopping a car, P(S) = P(uninsured motorist) = 0.14, P(S) is the same on each trial, and the trials are independent. To solve this problem, define the geometric random variable and write a probability statement for each part.

Example 5.18 Uninsured Drivers   The Insurance Research Council estimates that 14% of all drivers in the United States are uninsured. Suppose the State Police establish a checkpoint and randomly stop cars to inspect the driver's license, registration, and proof of insurance.

a. What is the probability that the third person stopped will be the first uninsured motorist?

b. What is the probability that it will take at least five cars before the first uninsured motorist is found?

SOLUTION

a. Let X be the number of the car in which the first uninsured motorist is found. X is a geometric random variable with $P(S) = 0.14 = p$.

The probability that the third motorist will be the first success:

$$P(X = 3) = (1 - p)^{3-1}p \qquad \text{Equation 5.9.}$$
$$= (1 - 0.14)^2(0.14) = (0.86)^2(0.14) = \boxed{0.1035} \qquad \text{Use } p = 0.14.$$

The probability that the first uninsured motorist will be in the third stopped car is 0.1035.

b. *At least five cars to find the first uninsured motorist* means the first success will occur on the fifth trial or greater.

The probability that at least five cars need to be stopped:

$P(X \geq 5)$
$\quad = 1 - P(X < 5)$ The Complement Rule.
$\quad = 1 - P(X \leq 4)$ The first value X takes on *less than* 5 is 4.
$\quad = 1 - \left[1 - (1 - p)^4 \right]$ Use Equation 5.11.
$\quad = 1 - \left[1 - (0.86)^4 \right]$ Use $p = 0.14$.
$\quad = 1 - 0.4530 = \boxed{0.5470}$ Expand and simplify.

> Remember, a success does *not* have to be a *good* thing.

The probability that it takes five or more cars to find the first uninsured motorist is 0.5470.

Figures 5.15 and 5.16 show technology solutions:

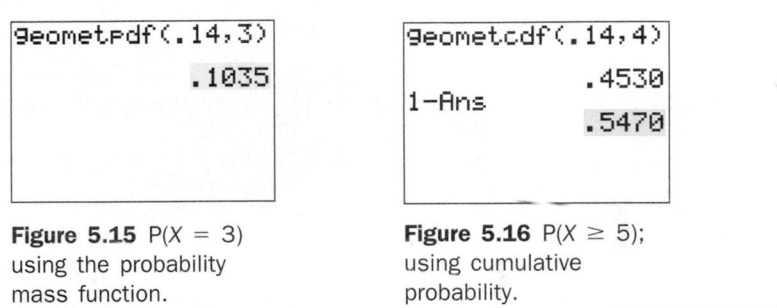

Figure 5.15 $P(X = 3)$ using the probability mass function.

Figure 5.16 $P(X \geq 5)$; using cumulative probability.

> The distribution is named after the French mathematician Simeon Denis Poisson (1781–1840).

The **Poisson probability distribution** has many practical applications and is often associated with *rare* events. A **Poisson random variable** is a count of the number of occurrences of a certain event in a given unit of time, space, volume, distance, etc. Examples are the number of arrivals to an emergency room in a certain 30-minute period, the number of asteroids that pass through Earth's orbit during a given year, or the number of bacteria in a milliliter of drinking water.

PROPERTIES OF A POISSON EXPERIMENT

1. The probability that a single event occurs in a given interval (of time, length, volume, etc.) is the same for all intervals.
2. The number of events that occur in any interval is independent of the number that occur in any other interval.

These properties are often referred to as a *Poisson process* and can be difficult to verify.

THE POISSON RANDOM VARIABLE

The **Poisson random variable** is a count of the number of times the specific event occurs during a given interval.

The Poisson distribution is completely determined by the mean, denoted by the Greek letter lambda, λ. Since the Poisson distribution is often used to count rare events, the mean number of events per interval is usually small. The probability distribution is given below.

THE POISSON PROBABILITY DISTRIBUTION

Suppose X is a Poisson random variable with mean λ. Then

$$p(x) = P(X = x) = \frac{e^{-\lambda}\lambda^x}{x!}, \quad x = 0, 1, 2, 3, \ldots \quad (5.12)$$

$$\mu = \lambda, \quad \sigma^2 = \lambda. \quad (5.13)$$

ILLUMINATING THE CONCEPTS

1. The Poisson random variable is discrete. The number of possible values is countably infinite: 0, 1, 2, 3, ...

2. The Poisson distribution is completely characterized by only one parameter, λ. The mean and the variance are both equal to the same value, λ.

3. Equation 5.12 is a valid probability distribution. All of the probabilities are between 0 and 1, and the sum of all the probabilities is 1 (another infinite series).

4. e in Equation 5.12 is the base of the natural logarithm. $e \approx 2.71828$ is an irrational number. Most calculators have this special constant built-in.

5. The denominator of Equation 5.12 contains $x!$ (x factorial).
Recall: $x! = x(x - 1)(x - 2) \ldots (3)(2)(1)$ and $0! = 1$.

6. Table II in the Appendix contains values for $P(X \leq x)$ (cumulative probability) for various values of λ. ■

Example 5.19 Clear the Runway The taxiways and runways of a major airport are carefully monitored to expedite takeoffs and landings and to prevent collisions. If a pedestrian or vehicle enters a radio-controlled surface at an airport without receiving permission (control tower clearance), this is called a deviation and incursion. In 2006, there were two serious deviations and incursions at the Los Angeles International Airport.[15] Suppose this is the mean number of deviations and incursions per year at LAX and a random year is selected.
a. Find the probability that exactly three serious deviations and incursions will occur.
b. Find the probability that at least five serious deviations and incursions will occur.
c. Find the probability that the number of deviations and incursions will be within one standard deviation of the mean.

SOLUTION

(Frhojdysz/Dreamstime.com)

Let X be the number of deviations and incursions per year at LAX. X has a Poisson distribution with $\lambda = 4.2$

a. The probability of *exactly three* means $P(X = 3)$.

$$P(X = 3) = \frac{e^{-2}2^3}{3!} = 0.1804 \qquad \text{Use Equation 5.12.}$$
$$= P(X \leq 3) - P(X \leq 2) \qquad \text{Or, convert to cumulative probability.}$$
$$= 0.8571 - 0.6767 = \boxed{0.1804} \qquad \text{Use Table II in the Appendix.}$$

b. *At least five* means 5 or more: $X \geq 5$.

$$
\begin{aligned}
P(X \geq 5) &= 1 - P(X < 5) \\
&= 1 - P(X \leq 4) \\
&= 1 - 0.9473 \\
&= \boxed{0.9473}
\end{aligned}
$$

The Complement Rule.
The first value X takes on *less than* 5 is 4.
Use Table II in the Appendix.

c. Within one standard deviation of the mean is the interval $(\mu - \sigma, \mu + \sigma)$.

$$
\mu = 2 = \sigma^2 \rightarrow \sigma = \sqrt{2} = 1.4142.
$$

$$
\begin{aligned}
P(\mu - \sigma &\leq X \leq \mu + \sigma) \\
&= P(2 - 1.4142 \leq X \leq 2 + 1.4142) \\
&= P(0.5858 \leq X \leq 3.4142) \\
&= P(1 \leq X \leq 3) \\
&= P(X \leq 3) - P(X \leq 0) \\
&= 0.8571 - 0.1353 \\
&= \boxed{0.7218}
\end{aligned}
$$

Use values for μ and σ.
Compute the difference and sum.
Use properties of the Poisson distribution.
Convert to cumulative probability.
Use Table II in the Appendix.
Compute the difference.

Figures 5.17 through 5.19 show technology solutions:

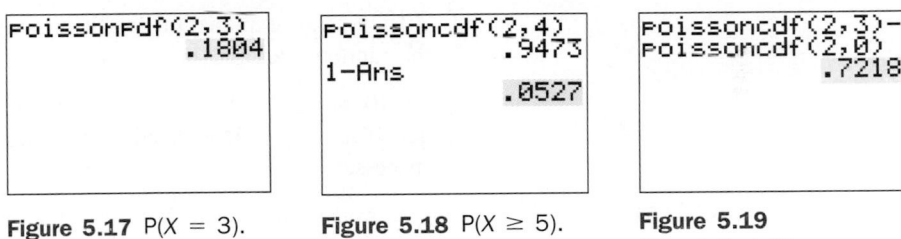

Figure 5.17 $P(X = 3)$.   **Figure 5.18** $P(X \geq 5)$.   **Figure 5.19** $P(1 \leq X \leq 3)$.

The **hypergeometric probability distribution** arises from an experiment in which there is sampling without replacement from a finite population. Each element in the population is labeled a success or failure. The **hypergeometric random variable** is a count of the number of successes in the sample. For example, consider a shipment of 12 automobile tires, of which 2 are defective, and a random sample of 4 tires. A hypergeometric random variable may be defined as a count of the number of *good* tires selected.

PROPERTIES OF A HYPERGEOMETRIC EXPERIMENT

1. The population consists of N objects, of which M are successes and $N - M$ are failures.
2. A sample of n objects is selected *without* replacement.
3. Each sample of size n is equally likely.

THE HYPERGEOMETRIC RANDOM VARIABLE

The **hypergeometric random variable** is a count of the number of successes in a random sample of size n.

The hypergeometric probability distribution is completely determined by n, N, and M. The probability of obtaining x successes is derived using many concepts introduced earlier: independence, the Multiplication Rule, equally likely outcomes, and combinations.

THE HYPERGEOMETRIC PROBABILITY DISTRIBUTION

Suppose X is a hypergeometric random variable characterized by sample size n, population size N, and number of successes M. Then

$$p(x) = P(x = x) = \frac{\binom{M}{x}\binom{N-M}{n-x}}{\binom{N}{n}} \tag{5.14}$$

$$\max(0, n - N + M) \le x \le \min(n, M)$$

$$\mu = n\frac{M}{N}, \quad \sigma^2 = \left(\frac{N-n}{N-1}\right)n\frac{M}{n}\left(1 - \frac{M}{N}\right) \tag{5.15}$$

💡 ILLUMINATING THE CONCEPTS

1. ◀ Here is an explanation for the strange restriction on the possible values for the random variable X.

   $\max(0, n - N + M) \le x$: x must be at least 0 or $n - N + M$, whichever is bigger. If $n - N + M$ is positive, it is impossible to obtain fewer than $n - N + M$ successes.

   $x \le \min(n, M)$: x can be at most n or M, whichever is smaller. The greatest number of successes possible is either n or the total number of successes in the population.

   Suppose $n = 5$, $N = 10$, and $M = 6$. Then:
   $\max(0, n - N + M) = \max(0, 5 - 10 + 6) = \max(0, 1) = 1$, and
   $\min(n, N) = \min(5, 10) = 5 \Rightarrow 1 \le x \le 5$

   It is impossible to obtain less than 1 success. Also, the greatest number of successes possible is 5. ▶

2. The hypergeometric random variable is discrete. All of the probabilities are between 0 and 1, and the probabilities do sum to 1.

3. Recall: $\binom{n}{r}$ is a *combination*. The number of combinations of size r is given by

   $$_nC_r = \binom{n}{r} = \frac{n!}{r!(n-r)!} \quad \blacksquare$$

Solution Trail

KEYWORDS

- 10 twin comforters.
- 2 stitched incorrectly.
- 4 selected at random.

⬇

TRANSLATION

- $N = 10$ (finite population).
- 2 failures, therefore $M = 8$ successes.
- Sample size $n = 4$.

⬇

CONCEPTS

Hypergeometric probability distribution.

⬇

VISION

The 4 comforters selected at random suggests sampling *without* replacement. Since there are 10 comforters to choose from, the population is finite. Each comforter selected is either a success (stitched correctly) or a failure. These clues suggest the number of successes (number of correctly stitched comforters) has a hypergeometric distribution.

Example 5.20 Torn Comforter A Target store has 10 Spider-Man twin comforters for sale. Two of the 10 have been stitched incorrectly at the factory and will split open when used. Suppose 4 of the comforters are randomly selected.
a. What is the probability that exactly 2 will be stitched correctly?
b. What is the probability that at least 3 will be stitched correctly?

SOLUTION

Let X be the number of successes in the sample. X has a hypergeometric distribution with $n = 4$, $N = 10$, and $M = 8$. Translate each question into a probability statement, convert to cumulative probability if necessary, and use Equation 5.14 and/or technology.

a. *Exactly* 2 means $X = 2$.

$$P(X = 2) = \frac{\binom{M}{x}\binom{N - M}{n - x}}{\binom{N}{n}} = \frac{\binom{8}{2}\binom{10 - 8}{4 - 2}}{\binom{10}{4}}$$

Use Equation 5.14.

$$= \frac{\binom{8}{2}\binom{2}{2}}{\binom{10}{4}}$$

In the numerator: from the 8 good comforters, choose 2; from the 2 bad comforters, choose 2. In the denominator: $\binom{10}{4}$ is the total number of ways to choose 4 comforters from 10.

$$= \frac{(28)(1)}{210} = 0.1333$$

Use the formula for a combination.

The probability of selecting exactly 2 correctly stitched comforters is 0.1333.

b. *At least* 3 means 3 or more. The maximum number of successes is $\min(n, M) = \min(4, 8) = 4$. In this case, 3 or more means 3 or 4.

$$P(X \geq 3) = P(X = 3) + P(X = 4)$$

Consider the values X can assume greater than or equal to 3.

$$= \frac{\binom{8}{3}\binom{2}{1}}{\binom{10}{4}} + \frac{\binom{8}{4}\binom{2}{0}}{\binom{10}{4}}$$

Use Equation 5.14.

$$= \frac{(56)(2)}{210} + \frac{(70)(1)}{210}$$

Use the formula for a combination.

$$= 0.5333 + 0.3333 = 0.8666$$

Note: This problem can also be solved using cumulative probability: $P(X \geq 3) = 1 - P(X \leq 2)$.

Figures 5.20 and 5.21 show technology solutions:

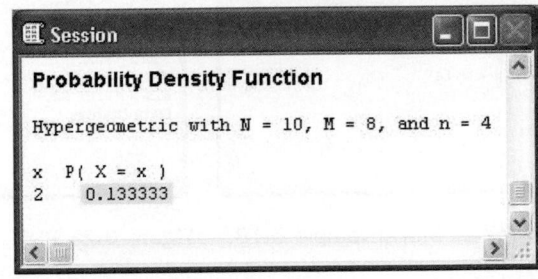

Figure 5.20 Minitab session window output using the Hypergeometric Distribution input window.

Figure 5.21 $P(X \geq 3)$ using the command language.

TECHNOLOGY CORNER

Procedure: Compute probabilities associated with a geometric, Poisson, or hypergeometric distribution.

Reconsider: Examples 5.18 (page 220), 5.19 (page 222), and 5.20 (page 224), solutions, and interpretations.

TI-84 Plus

There are built-in functions to compute cumulative probability and to evaluate the probability mass function associated with a geometric and Poisson random variable. Use the built-in function for combinations to compute probabilities associated with a hypergeometric random variable.

1. Suppose X is a geometric random variable with $P(S) = p$. Use the functions in the DISTR; DISTR menu. $P(X = x) = $ geometpdf(p,x). $P(X \leq x) = $ geometcdf(p,x). See Figures 5.15 and 5.16 (page 221).

2. Suppose X is a Poisson random variable with $\lambda = L$. Use the functions in the $\boxed{\text{DISTR}}$; DISTR menu. $P(X = x) = \texttt{poissonpdf(L,x)}$. $P(X \leq x) = \texttt{poissoncdf(L,x)}$. See Figures 5.17 and 5.18 (page 223).

3. Suppose X is a hypergeometric random variable with parameters n, N, and M. Use the built-in function for combinations, $\boxed{\text{MATH}}$; PRB; nCr, to compute probabilities associated with this random variable. Figure 5.22 shows the computations to find $P(X = 2)$ in Example 5.20. Figure 5.23 shows the computations to find $P(X = 3) + P(X = 4)$ in Example 5.20 using the calculator functions sum and seq.

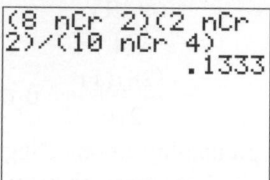

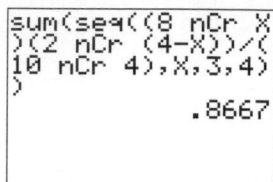

Figure 5.22 $P(X = 2)$ using the definition of the probability mass function.

Figure 5.23 $P(X = 3) + P(X = 4)$ using the sum and seq functions.

Minitab

There are built-in functions to compute cumulative probability and to evaluate the probability mass function associated with a geometric, Poisson, or hypergeometric random variable. These functions may be accessed through a graphical input window: Calc; Probability Distributions, or by using the command language.

1. Suppose X is a geometric random variable with $P(S) = p$. In a session window, use the commands PDF or CDF to evaluate the probability mass function or to compute cumulative probability. See Figures 5.24 and 5.25.

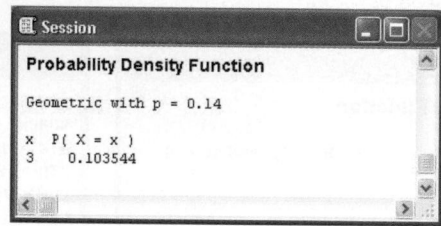

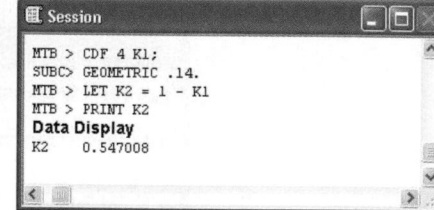

Figure 5.24 $P(X = 3)$ in Example 5.18.

Figure 5.25 $P(X \geq 5)$ in Example 5.18.

2. Suppose X is a Poisson random variable with $\lambda = L$. In a session window, use the commands PDF or CDF to evaluate the probability mass function or to compute cumulative probability. See Figures 5.26 and 5.27.

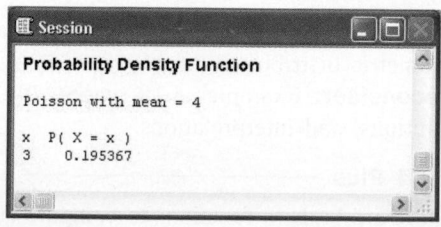

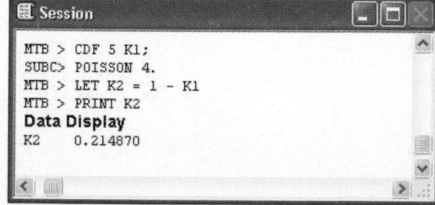

Figure 5.26 $P(X = 3)$ in Example 5.19.

Figure 5.27 $P(X \geq 5)$ in Example 5.19.

3. Suppose X is a hypergeometric random variable with parameters n, N, and M. In a session or Calc; Probability Distributions input window, evaluate the probability mass function or compute cumulative probability. See Figures 5.20 and 5.21 (page 225).

Excel

There are built-in functions to compute cumulative probability and to evaluate the probability mass function associated with a Poisson random variable. There is a built-in function to evaluate the probability mass function associated with a hypergeometric random variable. Use the function definitions to compute probabilities associated with a geometric random variable.

1. Suppose X is a geometric random variable with $P(S) = p$. Use the appropriate formula to find $P(X = x)$ or to find $P(X \leq x)$. See Figure 5.28.

	A	B
1	0.1035	= (1-0.14)^2*(0.14) = P(X = 3)
2	0.5470	= 1-(1-0.86^4) = P(X ≥ 5)

Figure 5.28 $P(X = 3)$ and $P(X \geq 5)$ in Example 5.18.

2. Suppose X is a Poisson random variable with $\lambda = L$. To evaluate the probability mass function, use the function POISSON with the last argument set to False. To compute cumulative probability, use the function POISSON with the last argument set to True. See Figure 5.29.

	A	B
1	0.1804	= POISSON(3,2,FALSE) = P(X = 3)
2	0.0527	= 1-POISSON(4,2,TRUE) = P(X ≥ 5)

Figure 5.29 $P(X = 3)$ and $P(X \geq 5)$ in Example 5.19.

3. Suppose X is a hypergeometric random variable with parameters n, N, and M. To evaluate the probability mass function, use the function HYPGEOMDIST. Sum the appropriate values to compute cumulative probability. See Figure 5.30.

	A	B
1	0.1333	= HYPGEOMDIST(2,4,8,10) = P(X = 2)
2	0.5333	= HYPGEOMDIST(3,4,8,10) = P(X = 3)
3	0.3333	= HYPGEOMDIST(4,4,8,10) = P(X = 4)
4	0.8667	= A2 + A3 = P(X ≥ 3)

Figure 5.30 $P(X = 2)$ and $P(X \geq 3)$ in Example 5.20.

SECTION 5.5 EXERCISES

Practice

5.71 Suppose X is a geometric random variable with probability of success 0.35. Find the following probabilities.
a. $P(X = 4)$.
b. $P(X \geq 3)$
c. $P(X \leq 2)$
d. $P(X \geq \mu)$

5.72 Suppose X is a geometric random variable with mean $\mu = 4$. Find the following probabilities.
a. $P(X = 1)$
b. $P(3 \leq X \leq 7)$
c. $P(X > \mu + 2\sigma)$

5.73 Suppose X is a Poisson random variable with $\lambda = 2$. Find the following probabilities.
a. $P(X = 0)$
b. $P(2 \leq X \leq 8)$
c. $P(X > 5)$
d. $P(X \leq 6)$

5.74 Suppose X is a Poisson random variable with $\lambda = 4.5$. Find the following probabilities.
a. $P(X > \mu)$
b. $P(X = 2)$
c. The probability X is either 4 or 5.
d. $P(X \leq \mu + 2\sigma)$

5.75 Suppose X is a hypergeometric random variable with $n = 5$, $N = 12$, and $M = 6$.
a. Find $P(X = 2)$.
b. Find $P(X = 5)$.
c. Find the mean, variance, and standard deviation of X.

5.76 Suppose X is a hypergeometric random variable with sample size 8, population size 16, and number of successes in the population 12.
a. List the possible values for X.
b. Find the mean, variance, and standard deviation of X.
c. Find $P(X = 5)$.
d. Find $P(X = 8)$.

Applications

5.77 Fuel Consumption and Cars The Idaho Transportation Department issued 229,236 class D licenses in 2007.[16] Suppose the probability of a person passing the Idaho road test for a Class D driver's license is 0.70. This probability remains the same for people who must take the road test several times in order to obtain a license. Suppose a randomly selected person is preparing to take the road test.
a. What is the probability that the person will obtain a license on the third try?
b. What is the probability that it will take the person at least two attempts in order to obtain a license?
c. What is the mean number of road-test attempts necessary to obtain a license?
d. Suppose 50 people preparing for the road test are selected at random. What is the expected total number of road tests needed for all 50 to obtain their licenses?

5.78 Psychology and Human Behavior During the second quarter of 2008, the mean number of bank robberies per day in the Middle Atlantic States was 2.06.[17] Suppose a day is selected at random.
a. What is the probability of exactly two bank robberies on that day?
b. What is the probability that there will be more than five robberies on that day?
c. Suppose two banking business days are randomly selected. What is the probability that there will be no robberies on both days?

5.79 Public Health and Nutrition While a heart murmur might be scary, most are not life threatening. According to the Marshfield Clinic (2008), heart murmurs are extremely common but most are harmless. The Cincinnati Children's Heart center reports that heart murmurs can be detected in approximately 90% of children ages 4 to 7.[18] Suppose a physician examines every child ages 4 to 7 who is admitted to a hospital.
a. What is the probability that the second child examined will be the first to have a heart murmur detected?
b. What is the probability that the seventh child examined will be the first to have a heart murmur detected?
c. What is the mean number of children examined before one will have a heart murmur detected?
d. What is the probability that at least five children will be examined before one will have a heart murmur detected?

5.80 Sports and Leisure During the 2007 season, the Rutgers men's soccer team had the highest mean number of yellow cards (warnings or cautions by the referee) per game, 1.95, in the Big East Conference.[19] Suppose one Rutgers men's soccer game is randomly selected.
a. What is the probability of exactly 2 yellow cards during the game?
b. What is the probability of fewer than 6 yellow cards during the game?
c. What is the probability of at least 3 yellow cards during the game?
d. Suppose there are between 2 and 10 (inclusive) yellow cards. What is the probability of more than 5 yellow cards?

5.81 Travel and Transportation The Canadian Pacific Railway has a joint union/management safety team and as a result has a superb safety record. The personal injury frequency for 2008 was approximately 1.4 injuries per 200,000 employee hours.[20] Suppose 200,000 Canadian Pacific Railway employee hours are randomly selected.
a. Find the probability that there are no more than 4 employee injuries.
b. Find the probability that the number of employee injuries is more than $\mu + 2\sigma$.
c. In order to win a quality-control award, the railroad system must record five consecutive 200,000-hour periods without any employee injuries. What is the probability of this happening?

5.82 Manufacturing and Product Development Representatives for Ariens lawn mowers claim that the probability the engine will start on any given pull is 0.90. Suppose the engine is properly primed and pulls are independent.
a. What is the probability that the engine starts in two pulls or fewer?
b. What is the mean number of pulls necessary in order to start the engine?
c. If the engine does not start within five pulls, a customer becomes frustrated. What is the probability that the customer will become frustrated?

5.83 Manufacturing and Product Development Flat-panel LCD displays in televisions and computer monitors often develop *dead pixels*, pixels that become locked in one state, for example, red at all times. Dell Computer Corporation maintains that dead pixels are a natural defect and 1 to 5 dead pixels in an LCD monitor is an acceptable number.[21] Suppose the mean number of dead pixels in a new Dell 17-inch LCD computer monitor is 2.5. One of these monitors is randomly selected and inspected for dead pixels.
a. What is the probability that there will be no dead pixels?
b. If the number of dead pixels is more than $\mu + 3\sigma$, the assembly line is automatically stopped and examined. What is the probability that the assembly line will be stopped?
c. What is the probability that the number of dead pixels will be within two standard deviations of the mean?

5.84 Technology and Internet In August 2008, Kaspersky Lab reported that approximately 80% of all email is junk mail, or spam.[22] Each morning a computer user turns on his computer and activates his email system.
a. What is the probability that the fourth email received will be the first piece of spam?
b. What is the mean number of emails until the first piece of spam arrives?
c. What is the probability that the first spam will arrive within the first three emails?
d. What is the probability that the first spam will arrive more than three standard deviations above the mean?

5.85 Business and Management Fifteen lobstermen have their boats anchored at a small pier along the New Hampshire coast.

Five of these lobstermen have been fined within the past year for commercial lobster-size violations. Suppose 4 lobstermen are selected at random.

a. What is the probability that exactly 2 have been fined for violations within the past year?

b. What is the probability that all 4 have been fined for violations within the past year?

c. What is the probability that at least 1 has been fined for violations within the past year?

5.86 Demographics and Population Statistics Buchtal, a manufacturer of ceramic tiles, reports 3.9 job-related accidents per year. Accident categories include trip, fall, struck by equipment, transportation, and handling. Suppose a year is selected at random.

a. What is the probability that there will be no job-related accidents?

b. What is the probability that the number of accidents that year will be between two and five (inclusive)?

c. If the number of accidents is more than three standard deviations above the mean, the company insurance carrier will raise the rates. What is the probability of an increase in the company's insurance bill?

Extended Applications

5.87 Marketing and Consumer Behavior The Snapple Beverage Group is sponsoring a conventional bottle cap sweepstakes game. Under each bottle cap there is a note saying either "You are not a winner," or the prize awarded. Suppose there are 20 of the game bottles on a shelf in the supermarket, and 2 of them are winners. A customer randomly selects 6 bottles from the shelf.

a. What is the probability of selecting no winning bottles?

b. What is the probability of selecting both winning bottles?

c. What is the mean number of winning bottles selected?

d. How many bottles would the customer have to purchase in order to expect 1 winning bottle?

5.88 Psychology and Human Behavior According to an article in the *UK News*, drinking four cups of tea per day may reduce the risk of a heart attack. However, the mean number of cups per day is three.[23] Suppose an adult in Britain is selected at random and the number of cups of tea that the person drinks during a day is recorded.

a. What is the probability that the person will have exactly two cups of tea?

b. What is the probability that the person will have at most five cups of tea?

c. Suppose the resident has eight cups of tea (the optimal number as suggested in the article). Is there any evidence to suggest the mean is different from three? Justify your answer.

5.89 Business and Management Cabela's online store notes that there are a variety of errors that may occur in customer orders received via the telephone. A recent audit revealed that the probability of some type of error in a telephone order is 0.20. In an attempt to correct these errors, a supervisor randomly selects telephone orders and carefully inspects each one.

a. What is the probability that the third telephone order selected will be the first to contain an error?

b. What is the probability that the supervisor will inspect between two and six (inclusive) telephone orders before finding an error?

c. What is the probability that the inspector will examine at least seven orders before finding an error?

d. What is the probability that the first error will be on the fourth telephone order or later?

e. Suppose the first four telephone orders contain no errors. What is the probability that the first error will be on the eighth order or later?

5.90 Economics and Finance The manager of San Lagos at Arrowhead Highlands, an apartment complex in Glendale, Arizona, collects the rent from each tenant on the first day of every month. Past records indicate that the mean number of tenants who do not pay the rent on time in any given month is 4.7. Consider the rent collection for the next month.

a. Find the probability that every tenant will pay the rent on time.

b. Find the probability that at least seven tenants will be late with their rent.

c. Suppose the number of delinquent rent payments in a month is independent of the number in every other month. What is the probability that at most three tenants will be late with their rent in two consecutive months?

5.91 Fuel Consumption and Cars According to a survey by GasBuddy Organization, Inc., approximately 15% of automobile owners wash their primary vehicle 10–12 times per year.[24] Suppose the mean number of times that people wash their primary vehicles per year is 4.

a. What is the probability that a randomly selected automobile owner does not wash his or her primary vehicle during a year? Once during the year? Twice during the year?

b. Suppose two automobile owners are selected at random. What is the probability that the total number of primary vehicle washes for the two owners is 0? 1? 2?

c. Suppose the mean number of times that people wash their primary vehicles during a 2-year period is 8. What is the probability that a randomly selected automobile owner does not wash his or her primary vehicle during a 2-year period? Once during the 2-year period? Twice during the 2-year period?

d. How do your answers in parts (b) and (c) compare? What property does this suggest about a Poisson random variable?

Challenge

5.92 Approaching Poisson Suppose X is a Poisson random variable with $\lambda = 2$. Let the related random variable Y have a probability distribution as given in the following table.

y	$P(Y = y)$
0	$P(X = 0) = 0.1353$
1	$P(X = 1) = 0.2707$
2	$P(X = 2) = 0.2707$
3	$P(X \geq 3) = 0.3233$

Find the expected value of Y.

Suppose the distribution of Y is changed slightly, at the right tail, as given in the following table.

y	$P(Y = y)$
0	$P(X = 0) = 0.1353$
1	$P(X = 1) = 0.2707$
2	$P(X = 2) = 0.2707$
3	$P(X = 3) = 0.1804$
4	$P(X \geq 4) = 0.1429$

Find the expected value of Y.

Suppose the distribution of Y is changed again, once more at the right tail.

y	$P(Y = y)$
0	$P(X = 0) = 0.1353$
1	$P(X = 1) = 0.2707$
2	$P(X = 2) = 0.2707$
3	$P(X = 3) = 0.1804$
4	$P(X = 4) = 0.0902$
5	$P(X \geq 5) = 0.0527$

Find the expected value of Y.

Continue in this manner. To what number is $E(Y)$ converging, and why does this make sense?

Chapter 5 Challenge Wrap-Up

A random sample of 50 people is selected, each is given a flu shot, and the chance of contracting the flu is assumed to be 1 in 4 (0.25). This information suggests a binomial experiment. There are $n = 50$ trials; each person either contracts the flu (a success) or does not; each person contracts, or does not contract, the flu independently of every other person; and the probability of contracting the flu is the same for each person.

The random variable X, the number of people who contract the flu, has a binomial distribution with $n = 50$ and $p = 0.25$: $X \sim B(50, 0.25)$. In order to check the claim, we follow the four-step inference procedure.

(Zuma/Newscome)

Claim: $p = 0.25 \rightarrow X \sim B(50, 0.25)$.

Experimental outcome: $x = 19$ people actually contracted the flu.

Likelihood: $\mu = np = (50)(0.25) = 12.5$. Since $x = 19$ is to the right of the mean, and we are checking for evidence to suggest $p > 0.25$, consider a right-tail probability.

$$
\begin{aligned}
P(X \geq 19) &= 1 - P(X < 19) && \text{The Complement Rule.}\\
&= 1 - P(X \leq 18) && \text{First value of } X \text{ less than 19 is 18.}\\
&= 1 - 0.9713 && \text{Use Table I in the Appendix.}\\
&= 0.0287
\end{aligned}
$$

A technology solution:

```
binomcdf(50,.25,
18)
            .9713
1-Ans
            .0287
```

Figure 5.31 $P(X \geq 19)$.

Conclusion: Since this probability is so small, there is evidence to suggest that the claim is false. If $p = 0.25$ the probability of 19 or more people contracting the flu is tiny. But it happened! Therefore, there is evidence to suggest that p is larger than 0.25 and that the vaccine is not as effective as advertised.

CHAPTER 5 SUMMARY

Concept	Page	Notation / Formula / Description
Random variable	184	A function that assigns a unique numerical value to each outcome in a sample space.
Discrete random variable	186	The set of all possible values is finite, or countably infinite.
Continuous random variable	186	The set of all possible values is an interval of numbers.
Probability distribution for a discrete random variable of the random variable	188, 194	A method for conveying all the possible values and the probability associated with each value.
Mean, or expected value, of a discrete random variable X	198	$\mu = E(X) = \sum_{\text{all } x} [x \cdot p(x)]$.
Variance of a discrete random variable X	200	$\sigma^2 = \text{Var}(X) = \sum_{\text{all } x} [(x - \mu)^2 \cdot p(x)]$. $= E[(X - \mu)^2] = E(X^2) - E(X)^2$ $= E(X^2) - \mu^2$

(continued)

Concept	Page	Notation / Formula / Description
Standard deviation of a discrete random variable X	200	$\sigma = \sqrt{\sigma^2}$, The positive square root of the variance.
Properties of a binomial experiment	206	1. n identical trials. 2. Each trial can result in only a success (S) or a failure (F). 3. Trials are independent. 4. Probability of a success is constant from trial to trial.
Binomial random variable	207	The number of successes in n trials.
Binomial probability distribution	208	If $X \sim \mathrm{B}(n, p)$ then $$p(x) = \binom{n}{x} p^x (1 - p)^{n-x}, \quad x = 0, 1, 2, 3, \dots, n$$ where $\mu = np$, $\sigma^2 = np(1 - p)$, and $\sigma = \sqrt{np(1 - p)}$.
Cumulative probability	209	$P(X \leq x)$
Geometric random variable	219	The number of trials necessary to realize the first success.
Geometric probability distribution	220	$$p(x) = (1 - p)^{x-1} p, \quad x = 1, 2, 3, \dots$$ where $\mu = \dfrac{1}{p}$, $\sigma^2 = \dfrac{1 - p}{p^2}$, and $\sigma = \sqrt{\dfrac{1 - p}{p^2}}$
Poisson random variable	221	A count of the number of times a specific event occurs during a given interval.
Poisson probability distribution	222	$$p(x) = \frac{e^{-\lambda} \lambda^x}{x!}, \quad x = 0, 1, 2, 3, \dots$$ where $\mu = \lambda$, $\sigma^2 = \lambda$, and $\sigma = \sqrt{\lambda}$.
Hypergeometric random variable		A count of the number of successes in a random sample of size n from a population of size N, and number of successes M.
Hypergeometric probability distribution	224	$$p(x) = \frac{\binom{M}{x}\binom{N - M}{n - x}}{\binom{N}{n}}, \quad \max(0, n - N + M) \leq x \leq \min(n, M)$$ where $\mu = n\dfrac{M}{N}$, $\sigma^2 = \left(\dfrac{N - n}{N - 1}\right) n\dfrac{M}{N}\left(1 - \dfrac{M}{N}\right)$, and $\sigma = \sqrt{\left(\dfrac{N - n}{N - 1}\right) n\dfrac{M}{N}\left(1 - \dfrac{M}{N}\right)}$.

CHAPTER 5 EXERCISES

Applications

5.93 Fuel Consumption and Cars In May 2008, the automobile enhanced emissions tests in Colorado became stricter, in order to address the ground-level ozone pollution along Colorado's Front Range. However, the Air Pollution Control Division suggested that approximately 93% of automobiles, 1982 and newer, will still pass the enhanced emissions tests.[25] Suppose 30 automobiles, 1982 or newer, in Colorado are selected at random.

a. Find the probability that exactly 28 automobiles will pass the enhanced emissions tests.

b. Find the probability that at least 25 automobiles will pass the enhanced emissions tests.

c. Suppose 30 automobiles are inspected in 2009 and only 20 pass the enhanced emissions tests. Is there any evidence to suggest that the proportion of automobiles that pass the enhanced emissions test has changed? Justify your answer.

5.94 Business and Management Bytes and Pieces, a computer company that offers forensic software and hardware, classifies all telephone calls to its technical support staff by the amount of time the customer is on hold. If the customer is on hold for no more than 60 seconds, then the call is classified as successful (actually this sounds like a miracle). The supervisor in technical support claims that 80% of all calls are successful. Suppose 25 calls to technical support are selected at random.

a. Find the mean, variance, and standard deviation of the number of successful calls.

b. Find the probability that at least 18 calls will be successful.

c. Suppose 21 calls are successful. Is there any evidence to suggest the supervisor's claim is false? Justify your answer.

5.95 Economics and Finance Fidelity National Financial conducted a survey to learn more about personal debt and financing. Let X be the number of credit cards a randomly selected adult in the United States carries regularly. The probability distribution for X is given in the table that follows.

x	1	2	3	4	5
$p(x)$	0.02	0.06	0.08	0.10	0.11

x	6	7	8	9	10
$p(x)$	0.16	0.23	0.15	0.07	0.02

(*Source*: CardWeb).

a. Find the mean, variance, and standard deviation of the number of credit cards carried.

b. Find the probability that a randomly selected individual carries more than seven credit cards.

c. Find the probability that the number of cards carried by a randomly selected individual is less than $\mu - \sigma$.

d. Suppose three people are selected at random. What is the probability that at least one person carries fewer than six credit cards?

5.96 Manufacturing and Product Development BlueCosmo has started to manufacture, sell, and launch communications satellites. The most common reason for a satellite to fail once in orbit is a problem related to opening and initiating the solar panels. Suppose the probability of a failure related to the solar panels is 0.08.

a. What is the probability that the fifth satellite launched will be the first to fail due to a solar-panel problem?

b. What is the expected number of satellites that will launch before the first one fails due to a solar-panel problem?

c. BlueCosmo is preparing an advertising campaign in which they claim to have had 20 successful launches in a row. What is the probability that the first failure due to a solar-panel problem will occur after the 20th launch?

5.97 Business and Management An easy-assembly, no-tools-required gas grill comes with detailed step-by-step instructions. Even though each grill is carefully packaged, there are often missing pieces. This can aggravate the customer and increase the cost to the producer, who must provide phone support and ship the missing parts. Suppose the mean number of missing pieces per packaged grill is 0.7, and one grill is randomly selected from the stockroom.

a. What is the probability that there will be no missing pieces in the package?

b. If there are more than five missing pieces, the producer identifies the packager and issues a warning. What is the probability of a warning being issued at the packaging plant?

c. Suppose three grills are randomly selected. What is the probability that each will have no more than one missing piece?

5.98 Psychology and Human Behavior According to the Gallup Poll 2008 annual crime survey, approximately 64% of Americans still support the death penalty.[26] In a pool of 24 potential jurors, 16 support the death penalty. Suppose there is a capital case on the court's docket and 12 jurors are selected at random.

a. What is the probability that fewer than 6 jurors support the death penalty?

b. What is the probability that at least 1 juror supports the death penalty?

c. What is the probability that all 12 jurors support the death penalty?

5.99 Sports and Leisure At the beaches in Wellington, New Zealand, there were approximately 1.8 rescues per hour during 2007.[27] Suppose an hour is selected at random.

a. What is the probability that no rescues are necessary?

b. What is the probability that between two and six (inclusive) rescues are necessary?

c. If there is evidence to suggest that the mean number of rescues per hour has increased, the supervisor may hire more lifeguards to adequately protect the public. Suppose there are nine rescues necessary in the randomly selected hour. Does the supervisor have justification for hiring more lifeguards? Justify your answer.

d. Although the beaches in Hawaii are known for their beauty, they can be dangerous too. At the most dangerous beach, Makapuu, there are 9.1 rescues per 10,000 swimmers and surfers.[28] Suppose 10,000 swimmers and surfers are selected at random. What is the probability that there will be no rescues?

5.100 Travel and Transportation According to the Independent Insurance Agents of America (IIAA) and College Parents of America (CPA), approximately 70% of college students have access to a car at school (either their own or their parents' car). Suppose 40 college students are selected at random.

a. What is the probability that exactly 25 students have access to a car?

b. What is the probability that at most 30 students have access to a car?

c. What is the probability that more than 33 students have access to a car?

d. Suppose the number of students who have access to a car is within two standard deviations of the mean. What is the probability that the actual number who have access to a car is within one standard deviation of the mean?

5.101 Business and Management Only 4 out of every 5 new small businesses survive their first year, according to a report on small business closings.[29] Suppose 35 new small businesses are selected at random.

a. Find the probability that at least 30 will survive their first year.

b. Find the probability that all 35 will survive their first year.

c. An economist estimates that for every small business that closes, 12 people lose their jobs and sign up for unemployment benefits. How many people from these 35 new small businesses may be expected to lose their jobs during the next year?

5.102 Psychology and Human Behavior The army emphasizes cleanliness and neatness in a military barracks. Each cadet is responsible for maintaining his or her area in top condition. Periodic inspections are held, and those receiving top scores are rewarded. Suppose the mean number of violations discovered per cadet during a barracks inspection is 2.7.

a. What is the probability that a randomly selected cadet will have exactly three violations during an inspection?

b. If a cadet has six or more violations, he or she is assigned to KP (Kitchen Patrol) duty for one week. What is the probability that a randomly selected cadet will be assigned to KP duty following a barracks inspection?

c. If every member of a 10-cadet unit has no violations, then each will receive a weekend pass. What is the probability of this happening following a barracks inspection?

5.103 Travel and Transportation In November 2005, a Boeing 777-200LR Worldliner broke the record for the longest nonstop flight by a commercial jet. Following this news, CNN conducted a survey in which they asked people to describe a 23-hour nonstop flight. Eighty-four percent of those who voted described the flight as a nightmare. Suppose 50 frequent air travelers are selected at random and each is asked to describe a 23-hour flight.

a. Find the mean, variance, and standard deviation of the number of people who describe this flight as a nightmare.

b. What is the probability that at least 40 people will describe the flight as a nightmare?

c. Find the probability that the number of people who describe this flight as a nightmare will be within two standard deviations of the mean.

5.104 Economics and Finance CNBC reported the results of a survey concerning personal job security of American workers.[30] At the time of survey, unemployment had risen to 6.1%, the highest level in 5 years. Despite this figure, it was reported that 10% of workers in Chicago, 6% of workers in Houston, and 7% of workers in Boston felt more secure about their jobs. Suppose attitudes have not changed and 30 workers from Chicago, 30 workers from Houston, and 30 workers from Boston are selected at random.

a. What is the probability that fewer than 7 workers from Chicago feel more secure about their jobs?

b. What is the probability that at least 5 workers from each city feel more secure about their jobs?

c. What is the probability that a total of exactly 2 workers from all three cities feel more secure about their jobs?

5.105 Public Policy and Political Science In a recent nationwide study, it was reported that U.S. adults continue to believe that big companies and lobbyists have too much power. In particular, 85% of those polled indicated that PACs (Political Action Committees) have too much influence in Washington.[31] Suppose 50 U.S. adults are selected at random.

a. What is the probability that at least 45 U.S. adults think PACs have too much power?

b. What is the probability that between 38 and 42 (inclusive) U.S. adults think PACs have too much power?

c. Suppose 35 U.S. adults think PACs have too much power. Is there any evidence to suggest that the poll results are wrong? Justify your answer.

5.106 Sports and Leisure The FOX television drama *House* stars an irreverent, sneaky, discourteous, but brilliant, diagnostic physician, played by Hugh Laurie. Each week a team of doctors is presented with a unique, usually life-threatening,

medical puzzle. The producers of the show have set the probability to be 0.85 that the team will make a correct diagnosis and the patient will live. Suppose 30 random episodes of *House* are selected.

a. Find the expected value, variance, and standard deviation of the number of patients who will live.

b. What is the probability that at least 23 patients will live?

c. Suppose the patient dies in 9 episodes. Is there any evidence to suggest that the proportion of patients who live has changed? Justify your answer.

Extended Applications

5.107 Discrete Uniform Random Variable Suppose X is a random variable with probability distribution given by

$$p(x) = \frac{1}{5}, \quad x = 1, 2, 3, 4, 5.$$

a. Find the mean, variance, and standard deviation of X.

b. Suppose $p(x) = 1/6, x = 1, 2, 3, 4, 5, 6$. Find the mean, variance, and standard deviation of X.

c. Suppose $p(x) = 1/n, x = 1, 2, 3, \ldots, n$. Find the mean, variance, and standard deviation of X in terms of n.

5.108 Marketing and Consumer Behavior According to an article in *TravelMole* (May 24, 2007), 34% of travelers claim that they have been inaccurately charged for an item in the minibar. The hotel manager in a Hyatt is going to conduct an experiment in order to determine whether more than 34% of hotel guests are inaccurately charged for something in the minibar. The manager has decided to select 25 guests at random. If the number of guests who are inaccurately charged is 11 or fewer, then no action will be taken. Otherwise, a new minibar price scanning system will be implemented.

a. Suppose the true proportion of guests who are inaccurately charged is 0.34. What is the probability that the new price scanning system will be implemented?

b. Suppose the true proportion of minibar price mistakes has risen to 0.40. What is the probability that no action will be taken? What if the true proportion is 0.50?

c. Suppose the decision rule is changed such that if the number of guests who are inaccurately charged is 12 or fewer, then no action will be taken. Answer parts (a) and (b) using this rule.

5.109 Marketing and Consumer Behavior Value City is running a sale in which customers may save as much as 40% on any purchase. Once a customer decides to make a purchase, he or she selects two sales prize tickets at random from a large bin at the front of the store. Each ticket has a percentage marked on it and the probability of selecting each ticket is given in the table below.

Percentage	10%	20%	30%	40%
Probability	0.50	0.35	0.10	0.05

The larger of the two percentages selected is used for the purchase.

a. Let X be the maximum of the two prize ticket percentages. Find the probability distribution for X.

b. Find the mean, variance, and standard deviation of X.

c. What is the probability that a customer will receive at least 20% off a purchase?

5.110 Sports and Leisure Over the past several years, the number of *older* skiers has risen steadily. A recent study indicated that the proportion of skiers aged 45 to 54 is approximately 19.9%.[32] Suppose this percentage remains the same for the next skiing season, and 40 skiers in the United States are selected at random.

a. What is the probability that exactly 12 skiers are 45 to 54?

b. Find the largest value c such that the probability of c or fewer skiers 45 to 54 is at most 0.2.

c. Suppose 16 skiers are 45 to 54. Is there any evidence to suggest that the proportion of skiers 45 to 54 has changed? Justify your answer.

Challenge

5.111 A Day on the Dock Two crews work on a receiving dock at a fabric manufacturing plant. The first crew unloads four shipments every day and the second crew unloads seven shipments every day. A supervisor records whether each shipment is complete (a success) or missing items (a failure).

Suppose X_1 is a binomial random variable representing the number of complete shipments for crew 1, with parameters $n_1 = 4$ and $p = 0.6$. Similarly, let X_2 be a binomial random variable representing the number of complete shipments for crew 2, with parameters $n_2 = 7$ and $p = 0.6$. Assume X_1 and X_2 are independent.

a. Use technology to generate a random observation for X_1 (the number of complete shipments for crew 1) and a random observation for X_2 (the number of complete shipments for crew 2). Add these two values to compute a random total number of complete shipments for crews 1 and 2.

Repeat this process to generate 1000 random total number of complete shipments for crews 1 and 2. Compute the relative frequency of occurrence of each observation.

Suppose Y is a binomial random variable with $n = 11$ and $p = 0.6$. Use technology to construct a table of probabilities for $Y = 0, 1, 2, 3, \ldots, 11$. Compare these probabilities with the relative frequencies obtained above.

b. Suppose a new receiving crew is added and it unloads five shipments each day. Let X_3 be a binomial random variable, representing the number of complete shipments for crew 3, with parameters $n_3 = 5$ and $p = 0.6$.

Use technology to generate random observations for X_1, X_2, and X_3. Add these three values to compute a random total number of complete shipments for crews 1, 2, and 3.

Repeat this process to generate 1000 random total number of complete shipments for crews 1, 2, and 3. Compute the relative frequency of occurrence of each observation.

Suppose Y is a binomial random variable with $n = 16$ and $p = 0.6$. Use technology to construct a table of probabilities for $Y = 0, 1, 2, 3, \ldots, 16$. Compare these probabilities with the relative frequencies obtained above.

c. Suppose another receiving crew is added and it unloads nine shipments each day. Let X_4 be a binomial random variable representing the number of complete shipments for crew 4, with parameters $n_2 = 9$ and $p = 0.6$. Let Y represent the total number of complete shipments for all four crews.

   (i) Find $P(Y = 15)$ (the probability of exactly 15 total complete shipments).

   (ii) Find $P(Y \leq 12)$.

   (iii) Find $P(Y > 16)$.

   (iv) How many total complete shipments can be expected?

Continuous Probability Distributions

6

Chapter 6 Challenge

When is the surf too rough for snorkeling?

Buck Island Reef, located off the coast of St. Croix in the U.S. Virgin Islands, is a national monument. It contains some of the most beautiful marine gardens in the Caribbean Sea. Two-thirds of the island is surrounded by an extraordinary barrier reef that is home to colorful fishes and even sea turtles. The island also has an overland nature trail and bright white coral sand beaches.

Concessionaires, or tour boats, approved by the National Park Service offer half- and full-day expeditions to Buck Island from St. Croix. Some provide snorkeling lessons and a guided tour along the underwater trail. The tours run daily, except when the surf is too rough. High seas make it difficult for the catamarans to navigate to the island and dangerous for amateur snorkelers to swim.

During the months of January through May, the mean height of waves off the coast of St. Croix each day is 7 feet with a standard deviation of 0.75 feet. The distribution of wave height is approximately normal. Suppose tours to Buck Island are canceled due to high waves on only 2% of all days in January through May. How high would the waves have to be in order for all tours to be canceled? The solution is presented in the Chapter Challenge Wrap-Up (page 282).

(Frhojdysz/Dreamstime.com)

Review
- Remember how to completely describe and compute probabilities associated with a discrete random variable.
- Recall the characteristics of and probability computations associated with the binomial, geometric, Poisson, and hypergeometric random variables.

Preview
- Learn how to completely describe a continuous random variable.
- Compute probabilities associated with a continuous random variable.
- Understand the characteristics of the normal distribution and compute probabilities involving a normal random variable.

CONTENTS

6.1 PROBABILITY DISTRIBUTIONS FOR CONTINUOUS RANDOM VARIABLES

Suppose X is a continuous random variable; X takes on any value in some interval of numbers. A **continuous probability distribution** completely describes the random variable and is used to compute probabilities associated with the random variable.

> **DEFINITION**
>
> A **probability distribution for a continuous random variable** X is given by a smooth curve called a **density curve** or **probability density function** (pdf). The curve is defined so that the probability that X takes on a value between a and b ($a < b$) is the area under the curve between a and b.

 ILLUMINATING THE CONCEPTS

1. Probability in a continuous world is *area under a curve*. Figures 6.1 to 6.3 illustrate the correspondence between the probability of an event (defined in terms of a continuous random variable) and the area under the density curve.

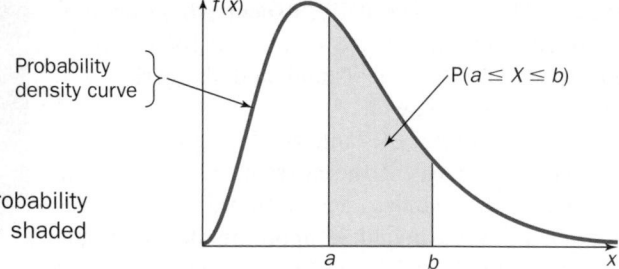

Figure 6.1 The probability is the area of the shaded region.

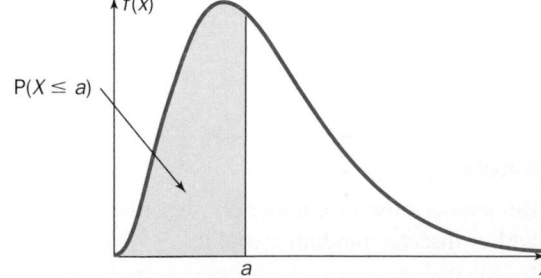

Figure 6.2 The shaded area is P($X \le a$).

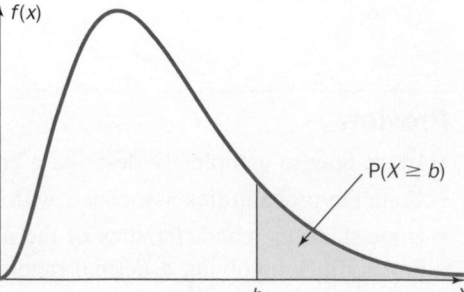

Figure 6.3 The shaded area is P($X \ge b$).

2. The density curve, or **probability density function**, is usually denoted by f. It is a *function* defined for *all* real numbers. $f(x)$ is *not* the probability that the random variable X equals the specific value x. Rather, the function f leads to, or conveys, probability through area.

3. The shape of the graph of a density function can vary considerably. However, a density function must satisfy the following two properties:

   a. f must be defined so that the total area under the curve is 1. The total probability associated with any random variable must be 1. $f(x)$, a specific value of the density function, *may* be greater than 1 (while the total area under the curve is still exactly 1).

   b. $f(x) \geq 0$ for all x. Therefore, the entire graph lies on or above the x-axis.

> Remember, $f(x)$ is *not* a probability. The density function *leads to* probability.

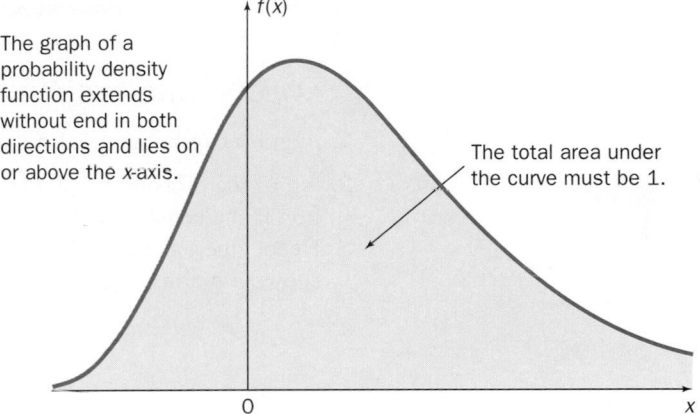

The graph of a probability density function extends without end in both directions and lies on or above the *x*-axis.

The total area under the curve must be 1.

Figure 6.4 A valid probability density function.

4. If X is a continuous random variable with density function f, the probability that X equals *any* one specific value is 0. That is, $P(X = a) = 0$ for any a. The reason is that there is no area under a single point.

> $P(X = a)$ translated: Find the area under the curve between a and a. This is asking for the area of a line segment. There is no second dimension. Hence the area is 0.

◀ This seems like a contradiction. Certainly we can observe specific values of X. Yet the probability of observing any single value is 0. Recall: Probability is a *limiting relative frequency*. There are an (uncountably) infinite number of values for any continuous random variable. Therefore, the limiting relative frequency of occurrence of any single value is 0.

Since there is no probability associated with a single point, the following four probabilities are all the same:

> This is not necessarily true for a discrete random variable.

$$P(a \leq X \leq b) = P(a < X \leq b) = P(a \leq X < b) = P(a < X < b). \quad (6.1)$$

In fact, we can remove as many single points as we want from any interval, and the probability will stay the same. The only reasonable probability questions concerning continuous random variables involve intervals. We can almost always sketch a graph to visualize these probabilities, or regions. ▶ ■

So, how do we find area under a curve, and therefore probability? In general this is a calculus question. Don't panic. We'll use a little geometry, tables, and technology to find the necessary area (probability).

The (continuous) **uniform distribution** provides a good opportunity to illustrate the connection between area under the curve and probability. For this random variable, the total probability, 1, is distributed evenly, or uniformly, between two values.

The task of computing probabilities associated with this random variable reduces to finding the area of a rectangle.

DEFINITION

The random variable X has a **uniform distribution** on the interval $[a, b]$ if

$$f(x) = \begin{cases} \dfrac{1}{b-a} & \text{if } a \leq x \leq b \\ 0 & \text{otherwise} \end{cases} \qquad -\infty < a < b < \infty \qquad (6.2)$$

$$\mu = \frac{a+b}{2}, \quad \sigma^2 = \frac{(b-a)^2}{12}. \qquad (6.3)$$

💡 **ILLUMINATING THE CONCEPTS**

1. a and b can be any real numbers, as long as a is less than b ($a < b$).

2. All of the probability (action) is between a and b. The probability density function is the constant $1/(b-a)$ between a and b, and zero outside of this interval. Hence, there is no area and no probability outside the interval $[a, b]$. Figure 6.5 shows a graph of the uniform probability density function.

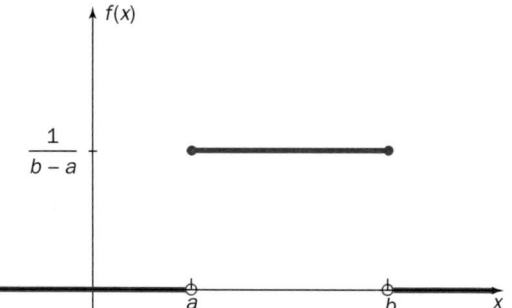

Figure 6.5 The graph of the probability density function for a uniform random variable.

3. Equation 6.2 is a valid probability density function because $f(x) \geq 0$ for all x, and the total area under the curve is 1. The area under the curve for $x < a$ is 0, and the area under the curve for $x > b$ is 0. Between a and b, the area under the curve is the area of a rectangle (Area = Width × Height). See Figure 6.6.

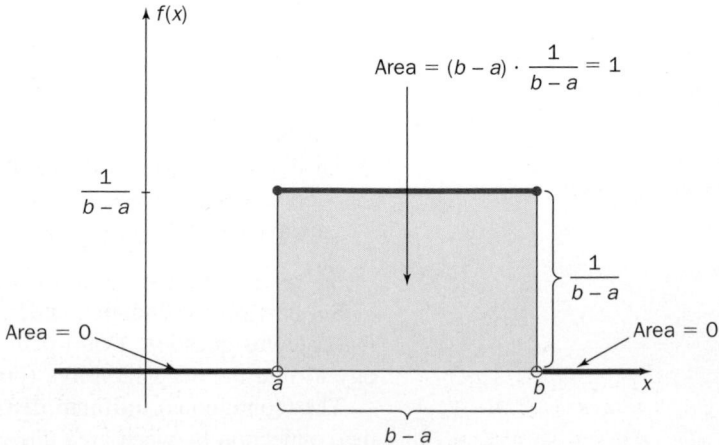

Figure 6.6 The total area under the curve is 1. Here, the density *curve* consists of three line segments. ■

The following example involves a uniform distribution and illustrates visualizing and calculating probabilities associated with a continuous random variable.

Example 6.1 Time to Fill a Prescription   Some people complain about the length of time it takes for a pharmacist to fill a prescription. It takes time because the pharmacist must follow several steps in order to prepare prescription medication. The pharmacist inspects a customer's medication history, looks for drug interactions, and considers possible allergic reactions. A careful audit at a Medicap Pharmacy has shown that the time it takes to fill a randomly selected prescription has a uniform distribution between 5 and 25 minutes. Suppose a prescription is selected at random.

a. Carefully sketch a graph of the probability density function.
b. Find the probability that it takes at most 10 minutes to fill the prescription.
c. Find the probability that it takes between 10 and 20 minutes to fill the prescription.
d. Find the mean time it takes to fill a prescription, the variance, and the standard deviation.

Solution Trail

KEYWORDS

- Uniform distribution.
- Between 5 and 25 minutes.

$\downarrow$

TRANSLATION

- Uniform random variable.
- $a = 5$, $b = 25$.

$\downarrow$

CONCEPTS

- Uniform probability distribution.
- Probability is area under the density curve.

$\downarrow$

VISION

The time it takes to fill a prescription is uniform between 5 and 25. As before, the two most important components for solving this kind of problem are the probability distribution and the probability statement. Use Equation 6.2 to draw the density function, and sketch a graph corresponding to each probability statement.

SOLUTION

a. Let X be the time it takes to fill a prescription. X is uniform between the times $a = 5$ and $b = 25$. Use Equation 6.2 to find

$$\frac{1}{b - a} = \frac{1}{25 - 5} = \frac{1}{20} = 0.050.$$

The probability density function is

$$f(x) = \begin{cases} 0.050 & \text{if } 5 \leq x \leq 25 \\ 0 & \text{otherwise} \end{cases}$$

Figure 6.7 is a graph of the probability density function.

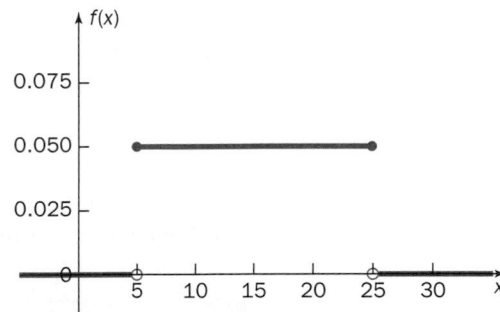

Figure 6.7 The graph of the probability density function for a uniform random variable on the interval $a = 5$ to $b = 25$.

b. We have the distribution of X. Translate the question in part (b) into a probability statement, sketch the region corresponding to the probability statement, and find the area of that region.

At most 10 means up to and including 10. We need the probability that X is less than or equal to 10: $P(X \leq 10)$. See Figure 6.8.

$$\begin{aligned} P(X \leq 10) &= \text{area under the density curve between 5 and 10} \\ &= \text{area of a rectangle} \\ &= \text{width} \times \text{height} \\ &= (5)(0.050) = 0.25 \end{aligned}$$

The probability that it takes at most 10 minutes is 0.25.

The probability statement $P(X \leq 10)$ simplifies to $P(5 \leq X \leq 10)$ in this case because there is no probability (area) for X less than 5.

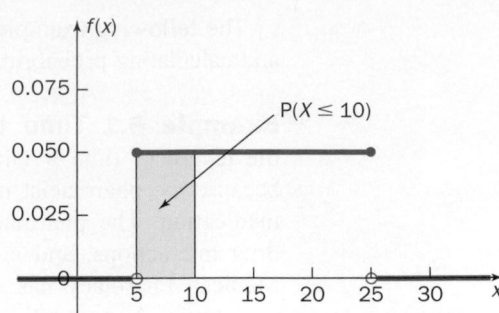

Figure 6.8 The area of the shaded region is $P(X \leq 10)$.

P(10 ≤ X ≤ 20)
  = P(10 < X ≤ 20)
  = P(10 ≤ X < 20)
  = P(10 < X < 20)

c. The probability that it takes between 10 and 20 minutes to fill the prescription in terms of the random variable X is $P(10 \leq X \leq 20)$. Even though the word *inclusive* is not used in the question, we chose to write the interval including the endpoints. It doesn't really matter! Remember: In a continuous world, single values contribute no probability and do not change the probability calculation.

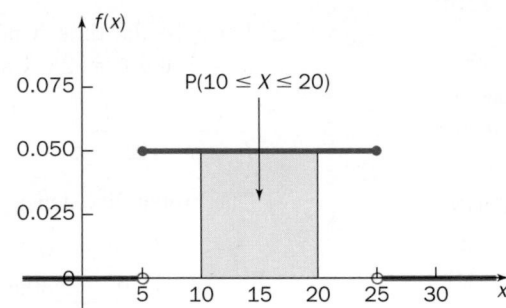

Figure 6.9 The area of the shaded region is $P(10 \leq X \leq 20)$.

$$P(10 \leq X \leq 20) = \text{area under the density curve between 10 and 20}$$
$$= \text{area of a rectangle}$$
$$= \text{width} \times \text{height}$$
$$= (10)(0.050) = 0.50$$

The probability that it takes between 10 and 20 minutes is 0.50.

d. Use Equation 6.3 to find the mean and variance.

$$\mu = \frac{a+b}{2} = \frac{5+25}{2} = \frac{30}{2} = 15$$

The mean time it takes to fill a prescription is 15 minutes.

Since the uniform distribution is *symmetric*, the mean is the middle of the distribution, and the mean is equal to the median.

$$\sigma^2 = \frac{(b-a)^2}{12} = \frac{(25-5)^2}{12} = \frac{20^2}{12} = \frac{400}{12} \approx 33.3$$
$$\sigma = \sqrt{\sigma^2} = \sqrt{33.3} \approx 5.8$$

Challenge: Find a length of time, t, such that 90% of prescriptions are filled within t minutes.

The standard deviation is approximately 5.8 minutes.

In order to find a probability associated with any continuous random variable, we will often rewrite the probability statement to use cumulative probability. From Chapter 5, cumulative probability means *accumulate* probability up to and including a fixed value. Find all the area under the density curve to the *left* of the fixed value. For a continuous random variable X, the cumulative probability up to x is $P(X \leq x)$. Figure 6.10 illustrates this cumulative probability.

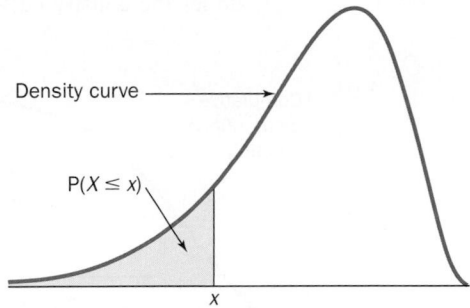

Figure 6.10 The shaded area is the cumulative probability $P(X \leq x)$.

It doesn't matter whether we use $\leq$ or $<$, since one point contributes no probability. However, for consistency and accuracy throughout this text, cumulative probability means *up to and including x*, so we use $\leq$ (not $<$).

Suppose X is a continuous random variable, and a and b are constants. Here are some typical probability statements involving X and equivalent expressions using cumulative probability. See Figures 6.11 and 6.12.

$$P(X \geq b) = 1 - P(X < b) \qquad \text{The Complement Rule.}$$
$$= 1 - \underbrace{P(X \leq b)}_{\text{cumulative probability}} \qquad \text{A single value contributes no probability.}$$

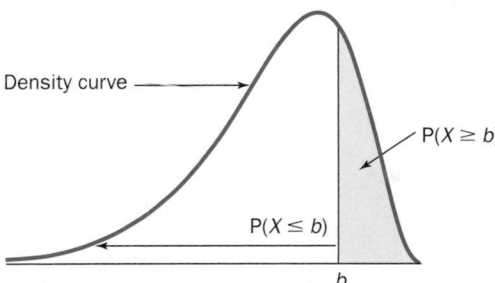

Figure 6.11 Use the Complement Rule to convert to cumulative probability.

$$P(a \leq X \leq b) = P(X \leq b) - P(X < a)$$
$$= P(X \leq b) - P(X \leq a) \qquad \text{A single value contributes no probability}$$

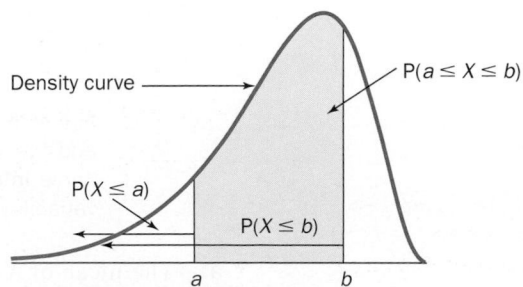

Figure 6.12 Subtract cumulative probabilities to obtain $P(a \leq X \leq b)$.

Find all the probability up to b, find all the probability up to a, and subtract. The difference is the probability that X lies in the interval from a to b. ●

Here is one more way to picture cumulative probability. As x moves from left to right, we accumulate more and more probability. As x increases, cumulative probability also increases. Imagine starting at an altitude of zero and walking up a (smooth) hill. At any point along the walk, measure the altitude. This distance is the cumulative probability. Figure 6.13 shows the relationship between the area under the density curve and the altitude.

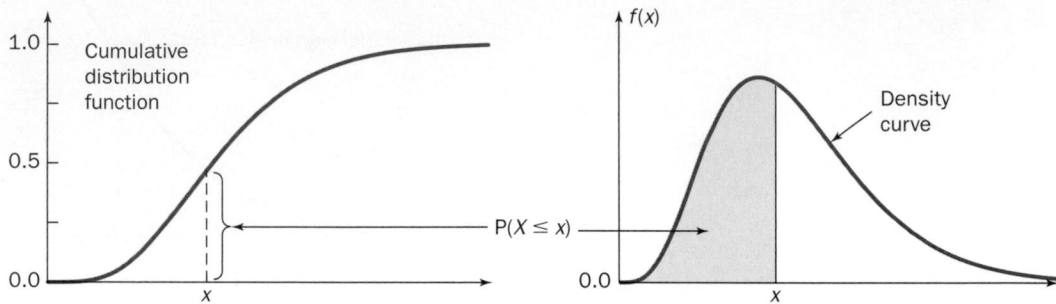

Figure 6.13 Picturing cumulative probability: The altitude is equal to the shaded area.

 ILLUMINATING THE CONCEPTS

Why is 1 the maximum value of the cumulative distribution function?

1. The graph on the left in Figure 6.13 is a graph of a **cumulative distribution function**. This function starts at 0 and is always increasing, until it reaches a maximum value of 1.

2. The mean, μ, and the variance, σ^2, for a continuous random variable are computed using calculus. Although we will not consider any of these calculations, we will interpret and use these values as usual. μ is a measure of the *center* of the distribution, and σ^2 (or σ) is a measure of the spread, or *variability*, of the distribution. Figure 6.14 shows density functions for the random variables X and Y.

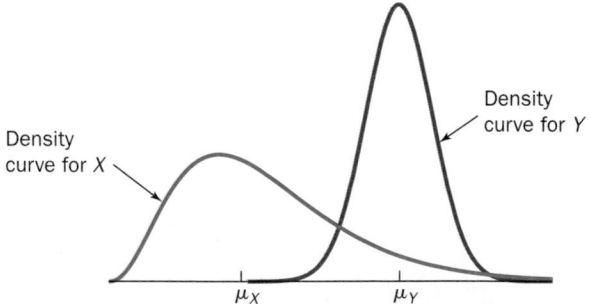

Figure 6.14 Density functions for X and Y. The mean and the variance (and standard deviation) convey the same information as before about center and variability.

a. The mean of X is less than the mean of Y, $\mu_X < \mu_Y$, because the *center* of the distribution of X is to the left of the *center* of the distribution of Y.

b. The standard deviation of X is greater than the standard deviation of Y, $\sigma_X > \sigma_Y$, because the distribution of X is more spread out, and thus has more variability, than the distribution of Y. ■

The following example illustrates the use of cumulative probability to compute probability associated with a continuous random variable.

Example 6.2 Keeping Good Time Each Citizen wristwatch is carefully tested for accuracy before it is packaged and shipped. If the watch gains or loses time during the 24-hour testing period, it is sent to a technician for adjustment. The time inconsistency (in seconds) is a random variable, X, with probability density function shown in Figure 6.15. A negative value of X indicates that the watch lost time, and a positive value indicates that the watch gained time.

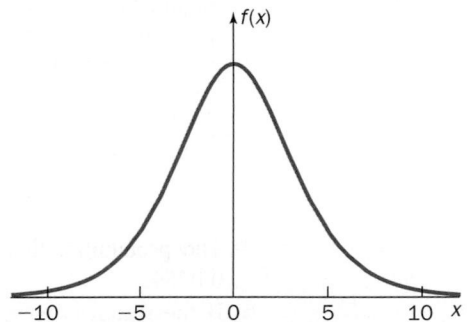

Figure 6.15 The probability density function for the time inconsistency of a wristwatch.

Cumulative probability for X is illustrated in Figure 6.16 and can be computed using the equation that follows. The cumulative probability (the area of the shaded region in Figure 6.16) is

Recall: e is the base of the natural logarithm; $e \approx 2.71828$. Most calculators have a specific key for e.

$$P(X \le x) = \frac{1}{1 + e^{-x/2}} \text{ for all } x. \tag{6.4}$$

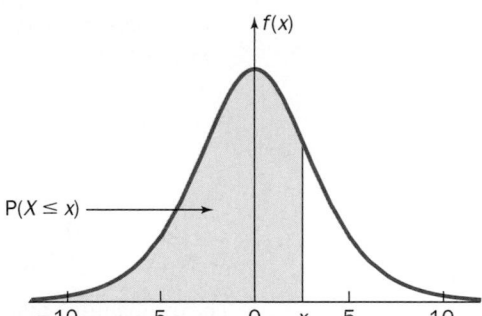

Figure 6.16 The cumulative probability for the time inconsistency of a wristwatch.

Suppose a watch is randomly selected.
a. What is the probability that the watch is 5 seconds slow or slower?
b. What is the probability that the watch is more than 10 seconds fast?
c. What is the probability that the watch is between 3 seconds slow and 3 seconds fast?

SOLUTION

a. If the watch is 5 seconds slow or slower, this means $X \le -5$. The probability $P(X \le -5)$ is cumulative probability already. The calculation and the graphical interpretation (Figure 6.17) follow.

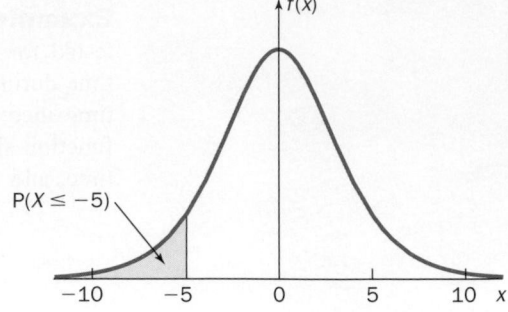

Figure 6.17 The cumulative probability for watch inconsistency, P(X ≤ 5).

$$P(X \le -5) = \frac{1}{1 + e^{-(-5)/2}}$$

Use Equation 6.4.

$$= 0.0759$$

The probability that a randomly selected watch is 5 seconds slow or slower is 0.0759.

b. If the watch is more than 10 seconds fast, this means $X > 10$. To compute the corresponding probability, use the Complement Rule to convert to an expression involving cumulative probability, and use Equation 6.4. See Figure 6.18.

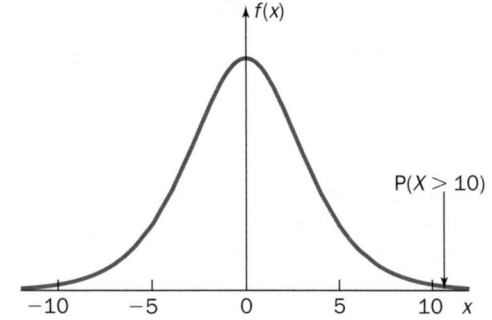

Figure 6.18 The probability for watch inconsistency, $P(X > 10)$.

$$P(X > 10) = 1 - P(X \le 10)$$

The Complement Rule.

$$= 1 - \frac{1}{1 + e^{-10/2}}$$

Use Equation 6.4.

$$= 1 - 0.9933$$

Simplify.

$$= 0.0067$$

The probability that a randomly selected watch is at least 10 seconds fast is 0.0067.

c. If the watch is between 3 seconds slow and 3 seconds fast, this means $-3 \le X \le 3$. To compute the corresponding probability, find the difference between two cumulative probabilities. See Figure 6.19.

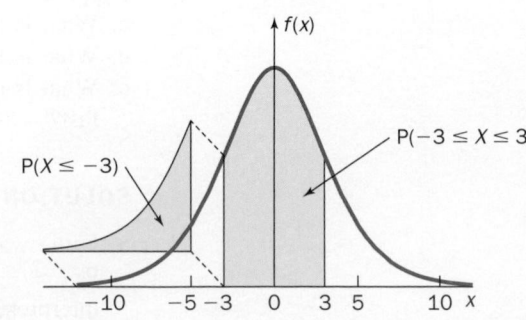

Figure 6.19 The probability for watch inconsistency, $P(-3 \le X \le 3)$.

$$P(-3 \le X \le 3) = P(X \le 3) - P(X < -3)$$
$$= P(X \le 3) - P(X \le -3) \quad \text{A single value contributes no probability.}$$
$$= \left(\frac{1}{1 + e^{-3/2}}\right) - \left(\frac{1}{1 + e^{-(-3)/2}}\right) \quad \text{Use Equation 6.4.}$$
$$= 0.8176 - 0.1824 = 0.6352 \quad \text{Simplify.}$$

The probability that a randomly selected watch is between 3 seconds slow and 3 seconds fast is 0.6352. ●

Sometimes a probability is given and we need to work *backward* in order to find a solution. Consider the following example.

(Cullenphoto/Dreamstime.com)

This is a *backward* problem since we know the probability (0.75) and need to find a starting point (*w*), a value of *X* that produces this probability.

Example 6.3 Eaglet Weights   Female bald eagles lay only two eggs per year. The eaglets usually hatch a few days apart and normally only one eaglet survives.[1] Suppose the weight of a randomly selected eaglet at birth, X, has a uniform distribution between 100 and 125 grams. Find the weight w such that 75% of all eaglets weigh at most w.

SOLUTION

STEP 1   Since X has a uniform distribution with $a = 100$ and $b = 125$, the probability density function is:

$$f(x) = \begin{cases} \dfrac{1}{25} = 0.04 & 100 \le x \le 125 \\ 0 & \text{otherwise} \end{cases}$$

We need to find the value of w such that $P(X \le w) = 0.75$.

STEP 2   From Figure 6.20, the probability that an eaglet weighs less than w is

$$P(X \le w) = \text{area under the density curve from 100 to } w$$
$$= \text{area of a rectangle}$$
$$= \text{width} \times \text{height}$$
$$= (w - 100)(0.04)$$

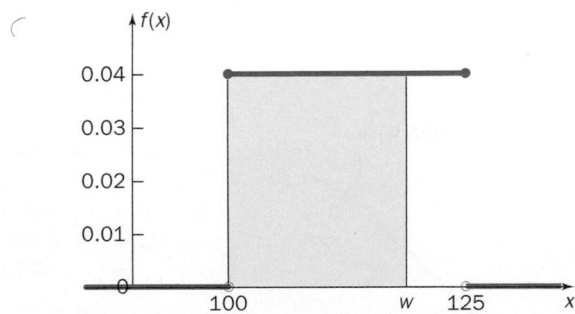

Figure 6.20 The area of the shaded region is $P(X \le w)$.

STEP 3   Set the expression for probability equal to 0.75 and solve for w.

$$(w - 100)(0.04) = 0.75$$
$$w - 100 = \frac{0.75}{0.04} = 18.75 \quad \text{Divide both sides by 0.04.}$$
$$w = 18.75 + 100.00 = 118.75 \quad \text{Add 100 to both sides.}$$

Seventy-five percent of all eaglets weigh at most 118.75 grams. ●

SECTION 6.1 EXERCISES

Practice

6.1 Suppose X is a uniform random variable with $a = 0$ and $b = 16$.
a. Carefully sketch a graph of the probability density function for X.
b. Find the mean, variance, and standard deviation of X.
c. Find $P(X \geq 4)$.
d. Find $P(2 \leq X < 12)$.
e. Find $P(X \leq 7)$.

6.2 Suppose X is a uniform random variable with $a = -5$ and $b = 25$.
a. Carefully sketch a graph of the probability density function for X.
b. Find the mean, variance, and standard deviation of X.
c. Find $P(-10 < X < -1)$.
d. Find $P(X > 0)$ and $P(X \geq 0)$.
e. Find $P(X \geq 20 \,|\, X \geq 10)$.

6.3 Suppose X is a uniform random variable with $a = 50$ and $b = 100$.
a. Find the mean, variance, and standard deviation of X.
b. Find $P(\mu - \sigma \leq X \leq \mu + \sigma)$.
c. Find $P(X \geq \mu + 2\sigma)$.
d. Find a value c such that $P(X \leq c) = 0.20$.

6.4 Suppose X is a uniform random variable with $a = 25$ and $b = 65$.
a. Find the mean, variance, and standard deviation of X.
b. Find the probability that X is more than two standard deviations from the mean.
c. Find a value c such that $P(X \geq c) = 0.40$.
d. Suppose two values of X are selected at random. What is the probability that both values are between 30 and 40?

6.5 Suppose X is a continuous random variable with probability density function given by

$$f(x) = \begin{cases} \dfrac{x}{8} & \text{if } 0 \leq x \leq 4 \\ 0 & \text{otherwise} \end{cases}$$

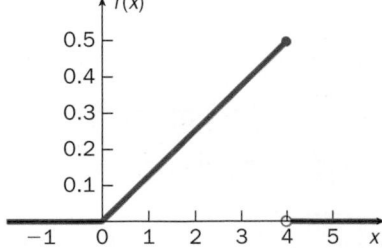

a. Find $P(X \leq 1)$.
b. Find $P(X > 3)$.
c. Find $P(X > 4)$.
d. Find $P(2 \leq X \leq 3)$.
e. Find $P(X \leq 2 \,|\, X \leq 3)$.
f. Find a value c such that $P(X \leq c) = 0.5$. Explain why c is *not* equal to 2.

Applications

6.6 Manufacturing and Product Development A Gold Canyon candle is designed to last nine hours. However, depending on the wind, air bubbles in the wax, the quality of the wax, and the number of times the candle is relit, the actual burning time (in hours) is a uniform random variable with $a = 6.5$ and $b = 10.5$. Suppose one of these candles is randomly selected.
a. Find the probability that the candle burns at least seven hours.
b. Find the probability that the candle burns at most eight hours.
c. Find the mean burning time and the probability that the burning time of a randomly selected candle will be within one standard deviation of the mean.
d. Find a time t such that 25% of all candles burn longer than t hours.

6.7 Sports and Leisure According to Major League Baseball rules, a baseball should weigh between 5 and 5.25 ounces and have a circumference of between 9 and 9.25 inches. Suppose the weight of a baseball (in ounces) has a uniform distribution with $a = 5.085$ and $b = 5.155$, and the circumference (in inches) has a uniform distribution with $a = 9.0$ and $b = 9.1$.
a. Find the probability that a randomly selected baseball has a weight greater than 5.14 ounces.
b. Find the probability that a randomly selected baseball has a circumference less than 9.03 inches.
c. Suppose the weight and the circumference are independent. Find the probability that a randomly selected baseball will have a weight between 5.11 and 5.13 ounces and a circumference between 9.04 and 9.06 inches.

6.8 Manufacturing and Product Development Premanufactured wooden roof trusses allow builders to complete projects faster and with lower on-site labor costs. The connector plates for trusses are made from Grade A steel and are hot-dip galvanized. The thickness of a truss connector (in inches) varies slightly and has a uniform distribution with $a = 0.036$ and $b = 0.050$.
a. If the manufacturer will only use connectors with a minimum thickness of 0.04 inches, what proportion of connectors is rejected?
b. Suppose a truss connector is selected at random. Find the probability that the truss connector has a thickness between 0.042 and 0.045.
c. Find the mean, variance, and standard deviation of the thickness of a truss connector.

6.9 Travel and Transportation When the Department of Transportation (DOT) repaints the center lines, edge lines, or no-passing-zone lines on a highway, epoxy paint is sometimes applied. This paint is more expensive than latex but lasts longer. If this paint splashes onto a vehicle, it has to be completely sanded off, and that area of the vehicle has to be repainted. The DOT has warned motorists that the drying time for this epoxy paint (in minutes) has a uniform distribution with $a = 30$ and $b = 60$. Suppose epoxy paint is applied to a small section of center line.
a. What is the probability that the paint will be dry within 45 minutes?

b. What is the probability that the paint will be dry between 40 and 50 minutes?

c. Find a value t such that the probability of the paint taking at least t minutes to dry is 0.75.

d. If the DOT road crew removes all of the cones on the center line 55 minutes after painting, what is the probability that the paint is still wet at this time?

6.10 Physical Sciences In Grafton, a rural area in Vermont, the distance (in meters) between telephone poles has a uniform distribution with $a = 40$ and $b = 65$. Suppose two consecutive telephone poles are selected at random.

a. What is the probability that the distance between the poles is less than 60 meters?

b. What is the probability that the distance between the poles is between 45 and 55 meters?

c. Any distance between poles greater than 50 meters is considered to be *environment friendly*. What is the probability that the distance is environment friendly?

Extended Applications

6.11 Psychology and Human Behavior Melatonin is widely used to help people with insomnia fall asleep. The probability density function for X, the time (in minutes) it takes to fall asleep after taking a melatonin tablet, is given as follows.

$$f(x) = \begin{cases} -0.005x + 0.1 & \text{if } 0 \leq x \leq 20 \\ 0 & \text{otherwise} \end{cases}$$

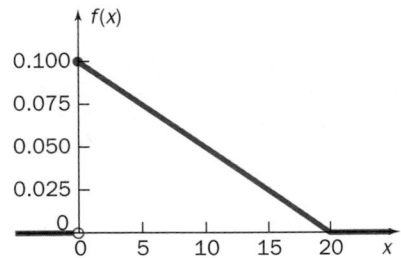

a. Verify that this is a valid probability density function.

b. If a randomly selected person takes a melatonin tablet at bedtime, what is the probability that he or she will fall asleep within 5 minutes?

c. What is the probability that the person will fall asleep between 5 and 10 minutes after taking a tablet?

d. Find a value t such that the probability of falling asleep within t minutes after taking a tablet is 0.50.

e. If it takes less than 15 minutes to fall asleep after taking a tablet, people consider the medication a success. Suppose 20 people are selected at random. What is the probability that exactly 15 people fall asleep successfully? What is the probability that at least 18 people fall asleep successfully? What is the probability that all 20 people fall asleep successfully?

6.12 Biology and Environmental Science The grasshopper sparrow lives in grassy areas and feeds mostly on insects. The territory size of this bird varies from 1 to 3 acres.[2] The probability density function for territory size is given below. Suppose a grasshopper sparrow is selected at random.

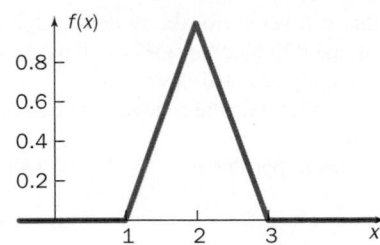

a. What is the probability that the bird has a territory size less than 2 acres?

b. What is the probability that the bird has a territory size less than 1.4 acres?

c. What is the probability that the bird has a territory size greater than 2.6 acres?

d. What is the probability that the bird has a territory size between 1.4 and 2.6 acres?

6.13 Marketing and Consumer Behavior Marini's candy store on the beach boardwalk in Santa Cruz sells candy in bulk. Customers can mix products from over 100 barrels. The probability distribution for the number of pounds of candy purchased by a randomly selected customer is shown below.

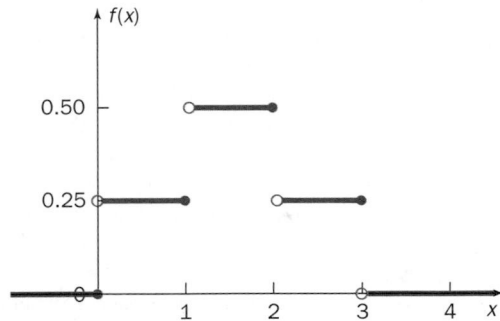

a. Verify that this is a valid probability density function.

b. Find the probability that the next customer buys at most 2 pounds of candy.

c. Find the probability that the next customer buys more than 1 pound of candy.

d. Suppose the next customer buys at most 1.5 pounds of candy. What is the probability that he or she buys at most 0.5 pounds of candy?

6.14 Economics and Finance On any given trading day, the fluctuation, or change, in the price (in dollars) of CalAmp stock, listed on the New York Stock Exchange, is between -2.00 and 2.00. Suppose the change in price is a random variable with probability density function shown as follows.

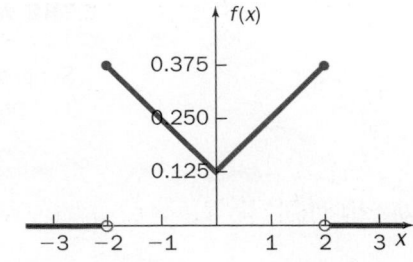

a. Verify that this is a valid probability density function.
b. What is the probability that the stock price increases by at least $1.00 on a randomly selected day?
c. What is the probability that the change in stock price is between -1.00 and 1.00?
d. Find a value c such that $P(-c \leq X \leq c) = 0.90$.

Challenge

6.15 Marketing and Consumer Behavior Dinner customers at the Red Iguana Restaurant often experience a long wait for a table. For a randomly selected customer who arrives at the restaurant between 6:00 P.M. and 7:00 P.M., the waiting time (in minutes) is a continuous random variable such that

$$P(X \leq x) = \begin{cases} 1 - e^{-0.05x} & \text{if } x \geq 0 \\ 0 & \text{otherwise} \end{cases}$$

Suppose a dinner customer is randomly selected.
a. What is the probability that the person must wait for a table at most 20 minutes?
b. What is the probability that the person must wait for a table over one half-hour?
c. What is the probability that the person must wait for a table between 15 and 30 minutes?

6.16 Psychology and Human Behavior Parents with children under age 16 often spend a lot of time during the day driving their kids to various places, for example, to and from after-school activities, music practice, sports practices and games, the library, and friends' homes. Suppose a family has k child(ren) under age 16 ($k = 1, 2, 3, 4, 5$), and let the random variable X_k be the time (in hours) spent *taxiing* during the day. X_k has a uniform distribution with $a = 0$ and $b = k$. For example, for a family with two children, X_2 has a uniform distribution with $a = 0$ and $b = 2$.
a. For a family with three children, what is the probability that parents will spend less than one hour driving kids on a randomly selected day?

b. For a family with four children, what is the mean number of hours spent driving kids? What is the probability that the driving time will be greater than two standard deviations from the mean?
c. For a family with five children under 16, find a time t such that the probability of driving kids more than t hours is 0.25.
d. Suppose five families are selected at random, the first with one child under 16, the second with two children under 16, etc. What is the probability that all five families drive less than 30 minutes on a randomly selected day? What is the probability that all five families drive more than 90 minutes on a randomly selected day?

6.17 Suppose X is a continuous random variable such that

$$P(X \leq x) = \begin{cases} 1 - e^{-x^2/8} & \text{if } x \geq 0 \\ 0 & \text{otherwise} \end{cases}$$

a. Find $P(X \leq 4)$.
b. Find $P(X > 2)$.
c. Find $P(1 \leq X \leq 3)$.
d. Find $P(X \leq 2 \mid X \leq 4)$.

6.18 Sports and Leisure A figure skating routine is designed to last 6 minutes. The amount of time (in minutes) less than or greater than 6 minutes is a random variable, X, with probability density function given by

$$f(x) = \begin{cases} \sqrt{\dfrac{2}{\pi} - x^2} & \text{if } -\sqrt{\dfrac{2}{\pi}} \leq x \leq \sqrt{\dfrac{2}{\pi}} \\ 0 & \text{otherwise} \end{cases}$$

If the value of X is negative, then the routine was under 6 minutes. If the value of X is positive, the routine went too long.
a. Carefully sketch a graph of the density function.
b. Find the probability that a randomly selected performance is within $1/\sqrt{\pi}$ minutes of 6. That is, find

$$P\left(-\frac{1}{\sqrt{\pi}} \leq X \leq \frac{1}{\sqrt{\pi}}\right).$$

6.2 THE NORMAL DISTRIBUTION

The **normal probability distribution** is very common and is the most important distribution in all of statistics. This *bell-shaped* density curve can be used to model many natural phenomena and is used extensively in statistical inference. Recall, a random variable is completely described by certain *parameters*, for example, a binomial random variable by n and p, and a Poisson random variable by λ. A normal distribution is completely characterized, or determined, by its mean μ and variance σ^2 (or by its mean μ and standard deviation σ).

THE NORMAL PROBABILITY DISTRIBUTION

Suppose that X is a normal random variable with mean μ and variance σ^2. The probability density function is given by

$$f(x) = \frac{1}{\sigma\sqrt{2\pi}} e^{-(x-\mu)^2/2\sigma^2}, \text{ and} \tag{6.5}$$

$$-\infty < x < \infty, \quad -\infty < \mu < \infty, \quad \sigma^2 > 0 \tag{6.6}$$

ILLUMINATING THE CONCEPTS

We've seen *e* before, in the Poisson distribution.

1. In this probability density function, *e* is the base of the natural logarithm; $e \approx 2.71828$. π is another constant, commonly used in trigonometry; $\pi \approx 3.14159$.

2. We use the shorthand notation $X \sim N(\mu, \sigma^2)$ to indicate that X is (distributed as) a normal random variable with mean μ and variance σ^2.

   For example, $X \sim N(5, 36)$ means that X is a normal random variable with mean $\mu = 5$ and variance $\sigma^2 = 36$ (and $\sigma = 6$).

3. Equation 6.6 means that x can be any real number (the density curve continues forever in both directions), the mean μ can be any real number (positive or negative or zero), and the variance can be any positive real number.

4. For *any* mean μ and variance σ^2, the density curve is symmetric about the mean μ, unimodal, and bell-shaped as shown in Figure 6.21.

To visualize what *bell-shaped* means, place a bell on a table and pass a plane (a piece of paper) *through* the bell perpendicular to the table. The intersection of the plane and the bell is a *bell-shaped curve*.

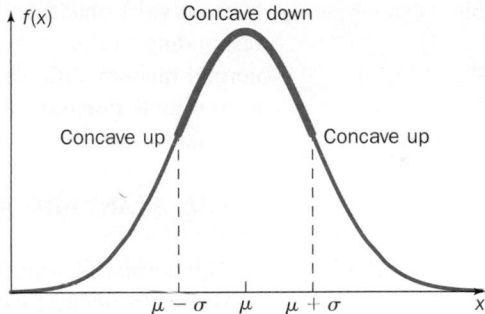

Figure 6.21 Graph of the probability density function for a normal random variable with mean μ and variance σ^2.

◀ The graph of the probability density function changes *concavity* at $x = \mu - \sigma$ and again at $x = \mu + \sigma$. ▶

The mean is equal to the median. (Why?)

◀ It can be shown (using calculus) that the total area under this density curve is 1 (even though it extends forever in both directions, getting closer and closer to the *x*-axis but never touching it). ▶

5. The mean μ is a *location parameter*, and the variance σ^2 determines the spread of the distribution. As the variance increases, the total area under the probability density function (1) is rearranged. The graph is compressed down and pushed out (on the tails). Figures 6.22 and 6.23 show the effects of μ and σ^2 on the location (center) and spread of the density curve.

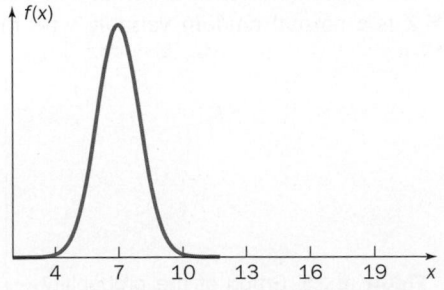

Figure 6.22 Normal probability density function with $\mu = 7$ and small σ^2.

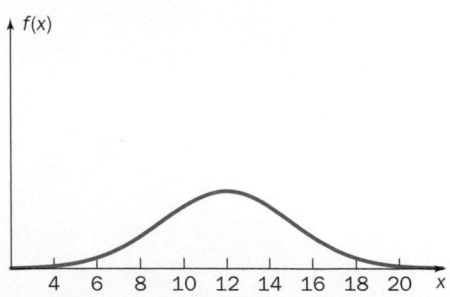

Figure 6.23 Normal probability density function with $\mu = 12$ and large σ^2. ∎

Suppose that X is a normal random variable with mean μ and variance σ^2: $X \sim \mathrm{N}(\mu, \sigma^2)$. The probability X lies in some interval; for example $[a, b]$ is the area under the density curve between a and b (Figure 6.24).

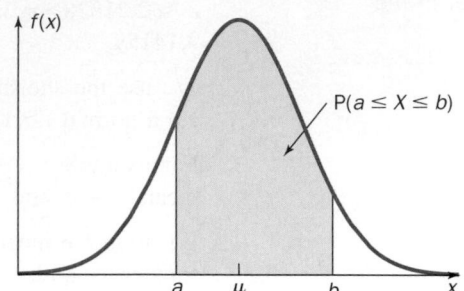

Figure 6.24 The shaded region corresponds to P($a \leq X \leq b$).

The shaded region in Figure 6.24 is not a a simple geometric figure; it's bounded by a curve! Consequently, there is no *nice* formula for the area of this region, corresponding to P($a \leq X \leq b$). However, a probability statement associated with *any* normal random variable can be *transformed* into an equivalent expression involving a **standard normal random variable** (defined below). Cumulative probabilities associated with this distribution are provided in the Appendix, Table III.

THE STANDARD NORMAL DISTRIBUTION

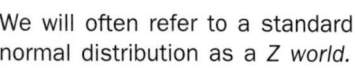

Let $\mu = 0$ and $\sigma = 1$ in Equation 6.5.

The normal distribution with $\mu = 0$ and $\sigma^2 = 1$ (and $\sigma = 1$) is called the **standard normal distribution**. A random variable that has a standard normal distribution is called a **standard normal random variable**, usually denoted Z. The probability density function for Z is given by

$$f(z) = \frac{1}{\sqrt{2\pi}} e^{-z^2/2}, \quad -\infty < z < \infty. \tag{6.7}$$

💡 ILLUMINATING THE CONCEPTS

We will often refer to a standard normal distribution as a *Z world*.

1. In Equation 6.7 the independent variable z is used to define the probability density function simply because the standard normal random variable is usually denoted by Z.

2. Figure 6.25 shows a graph of the probability density function for a standard normal random variable. The mean is $\mu = 0$ and the standard deviation is $\sigma = 1$. Note that most of the probability (area) is within three standard deviations of the mean, between -3 and 3. The shorthand notation $Z \sim \mathrm{N}(0, 1)$ means that Z is a normal random variable with mean that 0 and variance 1.

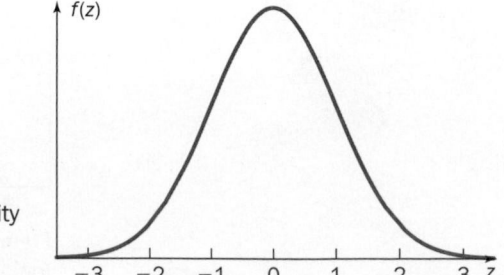

Figure 6.25 Graph of the probability density function for a standard normal random variable.

3. The **standard normal distribution** is not common, but it is used extensively as a *reference* distribution. Any probability statement involving any normal random variable can be transformed into an equivalent expression (with the same probability) involving a Z random variable. We will learn how to *standardize* shortly. Therefore, you need to become an expert at computing probabilities in the Z world. Probabilities associated with Z are computed using cumulative probability, as shown in Figure 6.26, which depicts the steps for computing probabilities associated with a normal random variable.

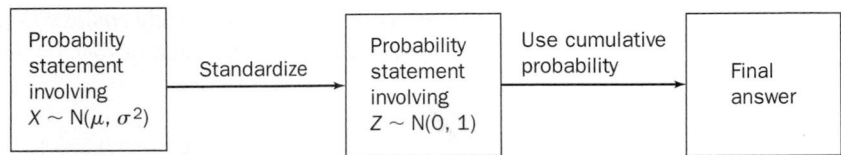

Figure 6.26 Strategy for computing a probability associated with any normal random variable. ■

Probabilities associated with a standard normal random variable, Z, are computed using cumulative probability. Table III in the Appendix contains values for $P(Z \le z)$ for selected values of z. Figure 6.27 shows the geometric region corresponding to $P(Z \le z)$, and Figure 6.28 illustrates the use of Table III. Locate the units and tenths digits in z along the left side of the table. Find the hundredths digit in z across the top row. The intersection of this row and column in the body of the table contains the cumulative probability.

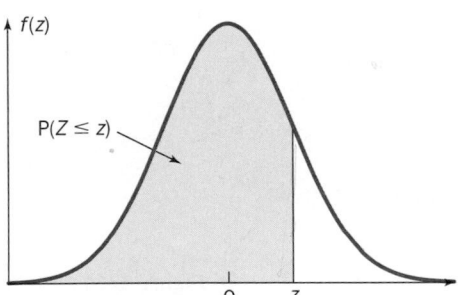

Figure 6.27 The shaded area under the standard normal density curve corresponds to $P(Z \le z)$.

z	0.00	0.01	0.02	0.03	0.04	0.05	0.06	0.07	0.08	0.09
⋮	⋮	⋮	⋮	⋮	⋮	⋮	⋮	⋮	⋮	⋮
1.0	0.8413	0.8438	0.8461	0.8485	0.8508	0.8531	0.8554	0.8577	0.8599	0.8621
1.1	0.8643	0.8665	0.8686	0.8708	0.8729	0.8749	0.8770	0.8790	0.8810	0.8830
1.2	0.8849	0.8869	0.8888	0.8907	0.8925	0.8944	0.8962	0.8980	0.8997	0.9015
1.3	0.9032	0.9049	0.9066	0.9082	0.9099	0.9115	0.9131	0.9147	0.9162	0.9177
1.4	0.9192	0.9207	0.9222	0.9236	0.9251	0.9265	0.9279	0.9292	0.9306	0.9319
⋮	⋮	⋮	⋮	⋮	⋮	⋮	⋮	⋮	⋮	⋮

Figure 6.28 $P(Z \le 1.23) = 0.8907$ in Table III in the Appendix.

The following example illustrates the use of Table III in the Appendix to find probabilities associated with Z.

Example 6.4 Probability Calculations Associated with the Standard Normal Distribution Use Table III in the Appendix to find each probability associated with the standard normal distribution.

a. $P(Z \leq 1.45)$.
b. $P(Z \geq -0.6)$.
c. $P(-1.25 \leq Z \leq 2.13)$.
d. Find the value b such that $P(Z \leq b) = 0.90$.

SOLUTION

a. This expression is already *cumulative probability.* Go directly to Table III (page T-7), and find the intersection of row 1.4 and column 0.05. See Figure 6.29.

A technology solution (a):

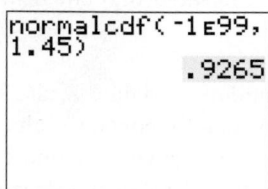

Figure 6.30 $P(Z \leq 1.45)$

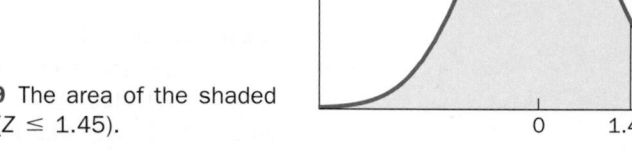

Figure 6.29 The area of the shaded region is $P(Z \leq 1.45)$.

$P(Z \leq 1.45) = \boxed{0.9265}$ Cumulative probability; use Table III in the Appendix.

b. This is a right-tail probability. Convert to cumulative probability and use Table III in the Appendix. See Figure 6.31.

A technology solution (b):

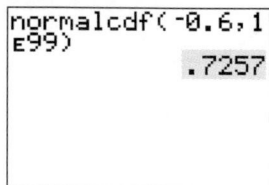

Figure 6.32 $P(Z \geq -0.6)$

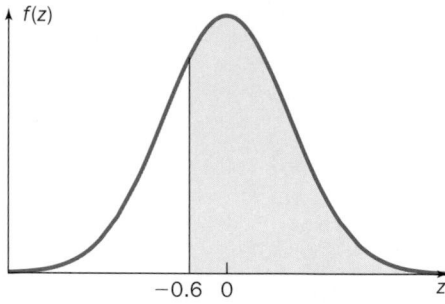

Figure 6.31 The area of the shaded region is $P(Z \geq -0.6)$.

$$
\begin{aligned}
P(Z \geq -0.6) &= 1 - P(Z < -0.6) &&\text{The Complement Rule.}\\
&= 1 - P(Z \leq -0.6) &&\text{One value doesn't matter.}\\
&= 1 - 0.2743 &&\text{Use Table III in the Appendix.}\\
&= \boxed{0.7257}
\end{aligned}
$$

c. Find all the probability up to 2.13, find all the probability up to -1.25, and subtract. The difference is the probability that Z lies in this interval. See Figure 6.33.

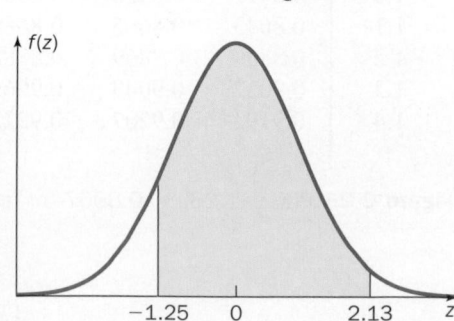

Figure 6.33 The area of the shaded region is $P(-1.25 \leq Z \leq 2.13)$.

A technology solution (c):

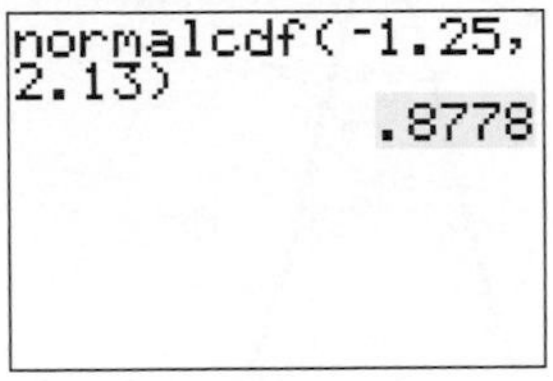

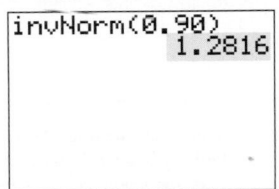

Figure 6.34
$P(-1.25 \leq Z \leq 2.13)$

A technology solution (d):

Figure 6.35 Inverse
cumulative probability.

$P(-1.25 \leq Z \leq 2.13)$
$= P(Z \leq 2.13) - P(Z < -1.25)$ Use cumulative probability.
$= P(Z \leq 2.13) - P(Z \leq -1.25)$ One value doesn't matter.
$= 0.9834 - 0.1056$ Use Table III in the Appendix.
$= 0.8778$

d. In this problem, we need to work *backward* in order to find the solution. This is an **inverse cumulative probability** problem. The cumulative probability is given. We need the value b such that the cumulative probability is 0.90. Search the body of Table III in the Appendix to find a cumulative probability as close to 0.90 as possible. Read the row and column entries to find b.

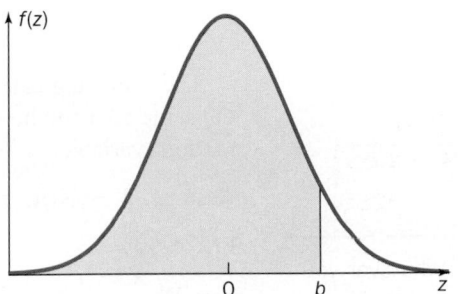

Figure 6.36 The area of the shaded region is $0.90 = P(Z \leq b)$.

In the body of Table III, the closest cumulative probability to 0.90 is 0.8997. This corresponds to $1.28 \approx b$.

Note: Linear interpolation can be used to find a more exact answer. The technology solutions presented use special inverse cumulative probability functions. ●

The following rule provides the connection between *any* normal random variable and the standard normal random variable.

STANDARDIZATION RULE

If X is a normal random variable with mean μ and variance σ^2, then a standard normal random variable is given by

$$Z = \frac{X - \mu}{\sigma} \qquad (6.8)$$

There are other types of standardization.
$Z = (X - \mu)/\sigma$ is the most common.

💡 **ILLUMINATING THE CONCEPTS**

1. The process of converting from X to Z is called **standardization**. Z is a *standardized* random variable.

2. Using this rule, any probability involving a normal random variable can be transformed into an equivalent expression involving a Z random variable. We can then convert to cumulative probability if necessary, and use Table III in the Appendix.

3. The rule above is illustrated in Figure 6.37, using cumulative probability.

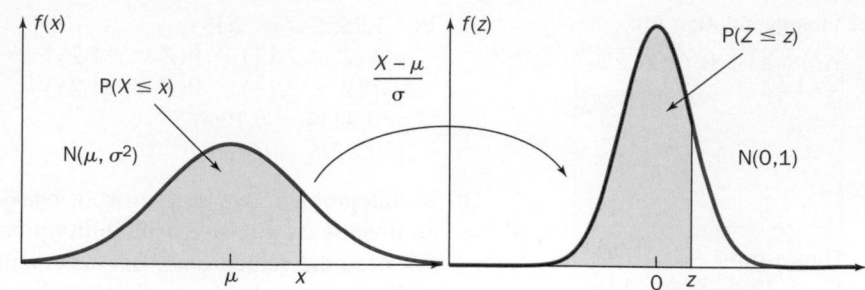

Figure 6.37 An illustration of standardization. The areas of the shaded regions are equal. ∎

The following calculation shows why the two shaded regions in Figure 6.37 have the same area and how to use the rule to compute probabilities involving any normal random variable.

Assume: $X \sim N(\mu, \sigma^2)$.

$P(X \le x)$	The original (cumulative) probability statement.
$= P\left(\dfrac{X - \mu}{\sigma} \le \dfrac{x - \mu}{\sigma}\right)$	Work *within* the probability statement. Subtract the mean of X and divide by the standard deviation of X on both sides of the inequality (standardize).
$= P(Z \le z)$	Apply the standardization rule *within* the probability statement. The expression with X is transformed into Z. The expression with x becomes some fixed value z. Use Table III in the Appendix to find this probability.

> Remember the rule: Whatever you do to one side of the inequality, you have to do to the other side.

The examples below involve normal random variables and standardization. The hardest part of these types of problems is (as before) (1) to define and identify the probability distribution, and (2) to write a probability statement. Given a probability statement involving a normal random variable, all we have to do is standardize and use cumulative probability. Even for backward problems (with a known probability), we still standardize and still use cumulative probability. Note that the technology solutions presented below do *not* require standardization.

Example 6.5 Probability Calculations Associated with a Normal Random Variable Suppose X is a normal random variable with mean 10 and variance 4: $X \sim N(10, 4)$, and $\sigma = \sqrt{4} = 2$.
a. Find $P(X > 12.5)$.
b. Find $P(9 \le X \le 10)$.
c. Find the value b such that $P(X \le b) = 0.75$.

SOLUTION

a. X is normal. We know the mean and standard deviation. Standardize and use cumulative probability associated with Z.

$$P(X > 12.5) = P\left(\frac{X - 10}{2} > \frac{12.5 - 10}{2}\right) \qquad \text{Standardize.}$$
$$= P(Z > 1.25) \qquad \text{Use Equation 6.8; simplify.}$$
$$= 1 - P(Z \le 1.25) \qquad \text{The Complement Rule.}$$
$$= 1 - 0.8944 = 0.1056 \qquad \text{Use Table III in the Appendix.}$$

Standardization illustrated part (b):

P(9 ≤ X ≤ 10)

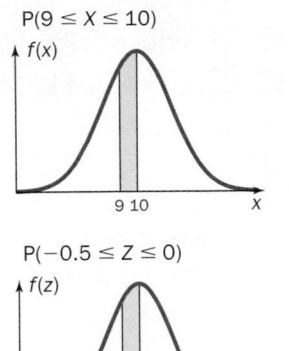

P(−0.5 ≤ Z ≤ 0)

Standardization illustrated part (c):

P(X ≤ b) = 0.75

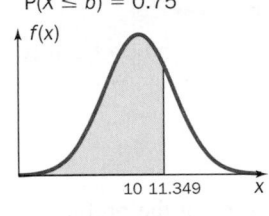

$$P\left(Z \leq \frac{b - 10}{2}\right)$$

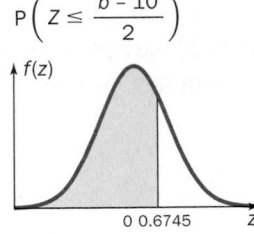

Recall, interpolation is a method of approximation. It is often used to estimate a value at a position between two given values in a table. Linear interpolation assumes that the two known values lie on a straight line.

0.75 is between the known cumulative probabilities 0.7486 and 0.7517. The known values are (0.7486, 0.67) and (0.7517, 0.68).

The approximate z value corresponding to the cumulative probability 0.75 is

0.67 + (0.01)(0.75 − 0.7486)/ (0.7517 − 0.7486) = 0.6745.

Figure 6.38 illustrates this solution.

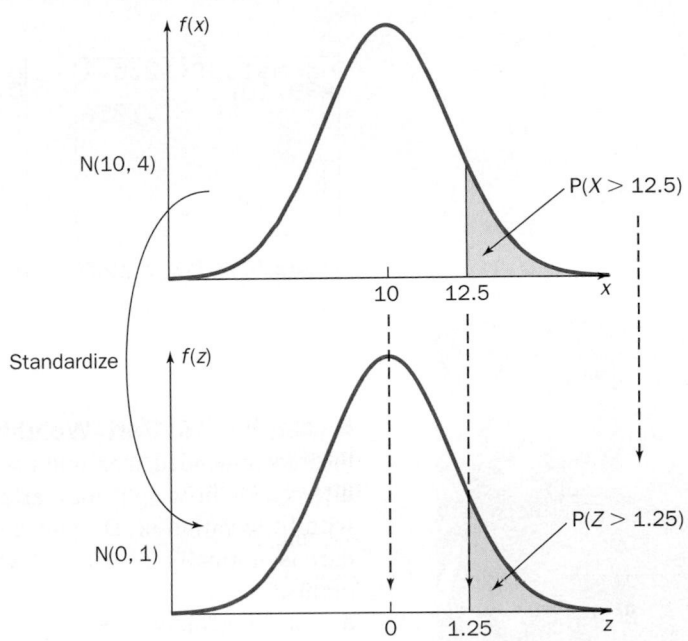

Figure 6.38 Example 6.5 (a) standardization illustrated: 10 is transformed to 0. 12.5 is transformed to 1.25. The areas of the shaded regions are the same.

b. Standardize again. Work within the probability statement to write an equivalent expression involving Z.

$$P(9 \leq X \leq 10) = P\left(\frac{9 - 10}{2} \leq \frac{X - 10}{2} \leq \frac{10 - 10}{2}\right) \quad \text{Standardize.}$$

$$= P(-0.5 \leq Z \leq 0) \quad \text{Use Equation 6.8; simplify.}$$

$$= P(Z \leq 0) - P(Z < -0.5) \quad \text{Use cumulative probability.}$$

$$= P(Z \leq 0) - P(Z \leq -0.5) \quad \text{One value doesn't matter.}$$

$$= 0.5000 - 0.3085 \quad \text{Use Table III in the Appendix.}$$

$$= 0.1915$$

c. Convert the expression into cumulative probability involving Z. Since the probability is already given, this is an inverse cumulative probability problem. Work backward in Appendix Table III.

$$P(X \leq b) = P\left(\frac{X - 10}{2} \leq \frac{b - 10}{2}\right) \quad \text{Standardize.}$$

$$= P\left(Z \leq \frac{b - 10}{2}\right) = 0.75 \quad \text{Use Equation 6.8.}$$

There is no other simplification within the probability statement. However, the resulting probability statement involves Z and is cumulative probability. Find a value in the body of Table III, as close to 0.75 as possible. Set the corresponding z equal to $\left(\dfrac{b - 10}{2}\right)$, and solve for b.

$$\frac{b - 10}{2} = 0.6745 \quad \text{Table III; interpolation.}$$

$$b - 10 = 1.349 \quad \text{Multiply both sides by 2.}$$

$$b = 11.349 \quad \text{Add 10 to both sides.}$$

Therefore, $P(X \leq 11.349) = 0.75$ and hence $b = 11.349$.

Figures 6.39 through 6.41 show technology solutions:

```
normalcdf(12.5,1
E99,10,2)
            .1056
```

```
normalcdf(9,10,1
0,2)
            .1915
```

```
invNorm(.75,10,2
)
         11.3490
```

Figure 6.39 P(X ≥ 12.5).

Figure 6.40
P(9 ≤ X ≤ 10).

Figure 6.41 Inverse
cumulative probability. ●

(iStockphoto)

Recall that [16.55, 17.45] means
16.55 ≤ X ≤ 17.45.

Solution Trail

KEYWORDS

- Normally distributed.
- Mean.
- Standard deviation.

↓

TRANSLATION

- Normal random variable.
- $\mu = 17$.
- $\sigma = 0.3$.

↓

CONCEPTS

- Normal probability distribution.
- Standardization.

↓

VISION

Define a normal random variable
and translate each question into a
probability statement. Standardize
and use cumulative probability
associated with Z if necessary.

Example 6.6 Dart Weights According to the National Dart Association, the recommended maximum weight for soft-tip darts is 18 grams. Since slight differences in weight may affect trajectory, tournament darts are manufactured with little variation. Despite rigorous quality control, the weight of a tournament dart is normally distributed with mean 17 grams and standard deviation 0.3 grams.

a. For a randomly selected tournament dart, find the probability that the weight is between 16.55 and 17.45.

b. Any dart over 18 grams is illegal in tournament play. Find the probability that a randomly selected dart used in a tournament is illegal.

SOLUTION

a. Let X be the weight of a tournament dart. The keywords in the problem suggest $X \sim N(17, 0.09)$, $\sigma = 0.3$.

Between 16.55 and 17.45 means in the interval [16.55, 17.45] (whether it is closed or open doesn't matter). Find the probability that X lies in this interval.

$$P(16.55 \leq X \leq 17.45)$$
$$= P\left(\frac{16.55 - 17}{0.3} \leq \frac{X - 17}{0.3} \leq \frac{17.45 - 17}{0.3}\right) \qquad \text{Standardize.}$$
$$= P(-1.50 \leq Z \leq 1.50) \qquad \text{Use Equation 6.8; simplify.}$$
$$= P(Z \leq 1.50) - P(Z \leq -1.50) \qquad \text{Use cumulative probability.}$$
$$= 0.9332 - 0.0668 \qquad \text{Use Table III in the Appendix.}$$
$$= 0.8664$$

The probability that a randomly selected tournament dart weighs between 16.55 and 17.45 grams is 0.8664.

b. A dart is overweight and illegal if the value of X is greater than 18 grams. Find $P(X > 18)$.

$$P(X > 18) = P\left(\frac{X - 17}{0.3} > \frac{18 - 17}{0.3}\right) \qquad \text{Standardize.}$$
$$= P(Z > 3.33) \qquad \text{Use Equation 6.8; simplify (and round off).}$$
$$= 1 - P(Z \leq 3.33) \qquad \text{Use cumulative probability.}$$
$$= 1 - 0.9996 \qquad \text{Use Table III in the Appendix.}$$
$$= 0.0004$$

The probability that a randomly selected tournament dart is illegal is 0.0004.

Figures 6.42 and 6.43 show technology solutions.

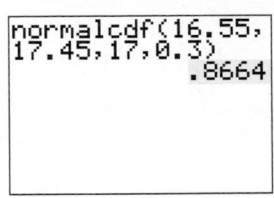

Figure 6.42
P($16.55 \leq X \leq 17.45$).

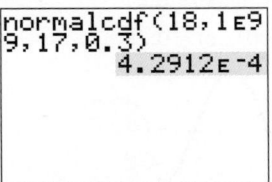

Figure 6.43 P($X > 18$).

Standardization illustrated:

P($X > 18$)

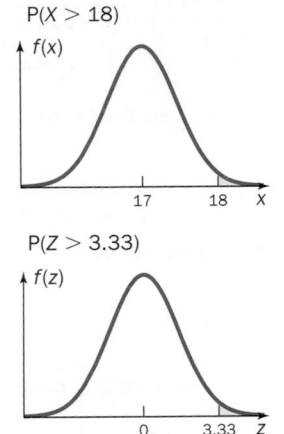

P($Z > 3.33$)

Solution Trail

KEYWORDS

- Normally distributed.
- Mean.
- Standard deviation.
- 95%.

↓

TRANSLATION

- Normal random variable.
- $\mu = 600$.
- $\sigma = 100$.
- Probability 0.95.

↓

CONCEPTS

- Normal probability distribution.
- Standardization.

↓

VISION

Define a random variable and translate the question into a probability statement. Since a probability is given (0.95), this suggests an inverse cumulative probability problem.

Example 6.7 Backpacks and Back Pain   Chronic back pain has become common in children who carry overfilled and overweight backpacks. Heavy school books, notebooks, calculators, and computer equipment, all crammed into a backpack and lugged around all day, increase the chance of neck and shoulder muscle spasms and lower back pain. Research has shown that the total weight carried is directly related to the volume of a backpack. The volume of a randomly selected backpack sold commercially is normally distributed with mean 600 in³ and standard deviation 100 in³. Find a symmetric interval about the mean volume, $[\mu - b, \mu + b]$, such that 95% of all backpack volumes lie in this interval.

SOLUTION

STEP 1   Let X be the volume (in in³) of a randomly selected backpack. The information given indicates that X is a normal random variable with mean $\mu = 600$ and standard deviation $\sigma = 100$: $X \sim N(600, 10000)$.

Find a symmetric interval about the mean such that 95% of all backpack volumes lie in this interval (find a value of b such that P($600 - b \leq X \leq 600 + b$) = 0.95). Figure 6.44 illustrates this probability statement.

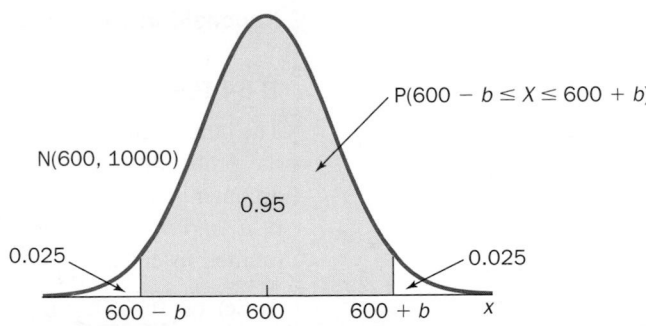

Figure 6.44 Visualizing the problem.

STEP 2   This problem reduces to finding the value for b. Since this question involves a normal random variable, we will certainly have to standardize. And since the probability is already given, this is a backward problem. In order to use Table III in the Appendix, we need a cumulative probability statement. We need another interpretation of Figure 6.44 involving cumulative probability *and* b. Here are two possibilities.

a. P($X \leq 600 - b$) = 0.025

The area (or probability) in the tails of the distribution is $1 - 0.95 = 0.05$ (the Complement Rule). Since the distribution is symmetric, the probability to the left of ($600 - b$) is $0.05/2 = 0.025$.

b. P($X \leq 600 + b$) = 0.975

The probability to the left of ($600 + b$) is $0.95 + 0.025 = 0.975$.

Standardization illustrated:

$P(X \leq 600 - b) = 0.25$

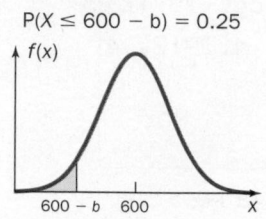

$P\left(Z \leq \dfrac{-b}{100}\right) = 0.025$

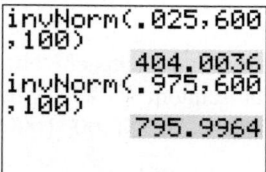

A technology solution:

```
invNorm(.025,600
,100)
          404.0036
invNorm(.975,600
,100)
          795.9964
```

Figure 6.45 Use inverse cumulative probability to find each endpoint.

 Try the Normal Density Curve statistical applet on the text web site.

STEP 3 We'll use the expression in (a).

$$P(X \leq 600 - b) = P\left(\frac{X - 600}{100} \leq \frac{(600 - b) - 600}{100}\right) = 0.025$$

Standardize.

$$= P\left(Z \leq \frac{-b}{100}\right)$$

Use Equation 6.8; simplify.

There is no further simplification within the probability statement. The resulting expression involves Z and is a cumulative probability. Find a value in the body of Table III as close to 0.025 as possible.

Set the corresponding z equal to $\dfrac{-b}{100}$, and solve for b.

$$\frac{-b}{100} = -1.96$$

Use Table III in the Appendix.

$$-b = -196.00$$

Multiply both sides by 100.

$$b = 196.00$$

Multiply both sides by -1.

The value of b is 196 and the symmetric interval about the mean is

$$P(600 - b \leq X \leq 600 + b) = P(600 - 196 \leq X \leq 600 + 196)$$
$$= P(\boxed{404} \leq X \leq \boxed{796}) = 0.95$$

95% of all backpacks have a volume between 404 and 796 in^3.

Technology can be used to find the endpoints of the interval without solving for b. ●

TECHNOLOGY CORNER

Procedure: Solve probability questions involving a normal random variable.
Reconsider: Example 6.6, page 258, solutions, and interpretations.

TI-84 Plus

The built-in function `normalcdf` is used to find (calculator) cumulative probability, the probability that X takes on a value between a and b. This function takes four arguments: a, b, μ, and σ. The built-in function `invNorm` takes three arguments: p, μ, and σ. This function returns a value x such that $P(X \leq x) = p$. If μ and σ are omitted in either function, the distribution is assumed to be standard normal.

1. Select DISTR; DISTR; `normalcdf`. Enter the left endpoint, 12.5, the right endpoint, 1E99 (calculator infinity), the mean, 10, and the standard deviation, 2. See Figure 6.39 (page 258).
2. Select DISTR; DISTR; `normalcdf`. Enter the left endpoint, 9, the right endpoint, 10, the mean, 10, and the standard deviation, 2. See Figure 6.40 (page 258).
3. Select DISTR; DISTR; `invNorm`. Enter the cumulative probability, 0.75, the mean, 10, and the standard deviation, 2. See Figure 6.41 (page 258).

Minitab

There are built-in functions to compute (strict) cumulative probability and inverse cumulative probability. These functions may be accessed through a graphical input window: Calc; Probability Distributions; Normal, or by using the command language.

1. In a session window, use the function CDF to find $P(X \leq 12.5)$. Use the complement rule to find $P(X > 12.5)$. See Figure 6.46.

2. Use the function CDF to find P($X \le$ 10). Store the result in K1. Use the function CDF to find P($X \le$ 9). Store the result in K2. Find the difference, and print the result. See Figure 6.47.

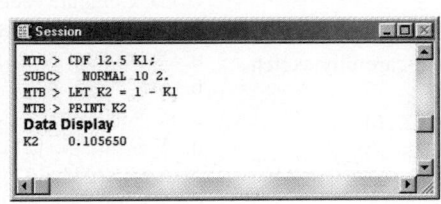

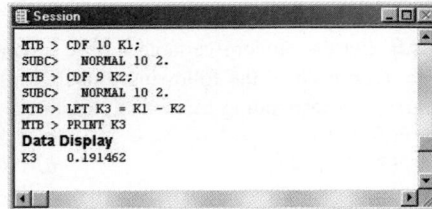

Figure 6.46 P($X >$ 12).

Figure 6.47 P(9 $\le X \le$ 10).

3. In a session window, use the function INVCDF, or select Calc; Probability Distributions; Normal. Choose Inverse cumulative probability, enter the Mean, 10, and the standard deviation, 2. Select Input constant (p), and enter 0.75. See Figure 6.48. The value of x is displayed in the session window. See Figure 6.49.

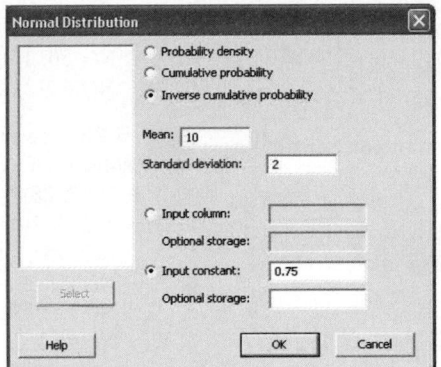

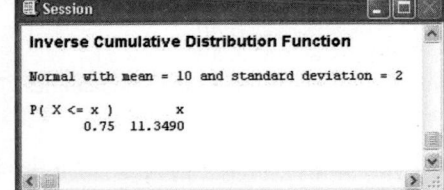

Figure 6.48 Graphical input window.

Figure 6.49 Inverse cumulative distribution function output.

Excel

There are built-in functions to compute (strict) cumulative probability associated with a standard normal random variable (NORMSDIST) and a normal random variable with mean $\mu \ne 0$ and/or standard deviation $\sigma \ne 1$ (NORMDIST). The functions NORMSINV and NORMINV are the corresponding inverse cumulative probability functions.

1. Use the function NORMDIST to find P($X \le$ 12.5). Use the complement rule to find P($X >$ 12.5). See Figure 6.50.
2. Use the function NORMDIST to find P($X \le$ 10). Use the function NORMDIST to find P($X \le$ 9). Find the difference: P(9 $\le X \le$ 10). See Figure 6.51.

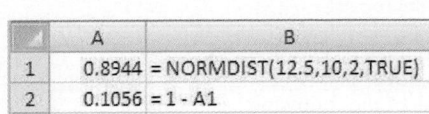

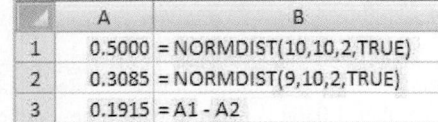

Figure 6.50 P($X >$ 12).

Figure 6.51 P(9 $\le X \le$ 10).

3. Use the function NORMINV. Enter the cumulative probability, 0.75, the mean, 10, and the standard deviation, 2. See Figure 6.52.

	A	B
1	11.3490	= NORMINV(0.75,10,2)

Figure 6.52 Inverse cumulative probability.

SECTION 6.2 EXERCISES

Practice

6.19 Let the random variable Z have a standard normal distribution. Find each of the following probabilities and carefully sketch a graph corresponding to each expression.

a. $P(Z \le 2.16)$ b. $P(Z < 2.16)$
c. $P(Z \le -0.47)$ d. $P(0.73 > Z)$
e. $P(-1.75 \ge Z)$ f. $P(-0.35 \le Z \le 0.65)$
g. $P(Z < 5)$ h. $P(Z \le -4)$
i. $P(Z \le 4)$

6.20 Let the random variable Z have a standard normal distribution. Find each of the following probabilities and carefully sketch a graph corresponding to each expression.

a. $P(-1.33 > Z)$ b. $P(Z < 2.35)$
c. $P(Z > 2.59)$ d. $P(-1.56 < Z < -0.56)$
e. $P(0.13 < Z < 2.44)$ f. $P(-0.05 < Z < 0.76)$
g. $P(Z \ge 2.67)$ h. $P(Z \le 1.42)$
i. $P(Z \le -2.00 \cup Z \ge 2.00)$ j. $P(-1.82 < Z \le -0.94)$

6.21 Let the random variable Z have a standard normal distribution. Find each of the following probabilities.

a. $P(-1.00 \le Z \le 1.00)$ b. $P(-2.00 \le Z \le 2.00)$
c. $P(-3.00 \le Z \le 3.00)$
Do you recognize these three probabilities? With which rule are they associated?

6.22 Let the random variable Z have a standard normal distribution. Solve each expression for b. Carefully sketch a graph corresponding to each probability statement.

a. $P(Z \le b) = 0.8686$ b. $P(Z < b) = 0.1867$
c. $P(Z < b) = 0.0016$ d. $P(Z \ge b) = 0.2643$
e. $P(Z > b) = 0.9382$ f. $P(Z \ge b) = 0.5000$
g. $P(b < Z) = 0.0192$ h. $P(b > Z) = 0.9938$
i. $P(-b < Z < b) = 0.7995$
j. $P(-b \le Z \le b) = 0.5527$

6.23 Let the random variable Z have a standard normal distribution. Solve each expression for b. Carefully sketch a graph corresponding to each probability statement.

a. $P(Z \le b) = 0.5100$ b. $P(Z > b) = 0.1080$
c. $P(Z \ge b) = 0.0500$ d. $P(Z \le b) = 0.0100$
e. $P(-b \le Z \le b) = 0.8000$
f. $P(-b < Z < b) = 0.6535$

6.24 Let the random variable Z have a standard normal distribution. Recall the definition for percentiles. Since $P(Z \le 1.0364) = 0.85$, then 1.0364 is the 85th percentile. Find each of the following percentiles for a standard normal distribution.

a. 10th   b. 27th   c. 85th
d. 40th   e. 49th   f. 61st

6.25 Let Z be a standard normal random variable and recall the calculations necessary in order to construct a box plot.

a. Find the first and the third quartile for a standard normal distribution.
b. Find the inner fences for a standard normal distribution.
c. Find the probability that Z is beyond the inner fences.

d. Find the outer fences for a standard normal distribution.
e. Find the probability that Z is beyond the outer fences.

6.26 Compute each probability and carefully sketch a graph corresponding to each expression.

a. $X \sim N(3, 0.0225)$, $P(X \le 3.25)$.
b. $X \sim N(52, 49)$, $P(X > 60)$.
c. $X \sim N(-7, 1)$, $P(X \le -4.5)$.
d. $X \sim N(235, 121)$, $P(X > 200)$.
e. $X \sim N(242, 132)$, $P(X \ge 350)$.
f. $X \sim N(1.17, 3.94)$, $P(X < -1.45)$.

APPLET   **6.27** Use the Normal Density Curve statistical applet to compute each probability and to carefully sketch a graph corresponding to each expression.

a. $X \sim N(3.7, 4.55)$, $P(3.0 \le X \le 4.0)$.
b. $X \sim N(62, 100)$, $P(50 < X < 70)$.
c. $X \sim N(32, 30)$, $P(X \ge 45)$.
d. $X \sim N(77, 0.01)$, $P(X < 76.95)$.
e. $X \sim N(-50, 16)$, $P(X < -55 \cup X > -45)$.
f. $X \sim N(7.6, 12)$, $P(8 \le X \le 9)$.

APPLET   **6.28** Use the Normal Density Curve statistical applet to solve each expression for b.

a. $X \sim N(17, 28)$, $P(X < b) = 0.75$.
b. $X \sim N(303, 70)$, $P(X \le b) = 0.05$.
c. $X \sim N(0, 25)$, $P(-b \le X \le b) = 0.90$.
d. $X \sim N(-12, 2)$, $P(X > b) = 0.35$.
e. $X \sim N(37, 2.25)$, $P(\mu - b \le X \le \mu + b) = 0.68$.
f. $X \sim N(26.35, 7.21)$, $P(X < b) = 0.11$.

6.29 Suppose that X is a normal random variable with mean 25 and standard deviation 6: $X \sim N(25, 36)$.

a. Find the first and third quartiles for X.
b. Find the inner fences for X.
c. Find the probability that X is beyond the inner fences.
d. Find the outer fences for X.
e. Find the probability that X is beyond the outer fences.

Applications

6.30 **Demographics and Population Statistics** The mean age of patients undergoing surgery in a study was 60.4 years with a standard deviation of 7.5 years.[3] Suppose the distribution of the age of patients is normal and one patient is selected at random.

a. Find the probability that the patient is less than 50 years old.
b. Find the probability that the patient is between 60 and 65 years old.
c. Find an age a such that 90% of all patients are less than a years old.

6.31 **Marketing and Consumer Behavior** According to The Wedding Report, Inc., the mean cost for a wedding in the United States is $28,732 (as of November 2008). Suppose the cost for a wedding is normally distributed, with a standard deviation of $1,500, and that a wedding is selected at random.

a. Find the probability that the wedding costs more than $32,000.
b. Find the probability that the wedding costs between $25,000 and $30,000.
c. Find the probability that the wedding costs less than $22,000.

6.32 Technology and Internet Social networking sites, for example, MySpace and Black Planet, have grown in popularity as users create web pages loaded with music, photographs, and profiles. Hitwise reported that the mean time spent by a user at Facebook during April 2008 was 20 minutes, 52 seconds. Suppose the distribution of time spent at Facebook per month is normal, the standard deviation is 4.5 minutes, and a Facebook user is selected at random.

a. Find the probability that the user spends less than 16 minutes per month at the site.

b. Find the probability that the user spends between 20 and 30 minutes per month at the site.

c. Find a value t such that only 2.5% of all Facebook users spend more than t minutes at Facebook per month.

6.33 Demographics and Population Statistics A company that manufactures military hats reports that the mean head circumference of all men in the military is 22.05 inches with a standard deviation of 0.68 inches.[4] Suppose the distribution of head circumference is approximately normal, and one man in the military is selected at random.

a. Find the probability that the man's head circumference is less than 21 inches.

b. Find the probability that the man's head circumference is between 22 and 23 inches.

c. If a man in the military has a head circumference greater than 24 inches, a surcharge is applied to the cost of the hat. What is the probability that a surcharge will be applied?

d. Find a symmetric interval about the mean such that 95% of all measurements lie in this interval.

6.34 Marketing and Consumer Behavior Movie trailers are designed to entice audiences by showing scenes from coming attractions. Several trailers are usually shown in a theater before the start of the main movie, and most are available via the Internet. The duration of a movie trailer is approximately normal with mean 150 seconds and standard deviation 30 seconds.

a. What is the probability that a randomly selected trailer lasts less than 1 minute?

b. Find the probability that a randomly selected trailer lasts between 2 minutes and 3 minutes, 15 seconds.

c. Any movie trailer that lasts beyond 4 minutes and 30 seconds is considered too long. What proportion of movie trailers are too long?

d. Find a symmetric interval about the mean such that 99% of all movie trailer durations lie in this interval.

6.35 Biology and Environmental Science The salinity, or salt content, in the ocean is expressed in parts per thousand (ppt). The number varies due to depth, rainfall, evaporation, river runoff, and ice formation. During November 2008, the mean salinity in a region of the tropical Pacific Ocean was approximately 34.95 ppt.[5] Suppose the distribution of salinity is normal and the standard deviation is 0.52 ppt, and suppose a random sample of ocean water from this region is obtained.

a. What is the probability that the salinity is more than 36 ppt?

b. What is the probability that the salinity is less than 33.5 ppt?

c. A certain species of fish can only survive if the salinity is between 33 ppt and 35 ppt. What is the probability that this species can survive in a randomly selected area?

d. Find a symmetric interval about the mean salinity such that 50% of all salinity levels lie in this interval. What are the endpoints of this interval called?

6.36 Manufacturing and Product Development Pressurized carbon dioxide is usually added to soda in order to alleviate the sweet taste of the drink and to stop some microbes from growing. While the amount of carbon dioxide varies, most sodas are bottled under high pressure. The internal pressure of a refrigerated can of a certain cola has a mean value of 276 kilopascals (kPa) at 16 °C.[6] Suppose the pressure is normally distributed, with a standard deviation 7 kPa. A can of this cola (at 16 °C) is randomly selected.

a. What is the probability that the internal pressure is less than 270 kPa?

b. What is the probability that the internal pressure is between 265 and 275 kPa?

c. If the internal pressure is more than 300 kPa when the can is opened, the liquid will spill out of the can. What is the probability of a spill?

d. If a person shakes the can for a minimum of 30 seconds, the mean internal pressure changes to 289 kPa (while the standard deviation remains the same). What is the probability that the liquid in a shaken can will spill out when the can is opened?

6.37 Public Health and Nutrition The NHANES report, 2005–2006, *What We Eat in America*, studied the nutrient intakes from food for individuals over 2 years old. For women age 20 and over, the mean amount of selenium consumed from food per day was 92.6 μg.[7] Assume the standard deviation is 18.4 μg and the amount of selenium consumed per day is normally distributed. Suppose a woman over age 20 is selected at random.

a. What is the probability that her daily intake of selenium from food will be at most 100 μg?

b. What is the probability that her daily intake of selenium from food will be between 60 and 75 μg?

c. Suppose the FDA has established 90 μg of selenium from food as the minimum daily intake. What is the probability that the woman selected will meet this minimum?

d. Find a value g such that 75% of all women in this age group have a daily intake of selenium from food of at least g grams.

6.38 Psychology and Human Behavior In many American families, both parents work outside the home, while children spend time at daycare centers or are cared for by other relatives. The mean amount of time fathers spend with their child(ren) 0–6 years old per day is 1.22 hours.[8] Suppose this time distribution is approximately normal with standard deviation 0.60 hours, and suppose a father with a child 0–6 years old is randomly selected.

a. What is the probability that the father spends at least 2 hours with his child on any given day?

b. What is the probability that the father spends between 1 and 1.5 hours with his child on any given day?

c. If the child sees his or her father for less than 2.5 hours per day, the parental bond is weakened. What is the probability that this special bond will be weakened on any given day?

Extended Applications

6.39 Sports and Leisure People in hot-air balloons usually fly just above the treetops at 200–500 feet. In populated areas, however, they usually stay at an altitude of at least 1000 feet. The amount of flying time possible in a hot-air balloon is dependent on many factors: number of propane burners, number of people in the balloon, and weather. Assume the time spent aloft is normally distributed with mean 1.5 hours and standard deviation 0.45 hours. Suppose a hot-air balloon flight is selected at random.

a. What is the probability that the flight time is between 1 and 2 hours?

b. What is the probability that the flight time is more than 1 hour and 15 minutes?

c. Find a value *t* such that 10% of all flights last less than *t* hours.

d. Suppose a person offering hot-air balloon rides charges $50 for each ride of at least 1 hour, and $1.00 for every minute after 1 hour. What proportion of rides costs more than $100?

6.40 Public Policy and Political Science In September 2008, Hurricane Ike flooded approximately 75% of the 24,000 properties in Galveston, Texas. As a result, the city's planning department was inundated with requests for permits to repair damage from the storm. The mean time waiting in line at the permit office was 2 hours.[9] Assume the time spent waiting in line is normally distributed with standard deviation 45 minutes. Suppose one person who waited in line for a permit is selected at random.

a. What is the probability that the waiting time was between 1 and 3 hours?

b. What is the probability that the waiting time was more than 4 hours?

c. What is the probability that the waiting time was less than 1 hour?

d. Suppose five people who waited in line were selected at random. What is the probability that exactly two of the five had a waiting time of less than 30 minutes?

6.41 Biology and Environmental Science Since 1990 the timber harvest in the national forests has steadily declined. In 2007 approximately 233,000 acres were cleared for timber.[10] While timber volume varies by site, the volume per hectare is normally distributed with mean 231 m^3 and standard deviation 8 m^3. Suppose a harvested hectare is selected at random.

a. What is the probability that the volume harvested is between 225 and 235 m^3?

b. What is the probability that the volume harvested is between 230 and 240 m^3?

c. Suppose the hectare has already produced 230 m^3 of timber. What is the probability that the volume harvested will be more than 240 m^3?

d. If a selected hectare produces less than 210 m^3 then the cost of the harvest is more than the lumber revenue, resulting in a loss. What is the probability of a loss?

6.42 Manufacturing and Product Development The AAA batteries used in many calculators are manufactured according to specifications regulating the diameter, height, rated capacity, weight, and volume. The volume of a randomly selected AAA battery is normally distributed with mean 3.85 cm^3 and standard deviation 0.01 cm^3.[11] Suppose an AAA battery is randomly selected.

a. What is the probability that the battery volume is between 3.84 and 3.87 cm^3?

b. What is the probability that the battery volume differs from the mean by more than 0.015 cm^3?

c. Find a value *v* such that only 10% of all batteries have volume less than *v*.

d. If a battery has volume 3.88 cm^3 or greater, it will not fit properly into the slot. For a calculator that requires four AAA batteries, what is the probability that all four batteries will fit?

6.43 Biology and Environmental Science Many backyard gardeners prefer Silver Queen Hybrid corn. This late-season variety is very sweet and has tender, white kernels. In some locations in the Northeast, gardeners have trouble harvesting this variety because it has a longer growing time. The temperature of the soil should be at least 65 °F before planting, and the growing time is approximately normal with mean 92 days and standard deviation 5 days.

a. What is the probability that a randomly selected seed will mature in less than 90 days?

b. What is the probability that a randomly selected seed will mature between 95 and 100 days?

c. Suppose a row in a backyard garden contains 12 plants. What is the probability that 4 will be ready for dinner by the 95th day?

d. Find a value *h* such that 99% of all plants are ready to be harvested within *h* days.

6.44 Manufacturing and Product Development Classical violin bows are made from different woods to accommodate musicians' preferences and demands. Some commonly used woods include snakewood, ironwood, hakia, and pernambuco. While the bows are carefully handcrafted, they vary slightly in weight. Suppose a bowmaker claims that the weight of his classical bows is normally distributed with mean 60 grams and standard deviation 3.2 grams.[12]

a. What is the probability that the weight of a randomly selected ironwood bow is between 58 and 62 grams?

b. Good musicians can detect an *unacceptable* bow weight (i.e., a weight that differs from the mean by more than two standard deviations). What is the probability that a bow weight is unacceptable?

c. Any manufactured bow that weighs more than 66 grams is reworked in order to decrease the weight. What is the probability that a randomly selected ironwood bow will need rework?

d. Suppose that the weight of a randomly selected bow is 55 grams. Is there any evidence to suggest that the mean weight is less than 60 grams? Justify your answer.

6.45 Medicine and Clinical Studies Repeated industrial tasks often cause work-related muscle disorders. Measurements of joint angles (of the shoulder and elbow, for example) required to complete a certain task can be used to predict future injuries. The shoulder joint angle required to fasten an aluminum door frame on an assembly line varies according to height, arm length, and location. The shoulder joint angle for this task is normally distributed with mean 23.7 degrees and standard deviation 1.9 degrees. Suppose an employee is randomly selected.

a. What is the probability that the employee's shoulder joint angle will be between 20 and 25 degrees?

b. What is the probability that the joint angle will be less than 18 degrees?

c. If the joint angle is more than 28 degrees, there is a good chance the employee will suffer from a muscle disorder. What is the probability that the employee will suffer from a muscle disorder?

d. If the joint angle is between 21.7 and 25.7 degrees, then management believes the ergonomics of the task are adequate. If five employees are randomly selected, what is the probability that four of the five have adequate ergonomics?

6.46 Biology and Environmental Science Many lakes are carefully monitored for pH concentrations, total phosphorus, chlorophyll, nitrogen, and total suspended solids. These data are used to characterize the condition of the lake and to chart year-to-year variability. According to the Lake Partner Program, Ontario Ministry of the Environment, Aberdeen Lake has a mean total phosphorus concentration of 14.6 mg/L and standard deviation 5.8 mg/L. Suppose the total phosphorus concentration is normally distributed, a day is selected at random, and a total phosphorus measure from Aberdeen Lake is obtained.

a. What is the probability that the total phosphorus is less than 13 mg/L?

b. What is the probability that the total phosphorus differs from the mean by more than 5 mg/L?

c. Suppose that the total phosphorus is less than 20 mg/L. What is the probability that it is less than 14 mg/L?

d. If the total phosphorus measurement is 27 mg/L, is there any evidence to suggest that the mean has increased?

6.47 Manufacturing and Product Development High-pressure washers have become popular for cleaning siding and windows. This equipment is available in various engine types and horsepower. Suppose the power rating (in horsepower, hp) for a residential pressure washer is normally distributed with mean 20 hp and standard deviation σ.

a. The probability that a randomly selected power rating is within 2.5 hp of the mean is 0.7229. Find the value of σ.

b. A leading consumer magazine advised its readers to purchase pressure washers with a power rating of 15 hp or above. What proportion of pressure washers has this rating?

c. If the power rating is above 26.5 hp, the pressure washer will crack, or even break, certain windows. What is the probability that a pressure washer could break a window?

6.48 Physical Sciences Hydroelectric projects are carefully monitored and their energy capability is predicted for several years into the future. Suppose the Klamath Hydro Project, located on the upper Klamath River in south-central Oregon, generates electricity according to a normal distribution. The Pacific Northwest Utilities Conference Committee claims that the mean electricity generated per year is 35 megawatts.

a. The probability that the Klamath Hydro Project generates less than 34 megawatts during any randomly selected year is 0.3540. Find the standard deviation.

b. Suppose the years are independent, and the hydro project will record a profit in a given year if it is able to generate at least 37.8 megawatts that year. What is the probability that the project will record a profit for four consecutive years?

c. Suppose that the electricity generated during a certain year is 33.5 megawatts. Is there any evidence to suggest that the claim by the Pacific Northwest Utilities Conference Committee is false? Justify your answer.

6.49 Manufacturing and Product Development Dining room chairs come in many different woods, styles, and shapes. The height of the seat of a randomly selected oak dining-room chair is approximately normal with mean 85 cm and standard deviation 1.88 cm.

a. Find a value h such that 99% of all dining-room chairs have height less than h.

b. Consumer testing indicates that any chair seat higher than 90 cm is uncomfortable for use when eating. What is the probability that a randomly selected dining-room chair is uncomfortable?

c. Find the first and third quartiles of the dining-room chair heights distribution.

d. There is some evidence to suggest that, after five years of use, the mean height of these chairs has decreased, due to wear, erosion, and humidity. Suppose that after five years, the probability that the height is more than 86 cm is 0.0718. Find the mean height after five years.

Challenge

6.50 Sports and Leisure The International Tennis Federation (ITF) establishes the specifications for tennis balls. The diameter of a tennis ball used in any tournament must be between 2.5 and 2.625 inches. Suppose that the diameter of a tennis ball is approximately normal with mean 2.5625 inches and standard deviation 0.04 inches.

a. What is the probability that a randomly selected tennis ball will meet ITF diameter specifications?

b. Suppose six tennis balls will be used in a tournament game. What is the probability that exactly one will not meet ITF diameter specifications?

6.3 CHECKING THE NORMALITY ASSUMPTION

Almost every inferential statistics procedure requires certain assumptions, for example, that observations are selected independently or that variances are equal (for analysis of variance). And many statistical techniques are valid *only* if the observations are from a normal distribution. If an inference procedure requires normality, and the population distribution is not normal, then the conclusions are worthless. Therefore, it seems reasonable to perform some kind of check for normality, to make sure there is no evidence to refute this assumption.

Until now we have been using the normal distribution as a model for describing the variability of a random variable X, and we have been assuming that we knew the values of the population mean μ and the population variance σ^2. If those values are unknown, the sample mean $\bar{x}$ and the sample standard deviation s can be used as estimates of the unknown parameters μ and σ. But we still cannot be sure that the normal distribution is an appropriate model to describe a particular set of observations. We need a way to check whether a set of observations does seem to come from a population with a normal distribution. There are four different methods we can use to look for any evidence of nonnormality. Three of them use techniques that we have seen before; the fourth one is a new method.

Given a set of observations, $x_1, x_2, x_3, \ldots, x_n$, the following four methods may be used to check for any evidence of nonnormality, for example, a distribution that is not bell-shaped, a skewed distribution, or a distribution with heavy tails.

1. Graphs

Construct a histogram, a stem-and-leaf plot, and/or a dot plot. Examine the shape of the distribution for any indications that the distribution is not bell-shaped. In a random sample, the distribution of the sample should be similar to the distribution of the population.

2. Backward Empirical Rule

To use the Empirical Rule to test for normality, find the mean, the standard deviation, and the three symmetric intervals about the mean $(\bar{x} - ks, \bar{x} + ks)$, $k = 1, 2, 3$. Compute the *actual* proportion of observations in each interval. If the actual proportions are close to 0.68, 0.95, and 0.997, then normality seems reasonable. Otherwise, there is evidence to suggest that the shape of the distribution is not normal.

3. *IQR/s*

Find the interquartile range, *IQR*, and standard deviation, s, for the sample, and compute the ratio *IQR/s*. If the data are approximately normal, then $IQR/s \approx 1.3$.

$P(Z \leq -0.6745) = 0.25$, and
$P(Z \leq 0.6745) = 0.75$.

Here is some justification for this ratio. Consider a standard normal random variable, Z ($\mu = 0$, $\sigma = 1$). The first quartile for Z is -0.6745 and the third quartile is 0.6745. The interquartile range divided by the standard deviation is $[0.6745 - (-0.6745)]/1 = 1.349$. In a random sample, the interquartile range should be close to the population interquartile range, and the standard deviation should be close to the population standard deviation. Since any normal distribution can be standardized, or compared to Z, $IQR/s \approx 1.3$.

4. Normal Probability Plot

A **normal probability plot** is a scatter plot of each observation versus its corresponding standardized normal score. For a normal distribution, the points will fall along a straight line.

The *standardized normal scores* are expected values. For example, in repeated samples of size n from the Z distribution, on average the smallest value is z_1, the next largest value is z_2, and so on; the largest value is z_n.

HOW TO CONSTRUCT A NORMAL PROBABILITY PLOT

Suppose $x_1, x_2, \ldots, x_n$ is a set of observations.
a. Order the observations from smallest to largest and let $x_{(1)}, x_{(2)}, \ldots, x_{(n)}$ represent the set of ordered observations.
b. Find the standardized normal scores for a sample of size n in Table IV in the Appendix: $z_1, z_2, \ldots, z_n$.
c. Plot the ordered pairs $(z_i, x_{(i)})$.

If the scatter plot is nonlinear, then there is evidence to suggest that the data did not come from a normal distribution. Most statistical software (the TI-84 and Minitab included) automatically computes the expected Z values. Table IV in the Appendix provides standardized normal scores for some values of n.

Most of the standardized normal scores are always between -2.0 and $+2.0$ since approximately 95% of all observations lie within two standard deviations of the mean.

Figures 6.53–6.56 are examples of normal probability plots.

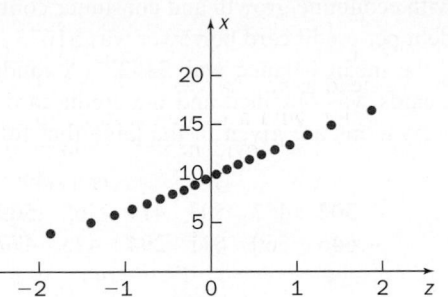

Figure 6.53 A normal probability plot. The points lie along an approximate straight line. There is no evidence of nonnormality.

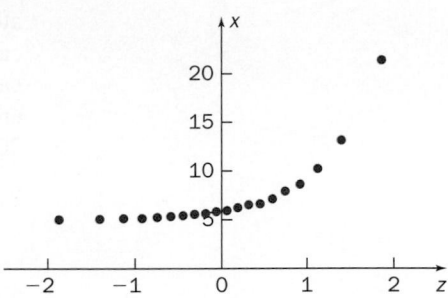

Figure 6.54 A normal probability plot. The curved graph suggests that the distribution is not normal and is skewed.

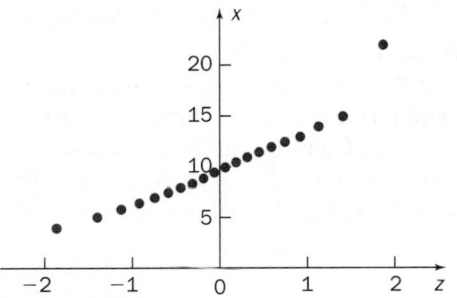

Figure 6.55 A normal probability plot. The plot suggests that the distribution is not normal and that the data set contains an outlier.

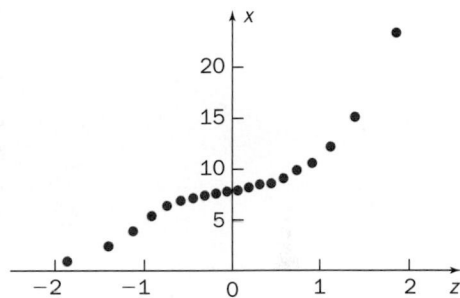

Figure 6.56 A normal probability plot. The plot suggests that the distribution is not normal and has *heavy tails*.

The data axis can be horizontal or vertical. To use a horizontal data axis, plot the points $(x_{(i)}, z_i)$. Figure 6.57 shows a normal probability plot with the data plotted on the vertical axis and Figure 6.58 shows a normal probability plot (using the same data and standardized normal scores) with data plotted on the horizontal axis.

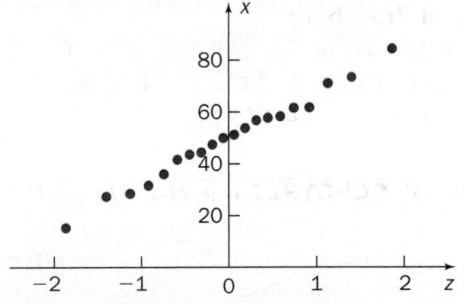

Figure 6.57 A normal probability plot with the data plotted on the vertical axis.

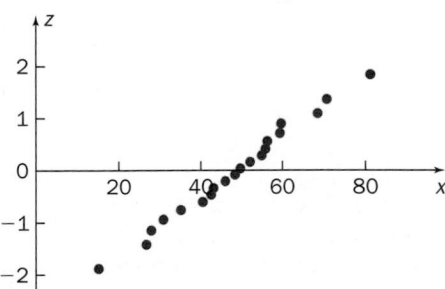

Figure 6.58 A normal probability plot with the data plotted on the horizontal axis.

Interpretation of a normal probability plot is very subjective. Even if the axes are reversed, we are still looking for the points to lie along a straight line.

All four methods can be used to check the normality assumption, and any one (or several) may suggest that the data did *not* come from a normal distribution. Since we are searching for evidence of nonnormality, even if we fail to reject the normality assumption in each test, we still *cannot* say with absolute certainty that the data came from a normal distribution.

Example 6.8 Credit Card Debt   Household credit card debt is often associated with economic growth and consumer confidence. In June 2008, the mean credit card debt per credit card borrower was $1673.[13] Among college students with credit cards, the mean balance was $452.[14] A random sample of college students with credit cards was obtained and the credit card balance was recorded for each. The 20 observations are given in the table that follows.

302	467	503	419	568	509	399	356	498	416
646	560	671	294	428	477	518	311	236	476

Is there any evidence to suggest that this distribution is not normally distributed?

SOLUTION

STEP 1   Figure 6.59 shows a frequency histogram for these data. There are no obvious outliers, and the distribution seems approximately normal.

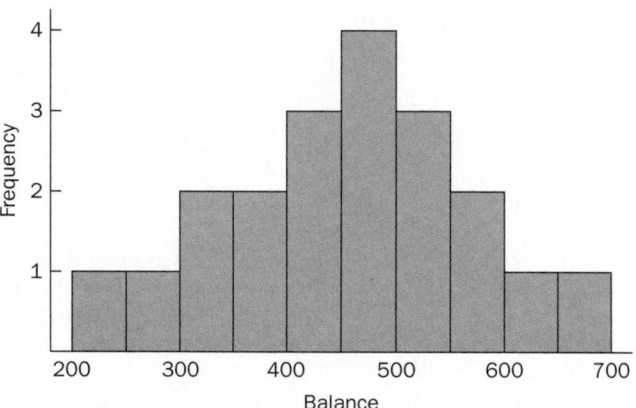

Figure 6.59 Frequency histogram for the credit-card balance data.

STEP 2   The sample mean and the sample standard deviation are $\bar{x} = 452.70$ and $s = 115.49$. The following table lists three symmetric intervals about the mean, the number of observations in each interval, and the proportion of observations in each interval (recall that $n = 20$).

Interval	Frequency	Proportion
$(\bar{x} - s, \bar{x} + s) = (337.21, 568.19)$	14	0.70
$(\bar{x} - 2s, \bar{x} + 2s) = (221.72, 683.68)$	20	1.00
$(\bar{x} - 3s, \bar{x} + 3s) = (106.23, 799.17)$	20	1.00

The actual proportions are *close* to those given by the Empirical Rule (0.68, 0.95, and 0.997).

STEP 3   The quartiles are $Q_1 = 377.50$, $Q_3 = 513.50$.

$IQR/s = (513.50 - 377.50)/115.50 = 1.1776$

This ratio is *close* to 1.3.

STEP 4   The table below lists each observation along with the corresponding normal score from Table IV in the Appendix.

Observation	Normal score	Observation	Normal score
236	−1.87	476	0.06
294	−1.40	477	0.19
302	−1.13	498	0.31
311	−0.92	503	0.45
356	−0.74	509	0.59
399	−0.59	518	0.74
416	−0.45	560	0.92
419	−0.31	568	1.13
428	−0.19	646	1.40
467	−0.06	671	1.87

Plot these points in order to obtain the normal probability plot, as shown in Figure 6.60.

The points lie along an approximate straight line.

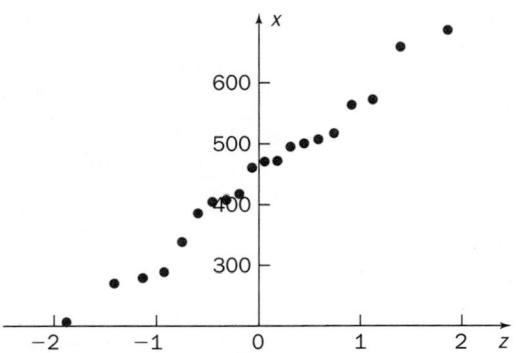

Figure 6.60 Normal probability plot for the credit-card balance data.

A technology solution:

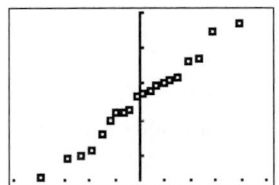

Figure 6.61 Normal probability plot.

The histogram, backward Empirical Rule, IQR/s, and normal probability plot show no significant evidence of nonnormality. Remember, this decision is very subjective.

Example 6.9 Chemotherapy Protocol A certain protocol for chemotherapy states that the total dose for patients under age 12 is no greater than 450 mg/m^2 within six months. A random sample of 30 patients undergoing this form of chemotherapy was obtained, and their medical records were examined to determine the total dose of the drug they received over the previous six months. The data are given in the following table.

354	350	351	352	353	358	361	364	371	376
377	378	387	396	399	402	406	408	412	424
427	430	432	437	440	441	443	446	447	449

(*Source:* National Cancer Institute clinical trials.)

Is there any evidence to suggest that the distribution of six-month total dosage is not normally distributed?

SOLUTION

STEP 1 Figure 6.62 shows a frequency histogram for these data. Although the graph seems symmetric, it is not bell-shaped. Most of the data are concentrated in the tails of the distribution. This suggests that the data are not from a normal distribution.

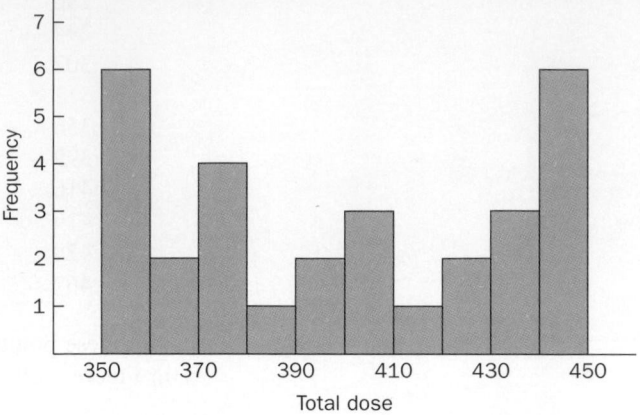

Figure 6.62 Frequency histogram for the cumulative chemotherapy dose data.

STEP 2 The sample mean is $\bar{x} = 399.03$ and the sample standard deviation is $s = 34.94$. The following table lists three symmetric intervals about the mean, the number of observations in each interval, and the proportion of observations in each interval (computed using $n = 30$).

Interval	Frequency	Proportion
$(\bar{x} - s, \bar{x} + s) = (364.09, 433.97)$	15	0.50
$(\bar{x} - 2s, \bar{x} + 2s) = (329.15, 468.91)$	30	1.00
$(\bar{x} - 3s, \bar{x} + 3s) = (294.21, 503.85)$	30	1.00

The first two proportions (0.50 and 1.00) are significantly different from those given by the Empirical Rule (0.68 and 0.95). This suggests that the population of total chemotherapy doses is not normal.

STEP 3 The quartiles are $Q_1 = 364.00$ and $Q_3 = 432.00$.

$IQR/s = (432.00 - 364.00)/34.94 = 1.9462$

Since this ratio is significantly different from 1.3, there is more evidence to suggest that the underlying population is not normal.

STEP 4 The following table lists each observation along with the corresponding normal score.

Observation	Normal score	Observation	Normal score	Observation	Normal score
350	−2.04	377	−0.38	427	0.47
351	−1.61	378	−0.29	430	0.57
352	−1.36	387	−0.21	432	0.67
353	−1.18	396	−0.12	437	0.78
354	−1.02	399	−0.04	440	0.89
358	−0.89	402	0.04	441	1.02
361	−0.78	406	0.12	443	1.18
364	−0.67	408	0.21	446	1.36
371	−0.57	412	0.29	447	1.61
376	−0.47	424	0.38	449	2.04

A technology solution:

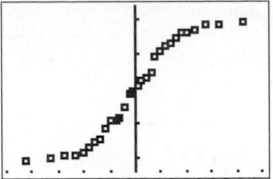

Figure 6.63 Normal probability plot.

The normal probability plot is shown in Figure 6.64.

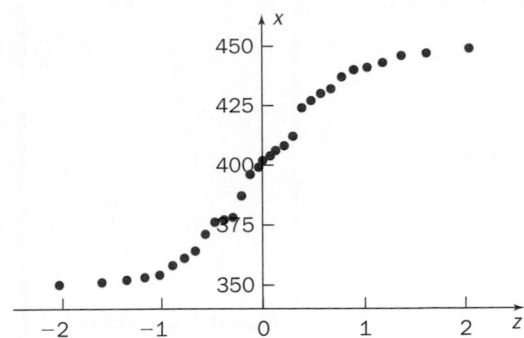

Figure 6.64 Normal probability plot for the chemotherapy dose data.

The points do not lie along a straight line. Each tail is flat, which makes the graph look "S-shaped." This suggests that the underlying population is not normal.

The histogram, backward Empirical Rule, IQR/s, and normal probability plot all indicate that this sample did not come from a normal population. ●

TECHNOLOGY CORNER

Procedure: Construct a normal probability plot.
Reconsider: Example 6.8, page 268, solution, and interpretations.

TI-84 Plus

A normal probability plot is one of the six built-in statistical plots.

1. Enter the data into list L1.
2. Choose STATPLOT ; STAT PLOTS; Plot1. Turn the plot On, select Type normal probability plot (the last graph icon), enter the Data List, L1, set the Data Axis to Y, and choose a Mark (for the points on the graph).
3. Enter appropriate window settings and press GRAPH to view the normal probability plot. See Figure 6.61, page 269.

Minitab

Use the built-in function NSCORES to compute the normal scores. Construct a scatter plot of the data versus the normal scores.

1. Enter the data into column C1.
2. Compute the normal scores in a session window and store the results in column C2: NSCORES C1 C2.
3. Construct a scatter plot:
   a. In a session window: PLOT C1*C2.
   b. In a graphical input window: Graph; Scatterplot. Select Simple and let the Y variable be the data column (C1) and the X variable be the normal scores column (C2). See Figure 6.65, page 272.

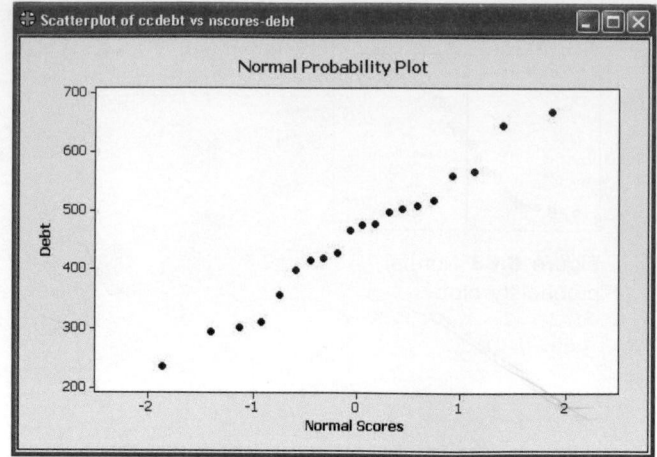

Figure 6.65 Normal probability plot.

Excel

There is a normal probability plot option in Data; Data Analysis; Regression. However, this option uses a different algorithm to compute the normal scores and the Regression function requires two input columns. To construct a normal probability plot, compute the normal scores using the formula that follows.

1. Enter the data into column C, in increasing order, and enter the numbers 1 to $n = 20$ in column A.
2. Set the cell B1 equal to NORMINV((A1−3/8)/(20+1/4),0,1). Copy this result and paste into the cells B2−B20. These are (approximately) the normal scores.
3. Highlight the data range B1:C20. Under the Insert tab, select Scatter; Scatter with only Markers. See Figure 6.66.

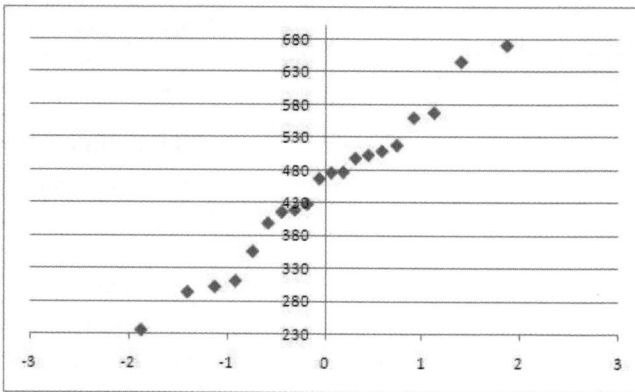

Figure 6.66 Normal probability plot.

SECTION 6.3 EXERCISES

Practice

6.51 Consider the following 20 observations.

15.4	13.9	14.9	16.2	16.6	15.4	17.2	18.5
19.3	13.0	16.5	20.2	16.4	15.3	18.5	17.9
15.5	17.4	16.3	14.3				

Construct a normal probability plot. Is there any evidence to suggest that the data are from a nonnormal population? Justify your answer.

6.52 Consider the following 20 observations.

52.0	52.1	58.8	88.0	49.9	18.7	43.1	47.6
90.0	49.8	54.8	35.1	56.1	53.2	76.5	45.4
34.1	19.5	58.7	25.7				

Construct a normal probability plot. Is there any evidence to suggest that the data are from a nonnormal population? Justify your answer.

6.53 Examine each of the following four normal probability plots. Is there any evidence to suggest that the data are from a nonnormal population?

a.

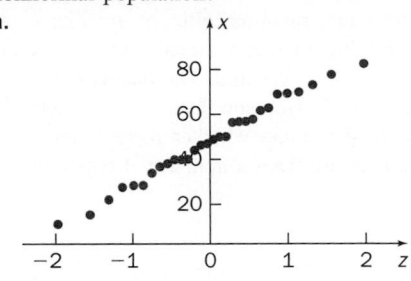

b.

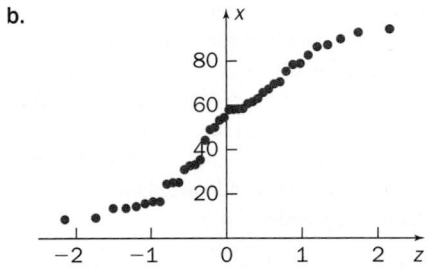

c.

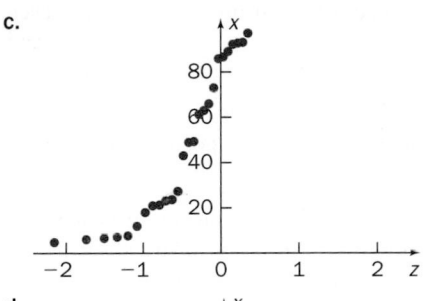

d.

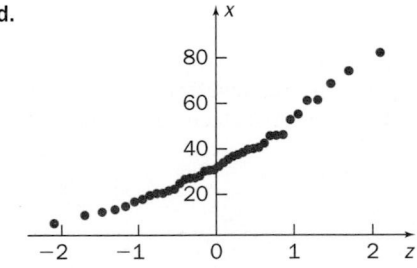

6.54 Consider the data on the data CD and book's web site. Use the four methods presented in this section to determine whether there is any evidence to suggest that the data are from a nonnormal population.

6.55 Consider the following data.

5.32	9.87	11.25	10.94	5.58
6.29	7.47	10.75	6.22	8.00

Use the four methods presented in this section to determine whether there is any evidence to suggest that the data are from a nonnormal population.

Applications

6.56 Biology and Environmental Science The Alaska River Forecast Center publishes summary information concerning river depth, precipitation, and ice thickness. Suppose the thickness of the ice (in feet) on Lake Hood near Anchorage during the first week of January for the past 20 years is given in the table below.

8.9	15.3	18.6	16.4	15.9	15.6	13.1	13.2
16.0	17.2	15.5	15.6	15.7	13.3	13.9	14.0
12.8	13.9	15.0	18.6				

(*Source:* National Weather Service.)

a. Construct a normal probability plot for these data. Is there any evidence to suggest nonnormality?
b. Use the backward Empirical Rule to check for evidence of nonnormality.

6.57 Education and Child Development Some people claim that children who practice yoga are more physically fit, self-confident, and self-aware. A random sample of preteens (ages 10–12) practicing yoga was obtained, and the amount of time they spent doing meditation or quiet breathing (in minutes) per day was recorded. The data are given on the data CD and book's web site. Use the four methods presented in this section to determine whether there is any evidence to suggest that the data are from a nonnormal population.

6.58 Sports and Leisure Many NBA players express themselves on the court through their sneakers. High-school basketball players often insist on wearing the same expensive sneakers worn by their favorite player. A random sample of sneakers worn by Notre Dame Prep (in Massachusetts) basketball players was obtained, and the retail price (in dollars) for each pair of sneakers they wore is given in the table below.

96.70	112.05	120.70	106.40	86.60
126.40	134.75	76.75	142.20	116.70

Use the four methods presented in this section to determine whether there is any evidence to suggest that the data are from a nonnormal population.

6.59 Public Policy and Political Science Bicycle paths are usually planned and constructed according to certain guidelines (such as the American Association of State Highway and Transportation Officials guidelines for construction). There are construction standards for width, offset from the road, maximum grade, and horizontal and vertical clearances. A random sample of bicycle-path widths (in feet) was obtained, and the data are given on the data CD and book's web site.

a. Find the sample mean and the sample standard deviation for these data.
b. Compute the intervals $(\bar{x} - s, \bar{x} + s), (\bar{x} - 2s, \bar{x} + 2s),$ $(\bar{x} - 3s, \bar{x} + 3s)$.
c. Find the proportion of observations in each interval in part (b). Is there any evidence to suggest that the data are from a nonnormal population?

6.60 Physical Sciences Near Earth Objects (NEOs) are comets and asteroids that have entered the Earth's *neighborhood*. The National Aeronautics and Space Administration maintains a list of NEOs deemed to have the potential of colliding with Earth. The diameter (in kilometers) of several of these objects is given on the data CD and book's web site.[15] Use the methods presented in this section to determine whether there is any evidence to suggest that the data are from a nonnormal population.

6.61 Sports and Leisure A typical round of golf takes approximately 3.5 hours (walking, without an electric cart). Many weekend golfers take more time due to lost golf balls, thinking about certain shots, and talking to other players. The manager at the Wawona Golf Course in Wawona, California, took a random sample of round times (in hours), and the data are reported in the following table.

3.86	4.92	4.15	3.83	4.34	4.56	4.24	4.33
4.36	4.09	4.30	4.23	4.28	4.63	4.34	3.73
4.66	4.40	4.45	4.42				

a. Construct a stem-and-leaf plot for these data.
b. Compute the ratio IQR/s.
c. Construct a normal probability plot for these data.
d. Is there any evidence to suggest that these data are from a nonnormal population? Justify your answer.

6.62 Marketing and Consumer Behavior The predominant acid in frozen concentrated orange juice (FCOJ) is citric acid, and the amount is usually given as a percentage. The Brix is also a percentage, and is a measure of the total soluble solids in FCOJ. The Brix/acid ratio is computed by simple division and 12 is considered an ideal ratio. A random sample of FCOJ was obtained from different sellers, and the Brix/acid ratio was measured for each. These measurements were used to construct the following normal probability plot.

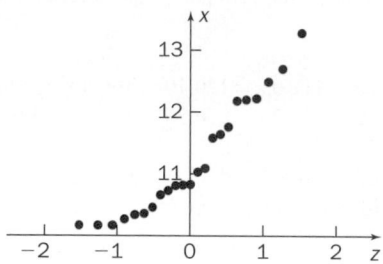

Does this plot suggest that the data are from a nonnormal distribution? Justify your answer.

6.63 Public Health and Nutrition The Food and Drug Administration recently issued warning letters to pharmacies for selling nicotine lollipops. While these products may help some people break the cigarette habit, nicotine lollipops are illegal. A random sample of nicotine lollipops was obtained, and the amount of nicotine in each was carefully measured. The data (in mg) are given on the data CD and book's web site. Use the four methods presented in this section to determine whether there is any evidence to suggest that the data are from a nonnormal population.

Challenge

6.64 Normal Scores Generate 500 random samples of size 10 from a standard normal distribution. Order each sample from smallest to largest. Find $\bar{x}_{(1)}$, the sample mean of the 500 smallest values from each sample. Consider the next largest value in each sample. Find $\bar{x}_{(2)}$, the sample mean of these 500 values. Continue in this manner to find $\bar{x}_{(3)}, \bar{x}_{(4)}, \ldots$, and $\bar{x}_{(10)}$, where $\bar{x}_{(i)}$ is the sample mean of the 500 ith largest values. Compare these 10 sample means, $\bar{x}_{(1)}, \bar{x}_{(2)}, \ldots, \bar{x}_{(10)}$ with the standardized normal scores for $n = 10$ in Table IV in the Appendix.

Try a similar procedure for $n = 20$. Generate 500 random samples of size 20 from a standard normal distribution. Order each sample from smallest to largest. Find $\bar{x}_{(1)}, \bar{x}_{(2)}, \ldots, \bar{x}_{(20)}$ and compare these values with the standardized normal scores for $n = 20$ in Table IV.

Explain why these sample means should be good estimates of the standardized normal scores.

Generate 500 random observations from a normal distribution with mean 50 and standard deviation 10. Arrange the observations in order from smallest to largest and denote this ordered list $x_{(1)}, x_{(2)}, \ldots, x_{(500)}$.

Form the ordered pairs $(x_{(1)}, 1/500), (x_{(2)}, 2/500), \ldots, (x_{(i)}, i/500), \ldots, (x_{(500)}, 1)$. Plot these points in a rectangular coordinate system and describe the shape of the graph. Which curve does this graph approximate? Why?

6.4 THE EXPONENTIAL DISTRIBUTION

There are many other common continuous distributions besides the normal distribution, for example, the t distribution, chi-square distribution, and F distribution. We will learn a little about each of these distributions in Chapters 8 and 9 as we study confidence intervals and hypothesis tests. This section presents the exponential distribution, which is related to several continuous distributions. Remember that probability in a continuous world is area under the curve, and where appropriate try to convert any probability statement into an equivalent expression involving *cumulative* probability.

The **exponential probability distribution** is often used to model the time to failure of an electronic part or the waiting time between events. This distribution is completely characterized by one parameter, λ.

THE EXPONENTIAL PROBABILITY DISTRIBUTION

Suppose X is an exponential random variable with parameter λ (with $\lambda > 0$). The probability density function is given by

$$f(x) = \begin{cases} \lambda e^{-\lambda x} & \text{if } x \geq 0 \\ 0 & \text{otherwise} \end{cases} \tag{6.9}$$

ILLUMINATING THE CONCEPTS

1. The symbol e in Equation 6.9 is the base of the natural logarithm ($e \approx 2.71828$). The constant e is also used in the Poisson distribution and the normal distribution.

2. If the exponential distribution is used to model the lifetime of a light bulb, machine, or even a human being, then λ represents the failure rate.

3. The exponential distribution has positive probability only for $x \geq 0$. Figures 6.67 and 6.68 show the graphs of a general probability density function for an exponential random variable and a probability density function with $\lambda = 2$.

Notice that $f(x) = \lambda$ when $x = 0$ (because $e^0 = 1$).

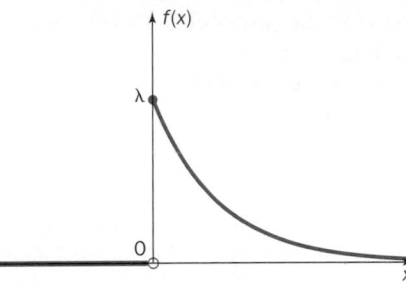

Figure 6.67 The probability density function for an exponential random variable with parameter λ.

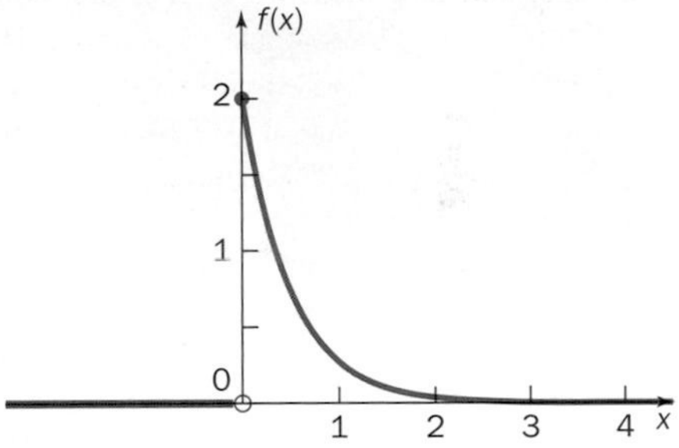

Figure 6.68 The probability density function for an exponential random variable with $\lambda = 2$.

4. The mean and variance for an exponential random variable, X, with parameter λ are

$$E(X) = \mu = \frac{1}{\lambda}, \quad \sigma^2 = \frac{1}{\lambda^2}. \tag{6.10}$$

Probabilities associated with an exponential random variable with parameter λ are computed using cumulative probability. We do not need a table for these calculations! Figure 6.69 illustrates cumulative probability associated with an exponential random variable. Remember, there is no area (or probability) for $x < 0$.

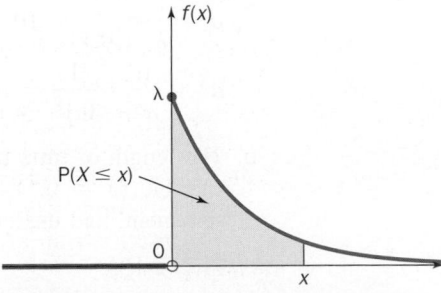

Figure 6.69 The probability density function for an exponential random variable. The shaded area corresponds to $P(X \leq x)$.

The formula for cumulative probability is given by

$$P(X \le x) = \begin{cases} 0 & \text{if } x < 0 \\ 1 - e^{-\lambda x} & \text{if } x \ge 0 \end{cases} \qquad (6.11)$$

If $x \ge 0$, the probability that X assumes a value greater than x is a right-tail probability given by

$$P(X > x) = 1 - P(X \le x) = 1 - (1 - e^{-\lambda x}) = 1 - 1 + e^{-\lambda x} = e^{-\lambda x}. \quad (6.12)$$

The following example illustrates the use of cumulative probability and the formula for right-tail probability to find probabilities associated with an exponential random variable.

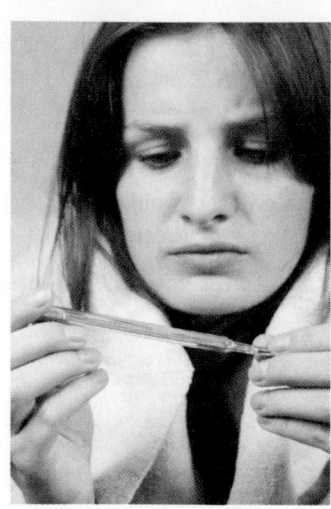

(Anyra/Dreamstime.com)

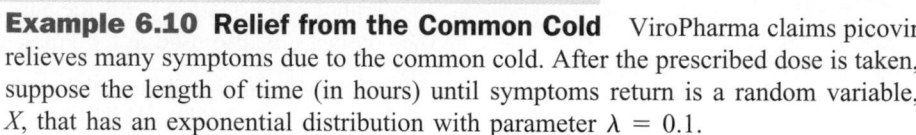
Example 6.10 Relief from the Common Cold ViroPharma claims picovir relieves many symptoms due to the common cold. After the prescribed dose is taken, suppose the length of time (in hours) until symptoms return is a random variable, X, that has an exponential distribution with parameter $\lambda = 0.1$.

a. Carefully sketch a graph of the probability density function for X. Find the mean, variance, and standard deviation of X.
b. What is the probability that the length of time until symptoms return is less than the mean?
c. What is the probability that the length of time until symptoms return is at least 12 hours?
d. What is the probability that the length of time until symptoms return is between 8 and 16 hours?

SOLUTION

a. Use Equation 6.9 to sketch the graph (Figure 6.70) and Equation 6.10 to compute the mean, variance, and standard deviation.

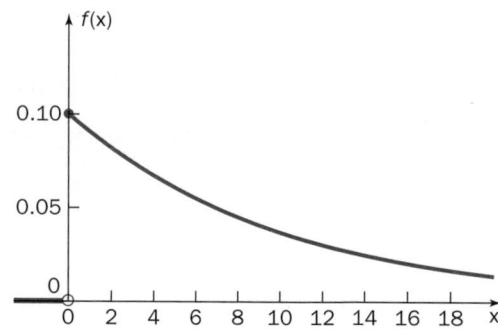

Figure 6.70 The probability density function for an exponential random variable with $\lambda = 0.1$.

$$\mu = \frac{1}{\lambda} = \frac{1}{0.1} = 10$$

$$\sigma^2 = \frac{1}{\lambda^2} = \frac{1}{0.1^2} = 100, \quad \sigma = \sqrt{\sigma^2} = \sqrt{100} = 10$$

b. The length of time until symptoms return is modeled by an exponential random variable ($\lambda = 0.1$). The mean is 10. Translate the question into a probability statement, and use cumulative probability where appropriate.

$P(X < 10)$ Translation to a probability statement.

$\quad = 1 - e^{-0.1(10)}$ Use the formula for cumulative probability.

$\quad = 1 - e^{-1}$ Simplify.

$\quad = 1 - 0.3679 = \boxed{0.6321}$

The probability that the length of time until symptoms return is less than the mean is 0.6321. Figure 6.71 illustrates this probability.

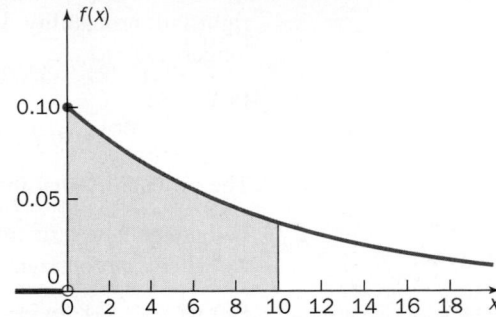

Figure 6.71 The shaded region represents the probability that the symptoms return in less than 10 hours.

c. Translate the question into a probability statement, and use cumulative probability where appropriate. *At least 12 hours* means 12 hours or greater.

$$P(X \geq 12)$$
Translation to a probability statement.

$$=e^{-0.1(12)}$$
Use the formula for a right-tail probability.

$$=e^{-1.2} = \boxed{0.3012}$$
Simplify.

The probability that the symptoms will return in 12 hours or more is 0.3012.

d. Translate the question into a probability statement, and use cumulative probability where appropriate.

$$P(8 \leq X \leq 16)$$
Translation to a probability statement.

$$=P(X \leq 16) - P(X < 8)$$
Use cumulative probability.

$$=\left[1 - e^{-0.1(16)}\right] - \left[1 - e^{-0.1(8)}\right]$$
Formula for cumulative probability.

$$=0.7981 - 0.5507 = \boxed{0.2474}$$

The probability that the symptoms will return in 8 to 16 hours is 0.2474.

Figures 6.72 through 6.74 show technology solutions.

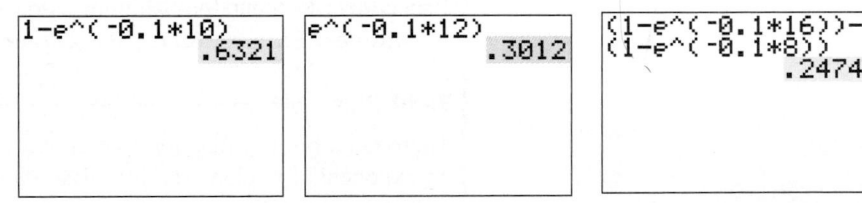

Figure 6.72
P(X < 10): use the formula for cumulative probability.

Figure 6.73
P(X ≥ 12): use the formula for right-tail probability.

Figure 6.74
P(8 ≤ X ≤ 16): compute each cumulative probability and subtract. ●

The following example illustrates an unusual property of the exponential distribution.

Example 6.11 Heat Pump Lifetime A Carrier XH heat pump, designed to heat a home in the winter and cool the home in the summer, lasts 16 years on average. The lifetime (in years) of this system can be modeled by an exponential random variable, X, with $\lambda = 0.0625$. Suppose a heat pump is selected at random.
a. What is the probability that the heat pump will last for at least 5 years?
b. Suppose the heat pump lasts for 5 years. What is the probability that it will last for at least another 5 years?

SOLUTION

a. Translate the question into a probability statement. Use cumulative probability or right-tail probability where appropriate. *At least 5 years* means 5 years or longer.

$$P(X \geq 5)$$
Translation to a probability statement.

$$= e^{-0.0625(5)} = 0.7316$$
Right-tail probability.

The probability that the heat pump lasts for at least 5 years is 0.7316.

b. Use the definition of conditional probability, and cumulative or right-tail probability where appropriate.

$$P(X \geq 5 + 5 \mid X \geq 5)$$
Translation to a probability statement.

$$= \frac{P[(X \geq 10) \cap (X \geq 5)]}{P(X \geq 5)}$$
Definition of conditional probability.

$$= \frac{P(X \geq 10)}{P(X \geq 5)}$$
$(X \geq 10) \cap (X \geq 5) = X \geq 10$.

$$= \frac{e^{-0.0625(10)}}{e^{-0.0625(5)}}$$
Right-tail probabilities.

$$= 0.7316$$

This *conditional* probability of lasting an additional 5 years is the same as the *unconditional* probability of lasting an initial 5 years. According to this model, the fact that it has lasted 5 years does not affect its probability of lasting an additional 5 years. This unrealistic result is called the *memoryless property* of an exponential random variable. At any point in time, the exponential random variable *forgets* or *ignores* what happened earlier: "The future is independent of the past." ●

---

b. Solution Trail

KEYWORDS

Suppose the heat pump lasts for 5 years.

↓

TRANSLATION

Given that the heat pump lasts 5 years.

↓

CONCEPTS

Conditional probability.

↓

VISION

In formulating the conditional probability statement, remember X is a continuous random variable. The events, stated in terms of X, must describe an interval. Given the heat pump lasts for 5 years means it lasts for *at least* 5 years.

TECHNOLOGY CORNER

Procedure: Compute probabilities associated with an exponential random variable.
Reconsider: Example 6.11, page 277, solutions, and interpretations.

TI-84 Plus

There is no built-in function to compute (cumulative) probabilities associated with an exponential random variable. Use the formulas for cumulative probability and right-tail probability.

1. $P(X < 10)$: use the formula for cumulative probability. See Figure 6.72, page 277.
2. $P(X > 12)$: use the formula for right-tail probability. See Figure 6.73, page 277.
3. $P(8 \leq X \leq 16)$: compute each cumulative probability and subtract. See Figure 6.74, page 277.

Minitab

There are built-in functions to compute (strict) cumulative probability and inverse cumulative probability. These functions may be accessed through a graphical input window or by using the command language.

1. In a session window, use the function CDF to find $P(X < 10)$, or select Calc; Probability Distributions; Exponential. Choose Cumulative probability and enter the Scale = μ = 10 and Threshold = 0. Select input constant (x) and enter 10. See Figure 6.75, page 279. The cumulative probability is displayed in the session window. See Figure 6.76, page 279.

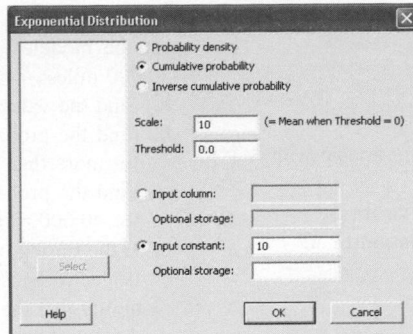

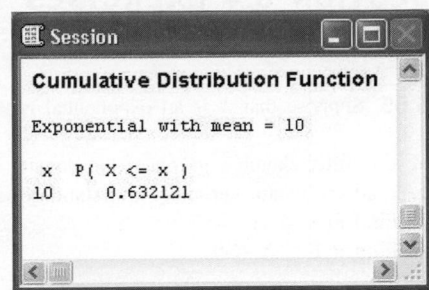

Figure 6.75 Graphical input window.

Figure 6.76 Cumulative distribution function output.

2. In a session window, use the function CDF to find $P(X \leq 12)$. Use the complement rule to find $P(X \geq 12)$. See Figure 6.77.

3. Use the function CDF to find $P(X \leq 16)$. Note that Minitab uses the mean to completely describe an exponential random variable. Store the result in K1. Use the function CDF to find $P(X \leq 8)$. Store the result in K2. Find the difference and print the result. See Figure 6.78.

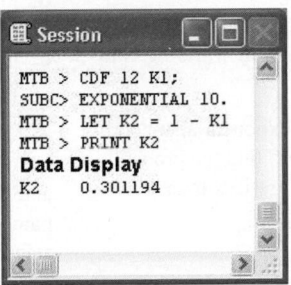

Figure 6.77 $P(X \geq 12)$.    **Figure 6.78** $P(8 \leq X \leq 16)$.

Excel

The built-in function EXPONDIST may be used to compute (strict) cumulative probability.

1. Use the function EXPONDIST to find $P(X < 10)$. The arguments for this function are x, λ, and the logical value True to compute cumulative probability. See Figure 6.79.

2. Use the function EXPONDIST to find $P(X \leq 12)$. Use the complement rule to find $P(X \geq 12)$. See Figure 6.80.

	A	B
1	0.6321	= EXPONDIST(10,0.1,TRUE)

Figure 6.79 $P(X < 10)$ Use the function for cumulative probability.

	A	B
1	0.6988	= EXPONDIST(12,0.1,TRUE)
2	0.3012	= 1 - A1

Figure 6.80 $P(X \geq 12)$: Use the function for cumulative probability and the complement rule.

3. Use the function EXPONDIST to find $P(X \leq 16)$. Use the function EXPONDIST to find $P(X \leq 8)$. Find the difference. See Figure 6.81.

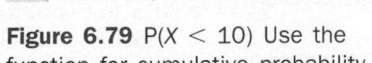

	A	B
1	0.7981	= EXPONDIST(16,0.1,TRUE)
2	0.5507	= EXPONDIST(8,0.1,TRUE)
3	0.2474	= A1 - A2

Figure 6.81 $P(8 \leq X \leq 16)$

SECTION 6.4 EXERCISES

Practice

6.65 Suppose that X is an exponential random variable with $\lambda = 0.2$.
a. Carefully sketch a graph of the density curve for X.
b. Find the mean, variance, and standard deviation of X.
c. Find $P(X < 3)$.
d. Find $P(1 < X < 9)$.

6.66 Suppose that X is an exponential random variable with $\lambda = 0.25$.
a. Carefully sketch a graph of the density curve for X.
b. Find $P(0.01 \leq X \leq 0.05)$.
c. Find $P(X > 0.06)$.

6.67 Suppose that X is an exponential random variable with $\lambda = 0.025$.
a. Find $P(X > 30)$ and illustrate this probability using an appropriate density curve.
b. Find $P(X > 20)$.
c. Find $P(X > 50 | X \geq 30)$.

6.68 Suppose that X is an exponential random variable and $P(X \leq 20) = 0.7981$. Find the value of λ.

6.69 Suppose that X_i ($i = 1, 2, 3, 4$) is an exponential random variable with $\lambda = i/10$. If the X_is are independent, find the probability that the values of all four random variables are less than 1.

Applications

6.70 **Public Policy and Political Science** More than 65,000 politically active Americans receive *The Weekly Standard* magazine each week. This magazine is hand-delivered to the most powerful men and women in government, politics, and the media. According to a survey conducted by magazine staff, the mean amount of money that subscribers spent during the past 12 months for personal or vacation travel is $6159. Suppose the distribution of the amount spent by a subscriber during the past 12 months on personal or vacation travel can be modeled by an exponential random variable.
a. Find the value of λ.
b. Find the probability that a randomly selected subscriber spent more than $8000 for personal or vacation travel during the past 12 months.
c. Find the probability that a randomly selected subscriber spent between $5000 and $10,000 for personal or vacation travel during the past 12 months.
d. If a subscriber spent less than $500 for personal or vacation travel during the past 12 months, the circulation department will no longer hand-deliver each issue. What is the probability that a randomly selected subscriber will no longer receive hand-delivered issues?

6.71 **Fuel Consumption and Cars** The purpose of an automobile timing belt is to provide a connection between the camshaft and the crankshaft. This allows the valves to open and close in sync with the pistons. Suppose the duration of a timing belt (in miles) can be modeled by an exponential random variable with mean 50,000 miles.
a. Find the value of λ.
b. Find the probability that a randomly selected timing belt lasts for more than 60,000 miles.
c. Find the probability that a randomly selected timing belt lasts for 40,000 to 80,000 miles.
d. If the timing belt on a new car breaks within 20,000 miles, the dealer will install a new belt free of charge. What is the probability that the dealer will be forced to install a new timing belt free of charge on a randomly selected car?

6.72 **Business and Management** Suppose the dentists at the Eastin Center for Modern Dentistry in Coeur D'Alene, Idaho, never see a patient on time. The amount of time (in minutes) past the appointment time that a random patient must wait to see a dentist has an exponential distribution with $\lambda = 0.01$. Suppose a patient is selected at random, and the appointment time has passed.
a. What is the probability that the dentist will see the patient within 5 minutes after the appointment time?
b. Find the probability that the dentist will see the patient between 10 and 20 minutes after the appointment time.
c. Find the probability that the patient will have to wait at least 30 minutes past the appointment time.

6.73 **Psychology and Human Behavior** In Columbia, South Carolina, on a randomly selected Friday night between 9:00 P.M. and 2:00 A.M., the time (in minutes) between calls to a 911 dispatcher has an exponential distribution with $\lambda = 0.125$. Suppose a Friday evening in Columbia is selected at random.
a. If a 911 call is taken at 10:07 P.M., what is the probability that the next call will occur before 10:30 P.M.?
b. If a 911 call is taken at 11:30 P.M., what is the probability that the next call will occur after 11:45 P.M.?
c. Find the mean, variance, and standard deviation for the time between 911 calls.
d. Find a value t such that if a 911 call is received between 9:00 P.M. and 2:00 A.M. then the probability that the next 911 call occurs more than t minutes later is 0.75.

6.74 **Public Policy and Political Science** During the Cold War there were frequent radio tests of alert warnings. The regular program was interrupted and replaced by a long high-pitched signal. An announcer would break in with a message similar to, "This has been a test of the emergency broadcasting system." Suppose the time (in hours) between these tests had an exponential distribution with mean 24 hours.
a. If a test occurred at 6:00 A.M., what is the probability that another one would occur before 6:00 P.M.?
b. If a test occurred at 10:00 P.M., what is the probability that another one would not occur until after 6:00 A.M.?
c. If a test occurred at 9:00 A.M., what is the probability that the next test would occur between 12:00 P.M. and 1:00 P.M.?

Extended Applications

6.75 **Public Health and Nutrition** The U.S. Public Health Service's Advisory Committee on Immunization Practices recommends a tetanus booster every 10 years. Suppose the lifetime (in years) of a tetanus booster shot has an exponential distribution with $\lambda = 0.05$.

a. Find the probability that a randomly selected tetanus booster shot is still protecting against tetanus after more than 10 years.

b. Suppose that a physician recommends a booster shot after a period when only 10% of all tetanus shots still have an effect. How long should the patient wait before getting another booster shot?

c. Suppose that two people independently receive tetanus shots. What is the probability that both shots still have an effect after more than 5 years?

6.76 Medicine and Clinical Studies A patient's blood cholesterol level is often checked during a routine physical examination. A blood sample is taken (from the finger or arm) and tested for total cholesterol and HDL cholesterol levels in milligrams per deciliter (mg/dL). If the total cholesterol is less than 200, then no action is taken. If a patient's total cholesterol is approximately 300, a new drug together with a strict diet is prescribed to reduce this number to a safe level. The amount of time (in days) it takes to reduce total cholesterol to a safe level has an exponential distribution with $\lambda = 1/15$. A patient with total cholesterol of 300 is randomly selected and placed on this drug-and-diet regimen.

a. Find the mean number of days until the total cholesterol level is safe (less than 200).

b. Find the probability that the total cholesterol level will be safe in a number of days within two standard deviations of the mean.

c. Find a value d such that 75% of all such patients have a safe cholesterol level with d days.

d. Suppose two patients are selected at random. Find the probability that this drug-and-diet regimen works for at least one of the patients within 10 days.

6.77 Marketing and Consumer Behavior A toy manufacturer routinely tests new toys in a controlled environment before deciding whether to actually market a toy. Research has shown that the amount of time (in minutes) a randomly selected child plays with a new toy has an exponential distribution with $\lambda = 0.05$. Suppose a new toy is presented for study.

a. What is the probability that a randomly selected child will play with the toy for at most 10 minutes?

b. What is the probability that a randomly selected child will play with the toy for between 5 and 20 minutes?

c. Suppose a child plays with the toy for at least 15 minutes. What is the probability that the child will play with the toy for another 20 minutes?

d. If four different children each independently play with a new toy for at least 25 minutes, then the toy is immediately brought to market. What is the probability of this happening for a newly designed toy?

6.78 Physical Sciences Four large water pumps supply Bellingham, Washington, with water. If water pump i ($i = 1, 2, 3, 4$) breaks down, then the time to repair it has an exponential distribution with $\lambda = 1/(2i)$ hours. Suppose the water pumps operate independently, and when breakdowns occur, the repair times are also independent.

a. Suppose water pump 1 breaks down. What is the probability that it will take more than 30 minutes to repair?

b. Answer part (a) for water pumps 2, 3, and 4.

c. Suppose all four water pumps break simultaneously. What is the probability that at least one of the four water pumps will not be repaired within 1 hour?

6.79 Public Health and Nutrition From the instant a fresh-baked chocolate chip cookie is taken out of the oven, the time (in minutes) the wonderful aroma lasts has an exponential distribution with $\lambda = 1/30$. Suppose a chocolate chip cookie is done baking and is taken out of the oven.

a. Find the mean, variance, and standard deviation for the time the aroma lasts.

b. What is the probability that the aroma lasts for at least 40 minutes?

c. What is the probability that the aroma lasts for between 30 and 50 minutes?

d. Find a value t such that the probability the aroma lasts for at most t minutes is 0.90.

e. Suppose a second batch of cookies is taken out of the oven 10 minutes after the first. What is the probability that there will be no aroma from either batch 35 minutes after the first batch was done?

Challenge

6.80 Memoryless Property Suppose X is an exponential random variable with parameter λ. If a and b are constants (> 0), confirm the *memoryless property* by showing $P(X \geq a) = P(X \geq a + b \mid X \geq b)$.

(Frhojdysz/Dreamstime.com)

Chapter 6 Challenge Wrap-Up

Each day, the height of waves off the coast of St. Croix is approximately normal with mean 7 feet and standard deviation 0.75 feet. Tours are canceled on only 2% of all days (due to high waves). Therefore, 98% of the days have waves with acceptable heights.

Let X be the height of waves. X is a normal random variable with mean $\mu = 7$ and standard deviation $\sigma = 0.75$: $X \sim N(7, 0.75^2)$. Find a height h such that $P(X > h) = 0.02$, or in terms of cumulative probability, $P(X \le h) = 0.98$. Since the probability is already known, this is an *inverse* cumulative probability problem. Convert the expression to a statement involving Z, and work backwards in Table III.

$$P(X \le h) = P\left(\frac{X - 7}{0.75} \le \frac{h - 7}{0.75}\right) \qquad \text{Standardize.}$$

$$= P\left(Z \le \frac{h - 7}{0.75}\right) = 0.98 \qquad \text{Use Equation 6.8.}$$

There is no other simplification within the probability statement. However, the resulting expression involves Z and is cumulative probability. Find a value in the body of Table III in the Appendix as close to 0.98 as possible. Set the corresponding z equal to $\left(\frac{h - 7}{0.75}\right)$, and solve for h.

$$\frac{h - 7}{0.75} = 2.0537 \qquad \text{Table III in the Appendix; interpolation.}$$

$$h - 7 = 1.5403 \qquad \text{Multiply both sides by 0.75.}$$

$$h = 8.5403 \qquad \text{Add 7 to both sides.}$$

If tours are canceled when the waves are at least 8.54 feet high, then approximately 2% of all trips to Buck Island will be postponed.

Figure 6.82 shows a technology solution.

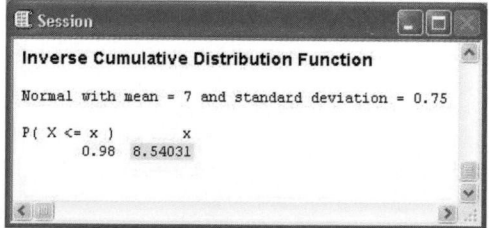

Figure 6.82 Minitab inverse cumulative distribution function.

CHAPTER 6 SUMMARY

Concept	Page	Notation / Formula / Description
Probability distribution for a continuous random variable	238	A smooth curve (density curve) defined such that the probability X takes on a value between a and b is the area under the curve between a and b.
Uniform distribution	240	If X has a uniform distribution on the interval $[a, b]$, then $$f(x) = \begin{cases} \dfrac{1}{b-a} & \text{if } a \le x \le b \\ 0 & \text{otherwise} \end{cases} \quad \mu = \frac{a+b}{2}, \quad \sigma^2 = \frac{(b-a)^2}{12}$$
Normal distribution	250	If X is a normal random variable with mean μ and variance σ^2, then the probability density function is given by $$f(x) = \frac{1}{\sigma\sqrt{2\pi}} e^{-(x-\mu)^2/2\sigma^2}, \quad -\infty < x < \infty, \ -\infty < \mu < \infty, \ \sigma^2 > 0$$
Standard normal distribution	252	A normal distribution with $\mu = 0$ and $\sigma^2 = 1$ is the standard normal distribution. The standard normal random variable is usually denoted by Z: $Z \sim N(0, 1)$ $-\infty < z < \infty$.
Standardization	255	If X is a normal random variable with mean μ and variance σ^2, then $Z = \dfrac{X - \mu}{\sigma}$ is a standard normal random variable.
Normal probability plot	266	A scatter plot of each observation versus its corresponding expected value from a Z distribution. For a normal distribution, the points will fall along a straight line.
Exponential distribution	275	If X is an exponential random variable with parameter λ, then the probability density function is given by $$f(x) = \begin{cases} \lambda e^{-\lambda x} & \text{if } x > 0 \\ 0 & \text{otherwise} \end{cases} \quad \mu = \frac{1}{\lambda}, \sigma^2 = \frac{1}{\lambda^2}$$

CHAPTER 6 EXERCISES

Applications

6.81 Technology and Internet In the advertising department of SBC Communications, Inc., all of the computers are part of a local area network connected to one main printer. When an advertising employee prints a document, the job is placed in a queue. It may take several minutes before the document begins to print due to the number of other print jobs and the complexity of the document. The time (in minutes) until the document starts to print has an exponential distribution with $\lambda = 0.40$. Suppose a randomly selected document is sent to the main printer.

a. What is the mean time until the document begins to print?
b. What is the probability that the document will begin to print within 30 seconds?
c. What is the probability that the document will take more than 5 minutes before starting to print?
d. Find a value t such that only 2% of all documents take at least t minutes before starting to print.

6.82 Public Health and Nutrition The amount of sodium in a randomly selected 8-ounce serving of chicken noodle soup has a normal distribution with mean $\mu = 1106$ mg and standard deviation $\sigma = 150$ mg.[16] Suppose an 8-ounce serving is randomly selected.

a. What is the probability that the amount of sodium is over 1000 mg?
b. What is the probability that the amount of sodium is between 800 and 1200 mg?
c. Find a symmetric interval about the mean such that 90% of all 8-ounce servings have sodium amounts in that interval.
d. Suppose a randomly selected 8-ounce serving has a sodium level of 750 mg. Is there any evidence to suggest that the mean sodium level reported above is false?

6.83 Public Health and Nutrition The Food and Drug Administration (FDA) reviews all advertisements for drugs to check for omissions regarding a drug's risk; inadequate, incorrect, or inconsistent labeling; misleading claims; unsupported comparative claims; and unapproved purposes. Lengthy legal reviews of advertisements have increased the total review time. Suppose the length of time between a request for a review and final approval has a normal distribution with mean $\mu = 21$ days and $\sigma = 4$ days. Consider a randomly selected advertisement submitted to the FDA for review.

a. What is the probability that the advertisement will be reviewed in fewer than 14 days?

b. What is the probability that the advertisement will be reviewed in 15 to 19 days?

c. Suppose the advertisement takes at least 20 days for review. What is the probability that it will take fewer than 30 days for review?

d. Suppose two independent advertisements are submitted for review simultaneously. What is the probability that both will take more than 30 days for review?

6.84 Marketing and Consumer Behavior Canister vacuums are often rated according to their ease of use, noise level, emissions, cleaning ability, and length of the power cord. The length (in feet) of the electric cord on a canister vacuum is a random variable, X, with a uniform distribution on the interval [20, 30]. Suppose a canister vacuum is selected at random.

a. Carefully sketch a graph of the probability density curve for the random variable X.

b. What is the probability that the power cord has a length less than 22 feet?

c. What is the probability that the power cord has a length greater than 26 feet?

d. Find a value f such that 75% of all power cords have length greater than f feet.

6.85 Fuel Consumption and Cars A random sample of the braking distance (in feet) for midsize cars with tire pressure 25 psi on dry asphalt at 60 mph is given on the data CD and book's web site.[17] Is there any evidence to suggest that the data are from a nonnormal population?

6.86 Business and Management The manager for the Wingate Inn claims that the time it takes to make a room reservation over the phone is approximately normally distributed with mean 4 minutes and standard deviation 45 seconds. Suppose a call placed to the inn in order to make a room reservation is selected at random.

a. What is the probability that it will take less than 3 minutes to make the reservation?

b. What is the probability that it will take between 3.5 and 4.5 minutes to make the reservation?

c. Find the first and the third quartile times.

d. Suppose it takes 7 minutes to make the reservation. Is there any evidence to suggest that the inn's claim ($\mu = 4$ minutes) is false? Justify your answer.

6.87 Manufacturing and Product Development Film for a 35-mm camera includes a celluloid base with various coatings necessary for manufacture and processing. The thickness of the celluloid base is approximately normal with mean 0.125 mm and standard deviation 0.025 mm. Suppose a roll of 35-mm film is randomly selected.

a. What is the probability that the thickness of the celluloid is between 0.10 and 0.17 mm?

b. What is the probability that the thickness of the celluloid is more than 0.20 mm?

c. If the thickness of the celluloid is less than 0.04 mm, the film cannot be processed. What is the probability that the film cannot be processed?

d. Suppose three independent rolls of film are selected at random. What is the probability that all three will have celluloid thickness greater than 0.15 mm?

6.88 Manufacturing and Product Development Oriental rugs are made from various wools and woven in several different countries. The pile height (in mm) of an oriental rug varies slightly and can be modeled by a uniform random variable on the interval [6, 10].

a. Carefully sketch a graph of the density function for pile height.

b. What is the probability that a randomly selected oriental rug will have pile height less than 7 mm?

c. What is the probability that a randomly selected oriental rug will have pile height between 8.5 and 9.5 mm?

d. Find a value h such that 90% of all oriental rugs have pile height less than h.

6.89 Technology and Internet The life expectancy of a laser-printer toner cartridge varies considerably according to the toner and drum type and how the cartridge is used. Printed pages containing lots of graphics require more toner, while pages with mostly text require considerably less toner. Page coverage is usually measured as a proportion (or percentage). For example, a typical text page has approximately 0.05 coverage. A random sample of printed pages from a office printer was obtained, and the page coverage was carefully measured. The data are given on the data CD and book's web site. Is there any evidence to suggest that the data are from a nonnormal population?

6.90 Sports and Leisure Jockeys in the United States and England work very hard to keep their weight down. Many participate in weight-loss programs, carefully monitor their diet, and exercise regularly. The weight of a jockey is approximately normal with mean 52 kg and standard deviation 1.2 kg. Suppose a jockey is randomly selected.

a. What is the probability that the jockey weighs more than 53 kg?

b. What is the probability that the jockey weighs between 50 and 54 kg?

c. Find a value w such that 80% of all jockeys weigh more than w.

d. Suppose the jockey selected weighs 57 kg. Is there any evidence to suggest that the claimed mean (52 kg) is wrong? Justify your answer.

6.91 Public Health and Nutrition The amount of caffeine in a cup of coffee varies considerably, even if it is brewed by the same person using the same brewing method and ingredients. Suppose the amount of caffeine in an 8-ounce cup of Maxwell House coffee is approximately normally distributed with mean 110 mg and standard deviation 7.75 mg.[18]

a. What is the probability that a randomly selected cup of Maxwell House coffee will have less than 100 mg of caffeine?

b. What is the probability that a randomly selected cup of Maxwell House coffee will have between 105 and 120 mg of caffeine?

c. A recent article in a medical journal suggests that an 8-ounce cup of coffee with more than 130 mg of caffeine could cause a person's heart to race. What is the probability that a randomly selected cup will have more than 130 mg of caffeine?

6.92 Manufacturing and Product Development The quality of a kitchen knife is often measured by the sharpness and total lifetime of the blade. One test for sharpness involves mounting the knife with the blade vertical and lowering a specially designed pack of paper onto the blade. The sharpness is measured by the depth of the cut. A larger depth indicates a sharper knife. Suppose the depth of the cut for a randomly selected knife has a normal distribution with mean 92 mm and standard deviation 21 mm.

a. What is the probability that a randomly selected kitchen knife has a sharpness measure less than 75 mm?

b. A kitchen knife with sharpness measure of at least 100 mm qualifies as a steak knife. What proportion of kitchen knives are steak knives?

c. The Kitchen Gadgets Association would like to set a maximum sharpness for butter knives. Find a value c such that 15% of all knives have sharpness below c.

d. Suppose a randomly selected kitchen knife has sharpness greater than 90 mm. What is the probability that it has sharpness greater than 100 mm?

6.93 Economics and Finance The National Automobile Dealers Association (NADA) monitors the number of new car loans, the duration of the loan, and the amount borrowed. According to the NADA, the average new car loan is $17,000. Suppose the amount of a new car loan is approximately normal with standard deviation $4,500.

a. What is the probability that a new car loan is for at most $10,000?

b. What is the probability that a new car loan is for $15,000 to $20,000?

c. Find a symmetric interval about the mean such that 95% of all new car loans fall in this interval.

d. Suppose the length (in months) of a new car loan is also approximately normal with mean 62 and standard deviation 4. If the length of the loan and the amount borrowed are independent, what is the probability that the loan will be for more than $25,000 and for fewer than 56 months?

6.94 Marketing and Consumer Behavior According to the National Retail Federation, the mean amount spent by each consumer on Valentine's Day gifts in 2008 was expected to be $123. A random sample of consumers was obtained, and each was asked for the amount (in dollars) spent on Valentine's Day gifts. The data are given on the data CD and book's web site. Is there any evidence to suggest that the data are from a nonnormal population?

6.95 Physical Sciences Large reservoirs of oil found underground are under very high pressure, which allows the oil to be pumped to the surface. All oil fields contain some water, and as water is pumped back into a well in order to maintain high pressure, the water content increases. Suppose the proportion of water in a randomly selected barrel of oil pumped to the surface has a normal distribution with mean 0.12 and standard deviation 0.025.

a. What is the probability that a randomly selected barrel of oil has a proportion of water less than 0.12?

b. What is the probability that a randomly selected barrel of oil has a proportion of water between 0.15 and 0.17?

c. The higher the proportion of water in oil, the more expensive it is to separate the oil from water. If the proportion of water is greater than 0.20, then the well is too expensive to operate and maintain. What is the probability that a randomly selected well is too expensive?

Extended Applications

6.96 Travel and Transportation In order to alleviate the concerns of American citizens, the Easy Crossing Council reported that there are minimal to no delays (less than 10 minutes) for vehicles entering Canada at Niagara's border crossing.[19] Suppose the time to cross the border at Niagara is a random variable, X, with probability density function given by

$$f(x) = \begin{cases} 0.02x & \text{if } 0 \leq x \leq 10 \\ 0 & \text{otherwise} \end{cases}$$

a. Carefully sketch a graph of the density function.

b. Find the probability that it takes less than 5 minutes for a randomly selected vehicle to cross the border at Niagara.

c. Find the probability that it takes more than 8 minutes for a randomly selected vehicle to cross the border at Niagara.

d. Find the probability that it takes between 2 and 6 minutes for a randomly selected vehicle to cross the border at Niagara.

e. Suppose it takes fewer than 2 minutes for a randomly selected vehicle to cross the border at Niagara. What is the probability that it takes less than 1 minute?

6.97 Public Health and Nutrition Meat or poultry classified as *lean* has less than 4 grams of saturated fat. Suppose the amount of saturated fat in a randomly selected piece of lean meat or poultry is a random variable, X, with probability density function given by

$$f(x) = \begin{cases} 0.1 & \text{if } 0 \leq x < 1 \\ 0.2 & \text{if } 1 \leq x < 2 \\ 0.3 & \text{if } 2 \leq x < 3 \\ 0.4 & \text{if } 3 \leq x < 4 \\ 0 & \text{otherwise} \end{cases}$$

a. Carefully sketch a graph of the density function.

b. What is the probability that a randomly selected piece of lean meat or poultry has less than 1.5 grams of saturated fat?

c. What is the probability that a randomly selected piece of lean meat or poultry has more than 3 grams of saturated fat?

d. What is the probability that a randomly selected piece of lean meat or poultry has between 2 and 4 grams of saturated fat?

e. Suppose a randomly selected piece of lean meat or poultry has at most 3 grams of saturated fat. What is the probability that it has at most 1 gram of saturated fat?

6.98 Manufacturing and Product Development Four people working independently on an assembly line all perform the same task. The time (in minutes) to complete this task for person i ($i = 1, 2, 3, 4$) has a uniform distribution on the interval $[0, i]$. Suppose each person begins the task at the same time.

a. What is the probability that person 2 takes less than 90 seconds to complete the task?

b. What is the mean completion time for each person?

c. What is the probability that all four people will complete the task in less than 30 seconds?

d. What is the probability that exactly one person will complete the task in less than 1 minute?

6.99 Probability Calculations Using a Density Function The probability density function for a random variable X is given by

$$f(x) = \begin{cases} -\dfrac{1}{4}x + \dfrac{1}{2} & \text{if } 0 \leq x < 2 \\ -\dfrac{1}{4}x + 1 & \text{if } 2 \leq x < 4 \\ 0 & \text{otherwise} \end{cases}$$

a. Carefully sketch a graph of the density function.

b. Find $P(X < 1)$.

c. Find $P(X \geq 3)$.

d. Find $P(1 < X < 3)$.

Sampling Distributions

Chapter 7 Challenge

How long does the average person spend on a treadmill?

A treadmill is often the center of an exercise regimen. This piece of equipment allows walkers, joggers, and runners to complete a serious workout indoors, even when the weather is nasty. Great for the cardiovascular system, many compact treadmills store easily yet feature a variety of options. Some make it easy to read, listen to music, or watch television while walking.

Most health clubs feature an array of treadmills for their members. Many users routinely use a treadmill as all or part of a typical workout. Walkers, runners, and joggers often gauge their workout by the amount of time spent on a treadmill. Suppose a certain health club carefully monitors the use of its treadmills. The manager claims that the mean amount of time spent on a treadmill is 35 minutes with standard deviation 3.5 minutes.

A random sample of 40 treadmill users was obtained. The sample mean time spent on the treadmill for these users was 32.75 minutes. Is there any evidence to sug-

(Tetra Images/Alamy)

gest that the true mean time spent on a treadmill is less than 35 minutes? The answer is in the Chapter Challenge Wrap-Up (page 317). The concepts presented in this chapter will allow us to find the distribution of the sample mean, to answer probability questions concerning this random variable, and to make inferences about a population mean. These ideas all rely on the Central Limit Theorem, the most important result in probability and statistics!

Review
- Recall that numbers like μ, σ^2, and p completely characterize, or describe, a population.
- These numbers are constant and usually unknown, but we would like to estimate these values, or draw a conclusion concerning these numbers.

Preview
- Understand and utilize methods to find a sampling distribution (for $\overline{X}$ and for $\widehat{P}$) and use it in statistical inference.
- Discover the Central Limit Theorem, and solve probability and inference problems using this result.

CONTENTS

7.1 STATISTICS, PARAMETERS, AND SAMPLING DISTRIBUTIONS

The terms **parameter** and **statistic** have been used in previous chapters in intuitive contexts. The following definitions distinguish between measures represented by symbols like μ, σ^2, and p, and the quantities used to estimate these values.

> **DEFINITION**
>
> A **parameter** is a numerical descriptive measure of a population.
> A **statistic** is any quantity computed from values in a sample.

 ILLUMINATING THE CONCEPTS

1. A parameter is a population quantity. It is used to describe some characteristic of a population. Usually we cannot measure a parameter; it is an unknown constant that we would like to estimate.

2. A statistic is *any* sample quantity. There are infinitely many quantities one could compute using the data in a sample. For example, $\bar{x}$ and $\tilde{x}$ are statistics, as is the sum of the smallest and the largest values divided by two. ■

The difference between a parameter and a statistic is key: parameters describe populations, and statistics describe samples. We use statistics to make inferences about parameters. Therefore, the properties of a statistic are important.

Example 7.1 Parameter versus Statistic   In each of the following statements, identify the **boldface** number as the value of a population parameter or a sample statistic.

a. In a recent survey of American males, **42**% said more money would make their lives better.
b. A spokesman for a large insurance agency reported that the proportion of all women with some form of life insurance is **0.32**.
c. The Department of Transportation recently reported that the mean age of all highway bridges in the United States is **13.8** years.
d. The manager of a large hotel located near Disney World indicated that 20 selected guests had a mean length of stay equal to **5.6** days.
e. In a psychology experiment, **65**% of tested individuals indicated that teal is a soothing color.

SOLUTION

a. 42% is a statistic. This number describes a characteristic of a *sample* of American males.
b. 0.32 is a parameter. This number describes a characteristic of the *entire* population of women.
c. 13.8 is a parameter. This number is a characteristic of *all* the highway bridges in the United States.
d. 5.6 is a statistic. This number describes a characteristic of the *sample* of 20 guests.
e. 65% is a statistic. This percentage describes a characteristic of the *sample* of subjects involved in the experiment.   ●

Suppose the mean is computed for a sample of size n from a population of interest. Denote this value $\bar{x}_1$. In a second sample of size n from the same population, let $\bar{x}_2$ be the mean. It is reasonable to expect $\bar{x}_1$ and $\bar{x}_2$ to be *close* but not equal. The important realization is that $\bar{x}_1$ will be *different* from $\bar{x}_2$. In fact, the sample mean will differ from sample to sample. This statistic is subject to sampling variability. Therefore, the sample mean, $\overline{X}$, is a random variable, and $\overline{X}$ has a mean, a variance, a standard deviation, and a probability distribution. This distribution is called a **sampling distribution**.

Any statistic is a random variable because it differs from sample to sample. One cannot predict the value of a statistic with absolute certainty. In order to make a reliable inference based on a specific statistic, we need to know the properties of the distribution of the statistic.

Statistics are random variables!

DEFINITION

The **sampling distribution** of a statistic is the probability distribution of the statistic.

 ILLUMINATING THE CONCEPTS

1. A sampling distribution (like any random variable) describes the long-run behavior of the statistic.

2. There are two techniques we can use to obtain, or find, a sampling distribution.
   a. Recall that to approximate the distribution of a population, we construct a histogram (or stem-and-leaf plot) using values from the population. If the sample is representative, then the histogram should be similar in shape, center, and spread to the population distribution.

      Similarly, to approximate the distribution of a statistic, we obtain (many) values of the statistic and construct a histogram. The resulting graph approximates the sampling distribution of the statistic.

      For example, here are the steps to approximate the sampling distribution of the mean of a sample of size $n = 10$ from a population. (1) Obtain several samples of size 10 from the population. (2) Compute the sample mean for each sample. (3) Construct a histogram using all the sample means. The histogram approximates the sampling distribution of the sample mean.

   b. In some cases, the exact sampling distribution of a statistic can be obtained. If the statistic is a discrete random variable, the sampling distribution includes all the values the statistic assumes and the associated probabilities. If the statistic is a continuous random variable, the sampling distribution consists of a probability density curve. ■

Example 7.2 Stuffed Animals Sarah has started a stuffed animal collection and currently has five koalas on her dresser. The plush toys vary in color, design, and size. These factors, along with different stuffing material, contribute to the various weights (in ounces) of the five koalas: 10.9, 14.5, 17.1, 18.1, and 17.6.

Suppose a sample of three koalas is selected and the (sample) median weight is computed. Find the sampling distribution for the (sample) median.

SOLUTION

The order of selection does not matter here. Therefore, this is a combination.

STEP 1 There are $\binom{5}{3} = 10$ ways to select three koalas from the five in the population. List all the possible samples, the computed value of the statistic for each sample, and the probability of selecting each sample. Use this table to construct the sampling distribution.

STEP 2 Here is the resulting table.

Sample	$\tilde{x}$	Probability	Sample	$\tilde{x}$	Probability
10.9, 14.5, 17.1	14.5	0.1	10.9, 18.1, 17.6	17.6	0.1
10.9, 14.5, 18.1	14.5	0.1	14.5, 17.1, 18.1	17.1	0.1
10.9, 14.5, 17.6	14.5	0.1	14.5, 17.1, 17.6	17.1	0.1
10.9, 17.1, 18.1	17.1	0.1	14.5, 18.1, 17.6	17.6	0.1
10.9, 17.1, 17.6	17.1	0.1	17.1, 18.1, 17.6	17.6	0.1

The median for each sample is the middle value. The probability of each sample is $1/10 = 0.1$, since we assume each sample is equally likely.

STEP 3 Out of all the possible samples, there are only three possible values for the sample median in this case. Sum the probabilities associated with each value. The probability distribution for the random variable $\tilde{X}$ lists the values (of $\tilde{X}$) and the associated probabilities.

$\tilde{x}$	14.5	17.1	17.6
$p(\tilde{x})$	0.3	0.4	0.3

●

Challenge: Find the sampling distribution for $\tilde{X}$ if sampling is done with replacement. Hint: There are $5 \times 5 \times 5 = 125$ possible samples.

💡 ILLUMINATING THE CONCEPTS

1. In the example above, $\tilde{X}$ is a discrete random variable. Using Equations 5.1, 5.3, and 5.4, the mean, variance, and standard deviation for $\tilde{X}$ are $\mu = 16.47$, $\sigma^2 = 1.71$, $\sigma = 1.31$.

2. The koalas were selected above *without* replacement. Suppose three koalas are selected *with* replacement. That is, select a stuffed animal, record the weight, and place it back on the dresser. The same one could be selected two or three times. This sampling scheme changes the probability distribution for $\tilde{X}$. ■

Example 7.3 Sick Days New employees at Whittmanhart are allowed up to 3 sick days during a calendar year.[1] Suppose the probability distribution for X, the number of sick days used by a new employee during a year, is given in the following table.

x	0	1	2	3
$p(x)$	0.40	0.35	0.20	0.05

Two new employees are independently selected at random, and the number of sick days is recorded for each. Consider the statistic M, the maximum number of sick days taken by either employee. Find the probability distribution for M.

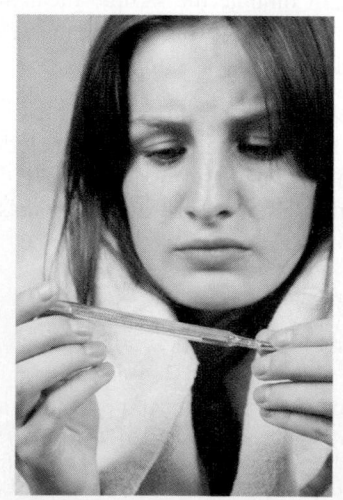

(Anyra/Dreamstime.com)

SOLUTION

STEP 1   A sample consists of two numbers. The first represents the sick days for employee 1, and the second denotes the sick days for employee 2. There are 16 possible samples. Using the Multiplication Rule, there are two slots to fill: $4 \times 4 = 16$. List all the possible samples, the computed value of the statistic for each sample, and the probability of selecting each sample.

STEP 2   The probability associated with each sample is computed using the independence assumption. For example, the probability that the first employee used one sick day *and* the second employee used two sick days is given by

Each *event* is given in terms of a random variable.

$$P(X = 1 \cap X = 2)$$
$$= P(X = 1) \cdot P(X = 2) \qquad \text{Independent events.}$$
$$= (0.35)(0.20) = 0.07 \qquad \text{Use the probability distribution for } X.$$

STEP 3   Use the table below to construct the sampling distribution.

Sample	*m*	Probability	Sample	*m*	Probability
0, 0	0	0.1600	2, 0	2	0.0800
0, 1	1	0.1400	2, 1	2	0.0700
0, 2	2	0.0800	2, 2	2	0.0400
0, 3	3	0.0200	2, 3	3	0.0100
1, 0	1	0.1400	3, 0	3	0.0200
1, 1	1	0.1225	3, 1	3	0.0175
1, 2	2	0.0700	3, 2	3	0.0100
1, 3	3	0.0175	3, 3	3	0.0025

The maximum for each sample is the largest of the two values.

STEP 4   There are four possible values for the discrete random variable *M*. Sum the probabilities associated with each value. The probability distribution is given in the table below.

Remember, a capital letter like *M* represents a random variable. The corresponding small letter *m* denotes a specific value that the random variable *M* can assume.

m	0	1	2	3
p(m)	0.1600	0.4025	0.3400	0.0975

Challenge: Find the mean, variance, and standard deviation for the random variable *M*.

In almost all observational studies it is assumed that the data are obtained from a **simple random sample**. Usually the sampling is done *without* replacement. Consider an exit poll or a study of the time spent each week on lawn care by homeowners. An individual is selected from the population and an observation is recorded. The individual is *not* placed back into the population. There is no chance the individual will be selected again.

If sampling is done without replacement, individual responses are dependent. However, if the population is *large enough* and the sample is small relative to the size of the population, then the responses are *almost independent*. Calculating probabilities assuming independence results in little loss of accuracy. As a rule of thumb, if the sample size is at most 5% of the total population, then successive observations can be considered independent. Even though sampling is done without replacement, the data are assumed to be part of a simple random sample.

Recall the following definition from Chapter 1.

> **DEFINITION**
>
> A (**simple**) **random sample** (SRS) of size n is a sample selected in such a way that every possible sample of size n has the same chance of being selected.

 Try the Simple Random Sample statistical applet on the text web site.

ILLUMINATING THE CONCEPTS

1. Given a finite population of size N and a sample of size n, the number of possible simple random samples is $\binom{N}{n}$.

2. A random sample consists of individuals or objects, and a variable is a characteristic of an individual or object. A value of the variable is obtained for each member of the random sample. We often refer to the *values* of the variable as the random sample rather than the *individuals*. For example, consider a study in which the amount of the trace element chromium is measured in coal from around the world. The random sample consists of pieces of coal (objects) and the values are the chromium measurements. However, it is common practice to say, "Consider the random sample of chromium measurements."

3. Unless stated otherwise, all data presented in this text are obtained from a simple random sample. ■

In the previous two examples, the *exact* sampling distribution of each statistic was determined. In the following example, a histogram is used to *approximate* a sampling distribution.

Note: The population mean is the sum of all the observations divided by $n = 20$: $\mu = 4.3$.

Example 7.4 Fat in Ice Cream The following table lists the total fat (in grams) in one cup of 20 popular brands of ice cream.

0	0	0	0	0	4	4	4	4	5
5	6	7	6	6	8	9	18	0	0

(*Source*: Center for Science in the Public Interest)

Find an approximate sampling distribution for the sample mean of five observations from this population.

SOLUTION

STEP 1 There are $\binom{20}{5} = 15504$ possible samples of size five. Instead of considering every one of these samples, select some (say 100) samples of size five, compute the mean for each sample, and construct a histogram of the sample means.

STEP 2 The table that follows lists the first few samples of size five and the mean for each sample.

Sample					$\bar{x}$	Sample					$\bar{x}$
8	4	6	18	0	7.2	0	0	4	0	0	0.8
6	4	6	4	6	5.2	8	0	9	4	0	4.2
0	5	6	0	9	4.0	18	4	5	4	8	7.8
		⋮			⋮			⋮			⋮

STEP 3 Figure 7.1 shows a histogram of the sample means for 100 samples of size five.

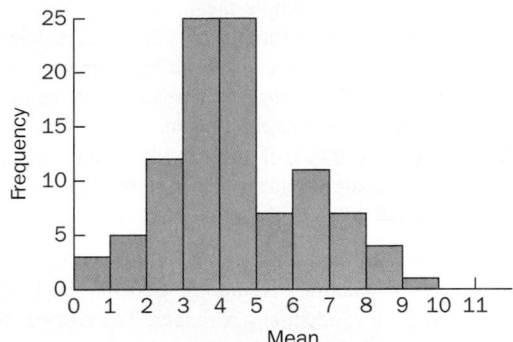

Figure 7.1 Histogram of sample means.

STEP 4 There are some very interesting, curious results here. Even though the population is not normally distributed, the shape of the sampling distribution appears to be approximately normal! In addition, the *center* of the sampling distribution of the mean is approximately the *population* mean ($\mu = 4.3$). Although the relationship between the original population parameters and the sampling distribution parameters is not clear, there is certainly *less* variability in the sampling distribution than in the population distribution. These three observations suggest the exact distribution of the sample mean, as discussed in the next section.

SECTION 7.1 EXERCISES

Applications

7.1 Parameter versus Statistic In each of the following statements, identify the **boldface** number as the value of a population parameter or a sample statistic.
a. A researcher in Boston conducted a study to investigate the prevalence of Alzheimer's disease in people over the age of 85. The results indicated that **47.2**% of the patients studied experienced symptoms consistent with the disease.
b. A brokerage firm reported that the mean dividend paid by all Fortune 500 companies for the year 2003 was $**0.62**.
c. A consumer magazine tested a random sample of 19 green teas for taste and health benefits. The mean cost per cup of tea was reported to be **14** cents.
d. A computer manufacturer claims that the mean lifetime of laptop batteries is **6.7** hours.
e. A new law requires that all propane tanks (used, for example, on barbecue grills) must have new, safer valves before they can be refilled. A gas company that refills tanks took a random sample of customers on a recent Saturday morning and reported the mean age of their propane tanks to be **3.45** years.

7.2 Parameter versus Statistic In each of the following statements, identify the **boldface** number as the value of a population parameter or a sample statistic.
a. A half-ton pickup truck is designed to safely carry a maximum of 1000 pounds. The Department of Transportation is concerned that drivers are hauling much heavier loads. The State Police randomly stopped 50 pickup trucks on an interstate highway and carefully weighed the contents in each truck bed. The mean weight was **1037** pounds.
b. A toy manufacturer issued a recall on a small wooden toy car because the wheels could break off and pose a choking hazard for small children. The proportion of buyers who took advantage of the recall was reported to be **0.45**.
c. A recent study indicated that more people are rising before 6:00 A.M. each weekday in order to prepare for work and to help children get ready for school. In a random sample of 500 adults, **42**% said they get up each weekday before 6:00 A.M.
d. In a random sample of dentists, **80**% recommended a certain product to help whiten teeth.
e. During a recent winter, the month of January was particularly cold. A power company in the state of Pennsylvania reported that the mean number of kilowatts used by each customer during January was **1346**.

7.3 Sports and Leisure The Sports Authority sells five different medicine balls with weights (in kg) given in the table below.

| 10 | 12 | 15 | 18 | 25 |

a. Find the population mean and the population median.
b. Suppose a random sample of size three is selected from this population without replacement. Find the sampling distribution of the sample mean. Find the mean, variance, and standard deviation for the sample mean.
c. Suppose a random sample of size three is selected from this population without replacement. Find the sampling distribution of the sample median. Find the mean, variance, and standard deviation for the sample median.
d. Compare the mean of the sample mean with the population mean. Compare the mean of the sample median with the population median.

7.4 Marketing and Consumer Behavior Many different manufacturers (for example, Kenmore, General Electric, Maytag, and Whirlpool) sell residential gas ranges. The cost (in dollars) of four gas ranges is given below.

| 529 | 664 | 709 | 800 |

(*Source*: NexTag.com)

a. Suppose a random sample of size two is selected from this population *without* replacement. Find the sampling distribution of the sample mean.
b. Suppose a random sample of size two is selected from this population *with* replacement. Find the sampling distribution of the sample mean.
c. How are these two distributions similar? How are they different?

7.5 Manufacturing and Product Development The coverage of a gallon of paint depends on the surface, type, and quality of the paint, and on the applicator (*Source*: p1m.com). A local hardware store stocks 30 different paints. The coverage (in square feet) for a gallon of each type is given on the data CD and book's web site.
a. Use a computer or calculator to draw 50 random samples of size 5, and compute the mean for each sample.
b. Construct a histogram of the sample means.
c. Use the histogram to approximate the sampling distribution of the mean. What is the approximate shape of the distribution? What is the approximate value of the mean of the sample mean?
d. Find the population mean. How does this compare with the approximate mean of the sampling distribution?

7.6 Demographics and Population Statistics For planning purposes, U.S. Bancorp has determined the probability distribution for the retirement age, X, of employees in the mortgage work group. The probability distribution for X is given in the table below.

x	64	65	66
$p(x)$	0.1	0.7	0.2

a. Find the mean of X.
b. Suppose two employees from this work group are selected at random. Find the exact probability distribution for the sample mean, $\overline{X}$.
c. Find the mean of $\overline{X}$. How does this compare with your answer in part (a)?

7.7 Public Health and Nutrition The Food and Drug Administration regulates the use of certain terms used on food labels, for example, reduced fat, low fat, and light. Lean meat must contain less than 4.5 grams of saturated fat per serving.[2] Suppose the probability distribution for the amount of saturated fat in a randomly selected serving of lean meat is given in the following table.

x	0	1	2	3	4
$p(x)$	0.50	0.25	0.10	0.10	0.05

a. Suppose two servings of lean meat are selected at random. Find the exact probability distribution for the sample median amount of saturated fat, $\tilde{X}$.
b. Find the mean, variance, and standard deviation of $\tilde{X}$.

7.8 Biology and Environmental Science In February 2007, New Zealand fishermen caught a colossal squid in the Ross Sea. The squid was estimated to weigh 990 pounds and was approximately 39 feet long.[3] Suppose five squid were captured in the same area with lengths (in feet) given in the following table.

| 11 | 15 | 23 | 27 | 11 |

A random sample of three of these squid is selected without replacement.
a. Find the sampling distribution of the sample mean $\overline{X}$.
b. Find the sampling distribution of the total length for all three squid T.

7.9 Education and Child Development The number of copies sold (in millions) for the all-time best-selling children's books are given in the table below.

| 9.9 | 9.7 | 7.1 | 7.0 | 6.8 |

Suppose two of these books are selected at random without replacement.
a. Find the sampling distribution of the sample mean.
b. Find the sampling distribution of the total number of books sold.

Extended Applications

7.10 Fuel Consumption and Cars An automobile manufacturer lists several specifications for every one of its cars. For example, the length, width, wheelbase, turning circle, curb weight, and interior room measurements are readily available to customers. There are 10 cars in the manufacturer's fleet, and the acceleration time from 0 to 60 mph (in seconds) for each car is given in the table that follows.

| 2.9 | 3.8 | 3.8 | 3.9 | 4.0 | 4.0 | 4.2 | 4.2 | 4.2 | 4.2 |

(*Source*: cars.com)

a. Use a computer or calculator to draw 30 random samples of size three, and compute the standard deviation for each sample.
b. Construct a histogram of the sample standard deviations.
c. Use the histogram to describe the shape of the sampling distribution.
d. Compute the *population* standard deviation. Find an approximate mean of the sampling distribution. How do these two numbers compare?

7.11 Travel and Transportation American Airlines offers limited first-class seating to passengers on a flight from Newark to Los Angeles. The probability distribution for the number of passengers in first-class on a randomly selected flight is a random variable, X. The probability distribution for X is given in the table below.

x	5	6	7	8
$p(x)$	0.50	0.30	0.15	0.05

a. Find the variance of X.
b. Suppose two American Airlines cross-country flights are randomly selected. Find the exact probability distribution for the sample variance, S^2.
c. Find the mean of S^2. How does this compare with your answer in part (a)?

7.12 Sports and Leisure The times (in seconds) for three men in the 100-meter hurdles from the 2008 U.S. Olympic Team Trials are given in the following table.

13.00	13.27	13.43

(*Source*: USA Track and Field, Inc.)

Suppose a random sample of two of these times is selected with replacement.
a. Find the sampling distribution of the minimum time.
b. Find the sampling distribution of the total time.

7.13 Manufacturing and Product Development The human transporter is a two-wheeled scooter that can travel up to 17 miles per hour. The weight of each scooter varies, depending on optional equipment. A small bicycle store has four scooters in stock, with weights (in pounds) given in the following table.

83	100	95	70

(*Source*: segway.com)

Suppose a random sample of two of these scooters is selected without replacement.
a. Find the sampling distribution for the maximum weight.
b. Find the sampling distribution for the total weight.

7.14 Psychology and Human Behavior Five people own and operate the Victrola Coffee Shop in Seattle. Each person is married, and the number of years each has been married is given in the following table.

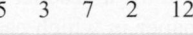

5	3	7	2	12

Suppose a random sample of two of the owners is selected with replacement. Let D be a statistic defined to be the absolute value of the difference in the number of years each has been married. For example, if 2 and 7 were selected, the value of D would be $|2 - 7| = |-5| = 5$. Find the sampling distribution of D.

 7.15 Simple Random Sample Consider the population consisting of the numbers 1 through 50.
a. Find the population mean μ.
b. Use the Applet to select 50 random samples of size 10 from this population. Find the sample mean for each sample.
c. Construct a histogram of the 50 sample means. Where is the histogram centered?

Challenge

7.16 Which Estimator Is Better? Each of the following graphs shows the probability distribution for two statistics that could be used to estimate a parameter θ (describing an underlying population). Select the statistic that would be a *better* estimator of θ and justify your answer.

a.

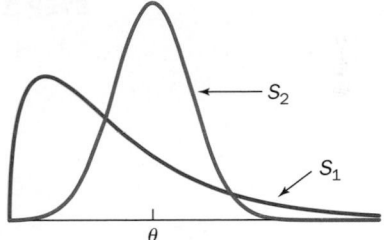

b.

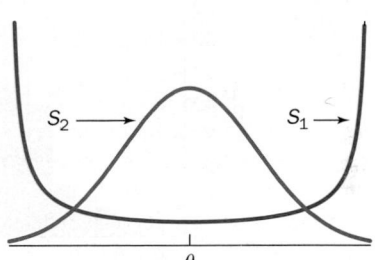

c.

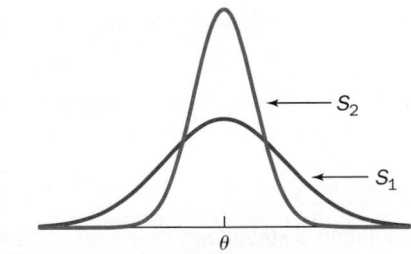

d.

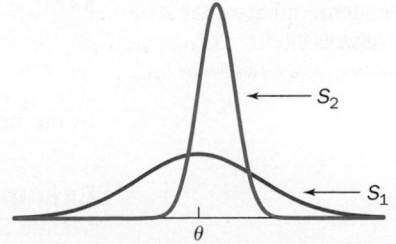

7.2 THE SAMPLING DISTRIBUTION OF THE SAMPLE MEAN

It seems reasonable to use a value of the sample mean, $\bar{x}$, in order to make an inference concerning the population mean μ. However, as discussed in the previous section, the sample mean varies from sample to sample. This **sampling variability** makes it difficult to know how far a specific $\bar{x}$ is from μ, or even whether the sample mean is an overestimate or underestimate. In this section, the sampling variability of the sample mean is completely characterized. The probability distribution of the sample mean, $\overline{X}$, can be used to make a sensible guess for the true value of the population mean.

In order to make a reliable estimate, we need to know the exact probability distribution of the sample mean. As the next few examples show, the distribution of $\overline{X}$ is related to n, the sample size, and to the parameters of the original, or underlying, population.

Example 7.5 Approximate Distribution of $\overline{X}$ Consider a population consisting of the numbers 1, 2, 3, . . . , 20. The population mean is $\mu = (1 + 2 + 3 + \ldots + 20) / 20 = 10.5$. Use frequency histograms to approximate the distribution of the mean, $\overline{X}$.

SOLUTION

STEP 1   Consider a random sample of n observations selected with replacement. For $n = 5$, five numbers are selected at random from the population, and the sample mean is computed. This procedure is repeated 500 times. A histogram of the resulting 500 sample means is shown in Figure 7.2. For $n = 10$, a similar procedure is followed. The resulting histogram of the sample means is shown in Figure 7.3.

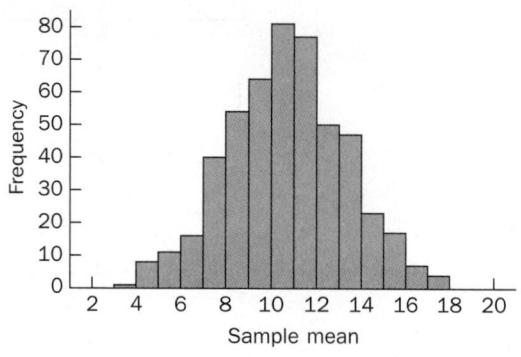

Figure 7.2 Histogram of sample means for $n = 5$.

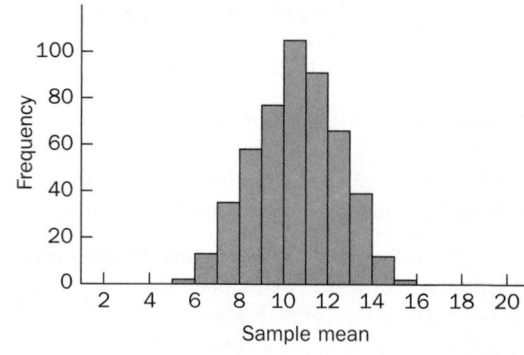

Figure 7.3 Histogram of sample means for $n = 10$.

STEP 2   These histograms suggest some very surprising results. Each distribution appears to be centered near the population mean 10.5. In addition, the shape of each distribution is approximately normal! Also notice that the variability of the sampling distribution decreases as n increases. The sampling distribution for $\overline{X}$ for $n = 10$ is more compact than that for $n = 5$.

STEP 3   This example implies that the distribution of the sample mean is approximately normal, with mean equal to the underlying, original, population mean and variance related to the sample size, n.   ●

The original population distribution is certainly not normal. It is amazing and perhaps counterintuitive that $\overline{X}$ has a normal distribution.

In the next example, the exact sampling distribution of the mean is obtained.

Example 7.6 Essay Sources Mary Dunn, a teacher at the Messalonskee Middle School, has assigned a short research paper in which each student must write about an endangered species. Ms. Dunn requires each student to correctly reference at least two sources. Past experience indicates that the number of sources cited is a random variable, X, with probability distribution given in the following table.

x	2	3	4
$p(x)$	0.5	0.3	0.2

Suppose a random sample of three student papers is selected.
a. Find the sampling distribution of $\overline{X}$, the sample mean number of sources cited.
b. Find the mean and the variance of the random variable X.
c. Find the mean and the variance of the random variable $\overline{X}$.
d. How do the results from (b) compare with the results from (c)?

SOLUTION

a. A sample consists of three numbers. The first represents the number of sources for student 1, the second for student 2, and the third for student 3. By the Multiplication Rule, there are $3 \times 3 \times 3 = 27$ possible samples. For each sample, list the sample mean and the probability of selecting that sample.

The probability associated with each sample is computed using independence. For example, the probability of observing the sample 2, 2, 4 is

$$P(X = 2 \cap X = 2 \cap X = 4)$$
$$= P(X = 2) \cdot P(X = 2) \cdot P(X = 4)$$
$$= (0.5)(0.5)(0.2) = 0.05$$

Intersection of three events.

Independence.

Given probabilities.

Use the table below to construct the sampling distribution.

Sample	$\overline{x}$	Probability	Sample	$\overline{x}$	Probability
2, 2, 2	2	0.125	3, 3, 4	10/3	0.018
2, 2, 3	7/3	0.075	3, 4, 2	3	0.030
2, 2, 4	8/3	0.050	3, 4, 3	10/3	0.018
2, 3, 2	7/3	0.075	3, 4, 4	11/3	0.012
2, 3, 3	8/3	0.045	4, 2, 2	8/3	0.050
2, 3, 4	3	0.030	4, 2, 3	3	0.030
2, 4, 2	8/3	0.050	4, 2, 4	10/3	0.020
2, 4, 3	3	0.030	4, 3, 2	3	0.030
2, 4, 4	10/3	0.020	4, 3, 3	10/3	0.018
3, 2, 2	7/3	0.075	4, 3, 4	11/3	0.012
3, 2, 3	8/3	0.045	4, 4, 2	10/3	0.020
3, 2, 4	3	0.030	4, 4, 3	11/3	0.012
3, 3, 2	8/3	0.045	4, 4, 4	4	0.008
3, 3, 3	3	0.027			

The probability distribution for $\overline{X}$ is given in the following table. It lists all the values $\overline{X}$ can assume and the corresponding probabilities.

$\overline{x}$	2	7/3	8/3	3	10/3	11/3	4
$p(\overline{x})$	0.125	0.225	0.285	0.207	0.114	0.036	0.008

b. The mean and variance for X are

$$E(X) = (2)(0.5) + (3)(0.3) + (4)(0.2) = 2.7$$
$$\text{Var}(X) = (2 - 2.7)^2(0.5) + (3 - 2.7)^2(0.3) + (4 - 2.7)^2(0.2) = 0.61$$

c. The mean and variance for $\overline{X}$ are

$$E(\overline{X}) = (2)(0.125) + (7/3)(0.225) + \cdots + (4)(0.008) = 2.7$$
$$\text{Var}(\overline{X}) = (2 - 2.7)^2(0.125) + \cdots + (4 - 2.7)^2(0.008) = 0.2033$$

d. Notice the following extraordinary relationships between the means and variances.

$$E(\overline{X}) = 2.7 = E(X)$$

The mean of the sample mean is equal to the original population mean!

$$\text{Var}(\overline{X}) = 0.2033 = 0.61/3 = \text{Var}(X)/n$$

The variance of $\overline{X}$ is equal to the original variance divided by the sample size. ●

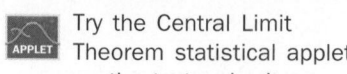

Try the Central Limit Theorem statistical applet on the text web site.

Here is one more approach to illustrate the very important connections between the distribution of the sample mean and the distribution of the original population.

Consider three original, or underlying, distributions: (1) the standard normal distribution, a normal distribution with mean 0 and standard deviation 1; (2) a uniform distribution with parameters $a = 0$ and $b = 1$; and (3) an exponential distribution with parameter $\lambda = 0.5$. The probability density function and the mean for each distribution are shown in the first row of Figure 7.4.

Consider the following process for each distribution. Select 500 samples of size $n = 2$ and compute the mean for each sample. Construct a histogram of the sample means. The resulting *smoothed* histogram is shown in the second row of Figure 7.4. Repeat this procedure for $n = 5$, 10, and 20, for each underlying distribution. These smoothed histograms are also given in Figure 7.4.

Notice the following incredible patterns.

1. If the underlying population is normal, the distribution of the sample mean appears to be normal, regardless of the sample size.

2. Even if the underlying population is *not* normal, the distribution of the sample mean becomes *more normal* as n increases.

3. The sampling distribution of the mean is centered at the mean of the underlying population.

4. As the sample size, n, increases, the variance of the distribution of the sample mean decreases.

The previous examples and observations lead to the following properties concerning the sample mean, $\overline{X}$.

PROPERTIES OF THE SAMPLE MEAN

Let $\overline{X}$ be the mean of observations in a random sample of size n drawn from a population with mean μ and variance σ^2.

1. The mean of $\overline{X}$ is equal to the mean of the underlying population. In symbols, $\mu_{\overline{X}} = \mu$.

2. The variance of $\overline{X}$ is equal to the variance of the underlying population divided by the sample size.

In symbols, $\sigma_{\overline{X}}^2 = \dfrac{\sigma^2}{n}$.

The standard deviation of $\overline{X}$ is $\sigma_{\overline{X}} = \sqrt{\dfrac{\sigma^2}{n}} = \dfrac{\sigma}{\sqrt{n}}$.

3. If the underlying population is distributed normally, then the distribution of $\overline{X}$ is also *exactly* normal for any sample size.

In symbols, $\overline{X} \sim N(\mu, \sigma^2/n)$.

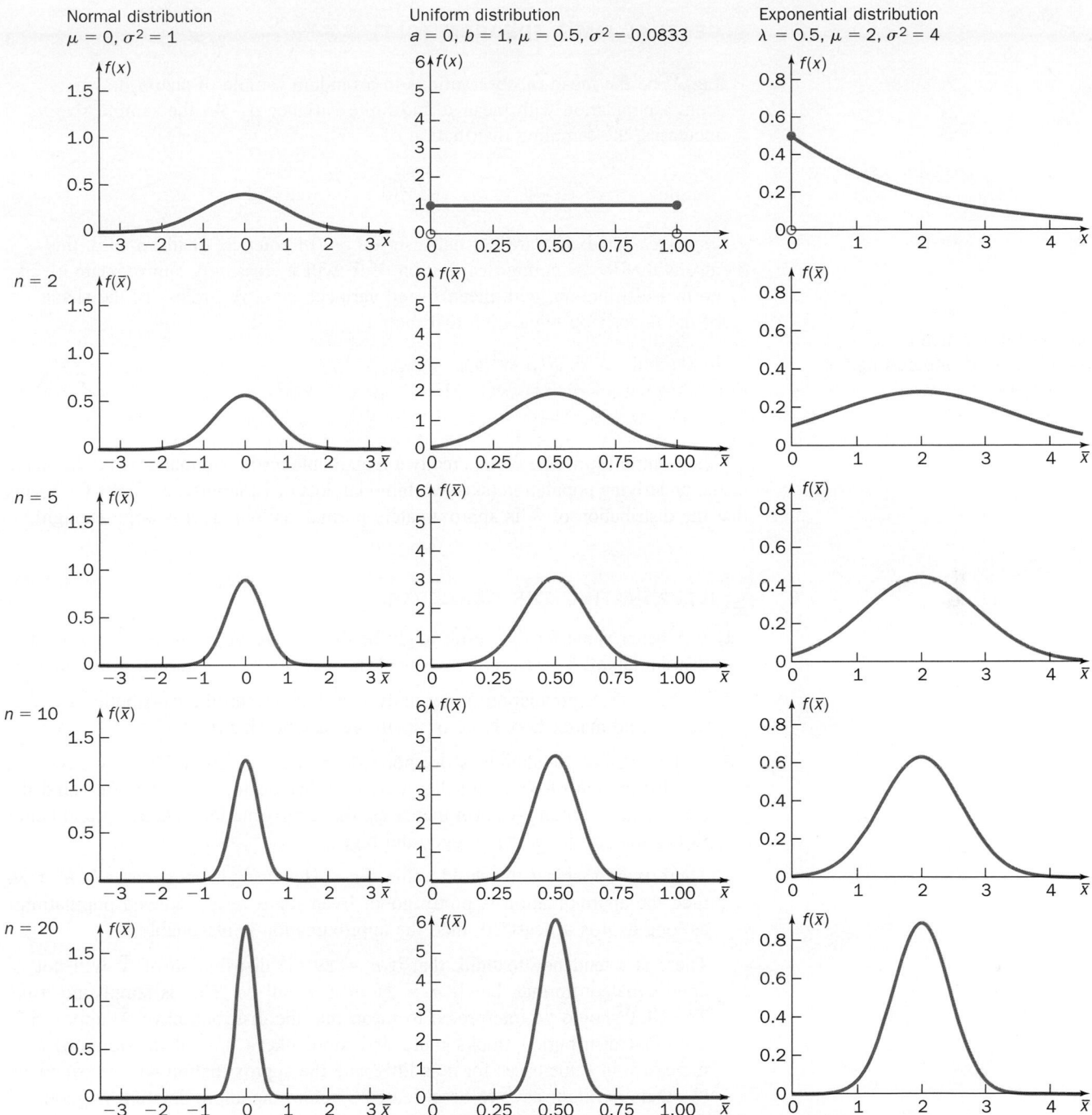

Figure 7.4 Three different original, or underlying, populations, and approximations to the distribution of the sample mean for various sample sizes *n*.

Since the mean of the sampling distribution of $\overline{X}$ is the same as the mean of the underlying distribution that we want to estimate, the sample mean is an **unbiased estimator** of the population mean μ.

Even if the underlying population is *not* normal, the previous examples suggest that the distribution of $\overline{X}$ *becomes more normal* as the sample size increases. This amazing result is the most important idea in all of statistics, the **Central Limit Theorem**.

CENTRAL LIMIT THEOREM (CLT)

Let $\overline{X}$ be the mean of observations in a random sample of size n drawn from a population with mean μ and finite variance σ^2. As the sample size n increases, the sampling distribution of

$$Z = \frac{\overline{X} - \mu}{\sigma/\sqrt{n}}$$

approaches the standard normal distribution. In practice, or informally, this means that the sampling distribution of $\overline{X}$ will increasingly approximate a normal distribution, with mean μ and variance σ^2/n, regardless of the shape of the underlying population distribution.

In symbols, $\overline{X} \stackrel{\cdot}{\sim} N(\mu, \sigma^2/n)$.

The symbol $\stackrel{\cdot}{\sim}$ means "is *approximately* distributed as."

The Central Limit Theorem is really a remarkable result. No matter what the shape of the underlying population (skewed, bimodal, lots of variability, etc.), the CLT says that the distribution of $\overline{X}$ is approximately normal, as long as n is large enough!

ILLUMINATING THE CONCEPTS

1. ◀ A better name for this result might be the *Normal Convergence Theorem*. The distribution of $\overline{X}$ *converges*, or gets closer and closer, to a normal distribution. ▶

2. If the original population is normally distributed, then the distribution of $\overline{X}$ is normal, no matter how large or small the sample size (n).

3. If the original population is *not* normal, the Central Limit Theorem says that the distribution of $\overline{X}$ approaches a normal distribution as n increases, and the approximation improves as n increases; the approximation gets better and better as the sample size n gets bigger and bigger.

   There is no magical threshold value for n. However, in most cases if $n \geq 30$ then the approximation is pretty good. Even for severely skewed populations, as long as n is at least 30, then the approximation is reasonable.

   There is a tendency to think that if $n = 29$ the distribution of $\overline{X}$ will not be approximately normal, but if $n = 31$ then it will be. This is simply not true. The CLT should be interpreted to mean that the distribution of $\overline{X}$ approaches a normal distribution (looks more and more like a normal distribution) as n increases. In some cases for n as little as 5 the approximation will be excellent. In others, n might have to be at least 26 before the approximation is good.

4. To compute a probability involving the sample mean, we treat $\overline{X}$ just like any other normal random variable: standardize and use cumulative probability where appropriate. We use the same method even if $\overline{X}$ is only *approximately* normal.

5. The expression for the variance of $\overline{X}$ mathematically confirms our observations regarding the spread of the sampling distribution. As n increases, the distribution of the sample mean becomes more compact. $\sigma_{\overline{X}}^2 = \sigma^2/n$ and σ^2 is constant. This fraction becomes smaller as n (the denominator) increases.

6. A more general version of the CLT includes a statement about the sum of independent observations, T. If n is sufficiently large, the distribution of T approaches a normal distribution with mean $n\mu$ and variance $n\sigma^2$.

   In symbols, $T \stackrel{\cdot}{\sim} N(n\mu, n\sigma^2)$. ■

(Visions of America, LLC/Alamy)

An even more general version of the CLT concludes, essentially, that any statistic that is a sum or a mean tends toward a normal distribution as *n* increases. This is useful in many inference problems because the appropriate statistic is often a sum or a mean. It is easy to compute probabilities associated with a normal statistic (random variable). Therefore, the likelihood, or tail probability, associated with an observed value of the statistic is a straightforward calculation.

In addition, the Central Limit Theorem helps to explain why so many real-world distributions are approximately normal. Almost any statistic can be decomposed into a sum of other variables. For example, the height of a tomato plant after six weeks might be directly related to, or might be the *sum* of, the effects of a large number of independent variables including the amount of water, fertilizer, and sunlight it has received. Therefore, the distribution of the height should be approximately normal. This single theorem explains empirical evidence that suggests almost every *measurement* distribution is approximately normal.

The following examples illustrate the properties of $\overline{X}$ and the Central Limit Theorem.

Solution Trail

KEYWORDS

- Normally distributed.
- Mean 50, standard deviation 4, 25 trips.
- Sample mean.

↓

TRANSLATION

- Normal, $\mu = 50$, $\sigma^2 = 16$, $n = 25$.
- $\overline{X}$.

↓

CONCEPTS

Properties of $\overline{X}$.

↓

VISION

These questions involve the sample mean $\overline{X}$. Since the underlying distribution is normal, the distribution of $\overline{X}$ is *exactly* normal. Find the mean, variance, and standard deviation of $\overline{X}$, translate each question into a probability statement, standardize, and use cumulative probability where appropriate.

Example 7.7 Green Line Time On the Massachusetts Bay Transportation Authority (MBTA) Green Line from Eliot to Lechmere in Boston, trolleys leave regularly throughout the day beginning at 5:00 A.M. Although the length of the trip and the number of stops are constant, the time taken for each trip varies due to weather, traffic, and time of day. According to the MBTA, the mean time taken for the trip is 50 minutes. Suppose the travel time is normally distributed with standard deviation $\sigma = 4$ minutes. A random sample of 25 trips is obtained, and the time for each is recorded.

a. Find the probability that the sample mean time will be less than 48 minutes.

b. Find the probability that the sample mean will be within 1 minute of the population mean (50 minutes).

SOLUTION

a. The underlying distribution is normal with $\mu = 50$ and $\sigma^2 = 16$. The sample size is $n = 25$. The sample mean is (exactly) normally distributed:

$$\overline{X} \sim N(\mu, \sigma^2/n) = N(50, 16/25); \quad \sigma_{\overline{X}} = \sqrt{16/25} = 4/5 = 0.80.$$

To solve part (a), begin with a probability statement involving the random variable $\overline{X}$. We need the probability that $\overline{X}$ will be less than 48 minutes.

$$P(\overline{X} < 48) = P\left(\frac{\overline{X} - 50}{0.80} < \frac{48 - 50}{0.80}\right) \quad \text{Standardize.}$$
$$= P(Z < -2.5) \quad \text{Use Equation 6.8; simplify.}$$
$$= 0.0062 \quad \text{Use Table III in the Appendix.}$$

The probability that the sample mean time for the 25 trips will be less than 48 minutes is 0.0062.

Figure 7.5 shows a technology solution.

Standardization illustrated:

P($\overline{X}$ < 48)

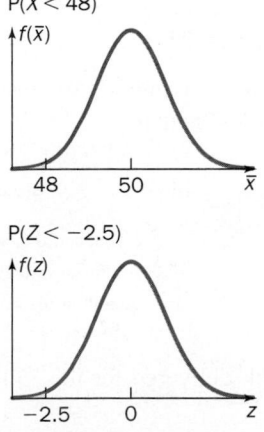

P(Z < −2.5)

Figure 7.5 Compute cumulative probability using <u>C</u>alc; Probability <u>D</u>istributions; <u>N</u>ormal.

```
▣ Session                                    ▢◻✕

Cumulative Distribution Function

Normal with mean = 50 and standard deviation = 0.8

  x   P( X <= x )
 48     0.0062097
```

Remember, one point in a continuous world (Z) contributes no probability: P(Z < −1.25) = P(Z ≤ −1.25).

Standardization illustrated:

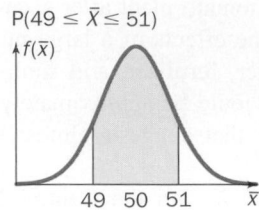

P(49 ≤ $\overline{X}$ ≤ 51)

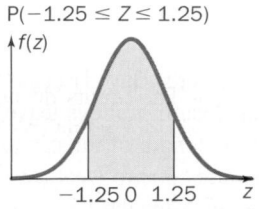

P(−1.25 ≤ Z ≤ 1.25)

b. *Within 1 minute of the population mean* means $49 \le \overline{X} \le 51$. Write the probability statement, and standardize again.

$$P(49 \le \overline{X} \le 51)$$
$$= P\left(\frac{49 - 50}{0.8} \le \frac{\overline{X} - 50}{0.8} \le \frac{51 - 50}{0.8}\right) \qquad \text{Standardize.}$$
$$= P(-1.25 \le Z \le 1.25) \qquad \text{Use Equation 6.8; simplify.}$$
$$= P(Z \le 1.25) - P(Z < -1.25) \qquad \text{Use cumulative probability.}$$
$$= 0.8944 - 0.1056 \qquad \text{Use Table III in the Appendix.}$$
$$= \boxed{0.7888}$$

The probability that the sample mean time for the 25 trips will be between 49 and 51 minutes is 0.7888.

Figure 7.6 shows a technology solution.

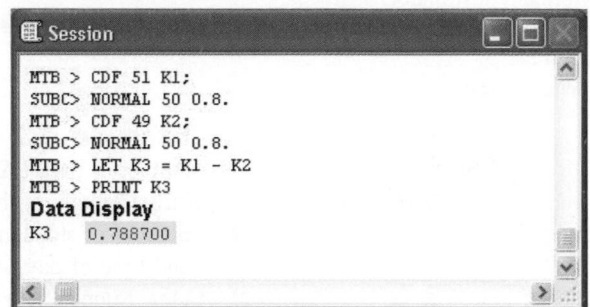

Figure 7.6 Use the command language to compute cumulative probabilities and to find the difference.

Example 7.8 Milk Deliveries In upstate New York, milk tanker trucks follow a daily routine, stopping at the same dairy farms every day. Farm output varies due to weather, time of year, number of cows, and other factors. From years of recorded data, the mean amount of milk delivered for processing by a truck is 7750 liters, with a standard deviation of 150 liters. Suppose 36 trucks are randomly selected.
a. Find the probability that the sample mean amount of milk delivered for the 36 trucks is more than 7800 liters.
b. Find a value m such that the probability that the sample mean is less than m is 0.1.

SOLUTION

a. The underlying distribution has $\mu = 7750$ and $\sigma = 150$. The sample size is $n = 36$. By the Central Limit Theorem, the distribution of $\overline{X}$ is approximately normal:

$$\overline{X} \overset{\cdot}{\sim} N(\mu, \sigma^2/n) = N(7750, 150^2/36) = N(7750, 625).$$

The standard deviation of $\overline{X}$ is

$$\sigma_{\overline{X}} = \sqrt{\sigma^2/n} = \sqrt{625} = 25 \text{ (or } \sigma_{\overline{X}} = \sigma/\sqrt{n} = 150/6 = 25).$$

To solve part (a), start with a probability statement involving $\overline{X}$. We need the probability that the sample mean will be *more than* 7800 liters.

$$P(\overline{X} > 7800) = P\left(\frac{\overline{X} - 7750}{25} > \frac{7800 - 7750}{25}\right) \qquad \text{Standardize.}$$
$$= P(Z > 2) \qquad \text{Use Equation 6.8; simplify.}$$
$$= 1 - P(Z \le 2) \qquad \text{Use cumulative probability.}$$
$$= 1 - 0.9772 \qquad \text{Use Table III in the Appendix.}$$
$$= \boxed{0.0228}$$

The probability that the sample mean amount of milk delivered for the 36 trucks will be greater than 7800 is 0.0228.

a. Solution Trail

KEYWORDS

- Mean 7750, standard deviation 150, 36 trucks.
- Sample mean.

TRANSLATION

- $\mu = 7750$, $\sigma = 150$, $n = 36$.
- $\overline{X}$.

CONCEPTS

Central Limit Theorem.

VISION

These questions involve the sample mean $\overline{X}$. The exact distribution of the underlying population is not known. The sample size is large: $n = 36$ (≥ 30). Therefore, the Central Limit Theorem can be applied. Find the approximate distribution of $\overline{X}$, translate each question into a probability statement, standardize, and use cumulative probability where appropriate.

Standardization illustrated:

$P(\overline{X} > 7800)$

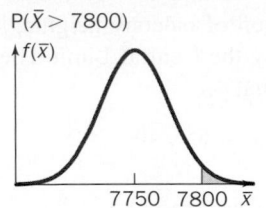

Standardization illustrated:

$P(\overline{X} < m) = 0.1$

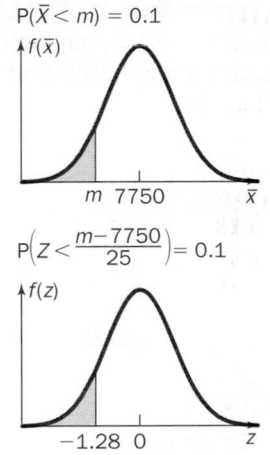

b. To solve part (b), translate the question into a probability statement. Convert the expression into cumulative probability involving Z. Since the probability is already given, this is an inverse cumulative probability problem. Work backward in Table III.

$$P(\overline{X} < m) = P\left(\frac{\overline{X} - 7750}{25} < \frac{m - 7750}{25}\right) \qquad \text{Standardize.}$$

$$= P\left(Z < \frac{m - 7750}{25}\right) = 0.1 \qquad \text{Use Equation 6.8.}$$

There is no further simplification within the probability statement. However, the resulting probability statement involves Z and is cumulative probability. Find a value in the body of Table III, as close to 0.1 as possible. Set the corresponding z value equal to $\left(\dfrac{m - 7750}{25}\right)$, and solve for m.

$$\frac{m - 7750}{25} = -1.28 \qquad \text{Table III in the Appendix.}$$

$$m - 7750 = -32 \qquad \text{Multiply both sides by 25.}$$

$$m = \boxed{7718} \qquad \text{Add 7750 to both sides.}$$

The probability that the sample mean will be less than $m = 7718$ is (approximately) 0.1.

Figures 7.7 and 7.8 show technology solutions.

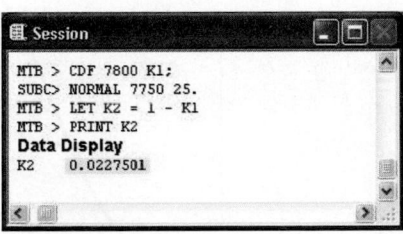

Figure 7.7 Find the cumulative probability and use the complement rule.

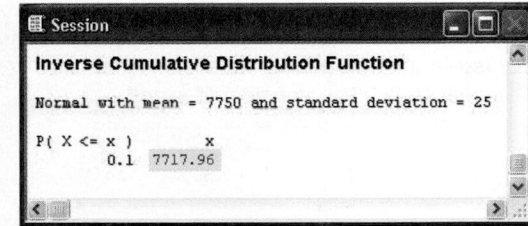

Figure 7.8 Use the inverse cumulative distribution function.

The last example in this section involves an inference using the random variable $\overline{X}$.

Example 7.9 Building Regulations The Building Authority (BA) in Oakland, California, monitors thousands of regulations concerning structural integrity and interior restrictions. One regulation subject to increased scrutiny is the height of balconies on commercial buildings. For a variety of safety reasons, a balcony cannot be more than 15 meters above the ground. Suppose the population standard deviation for the height of balconies (in meters) is $\sigma = 5$. A BA Inspector selects a random sample of 40 balconies from buildings around the city and finds the sample mean height: $\overline{x} = 15.5$. Is there any evidence to suggest that the population mean height of balconies around the city is greater than 15 meters?

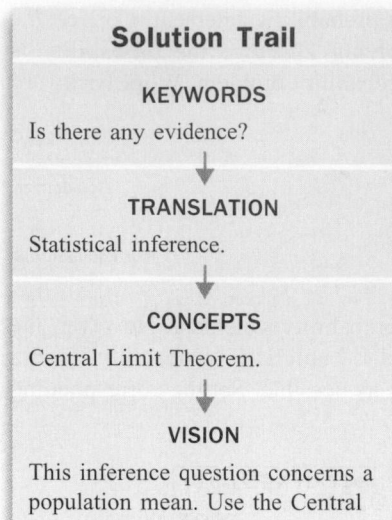

Solution Trail

KEYWORDS

Is there any evidence?

↓

TRANSLATION

Statistical inference.

↓

CONCEPTS

Central Limit Theorem.

↓

VISION

This inference question concerns a population mean. Use the Central Limit Theorem and the distribution of $\overline{X}$ to determine the likelihood of an observed sample mean of $\overline{x} = 15.5$ or greater. Use the four-step procedure outlined in Chapter 1. If the probability is small, there is evidence to suggest that the claim (population mean $\mu = 15$) is wrong.

In statistical inference problems, usually any probability less than or equal to 0.05 is considered small.

SOLUTION

STEP 1 Assume the underlying distribution has $\mu = 15$ and $\sigma = 5$. We do not know the shape of the underlying distribution of balcony heights, but the sample size is large: $n = 40$ (≥ 30). By the Central Limit Theorem, the distribution of $\overline{X}$ is approximately normal.

STEP 2 **Claim:** $\mu = 15 \Rightarrow \overline{X} \sim N(15, 5^2/40)$, $\sigma_{\overline{X}} = 5/\sqrt{40}$

Experiment: $\overline{x} = 15.5$

Likelihood: Since the BA Inspector is looking for evidence that the mean height is *greater* than 15, find the right-tail probability (to be conservative).

$$
\begin{aligned}
P(\overline{X} \geq 15.5) &= P\left(\frac{\overline{X} - 15}{5/\sqrt{40}} \geq \frac{15.5 - 15}{5/\sqrt{40}}\right) && \text{Standardize.} \\
&= P(Z \geq 0.63) && \text{Use Equation 6.8; simplify.} \\
&= 1 - P(Z < 0.63) && \text{The complement Rule; Use cumulative probability.} \\
&= 1 - 0.7357 && \text{Table III in the Appendix.} \\
&= 0.2643
\end{aligned}
$$

Conclusion: Since this probability is large (> 0.05), there is no evidence to suggest that the mean balcony height is greater than 15. Even if the mean height is 15, an observation of $\overline{x} = 15.5$ is not very unlikely, and so we cannot doubt that the mean is really 15. Figure 7.9 illustrates the distribution of $\overline{X}$ and depicts the right-tail probability.

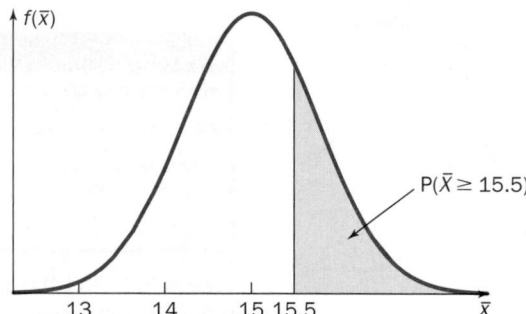

Figure 7.9 The distribution of $\overline{X}$. The shaded region represents the right-tail probability $P(\overline{X} \geq 15.5)$.

Figure 7.10 shows a technology solution.

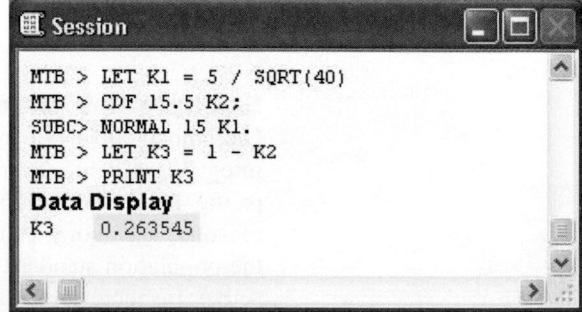

Figure 7.10 Find cumulative probability and use the complement rule.

TECHNOLOGY CORNER

Procedure: Solve probability questions involving the sample mean $\overline{X}$.
Reconsider: Example 7.7, page 301, solutions, and interpretations.

TI-84 Plus

If $\overline{X}$ is normal or approximately normal, use the built-in function `normalcdf` to find (calculator) cumulative probability and the built-in function `invNorm` to solve inverse cumulative probability problems.

1. Select $\boxed{\text{DISTR}}$; DISTR; `normalcdf`. Enter the left endpoint, –1E99 (calculator negative infinity); the right endpoint, 48; the mean, 50; and the standard deviation, 0.8. See Figure 7.11.
2. Select $\boxed{\text{DISTR}}$; DISTR; `normalcdf`. Enter the left endpoint, 49, the right endpoint, 51, the mean, 50, and the standard deviation, 0.8. See Figure 7.12.

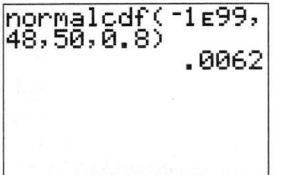

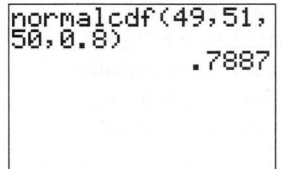

Figure 7.11 Find cumulative probability.

Figure 7.12 Find (calculator) cumulative probability.

Minitab

As in Chapter 6 for normal random variables, use the built-in functions to find cumulative probability or inverse cumulative probability. These functions may be accessed through a graphical input window: <u>C</u>alc; Probability <u>D</u>istributions; <u>N</u>ormal, or by using the command language.

1. In a session window, use the function CDF to find cumulative probability, or select <u>C</u>alc; Probability <u>D</u>istributions; <u>N</u>ormal. Choose Cumulative probability; enter the Mean, 50; and enter the Standard deviation, 0.8. Select Input constant and enter 48. See Figure 7.5, page 301.
2. In a session window, use the function CDF to find $P(\overline{X} \le 51)$. Store the result in K1. Use the function CDF to find $P(\overline{X} \le 49)$. Store the result in K2. Find the difference, and print the result. See Figure 7.6, page 302.

Excel

As in Chapter 6 for normal random variables, use the built-in functions to compute (strict) cumulative probability associated with a standard normal random variable (NORMSDIST) and a normal random variable with mean $\mu \neq 0$ and/or standard deviation $\sigma \neq 1$ (NORMDIST). The functions NORMSINV and NORMINV are the corresponding inverse cumulative probability functions.

1. Use the function NORMDIST to find $P(\overline{X} < 48)$. See Figure 7.13.
2. Use the function NORMDIST to find $P(\overline{X} \le 51)$. Use the function NORMDIST to find $P(\overline{X} \le 49)$. Find the difference: $P(49 \le \overline{X} \le 51)$. See Figure 7.14.

	A	B
1	0.8944	= NORMDIST(51,50,0.8,TRUE)
2	0.1056	= NORMDIST(49,50,0.8,TRUE)
3	0.7887	= A1 - A2

	A	B
1	0.0062	= NORMDIST(48,50,0.8,TRUE)

Figure 7.13 $P(\overline{X} < 48)$: cumulative probability.

Figure 7.14 Calculations for $P(49 \le \overline{X} \le 51)$.

SECTION 7.2 EXERCISES

Note: To find the distribution of $\overline{X}$, or any random variable, you need to describe the distribution, by name if possible, and provide the values of the parameters that characterize the distribution. For example, $\overline{X}$ is normally distributed with mean $\mu = 27$ and variance $\sigma_{\overline{X}}^2 = 0.35$. $\overline{X} \sim N(27, 0.35)$ completely describes the distribution of $\overline{X}$.

Practice

7.17 Consider a normally distributed population with mean $\mu = 10$ and standard deviation $\sigma = 2.5$. Suppose a random sample of size n is selected from this population. Find the distribution of $\overline{X}$ and the indicated probability in each of the following cases.
a. $n = 7$, $P(\overline{X} \le 9)$.
b. $n = 12$, $P(\overline{X} > 11.5)$.
c. $n = 15$, $P(9.5 \le \overline{X} \le 10.5)$.
d. $n = 25$, $P(\overline{X} \ge 10.25)$.
e. $n = 100$, $P(\overline{X} \le 9.8 \cup \overline{X} \ge 10.2)$.

7.18 Suppose X is a normal random variable with mean $\mu = 17.5$ and standard deviation $\sigma = 6$. A random sample of size $n = 24$ is selected from this population.
a. Find the distribution of $\overline{X}$.
b. Carefully sketch a graph of the probability density functions for X and $\overline{X}$ on the same coordinate axes.
c. Find $P(X \le 14)$ and $P(\overline{X} < 14)$.
d. Find $P(15 < X < 19)$ and $P(15 < \overline{X} < 19)$.

7.19 Suppose X is a random variable with mean $\mu = 50$ and variance $\sigma^2 = 49$. A random sample of size $n = 38$ is selected from this population.
a. Find the approximate distribution of $\overline{X}$. Why is the Central Limit Theorem necessary here?
b. Find $P(\overline{X} < 49)$.
c. Find $P(\overline{X} \ge 52)$.
d. Find $P(49.5 \le \overline{X} \le 51.5)$.
e. Find a value c such that $P(\overline{X} > c) = 0.15$.

7.20 Suppose X is a random variable with mean $\mu = 1000$ and standard deviation $\sigma = 100$. A random sample of size $n = 36$ is selected from this population.
a. Find the approximate distribution of $\overline{X}$. Carefully sketch a graph of the probability density function.
b. Find $P(\overline{X} > 975)$.
c. Find $P(\overline{X} \le 1030)$.
d. Find $P(\mu_{\overline{X}} - \sigma_{\overline{X}} \le \overline{X} \le \mu_{\overline{X}} + \sigma_{\overline{X}})$.
e. Find a symmetric interval about the mean $\mu_{\overline{X}}$ such that $P(\mu_{\overline{X}} - c \le \overline{X} \le \mu_{\overline{X}} + c) = 0.95$.

7.21 Suppose X is a random variable with mean $\mu = 30$ and standard deviation $\sigma = 50$. A random sample of size $n = 40$ is selected from this population.
a. Find the approximate distribution of $\overline{X}$. Carefully sketch a graph of the probability density function.
b. Find $P(\overline{X} \ge 38)$.
c. Find $P(20 \le \overline{X} \le 40)$.
d. Find $P(\overline{X} < 15)$.
e. Find a value of c such that $P(\overline{X} \le c) = 0.001$.

7.22 The following figure shows the graphs of the probability density functions for the random variable X, the random variable $\overline{X}$ for $n = 5$, and the random variable $\overline{X}$ for $n = 15$.

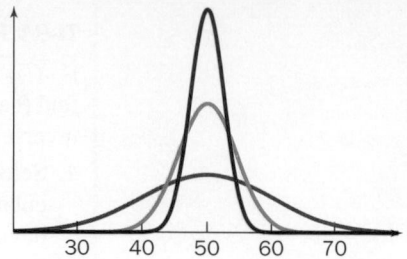

Identify each probability density function.

7.23 The following figure shows graphs of the probability density function for the random variable X and the approximate density functions for the random variable $\overline{X}$ for $n = 5$ and the random variable $\overline{X}$ for $n = 15$.

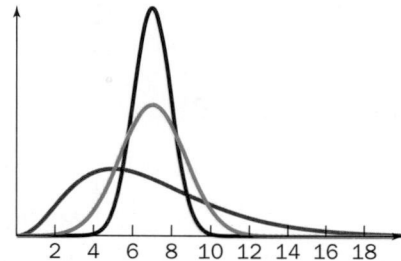

Identify each probability density function.

Applications

7.24 Biology and Environmental Science In a survey conducted by the Big Creek State Marine Reserve in California, the length of various species of fish caught at two locations were recorded. Suppose the length of a yellowtail caught at the Santa Lucia Kelp Bed is normally distributed with mean 38 inches and standard deviation 2.9 inches.[4]
a. Suppose 1 yellowtail is selected at random. What is the probability that the length of the yellowtail is more than 41 inches?
b. Suppose 32 yellowtails are selected at random. What is the probability that the sample mean length is greater than 41 inches?
c. What is the probability that the sample mean length is less than 38.5 inches?
d. A respected fishing guide claims that most yellowtails caught in this area are between 37 and 40 inches. What is the probability that the sample mean will be within this length interval?

7.25 Public Policy and Political Science In certain hurricane-prone areas of the United States, concrete columns used in construction must meet specific building codes. The minimum diameter for a cylindrical column is 8 inches. Suppose the mean diameter for all columns is 8.25 inches with standard deviation 0.1 inch. A building inspector randomly selects 35 columns and measures the diameter of each.
a. Find the approximate distribution of $\overline{X}$. Carefully sketch a graph of the probability density function.

b. What is the probability that the sample mean diameter for the 35 columns will be greater than 8 inches?

c. What is the probability that the sample mean diameter for the 35 columns will be between 8.2 and 8.4 inches?

d. Suppose that the standard deviation is 0.15 inch. Answer parts (a), (b), and (c) using this value of σ.

7.26 Manufacturing and Product Development A large part of the luggage market is made up of overnight bags. These bags vary by weight, exterior appearance, material, and size. Suppose the volume of overnight bags is normally distributed with mean $\mu = 1750$ cubic inches and standard deviation $\sigma = 250$ cubic inches. A random sample of 15 overnight bags is selected, and the volume of each is found.

a. Find the distribution of $\overline{X}$.

b. What is the probability that the sample mean volume is more than 1800 cubic inches?

c. What is the probability that the sample mean volume is within 100 cubic inches of 1750?

d. Find a symmetric interval about 1750 such that 95% of all values of the sample mean volume lie in this interval.

7.27 Biology and Environmental Science The Environmental Protection Agency (EPA) is concerned about pollution caused by factories that burn sulfur-rich fuel. In order to decrease the impact on the environment, factory chimneys must be high enough to allow pollutants to dissipate over a larger area. Assume that the mean height of chimneys in these factories is 100 meters (an EPA-acceptable height) with standard deviation 12 meters. A random sample of 40 chimney heights is obtained.

a. What is the probability that the sample mean height for the 40 chimneys is greater than 102 meters?

b. What is the probability that the sample mean height is between 101 and 103 meters?

c. Suppose the sample mean is 98.5 meters. Is there any evidence to suggest that the true mean height for chimneys is less than 100 meters? Justify your answer.

7.28 Manufacturing and Product Development The manager at an HEB grocery store in Corpus Christi, Texas, suspects that a supplier is systematically underfilling 12-ounce bags of potato chips. To check the manufacturer's claim, a random sample of 100 bags of potato chips is obtained, and each bag is carefully weighed. The sample mean is 11.9 ounces. Assume $\sigma = 0.3$ ounces.

a. Find the probability that the sample mean is 11.9 ounces or less.

b. How can this probability in part (a) be so small when 11.9 is so *close* to the population mean $\mu = 12$?

c. From your answer to part (a), is there any evidence to suggest that the mean weight of bags of potato chips is less than 12 ounces?

7.29 Psychology and Human Behavior In attempting to flee from police, criminals sometimes barricade themselves inside a building and create a police standoff. The criminal is usually armed with a dangerous weapon and may hold hostages. Suppose the mean length of a police standoff, ending with some sort of resolution, is 6.5 hours with standard deviation 4 hours. A random sample of 35 police standoffs is selected.

a. Find the distribution of the sample mean.

b. What is the probability that the sample mean for the 35 police standoffs will be greater than 7 hours?

c. A new psychological technique was used in negotiations with the criminals involved in the 35 police standoffs. Suppose the sample mean for the 35 police standoffs is $\overline{x} = 5.1$ hours. Is there any evidence to suggest that the mean police standoff time is lower when this new technique is used?

7.30 Biology and Environmental Science During winter in the Northern Hemisphere, the mean jet-stream wind speed at altitudes of 7 to 8 miles is 75 miles per hour.[5] Suppose the standard deviation is 20 mph and 30 winter days are selected at random.

a. What is the probability that the sample mean jet-stream wind speed will be less than 70 mph?

b. What is the probability that the sample mean jet-stream wind speed will be greater than 83 mph?

c. Find a symmetric interval about the mean, 75, such that the probability that the sample mean lies in this interval is 0.90.

7.31 Sports and Leisure At the 2005 Winston Cup NASCAR race at the Dover International Speedway, the mean speed per lap was 115.054 miles per hour (mph).[6] Suppose 50 car laps are selected at random and the standard deviation for speed is $\sigma = 30$ mph.

a. Find the probability that the sample mean speed for the 50 cars is less than 110 mph.

b. Find the probability that the sample mean speed is between 115 and 120 mph.

c. Find the probability that the sample mean speed is greater than 125 mph.

7.32 Public Health and Nutrition Typhoid fever is more common in developing countries with poor sanitation and greater incidence of food contamination. The mean duration of this illness is approximately 4 weeks.[7] Periodically, a team of doctors randomly selects 10 patients with typhoid fever and carefully measures the duration of the illness. If the mean duration is greater than 4.5 weeks, then a health alert is issued. Suppose the duration of this illness is normally distributed with standard deviation 1 week.

a. Find the probability that the sample mean duration is less than 4.5 weeks.

b. Find the probability that a health alert is issued.

c. Suppose the true mean duration has increased to 4.3 weeks. What is the probability that a health alert will be issued?

7.33 Biology and Environmental Science There are many regulations for catching lobsters off the coast of New England, including required permits, allowable gear, and size prohibitions. The Massachusetts Division of Marine Fisheries requires a minimum carapace length measured from a rear eye socket to the center line of the body shell. Any lobster measuring less than 3.25 inches must be returned to the ocean. For all lobsters caught, suppose the carapace length is normally distributed with mean 4.125 inches and standard deviation 1 inch. A random sample of 15 lobsters is obtained.

a. Find the distribution of the sample mean carapace length.

b. What is the probability that the sample mean carapace length is more than 4.5 inches?

c. What is the probability that the sample mean carapace length is between 3.375 and 4.125 inches?

d. If the sample mean carapace length is less than 3.5 inches, a lobsterman will look for other places to set his traps. What is the probability that a lobsterman will be looking for a different location?

7.34 Sports and Leisure One measure of an athlete's ability is the height of his/her vertical leap. Many professional basketball players are known for their remarkable vertical leaps, which lead to amazing dunks. The mean vertical leap of all NBA players is 28 inches.[8] Suppose the standard deviation is 7 inches and 36 NBA players are selected at random.

a. What is the probability that the mean vertical leap for the 36 players will be less than 26 inches?

b. What is the probability that the mean vertical leap for the 36 players will be between 27.5 and 28.5 inches?

c. A high-priced athletic trainer has been hired to work with a group of NBA players to improve their hip flexibility, which should improve their vertical leap. After 1 month of training, the mean vertical leap for 50 of these players selected at random was 29.75 inches. Is there any evidence to suggest that this flexibility program has increased the mean vertical leap (from $\mu = 28$ inches)?

7.35 Physical Sciences The thickness of the ozone layer surrounding the Earth varies according to season and location. A measurement station in Boulder, Colorado, has determined that the mean thickness of the ozone layer above the United States is approximately 320 Dobson units (DU) with standard deviation 35 DU. Suppose 49 days are selected at random, and the thickness of the ozone layer is measured each day.[9]

a. What is the probability that the sample mean ozone-layer thickness is more than 330 DU?

b. What is the probability that the sample mean ozone layer thickness is between 315 and 325 DU?

c. Many researchers suggest that increased pollution is causing the thickness of the ozone layer to decrease. Suppose the sample mean for the 49 days is 310 DU. Is there any evidence to suggest that the mean ozone layer thickness is less than 320 DU?

Extended Applications

7.36 Public Health and Nutrition Recent research studies suggest that lycopene, found in tomatoes, may reduce the risk of certain cancers. According to the USDA/NCC Carotenoid Database for U.S. Foods, the mean amount of lycopene in a medium-sized fresh tomato is 3.7 milligrams (mg). Suppose the amount of lycopene is a normal random variable with standard deviation 1 mg. Consider the sample mean based on 5 observations, $\overline{X}_5$, and the sample mean based on 20 observations, $\overline{X}_{20}$.

a. Find the distributions for $\overline{X}_5$ and $\overline{X}_{20}$.

b. Carefully sketch the probability density functions for X, $\overline{X}_5$, and $\overline{X}_{20}$ on the same coordinate axes.

c. Find $P(X < 3)$, $P(\overline{X}_5 < 3)$, and $P(\overline{X}_{20} < 3)$.

d. Find $P(3.6 \leq X \leq 3.8)$, $P(3.6 \leq \overline{X}_5 \leq 3.8)$, and $P(3.6 \leq \overline{X}_{20} \leq 3.8)$.

7.37 Sports and Leisure A health club recently added tanning booths in addition to a weight room, pool, and racquetball courts. For insurance purposes, an employee maintains careful records of the length of time each member spends in a tanning booth. The mean length of time spent tanning is 15 minutes with standard deviation 2 minutes. Consider a random sample of 35 members who use a tanning booth.

a. Find the distribution of the total time spent tanning by the 35 members, T.

b. Find the probability that the total time spent tanning is between 8 and 9 hours.

c. If the total time spent tanning is more than 9.2 hours, an employee must work overtime. What is the probability that the employee works overtime on a day when 35 members tan?

d. Find a value t such that $P(T \geq t) = 0.01$.

7.38 Sports and Leisure The Tonight Show with Johnny Carson lasted for 30 years and featured movie stars, comedians, animal acts, and Carnac the Magnificent. The first part of the show was reserved for Johnny's monologue. The mean length of a monologue was 12 minutes with standard deviation 45 seconds. Suppose 40 shows are selected at random and the length of each monologue is recorded.

a. Find the distribution of the total monologue time for the 40 shows, T.

b. Find $P(T > 500)$.

c. Find $P(470 \leq T \leq 490)$.

d. After Johnny Carson retired, his company started to sell DVDs of his monologues. Suppose one DVD holds eight hours. What is the probability that the company will be able to fit 40 randomly selected monologues onto one DVD?

7.39 Manufacturing and Product Development A (destructive) tensile test is standard procedure for testing a cross-wire weld. Suppose a certain weld is designed to withstand a force with mean 0.8 kN. In order to check the quality of welds, a manufacturer randomly selects 25 welds (every hour) and performs a tensile test on each weld. If the mean force required (for the 25 welds) is between 0.75 and 0.85 kN, the process is allowed to continue. Otherwise, it is shut down. Suppose the force distribution is normal with standard deviation 0.1 kN.

a. If the true mean is 0.8 kN, what is the probability that the process is shut down? This probability represents the chance of making one kind of error.

b. If the true mean is 0.82 kN, what is the probability that the process is allowed to continue? What if $\mu = 0.84$ kN? These probabilities represent the chance of making a different kind of error.

c. Suppose the process is allowed to continue if the mean force required is between 0.76 and 0.84 kN. Answer parts (a) and (b) with this new interval.

7.40 Manufacturing and Product Development Shetland wool is considered to be some of the finest in the world because it is soft, durable, and easy to spin. The mean fleece weight from a typical sheep is 3.25 pounds with a standard deviation of 0.4 pounds. Suppose a farmer has 100 sheep ready to be sheared.

a. Find the distribution for the total weight of fleece from the 100 sheep, T.

b. What is the probability that the total fleece weight is less than 323 pounds?

c. What is the probability that the total fleece weight is between 330 and 340 pounds?

d. Find a value t such that $P(T \geq t) = 0.15$.

Challenge

7.41 Fuel Consumption and Cars Companies receiving large shipments of raw materials or of any product often use a specific plan for accepting the entire shipment. An *acceptance sampling plan* usually includes the sample size for close inspection, the *acceptance* criterion, and the *rejection* criterion. For a given plan, the *operating characteristic* (OC) curve shows the probability of accepting the entire lot as a function of the actual quality level.

Standard clip-on weights for steel rims on automobiles (used when tires are balanced) are available in 0.25-ounce to 6-ounce sizes. Suppose an automobile garage receives a large shipment of 2-ounce weights. A garage mechanic will select a random sample of 30 weights and weigh each one on a precise scale. If the sample mean is within 0.05 ounces of the printed weight (of 2 ounces), then the shipment is accepted. Otherwise, the entire shipment is rejected and returned to the manufacturer. Suppose the population standard deviation is 0.13 ounces.

a. Find the probability of accepting the entire shipment if the true population mean weight is 1.86, 1.88, 1.90, 1.92, 1.94, 1.96, 1.98, 2.00, 2.02, 2.04, 2.06, 2.08, 2.10, 2.12, or 2.14 ounces.

b. Plot the probability of accepting the entire shipment versus the true population mean weight. The resulting graph is the OC curve for the given acceptance sampling plan.

7.42 Normal Approximation to the Binomial Distribution Consider a binomial experiment with n trials and probability of success p. If we assign a 0 to each failure and a 1 to each success then the binomial random variable can be defined as a sum. By the Central Limit Theorem, as n increases, the distribution of X (the sum) approaches a normal distribution with mean np and variance $np(1 - p)$. Suppose X is a binomial random variable with $n = 30$ and probability of success $p = 0.5$.

a. Construct a probability histogram for the binomial random variable X. Find $P(12 \leq X \leq 16)$ using the binomial distribution.

b. Find the approximate normal distribution for X. Find $P(12 \leq X \leq 16)$ using the normal distribution for X.

c. Compare the probabilities found in parts (a) and (b).

d. Find $P(11.5 \leq X \leq 16.5)$ using the normal distribution for X. Compare this answer with the probability in part (a). Why do you think this is a much better approximation?

7.43 Manufacturing and Product Development A hardware store has 20 interior plantation shutter sets in stock. The width (in inches) of each set is given in the following table.

30	30	28	34	36	28	34	36	24	35
28	30	32	30	44	34	22	32	20	30

a. Find the (population) mean width of the 20 shutters.

b. Use technology to generate at least 100 samples of size five (without replacement). Find the sample mean width for each sample.

c. Construct a histogram of the sample means found in part (a). Describe the distribution.

7.3 THE DISTRIBUTION OF THE SAMPLE PROPORTION

In Section 7.2 the sampling distribution of the sample mean was introduced. The Central Limit Theorem helps if the underlying population is not normal. Knowing the distribution of $\overline{X}$, we can use the sample mean to make an inference about the population mean μ. Similarly, we are often interested in drawing a conclusion about the population proportion p (the probability of a success). For example, a politician may want to estimate the proportion of voters in a district in favor of a certain highway bill, or a quality-control supervisor may need to estimate the true proportion of defective parts in a large shipment.

It seems reasonable to use a value of the sample proportion, $\hat{p}$, to make an inference concerning the population proportion p. Therefore, a knowledge of the **sampling distribution of the sample proportion** is necessary. We need to completely characterize the variability of this statistic.

Consider a sample of n individuals or objects (or trials) and let X be the number of successes in the sample. The sample proportion is defined to be

$$\widehat{P} = \frac{X}{n} = \frac{\text{The number of successes in the sample}}{\text{The sample size}}. \tag{7.1}$$

The sample proportion is simply the proportion of successes in the sample, or a relative frequency of success.

An approach similar to the one in Section 7.2 can be used to approximate the distribution of $\widehat{P}$: generate lots of sample proportions, construct a histogram, and try to characterize the distribution in terms of shape, center, and variability. The sampling distribution is summarized as follows.

As *n* increases, the distribution of $\hat{P}$ approaches a normal distribution. There is no threshold value for *n*. The larger the value of *n* is and the closer *p* is to 0.5, the better the approximation.

THE SAMPLING DISTRIBUTION OF $\hat{P}$

Let $\hat{P}$ be the sample proportion of successes in a sample of size *n* from a population with true proportion of success *p*.

1. The mean of $\hat{P}$ is the true population proportion.
   In symbols: $\mu_{\hat{P}} = p$.

2. The variance of $\hat{P}$ is $\sigma_{\hat{P}}^2 = \dfrac{p(1-p)}{n}$.

   The standard deviation of $\hat{P}$ is $\sigma_{\hat{P}} = \sqrt{\dfrac{p(1-p)}{n}}$.

3. If *n* is large and both $np \geq 5$ and $n(1-p) \geq 5$, then the distribution of $\hat{P}$ is approximately normal.
   In symbols: $\hat{P} \stackrel{\cdot}{\sim} N(p, p(1-p)/n)$.

 ILLUMINATING THE CONCEPTS

1. It may be a little surprising to learn that $\hat{P}$ is approximately normal. However, the sample proportion can be written as a sample mean. Assign 0 to each failure and 1 to each success. *X*, the total number of successes, is a sum (of 0s and 1s). Therefore, the sample proportion, *X/n*, is really a sample mean, and the Central Limit Theorem says that the sample mean is approximately normal.

2. Since the mean of $\hat{P}$ is the true population proportion, $\hat{P}$ is an unbiased estimator for *p*.

3. A *large* sample isn't enough for normality. The two products *np* and $n(1-p)$ must *both* be greater than or equal to 5. These are called the *nonskewness criteria*. They guarantee that the distribution of $\hat{P}$ is approximately symmetric (i.e., centered far enough away from 0 or 1).

4. To compute a probability involving the sample proportion, treat $\hat{P}$ like any other normal random variable: standardize and use cumulative probability where appropriate. ∎

The following example illustrates the properties of $\hat{P}$ and the technique for computing probabilities associated with this random variable.

(AFP/Getty Images)

Example 7.10 International Visitors
Foreign visitors to the United States are carefully monitored by security organizations and the tourism industry. During March 2008, there were 4.7 million international visitors to the United States. Forty percent of all Western European visitors were from the United Kingdom (UK).[10] Suppose 120 March visitors from Western Europe are selected at random and the number of visitors from the UK are determined.

a. Find the distribution of the sample proportion of visitors from the UK, $\hat{P}$. Carefully sketch the probability density function for this random variable.

b. What is the probability that the sample proportion (for the 120 visitors selected) is greater than 0.50?

c. Find the probability that the sample proportion will be between 0.32 and 0.37.

Solution Trail

KEYWORDS

- 120 visitors, proportion is 0.40.
- Sample proportion.

↓

TRANSLATION

- $n = 120$ trials, probability of a success (a UK visitor) is $p = 0.40$.
- $\hat{P}$.

↓

CONCEPTS

- Distribution of $\hat{P}$.

↓

VISION

This question involves the random variable $\hat{P}$. Use the properties of $\hat{P}$ to find the distribution, translate each question into a probability statement, and write an equivalent probability statement involving Z.

Standardization illustrated:

$P(0.32 \leq \hat{P} \leq 0.37)$

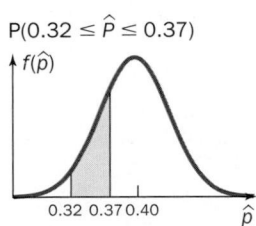

$P(-1.79 \leq Z \leq -0.67)$

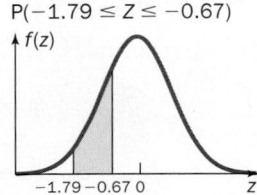

SOLUTION

a. For $n = 120$ and $p = 0.40$, check the nonskewness criteria.

$$np = (120)(0.40) = 48 \geq 5 \quad \text{and} \quad n(1-p) = (120)(0.60) = 72 \geq 5.$$

Both inequalities are satisfied. The distribution of $\hat{P}$ is approximately normal with

$$\mu_{\hat{p}} = p = 0.40 \quad \text{and} \quad \sigma_{\hat{p}}^2 = \frac{p(1-p)}{n} = \frac{(0.40)(0.60)}{120} = 0.002.$$

In symbols, $\hat{P} \stackrel{.}{\sim} N(0.40, 0.002) \qquad \sigma_{\hat{p}} = \sqrt{0.002}$.

Figure 7.15 shows a graph of the probability density function and the associated probability.

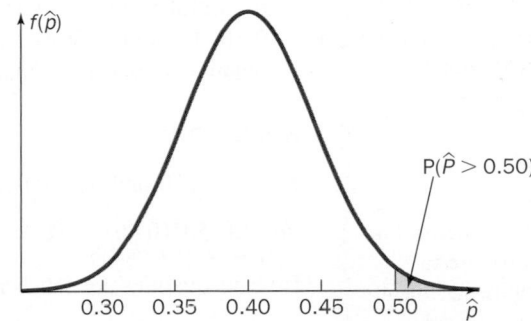

Figure 7.15 The probability density function for $\hat{P}$. The shaded area represents $P(\hat{P} > 0.50)$.

b. $P(\hat{P} > 0.50)$ Probability that the sample proportion is greater than 0.50.

$$= P\left(\frac{\hat{P} - 0.40}{\sqrt{0.002}} > \frac{0.50 - 0.40}{\sqrt{0.002}}\right) \qquad \text{Standardize.}$$

$$= P(Z > 2.24) \qquad \text{Use Equation 6.8; simplify.}$$

$$= 1 - P(Z \leq 2.24) \qquad \text{Use cumulative probability.}$$

$$= 1 - 0.9875 \qquad \text{Use Table III in the Appendix.}$$

$$= 0.0125$$

The probability that the sample proportion is greater than 0.50 is 0.0125.

c. $P(0.32 \leq \hat{P} \leq 0.37)$ Probability that the sample proportion is between 0.32 and 0.37.

$$= P\left(\frac{0.32 - 0.40}{\sqrt{0.002}} \leq \frac{\hat{P} - 0.40}{\sqrt{0.002}} \leq \frac{0.37 - 0.40}{\sqrt{0.002}}\right) \qquad \text{Standardize.}$$

$$= P(-1.79 \leq Z \leq -0.67) \qquad \text{Use Equation 6.8; simplify.}$$

$$= P(Z \leq -0.67) - P(Z \leq -1.79) \qquad \text{Use cumulative probability.}$$

$$= 0.2514 - 0.0367 \qquad \text{Use Table III in the Appendix.}$$

$$= 0.2147$$

The probability that the sample proportion is between 0.32 and 0.37 is 0.2147.

Figures 7.16 and 7.17 show technology solutions to parts (b) and (c).

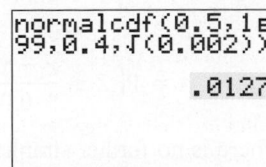

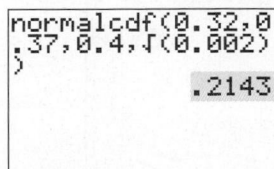

Figure 7.16
$P(\hat{P} > 0.50)$: right-tail probability.

Figure 7.17
$P(0.32 \leq \hat{P} \leq 0.37)$.

Another way to ask this question: Find the third quartile of the $\hat{P}$ distribution.

The following example illustrates an inverse cumulative probability problem and an inference associated with the random variable $\hat{P}$.

Example 7.11 Too Many Regulations Company executives often complain that there are too many government regulations. Stifling rules and endless bureaucracy may limit creativity, new product research, and corporate profits. Suppose that 60% of all CEOs believe there are too many government regulations. A random sample of 150 CEOs is obtained, and each is asked whether he or she believes that there are too many government regulations.

a. Find a value r such that the probability that the sample proportion is greater than r is 0.25.

b. In recent years, big business has lobbied politicians to relax regulations in order to stimulate the economy. Suppose the sample proportion for the 150 CEOs is 0.56. Is there any evidence to suggest that the true proportion of CEOs who believe there are too many regulations has decreased?

SOLUTION

For $n = 150$ and $p = 0.60$, check the nonskewness criteria.

$$np = (150)(0.60) = 90 \geq 5 \quad \text{and} \quad n(1 - p) = (150)(0.40) = 60 \geq 5.$$

Both inequalities are satisfied. The distribution of $\hat{P}$ is approximately normal with

$$\mu_{\hat{p}} = p = 0.60 \quad \text{and} \quad \sigma_{\hat{p}}^2 = \frac{p(1 - p)}{n} = \frac{(0.60)(0.40)}{150} = 0.0016.$$

In symbols, $\hat{P} \sim N(0.60, 0.0016)$ $\sigma_{\hat{p}} = \sqrt{0.0016} = 0.04$.

a. Find a value r such that $P(\hat{P} > r) = 0.25$. Figure 7.18 illustrates this probability statement.

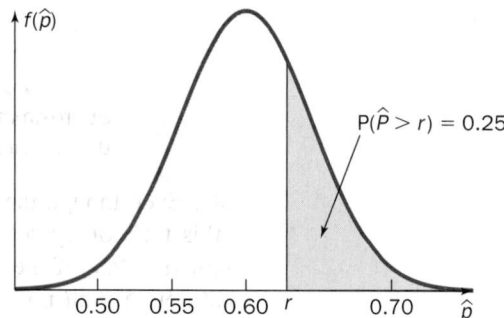

Figure 7.18 The shaded area represents $P(\hat{P} > r)$.

A technology solution:

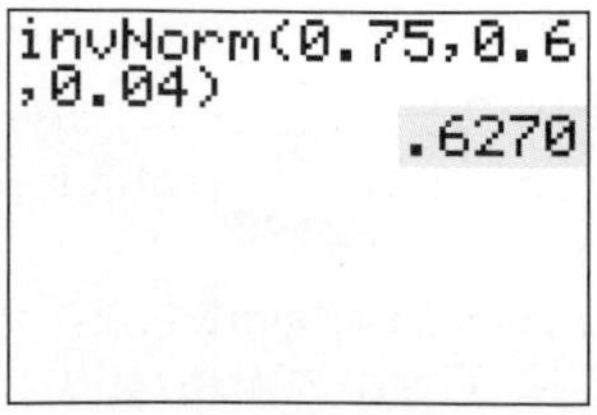

```
invNorm(0.75,0.6
,0.04)
            .6270
```

Figure 7.19 Use `invNorm` to solve an inverse cumulative probability problem.

To find the value r, write an equivalent expression involving cumulative probability. Standardize and work backward in Table III.

$$P(\hat{P} \leq r) = P\left(\frac{\hat{P} - 0.60}{0.04} \leq \frac{r - 0.60}{0.04}\right) \qquad \text{Standardize.}$$

$$= P\left(Z \leq \frac{r - 0.60}{0.04}\right) = 0.75 \qquad \text{Use Equation 6.8.}$$

There is no further simplification, but the resulting probability statement involves Z and is cumulative probability. Find a value in the body of Table III as close to 0.75 as possible. Set the corresponding z value equal to $\left(\dfrac{r - 0.60}{0.04}\right)$, and solve for r.

Standardization illustrated:

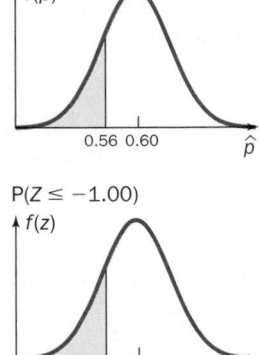

Remember, the TI-84 Plus, and technology solutions in general, are more accurate. In the *by-hand* solution, the z values are rounded to two decimal places, and values in Table III are rounded to four decimal places.

$$\frac{r - 0.60}{0.04} = 0.675 \qquad \text{Table III in the Appendix; interpolation.}$$

$$r - 0.60 = 0.027 \qquad \text{Multiply both sides by 0.04.}$$

$$r = 0.627 \qquad \text{Add 0.60 to both sides.}$$

The probability that the sample proportion is greater than 0.627 is 0.25.

b. Follow the usual four-step inference procedure and consider a tail probability as a measure of likelihood.

Claim: 60% of all CEOs believe that there are too many government regulations. $p = 0.60 \Rightarrow \hat{P} \sim N(0.60, 0.04)$.

Experimental outcome: The sample proportion of CEOs who believe that there are too many government regulations is $\hat{p} = 0.56$.

Likelihood: Since $\hat{p} = 0.56$ is to the left of the mean ($p = 0.60$), consider a left-tail probability as a measure of likelihood.

$$P(\hat{P} \le 0.56) = P\left(\frac{\hat{P} - 0.60}{0.04} \le \frac{0.56 - 0.60}{0.04}\right) \qquad \text{Standardize.}$$

$$= P(Z \le -1.00) \qquad \text{Use Equation 6.8; simplify.}$$

$$= 0.1587 \qquad \text{Table III in the Appendix.}$$

In statistical inference problems, any probability less than or equal to 0.05 is considered small.

Conclusion: Since this probability is larger than 0.05, it is reasonable to observe a sample proportion of 0.56 or smaller. There is no evidence to suggest that the claim of $p = 0.60$ is wrong.

A technology solution:

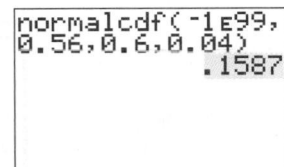

Figure 7.20 Since $\hat{P}$ is approximately normal, use `normalcdf`.

Note: Even though the sample proportion cannot be less than 0, −1E99 was used in this technology solution because a normal distribution is defined for all real numbers. Most of the time, when the approximation is good, using 0 (or 1 as a right bound) will not change the probability. ●

TECHNOLOGY CORNER

Procedure: Solve probability questions involving the sample proportion $\hat{P}$.

Reconsider: Example 7.11, page 312, solutions, and interpretations.

TI-84 Plus

If $\hat{P}$ is approximately normal, use the built-in function `normalcdf` to find (calculator) cumulative probability and the built-in function `invNorm` to solve inverse cumulative probability problems.

1. Select DISTR; DISTR; `invNorm`. Enter the cumulative probability, 0.75, the mean, 0.6, and the standard deviation, 0.04. See Figure 7.19, page 312.
2. Select DISTR ; DISTR; `normalcdf`. Enter the left endpoint, −1E99; the right endpoint, 0.56; the mean, 0.6; and the standard deviation, 0.04. See Figure 7.20.

Remember, it is very important for the sample to be selected *randomly* from the underlying population. Otherwise, the results are not valid.

Minitab

As in Chapter 6 for normal random variables, use the built-in functions to find cumulative probability or inverse cumulative probability. These functions may be accessed through a graphical input window: <u>C</u>alc; Probability <u>D</u>istributions; <u>N</u>ormal, or by using the command language.

1. In a session window, use the function INVCDF to solve inverse cumulative probability problems. See Figure 7.21.
2. In a session window, use the function CDF to find cumulative probability. See Figure 7.22.

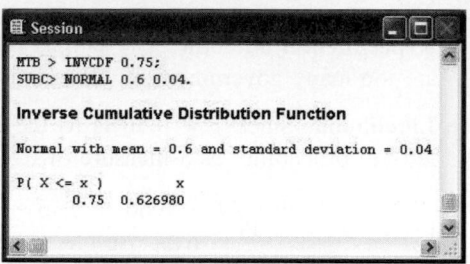

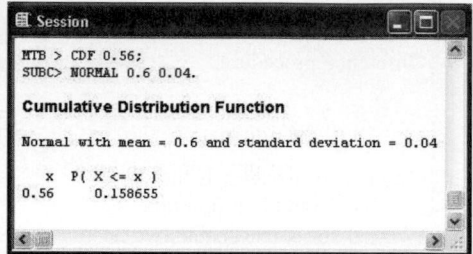

Figure 7.21 Use INVCDF to solve inverse cumulative probability problems.

Figure 7.22 Use CDF to find cumulative probability.

Excel

As in Chapter 6 for normal random variables, use the built-in function NORMDIST to compute (strict) cumulative probability associated with a normal random variable and the function NORMINV to solve inverse cumulative probability problems.

1. Use the function NORMINV to find the value r such that $P(\hat{P} \le r) = 0.75$. See Figure 7.23.
2. Use the function NORMDIST to find cumulative probability. See Figure 7.24.

Figure 7.23 Use NORMINV to solve inverse cumulative probability problems.

Figure 7.24 Use NORMDIST to find cumulative probability.

SECTION 7.3 EXERCISES

Practice

7.44 Suppose a random sample of size n is obtained. In each problem below, check the nonskewness criteria, and find the distribution of the sample proportion $\hat{P}$.
a. $n = 100$, $p = 0.25$.
b. $n = 150$, $p = 0.90$.
c. $n = 100$, $p = 0.75$.
d. $n = 1000$, $p = 0.85$.
e. $n = 5000$, $p = 0.006$.

7.45 Suppose a random sample of size $n = 200$ is obtained from a population with probability of success $p = 0.40$. Find each of the following probabilities.
a. $P(\hat{P} \le 0.37)$.
b. $P(\hat{P} > 0.45)$.

c. $P(0.38 \le \hat{P} \le 0.42)$.
d. $P(\hat{P} < 0.33)$ or $P(\hat{P} > 0.47)$.

7.46 Suppose a random sample of size $n = 500$ is obtained from a population with probability of success $p = 0.50$. Find each of the following probabilities.
a. $P(0.52 > \hat{P})$.
b. $P(\hat{P} \ge 0.47)$.
c. $P(0.44 \le \hat{P} < 0.49)$.
d. $P(0.45 \le \hat{P} \le 0.55)$.

7.47 Suppose a random sample of size $n = 80$ is obtained from a population with probability of success $p = 0.35$.
a. Find a value a such that $P(\hat{P} \le a) = 0.10$.
b. Find a value b such that $P(\hat{P} > b) = 0.01$.
c. Find a value c such that
$$P(0.35 - c \le \hat{P} \le 0.35 + c) = 0.95.$$

7.48 Suppose a random sample of size $n = 1000$ is obtained from a population with probability of success $p = 0.25$.

a. Find a value a such that $P(\hat{P} \le a) = 0.05$.

b. Find a value b such that $P(\hat{P} > b) = 0.005$.

c. Find a value c such that $P(0.25 - c \le \hat{P} \le 0.25 + c) = 0.99$.

d. Find the quartiles of the distribution of $\hat{P}$.

Applications

7.49 Demographics and Population Statistics Approximately 70% of high school students in the United States graduate on time.[11] Suppose a random sample of 250 high school students who were scheduled to graduate in 2009 is obtained, and school records are checked to determine whether or not each student graduated on time.

a. Find the distribution of the sample proportion of high school students who graduated on time.

b. Find the probability that the sample proportion is less than 0.66.

c. Find the probability that the sample proportion is more than 0.71.

d. Find the probability that the sample proportion is between 0.68 and 0.78.

7.50 Sports and Leisure During 2008, the most difficult golf course on the PGA tour, determined by driving accuracy, was Oakland Hills CC. During the PGA Championship tournament there, the proportion of all shots that hit the fairway off the tee was 0.48.[12] Suppose 400 tee shots from the tournament are selected at random.

a. Find the distribution of the sample proportion of tee shots that hit the fairway.

b. Find the probability that the sample proportion is less than 0.45.

c. Find the probability that the sample proportion is more than 0.52.

d. Find a value t such that the probability that the sample proportion is less than t is 0.01.

7.51 Psychology and Human Behavior A recent survey indicated that one-third of all Americans read at least 10 books during a year. One of the most popular genres is mystery, thriller, and crime (MTC). Approximately 48% of those Americans who read at least one book during a year read an MTC book.[13] Suppose 120 Americans who have read at least 1 book during the past year are selected at random.

a. Find the distribution of the sample proportion of MTC readers.

b. Find the probability that the sample proportion is more than 0.50.

c. Find the probability that the sample proportion is less than 0.40.

d. Find a symmetric interval about the mean ($p = 0.48$) such that the probability that the sample proportion is in this interval is 0.90.

7.52 Sports and Leisure People who bet money in a casino are classified as winners if they win more than they lose. Casino operators in Atlantic City, New Jersey, believe that the proportion of all players who *go home* a winner is 0.46. Suppose 75 Atlantic City gamblers are selected at random.

a. Find the sampling distribution of the proportion of gamblers who go home winners.

b. Find the probability that the sample proportion is less than 0.40.

c. Find the probability that the sample proportion is more than 0.45.

d. Find a symmetric interval about the mean ($p = 0.46$) such that the probability that the sample proportion is in this interval is 0.99.

7.53 Marketing and Consumer Behavior Instead of risking a trip to the local mall, many consumers purchase their Christmas gifts on the Internet. Of those people who buy Christmas gifts online, 36% purchase toys and games.[14] Suppose 200 customers who purchased Christmas gifts online are randomly selected.

a. Find the probability that the sample proportion of customers who purchased toys and games online is less than 0.35.

b. Find the probability that the sample proportion of customers who purchased toys and games online is more than 0.40.

c. Find a value t such that $P(\hat{P} < t) = 0.95$.

7.54 Business and Management The depressed economy and tight credit market in 2008 brought attention to the huge bonuses paid to some company executives. A survey indicated that 23% of U.S. financial services professionals expected lower bonuses in 2008.[15] Suppose a random sample of 90 financial services professionals is obtained.

a. Find the probability that the sample proportion of financial service professionals who expect a lower bonus is less than 0.20.

b. Find the probability that the sample proportion of financial service professionals who expect a lower bonus is between 0.25 and 0.30.

c. Suppose the actual sample proportion of financial service professionals who expect a lower bonus was 0.10. Is there any evidence to suggest that the claim is wrong? Justify your answer.

7.55 Education and Child Development Admissions offices at universities and colleges keep careful records of acceptance and yield rates. The yield rate is the percentage of admitted students who decide to accept an offer of admission. Historical data help the officers to plan for the incoming classes and guide some admission decisions. The proportion of applicants accepted for admission to Ramapo College of New Jersey is 0.40.[16] One hundred Ramapo College applicants are randomly selected.

a. Find the probability that the sample proportion of accepted applicants is less than 0.42.

b. Find the probability that the sample proportion of accepted applicants is between 0.30 and 0.35.

c. Suppose the actual sample proportion of accepted applicants this year is 0.47. Is there any evidence to suggest that the admissions office's acceptance rate has increased? Justify your answer.

7.56 Business and Management A certain philanthropic organization funds 1 out of every 10 grant proposals. Suppose a random sample of 300 grant proposals is obtained.

a. Find the probability that the sample proportion of funded grant proposals is less than 0.075.

b. Find the probability that the sample proportion of funded grant proposals is between 0.11 and 0.15.

c. If the funding rate increases, the board of directors may become concerned about resources being depleted. Suppose the actual sample proportion of funded grant proposals is 0.16. Is there any evidence to suggest that the funding rate has increased? Justify your answer.

7.57 Medicine and Clinical Studies According to the web site PeanutAllergy.com (2008), the proportion of all Americans who have a peanut allergy is approximately 0.005. Suppose 1000 Americans are randomly selected and each is tested for an allergic reaction to peanuts.

a. Find the distribution for the sample proportion of Americans who have a peanut allergy.

b. Find the probability that the sample proportion is less than 0.002.

c. Find the probability that the sample proportion is more than 0.010.

d. Find a value a such that $P(\widehat{P} \geq a) = 0.80$.

7.58 Economics and Finance A survey from December 2007 indicated that very few Americans planned to refinance their homes during 2008. However, 42% said they intended to reduce (pay down) their level of debt.[17] A random sample of 225 Americans was obtained and they were all asked if they planned to reduce their level of debt.

a. Find the probability that the sample proportion of Americans who intend to reduce their debt is less than 0.40.

b. Find the probability that the sample proportion is between 0.45 and 0.55.

c. Find a value r such that $P(\widehat{P} \geq r) = 0.01$.

7.59 Biology and Environmental Science Ships dumping garbage and ordinary beach goers contribute to the increasing amount of trash that washes onto the shore and collects in the oceans. In 2007 the International Coastal Cleanup collected more than 6 million pounds of garbage from the world's oceans. Common debris items collected included food wrappers, caps, and beverage bottles. Cigarettes accounted for 27.2% of all debris items.[18] A random sample of 430 debris items from a cleanup along a Texas beach was obtained.

a. Find the probability that the proportion of cigarette debris items is greater than 0.31.

b. Find the probability that the proportion of cigarette debris items is between 0.25 and 0.30.

c. Suppose 115 of the debris items are cigarettes. Is there any evidence to suggest that the true proportion of cigarette debris items is different from 0.272? Justify your answer.

Extended Applications

7.60 Manufacturing and Product Development Low-quality coffee shipped from various locations in Europe tends to contain a high proportion of *defective* beans (beans composed of foreign matter; moldy, black, unripe, or fermented beans; or those known as stinkers). Suppose a shipper claims that the proportion of defective beans is 0.07. A U.S. packaging company received a huge shipment of coffee beans and randomly selected 1000 beans. If the sample proportion of defective beans is more than 0.09, then the entire shipment will be returned to the supplier in Europe.

a. If the true proportion of defective beans is 0.07 (as claimed), what is the probability that the shipment will be sent back?

b. If the true proportion of defective beans is 0.08, what is the probability that the shipment will be accepted?

7.61 Manufacturing and Product Development McGuckin Hardware in Boulder, Colorado, routinely receives shipments of 4×8 foot sheets of 0.5-inch-thick plywood. A plywood sheet may contain defects, for example, a knot, a split, or a deviation in wood structure. Defective sheets decrease profits because they are sold at a reduced price. The manufacturer claims that the proportion of all plywood sheets that are defective is 0.05. Suppose a large shipment of plywood sheets is received and 200 are randomly selected for inspection. If the sample proportion of defective sheets is more than 0.09, then the entire shipment will be sent back to the supplier.

a. If the true proportion of defective plywood sheets is 0.05, what is the probability that the entire shipment will be sent back?

b. If the true proportion of defective plywood sheets is 0.03, what is the probability that the entire shipment will be sent back?

c. If the true proportion of defective plywood sheets is 0.10, what is the probability that the shipment will be accepted?

Challenge

7.62 Manufacturing and Product Development Suppose a company is receiving a large shipment of peel-and-stick vinyl floor tiles. The manufacturer claims that the proportion of defective floor tiles is 0.05. The company will select a random sample of 200 floor tiles, carefully inspect each, and determine whether each is defective. The acceptance sampling plan is to accept the entire shipment if the sample proportion of defective tiles is 0.08 or less; otherwise, reject the entire shipment.

a. Find the probability of accepting the entire shipment if the true proportion of defective floor tiles is 0.01, 0.02, 0.03, 0.04, 0.05, 0.06, 0.07, 0.08, 0.09, 0.10, or 0.15.

b. Plot the probability of accepting the entire shipment (on the y-axis) versus the true proportion of defective floor tiles (on the x-axis). The resulting graph is the OC curve for the given acceptance sampling plan.

7.63 Maximum Variance For a fixed sample size n, find the value of p that maximizes the variance of the sample proportion,
$$\sigma_{\widehat{P}}^2 = \frac{p(1-p)}{n}.$$

Hint: Compute the value of the variance for several different values of p. Plot the variance (on the y-axis) versus the value of p (on the x-axis).

Chapter 7 Challenge Wrap-Up

The manager of the health club claims that the mean amount of time spent on a treadmill is 35 minutes with standard deviation 3.5 minutes. In a random sample of 40 treadmill users, the sample mean time spent was 32.75 minutes. The distribution of time spent on a treadmill is unknown, but $n = 40 \geq 30$. Therefore, by the Central Limit Theorem the distribution of $\overline{X}$ is approximately normal with mean 35 minutes and standard deviation $3.5/\sqrt{40} \approx 0.5534$ minutes.

Use the four-step inference procedure outlined in Chapter 1 to determine whether there is any evidence to suggest that the true mean time spent on a treadmill is less than 35 minutes.

(Tetra Image/Alamy)

Claim: $\mu = 35 \Rightarrow \overline{X} \stackrel{\cdot}{\sim} N(35, 3.5^2/40)$, $\sigma_{\overline{X}} = 3.5/\sqrt{40} \approx 0.5534$.

Experiment: $\overline{x} = 32.75$.

Likelihood: Since we are looking for evidence to suggest that the true mean is less than 35 minutes, we find the left-tail probability (to be conservative).

$$
\begin{aligned}
P(\overline{X} \leq 32.75) &= P\left(\frac{\overline{X} - 35}{3.5/\sqrt{40}} \leq \frac{32.75 - 35}{3.5/\sqrt{40}}\right) && \text{Standardize.} \\
&= P(Z \leq -4.07) && \text{Use Equation 6.8; simplify.} \\
&\approx 0.0000 \,(\,0.00002\,) && \text{Table III in the Appendix.}
\end{aligned}
$$

Conclusion: Since this probability is small (≤ 0.05), there is evidence to suggest that the mean time spent on a treadmill is less than 35 minutes. Figure 7.25 illustrates the distribution of the sample mean and the left-tail probability. The technology solution follows in Figure 7.26.

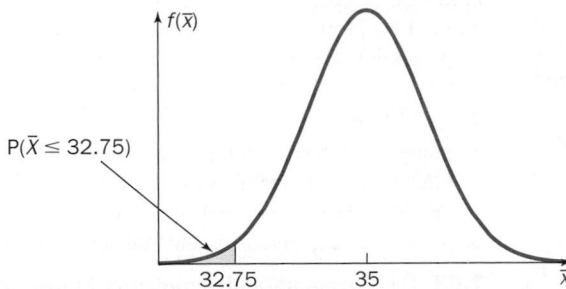

Figure 7.25 The distribution of the sample mean. The shaded area represents $P(\overline{X} \leq 32.75)$.

A technology solution:

	A	B
1	0.00002	= NORMDIST(32.75,35,3.5/SQRT(40),TRUE)

Figure 7.26 Use the function NORMDIST to find cumulative probability.

CHAPTER 7 SUMMARY

Concept	Page	Notation / Formula / Description
Parameter	288	A numerical descriptive measure of a population.
Statistic	288	Any quantity computed from values in a sample.
Sampling distribution	289	The probability distribution of a statistic.
Properties of the sample mean $\overline{X}$	298	$\mu_{\overline{X}} = \mu$, $\sigma_{\overline{X}}^2 = \dfrac{\sigma^2}{n}$. If the underlying population is normal, then $\overline{X}$ is normal.
Central Limit Theorem	300	As the sample size n increases, the sampling distribution of $\overline{X}$ will increasingly approximate a normal distribution, with mean μ and variance σ^2/n, regardless of the shape of the underlying population distribution.
Distribution of the sample proportion of successes $\hat{P}$	310	$\mu_{\hat{p}} = p$, $\sigma_{\hat{p}}^2 = \dfrac{p(1-p)}{n}$. If n is large and both $np \geq 5$ and $n(1-p) \geq 5$, then $\hat{P}$ is approximately normal.

CHAPTER 7 EXERCISES

Applications

7.64 Manufacturing and Product Development Up until the early 1900s, people who colored their hair used only herbs and natural dyes. Today, hair coloring products contain two main ingredients: hydrogen peroxide and ammonia. The makers of a Clairol hair-color product claim that the mean amount of hydrogen peroxide in each bottle is 0.10 mg/m^3. Assume that the standard deviation is 0.05 mg/m^3. A random sample of these hair-color products was obtained, and the amount of hydrogen peroxide in each bottle was measured. The resulting data (in mg/m^3) are given on the data CD and book's web site.
a. Suppose that the manufacturer's claim is true. Find the distribution of the sample mean.
b. Is there any evidence to suggest that the manufacturer is including too much hydrogen peroxide in the product? Justify your answer.

7.65 Physical Sciences A floor slip tester is used to measure the safety of a floor by comparing the measured coefficient of static friction with accepted standards and guidelines. There are several factors that can affect floor safety, for example, dampness, polishes, and maintenance chemicals. A marble floor is consider *safe* if the coefficient of static friction is no greater than 0.5. A random sample of 50 rainy days was selected, and the coefficient of static friction for the marble floor was measured on each day. The resulting sample mean was 0.6. Is there any evidence to suggest that the marble floor is unsafe on rainy days? Assume the underlying population standard deviation is 0.2 and justify your answer.

7.66 Marketing and Consumer Behavior In the fresh fruits and vegetables section of a Kroger grocery store, customers can purchase any desired amount (by placing the food in a plastic bag to be weighed and/or priced at the checkout line). The probability distribution for the number of cucumbers purchased is given in the table that follows.

x	1	2	3	4	5
$p(x)$	0.10	0.50	0.20	0.15	0.05

a. Suppose two customers who purchase cucumbers are selected at random. Find the exact probability distribution for the sample mean number of cucumbers purchased, $\overline{X}$.
b. Find the mean, variance, and standard deviation of $\overline{X}$.

7.67 Marketing and Consumer Behavior A drive-in movie theater charges viewers by the carload, but keeps careful records of the number of people in each car. The probability distribution for the number of people in each car entering the drive-in is given in the table below.

x	1	2	3	4	5	6
$p(x)$	0.02	0.30	0.10	0.30	0.20	0.08

a. Suppose two cars entering the drive-in are selected at random. Find the exact probability distribution for the maximum number of people in either one of the cars, M.
b. Find the mean, variance, and standard deviation of M.

7.68 Demographics and Population Statistics Many clubs and companies around the country offer hot-air balloon rides. Most impose strict safety regulations and take at most four adults at one time, plus a pilot. The mean weight for an adult male in the United States is 190 pounds.[19] Suppose the distribution is normal with a standard deviation of 30 pounds.
a. If a hot-air balloon pilot (an adult male) takes three other adult males for a ride, what is the distribution for the total weight aboard, T? Carefully sketch the probability distribution for T.
b. If the total weight is less than 750 pounds, the pilot will have enough fuel to extend the ride by a few minutes. What is the probability of an extended ride?

c. If the total weight is over 1000 pounds, then the balloon will not be able to take off. What is the probability that the balloon will not be able to take off?

7.69 Biology and Environmental Science A typical houseplant produces oxygen (from carbon dioxide) in varying amounts, depending on the amount of light and water. Suppose a medium Norfolk Island pine produces 7.5 ml per hour of oxygen when exposed to normal sunlight. Thirty-five Norfolk Island pines are selected at random, and the oxygen output is carefully measured for each. Assume $\sigma = 1.75$ ml/h.

a. What is the probability that the sample mean oxygen produced is less than 7 ml/h?

b. What is the probability that the sample mean oxygen produced is between 7.25 and 7.5 ml/h?

c. Suppose each plant is exposed to a new high-intensity grow light and the sample mean oxygen produced for the 35 plants is 8.1 ml/h. Is there any evidence that the new lamp has increased oxygen output?

d. Answer parts (a), (b), and (c) if $\sigma = 3.75$ ml/h.

7.70 Marketing and Consumer Behavior Seafood restaurants along the coast of New England offer a variety of entrees, but lobster is the most popular meal. At Newick's Lobster Home in Dover, New Hampshire, 37% of all diners order lobster. Suppose 120 customers are selected at random, and the meal ordered by each is recorded.

a. Find the distribution of the sample proportion of diners who order lobster, and carefully sketch the probability distribution.

b. Find the probability that the sample proportion of diners who order lobster is less than 0.30.

c. Find the probability that the sample proportion of diners who order lobster is between 0.35 and 0.40.

d. The manager of Newick's is concerned that more customers might be ordering lobster. This would require a change in restaurant ordering and a shift in kitchen staff. Suppose the actual proportion of diners who order lobster is 0.42. Is there any evidence to suggest that the proportion of diners who order lobster has increased? Justify your answer.

7.71 Physical Sciences Tropical rainforests cover approximately 6% of the Earth's surface. The temperature in a typical rainforest ranges from 68 to 93°F, and the humidity is usually between 77% and 88%.[20] The mean amount of rain per year in any one rainforest is $\mu = 155$ inches with standard deviation $\sigma = 35$ inches. Suppose 30 rainforests are selected at random and the amount of rain per year is recorded for each.

a. What is the probability that the mean rainfall per year for the 30 rainforests is more than 170 inches?

b. What is the probability that the mean rainfall is between 140 and 150 inches?

c. Find a value r such that the probability that the mean rainfall is less than r is 0.001.

7.72 Technology and Internet An office manager has several computers running distributed programs. Because of the demands on the system, the machines may crash at various times during the day and require a hard reset. The probability distribution for the number of times a randomly selected machine crashes during a day, X, is given in the following table.

x	0	1	2	3	4	5
$p(x)$	0.50	0.30	0.10	0.07	0.02	0.01

a. Find the mean, variance, and standard deviation for the number of crashes by a single machine during a day.

b. Suppose $n = 2$ machines are selected at random. Find the sampling distribution of the statistic T, the total number of crashes for the two machines.

c. Find the mean, variance, and standard deviation of T.

d. Verify the relationships $\mu_T = 2\mu_X$ and $\sigma_T^2 = 2\sigma_X^2$.

7.73 Parameter versus Statistic In each of the following statements, identify the **boldface** number as the value of a population parameter or a sample statistic.

a. Some political observers claim that state senators spend too much time addressing colleagues about pending legislation. The mean length for a random sample of speeches was **23.7** minutes.

b. A consulting firm prepared a report from census data and concluded that the proportion of single-family homes in a certain county is **0.52**.

c. In a random sample of adults, the mean number of blinks per day was **22,037**.

d. A ballet instructor found the variance of the number of round-trip miles (to practice) traveled by all students was **150.76**.

e. A random sample of people who snowboard at least five times per year found the mean number of injuries per person per year to be **3.4**.

7.74 Medicine and Clinical Studies The stirrup bone is the smallest bone in the human body, with a mean length of 3 mm.[21] Suppose the standard deviation of the length is 0.16 mm. This bone, located in the ear, is part of a leverage system that can affect hearing. There is some speculation that high noise levels can affect the development of this bone and inhibit hearing. A random sample was obtained of adults who have lived in a large city their entire lives, and the length of the right-ear stirrup bone was recorded for each. The data (in mm) are given on the data CD and book's web site.

a. Find the distribution of the sample mean. Carefully sketch the probability density function.

b. Is there any evidence to suggest that the length of the stirrup bone for lifetime city adults is different from 3 mm? Justify your answer.

7.75 Sampling Distribution Each of the following graphs shows the probability distribution for an underlying distribution and for the sampling distribution of the sample mean of n observations (drawn from the underlying distribution). Identify each probability distribution function.

a.

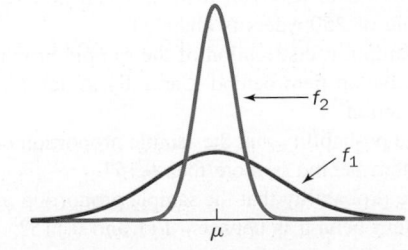

b.

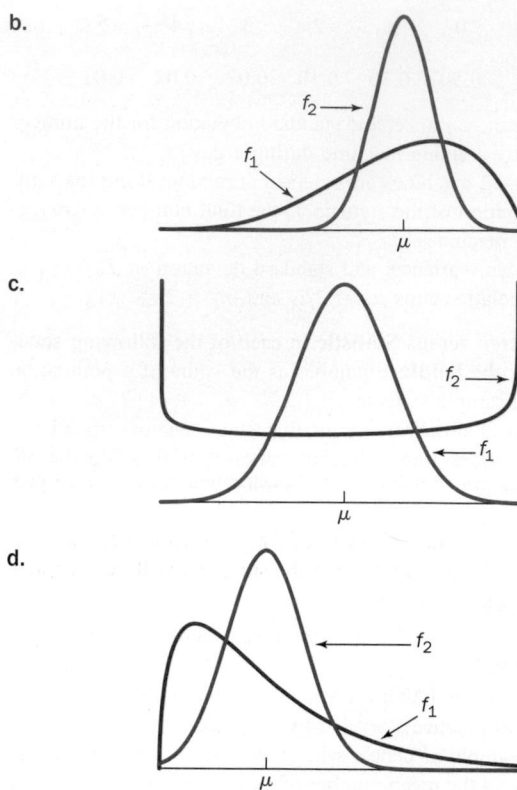

c.

d.

7.76 Public Health and Nutrition A manufacturer claims that the amount of nicotine in a certain brand of cigarette is normally distributed with mean 0.9 mg and standard deviation 0.35 mg. A random sample of these cigarettes was obtained and the nicotine content in each was carefully measured. The resulting data (in mg) are given in the following table.

1.06	2.03	0.67	0.40	1.04
0.76	1.12	1.02	1.26	1.02

(*Source*: Federal Trade Commission.)

a. Suppose that the manufacturer's claim is true. Find the distribution of the total amount of nicotine in these cigarettes.
b. Use the total amount of nicotine in these cigarettes to determine whether there is any evidence to suggest that the manufacturer's claim is false.

7.77 Travel and Transportation The Greenline Taxi Company in New York City keeps careful records of items left behind by riders. Most items are claimed, but many remain in a lost-and-found area at the company's headquarters. Records indicate that the proportion of riders who leave an item in a taxi is 0.12. Suppose a random sample of 250 riders is obtained.

a. Find the sampling distribution of the sample proportion of riders who leave an item behind. Carefully sketch the probability density function.
b. What is the probability that the sample proportion of riders who leave an item behind is more than 0.15?
c. What is the probability that the sample proportion of riders who leave an item behind is between 0.11 and 0.115?

d. Some people speculate that in bad economic times, riders (and people in general) are more careful with their belongings. Suppose this sample was obtained during a recession, and the sample proportion of riders who left an item behind was 0.09. Is there any evidence to suggest that the true proportion is different from 0.12?

7.78 Manufacturing and Product Development Eyedrops are used by many people to soothe and relieve irritation and to lubricate their eyes. A manufacturer claims that the amount of dextran, which helps to prolong the effect of eyedrops, contained in its eyedrops product is 70%. Thirty-six randomly selected bottles of these eye drops were obtained and the sample mean amount of dextran was 68.25%. Assume the population standard deviation is $\sigma = 5\%$.

a. Find the sampling distribution of the sample mean.
b. Find the probability that the amount of dextran is more than 71%.
c. Does the sample mean found suggest that the manufacturer's claim is wrong? Justify your answer.

7.79 Public Health and Nutrition Individuals who belong to a health maintenance organization (HMO) usually share the cost of certain medical services, for example, paying $10 for an office visit. An insurance company study indicates that the proportion of all people who belong to an HMO who have a copayment of more than $10 for an office visit is 0.65. Suppose 1000 people who belong to an HMO are randomly selected, and the office copayment amount is recorded.

a. Find the distribution of the sample proportion of people who have a copayment of more than $10 for an office visit. Verify the *nonskewness criteria*.
b. Find the probability that the sample proportion is more than 0.66.
c. Find the probability that the sample proportion is between 0.64 and 0.67.
d. Find a value h such that the probability that the sample proportion is less than h is 0.01.

Extended Applications

7.80 Physical Sciences Carlinville, Illinois, has just started an ambitious recycling program. Special trucks collect recyclable products once a week, sorted into barrels of paper, glass, and plastic. The amount of glass recycled per week by a single household is normally distributed with mean $\mu = 27$ pounds and standard deviation $\sigma = 7$ pounds. Suppose 12 households are randomly selected.

a. What is the probability that the total glass collected for the 12 homes, T, will be less than 350 pounds?
b. Find a value g such that $P(T \geq g) = 0.05$.
c. If the total glass collected for the 12 homes is more than 400 pounds, the recycling plant will make a profit. Find a value of μ such that the probability of making a profit is 0.10.

7.81 Mail Delivery During 2008, a special Canadian review committee was considering the issue of postal deregulation. A poll indicated that 69% of the public was against mail delivery by private companies, that is, against deregulation.[22] Suppose 250

Canadians are selected at random and each is asked whether they favor postal deregulation.

a. What is the probability that the sample proportion of Canadians who favor deregulation is between 0.20 and 0.30?

b. What is the probability that the sample proportion is more than 0.35?

c. Find a value of the sample size n such that the probability that the sample proportion is less than 0.28 is 0.05.

7.82 Manufacturing and Product Development A manufacturer of ice pops fills each plastic container with a fruity liquid, leaving enough room so that consumers can freeze the product. A filling machine is set so that the amount of liquid in each ice pop is normally distributed with mean 8.00 ounces and standard deviation 0.25 ounces. Suppose 16 ice pops are randomly selected.

a. Find the probability distribution of the sample mean number of ounces in each container, $\overline{X}$.

b. Find the probability that the sample mean is less than 7.9 ounces.

c. Find the probability that the sample mean is more than 8.15 ounces.

d. Suppose the filling machine operator can fine-tune the process by controlling the standard deviation of the fill. Find a value for σ such that the probability that the sample mean is more than 8.05 ounces is 0.05.

7.83 Sports and Leisure A large sporting goods company has just received a shipment of 100,000 table tennis balls. USA Table Tennis tournament regulations specify that the diameter of the ball must be 40 millimeters (mm). Suppose the distribution of the diameter is normal with standard deviation 0.4 mm. Twenty-five table tennis balls will be selected at random, and the diameter of each will be carefully measured. If the mean diameter is within 0.2 mm of 40 mm, then the shipment is accepted. Otherwise the entire shipment is returned to the manufacturer.

a. Suppose the true mean diameter of the table tennis balls is 40 mm. What is the probability that the entire shipment will be sent back to the manufacturer?

b. Suppose the true mean diameter of the table tennis balls is 40.4 mm. What is the probability that the shipment will be accepted by the sporting goods store?

c. Suppose the true mean diameter of the table tennis balls is 39.4 mm. What is the probability that the shipment will be accepted by the sporting goods store?

Confidence Intervals Based on a Single Sample

8

Chapter 8 Challenge

Is a typical train car loaded with coal overweight?

Coal is still a major source of energy in the United States and is used to generate electricity in many power plants. This non-renewable energy source is usually mined in one of two ways: underground mining (if the coal is at least 200 feet deep) or surface mining. New mining technology that has improved safety includes robots, lasers, computers, and sophisticated air filters.

Once coal is taken from a mine, it is crushed into small pieces and transported by conveyors to dump trucks or train coal-hopper cars. A tipple is used to load each coal car by gravity, and after each car is loaded, it is weighed using a computerized scale. The weight of the empty car, or the tare weight, is subtracted in order to obtain the exact weight of the coal in each car.

(Kenneth Sponster/iStock photo)

Each car can hold up to 120 tons of coal, and 150 cars may be joined together to form a coal train. The Department of Transportation (DOT) is concerned that some operators may be overloading coal cars, causing increased wear on the rail system and posing greater accident risk. The DOT would like to predict the weight of coal carried by a typical train car. Suppose 15 loaded coal cars are selected at random, and the exact weight of the coal in each is obtained. The summary statistics are $\bar{x} = 105.35$ tons and $s = 6.30$ tons. The concepts presented in this chapter will be used to construct an interval in which we are fairly certain the true mean coal weight per train car lies. This interval of numbers is given in the Chapter Challenge Wrap-Up (page 373) and can also be used to make an inference concerning the population mean.

Review

- Remember numerical summary measures like the sample mean, sample variance, and sample proportion.
- Recall that the sampling distribution of a statistic is the probability distribution of the statistic.

Preview

- Learn the properties of point estimators and understand what makes an estimator good.
- Construct confidence intervals (CIs) for various parameters.

8.1 POINT ESTIMATION

A **point estimate** of a population parameter is a single number computed from a sample that serves as a guess for the parameter. Using the terminology introduced in Chapter 7,

1. An **estimator** is a statistic of interest and is, therefore, a random variable. An estimator has a distribution, a mean, a variance, and a standard deviation.

2. An *estimate* is simply a specific value of an estimator.

DEFINITION

An **estimator** (statistic) is a rule used to produce a point estimate of a population parameter.

Suppose we need to estimate a population parameter θ and there are many different statistics (rules) available. Which one should we use? Figure 8.1 shows the sampling distributions of three different statistics for estimating θ. These graphs suggest some properties of a *good* statistic.

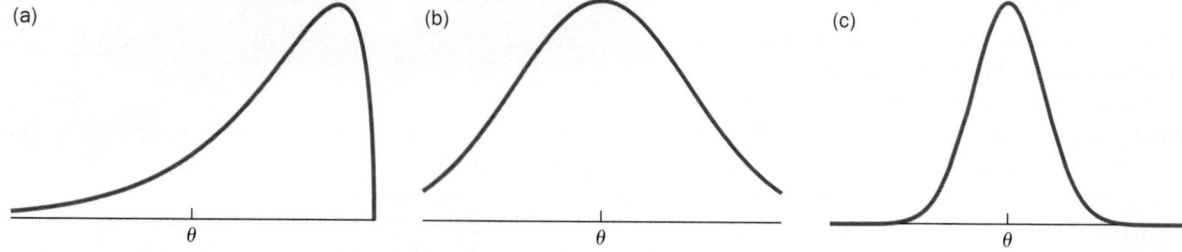

Figure 8.1 The sampling distributions for three different statistics for estimating θ: (a) a skewed statistic, (b) a statistic with large variance, and (c) a statistic with small variance. The horizontal axis represents all possible values of $\widehat{\theta}$.

The statistic in (a) is unlikely to produce a value close to θ. The sampling distribution is skewed to the left, and most of the values of the statistic are to the right of θ. The statistic in (b) is centered at θ. On average (in the long run), this statistic will produce θ (that's good!). However, the statistic has large variance (that's bad!). Even though the sampling distribution is centered at the true value of the population parameter, specific estimates will probably be *far away* from θ. The statistic in (c) exhibits two very desirable properties. It is centered at the true value of the population parameter, and it has small variance. These observations suggest two rules for selecting a statistic.

DEFINITION

A statistic $\widehat{\theta}$ is an **unbiased estimator** of a population parameter θ if $E(\widehat{\theta}) = \theta$, the mean of $\widehat{\theta}$ is θ.

If $E(\widehat{\theta}) \neq \theta$, then the statistic $\widehat{\theta}$ is a **biased estimator** of θ.

Figure 8.2 (a) illustrates the sampling distribution of an **unbiased estimator** for θ. The distribution of the statistic is centered at θ; the value of the statistic is, on average, θ. Figure 8.2 (b) shows the sampling distribution of a **biased estimator** for θ. On average, the value of this statistic is greater than θ.

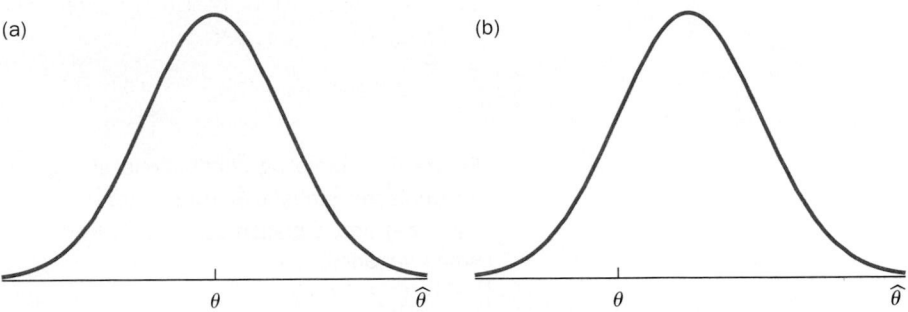

Figure 8.2 The sampling distribution for (a) an unbiased estimator for θ, and (b) a biased estimator for θ. The horizontal axis represents all possible values of $\hat{\theta}$.

We have already worked with several unbiased estimators in previous chapters:

1. The sample mean, $\overline{X}$, is an unbiased statistic for estimating the population mean μ, because $E(\overline{X}) = \mu$.

2. The sample proportion, $\hat{P}$, is an unbiased statistic for estimating the population proportion p, because $E(\hat{P}) = p$.

3. The sample variance, S^2, is an unbiased statistic for estimating the population variance σ^2, because $E(S^2) = \sigma^2$.

Even though S^2 is an unbiased estimator for σ^2, the sample standard deviation S is a *biased* estimator for the population standard deviation σ.

The expected-value operation does not pass freely through the square-root symbol.

$$E(S) = E(\sqrt{S^2}) \neq \sqrt{E(S^2)} = \sqrt{\sigma^2} = \sigma.$$

And even though S is biased, it is still important in statistical inference.

If there are several statistics available, it seems reasonable to use one that is unbiased. Therefore, the first rule for choosing a statistic is that the sampling distribution should be centered at θ; the estimator should be unbiased.

The second rule for choosing a statistic is that of all unbiased statistics, the best statistic to use is the one with the smallest variance. The point estimate produced using this statistic will, on average, be *close* to the true value of the population parameter. Figure 8.3 illustrates this second rule for choosing a statistic.

Figure 8.3 Sampling distributions of two unbiased statistics for estimating θ. Use the statistic with smaller variance.

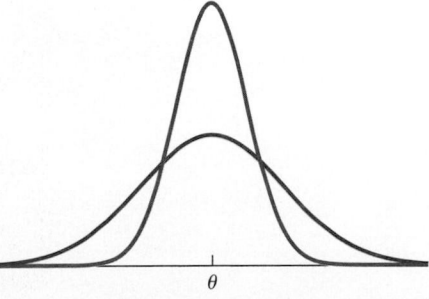

Suppose there are two statistics to choose from, $\hat{\theta}_1$ and $\hat{\theta}_2$, for estimating θ, as shown in Figure 8.4. The statistic $\hat{\theta}_1$ is unbiased but has large variance; $\hat{\theta}_2$ is slightly biased but has small variance. The choice of an estimator is a difficult decision, and there is no definitive answer.

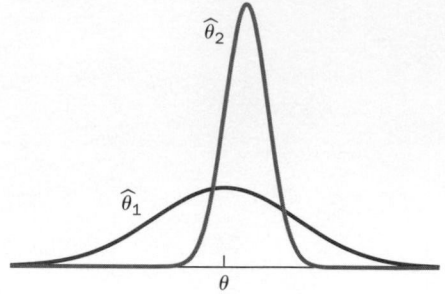

Figure 8.4 Sampling distributions of an unbiased statistic $\hat{\theta}_1$ (with large variance) and a biased statistic $\hat{\theta}_2$ (with small variance).

For a given parameter θ, an MVUE may not exist.

Suppose we need to estimate the population parameter θ, and there are several unbiased statistics from which to choose. If one of these statistics has the smallest possible variance, it is called the MVUE (minimum-variance unbiased estimator). If the underlying population is normal, the sample mean, $\overline{X}$, is the MVUE for estimating μ. So if the population is normal, the sample mean is a *really good* statistic to use for estimating μ. $\overline{X}$ is unbiased, and it has the smallest variance of all possible unbiased estimators for μ.

Example 8.1 Storing Petroleum Solvents A dry-cleaning company must comply with many government regulations, for example, the Federal Clean Water Act and the Oil Pollution Act. Some regulations are based on the above-ground storage capacity of petroleum solvent. A random sample of dry-cleaning companies in Tacoma, Washington, was obtained. The petroleum-solvent storage capacity (in gallons) for each is given in the table below.

770	875	850	1000	830	980
800	950	940	1125	925	1100

Find point estimates for the population mean petroleum-solvent storage capacity and for the population median petroleum-solvent storage capacity.

SOLUTION

STEP 1 Use the sample mean to estimate the population mean.

$$\bar{x} = \frac{1}{12}(770 + 875 + \cdots + 925 + 1100)$$

$$= \frac{1}{12}(11145) = 928.75$$

A point estimate of the population mean petroleum-solvent storage capacity is 928.75 gallons.

STEP 2 Use the sample median to estimate the population median.

Order the observations from smallest to largest. The sample median is the *middle* value. Since there are $n = 12$ observations, the middle value is in position 6.5 (i.e., the mean of the values in positions 6 and 7).

A technology solution:

```
mean(L1)
            928.7500
median(L1)
            932.5000
```

Figure 8.5 The sample mean and the sample median computed using ⌊LIST⌋; MATH functions.

The observations in order are as follows:

$$770 \quad 800 \quad 830 \quad 850 \quad 875 \quad 925 \quad 940 \quad 950 \quad 980 \quad 1000 \quad 1100 \quad 1125$$

$$\uparrow$$

$$\tilde{x} = \frac{1}{2}(925 + 940) = 932.50$$

A point estimate of the population median petroleum-solvent storage capacity is 932.50 gallons.

TECHNOLOGY CORNER

Procedure: Compute point estimates.
Reconsider: Example 8.1, page 326, solution, and interpretations.

TI-84 Plus

There are several built-in functions to compute summary statistics. ⌊STAT⌋; CALC; 1-Var Stats may be used to find several summary statistics at once. The ⌊LIST⌋; MATH menu contains several functions that return single point estimates.

1. Enter the data into list L1.
2. Use ⌊LIST⌋; MATH; mean to find the sample mean and ⌊LIST⌋; MATH; median to find the sample median. See Figure 8.5.

Minitab

There are several built-in functions to compute summary statistics in a session window, using a graphical input window, or using the Calculator.

1. Enter the data into column C1.
2. Select Stat; Basic Statistics; Display Descriptive Statistics.
3. Enter C1 in the Variables input window. Use the Statistics options button to select the desired estimates. See Figure 8.6.

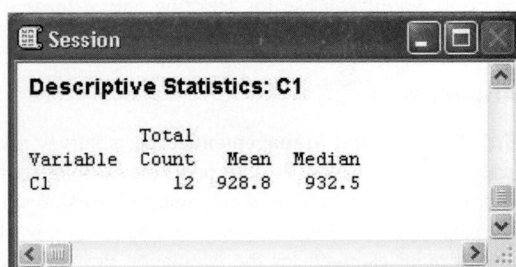

Figure 8.6 Descriptive statistics, selected in a graphical input window.

Session

Descriptive Statistics: C1

Variable	Total Count	Mean	Median
C1	12	928.8	932.5

Excel

There are several built-in functions to compute summary statistics. Under the Data tab, Data Analysis; Descriptive Statistics returns several summary statistics at once.

1. Enter the data into column A.
2. Use the function AVERAGE to find the sample mean and the function MEDIAN to find the sample median. See Figure 8.7.

Figure 8.7 Built-in functions to find the sample mean and sample median.

B	C
928.75	=AVERAGE(A1:A12)
932.50	=MEDIAN(A1:A12)

SECTION 8.1 EXERCISES

Practice

8.1 The graph below shows the probability density functions for three different statistics that could be used to estimate a population parameter θ. Which statistic would you use and why?

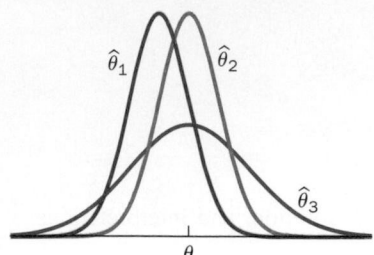

8.2 The graph below shows the probability density functions for three different statistics that could be used to estimate a population parameter θ. Which statistic would you use and why?

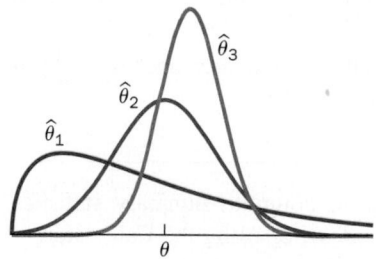

8.3 Why is an unbiased estimator for θ better than a biased estimator for θ?

8.4 Suppose there are several unbiased estimators for θ. What criterion would you use for selecting one of these estimators, and why?

Applications

8.5 Business and Management A recent survey asked employees in technology-related jobs what perks they would like their employer to provide. Out of 1200 randomly selected workers who were polled, 975 said they would like ongoing training paid for by their employer. Find a point estimate for the population proportion of all workers in technology-related jobs who would like ongoing training.

8.6 Biology and Environmental Science A *hand* is a traditional unit used to measure the height of a horse. One hand is 4 inches, and the height of a horse is measured from the ground to the horse's shoulder. A random sample of horses sold at auctions around the country revealed the following heights (in hands).

15.8	13.7	11.0	17.1	19.3	14.6	14.4	13.8
18.7	16.5	12.2	17.9	16.8	16.3	12.8	18.7
10.4	15.6	11.7	12.8				

a. Find a point estimate for the population mean height of all horses sold at auctions.

b. Find a point estimate for the population median height of all horses sold at auctions.

c. Find a point estimate for the population variance of the height of all horses sold at auctions.

d. Any horse with a height less than 14.5 hands is considered to be a pony. Find a point estimate for the true proportion of ponies sold at auctions.

8.7 Demographics and Population Statistics A survey of 500 randomly selected American families revealed that 90 speak a non-English language at home. Find a point estimate for the proportion of all American families that speak a non-English language at home.[1]

8.8 Biology and Environmental Science Cowrie shells have been used as money in many parts of the world, for example, China and Africa. Today, they are used in decorations and jewelry, and cowrie shell bracelets and necklaces are popular near seaside resorts. A jeweler recently purchased several hundred cowrie shells to use in making earrings. A random sample of the finished earrings was obtained, and the weight (in grams) of each is given in the table below.

7.3	7.2	7.2	7.9	7.3	7.3	7.0	7.0	7.4	7.4
7.4	7.2	7.5	7.7	7.1	7.0	7.2	7.2	7.7	6.9
7.4	7.8	7.7	7.4	7.5	7.7	7.5	7.3	7.6	7.3

(*Source*: africashowcase.com.)

a. Find a point estimate for the first quartile and a point estimate for the third quartile.

b. The smallest (in weight) 20% of all cowrie shell earrings are sold at a discount. Find a point estimate for the 20th percentile of the cowrie shell earring weight distribution.

8.9 Manufacturing and Product Development A company that manufactures a centrifugal pump for golf-course sprayers would like to rate the pressure developed by this unit. Thirty pumps were randomly selected and tested. The pressure (in psi) developed by each unit is given on the data CD and book's web site.

a. Find a point estimate for the minimum pressure developed by this pump.

b. Find a point estimate for the maximum pressure developed by this pump.

c. Use your answers to parts (a) and (b) to construct an *interval* estimate for the pressure developed by this pump.

8.10 Sports and Leisure *USA Weekend* readers were asked, "Of the 70,000 American movies ever made, which is most beloved?" Of the 54,234 readers polled, 8,134 said *Titanic* was their most beloved movie, and 6,508 named *The Godfather*.[2]

a. Find a point estimate for the proportion of all *USA Weekend* readers whose favorite movie is *Titanic*.

b. Find a point estimate for the proportion of all *USA Weekend* readers whose favorite movie is *The Godfather*.

c. Let p_d denote the difference in population proportions between all *USA Weekend* readers whose favorite movie is *Titanic* and all *USA Weekend* readers whose favorite movie is *The Godfather*. Find a point estimate for p_d.

8.11 Manufacturing and Product Development Film from 35-mm cameras is processed at a local store using a special automated machine. The film must be precisely loaded in this machine and several adjustments are needed to ensure proper alignment. Due to human error and extensive use, the machine may cut off the left portion of a negative and cause the finished print to appear off-center. This printing error occurs frequently, though most customers rarely notice a problem.

A random sample of negatives and their corresponding prints was obtained. The length (in mm) of the portion of the negative cut off

from the left was measured on each print, and the data are given on the data CD and book's web site.

a. Find a point estimate for the mean length cut off from the left of all negatives.

b. Find a point estimate for the variance of the length cut off from the left of all negatives.

c. Find a point estimate for the first quartile and a point estimate for the third quartile.

8.2   A CONFIDENCE INTERVAL FOR A POPULATION MEAN WHEN σ IS KNOWN

In Section 8.1, we discovered that a *good* estimator is unbiased and has small variance. An estimator produces only a single value that serves as a best guess for a population parameter. In this section, we use this single value to produce a **confidence interval**. This interval of values is constructed so that we can be reasonably sure that the true value of the population parameter lies in this interval.

> **DEFINITION**
>
> A **confidence interval** (CI) for a population parameter is an interval of values constructed so that, with a specified degree of confidence, the value of the population parameter lies in this interval.
>
> The **confidence coefficient** is the probability that the CI encloses the population parameter in repeated samplings.
>
> The **confidence level** is the confidence coefficient expressed as a percentage.

 ILLUMINATING THE CONCEPTS

1. A confidence interval is usually expressed as an *open* interval, for example, (10.5, 15.8). 10.5 is the left endpoint, or lowerbound, and 15.8 is the right endpoint, or upperbound. The interval extends all the way to, but does not include, the endpoints.

2. Typical **confidence coefficients** are 0.95 and 0.99.

3. Typical **confidence levels** are, therefore, 95% and 99%. ∎

The following *steps* provide background for the construction of a confidence interval for a population mean μ.

1. Suppose either (a) the underlying population is normal, or (b) the sample size n is large, or both, and the population standard deviation σ is known.

2. Using the properties of $\overline{X}$ and the CLT (if necessary), the sample mean $\overline{X}$ is (approximately) normal with mean μ and variance σ^2/n.

   In symbols, $\overline{X} \sim N(\mu, \sigma^2/n)$.

3. Using the Empirical Rule, approximately 95% of all values of the sample mean lie within two standard deviations of the mean. Figure 8.8 shows an example of a single value, or point estimate, $\bar{x}$, within two standard deviations of the mean.

4. Even though we know the distribution of $\overline{X}$ is centered at μ, we do not know the true value of μ. In order to *capture* μ it seems reasonable to *step* two standard deviations from an estimate $\bar{x}$ in both directions. The resulting (rough) 95% confidence interval $(\bar{x} - 2\sigma/\sqrt{n}, \bar{x} + 2\sigma/\sqrt{n})$ is illustrated in Figure 8.9.

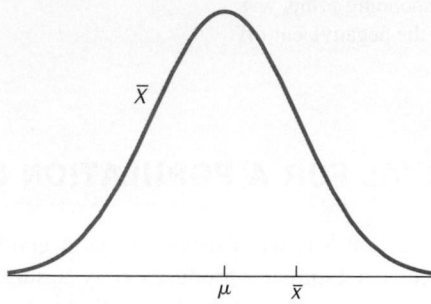

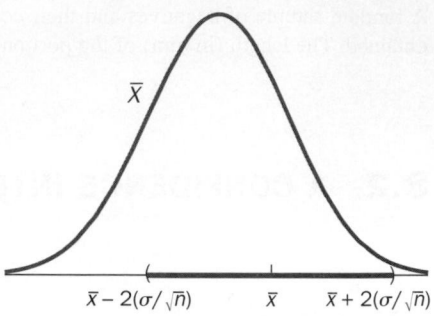

Figure 8.8 The sampling distribution of $\overline{X}$ and a typical value $\bar{x}$.

Figure 8.9 A rough 95% confidence interval that probably captures the true value μ.

In order to construct a more accurate 95% confidence interval for μ, using the same assumptions, begin with the standardized random variable Z.

1. $X \sim \mathrm{N}(\mu, \sigma^2/n) \longrightarrow Z = \dfrac{\overline{X} - \mu}{\sigma/\sqrt{n}} \sim \mathrm{N}(0, 1)$

2. Find a symmetric interval about 0 such that the probability that Z lies in this interval is 0.95.

$$P(-1.96 < Z < 1.96) = 0.95$$

Use Table III; see Figure 8.10.

Figure 8.10 A symmetric interval about 0 such that $P(-1.96 < Z < 1.96) = 0.95$.

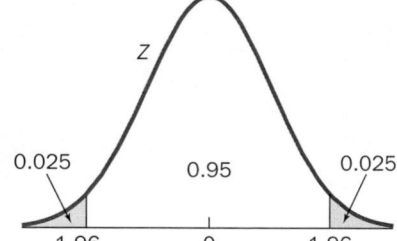

3. Substitute for Z and manipulate the interval inside the probability statement so that μ is *caught* in the middle.

$P\left(-1.96 < \dfrac{\overline{X} - \mu}{\sigma/\sqrt{n}} < 1.96\right) = 0.95$ Substitute for Z.

$P\left((-1.96)\dfrac{\sigma}{\sqrt{n}} < \overline{X} - \mu < (1.96)\dfrac{\sigma}{\sqrt{n}}\right) = 0.95$ Multiply all three parts by $\sigma/\sqrt{n}$.

$P\left(-\overline{X} - (1.96)\dfrac{\sigma}{\sqrt{n}} < -\mu < -\overline{X} + (1.96)\dfrac{\sigma}{\sqrt{n}}\right) = 0.95$ Subtract $\overline{X}$ from all three parts.

$P\left(\overline{X} + 1.96\dfrac{\sigma}{\sqrt{n}} > \mu > \overline{X} - 1.96\dfrac{\sigma}{\sqrt{n}}\right) = 0.95$ Multiply all three parts by -1. Multiplying by -1 changes the directions of the inequalities.

$P\left(\overline{X} - 1.96\dfrac{\sigma}{\sqrt{n}} < \mu < \overline{X} + 1.96\dfrac{\sigma}{\sqrt{n}}\right) = 0.95$ Rewrite expressions in increasing order. ▶

The last expression includes a formula for a 95% confidence interval for μ: step exactly 1.96 (not 2) standard deviations from a specific value $\bar{x}$ in both directions. This interval is shown in Figure 8.11.

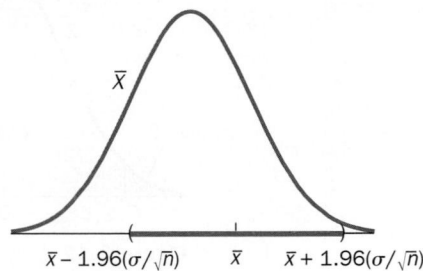

$$\bar{x} - 1.96(\sigma/\sqrt{n}) \qquad \bar{x} \qquad \bar{x} + 1.96(\sigma/\sqrt{n})$$

Figure 8.11 An exact 95% confidence interval for μ.

The last step is to find a more general $100(1 - \alpha)\%$ confidence interval for μ, using the same assumptions. α is usually small. For example, if $\alpha = 0.05$, the resulting confidence level is $100(1 - 0.05)\% = 95\%$. Usually the confidence level is given, and we need to work backward to find α. The following definition is necessary in order to construct a more general confidence interval.

In this definition, the subscript on z could be *any variable* or letter. For example, $P(Z \geq z_{\text{©}}) = c$.

DEFINITION

$z_{\alpha/2}$ is a **critical value**. It is a value on the measurement axis in a **standard normal distribution** such that $P(Z \geq z_{\alpha/2}) = \alpha/2$.

💡 **ILLUMINATING THE CONCEPTS**

1. $z_{\alpha/2}$ is simply a z value such that there is $\alpha/2$ of the area (probability) to the right of $z_{\alpha/2}$. $-z_{\alpha/2}$ is just the negative critical value.

2. **Critical values** are *always* defined in terms of right-tail probability.

3. z critical values are easy to find using the Complement Rule and working backward. For example,

   $P(Z \geq z_{\alpha/2}) = \alpha/2$ Definition of critical value.
   $P(Z \leq z_{\alpha/2}) = 1 - \alpha/2$ The Complement Rule.

   Work backward in Table III to find $z_{\alpha/2}$. ■

To find a general confidence interval for μ, start once again in the Z world. Find a symmetric interval about 0 such that the probability that Z lies in this interval is $1 - \alpha$ (Figure 8.12).

$$P(-z_{\alpha/2} < Z < z_{\alpha/2}) = 1 - \alpha \tag{8.1}$$

Manipulate Equation 8.1 to obtain the probability statement

$$P\left(\overline{X} - z_{\alpha/2} \frac{\sigma}{\sqrt{n}} < \mu < \overline{X} + z_{\alpha/2} \frac{\sigma}{\sqrt{n}}\right) = 1 - \alpha.$$

Figure 8.13 illustrates this interval for a specific value $\bar{x}$.

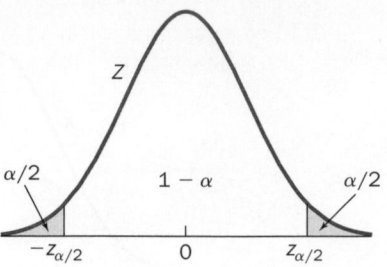

Figure 8.12 A symmetric interval about 0 such that $P(-z_{\alpha/2} < Z < z_{\alpha/2}) = 1 - \alpha$.

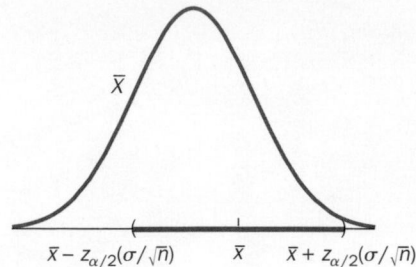

Figure 8.13 A $100(1 - \alpha)$% confidence interval for μ.

These derivations lead to the following general result.

> **HOW TO FIND A $100(1 - \alpha)$% CONFIDENCE INTERVAL FOR A POPULATION MEAN WHEN σ IS KNOWN**
>
> Given a random sample of size n from a population with mean μ, if
>
> **1.** the underlying population distribution is normal and/or n is large, and
> **2.** the population standard deviation σ is known,
>
> then a $100(1 - \alpha)$% confidence interval for μ has as endpoints the values
>
> $$\bar{x} \pm z_{\alpha/2} \frac{\sigma}{\sqrt{n}}. \tag{8.2}$$

Reminder: a *large* sample usually means $n \geq 30$.

ILLUMINATING THE CONCEPTS

1. Equation 8.2 can be used *only* if σ is known.

2. If n is large and σ is unknown, some statisticians suggest using the sample standard deviation s in Equation 8.2 in place of σ. This produces an *approximate* confidence interval for μ. The next section presents an *exact* confidence interval for μ when σ is unknown.

3. As the confidence coefficient increases (with σ and n constant), the critical value $z_{\alpha/2}$ increases. Therefore, the confidence interval is wider. ■

Example 8.2 Tire Weight   The total weight of a filled tire can dramatically affect the performance and safety of an automobile. Some transportation officials argue that mechanics should check the tire weights of every vehicle as part of an annual inspection. Suppose the weight of a 185/60/14 filled tire is normally distributed with standard deviation 1.25 pounds. In a random sample of 15 filled tires, the sample mean weight was $\bar{x} = 18.75$ pounds. Find a 95% confidence interval for the true mean weight of 185/60/14 tires.

Solution Trail

KEYWORDS

- 95% CI for the true mean.
- Normally distributed.
- Standard deviation 1.25 pounds.

↓

TRANSLATION

- 95% CI for μ.
- Underlying population is normal.
- $\sigma = 1.25$.

↓

CONCEPTS

A $100(1 - \alpha)$% CI for a population mean when σ is known.

↓

VISION

We need a 95% CI for μ. The population is normal and σ is known. Find the appropriate critical value and use Equation 8.2.

SOLUTION

STEP 1 $\bar{x} = 18.75,$ $\sigma = 1.25,$ $n = 15.$ Given.

$1 - \alpha = 0.95 \longrightarrow \alpha = 0.05 \longrightarrow \alpha/2 = 0.025$ Find $\alpha/2$.

$P(Z \geq z_{\alpha/2}) = P(Z \geq z_{0.025}) = 0.025$ Definition of critical value.

$P(Z \leq z_{0.025}) = 1 - 0.025 = 0.975$ The Complement Rule.

$z_{0.025} = 1.96$ Use Table III.

STEP 2   Use Equation 8.2.

$$\bar{x} \pm z_{\alpha/2} \frac{\sigma}{\sqrt{n}}$$ Equation 8.2.

$$= \bar{x} \pm z_{0.025} \frac{\sigma}{\sqrt{n}}$$ Use the value of α.

$$= 18.75 \pm (1.96) \frac{1.25}{\sqrt{15}}$$ Use summary statistics and values for σ and $z_{0.025}$.

$$= 18.75 \pm 0.63$$ Simplify.

$$= (18.12, 19.38)$$ Compute endpoints.

$(18.12, 19.38)$ is a 95% confidence interval for the true mean weight (in pounds) of 185/60/14 tires.

Figures 8.14 and 8.15 together show a technology solution.

Figure 8.14 ZInterval input screen.

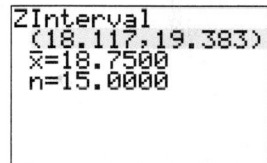

Figure 8.15 Resulting confidence interval.

There are two important ideas to remember when constructing a confidence interval.

1. The population parameter, in this case μ, is *fixed*. The confidence interval *varies* from sample to sample. It is correct to say, "We are 95% confident that the interval *captures* the true mean μ." It is incorrect to imply that the interval is fixed and the parameter μ varies.

2. The confidence coefficient, a probability, is a long-run limiting relative frequency. In repeated samples, the proportion of confidence intervals that capture the true value of μ approaches 0.95. Figure 8.16 illustrates this concept. We cannot be certain about any one specific confidence interval. The confidence is in the long-run process.

In the following example, actual data are presented rather than summary statistics. Equation 8.2 is used again to construct a confidence interval for a population mean.

Example 8.3 Airbag Safety   According to the National Highway Traffic Safety Administration, 284 people died between 1990 and 2007 as a result of an injury caused by an airbag. This safety device is designed to explode out of the dashboard or a side door panel in the event of a collision. An airbag protects the driver and passenger from impact with the windshield, steering column, and dashboard. Newer airbags open at a speed proportional to the speed of an impact. Older models open

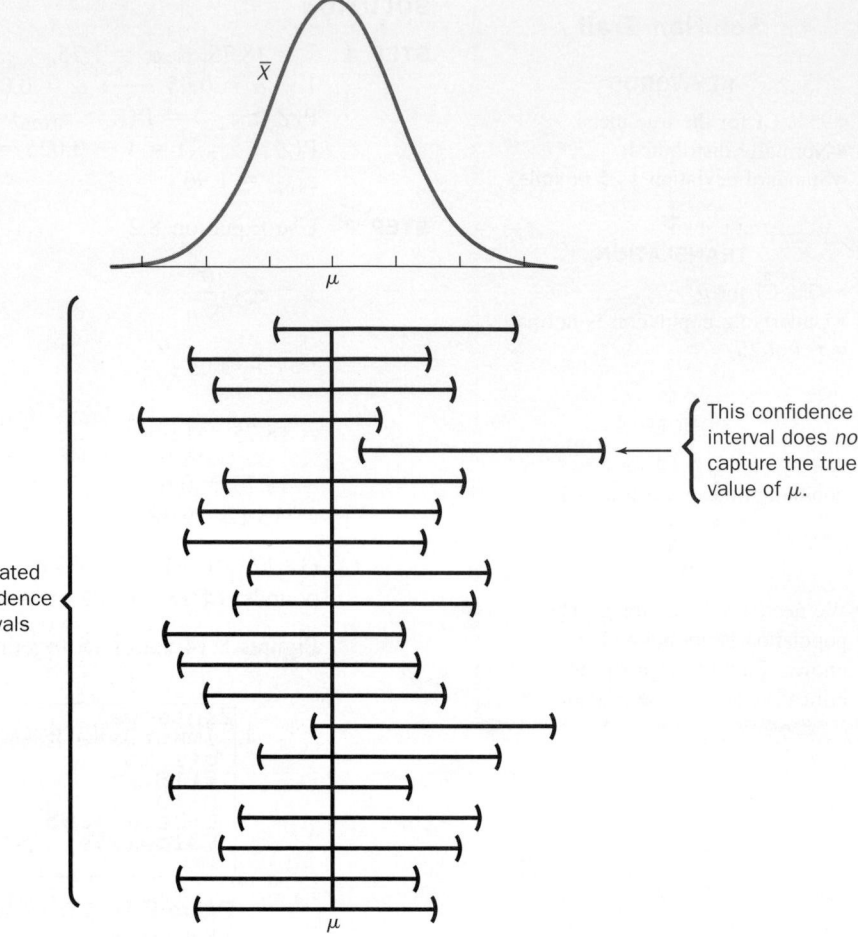

Figure 8.16 An illustration of the meaning of confidence coefficient. In repeated CIs, the proportion of all 95% confidence intervals that capture the true value of μ is 0.95.

Try the Confidence Intervals statistical applet on the text web site.

---

a. Solution Trail

KEYWORDS

- 99% CI for the mean.
- $\sigma = 10$ mph.

↓

TRANSLATION

- 99% CI for μ.
- Known standard deviation.

↓

CONCEPTS

A $100(1 - \alpha)\%$ CI for a population mean when σ is known.

↓

VISION

Construct a 99% CI for μ. No information is given about the shape of the underlying population distribution. However, $n = 40$ (≥ 30), so the CLT applies, and we can use Equation 8.2.

---

at a constant speed, up to 200 mph, equal to the force of a heavyweight knockout punch. A random sample of older-model airbags was obtained and tested. The opening speed (in mph) of each is given in the table below. Assume $\sigma = 10$ mph.

169	165	174	176	169	170	180	180	173	189
166	181	188	172	188	164	180	174	176	167
182	168	172	177	169	158	171	164	183	175
177	193	177	173	172	169	167	167	168	172

a. Find a 99% confidence interval for the mean opening speed of older-model airbags.

b. Using the confidence interval in part (a), is there any evidence to suggest that the population mean opening speed in older-model airbags is more than 175 mph?

SOLUTION

a. $\sigma = 10, \quad n = 40.$ *Given.*

$$\bar{x} = \frac{1}{40}(169 + \cdots + 172) = 173.88$$ *Compute the sample mean.*

$$1 - \alpha = 0.99 \longrightarrow \alpha = 0.01 \longrightarrow \alpha/2 = 0.005$$ *Find $\alpha/2$.*

$$P(Z \leq z_{\alpha/2}) = P(Z \geq z_{0.005}) = 0.005$$ *Definition of critical value.*

$$P(Z \leq z_{0.005}) = 1 - 0.005 = 0.995$$ *The Complement Rule.*

$$z_{0.005} = 2.58$$ *Use Table III.*

Use Equation 8.2.

$$\bar{x} \pm z_{\alpha/2} \frac{\sigma}{\sqrt{n}}$$ 　Equation 8.2.

$$= \bar{x} \pm z_{0.005} \frac{\sigma}{\sqrt{n}}$$ 　Use the value of α.

$$= 173.88 \pm (2.58) \frac{10}{\sqrt{40}}$$ 　Use summary statistics and values for σ and $z_{0.005}$.

$$= 173.88 \pm 4.08$$ 　Simplify.

$$= (169.80, 177.96)$$ 　Compute endpoints.

(169.80, 177.96) is a 99% confidence interval for the true mean opening speed (in mph) of older-model airbags.

Figure 8.17 shows a technology solution.

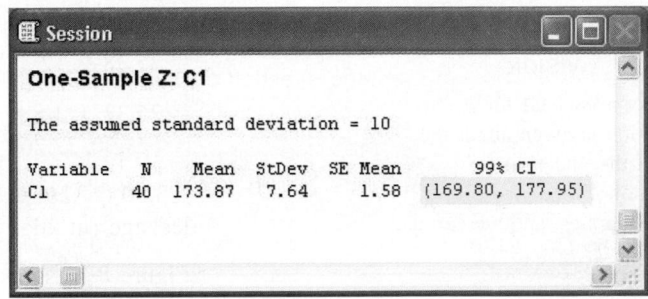

Figure 8.17 Minitab 1-Sample $\underline{Z}$ confidence interval.

b. **Claim**: $\mu = 175$.

Experiment: $\bar{x} = 173.88$.

Likelihood: The likelihood in this case is expressed as a 99% confidence interval, an interval of likely values for μ: (169.80, 177.96), from part (a).

Conclusion: Since this CI includes 175, there is no evidence to suggest that μ is different from 175.　　●

Here is one more example involving actual data.

b. Solution Trail

KEYWORDS

Is there any evidence?

↓

TRANSLATION

Inference procedure.

↓

CONCEPTS

A $100(1 - \alpha)$% CI for a population mean when σ is known.

↓

VISION

A 99% CI for μ is an interval in which we are 99% confident the true value of μ lies. If the CI in part (a) captures, or includes, 175, then there is no evidence to suggest that μ is different from 175. Use the four-step inference procedure.

Example 8.4 Wasting Electricity　A typical Canadian home leaks approximately 389 to 513 kilowatt-hours of electricity per year.[3] This waste is caused by electronic devices that are turned off but remain in standby mode, for example, television sets, computer printers, garage door openers, cordless drills, and audio systems. A voluntary Energy Star program allows manufacturers to display a federal Energy Star label if a TV uses at most 3 watts of power in standby mode. A random sample of 50 homes from around the country was obtained, and each was carefully measured for electricity leaks. The monthly waste (in kilowatts) for each home is given in the table that follows. Assume $\sigma = 5.7$ kilowatts.

31	28	25	21	24	21	27	28	19	18
30	23	30	32	27	33	24	22	15	19
29	25	21	23	32	27	23	15	29	34
27	24	29	22	23	32	30	19	32	35
37	25	29	32	22	22	27	19	10	18

Find a 98% confidence interval for the true mean electricity leakage per home per month.

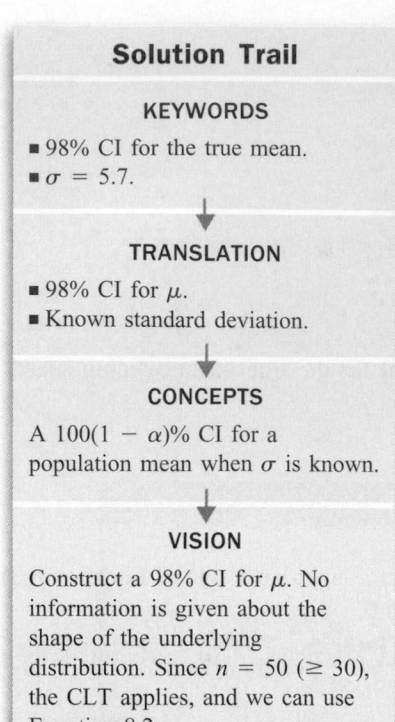

Solution Trail

KEYWORDS

- 98% CI for the true mean.
- $\sigma = 5.7$.

↓

TRANSLATION

- 98% CI for μ.
- Known standard deviation.

↓

CONCEPTS

A $100(1 - \alpha)\%$ CI for a population mean when σ is known.

↓

VISION

Construct a 98% CI for μ. No information is given about the shape of the underlying distribution. Since $n = 50$ (≥ 30), the CLT applies, and we can use Equation 8.2.

SOLUTION

STEP 1 $\sigma = 5.7,\quad n = 50.$ *Given.*

$$\bar{x} = \frac{1}{50}(31 + \cdots + 18) = 25.38$$ *Compute the sample mean.*

$1 - \alpha = 0.98 \longrightarrow \alpha = 0.02 \longrightarrow \alpha/2 = 0.01$ *Find $\alpha/2$.*

$P(Z \geq z_{\alpha/2}) = P(Z \geq z_{0.01}) = 0.02$ *Definition of critical value.*

$P(Z \leq z_{0.01}) = 1 - 0.01 = 0.99$ *The Complement Rule.*

$z_{0.01} = 2.33$ *Use Table III.*

STEP 2 Use Equation 8.2.

$$\bar{x} \pm z_{\alpha/2}\frac{\sigma}{\sqrt{n}}$$ *Equation 8.2.*

$$= \bar{x} \pm z_{0.01}\frac{\sigma}{\sqrt{n}}$$ *Use the value of α.*

$$= 25.38 \pm (2.33)\frac{5.7}{\sqrt{50}}$$ *Use summary statistics and values for σ and $z_{0.01}$.*

$$= 25.38 \pm 1.88$$ *Simplify.*

$$= (23.50, 27.26)$$ *Compute endpoints.*

$(23.50, 27.26)$ is a 98% confidence interval for the true mean electricity leakage (in kilowatts) in an American home per month.

Figure 8.18 shows a technology solution.

B	C	D
25.3800	= AVERAGE(A1:A50)	The sample mean
1.8753	= CONFIDENCE(0.02,5.7,50)	The step
23.5047	= B1-B2	Left endpoint
27.2553	= B1+B2	Right endpoint

Figure 8.18 Excel calculations to construct a confidence interval.

Many confidence intervals have endpoint formulas that are of the same general form. Suppose the statistic $\hat{\theta}$ is used to estimate the population parameter θ. The general form of a confidence interval is

(Point estimate using $\hat{\theta}$) $\pm$ (Critical value) • [(Estimate of) standard deviation of $\hat{\theta}$].

The critical value may be from the standard normal distribution or some other reference distribution. If the actual standard deviation of the statistic is not known (a likely situation), then an estimate is used.

The (two-sided) confidence interval derived above is the most common. However, there are other two-sided confidence intervals, and also one-sided confidence intervals consisting of only a lower bound or only an upper bound. The derivation of several one-sided confidence intervals is outlined in several Challenge exercises.

Most of the time, as in the examples above, there is no control over the sample size. The data or summary statistics are presented and a confidence interval is constructed. However, a certain desired accuracy may be expressed by the width of a confidence interval. Given σ and the confidence level, a sample size n can be computed such that the resulting confidence interval has a certain desired width.

The error of estimation is the step, the distance from the sample mean to the left and to the right endpoint of the confidence interval. We would like to find a sample size n so that the step is no bigger than B. Therefore, B is called a bound on the error of estimation.

Suppose n is large (but unknown), σ is known, and the confidence level is $100(1 - \alpha)\%$. If the desired width is W, let $B = W/2$, called the **bound on the error of estimation**. Half the width of the confidence interval is given by the step (in each direction) from $\bar{x}$. The endpoints of the confidence interval for μ are (from Equation 8.2)

$$\bar{x} \pm \underbrace{z_{\alpha/2}\frac{\sigma}{\sqrt{n}}}_{B \,=\, \text{bound}}.$$

Let $B = z_{\alpha/2}(\sigma/\sqrt{n})$, and solve for n. The resulting formula for n is given by

$$n = \left[\frac{\sigma z_{\alpha/2}}{B}\right]^2. \tag{8.3}$$

In symbols:
$\sigma \uparrow \;\Rightarrow n \uparrow$
$z_{\alpha/2} \uparrow \;\Rightarrow n \uparrow$
$B \downarrow \;\Rightarrow n \uparrow$

💡 ILLUMINATING THE CONCEPTS

1. ◀ Consider the effect on n as one value (on the right-hand side of Equation 8.3) changes while the other two remain constant.

   As σ increases (with $z_{\alpha/2}$ and B constant), the fraction becomes larger, the square becomes larger, and n becomes larger. This result makes sense, since a larger underlying variance would require a larger sample size to preserve a certain bound on the error of estimation.

   As $z_{\alpha/2}$ increases (equivalently, as the confidence level increases, with σ and B constant), again the fraction becomes larger, the square becomes larger, and n becomes larger. More confidence requires a larger n to maintain the same bound on the error of estimation.

   As B decreases (with σ and $z_{\alpha/2}$ constant), the denominator becomes smaller, which makes the fraction larger, the square larger, and n larger. A smaller bound on the error of estimation (i.e., a smaller CI) means more information (a larger sample size) is needed to maintain the same confidence level. ▶

2. When Equation 8.3 is applied, it is likely that the value of n will *not* be an integer. *Always* round up. This guarantees a $100(1 - \alpha)\%$ confidence interval with a bound on the error of estimation of at most B. Any larger sample size will also be sufficient.

3. It is unlikely that σ, needed in Equation 8.3, is known. However, often an experimenter can make a very good guess at σ from previous experience or a small preliminary study. ■

The following example illustrates the use of Equation 8.3.

(NPS Photo by R. Robinson)

Example 8.5 Geyser Height   Yellowstone National Park, which opened in 1872, includes magnificent scenery, historic sites, attractions of scientific interest, and recreational areas. The Beehive Geyser is one of the gushing geysers in Yellowstone and is a popular tourist and scientific attraction. During the summertime, its eruptions are eight hours to a few days apart, and it usually erupts at least once every day.[4] A National Park Service researcher would like to find a 95% confidence interval for the mean height of the Beehive Geyser eruption with a bound on the error of estimation of 5 feet. Previous experience suggests that the population standard deviation for the height is approximately 15 feet. How large a sample is necessary in order to achieve this accuracy?

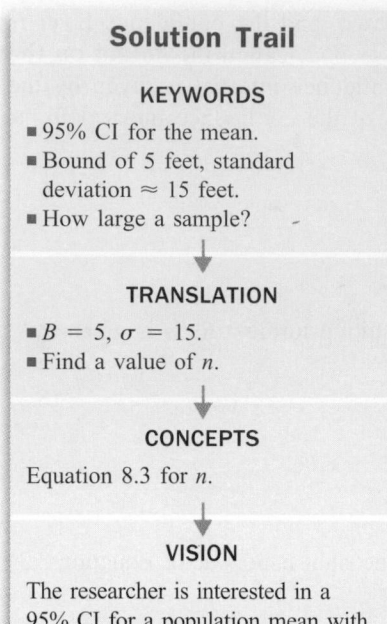

Solution Trail

KEYWORDS

- 95% CI for the mean.
- Bound of 5 feet, standard deviation ≈ 15 feet.
- How large a sample?

↓

TRANSLATION

- $B = 5$, $\sigma = 15$.
- Find a value of n.

↓

CONCEPTS

Equation 8.3 for n.

↓

VISION

The researcher is interested in a 95% CI for a population mean with a certain bound on the error of estimation. There is a good estimate for the population standard deviation. Find the appropriate critical value, and use Equation 8.3 to determine a lower bound for the sample size.

SOLUTION

STEP 1 $\sigma = 15$, $B = 5$.   Given.

$1 - \alpha = 0.95 \longrightarrow \alpha/2 = 0.025 \longrightarrow z_{0.025} = 1.96$ Find the critical value.

STEP 2   Use Equation 8.3.

$$n = \left[\frac{\sigma z_{\alpha/2}}{B}\right]^2$$ Equation 8.3.

$$= \left[\frac{(15)(1.96)}{5}\right]^2$$ Substitute given values.

$$= [5.88]^2 = 34.57$$ Compute.

The necessary sample size is $n \geq 35$ (always round up). This will guarantee that a 95% confidence interval for the population mean height will have a bound on the error of estimation no greater than five. ●

TECHNOLOGY CORNER

Procedure: Construct a confidence interval for a population mean when σ is known.
Reconsider: Example 8.2, page 332, solution, and interpretation.

TI-84 Plus

Use the built-in function `ZInterval`. Input is either summary statistics or a list containing data.

1. Select $\boxed{\text{STAT}}$; TESTS; `ZInterval`.
2. Highlight `Stats` (for summary statistics). Enter values for σ, $\bar{x}$, n, and the `C-Level` (the confidence coefficient). See Figure 8.14, page 333.
3. Highlight `Calculate` and press $\boxed{\text{ENTER}}$. See Figure 8.15, page 333.

Minitab

Use the built-in function 1-Sample Z. Input is either summary statistics or a column containing data.

1. Select Stat; Basic Statistics; 1-Sample Z.
2. Choose Summarized data and enter the Sample size, Mean, and Standard deviation.
3. Select the Option button and enter the confidence level. The results are displayed in the session window. See Figure 8.19.

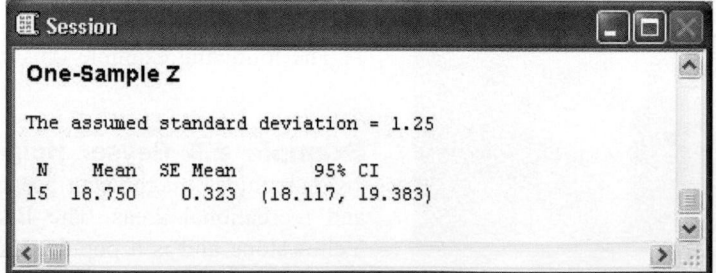

Figure 8.19 Confidence interval for the population mean.

Excel

There is no built-in function to automatically compute the endpoints of a confidence interval for a population mean when σ is known. Use the function `CONFIDENCE` to find the step ($z_{\alpha/2} \cdot \sigma/\sqrt{n}$) in each direction from the mean, and the function AVERAGE to find the sample mean, if necessary.

1. Use the command CONFIDENCE to find the step. The arguments are α, σ, and n.
2. Compute the left and right endpoints of the confidence interval. See Figure 8.20.

	A	B	C
1	0.6326	=CONFIDENCE(0.05,1.25,15)	The step
2	18.1174	=18.75-A1	Left endpoint
3	19.3826	=18.75+A1	Right endpoint

Figure 8.20 Calculations to construct the confidence interval.

SECTION 8.2 EXERCISES

Practice

8.12 Find each of the following critical values.

a. $z_{0.10}$ b. $z_{0.05}$ c. $z_{0.025}$ d. $z_{0.01}$
e. $z_{0.005}$ f. $z_{0.001}$ g. $z_{0.0005}$ h. $z_{0.0001}$

8.13 In each of the following problems, the sample mean, the sample size, the population standard deviation, and the confidence level are given. Assume the underlying population is normally distributed. Find the associated confidence interval for the population mean.

a. $\bar{x} = 15.6$, $n = 12$, $\sigma = 3.7$, 95%.
b. $\bar{x} = 6322$, $n = 17$, $\sigma = 225$, 90%.
c. $\bar{x} = -45.78$, $n = 9$, $\sigma = 12.35$, 80%.
d. $\bar{x} = 0.0795$, $n = 24$, $\sigma = 0.006$, 99%.
e. $\bar{x} = 37.68$, $n = 27$, $\sigma = 2.2$, 99.9%.

8.14 In each of the following problems, the sample mean, the sample size, the population standard deviation, and the confidence level are given. Find the associated confidence interval for the population mean.

a. $\bar{x} = 17.6$, $n = 32$, $\sigma = 10.27$, 95%.
b. $\bar{x} = 136.8$, $n = 45$, $\sigma = 25.44$, 99%.
c. $\bar{x} = 335.7$, $n = 65$, $\sigma = 125.3$, 90%.
d. $\bar{x} = -6.7$, $n = 52$, $\sigma = 2.25$, 98%.
e. $\bar{x} = 20.11$, $n = 37$, $\sigma = 1.76$, 99.99%.

8.15 Two statisticians are given the same data from an experiment. Each uses these data to construct a confidence interval for the true population mean μ. The resulting CIs are (8.55, 10.85) and (8.40, 11.0).
a. What is the value of the sample mean $\bar{x}$?
b. One CI has a confidence level of 95%, and the other has a confidence level of 99.9%. Match the confidence level with the confidence interval, and justify your answer.

8.16 In each of the following problems, the population standard deviation, the bound on the error of estimation, and the confidence level are given. Find a value for the sample size n necessary to satisfy these requirements.

a. $\sigma = 7.9$, $B = 2.5$, 95%.
b. $\sigma = 10.77$, $B = 5$, 99%.
c. $\sigma = 0.55$, $B = 0.001$, 98%.
d. $\sigma = 35.97$, $B = 3.5$, 95%.
e. $\sigma = 55$, $B = 2$, 99.9%.

Applications

8.17 Physical Sciences A large number of meteorites composed of iron, chondrite, or achondrite fall onto the Earth every day. However, most are very small when they hit the surface and are unreported. In a random sample of 45 meteorite falls and finds, the mean weight was 106.14 kg. Assume $\sigma = 35$ grams. Find a 95% confidence interval for the mean weight of all meteorite finds.[5]

8.18 Public Health and Nutrition After a ham is cured, it may be smoked to add flavor or to ensure that it lasts longer. Typical grocery-store hams are smoked for a short period of time, whereas gourmet hams are usually smoked for at least one month. A random sample of 36 grocery-store hams was obtained, and the length of the smoking time was recorded for each. The mean was $\bar{x} = 140$ hours. Assume $\sigma = 8$ hours.
a. Find a 99% confidence interval for the mean amount of time a grocery-store ham is smoked.
b. What assumptions did you make in order to construct the confidence interval in part (a)?

8.19 Biology and Environmental Science The production of cultivated edible mushrooms increased 1000% from 1965 to 1991. Mainland China is the top producer, and the United States grows approximately 16% of the worldwide total.[6] The most popular cultivated mushroom is *Agaricus bisporus*. A random sample of these mushrooms was obtained, and the diameter of each (in inches) was measured. The data are given on the data CD and book's web site. Assume $\sigma = 0.34$.

a. Find a 95% confidence interval for the mean diameter of this specialty mushroom.

b. Find a 99% confidence interval for the mean diameter of this specialty mushroom.

c. Explain why the interval in part (b) is wider.

8.20 Marketing and Consumer Behavior The owner of a small tailor shop keeps careful records of all alterations. Two common alterations to men's clothing include lengthening the inseam on pants and letting out a sports coat around the waist. A random sample of each type of alteration was obtained, and the summary statistics are given in the table below. Measurements are in inches.

Alteration	Sample size	Sample mean	Assumed σ
Pants inseam	33	0.74	0.22
Sports coat waist	42	1.05	0.37

a. Find a 95% confidence interval for the mean alteration of the inseams on men's pants.

b. How large a sample is necessary in order for the bound on the error of estimation to be 0.05 for the confidence interval in part (a)?

c. Find a 90% confidence interval for the mean alteration of the waists of sports coats.

d. How large a sample is necessary in order for the bound on the error of estimation to be 0.07 for the confidence interval in part (c)?

8.21 Physical Sciences An iceberg consists of frozen freshwater that has broken away from a glacier. Icebergs extend above and below the surface of the ocean and are usually white, blue, or green. The distribution of iceberg keel depths (the distance the iceberg extends below the water line) in the Antarctic is multimodal, with three distinct peaks: 140–200 m, 250–300 m, and 500–600 m.[7] Suppose a team of scientists aboard a research vessel operating in the Antarctic observed a random sample of 49 icebergs. The keel depth was carefully measured for each and the sample mean was $\bar{x} = 260$ meters. Assume $\sigma = 25$ meters.

a. Find a 99% confidence interval for the mean keel depth of icebergs in the Antarctic.

b. Some experts suggest that global warming has increased the number of icebergs with greater keel depth. Using the interval in part (a), is there any evidence to suggest that the mean depth of icebergs is more than 200 meters? Justify your answer.

8.22 Travel and Transportation Lighthouses are constructed to guide ships traveling in rocky waters and to allow sailing at night. There are many ways to describe the size of a lighthouse. However, the height is usually measured from the base of the tower to the top of the ventilator ball. A random sample of 18 lighthouses in France and England was obtained, and the height of each was recorded. The sample mean was $\bar{x} = 33.75$ meters. Assume the distribution of lighthouse heights is normal and $\sigma = 5.4$ meters.

a. Find a 95% confidence interval for the mean height of all lighthouses in France and England.

b. How large a sample is necessary to ensure that the width of the resulting 95% confidence interval is 2 meters?

c. Why is the confidence interval constructed in part (a) valid even though the sample size $n = 18$ is less than 30?

8.23 Biology and Environmental Science Koalas are not really bears but are related to wombats and kangaroos. These cuddly creatures number only 2000–8000 in the wild and grow to weigh 15–30 pounds.[8] A random sample of 35 koalas was obtained, and each was carefully weighed. The mean weight was $\bar{x} = 20.75$ pounds. Assume $\sigma = 3.05$ pounds.

a. Find a 99% confidence interval for the mean weight of all koalas.

b. Researchers believe that if the true mean weight is less than 23 pounds, these animals may be suffering from malnutrition. Is there any evidence to suggest that the koala population is suffering from malnutrition? Justify your answer.

8.24 Technology and Internet Many companies are increasingly concerned about computer security and employees using their computers for personal use. A random sample of 50 employees of Liberty Mutual Insurance Company was obtained, and each was asked whether they use the Internet for nonbusiness (personal) purposes during the day. The number of hours for each employee is given on the data CD and book's web site. Assume $\sigma = 0.5$ hours.

a. Find an 95% confidence interval for the mean number of hours all employees at Liberty Mutual use the Internet for personal reasons.

b. Using the confidence interval in part (a), is there any evidence to suggest that the mean time spent using the Internet for personal reasons is more than 1 hour? Justify your answer.

c. How large a sample is necessary in order for the bound on the error of estimation to be 0.1 hours?

d. Do you think that the underlying distribution of time spent on the Internet for personal use is normal? Explain your answer.

8.25 Sports and Leisure A random sample of professional wrestlers was obtained, and the annual salary (in dollars) for each was recorded. The summary statistics were $\bar{x} = 47,500$ and $n = 18$. Assume the distribution of annual salary is normal with $\sigma = 8,500$.

a. Find a 90% confidence interval for the true mean annual salary for all professional wrestlers.

b. How large a sample is necessary in order for the bound on the error of estimation to be 3000?

c. How large a sample is necessary in order for the bound on the error of estimation to be 1000?

8.26 Education and Child Development Due to a decrease in state funding, rising health care costs, and the cost of technology, the tuition at state colleges and universities has risen rapidly over the last decade. According to the College Board, the mean tuition and fees at state colleges for the 2008–2009 academic year was $6585.[9] Suppose a random sample of 12 New Jersey state colleges was obtained and the mean tuition and fees for 2008–2009 was $6,353. Assume the tuition distribution is normal and $\sigma = 225$.

a. Find a 99% confidence interval for the true mean tuition and fees at New Jersey state colleges and universities.

b. Suppose the governor of New Jersey believes that the majority of New Jersey residents cannot afford to attend a state college or university if the tuition is greater than $6200. Is there any evidence to suggest that the true mean tuition and fees for New Jersey state colleges is greater than $6200? Justify your answer.

8.27 Public Policy and Political Science Many chimney fires are caused by a buildup of creosote, a highly flammable material that forms in flues due to the condensation of certain gases. In an effort to promote fireplace and wood stove safety, a town in Vermont has started a new inspection program. Forty homes with wood stoves were selected at random, and the amount of creosote buildup 1 foot from the top of the chimney was carefully measured. The resulting sample mean thickness was $\bar{x} = 0.131$ inches. Assume $\sigma = 0.02$ inches.

a. Find a 95% confidence interval for the true mean thickness of creosote buildup 1 foot from the top of chimneys.

b. One-eighth of an inch of creosote buildup is considered safe. If there is evidence to suggest that the true mean thickness is greater than 1/8 inch, the town will launch an extensive safety education program. Using the interval constructed in part (a), should the town stress greater safety? Justify your answer.

8.28 Physical Sciences According to the National Weather Service, at any given moment of any day, there are approximately 2000 thunderstorms occurring worldwide. Many of these storms include lightning strikes. Sensitive electronic equipment is used to record the number of lightning strikes worldwide every day. Twelve days were selected at random, and the number of lightning strikes on each day was recorded. The sample mean was $\bar{x} = 8.6$ million. Assume the distribution of the number of lightning strikes per day is normal with $\sigma = 0.35$ million.

a. Find a 99% confidence interval for the mean number of lightning strikes per day worldwide.

b. Do you think the normality assumption in this problem is reasonable? Why or why not?

8.29 Biology and Environmental Science The Australian dragonfly, the world's fastest insect, has been clocked at 36 miles per hour over short distances.[10] A dragonfly's wings, two on each side, are the reason for this speed. In a recent study, 45 randomly selected dragonflies were captured, and the wingspan on each was measured. The sample mean was $\bar{x} = 6.6$ cm. Assume $\sigma = 0.5$ cm.

a. Find a 95% confidence interval for the true mean wingspan for all dragonflies.

b. Suppose $\sigma = 0.25$ cm. Find a 95% confidence interval for the true mean wingspan for all dragonflies.

c. Write a sentence to explain the meaning of each confidence interval in relationship to the true mean wingspan for all dragonflies.

Extended Applications

8.30 Fuel Consumption and Cars The United States is the third largest producer of oil, behind Saudi Arabia and the former Soviet Union. A random sample of total daily production was obtained for the United States and for Saudi Arabia (in millions of barrels). The data are given on the data CD and book's web site and are based on information in the BP Statistical Review of World Energy, June 2008. Assume $\sigma = 0.8$ for United States daily production and $\sigma = 0.6$ for Saudi Arabia.

a. Find a 95% confidence interval for the United States' daily oil production in millions of barrels.

b. Find a 95% confidence interval for Saudi Arabia's daily oil production in millions of barrels.

c. Using the confidence intervals in parts (a) and (b), is there any evidence to suggest that the mean daily oil production for the United States and Saudi Arabia are different?

8.31 Economics and Finance Representative Richard Baker has become concerned about the disparity in home prices in Louisiana's sixth district (Monroe). A random sample of homes sold within the past year in each of two parishes was obtained. The summary statistics are given in the table below. Prices are in dollars.

Parish	Sample size	Sample mean	Assumed σ
East Feliciana	30	125,200	5,750
Iberville	36	155,900	25,390

a. Find a 99% confidence interval for the mean selling price of all homes in East Feliciana Parish.

b. Find a 99% confidence interval for the mean selling price of all homes in Iberville Parish.

c. Using the confidence intervals in parts (a) and (b), is there any evidence to suggest that the mean selling price is different for the two parishes? Justify your answer.

8.32 Sports and Leisure There are three main types of exercises: range-of-motion (flexibility), strengthening, and endurance. A random sample of people who exercise regularly was obtained. The type of exercise and length (in minutes) of each workout was recorded. The table below summarizes the information obtained.

Exercise type	Sample size	Sample mean	Assumed σ
Range-of-motion	65	25.2	5.2
Strengthening	32	73.6	10.7
Endurance	40	82.2	12.5

a. Construct a 99% confidence interval for the mean workout length for each of the three types of exercises.

b. A group of exercise scientists claims that the length of a workout for range-of-motion exercises is the same as for the other two types. Based on the confidence intervals in part (a), is there any evidence to refute this claim?

8.33 Psychology and Human Behavior A random sample of male college athletes was obtained, and their coping skills levels were measured using an extensive psychological profile. Each athlete was also classified by sport. The table below summarizes the information obtained.

Sport	Sample size	Sample mean	Assumed σ
Football	35	65.77	14.07
Basketball	30	53.90	12.50
Hockey	32	68.45	10.25

a. Find a 95% confidence interval for the true mean coping skills level for all male athletes in each sport.

b. Use your answers to part (a) to determine whether there is any evidence to suggest that the mean coping skills level for male basketball players is different from that of male football players. Justify your answer.

c. How large a sample is necessary in order for the bound on the error of estimation for each confidence interval in part (a) to be 2?

8.34 Sports and Leisure Two competing ski slopes in Colorado advertise their powder base each day in an effort to attract more business. A random sample of the powder base depth was obtained for each ski resort on days during a recent winter. The depth (in inches) on each day is given on the data CD and book's web site. Assume $\sigma_B = 2.5$ and $\sigma_V = 2.7$.

a. Find a 99% confidence interval for the true mean powder depth at Breckenridge Ski Resort.

b. Find a 99% confidence interval for the true mean powder depth at Vail Ski Resort.

c. Use the results in parts (a) and (b) to determine whether there is any evidence to suggest a difference in the true mean powder depths at these two ski resorts.

d. What assumptions were necessary in order to construct the confidence intervals in parts (a) and (b)?

8.35 Public Health and Nutrition The nutritional value of every food product sold in the United States is listed on the package in terms of protein, fat, etc. A researcher randomly selected 50 one-ounce samples of various nuts and carefully measured the amount of protein (in grams) in each sample. The sample mean amount of protein and the assumed standard deviation for each type of nut are given in the table below.

Nut	Sample mean	Assumed σ
Cashew	5.17	0.40
Filbert	4.24	0.60
Pecan	2.60	0.95

(*Source*: USDA National Nutrient Database for Standard Reference.)

a. Find a 95% confidence interval for the true mean amount of protein in each type of nut. Based on these intervals, is there any evidence to suggest that the mean amount of protein is different for cashews and pecans? How about filberts and pecans?

b. Answer part (a) using a sample size for each nut of 18.

Challenge

8.36 One-Sided Confidence Intervals Given a random sample of size n from a population with mean μ, assume that the underlying population is normal or n is large and that the population standard deviation σ is known.

a. Manipulate the probability statement

$$P\left(\frac{\bar{X} - \mu}{\sigma/\sqrt{n}} > -z_\alpha\right) = 1 - \alpha$$

to find a *one-sided* $100(1 - \alpha)\%$ confidence interval for μ. That is, find an upper bound for the population mean μ.

b. Manipulate a similar probability statement to find a one-sided $100(1 - \alpha)\%$ confidence interval for μ bounded below.

c. First-class mail in the United States includes personal correspondence and all kinds of bills. Each piece of first-class mail must weigh less than 13 ounces. In a random sample of 25 first-class letters, the sample mean weight was $\bar{x} = 2.2$ ounces. Assume the population is normal and $\sigma = 0.75$. Find a 95% one-sided confidence interval bounded above for the population mean weight of first-class letters.

8.37 The "Best" CI There are actually *infinitely* many different confidence intervals for a population mean μ. For example, suppose in a random sample of size n from a population with mean μ, the underlying population is normal or n is large, and the population standard deviation σ is known. One could start with a probability statement of the form

$$P\left(-z_{3\alpha/4} < \frac{\bar{X} - \mu}{\sigma/\sqrt{n}} < z_{\alpha/4}\right) = 1 - \alpha$$

to find a $100(1 - \alpha)\%$ confidence interval for μ. Why is the traditional two-sided confidence interval (presented in this chapter) the *best* CI for a parameter?

APPLET **8.38 The Real Meaning of a 95% CI**

a. Use the Confidence Intervals applet to generate a random sample of size 50 from a normal population with mean μ. Record the percentage of CIs that capture the true population mean μ (Percent hit).

b. Press the Reset button, and repeat this process an additional 50 times.

c. Construct a histogram of the Percent hits. Describe the shape of the histogram and the center.

d. Press the Reset button. Generate a random sample of size 50 from a normal population with mean μ. Record the percentage of CIs that capture the true population mean.

e. Press the Sample 50 button again. Record the new Percent hit. Continue in this manner until the Total is 2000.

f. Create a scatter plot of Percent hit versus Total. Describe the pattern in this plot.

8.3 A CONFIDENCE INTERVAL FOR A POPULATION MEAN WHEN σ IS UNKNOWN

In Section 8.2, a confidence interval for a population mean μ was presented. However, Equation 8.2, which is based on a standard normal, or Z, distribution, is valid only if σ is known (and either the underlying population is normal or the sample size is large). It is unrealistic to assume that the population standard deviation is known. A more practical and useful approach to constructing a confidence interval for μ is presented in this section.

Recall, if σ is known and either the underlying population is normal or the sample size is large, then the expression

$$Z = \frac{\overline{X} - \mu}{\sigma/\sqrt{n}}$$

has a standard normal distribution. This expression was used to derive the confidence interval in Section 8.2. Note, only one component, $\overline{X}$, in this expression contributes to the variability.

If σ is unknown, a similar *standardization* is used,

$$T = \frac{\overline{X} - \mu}{S/\sqrt{n}}. \tag{8.4}$$

However, the distribution of this random variable is not normal. It is reasonable to believe the distribution of T is centered at 0, but there is more variability in this expression, with contributions from two sources, $\overline{X}$ and S. The most common confidence interval for a population mean μ is based on Equation 8.4 and the *t distribution* introduced below.

The t distribution is closely related to the normal distribution. While most continuous random variables are defined by a probability density function, it is only important here to understand the properties of a t distribution.

The symbol T is used in notation to represent both the sample total (in Chapter 7) and a random variable with a t distribution. The context in which the notation is used implies the appropriate concept.

Okay, here is the probability density function for a t random variable:

$$f(x) = \frac{1}{\sqrt{\pi\nu}} \frac{\Gamma\left(\frac{\nu+1}{2}\right)}{\Gamma\left(\frac{\nu}{2}\right)} \left(1 + \frac{x^2}{\nu}\right)^{-\frac{\nu+1}{2}}$$

$-\infty < x < \infty, \nu = \text{df}.$

The t distribution was derived by William Gosset in 1908. Working for Guinness Breweries, he published his result using the pseudonym *Student*. For this reason, the distribution is still often called "Student's t distribution."

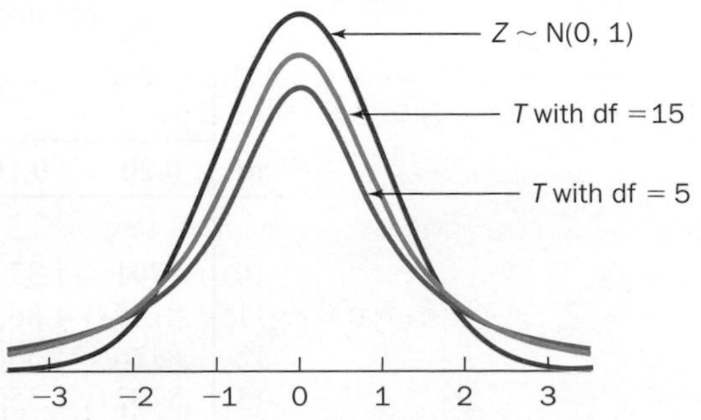

PROPERTIES OF A t DISTRIBUTION

1. A t distribution is completely determined (characterized) by only one parameter, df, called the number of degrees of freedom. df must be a positive integer, df = 1, 2, 3, 4, . . ., and there is a different t distribution corresponding to each value of df.
2. If T (a random variable) has a t distribution with df degrees of freedom,

$$\mu_T = 0 \quad \text{and} \quad \sigma_T^2 = \frac{\text{df}}{\text{df} - 2}, \quad (\text{df} \geq 3). \tag{8.5}$$

3. The density curve for every t distribution is bell-shaped and centered at 0, but more spread out than the density curve for a standard normal random variable, Z. As df increases, the density curve for T becomes more compact and closer to the density curve for Z. See Figure 8.21.

Figure 8.21 A comparison of density curves for two t distributions with the density curve for a standard normal random variable.

In order to construct a confidence interval for μ based on a t distribution, the following definition for a *t* **critical value** is necessary

DEFINITION

t_α is a **critical value** related to a ***t* distribution** (a ***t* critical value**) with df degrees of freedom. If T has a t distribution with df degrees of freedom, then $P(T \geq t_\alpha) = \alpha$.

ILLUMINATING THE CONCEPTS

Remember, the α in t_α is simply a *placeholder*. Any symbol could be used here to represent a right-tail probability.
$P(T \geq t_{@}) = \alpha$

1. For any ***t* distribution**, t_α is simply a t value (a value on the measurement axis) such that there is α of the area (probability) to the right of t_α. $-t_\alpha$ is the negative **critical value**. Because the t distribution is symmetric, $P(T \leq -t_\alpha) = P(T \geq t_\alpha) = \alpha$. See Figure 8.22.
2. Remember, critical values are always defined in terms of right-tail probability.

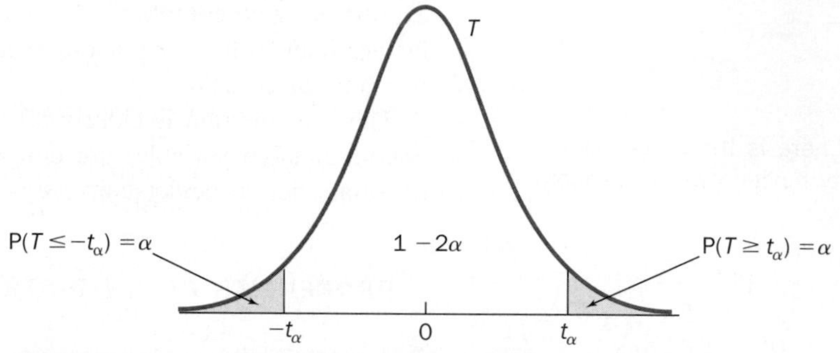

Figure 8.22 An illustration of the definition of a t critical value. ■

Table V (see the Appendix) presents selected critical values associated with various t distributions. Right-tail probabilities are in the top row and degrees of freedom are listed in the left column. In the body of the table, t_α is at the intersection of the α column and the df row. The following example illustrates the use of this table for finding critical values associated with a t distribution.

Example 8.6 Critical Value Look-Ups Find each critical value: **a.** $t_{0.05}$ with df = 12, **b.** $t_{0.01}$ with df = 21.

SOLUTION

a. Using the following portion of Table V, find the intersection of the $\alpha = 0.05$ column and the df = 12 row.

	α								
df	**0.20**	**0.10**	**0.05**	**0.025**	**0.01**	**0.005**	**0.001**	**0.0005**	**0.0001**
⋮	⋮	⋮	⋮	⋮	⋮	⋮	⋮	⋮	⋮
10	.8791	1.3722	1.8125	2.2281	2.7638	3.1693	4.1437	4.5869	5.6938
11	.8755	1.3634	1.7959	2.2010	2.7181	3.1058	4.0247	4.4370	5.4528
12	.8726	1.3562	1.7823	2.1788	2.6810	3.0545	3.9296	4.3178	5.2633
13	.8702	1.3502	1.7709	2.1604	2.6503	3.0123	3.8520	4.2208	5.1106
14	.8681	1.3450	1.7613	2.1448	2.6245	2.9768	3.7874	4.1405	4.9850
⋮	⋮	⋮	⋮	⋮	⋮	⋮	⋮	⋮	⋮

Therefore, $t_{0.05} = 1.7823$, and if T has a t distribution with df = 12, then $P(T \geq 1.7823) = 0.05$, as illustrated in Figure 8.23.

b. Similarly, using Table V, find the intersection of the $\alpha = 0.01$ column and the df = 21 row.

$t_{0.01} = 2.5176$ and if T has a t distribution with df = 21, then $P(T \geq 2.5176) = 0.01$, as illustrated in Figure 8.24.

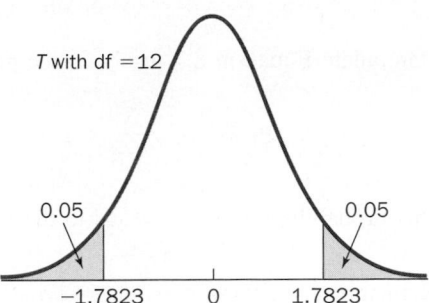

Figure 8.23 Visualization of $t_{0.05} = 1.7823$.

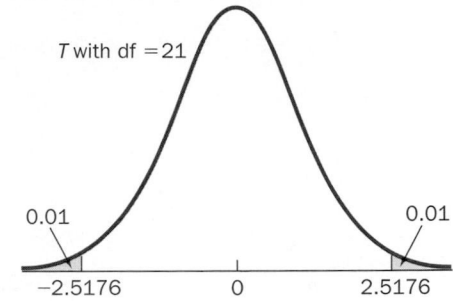

Figure 8.24 Visualization of $t_{0.01} = 2.5176$.

Figure 8.25 shows a technology solution.

Figure 8.25 Use the Inverse Cumulative Distribution Function to find critical values associated with a t distribution.

```
Session                                    [_][□][X]

Inverse Cumulative Distribution Function

Student's t distribution with 21 DF

P( X <= x )        x
        0.99    2.51765
```

 ILLUMINATING THE CONCEPTS

1. Table V is *very* limited. However, a calculator or computer can find almost any critical value needed.

2. In Table V, as df increases, the t critical values approach the corresponding Z critical values. This numerical observation is analogous to the graphical comparison in Figure 8.21. The density curve for T approaches the density curve for Z as df increases. ■

Even for df = 40, $t_\alpha \approx z_\alpha$.

◀ A confidence interval for a population mean when σ is unknown is based on the following theorem.

8.1 THEOREM

Let $\overline{X}$ be the mean of a random sample of size n from a normal distribution with mean μ. The random variable

$$T = \frac{\overline{X} - \mu}{S/\sqrt{n}}$$

(8.6)

Even though n is in the denominator, $n - 1$ degrees of freedom is correct!

has a t distribution with $n - 1$ degrees of freedom.

This theorem is used to construct a confidence interval for μ. As in Section 8.2, start in the appropriate t world. Find a symmetric interval about 0 such that the probability T lies in this interval is $1 - \alpha$.

$$P(-t_{\alpha/2} < T < t_{\alpha/2}) = 1 - \alpha$$

$$P\left(-t_{\alpha/2} < \frac{\overline{X} - \mu}{S/\sqrt{n}} < t_{\alpha/2}\right) = 1 - \alpha \qquad (8.7)$$

Manipulate Equation 8.7 to obtain the probability statement

$$P\left(\overline{X} - t_{\alpha/2}\frac{S}{\sqrt{n}} < \mu < \overline{X} + t_{\alpha/2}\frac{S}{\sqrt{n}}\right) = 1 - \alpha. \qquad (8.8)$$

This probability statement leads to the following general result. ▶

HOW TO FIND A 100(1 − α)% CONFIDENCE INTERVAL FOR A POPULATION MEAN WHEN σ IS UNKNOWN

Given a random sample of size n with sample standard deviation s from a population with mean μ, if the underlying population distribution is normal, a 100(1 − α)% confidence interval for μ has as endpoints the values

$$\overline{x} \pm t_{\alpha/2}\frac{s}{\sqrt{n}} \qquad (8.9)$$

where $t_{\alpha/2}$ is based on $n - 1$ degrees of freedom.

ILLUMINATING THE CONCEPTS

1. Equation 8.9 can be used with *any* sample size n (≥ 2) and produces an *exact* (not an approximate) confidence interval for μ.

2. This confidence interval for μ (Equation 8.9) is valid *only* if the underlying population is normal. ■

(Emilio Ereza/AgeFotostock)

Example 8.7 Orange Blossom Perfume Distillation is a process for separating and collecting substances according to their reaction to heat. When heat is applied to a mixture, the substance that evaporates and is collected as it cools is the distillate. The unevaporated portion of the mixture is the residue. Oil obtained from orange blossoms through distillation is used in perfume. Suppose the oil yield is normally distributed. In a random sample of 11 distillations, the sample mean oil yield was $\overline{x} = 980.2$ grams with standard deviation $s = 27.6$ grams. Find a 95% confidence interval for the true mean oil yield per batch.

SOLUTION

STEP 1 $\overline{x} = 980.2, \quad s = 27.6, \quad n = 11.$ Given.

$1 - \alpha = 0.95 \longrightarrow \alpha = 0.05 \longrightarrow \alpha/2 = 0.025$ Find $\alpha/2$.

$t_{\alpha/2} = t_{0.025} = 2.2281$ Use Table V with df = 10.

Solution Trail

KEYWORDS

- 95% CI for the true mean.
- Normally distributed.
- $s = 27.6$.

TRANSLATION

- 95% CI for μ.
- Underlying population is normal.
- σ is unknown.

CONCEPTS

A $100(1 - \alpha)\%$ CI for a population mean when σ is unknown.

VISION

We need a 95% CI for μ. The population is normal, and σ is unknown. Find the appropriate t critical value, and use Equation 8.9.

STEP 2 Use Equation 8.9.

$$\bar{x} \pm t_{\alpha/2} \frac{s}{\sqrt{n}} \qquad \text{Equation 8.9.}$$

$$= \bar{x} \pm t_{0.025} \frac{s}{\sqrt{n}} \qquad \text{Use the value of } \alpha.$$

$$= 980.2 \pm (2.2281) \frac{27.6}{\sqrt{11}} \qquad \text{Use summary statistics and values for } s \text{ and } t_{0.025}.$$

$$= 980.2 \pm 18.54 \qquad \text{Simplify.}$$

$$= (961.66, 998.74) \qquad \text{Compute endpoints.}$$

$(961.66, 998.74)$ is a 95% confidence interval for the true mean oil yield (in grams) through distillation in a typical batch.

Figures 8.26 and 8.27 together show a technology solution.

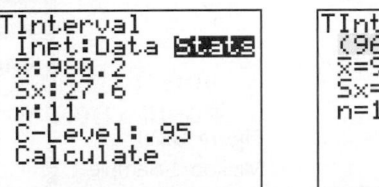

Figure 8.26 TInterval input screen.

Figure 8.27 Resulting confidence interval.

The following example illustrates the same process using actual data rather than summary statistics. The resulting confidence interval is used to draw a conclusion about the value of the population mean.

Example 8.8 Impact Force A SPIT 326 electro Pneumatic hammer has an advertised impact force of 2.2 joules.[11] In a random sample of 23 hammers, the impact force for each tool was carefully measured (in joules) and is given in the following table.

2.16	1.69	2.30	2.08	1.72	2.17	2.25	2.06
2.00	2.29	2.15	2.49	2.12	2.17	1.93	2.39
2.22	2.26	2.14	1.92	2.06	2.09	2.08	

a. Find a 99% confidence interval for the true mean impact force for this type of pneumatic hammer. Assume the underlying population is normal.
b. Using the confidence interval constructed in part (a), is there any evidence to suggest that the true mean impact force is different from 2.2 joules as advertised? Justify your answer.

SOLUTION

a. Solution Trail

KEYWORDS

- 99% CI for the true mean.
- Underlying population is normal.

TRANSLATION

- 99% CI for μ.
- σ is unknown.

CONCEPTS

A $100(1 - \alpha)\%$ CI for a population mean when σ is unknown.

VISION

Construct a 99% CI for μ. The underlying distribution is assumed to be normal, and σ is unknown. Use Equation 8.9.

a. $n = 23$ Sample size.

$$\bar{x} = \frac{1}{23}(2.16 + \cdots + 2.08) = 2.12 \qquad \text{Compute the sample mean.}$$

$$s^2 = \frac{1}{23}[(2.16 - 2.12)^2 + \cdots + (2.08 - 2.12)^2] = 0.0351$$

$$s = \sqrt{0.0351} = 0.1873 \qquad \text{Compute the sample standard deviation.}$$

$$1 - \alpha = 0.99 \longrightarrow \alpha = 0.01 \longrightarrow \alpha/2 = 0.005 \qquad \text{Find } \alpha/2.$$

$$t_{\alpha/2} = t_{0.005} = 2.8187 \qquad \text{Use Table V with df} = 22.$$

Use Equation 8.9.

$$\bar{x} \pm t_{\alpha/2}\frac{s}{\sqrt{n}}$$

Equation 8.9.

$$= \bar{x} \pm t_{0.005}\frac{s}{\sqrt{n}}$$

Use the value of α.

$$= 2.12 \pm (2.8187)\frac{0.1873}{\sqrt{23}}$$

Use summary statistics and value for $t_{0.005}$.

$$= 2.12 \pm 0.110$$

Simplify.

$$= (2.01,\ 2.23)$$

Compute endpoints.

$(2.01, 2.23)$ is a 99% confidence interval for the true mean impact force (in joules) of this type of pneumatic hammer.

Figure 8.28 shows a technology solution.

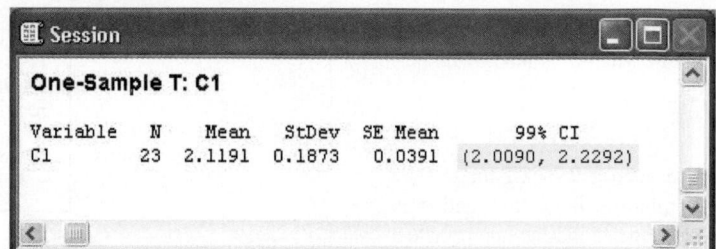

Figure 8.28
Minitab 1-Sample t confidence interval.

One-Sample T: C1

Variable	N	Mean	StDev	SE Mean	99% CI
C1	23	2.1191	0.1873	0.0391	(2.0090, 2.2292)

b. Claim: $\mu = 2.2$.

Experiment: $\bar{x} = 2.12$.

Likelihood: The likelihood is expressed as a 99% confidence interval, an interval of likely values for μ: $(2.01, 2.23)$, from part (a).

Conclusion: The advertised impact force, 2.2 joules, is included in this confidence interval. There is no evidence to suggest that μ is different from 2.2. ●

💡 **ILLUMINATING THE CONCEPTS**

1. Technology solutions can find the appropriate t critical value for any sample size and any confidence level. If you must use Table V to find a critical value for a df and/or a value of α not listed, use linear interpolation.

   Recall, interpolation is a method of approximation. It is often used to estimate a value at a position between two given values in a table. Linear interpolation assumes the two known values lie on a straight line; it was discussed in Chapter 6.

2. Sample size calculation is more complicated in this case. σ is unknown and the critical value, $t_{\alpha/2}$, depends on n. Equation 8.3 is often used with an estimate for σ and the assumption that $z_{\alpha/2} \approx t_{\alpha/2}$. ■

TECHNOLOGY CORNER

Procedure: Construct a confidence interval for a population mean when σ is unknown.
Reconsider: Example 8.8, page 347, solution, and interpretation.

TI-84 Plus

Use the built-in function `TInterval`. Input is either summary statistics or a list containing data.
1. Enter the data into list L1.
2. Select STATS; TESTS; TInterval.

---

b. Solution Trail

KEYWORDS

Is there any evidence?

↓

TRANSLATION

Inference procedure.

↓

CONCEPTS

A $100(1 - \alpha)$% CI for a population mean when σ is unknown.

↓

VISION

The CI constructed in part a is an interval in which we are 99% confident the true value of μ lies. If 2.2 lies in this interval, there is no evidence to suggest μ is different from 2.2. Use the four-step inference procedure.

3. Highlight `Data`. Enter the name of the list `L1`, set the frequency to 1, and enter the confidence level. See Figure 8.29.

4. Highlight `Calculate` and press $\boxed{\text{Enter}}$. The confidence interval and summary statistics are displayed on the Home Screen. See Figure 8.30.

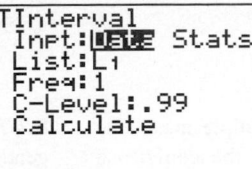

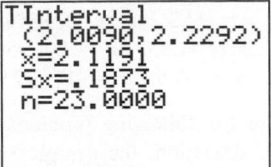

Figure 8.29
`TInterval` input screen.

Figure 8.30 Resulting confidence interval.

Minitab

Use the built-in function 1-Sample t. Input is either summary statistics or a column containing data.

1. Enter the data into column C1.

2. Select Stat; Basic Statistics; 1-Sample t.

3. Choose Samples in columns and enter C1 in the entry window.

4. Select the Options button and enter the confidence level. The results are displayed in a session window. See Figure 8.28, page 348.

Excel

There is no built-in function to compute the endpoints of a confidence interval for a population mean when σ is unknown. Use the function TINV to find the critical value and the functions AVERAGE and STDEV if necessary.

1. Enter the data into column A.

2. Compute the sample mean and the sample standard deviation.

3. Find the critical value using the function TINV. The arguments are α and $n - 1$.

4. Compute the left and right endpoints of the confidence interval. See Figure 8.31.

B	C	D
2.1191	= AVERAGE(A1:A23)	Sample mean
0.1873	= STDEV(A1:A23)	Sample standard deviation
2.8188	= TINV(0.01,22)	Critical value
2.0090	= B1-B3*B2/SQRT(23)	Left endpoint
2.2292	= B1+B3*B2/SQRT(23)	Right endpoint

Figure 8.31 Calculations to construct the confidence interval.

SECTION 8.3 EXERCISES

Practice

8.39 Find each of the following critical values.
a. $t_{0.10}$ with df = 5   **b.** $t_{0.20}$ with df = 24
c. $t_{0.005}$ with df = 19   **d.** $t_{0.025}$ with df = 7
e. $t_{0.005}$ with df = 15   **f.** $t_{0.001}$ with df = 6
g. $t_{0.0005}$ with df = 23 **h.** $t_{0.0001}$ with df = 3

8.40 In each of the following problems, the sample size and the confidence level are given. Assume σ is unknown. Find the appropriate t critical value for use in constructing a confidence interval for the population mean.

a. $n = 15$, 95%   **b.** $n = 21$, 98%
c. $n = 31$, 99%   **d.** $n = 17$, 99.9%
e. $n = 12$, 99.99% **f.** $n = 4$, 98%

8.41 Use Table V and linear interpolation (if necessary) to approximate each of the following critical values. Verify each approximation using technology.
a. $t_{0.15}$ with df = 10   **b.** $t_{0.07}$ with df = 23
c. $t_{0.0025}$ with df = 20 **d.** $t_{0.01}$ with df = 35
e. $t_{0.005}$ with df = 42   **f.** $t_{0.025}$ with df = 75
g. $t_{0.02}$ with df = 45   **h.** $t_{0.003}$ with df = 52

8.42 In each of the following problems, the sample mean, the sample standard deviation, the sample size, and the confidence level are given. Assume the underlying population is normally distributed. Find the associated confidence interval for the population mean.

a. $\bar{x} = 211.2$, $s = 44.37$, $n = 27$, 95%.
b. $\bar{x} = 74.42$, $s = 31.8$, $n = 10$, 98%.
c. $\bar{x} = 138.9$, $s = 22.3$, $n = 28$, 99%.
d. $\bar{x} = -28.3$, $s = 41.33$, $n = 20$, 95%.
e. $\bar{x} = 1014.5$, $s = 67.9$, $n = 17$, 99.9%.

8.43 In each of the following problems, the sample mean, the sample standard deviation, the sample size, and the confidence level are given. Assume the underlying population is normally distributed. Find the associated confidence interval for the population mean.

a. $\bar{x} = 0.234$, $s = 0.081$, $n = 16$, 95%.
b. $\bar{x} = 259.6$, $s = 76.9$, $n = 26$, 99%.
c. $\bar{x} = 22.85$, $s = 7.19$, $n = 27$, 99%.
d. $\bar{x} = 380.9$, $s = 28.4$, $n = 21$, 95%.
e. $\bar{x} = 88.1$, $s = 17.45$, $n = 19$, 99.9%.

8.44 In each of the following problems, put the values in order from smallest to largest. (*Note:* There is no need to use a table or technology here. Use your knowledge of the t distribution and the Z distribution.)

a. $t_{0.01}$ df = 5, $t_{0.01}$ df = 27, $z_{0.01}$, $t_{0.01}$ df = 17.
b. $t_{0.025}$ df = 13, $t_{0.025}$ df = 11, $t_{0.025}$ df = 45, $z_{0.025}$.
c. $t_{0.05}$ df = 15, $t_{0.001}$ df = 15, $t_{0.02}$ df = 15, $t_{0.025}$ df = 15.
d. $t_{0.0001}$ df = 21, $t_{0.1}$ df = 21, $t_{0.005}$ df = 21, $t_{0.05}$ df = 21.
e. $t_{0.10}$ df = 6, $t_{0.001}$ df = 26, $t_{0.05}$ df = 17, $z_{0.0001}$.

Applications

8.45 Manufacturing and Product Development During the manufacture of certain commercial windows and doors, hot steel ingots are passed through a rolling mill and flattened to a prescribed thickness. The machinery is set to produce a steel section 0.25 inches thick. Fourteen steel sections were selected at random and the thickness of each was recorded. The data are given in the table that follows.

0.213	0.298	0.236	0.324	0.254	0.271	0.204
0.252	0.307	0.297	0.301	0.291	0.222	0.246

Assume the underlying distribution of section thickness is normal.
a. Find a 99% confidence interval for the true mean steel section thickness.
b. As the rollers erode, the machine begins to produce steel sections that are too thick. Using your answer to part (a), is there any evidence to suggest that the true mean steel section thickness is more than 0.25 inches? Justify your answer.
c. Use the methods described in Section 6.3 to check for any evidence of nonnormality.

8.46 Manufacturing and Product Development A typical washing machine has several different cycles, including soak, wash, and rinse. The energy consumption of a washing machine is linked to the length of each cycle. A random sample of 21 washing machines was obtained, and the length (in minutes) of each wash cycle was recorded. The summary statistics are $\bar{x} = 37.8$

and $s = 5.9$. Assume the underlying distribution of main wash cycle times is normal.
a. Find a 90% confidence interval for the true mean wash cycle time.
b. Interpret the confidence interval found in part (a).
c. Suppose a 95% confidence interval for the true mean wash cycle time is constructed. Would this interval be smaller or larger than the interval in part (a)? Justify your answer.

8.47 Biology and Environmental Science Manatees are very gentle, slow-moving creatures found in coastal waterways. They spend most of their time eating vegetation, resting, and swimming. Conservationists are concerned that pollution has decreased the amount of vegetation available for manatees. In a recent research study, 12 randomly selected adult manatees were captured, carefully weighed, and released. Each weight (in pounds) is given in the table that follows.[12]

956	1012	954	988	973	1048
1075	1064	856	1026	1031	1064

a. Assuming the underlying distribution of manatee weights is normal, find a 95% confidence interval for the true mean weight of adult manatees.
b. In a similar study 10 years ago, researchers concluded that the true mean weight of adult manatees was approximately 1000 pounds. Using your answer to part (a), is there any evidence to suggest that the true mean weight of adult manatees is less than 1000 pounds now? Justify your answer.

8.48 Physical Sciences The Earth is structured in layers: crust, mantle, and core. A recent study was conducted to estimate the mean depth of the upper mantle in a specific farming region in California. Twenty-six sample sites were selected at random, and the depth of the upper mantle was measured using changes in seismic velocity and density. The summary statistics are $\bar{x} = 127.5$ km and $s = 21.3$ km. Suppose the depth of the upper mantle is normally distributed. Find a 95% confidence interval for the true mean depth of the upper mantle in this farming region, and interpret your result.

8.49 Fuel Consumption and Cars In many areas, newspaper carriers deliver morning papers using their automobiles because the routes are too long to walk. In a random sample of 28 carriers who use their automobiles, the sample mean route length was $\bar{x} = 16.7$ miles with $s = 3.4$.
a. If the distribution of route lengths is normal, find a 95% confidence interval for the true mean route length of newspaper carriers who use their automobiles.
b. If the mean length of the routes is over 20 miles, the circulation department becomes concerned that papers will not be delivered by 7:00 A.M. Using your answer to part (a), is there any evidence to suggest that the true mean route length is over 20 miles? Justify your answer.

8.50 Sports and Leisure According to the International Federation of Competitive Eating, Joey Chestnut, from San Jose, CA, is the number one ranked eater in the world. On July 4, 2008, he won the Nathans Famous 4th of July International Hot Dog-Eating contest by eating 64 hot dogs and buns. The hot dogs are carefully prepared and weighed by official judges. The weights (in ounces) of 23 randomly selected hot dogs are given in the table below.

2.98	2.89	3.09	3.12	3.09	3.07	3.08	3.01
3.16	2.99	2.89	3.05	2.92	3.02	2.89	3.00
3.07	3.20	3.01	2.92	2.92	2.90	3.09	

a. Use the methods described in Section 6.3 to determine whether there is any evidence to suggest that the hot dog weight data are from a nonnormal distribution.
b. Construct a 99% confidence interval for the true mean hot dog weight.
c. Using your answer to part (b), is there any evidence to suggest that the true mean hot dog weight is different from 3 ounces? Justify your answer.

8.51 Technology and Internet A computer supply store sells a wide variety of generic replacement ink cartridges for printers. A consumer group is concerned that the cartridges may not contain the specified amount of ink (30 ml). A random sample of 17 black replacement cartridges was obtained, and the amount of ink (in ml) in each is given in the table below.

30.27	29.70	29.35	29.08	29.74	29.26	29.50
29.12	29.68	28.54	30.01	29.87	30.61	29.80
29.33	29.21	28.84				

Assume the underlying distribution of ink amount is normal.
a. Find a 95% confidence interval for the true mean amount of ink in each black cartridge.
b. Is there any evidence to suggest that the cartridges are underfilled? Justify your answer.

8.52 Travel and Transportation Some municipal managers complain that a city bus route is the most difficult service to maintain. Traffic jams and roadwork are often the cause of unreliable service. In an effort to analyze current city bus transportation in Atlanta, a random sample of routes and stops was obtained. The number of minutes the bus was late (compared with the posted route times) was recorded. The resulting data are given in the table that follows. A value of 0.00 means the bus arrived on time.

10.29	0.00	0.00	9.96	2.10	0.00	9.52
1.83	2.37	1.47	0.00	7.98	5.19	0.40
3.21	0.00	6.91	0.00	5.02		

a. Assume the underlying distribution of late times is normal. Find a 90% confidence interval for the true mean number of minutes a city bus in Atlanta is late.
b. Use the methods described in Section 6.3 to determine whether there is there any evidence to suggest that the distribution of late times is nonnormal.
c. If city buses in Atlanta consistently arrive more than 5 minutes late, the mayor receives a large number of phone calls from irate citizens. Using your answer to part (a), is there any evidence to suggest that the mayor will be receiving nasty phone calls? Justify your answer.

8.53 Psychology and Human Behavior According to the U.S. Fire Administration, approximately 25,000 fires are caused by fireworks each year in the United States. Despite numerous public warnings against the use of fireworks, the home property damage due to these fires is enormous. In a random sample of 25 fires due to fireworks, the resulting mean property damage (in dollars) was $\bar{x} = 860.75$ with a standard deviation of $s = 350.50$. Assume the underlying distribution of property damage due to these fires is normal.
a. Find a 99% confidence interval for the true mean property damage due to a fire caused by fireworks.
b. Many homeowner policies include a $500 deductible for all claims. Is there any evidence to suggest that the true mean property damage due to a fire caused by fireworks is greater than $500? Justify your answer.

8.54 Sports and Leisure The various areas of a golf course have grass cut to different lengths. For example, the greens are usually cut to the shortest length while the primary rough is cut to the greatest length. For each area classification, a random sample of golf courses in the United States was obtained. The length (in inches) of the grass in the corresponding area was carefully measured, and the resulting data are given in the table that follows.

Course area	Sample size	Sample mean	Sample standard deviation
Tees	15	0.380	0.05
Fairways	18	0.485	0.07
Greens	22	0.115	0.02
Primary rough	15	5.120	1.25
Intermediate rough	26	1.320	0.35

a. Assuming normality, find a 95% confidence interval for the true mean length of the grass for each of the five areas of a golf course.
b. The sample size for tees and primary rough is the same. Why is the 95% confidence interval for the primary rough mean length wider?
c. The Professional Golfers' Association specifies that fairways should be cut to half an inch. Is there any evidence to suggest that the true mean length of grass in fairways at U.S. courses is different from 0.5 inch? Justify your answer.

8.55 Biology and Environmental Science On June 23, 2003, a huge, unidentified sea creature was found washed ashore on a beach in Chile. The animal was described as a 40-foot-long mass of rotting gray flesh that scientists estimated to weigh about 13 tons.[13] Some scientists believed that the creature was a giant octopus, despite the absence of obvious tentacles. The largest octopi are found along the northern Pacific coast of the United States. A team of researchers recently obtained a random sample of octopi living in the northern Pacific Ocean. The length (in feet) of each specimen is given on the data CD and book's web site. Assuming the distribution of octopus lengths is normal, find a 99% confidence interval for the true mean length of octopi living in the northern Pacific Ocean.

8.56 Economics and Finance Along with gold and silver, platinum is considered a precious metal and is traded on the commodities market. A random sample of the price of platinum (in dollars per troy ounce) was obtained from jewelers around the world. The resulting summary statistics were $\bar{x} = 664.50$, $s = 5.25$, and $n = 55$. Assume the underlying distribution of the price is normal.

a. Find a 95% confidence interval for the current true mean price per troy ounce of platinum.

b. One month ago, a brokerage firm reported the true mean price of platinum to be $660.79 per troy ounce, with a recommendation to buy. Is there any evidence to suggest that the true mean price has increased? Justify your answer.

Extended Applications

8.57 Psychology and Human Behavior The ambient temperature in which humans are comfortable varies with culture, activity, metabolic rate, psychological state, environment, and season. For most people in the United States, the *comfort zone* is 68 to 78°F. During a recent winter, a random sample of homeowners was selected from two different parts of the country. The thermostat temperature setting (in °F) was recorded for each home, and the summary statistics are given in the following table.

Region	Sample size	Sample mean	Sample standard deviation
New England	14	70.2	2.75
South	11	72.1	1.55

a. Find a 95% confidence interval for the true mean thermostat setting for New England homeowners during winter.

b. Find a 95% confidence interval for the true mean thermostat setting for Southern homeowners during winter.

c. What assumption(s) did you make in constructing these two confidence intervals?

d. Using your answers to parts (a) and (b), is there any evidence to suggest that the New England mean thermostat setting is different from the Southern thermostat setting during winter? Justify your answer.

8.58 Public Policy and Political Science Juvenile courts in all states maintain careful records of cases, including demographics, charges, and dispositions. One variable of interest for repeat offenders is the number of days since the last offense (or arrest). A random sample of repeat offenders appearing in juvenile court was obtained for three states. The number of days since the last offense was recorded for each, and the summary statistics are given in the following table.

State	Sample size	Sample mean	Sample standard deviation
Ohio	17	180.6	37.8
California	29	162.7	25.2
Massachusetts	22	115.3	17.6

a. Assume the underlying distribution of days since the last offense is normal for each state. Find a 99% confidence interval for the true mean number of days since the last offense for repeat offenders in each state.

b. Using your answers to part (a), is there any evidence to suggest that the true mean number of days since the last offense in Ohio and California is different? How about California and Massachusetts? Justify each answer.

8.59 Medicine and Clinical Studies Arrhythmia, or an irregular heart beat, may be caused by heart disease or by environmental factors such as stress, caffeine, tobacco, or even cold medicine. One type of arrhythmia is atrial flutter, in which the heart beats very fast, at over 250 beats per minute. In a recent research study, randomly selected patients identified to have atrial flutter were carefully monitored. The number of beats per minute for the most recent flutter for each patient, by gender, is given on the data CD and book's web site. Assume the underlying distribution of beats per minute in atrial flutter patients is normal.

a. Find a 98% confidence interval for the true mean beats per minute in male atrial flutter patients.

b. Find a 98% confidence interval for the true mean beats per minute in female atrial flutter patients.

c. Using your answers to parts (a) and (b), is there any evidence to suggest that the true mean beats per minute in atrial flutter patients is different for men and women? Justify your answer.

d. Do you think the normality assumption is reasonable in this case? Why or why not?

8.60 Sports and Leisure The Wimbledon Tennis Tournament, where women players curtsy to the Queen and fans eat strawberries and cream, finishes in the first week of July. Each match is played on a grass court, as opposed to a clay or hard court. Some people believe that the grass court is more challenging and speeds up the game. Random samples of match lengths (in minutes) from Wimbledon and from the U.S. Clay Court Championships were obtained. The data are summarized in the following table.

Court type	Sample size	Sample mean	Sample standard deviation
Grass	18	65.7	25.3
Clay	12	83.2	35.8

Assume the underlying distribution of match lengths is normal on grass and on clay.

a. Find a 99% confidence interval for the true mean match length for grass courts at Wimbledon.

b. Find a 99% confidence interval for the true mean match length for clay courts at the U.S. Clay Court Championships.

c. Is there any evidence to suggest that the mean match time for grass courts and clay courts is different? Justify your answer.

8.61 Public Health and Nutrition During a medical emergency, people who dial 911 expect an ambulance to arrive quickly and personnel to provide vital care. Health insurance companies believe there is a marked difference in the response time for rural areas versus cities due to differences in coverage area and number of qualified paramedics. Random samples of ambulance response times (in minutes) were obtained for rural areas and cities. The data are given on the data CD and book's web site.

a. Assuming normality, find a 99% confidence interval for the true mean response time for rural-area ambulances.

b. Assuming normality, find a 99% confidence interval for the true mean response time for city ambulances.

c. Use the methods described in Section 6.3 to determine whether there is any evidence to suggest that either response time distribution is nonnormal.

d. Is there any evidence to suggest that the mean response time is different for rural areas and cities? Justify your answer.

8.62 Medicine and Clinical Studies The rising cost of malpractice insurance has caused many physicians to leave the field of medicine. Others have gave on strike, stopping all risky surgeries. The yearly premium varies by state and specialty. According to a survey conducted for Medical Economics, OB/GYNs pay approximately $55,000 a year in insurance premiums, pediatricians $12,500, and primary care physicians $17,500.[14] In order to confirm this report, a random sample of malpractice insurance premiums was obtained for physicians in various fields. The resulting data are given in the following table.

Physician specialty	Sample size	Sample mean	Sample standard deviation
OB/GYN	10	57,500	5,175
Pediatricians	15	14,590	2,789
Primary care	23	13,505	1,535

Assume each underlying distribution of malpractice-insurance premium is normal.

a. Find a 95% confidence interval for the true mean malpractice-insurance premium for each type of physician.

b. Is there any evidence that any of the true mean medical malpractice insurance premiums are different from those reported by *Medical Economics*? Justify your answer.

8.63 Sports and Leisure The Jackpine Gypsies Motorcycle Club holds the Sturgis Rally in the Black Hills of South Dakota every summer. There are motocross, hill climb, and short track races, and plenty of parties and bikes. In an attempt to learn more about the approximately 1 million participants, a random sample of bike enthusiasts was obtained. Some of the summary data are given in the table that follows.

Variable	Sample size	Sample mean	Sample standard deviation
Age (males, years)	60	38.9	7.9
Age (females, years)	40	35.6	4.5
Distance traveled (miles)	75	257.5	56.8

Assume the underlying distributions are normal.

a. Find a 95% confidence interval for the true mean age of men and of women attending the rally.

b. Is there any evidence to suggest that the mean age for men is different from the mean age for women? Justify your answer.

c. Find a 99% confidence interval for the true mean distance traveled to the rally. Interpret this result.

Challenge

8.64 Sports and Leisure Given a random sample of size n from a normal population with mean μ (and σ unknown):

a. Use the method outlined in Exercise 8.36 to find a one-sided $100(1 - \alpha)\%$ confidence interval for μ bounded above and another bounded below.

b. Steeplechase horse races are run on courses that include obstacles like brush fences, stone walls, timber walls, and water jumps. A random sample of 17 winning times (in seconds) in the 2-3/8-mile Saratoga Steeplechase Race was obtained. The sample mean was $\bar{x} = 259.79$ with a standard deviation of $s = 7.5$. Assume the population distribution of winning times is normal, and find a one-sided 99% confidence interval, bounded below, for the true mean winning time.

8.4 A LARGE-SAMPLE CONFIDENCE INTERVAL FOR A POPULATION PROPORTION

Many surveys and experiments are conducted in order to estimate a population proportion, the true fraction of individuals or objects that exhibit a specific characteristic. For example, pollsters routinely estimate the proportion of Americans who favor a particular political candidate. Food companies use randomly selected consumers to test-market new products, because the true proportion of shoppers who will purchase a product is important for predicting profit. And insurance agencies constantly analyze data to estimate the proportion of 18-year-old drivers who will be in an accident during the next year.

Let p = the true population proportion, the fraction of individuals or objects with a specific characteristic (the probability of a success). As in Section 7.3, it is reasonable to use a value of the sample proportion, $\hat{p}$, in order to construct a confidence interval for p. In a sample of n individuals or objects, let X be the number of individuals with the characteristic (or the number of successes in the sample). Recall that the sample proportion is the proportion of individuals with a specific characteristic in the sample, or a relative frequency.

$$\hat{P} = \frac{X}{n} = \frac{\text{The number of individuals with the characteristic}}{\text{The sample size}} \qquad (8.10)$$

From Section 7.3, if n is large and both $np \geq 5$ and $n(1 - p) \geq 5$, then the random variable $\widehat{P}$ is approximately normal with mean p and variance $p(1 - p)/n$: $\widehat{P} \stackrel{.}{\sim} N(p, p(1 - p)/n)$. Since $\widehat{P}$ is approximately normal, standardize to obtain

$$Z = \frac{\widehat{P} - p}{\sqrt{\dfrac{p(1 - p)}{n}}} \stackrel{.}{\sim} N(0, 1) \tag{8.11}$$

◀ As in Section 8.2, start with an appropriate probability statement involving the random variable Z in Equation 8.11. Find a symmetric interval about 0 such that the probability that Z lies in this interval is $1 - \alpha$.

$$P(-z_{\alpha/2} < Z < z_{\alpha/2}) = 1 - \alpha$$

$$P\left(-z_{\alpha/2} < \frac{\widehat{P} - p}{\sqrt{\dfrac{p(1 - p)}{n}}} < z_{\alpha/2} \right) = 1 - \alpha \tag{8.12}$$

Trying to sandwich p in Equation 8.12 is a little tricky because p appears in both the numerator and the denominator (inside a square root). Instead, since n is large, we usually use the sample proportion, $\hat{p}$, as a good estimate of *p in the denominator*. Manipulating the inequality (inside the probability statement) leads to the following general result. ▶

A Challenge exercise in this section asks you to find a confidence interval for p without using $\hat{p}$ in the denominator.

Use $\hat{p}$ as an estimate of p to check the *nonskewness criteria*.

> ### HOW TO FIND A LARGE-SAMPLE $100(1 - \alpha)\%$ CONFIDENCE INTERVAL FOR A POPULATION PROPORTION
>
> Given a random sample of size n, if n is large and both $n\hat{p} \geq 5$ and $n(1 - \hat{p}) \geq 5$, then a large-sample $100(1 - \alpha)\%$ confidence interval for p, the true population proportion, has as endpoints the values
>
> $$\hat{p} \pm z_{\alpha/2} \sqrt{\frac{\hat{p}(1 - \hat{p})}{n}}. \tag{8.13}$$

---

a. Solution Trail

KEYWORDS

- 95% CI for the true proportion.
- 1100 adults.
- 319 with hypertension.

↓

TRANSLATION

- 95% CI for p.
- $n = 1100$.
- $x = 319$.

↓

CONCEPTS

A large-sample $100(1 - \alpha)\%$ CI for a population proportion.

↓

VISION

We need a 95% CI for p. Check the nonskewness criteria to ensure the distribution of $\widehat{P}$ is approximately normal. Find the appropriate critical value, and use Equation 8.13.

Example 8.9 Americans with Hypertension High blood pressure, or hypertension, occurs when the force of blood against the artery walls is too strong. This added pressure can cause damage to the arteries, heart, and kidneys and could lead to a stroke. In a recent survey, 1100 adult Americans were randomly selected and examined for high blood pressure. The number of patients classified with hypertension is 319.

a. Find a 95% confidence interval for the true proportion of adult Americans with hypertension.

b. The U.S. Department of Health and Human Services has a target of 16% for hypertension prevalence by the year 2010. Is there any evidence to suggest that the prevalence of hypertension is different from the target value? Justify your answer.

SOLUTION

a. The following information is given:
Sample size: $n = 1100$.
Number of people with the specific characteristic, hypertension: $x = 319$.

Compute the sample proportion and check the nonskewness criteria.

$$\hat{p} = \frac{x}{n} = \frac{319}{1100} = 0.29.$$
$$n\hat{p} = (1100)(0.29) = 319 \geq 5$$
$$n(1 - \hat{p}) = (1100)(0.71) = 781 \geq 5$$

Since both inequalities are satisfied, $\hat{P}$ is approximately normal, and Equation 8.13 can be used to construct a confidence interval for p.

$$1 - \alpha = 0.95 \longrightarrow \alpha = 0.05 \longrightarrow \alpha/2 = 0.025$$ Find $\alpha/2$.
$$z_{\alpha/2} = z_{0.025} = 1.96$$ Use Table III.

Use Equation 8.13.

$$\hat{p} \pm z_{\alpha/2}\sqrt{\frac{\hat{p}(1 - \hat{p})}{n}}$$ Equation 8.13.

$$= \hat{p} \pm z_{0.025}\sqrt{\frac{\hat{p}(1 - \hat{p})}{n}}$$ Use the value of α.

$$= 0.29 \pm 1.96\sqrt{\frac{(0.29)(0.71)}{1100}}$$ Use the values for $z_{0.025}$ and $\hat{p}$.

$$= 0.29 \pm 0.0268$$ Simplify.

$$= (0.2632, 0.3168)$$ Compute endpoints.

$(0.2632, 0.3168)$ is a 95% confidence interval for the true proportion, p, of adult Americans with hypertension.

Figures 8.32 and 8.33 together show a technology solution.

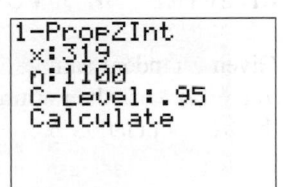

Figure 8.32 1-PropZInt input screen.

Figure 8.33 Resulting confidence interval.

b. **Claim**: $p = 0.16$.

Experiment: $\hat{p} = 0.29$.

Likelihood: The likelihood is expressed as a 95% confidence interval, an interval of likely values for p: $(0.2632, 0.3168)$.

Conclusion: The target value, 0.16, does not lie in the confidence interval. There is evidence to suggest that p is different from 0.16. ●

Here is one more example involving a confidence interval for a population proportion.

Example 8.10 Solar Flare Study Solar flares on the sun are violent explosions that release enormous amounts of energy and radiation. These emissions can harm spacecraft and satellites in the Earth's orbit and disrupt communication on the ground. In an effort to plan for these possible disruptions, a study was conducted over several years. On 86 of 350 randomly selected days there was at least one solar flare. Find a 99% confidence interval for the true proportion of days on which there is at least one solar flare.

---

b. Solution Trail

KEYWORDS

Is there any evidence.

↓

TRANSLATION

Inference procedure.

↓

CONCEPTS

A large-sample $100(1 - \alpha)\%$ CI for a population proportion.

↓

VISION

The CI constructed in part (a) is an interval in which we are 95% confident the true value of p lies. If 0.16 lies in this interval, there is no evidence to suggest p is different from 0.16. Use the four-step inference procedure.

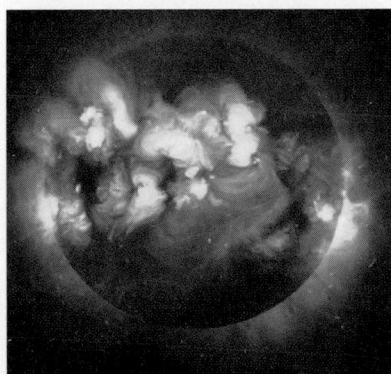

(NASA)

SOLUTION

STEP 1 $n = 350$, $x = 86$.   Given.

STEP 2 The sample proportion: $\hat{p} = \dfrac{86}{350} = 0.2457$.

$n\hat{p} = (350)(0.2457) = 86 \geq 5$
$n(1 - \hat{p}) = (350)(0.7543) = 264 \geq 5$
Both inequalities are satisfied. $\hat{P}$ is approximately normal.

STEP 3 $1 - \alpha = 0.99 \longrightarrow \alpha = 0.01 \longrightarrow \alpha/2 = 0.005$ Find $\alpha/2$.

$z_{\alpha/2} = z_{0.005} = 2.5758$ Use Table III.

STEP 4 Use Equation 8.13.

$$\hat{p} \pm z_{\alpha/2}\sqrt{\frac{\hat{p}(1 - \hat{p})}{n}}$$ Equation 8.13.

$$= \hat{p} \pm z_{0.005}\sqrt{\frac{\hat{p}(1 - \hat{p})}{n}}$$ Use the value of α.

$$= 0.2457 \pm 2.5758\sqrt{\frac{(0.2457)(0.7543)}{350}}$$ Use the values for $z_{0.005}$ and $\hat{p}$

$$= 0.2457 \pm 0.0593$$ Simplify.

$$= (0.1864, 0.3050)$$ Compute endpoints.

(0.1864, 0.3050) is a 99% confidence interval for the true proportion, p, of days on which there is at least one solar flare.

Figure 8.34 shows a technology solution.

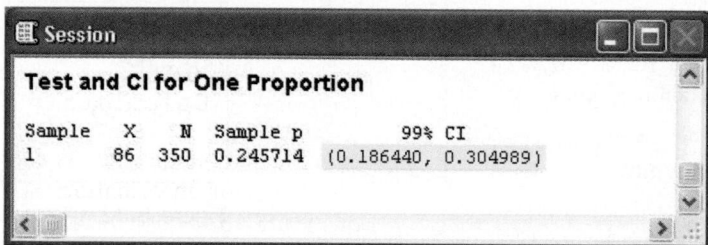

Figure 8.34 Minitab 1-Proportion confidence interval using the normal distribution. ●

Recall from Section 8.2: A sample size can be determined such that the resulting confidence interval for μ has certain properties. Similarly, given the confidence level, a sample size n can be computed such that the resulting confidence interval for p has a desired width.

Suppose n is large, the confidence level is $100(1 - \alpha)\%$, and B is the bound on the error of estimation. Half the width of the confidence interval is the step in each direction from $\hat{p}$. Therefore, the bound on the error of estimation is

$$B = z_{\alpha/2} \cdot \sqrt{\frac{\hat{p}(1 - \hat{p})}{n}}. \qquad (8.14)$$

Solving for n yields

$$n = \hat{p}(1 - \hat{p})\left[\frac{z_{\alpha/2}}{B}\right]^2. \qquad (8.15)$$

Look carefully at this result. While the mathematics is correct, there is a real problem with Equation 8.15. The critical value, $z_{\alpha/2}$, and the bound on the error of estimation, B, can be specified. But $\hat{p}$ *is unknown*. We do not know $\hat{p}$ until we have the sample size n (and x).

There are two solutions to this problem.

1. Use a reasonable estimate for $\hat{p}$ from previous experience. Researchers often conduct many similar experiments over time and can make very realistic guesses for $\hat{p}$.

2. If no prior information is available, use $\hat{p} = 0.5$ in Equation 8.15. This produces a very *conservative,* large value for n.

For given $z_{\alpha/2}$ and B, Equation 8.15 is greatest (maximized) when $\hat{p} = 0.5$.

In either case, it is unlikely that the value of n will be an integer. *Always* round up. This guarantees that the resulting confidence interval will have a bound on the error of estimation of at most B. Any larger sample size will also suffice. The following example illustrates the use of Equation 8.15.

Example 8.11 Telephone Pole Replacement   Each year a regional Bell telephone company inspects telephone poles and replaces those that are defective (cracked, rotted, etc.). In order to effectively allocate resources, the company plans to estimate the proportion of defective poles. A 95% confidence interval for p with bound on the error of estimation 0.02 is needed. How large a sample size is necessary in each of the following cases?
a. Prior experience suggests $\hat{p} \approx 0.1$.
b. There is no prior information regarding the proportion of defective poles.

SOLUTION

a. Use $\hat{p} = 0.1$ and $B = 0.02$ in Equation 8.15.

$$1 - \alpha = 0.95 \longrightarrow \alpha/2 = 0.05 \longrightarrow z_{0.025} = 1.96 \qquad \text{Find the critical value.}$$

$$n = \hat{p}(1 - \hat{p})\left[\frac{z_{\alpha/2}}{B}\right]^2 \qquad \text{Equation 8.15.}$$

$$= (0.1)(0.9)\left[\frac{1.96}{0.02}\right]^2 \qquad \text{Substitute given values.}$$

$$= (0.09)[98]^2 = 864.36 \qquad \text{Compute.}$$

The necessary sample size is $n \geq 865$.

b. With no prior information, use $\hat{p} = 0.5$, $B = 0.02$, and $z_{0.025} = 1.96$ in Equation 8.15.

$$n = \hat{p}(1 - \hat{p})\left[\frac{z_{\alpha/2}}{B}\right]^2 \qquad \text{Equation 8.15.}$$

$$= (0.5)(0.5)\left[\frac{1.96}{0.02}\right]^2 \qquad \text{Substitute given values.}$$

$$= (0.25)[98]^2 = 2401 \qquad \text{Compute.}$$

The necessary sample size in this case is $n \geq 2401$.   ●

TECHNOLOGY CORNER

Procedure: Construct a confidence interval for a population proportion.
Reconsider: Example 8.10, page 355, solution, and interpretation.

TI-84 Plus

Use the built-in function 1-PropZInt.

1. Select STATS; TESTS; 1-PropZInt.
2. Enter values for x, the number of successes; n, the total number of trials; and C-Level, the confidence coefficient. See Figure 8.35.
3. Highlight Calculate and press ENTER. The confidence interval and summary statistics are displayed on the Home Screen. See Figure 8.36.

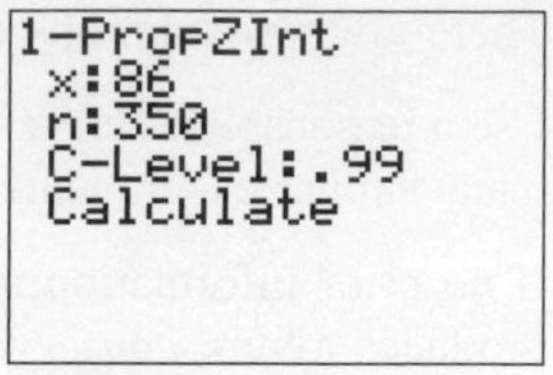

```
1-PropZInt
 x:86
 n:350
 C-Level:.99
 Calculate
```

```
1-PropZInt
 (.1864,.3050)
 p̂=.2457
 n=350.0000
```

Figure 8.35
1-PropZInt input
screen.

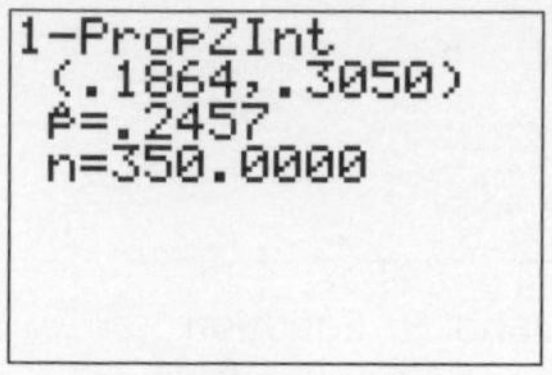

Figure 8.36 Resulting
confidence interval.

Minitab

Use the built-in function 1 Proportion. Input is either a sample in a column or summarized data.
1. Select Stat; Basic Statistics; 1 Proportion.
2. Choose Summarized data and enter the Number of events (the number of successes), and the Number of trials.
3. Select the Options button. Enter the Confidence level and check the box for Use test and interval using normal distribution. The results are displayed in a session window. See Figure 8.34, page 356.

Excel

There is no built-in function to compute the endpoints of a confidence interval for a population proportion. Use the function NORMSINV to find the critical value.
1. Enter values for x, the number of successes, n, the number of trials, and α.
2. Use the function NORMSINV to find the critical value.
3. Compute the left and right endpoints of the confidence interval. See Figure 8.37.

	A	B	C
1	86	= x	Number of successess
2	350	= n	Number of trials
3	0.2457	= A1/A2	Sample proportion
4	0.01	= alpha	Alpha
5	2.5758	= NORMSINV(1-A4/2)	Critical value
6	0.1864	= A3-A5*SQRT(A3*(1-A3)/A2)	Left endpoint
7	0.3050	= A3+A5*SQRT(A3*(1-A3)/A2)	Right endpoint

Figure 8.37 Calculations to construct the confidence interval.

SECTION 8.4 EXERCISES

Practice

8.65 In each of the following problems, the sample size and the number of individuals or objects with a specified characteristic are given. Check the nonskewness criteria to determine whether the distribution of $\hat{P}$ is approximately normal.
a. $n = 105, x = 85$ b. $n = 1750, x = 1645$
c. $n = 225, x = 220$ d. $n = 183, x = 3$
e. $n = 377, x = 350$ f. $n = 480, x = 478$

8.66 In each of the following problems, the sample size, the number of individuals or objects with a specified characteristic, and the confidence level are given. Find the associated confidence interval for the population proportion.

a. $n = 150,$ $x = 70,$ 95%.
b. $n = 225,$ $x = 65,$ 98%.
c. $n = 500,$ $x = 468,$ 90%.
d. $n = 95,$ $x = 63,$ 99%.
e. $n = 2450,$ $x = 986,$ 99.9%.

8.67 In each of the following problems, the sample size, the number of individuals or objects with a specified characteristic, and the confidence level are given. Find the associated confidence interval for the population proportion.
a. $n = 1336,$ $x = 1001,$ 99%. b. $n = 775,$ $x = 680,$ 95%.
c. $n = 85,$ $x = 41,$ 95%. d. $n = 335,$ $x = 290,$ 98%.
e. $n = 566,$ $x = 47,$ 99.9%.

8.68 In each of the following problems, the confidence level, the bound on the error of estimation, and an estimate for $\hat{p}$ are given.

Find the sample size necessary to produce a confidence interval for p with this bound and confidence.
a. 95%, $B = 0.05$, $\hat{p} \approx 0.45$.
b. 98%, $B = 0.07$, $\hat{p} \approx 0.32$.
c. 99%, $B = 0.10$, $\hat{p} \approx 0.14$.
d. 90%, $B = 0.001$, $\hat{p} \approx 0.057$.
e. 99.9%, $B = 0.03$, $\hat{p} \approx 0.22$.

8.69 In each of the following problems, the confidence level and the bound on the error of estimation are given. Find the sample size necessary to produce a confidence interval for p with this bound and confidence level.
a. 99%, $B = 0.06$.
b. 95%, $B = 0.10$.
c. 98%, $B = 0.002$.
d. 90%, $B = 0.05$.
e. 99.9%, $B = 0.2$.

8.70 There are three factors that affect the width of a large-sample confidence interval for p: the confidence level, the sample size, and the sample proportion, $\hat{p}$. In each of the following problems, determine whether the width of the resulting confidence interval for p increases or decreases.
a. Confidence level and $\hat{p}$ are constant, and n increases.
b. Sample size and $\hat{p}$ are constant, and confidence level decreases.
c. Sample size and confidence level are constant, and $\hat{p}$ (> 0.5) increases.

8.71 There are three factors that affect the size of the sample necessary to produce a confidence interval for p with certain properties: the confidence level, the bound on the error of estimation, and the estimate of $\hat{p}$. In each of the following problems, determine whether the necessary sample size increases or decreases.
a. Confidence level and estimate of $\hat{p}$ are constant, and B is smaller.
b. B and estimate of $\hat{p}$ are constant, and confidence level increases.
c. B and confidence level are constant, and estimate of $\hat{p}$ is closer to 0.
d. B and confidence level are constant, and estimate of $\hat{p}$ is closer to 1.

Applications

8.72 Public Health and Nutrition Good vision is very important for effective learning, especially for children in elementary school. Some optometrists argue that as many as 25% of all elementary school children may have impaired vision.[15] A random sample of 256 elementary school children was obtained, and each was given a vision test. Fifty were found to have some form of impaired vision.
a. Find a 95% confidence interval for the true proportion of elementary school children with impaired vision.
b. Is there any evidence to suggest that the optometrists' claim of $p = 0.25$ is wrong? Justify your answer.

8.73 Public Policy and Political Science A new standard for 30-, 40-, and 50-gallon residential gas water heaters includes a flame arrester. This device helps to prevent flashback fires from flammable liquid vapor nearby. The Consumer Product Safety Commission would like to estimate the proportion of homes affected by this safety standard.

a. In a random sample of 575 homes, 235 had gas water heaters. Find a 90% confidence interval for the true proportion of homes with gas water heaters.
b. Prior research suggests that the proportion of homes with gas water heaters is approximately 0.40. How large a sample is necessary in order for the bound on the error of estimation to be 0.03 for a 95% confidence interval?

8.74 Marketing and Consumer Behavior A successful company usually has high brand name and logo recognition among consumers. For example, Coca-Cola products are available to 98% of all people in the world, and therefore may have the highest logo recognition of any company. A software firm developing a product would like to estimate the proportion of people who recognize the Linux penguin logo. Of the 952 randomly selected consumers surveyed, 132 could identify the product associated with the penguin.
a. Is the distribution of the sample proportion, $\hat{P}$, approximately normal? Justify your answer.
b. Find a 95% confidence interval for the true proportion of consumers who recognize the Linux penguin.
c. The company will market a Linux version of their new software if the true proportion of people who recognize the logo is greater than 0.10. Is there any evidence to suggest that the true proportion of people who recognize the logo is greater than 0.10? Justify your answer.

8.75 Travel and Transportation In order to advertise appropriate vacation packages, Best Bets Travel would like to learn more about families planning overseas trips. In a random sample of 125 families planning a trip to Europe, 15 indicated France as their travel destination.
a. For those families planning vacations to Europe, find a 98% confidence interval for the true proportion traveling to France.
b. Suppose no prior estimate of $\hat{p}$ is known. How large a sample is necessary in order for the bound on the error of estimation to be 0.05 with the confidence level in part (a)?

8.76 Public Policy and Political Science In a recent survey, 650 randomly selected Americans were asked several questions regarding television broadcasting and programming. The table below lists three proposals and the number of Americans who responded in favor of each.

Proposal	Number in favor
To limit the number of commercials shown during each children's show	566
To provide more adult-education programs	530
To make children's shows commercial free	468

a. Find a 95% confidence interval for the true proportion of Americans in favor of limiting the number of commercials shown during each children's show.
b. Find a 95% confidence interval for the true proportion of Americans in favor of providing more adult-education programs.
c. Find a 95% confidence interval for the true proportion of Americans in favor of making all children's programming commercial free.
d. Which of the above confidence intervals is the narrowest? Why?

8.77 Psychology and Human Behavior In a Harris Interactive Poll conducted between November 10 and 17, 2008, concerning the religious beliefs of Americans, adults were asked questions concerning ghosts, astrology, and reincarnation. Of the 2126 adults who responded, 936 believe in ghosts, 659 believe in astrology, and 510 believe in reincarnation.

a. Find a 95% confidence interval for the true proportion of American adults who believe in ghosts.

b. Find a 95% confidence interval for the true proportion of American adults who believe in astrology.

c. Find a 95% confidence interval for the true proportion of American adults who believe in reincarnation.

8.78 Travel and Transportation Americans are extremely mobile, moving due to a change in job, wanting a different climate, or wanting to be closer to (or farther from) family. In a random sample of 575 Americans, 76 indicated that they had moved within the past year.[16]

a. Find a 90% confidence interval for the true proportion of Americans who moved within the past year.

b. Suppose no prior estimate of $\hat{p}$ is known. How large a sample is necessary in order for the bound on the error of estimation to be 0.02 with confidence level 90%?

8.79 Economics and Finance When unemployment rises, high school students face more competition from college students and adult workers for summer jobs. In a random sample of 188 high school students looking for summer work, 61 said they were able to find a job.

a. Find a 95% confidence interval for the true proportion of high school students who were able to find a summer job.

b. In a similar study the previous year, the sample proportion of high school students who were able to find a job was 0.25. Use this estimate to find the sample size necessary in order for the bound on the error of estimation to be 0.025 with confidence level 95%.

8.80 Marketing and Consumer Behavior Most walk-behind lawn mowers have three options for disposal of grass clippings: bagging, mulching, or side discharge. The manager at an Aubuchon Hardware store conducted a survey to determine which disposal method is most common. The results are given in the table below, classified by area mowed. A small area is less than 0.5 acres, a medium area is 0.5 to 1 acre, and a large area is over 1 acre.

| Area | Sample size | Disposal method | | |
		Bagging	**Mulching**	**Side discharge**
Small	125	85	35	5
Medium	157	70	40	47
Large	144	42	45	57

a. For people with small yards, find a 95% confidence interval for the true proportion who dispose of grass clippings by bagging.

b. For people with medium yards, find a 95% confidence interval for the true proportion who dispose of grass clippings by mulching.

c. For people with large yards, find a 95% confidence interval for the true proportion who dispose of grass clippings by side discharge.

8.81 Business and Management A United States textile company is interested in the proportion of orders shipped to another country. In a random sample of 1560 clothing orders placed to American companies, 500 were exported.

a. Find a 99% confidence interval for the true proportion of clothing orders shipped to other countries.

b. In the previous year, the true proportion of clothing orders shipped to other countries was believed to be 0.30. Using your answer to part (a), is there any evidence to suggest that the true proportion has changed? Justify your answer.

8.82 Business and Management Many dairy farms have experienced bankruptcy over the past decade due to the wild fluctuations in conventional milk prices. However, organic farms, those that do not treat cows with antibiotics or hormones and that use feed grown without chemicals, have remained solvent and even expanded. In a random sample of 1400 New England dairy farms, 90 are certified as organic.

a. Find a 99% confidence interval for the true proportion of New England dairy farms certified as organic.

b. Five years ago, an extensive census reported that 3% of all New England dairy farms were organic. Is there any evidence to suggest that this proportion has changed? Justify your answer.

8.83 Marketing and Consumer Behavior In a recent survey of home-buyer preferences, consumers were asked about desired characteristics, such as a wood-burning fireplace, a den/library, and flooring. In particular, each person was asked whether a separate dining room is essential. The sample size and the number who responded *Yes* to this question are given in the following table by geographic region.

Geographic region	Sample size	Number who responded *Yes*
Northeast	225	180
Midwest	276	224
South Central	301	232
South Atlantic	454	377
West	366	304

a. Find a 99% confidence interval for the true proportion of home buyers in each geographic region who believe that a separate dining room is essential.

b. Which confidence interval in part (a) is the largest? Why?

8.84 Sports and Leisure Coin collecting is common among historians and ordinary consumers. Many coins are extraordinary works of art, play an important role in the growth and development of a nation, and may dramatically increase in value. The 50 State Quarters Program started in 1999, with five new coins released each year. In an effort to assess public interest, *The Washington Post* conducted a survey to estimate the proportion of Americans attempting to amass a collection of all 50 state quarters. In a random sample of 2350 Americans, 1050 said they were trying to accumulate all 50 state quarters.

a. Find a 99% confidence interval for the true proportion of Americans who are attempting to collect all 50 state quarters.

b. The U.S. Mint claims that 46% of all Americans are trying to collect all 50 state quarters.[17] Is there any evidence to suggest that this claim is false? Justify your answer.

Extended Applications

8.85 Demographics and Population Statistics In an articles in the *Journal of Gambling Issues*, the authors present survey results concerning gambling-related problems among older adults (55 and over) in Ontario, Canada.[18] It was reported that 5.4% of secondary school graduates of the adults sampled, 3.8% of adults with other post-secondary education, and 4.3% of adults with a post-secondary education experienced a gambling problem. A random sample of adults 55 and over living in Ontario was obtained and asked whether they experience a gambling problem. The resulting data are given in the table below, by education level.

Sample size	Number with gambling problem (x)	Education level
257	20	Secondary school
380	27	Other post-secondary
305	11	Post-secondary

a. Find a 99% confidence interval for the true proportion of adults in Ontario who experience a gambling problem for each education level.

b. Is there any evidence to suggest that the true proportion of *post-secondary* adults who experience a gambling problem is different from that in either of the other two education levels? Justify your answer.

8.86 Psychology and Human Behavior The Drivers Technology Association in the United Kingdom recently studied the behavior of drivers with and without radar detectors. A random sample of 550 users and 562 nonusers was obtained. In the past three years, 108 users and 68 nonusers have had an accident.

a. Find a 99% confidence interval for the true proportion of radar-detector users who have had an accident in the past three years.

b. Find a 99% confidence interval for the true proportion of radar-detector nonusers who have had an accident in the past three years.

c. Is there any evidence to suggest that the two true proportions are different? Justify your answer.

8.87 Psychology and Human Behavior Tattoos and body piercings have become more popular in recent years, especially among college students. Arms, hips, and ankles are the most common sites for a tattoo, and many students have more than one tattoo. In a survey released by the Pew Research Center (January 9, 2007), approximately 36% of Gen Nexters and 40% of Gen Xers currently have or have had a tattoo. Suppose 481 Gen Nexters and 1028 Gen Xers participated in the survey.

a. Find a 95% confidence interval for the true proportion of Gen Nexters who have or have had a tattoo.

b. Find a 95% confidence interval for the true proportion of Gen Xers who have or have had a tattoo.

c. Is there any evidence to suggest that the two true proportions are different? Justify your answer.

8.88 Economics and Finance The first question on IRS Income Tax Form 1040 asks each taxpayer whether he or she wants $3 to go to the Presidential Election Campaign Fund. A large Washington political-action committee obtained a random sample of registered voters and asked them if they checked *Yes* on this question on the past year's return. The results are given in the table below, by party affiliation.

Political party	Sample size	Number who checked *Yes*
Democrat	237	70
Republican	388	184
Independent	155	23

a. Find a 95% confidence interval for the true proportion of registered voters in each political party who checked *Yes* on the Presidential Election Campaign Fund question.

b. Is there any evidence to suggest that the proportion of Independent voters who checked *Yes* is different from either the proportion of Democrat or Republican voters who checked *Yes*? Justify your answer.

c. Suppose the results obtained in this study are preliminary and a larger survey is planned. How large a sample is necessary for each political party in order for the bound on the error of estimation to be 0.02 with confidence level 95%?

8.89 Medicine and Clinical Studies A new prescription medication is designed to ease the pain of arthritis. In a clinical trial, both treatment and placebo groups were studied in order to determine whether there are any adverse reactions. The table that follows lists the number of patients in each group who experienced each adverse reaction. Assume each group represents a random sample.

Adverse reaction	Treatment group ($n = 465$)	Placebo group ($n = 154$)
Headache	61	15
Rash	31	9

a. Find a 95% confidence interval for the true proportion of people who suffered a headache in each of the groups.

b. Is there any evidence to suggest that the true proportion of people who suffered a headache is different for the two groups? Justify your answer.

c. Find a 98% confidence interval for the true proportion of people who experienced a rash in each of the groups.

d. Is there any evidence to suggest that the true proportion of people who experienced a rash is different for the two groups? Justify your answer.

Challenge

8.90 Technology and Internet Given a random sample of size n, suppose n is large and both $n\hat{p} \geq 5$ and $n(1 - \hat{p}) \geq 5$.

a. Use the method outlined in Exercise 8.36 to find a one-sided $100(1 - \alpha)\%$ confidence interval for p bounded above and another bounded below.

b. Many Internet users download and share illegal copies of songs (and movies). A random sample of 260 Internet users who regularly download music was obtained, and 171 indicated that they did not care if they were violating copyright laws. Find a one-sided 95% confidence interval, bounded above, for the population proportion of Internet users who download music who do not care about violating copyright laws.

8.91 The Wilson CI Consider the probability statement used to construct a confidence interval for a population proportion (Equation 8.12):

$$P\left(-z_{\alpha/2} < \frac{\hat{P} - p}{\sqrt{\dfrac{p(1 - p)}{n}}} < z_{\alpha/2} \right) = 1 - \alpha.$$

Manipulate the inequality (without substituting $\hat{p}$ for p in the denominator) to obtain a $100(1 - \alpha)\%$ confidence interval for p (the Wilson interval, with endpoints given as follows).

$$\frac{n\hat{p} + z_{\alpha/2}^2/2}{n + z_{\alpha/2}^2} \pm \frac{z_{\alpha/2}\sqrt{n}}{n + z_{\alpha/2}^2}\sqrt{\hat{p}(1 - \hat{p}) + z_{\alpha/2}^2/(4n)} \quad (8.16)$$

a. In a random sample of size 100, suppose the sample proportion is $\hat{p} = 0.60$. Find a 95% confidence interval for p using

Equation 8.13. Find a 95% confidence interval for p using the Wilson interval (Equation 8.16). Which is wider? Why?

b. Let $n = 120, 140, 160, \ldots, 500$. For each value of n, let $\hat{p} = 0.60$ and compute both CIs for p. What happens to the Wilson CI as n increases? Why?

8.92 Necessary Sample Size In order to find the sample size necessary to construct a $100(1 - \alpha)\%$ confidence interval for a population proportion with bound on the error of estimation B, Equation 8.15 is used:

$$n = \hat{p}(1 - \hat{p})\left[\frac{z_{\alpha/2}}{B}\right]^2.$$

For a 95% confidence interval and $B = 0.10$, let $\hat{p} = 0.05, 0.10, 0.15, \ldots, 0.95$, and find the sample size necessary for each value of $\hat{p}$. Plot the values of n versus $\hat{p}$ (n on the vertical axis and $\hat{p}$ on the horizontal axis). Describe the pattern. When is n largest?

8.5 A CONFIDENCE INTERVAL FOR A POPULATION VARIANCE

Many real-world problems involve estimation of variability to find out whether measurements are clustered around a central value or spread out over a wide range. For example, quality-control specialists continuously monitor production processes for increases or changes in range or variability, and most scientists agree that increased quantities of carbon dioxide in the atmosphere have contributed to greater climate variability. It seems reasonable to use the sample variance, S^2, as an estimator for the population variance, σ^2, a measure of variability. A confidence interval for a population variance, σ^2, is based on S^2, a new *standardization*, and a **chi-square distribution** introduced below.

S^2 is a *good* estimator for σ^2. It is *unbiased*: $E(S^2) = \sigma^2$.

A **chi-square** (abbreviated χ^2) **distribution** has positive probability only for nonnegative values. The probability density function for a χ^2 random variable (details given in the margin) is 0 for $x < 0$. Focus on the properties of a chi-square distribution and the method for finding critical values associated with this distribution.

The probability density function for a chi-square random variable, X, with df $= \nu$ is given by

$$f(x) = \begin{cases} \dfrac{e^{-x/2}x^{(\nu/2)-1}}{2^{\nu/2}\Gamma(\nu/2)} & x \geq 0 \\ 0 & \text{otherwise} \end{cases}$$

PROPERTIES OF A CHI-SQUARE DISTRIBUTION

1. A chi-square distribution is completely determined by one parameter, the number of degrees of freedom (df). The degrees of freedom must be a positive integer, df $= 1, 2, 3, 4, \ldots$. There is a different chi-square distribution corresponding to each value of df.

2. If X has a chi-square distribution with df degrees of freedom, denoted $X \sim \chi^2$ with df, then

$$\mu_X = \text{df} \quad \text{and} \quad \sigma_X^2 = 2\,(\text{df}). \quad (8.17)$$

The mean of X is df, the number of degrees of freedom, and the variance is 2 (df), twice the number of degrees of freedom.

3. Suppose $X \sim \chi^2$ with df degree of the freedom. The density curve for X is positively skewed (*not* symmetric), and as x increases it gets closer and closer to the x-axis but never touches it. As df increases, the density curve becomes flatter and actually looks more normal. See Figure 8.38.

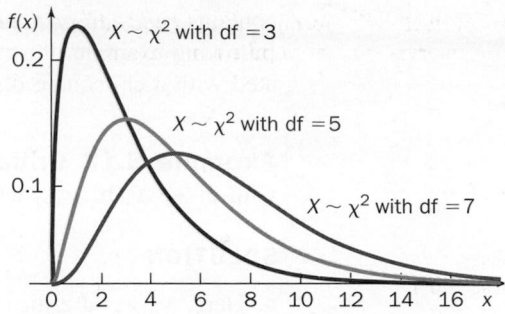

Figure 8.38 Density curves for several chi-square distributions.

The definition and notation for a χ^2 **critical value** is analogous to those for a Z and a t critical value.

DEFINITION

χ_α^2 is a **critical value** related to a **chi-square distribution** (a χ^2 **critical value**) with df degrees of freedom. If $X \sim \chi^2$ with df degrees of freedom, then $P(X \geq \chi_\alpha^2) = \alpha$.

 ILLUMINATING THE CONCEPTS

1. χ_α^2 is a value on the measurement axis in a chi-square world with df degrees of freedom such that there is α of the area (probability) to the right of χ_α^2. There is *no* symmetry in chi-square critical values (as there was for Z and t critical values).

2. **Critical values** are always defined in terms of right-tail probability. The notation is just a little trickier here because a chi-square distribution is *not* symmetric. It will be necessary to find critical values denoted $\chi_{1-\alpha}^2$ (with *large* values for $1 - \alpha$). By definition, $P(X \geq \chi_{1-\alpha}^2) = 1 - \alpha$, and by the Complement Rule, $P(X \leq \chi_{1-\alpha}^2) = \alpha$. See Figure 8.39.

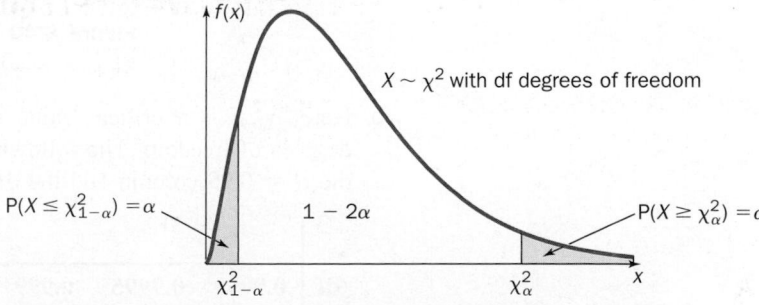

Figure 8.39 An illustration of chi-square critical values. ■

Table VI (see the Appendix) presents selected critical values associated with various chi-square distributions. Right-tail probabilities are in the top row and degrees of freedom are listed in the left column. In the body of the table, χ_α^2 is at the intersection of the α column and the df row. The first part of this table presents *left-tail critical values* corresponding to *large* right-tail probabilities. The second half contains *right-tail critical values* corresponding to *small* right-tail probabilities. The following example illustrates the use of this table for finding critical values associated with a chi-square distribution.

Example 8.12 Critical Value Look Ups Find each critical value: **a.** $\chi^2_{0.05}$ with df $= 10$, **b.** $\chi^2_{0.99}$ with df $= 7$.

SOLUTION

a. Here, $\chi^2_{0.05}$ is a critical value related to a chi-square distribution with 10 degrees of freedom. By definition, if $X \sim \chi^2$ with df $= 10$, then $P(X \geq \chi^2_{0.05}) = 0.05$. The following portion of Table VI shows the intersection of the $\alpha = 0.05$ column and the df $= 10$ row.

df	0.10	0.05	0.025	0.01	0.005	0.001	0.0005	0.0001
⋮	⋮	⋮	⋮	⋮	⋮	⋮	⋮	⋮
8	13.3616	15.5073	17.5345	20.0902	21.9550	26.1245	27.8680	31.8276
9	14.6837	16.9190	19.0228	21.6660	23.5894	27.8772	29.6658	33.7199
10	15.9872	18.3070	20.4832	23.2093	25.1882	29.5883	31.4198	35.5640
11	17.2750	19.6751	21.9200	24.7250	26.7568	31.2641	33.1366	37.3670
12	18.5493	21.0261	23.3367	26.2170	28.2995	32.9095	34.8213	39.1344
⋮	⋮	⋮	⋮	⋮	⋮	⋮	⋮	⋮

The table is headed by α.

Therefore, $\chi^2_{0.05} = 18.3070$, and if $X \sim \chi^2$ with df $= 10$, then $P(X \geq 18.3070) = 0.05$, as illustrated in Figure 8.40.

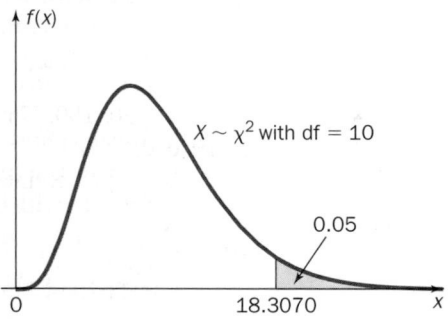

Figure 8.40 Visualization of $\chi^2_{0.05} = 18.3070$.

b. Here, $\chi^2_{0.99}$ is a critical value related to a chi-square distribution with seven degrees of freedom. The following portion of Table VI shows the intersection of the $\alpha = 0.99$ column and the df $= 7$ row.

df	0.9999	0.9995	0.999	0.995	0.99	0.975	0.95	0.90
⋮	⋮	⋮	⋮	⋮	⋮	⋮	⋮	⋮
5	0.0822	0.1581	0.2102	0.4117	0.5543	0.8312	1.1455	1.6103
6	0.1724	0.2994	0.3811	0.6757	0.8721	1.2373	1.6354	2.2041
7	0.3000	0.4849	0.5985	0.9893	1.2390	1.6899	2.1673	2.8331
8	0.4636	0.7104	0.8571	1.3444	1.6465	2.1797	2.7326	3.4895
9	0.6608	0.9717	1.1519	1.7349	2.0879	2.7004	3.3251	4.1682

The table is headed by α.

$\chi^2_{0.99} = \boxed{1.2390}$ and if $X \sim \chi^2$ with df $= 7$ then $P(X \geq 1.2390) = 0.99$ and $P(X \leq 1.2390) = 0.01$, as illustrated in Figure 8.41.

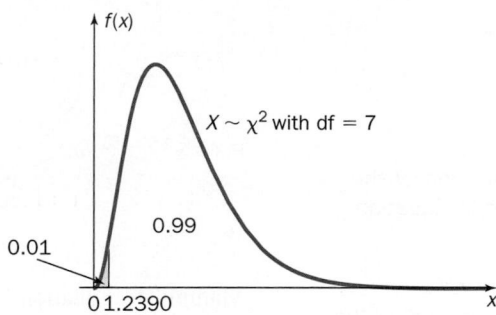

Figure 8.41 Visualization of $\chi^2_{0.99} = 1.2390$.

Figure 8.42 shows a technology solution.

Figure 8.42 Use the Inverse Cumulative Distribution Function to find critical values associated with a chi-square distribution.

Table VI is limited. There are only a handful of values for α, and $\nu = 1$–40, 50, 60, 70, 80, 90, 100. However, a calculator or computer can find almost any critical value needed.

A confidence interval for a population variance is based on the following theorem.

8.2 THEOREM

Let S^2 be the sample variance of a random sample of size n from a normal distribution with variance σ^2. The random variable

$$X = \frac{(n-1)S^2}{\sigma^2} \qquad (8.18)$$

has a chi-square distribution with $n - 1$ degrees of freedom.

This is another kind of *standardization*, a transformation to a chi-square distribution.

This theorem is used to construct a confidence interval for σ^2. ◀ Let $X \sim \chi^2$ with df $= n - 1$ and find an interval that captures $1 - \alpha$ in the *middle* of this chi-square distribution (see Figure 8.43).

$$P(\chi^2_{1-\alpha/2} < X < \chi^2_{\alpha/2}) = 1 - \alpha \qquad (8.19)$$

$$P\left(\chi^2_{1-\alpha/2} < \frac{(n-1)S^2}{\sigma^2} < \chi^2_{\alpha/2}\right) = 1 - \alpha \qquad (8.20)$$

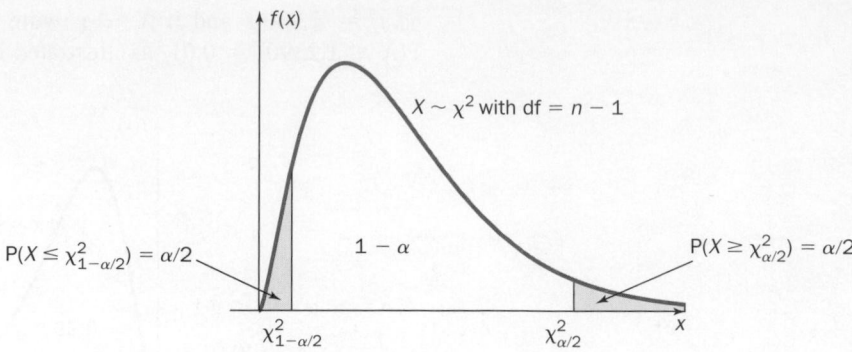

Figure 8.43 An illustration of the probability statement in Equation 8.20.

Manipulate Equation 8.20 to obtain the probability statement

$$P\left(\frac{(n-1)S^2}{\chi^2_{\alpha/2}} < \sigma^2 < \frac{(n-1)S^2}{\chi^2_{1-\alpha/2}}\right) = 1 - \alpha. \quad (8.21)$$

This probability statement leads to the following general result. ▶

Manipulate here means, inside the probability statement, divide each term by $(n-1)S^2$ and take the reciprocal of each term (change the direction of the inequality).

The critical value in the right tail of the chi-square distribution is part of the expression for the left endpoint, and the critical value in the left tail of the chi-square distribution is part of the expression for the right endpoint.

HOW TO FIND A 100(1 − α)% CONFIDENCE INTERVAL FOR A POPULATION VARIANCE

Given a random sample of size n from a population with variance σ^2, if the underlying population is normal, a $100(1 - \alpha)\%$ confidence interval for σ^2 is given by

$$\left(\frac{(n-1)s^2}{\chi^2_{\alpha/2}}, \frac{(n-1)s^2}{\chi^2_{1-\alpha/2}}\right) \quad (8.22)$$

with critical values based on df = $n - 1$.

💡 **ILLUMINATING THE CONCEPTS**

1. This confidence interval for σ^2 is valid *only* if the underlying population is normal.

2. Take the square root of each endpoint of Equation 8.22 in order to find a $100(1 - \alpha)\%$ confidence interval for the population standard deviation σ. ■

The following example illustrates the use of Equation 8.22 to construct a confidence interval for σ^2 and to answer an inference question.

Example 8.13 Kiln-Fired Dishes Earthenware dishes are made from clay and are fired, or exposed to heat, in a large kiln. Large fluctuations in the kiln temperature can cause cracks, bumps, or other flaws (and increase cost). With the kiln set at 800°C, a random sample of 19 temperature measurements (in °C) was obtained. The sample variance was $s^2 = 17.55$.

a. Find a 95% confidence interval for the true population variance in temperature of the kiln when it is set to 800°C. Assume that the underlying distribution is normal.

b. Quality-control engineers have determined that the maximum variance in temperature during firing should be 16°C. Using the confidence interval constructed in part (a), is there any evidence to suggest that the true temperature variance is greater than 16°C? Justify your answer.

(Xavier Gonzalez/FeaturePics)

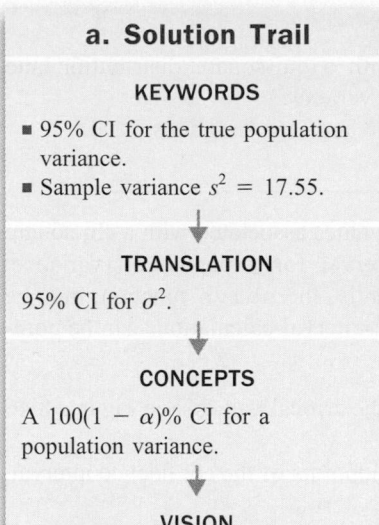

a. Solution Trail

KEYWORDS

- 95% CI for the true population variance.
- Sample variance $s^2 = 17.55$.

↓

TRANSLATION

95% CI for σ^2.

↓

CONCEPTS

A $100(1 - \alpha)$% CI for a population variance.

↓

VISION

The underlying distribution is assumed to be normal. Find the appropriate chi-square critical value, and use Equation 8.22 to construct a 95% CI for σ^2.

SOLUTION

a. $s^2 = 17.55$, $n = 19$, df $= 19 - 1 = 18$. Given.

$1 - \alpha = 0.95 \longrightarrow \alpha = 0.05 \longrightarrow \alpha/2 = 0.025$ Find $\alpha/2$.

$\chi^2_{\alpha/2} = \chi^2_{0.025} = 31.5264$

$\chi^2_{1-\alpha/2} = \chi^2_{0.975} = 8.2307$ Use Table VI with df $= 18$ to find the critical values.

Use Equation 8.22.

$$\frac{(n-1)s^2}{\chi^2_{\alpha/2}} = \frac{(n-1)s^2}{\chi^2_{0.025}} = \frac{(18)(17.55)}{31.5264} = 10.0202 \qquad \text{Left endpoint.}$$

$$\frac{(n-1)s^2}{\chi^2_{1-\alpha/2}} = \frac{(n-1)s^2}{\chi^2_{0.975}} = \frac{(18)(17.55)}{8.2307} = 38.3807 \qquad \text{Right endpoint.}$$

$(10.0202, 38.3807)$ is a 95% confidence interval for the true population variance in temperature when the kiln is set to $800°C$.

$(\sqrt{10.0202}, \sqrt{38.3807}) = (3.1655, 6.1952)$ is a 95% confidence interval for the population standard deviation.

Figure 8.44 shows a technology solution.

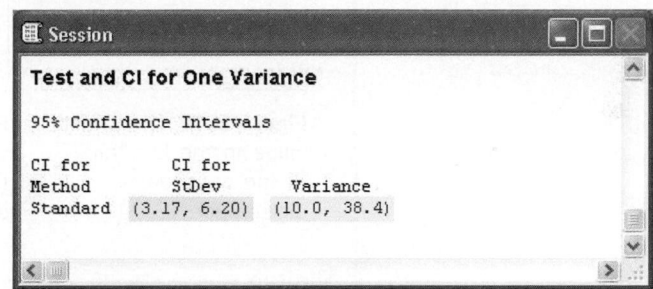

Figure 8.44 Confidence intervals for the population standard deviation and the population variance.

b. Solution Trail

KEYWORDS

Is there any evidence?

↓

TRANSLATION

Inference procedure.

↓

CONCEPTS

A $100(1 - \alpha)$% CI for a population variance.

↓

VISION

The CI constructed in part a is an interval in which we are 95% confident the true value of σ^2 lies. If 16 lies in this interval, there is no evidence to suggest σ^2 is different from 16. Use the four-step inference procedure.

b. Claim: $\sigma^2 = 16$.

Experiment: $s^2 = 17.55$.

Likelihood: The likelihood is expressed as a 95% confidence interval, an interval of likely values for σ^2: $(10.0202, 38.3807)$, from part (a).

Conclusion: The required kiln temperature variance, $16°C$, is included in this confidence interval. There is no evidence to suggest that σ^2 is different from 16. ●

The confidence interval for σ^2 is *not* symmetric about the point estimate s^2 and is quite large (see Figure 8.45). The confidence interval will be *narrow* only for very large values of n.

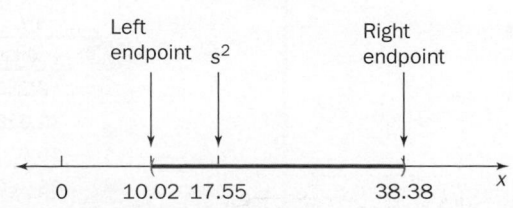

Figure 8.45 A 95% confidence interval for the population variance σ^2.

TECHNOLOGY CORNER

Procedure: Find critical values associated with a chi-square distribution and construct a confidence interval for a population variance.
Reconsider: Example 8.13, page 366, solutions, and interpretations.

TI-84 Plus

There is no built-in function to compute critical values associated with a chi-square distribution nor to construct a confidence interval for a population variance. However, the `EQUATION SOLVER`, or equivalently, the `solve` function, may be used to find critical values, and other confidence-interval calculations can be completed on the Home Screen.

1. Use the `solve` function with χ^2 `cdf` to find the critical values. See Figures 8.46 and 8.47.
2. On the Home Screen, find the left and right endpoints of the confidence interval. See Figure 8.48.

```
solve(χ²cdf(0,X,
18)-0.025,X,5,{0
,10})→L
         8.2307
```

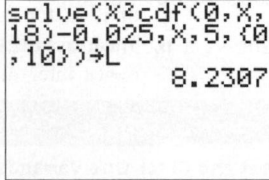

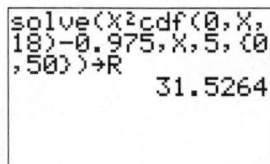

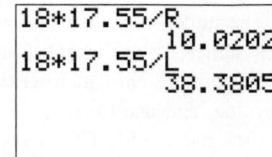

Figure 8.46 The critical value in the left tail of the chi-square distribution.

```
solve(χ²cdf(0,X,
18)-0.975,X,5,{0
,50})→R
        31.5264
```

Figure 8.47 The critical value in the right tail of the chi-square distribution.

```
18*17.55/R
        10.0202
18*17.55/L
        38.3805
```

Figure 8.48 Calculations to find the left and right endpoint of the confidence interval.

Minitab

Use the built-in function 1 Variance. Input is either summary statistics or a column containing data.

1. Select Stat; Basic Statistics; 1 Variance.
2. Choose Enter variance from the pull-down menu and Summarized data. Enter the Sample size and the Sample variance.
3. Select the Options button and enter the Confidence level. The results (a confidence interval for the population standard deviation and for the population variance) are displayed in a session window. See Figure 8.44, page 367.

Excel

There is a built-in function to find critical values. Use these values in the appropriate formula to construct a confidence interval for a population variance.

1. Use the function `CHIINV` to find the critical values. The arguments are cumulative probability and degrees of freedom.
2. Compute the left and right endpoints of the confidence interval. See Figure 8.49.

	A	B	C
1	8.2307	=CHIINV(0.975,18)	Left tail critical value
2	31.5264	=CHIINV(0.025,18)	Right tail critical value
3	10.0202	=18*17.55/A2	Left endpoint
4	38.3805	=18*17.55/A1	Right endpoint

Figure 8.49 Calculations to construct the confidence interval.

SECTION 8.5 EXERCISES

Practice

8.93 Find each of the following critical values.

a. $\chi^2_{0.10}$ for df = 5
b. $\chi^2_{0.001}$ for df = 31
c. $\chi^2_{0.05}$ for df = 16
d. $\chi^2_{0.025}$ for df = 21
e. $\chi^2_{0.99}$ for df = 11
f. $\chi^2_{0.95}$ for df = 15
g. $\chi^2_{0.975}$ for df = 23
h. $\chi^2_{0.995}$ for df = 9

8.94 Find each of the following critical values.

a. $\chi^2_{0.90}$ for df = 12
b. $\chi^2_{0.01}$ for df = 15
c. $\chi^2_{0.9999}$ for df = 22
d. $\chi^2_{0.005}$ for df = 3
e. $\chi^2_{0.9995}$ for df = 19
f. $\chi^2_{0.005}$ for df = 34
g. $\chi^2_{0.999}$ for df = 26
h. $\chi^2_{0.0001}$ for df = 40

8.95 In each of the following problems, the sample size and the confidence level are given. Find the appropriate chi-square critical values for use in constructing a confidence interval for the population variance.

a. $n = 22$, 95%
b. $n = 37$, 99%
c. $n = 11$, 98%
d. $n = 31$, 90%
e. $n = 5$, 95%
f. $n = 37$, 99.9%

8.96 In each of the following problems, the sample variance, the sample size, and the confidence level are given. Assume the underlying population is normally distributed. Find the associated confidence interval for the population variance.

a. $s^2 = 5.65$, $n = 35$, 95%.
b. $s^2 = 45.62$, $n = 26$, 98%.
c. $s^2 = 50.41$, $n = 6$, 80%.
d. $s^2 = 7.68$, $n = 37$, 90%.
e. $s^2 = 32.22$, $n = 5$, 99%.
f. $s^2 = 70.67$, $n = 28$, 99.9%.

8.97 In each of the following problems, the sample variance, the sample size, and the confidence level are given. Assume the underlying population is normally distributed. Find the associated confidence intervals for the population variance and the population standard deviation.

a. $s^2 = 3.08$, $n = 14$, 99%.
b. $s^2 = 64.10$, $n = 11$, 95%.
c. $s^2 = 59.07$, $n = 6$, 80%.
d. $s^2 = 7.35$, $n = 27$, 99.98%.
e. $s^2 = 31.38$, $n = 22$, 95%.
f. $s^2 = 12.39$, $n = 18$, 99%.

8.98 Use Table VI and linear interpolation to approximate each critical value. Verify each approximation using technology.

a. $\chi^2_{0.05}$ for df = 45
b. $\chi^2_{0.005}$ for df = 65
c. $\chi^2_{0.01}$ for df = 72
d. $\chi^2_{0.025}$ for df = 56
e. $\chi^2_{0.95}$ for df = 85
f. $\chi^2_{0.999}$ for df = 52
g. $\chi^2_{0.90}$ for df = 66
h. $\chi^2_{0.975}$ for df = 75

Applications

8.99 Business and Management A personnel manager at Inserra Supermarkets is concerned about the pattern of overtime hours claimed by employees. While there is a special budget to pay for overtime hours, large fluctuations may cause cash-flow problems.

In a random sample of 25 employees, the sample variance for the number of overtime hours claimed was 4.25. Assume the distribution of overtime hours is normal.

a. Find a 95% confidence interval for the population variance of overtime hours.
b. Find a 95% confidence interval for the population standard deviation of overtime hours.

8.100 Medicine and Clinical Studies One measure of lung function is the forced vital capacity (FVC), the volume of air expelled under certain conditions. The FVC varies with age, gender, and height and is affected by respiratory disease. In a study published in the *American Journal of Epidemiology*, the authors concluded that dietary fiber improves lung function.[19] Suppose a random sample of adults on a high fiber diet was obtained and the FVC for each was carefully measured. The data are given on the data CD and book's web site.

a. Find a 99% confidence interval for the true population variance of FVC for this population of adults.
b. What assumption(s) did you make in constructing this confidence interval?
c. A physician involved in this study claims that the variance for FVC is $\sigma^2 = 100$. Is there any evidence to refute this claim? Justify your answer.

8.101 Technology and Internet Everyone who owns or has access to a computer probably has several passwords for various programs. These codes are supposed to be a secret, random combination of characters. However, some psychologists believe passwords are predictable based on personality traits, and many computer hackers claim that any password can be *acquired*. In order to study computer security at Spectaguard Acquisition, a special program was developed and used to systematically try various passwords for randomly selected user accounts. The time (in hours) needed to obtain each password is given in the table below.

1.88	1.71	2.09	6.60	2.28	3.52	2.64	4.94
1.78	4.13	2.55	1.66	0.28	4.02	4.47	0.68
5.67	0.03						

Assume the distribution of times is normal.

a. Find a 98% confidence interval for the true population variance of the time needed to acquire some one's password.
b. Use the methods described in Section 6.3 to check for any evidence of nonnormality.

8.102 Physical Sciences St. Simons Island, Georgia, is a popular vacation destination. Attractions include a salt marsh tour, a dolphin watch, a golf course, and the art and antiques trail. In a sample of high tide observations (in feet) from March 2005, $\bar{x} = 6.88$, $s = 0.597$, and $n = 30$.[20]

a. Assuming normality, find a 99% confidence interval for the true variance in high-tide level.
b. Historical evidence suggests that the variance in high tide is 0.50. Is there any evidence to suggest that the variance in high tide has decreased? Justify your answer.

8.103 Physical Sciences July temperatures in Phoenix, Arizona, routinely soar above 100°F, but residents claim it's a "dry" heat.

The low humidity makes outside physical activity easier and attracts tourists, especially golfers, and retirees. The following table contains relative humidity data (percentages) from the Arizona Regional Hourly Weather Roundup.

14	19	13	16	17	18	14	16	14	16
17	16	19	16	17	17	17	18	16	16
17	17	19	15	22	23	16	13	15	13

(*Source*: National Weather Service, Phoenix, Arizona.)

a. Find a 95% confidence interval for the true variance in relative humidity in Phoenix.

b. Is there any evidence to suggest that the data are from a non-normal distribution? Justify your answer.

8.104 Manufacturing and Product Development In residential and commercial buildings, the thickness of window glass varies according to the width of the window. Wider windows require thicker window panes. As part of quality control, a manufacturer randomly samples windows and carefully measures the width of each (in millimeters). Consider the data in the following table.

Window width	Sample size	Sample mean	Sample variance
< 250	15	6.02	0.003
250–1100	18	7.95	0.04
1100–2250	22	10.12	0.5

a. Find a 99% confidence interval for the true variance of window thickness for each window width category. Assume normality.

b. Suppose a process is considered to be "in control" if the variance is 0.02 mm (or less). Is there any evidence to suggest that any of the three processes is out of control?

8.105 Fuel Consumption and Cars New automobiles contain lots of electronic equipment which require more power to operate. The Honda Insight is designed so that the motor produces 10 kW of power at 3000 rpm (*Source*: Eartheasy.com). Large variations in the power output will cause electrical problems throughout the automobile. To check the variability, 17 cars were selected at random, and the power output (in kW) at 3000 rpm was measured. The sample variance was 0.55. Assume normality.

a. Find a 90% confidence interval for the true population variance in power output at 3000 rpm.

b. Find a 90% confidence interval for the true population standard deviation in power output at 3000 rpm.

c. If the standard deviation is greater than 1, there is a good chance of electrical failure. Is there any evidence to suggest that the standard deviation is greater than 1? Justify your answer.

8.106 Sports and Leisure Perhaps the most grueling parts of the Tour de France bicycling championship are the treacherous mountain stages. In a random sample of 25 riders on Stage 15 of the 2008 Tour de France, Lézatsur-Léze to Saint-Lary Soulan (Pla d'Adet), times were measured (in minutes) and the sample variance was 27.42.[21] Assuming normality, find a 99% confidence

interval for the true population variance in times for Stage 15 of the Tour de France.

8.107 Sports and Leisure During the National Football League preseason, coaches appraise new players and tested veterans. Suppose a rookie is trying to unseat a veteran for the starting tailback position. A random sample of preseason runs is selected for each player, and the number of yards gained on each play is recorded. The data are given on the data CD and book's web site.

a. Find a 99% confidence interval for the variance of yardage gained per run for the veteran tailback.

b. Find a 99% confidence interval for the variance of yardage gained per run for the rookie tailback.

c. Suppose the coach will start whichever tailback is more consistent and dependable. Which player do you think will get the starting position, and why?

8.108 Medicine and Clinical Studies The thickness of the carotid artery wall in a patient can be measured using a noninvasive ultrasound technique. Some research studies suggest that thicker walls are associated with a higher occurrence of cardiovascular disease. Twenty-nine patients with coronary artery disease were randomly selected, and the carotid artery wall thickness (in mm) was measured for each. The sample variance was 0.0225.[22]

a. Find a 95% confidence interval for the population variance in carotid artery wall thickness. Assume normality.

b. Some physicians believe that the variance in wall thickness is greater in patients with coronary artery disease. Is there any evidence to suggest that the population variance is greater than 0.06—the assumed variance in wall thickness for healthy adults?

8.109 Physical Sciences The water table depth is important for agriculture and for the development and maintenance of residential and commercial areas. During 2004, 102 wells near La Junta, Colorado, were monitored for water table depth and salinity. A sample of the water table depths (in meters) is given in the following table.

3.51	3.76	3.55	3.04	3.52	3.45	3.30	3.39
3.50	3.27	2.90	3.44	3.49	2.99	3.64	3.44
3.14	3.00	3.36	3.73				

(*Source*: Colorado AES Projects 2005–2006.)

a. Find a 98% confidence interval for the population variance in water-table depth at this site.

b. Find a 98% confidence interval for the population standard deviation in water-table depth at this site.

c. Use the methods described in Section 6.3 to test for any evidence of nonnormality.

8.110 Fuel Consumption and Cars One measure of a vehicle's front-end alignment is the caster angle—the relationship between the upper and lower ball joints. The specifications for a certain sports car include a caster angle of 2.8° with variance 0.40. Twenty-two new sports cars were randomly selected and the caster angle was measured for each. The sample variance was 0.55.

a. Find a 95% confidence interval for the population variance in caster angle. Assume normality.

b. Is there any evidence to suggest that the population variance in caster angle is different from 0.40? Justify your answer.

8.111 Business and Management Even though the budget for a movie may be close to 200 million dollars, worldwide distribution can provide a huge net income for a studio. The following table includes a sample of movies, the budget (in dollars) for each, and the U.S. gross income (in dollars) for each.

Movie	Budget	Gross
Titanic	200,000,000	600,788,188
Waterworld	175,000,000	88,246,220
Batman Begins	150,000,000	205,085,404
Troy	150,000,000	133,298,577
Die Another Day	142,000,000	160,932,247
Men in Black 11	140,000,000	190,418,803
Lethal Weapon 4	140,000,000	130,444,603
Spider-Man	139,000,000	403,706,375
Stealth	138,000,000	31,704,416
The Blair Witch Project	35,000	140,539,099
Pi	68,000	3,221,152
Mad Max	200,000	8,750,000
Super Size Me	65,000	11,529,368
Hulk	137,000,000	132,160,047
Armageddon	140,000,000	201,578,182

(*Source*: The Numbers, Box Office Data.)

a. Find a 95% confidence interval for the population variance in gross income.

b. Compute the ratio of U.S. revenue to budget (ratio = gross/budget) for each movie. Find a 95% confidence interval for the population variance in ratio of revenue to budget.

c. If the variance in the ratio is greater than 1, it is considered *risky* to invest in a new film. Is there any evidence to suggest that investing in a new movie is risky? Justify your answer.

8.112 Manufacturing and Product Development Silver bars used in trading on the bullion market are manufactured to certain height, width, and length specifications. In addition, there is variation in the silver content. Five randomly selected silver bars were carefully analyzed for silver content (in ounces). The sample variance was 65.5. Assume the distribution of silver content is normal, and find a 99% confidence interval for the population variance in silver content.

8.113 Psychology and Human Behavior The Great Wall of China was started in 214 B.C. and designed as a defense against nomadic tribes. This 1500-mile long wall, built from earth and stone, is visible from space and is one of the most famous structures in the world. A random sample of midpoints between guard towers along the Great Wall was obtained, and the height of the wall at each point was measured (in feet). The data are given in the following table.

26.9	22.2	21.8	26.6	19.2	29.3	21.9	19.0
23.5	26.7	18.3	27.7	18.8	25.0	24.9	21.1
24.1	24.3	19.5	20.8	27.1			

(*Source*: www.infoplease.com and www.teachingtools.com.)

a. Find a 95% confidence interval for the population variance in height at the midpoints between guard towers of the Great Wall.

b. Using the confidence interval constructed in part (a), is there any evidence to suggest that the true population variance in the height at midpoints between guard towers is less than 12 ft^2? Justify your answer.

Extended Applications

8.114 Medicine and Clinical Studies Many researchers believe that moderate physical activity, like walking, will help prevent weight gain. In order to study this claim, doctors in Colorado randomly selected patients to wear a pedometer to measure the distance walked (in miles) per week. The table below presents data for men and women.

	Sample size	Sample variance
Men	32	5.75
Women	28	7.66

a. Find a 95% confidence interval for the true variance in distance walked per week for men.

b. Find a 95% confidence interval for the true variance in distance walked per week for women.

c. Is there any evidence to suggest that the true variance is different for men and women? Justify your answer.

d. What assumption(s) did you make in constructing the confidence intervals above?

8.115 Biology and Environmental Science Rice is a staple food for much of the world's population. The height of a rice plant (usually 0.4–5 mm) depends on the rice variety and the environment. Some corporations are attempting to genetically engineer rice plants to increase their resistance to disease and to decrease the variability in plant height. In a controlled experiment, plant heights were measured; 30 natural rice plants had a sample variance in height of 1.5 mm, and 22 genetically engineered rice plants had a sample variance in height of 0.89 mm. Assume normality.

a. Find a 95% confidence interval for the true variance in height for natural rice plants.

b. Find a 95% confidence interval for the true variance in height for genetically engineered rice plants.

c. Is there any evidence to suggest that the population variance is different for natural and genetically engineered rice plants?

8.116 Economics and Finance The rate for a home mortgage varies from lender to lender. Consumer advocates advise buyers to shop around for the best available offer. The data on the data CD and book's web site represent a random sample of the interest rate (as a percentage) on a $100,000, 0 points, 30-year fixed rate mortgage, at banks in the Columbus, Ohio, area.[23]

a. Find a 95% confidence interval for the population variance in mortgage interest rates.

b. Find a 95% confidence interval for the population standard deviation in mortgage interest rates.

c. Suppose the population mean interest rate is 5.8. Use the left endpoint of the CI in part (b) as an estimate for σ. Assuming normality, find the probability that a randomly selected interest rate is less than 5.500. Use the right endpoint of the CI and find the probability that a randomly selected interest rate is less than 5.500.

8.117 Physical Sciences A microwave radiometer is used to measure the column water vapor and the infrared brightness (IB) temperatures in clouds. Clouds were randomly selected, and a weather station in Coffeyville, Kansas, collected the following summary statistics.

Cloud type	Column water vapor (cm) Sample size	Column water vapor (cm) Sample variance	IB Temperature (°C) Sample size	IB Temperature (°C) Sample variance
Cirrus	11	0.06	17	201.7
Cumulus	21	0.08	28	225.6

a. Find a 99% confidence interval for the population variance in column water vapor and in infrared brightness temperature for cirrus clouds.
b. Find a 99% confidence interval for the population variance in column water vapor and in infrared brightness temperature for cumulus clouds.
c. Is there any evidence to suggest that the variance in column water vapor or infrared brightness temperature is different in cirrus and cumulus clouds? Justify your answers.

8.118 Medicine and Clinical Studies A new medicinal spray has been developed to help ease the itch and burn associated with poison ivy and poison oak. People who suffer from these conditions want immediate relief. A research study was conducted to measure the time (in minutes) from application of this spray to relief from itching. The following summary statistics were reported for a random sample of children and adults.

Population	Sample size	Sample variance
Children	11	1.57
Adults	25	2.38

a. Find a 95% confidence interval for the population variance in time to relief for children.
b. Find a 95% confidence interval for the population variance in time to relief for adults.
c. Is there any evidence to suggest that the variance in time to relief is different in children and adults? Justify your answer.

Challenge

8.119 Sports and Leisure Given a random sample of size n from a normal population with variance σ^2:
a. Use the method outlined in Exercise 8.36 to find a one-sided $100(1 - \alpha)\%$ confidence interval for σ^2 bounded above and another bounded below.
b. A random sample of soccer stadiums from around the world was obtained, and the seating capacity for each is given on the data CD and book's web site. Find a one-sided 95% confidence interval, bounded above, for the population variance in seating capacity for soccer stadiums.

8.120 Normal Approximation to the Chi-Square Distribution Given a random sample of size n from a normal population with variance σ^2, a $100(1 - \alpha)\%$ confidence interval for σ^2 (using a chi-square distribution) is given by Equation 8.22:

$$\frac{(n - 1)s^2}{\chi^2_{\alpha/2}} < \sigma^2 < \frac{(n - 1)s^2}{\chi^2_{1 - \alpha/2}}.$$

If n is large ($n > 30$), then the chi-square random variable $(n - 1)S^2/\sigma^2$ is approximately normal with mean $n - 1$ and variance $2(n - 1)$. Therefore, for large n,

$$P\left(-z_{\alpha/2} < \frac{\frac{(n - 1)S^2}{\sigma^2} - (n - 1)}{\sqrt{2(n - 1)}} < z_{\alpha/2}\right) = 1 - \alpha. \quad (8.23)$$

a. Manipulate Equation 8.23 to obtain an approximate $100(1 - \alpha)\%$ confidence interval for a population variance.
b. The thickness of pavement on roads and highways depends on the predicted weight and volume of vehicular traffic. The thickness of the pavement (in mm) was measured at 51 random locations along Route 95. The sample variance was $s^2 = 16.25$.
   i. Find a 95% confidence interval for the population variance in pavement thickness using Equation 8.22.
   ii. Find an approximate 95% confidence interval for the population variance in pavement thickness using the equation derived in part (a).
   iii. Which of these two intervals is wider? Why?

Chapter 8 Challenge Wrap-Up

For the 15 randomly selected coal cars, the mean weight of the coal was $\bar{x} = 105.35$ tons and the standard deviation was $s = 6.3$ tons. In order to verify tipple loading operations, construct a 95% confidence interval for the true mean coal weight. Assume that the distribution of coal weight is normal. Since the population standard deviation is unknown, find the appropriate t critical value and use Equation 8.9.

(Kenneth Sponster/iStock photo)

$\bar{x} = 105.35, \quad s = 6.3, \quad n = 15, \quad df = 15 - 1 = 14.$ Given.

$1 - \alpha = 0.95 \longrightarrow \alpha = 0.05 \longrightarrow \alpha/2 = 0.025$ Find $\alpha/2$.

$t_{\alpha/2} = t_{0.025} = 2.1448$ Use Table V with df = 14 to find the critical value.

Use Equation 8.9.

$$\bar{x} \pm t_{\alpha/2}\frac{s}{\sqrt{n}}$$ Equation 8.9.

$$= \bar{x} \pm t_{0.025}\frac{s}{\sqrt{n}}$$ Use the value of α.

$$= 105.35 \pm 2.1448\frac{6.3}{\sqrt{15}}$$ Use summary statistics, values for s, $t_{0.025}$, and n.

$$= 105.35 \pm 3.49$$ Simplify.

$$= (101.86, 108.84)$$ Compute endpoints.

(101.86, 108.84) is a 95% confidence interval for the true mean weight (in tons) of coal carried in railroad cars.

Figures 8.50 and 8.51 together show a technology solution.

```
TInterval
 Inpt:Data Stats
 x:105.35
 Sx:6.3
 n:15
 C-Level:.95
 Calculate
```

```
TInterval
 (101.86,108.84)
 x̄=105.3500
 Sx=6.3000
 n=15.0000
```

Figure 8.50 TInverval input screen.

Figure 8.51 Resulting confidence interval.

Since the entire confidence interval is less than 120 tons, there is evidence to suggest that the true mean weight of train coal cars is less than this maximum weight. The DOT is pleased with tipple operators.

CHAPTER 8 SUMMARY

Concept	Page	Notation/Formula/Description
Estimator	324	A statistic, or rule, used to produce a point estimate of a population parameter.
Unbiased estimator	324	A statistic $\widehat{\theta}$ is an unbiased estimator of θ if $E(\widehat{\theta}) = \theta$.
Biased estimator	324	A statistic $\widehat{\theta}$ is a biased estimator of θ if $E(\widehat{\theta}) \neq \theta$.
MVUE	326	Minimum variance unbiased estimator.
Confidence interval (CI)	329	An interval of values constructed so that with a specified degree of confidence, the value of the population parameter lies in this interval.
Confidence coefficient	329	The probability that the confidence interval encloses the population parameter in repeated samplings.
Confidence level	329	The confidence coefficient expressed as a percentage.
z_α	331	z critical value; a value such that $P(Z \geq z_\alpha) = \alpha$.
t_α	344	t critical value; a value such that $P(T \geq t_\alpha) = \alpha$ where T has a t distribution with df degrees of freedom.
χ_α^2	363	Chi-square critical value; a value such that $P(X \geq \chi_\alpha^2) = \alpha$ where X has a chi-square distribution with df degrees of freedom.

Summary of confidence intervals.

Parameter	Assumptions	$100(1 - \alpha)$% confidence interval
μ	n large, σ known, or normality, σ known	$\bar{x} \pm z_{\alpha/2} \dfrac{\sigma}{\sqrt{n}}$
μ	normality, σ unknown	$\bar{x} \pm t_{\alpha/2} \dfrac{s}{\sqrt{n}}$
p	n large, nonskewness	$\widehat{p} \pm z_{\alpha/2} \sqrt{\dfrac{\widehat{p}(1 - \widehat{p})}{n}}$
σ^2	normality	$\left(\dfrac{(n-1)s^2}{\chi_{\alpha/2}^2}, \dfrac{(n-1)s^2}{\chi_{1-\alpha/2}^2} \right)$

Common sample-size calculations.
B = Bound on the error of estimation, half the width of confidence interval.

Parameter	Estimate	Sample size
μ	$\bar{x}$	$n = \left[\dfrac{\sigma z_{\alpha/2}}{B} \right]^2$
p	$\widehat{p}$	$n = \widehat{p}(1 - \widehat{p}) \left[\dfrac{z_{\alpha/2}}{B} \right]^2$

CHAPTER 8 EXERCISES

Applications

8.121 Medicine and Clinical Studies Most patients in need of a kidney transplant are placed on a dialysis machine. There is some evidence to suggest that the longer patients remain on dialysis, the worse they fare following a transplant. A random sample of kidney transplant patients was obtained, and the wait time (in months) was recorded for each. The data are given in the following table.

37.3	38.1	23.6	46.2	32.7	19.1	41.0	33.3
30.6	49.1	48.3	33.1	40.0	25.6	19.4	43.7
50.1	33.7	36.5	39.4	34.7	48.0	42.3	23.5

(*Source*: American Kidney Fund.)

a. Find a 95% confidence interval for the true mean wait time for a kidney transplant.
b. Find a 95% confidence interval for the true variance in wait time for a kidney transplant.
c. Construct a normal probability plot for the wait-time data. Is there any evidence to suggest non-normality?

8.122 Biology and Environmental Science The Air Quality Index (AQI) is a measure developed by the Environmental Protection Agency to indicate the severity of air pollution. The AQI ranges from 0–50 (good air quality) to 500 (hazardous air quality) and is calculated for five major pollutants. In a random sample of 35 days in December 2008, the AQI for ozone was carefully measured in the Dallas–Fort Worth area. The sample mean AQI was $\bar{x} = 32.86$. Assume the population standard deviation is $\sigma = 6.7$.[24]

a. Find a 98% confidence interval for the true mean ozone AQI for Dallas–Fort Worth.
b. Find the sample size necessary to construct a 98% confidence interval for the true mean ozone AQI with a bound on the error of estimation of $B = 2$.

8.123 Marketing and Consumer Behavior Parrot Jungle Island is a roadside attraction in Miami, Florida, featuring tropical birds, crocodiles, and over 2000 varieties of plants and flowers. A new advertising campaign was developed to attract more out-of-state visitors. In a random sample of 270 visitors, 189 were area residents.

a. Find the sample proportion of visitors who were area residents. Check the nonskewness criteria.
b. Find a 99% confidence interval for the true proportion of visitors who were area residents.
c. Suppose that you have no prior knowledge of the proportion of visitors who were area residents. Find the sample size necessary for a 99% confidence interval with a bound on the error of estimation of 0.05.

8.124 Sports and Leisure Most major league baseball parks have a device to measure the speed of every pitch. The results are often displayed on a scoreboard and tracked by coaches. A random sample of pitches made during the first inning of baseball games

around the country was selected. The speed of each pitch (in mph) is given in the table below.

85	90	82	86	83	88	87	90	92	84	92
90	89	90	89	89	84	87	92	89	86	89

a. Assume normality. Find a 95% confidence interval for the population mean speed of pitches in the first inning of major league baseball games.
b. Find a 99% confidence interval for the population variance in speed of pitches in the first inning of major league baseball games.
c. The mean speed of all pitches made in the first inning of games during the previous season was 90.225. Is there any evidence to suggest that the mean speed has changed?

8.125 Economics and Finance ATM machines have made it easier and quicker for customers to check account balances, transfer money, and obtain cash. Many banks are now considering the next generation of ATM machines that will feature news headlines, full-motion video, and tickets to events. In a random sample of 500 customers, 280 said that ATM machines should offer postage stamps.

a. Find the sample proportion of customers who believe that ATM machines should offer postage stamps, and check the nonskewness criteria.
b. Find a 99% confidence interval for the true proportion of customers who believe that ATM machines should offer postage stamps.
c. A bank official claims that the proportion of customers who believe ATM machines should offer postage stamps is 0.60. Is there any evidence to refute this claim?

8.126 Manufacturing and Product Development A 0.2-kiloton bunker buster missile is designed to destroy enemy bunkers approximately 70 feet deep. In order to prevent fallout, the missile must penetrate 120 feet below the ground. In tests conducted by the military, eight missiles were fired and the penetration depth (in feet) of each was recorded. The data are given in the following table.

122	117	119	119	121	124	119	120

a. Assume normality. Find a 99% confidence interval for the population mean penetration depth.
b. Is there any evidence to suggest that the population mean penetration depth is less than 120 feet?

8.127 Medicine and Clinical Studies Passengers on long airline flights may develop deep vein thrombosis (DVT), potentially dangerous blood clots. While most blood clots dissolve naturally in the blood stream, travelers on long flights have three times the risk of developing a blood clot. In a research study of people who traveled frequently as part of their job, 22 of 8755 developed a blood clot within 8 weeks of taking a long flight.[25]

a. Find a 90% confidence interval for the true proportion of passengers on long flights who develop DVT within 8 weeks of a long flight.
b. Is there any evidence to suggest that this proportion is greater than 0.5%? Justify your answer.

8.128 Sports and Leisure According to Golf Washington, there was a net loss of golf courses in 2007 and the number of rounds played in the United States was 0.5% less than in 2006. The latest survey by the National Golf Foundation suggests that 7.4% of all adult golfers play left handed. In a random sample of 1250 golfers from around the country, 101 indicated that they play left handed.
a. Find the sample proportion of golfers who play left handed. Check the nonskewness criteria.
b. Find a 95% confidence interval for the population proportion of golfers who play left handed.
c. Find the sample size necessary to construct a 95% confidence interval with a bound on the error of estimation of 0.02 (assuming $p = .074$).

8.129 Marketing and Consumer Behavior The U.S. per capita consumption of sweeteners, in grams per day, (other than refined sugar and high fructose corn syrup) was 16.0 in 2007.[26] Suppose a random sample of adults was selected, and each was monitored for sweetener intake for 1 day. The consumption of sweeteners (in grams) for each is given on the data CD and book's web site.
a. Find a 99% confidence interval for the population mean sweetener amount consumed per day.
b. Find a 99% confidence interval for the population variance in sweetener amount consumed per day.
c. What assumptions did you make in constructing these two confidence intervals?
d. A physicians' group believes that consumers should ingest no more than 15 grams of sweetener per day. Is there any evidence to suggest that the mean sweetener amount consumed per day is over this healthy limit?

8.130 Business and Management Many businesses rent limousines to chauffeur important clients to and from airports, hotels, or offices. A random sample of the cost (in dollars) of renting a limousine for the entire day in Cincinnati was obtained. For $n = 25$, the mean was $\bar{x} = 410.25$ and the standard deviation was $s = 35.07$. Assume the cost distribution is normal.
a. Find a 95% confidence interval for the population mean cost of renting a limousine for the entire day.
b. Find a 95% confidence interval for the population variance in the cost of renting a limousine for the entire day.
c. Find a 95% confidence interval for the standard deviation in the cost of renting a limousine for the entire day.

8.131 Psychology and Human Behavior The Christmas holiday shopping season is the height of shoplifting and employee theft. Retailers lose approximately 30 billion dollars per year due to these crimes. A random sample of 17 (apprehended) shoplifters was obtained, and the mean value of merchandise taken was $128.00.[27] Assume the distribution of the amount stolen is normal and that $s = 75.50$.
a. Find a 99% confidence interval for the true mean amount of merchandise stolen by a shoplifter.
b. Suppose Sears assumes a certain amount of merchandise will be lost to shoplifters and that each shoplifter takes approximately $100 worth of merchandise. Is there any evidence to suggest that the mean amount of merchandise lost to shoplifters is greater than $100?
c. Find a 99% confidence interval for the variance in the amount of merchandise lost to shoplifters.

d. A large variance in the amount of merchandise lost to shoplifters suggests that store security personnel need to make more frequent rounds. Is there any evidence to suggest that the population variance in the amount of merchandise lost to shoplifters is greater than $2500?

8.132 Physical Sciences The Quabbin reservoir in Massachusetts is 18 miles long and has a capacity of 412 billion gallons. This reservoir serves approximately 2.5 million people, with a daily yield of 300 million gallons of water. The depth of the reservoir is carefully monitored and measured at certain locations and times each day. In a random sample of 18 summer days, the depth (in feet) was measured at location S-1 at noon. The mean depth was 75.4. Assume the population standard deviation is $\sigma = 7.58$.
a. Find a 99% confidence interval for the depth of the reservoir at this location.
b. Historical records indicate that the mean depth at this location for the previous 10 years is 78.4 feet. Is there any evidence to suggest that this population mean has changed?

8.133 Public Health and Nutrition Rising costs have caused many Americans to do without health insurance. A random sample of adults of various ethnicities was obtained, and each was asked whether they had health insurance. The resulting data are given in the table below.

Group	Sample size	Number without health insurance
Caucasian	1220	166
African American	1080	205
Hispanic	1156	384

a. Find a 95% confidence interval for the proportion of adults without health insurance for each group.
b. A government publication claims that the proportion of people without health insurance for the general population is 0.146. Using the confidence intervals in part (a), is there evidence that any group has a different population proportion without health insurance?

Extended Applications

8.134 Medicine and Clinical Studies The drugs Ritalin and Adderall are designed to stimulate the central nervous system and are widely prescribed for children diagnosed with attention-deficit disorder (ADD). Suppose a random sample of girls and boys in 12th grade was obtained and the following results were obtained.

Group	Sample size	Number taking Ritalin/Adderall
Girls	375	24
Boys	480	39

(*Source*: Indiana Youth Survey, 2008)

a. Find a 95% confidence interval for the true proportion of 12th-grade girls who are taking Ritalin/Adderall.

b. Find a 95% confidence interval for the true proportion of 12th-grade boys who are taking Ritalin/Adderall.

c. Is there any evidence to suggest that the proportion of girls in this age group taking Ritalin/Adderall is different from the proportion of boys? Justify your answer.

8.135 Public Health and Nutrition Tannin is a general term for certain nonvolatile phenolic substances in many fruits that provide an astringent sensation, in apple cider for example. There is some evidence to suggest that the tannin level in apples is affected by the fertilizer regimen. A random sample of apples was obtained from both fertilized and unfertilized trees. The percentage of tannin was measured in each apple, and the data (sample size, sample mean, and sample standard deviation) are reported in the table below.

Apples	Sample size	Sample mean	Sample standard deviation
Fertilized trees	48	0.30	0.058
Unfertilized trees	55	0.35	0.077

a. Find a 95% confidence interval for the true mean tannin percentage in apples from fertilized trees.

b. Find a 95% confidence interval for the true mean tannin percentage in apples from unfertilized trees.

c. Is there any evidence to suggest that the percentage of tannin is different in apples from fertilized and unfertilized trees?

8.136 Sports and Leisure Most high schools across the country offer a wide variety of sports for both men and women. However, the participation rate has historically been higher for boys. A random sample of high school students from across the country was obtained. The students were asked whether they participate in high school sports, and the results are given in the following table.

Group	Sample size	Number participating
Boys	1250	583
Girls	1475	494

a. Find a 95% confidence interval for the true proportion of boys participating in a high school sport.

b. Find a 95% confidence interval for the true proportion of girls participating in a high school sport.

c. Is there any evidence to suggest that the proportion of boys and girls participating in a high school sport is different? Justify your answer.

8.137 Public Health and Nutrition Elevated levels of mercury can cause hair loss, fatigue, and memory lapses. The concentration of mercury in a person's blood can be greatly affected by diet, specifically, the amount of fish consumed. Random samples of adults who eat fish two to three times per week and who never eat fish were obtained. The mercury concentration in the blood of each person was measured (in micrograms per liter of blood), and the summary data are reported in the following table.

Group	Sample size	Sample mean	Sample standard deviation
Fish	18	4.662	0.298
No fish	27	2.079	0.309

Assume both distributions are normal.

a. Find a 95% confidence interval for the population mean mercury concentration in the blood of adults who eat fish regularly.

b. Find a 95% confidence interval for the population mean mercury concentration in the blood of adults who never eat fish.

c. The safe level of mercury (set by the U.S. Environmental Protection Agency) is 5 micrograms per liter of blood. Is there any evidence to suggest that either group is over the safe limit of mercury concentration?

d. Use the confidence intervals in parts (a) and (b) to determine whether the two groups have different mean mercury concentration levels.

8.138 Biology and Environmental Science There is growing concern that Americans are generating too much trash. More zoning permits are being sought for landfills in rural areas, and trash haulers move garbage out-of-state and even out to the ocean. In a recent study, 65 households were randomly selected, and the amount of trash (in pounds) generated by each in 1 week was recorded. The sample mean was $\bar{x} = 52.3$ with a standard deviation of $s = 10.75$. Assume normality.

a. Find a 98% confidence interval for the true mean amount of trash generated by an American household per week.

b. Find an *approximate* 98% confidence interval for the true mean amount of trash generated by an American household per week (based on a Z distribution rather than a t distribution). Compare your answers in parts (a) and (b). How do they differ?

c. Find a 95% confidence interval for the variance in the amount of trash generated by an American household per week.

8.139 Medicine and Clinical Studies In a major health study, a random sample of adult males was obtained and each was tested for symptoms of heart disease yearly for a decade. Subjects were divided into those who regularly donated blood and those who did not. The results are given in the following table.

Group	Sample size	Number with heart disease
Donate blood	145	51
Do not donate blood	527	210

a. Find a 95% confidence interval for the population proportion of males who donate blood who have heart disease.

b. Find a 95% confidence interval for the population proportion of males who do not donate blood who have heart disease.

c. Is there any evidence to suggest that the proportion of males with heart disease is different for those who donate blood and those who do not?

8.140 Manufacturing and Product Development Teflon-coated pots and pans are designed to be non-stick and to make cooking and cleanup easier. A recent study compared the thickness of the Teflon layer in new, factory-coated pans with recoated pans. A random sample of pans was obtained for each group, and the Teflon thickness was carefully measured (inches). The data are given on the data CD and book's web site.

a. Find a 95% confidence interval for the population mean Teflon thickness in factory-coated pans.

b. Find a 95% confidence interval for the population mean Teflon thickness in recoated pans.

c. Is there any evidence to suggest that these two population means are different? Justify your answer.

Hypothesis Tests Based on a Single Sample

9

Chapter 9 Challenge

Should city officials redesign St. Mark's Trail?

The use of trails in national parks and forests has sparked controversy among off-road vehicle (ORV) enthusiasts, hikers, and organizations like The Wilderness Society and the Sierra Club. Some parks have banned the use of ORVs on certain trails, and newspapers and politicians have been barraged with letters and pleas from all parties. Primary issues are air, water, and noise pollution; possible vegetation damage caused by four-wheeled all-terrain vehicles (ATVs), and the impact of these vehicles on the recreational experience of other trail users.

Hiking, bicycling, and backpacking have gained in popularity, and ATV, motorcycle, and snowmobile users have started to make extensive use of man-made trails. Government land managers develop new trails and maintenance plans on the basis of trail users' opinions. As reported by the Iowa Department of Transportation, a study conducted for the National Park Service (NPS) considered adjacent landowner perceptions and the economic impacts of three trails. According to the NPS study, 65% of the respondents use the Heritage Trail in Iowa for bicycling, 81% use St. Mark's Trail in Tallahassee, Florida, for bicycling, and 63% use the Lafayette/Morage Trail near San Francisco for walking. Suppose city officials will redesign St. Mark's Trail if there is conclusive statistical evidence that more than 75% of the respondents use the trail for bicycling. The statistical test and resulting decision are presented in the Chapter Challenge Wrap-Up (page 439).

(Mike Powell/AgeFotostock)

Review

- Recall the point estimates for the parameters μ, p, and σ^2.
- Remember how to construct and interpret confidence intervals.
- Think about the concept of a sampling distribution for a statistic and the process of standardization.

Preview

- Use the available information in a sample to make a specific decision about a population parameter.
- Understand the formal decision process and learn the four-part hypothesis test procedure.
- Conduct formal hypothesis tests concerning the population parameters μ, p, and σ^2.

9.1 THE PARTS OF A HYPOTHESIS TEST AND CHOOSING THE ALTERNATIVE HYPOTHESIS

You have probably heard the word **hypothesis** used in many different contexts. An engineer might have a hypothesis concerning gas mileage on a certain car. She might claim that a new engine design will significantly improve the miles-per-gallon rating. Or an agronomist might hypothesize that a special combination of nutrients will significantly increase the growth rate of yellow corn.

> **DEFINITION**
>
> In statistics, a **hypothesis** is a declaration, or claim, in the form of a mathematical statement, about the value of a specific population parameter (or about the values of several population characteristics).

Here are some examples of statistical hypotheses.

1. $\mu = 14.5$

   where μ is the population mean time (in minutes) it takes for an adult's pupils to dilate after treatment with phenylephrine.

2. $p > 0.75$

   where p is the population proportion of all voters in a district who favor property tax reform.

3. $\sigma^2 \neq 30.5$

   where σ^2 is the population variance in the amount (in gallons) of coal tar in a 5-gallon bucket.

A hypothesis is a claim about a *population* parameter, **not** about a sample statistic. For example, $\mu = 5$ and $p = 0.27$ are valid hypotheses, but $\bar{x} = 27$ and $s = 32.5$ are not.

There are four parts to every hypothesis test, and it is important to identify each part in every test.

H_0 is read as "*H* sub zero," and H_a is read as "*H* sub a."

> **FOUR PARTS OF A HYPOTHESIS TEST**
>
> 1. The **null hypothesis**, denoted H_0.
>   This is the claim (about a population parameter) assumed to be true, what is believed to be true, or the hypothesis to be tested. Sometimes referred to as the *no change* hypothesis, this claim usually represents the status quo or existing state. There is an implied inequality in H_0, however, the null hypothesis is written in terms of a single value (with an equal sign), for example, $\theta = 5$. While it may seem strange, we usually try to *reject* the null hypothesis.
> 2. The **alternative hypothesis**, denoted H_a.
>   This statement identifies other possible values of the population parameter, or simply a possibility not included in the null hypothesis. H_a indicates the possible values of the parameter if H_0 is false. Experiments are often designed to determine whether there is evidence in favor of H_a. The alternative hypothesis represents change in the current standard or existing state.

3. The **test statistic**, denoted TS.
   This statistic is a rule related to the null hypothesis that involves the information in a sample. The *value* of the test statistic will be used to determine which hypothesis is more likely to be true, H_0 or H_a.
4. The **rejection region** or **critical region**, denoted RR or CR.
   This is an interval or set of numbers specified such that if the value of the test statistic lies in the rejection region, then the null hypothesis is rejected. There is also a corresponding *nonrejection region*; if the value of the test statistic lies in this set, then we *cannot reject H_0*.

ILLUMINATING THE CONCEPTS

1. The test of a statistical hypothesis is a procedure by which we decide whether or not there is evidence to suggest that the **alternative hypothesis**, H_a, is true. The ultimate objective of a hypothesis test is to use the information in a sample to decide which hypothesis is more likely to be true, H_0 or H_a. Usually, we are trying to reject the **null hypothesis**.

2. The **rejection region** and the nonrejection region divide the world (values of the **test statistic**) into parts. Figure 9.1 illustrates this concept in terms of a parameter θ. The cutoff point, or dividing line, is determined by considering likely values for the test statistic if H_0 is true and is included in one of the regions.

   The value of θ must lie in one of the regions. In Section 9.3, we'll see how easy it is to specify a rejection region.

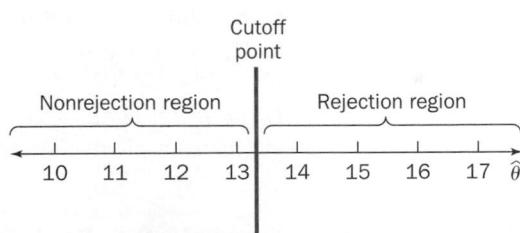

Figure 9.1 An illustration of a rejection region and a nonrejection region.

3. The hypothesis test procedure is very prescriptive. Once the four parts are identified, the sample data are used to compute a value of the test statistic. There are only two possible conclusions:
   a. If the value of the test statistic lies in the rejection region, then we reject H_0. There is evidence to suggest that the alternative hypothesis is true.
   b. If the value of the test statistic does not lie in the rejection region, then we cannot reject H_0. There is no evidence to suggest that the alternative hypothesis is true.

   We never say that we *accept H_0*. A hypothesis test is designed to prove the alternative hypothesis. If there is no evidence in favor of H_a, this does *not* imply H_0 is true.

   A hypothesis test can only provide support in favor of H_a. If the value of the test statistic lies in the rejection region, reject the null hypothesis. There is evidence to suggest that H_a is true. If the value of the test statistic does not lie in the rejection region, do not reject H_0. There is no evidence to suggest that H_a is true. Watch the wording: We **never** *accept* the null hypothesis. Rather, we say that the value of the test statistic *does not lie in the rejection region*.

4. This formal hypothesis test procedure is analogous to the four-step inference procedure used in the previous chapters. In fact, many of the concepts presented in earlier chapters are combined here to produce this traditional, well-established hypothesis test procedure.

The **Claim** corresponds to H_0, a claim about a population parameter. The **Experiment** is equivalent to a value of the test statistic. **Likelihood** is expressed in terms of the nonrejection region (likely values of the test statistic) and the rejection region (unlikely values of the test statistic). The **Conclusion** is completely determined by the region in which the value of the test statistic lies. If the value is in the rejection region, we reject H_0; otherwise, we cannot reject H_0. ∎

Suppose a hypothesis test is conducted concerning the population parameter θ, and θ_0 is a specific value of θ. The null hypothesis is always stated in terms of a single value. There are only three possible alternative hypotheses:

$$H_0 : \theta = \theta_0$$
$$H_a : \left.\begin{array}{l} \theta > \theta_0 \\ \theta < \theta_0 \end{array}\right\} \text{ one-sided alternatives}$$
$$\theta \neq \theta_0 \text{ } \} \text{ two-sided alternative}$$

There are two one-sided alternatives and one two-sided alternative.

Only one alternative hypothesis is selected. H_a answers the question, "What is the experimenter trying to prove, or detect, about θ?" It takes a little practice to decide which H_a is appropriate. The same specific value of the parameter (θ_0, above) always appears in H_0 and H_a.

A valid set of null and alternative hypotheses must include a statement similar to H_0 above and the relevant alternative, one of the three given above. For example, $H_0: \mu = 17$; $H_a: \mu < 17$ is a valid set of null and alternative hypotheses. $H_0: \tilde{\mu} \neq 25$; $H_a: \tilde{\mu} = 26$ is not. The following examples focus on identifying H_0 and the relevant alternative hypothesis.

Example 9.1 Back Pain Millions of Americans suffer from back pain, and some experience disc problems so severe that simple tasks, such as driving, sitting in a chair, or even sleeping are painful. The traditional remedy for a damaged disc is surgery: spinal fusion. Historical records indicate that 65% of all patients who endure this costly, complicated surgery experience reduced pain and greater mobility. A new treatment (IDET, intradiscal electrothermal annuloplasty) has been developed, and researchers claim that this procedure is more effective, cheaper, and less painful. An experiment is conducted to determine whether IDET is more effective than spinal fusion. What null and alternative hypotheses should be used?

SOLUTION

STEP 1   The claim assumed to be true involves the population proportion of patients who experience reduced pain and greater mobility with the existing surgical technique. It is assumed that 65% of all patients experience relief.

Therefore, the null hypothesis, stated in terms of a single value for p, is

$$H_0: p = 0.65.$$

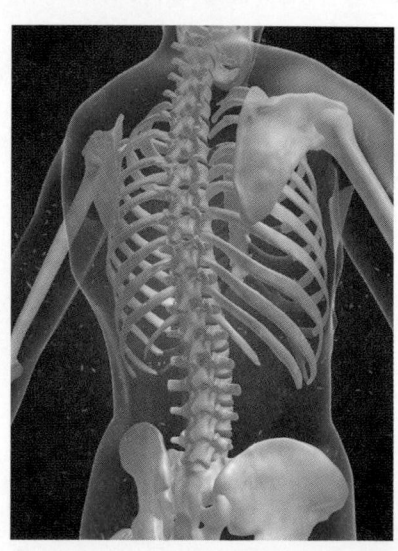

(Eraxion/Dreamstime.com)

STEP 2   The existing proportion of patients who experience relief is 0.65.
The experiment is designed to detect an increase in this proportion
(i.e., to answer the question, "Is the new procedure more effective?").
Researchers hope to find evidence that the proportion of patients who
experience relief with IDET is greater than 0.65. Therefore, the alternative
hypothesis is

H_a: $p > 0.65$.

Example 9.2 The Cost of Divorce   Although the divorce rate has stabilized
over the past few years, the institution of marriage has changed dramatically. A
divorce procedure can be very messy, time consuming, contentious, and costly.
Historical records indicate that the mean cost of an uncontested divorce processed
through attorneys is $1200.[1] A new online service offers an uncontested divorce
service with a cost structure that may save couples money. An observational study
is conducted to determine whether the mean cost of an online divorce is less than
that of a lawyer-processed divorce. State the appropriate null and alternative
hypotheses.

SOLUTION

STEP 1   The claim assumed to be true involves the population mean cost (in
dollars) of an uncontested divorce processed through an attorney.
The null hypothesis, the existing standard, is given in terms of the
parameter μ.

H_0: $\mu = 1200$.

STEP 2   The investigator is searching for evidence in favor of a *lower* mean cost
for an online divorce. The experiment is designed with the hope of
finding evidence that the mean cost for an online divorce is less than
the mean cost using lawyers. Therefore,

H_a: $\mu < 1200$.

 Try the Test of Significance
statistical applet on the text
web site.

Example 9.3 Recycled Paper   The thickness (measured in inches) of recy-
cled printer paper is important, because sheets that are too thick will clog the
machine, and paper that is too thin will rip and bleed toner. The variance in thickness for 20-lb printer paper at a manufacturing plant is known to be 0.0007. A new
process is developed using more recycled fiber, and an experiment is conducted to
detect any difference in the variance in paper thickness. State the appropriate null
and alternative hypotheses.

SOLUTION

STEP 1   The assumption, or existing state, involves the population variance in
thickness of recycled printer paper. The null hypothesis is given in
terms of σ^2.

H_0: $\sigma^2 = 0.0007$.

STEP 2   The experiment is designed to detect *any* difference in the population
variance. This suggests a two-sided alternative.

H_a: $\sigma^2 \neq 0.0007$.

SECTION 9.1 EXERCISES

Practice

9.1 Determine whether each of the following statements is a valid hypothesis. Classify each valid hypothesis as a null hypothesis or an alternative hypothesis. Justify your answers.

a. $\mu = 0.355$ b. $\hat{p} < 0.42$ c. $s > 3.5$
d. $\bar{x} \neq 16$ e. $\mu > 22.66$ f. $p \neq 0.15$
g. $\tilde{x} < 47.5$ h. $\tilde{\mu} = 12$

9.2 Determine whether each of the following statements is a valid hypothesis. Classify each valid hypothesis as a null hypothesis or an alternative hypothesis. Justify your answers.

a. $\sigma^2 = 49.55$ b. $IQR = 25$ c. $\mu \neq 17$
d. $\bar{y} = 100.7$ e. $Q_3 = 7.65$ f. $\mu < 33.79$
g. $\sigma \neq 8.95$ h. $p = 0.77$

9.3 Determine whether each pair of statements is a valid set of null and alternative hypotheses. Justify your answers.

a. $H_0: p = 0.55$; $H_a: p < 0.55$
b. $H_0: \mu = 9.7$; $H_a: \mu \geq 9.7$
c. $H_0: \sigma^2 = 98.6$; $H_a: \sigma^2 = 101$
d. $H_0: \tilde{\mu} = 38.9$; $H_a: \tilde{\mu} < 38.9$

9.4 Determine whether each pair of statements is a valid set of null and alternative hypotheses. Justify your answers.

a. $H_0: \mu = 30$; $H_a: \mu \neq 30$
b. $H_0: \sigma = 3.55$; $H_a: \sigma > 3.55$
c. $H_0: p \leq 0.32$; $H_a: p > 0.32$
d. $H_0: \bar{x} = 78.5$; $H_a: \bar{x} \neq 78.5$

9.5 Determine whether each pair of statements is a valid set of null and alternative hypotheses. Justify your answers.

a. $H_0: p = 0.50$; $H_a: p \neq 0.50$
b. $H_0: \mu = 25.6$; $H_a: \mu < 25.6$
c. $H_0: \mu < 35.9$: $H_a: \mu \geq 35.9$
d. $H_0: \sigma^2 = 95$; $H_a: \sigma^2 > 95$

9.6 In each of the following problems, a conclusion to a hypothesis test is presented. Determine whether each statement is permissible.

a. The value of the test statistic does not lie in the rejection region. Therefore, we accept the null hypothesis.
b. The value of the test statistic lies in the rejection region. Therefore, there is evidence to suggest that the null hypothesis is not true.
c. The value of the test statistic does not lie in the rejection region. Therefore, there is evidence to suggest that the null hypothesis is true.
d. The value of the test statistic lies in the non-rejection region. Therefore, there is no evidence to suggest that the alternative hypothesis is true.
e. The value of the test statistic does not lie in the rejection region. Therefore, there is no evidence to suggest that the alternative hypothesis is true.
f. The value of the test statistic lies in the rejection region. Therefore, there is evidence to suggest that the alternative hypothesis is true.

Applications

9.7 Education and Child Development The College Board reported that the mean cumulative SAT score (Critical Reading, Mathematics, and Writing) for 2007 college-bound seniors was 1511.[2] Officials from the Pennsylvania State System of Higher Education would like to know whether the students enrolled for fall 2008 classes have a mean cumulative SAT score greater than 1511. A random sample of students from across the system was obtained. Each cumulative SAT score was recorded, and the mean cumulative score was recorded. State the null and alternative hypotheses.

9.8 Marketing and Consumer Behavior A marketing research study conducted in 2009 indicated that 11% of all households in the United States have at least one telescope. During November 2008, the moon, Venus, and Jupiter were clustered together in the night sky. A company that manufactures telescopes believes the increased media attention caused more people to buy telescopes in order to see this spectacular sight. State the null and alternative hypotheses in terms of the population proportion of households that have at least one telescope.

9.9 Biology and Environmental Science During the summer months, wildfires in the western United States pose a great hazard to people, residential and commercial buildings, and animals. Records indicate that the mean number of acres burned during a wildfire is 17,060. The most recent summer was unusually wet, and fire-fighting officials would like to know whether the mean number of acres burned during wildfires was any less than normal. State the null and alternative hypotheses.

9.10 Fuel Consumption and Cars In 2007, there were 8225 passenger vehicle occupant deaths as a result of side-impact automobile accidents (*Source:* Insurance Institute for Highway Safety). Side air bags that protect the head, chest, and abdomen lower the risk of death in such crashes. However, these safety devices still are not standard in all vehicles, because of cost and because many drivers fear that side air bags pose a threat to small children. Only 7% of 2007 pickups offered side air bags.[3] A large medical insurance company would like to know whether the proportion of 2008 model pickups with side air bags is different from 0.07. State the null and alternative hypotheses.

9.11 Sports and Leisure The stereotypical videogame player is a male teenager. However, recent studies suggest that more women over 18 than young boys regularly play video games.[4] A software company has decided to develop and market a sophisticated stock market video game if the mean age of all video-game players is greater than 25. A random sample of players will be obtained, and the resulting data will be used to test the relevant hypothesis. Let μ represent the mean age of all video-game players. Which of the following sets of null and alternative hypotheses is appropriate? Justify your answer.

a. $H_0: \mu = 25$ versus $H_a: \mu > 25$.
b. $H_0: \mu = 25$ versus $H_a: \mu < 25$.
c. $H_0: \mu = 25$ versus $H_a: \mu \neq 25$.

9.12 Physical Sciences There are approximately 16 fiber-optic lines under the ocean off the Florida coast. These lines carry telephone and Internet communications between Florida and Europe, Latin

America, and the Caribbean. During storms, these lines sway dramatically, often damage coral, and get caught in anchors. Technical reports of the distance the line swayed indicate that the variance in sway during storms is 32 feet. A study will be conducted to determine whether a new, heavier cable housing will decrease the variance in sway. State the relevant null and alternative hypotheses in terms of σ^2, the variance in cable sway.

9.13 Technology and Internet Many limousines now offer the latest in high-tech gadgets. DVD players, satellite radio, and wireless Internet connections are some of the high-priced accessories. Suppose Boston Coach will install wireless Internet in all of its vehicles if more than 75% of all clients prefer this service. A random sample of clients will be selected, and the resulting information will be used to test the relevant hypothesis. Let p represent the proportion of all clients who prefer wireless Internet in a limousine. State the null and alternative hypotheses in terms of p.

9.14 Sports and Leisure The mean duration of a Major League baseball game from the first pitch to the last out over the first seven weeks of the 2008 season was 2 hours and 51 minutes.[5] During the second half of the season, major league umpires tried various techniques to speed up the game. A random sample of second-half games will be selected, and the resulting game durations will be used to determine whether games take less time to complete. Let μ be the mean duration of a baseball game. State the relevant null and alternative hypotheses in terms of μ.

9.15 Travel and Transportation DATTCO, a school bus company in New Britain, Connecticut, will install seat belts on all buses if more than 50% of all parents favor this change. A random sample of parents in the school district will be obtained, and the responses will be used to test the relevant hypothesis. If p is the true proportion of parents who favor the seat belt installation, which of the following sets of null and alternative hypotheses is appropriate? Justify your answer.
a. $H_0: p = 0.50$ versus $H_a: p \neq 0.50$.
b. $H_0: p = 0.50$ versus $H_a: p < 0.50$.
c. $H_0: p = 0.50$ versus $H_a: p > 0.50$.

9.16 Public Policy and Political Science Suppose Dan Kotowski, a state senator from Mount Prospect, Illinois, is considering a run for governor. He will enter the campaign if there is evidence to suggest that more than 65% of all state residents favor his candidacy. A random sample of likely voters is obtained, and the resulting survey data will be used to test the relevant hypothesis. Let p be the proportion of all voters who favor his candidacy. State the null and alternative hypotheses in terms of p.

9.17 Economics and Finance Workers' Compensation officials will lay off investigators if there is evidence to suggest that the proportion of fraudulent claims is less than 0.35. A random sample of claims will be selected and carefully investigated, and the data will be used to test the relevant hypothesis. State the null and alternative hypotheses in terms of p, the proportion of fraudulent workers' compensation claims.

9.18 Education and Child Development Students at Stetson University plan to ask administrators to build more on-campus housing on the DeLand campus if there is any evidence that the median monthly rent for off-campus housing is more than $350 per person. A random sample of students living off-campus will be obtained, and the resulting rent data will be used to test the relevant hypothesis. Let $\tilde{\mu}$ be the median monthly rent for students. State the null and alternative hypotheses in terms of the population median $\tilde{\mu}$.

9.19 Medicine and Clinical Studies A new surgical procedure has been developed to remove cataracts; it involves a smaller incision. This more expensive procedure will be implemented at a major hospital only if there is evidence that the standard deviation in time to recovery is less than seven days. A random sample of patients will receive the new procedure, and the resulting recovery times will be used to test the relevant hypothesis. Let σ be the population standard deviation in recovery time. State the null and alternative hypotheses in terms of σ.

9.20 Public Policy and Political Science Officials in Dexter, a small town in upstate New York, have decided to install more fire hydrants in order to decrease insurance rates for many businesses. Prior to the new installations, the mean distance to a fire hydrant for downtown buildings was 525 feet. After the new installations, a random sample of downtown buildings will be obtained, and the distance to the nearest fire hydrant will be recorded. The data will be used to determine whether there is evidence that the mean distance to a fire hydrant has decreased. State the null and alternative hypotheses in terms of μ, the population mean distance to a fire hydrant.

9.21 Public Policy and Political Science The Health Insurance Portability and Accountability Act of 1996 (HIPAA) was designed to protect personal privacy. However, this law has created mountains of paperwork and may even increase the cost of medical care. The federal government has decided to consider repeal of this law if more than 60% of all hospitals are experiencing increased costs due to this regulation. A random sample of hospitals will be obtained, and the resulting information will be used to test the relevant hypothesis. State the null and alternative hypotheses in terms of p, the population proportion of hospitals experiencing increased costs due to HIPAA.

9.22 Economics and Finance The first phase of a tax amnesty program in California ended in 2005. This program was designed for businesses and residents who underpaid their taxes or did not pay any taxes at all. These residents could pay their taxes in full before May 31 and would then be removed from the delinquent list. Suppose this new plan will be enacted again in future years if there is evidence to suggest that the mean payment is greater than $1235. State the null and alternative hypotheses in terms of μ, the population mean payment for residents taking advantage of the amnesty program.[6]

9.23 Public Policy and Political Science The Sparks City Council in Nevada will build a new dock at the Sparks Marina if more than 80% of the residents favor the plan. A random sample of residents will be obtained, and their responses will be used to test the relevant hypothesis. Let p be the population proportion of residents who favor a new marina. Which of the following sets of null and alternative hypotheses is appropriate? Justify your answer.
a. $H_0: p = 0.80$ versus $H_a: p \neq 0.80$.
b. $H_0: p = 0.80$ versus $H_a: p < 0.80$.
c. $H_0: p = 0.80$ versus $H_a: p > 0.80$.

9.24 Business and Management The Savannah Sugar Refinery in Louisiana will invest in new energy-saving devices if there is evidence to suggest that the mean amount of energy (in kilowatts) used per day will decrease from 1925. Suppose μ is the mean amount of energy used per day, and an experiment is designed to test the new equipment. State the null and alternative hypotheses in terms of μ.

9.25 Economics and Finance Suppose monetary policy will be set by the Federal Reserve Board by examining the median consumer price, $\tilde{\mu}$, for a large fixed set of common commodities. The Board will raise the interest rate if there is evidence that the median price is less than $125.50. A random sample of counties will be obtained, and the cost of these goods in each county will be used to test the relevant hypothesis. State H_0 and H_a in terms of $\tilde{\mu}$.

9.2 HYPOTHESIS TEST ERRORS

To conduct a hypothesis test, we use the information in a sample in order to reach a decision about the value of a population parameter. The sample data lead to a value of the test statistic and the ultimate decision (reject or do not reject H_0). However, there is always a chance of making a mistake (the wrong decision). Even a simple random sample is only a (usually small) portion of the entire population. This limited information could lead to the wrong conclusion about the population parameter. To fully understand the structure of a hypothesis test, we need to examine what could possibly go wrong.

The water in Lake Jean, located in Ricketts Glen State Park, Pennsylvania, is clear and cool, sometimes *very* cool. Park officials have decided to post an advisory warning for swimmers if there is any evidence to suggest that the mean water temperature, μ, is less than 62 °F. Otherwise, the lifeguards will post no signs and allow swimming as usual. Each day, lifeguards measure the temperature of the water (in °F) at 15 randomly selected locations around the lake. They use the sample mean water temperature, $\bar{x}$, to test the hypotheses

> In this hypothesis test, $\bar{x}$ is the test statistic.

$$H_0: \mu = 62 \quad \text{versus} \quad H_a: \mu < 62.$$

The lifeguards have decided to use a *cutoff point* of 61.5 °F. They cannot measure the temperature of the water at every location in the lake (and hence compute the population mean) and remember, the sample mean varies around the population mean. The lifeguards have decided that a sample mean of $\bar{x} \le 61.5$ is far enough away from 62 that it cannot be attributed to ordinary variation about the population mean.

The rejection region is any value of $\bar{x}$ less than or equal to 61.5 (see Figure 9.2). If the value of $\bar{x}$ is 61.5 or less, then H_0 is rejected, and an advisory warning is posted. If the value of $\bar{x}$ is greater than 61.5, the lifeguards cannot reject H_0. There is no evidence to suggest that the mean water temperature is less than 62 °F.

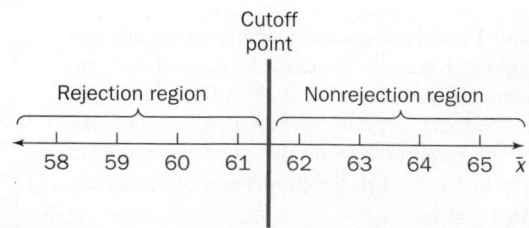

Figure 9.2 The rejection region and nonrejection region for the Lake Jean water temperature hypothesis test.

Here is what can happen when the sample data are collected and the hypothesis is tested.

1. Suppose the true population mean μ is 62 °F or greater; the water is warm enough for swimming.
   a. If $\bar{x} > 61.5$ °F, then there is no evidence to reject H_0. This conclusion is *correct*.
   b. If $\bar{x} \leq 61.5$ °F, then there is evidence to reject H_0. An advisory warning is posted. But this conclusion is *incorrect*. The water really is warm enough for swimming.
2. Suppose the true population mean μ is less than 62 °F; the water is too cold for swimming.
   a. If $\bar{x} > 61.5$ °F, then there is no evidence to reject H_0, so the lifeguards decide to allow swimming. But this conclusion is *incorrect*. The water is really too cold for swimming.
   b. If $\bar{x} \leq 61.5$ °F, then there is evidence to reject H_0. An advisory warning is posted. This conclusion is *correct*.

This example illustrates the two possible errors in a hypothesis test.

> This is all very theoretical. *You never really know whether you made a mistake in a hypothesis test.*

DEFINITION

1. The value of the test statistic may lie in the rejection region, but the null hypothesis is true. If we reject H_0 when H_0 is really true, this is called a **type I error**. The probability of a type I error is called the *significance level* of the hypothesis test and is denoted by α: P(type I error) = α.

2. The value of the test statistic may not lie in the rejection region, but the alternative hypothesis is true. If we do not reject the null hypothesis when H_a is really true, this is called a **type II error**. The probability of a type II error is denoted by β: P(type II error) = β.

The following table illustrates the decisions and errors in a hypothesis test.

		Decision	
		Reject H_0	Do not reject H_0
Truth	H_0	Type I error	Correct decision
	H_a	Correct decision	Type II error

Each example below describes a specific hypothesis test. The **type I error** and the **type II error** are described in context, and the real-world consequences of a wrong decision are given.

(Robert Deal/iStockphoto)

Example 9.4 Top Quality Control   The National Institute of Standards and Technology (NIST) reported that the Ames Rubber Corporation (a Malcolm Baldridge National Quality Award Winner) is a benchmark producer of fuse rollers for copiers. NIST claims the proportion of defective fuse rollers manufactured by this company is 0.000011. Suppose Xerox receives a large shipment of fuse rollers for use in its copiers. The manager knows that some are defective but does not have the time or resources to inspect each fuse roller. Xerox is willing to accept 0.000011 defective

There are two possible truth assumptions, as indicated in the decision table: H_0 is true or H_a is true.

fuse rollers or any smaller proportion, of course. A random sample of fuse rollers will be obtained and carefully inspected. The information in the sample will be used to test the hypotheses

$$H_0: p = 0.000011 \quad \text{versus} \quad H_a: p > 0.000011.$$

If H_0 is rejected, the entire shipment of fuse rollers is sent back to the Ames Rubber Corporation. Discuss the consequences of the decision to reject or not to reject for each truth assumption.

SOLUTION

STEP 1   Suppose H_0 is true; the proportion of defective fuse rollers is really 0.000011 (or less).
  a. If H_0 is rejected, the entire fuse roller shipment is sent back. This is a type I error and in this case Ames Rubber Corporation is not happy. The fuse rollers were of acceptable quality, but the entire shipment is returned.
  b. If H_0 is not rejected, then the fuse roller shipment is accepted. Everyone is happy in this case. The hypothesis test indicated no evidence of a higher proportion of defective fuse rollers.

STEP 2   Suppose H_a is true; the proportion of fuse rollers is *greater than* 0.000011.
  a. If H_0 is rejected, the entire fuse roller shipment is returned. This is the correct conclusion. The hypothesis test indicated the fuse roller shipment contained a higher proportion of defectives than the minimum acceptable level, and Xerox is glad to return a bad batch.
  b. If H_0 is not rejected, then the fuse roller shipment is accepted. This is a type II error and in this case Xerox is not happy. Too many fuse rollers are defective, yet Xerox is accepting a shipment of poor quality.   ●

Example 9.5 Take Two Aspirins   *FDA Consumer Magazine* reported that Americans take 80 million aspirin tablets each day. Surprisingly, most are taken to reduce the risk of heart disease, not for everyday aches and pains. Suppose Walgreens receives a large shipment of Bayer aspirin and each tablet supposedly contains 325 mg of acetylsalicylic acid, the active ingredient in aspirin. A random sample of tablets will be obtained and chemically analyzed. The information in the sample will be used to test the hypotheses

$$H_0: \mu = 325 \quad \text{versus} \quad H_a: \mu \neq 325$$

where μ is the mean amount of acetylsalicylic acid in each tablet. If H_0 is rejected, the entire aspirin shipment will be returned to the Bayer corporation. Discuss the consequences of the decision to reject or not to reject for each truth assumption.

SOLUTION

STEP 1   Suppose H_0 is true; the mean amount of the active ingredient is 325 mg.
  a. If H_0 is rejected, the aspirin shipment is returned to Bayer. This is a type I error and Bayer would not like this decision. The aspirin has the specified amount of acetylsalicylic acid, but the entire shipment is returned.
  b. If H_0 is not rejected, then the aspirin bottles are placed on the shelves at Walgreens. Everyone is happy. The aspirin contains the correct amount of the active ingredient, and the hypothesis test did not indicate otherwise.

STEP 2 Suppose H_a is true; the mean amount of active ingredient is not 325 mg.

    a. If H_0 is rejected, the aspirin shipment is returned to Bayer. This is the correct conclusion. The hypothesis test indicated that the mean amount of acetylsalicylic acid is different from 325 mg, and Walgreens is happy to return a flawed shipment.

    b. If H_0 is not rejected, the aspirin bottles are placed on the shelves at Walgreens, ready for sale. This is a type II error, and in this case Walgreens customers suffer. The mean amount of the active ingredient is not as specified, yet the aspirin is placed in store stock and sold. ●

Example 9.6 Plastic-Sheet Thickness The Kleerdex Company produces KYDEX plastic sheet products in a wide range of textures, patterns, and colors. Some applications include bus seats, clean-room walls and ceilings, and aircraft interior components. The plastic sheets are designed to have a certain thickness, but the focus on the assembly line is on the variance in thickness. Each hour, a random sample of plastic sheets is obtained and the thickness of each sheet is carefully measured (in inches). The sample variance is used to test the hypotheses

$$H_0: \sigma^2 = 0.2 \quad \text{versus} \quad H_a: \sigma^2 > 0.2$$

where σ^2 is the population variance in plastic sheet thickness. If H_0 is rejected, the entire assembly line is shut down for inspection. Discuss the consequences of the decision to reject or not to reject for each truth assumption.

> Remember, H_0 is stated in terms of an equality, but there is an implied inequality in the null hypothesis. The assembly line will be stopped only if there is evidence that the population variance is greater than 0.2. The status quo or no change state is really $\sigma^2 \leq 0.2$.

SOLUTION

STEP 1 Suppose H_0 is true; the population variance is 0.2 (or less).

    a. If H_0 is rejected, the assembly line is shut down. This is a type I error. The null hypothesis is true, but H_0 is rejected. This error is bad for Kleerdex. The line has been shut down unnecessarily, and production time is lost.

    b. If H_0 is not rejected, then the assembly line continues to manufacture plastic sheets. This is a correct decision.

STEP 2 Suppose H_a is true; the population variance in thickness is greater than 0.2.

    a. If H_0 is rejected, the correct decision has been made. The variance is too high and the hypothesis test procedure suggested that the assembly line should be shut down for inspection.

    b. If H_0 is not rejected, the assembly line continues to hum along. This is an incorrect decision, a type II error. H_a is true but the null hypothesis is not rejected. In this case, there is a good chance the plastic sheets are being made with too much variability in thickness. ●

> A *perfect* hypothesis test would have the probability of a type I error and the probability of a type II error both equal to 0.

The *efficiency* or *goodness* of a hypothesis test is often measured in terms of α and β. Intuitively, an effective test should make the probability of both types of errors small. In fact, since no one wants to make a mistake, it would be ideal to have $\alpha = \beta = 0$. However, this is impossible, since any decision in a hypothesis test is made using information in a *sample*. We can never examine the entire population, so there is always a chance of making a mistake.

> α and β are inversely related, but not by any set formula. We know only that as α decreases, β increases, and as β decreases, α increases.

◀ In a hypothesis test, we usually control the value of α. We do not have any direct control over β, but can often compute the probability of a type II error for a specific alternative value of the parameter. It seems reasonable to set α as small as possible, but for a fixed sample size, α and β are inversely related. For a fixed n, making α smaller forces β to be larger. The only way to decrease both α and β (simultaneously) is to increase the sample size. More information (larger n) means a smaller chance of making a mistake. The most common values for α are 0.05 and 0.01. The probability of a type II error actually *depends* on the specific alternative value of the population parameter being tested. Therefore, the probability of a type II error is usually written as a *function* of the population parameter. For example, $\beta(\mu_a)$ represents the probability of a type II error if the true value of μ is μ_a.

Consider a hypothesis test with $H_0: \theta = \theta_0$ and $H_a: \theta \neq \theta_0$. As the alternative value of the parameter, θ_a, moves farther away from the hypothesized value, θ_0, the probability of a type II error, $\beta(\theta_a)$, decreases. This seems reasonable because for values of θ_a far away from θ_0, we have a better chance of detecting the difference between the hypothesized value and the alternative value of the population parameter.

Later in this chapter, we will visualize and learn how to compute the probability of a type II error for a specific alternative value of the population characteristic. ▶

SECTION 9.2 EXERCISES

Practice

9.26 Consider a hypothesis test with

$$H_0: \mu = 180 \quad \text{and} \quad H_a: \mu < 180.$$

Determine whether each of the following decisions is correct or in error. Identify each error as type I or type II.
a. The true value of μ is 180 and H_0 is rejected.
b. The true value of μ is 179 and H_0 is rejected.
c. The true value of μ is 160 and H_0 is not rejected.
d. The true value of μ is 182 and H_0 is rejected.

9.27 Consider a hypothesis test with

$$H_0: p = 0.44 \quad \text{and} \quad H_a: p \neq 0.44.$$

Determine whether each of the following decisions is correct or in error. Identify each error as type I or type II.
a. The true value of p is 0.44 and H_0 is not rejected.
b. The true value of p is 0.41 and H_0 is not rejected.
c. The true value of p is 0.45 and H_0 is not rejected.
d. The true value of p is 0.42 and H_0 is rejected.

9.28 Consider a hypothesis test with

$$H_0: \sigma^2 = 26.5 \quad \text{and} \quad H_a: \sigma^2 > 26.5.$$

Determine whether each of the following decisions is correct or in error. Identify each error as type I or type II.
a. The true value of σ^2 is 26.0 and H_0 is rejected.
b. The true value of σ^2 is 27.0 and H_0 is not rejected.
c. The true value of σ^2 is 26.4 and H_0 is not rejected.
d. The true value of σ^2 is 26.5 and H_0 is not rejected.

9.29 Recall, the probability of a type II error depends on the alternative specific value of the population parameter. Consider a hypothesis test with

$$H_0: \mu = 10 \quad \text{and} \quad H_a: \mu > 10.$$

a. For a fixed sample size, how do $\beta(11)$ and $\beta(15)$ compare? Are these two values approximately the same or different? If they are different, which is smaller, and why?

b. What happens to $\beta(\mu_a)$ as μ_a increases (gets further and further away from 10)?

9.30 Why is there always a chance of making a mistake (an incorrect decision) in any hypothesis test?

9.31 Since we usually control the value of α, we could simply set the probability of a type I error to a very small value, say 0.0001. What's wrong with this strategy?

Applications

9.32 Travel and Transportation Traffic on Route 95 in New Hampshire and Maine is slowed by accidents, bad weather, and, of course, toll booths. Department of Transportation (DOT) officials have decided to conduct a hypothesis test and will add additional toll booths if there is evidence to suggest that the mean number of cars passing through the Dover Toll Booth Plaza is more than 40,000 vehicles per day.[7]
a. Write the null and alternative hypotheses about μ, the mean number of cars passing through the Dover Plaza.
b. For the hypotheses in part (a), describe the type I and type II errors.
c. If a type I error is committed, who is more angry, the DOT officials or drivers, and why?
d. If a type II error is committed, who is more angry, the DOT officials or drivers, and why?

9.33 Public Policy and Political Science The Avenal State Prison in California has thousands of files on former inmates comprising an estimated 20 million pages of documents. A proposal has been made to archive many of these old files to compact discs. The state will release money for this project only if prison officials present evidence to suggest that the true mean age of all files is greater than 10 years. A random sample of files will be obtained, and the information in the sample will be used to test the hypotheses

$$H_0: \mu = 10 \quad \text{versus} \quad H_a: \mu > 10.$$

a. Describe a type I error and a type II error in this context.
b. From the warden's perspective, which error is more serious? Why?
c. From a state senator's point of view, which error is more serious? Why?

9.34 Physical Sciences The annual Waikiki Roughwater Swim contest is held over a 2.4-mile course and ends near the Hilton Rainbow Tower. The 2008 winner was Trent Grimsey in 47 minutes, 59 seconds.[8] Before the race, a random sample of the water current velocity (in knots) along the race course is obtained, and the resulting information is used to determine whether the race should be canceled. A mean current velocity, μ, of more than 0.65 knots is considered unsafe.
a. State the null and alternative hypotheses.
b. Describe type I and type II errors in this context.
c. Which error is more serious for the swimmers? Why?
d. Which error is more serious for the race organizers? Why?

9.35 Education and Child Development A Harris Interactive Poll for Tylenol PM indicated that 8% of American students have missed a test because they overslept. Suppose officials at a state college decide to apply a new academic policy regarding exams. If a student arrives late for a test, then the student receives a 0 for the test and cannot take a make-up test at a later time.
a. What hypotheses should be tested, in terms of p, in order for college officials to prove that the new academic policy is causing fewer students to be late for exams?
b. Which error, type I or type II, is more serious for college officials? Why?
c. Which error, type I or type II, is more serious for students? Why?

9.36 Medicine and Clinical Studies Recently, scientists have speculated that eating a moderate amount of dark chocolate may increase the level of antioxidants, compounds that protect us against free radicals, which can cause heart disease and cancer. Levels of antioxidants in the bloodstream have been shown to increase significantly after consumption of dark chocolate.[9] A study was designed, and the antioxidant concentration in patients who ate dark chocolate regularly was measured. This information was used to test the hypotheses

$$H_0: \mu = 0.4 \quad \text{versus} \quad H_a: \mu > 0.4$$

where μ is the true mean percentage of antioxidants in the bloodstream.
a. Describe a type I error and a type II error in this context.
b. Hershey Food Corporation, maker of Hershey's chocolates, is very interested in the results of this study. Which error is more serious for this company? Why?

9.37 Public Policy and Political Science The Dallas city council is going to consider a zoning variance for the Estates on Frankford apartment complex so that the developer may extend the structure closer to the nearest road. Some council members are concerned about safety, but will vote for the measure if more than 60% of all town residents favor the variance. A random sample of residents will be obtained and asked whether they favor the zoning variance. Let p be the true proportion of residents who favor the extended structure.
a. State the null and alternative hypotheses in terms of p.
b. Describe a type I error and a type II error in this context.
c. Which error is more serious for the developer? Why?
d. Which error is more serious for the city council members? Why?

9.38 Business and Management Many corporate executives are concerned that employees are using their computers for personal matters rather than business. However, Instant Messaging (IM) is fast becoming an acceptable informal means of communication. Some companies are relying on this technology and integrating this software with other management systems. Shaw Pittman, a law firm in Washington, D.C., has decided to use IM for communication with clients only if more than 60% of all employees are already familiar with and regularly use this kind of software (at home, of course). Let p be the true proportion of employees who regularly use IM.
a. State the null and alternative hypotheses.
b. Describe a type I error and a type II error in this context.

9.39 Economics and Finance A shadow chief secretary to the treasury claims that first-time home buyers in the South Downs area (near Worthing in Sussex, England) are paying (on average) more than £1,367 in stamp duty. The shadow chief secretary believes this steep tax is prohibiting families from buying a first home in this area. A random sample of first-time home buyers will be obtained, and the stamp duty paid by each family will be recorded. These data will be used to determine whether there is any evidence that the mean stamp duty for first-time home buyers, μ, is more than £1,367.
a. State the null and alternative hypotheses in terms of μ.
b. Describe a type I error and a type II error in this context.

9.40 Public Health and Nutrition A recent study of healthcare systems in Europe revealed that patients in Switzerland pay the highest premiums and believe their system is the best in the world. However, a citizens group claims that more than half of all patients in Switzerland perceive a lack of consultation regarding medical treatment. A random sample of patients in Switzerland will be obtained, and each will be asked to complete a survey following treatment. The information will be used to determine whether there is any evidence that the proportion of patients who perceive a lack of consultation, p, is greater than 0.50. If there is evidence that p is greater than 0.50, physicians' malpractice insurance premiums will be increased.
a. State the null and alternative hypotheses in terms of p.
b. Describe a type I and a type II error in this context.
c. Which error is more serious for physicians? Why?
d. Which error is more serious for patients? Why?

9.41 Medicine and Clinical Studies An electroencephalogram (EEG) is often used to measure brain, or neural, activity expressed as electrical voltage. Research suggests that transcendental meditation (TM) produces a simplified state of rest and relaxation resulting in an EEG with smaller than normal variance. Suppose normal brain activity has variance $\sigma^2 = 15$ volts^2. A random sample of patients who can achieve this state of consciousness (TM) will be selected, and their brain activity will be measured during TM. The resulting information will be used to test the research theory.
a. State the null and alternative hypotheses in terms of σ^2.
b. Describe a type I and a type II error in this context.
c. If there is evidence to suggest that TM decreases the variance in brain activity, then the National Science Foundation (NSF) will commit more money for TM research. Which error is more serious for the NSF? Why? Which error is more serious for TM researchers? Why?

Extended Applications

9.42 Marketing and Consumer Behavior Many families have decided to use a satellite dish instead of traditional cable TV. Satellite-dish companies offer a wide variety of premium channels and popular sports packages. The local cable TV provider in Iowa City, Iowa, Mediacom, is concerned about losing market share and is going to conduct a hypothesis test to determine whether more advertising is needed. A random sample of homes in the city will be obtained, and the data will be used to determine whether there is any evidence that the true proportion of homes with satellite dishes is greater than 0.25.

a. State the null and alternative hypotheses in terms of p.

b. Describe type I and type II errors in this context.

c. For a fixed sample size, which is smaller, $\beta(0.27)$ or $\beta(0.35)$? Justify your answer.

9.43 Sports and Leisure Department of Natural Resources personnel enforce hunting and fishing regulations and conduct routine safety checks on recreational boats. Past experience indicates that 15% of all boats inspected have at least one safety violation. Recent accidents on lakes in South Carolina have prompted calls for more extensive inspections. A random sample of recreational boats will be obtained and inspected. If there is evidence that the true proportion of boats with safety violations, p, is more than 15%, a methodical inspection of every boat launched at popular sites will be started.

a. State the null and alternative hypotheses in terms of p.

b. Describe a type I error and a type II error in this context.

c. What happens to the probability of a type I error as the value of p approaches 0.15 from the right (that is, 0.20, 0.19, . . .)?

9.44 Physical Sciences Civil engineers are going to test a highway bridge outside of Washington, D.C., for structural integrity. A random sample of locations along the bridge will be selected, and the concrete stress (in pounds per square inch) will be measured. These data will be used to determine the safety of the bridge.

a. Suppose 6400 psi is considered a safe concrete stress level. Which pair of hypotheses should be tested?

$$H_0: \mu = 6400 \quad \text{versus} \quad H_a: \mu > 6400, \text{or}$$
$$H_0: \mu = 6400 \quad \text{versus} \quad H_a: \mu < 6400.$$

b. If you regularly drive over the bridge being tested, would you prefer $\alpha = 0.1$ or $\alpha = 0.01$? Why?

9.3   HYPOTHESIS TESTS CONCERNING A POPULATION MEAN WHEN σ IS KNOWN

In the previous sections and chapters, all of the statistics tools have been provided to construct a formal hypothesis test. Recall, every hypothesis test consists of four parts, and the null hypothesis, an equality stated in terms of a population parameter, represents the current state (i.e., what is assumed to be true). The following example is used to develop a hypothesis test about a population mean μ when σ is known.

Example 9.7 Patient Triage Hospital emergency rooms across the country are experiencing shortages of doctors and nurses and have too few beds. These constraints make it difficult to treat patients in a timely manner. St. Paul's Hospital in Providence, Rhode Island, which treats approximately 50,000 patients in its emergency room each year, decided to address this issue by moving into the waiting room to treat patients, similar to the way a MASH unit operates. Prior to this experiment, the mean time to treat *very ill* patients (as opposed to critically ill patients or those with a minor injury) entering the emergency room was 20 minutes (with standard deviation $\sigma = 5.0$ minutes). During the waiting room experiment, a random sample of 36 very ill patients was selected and the time to treatment for each was recorded. The sample mean time was $\bar{x} = 16.1$ minutes. Conduct a hypothesis test to determine whether there is any evidence to suggest that the waiting room experiment has reduced the mean time to treatment for very ill patients. Use $\alpha = 0.05$.

An Army MASH unit is a mobile surgical hospital, usually deployed close to battle lines. It is designed to provide early surgical procedures and to stabilize critically wounded soldiers.

SOLUTION

STEP 1   Let μ be the mean time to treatment for very ill patients entering the emergency room. The current state, or what is assumed to be true, involves time to treatment prior to the waiting room experiment. Therefore, the null hypothesis is

$$H_0: \mu = 20.$$

STEP 2   The chairman of the Department of Emergency Medicine for Providence Health Care hopes doctors going directly to patients in the waiting room will *decrease* time to treatment for very ill patients. The alternative hypothesis is

$$H_a: \mu < 20.$$

STEP 3   We need to know whether 16.1 minutes is a reasonable observation under the null hypothesis, or whether it is *too far away* from the assumed mean, 20 minutes. Since H_0 is assumed to be true, and $n = 36 \geq 30$, by the Central Limit Theorem, $\overline{X}$ is approximately normal with mean $\mu = 20$ and standard deviation $\sigma/\sqrt{n}$. Although $\overline{X}$ could be used as the test statistic, it's easier to identify unlikely values of a standard normal random variable. Therefore, the appropriate test statistic is obtained by standardizing:

$$\text{TS:} \, Z = \frac{\overline{X} - 20}{\sigma/\sqrt{n}}$$

If H_0 is true, then Z is approximately standard normal.

STEP 4   Since the alternative hypothesis is one-sided and left-tailed, unusual values of the test statistic, Z, are in the left tail of the distribution. We should reject the null hypothesis if the value of Z is to the left of some cutoff value, or endpoint. The cutoff value is determined so that the probability of a type I error is 0.05. Select the cutoff value such that if H_0 is true,

$$P(Z \leq \text{cutoff value}) = 0.05.$$

Using the definition of a Z critical value (from Section 8.2),

$$P(Z \leq -z_{0.05}) = 0.05 \quad \text{and} \quad -z_{0.05} = -1.6449. \qquad \text{Table III in the Appendix.}$$

The rejection region is written as

> In interval notation, the rejection region is $(-\infty, -1.6449]$.

$$\text{RR:} \, Z \leq -z_{0.05} = -1.6449.$$

If the value of Z is less than or equal to the critical value -1.6449, then the observed value of $\overline{X}$ is considered rare, and we reject the null hypothesis. Figure 9.3 illustrates the critical value and the rejection region for this hypothesis test.

> The critical value divides the z measurement axis into two parts: the rejection region and the nonrejection region.

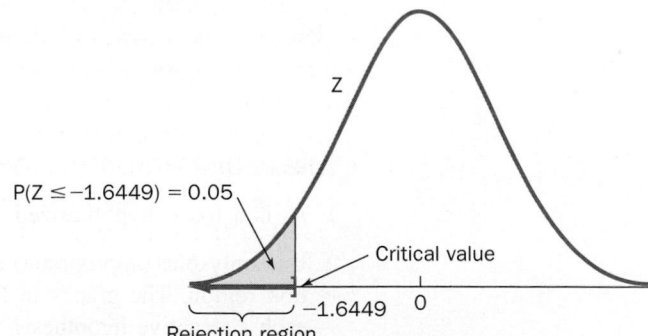

Figure 9.3 The critical value and the rejection region for the emergency room hypothesis test.

STEP 5   The value of the test statistic is

> The small letters here (z and $\bar{x}$) represent actual values.

$$z = \frac{\bar{x} - 20}{\sigma/\sqrt{n}} = \frac{16.1 - 20}{5.0/\sqrt{36}} = -4.68.$$

The value $\bar{x} = 16.1$ is 4.68 standard deviations to the left of the mean—a very unusual observation if the null hypothesis is really true. This means that either 16.1 is an incredibly lucky observation or the assumption ($\mu = 20$) is wrong. As usual, we discount the lucky possibility.

More formally, since $z = -4.68$ (≤ -1.6449) lies in the rejection region, we reject the null hypothesis at the $\alpha = 0.05$ level of significance. There is evidence to suggest that the true population mean time to treatment, μ, is less than 20 minutes. ●

In any hypothesis test, as in the example above, we assume H_0 is true and consider the likelihood of the sample outcome (expressed as a single value of the test statistic).

1. If the value of the test statistic is reasonable under the null hypothesis, then we cannot reject H_0.

2. If the value of the test statistic is unlikely under the null hypothesis, then we reject H_0.

As presented in Section 9.1, there are three possible alternative hypotheses: one two-sided alternative and two one-sided alternatives. The rejection region depends on the alternative hypothesis. The general procedure for a hypothesis test concerning μ is summarized below.

Use this as a *template* for a hypothesis test about a population mean when σ is known.

HYPOTHESIS TESTS CONCERNING A POPULATION MEAN WHEN σ IS KNOWN

Given a random sample of size n from a population with mean μ, assume

1. The underlying population is normal or n is large, and
2. The population standard deviation σ is known.

A hypothesis test about the population mean μ with significance level α has the form:

$H_0: \mu = \mu_0$

$H_a: \mu > \mu_0, \quad \mu < \mu_0, \quad \text{or} \quad \mu \neq \mu_0$

$\text{TS: } Z = \dfrac{\bar{X} - \mu_0}{\sigma/\sqrt{n}}$

$\text{RR: } Z \geq z_\alpha, \quad Z \leq -z_\alpha, \quad \text{or} \quad |Z| \geq z_{\alpha/2}$

The rejection region *always* includes the endpoint of the (infinite) interval.

 ILLUMINATING THE CONCEPTS

1. μ_0 is a fixed, hypothesized value of the population mean μ.

2. Use only one (appropriate) alternative hypothesis and the corresponding rejection region. The graphs in Figures 9.4 to 9.6 illustrate the rejection region for each alternative hypothesis.

3. For a two-sided alternative hypothesis, the rejection region $|Z| \geq z_{\alpha/2}$ (written using the absolute value of Z) is simply a shorthand way to write $Z \geq z_{\alpha/2}$ or $Z \leq -z_{\alpha/2}$.

4. For a given significance level, the corresponding critical value is found using Table III in the Appendix, backward. This procedure was presented in Chapter 8. Common values for α are 0.05, 0.025, and 0.01.

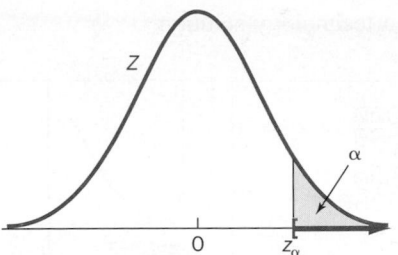

Figure 9.4 Rejection region for $H_a: \mu > \mu_0$.

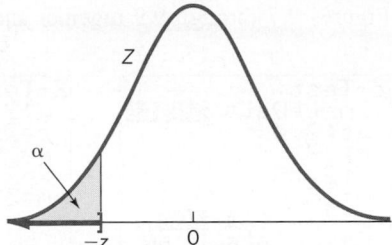

Figure 9.5 Rejection region for $H_a: \mu < \mu_0$.

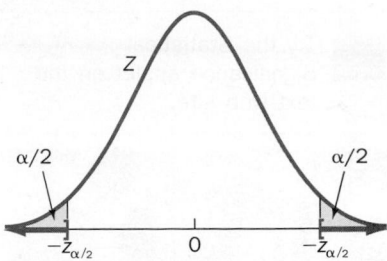

Figure 9.6 Rejection region for $H_a: \mu \neq \mu_0$.

5. The hypothesis test procedure described above can be used *only* if σ is known. If n is large and σ is unknown, some statisticians suggest using the sample standard deviation s in place of σ. This produces an *approximate* test statistic. Section 9.5 presents an *exact* test procedure concerning μ when σ is unknown. ■

The following examples illustrate this hypothesis test procedure.

Example 9.8 Automobile Inventories The collapse of automobile sales during the last half of 2008 forced General Motors and Chrysler to ask the federal government for a bailout in the form of loans. One measure of automobile sales is the inventory-to-sales ratio, a measure of the time, in months, necessary to sell all cars on a dealer's lot if sales were to remain at the current level. In December 2008 this ratio had risen to 4.2, while the long-term mean is 2.4 months.[10] Suppose a random sample of 35 car dealers was obtained in March 2009, and the mean inventory-to-sales ratio was 2.75. Is there any evidence to suggest that the current mean inventory-to-sales ratio is greater than the long-term mean? Assume that the population standard deviation is 0.70 and use $\alpha = 0.025$.

SOLUTION

STEP 1 The current state, or assumed mean, is $\mu = 2.4 \ (= \mu_0)$.

The sample size is $n = 35$; $\sigma = 0.70$ and $\alpha = 0.025$.

We are trying to prove that the current inventory-to-sales ratio is *greater* than the long-term mean. Therefore, the alternative hypothesis is one-sided, right-tailed.

STEP 2 The four parts of the hypothesis test are

$H_0: \mu = 2.4$

$H_a: \mu > 2.4$

$\text{TS: } Z = \dfrac{\overline{X} - \mu_0}{\sigma/\sqrt{n}}$

$\text{RR: } Z \geq z_\alpha = z_{0.025} = 1.96$

STEP 3 The value of the test statistic is

$$z = \frac{\overline{x} - \mu_0}{\sigma/\sqrt{n}} = \frac{2.75 - 2.4}{0.70/\sqrt{35}} = 2.958 \geq 1.96$$

STEP 4 Since 2.958 lies in the rejection region, we reject the null hypothesis at the $\alpha = 0.025$ level. There is evidence to suggest that the current mean inventory-to-sales ratio is greater than 2.4 months.

Solution Trail

KEYWORDS

- Is there any evidence?
- Greater than the long term mean.
- Standard deviation 0.70.

TRANSLATION

- Conduct a one-sided, right-tailed test about μ.
- $\mu_0 = 2.4$.
- $\sigma = 0.70$.

CONCEPTS

Hypothesis test concerning a population mean when σ is known.

VISION

Use the template for a one-sided, right-tailed test about μ. The underlying population distribution is unknown, but n is large and σ is known. Determine the appropriate alternative hypothesis and the corresponding rejection region, find the value of the test statistic, and draw a conclusion.

Try the Statistical Significance applet on the text web site.

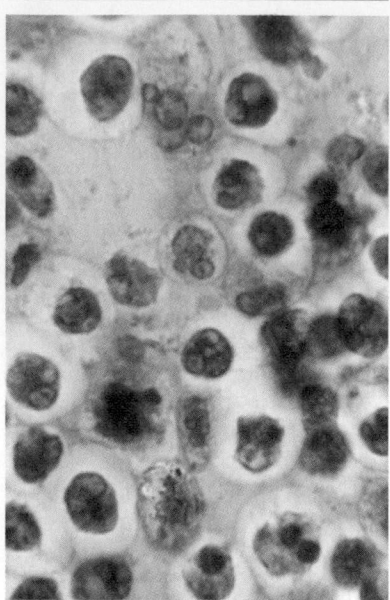

(CDC)

Figures 9.7 through 9.9 together show a technology solution.

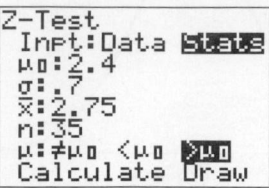

Figure 9.7 TI-84 Plus Z-Test input screen.

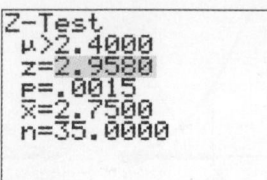

Figure 9.8 TI-84 Plus Z-Test Calculate results.

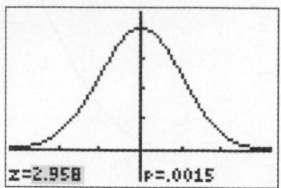

Figure 9.9 TI-84 Plus Z-Test Draw results.

Example 9.9 Natural Defense White blood cells are the body's natural defense mechanism against disease and infection. The mean white blood cell count in healthy adults, measured as part of a CBC (complete blood count), is approximately $7.5 \times 10^3/\mu\text{l}$.[11] A company developing a new drug to treat arthritis pain must check for any side effects. A random sample of patients using the new drug was selected, and the white blood cell count of each patient was measured. The results are given in the following table ($\times 10^3/\mu\text{l}$).

6.50	8.69	6.85	6.76	6.58	8.84	8.44
8.28	7.65	6.95	10.12	8.74	8.00	8.84
7.93	7.65	7.00	6.70	9.20	6.45	7.66

Assume the distribution of white blood cell counts is normal and $\sigma = 1.1$. Conduct a hypothesis test to determine whether there is any change in the mean white blood cell count due to the arthritis drug. Use $\alpha = 0.01$.

SOLUTION

STEP 1 The assumed mean is $\mu = 7.5$ ($= \mu_0$); the sample size is $n = 21$; $\sigma = 1.1$; and $\alpha = 0.01$.

The company is looking for *any* change in the mean white blood cell count. Therefore, the relevant alternative hypothesis is two-sided.

The sample size is small ($n < 30$), but the population is assumed to be normal. The hypothesis test concerning a population mean when σ is known can be used.

STEP 2 The four parts of the hypothesis test are

$H_0: \mu = 7.5$
$H_a: \mu \neq 7.5$
TS: $Z = \dfrac{\bar{X} - \mu_0}{\sigma/\sqrt{n}}$
RR: $|Z| \geq z_{\alpha/2} = z_{0.005} = 2.5758$ ($Z \leq -2.5758$ or $Z \geq 2.5758$)

STEP 3 The sample mean is

$$\bar{x} = \frac{1}{21}(6.50 + 8.69 + \cdots + 7.66) = 7.8014$$

The value of the test statistic is

$$z = \frac{\bar{x} - \mu_0}{\sigma/\sqrt{n}} = \frac{7.8014 - 7.5}{1.1/\sqrt{21}} = 1.2556$$

A technology solution:

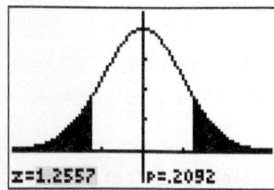

Figure 9.10 TI-84 Plus Z-Test input screen.

```
Z-Test
 µ≠7.5000
 z=1.2557
 P=.2092
 x̄=7.8014
 Sx=1.0368
 n=21.0000
```

Figure 9.11 TI-84 Plus Z-Test Calculate results.

Figure 9.12 TI-84 Plus Z-Test Draw results.

STEP 4   The value of the test statistic, $z = 1.2556$, does *not* lie in the rejection region—we do not reject the null hypothesis at the $\alpha = 0.01$ level of significance. There is no evidence to suggest that the new arthritis drug has changed the mean white blood cell count.   ●

◀ Recall, the probability of a type II error depends on the alternative value of the parameter under investigation. The example below presents a method for computing β in a hypothesis test about a population mean μ when σ is known. ▶

Example 9.10 Watermelon Weight   The mean weight of a Dixielee watermelon is 22.0 pounds. Dixielee watermelons that are too light are dry and are less flavorful. When a large grocery chain receives a shipment of these watermelons for sale, the produce supervisor randomly selects 36 watermelons and weighs each. A hypothesis test is conducted to determine whether there is any evidence that the mean weight of the watermelons is less than 22.0 pounds. Let $\alpha = 0.025$, assume $\sigma = 3.0$, and find the probability of a type II error if the true mean weight of the watermelon is 20.5 pounds.[12]

SOLUTION

STEP 1   The assumed mean is $\mu = 22.0 \, (= \mu_0)$; $n = 36$, $\sigma = 3.0$, and $\alpha = 0.025$.

The supervisor of produce is looking for evidence of a smaller mean weight. This is a one-sided, left-tailed test.

The underlying population distribution is unknown, but the sample size is large ($n = 36 \geq 30$). The hypothesis test concerning a population mean μ when σ is known is relevant.

STEP 2   The four parts of the hypothesis test are

$H_0: \mu = 22.0$

$H_a: \mu < 22.0$

$$\text{TS: } Z = \frac{\overline{X} - \mu_0}{\sigma/\sqrt{n}} = \frac{\overline{X} - 22.0}{3.0/\sqrt{36}}$$

RR: $Z \leq -z_\alpha = -z_{0.025} = -1.96$

STEP 3   To compute the probability of a type II error, you must visualize the rejection region in terms of $\overline{X}$. Start by writing the definition of a type I error, and work backward to the $\overline{X}$ world.

$P(Z \leq -1.96) = 0.025$ Definition of type I error.

$$P\left(\frac{\overline{X} - 22.0}{3.0/\sqrt{36}} \leq -1.96\right) = 0.025$$ Use the definition of the test statistic.

$P(\overline{X} \leq 21.02) = 0.025$ Isolate $\overline{X}$. Within the probability expression, multiply both sides by $3/\sqrt{36} = 0.5$, and add 22.0 to both sides.

$21.02 = \overline{X}_c$ is the *critical value in the $\overline{X}$ world* (see Figure 9.13). If the sample mean weight of the watermelons is less than or equal to 21.02, we reject the null hypothesis.

STEP 4 $\beta(20.5)$ is the probability of not rejecting the null hypothesis if the real mean is 20.5 (often denoted $\mu_a = 20.5$; the a in the subscript stands for **alternative mean**). Write a probability statement for $\beta(20.5)$ using the critical value 21.02, and standardize.

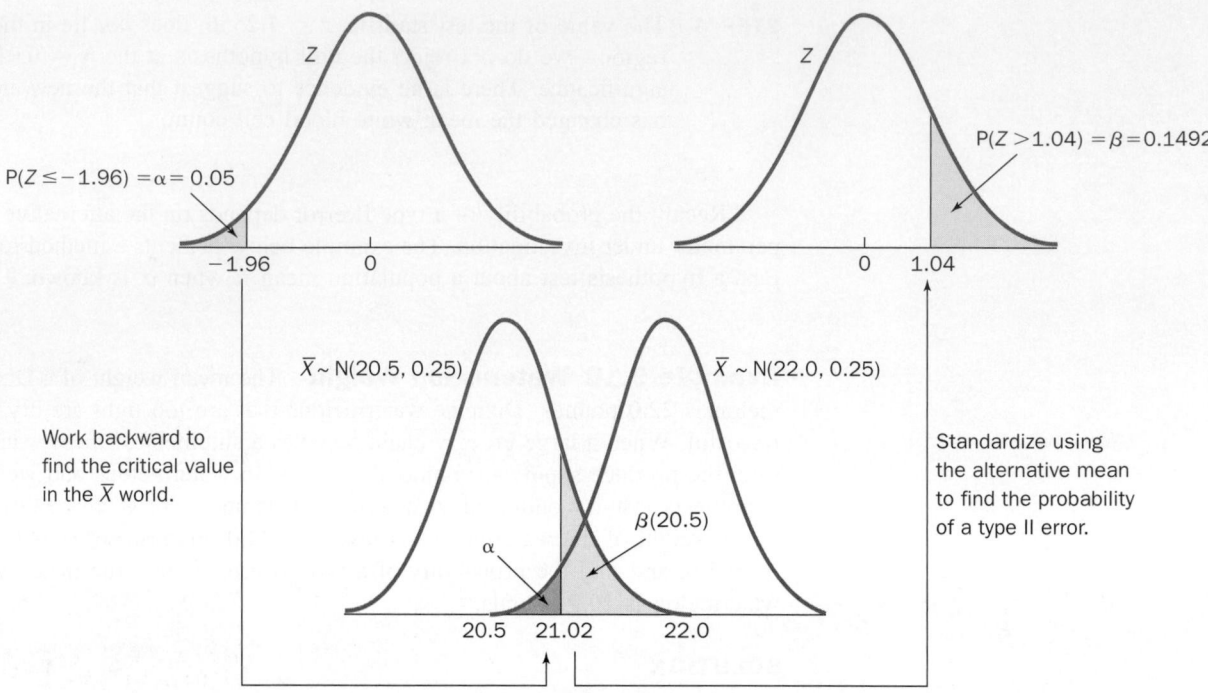

Figure 9.13   Visualization of the calculations for $\beta(20.5)$, the probability of a type II error for $\mu_a = 20.5$.

 Try the Power of a Test
statistical applet.

$$\beta(\mu_a) = P(\overline{X} > \overline{X}_c)$$
$$\beta(20.5) = P(\overline{X} > 21.02)$$

Definition of a type II error.
Use values for μ_a and $\overline{X}_c$.

$$= P\left(\frac{\overline{X} - 20.5}{3.0/\sqrt{36}} > \frac{21.02 - 20.5}{3.0/\sqrt{36}}\right)$$

Standardize using $\mu_a = 20.5$.
Assume $\sigma = 3.0$ is unchanged
even if the true mean is 20.5.

$$= P(Z > 1.04)$$

Equation 6.8.

$$= 1 - P(Z \le 1.04)$$

Use cumulative probability.

$$= 1 - 0.8508 = 0.1492$$

Use Table III in the Appendix.

If the true mean weight is really 20.5 pounds, the probability of a type II error is 0.1492, at the $\alpha = 0.025$ significance level.   ●

💡 ILLUMINATING THE CONCEPTS

1. By referring to Figure 9.13, we can visualize the inverse relationship between α and β. For example, if α increases, $\overline{X}_c$ moves to the right, and the area of the region corresponding to β decreases. Problem 9.55 in the Exercises is designed to explore and confirm this concept numerically.

2. The logic and method for finding the probability of a type II error for the other alternative hypotheses is the same:
   a. Work backward to find $\overline{X}_c$.
   b. Write a probability statement for $\beta(\mu_a)$ in terms of $\overline{X}$ and $\overline{X}_c$.
   c. Standardize to find the probability.

A little educational philosophy
sneaks in here.

3. Rather than trying to memorize a formula for the probability of a type II error corresponding to each alternative hypothesis, use the definition, a drawing, and the three-step procedure described above. This approach will help when we consider the hypothesis test procedure concerning a population proportion. ■

TECHNOLOGY CORNER

Procedure: Hypothesis test concerning a population mean when σ is known.
Reconsider: Example 9.9, page 396, solution, and interpretations.

TI-84 Plus

The input for the TI-84 Plus Z-Test function is either data in a list or summary statistics.

1. Enter the data into list L1.
2. Select $\boxed{\text{STAT}}$; TESTS; Z-Test.
3. The data are given in this example—highlight Data. Enter μ_0, σ, the name of the list containing the data, the frequency of occurrence of each observation, and highlight the appropriate alternative hypothesis. See Figure 9.10, page 396.
4. In the last row of the input screen, select Calculate and press $\boxed{\text{ENTER}}$.
5. The p value associated with the hypothesis test is given on the Z-Test output screen. This topic is covered in Section 9.4. See Figure 9.11, page 396.
6. The Z-Test input screen includes an option to Draw, or visualize, the results of the hypothesis test. The calculator will automatically determine a suitable WINDOW, sketch a standard normal density curve, display the computed z and p values, and shade the area under the curve corresponding to the p value. See Figure 9.12, page 396.

Minitab

The input is either data in a column (Samples in columns) or summary statistics (Summarized data).

1. Enter the data into column C1.
2. Choose <u>S</u>tat; <u>B</u>asic Statistics; 1-Sample <u>Z</u>.
3. In this example, select Samples in columns and enter C1. Enter the Standard deviation, check the box to Perform a hypothesis test, and enter the Hypothesized mean. See Figure 9.14.
4. Choose Options. Enter a Confidence level and select the Alternative hypothesis. See Figure 9.15.
5. Summary statistics, the confidence interval, the value of the test statistic, and the p value are displayed in the Session window. See Figure 9.16

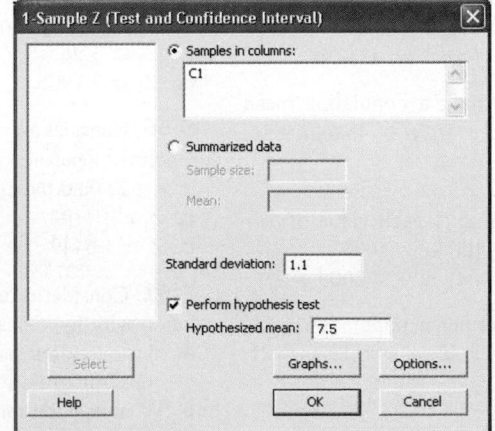

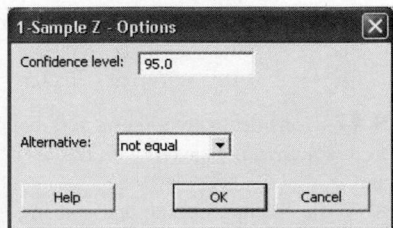

Figure 9.14 1-Sample Z input screen.

Figure 9.15 1-Sample Z Options screen.

```
Session                                              _ □ ✕

One-Sample Z: C1

Test of mu = 7.5 vs not = 7.5
The assumed standard deviation = 1.1


Variable   N   Mean  StDev  SE Mean      95% CI        Z     P
C1        21  7.801  1.037   0.240   (7.331, 8.272)  1.26  0.209
```

Figure 9.16 Hypothesis test results.

Excel

Excel can be used to compute the sample mean and the value of the test statistic. In the next Section, the Excel function ZTEST is used to compute the p value associated with a hypothesis test concerning a population mean when σ is known.

1. Enter the data into column A.
2. Compute the sample mean using the Excel command AVERAGE.
3. Compute the value of the test statistic. See Figure 9.17.

	B	C	D
1	7.8014	=AVERAGE(A1:A21)	The sample mean
2	1.2557	=(B1-7.5)/(1.1/SQRT(21))	The value of z

Figure 9.17 Excel computations.

SECTION 9.3 EXERCISES

Practice

9.45 Consider a hypothesis test concerning a population mean with H_0: $\mu = 170$, H_a: $\mu < 170$, $n = 38$, and $\sigma = 15$.
a. Write the appropriate test statistic.
b. Write the rejection region corresponding to each value of α.
   (i) $\alpha = 0.01$ (ii) $\alpha = 0.025$ (iii) $\alpha = 0.05$
   (iv) $\alpha = 0.10$ (v) $\alpha = 0.001$ (vi) $\alpha = 0.0001$

9.46 Consider a hypothesis test concerning a population mean from a normal population with H_0: $\mu = 45.6$, H_a: $\mu > 45.6$, $n = 16$, and $\sigma = 15$.
a. Write the appropriate test statistic.
b. Write the rejection region corresponding to each value of α.
   (i) $\alpha = 0.01$ (ii) $\alpha = 0.025$ (iii) $\alpha = 0.05$
   (iv) $\alpha = 0.10$ (v) $\alpha = 0.005$ (vi) $\alpha = 0.0005$

9.47 Consider a hypothesis test concerning a population mean from a normal population with H_0: $\mu = -11$, H_a: $\mu \neq -11$, $n = 21$, and $\sigma = 4.5$.
a. Write the appropriate test statistic.
b. Write the rejection region corresponding to each value of α.
   (i) $\alpha = 0.01$ (ii) $\alpha = 0.2$ (iii) $\alpha = 0.05$
   (iv) $\alpha = 0.10$ (v) $\alpha = 0.001$ (vi) $\alpha = 0.0002$

9.48 Consider a hypothesis test concerning a population mean with H_0: $\mu = 3.55$, H_a: $\mu < 3.55$, $n = 49$, and $\sigma = 6.2$. Find the significance level (α) for each rejection region.
a. $Z \leq -1.6449$ b. $Z \leq -2.5758$ c. $Z \leq -2.0537$
d. $Z \leq -2.3263$ e. $Z \leq -3.0902$ f. $Z \leq -3.7190$

9.49 Consider a hypothesis test concerning a population mean with H_0: $\mu = 7.6$, H_a: $\mu \neq 7.6$, $n = 37$, and $\sigma = 4.506$. Find the significance level (α) for each rejection region.
a. $|Z| \geq 1.96$ b. $|Z| \geq 1.6449$ c. $|Z| \geq 2.8070$
d. $|Z| \geq 3.2905$ e. $|Z| \geq 1.2816$ f. $|Z| \geq 2.3263$

9.50 Consider a hypothesis test concerning a population mean from a normal population with H_0: $\mu = 98.6$, H_a: $\mu > 98.6$, $n = 10$, and $\sigma = 1.2$. Find the significance level (α) for each rejection region.
a. $Z \geq 3.7190$ b. $Z \geq 0.8416$ c. $Z \geq 2.3263$
d. $Z \geq 1.6449$ e. $Z \geq 3.2905$ f. $Z \geq 2.8782$

9.51 Consider a random sample of size 25 from a normal population with hypothesized mean 212 and $\sigma = 2.88$.
a. Write the four parts for a one-sided, right-tailed hypothesis test concerning the population mean with $\alpha = 0.01$.
b. What assumptions must be made in order for the test in part (a) to be appropriate?
c. Suppose the sample mean is $\bar{x} = 213.5$. Find the value of the test statistic, and draw a conclusion about the population mean.

9.52 Consider a random sample of size 32, given on the data CD and book's web site, from a population with hypothesized mean 3.14 and $\sigma = 6.8$.

a. Write the four parts for a one-sided, left-tailed hypothesis test concerning the population mean with $\alpha = 0.001$.

b. What assumptions are made in order for the test in part (a) to be appropriate?

c. Compute the sample mean, and find the value of the test statistic. Draw a conclusion about the population mean.

9.53 Consider a random sample of size 48 from a population with hypothesized mean 365.25 and $\sigma = 22.3$.

a. Write the four parts for a two-sided hypothesis test concerning the population mean with $\alpha = 0.05$.

b. What assumptions are necessary in order for the test in part (a) to be appropriate?

c. Suppose the sample mean is $\bar{x} = 360.0$. Find the value of the test statistic, and draw a conclusion about the population mean.

9.54 Consider a one-sided, right-tailed hypothesis test concerning the mean, μ_0, from a normal population with σ known, sample size n, and $\alpha = 0.05$. Explain the error in each of the following statements.

a. The rejection region is RR: $Z \le 1.96$.

b. The test statistic is $Z = \dfrac{\mu_0 - \bar{X}}{\sigma/\sqrt{n}}$.

c. The value of the test statistic does not lie in the rejection region. We accept the null hypothesis and conclude that $\mu = \mu_0$.

d. The null hypothesis is $H_0 : \mu > \mu_0$.

e. The probability of a type II if error if $\mu = \mu_a$ is $1 - \alpha = 1 - 0.05 = 0.95$.

Applications

9.55 Business and Management The mean income per year (in dollars) of employees who produce internal and external newsletters and magazines for corporations was reported to be 51,500. These editors and designers work on corporate publications but not on marketing materials. Due to poor economic conditions and oversupply, these corporate communications workers may be experiencing a decrease in salary. A random sample of 38 corporate communications workers revealed their mean salary was $\bar{x} = 49,762$. Conduct a hypothesis test to determine whether there is any evidence to suggest that the mean income per year of corporate communications workers has decreased. Assume $\sigma = 3750$ and use $\alpha = 0.01$.

9.56 Physical Sciences The amount of lava that flows out of an erupting volcano varies widely. The smallest amount measured came from an eruption in Iceland in 1977, and the largest lava flow has been recorded from Kilauea in Hawaii. Historical evidence suggests that the mean lava flow from a volcanic eruption is 10 million cubic meters (m^3). A random sample of the amount of lava from volcano eruptions during 2005 is given on the data CD and book's web site (in millions of cubic meters). Is there any evidence to suggest the mean lava flow has changed? Assume $\sigma = 3.1$ and use $\alpha = 0.05$.[13]

9.57 Marketing and Consumer Behavior After analyzing a database of over 2 million telephone calls, CogniWorld Long Distance reported that the mean length of all international calls was 295 seconds. In an effort to increase the length of international calls and boost profits, the company conducted an extensive advertising campaign. A few months after the ads first appeared, a random sample of 48 calls was obtained. The sample mean was $\bar{x} = 306.3$ seconds. Assume that the population standard deviation is 52 seconds. Is there any evidence to suggest that the advertising campaign has been successful? Use $\alpha = 0.01$.

9.58 Physical Sciences The United States Geological Survey collects water-level measurement data from various locations. Data from the La Pine Basin, Oregon, are being used to study the changes in the water table. Twelve days were randomly selected, and the water table in feet below land surface on each day is given in the following table.

12.30	12.29	12.30	12.35	12.37	12.40
12.38	12.37	12.38	12.42	12.46	12.52

Is there any evidence to suggest that the mean water table is less than 12.4 feet? Assume $\sigma = 0.07$ and use $\alpha = 0.025$.

9.59 Sports and Leisure The length of a motorboat is traditionally measured in two ways. The distance along the centerline from the outside of the front hull to the rear is the length overall (LOA). The length of waterline, or load waterline (LWL), is the length of the boat on the line where the boat meets the water. Members of the Community Association in the lakeside city of Vermilion, Ohio, are concerned that residents are using bigger boats, contributing to more noise and water pollution. Past registration records indicate that the mean LOA for boats allowed on the lake is 35 feet. A random sample of 41 boats on the lake was obtained, and each LOA was carefully measured. The sample mean LOA was $\bar{x} = 36.22$ feet. Assume $\sigma^2 = 5.7$ ft^2.

a. Is there any evidence to suggest that the mean LOA has increased? Use $\alpha = 0.01$.

b. Does your answer to part (a) change if $\alpha = 0.1$? Why or why not?

9.60 Education and Child Development Suppose a typical dorm room at Wake Forest University is 166 square feet. Eddie Hull, president of the Association of College and University Housing Officers International, explained that a newly constructed standard double room would be about 220 to 240 square feet. A random sample of standard double dorm rooms at private schools across the country was obtained, and the area of each was measured. The data are given on the data CD and book's web site. Conduct a hypothesis test to determine whether there is any evidence that standard double dorm rooms at private schools have a mean area less than 220 square feet. Use $\alpha = 0.05$ and assume $\sigma = 15.7$.

9.61 Public Health and Nutrition The mean daily energy requirement for eight-year-old boys is 2200 calories. An education researcher believes that many students in this group do poorly in school due to an inadequate diet and not having enough energy. A random sample of academically at-risk eight-year-old boys was obtained, and their caloric intake was carefully measured for 1 day. The summary statistics were $n = 37$, $\bar{x} = 2089$. Assume $\sigma = 358$.

a. Is there any evidence that this group of students has a mean caloric intake below the daily energy requirement? Use $\alpha = 0.05$.

b. How would your answer to part (a) change if $\alpha = 0.01$?

9.62 Biology and Environmental Science An average lawn has a mean of 21 blades of grass per square inch. A garden store sells an expensive fertilizer designed to transform an average lawn into a lush, thick carpet within three weeks. To test the claim, a random sample of average lawn plots was obtained. Each was treated with the fertilizer according to the instructions on the package. Three weeks later the density of each plot was measured by counting the blades of grass per square inch. The summary statistics were: $n = 32$, $\bar{x} = 22.4$. Is there any evidence to suggest that the fertilizer improves the thickness of an average lawn? Assume $\sigma = 2.7$ and use $\alpha = 0.005$.

9.63 Physical Sciences The Australian Bureau of Meteorology collects weather information each day for several locations across Australia. Since 1939, the mean air temperature at 9:00 A.M. during July at the Canberra Airport is 3.9 °C.[14] In a random sample of 10 days during July 2008, the sample mean daily temperature at 9:00 A.M. was 4.65 °C.

a. Suppose the underlying population is normal and the standard deviation for daily temperature is $\sigma = 2.3$. Is there any evidence to suggest that the mean daily temperature in 2008 was higher than the long-term mean? Use $\alpha = 0.01$.

b. Suppose the standard deviation for daily temperature is $\sigma = 1.3$. Does your conclusion change?

9.64 Public Policy and Political Science The U.S. Department of Health and Human Services defines "response time" as the time from receipt of a report of child neglect to the initial investigation. Once a call to the agency is received and logged in, the clock starts. Face-to-face contact with the alleged victim marks the initial investigation. There is some concern that workload and poor organization have contributed to a significant increase in the long-term mean response time of 14.0 hours. A random sample of cases was selected, and the response time for each is given on the data CD and book's web site. Assume $\sigma = 3.2$ and conduct a hypothesis test to determine whether there is any evidence that the mean response time has increased. Use $\alpha = 0.001$.

9.65 Manufacturing and Product Development A major cause of injuries in highway work zones is weak construction barriers. One federal government road-barrier specification involves the velocity of a front-seat passenger immediately following impact. This impact velocity must be less than 12 m/s. TSS GmbH, a German company, has just developed a new high-impact road barrier with special shock-absorbing material. In controlled tests the impact velocity was measured for 12 randomly selected crashes. The sample mean was 11.85 m/s. Assume that the distribution of impact velocities is normal and $\sigma = 0.26$.

a. Conduct a hypothesis test to determine whether there is sufficient evidence to suggest that the true mean impact velocity is less than 12 m/s. Use $\alpha = 0.05$.

b. What is your conclusion if $\alpha = 0.01$?

9.66 Fuel Consumption and Cars The National Highway Traffic Safety Administration conducts crash tests on automobiles. The maximum crush distance (in mm) is the damage penetration regardless of its location. The maximum crush distances for 2005 automobiles are given on the data CD and book's web site. Conduct a hypothesis test to determine whether there is evidence to suggest that the true mean maximum crush distance for 2005 automobiles is less than 450 mm. Assume $\sigma = 146$ and use $\alpha = 0.05$.[15]

9.67 Manufacturing and Product Development The Brunton Optimus Crux Camping Stove, a lightweight camping stove, is designed to accept a triblend fuel made of propane, iso-butane, and butane. The cartridges for this stove are sold with camping equipment, and the label indicates 225 grams of fuel. To check this claim, 52 cartridges were randomly selected and the amount of fuel in each was carefully measured. The sample mean was $\bar{x} = 224.2$ grams.

a. Is there any evidence that the true mean amount of fuel in these cartridges is less than the advertised amount? Assume $\sigma = 3.6$ and use $\alpha = 0.05$.

b. What assumptions did you make in order to conduct the hypothesis test in part (a)?

9.68 Public Health and Nutrition Egg Beaters™ are egg substitutes with no fat or cholesterol but with the same amount of protein as a regular egg (6 grams).[16] In order to check the protein claim, 18 Egg Beaters™ were randomly selected and the amount of protein in each was measured. The resulting data are given in the following table.

6.54	6.83	5.45	5.41	6.51	7.46	5.95
4.83	5.64	6.71	6.62	5.11	7.26	5.40
4.01	7.26	5.84	6.49			

Is there any evidence to suggest the mean amount of protein in an Egg Beater™ is less than 6 grams? Assume the distribution of protein is normal and $\sigma = 0.95$. Use $\alpha = 0.02$.

9.69 Public Policy and Political Science Residential mailboxes in Des Moines, Iowa, should be installed such that the bottom of the mailbox is 42 inches above the ground. This rule is designed for safety and to accommodate short mail carriers. A random sample of 75 mailboxes in a large city was selected. The height of each was carefully measured, and the sample mean was 43.22 inches. Assume $\sigma = 7.6$ inches and use $\alpha = 0.05$. Is there any evidence to suggest that the true mean height of mail boxes in Des Moines is different from 42 inches?

9.70 Biology and Environmental Science Sweet corn is a backyard barbecue favorite, especially around the Fourth of July, and almost every state produces some variety. The data in the following table represent a random sample of 16 states' fresh sweet corn production (in 1000 cwt) for 2007.

4608	1302	320	280	5015	4125	726	459
168	216	400	723	102	675	3300	740

(*Source*: National Agricultural Statistics Service, USDA and State Departments of Agriculture.)

a. Is there any evidence to suggest that the true mean sweet corn production per state is greater than 1400 (1000 cwt)? Assume the underlying population is normal, and use $\sigma = 1700$ and $\alpha = 0.01$.

b. Is there any evidence to suggest that the true mean sweet corn production per state is greater than 900 (1000 cwt)? Why do you suppose your conclusion does not change even though 900 is so far away from the sample mean?

Extended Applications

9.71 The four parts of a hypothesis test concerning a population mean from a normal population are shown below.

$H_0: \mu = 50$

$H_a: \mu > 50$

TS: $Z = \dfrac{\overline{X} - \mu_0}{\sigma/\sqrt{n}}$

RR: $Z \geq z_\alpha$

Assume that the sample size is $n = 25$; $\sigma = 7.5$ and $\alpha = 0.01$.
a. Find the probability of a type II error for the alternative mean $\mu_a = 54$; that is, find $\beta(54)$.
b. Find $\beta(55)$ and $\beta(56)$.
c. Repeat parts (a) and (b) for $\alpha = 0.025$.

9.72 Fuel Consumption and Cars Biodiesel fuels are made from vegetable oils or animal fats and may be used instead of conventional diesel fuel. One advantage to using biodiesel is a possible decrease in regulated emissions, specifically total hydrocarbons (HC). Using the heavy-duty transient Federal Test Procedure (FTP), the mean HC emission is 0.23 g/hp-hr (grams per horsepower-hour) with standard deviation $\sigma = 0.07$. A random sample of heavy-duty engines was obtained, and each was tested with biodiesel fuel. The resulting HC measurements are given in the following table.

0.17	0.19	0.10	0.21	0.15	0.23	0.06	0.21
0.18	0.01	0.24	0.29	0.23	0.14	0.18	0.10
0.02	0.16	0.13	0.10	0.14	0.30	0.24	0.24
0.18	0.06						

a. Suppose the underlying population is normal. Is there any evidence to suggest that the use of biodiesel has decreased the mean level of HC emissions? Use $\alpha = 0.01$.
b. Suppose a company is thinking about building a \$20 million biodiesel fuel production facility. The company will invest the money only if there is overwhelming evidence in favor of decreased HC emissions. Which error (type I or type II) is more important to the fuel company? Would the company prefer a smaller or a larger significance level? Justify your answers.[17]

9.73 Manufacturing and Product Development The left outside panel of an apartment-size, frost-free refrigerator is designed to have a width of $23\frac{5}{8}$ inches. Each hour, 10 such panels are randomly selected from the assembly line and carefully measured. If there is any evidence that the mean width is different from $23\frac{5}{8}$ inches, the assembly line is shut down for cleaning and inspection. Suppose the distribution of panel widths is normal and $\sigma = 0.15$.
a. During a specific hour of operation, $\overline{x} = 23.7$ inches. Should the assembly line be shut down? Use $\alpha = 0.05$.
b. Using $\alpha = 0.05$, find the critical values in the $\overline{X}$ world. (*Note*: There are two critical values, since this is a two-sided hypothesis test.)

9.74 Manufacturing and Product Development A manufacturer claims the weight of a package of its unsalted pretzels is (at least) 15.5 ounces. In order to test this claim, a consumer group randomly selected 250 packages and carefully weighed the contents of each. The sample mean was $\overline{x} = 15.45$ ounces.
a. Conduct a one-sided, left-tailed hypothesis test to see whether there is any evidence that the true mean weight of the pretzel packages is less than 15.5 ounces. Use $\alpha = 0.01$ and assume $\sigma = 0.26$.
b. If your conclusion in part (a) is to reject the null hypothesis, explain how this conclusion is possible even though the sample mean, 15.45, is so close to the hypothesized mean, 15.5. If your conclusion in part (a) is to not reject H_0, check those calculations one more time.

9.75 Physical Sciences Suppose the historical mean thickness of the ice just off the coast of Barrow, Alaska, is 3.32 meters. Many scientists are concerned about changes in the ice due to global warming and ice streams that have slowed or halted. A random sample of the ice thickness near Barrow during 2008 was obtained, and the data are given on the data CD and book's web site.[18] Assume $\sigma = 2.8$ meters.
a. Conduct a hypothesis test to determine whether there is any change in the mean ice thickness. Use $\alpha = 0.01$.
b. Find the probability of a type II error if the true ice mean thickness is 3.0; that is, find $\beta(3.0)$.
c. Carefully sketch a graph illustrating the probability found in part (b) using density curves for $\overline{X}$.
d. Find $\beta(2.8)$.

9.76 Marketing and Consumer Behavior The water treatment facility in Lake Havasu City, Arizona, recently proposed changes to all user fees. There is a treatment capacity fee, a connection fee, and a monthly minimum fee. All prices have been set assuming that the mean monthly usage is 714 cubic feet per household. In order to check this claim, a random sample of 16 homes was obtained and the monthly usage for each home was recorded. The sample mean was 601.2 cubic feet. Assume the distribution of monthly water usage per household is normal with standard deviation 183 cubic feet.
a. Conduct a two-sided hypothesis test to determine whether there is any evidence that the mean monthly water usage is different from 714 cubic feet. Use $\alpha = 0.01$.
b. If your conclusion in part (a) is to not reject the null hypothesis, explain how this conclusion is possible even though the sample mean, 601.2, is so far away from the hypothesized mean, 714. If your conclusion in part (a) is to reject the null hypothesis, try checking your calculations one more time.

9.77 Marketing and Consumer Behavior The Carpet Corner in Gladstone, Missouri, offers a variety of carpets, wood floors, and tiles. Sales records indicate that the mean amount of carpet installed in a wall-to-wall carpeted residential home by crews from this store is 1250 square feet. The store manager, Frank Vida, believes that when there is a sale, customers translate the savings into carpeting a larger area. A random sample of 45 wall-to-wall carpet purchases was selected during a sale. The sample mean was $\overline{x} = 1305$ square feet and assume $\sigma = 155$ square feet.
a. Conduct a one-sided, right-tailed test of $H_0: \mu = 1250$ versus $H_a: \mu > 1250$. Is there any evidence to suggest that the true mean square footage is larger during a sale? Use $\alpha = 0.01$.
b. Find the probability of a type II error if the true mean square footage during a sale is $\mu_a = 1330$; that is, find $\beta(1330)$.

c. Carefully sketch a graph that includes the distribution of $\overline{X}$ if $\mu = 1250$ and $\mu_a = 1330$. Shade in the areas that correspond to the probability of a type I error (0.01) and to the probability of a type II error found in part (b).

9.78 Physical Sciences Kiln-dried solid grade A teak wood should have a moisture content of no more than 12%. A furniture company recently purchased a large shipment of this wood to construct dining room sets. Thirty-seven pieces of teak wood were randomly selected and carefully measured for moisture content. The sample mean moisture content was $\overline{x} = 12.3\%$. Assume $\sigma = 1.25$.

a. Is there any evidence to suggest that the true population mean moisture content is greater than 12%? Use $\alpha = 0.01$.

b. Find the probability of a type II error if the true population mean moisture content is 12.2%; that is, find $\beta(12.2)$.

9.79 Fuel Consumption and Cars The joint venture ContiTech-Jiebao Power Transmission Systems, Ltd. produces many makes and models of automobile drive-belts. One popular poly V-belt is designed to have length 1050 mm. In order to ensure that the belts meet this specification, 18 are randomly selected from the assembly line every hour and carefully measured. If there is any evidence that the population mean length is different from 1050 mm, the entire line is shut down for inspection.

a. Assume that the distribution of belt lengths is normal and $\sigma = 3.7$ mm. In a two-sided hypothesis test of $H_0: \mu = 1050$ versus $H_a: \mu \neq 1050$, find the critical values in the $\overline{X}$ world. Use $\alpha = 0.01$.

b. In a sample of 18 belts, suppose $\overline{x} = 1049$. Should the assembly line be shut down? Justify your answer using the Z distribution and the appropriate $\overline{X}$ distribution.

Challenge

9.80 The Power of a Test The probability of a type II error, β, represents the likelihood of accepting the null hypothesis when the alternative is true. The *power* of a statistical test is $\pi = 1 - \beta$. This is the probability of (correctly) rejecting the null hypothesis, of detecting a difference in the hypothesized value of the population parameter. Try the Power statistical applet on the text web site.

A hot torsion test is used to determine the workability of a metal. Suppose a carbon steel rod is designed to fail (break) with mean axial load of 800 N (under certain temperature and speed conditions). A random sample of 25 rods is obtained, and the axial load failure is measured for each. Suppose the underlying distribution of axial load is normal and $\sigma = 50$.

a. Consider a hypothesis test of $H_0: \mu = 800$ versus $H_a: \mu < 800$ with $\alpha = 0.025$. Find the probability of a type II error and the power of this test for $\mu = 775$; that is, find $\beta(775)$ and $\pi(775)$.

b. Use a calculator or computer to find the power, $\pi(\mu_a)$, for $\mu_a = 730, 735, 740, \ldots, 800$.

c. Use the values from part (b) to carefully sketch a plot of $\pi(\mu_a)$ versus μ_a. The resulting plot is called a *power curve*.

9.4 *p* VALUES

The last piece of a hypothesis test, the rejection region, establishes a firm decision rule based on the value of the test statistic. If the value of the test statistic lies in the rejection region, we reject H_0. If not, we cannot reject the null hypothesis. Using a fixed rejection region associated with a specific value of α (the significance level) can lead to a peculiar dilemma. Consider the following example.

Example 9.11 Sleep Deprivation   *Healthday* published an article in which research suggests that many women suffer from sleep deprivation. For reasons such as household responsibilities, job stress, or even insomnia, women reportedly do not get enough sleep. Suppose women (and men) need 7.0 hours of sleep per night. A random sample of 32 women was obtained, and the sleeping time for each was recorded. The sample mean was $\overline{x} = 6.8$ hours and assume $\sigma = 0.67$ hours. Is there any evidence to suggest that the mean sleeping time for women is less than 7 hours?

SOLUTION

STEP 1   The assumed mean is $\mu = 7.0 (= \mu_0)$; $n = 32$ and $\sigma = 0.67$.

We are searching for any evidence to suggest that the mean sleeping time for women is less than 7.0 hours. This is a one-sided, left-tailed test. The underlying population distribution is unknown, but the sample size is large ($n = 32 \geq 30$). The hypothesis test concerning a population mean μ when σ is known is relevant.

STEP 2   No value of α is given. The first three parts of the hypothesis test are

$$H_0: \mu = 7$$
$$H_a: \mu < 7$$
$$\text{TS: } Z = \frac{\overline{X} - \mu_0}{\sigma/\sqrt{n}} = \frac{\overline{X} - 7}{0.67/\sqrt{32}}$$

The value of the test statistic is $z = \dfrac{6.8 - 7.0}{0.67/\sqrt{32}} = -1.6886$.

STEP 3   The following table shows the resulting rejection region and conclusion for various values of α.

α	**Rejection region**	**Conclusion**
0.10	$Z \leq -z_{0.10} = -1.2816$	$z = -1.6886$ lies in the rejection region. We **reject** the null hypothesis at the $\alpha = 0.10$ level.
0.05	$Z \leq -z_{0.05} = -1.6449$	$z = -1.6886$ lies in the rejection region. We **reject** the null hypothesis at the $\alpha = 0.05$ level.
0.025	$Z \leq -z_{0.025} = -1.9600$	$z = -1.6886$ does *not* lie in the rejection region. We **cannot reject** the null hypothesis at the $\alpha = 0.025$ level.
0.01	$Z \leq -z_{0.01} = -2.3263$	$z = -1.6886$ does *not* lie in the rejection region. We **cannot reject** the null hypothesis at the $\alpha = 0.01$ level.

STEP 4   In this example, the decision to reject or not reject H_0 depends on the value of α. In order to avoid this dilemma, a different method for reporting the result of a hypothesis test is often used. This technique involves computing a tail probability, or **_p_ value**. ●

Most statistics software packages report the *p* value associated with a hypothesis test.

DEFINITION

The **_p_ value**, denoted p, for a hypothesis test is the smallest significance level (value of α) for which the null hypothesis, H_0, can be rejected.

The symbol *p* is used to represent both the population proportion of successes and the *p* value. The context in which the notation is used implies the appropriate concept.

The p value is simply a tail probability, and the tail is determined by the alternative hypothesis. Consider a hypothesis test concerning a population mean μ when σ is known. Suppose either the underlying population is normal or n is large, and let z be the value of the test statistic. The following table presents the probability definition of a p value for each alternative hypothesis.

Alternative hypothesis	**Probability definition**	
$H_a: \mu > \mu_0$	$p = P(Z \geq z)$	Figure 9.18
$H_a: \mu < \mu_0$	$p = P(Z \leq z)$	Figure 9.19
$H_a: \mu \neq \mu_0$	$p/2 = P(Z \geq z)$ if $z \geq 0$	Figure 9.20
	$p/2 = P(Z \leq z)$ if $z < 0$	Figure 9.21

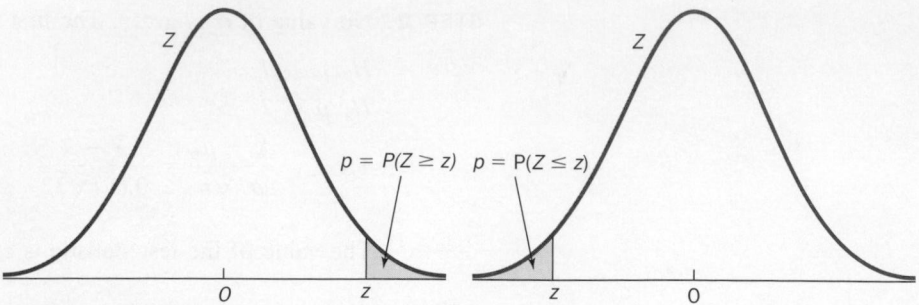

Figure 9.18 Illustration of the p value for H_a: $\mu > \mu_0$.

Figure 9.19 Illustration of the p value for H_a: $\mu < \mu_0$.

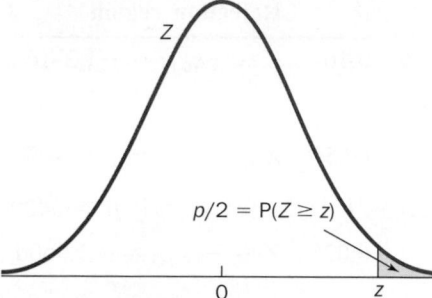

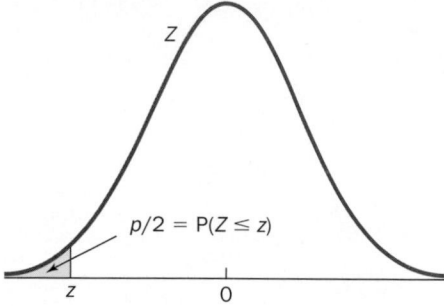

Figure 9.20 Illustration of the p value for H_a: $\mu \neq \mu_0$, $z \geq 0$.

Figure 9.21 Illustration of the p value for H_a: $\mu \neq \mu_0$, $z < 0$.

The p value conveys the strength of the evidence in favor of the alternative hypothesis. For a small p value, the value of the test statistic lies far out in a tail of the distribution. The smaller the p value, the more unlikely the value of the test statistic, and the more evidence in favor of H_a. *Small* values of p are usually $p \leq 0.05$.

If α is the significance level of the hypothesis test, the conclusion is usually written in one of two ways.

1. If $p \leq \alpha$: Reject the null hypothesis. There is evidence to suggest that H_a is true at the $p =$ (observed p value) level of significance.

2. If $p > \alpha$: Do not reject the null hypothesis. There is no evidence to suggest that the null hypothesis is false.

Consider a one-sided, right-tailed hypothesis test concerning a population mean μ when σ is known (H_a: $\mu > \mu_0$). Suppose either the underlying population is normal or n is large, α is the significance level, and let z be the value of the test statistic. Recall the definition of a critical value: $\alpha = P(Z \geq z_\alpha)$. Figures 9.22 and 9.23 demonstrate

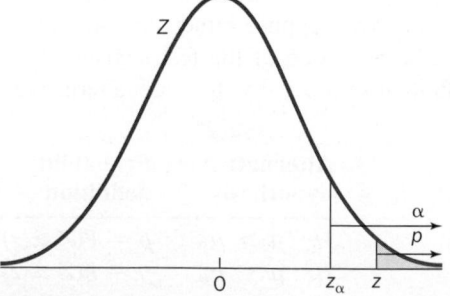

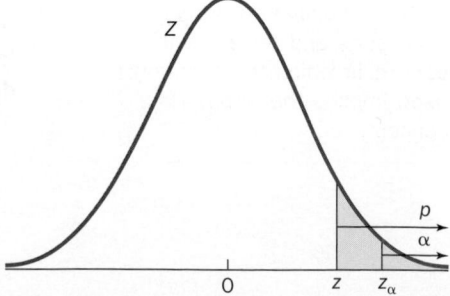

Figure 9.22 If $p \leq \alpha$, then $z \geq z_\alpha$. Conclusion: Reject H_0.

Figure 9.23 If $p > \alpha$, then $z < z_\alpha$. Conclusion: Do not reject H_0.

the relationship between the *p* value and a fixed significance level. These figures also illustrate the definition of a *p* value; that is, the *smallest* significance level for which H_0 can be rejected. If the significance level (α) is less than *p*, then $z < z_\alpha$, and we cannot reject H_0.

Example 9.12 Natural Gas Consumption The Arkansas Western Gas Company (AWGC) recently filed a rate increase request with the Public Service Commission. The rate increase was calculated based on a mean consumption of 6400 cubic feet of natural gas per month per customer. In order to check this usage figure, the Arkansas Attorney General's office obtained a random sample of 21 AWGC customers and recorded the monthly usage for each. The sample mean was $\bar{x} = 6575$ ft³. Assume that the distribution of natural gas consumption is normal with $\sigma = 450$. Conduct a one-sided, right-tailed hypothesis test with $\alpha = 0.05$ and compute the *p* value. Is there any evidence to suggest that the true mean usage is greater than 6400 ft³?

SOLUTION

STEP 1 The assumed mean is $\mu = 6400 \ (= \mu_0)$; $n = 21$ and $\sigma = 450$.

This is a one-sided, right-tailed test. The underlying population is assumed to be normal. The hypothesis test concerning a population mean μ when σ is known is relevant.

STEP 2 The value of the test statistic will be used to compute a *p* value. The first three parts of the hypothesis test are

$H_0: \mu = 6400$

$H_a: \mu > 6400$

$$\text{TS: } Z = \frac{\bar{X} - \mu_0}{\sigma/\sqrt{n}} = \frac{\bar{X} - 6400}{450/\sqrt{21}}$$

The value of the test statistic is $z = \dfrac{6575 - 6400}{450/\sqrt{21}} = \boxed{1.7821}$.

STEP 3 The *p* value is a right-tail probability (see Figure 9.27).

$p = P(Z \geq 1.7821)$ *Definition of p value for* $H_a: \mu > \mu_0$.

$= 1 - P(Z \leq 1.7812)$ *The Complement Rule.*

$= 1 - 0.9626$ *Use Table III in the Appendix.*

$= \boxed{0.0374}$

A technology solution:

Figure 9.24 TI-84 Plus Z-Test input screen.

```
Z-Test
 μ>6400.0000
 z=1.7821
 p=.0374
 x̄=6575.0000
 n=21.0000
```

Figure 9.25 TI-84 Plus Z-Test Calculate results.

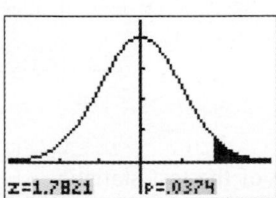

Figure 9.26 TI-84 Plus Z-Test Draw results.

Figure 9.27 Illustration of the *p* value for the Arkansas Western Gas Company example.

Since $p = 0.0374 \leq 0.05 \ (= \alpha)$, we reject the null hypothesis. There is evidence to suggest that the true mean consumption of natural gas is more than 6400 ft³ per month per customer at the $p = 0.0374$ level of significance.

The following example involves a two-sided hypothesis test.

(Dan Lamont/Corbis)

Example 9.13 Pollock Weight Alaska pollock is a whitefish known for its exceptional flavor and nutritional content. The historical mean weight per fish caught is 1.75 pounds, and over 3 million metric tons of pollock are caught each year. The Alaska Federation of Fisheries closely monitors this industry and watches for any changes in the fish population that may be attributable to weather, industry, or pollution. A random sample of Alaska pollock was obtained, and the weight (in pounds) of each fish caught is given in the following table.

1.80	1.69	1.53	1.64	1.83	1.76	1.56	1.58	1.95	1.72
1.62	1.89	1.66	1.76	1.69	1.83	1.79	1.64	1.81	1.60
1.60	1.71	1.79	1.83	1.84	1.73	1.71	1.66	1.69	1.67
1.79	1.68	1.77	1.39	1.74	1.90	1.68	1.88	1.75	1.84
1.74	1.78	1.74	1.61	1.60					

(*Source*: Alaska Fisheries Science Center).

Assume $\sigma = 0.12$ pounds. Is there any evidence that the true mean weight of Alaska pollock is different from 1.75 pounds? Use $\alpha = 0.05$ and the *p* value associated with the test statistic to justify your answer.

SOLUTION

STEP 1 The assumed mean is $\mu = 1.75$ (= μ_0); $n = 45$ and $\sigma = 0.12$.

The phrase "any difference" means this is a two-sided test. The distribution of the weight of Alaska pollock is unknown, but the sample size is large ($n = 45 \geq 30$). The hypothesis test concerning a population mean μ when σ is known is relevant.

STEP 2 The first three parts of the hypothesis test are

$H_0: \mu = 1.75$
$H_a: \mu \neq 1.75$
$$\text{TS: } Z = \frac{\bar{X} - \mu_0}{\sigma/\sqrt{n}} = \frac{\bar{X} - 1.75}{0.12/\sqrt{45}}$$

STEP 3 The sample mean is

$$\bar{x} = \frac{1}{45}(1.80 + 1.69 + \cdots + 1.60) = 1.7216$$

The value of the test statistic is

$$z = \frac{\bar{x} - \mu_0}{\sigma/\sqrt{n}} = \frac{1.7216 - 1.75}{0.12/\sqrt{45}} = -1.5876$$

STEP 4 Since this is a two-sided test and the value of the test statistic ($z = -1.5876$) is negative, $p/2$ is a left-tail probability (see Figure 9.28).

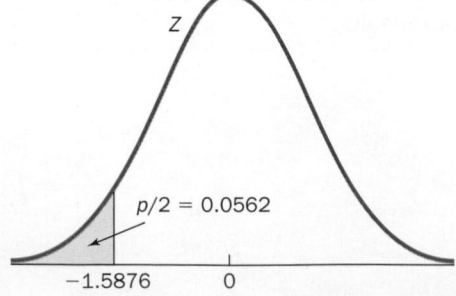

Figure 9.28 Illustration of the *p* value for the Alaska pollock example.

$p/2 = 0.0562$

-1.5876 0

A technology solution:

Figure 9.29 TI-84 Plus Z-Test input screen.

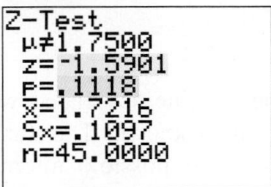

Figure 9.30 TI-84 Plus Z-Test Calculate results.

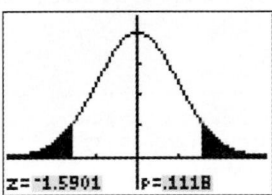

Figure 9.31 TI-84 Plus Z-Test Draw results.

$$p/2 = P(Z \le -1.5876)$$
$$= 0.0562$$
$$p = 2(0.0562) = \boxed{0.1124}$$

Definition of *p* value for $H_a: \mu \ne \mu_0$.

Use Table III in the Appendix.

Solve for *p*.

Since $p = 0.1124 > 0.05 \, (= \alpha)$, we do not reject the null hypothesis. There is no evidence to suggest that the mean weight of Alaska pollock is different from 1.75 pounds. ●

TECHNOLOGY CORNER

Procedure: Computing the *p* value in a hypothesis test concerning a population mean when σ is known.

Reconsider: Example 9.13, page 408, solution, and interpretations.

TI-84 Plus

The input is either data in a list or summary statistics.

1. Enter the data into list L1.
2. Select [STAT]; TESTS; Z-Test.
3. Highlight Data. Enter μ_0, σ, the name of the list containing the data, and the frequency of occurrence of each observation, and highlight the appropriate alternative hypothesis. See Figure 9.29.
4. In the last row of the input screen, select Calculate and press [ENTER].
5. The *p* value associated with the hypothesis test is given on the Z-Test output screen. See Figure 9.30.
6. The Draw results are shown in Figure 9.31.

Minitab

The input is either Samples in columns or Summarized data.

1. Enter the data into column C1.
2. Choose <u>S</u>tat; <u>B</u>asic Statistics; 1-Sample <u>Z</u>.
3. Select Samples in columns and enter C1. Enter the Standard deviation, check the box to Perform a hypothesis test, and enter the Hypothesized mean. See Figure 9.32
4. Choose Options. Enter a Confidence level and select the Alternative hypothesis. See Figure 9.33
5. Summary statistics, the confidence interval, the value of the test statistic, and the *p* value are displayed in the Session window. See Figure 9.34

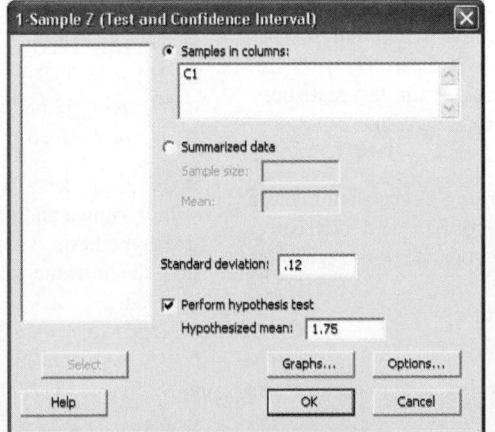

Figure 9.32 1-Sample Z input screen.

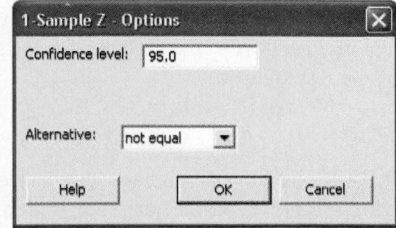

Figure 9.33 1-Sample Z Options screen.

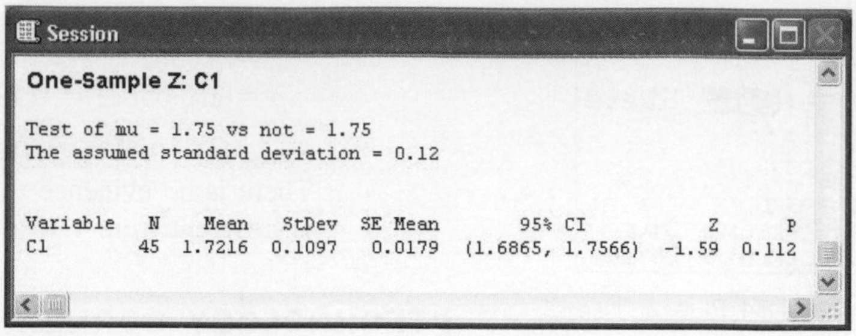

Figure 9.34 Hypothesis test results.

Excel

The Excel function ZTEST is used to compute the p value associated with a hypothesis test concerning a population mean when σ is known. This function returns the right-tail probability for each of the three possible alternative hypotheses.

1. Enter the data into column A.
2. Use ZTEST to compute the right-tail probability. The format is ZTEST (column, μ_0, σ).
3. Compute the appropriate p value. See Figure 9.35

	B	C	D
1	0.9441	=ZTEST(A1:A45,1.75,0.12)	The right-tail probability
2	0.1118	=2*(1-b1)	The p value

Figure 9.35 Ztest results and p value computation.

SECTION 9.4 EXERCISES

Practice

9.81 For each p value and significance level, determine whether the null hypothesis would be rejected or not rejected.
a. $p = 0.067$, $\alpha = 0.05$. **b.** $p = 0.0043$, $\alpha = 0.01$.
c. $p = 0.159$, $\alpha = 0.05$. **d.** $p = 0.026$, $\alpha = 0.025$.
e. $p = 0.001$, $\alpha = 0.05$. **f.** $p = 0.177$, $\alpha = 0.01$.

9.82 Consider a hypothesis test concerning a population mean with σ known, n large, and alternative hypothesis H_a: $\mu > \mu_0$. Find the p value associated with each value of the test statistic.
a. $z = 1.87$ **b.** $z = 2.55$ **c.** $z = 1.20$
d. $z = 0.57$ **e.** $z = 3.88$ **f.** $z = -1.14$

9.83 Consider a hypothesis test concerning a population mean with σ known, underlying population normal, and alternative hypothesis H_a: $\mu < \mu_0$. Find the p value associated with each value of the test statistic.
a. $z = -2.05$ **b.** $z = -1.43$ **c.** $z = -3.22$
d. $z = -0.67$ **e.** $z = -4.58$ **f.** $z = 0.25$

9.84 Consider a hypothesis test concerning a population mean with σ known, underlying population normal, and alternative hypothesis H_a: $\mu \neq \mu_0$. Find the p value associated with each value of the test statistic.

a. $z = -1.77$ **b.** $z = 1.43$
c. $z = 2.58$ **d.** $z = -0.37$
e. $z = 3.58$ **f.** $z = 0.85$

9.85 Consider a hypothesis test concerning a population mean with σ known and n large. For each alternative hypothesis, value of the test statistic, and significance level, find the p value and determine whether H_0 is rejected or not rejected.
a. H_a: $\mu > 12.5$, $z = 1.43$, $\alpha = 0.001$.
b. H_a: $\mu < -0.56$, $z = -2.05$, $\alpha = 0.05$.
c. H_a: $\mu \neq 1200$, $z = 1.75$, $\alpha = 0.10$.
d. H_a: $\mu > 37.7$, $z = 3.11$, $\alpha = 0.01$.
e. H_a: $\mu < 52.68$, $z = -1.16$, $\alpha = 0.025$.
f. H_a: $\mu \neq 46.68$, $z = -2.35$, $\alpha = 0.001$.

9.86 Consider a hypothesis test concerning a population mean with σ known and underlying population normal. For each alternative hypothesis, value of the test statistic, and significance level, find the p value and determine whether H_0 is rejected or not rejected.
a. H_a: $\mu > 3.14$, $z = 2.52$, $\alpha = 0.05$.
b. H_a: $\mu > 9.80$, $z = 1.39$, $\alpha = 0.05$.
c. H_a: $\mu < 186,000$, $z = -2.28$, $\alpha = 0.01$.
d. H_a: $\mu < 4.135$, $z = 0.17$, $\alpha = 0.01$.
e. H_a: $\mu \neq 1.62$, $z = -1.63$, $\alpha = 0.001$.
f. H_a: $\mu \neq 0.671$, $z = 2.96$, $\alpha = 0.001$.

9.87 Consider a hypothesis test concerning a population mean with σ known and underlying population normal. For each hypothesis test find the *p* value and determine whether H_0 is rejected or not rejected.

	H_0	H_a	$\bar{x}$	σ	n	α
a.	$\mu = 10$	$\mu > 10$	11.50	7.56	18	0.05
b.	$\mu = 2.718$	$\mu < 2.718$	2.60	0.56	21	0.01
c.	$\mu = 57.72$	$\mu \neq 57.72$	56.42	1.58	14	0.01
d.	$\mu = -16.18$	$\mu > -16.18$	2.35	21.23	8	0.001
e.	$\mu = 273$	$\mu < 273$	275.80	17.80	15	0.05
f.	$\mu = 6.63$	$\mu \neq 6.63$	7.17	1.08	27	0.05

Applications

9.88 Public Health and Nutrition High levels of dioxin, ingested through food and air, can cause diabetes, immune disorders, and cancer. The EPA has set a maximum acceptable level of dioxin in blood at 10 ppt (parts per trillion). Residents living near an old incinerator known to produce dioxin air pollution have complained of many health problems. A random sample of people living within a 10-mile radius of the facility was obtained, and the dioxin level (in ppt) in the blood of each person was measured. The data are given on the data CD and book's web site. Is there any evidence to suggest that the mean dioxin level in the blood of residents living near the incinerator is greater than 10 ppt? Assume $\sigma = 2.66$, use $\alpha = 0.05$, and compute the *p* value.[19]

9.89 Sports and Leisure The National Football League rates each quarterback after every game according to a formula that involves pass attempts, passing yards, touchdown passes, and interceptions. Suppose the mean quarterback rating per game is 87.6 with $\sigma = 15.7$. There is some speculation that rule changes benefiting receivers have also increased quarterback ratings. In a random sample of 43 quarterback game ratings, the sample mean was $\bar{x} = 92.8$. Is there any evidence to suggest that the true mean quarterback rating has increased? Use $\alpha = 0.05$ and compute the *p* value.[20]

9.90 Medicine and Clinical Studies The active ingredient in antiseptic liquid bandages is 8-hydroxyquinoline (8h). This chemical can cause harm through simple skin contact, swallowing, or inhalation. Suppose the list of ingredients on a 3M Nexcare liquid bandage indicates that the volume of 8h is 1%. A random sample of eight bottles of this product was obtained, and each was analyzed to obtain the volume of 8h. The sample mean was $\bar{x} = 1.025\%$. Assume that the underlying population is normal and $\sigma = 0.04$. Is there any evidence to suggest that the true mean percentage of 8h is different from 1%? Use $\alpha = 0.01$ and compute the *p* value.

9.91 Fuel Consumption and Cars Jack Keef General Motors, an automobile dealer in Lincoln, Nebraska, advertised a $9.99 forty-point safety checkup for any car and claimed that the service department would finish the job in less than 30 minutes. A local newspaper reporter selected a random sample of 58 customers who took advantage of the dealer's offer and found that the sample mean time to complete the safety checkup was $\bar{x} = 32.2$ minutes. Assume $\sigma = 5.7$ and use $\alpha = 0.01$.

a. Is there any evidence to suggest that the true mean time to complete the safety checkup is greater than 30 minutes?

b. Find the *p* value for the hypothesis test in part (a).

9.92 Business and Management Although gold has some practical uses (for example in electronics and dentistry), it is primarily used in jewelry. Gold is valuable because it is rare and cannot be manufactured. A random sample of gold mining companies was obtained from The Gold Report and the third quarter 2008 production (in thousands of ounces) was recorded for each.

1945	1277	1265	798	307	37
557	485	250	173	300	

Suppose historical records indicate that the mean production per company during the third quarter is 640 (thousands of ounces). Jewelers are concerned that if production increases, the price of gold will fall. Assume the underlying population is normal and $\sigma = 105$. Is there any evidence to suggest that the mean third quarter production per company is increasing? Use $\alpha = 0.05$ and compute the *p* value to draw a conclusion.

9.93 Manufacturing and Product Development Nitterhouse Masonry Products, LLC in Chambersburg, Pennsylvania, produces architectural concrete masonry products. The Dover is the largest block in a certain collection, is used primarily for residential retaining walls, and is manufactured to weigh 40 pounds. A quality-control inspector for the company randomly selected eight blocks and the weight (in pounds) of each is given in the following table.

43.4	39.8	42.3	42.6	41.0	39.8	39.6	38.8

Assume that the underlying population is normal and $\sigma = 2$ pounds. Is there any evidence to suggest that the true mean weight of the blocks is different from 40 pounds? Use $\alpha = 0.05$ and compute the *p* value to draw a conclusion.[21]

9.94 Travel and Transportation The technical specifications for a Boeing 767-200 indicate that a fully loaded plane (gross weight 300,000 pounds) can take off using only 5700 feet of runway.[22] In order to check this claim, officials from the FAA randomly selected 767-200 flights and measured the length (in feet) of the runway needed for takeoff. The data are given on the data CD and book's web site. Assume $\sigma = 365$ and use $\alpha = 0.05$. Is there any evidence to suggest that the true mean runway length needed for takeoff of a Boeing 767-200 is greater than 5700 feet? Use the *p* value to draw a conclusion.

9.95 Manufacturing and Product Development According to U.S. Inspect, roofing shingles weighed approximately 240 pounds per square foot prior to 1973 and modern shingles weigh approximately 190 pounds per square foot, depending on the material and the manufacturer. Suppose a brand of recycled synthetic shingles is advertised to weigh 185 pounds per square foot. A random sample of these shingles was obtained and the weight of each is given in the following table.

182.6	184.8	187.4	186.4	180.4	187.3
185.1	181.4	183.5	187.6	182.7	186.6
185.4	184.0	185.1	187.7	186.7	185.4
183.4	188.1	182.5	186.3	185.2	187.8
186.5	187.8	182.7	188.8	187.2	189.7
184.5	183.7				

Assume $\sigma = 2.7$. Is there any evidence to suggest that the true mean weight of these recycled shingles is different from 185 lbs/ft^2? Use $\alpha = 0.05$ and compute the p value to draw a conclusion.

9.96 Manufacturing and Product Development The Wood Flooring Manufacturers Association claims that a new product restores a high-gloss finish to old wood floors. A gloss meter is used to measure the shine of a wood floor and a high-gloss finish should have a rating of at least 80 units. In order to check the manufacturer's claim, the sealer and finish were applied to 35 randomly selected wood floors and the gloss was measured after each application dried. The sample mean gloss was $\bar{x} = 78.6$. Assume $\sigma = 3.45$. Is there any evidence to suggest that the manufacturer's claim is false? Use $\alpha = 0.01$ and compute the p value to draw a conclusion.

Extended Applications

9.97 Manufacturing and Product Development Arizona State procurement specifications for instant-dry highway paint indicate that the paint must dry in less than 60 seconds. A random sample of paint was obtained from a potential seller, and each batch was carefully tested. The drying time (in seconds) associated with each batch is given on the data CD and book's web site. Assume $\sigma = 8.0$ and use $\alpha = 0.05$.
a. Is there any evidence to suggest that the mean drying time for this highway paint is less than 60 seconds? Use the p value to draw a conclusion.
b. Using the results from part (a), do you believe there is overwhelming evidence to suggest the paint dries in less than 60 seconds? Why or why not?

c. Carefully sketch a graph to illustrate the p value in this hypothesis test.

9.98 Biology and Environmental Science A large farm selling round fescue hay bales claims that the mean weight per bale is 1600 pounds. In order to check this claim, personnel from the local extension service randomly selected 25 round bales and carefully weighed each. The sample mean was $\bar{x} = 1595.6$ pounds. Assume the underlying population is normal and $\sigma = 23$ pounds.
a. Is there any evidence to suggest that the true mean weight of these round hay bales is less than 1600 pounds? Use $\alpha = 0.01$ and compute the p value to draw a conclusion.
b. Carefully sketch a graph to illustrate the p value in this hypothesis test.

9.99 Public Policy and Political Science A technical panel investigating residential smoke alarms recommends that the sound emitted by the device (during a fire) should be at least 85 decibels at 10 feet away. In a random sample of 60 smoke alarms from homes in a large city, the sample mean decibel level (at 10 feet away) was $\bar{x} = 84.88$. Assume $\sigma = 5.6$.
a. Is there any evidence that the true mean decibel level produced by smoke alarms in this city is less than 85? Use $\alpha = 0.01$ and compute the p value to draw a conclusion.
b. What is the smallest significance level at which this hypothesis test would be significant?

9.5 HYPOTHESIS TESTS CONCERNING A POPULATION MEAN WHEN σ IS UNKNOWN

In Section 9.3, hypothesis tests concerning a population mean μ were developed. However, those tests are based on a standard normal, or Z, distribution and are valid only if σ is known (and the underlying population is normal, the sample size is large, or both). It is unrealistic to assume that the population standard deviation is known. As with confidence intervals, we assume that the underlying population is normal. The hypothesis test procedure is based on the following result (from Section 8.3).

9.1 THEOREM

Given a random sample of size n from a normal distribution with mean μ, the random variable

$$T = \frac{\bar{X} - \mu}{S/\sqrt{n}}$$

has a t distribution with $n - 1$ degrees of freedom (df).

Remember, this is simply another kind of standardization, and the distribution of T is characterized by $n - 1$ (not n) degrees of freedom. The general procedure for a hypothesis test concerning μ is summarized as follows.

> This is the *template* for a hypothesis test about a population mean when σ is unknown.

HYPOTHESIS TESTS CONCERNING A POPULATION MEAN WHEN σ IS UNKNOWN

Given a random sample of size n from a normal population with mean μ, a hypothesis test concerning the population mean μ with significance level α has the form

$H_0: \mu = \mu_0$

$H_a: \mu > \mu_0, \quad \mu < \mu_0, \quad \text{or} \quad \mu \neq \mu_0$

TS: $T = \dfrac{\bar{X} - \mu_0}{S/\sqrt{n}}$

RR: $T \geq t_\alpha, \quad T \leq -t_\alpha, \quad \text{or} \quad |T| \geq t_{\alpha/2}, \quad \text{df} = n - 1$

> Remember, use only one (appropriate) alternative hypothesis and the corresponding rejection region.

ILLUMINATING THE CONCEPTS

1. This procedure is often called a small-sample test (concerning a population mean), or simply a t test. As long as the underlying population is normal, this test is valid (and exact) for *any* sample size n (large or small).

2. Table V in the Appendix presents selected critical values associated with various t distributions. These critical values were used to construct confidence intervals in Section 8.3.

3. Figures 9.36, 9.37, and 9.38 illustrate the rejection region for each alternative hypothesis.

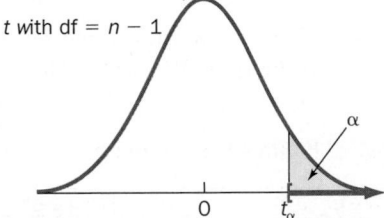

Figure 9.36 Rejection region for $H_a: \mu > \mu_0$.

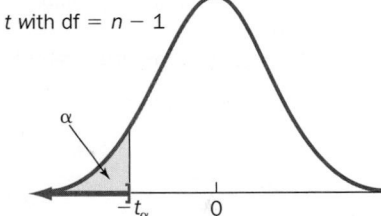

Figure 9.37 Rejection region for $H_a: \mu < \mu_0$.

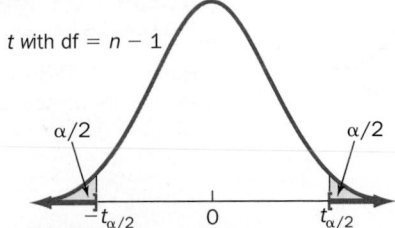

Figure 9.38 Rejection region for $H_a: \mu \neq \mu_0$. ■

The following example illustrates this hypothesis test procedure.

Example 9.14 Aerobic Physical Fitness   The VO_2 max is a measure of aerobic physical fitness and indicates the maximum amount of oxygen a person can use. When adjusted for body weight, the units of VO_2 max are milliliters per kilogram per minute (ml/kg/min). The mean VO_2 max for all men 16–24 years old is

Solution Trail

KEYWORDS

- Is there any evidence?
- Mean VO₂ max.
- Greater than 55.
- s = 4.8.
- Population is normal.

↓

TRANSLATION

Conduct a one-sided, right-tailed test about μ.

↓

CONCEPTS

Hypothesis test concerning a population mean when σ is unknown.

↓

VISION

Use the template for a one-sided, right-tailed t test about μ. The underlying population is assumed to be normal, and σ is unknown. Determine the appropriate alternative hypothesis and the corresponding rejection region, find the value of the test statistic, and draw a conclusion.

55 ml/kg/min.[23] A random sample of 12 men in this age group who exercise regularly was obtained. Each was measured for VO₂ max. The summary statistics were $\bar{x} = 58.6$ ml/kg/min and $s = 4.8$ ml/kg/min. Is there any evidence to suggest that the true mean VO₂ max for men in this age group who exercise regularly is greater than 55? Assume that the underlying population is normal, and use $\alpha = 0.05$.

SOLUTION

STEP 1 The assumed mean is $\mu = 55$ ($= \mu_0$); $n = 12$, $s = 4.8$, and $\alpha = 0.05$.

We are looking for evidence to suggest that the VO₂ max is *greater* in this group of men. The relevant alternative hypothesis is one-sided and right-tailed.

The underlying population is assumed to be normal, but σ is unknown. A t test is appropriate.

STEP 2 The four parts of the hypothesis test are

$H_0: \mu = 55$

$H_a: \mu > 55$

$$\text{TS: } T = \frac{\bar{X} - \mu_0}{S/\sqrt{n}}$$

RR: $T \geq t_\alpha = t_{0.05} = 1.7959$ df = 12 − 1 = 11

STEP 3 The value of the test statistic is

$$t = \frac{\bar{x} - \mu_0}{s/\sqrt{n}} = \frac{58.6 - 55}{4.8/\sqrt{12}} = 2.5981 \geq 1.7959$$

STEP 4 Since 2.5981 lies in the rejection region, we reject the null hypothesis at the $\alpha = 0.05$ level. There is evidence to suggest that the mean VO₂ max is greater than 55 in this group of men.

Figures 9.39 through 9.41 together show a technology solution.

p value illustration:
$p = P(T \geq 2.5981)$
$= 0.0124 \leq 0.05 = \alpha$

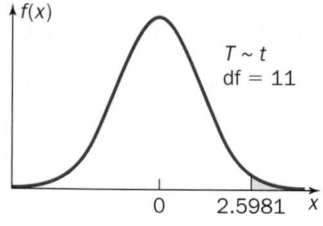

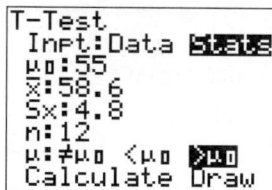

Figure 9.39 TI-84 Plus T-Test input screen.

Figure 9.40 TI-84 Plus T-Test Calculate results.

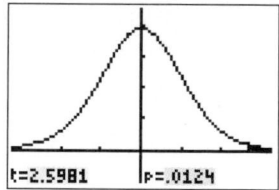

Figure 9.41 TI-84 Plus T-Test Draw results.

Example 9.15 Water Park Water Most water parks use a recycling system so that the only water loss is due to evaporation and splashing. Despite a sophisticated recycling system, the Schlitterbahn Waterpark resort in New Braunfels, Texas, has informed the city water department that it needs an additional 250,000 gallons of water per day. The city water department selected a random sample of days, and the park's water usage (in thousands of gallons) on each day is given in the following table.

140.1	270.1	234.3	242.3	233.3	337.4	248.6	210.4
244.6	269.6	292.3	263.0	229.0	264.0	260.5	236.9
244.1	201.0	205.7	219.0	303.5			

(Mayangsari/Dreamstime.com)

Is there any evidence to suggest that the mean water usage is different from 250,000 gallons? Assume that the underlying population is normal and use $\alpha = 0.01$.

SOLUTION

STEP 1   The assumed mean is 250 ($= \mu_0$) thousand gallons, $n = 21$, and $\alpha = 0.01$.

The water department is looking for water usage *different* from 250,000. Therefore, the relevant alternative hypothesis is two-sided.

The underlying population is assumed to be normal, and σ is unknown. A t test is appropriate.

STEP 2   The four parts of the hypothesis test are

$H_0: \mu = 250$

$H_a: \mu \neq 250$

TS: $T = \dfrac{\overline{X} - \mu_0}{S/\sqrt{n}}$

RR: $|T| \geq t_{\alpha/2} = t_{0.005} = 2.8453$ ($T \leq -2.8453$ or $T \geq 2.8453$)   df = 20

STEP 3   The sample mean is

$$\overline{x} = \frac{1}{21}(140.1 + 270.1 + \cdots + 303.5) = 245.2238$$

Find the sample standard deviation.

$$s^2 = \frac{1}{20}\left(1{,}296{,}137.59 - \frac{1}{21}(5149.7)^2\right) = 1665.4269$$
$$s = \sqrt{1665.4269} = 40.8096$$

The value of the test statistic is

$$t = \frac{\overline{x} - \mu_0}{s/\sqrt{n}} = \frac{245.2238 - 250}{40.8096/\sqrt{21}} = \boxed{-0.5363}$$

STEP 4   The value of the test statistic, $t = -0.5363$, does *not* lie in the rejection region. We cannot reject the null hypothesis. There is no evidence to suggest park water usage is different from 250,000 gallons per day.   ●

The table listing critical values for various t distributions is very limited. Using this table, the best we can do is *bound* the p value associated with a hypothesis test.

A technology solution:

```
T-Test
 Inpt:DATA Stats
 μ0:250
 List:L1
 Freq:1
 μ:≠μ0 <μ0 >μ0
 Calculate Draw
```

Figure 9.42 TI-84 Plus T-Test input screen.

```
T-Test
 μ≠250.0000
 t=-.5363
 p=.5976
 x̄=245.2238
 Sx=40.8096
 n=21.0000
```

Figure 9.43 TI-84 Plus T-Test results.

p value illustration:
$P = 2P(T \leq -0.5363)$
$= \boxed{0.5976} > 0.01 = \alpha$

$f(x)$

$T \sim t$
df = 20

-0.5363 0 0.5363 x

HOW TO BOUND THE p VALUE FOR A t TEST

Suppose t is the value of the test statistic in a one-sided hypothesis test.

1. Select the row in Table V in the Appendix that corresponds to $n - 1$, the number of degrees of freedom associated with the test.
2. Place $|t|$ in this ordered list of critical values.
3. Compute p:
   a. If $|t|$ is between two critical values in the $n - 1$ row, then the p value is bounded by the corresponding significance levels.
   b. If $|t|$ is greater than the largest critical value in the $n - 1$ row, then $p < 0.0001$ (the smallest significance level in the table).
   c. If $|t|$ is less than the smallest critical value in the $n - 1$ row, then $p > 0.20$ (the largest significance level in the table).

If $t < 0$ for a right-tailed test, or $t > 0$ for a left-tailed test, then $p > 0.5$. If the hypothesis test is two-sided, this method produces a bound on $p/2$.

Example 9.16 Delivery Time   A certain daily delivery route for Hostess breads and snack cakes includes eight grocery stores and four convenience stores. The historical mean time to complete the deliveries to all the stores and return to the distribution center is 6.5 hours. A new driver has been assigned to this route, and a random sample of his route completion times (in hours) was obtained. The data are given in the following table.

$$6.61 \quad 6.25 \quad 6.40 \quad 6.57 \quad 6.35 \quad 5.95 \quad 6.53 \quad 6.29$$

Assume that the underlying population is normal.

a. Is there any evidence to suggest that the new driver has been able to shorten the route completion time? Use $\alpha = 0.01$.

b. Find bounds on the p value associated with this hypothesis test.

SOLUTION

STEP 1   The assumed mean is 6.5 ($= \mu_0$); $n = 8$ and $\alpha = 0.01$.

We would like to know whether the new driver has been able to shorten the mean delivery time. The relevant alternative hypothesis is one-sided and left-tailed.

The underlying population is assumed to be normal, and σ is unknown. A t test is appropriate.

STEP 2   The four parts of the hypothesis test are

$H_0: \mu = 6.5$

$H_a: \mu < 6.5$

TS: $T = \dfrac{\overline{X} - \mu_0}{S/\sqrt{n}}$

RR: $T \leq -t_\alpha = -t_{0.01} = -2.9980$ df = 8 − 1 = 7

STEP 3   The sample mean is

$$\overline{x} = \frac{1}{8}(6.61 + 6.25 + \cdots + 6.29) = 6.3688$$

Find the sample standard deviation.

$$s^2 = \frac{1}{7}\left(324.8095 - \frac{1}{8}(50.95)^2\right) = 0.0460$$

$$s = \sqrt{0.460} = 0.2144$$

The value of the test statistic is

$$t = \frac{\overline{x} - \mu_0}{s/\sqrt{n}} = \frac{6.3688 - 6.5}{0.2144/\sqrt{8}} = -1.7308$$

The value of the test statistic, $t = -1.7308$, does *not* lie in the rejection region. We cannot reject the null hypothesis. There is no evidence to suggest that the new driver has been able to shorten the mean delivery time for this route.

STEP 4 $|t| = |-1.7308| = 1.7308$

In Table V in the Appendix, row $n - 1 = 8 - 1 = 7$, place 1.7308 in the ordered list of critical values.

$$1.4149 \leq 1.7308 \leq 1.8946$$

$$t_{0.10} \leq 1.7308 \leq t_{0.05}$$

Therefore, $0.05 \leq \quad p \quad \leq 0.10$

See Figure 9.44. The p value associated with this hypothesis test is between 0.05 and 0.10. Since the smallest possible value, 0.05, is greater than the significance level, 0.01, we cannot reject the null hypothesis.

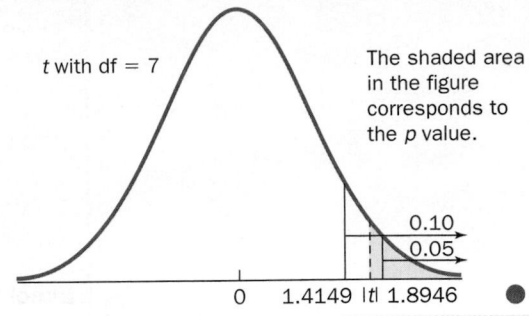

t with df = 7

The shaded area in the figure corresponds to the p value.

0.10
0.05

0   1.4149 |t| 1.8946

Figure 9.44 Visualization of the bounds on the p value.

A technology solution:

Figure 9.45 TI-84 Plus T-Test input screen.

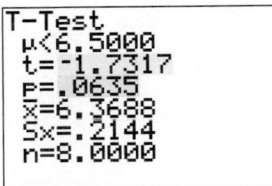

Figure 9.46 TI-84 Plus T-Test Calculate results.

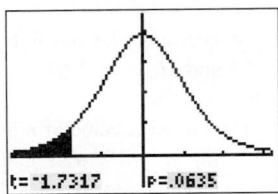

Figure 9.47 TI-84 Plus T-Test Draw results.

TECHNOLOGY CORNER

Procedure: Hypothesis test concerning a population mean when σ is unknown.
Reconsider: Example 9.16, page 416, solution, and interpretations.

TI-84 Plus

The input is either data in a list or summary statistics.

1. Enter the data into list L1.
2. Select $\boxed{\text{STAT}}$; TESTS; T-Test.
3. Highlight Data. Enter μ_0, the name of the list containing the data, and the frequency of occurrence of each observation, and highlight the appropriate alternative hypothesis. See Figure 9.45.
4. In the last row of the input screen, select Calculate and press $\boxed{\text{ENTER}}$.
5. The alternative hypothesis, value of the test statistic, p value, and summary statistics are displayed on the output screen. See Figure 9.46.
6. The Draw results are shown in Figure 9.47.

Minitab

The input is either Samples in columns or Summarized data.

1. Enter the data into Column C1.
2. Choose Stat; Basic Statistics; 1-Sample t.
3. Select Samples in Columns and enter C1. Check the box to Perform a hypothesis test, and enter the Hypothesized mean. See Figure 9.48.
4. Choose Options. Enter a confidence level and select the Alternative hypothesis. See Figure 9.49.

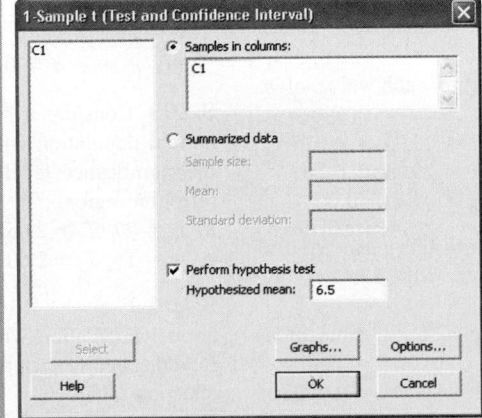

Figure 9.48 1-Sample t input screen.

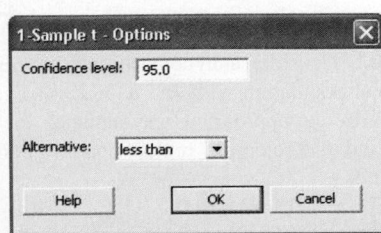

Figure 9.49 1-Sample t Options screen.

```
Session                                          ─ □ ×

One-Sample T: C1

Test of mu = 6.5 vs < 6.5


                                      95% Upper
Variable  N   Mean   StDev  SE Mean    Bound     T      P
C1        8  6.3688  0.2144  0.0758    6.5123  -1.73  0.063
```

Figure 9.50 Hypothesis test results.

Excel

Excel can be used to compute the sample mean, the sample standard deviation, and the value of the test statistic. The built-in Excel function TDIST can be used to find the p value associated with the hypothesis test.

1. Enter the data into column A.
2. Compute the sample mean using the Excel command AVERAGE.
3. Compute the sample standard deviation using the Excel command STDEV.
4. Compute the value of the test statistic.
5. Use the function TDIST to find the p value. See Figure 9.51.

	B	C	D
1	6.3688	=AVERAGE(A1:A8)	The sample mean
2	0.2144	=STDEV(A1:A8)	The sample standard deviation
3	8	=COUNT(A1:A8)	The sample size
4	-1.7317	=(B1-6.5)/(B2/SQRT(B3))	The value of t
5	0.0635	=TDIST(ABS(B4),B3-1,1)	The p value

Figure 9.51 Excel computations.

SECTION 9.5 EXERCISES

Practice

9.100 Consider a hypothesis test concerning a population mean from a normal population with H_0: $\mu = 10.5$ and H_a: $\mu > 10.5$.
a. Write the appropriate test statistic.
b. Find the rejection region corresponding to each value of n and α.

(i) $n = 6$, $\alpha = 0.01$ (ii) $n = 23$, $\alpha = 0.025$
(iii) $n = 17$, $\alpha = 0.05$ (iv) $n = 29$, $\alpha = 0.10$
(v) $n = 10$, $\alpha = 0.001$ (vi) $n = 9$, $\alpha = 0.0001$

9.101 Consider a hypothesis test concerning the mean from a normal population with H_0: $\mu = 22.41$ and H_a: $\mu < 22.41$.
a. Write the appropriate test statistic.
b. Find the rejection region corresponding to each value of n and α.

(i) $n = 15$, $\alpha = 0.01$ (ii) $n = 11$, $\alpha = 0.0005$
(iii) $n = 21$, $\alpha = 0.05$ (iv) $n = 24$, $\alpha = 0.10$
(v) $n = 5$, $\alpha = 0.001$ (vi) $n = 31$, $\alpha = 0.0001$

9.102 Consider a hypothesis test concerning the mean from a normal population with H_0: $\mu = 1.67$ and H_a: $\mu \neq 1.67$.
a. Write the appropriate test statistic.
b. Find the rejection region corresponding to each value of n and α.

(i) $n = 12$, $\alpha = 0.01$ (ii) $n = 19$, $\alpha = 0.20$
(iii) $n = 26$, $\alpha = 0.05$ (iv) $n = 28$, $\alpha = 0.10$
(v) $n = 7$, $\alpha = 0.001$ (vi) $n = 56$, $\alpha = 0.02$

9.103 Consider a hypothesis test concerning the mean from a normal population with H_0: $\mu = 0.082$ and H_a: $\mu > 0.082$. Find the significance level (α) corresponding to each value of n and rejection region.

a. $n = 27$, $T \geq 2.0555$ **b.** $n = 9$, $T \geq 4.5008$
c. $n = 16$, $T \geq 2.6025$ **d.** $n = 20$, $T \geq 3.5794$

9.104 Consider a hypothesis test concerning the mean from a normal population with H_0: $\mu = -15.76$ and H_a: $\mu < -15.76$. Find the significance level (α) corresponding to each value of n and rejection region.

a. $n = 23$, $T \leq -1.3212$ **b.** $n = 4$, $T \leq -10.2145$
c. $n = 31$, $T \leq -2.7500$ **d.** $n = 19$, $T \leq -2.5524$

9.105 Consider a hypothesis test concerning the mean from a normal population with H_0: $\mu = 5128$ and H_a: $\mu \neq 5128$. Find the significance level (α) corresponding to each value of n and rejection region.

a. $n = 30$, $|T| \geq 1.6991$ **b.** $n = 6$, $|T| \geq 4.0321$
c. $n = 17$, $|T| \geq 3.6862$ **d.** $n = 14$, $|T| \geq 5.1106$

9.106 Consider a hypothesis test concerning the mean from a normal population with H_0: $\mu = 2.53$ and H_a: $\mu > 2.53$. Find bounds on the p value for each value of n and test statistic t.

a. $n = 24$, $t = 2.35$ **b.** $n = 3$, $t = 8.55$
c. $n = 12$, $t = 5.68$ **d.** $n = 8$, $t = 1.52$

9.107 Consider a hypothesis test concerning the mean from a normal population with H_0: $\mu = 6.28$ and H_a: $\mu < 6.28$. Find bounds on the p value for each value of n and test statistic t.

a. $n = 25$, $t = -1.97$ **b.** $n = 13$, $t = -0.63$
c. $n = 18$, $t = -2.28$ **d.** $n = 27$, $t = -3.58$

9.108 Consider a hypothesis test concerning the mean from a normal population with H_0: $\mu = 1.414$ and H_a: $\mu \neq 1.414$. Find bounds on the p value for each value of n and test statistic t.

a. $n = 10$, $t = 2.04$ **b.** $n = 23$, $t = -3.14$
c. $n = 14$, $t = -5.52$ **d.** $n = 11$, $t = 1.75$

9.109 Consider a random sample of size 20 from a normal population with hypothesized mean 1.618.
a. Write the four parts for a one-sided, left-tailed hypothesis test concerning the population mean with $\alpha = 0.05$.
b. Suppose $\bar{x} = 1.5$ and $s = 0.45$. Find the value of the test statistic, and draw a conclusion about the population mean.
c. Find the p value associated with this hypothesis test.

9.110 Consider a random sample of 11 observations, given in the following table, from a normal population with hypothesized mean 57.71.

59.94	58.93	59.41	60.66	59.00	60.98
58.85	55.21	59.02	61.14	59.25	

a. Write the four parts for a one-sided, right-tailed hypothesis test concerning the population mean with $\alpha = 0.01$.
b. Compute the sample mean, the sample standard deviation, and the value of the test statistic. Draw a conclusion about the population mean.
c. Find the p value associated with this hypothesis test.

9.111 Consider a random sample of size 28 from a normal population with hypothesized mean $\mu_0 = 9.96$.
a. Write the four parts for a two-sided hypothesis test concerning the population mean with $\alpha = 0.002$.
b. Suppose $\bar{x} = 9.04$ and $s = 1.20$. Find the value of the test statistic, and draw a conclusion about the population mean.
c. Find the p value associated with this hypothesis test.

9.112 Consider a two-sided hypothesis test concerning the mean μ_0 from a normal population, with sample size $n = 25$, standard deviation s, and $\alpha = 0.05$. Explain the error in each of the following statements.

a. The test statistic is $T = \dfrac{\bar{X} - \mu_0}{\sigma/\sqrt{25}}$.

b. The test statistic is $T = \dfrac{\bar{X} - \mu_0}{S/\sqrt{24}}$.

c. The rejection region is RR: $T \geq 1.7109$.

d. If the value of the test statistic is $t = 2.6732$, then $p \leq 0.005$.

Applications

9.113 Biology and Environmental Science Hunting is a very popular sport in Vermont, and, in 2008, 1255 moose-hunting permits were randomly issued to the more than 13,000 people who applied.[24] While almost three-quarters of the state is open to moose hunting, the best hunting is at Island Pond, where the mean weight of a bull moose taken during the last hunting season was 871 pounds. Suppose a random sample of 12 moose taken from Island Pond during this hunting season was obtained. The summary statistics were $\bar{x} = 885.7$ and $s = 52$. Conduct a hypothesis test to determine whether there is any evidence that the mean weight of moose taken from Island Pond this year is greater than it was last year. Assume the distribution of moose weight is normal and use $\alpha = 0.05$.

9.114 Travel and Transportation During a routine commercial airline flight, some of the time in the aircraft is spent waiting to take off and taxiing to an arrival gate. However, airlines also keep careful records of the actual airborne time of each flight. A random sample of the airborne times (in minutes) of United Airlines flights from Boston to Denver during 2005 was obtained and the data are given in the following table.

235	250	247	241	240	246	231	233
230	255	228	246	247	238	223	

The airborne time for this flight is affected by the prevailing west-to-east winds and normally takes 245 minutes. Is there any evidence to suggest that the true mean airborne time is less than this 245 minutes? Assume that the underlying distribution is normal and use $\alpha = 0.025$.[25]

9.115 Biology and Environmental Science The historical mean yield of soybeans from farms in Indiana is 31.9 bushels per acre. Following a recent dry summer, a random sample of 26 farms from across the state was obtained. The mean yield in bushels per acre was $\bar{x} = 30.088$ with $s = 4.433$. Is there any evidence to suggest that the lack of rain adversely affected the soybean yield in Indiana? Assume that the underlying population is normal and use $\alpha = 0.01$.

9.116 Demographics and Population Statistics According to the Automotive Coalition for Traffic Safety, the mean penalty, including fines and court costs, for a stop-sign violation is $86.99.[26] In order to check this claim, a student randomly selected 23 stop-sign violations from court records and found that the sample mean cost was $\bar{x} = \$95.65$ and $s = \$11.25$. Is there any evidence to suggest that the mean penalty for a stop-sign violation is different from $86.99? Use $\alpha = 0.01$.

9.117 Business and Management The 40-hour workweek did not become an American standard until 1940. Today, many white-collar employees work more than 40 hours per week because management demands longer hours or offers large monetary incentives.

A random sample of white-collar employees was obtained, and the number of hours each worked during the last week are given in the following table.

44.7	42.0	45.8	43.0	42.8	50.9	47.0
41.9	49.3	45.6	45.7	39.4	39.0	44.4

Is there any evidence that the true mean number of hours worked by white-collar employees is greater than 40? Use $\alpha = 0.01$. What assumption(s) did you make in order to complete this hypothesis test?

9.118 Marketing and Consumer Behavior Kennebunkport, Maine, is a popular tourist town, but businesses suffer during the winter months, especially January. The mean hotel room occupation rate per day, a measure of tourist activity, is 23.1% during winter months. A new advertising campaign was launched in order to attract more tourists to this town during the winter. Following the campaign, a random sample of nine winter days was selected. The mean hotel room occupation rate was $\bar{x} = 24.6\%$ and $s = 2.1\%$. Is there any evidence to suggest that the mean hotel room occupation rate has increased? Assume normality and use $\alpha = 0.01$. If this test is significant, can you conclude that the ad campaign caused the increase?

9.119 Technology and Internet A 42-inch, large-screen, plasma TV is designed to consume only 350 watts and therefore produce less heat. In order to check this specification, a random sample of seven TVs was obtained and the power consumption was measured for each. The sample mean was $\bar{x} = 353.8$ watts and $s = 5.6$ watts. Is there any evidence to suggest that the mean power consumption for this model plasma TV does not meet the design specification? Assume normality and use $\alpha = 0.05$.

9.120 Business and Management The mean pay in 2008 for CEOs at the 500 biggest U.S. companies was 12.8 million dollars.[27] Public outcry and stockholder complaints have forced compensation committees to reconsider senior executive salaries. A random sample of CEOs was obtained, and the total pay (in millions of dollars) for each is given in the following table.

10.1	4.3	13.8	5.1	13.0	21.2	4.5
10.5	17.6	10.1	6.5	9.9	13.3	20.4

Is there any evidence to suggest that the mean pay for CEOs has decreased? Use $\alpha = 0.05$ and assume normality.

9.121 Public Health and Nutrition The dietary supplement coral calcium has been advertised to provide pain relief for patients with multiple sclerosis and other diseases. Unfortunately, this product is known to contain toxic minerals such as lead. In 2003 the FTC and FDA took action to stop deceptive and misleading advertising regarding coral calcium (*Source:* Federal Trade Commission for the Consumer). The Centers for Disease Control and Prevention recommend a maximum daily allowance of 25 micrograms (mcg) of lead. A random sample of coral calcium pills was obtained and the lead content (in mcg) in each is given on the data CD and book's web site. Assume that the distribution of lead content is normal and use $\alpha = 0.01$.

a. Is there any evidence to suggest that the true mean lead content in coral calcium is greater than the maximum daily allowance set by the CDC? Approximate the critical value in this test.
b. Find bounds on the p value associated with this test.

9.122 Public Health and Nutrition Medicare Part D is a federal program designed to help senior citizens pay for prescription drugs. A random sample of monthly premiums (in dollars) for 2009 Medicare Part D plans in Florida was obtained and the data are given in the following table.

16.70	21.20	23.60	12.10	63.00	31.60	27.00
32.50	31.50	35.10	30.50	22.00	37.80	56.70
56.50	67.70	61.00	45.50			

(*Source*: Medicare-PartD.com.)

Assume that the distribution of monthly Medicare Part D premiums is normal and use $\alpha = 0.05$.

a. Suppose Congress has decided to examine Medicare Part D plans if the premiums are too high. Is there any evidence to suggest that the true mean monthly benchmark premium for stand-alone Medicare drug plans is greater than $31.00?
b. Find bounds on the p value associated with this test.

9.123 Marketing and Consumer Behavior As the use of the Internet has increased, companies are spending more money advertising online. A sample of leading national advertisers was obtained, and the Internet advertising spending (in millions of dollars) for the year was recorded for each. The data are given in the following table.

Company	Spending
Procter & Gamble Co.	80.6
Verizon Communications	189.1
Time Warner	98.4
Walt Disney Co.	141.4
General Electric Co.	78.9
Toyota Motor Corp.	57.3
Sony Corp.	71.5
Kraft Foods	36.3
Macy's	11.1
PepsiCo	28.1
Pfizer	24.2
McDonald's Corp.	27.2

(*Source: Advertising Age*, June 2007.)

Suppose the mean spending on Internet advertising for these companies the previous year was $55.5 million. Is there any evidence to suggest that the true mean spending on Internet advertising has increased? Use $\alpha = 0.05$.

Extended Applications

9.124 Demographics and Population Statistics Police reports and insurance claims indicate that the mean loss during a residential break-in is $1381. Actuaries working for Eagle Pacific Insurance Company need to check whether the mean loss has changed during the past year in order to determine new policy rates. A random sample of 17 residential break-ins was obtained, and the loss due to each burglary was recorded. The sample mean loss (in dollars) was 1857 and the standard deviation was $s = 786$. Assume the loss distribution is normal and use $\alpha = 0.05$.

a. Is there any evidence to suggest that the true mean loss as a result of a residential break-in has changed?

b. Find bounds on the p value associated with this hypothesis test.

c. Carefully sketch a graph illustrating the p value and the value of the test statistic.

9.125 Physical Sciences The Hubbard Brook Experimental Forest is an area located in the White Mountain National Forest, New Hampshire. A research program established by the USDA Forest Service collects data associated with vegetation, the soil, and Mirror Lake. The chemistry of Mirror Lake outlet streamwater is carefully analyzed in weekly samples. A random sample of the calcium concentrations (in mg/L) in the outlet stream is given in the following table.

2.21	1.87	1.94	2.07	1.33	1.86	1.99
2.02	1.65	1.84	2.19	2.37	1.87	2.21
2.19	2.27	2.28	2.27	2.25	2.21	2.41

a. Is there any evidence to suggest that the true mean calcium concentration is different from 2 mg/L? Assume that the distribution is normal and use $\alpha = 0.01$.

b. Find bounds on the p value associated with this hypothesis test.

(The data used in this exercise were obtained by scientists of the Hubbard Brook Ecosystems Study; this exercise has not been reviewed by those scientists. The Hubbard Brook Experimental Forest is operated and maintained by the Northeastern Research Station, U.S. Department of Agriculture, Newtown Square, Pennsylvania.)

9.126 Fuel Consumption and Cars Coalbed methane is an important source of natural gas in the United States. The Powder River coalfield has approximately 2500 wells, each producing 159,350 cubic feet of gas per day. In order to maintain sufficient storage facilities, the mean well output is carefully monitored. A random sample of 11 wells was obtained, and the daily methane output for each was recorded. The sample mean was $\bar{x} = 163,288$ cubic feet and the standard deviation was $s = 8792$ cubic feet. Assume that the distribution of methane output per well per day is normal, and use $\alpha = 0.01$.

a. Is there any evidence to suggest that the mean methane output per well per day has increased?

b. Find bounds on the p value associated with this hypothesis test.

9.127 Marketing and Consumer Behavior The manager of a Piggly Wiggly grocery store claims a membership card will save consumers (through automatic discounts and extra coupons) at least $15.00 per week. In order to check this claim, a random sample of 11 shoppers with membership cards was obtained and their weekly grocery bills were inspected. The sample mean savings was $\bar{x} = \$14.35$ and $s = \$3.75$. Assume that the distribution of savings is normal.

a. Is there any evidence to refute the manager's claim? Use $\alpha = 0.025$.

b. If you rejected the null hypothesis, how can you explain this conclusion when $14.35 is so close to $15.00? If you did not reject the null hypothesis, how can you explain this conclusion when $14.35 is certainly less than $15.00?

c. Find bounds on the p value associated with this hypothesis test.

9.128 Physical Sciences In 2008, the U.S. Army Corps of Engineers dredged approximately 25 million cubic yards of material.[28] Suppose they dredged approximately 1.2 million cubic yards of material per site in 2007. While some of this material can be used on beaches or as wetland nourishment, it has become more difficult to dispose of such large quantities. A random sample of 12 dredging contracts from 2008 was selected, and the amount of disposed material was recorded for each site. The summary statistics were $\bar{x} = 0.810$ million cubic yards and $s = 0.626$ million cubic yards.

a. Is there any evidence to suggest that the true mean amount of material disposed of from dredging has decreased? Use $\alpha = 0.05$. What assumption(s) did you make in order to conduct this hypothesis test?

b. Find bounds on the p value associated with this hypothesis test.

9.129 Biology and Environmental Science The monthly production of Iowa dairy cows is approximately 1659 pounds of milk per cow.[29] In order to assess the effect of a prolonged heat wave on milk production, a random sample of Iowa dairy cows was obtained and the (summer) monthly milk production (in pounds) for each is given in the following table.

1665	1616	1645	1649	1644	1622	1657	1656
1625	1673	1658	1637	1675	1668	1659	1658
1643	1658	1692	1657	1651	1650	1648	1662

a. Is there any evidence to suggest that the true mean monthly milk production per cow has decreased? Use $\alpha = 0.05$. If the test is significant, do you believe that the heat wave caused this decrease?

b. Find bounds on the p value associated with this hypothesis test.

9.130 Biology and Environmental Science A quahog is a chewy Atlantic hard-shell clam with a blue-gray shell. Quahogs have different names, depending on size; for example, they are called littleneck clams if the width is under 5 centimeters (across the shell). The historical mean width of a littleneck is 4.75 cm. Following a recent oil spill off the coast of Maine, a random sample of 46 littlenecks was obtained and the width of each was recorded. The summary statistics were $\bar{x} = 4.66$ cm and $s = 0.25$ cm.

a. Use a one-sided, left-tailed hypothesis test to show that there is evidence to suggest that the mean width of littlenecks has decreased. Assume the population of littleneck widths is normal and use $\alpha = 0.01$.

b. Explain why there is statistical evidence that the population mean has decreased even though the sample mean (4.66) is so close to the historical mean (4.75).

c. Find bounds on the p value associated with this hypothesis test.

9.131 Physical Sciences The Palmer Drought Severity Index (PDSI) is a measure of prolonged abnormal dryness or wetness. It indicates general conditions and is not affected by local variations. An index value of -4 indicates extreme drought, and $+4$ represents very wet conditions. A random sample of the PDSI for selected regions in the central United States for the week ending July 9, 2005 was obtained, and the data are given in the following table.[30]

-0.91	-2.99	0.84	4.45	-2.89	2.83	1.41
0.79	-2.85	-6.64	2.57	-1.93	-3.67	1.59
-0.53	-1.67	-1.82	0.70	-0.75	3.06	

a. Is there any evidence to suggest that the true mean PDSI is different from 0? Assume that the underlying distribution is normal and use $\alpha = 0.01$. Find bounds on the p value associated with this test.

b. Explain your conclusion in part (a) in terms of farming conditions and reservoir levels.

9.132 Physical Sciences In late October 2008, Texas and Oklahoma were struck by several minor earthquakes. Although there was no serious damage, many residents felt the tremors and called police. A random sample of earthquakes around the world during December 2008 was obtained, and the magnitude of each is given in the following table.

5.9	4.7	2.7	3.4	3.2	4.4	4.3	2.8	2.5
5.1	2.5	5.0	3.5	2.7	5.2	4.7	5.5	5.3
2.6	3.3	3.6	5.4	4.9	5.1			

(*Source*: United States Geological Survey Earthquake Hazards Program.)

a. An earthquake with magnitude of 1.0 to 4.0 is considered weak and generally causes minor or no damage. Is there any evidence to suggest that the true mean magnitude of earthquakes is greater than 4.0? Assume that the underlying distribution is normal and use $\alpha = 0.05$.

b. Find bounds on the p value associated with this test.

9.133 Public Health and Nutrition Fluoride is added to city water and many dental products in order to help prevent tooth decay.[31] Sodium fluoride is a common additive in toothpaste, and the concentration is usually measured in parts per million by weight (ppmF). The manufacturer of Pepsodent brand toothpaste claims that the concentration of fluoride in every tube is 1000 ppmF. A random sample of toothpaste tubes was obtained, and the fluoride concentration in each is given on the data CD and book's web site. Some research suggests that high concentrations of fluoride can be toxic and may cause brain damage, immune disorders, and changes in bone structure and strength.

a. Test the relevant hypothesis concerning the mean fluoride concentration per tube of toothpaste. Assume normality and use $\alpha = 0.05$.

b. Find bounds on the p value associated with this hypothesis test.

9.134 Public Policy and Political Science During 2008, the mean monthly benefit per person participating in the federal food-stamp program was $109.93.[32] Due to the poor economy, more

people are using this support program to purchase food. A random sample of 15 people using food stamps was obtained, and the summary statistics for monthly benefits were $\bar{x} = \$112.97$, $s = \$9.50$. Assume that the distribution of monthly benefit per person is normal and use $\alpha = 0.05$.

a. Is there any evidence that the true mean monthly benefit per person has increased?

b. Suppose the sample size is $n = 20$ (with the same $\bar{x}$ and s). Is there any evidence that the true mean monthly benefit per person has increased?

c. Find the smallest value of n (with the same $\bar{x}$ and s) such that the hypothesis test is significant (at the $\alpha = 0.05$ level).

9.135 Economics and Finance In a Collegiate Case Study, *USA Today* reported that in the year 2000 the mean credit card debt for an undergraduate was $2748. In spring 2008, a random sample of undergraduates was obtained, and the credit card debt of each was recorded. The summary statistics were $\bar{x} = \$3022$, $s = \$457$, and $n = 11$. Assume that the distribution of credit card debt is normal and use $\alpha = 0.01$.

a. Is there any evidence to suggest that the true mean credit card debt for undergraduates has increased?

b. Find bounds on the p value associated with this test.

c. How large a sample size would be necessary for this test to be significant (assuming $\bar{x}$ and s remain the same)?

Challenge

9.136 Psychology and Human Behavior A small-sample (exact) hypothesis test concerning the mean of a Poisson random variable, λ, is constructed as follows. Suppose the random variable X is a count modeled by a Poisson random variable. The four parts of the hypothesis test are

$H_0: \lambda = \lambda_0$

$H_a: \lambda > \lambda_0$, $\quad \lambda < \lambda_0$, $\quad$ or $\quad \lambda \neq \lambda_0$

TS: X = the number of events that occur in the specified *interval*

RR: $X \geq x_\alpha$, $\quad X \leq x'_\alpha$, $\quad$ or $\quad X \leq x_{\alpha/2}$ $\quad$ or $\quad X \geq x_{\alpha/2}$

The critical values x_α, x'_α, $x_{\alpha/2}$, and $x'_{\alpha/2}$ are obtained from the Poisson distribution with parameter λ_0 to yield the desired significance level α.

At a large hotel in Los Angeles, on average four people per day forget to take their room key and, therefore, lock themselves out. In an effort to lower the number of these service calls, the hotel staff has placed special signs on the back of every hotel room door reminding visitors to take their key. On a randomly selected day, two people were locked out of their rooms.

a. Write the four parts of a small-sample, one-sided, left-tailed test concerning the mean number of people locked out of their room, based on a Poisson distribution. Use $\alpha = 0.05$.

b. Is there any evidence to suggest that the mean has decreased?

c. Find the p value associated with this hypothesis test.

9.137 Public Policy and Political Science A large-sample test concerning the mean of a Poisson random variable is based on a normal approximation. If X has a Poisson distribution with (large) mean λ, then X is approximately normal with mean λ and standard deviation $\sqrt{\lambda}$.

The four parts of the large-sample hypothesis test are

$H_0: \lambda = \lambda_0$

$H_a: \lambda > \lambda_0, \quad \lambda < \lambda_0, \quad$ or $\quad \lambda \neq \lambda_0$

$\text{TS: } Z = \dfrac{X - \lambda_0}{\sqrt{\lambda_0}}$

X = the number of events that occur in the specified *interval*.

$\text{RR: } Z \geq z_\alpha, \quad Z \leq -z_\alpha, \quad$ or $\quad |Z| \geq z_{\alpha/2}$

Postal workers who deliver mail are at high risk for dog bites. The number of dog bites to postal employees in the United States per (fiscal) year peaked during the mid-1980s, but due to increased public awareness and employee training the number steadily decreased to fewer than 3000 per year. However, in fiscal year 2002, the number of dog bites began to rise again, especially during the "dog days" of summer. In 2007 the mean number of dog bites to postal workers per day was 11.[33] A random day in July 2008 was obtained and the number of dog bites to postal workers was 18.

a. Write the four parts of a large-sample, one-sided, right-tailed test concerning the mean number of dog bites to postal employees per year, based on a normal approximation to the Poisson distribution. Use $\alpha = 0.01$.

b. Is there any evidence to suggest that the mean number of dog bites to postal employees per year has increased?

c. Find the p value associated with this hypothesis test.

9.6   LARGE-SAMPLE HYPOTHESIS TESTS CONCERNING A POPULATION PROPORTION

Many experiments and observational studies are conducted in order to draw a conclusion about a population proportion. For example, a quality-control inspector may need to decide whether the proportion of defective products in a delivery is greater than 0.05. Based on a random sample, a decision is made to either accept the delivery or send the entire shipment back. Medical researchers routinely assess the proportion of patients who recover from various illnesses. This information may be used to determine whether there is evidence that a new drug performs better than an existing treatment, or whether the proportion of patients who recover is greater than some threshold value.

These decisions involve p, the true population proportion, the fraction of individuals or objects with a specific characteristic, or the probability of a success. As in Sections 7.3 and 8.4, it is reasonable to use the sample proportion, $\hat{p}$, as an estimate of the population proportion p. In a sample of n individuals or objects, let X be the number of individuals with the relevant characteristic. Recall the definition of the random variable $\hat{P}$, the sample proportion.

$$\hat{P} = \frac{X}{n} = \frac{\text{The number of individuals with the characteristic}}{\text{The sample size}}. \tag{9.1}$$

From Section 7.3, if n is large and both $np \geq 5$ and $n(1 - p) \geq 5$, then the random variable $\hat{P}$ is approximately normal with mean p and variance $p(1 - p)/n$: $\hat{P} \sim N(p, p(1 - p)/n)$. These results concerning the distribution of $\hat{P}$ are used to construct a general procedure for a hypothesis test concerning p.

This is the template for a hypothesis test about a population proportion. The hypothesized value is p_0.

As usual, use only one (appropriate) alternative hypothesis and the corresponding rejection region.

LARGE-SAMPLE HYPOTHESIS TESTS CONCERNING A POPULATION PROPORTION

Given a random sample of size n, a large-sample hypothesis concerning the population proportion p with significance level α has the form

$H_0: p = p_0$

$H_a: p > p_0, \quad p < p_0, \quad$ or $\quad p \neq p_0$

$\text{TS: } Z = \dfrac{\hat{P} - p_0}{\sqrt{\dfrac{p_0(1 - p_0)}{n}}}$

$\text{RR: } Z \geq z_\alpha, \quad Z \leq -z_\alpha, \quad$ or $\quad |Z| \geq z_{\alpha/2}$

1. This test is valid as long as $np_0 \geq 5$ and $n(1 - p_0) \geq 5$ (the nonskewness criteria).

2. The critical values for this test are from the standard normal distribution, Z (as in a hypothesis test about a population mean when σ is known). ■

The following example illustrates this hypothesis test procedure.

Solution Trail

KEYWORDS

- Is there any evidence?
- Increased the proportion.
- 2500 cardholders.
- 902 use on a regular basis.
- 35%.

↓

TRANSLATION

- Conduct a one-sided, right-tailed test about p.
- $n = 2500$.
- $x = 902$.
- $p_0 = 0.35$.

↓

CONCEPTS

Large-sample hypothesis test concerning a population proportion.

↓

VISION

Check the nonskewness criteria. Use the template for a one-sided, right-tailed test about p. Use $\alpha = 0.01$ to find the critical value, compute the value of the test statistic, and draw a conclusion.

Example 9.17 Don't Leave Home without It In 2007, a survey by Brookfield Research/American Express indicated that 35% of American Express cardholders in Canada use only American Express (AE) and no other major credit card on a regular basis. An advertising campaign was developed and implemented in order to entice more credit cardholders to carry (and use) only an AE card. Following the promotion, a random sample of 2500 cardholders was obtained and 902 indicated that they use only AE on a regular basis. Is there any evidence that the proportion of cardholders who use only AE on a regular basis has increased? Use $\alpha = 0.01$.

SOLUTION

STEP 1 The given information is as follows:
Sample size: $n = 2500$.
Number of people with the specific characteristic, carrying an AE card: $x = 902$.
The sample proportion: $\hat{p} = x/n = 902/2500 = 0.3608$
The assumed value of the population proportion is 0.35 ($= p_0$), and the significance level is $\alpha = 0.01$.

STEP 2 Check the nonskewness criteria.

$$np_0 = (2500)(0.35) = 875 \geq 5$$
$$n(1 - p_0) = (2500)(0.65) = 1625 \geq 5$$

Since both inequalities are satisfied, $\hat{P}$ is approximately normal, and the large-sample hypothesis test concerning a population proportion can be used.

STEP 3 The four parts of the hypothesis test are

$H_0: p = 0.35$

$H_a: p > 0.35$

$$\text{TS: } Z = \frac{\hat{P} - p_0}{\sqrt{\dfrac{p_0(1 - p_0)}{n}}}$$

RR: $Z \geq z_\alpha = z_{0.01} = 2.3263$

STEP 4 The value of the test statistic is

$$z = \frac{\hat{p} - p_0}{\sqrt{\dfrac{p_0(1 - p_0)}{n}}} = \frac{0.3608 - 0.35}{\sqrt{\dfrac{(0.35)(0.65)}{2500}}} = 1.1321$$

STEP 5 The value of the test statistic does *not* lie in the rejection region. We do not reject the null hypothesis. There is no evidence to suggest that the true population proportion of cardholders is greater than 0.35. There is no evidence to suggest that the true proportion of AE cardholders has increased.

---

p value illustration:
$p = \text{P}(Z \geq 1.1321)$
$= 0.1288 > 0.01 = \alpha$

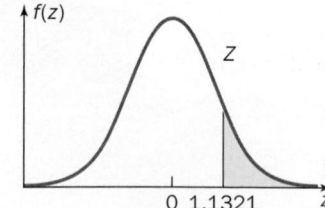

Figures 9.52 through 9.54 together show a technology solution.

Just a reminder: the p in H_0 and H_a represents the population proportion. The symbol p is also used later in this problem to represent the p value associated with the hypothesis test.

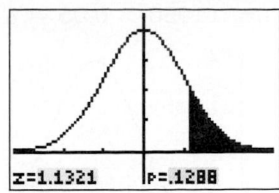

```
1-PropZTest
p0:.35
x:902
n:2500
prop≠p0 <p0 >p0
Calculate Draw
```

Figure 9.52 TI-84 Plus 1-PropZTest input screen.

```
1-PropZTest
prop>.3500
z=1.1321
p=.1288
p̂=.3608
n=2500.0000
```

Figure 9.53 TI-84 Plus 1-PropZTest Calculate results.

```
z=1.1321    p=.1288
```

Figure 9.54 TI-84 Plus 1-PropZTest Draw results.

Since a large-sample test concerning a population proportion is based on a standard normal distribution, Z, the p value is computed the same way as in a hypothesis test concerning a population mean when σ is known (which is also based on a standard normal distribution). The following example includes a p value computation.

Example 9.18 Gender Gap in Business Schools A poll of Canadian business executives indicated that most were not aware of the gender inequity in Canadian business schools. Approximately 25% of students enrolled in Canadian MBA programs are female.[34] A random sample of 350 students enrolled in Canadian MBA programs was obtained and 77 were female.

a. Is there any evidence to suggest that the proportion of female students enrolled in Canadian MBA programs is different from 0.25? Use $\alpha = 0.05$.
b. Find the p value associated with this hypothesis test.

SOLUTION

a. The sample size is $n = 350$ and $x = 77$.
The sample proportion is $\hat{p} = x/n = 77/350 = 0.22$
The assumed value of the population proportion is $p_0 = 0.25$ and $\alpha = 0.05$.

Check the nonskewness criteria:

$np_0 = (350)(0.25) = 87.5 \geq 5$
$n(1 - p_0) = (350)(0.75) = 262.5 \geq 5$

Both inequalities are satisfied. $\hat{P}$ is approximately normal, and the large-sample hypothesis test concerning a population proportion can be used.

Since we are looking for *any* difference from $p_0 = 0.25$, this is a two-sided hypothesis test:

$H_0: p = 0.25$
$H_a: p \neq 0.25$

$$\text{TS: } Z = \frac{\hat{P} - p_0}{\sqrt{\dfrac{p_0(1 - p_0)}{n}}}$$

RR: $|Z| \geq z_{\alpha/2} = z_{0.025} = 1.96$ ($Z \leq -1.96$ or $Z \geq 1.96$)

The value of the test statistic is

$$z = \frac{\hat{p} - p_0}{\sqrt{\dfrac{p_0(1 - p_0)}{n}}} = \frac{0.22 - 0.25}{\sqrt{\dfrac{(0.25)(0.75)}{350}}} = -1.2961$$

A technology solution:

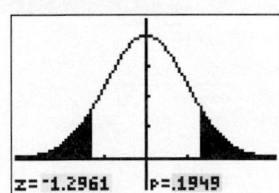

```
1-PropZTest
p0:.25
x:77
n:350
prop≠p0 <p0 >p0
Calculate Draw
```

Figure 9.55 TI-84 Plus 1-PropZTest input screen.

```
1-PropZTest
prop≠.2500
z=-1.2961
p=.1949
p̂=.2200
n=350.0000
```

Figure 9.56 TI-84 Plus 1-PropZTest Calculate results.

```
z=-1.2961    p=.1949
```

Figure 9.57 TI-84 Plus 1-PropZTest Draw results.

p value illustration:
$$p = 2P(Z \le -1.2961)$$
$$= 0.1950 > 0.05 = \alpha$$

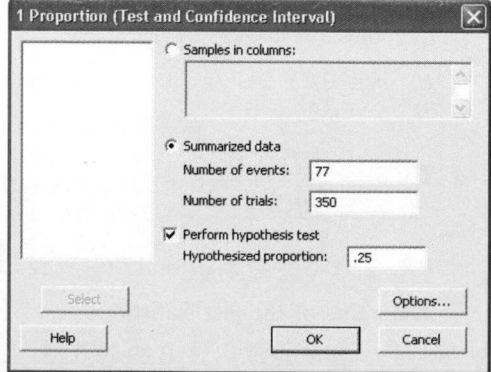

The value of the test statistic does *not* lie in the rejection region. We do not reject the null hypothesis. There is no evidence to suggest that the proportion of females in MBA programs has changed.

b. Since this is a two-sided test and the value of the test statistic ($z = -1.2961$) is negative, $p/2$ is a left-tail probability.

$$p/2 = P(Z \le -1.2961)$$
$$= 0.0975$$
$$p = 2(0.0975) = \boxed{0.1950} \ (> 0.05 = \alpha)$$

Definition of p value for H_a: $p \ne p_0$.

Use Table III in the Appendix.

Solve for p.

Since $p > \alpha$, we do not reject the null hypothesis. ●

TECHNOLOGY CORNER

Procedure: Hypothesis test concerning a population proportion.
Reconsider: Example 9.18, page 425, solution, and interpretations.

TI-84 Plus

The input is the number of successes and the number of trials.
1. Select $\boxed{\text{STAT}}$; TESTS; 1-PropZTest.
2. Enter p_0, the hypothesized proportion, x, the number of successes, and n, the sample size.
3. Choose the appropriate alternative hypothesis and Calculate. Press $\boxed{\text{ENTER}}$. See Figure 9.55, page 425.
4. The 1-PropZTest output includes the alternative hypothesis, the value of the test statistic, the p value, the sample proportion, and the sample size. See Figure 9.56, page 425.
5. The Draw results are shown in Figure 9.57, page 425.

Minitab

The input may be dichotomous data in a column or summarized data.
1. Select Stat; Basic Statistics; 1 Proportion.
2. Choose Summarized data. Enter the Number of events and the Number of trials. Check the box to Perform a hypothesis test, and enter the Hypothesized proportion. See Figure 9.58
3. Choose Options. Enter a Confidence level, select the Alternative hypothesis, and check the box to Use test and interval based on normal distribution. See Figure 9.59.
4. Summary statistics, the confidence interval, the value of the test statistic, and the p value are displayed in the Session window. See Figure 9.60

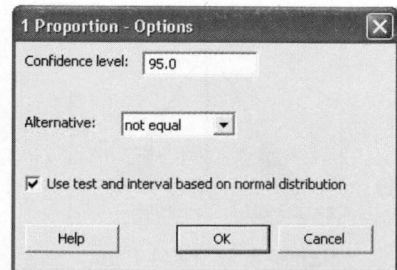

Figure 9.58 1 Proportion (Test and Confidence Interval) input screen.

Figure 9.59 1 Proportion Options screen.

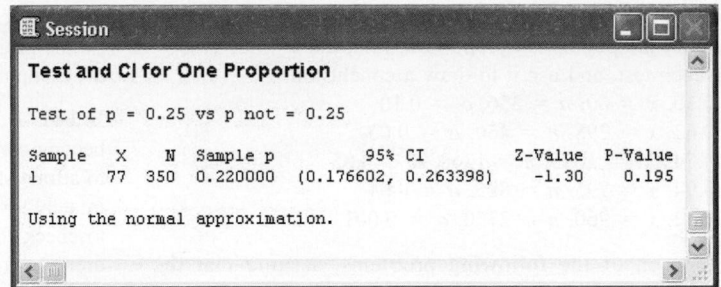

Figure 9.60 Hypothesis test results.

Excel

There is no built-in function to conduct a hypothesis test concerning a population proportion. However, Excel can be used to compute the value of the test statistic and the p value.

1. Enter p_0, the hypothesized proportion; x, the number of successes; and n, the sample size.

2. Compute the sample proportion, the value of the test statistic, and the p value. See Figure 9.61.

	A	B	C
1	0.2500	$= p\_0$	The hypothesized proportion
2	77	$= x$	The number of successes
3	350	$= n$	The sample size
4	0.2200	$= A2/A3$	The sample proportion
5	-1.2961	$=(A4-A1)/SQRT(A1*(1-A1)/A3)$	The value of z
6	0.1949	$=2*NORMDIST(A5,0,1,1)$	The p value

Figure 9.61 Excel computations.

SECTION 9.6 EXERCISES

Practice

9.138 In each of the following problems, the sample size and p_0 are given for a large-sample hypothesis test concerning a population proportion. Check the two nonskewness inequalities and determine whether or not this test is appropriate.

a. $n = 276$, $p_0 = 0.30$ **b.** $n = 1158$, $p_0 = 0.60$
c. $n = 645$, $p_0 = 0.03$ **d.** $n = 159$, $p_0 = 0.97$
e. $n = 322$, $p_0 = 0.38$ **f.** $n = 443$, $p_0 = 0.82$

9.139 In each of the following problems, assume that the null hypothesis is H_0: $p = p_0$ and the alternative hypothesis is H_a: $p > p_0$. Use the values of p_0, x, n, and α to conduct a large-sample hypothesis test about the population proportion.

a. $p_0 = 0.34$, $x = 121$, $n = 348$, $\alpha = 0.05$
b. $p_0 = 0.67$, $x = 200$, $n = 281$, $\alpha = 0.10$
c. $p_0 = 0.488$, $x = 535$, $n = 1020$, $\alpha = 0.01$
d. $p_0 = 0.02$, $x = 52$, $n = 2366$, $\alpha = 0.025$
e. $p_0 = 0.85$, $x = 570$, $n = 667$, $\alpha = 0.01$

9.140 In each of the following problems, assume that the null hypothesis is H_0: $p = p_0$ and the alternative hypothesis is H_a: $p < p_0$. Use the values of p_0, x, n, and α to conduct a large-sample hypothesis test about the population proportion.

a. $p_0 = 0.14$, $x = 40$, $n = 317$, $\alpha = 0.01$
b. $p_0 = 0.275$, $x = 98$, $n = 404$, $\alpha = 0.05$
c. $p_0 = 0.52$, $x = 250$, $n = 546$, $\alpha = 0.025$
d. $p_0 = 0.78$, $x = 2710$, $n = 3580$, $\alpha = 0.001$
e. $p_0 = 0.605$, $x = 1102$, $n = 1862$, $\alpha = 0.05$

9.141 In each of the following problems, assume that the null hypothesis is H_0: $p = p_0$ and the alternative hypothesis is H_a: $p \neq p_0$. Use the values of p_0, x, n, and α to conduct a large-sample hypothesis test about the population proportion.

a. $p_0 = 0.28$, $x = 88$, $n = 377$, $\alpha = 0.025$
b. $p_0 = 0.46$, $x = 130$, $n = 243$, $\alpha = 0.02$
c. $p_0 = 0.337$, $x = 120$, $n = 414$, $\alpha = 0.05$
d. $p_0 = 0.71$, $x = 865$, $n = 1250$, $\alpha = 0.005$
e. $p_0 = 0.93$, $x = 1515$, $n = 1600$, $\alpha = 0.01$

9.142 In each of the following problems, assume that the null hypothesis is H_0: $p = p_0$ and the alternative hypothesis is H_a: $p > p_0$.

Use the values of p_0, x, n, and α to conduct a large-sample hypothesis test about the population proportion. Find the p value associated with each test, and use it to draw a conclusion.

a. $p_0 = 0.15$, $x = 60$, $n = 356$, $\alpha = 0.10$
b. $p_0 = 0.62$, $x = 298$, $n = 450$, $\alpha = 0.05$
c. $p_0 = 0.743$, $x = 1035$, $n = 1360$, $\alpha = 0.05$
d. $p_0 = 0.94$, $x = 795$, $n = 825$, $\alpha = 0.01$
e. $p_0 = 0.32$, $x = 960$, $n = 2750$, $\alpha = 0.001$

9.143 In each of the following problems, assume that the null hypothesis is $H_0: p = p_0$ and the alternative hypothesis is $H_a: p < p_0$. Use the values of p_0, x, n, and α to conduct a large-sample hypothesis test about the population proportion. Find the p value associated with each test, and use it to draw a conclusion.

a. $p_0 = 0.54$, $x = 145$, $n = 301$, $\alpha = 0.05$
b. $p_0 = 0.39$, $x = 180$, $n = 460$, $\alpha = 0.005$
c. $p_0 = 0.07$, $x = 35$, $n = 566$, $\alpha = 0.01$
d. $p_0 = 0.64$, $x = 449$, $n = 747$, $\alpha = 0.025$
e. $p_0 = 0.47$, $x = 395$, $n = 925$, $\alpha = 0.005$

9.144 In each of the following problems, assume that the null hypothesis is $H_0: p = p_0$ and the alternative hypothesis is $H_a: p \neq p_0$. Use the values of p_0, x, n, and α to conduct a large-sample hypothesis test about the population proportion. Find the p value associated with each test, and use it to draw a conclusion.

a. $p_0 = 0.50$, $x = 418$, $n = 882$, $\alpha = 0.01$
b. $p_0 = 0.19$, $x = 158$, $n = 700$, $\alpha = 0.05$
c. $p_0 = 0.90$, $x = 445$, $n = 520$, $\alpha = 0.001$
d. $p_0 = 0.75$, $x = 1095$, $n = 1400$, $\alpha = 0.025$
e. $p_0 = 0.45$, $x = 2386$, $n = 5525$, $\alpha = 0.01$

Applications

9.145 Public Health and Nutrition In a recent research study concerning nutrition, it was reported that only 2% of children in the United States ages 6–10 eat the recommended number of servings for all five major food groups each day. A certain state conducted a nutrition campaign through elementary schools, newspapers, and television. Following the campaign, a random sample of 500 children in this age group was obtained, and 16 were found to eat the recommended number of servings each day.

a. Identify n, x, and p_0.
b. Check the nonskewness criteria. Is a large-sample hypothesis test about p appropriate?
c. Conduct the appropriate hypothesis test to determine whether the proportion of children who eat the recommended number of servings each day has increased. Use $\alpha = 0.05$.
d. Compute the p value associated with this test.

9.146 Marketing and Consumer Behavior It is claimed that the proportion of rentals (as opposed to owner-occupied housing) in Murfreesboro, Tennessee, is approximately 0.30. Recently, the Middle Tennessee State University off-campus housing office has had several complaints because students have not been able to find many rental properties available. A random sample of 60 dwellings in town is obtained, and 15 are rentals.

a. Identify n, x, and p_0.
b. Check the nonskewness criteria. Is a large-sample hypothesis test about p appropriate?

c. Conduct the appropriate hypothesis test to determine whether the proportion of rentals has decreased. Use $\alpha = 0.025$.
d. Compute the p value associated with this test.

9.147 Education and Child Development Home-schooling has become very popular in the United States, and many colleges try to attract students from this group. Evidence suggests that approximately 90% of all homeschooled children attend college. In order to check this claim, a random sample of 225 homeschooled children was obtained and 189 of them were found to have attended college.

a. Identify n, x, and p_0.
b. Check the nonskewness criteria. Is a large-sample hypothesis test about p appropriate?
c. Conduct the appropriate test to determine whether the proportion of homeschooled children who attend college is different from 0.90. Use $\alpha = 0.05$.
d. Compute the p value associated with this test.

9.148 Public Policy and Political Science It has always been difficult to attract general physicians to rural areas. Small-town doctors do not have the opportunity to take much time off, and the salary is relatively low. However, the quality of life is often appealing. Past records indicate that 52% of all physicians in rural areas leave after one year. The federal government decided to offer more incentives for doctors to stay in rural areas, for example, loan forgiveness and housing allowances. A few years after this program was implemented, a random sample of rural-area physician positions showed that 62 of 130 left after 1 year.

a. Identify n, x, and p_0.
b. Check the nonskewness criteria. Is a large-sample hypothesis test about p appropriate?
c. Conduct the appropriate test to determine whether the the turnover rate of rural doctors has decreased. Use $\alpha = 0.01$.
d. Compute the p value associated with this test.

9.149 Marketing and Consumer Behavior A small part of the federal budget is prepared by estimating the proportion of veterans who will take advantage of the education benefit. (The Montgomery GI Bill provides money for college, technical, or vocational school for active-duty personnel and veterans.) Previous records indicate that approximately 30% of all veterans take advantage of this allowance. A random sample of 470 veterans was obtained, and 178 indicated their intention to use the education benefit within the next year. Is there any evidence to suggest that the proportion of veterans using this benefit has increased? Use $\alpha = 0.001$.

9.150 Psychology and Human Behavior In a survey by Evans Data Corporation, it was reported that 49% of software developers were still writing applications for Windows XP.[35] In order to see whether the targeted operating system has changed, a random sample of 2500 software developers was obtained, and this time 1196 said they were writing applications for Windows XP. Is there any evidence to suggest that the true proportion of software developers writing code for Windows XP has changed? Use $\alpha = 0.05$ and use the p value to draw a conclusion.

9.151 Demographics and Population Statistics Beyond.com, Inc. conducted a survey to determine how professionals were coping with high gas prices. It was reported that 79% were trying to find a job closer to home.[36] In a random sample of 321 professionals,

245 indicated that they were searching for a job closer to home. Is there any evidence to suggest that the proportion of professionals searching for a job closer to home has decreased? Use $\alpha = 0.01$.

9.152 Psychology and Human Behavior In 2008, approximately 25% of all Baby Boomers volunteered their time to some charitable organization.[37] In order to determine whether this percentage has changed, a random sample of 1500 Baby Boomers was obtained, and 405 said they volunteer their time. Is there any evidence to suggest that the proportion of Baby Boomers who volunteer their time has changed? Use $\alpha = 0.01$.

9.153 Public Policy and Political Science Ron Littlefield is considering a campaign for mayor of Chattanooga. He will enter the race if there is evidence to suggest that fewer than 40% of all residents are satisfied with the local government. A random sample of 375 residents was obtained, and 127 indicated that they were satisfied with the local government. Do you think this politician will enter the race for mayor? Justify your answer. (Use $\alpha = 0.01$.)

9.154 Medicine and Clinical Studies The American Heart Association reports that 1.1 million people suffer a heart attack each year and 47% of them die.[38] Early detection techniques, better diagnostic procedures, and specially trained personnel may save more of these victims. In a random sample of 688 heart-attack victims, 288 could not be saved. Is there any evidence to suggest that the proportion of people who die from heart attacks has decreased? Let $\alpha = 0.01$, and compute the p value associated with this hypothesis test to draw a conclusion.

9.155 Public Policy and Political Science A local planning board must consider whether to require all new housing projects with five or more apartments to designate some of the units as rent-controlled. A random sample of 100 apartments in Cheyenne, Wyoming, was obtained, and 12 were found to be rent-controlled. Is there any evidence to suggest that the proportion of rent-controlled apartments is less than 10%? Let $\alpha = 0.05$, and use the p value associated with this hypothesis test to draw a conclusion.

9.156 Travel and Transportation In a 2008 summary by The Road Information Program (TRIP) it was reported that 21% of the bridges in Iowa were structurally deficient. Only Pennsylvania, Oklahoma, and Rhode Island had a higher percentage. Following increased attention to this problem, a random sample of 740 bridges was obtained in 2009, and 131 were determined to be structurally deficient. Is there any evidence to suggest that the Department of Transportation in Iowa has been able to decrease the proportion of structurally deficient bridges? Use $\alpha = 0.05$.

9.157 Demographics and Population Statistics During the 2007–2008 school year, approximately 64% of principals in Florida public schools were female.[39] In a 2008–2009 school year survey, 231 of 328 randomly selected principals in Florida public schools were female. Is there any evidence to suggest that the proportion of female principals in Florida public schools has changed? Use $\alpha = 0.05$.

Extended Applications

9.158 Education and Child Development College study-abroad programs have great educational value, compel students to become more globally aware, and invite participants to learn about differ-

ent nations and cultures. Some colleges sponsor several programs and have high participation rates. However, nationally only 1.16% of all college students participate in a study-abroad program.[40] Recent world events may have made more students leery of travel and life in another country. In a random sample of 1200 graduating college students, 11 said they had participated in a study-abroad program.

a. Check the nonskewness criteria. What part of this problem is an added warning to check the nonskewness criteria?

b. Is there any evidence to suggest that the proportion of students participating in these programs has decreased? Use $\alpha = 0.05$.

9.159 Physical Sciences Interstate Batteries guarantees that their lawn tractor battery will last at least three years. To increase sales, the company is planning to offer a new replacement warranty, as long as 95% of all batteries do indeed last at least three years. A random sample of 200 customers were contacted, and 183 had batteries that lasted at least three years.

a. Is there any evidence to suggest that the proportion of tractor batteries that last at least three years is less than 0.95? Use $\alpha = 0.05$.

b. Find the p value associated with this hypothesis test.

c. Based on your results, would you recommend implementing the new replacement warranty? Justify your answer.

9.160 Medicine and Clinical Studies A certain puzzle task is designed to measure spatial reasoning performance. Forty-five percent of all people attempting the puzzle complete the task within the allotted time. A researcher decided to test the theory that classical music increases brain activity and improves the ability to perform such tasks. A random sample of 400 people was obtained, and each listened to 15 minutes of classical music, then attempted the task. Two hundred eleven completed the task within the allotted time.

a. Is there any evidence to suggest that the proportion of people who complete the puzzle within the allotted time has increased? Use $\alpha = 0.001$.

b. Compute the p value associated with this test.

c. Carefully sketch a graph indicating the critical value from part (a) and the p value from part (b).

Challenge

9.161 Public Policy and Political Science The high cost of medication has forced some people to choose between prescribed drugs and other necessities such as food, heat, or electricity. A survey concluded that 29% of U.S. adults had not filled at least one prescription in the past two years because of cost.[41] Following a recent advertising campaign urging adults to fill all prescriptions, physicians' groups hope this percentage has decreased. A random sample of 1000 adults who received prescriptions was obtained, and 260 said they had not filled a prescription because of cost.

a. Is there any evidence to suggest that the proportion of U.S. adults who have not filled a prescription because of cost has decreased? Use $\alpha = 0.01$.

b. Find the probability of a type II error in this hypothesis test for $p_a = 0.26$; that is, find $\beta(0.26)$.

c. Find a value of p_a such that the probability of a type II error is 0.1; that is, solve $\beta(p_a) = 0.1$ for p_a.

d. Find the sample size necessary such that the probability of a type II error for $p_a = 0.24$ is 0.025; that is, solve $\beta(0.24) = 0.025$ for n.

9.162 Demographics and Population Statistics A large-sample hypothesis test concerning a population proportion is based upon a normal approximation and the nonskewness criteria. If n is small (and the nonskewness criteria fails), the test statistic is an exact hypothesis test concerning p and is based on the number of successes in the sample, X. If $H_0: p = p_0$ is true, then X has a binomial distribution with n trials and probability of success p_0.
The four parts of the hypothesis test are

$H_0: p = p_0$

$H_a: p > p_0, \quad p < p_0, \quad \text{or} \quad p \neq p_0$

TS: X = the number of successes in n trials

RR: $X \geq x_\alpha, \quad X \leq x'_\alpha, \quad \text{or} \quad X \leq x_{\alpha/2} \quad \text{or} \quad X \geq x'_{\alpha/2}$

The critical values $x_\alpha, x'_\alpha, x_{\alpha/2}$, and $x'_{\alpha/2}$ are obtained from the binomial distribution with parameters n and p_0 to yield the desired significance level α. For example, in a one-sided, right-tailed test, the critical value x'_α is found such that $P(X \geq x'_\alpha) \approx \alpha$.

The Josephson Institute released results from a survey indicating that 63% of students in the Northeast admitted to cheating.[42] A sociologist conducting research believes that this percentage is much lower. In a random sample of 25 Northeast students, 12 admitted to cheating.

a. Write the four parts of a small-sample (exact) one-sided, left-tailed test concerning the population proportion of Northeast students who admit to cheating, based on a binomial distribution. Use $\alpha = 0.01$.

b. Is there any evidence to suggest that the true proportion is less than 0.63?

c. Find the p value associated with this hypothesis test.

9.7 HYPOTHESIS TESTS CONCERNING A POPULATION VARIANCE OR STANDARD DEVIATION

Many real-world, practical decisions involve variability, or a population variance. For example, road inspectors make sure that asphalt pavement mix meets design specifications and that variability in asphalt properties is small. This improves the quality, safety, and lifetime of the road surface. The dose of an anticancer drug usually is determined by the patient's body surface area. However, there is still large variability in drug exposure, the amount of the drug absorbed into the bloodstream. Researchers continue to search for better methods of calculating the dose in order to decrease the variability in exposure from patient to patient.

As in Section 8.5, the sample variance, S^2, is used as an estimator for the population variance, σ^2. The hypothesis test procedure is based on Theorem 8.2, a standardization, or transformation, to a chi-square random variable. Recall, if S^2 is the sample variance of a random sample of size n from a normal distribution with variance σ^2, then the random variable $(n - 1)S^2/\sigma^2$ has a chi-square distribution with $n - 1$ degrees of freedom. This result is used to construct a general procedure for a hypothesis test concerning σ^2.

This is the template for a hypothesis test about a population variance. The hypothesized value is σ_0^2.

Use only one (appropriate) alternative hypothesis and the corresponding rejection region. Remember, the two-sided alternative rejection region *cannot* be written with an absolute value symbol. A chi-square distribution is not symmetric (about zero).

HYPOTHESIS TEST CONCERNING A POPULATION VARIANCE

Given a random sample of size n from a normal population with variance σ^2, a hypothesis test concerning the population variance σ^2 with significance level α has the form

$H_0: \sigma^2 = \sigma_0^2$

$H_a: \sigma^2 > \sigma_0^2, \quad \sigma^2 < \sigma_0^2, \quad \text{or} \quad \sigma^2 \neq \sigma_0^2$

TS: $X^2 = \dfrac{(n - 1)S^2}{\sigma_0^2}$

RR: $X^2 \geq \chi_\alpha^2, \quad X^2 \leq \chi_{1-\alpha}^2, \quad \text{or} \quad X^2 \leq \chi_{1-\alpha/2}^2 \quad \text{or} \quad X^2 \geq \chi_{\alpha/2}^2, \text{ (df} = n - 1)$

ILLUMINATING THE CONCEPTS

1. X (a Greek capital chi) is a random variable. χ (a Greek small chi) is a specific value.

2. This test is valid for any sample size, as long as the underlying population is normal.

3. Table VI in the Appendix presents selected critical values associated with various chi-square distributions. These critical values were used to construct confidence intervals in Section 8.5.

4. Figures 9.62, 9.63, and 9.64 illustrate the rejection region for each alternative hypothesis.

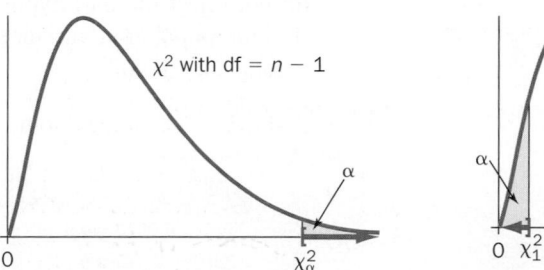

Figure 9.62 Rejection region for H_a: $\sigma^2 > \sigma_0^2$.

Figure 9.63 Rejection region for H_a: $\sigma^2 < \sigma_0^2$.

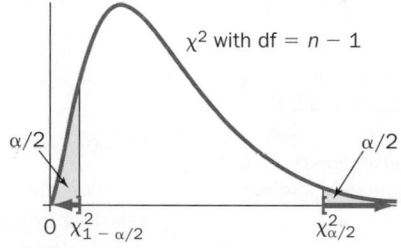

Figure 9.64 Rejection region for $\sigma^2 \neq \sigma_0^2$ ■

The following example illustrates this hypothesis test procedure.

(Robert Harding Picture Library Ltd / Alamy)

Example 9.19 Permafrost Measurements   One measure of permafrost is the volumetric liquid water content (VWC), a unitless quantity. This measurement is used to study climate changes in the Arctic ice cap, greenhouse gases, and vegetation. Along the northern part of the Trans-Alaskan Pipeline, variation in permafrost was an important consideration in planning and construction. Suppose a random sample of 15 locations was obtained, and the VWC was measured at each location during the winter months. The sample mean was $\bar{x} = 0.225$ and the sample variance was $s^2 = 0.0025$. Is there any evidence to suggest that the population variance is greater than 0.002? Use $\alpha = 0.05$.

SOLUTION

STEP 1   The given information:
Sample size: $n = 15$.
The sample variance: $s^2 = 0.0025$.
The assumed value of the population variance is 0.002 ($= \sigma_0^2$), and the significance level is $\alpha = 0.05$.

STEP 2 The four parts of the hypothesis test are

$$H_0: \sigma^2 = 0.002$$
$$H_a: \sigma^2 > 0.002$$
$$\text{TS: } X^2 = \frac{(n-1)S^2}{\sigma_0^2}$$
$$\text{RR: } X^2 \geq \chi_\alpha^2 = \chi_{0.05}^2 = 23.6848 \qquad df = 14$$

STEP 3 The value of the test statistic is

$$\chi^2 = \frac{(n-1)s^2}{\sigma_0^2} = \frac{(14)(0.0025)}{0.002} = 17.5$$

STEP 4 The value of the test statistic does *not* lie in the rejection region. We do not reject the null hypothesis. There is no evidence to suggest that the true population variance in volumetric liquid water content is greater than 0.002.

Figure 9.65 shows a technology solution.

	A	B	C
1	0.0020	= σ_0^2	The hypothesized variance
2	15	= n	The sample size
3	0.0025	= s^2	The sample variance
4	17.5000	=(A2-1)*A3/A1	The value of chi-square

Figure 9.65 Excel hypothesis test calculations.

 ILLUMINATING THE CONCEPTS

1. Neither the TI-84 Plus nor Excel has built-in commands to conduct a hypothesis test concerning a population variance. However, each can be used to find the appropriate critical value and associated p value.

2. The table listing critical values for various chi-square distributions is very limited. Using this table, the best we can do is bound the p value associated with a hypothesis test (as for a t test). ■

Example 9.20 Diesel Fuel Use Reports completed by independent truck drivers indicate that the mean amount of diesel fuel purchased per week is 346 gallons with a standard deviation of 23.6 gallons. Truck-stop managers use this information to plan work schedules and to order gasoline supplies. A random sample of independent truck drivers was obtained, and their weekly gas purchases (in gallons) are given in the following table.

339	348	324	343	365	355	347	343	343	358
328	326	343	358	340	331	326	355	342	349

a. Is there any evidence to suggest that the true population variance in diesel fuel purchased per week is different from 556.96 gallons^2? Assume normality and use $\alpha = 0.01$.

b. Find bounds on the p value associated with this hypothesis test.

The initial bounds are on $p/2$ since the hypothesis test is two-sided.

Solution Trail

KEYWORDS

- Is there any evidence?
- Population variance.
- Different from 556.96.
- Weekly gas purchases are given.

TRANSLATION

- Conduct a two-sided test about σ^2.
- Find n and s^2.

CONCEPTS

Hypothesis test concerning a population variance.

VISION

We can assume that the underlying distribution is normal. Use the template for a two-sided test about σ^2. Use $\alpha = 0.01$ to find the critical values. Compute the value of the test statistic, and draw a conclusion. Use Table VI in the Appendix to find bounds on the p value.

SOLUTION

STEP 1 The assumed population variance is $23.6^2 = 556.96 \ (= \sigma_0^2)$; $n = 20$ and $\alpha = 0.01$.

We would like to know whether the population variance is different from 556.96. The relevant alternative hypothesis is two-sided.

The underlying population is assumed to be normal. A hypothesis test concerning a population variance can be used.

STEP 2 The four parts of the hypothesis test are

$H_0: \sigma^2 = 556.96$

$H_a: \sigma^2 \neq 556.96$

TS: $X^2 = \dfrac{(n-1)S^2}{\sigma_0^2}$

RR: $X^2 \leq \chi_{1-\alpha/2}^2 = \chi_{0.995}^2 = 6.8440$ or

$ X^2 \geq \chi_{\alpha/2}^2 = \chi_{0.005}^2 = 38.5823$ df = 19

STEP 3 The sample variance is

$$s^2 = \frac{1}{n-1}\left[\sum x_i^2 - \frac{1}{n}\left(\sum x_i\right)^2\right]$$ Computational formula for sample variance.

$$= \frac{1}{19}\left(2{,}357{,}671 - \frac{1}{20}(6863)^2\right) = 138.56$$ Use given data.

The value of the test statistic is

$$\chi^2 = \frac{(n-1)s^2}{\sigma_0^2} = \frac{(19)(138.56)}{556.96} = 4.7268$$

The value of the test statistic lies in the rejection region ($\chi^2 = 4.7268 \leq 6.8840$). We reject the null hypothesis. There is evidence to suggest that the true population variance is different from 556.96.

STEP 4 $\chi^2 = 4.7268$.

In Table VI in the Appendix, row $n - 1 = 20 - 1 = 19$, place 4.7268 in the ordered list of critical values.

$$3.9683 \leq 4.7268 \leq 4.9123$$
$$\chi_{0.9999}^2 \leq 4.7268 \leq \chi_{0.9995}^2$$
$$\chi_{1-0.0001}^2 \leq 4.7268 \leq \chi_{1-0.0005}^2$$

Therefore, $0.0001 \leq p/2 \leq 0.0005$

And $0.0002 \leq p \leq 0.0010$ See Figure 9.66.

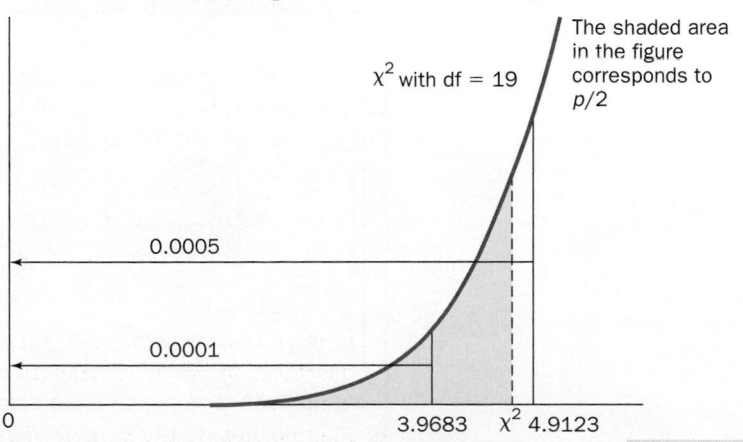

Figure 9.66 The left tail of a chi-square distribution with 19 degrees of freedom.

TECHNOLOGY CORNER

Procedure: Hypothesis test concerning a population variance or standard deviation.
Reconsider: Example 9.20, page 432, solution, and interpretations.

TI-84 Plus

There is no built-in function to conduct a hypothesis test concerning a population variance or standard deviation. However, the calculator may be used to compute summary statistics, the test statistic, and the p value.

1. Enter the data into list L1.
2. Use $\boxed{\text{LIST}}$; MATH; stdDev to compute the sample standard deviation.
3. Compute the value of the test statistic. See Figure 9.67.
4. Use $\boxed{\text{DISTR}}$; DISTR; χ^2cdf to compute the p value. See Figure 9.68.

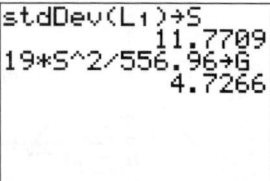

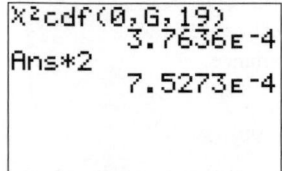

Figure 9.67 Standard deviation and test statistic calculations.

Figure 9.68 p value calculations.

Minitab

The input may be either Samples in columns or Summarized data, and the test may be conducted in terms of the variance or standard deviation.

1. Enter the data into column C1.
2. Select Stat; Basic Statistics; 1 Variance.
3. Choose Enter variance from the drop down menu. Choose Samples in columns and enter C1. Check the box to Perform a hypothesis test, and enter the Hypothesized variance. See Figure 9.69.
4. Choose Options. Enter a Confidence level and select the Alternative hypothesis. See Figure 9.70.
5. Summary statistics, the confidence interval, the value of the test statistic, and the p value are displayed in the Session window. See Figure 9.71.

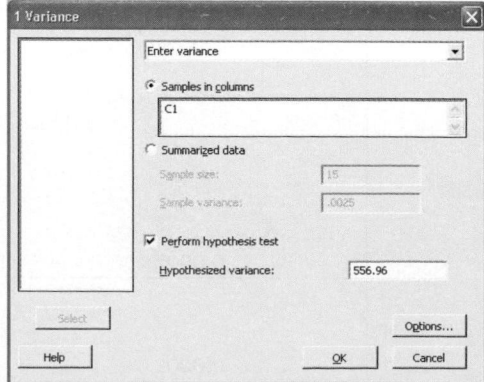

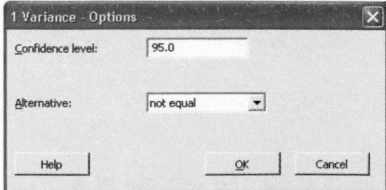

Figure 9.69 1 Variance input screen.

Figure 9.70 1 Variance Options screen.

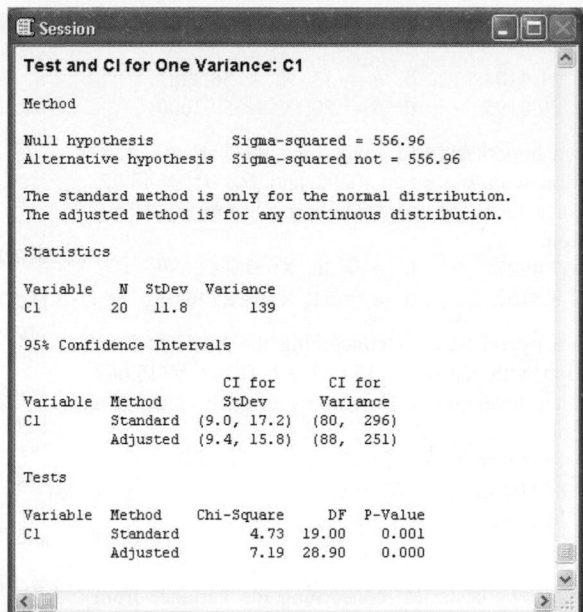

Figure 9.71 Hypothesis test results.

Excel

There is no built-in function to conduct a hypothesis test concerning a population variance or standard deviation. However, Excel may be used to compute summary statistics, the test statistic, and the p value.

1. Enter the data into column A.
2. Compute the sample variance and the value of the test statistic.
3. Use CHDIST to calculate the p value. See Figure 9.72.

	B	C	D
1	556.9600	= σ_0^2	The hypothesized variance
2	20	= COUNT(A1:A20)	The sample size
3	138.5553	= VAR(A1:A20)	The sample variance
4	4.7266	= (B2-1)*B3/B1	The value of chi-square
5	0.0008	= 2*(1-CHIDIST(B4,B2-1))	The p value

Figure 9.72 Excel computations.

SECTION 9.7 EXERCISES

Practice

9.163 Consider a hypothesis test concerning a population variance from a normal population with H_0: $\sigma^2 = 27.2$ and H_a: $\sigma^2 > 27.2$.
a. Write the appropriate test statistic.
b. Find the rejection region corresponding to each value of n and α.
 (i) $n = 12$, $\alpha = 0.05$ (ii) $n = 19$, $\alpha = 0.025$
 (iii) $n = 23$, $\alpha = 0.01$ (iv) $n = 29$, $\alpha = 0.10$
 (v) $n = 6$, $\alpha = 0.001$ (vi) $n = 15$, $\alpha = 0.0001$

9.164 Consider a hypothesis test concerning a population variance from a normal population with H_0: $\sigma^2 = 352.98$ and H_a: $\sigma^2 < 352.98$.

a. Write the appropriate test statistic.
b. Find the rejection region corresponding to each value of n and α.
 (i) $n = 17$, $\alpha = 0.001$ (ii) $n = 13$, $\alpha = 0.05$
 (iii) $n = 37$, $\alpha = 0.0001$ (iv) $n = 27$, $\alpha = 0.10$
 (v) $n = 33$, $\alpha = 0.05$ (vi) $n = 8$, $\alpha = 0.005$

9.165 Consider a hypothesis test concerning a population variance from a normal population with H_0: $\sigma^2 = 43.8$ and H_a: $\sigma^2 \neq 43.8$.
a. Write the appropriate test statistic.
b. Find the rejection region corresponding to each value of n and α.
 (i) $n = 40$, $\alpha = 0.05$ (ii) $n = 31$, $\alpha = 0.01$
 (iii) $n = 22$, $\alpha = 0.001$ (iv) $n = 16$, $\alpha = 0.02$
 (v) $n = 29$, $\alpha = 0.002$ (vi) $n = 5$, $\alpha = 0.20$

9.166 Consider a hypothesis test concerning the variance from a normal population with H_0: $\sigma^2 = 1.28$ and H_a: $\sigma^2 > 1.28$. Find the significance level (α) corresponding to each value of n and rejection region.

a. $n = 21$, $X^2 \geq 31.4104$ b. $n = 33$, $X^2 \geq 56.3281$
c. $n = 14$, $X^2 \geq 29.8195$ d. $n = 29$, $X^2 \geq 59.3000$

9.167 Consider a hypothesis test concerning the variance from a normal population with H_0: $\sigma^2 = 48.92$ and H_a: $\sigma^2 < 48.92$. Find the significance level (α) corresponding to each value of n and rejection region.

a. $n = 25$, $X^2 \leq 7.4527$ b. $n = 36$, $X^2 \leq 18.5089$
c. $n = 12$, $X^2 \leq 3.8157$ d. $n = 51$, $X^2 \leq 27.9907$

9.168 Consider a hypothesis test concerning the variance from a normal population with H_0: $\sigma^2 = 15.667$ and H_a: $\sigma^2 \neq 15.667$ Find the significance level (α) corresponding to each value of n and rejection region.

a. $n = 9$, $X^2 \leq 1.3444$ or $X^2 \geq 21.9550$
b. $n = 21$, $X^2 \leq 8.2604$ or $X^2 \geq 37.5662$
c. $n = 38$, $X^2 \leq 15.0202$ or $X^2 \geq 73.3512$
d. $n = 18$, $X^2 \leq 7.5642$ or $X^2 \geq 30.1910$

9.169 Consider a hypothesis test concerning the variance from a normal population with H_0: $\sigma^2 = 11.4$ and H_a: $\sigma^2 > 11.4$. Find bounds on the p value for each value of n and test statistic.

a. $n = 32$, $\chi^2 = 50.05$ b. $n = 7$, $\chi^2 = 11.99$
c. $n = 19$, $\chi^2 = 38.62$ d. $n = 24$, $\chi^2 = 60.15$

9.170 Consider a hypothesis test concerning the variance from a normal population with H_0: $\sigma^2 = 404.7$ and H_a: $\sigma^2 < 404.7$. Find bounds on the p value for each value of n and test statistic.

a. $n = 11$, $\chi^2 = 1.36$ b. $n = 17$, $\chi^2 = 1.97$
c. $n = 31$, $\chi^2 = 14.05$ d. $n = 51$, $\chi^2 = 32.85$

9.171 Consider a hypothesis test concerning the variance from a normal population with H_0: $\sigma^2 = 232$ and H_a: $\sigma^2 \neq 232$. Find bounds on the p value for each value of n and test statistic.

a. $n = 25$, $\chi^2 = 41.67$ b. $n = 28$, $\chi^2 = 8.12$
c. $n = 5$, $\chi^2 = 0.5005$ d. $n = 16$, $\chi^2 = 38.88$

9.172 Consider a random sample of size 21 from a normal population with hypothesized variance 16.7.

a. Write the four parts for a one-sided, right-tailed hypothesis test concerning the population variance with $\alpha = 0.01$.
b. Suppose $s^2 = 28$. Find the value of the test statistic, and draw a conclusion about the population variance.
c. Find bounds on the p value associated with this hypothesis test, and carefully sketch a graph to illustrate this value.

9.173 Consider a random sample of 16 observations, given in the following table, from a normal population with hypothesized variance 36.8.

233.1	226.1	220.3	247.6	232.9	232.8	235.9
232.4	249.4	207.4	231.8	232.1	220.7	229.6
242.5	229.3					

a. Write the four parts for a two-sided hypothesis test concerning the population variance with $\alpha = 0.05$.
b. Compute the sample variance and the value of the test statistic. Draw a conclusion about the population variance.
c. Find bounds on the p value associated with this hypothesis test.

9.174 Consider a random sample of size 40 from a normal population with hypothesized variance 75.6.

a. Write the four parts for a one-sided, left-tailed hypothesis test concerning the population variance with $\alpha = 0.001$.
b. Suppose $s^2 = 48.5$. Find the value of the test statistic, and draw a conclusion about the population variance.
c. Find bounds on the p value associated with this hypothesis test.

Applications

9.175 **Physical Sciences** A water-droplet generator is used to produce simulated rain or fog, and, for example, to test emission drift rates from nuclear power plant cooling towers. A piezoelectric water-droplet generator is designed to produce 10 microliter (μl) mist with a variance of 0.25. A random sample of 35 water drops was obtained, the amount of water in each (in μl) was measured, and the summary statistics were $\bar{x} = 10.004$ and $s = 0.558$. Is there any evidence to suggest that the variance is larger than specified? Assume normality and use $\alpha = 0.05$.

9.176 **Biology and Environmental Science** The Snow Water Equivalent (SWE) is a measure used to estimate the amount of water contained in a snowpack. This data may be used to forecast the height of streams and rivers. A random sample of SWE measurements (in inches) from sites around Colorado during February 2008 was obtained and the data are given in the following table.

3.5	2.5	3.5	2.5	4.3	6.6	3.1	8.7	12.8	17.1
2.6	6.0	6.4	6.7	12.9	10.7	18.1	4.6	5.2	14.8
4.4	9.6	16.7	7.1	8.8	3.6	1.1	5.1	11.9	3.8

(*Source*: Natural Resources Conservation Service.)

Is there any evidence to suggest that the variance in SWE is less than 25 in^2? Assume normality and use $\alpha = 0.05$.

9.177 **Manufacturing and Product Development** A manufactured glass rod is tested for the stress required for fracture. Experiments suggest that a flame-polished rod has a higher mean fracture stress than a rod with an abraded surface. In both cases, the variance in stress is designed to be at most 324 MPa^2. A random sample of 12 polished glass rods was obtained, and the fracture stress for each was measured. The sample standard deviation in fracture stress was $s = 21.56$. Is there any evidence to refute the manufacturer's claim? Assume that the underlying distribution is normal and use $\alpha = 0.01$.

9.178 **Public Health and Nutrition** The Italian Peoples Bakery in Pennington, New Jersey, advertises an original-recipe cream puff packed with 3 ounces of filling. In order to produce a consistent product, the variance in cream filling is carefully monitored. In a random sample of 23 cream puffs, the variance in filling was $s^2 = 0.105$ ounces^2. Is there any evidence to suggest that the variance in filling is greater than 0.09 ounces^2? Assume normality and use $\alpha = 0.025$.

9.179 **Manufacturing and Product Development** In an effort to improve efficiency, Samuel Adams brewing company would like to produce yeast slurry with variance no greater than 62.5. A new delivery pump was installed, and a random sample of 10 slurry mixtures was obtained. Each mixture was measured in billion cells/ml and the sample variance was $s^2 = 70.1$. Is there any evidence to suggest that the variance is greater than the desired value? Assume normality and use $\alpha = 0.01$.

9.180 Sports and Leisure The specified weight of a 22-mm die used at casinos in Las Vegas, Nevada, is 10.4 grams (g). The variability in die weight must be negligible (at most 0.04 g^2) in order for patrons to believe that games involving dice are fair. A random sample of 25 new 22-mm dice was obtained and each die was carefully weighed. The sample mean weight was $\bar{x} = 10.38$ g and the sample standard deviation was $s = 0.244$ g. Conduct the relevant hypothesis test to determine whether there is any evidence that the variance in die weight is greater than 0.04. Use $\alpha = 0.01$ and assume normality.

9.181 Physical Sciences Some scientists believe that urban areas are simply hotter than rural areas because there is less vegetation and more concrete. The paper "Urban Landscaping and the Heat Island Effect in Mérida, Mexico" examined this issue and presented data regarding house size, lot size, percent shade, and housing characteristics for homes in Mérida.[43] In a random sample of 62 homes, the standard deviation of the lot size was $s = 191.184$ square meters. Is there any evidence that the standard deviation in lot size is less than 200? Assume normality and use $\alpha = 0.05$.

9.182 Business and Management Many farmers craft unique corn mazes in their fields after the growing season. These huge puzzles challenge both children and adults and increase revenue for farmers. Suppose a farmer has constructed a corn maze so that the mean time for completion is 1.5 hours with variance 0.57 hours^2. A random sample of 18 people entering the maze was selected, and the time (in hours) to complete the puzzle was recorded for each. The sample standard deviation was $s = 0.34$. Is there any evidence that the completion-time variance is different from the intended variance? Use $\alpha = 0.005$.

9.183 Sports and Leisure Bull riding has become a very popular rodeo sport—the action is fast and dangerous. A bull ride is scored by two judges; one actually rates the bull, while the other evaluates the rider. The variance in bull-riding time tends to be large, approximately 22.5 seconds^2. A random sample of 37 bull rides was selected over an entire rodeo season. The sample variance in riding times was $s^2 = 15.6$ seconds^2. Smaller variability in bull-riding time translates to a more monotonous, unexciting rodeo. Is there any evidence that bull riding has become less exciting? Assume normality and use $\alpha = 0.05$.

Extended Applications

9.184 Physical Sciences Researchers monitoring a certain area of the ocean are studying the concentration of chlorophyll. While there is some seasonal fluctuation, stable ecosystems tend to vary little. A random sample of surface water was obtained from vials filled at various times during the year, and the chlorophyll concentrations (in mg/m^3) are given on the data CD and book's web site.[44]
a. Is there any evidence to suggest that the variance in chlorophyll concentration is greater than 49? Assume normality and use $\alpha = 0.005$.
b. Find bounds on the p value associated with this hypothesis test, and carefully sketch a graph to illustrate this value.

9.185 Manufacturing and Product Development Bittersharp apple cider is made from apples that contain more than 0.45% malic acid. It is important for the variance in malic acid content to be small, since large variability may cause the resulting cider to be too sharp. In preparation for making a batch of cider, a random sample of 20 Foxwhelp Bittersharp apples was obtained, and the malic acid content in each was measured. The sample variance was $s^2 = 0.42$.
a. Is there any evidence to suggest that the population variance in malic acid is greater than 0.36? Assume that the underlying distribution is normal and use $\alpha = 0.05$.
b. Find bounds on the p value associated with this hypothesis test.

9.186 Biology and Environmental Science Scientists studying the effect of aerosols on the environment measure the light absorption coefficient (in M/m, square meters of absorbing cross section per cubic meter of air), an indication of how much sunlight is blocked from Earth. A random sample of 32 days in 2007 was selected and the light absorption coefficient from the Trinidad Head Station was recorded. The sample variance was $s^2 = 9.511$.[45]
a. Is there any evidence to suggest that the variance in the light absorption coefficient is different from 11.5? Assume normality and use $\alpha = 0.02$.
b. Find bounds on the p value associated with this hypothesis test.

9.187 Manufacturing and Product Development Cell phone companies attract customers by offering a wide coverage area and a consistent, strong signal. Signal strength is measured in dBm, decibels above or below 1.0 milliwatt. A random sample of locations in a company's coverage area was obtained, and the signal strength (in dBm) at each spot is given on the data CD and book's web site.
a. Suppose that a consistent signal has a variance of at most 230. Is there any evidence of an inconsistent signal in this company's coverage area? Use $\alpha = 0.05$ and assume normality.
b. Find bounds on the p value associated with this hypothesis test.

9.188 Manufacturing and Product Development The Estes Star Dart model rocket weighs only 1 ounce and can reach altitudes of over 2000 feet.[46] Suppose the company would like the standard deviation in height to be approximately 50 feet. If the standard deviation is any larger, then more rockets are lost (because they soar too high and drift away) or customers become angry because launches do not achieve the advertised height. A random sample of rockets was obtained, and the height (in feet) achieved on each launch is given on the data CD and book's web site.
a. Is there any evidence to suggest that the population variance in height is greater than the company's desired value? Assume normality and use $\alpha = 0.01$.
b. Find bounds on the p value for this hypothesis test.

9.189 Biology and Environmental Science A monarch butterfly may fly as many as 2000 miles during migration to Mexico. These amazing creatures vary dramatically in weight, color, and even flying behavior. Previous research suggests that the mean wingspan for a monarch butterfly is 50 mm with a standard deviation of 2.75 mm. There is some speculation that changes in the environment (for example, pollution and climate) have caused greater variability in the wingspan. A random sample of monarch butterflies was obtained, and the wingspan of each (in mm) is given in the following table.

49.3	52.4	43.3	51.2	55.1	48.4	41.6	54.9
51.9	51.6	47.0	50.7	47.2	55.7	50.3	48.3
52.9	47.8	49.4	56.8	47.0	51.5		

(*Source*: Monarch Lab, University of Minnesota.)

a. Is there any evidence that the variability in wingspan of the monarch butterfly has increased? Assume normality and use $\alpha = 0.05$.

b. Find bounds on the p value associated with this hypothesis test.

9.190 **Manufacturing and Product Development** Oxford Paper Company manufactures boxboard for folding cartons made entirely from recycled material. One specific product is designed to have thickness 375 micrometers with standard deviation 7 micrometers. A random sample of 21 pieces of boxboard was obtained from the assembly line.

a. Assume that the thickness distribution is normal and let $\alpha = 0.05$. Consider a two-sided hypothesis test to determine whether the boxboard is being manufactured with the designed variance in thickness.

b. Find the critical values in terms of the sample variance. Work backward to determine two values such that if $S^2 \leq s_L^2$ or $S^2 \geq s_H^2$ then the null hypothesis is rejected.

c. Suppose $s^2 = 56$. Use the critical values from part (b) to draw a conclusion about the population variance.

d. Suppose $s^2 = 15.6$. Use the critical values from part (b) to draw a conclusion about the population variance.

Challenge

9.191 **Economics and Finance** An approximate test concerning a population variance is based on a normal approximation. Recall, if S^2 is the sample variance of a random sample of size n from a normal distribution with variance σ^2, the random variable $X = (n - 1)S^2/\sigma^2$ has a chi-square distribution with $n - 1$ degrees of freedom. If n is large, the random variable X is approximately normal with mean $n - 1$ and variance $2(n - 1)$. An approximate hypothesis test is based on standardizing X to a Z random variable.

The four parts of the hypothesis test are

$H_0: \sigma^2 = \sigma_0^2$.

$H_a: \sigma^2 > \sigma_0^2, \quad \sigma^2 < \sigma_0^2, \quad \text{or} \quad \sigma^2 \neq \sigma_0^2$

TS: $Z = \dfrac{S^2 - \sigma_0^2}{\sqrt{2}\sigma_0^2/\sqrt{n - 1}}$

RR: $Z \geq z_\alpha, \quad Z \leq -z_\alpha, \quad \text{or} \quad |Z| \geq z_{\alpha/2}$

The historic variance in the exchange rate for the Japanese yen against the U.S. dollar is approximately 1.56. A random sample of 48 closing exchange rates was obtained, and the summary statistics were $\bar{x} = 103.74$ and $s^2 = 2.2$.

a. Write the four parts of a large-sample, two-sided test concerning the variance in exchange rate, using a normal approximation. Use $\alpha = 0.05$.

b. Is there any evidence to suggest that the variance in exchange rate has changed?

c. Find the p value associated with this hypothesis test.

d. Conduct an exact hypothesis test based on the chi-square distribution. Compute the p value, and compare it with your answer to part (c).

Chapter 9 Challenge Wrap-Up

In the National Park Service survey, suppose 324 (of 400 randomly selected people on St. Mark's Trail) were bicycling. The trail will be redesigned if there is evidence that more than 75% of *all* trail users are bicyclers. A one-sided, right-tailed hypothesis test will be used to determine whether there is any evidence that the true proportion of bicyclers is greater than 0.75.

(Mike Powell/AgeFotostock)

SOLUTION

STEP 1   The given information:
Sample size: $n = 400$.
Number of people with the specific characteristic, bicyclers: $x = 324$.
The sample proportion: $\hat{p} = 324/400 = 0.81$
The assumed value of the population proportion is $p_0 = 0.75$.
Use the significance level $\alpha = 0.05$.

STEP 2   Check the nonskewness criteria.

$np_0 = (400)(0.75) = 300 \geq 5$
$n(1 - p_0) = (400)(0.25) = 100 \geq 5$

Since both inequalities are satisfied, $\hat{P}$ is approximately normal, and the large-sample hypothesis test concerning a population proportion can be used.

STEP 3   The four parts of the hypothesis test are

$H_0: p = 0.75$
$H_a: p > 0.75$
TS: $Z = \dfrac{\hat{P} - p_0}{\sqrt{\dfrac{p_0(1 - p_0)}{n}}}$

RR: $Z \geq z_\alpha = z_{0.05} = 1.6449$

STEP 4   The value of the test statistic is

$$z = \frac{\hat{p} - p_0}{\sqrt{\dfrac{p_0(1 - p_0)}{n}}} = \frac{0.81 - 0.75}{\sqrt{\dfrac{(0.75)(0.25)}{400}}} = 2.7713 \geq 1.6449$$

STEP 5   The value of the test statistic lies in the rejection region. We reject the null hypothesis at the $\alpha = 0.05$ significance level. There is evidence to suggest that more than 75% of all trail users are bicyclers. City officials should begin the redesign program.

STEP 6   The p value is a right-tail probability.

$p = P(Z \geq 2.7713)$ *Definition of p value for $H_a: p > p_0$.*
$\quad = 1 = P(Z \leq 2.7713)$ *The Complement Rule.*
$\quad = 1 - 0.9972 = 0.0028 \ (\leq 0.05 = \alpha)$ *Use Table III in the Appendix, interpolate.*

p value illustration:
$p = P(Z \geq 2.7713)$
$\quad = 0.0028 \leq 0.05 = \alpha$

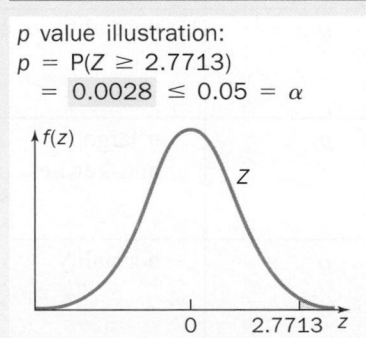

439

Figure 9.73 through 9.75 together show a technology solution.

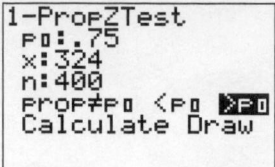

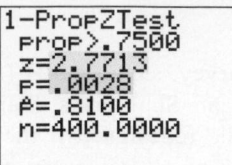

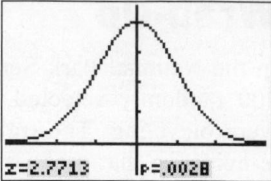

Figure 9.73
1-PropZTest input screen.

Figure 9.74
1-PropZTest Calculate results.

Figure 9.75
1-PropZTest Draw results.

CHAPTER 9 SUMMARY

Concept	Page	Notation / Formula / Description
Hypothesis	380	A claim about the value of a specific population parameter.
Null hypothesis	380	H_0, the claim assumed to be true, or the hypothesis to be tested.
Alternative hypothesis	380	H_a, the possible values of the parameter if H_0 is false.
Test statistic	381	A rule related to the null hypothesis, involving information in the sample. The value of the test statistic is used to determine which hypothesis, H_0 or H_a, is more likely.
Rejection region	381	An interval or set of numbers determined such that if the value of the test statistic lies in the rejection region, then the null hypothesis is rejected.
One-sided alternatives	382	H_a: $\theta > \theta_0$ (right-tailed), H_a: $\theta < \theta_0$ (left-tailed).
Two-sided alternative	382	H_a: $\theta \neq \theta_0$.
Type I error	387	H_0 is rejected (because the value of the test statistic lies in the rejection region), but H_0 is really true.
Type II error	387	H_0 is not rejected (because the value of the test statistic does not lie in the rejection region), but H_a is really true.
Significance level	387	The probability of a type I error, denoted by P(type I error) $= \alpha$.
p value	405	The smallest significance level for which the null hypothesis can be rejected.

Summary of hypothesis tests

Parameter	Assumptions	Alternative hypothesis	Test statistic	Rejection region
μ	n large, σ known, or normality, σ known	$\mu > \mu_0$ $\mu < \mu_0$ $\mu \neq \mu_0$	$Z = \dfrac{\overline{X} - \mu_0}{\sigma/\sqrt{n}}$	$Z \geq z_\alpha$ $Z \leq -z_\alpha$ $|Z| \geq z_{\alpha/2}$
μ	normality, σ unknown	$\mu > \mu_0$ $\mu < \mu_0$ $\mu \neq \mu_0$	$T = \dfrac{\overline{X} - \mu_0}{S/\sqrt{n}}$	$T \geq t_\alpha$ $T \leq -t_\alpha$ $|T| \geq t_{\alpha/2}$ (df $= n - 1$)
p	n large, nonskewness criteria	$p > p_0$ $p < p_0$ $p \neq p_0$	$Z = \dfrac{\hat{P} - p_0}{\sqrt{\dfrac{p_0(1 - p_0)}{n}}}$	$Z \geq z_\alpha$ $Z \leq -z_\alpha$ $|Z| \geq z_{\alpha/2}$
σ^2	normality	$\sigma^2 > \sigma_0^2$ $\sigma^2 < \sigma_0^2$ $\sigma^2 \neq \sigma_0^2$	$X^2 = \dfrac{(n - 1)S^2}{\sigma_0^2}$	$X^2 \geq \chi_\alpha^2$ $X^2 \leq \chi_{1-\alpha}^2$ $X^2 \leq \chi_{1-\alpha/2}^2$ or $X^2 \geq \chi_{\alpha/2}^2$ (df $= n - 1$)

CHAPTER 9 EXERCISES

Applications

9.192 Manufacturing and Product Development Vacuum-packed coffee stays fresh longer, but the package tends to be bumpy and unappealing. In a nitrogen-flushed package, all the oxygen is pushed out by heavier nitrogen, producing a smoother, more attractive package that also stays fresh. A machine used to produce a nitrogen-flushed package should dispense 1.6 moles of nitrogen for a 1-pound package of coffee. In a random sample of 23 packings, the sample mean amount of nitrogen dispensed was $\bar{x} = 1.78$ moles. Assume the underlying distribution is normal, with $\sigma = 0.5$ moles, and use $\alpha = 0.01$.

a. Is there any evidence to suggest that the machine is malfunctioning?

b. Compute the p value associated with this hypothesis test.

9.193 Manufacturing and Product Development Shanghai Creative Material Co., Ltd., makes plating tape for use on circuit boards, designed to be 4 mil thick. (A mil is equal to 0.001 inch: a milli-inch.) As part of quality control, every hour 40 random samples of tape are carefully measured. If there is any evidence that the thickness is different from 4 mil, then the entire process is stopped and the machinery is checked and cleaned. Assume $\sigma = 0.05$ mil and use $\alpha = 0.01$.

a. Suppose the sample mean is $\bar{x} = 4.014$. Should the process be stopped?

b. Suppose the sample mean is $\bar{x} = 3.979$. Should the process be stopped?

9.194 Biology and Environmental Science The population of British buzzards has grown considerably in the past few years, especially in Cheshire, Northamptonshire, and Lothians. The mean wingspan of this bird is approximately 120.5 cm.[47] In order to check this claim, a random sample of British buzzards was selected, and the wingspan for each is given on the data CD and book's web site.

a. Is there any evidence to suggest that the mean wingspan for a British buzzard is less than 120.5 cm? Assume $\sigma = 5.0$ cm and use $\alpha = 0.05$.

b. Compute the p value for this hypothesis test.

9.195 Biology and Environmental Science A special water channel has been constructed to connect to the Campbell River in British Columbia, in order to provide more area for chinook salmon spawning. The plans called for the channel to be 23 m wide. After construction was completed, a random sample of 18 locations along the channel was selected, and the width at each location was measured. The sample mean was $\bar{x} = 24.6$. Assume that the distribution of channel widths is normal and $\sigma = 2.28$ m.

a. Is there any evidence to suggest that the mean width of the channel is greater than 23 m? Use $\alpha = 0.01$.

b. Find the p value associated with this hypothesis test.

9.196 Manufacturing and Product Development A piston in a particular 12-cylinder diesel engine is manufactured to have diameter 13 mm. Any larger or smaller diameter will cause immediate, costly damage to the engine. Every half hour, 10 finished pistons are selected and the diameter of each (in mm) is carefully measured. If there is any evidence that the mean diameter is different from 13 mm, the manufacturing process is stopped and the

machinery is inspected. The quality-control inspector uses a significance level of $\alpha = 0.05$.

a. Suppose the sample mean is $\bar{x} = 12.89$ and the sample standard deviation is $s = 0.96$. Should the manufacturing process be stopped?

b. Suppose the sample mean is $\bar{x} = 13.04$ and the sample standard deviation is $s = 0.045$. Should the manufacturing process be stopped?

c. If you were buying these pistons, would you like the manufacturer to use a smaller or a larger significance level? Justify your answer.

9.197 Economics and Finance The Florida Energy Factor (FEF) is used to compare the efficiency of domestic solar hot water heaters and to determine certain building construction credits. Solar water heaters with a high FEF are more cost efficient. A random sample of approved heaters was obtained, and the North Region FEF for each is given in the following table.

3.3	3.2	1.8	2.1	4.2	3.9	4.6	6.3	1.9
1.9	4.9	6.7	5.2	3.1	1.4			

(*Source*: Florida Solar Energy Center Systems Approval Office.)

Is there any evidence to suggest that the true population mean FEF is greater than 3.0? Use $\alpha = 0.05$.

9.198 Business and Management According to a report issued by the management consulting firm McKinsey & Company, 20% of all companies require genetic or family medical history information from employees or job applicants. A random sample of 1500 companies was obtained, and 345 said they require this information from employees or job applicants.

a. Identify p_0 and n, and compute $\hat{p}$.

b. Check the nonskewness criteria.

c. Is there any evidence to suggest that the true proportion of companies that require this information is greater than 0.20? Use $\alpha = 0.01$.

d. Compute the p value associated with this hypothesis test.

9.199 Public Policy and Political Science A candidate for district attorney claims that 75% of all residents in Elko County, Nevada, favor granting police additional powers to tap telephone lines. In order to check this claim, a random sample of 560 residents was selected, and 392 said they were in favor of more phone taps.

a. Is there any evidence to suggest that the true proportion of residents who favor additional power to tap phones is less than 0.75? Use $\alpha = 0.01$.

b. Compute the p value associated with this hypothesis test.

9.200 Psychology and Human Behavior Although only a small proportion of all young men and women join a gang, it has been reported that gang members are responsible for the majority of youth violence. The proportion of men and women who join a gang is dependent on the city, the neighborhood, and the quality of local law enforcement. In a report issued by the attorney general of Florida, 8.2% of middle and high school students indicated that they belonged to a gang at some time.[48] In a new survey, 21 of 305 randomly selected middle and high school students in Florida said they were gang members at some time.

a. Is there any evidence to suggest that the proportion of youth gang members has changed? Use $\alpha = 0.05$.

b. Find the p value associated with this hypothesis test.

9.201 Manufacturing and Product Development Approximately 128 million counterfeit and pirated products entered the European Union (EU) in 2006. The Indian News reported that 80% of all counterfeit or pirated goods that enter the EU are from China.[49] Suppose in a 2009 survey of 3050 randomly selected counterfeit products brought into the EU, 2356 were from China. Is there any evidence to suggest that the proportion of counterfeit goods from China has decreased? Use $\alpha = 0.001$.

9.202 Fuel Consumption and Cars *USA Today* reported that the most annoying driving habit is not signaling. In a recent survey of 425 randomly selected drivers, 375 indicated that not signaling was an annoying driving habit. Is there any evidence to suggest that the true proportion of drivers who believe that not signaling is an annoying habit is different from 0.89? Use $\alpha = 0.01$.

9.203 Medicine and Clinical Studies The diameter of a virus is approximately 0.3 μm (micrometers). There is some speculation that new virus strains exhibit greater variability in diameter. A medical research lab obtained a random sample of 15 new virus strains and measured the diameter (in μm) of each. The sample mean diameter was 0.323 and the sample variance was $s^2 = 0.0026$. Is there any evidence to suggest that the true population variation in diameter of viruses has increased from 0.0015? Assume normality and use $\alpha = 0.05$.

9.204 Manufacturing and Product Development The radial shrinkage in paper birch wood from green to oven dry is approximately 6.3%. A new process has been developed in order to decrease the variability in shrinkage. In a random sample of 21 pieces of paper birch, the sample variance in shrinkage was 0.39. Is there any evidence to suggest that the population variance in shrinkage is less than 0.50? Assume normality and use $\alpha = 0.10$.

9.205 Medicine and Clinical Studies The normal blood platelet count for an adult ranges from 150,000 to 400,000, with a standard deviation of approximately 62,500. A researcher has speculated that increased exposure to pollutants has increased the variability in numbers of blood platelets. A random sample of 37 adults was obtained, and the blood platelet count was measured in each person. The sample standard deviation was $s = 65,268$. Is there any evidence to suggest that the population variance in blood platelet count has increased? Assume normality and use $\alpha = 0.001$.[50]

9.206 Public Health and Nutrition The fruitcake, a traditional holiday gift, has been the subject of many jokes and is often thrown away without even a nibble. *Tonight Show* host Johnny Carson used to say there was really only one fruitcake in the entire United States, and it was never really eaten by anyone, just passed as a gift from one family to another. Fruitcake is very dense, and recently fruitcake bakers have petitioned the FDA to adjust the serving size to 1.5 ounces, or about 160 calories.[51] A random sample of fruitcake servings was obtained, and the serving size of each was carefully measured (in ounces). The data are given in the following table.

1.52	1.61	1.42	1.54	1.62	1.43	1.62	1.60
1.76	1.54	1.56	1.72	1.53	1.59	1.68	1.54
1.45	1.62	1.45	1.76	1.70	1.37	1.58	

Suppose the FDA had adopted the new fruitcake serving size recommendations and the underlying distribution is normal. Is there any evidence to suggest that the true mean fruitcake serving size is greater than 1.5 ounces? Use $\alpha = 0.01$.

Extended Applications

9.207 Physical Sciences Workers at the Daivik diamond mine in northern Canada extract approximately 1800 DMT (dry metric tons) of ore per day, which is sifted and examined for diamonds. New machinery has just been installed that is designed to increase the amount of ore extracted per day. A random sample of 36 days was selected, and the amount of ore extracted each day was recorded. The sample mean was $\bar{x} = 1852$ DMT. Assume $\sigma = 202$ DMT.

a. Is there any evidence to suggest that the new machinery has improved production? Use $\alpha = 0.05$.

b. What is the probability of a type II error if the true mean amount of ore extracted has changed to 1875 DMT; that is, what is $\beta(1875)$? Find the probability of a type II error if the true mean is 1925 DMT.

9.208 Technology and Internet A chemical made to etch integrated circuits is rated at 650 Å/min (angstroms per minute). In order to check this claim, a random sample of 17 integrated circuits was selected, and each was etched using this chemical. The etch rates (in Å/min) are given in the following table.[52]

635	629	665	649	607	660	707	669	660
653	686	617	656	615	620	593	686	

a. Is there any evidence to suggest that the mean etch rate is different from 650 Å/min? Use $\alpha = 0.05$.

b. What assumption(s) did you make in order to conduct the hypothesis test in part (a).

c. Find bounds on the p value for the hypothesis test in part (a).

9.209 Physical Sciences The design specifications for a new gymnasium at a local high school call for the lights to produce at least 40 footcandles (a measure of brightness). Before the new facility was opened to students, the contractor collected a sample of 23 brightness measurements at random locations in the gym. The sample mean was $\bar{x} = 38.63$ footcandles and the sample standard deviation was $s = 5.6$ footcandles.

a. Is there any evidence to suggest that the mean brightness in the gym is less than the design specification? Assume normality and use $\alpha = 0.01$.

b. Find bounds on the p value associated with this hypothesis test.

9.210 Business and Management The largest aboveground storage tanks in Bayonne, New Jersey, have the capacity to hold 6,000,000 gallons of oil. For safety reasons, managers prefer the mean amount stored at any given time to be no greater than 4,500,000 gallons. A random sample of nine large tanks was selected, and the amount of oil stored in each (in gallons) was recorded. The summary statistics were $\bar{x} = 4,675,250$ and $s = 482,556$.

a. Is there any evidence to suggest that the mean amount of oil stored in the large tanks is above the safety level? Assume normality and use $\alpha = 0.025$.

b. Find bounds on the p value associated with this hypothesis test.

9.211 Public Health and Nutrition Cast iron is an extremely durable cookware material, good for searing and blackening foods. However, a cast-iron pan can be full of bacteria and can lend unwanted flavors to food. A company trying to promote alternative ceramic cookware asked members of a community to bring their favorite cast-iron cookware to their store for bacteria testing. The company manager claimed that at least 60% of all cast-iron cookware contains harmful bacteria. A random sample of 120 pans was selected, and 57 were found to contain harmful bacteria.

a. Is there any evidence to refute the manager's claim? Use $\alpha = 0.01$.
b. Find the p value associated with this hypothesis test.
c. Do you believe the sample is really random? Why or why not?

9.212 Marketing and Consumer Behavior An assisted-living home is an alternative to a nursing home and a bridge between a skilled-care facility and a patient's residence. A recent report indicated that 92% of all patients in assisted-living homes are satisfied with the facility and the care. An insurance company believes that this percentage is actually much lower (due to health-care violations and strict for-profit motives). A random sample of 5000 patients in assisted-living homes around the country was obtained, and 4576 said they were satisfied with the facility and the care.

a. Is there any evidence to suggest that the true proportion of assisted-living patients who are satisfied is less than 0.92? Use $\alpha = 0.01$.
b. Find the p value associated with this hypothesis test.
c. Carefully sketch a graph illustrating the critical value, significance level, value of the test statistic, and p value.

9.213 Public Health and Nutrition One measure of the quality of pasta is the b value, or yellow alkaline brightness. Consumers usually associate a brighter yellow color with higher quality. The following b values were reported for Desert Durum Variety pasta.

39.7	37.4	38.7	43.0	42.0	42.0	41.6	42.5
46.8	45.7	39.9	40.8	39.5	39.1	41.1	

(*Source*: California Wheat Commission.)

a. Suppose the company has designed a new milling process to increase product consistency. Is there any evidence to suggest that the standard deviation in b value has decreased from 3.1? Use $\alpha = 0.05$.
b. Find bounds on the p value associated with this hypothesis test.

Confidence Intervals and Hypothesis Tests Based on Two Samples or Treatments

10

Chapter 10 Challenge

Do children appreciate art?

Museum directors face a growing concern that young people have neither an interest in nor appreciation for the arts. Many art museums around the country are noticing a decrease in visitors (especially first-time visitors), fewer donations, and less public awareness. A trip to an art museum is generally no longer a planned family activity. In a recent study of visitor experiences and satisfaction at three Smithsonian art museums, 34% of all first-time visitors had simply wandered in to the facility.

Despite traditional collections that include paintings by famous artists, like "The Artist's Garden at Giverny" by French impressionist Claude Monet, special exhibitions, online tours, educational programs, and activities designed for children, attendance is down. While most art museums are publicly funded, government assistance is decreasing and grants are becoming scarce. Rather than raise ticket prices, many art museums are adopting private-sector practices, for example, museum gift shops, corporate sponsorship, and specialty restaurants.

As part of a profile of participation in culture in New York City, random samples were obtained of 125 adults (ages 18 and over) and 150 children (under age 18). Fifty-four adults and 89 children said they had visited an art museum within the past year.[1] The hypothesis test procedures presented in this chapter will be used to compare parameters (means or variances, for example) from two different populations. In this case, we will compare the population proportion of adults with the population proportion of children who have visited an art museum within the past year. The results are presented in the Chapter Challenge Wrap-Up (page 506). These tests are constructed using methods of *standardization* similar to those in Chapter 9.

(Manchan/Getty Images)

Review
- Recall the formal, four-part hypothesis test process.
- Remember the specific inference procedures concerning a single population parameter: μ, p, or σ^2.

Preview
- Adapt and extend the single-sample hypothesis test procedures.
- Construct confidence intervals to estimate the difference in two population parameters.
- Conduct hypothesis tests to compare two population parameters.

NOTATION

In order to conduct a hypothesis test to compare two (similar) population parameters, we will simply modify the single-sample procedures presented in the previous chapter. Perhaps the most tricky aspect of these procedures is the notation. The following table summarizes the notation used to represent similar parameters associated with two different populations.

	Population parameters			
	Mean	**Variance**	**Standard deviation**	**Proportion**
Population 1	μ_1	σ_1^2	σ_1	p_1
Population 2	μ_2	σ_2^2	σ_2	p_2

The following table summarizes the notation used to represent *values* of summary statistics associated with samples from two different populations.

	Sample statistics				
	Sample size	**Mean**	**Variance**	**Standard deviation**	**Proportion**
Sample from population 1	n_1	$\bar{x}_1$	s_1^2	s_1	$\hat{p}_1$
Sample from population 2	n_2	$\bar{x}_2$	s_2^2	s_2	$\hat{p}_2$

Note: We do not necessarily use every summary statistic associated with a population in every problem. For example, we may only need the sample size and proportion in one case, but use the sample size, mean, and standard deviation in another problem.

In order to compare two population parameters to see whether there is any evidence that they are different, we often consider a difference. For example, to compare two population means, μ_1 and μ_2, we consider the difference $\mu_1 - \mu_2$. In searching for evidence that p_1 is bigger than p_2, we look at the difference $p_1 - p_2$.

There are two reasons to consider a difference.

1. A typical relationship between two population parameters can be written in terms of a difference. For example, suppose we need to compare the means from two populations, μ_1 and μ_2.

Standard notation		**Difference notation**
$\mu_1 = \mu_2$	is equivalent to	$\mu_1 - \mu_2 = 0$
$\mu_1 > \mu_2$	is equivalent to	$\mu_1 - \mu_2 > 0$
$\mu_1 < \mu_2$	is equivalent to	$\mu_1 - \mu_2 < 0$

Therefore, a statistical test with null hypothesis $H_0: \mu_1 - \mu_2 = 0$ corresponds to a test of $H_0: \mu_1 = \mu_2$. And $H_a: \mu_1 - \mu_2 > 0$ is equivalent to $H_a: \mu_1 > \mu_2$. The hypothesized difference between the two means may be nonzero. The null hypothesis $H_0: \mu_1 = \mu_2 + 5$ written using a difference is equivalent to $H_0: \mu_1 - \mu_2 = 5$.

2. In addition, a difference (for example, $\mu_1 - \mu_2$) is itself a *single* population parameter. A natural, intuitive statistic, $\bar{X}_1 - \bar{X}_2$, may be used to estimate the value of this parameter. The properties of $\bar{X}_1 - \bar{X}_2$ will be used to develop a test statistic.

As in the statistical tests presented in the last chapter, in any two-sample hypothesis test we usually make certain assumptions. The assumptions associated with the hypothesis tests in this chapter include a statement concerning the selection of individuals or objects from *two* different populations.

> **DEFINITION**
>
> 1. Two samples are **independent** if the process of selecting individuals or objects in sample 1 has no effect on, or no relation to, the selection of individuals or objects in sample 2. If the samples are not independent, they are **dependent**.
>
> 2. A **paired** data set is the result of matching each individual or object in sample 1 with a *similar* individual or object in sample 2. A common experiment in which paired data are obtained involves a *before* and *after* measurement on each individual or object. Each *before* observation is matched, or paired, with an *after* observation.

Similar means the individuals or objects share some common, fundamental characteristic. They may even be the same individual or object!

The notation, the idea of using differences, and the extra assumptions are all used in the following sections to construct hypothesis tests for comparing various characteristics of two populations.

10.1 COMPARING TWO POPULATION MEANS USING INDEPENDENT SAMPLES WHEN POPULATION VARIANCES ARE KNOWN

As in Chapters 8 and 9, the first hypothesis test presented here, for comparing two population means, is instructive but not very realistic. Since $\overline{X}_1$ is a good estimator for μ_1 and $\overline{X}_2$ is a good estimator for μ_2, it is reasonable to use the estimator $\overline{X}_1 - \overline{X}_2$ to estimate the parameter $\mu_1 - \mu_2$. In order to develop a hypothesis test, we need to know the properties of the estimator, or the distribution of the random variable, $\overline{X}_1 - \overline{X}_2$.

> **PROPERTIES OF $\overline{X}_1 - \overline{X}_2$**
>
> Suppose that
>
> 1. $\overline{X}_1$ is the mean of a random sample of size n_1 from a population with mean μ_1 and variance σ_1^2,
> 2. $\overline{X}_2$ is the mean of a random sample of size n_2 from a population with mean μ_2 and variance σ_2^2, and
> 3. The samples are independent.
>
> If the distributions of both populations are normal, then the random variable $\overline{X}_1 - \overline{X}_2$ has the following properties.
>
> 1. $\mathrm{E}(\overline{X}_1 - \overline{X}_2) = \mu_{\overline{X}_1 - \overline{X}_2} = \mu_1 - \mu_2$
> $\overline{X}_1 - \overline{X}_2$ is an unbiased estimator of the parameter $\mu_1 - \mu_2$. The distribution is centered at $\mu_1 - \mu_2$.
>
> 2. $\mathrm{Var}(\overline{X}_1 - \overline{X}_2) = \sigma_{\overline{X}_1 - \overline{X}_2}^2 = \dfrac{\sigma_1^2}{n_1} + \dfrac{\sigma_2^2}{n_2}$ and the standard deviation is
> $$\sigma_{\overline{X}_1 - \overline{X}_2} = \sqrt{\dfrac{\sigma_1^2}{n_1} + \dfrac{\sigma_2^2}{n_2}}$$
>
> 3. The distribution of $\overline{X}_1 - \overline{X}_2$ is normal.
> If the underlying distributions are not known, but both n_1 and n_2 are large, then $\overline{X}_1 - \overline{X}_2$ is approximately normal (by the Central Limit Theorem).

Can you see the standardization coming?

Since the distribution of $\overline{X}_1 - \overline{X}_2$ is (approximately) normal, the usual standardization can be used to obtain a Z random variable. The resulting hypothesis test has a very typical form.

HYPOTHESIS TESTS CONCERNING TWO POPULATION MEANS WHEN POPULATION VARIANCES ARE KNOWN

Given two independent random samples, the first of size n_1 from a population with mean μ_1 and the second of size n_2 from a population with mean μ_2, assume

1. The underlying populations are normal and/or both sample sizes are large, and
2. The population variances, σ_1^2 and σ_2^2, are known.

A hypothesis test concerning two population means, in terms of the difference in means $\mu_1 - \mu_2$, with significance level α, has the following form:

$H_0: \mu_1 - \mu_2 = \Delta_0$

$H_a: \mu_1 - \mu_2 > \Delta_0, \quad \mu_1 - \mu_2 < \Delta_0, \quad \text{or} \quad \mu_1 - \mu_2 \neq \Delta_0$

$$\text{TS: } Z = \frac{(\overline{X}_1 - \overline{X}_2) - \Delta_0}{\sqrt{\dfrac{\sigma_1^2}{n_1} + \dfrac{\sigma_2^2}{n_2}}}$$

$\text{RR: } Z \geq z_\alpha, \quad Z \leq -z_\alpha, \quad \text{or} \quad |Z| \geq z_{\alpha/2}$

For reference, we'll call these the two-sample Z test assumptions.

This is the template for a hypothesis test concerning two population means when variances are known, sometimes called a two-sample Z test.

Δ is the uppercase Greek letter delta.

💡 ILLUMINATING THE CONCEPTS

1. The value Δ_0 is the fixed, hypothesized difference in means. Usually $\Delta_0 = 0$ (i.e., the means are assumed equal). The null hypothesis is then $H_0: \mu_1 - \mu_2 = 0$, which is equivalent to $H_0: \mu_1 = \mu_2$. However, Δ_0 may be some nonzero value. For example, two population means may historically differ by 12 so that $H_0: \mu_1 - \mu_2 = 12 \ (= \Delta_0)$. We may want to conduct a test to see whether there is any change in this difference, with $H_a: \mu_1 - \mu_2 \neq 12$.

2. Just a reminder: use only one (appropriate) alternative hypothesis and the corresponding rejection region. The z critical values are from the standard normal distribution.

3. This hypothesis test procedure can be used *only* if both population variances are known. If they are unknown but both sample sizes are large, some statisticians substitute s_1^2 for σ_1^2 and s_2^2 for σ_2^2. This produces an *approximate* test statistic. Section 10.2 presents an *exact* test procedure for comparing population means (under certain assumptions) when the population variances are unknown. ∎

The following example illustrates this hypothesis test procedure.

(Robert Brown/Dreamstime.com)

Example 10.1 Cast Crush Strength Most casts formed to help protect a broken limb are made from plaster of Paris and cotton wool. Another common type of cast is fabricated from polyurethane and fiberglass. This combination of materials is quick-setting and light. The most important aspect of any cast, however, is the crush resistance or crush strength. Independent random samples of each type of cast were obtained, and the crush resistance was measured (in newtons) for each cast one week after fabrication. The summary statistics and known variances are given in the following table.

Cast	Sample size	Sample mean	Population variance
Plaster of Paris (1)	$n_1 = 16$	$\bar{x}_1 = 210.5$	$\sigma_1^2 = 416.16$
Polyurethane (2)	$n_2 = 25$	$\bar{x}_2 = 225.1$	$\sigma_2^2 = 96.04$

Is there any evidence to suggest that the mean crush strength of plaster of Paris casts is less than the mean crush strength of polyurethane casts? Use $\alpha = 0.05$ and assume that each underlying distribution of crush strength is normal.

SOLUTION

STEP 1 Arbitrarily, let plaster of Paris casts be population 1, and polyurethane casts be population 2.

The current state, or assumption, is that the two population mean crush strengths are equal: $\mu_1 = \mu_2 \Rightarrow \mu_1 - \mu_2 = 0 \ (= \Delta_0)$.

The sample sizes, sample means, and population variances are given.

We are trying to find evidence that plaster of Paris casts have a *smaller* mean crush strength: $\mu_1 < \mu_2$, which is the same as $\mu_1 - \mu_2 < 0$. Therefore, the alternative hypothesis is one-sided, left-tailed.

STEP 2 The four parts of the hypothesis test are

$H_0: \mu_1 - \mu_2 = 0$

$H_a: \mu_1 - \mu_2 < 0$

$$\text{TS: } Z = \frac{(\bar{X}_1 - \bar{X}_2) - 0}{\sqrt{\dfrac{\sigma_1^2}{n_1} + \dfrac{\sigma_2^2}{n_2}}}$$

RR: $Z \le -z_\alpha = -z_{0.05} = -1.6449$

STEP 3 The value of the test statistic is

$$z = \frac{(\bar{x}_1 - \bar{x}_2) - 0}{\sqrt{\dfrac{\sigma_1^2}{n_1} + \dfrac{\sigma_2^2}{n_2}}} = \frac{210.5 - 225.1}{\sqrt{\dfrac{416.16}{16} + \dfrac{96.04}{25}}} = -2.6722 \ (\le -1.6449)$$

STEP 4 Since -2.6722 lies in the rejection region, we reject the null hypothesis at the $\alpha = 0.05$ significance level. There is evidence to suggest that the mean crush strength for plaster of Paris casts is less than the mean crush strength for polyurethane casts.

The p value for this hypothesis test is

$p = P(Z \le -2.6722) = 0.0038 \ (\le 0.05)$ Use Table III in the Appendix.

Since $p \le \alpha$, we reject the null hypothesis.

Solution Trail

KEYWORDS

- Is there any evidence?
- Less than.
- Known variances.
- Independent random samples.
- Each underlying distribution is normal.

$\downarrow$

TRANSLATION

Conduct a one-sided, left-tailed test to compare μ_1 and μ_2.

$\downarrow$

CONCEPTS

Hypothesis test concerning two population means when variances are known.

$\downarrow$

VISION

Use the template for this hypothesis test. The samples are random and independent, the underlying populations are normal, and the population variances are known. Use a one-sided alternative hypothesis and the corresponding rejection region, find the value of the test statistic, and draw a conclusion.

p value illustration:

$p = P(Z \le -2.6722)$

$= 0.0038 \le 0.05 = \alpha$

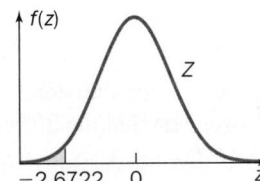

Figures 10.1 through 10.3 together show a technology solution.

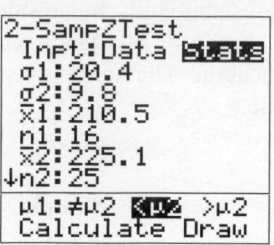

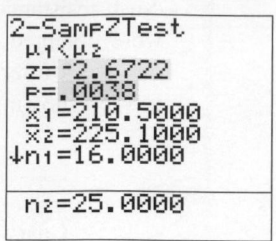

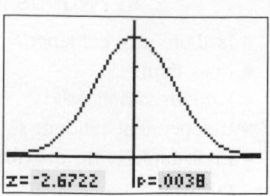

Figure 10.1
2-SampZTest input screen.

Figure 10.2 Hypothesis test results.

Figure 10.3
2-SampZTest Draw results. ●

The following example involves a hypothesis test with a nonzero value for the hypothesized difference in means, Δ_0.

Example 10.2 Low-Carb Ice Cream   Low-carbohydrate foods have become more popular as many Americans try to avoid this sugar and starch combination that they believe causes weight gain. An advertisement for a low-carb ice cream claims that the product has 16 fewer grams of carbohydrates per serving than the leading store brand. In order to check this claim, independent random samples of each type of ice cream were obtained, and the amount of carbohydrates in each serving was measured. The data are given (in grams) in the following tables.

Store brand (1)

15.4	20.4	21.0	24.3	23.3	18.7	19.8	22.5	18.9	22.8
25.4	25.1	20.3	24.1	16.6	22.6	22.1	19.4	16.6	24.4
17.8	18.6	14.9	24.6	19.1	17.9	18.7	20.1	26.3	18.4
21.8	17.1	21.5	19.6	22.9	22.2	21.5	18.3		

Low-carb brand (2)

3.7	3.9	4.5	4.3	3.2	3.6	3.7	3.6	3.7	4.0
4.1	3.1	4.3	3.4	3.4	3.5	4.4	4.9	3.7	3.8
4.1	4.7	3.7	4.2	3.1	4.4	4.2	3.4	4.8	3.6
3.2	3.4	4.2	3.0	3.9					

The variance in carbohydrates per serving is known to be 8.5 for the store brand and 0.253 for the low-carb brand. Is there any evidence to suggest that the difference in population means of carbohydrates per serving is not 16 grams? Use $\alpha = 0.01$.

SOLUTION

STEP 1   The hypothesized difference is $\mu_1 - \mu_2 = 16 \ (= \Delta_0)$.
The sample sizes are $n_1 = 38$ and $n_2 = 35$.
The known population variances are $\sigma_1^2 = 8.5$ and $\sigma_2^2 = 0.253$.
The significance level is $\alpha = 0.01$.

We are testing for *any* difference in population means other than 16 grams of carbohydrates. This is a two-sided test.

The samples are random and independent, and the population variances are known. The underlying population distributions are unknown, but both sample sizes are large (≥ 30). A hypothesis test concerning two population means when variances are known is relevant.

STEP 2 The four parts of the hypothesis test are

$H_0: \mu_1 - \mu_2 = 16$

$H_a: \mu_1 - \mu_2 \neq 16$

TS: $Z = \dfrac{(\bar{X}_1 - \bar{X}_2) - 16}{\sqrt{\dfrac{\sigma_1^2}{n_1} + \dfrac{\sigma_2^2}{n_2}}}$

RR: $|Z| \geq z_{\alpha/2} = z_{0.005} = 2.5758$

STEP 3 The sample means are

$$\bar{x}_1 = \frac{1}{38}(15.4 + 20.4 + \cdots + 18.3) = 20.6579$$

$$\bar{x}_2 = \frac{1}{35}(3.7 + 3.9 + \cdots + 3.9) = 3.8486$$

The value of the test statistic is

$$z = \frac{(\bar{x}_1 - \bar{x}_2) - 16}{\sqrt{\dfrac{\sigma_1^2}{n_1} + \dfrac{\sigma_2^2}{n_2}}} = \frac{(20.6579 - 3.8486) - 16}{\sqrt{\dfrac{8.5}{38} + \dfrac{0.253}{35}}} = 1.6842$$

STEP 4 The value of the test statistic, $z = 1.6842$, does not lie in the rejection region. We do not reject the null hypothesis. There is no evidence to suggest that the difference in population mean carbohydrates is different from 16 grams at the $\alpha = 0.01$ significance level.

Since this is a two-sided test and the value of the test statistic is positive, $p/2$ is a right-tail probability.

$p/2 = P(Z \geq 1.6842)$ — Definition of p value for a two-sided test.

$\quad = 1 - P(Z \leq 1.6842)$ — The Complement Rule.

$\quad = 1 - 0.9539 = 0.0461$ — Use Table III in the Appendix.

$p = 2(0.0461) = 0.0922$ — Solve for p.

Since $p = 0.0922 > 0.01$ ($= \alpha$), we do not reject the null hypothesis.

Figure 10.4 shows a technology solution.

p value illustration:

$p = 2P(Z \geq 1.6842)$

$\quad = 0.0922 > 0.01 = \alpha$

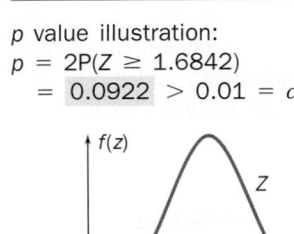

z-Test: Two Sample for Means		
	Variable 1	Variable 2
Mean	20.6579	3.8486
Known Variance	8.5	0.253
Observations	38	35
Hypothesized Mean Difference	16	
z	1.6842	
P(Z<=z) one-tail	0.0461	
z Critical one-tail	2.3263	
P(Z<=z) two-tail	0.0921	
z Critical two-tail	2.5758	

Figure 10.4 Hypothesis test results.

Given the two-sample Z test assumptions and the properties of the random variable $\bar{X}_1 - \bar{X}_2$, we can construct a confidence interval (CI) for the (difference) parameter $\mu_1 - \mu_2$. ◀ As usual, to find a general CI, start with an appropriate symmetric interval about 0 such that the probability that Z lies in this interval is $1 - \alpha$.

$$P\left(-z_{\alpha/2} < \underbrace{\frac{(\bar{X}_1 - \bar{X}_2) - (\mu_1 - \mu_2)}{\sqrt{\dfrac{\sigma_1^2}{n_1} + \dfrac{\sigma_2^2}{n_2}}}}_{z} < z_{\alpha/2}\right) = 1 - \alpha \qquad (10.1)$$

Manipulate the inequality in Equation 10.1 to *sandwich* the parameter $\mu_1 - \mu_2$. We obtain the following probability statement.

$$P\left[(\bar{X}_1 - \bar{X}_2) - z_{\alpha/2}\sqrt{\frac{\sigma_1^2}{n_1} + \frac{\sigma_2^2}{n_2}} < \mu_1 - \mu_2 < (\bar{X}_1 - \bar{X}_2) + z_{\alpha/2}\sqrt{\frac{\sigma_1^2}{n_1} + \frac{\sigma_2^2}{n_2}}\right] = 1 - \alpha.$$

This leads to the following general result. ▶

HOW TO FIND A $100(1 - \alpha)\%$ CONFIDENCE INTERVAL FOR $\mu_1 - \mu_2$ WHEN VARIANCES ARE KNOWN

Given the two-sample Z test assumptions, a $100(1 - \alpha)\%$ confidence interval for $\mu_1 - \mu_2$ has as endpoints the values

$$(\bar{x}_1 - \bar{x}_2) \pm z_{\alpha/2}\sqrt{\frac{\sigma_1^2}{n_1} + \frac{\sigma_2^2}{n_2}} \qquad (10.2)$$

(Tracy Hombrook/Dreamstime.com)

Example 10.3 Pizza Stone Weights Pizza stones designed for home use help cooks produce baked goods with brick-oven qualities, for example, a crusty loaf of bread or crispy-crust pizza. However, pizza stones can be very heavy and can also take up a lot of space in a traditional residential oven. Independent random samples of two similar types of round pizza stones were obtained, and the weight (in pounds) of each was recorded. The summary statistics and known variances are given in the following table.

Pizza stone	Sample size	Sample mean	Population variance
Kitchen Depot (1)	$n_1 = 35$	$\bar{x}_1 = 6.21$	$\sigma_1^2 = 2.1$
Head Chef (2)	$n_2 = 31$	$\bar{x}_2 = 7.08$	$\sigma_2^2 = 3.5$

Find a 95% confidence interval for the difference in population mean pizza-stone weights.

SOLUTION

STEP 1 Sample sizes, sample means, and known variances are given.

The underlying weight distributions are unknown, but the sample sizes are both large (≥ 30).

$1 - \alpha = 0.95 \Rightarrow \alpha = 0.05 \Rightarrow \alpha/2 = 0.025$ Find $\alpha/2$.

$z_{\alpha/2} = z_{0.025} = 1.960$ Find the z critical value.

STEP 2 Use Equation 10.2.

$$(\bar{x}_1 - \bar{x}_2) \pm z_{\alpha/2}\sqrt{\frac{\sigma_1^2}{n_1} + \frac{\sigma_2^2}{n_2}}$$ Equation 10.2.

$$= (6.21 - 7.08) \pm (1.96)\sqrt{\frac{2.1}{35} + \frac{3.5}{31}}$$ Use summary statistics and critical value.

$$= -0.87 \pm 0.8150$$ Simplify.

$$= (-1.6850, -0.0550)$$ Compute endpoints.

$(-1.6850, -0.0550)$ is a 95% confidence interval for the difference in population mean weights (in pounds) of the pizza stones, $\mu_1 - \mu_2$. This interval represents a set of very plausible values for the difference in population mean weights.

Figures 10.5 and 10.6 together show a technology solution.

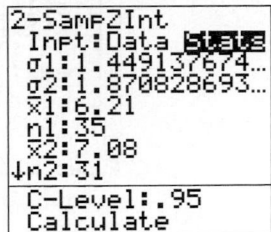

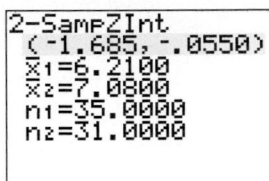

Figure 10.5
2-SampZInt input
screen.

Figure 10.6 Resulting
confidence interval.

TECHNOLOGY CORNER

Procedure: Hypothesis tests and confidence intervals concerning two population means when the population variances are known.
Reconsider: Example 10.2, page 450, solution, and interpretation.

TI-84 Plus

Use the built-in functions 2-SampZTest and 2-SampZInt. Input is either summary statistics or data in lists.

1. Enter the store brand data into list L1 and the low-carb brand data into list L2.
2. Subtract $\Delta_0 = 16$ from each observation in list L1 and store the results in list L1.
3. Select STAT; TESTS; 2-SampZTest. Highlight Data. Enter σ_1, σ_2, List1, and List2. Set each frequency to 1. Highlight the alternative hypothesis. See Figure 10.7.
4. Highlight Calculate and press ENTER. The results are displayed on the Home Screen. See Figure 10.8. The Draw results are shown in Figure 10.9.
5. To construct a confidence interval for the difference in population means, use the original data in list L1.

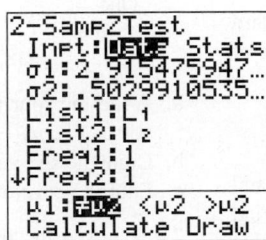

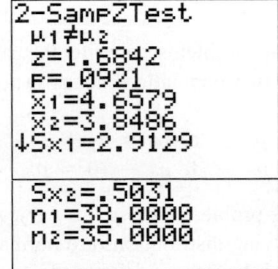

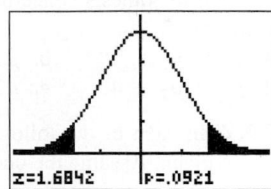

Figure 10.7
2-SampZTest input
screen.

Figure 10.8 Hypothesis
test results.

Figure 10.9
2-SampZTest Draw
results.

6. Select STAT; TESTS; 2-SampZInt. Highlight Data. Enter σ_1, σ_2, List1, and List2. Set each frequency to 1 and enter the C-Level. See Figure 10.10.

7. Highlight `Calculate` and press ENTER . The resulting confidence interval is displayed on the Home Screen. See Figure 10.11.

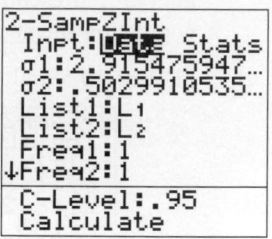

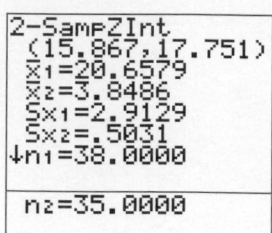

Figure 10.10
`2-SampZInt` input
screen.

Figure 10.11 Resulting
95% confidence interval.

Minitab

There is no built-in function to conduct hypothesis tests and construct confidence intervals concerning two population means when the population variances are known. Remember, this is an instructive situation, not very realistic. It is pretty unlikely that we would know the population variances and not the population means.

Excel

Use the built-in function z-test: Two Sample for Means.

1. Enter the store brand data into column A and the low-carb brand data into column B.
2. Subtract $\Delta_0 = 16$ from each observation in column A and store the results in column A.
3. Under the Data tab, select Data Analysis; z-test: Two Sample for Means.
4. Enter the Variable 1 Range, the Variable 2 Range, and the known population variances. Choose an Output option and click OK.
5. Summary statistics, along with the value of the test statistic, critical values, and p values are displayed. See Figure 10.4, page 451.

SECTION 10.1 EXERCISES

Practice

10.1 In each of the following problems, rewrite the standard notation hypothesis concerning two population means in terms of a difference, $\mu_1 - \mu_2$.

a. $\mu_1 = \mu_2$ b. $\mu_1 < \mu_2$ c. $\mu_1 \neq \mu_2 + 7$
d. $\mu_1 > \mu_2 - 4$ e. $\mu_1 \neq \mu_2$ f. $\mu_1 - 10 = \mu_2$

10.2 In each of the following problems, $\mu_1, \mu_2, \sigma_1, \sigma_2, n_1$, and n_2 are given. Assume the underlying distributions are normal. Find the mean, variance, and standard deviation of the random variable $\bar{X}_1 - \bar{X}_2$, and carefully sketch the probability density function.

a. $\mu_1 = 12, \mu_2 = 9, \sigma_1 = 3, \sigma_2 = 7, n_1 = 15, n_2 = 11$.
b. $\mu_1 = 25.6, \mu_2 = 37.8, \sigma_1 = 7.5, \sigma_2 = 10.5, n_1 = 10, n_2 = 25$.
c. $\mu_1 = 125.3, \mu_2 = 250.6, \sigma_1 = 15.6, \sigma_2 = 25.6, n_1 = 8, n_2 = 12$.
d. $\mu_1 = 3.1, \mu_2 = 2.2, \sigma_1 = 0.50, \sigma_2 = 0.75, n_1 = 21, n_2 = 21$.

10.3 Given the two-sample Z test assumptions, consider the following table of sample sizes, sample means, and known standard deviations.

Group	Sample size	Sample mean	Population standard deviation
One	18	17.5	1.5
Two	26	16.2	2.6

a. Write the four parts of a hypothesis test of $H_0: \mu_1 - \mu_2 = 0$ versus $H_a: \mu_1 - \mu_2 > 0$. Use $\alpha = 0.05$.
b. Compute the value of the test statistic and draw a conclusion.
c. Find the p value associated with this hypothesis test.

10.4 Given the two-sample Z test assumptions, consider the following table of sample sizes, sample means, and known variances.

Group	Sample size	Sample mean	Population variance
One	25	186	14.7
Two	24	190	23.8

a. Write the four parts of a hypothesis test of H_0: $\mu_1 - \mu_2 = 2$ versus H_a: $\mu_1 - \mu_2 < 2$. Use $\alpha = 0.01$.

b. Compute the value of the test statistic and draw a conclusion.

c. Carefully sketch a graph to illustrate the p value associated with this hypothesis test. Compute the p value.

10.5 Given the two-sample Z test assumptions, consider the following table of sample sizes, sample means, and known variances.

Group	Sample size	Sample mean	Population variance
One	37	1025.6	225.3
Two	42	1031.3	107.6

a. Write the four parts of a hypothesis test of H_0: $\mu_1 - \mu_2 = 0$ versus H_a: $\mu_1 - \mu_2 \neq 0$. Use $\alpha = 0.001$.

b. Compute the value of the test statistic and draw a conclusion.

c. Is the normality assumption necessary in order to conduct this hypothesis test? Justify your answer.

10.6 Two random samples were obtained independently and the resulting data are given on the data CD and book's web site. Assume both populations are normal with $\sigma_1 = 8$ and $\sigma_2 = 12$.

a. Find a 95% confidence interval for the true difference in means, $\mu_1 - \mu_2$.

b. Using the confidence interval in part (a), is there any evidence to suggest that the two population means are different? Justify your answer.

10.7 Suppose a random sample of size 15 is taken from a normal population with mean 25 and standard deviation 5, and a second, independent random sample of size 21 is taken from a normal population with mean 10 and standard deviation 4.

a. Describe the distribution of the difference in sample means, $\overline{X}_1 - \overline{X}_2$ (in terms of type of distribution, mean, variance, and standard deviation).

b. Carefully sketch the probability distribution for $\overline{X}_1 - \overline{X}_2$.

c. Find $P(\overline{X}_1 - \overline{X}_2 \geq 17)$.

d. Find $P(13.5 < \overline{X}_1 - \overline{X}_2 < 14.5)$.

e. Find $P(\overline{X}_1 < \overline{X}_2 + 14)$.

Applications

10.8 Manufacturing and Product Development The efficiency of an electric toothbrush is often judged by the rotation speed, in revolutions per minute (rpm). Two brands were selected for comparison, and independent random samples of each electric toothbrush were obtained. The rotation speed for each toothbrush was measured, and the summary statistics are given in the following table.

Electric toothbrush	Sample size	Sample mean	Population variance
Sonicare Elite	23	7992.2	1260.25
Oral-B	25	7988.2	1697.44

a. Is there any evidence to suggest that the Sonicare Elite has a greater population mean rotation speed than the Oral-B? Assume normality and use $\alpha = 0.05$.

b. Find the p value associated with this hypothesis test.

10.9 Business and Management Gift cards have become a popular present. Retailers like these cards because they are easier to process than paper gift certificates and more difficult to forge. Customers appreciate the convenience; they make great stocking stuffers and are easy to mail. Independent random samples of gift cards from two merchants were obtained and the purchased value (in dollars) of each was recorded. The summary statistics and known variances are given in the following table.

Store	Sample size	Sample mean	Population variance
Nordstrom	41	24.07	16.81
Macy's	38	26.61	10.24

Is there any evidence to suggest that the true mean purchased value is different between a Nordstrom's and a Macy's gift card? Use $\alpha = 0.01$.

10.10 Manufacturing and Product Development The energy rating, water consumption, and noise level of an electric dishwasher are all important selling features. Suppose the makers of the Hotpoint DF55 claim that this model has a lower noise-level rating than any other comparable dishwasher. Independent random samples of the Hotpoint DF55 and of a similar Maytag dishwasher were obtained. The noise level (in decibels) was measured for each, and the data are given on the data CD and book's web site. Is there any evidence to suggest that the population mean noise level for the Hotpoint dishwasher is less than the population mean noise level for the Maytag? Assume the underlying distributions are normal, with $\sigma_1 = 3.75$ and $\sigma_2 = 4.14$. Use $\alpha = 0.05$.

10.11 Technology and Internet A consumer organization conducted an experiment to compare the power–output rating (in watts) of two similar receivers (as part of a home audio system). Independent random samples were obtained, and the summary statistics and known variances are given in the following table.

Receiver	Sample size	Sample mean	Population variance
Pioneer (1)	16	99.79	1.5625
Sony (2)	21	100.36	3.9024

a. Assume the underlying distributions are normal and find a 95% confidence interval for the true difference in population mean power–output ratings.

b. Use the confidence interval to determine whether there is any evidence that the mean power–output ratings for the two brands differ.

10.12 Marketing and Consumer Behavior A new advertising program involves placing plain, old-fashioned text and graphics on the back of taxi front seats. The theory is that riders give their undivided attention to the ad during the entire trip. Independent random samples of taxi ride times (in minutes) in two cities were obtained. The data are given on the data CD and book's web site.[2] Is there any evidence to suggest that the mean taxi ride time is different in San Diego and Phoenix? Assume normality, with $\sigma_1 = 6.2$ and $\sigma_2 = 4.9$, and use $\alpha = 0.01$.

10.13 Manufacturing and Product Development The total weight (with the case) of a portable sewing machine is an important consideration. Suppose Singer claims to have the lightest machine by 5 pounds. Independent random samples of a Singer machine and a comparable Simplicity machine were obtained, and the weight (in pounds) of each was recorded. The summary statistics and known variances are given in the following table.

Sewing machine	Sample size	Sample mean	Population variance
Simplicity	42	17.99	2.89
Singer	38	13.26	2.25

a. Is there any evidence to refute the claim made by Singer? Use $\alpha = 0.01$.
b. Find the p value associated with this hypothesis test.
c. Is the normality assumption necessary in this problem? Why or why not?

10.14 Medicine and Clinical Studies Many people consume protein shakes to help build muscle mass and eliminate body fat. In a recent study, the amount of protein in two competing drinks was compared. Independent random samples were obtained, and the protein content (in grams) in each drink was measured. The summary statistics and known variances are given in the following table.

Protein drink	Sample size	Sample mean	Population variance
Met-Rx	12	39.38	5.06
Pure Gro	24	39.01	6.01

Is there any evidence to suggest that the mean amount of protein is different in these two products? Use $\alpha = 0.01$ and assume normality.

10.15 Technology and Internet Headphone manufacturers usually specify a sensitivity rating, determined by measuring the sound pressure level (SPL) 1 meter away. Independent random samples of two types of headphones were obtained, and the sensitivity of each (in dB SPL for 1 mW input) was measured. The summary statistics and known variances are given in the following table.

Headphone	Sample size	Sample mean	Population variance
Jensen	34	102.9	4.41
Sennheiser	35	97.6	14.44

a. Find a 99% confidence interval for the true mean difference in sensitivity, $\mu_1 - \mu_2$.
b. Using the interval in part (a), is there any evidence to suggest that the Jensen sensitivity is greater than the Sennheiser sensitivity? Justify your answer.

10.16 Travel and Transportation The recommended tire pressure for an off-road bicycle depends on the weight of the rider. As you would expect, the greater the weight, the greater the recommended tire pressure. At a well-used bicycle trail in the Black River State Forest in Wisconsin, independent random samples were obtained from two different weight groups. The front tire pressure (in psi) was measured for each person, and the summary statistics and known variances are given in the following table.

Weight group	Sample size	Sample mean	Population variance
$\approx$ 150 pounds	18	38.91	2.25
$\approx$ 180 pounds	23	41.99	6.25

Is there any evidence to suggest that the difference between the 180-pound riders' mean tire pressure and the 150-pound riders' mean tire pressure is greater than 3 psi? Assume normality and use $\alpha = 0.05$.

10.17 Technology and Internet A new cordless phone manufactured by Motorola is advertised to have the longest standby time of any cordless phone currently available. A consumer group selected a comparable Uniden phone and obtained independent random samples of both types. Each battery was fully charged, no calls were received or made during the test period, and the standby time (time until the first low-battery beep, in hours) was measured for each phone. The summary statistics for Motorola were $n_1 = 40$, $\bar{x}_1 = 73.41$, and $\sigma_1^2 = 13.69$, and for Uniden were $n_2 = 40$, $\bar{x}_2 = 70.49$, and $\sigma_2^2 = 21.16$.

a. Is there any evidence to suggest that the population mean standby time is greater for the Motorola phone than for the Uniden phone? Use $\alpha = 0.01$.
b. Find the p value associated with this hypothesis test.

10.18 Medicine and Clinical Studies The time it takes for general anesthesia to work (time to induction) is an important consideration during an emergency and for scheduled surgeries. Recently, a study compared the mean induction time of similar drugs administered via inhalation and intravenously. Independent random samples of patients requiring general anesthesia were obtained, and the induction times (in minutes) were measured. The data are given on the data CD and book's web site. Assume the variance in induction time for inhalation administration is 0.0625 and for intravenous administration is 0.1225. Is there any evidence to suggest that the mean time to induction is less for intravenous administration than for inhalation administration? Use $\alpha = 0.05$.

Extended Applications

10.19 Public Health and Nutrition Magnesium is used by every cell in your body, is required for over 300 biochemical reactions, and helps muscles and nerves function properly. According to the USDA National Nutritional Database, half a cup of vegetarian baked beans and 1 medium baked potato without the skin contain the same amount of magnesium (40 milligrams). In order to check this claim, independent random samples of baked beans and potatoes were obtained, and the amount of magnesium in each serving was recorded (in milligrams). The summary statistics and known variances are given in the following table.

Food	Sample size	Sample mean	Population variance
Vegetarian baked beans (1)	18	39.58	2.47
Medium potato (2)	18	40.12	0.87

a. Assume the underlying distributions are normal. Is there any evidence to refute the claim? Use $\alpha = 0.01$.
b. Suppose that instead the sample sizes are $n_1 = n_2 = 38$. Now, is there any evidence to refute the claim? Find the p value for this hypothesis test.
c. How large would the sample sizes $(n_1 = n_2)$ have to be in order for the hypothesis test to be significant at the $\alpha = 0.01$ level?

10.20 Physical Sciences The manufacturer of a Kenmore residential stove can order parts from two different suppliers, The Repair Clinic and The Parts Pros. The small burners, or elements, are designed to produce 7.4 kW at 240 volts. In order to decide which supplier to use, independent random samples from each supplier were obtained, and each element's output (in kW) was carefully measured. For The Repair Clinic: $n_1 = 12$, $\bar{x}_1 = 7.361$, and $\sigma_1^2 = 0.81$; for The Parts Pros: $n_1 = 15$, $\bar{x}_2 = 7.307$, and $\sigma_2^2 = 0.64$. Assume the underlying distributions are normal and use $\alpha = 0.05$ for the following.
a. Is there any evidence to suggest that the population mean output of elements from The Repair Clinic is different from 7.4?
b. Is there any evidence to suggest that the population mean output of elements from The Parts Pros is different from 7.4?
c. Is there any evidence to suggest that μ_1 is different from μ_2?
d. Using the results from parts (a), (b), and (c), which supplier should the manufacturer use?

10.21 Economics and Finance Wages for plumbers in the United States may vary dramatically, depending on the location, union versus nonunion, and experienced versus inexperienced. Independent random samples of plumbers from two cities were obtained, and each hourly wage (in dollars) was recorded. The summary statistics are given in the following table.

City	Sample size	Sample mean	Population variance
Utica, New York	60	29.21	392.04
Atlanta, Georgia	75	33.90	243.36

a. Is there any evidence to suggest that the mean hourly wage for a plumber in Utica is less than the mean hourly wage for a plumber in Atlanta? Use $\alpha = 0.01$.
b. Find the p value associated with this hypothesis test.
c. The two sample means, 29.21 and 33.90, seem far apart. What factor(s) contribute to an insignificant hypothesis test result?

Challenge

10.22 Sample Size Calculation Suppose that a $100(1 - \alpha)\%$ confidence interval is needed for the difference in two population means, $\mu_1 - \mu_2$. In addition, suppose that the underlying populations are normal, the population variances σ_1^2 and σ_2^2 are known, and the sample sizes are equal, $n_1 = n_2 = n$.
a. Find an expression for the sample size necessary (from each population) in order for the resulting confidence interval to have a bound on the error of estimation B (half the width of the confidence interval).
b. How large a sample size is necessary if $\sigma_1 = 12.7$, $\sigma_2 = 9.5$, $B = 5$, and the confidence level is 95%?
c. Use the sample size in part (b) with $\bar{x}_1 = 57.3$ and $\bar{x}_2 = 48.6$ to construct a 95% confidence interval for $\mu_1 - \mu_2$. Compute the exact bound on the error of estimation. How does this compare with $B = 5$?

10.2 COMPARING TWO POPULATION MEANS USING INDEPENDENT SAMPLES FROM NORMAL POPULATIONS

In Section 10.1, the hypothesis tests concerning the difference between two population means (or for comparing two population means) were based on the standard normal, or Z, distribution. These tests are valid *only* if both population variances are known (and with normality and/or large samples, and independent random samples). It is unrealistic to assume that the population variances are known. As in Chapter 9, we will assume that the underlying populations are normal. But there is one additional assumption necessary in order to construct a similar *two-sample t* test.

Suppose that

For reference, these are the two-sample *t* test assumptions.

1. $\bar{X}_1$ is the mean of a random sample of size n_1 from a normal population with mean μ_1,
2. $\bar{X}_2$ is the mean of a random sample of size n_2 from a normal population with mean μ_2,
3. The samples are independent, and
4. The two population variances are *unknown* but *equal*. The common variance is denoted $\sigma^2 (= \sigma_1^2 = \sigma_2^2)$.

A test of equality of population variances will be discussed in Section 10.5.

The last assumption is new and implies that we are comparing populations with the same variability. If we do not assume equal variances, there is no *nice* test procedure. More on this later.

PROPERTIES OF $\overline{X}_1 - \overline{X}_2$

If the two-sample t test assumptions are true, then the estimator $\overline{X}_1 - \overline{X}_2$ has the following properties.

1. $E(\overline{X}_1 - \overline{X}_2) = \mu_{\overline{X}_1 - \overline{X}_2} = \mu_1 - \mu_2$
   $\overline{X}_1 - \overline{X}_2$ is still an unbiased estimator of the parameter $\mu_1 - \mu_2$.

2. $\text{Var}(\overline{X}_1 - \overline{X}_2) = \sigma^2_{\overline{X}_1 - \overline{X}_2} = \dfrac{\sigma_1^2}{n_1} + \dfrac{\sigma_2^2}{n_2} = \dfrac{\sigma^2}{n_1} + \dfrac{\sigma^2}{n_2} = \sigma^2\left(\dfrac{1}{n_1} + \dfrac{1}{n_2}\right)$

   and the standard deviation is $\sigma_{\overline{X}_1 - \overline{X}_2} = \sqrt{\sigma^2\left(\dfrac{1}{n_1} + \dfrac{1}{n_2}\right)}$.

3. Since both underlying populations are normal, the distribution of $\overline{X}_1 - \overline{X}_2$ is also normal.

In the previous section, we used the known population variances, standardized, and constructed a test based on the Z distribution. Here, an estimate of the common variance σ^2 is necessary. The appropriate *standardization* results in a t distribution.

S_1^2 and S_2^2 are separate estimators for the common variance, but using only one of these means ignoring additional, useful information. Since σ^2 is the variance for both underlying populations, an estimator for this common variance should depend on both samples. However, it also seems reasonable for the estimator to rely more on the larger sample. Therefore an estimate of the common variance uses both S_1^2 and S_2^2 in a *weighted average*.

DEFINITION

The **pooled estimator** for the common variance σ^2, denoted S_p^2, is

$$S_p^2 = \frac{(n_1 - 1)S_1^2 + (n_2 - 1)S_2^2}{n_1 + n_2 - 2} \tag{10.3}$$

$$= \left(\frac{n_1 - 1}{n_1 + n_2 - 2}\right)S_1^2 + \left(\frac{n_2 - 1}{n_1 + n_2 - 2}\right)S_2^2.$$

The pooled estimator for the common standard deviation σ is $S_p = \sqrt{S_p^2}$.

💡 **ILLUMINATING THE CONCEPTS**

λ is the lowercase Greek letter lambda and represents a constant.

1. S_p^2 is indeed a weighted average. This estimator can be written in the form

$$S_p^2 = \lambda S_1^2 + (1 - \lambda)S_2^2 \quad \text{where} \quad 0 \le \lambda \le 1.$$

If $n_1 = n_2$, then $\lambda = 1/2$ and $S_p^2 = (1/2)S_1^2 + (1/2)S_2^2$. If $n_1 \ne n_2$, then more *weight* is given to the larger sample.

2. The constants in Equation 10.3 are related to degrees of freedom. S_1^2 contributes $n_1 - 1$ degrees of freedom and S_2^2 contributes $n_2 - 1$ degrees of freedom. Consequently, there are a total of $(n_1 - 1) + (n_2 - 1) = n_1 + n_2 - 2$ degrees of freedom associated with the estimator S_p^2. ∎

The hypothesis test procedure is based on the following theorem.

THEOREM

If the two-sample t test assumptions are true, then the random variable

$$T = \frac{(\overline{X}_1 - \overline{X}_2) - (\mu_1 - \mu_2)}{\sqrt{S_p^2\left(\dfrac{1}{n_1} + \dfrac{1}{n_2}\right)}}$$

has a t distribution with $n_1 + n_2 - 2$ degrees of freedom.

As in a two-sample Z test, the null and alternative hypotheses are stated in terms of the difference $\mu_1 - \mu_2$. The critical values are from the appropriate t distribution.

This is the template for a hypothesis test concerning two population means when variances are unknown but equal: a two-sample t test, with pooled variance.

HYPOTHESIS TESTS CONCERNING TWO POPULATION MEANS WHEN VARIANCES ARE UNKNOWN BUT EQUAL

Given the two-sample t test assumptions, a hypothesis test concerning two population means in terms of the difference in means $\mu_1 - \mu_2$, with significance level α, has the form:

$H_0: \mu_1 - \mu_2 = \Delta_0$

$H_a: \mu_1 - \mu_2 > \Delta_0, \quad \mu_1 - \mu_2 < \Delta_0, \quad$ or $\quad \mu_1 - \mu_2 \neq \Delta_0$

TS: $T = \dfrac{(\overline{X}_1 - \overline{X}_2) - \Delta_0}{\sqrt{S_p^2\left(\dfrac{1}{n_1} + \dfrac{1}{n_2}\right)}}$

RR: $T \geq t_\alpha, \quad T \leq -t_\alpha, \quad$ or $\quad |T| \geq t_{\alpha/2}, \quad$ df $= n_1 + n_2 - 2$

Example 10.4 Dissolved Metals   Water draining through abandoned mine tunnels collects minerals and bacteria. Some resulting streams can have a pH between 0 and 1, equivalent to battery acid. In addition, the concentration of dissolved metals can be dangerously high. Suppose the trace-metal concentrations in two streams in Montana were investigated. Independent random 1-liter samples of water were obtained from streams near each mine, and the amount of zinc (in micrograms per liter) in each sample was measured. The resulting summary statistics are given in the following table.

Location	Sample size	Sample mean	Sample variance
Jack Creek (1)	15	993.1	650.25
Cataract Creek (2)	17	968.6	1062.76

(*Source*: U.S. Geological Survey.)

Solution Trail

KEYWORDS

- Is there any evidence?
- Difference in population mean.
- Underlying distributions are normal, with equal variances.

↓

TRANSLATION

- Conduct a two-sided test to compare μ_1 and μ_2.
- Variances are unknown but assumed equal.

↓

CONCEPTS

Hypothesis test concerning two population means when variances are unknown but equal.

↓

VISION

Use the template for this hypothesis test. The samples are random and independent, the underlying distributions are normal, and the population variances are unknown but assumed equal. Use the two-sided alternative hypothesis and the corresponding rejection region, find the pooled estimate of the common variance, compute the value of the test statistic, and draw a conclusion.

a. Is there any evidence to suggest that there is a difference in the population mean amount of zinc per liter between the Jack Creek and Cataract Creek regions? (A difference may affect the cleanup schedule.) Use $\alpha = 0.05$ and assume the underlying distributions are normal, with equal variances.
b. Find bounds on the p value associated with this hypothesis test.

SOLUTION

STEP 1 Let the Jack Creek water be population 1 and the Cataract Creek water be population 2.

The null hypothesis is that the two population means are equal, with the same zinc level: $\mu_1 = \mu_2 \Rightarrow \mu_1 - \mu_2 = 0 \, (= \Delta_0)$.

The summary statistics are given, and the samples were obtained independently. The population variances are unknown but assumed equal. A two-sample t test is relevant.

Since we are looking for *any* difference in population means, this is a two-sided test.

STEP 2 The four parts of the hypothesis test are

$H_0: \mu_1 - \mu_2 = 0$

$H_a: \mu_1 - \mu_2 \neq 0$

$$\text{TS: } T = \frac{(\overline{X}_1 - \overline{X}_2) - 0}{\sqrt{S_p^2 \left(\dfrac{1}{n_1} + \dfrac{1}{n_2} \right)}}$$

RR: $|T| \geq t_{\alpha/2} = t_{0.025} = 2.0423$ df = 15 + 17 − 2 = 30

STEP 3 The pooled estimate of the common population variance is

$$s_p^2 = \frac{(n_1 - 1)s_1^2 + (n_2 - 1)s_2^2}{n_1 + n_2 - 2} = \frac{(14)(650.25) + (16)(1062.76)}{30}$$

$$= 870.2553$$

The value of the test statistic is

$$t = \frac{(\overline{x}_1 - \overline{x}_2) - 0}{\sqrt{s_p^2 \left(\dfrac{1}{n_1} + \dfrac{1}{n_2} \right)}} = \frac{993.1 - 968.6}{\sqrt{(870.2553)\left(\dfrac{1}{15} + \dfrac{1}{17} \right)}} = 2.3444 \,\, (\geq 2.0423)$$

The value of the test statistic, $t = 2.3444$, lies in the rejection region, hence we reject the null hypothesis at the $\alpha = 0.05$ significance level. There is evidence to suggest that the mean amount of zinc per liter is different in the water near Jack Creek and Cataract Creek.

STEP 4 Recall, due to the nature of the table of critical values for t distributions, we can only bound the p value.

$|t| = |2.3444| = 2.3444$

In Table V in the Appendix, row $n_1 + n_2 - 2 = 15 + 17 - 2 = 30$, place 2.3444 in the ordered list of critical values.

$2.0423 \leq 2.3444 \leq 2.4573$

$\quad\, t_{0.025} \quad\quad\, \leq 2.3444 \leq t_{0.01}$

Therefore, $\quad 0.01 \quad \leq \quad p/2 \quad \leq 0.025$

And, $\quad\quad\quad 0.02 \quad \leq \quad \boxed{p} \quad\,\, \leq 0.05$

p value illustration:

$p = 2P(T \geq 2.3444)$

$\quad = 0.0259 \leq 0.05 = \alpha$

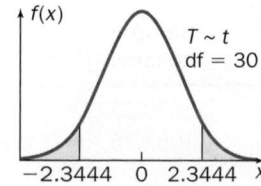

$T \sim t$
df = 30

$-2.3444 \quad 0 \quad 2.3444 \quad x$

Figures 10.12 through 10.14 together show a technology solution.

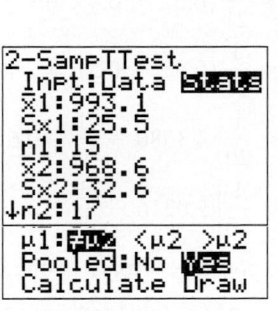

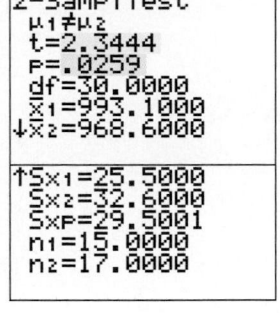

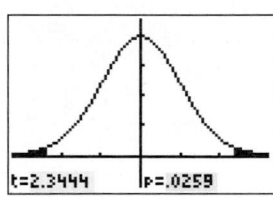

Figure 10.12
2-SampTTest input
screen.

Figure 10.13
Hypothesis test results.

Figure 10.14
2-SampTTest Draw
results.

Why do you suppose a small
significance level is important
here?

Solution Trail

KEYWORDS

- Is there any evidence?
- Smaller population mean
- Populations are normal, with
  equal variances.

↓

TRANSLATION

- Conduct a one-sided test to
  compare μ_1 and μ_2.
- Variances are unknown but
  assumed equal.

↓

CONCEPTS

Hypothesis test concerning two
population means when variances
are unknown but equal.

↓

VISION

Use the template for this hypothesis
test. The samples are random and
independent, the underlying
distributions are normal, and the
population variances are unknown
but assumed equal. Use a
one-sided alternative hypothesis
and the corresponding rejection
region, find the pooled estimate of
the common variance, compute the
value of the test statistic, and draw
a conclusion.

Example 10.5 Weight of Aluminum Cans

Aluminum cans are made from huge solid ingots pressed under high-pressure rollers and are cut like cookies from thin sheets. Aluminum is ideal for cans because it is lightweight, strong, and recyclable. A company claims that a new manufacturing process decreases the amount of aluminum needed to make a can, and therefore, decreases the weight. Independent random samples of aluminum cans made by old and new processes were obtained, and the weight (in ounces) of each is given in the following tables.

Old process (1)

0.52	0.49	0.47	0.47	0.48	0.52	0.55	0.49	0.52	0.50	0.50
0.50	0.51	0.51	0.50	0.53	0.49	0.51	0.52	0.51	0.51	

New process (2)

0.51	0.51	0.50	0.48	0.47	0.49	0.46	0.46	0.52	0.50	0.48
0.51	0.50	0.48	0.51	0.44	0.48	0.47	0.50	0.51	0.48	

Is there any evidence that the new-process aluminum cans have a smaller population mean weight? Assume the populations are normal, with equal variances, and use $\alpha = 0.01$.

SOLUTION

STEP 1 The null hypothesis is that the two mean weights are the same: $\mu_1 - \mu_2 = 0$. We are looking for evidence that the new-process cans have a smaller mean weight. The alternative hypothesis is $\mu_1 - \mu_2 > 0$.

The underlying populations are assumed normal with equal variances, and the samples were obtained independently. A two-sample t test is relevant.

STEP 2 The four parts of the hypothesis test are

$$H_0: \mu_1 - \mu_2 = 0$$
$$H_a: \mu_1 - \mu_2 > 0$$
$$\text{TS: } T = \frac{(\overline{X}_1 - \overline{X}_2) - 0}{\sqrt{S_p^2\left(\frac{1}{n_1} + \frac{1}{n_2}\right)}}$$

RR: $T \geq t_\alpha = t_{0.01} = 2.4233$ df = 21 + 21 − 2 = 40

STEP 3 The summary statistics are

$$\bar{x}_1 = \frac{1}{21}(0.52 + 0.49 + \cdots + 0.51) = 0.5048$$

$$\bar{x}_2 = \frac{1}{21}(0.51 + 0.51 + \cdots + 0.48) = 0.4886$$

$$s_1^2 = \frac{1}{20}\left[5.3580 - \frac{1}{21}(10.6)^2\right] = 0.0003762$$

$$s_2^2 = \frac{1}{20}\left[5.0216 - \frac{1}{21}(10.26)^2\right] = 0.0004429$$

The pooled estimate of the common population variances is

$$s_p^2 = \frac{(20)(0.0003762) + (20)(0.0004429)}{40} = 0.0004095$$

The value of the test statistic is

$$t = \frac{(\bar{x}_1 - \bar{x}_2) - 0}{\sqrt{s_p^2\left(\dfrac{1}{n_1} + \dfrac{1}{n_2}\right)}} = \frac{0.5048 - 0.4886}{\sqrt{(0.0004095)\left(\dfrac{1}{21} + \dfrac{1}{21}\right)}} = 2.5941$$

STEP 4 The value of the test statistic lies in the rejection region ($t = 2.5941 \geq 2.4233$). We reject the null hypothesis at the $\alpha = 0.01$ significance level. There is evidence to suggest that new-process aluminum cans have a smaller mean weight.

Figure 10.5 shows a technology solution.

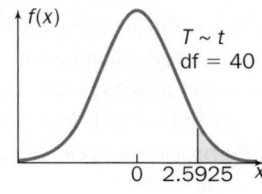

p value illustration:
$p = P(T \geq 2.5941)$
  $= 0.0066 \leq 0.01 = \alpha$

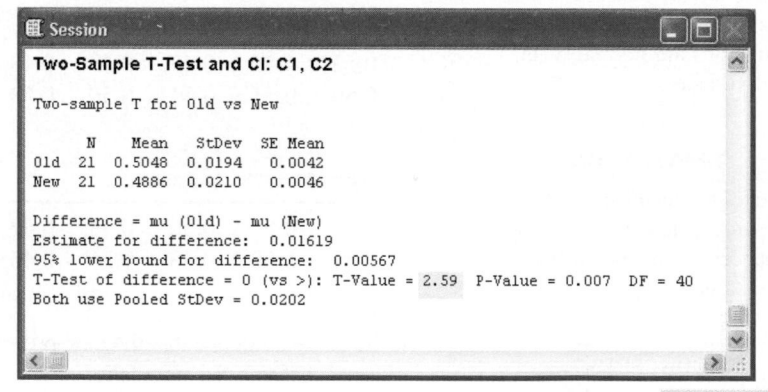

Figure 10.15 Hypothesis test (and confidence interval) results.

This methodology has been used several times, beginning in Chapter 8.

Using the assumptions presented in this section and the technique presented in Section 10.1, a confidence interval for $\mu_1 - \mu_2$ can be derived. Start with a symmetric interval about 0 such that the probability T lies in this interval is $1 - \alpha$. Manipulate the inequality to *sandwich* the parameter $\mu_1 - \mu_2$.

HOW TO FIND A $100(1 - \alpha)$% CONFIDENCE INTERVAL FOR $\mu_1 - \mu_2$ WHEN VARIANCES ARE UNKNOWN BUT EQUAL

Given the two-sample t test assumptions, a $100(1 - \alpha)$% confidence interval for $\mu_1 - \mu_2$ has as endpoints the values

$$(\bar{x}_1 - \bar{x}_2) \pm t_{\alpha/2}\sqrt{s_p^2\left(\frac{1}{n_1} + \frac{1}{n_2}\right)} \qquad (10.4)$$

with the critical value based on df $= n_1 + n_2 - 2$

Example 10.6 Dry Density of Wood   There are many different properties of wood, for example, thermal conductivity, vapor permeance, and air permeability. The dry density of wood is often used as a measure of shear strength. In a recent study, the strengths of plywood and oriented strand board (OSB) were compared. Independent random samples of each type of wood were obtained, and the dry density of each piece (in kg/m³) was measured. The summary statistics are given in the following table.

Wood type	Sample size	Sample mean	Sample standard deviation
Plywood (1)	12	581.8	5.56
OSB (2)	15	605.2	7.23

Assume the populations are normal and the variances are equal. Find a 99% confidence interval for the difference in population mean dry densities.[3]

SOLUTION

STEP 1   The summary statistics are given, the underlying distributions are assumed normal, and the population variances are assumed equal. Equation 10.4 can be used to construct a confidence interval for the difference $\mu_1 - \mu_2$.

$$1 - \alpha = 0.99 \quad \Rightarrow \quad \alpha = 0.01 \quad \Rightarrow \quad \alpha/2 = 0.005 \qquad \text{Find } \alpha/2.$$

$$t_{\alpha/2} = t_{0.005} = 2.7874 \qquad \text{Find the } t \text{ critical value, df} = 12 + 15 - 2 = 25.$$

STEP 2   Find the pooled estimate of the common variance.

$$s_p^2 = \frac{(n_1 - 1)s_1^2 + (n_2 - 1)s_2^2}{n_1 + n_2 - 2} = \frac{(11)(5.56)^2 + (14)(7.23)^2}{25} = 42.8748$$

STEP 3   Use Equation 10.4.

$$(\bar{x}_1 - \bar{x}_2) \pm t_{\alpha/2} \cdot \sqrt{s_p^2 \left(\frac{1}{n_1} + \frac{1}{n_2} \right)} \qquad \text{Equation 10.4.}$$

$$= (581.8 - 605.2) \pm (2.7874)\sqrt{(42.8748)\left(\frac{1}{12} + \frac{1}{15} \right)}$$

Use summary statistics and critical values.

$$= -23.4 \pm 7.0688 \qquad \text{Simplify.}$$

$$= (-30.4688, -16.3312) \qquad \text{Compute endpoints.}$$

$(-30.4688, -16.3312)$ is a 99% confidence interval for the difference (in kg/m³) in population mean dry densities, $\mu_1 - \mu_2$. Note that since 0 is not included in, or captured by, this interval, there is evidence to suggest that the mean dry densities are different. And since both endpoints of the CI are negative, this implies that the mean for OSB is larger than the mean for plywood.

Figures 10.16 and 10.17 together show a technology solution.

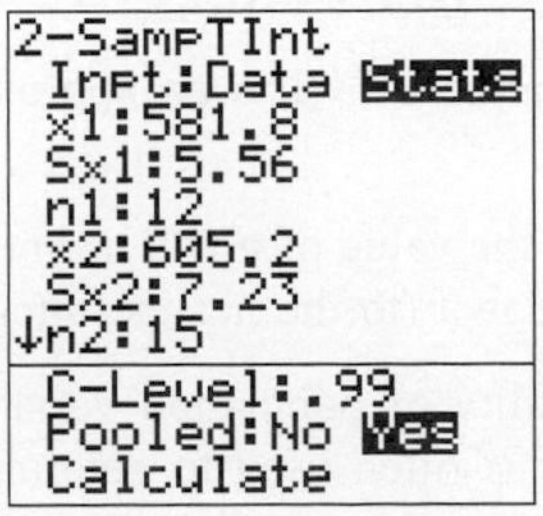

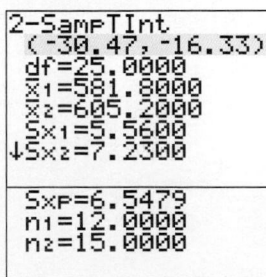

Figure 10.16
2-SampTTnt input screen.

Figure 10.17
Resulting 99% confidence interval.

The underlying distributions might only be approximately normal, or the population variances might not be exactly the same.

The hypothesis test procedure and the confidence interval formula presented in this section are *robust*. That is, if the assumptions aren't entirely true, the hypothesis test and the confidence interval are still very reliable. And even if the population variances are very different, as long as the underlying populations are normal and $n_1 = n_2$, the results are also very reliable.

If the underlying populations are normal, the population variances are unequal, and the sample sizes are different, there is no *nice* test procedure concerning $\mu_1 - \mu_2$ (or confidence interval for $\mu_1 - \mu_2$). It is reasonable to use each sample variance as an approximation for the corresponding population variance. However, the resulting logical standardization produces only an *approximate* test statistic. If the sample sizes are small and the underlying populations are not normal, then a nonparametric test must be used.

Nice means a reasonable *standardization* to produce a common random variable.

This is the template for an *approximate* two-sample *t* test.

HYPOTHESIS TESTS AND CONFIDENCE INTERVAL CONCERNING TWO POPULATION MEANS WHEN VARIANCES ARE UNKNOWN AND UNEQUAL

Given the *modified* two-sample *t* test assumptions (population variances unknown and assumed unequal), an *approximate* hypothesis test concerning two population means in terms of the difference, $\mu_1 - \mu_2$, with significance level α, has the following form.

$H_0: \mu_1 - \mu_2 = \Delta_0$

$H_a: \mu_1 - \mu_2 > \Delta_0, \quad \mu_1 - \mu_2 < \Delta_0, \quad \text{or} \quad \mu_1 - \mu_2 \neq \Delta_0$

$$\text{TS: } T' = \frac{(\overline{X}_1 - \overline{X}_2) - \Delta_0}{\sqrt{\dfrac{S_1^2}{n_1} + \dfrac{S_2^2}{n_2}}}$$

$\text{RR: } T' \geq t_\alpha, \quad T' \leq -t_\alpha, \quad \text{or} \quad |T'| \geq t_{\alpha/2}, \quad \text{df} = \nu$

The formula for ν is the Satterthwaite approximation for the number of degrees of freedom.

$$\text{where} \quad \nu \approx \frac{\left(\dfrac{s_1^2}{n_1} + \dfrac{s_2^2}{n_2}\right)^2}{\dfrac{(s_1^2/n_1)^2}{n_1 - 1} + \dfrac{(s_2^2/n_2)^2}{n_2 - 1}}$$

An approximate $100(1 - \alpha)\%$ confidence interval for $\mu_1 - \mu_2$, has as endpoints the values

$$(\overline{x}_1 - \overline{x}_2) \pm t_{\alpha/2} \sqrt{\frac{s_1^2}{n_1} + \frac{s_2^2}{n_2}}. \tag{10.5}$$

ILLUMINATING THE CONCEPTS

1. The random variable T' has an approximate *t* distribution with ν degrees of freedom.

2. It is likely that the value of ν will *not* be an integer. In order to be conservative, always round down (to the nearest integer).

3. A test for equality of population variances is presented in Section 10.5. This hypothesis test is often used to determine whether it is reasonable to assume equal population variances. ∎

(Long Ha/iStockphoto)

Example 10.7 Poker Chip Weights Clay-composite poker chips used in Las Vegas and Atlantic City weigh between 8.5 and 10 grams each, and last between three and six years. In a recent study of poker-chip weights, a casino obtained independent random samples of $100 and $500 chips. The weight of each chip (in grams) is given in the following tables.

$100 chips (1)

9.17	9.21	9.25	9.29	9.16	9.08	9.39	9.23	9.15	9.14
9.34	9.26	9.08	9.11						

$500 chips (2)

9.37	9.98	9.04	8.74	9.58	9.45	9.08	9.96	9.69

Is there any evidence to suggest that the population mean weight of $100 chips is different from that of $500 chips? Assume both populations are normal, and use $\alpha = 0.05$.

SOLUTION

STEP 1 The null hypothesis is that the two mean weights are the same, and the alternative is two-sided. The underlying populations are assumed normal and the samples were obtained independently. However, there is no assumption of equal variances. The approximate two-sample t test is appropriate.

STEP 2 The summary statistics are

$$\bar{x}_1 = \frac{1}{14}(9.17 + 9.21 + \cdots + 9.11) = 9.2043$$

$$\bar{x}_2 = \frac{1}{9}(9.37 + 9.98 + \cdots + 9.69) = 9.4322$$

$$s_1^2 = \frac{1}{13}\left[1186.1804 - \frac{1}{14}(128.86)^2\right] = 0.008934$$

$$s_2^2 = \frac{1}{8}\left[802.1295 - \frac{1}{9}(84.89)^2\right] = 0.1785$$

The approximate degrees of freedom are

$$\nu \approx \frac{\left(\dfrac{0.008934}{14} + \dfrac{0.1785}{9}\right)^2}{\dfrac{(0.008934/14)^2}{13} + \dfrac{(0.1785/9)^2}{8}} = 8.5177$$

We round ν down to 8.

STEP 3 The four parts of the hypothesis test are

$H_0\colon \mu_1 - \mu_2 = 0$
$H_a\colon \mu_1 - \mu_2 \neq 0$
$$\text{TS: } T' = \frac{(\bar{X}_1 - \bar{X}_2) - 0}{\sqrt{\dfrac{S_1^2}{n_1} + \dfrac{S_2^2}{n_2}}}$$
RR: $|T'| \geq t_{\alpha/2} = t_{0.025} = 2.3060$ df = 8

Solution Trail

KEYWORDS

- Is there any evidence?
- Population mean is different.
- Both populations are normal.

↓

TRANSLATION

- Conduct a two-sided test to compare μ_1 and μ_2.
- Variances are unknown and assumed unequal.

↓

CONCEPTS

Hypothesis test concerning two population means when variances are unknown and unequal.

↓

VISION

Use the template for this hypothesis test. The samples are random and independent, the underlying distributions are normal, and the population variances are unknown and unequal. Use the two-sided alternative hypothesis and the corresponding rejection region, find the value of the test statistic, and draw a conclusion.

p value illustration:
$$p = 2P(T \geq 1.5928)$$
$$= 0.1475 > 0.05 = \alpha$$

STEP 4   The value of the test statistic is

$$t' = \frac{(\bar{x}_1 - \bar{x}_2) - 0}{\sqrt{\dfrac{s_1^2}{n_1} + \dfrac{s_2^2}{n_2}}} = \frac{9.2043 - 9.4322}{\sqrt{\dfrac{0.008934}{14} + \dfrac{0.1785}{9}}} = -1.5928$$

The value of the test statistic does not lie in the rejection region. We do not reject the null hypothesis at the $\alpha = 0.05$ significance level. There is no evidence to suggest that the mean weight of \$100 chips is different from the mean weight of \$500 chips.

Figure 10.18 shows a technology solution.

D	E	F
t-Test: Two-Sample Assuming Unequal Variances		
	Variable 1	Variable 2
Mean	9.2043	9.4322
Variance	0.0089	0.1785
Observations	14	9
Hypothesized Mean Difference	0	
df	9	
t Stat	-1.5930	
P(T<=t) one-tail	0.0728	
t Critical one-tail	1.8331	
P(T<=t) two-tail	0.1456	
t Critical two-tail	2.2622	

Figure 10.18 Hypothesis test results.

TECHNOLOGY CORNER

Procedure: Hypothesis tests and confidence intervals concerning two population means when the population variances are unknown.
Reconsider: Example 10.5, page 461, solution, and interpretation.

TI-84 Plus

Use the built-in functions 2-SampTTest and 2-SampTInt. Input is either summary statistics or data in lists.

1. Enter the old process data into list L1 and the new process data into list L2.
2. Select STAT; TESTS; 2-SampTTest. Highlight Data. Enter List1, List2, and set each frequency to 1. Highlight the alternative hypothesis and Yes for Pooled. See Figure 10.19.
3. Highlight Calculate and press ENTER. The results are displayed on the Home Screen. See Figure 10.20.
4. The Draw results are shown in Figure 10.21.
5. Use the function STATS; TESTS; 2-SampTInt to construct a confidence interval for the difference of two population means.

Minitab

Use the built-in function 2-Sample t to conduct a hypothesis test and to construct a confidence interval. Input is either data in one or two columns (data in one column requires a subscript, or group identifying, column) or summarized data.

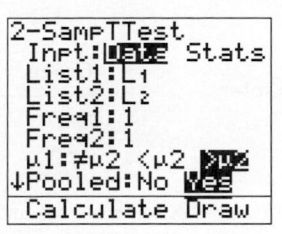

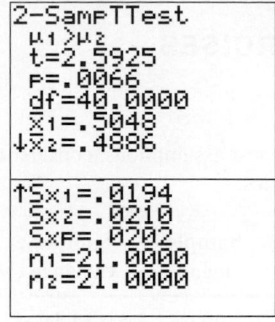

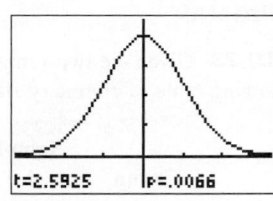

Figure 10.19
2-SampTTest input
screen.

Figure 10.20
Hypothesis test results.

Figure 10.21
2-SampTTest Draw
results.

1. Enter the old process data into column C1 and the new process data into column C2.
2. Select Stat; Basic Statistics; 2-Sample t.
3. Choose Samples in different columns and enter C1 in the First input window and C2 in the Second input window. Check the Assume equal variances box.
4. Choose the Options button. Enter a Confidence level, the hypothesized Test difference, and choose the appropriate Alternative.
5. The hypothesis test results and confidence interval are displayed in a session window. See Figure 10.15, page 462.

Excel

The Data Analysis toolkit contains two functions for comparing population means, assuming equal variances and assuming unequal variances. Use the appropriate formula and ordinary spreadsheet calculations to find the endpoints of a confidence interval.

1. Enter the old process data into column A and the new process data into column B.
2. Under the Data tab, select Data Analysis; t-Test: Two-Sample Assuming Equal Variances.
3. Enter the Variable 1 Range, Variable 2 Range, and the Hypothesized Mean Difference. Choose an Output option and click OK.
4. Summary statistics along with the value of the test statistic, critical values, and p values are displayed. See Figure 10.22.

D	E	F
t-Test: Two-Sample Assuming Equal Variances		
	Variable 1	Variable 2
Mean	0.5048	0.4886
Variance	0.0004	0.0004
Observations	21	21
Pooled Variance	0.0004	
Hypothesized Mean Difference	0	
df	40	
t Stat	2.5925	
P(T<=t) one-tail	0.0066	
t Critical one-tail	1.6839	
P(T<=t) two-tail	0.0132	
t Critical two-tail	2.0211	

Figure 10.22 Hypothesis test results.

SECTION 10.2 EXERCISES

Practice

10.23 Given the two-sample t test assumptions, consider the following table of summary statistics.

Group	Sample size	Sample mean	Sample variance
One	14	49.6	134.56
Two	16	50.2	243.36

a. Conduct a hypothesis test of $H_0: \mu_1 - \mu_2 = 0$ versus H_a: $\mu_1 - \mu_2 < 0$. Use $\alpha = 0.05$.
b. Find bounds on the p value associated with this test.

10.24 Given the two-sample t test assumptions, consider the following table of summary statistics.

Group	Sample size	Sample mean	Sample standard deviation
One	10	156.5	26.5
Two	11	132.6	21.5

a. Conduct a hypothesis test of $H_0: \mu_1 - \mu_2 = 0$ versus H_a: $\mu_1 - \mu_2 > 0$. Use $\alpha = 0.01$.
b. Find bounds on the p value associated with this test.

10.25 Given the two-sample t test assumptions, consider the independent random samples on the data CD and book's web site from two different populations.
a. Conduct a hypothesis test of $H_0: \mu_1 - \mu_2 = 0$ versus H_a: $\mu_1 - \mu_2 \neq 0$. Use $\alpha = 0.05$.
b. Find bounds on the p value associated with this test.

10.26 Given the two-sample t test assumptions, consider the following table of summary statistics.

Group	Sample size	Sample mean	Sample standard deviation
One	23	49.03	9.24
Two	23	49.57	8.15

a. Find a 95% confidence interval for the difference in population means, $\mu_1 - \mu_2$.
b. Using the confidence interval in part (a), is there any evidence to suggest that the two population means are different? Justify your answer.

10.27 In each of the following problems, n_1, n_2, s_1, and s_2 are given. Assume normal underlying distributions, independent random samples, and unknown, *unequal* variances. Find the approximate number of degrees of freedom, ν, in the critical value of an approximate two-sample t test.
a. $n_1 = 12$, $n_2 = 15$, $s_1 = 11.7$, $s_2 = 16.7$
b. $n_1 = 8$, $n_2 = 23$, $s_1 = 5.46$, $s_2 = 6.78$
c. $n_1 = 18$, $n_2 = 26$, $s_1 = 57.8$, $s_2 = 49.9$
d. $n_1 = 32$, $n_2 = 34$, $s_1 = 5.51$, $s_2 = 5.03$

10.28 Consider the following table of summary statistics.

Group	Sample size	Sample mean	Sample variance
One	8	173.9	320.41
Two	9	150.3	655.36

Assume normal underlying distributions, independent random samples, and unknown, unequal variances.
a. Conduct a hypothesis test of $H_0: \mu_1 - \mu_2 = 0$ versus H_a: $\mu_1 - \mu_2 > 0$. Use $\alpha = 0.05$.
b. Find bounds on the p value associated with this test.

10.29 Consider the following table of summary statistics.

Group	Sample size	Sample mean	Sample standard deviation
One	7	76.83	3.30
Two	16	66.80	14.00

Assume the underlying distributions are normal and the random samples were obtained independently.
a. Suppose that the population variances are assumed equal. Conduct a hypothesis test of $H_0: \mu_1 - \mu_2 = 0$ versus H_a: $\mu_1 - \mu_2 \neq 0$. Use $\alpha = 0.05$.
b. Suppose that the population variances are assumed unequal. Conduct a hypothesis test of $H_0: \mu_1 - \mu_2 = 0$ versus H_a: $\mu_1 - \mu_2 \neq 0$. Use $\alpha = 0.05$.
c. Which of the two tests do you think is more appropriate here? Justify your answer.

10.30 Independent random samples from two normal populations were obtained. The summary statistics were $n_1 = 17$, $\bar{x}_1 = 32.3$, $s_1 = 12.9$, $n_2 = 19$, $\bar{x}_2 = 43.8$, and $s_2 = 14.9$.
a. Assume the population variances are unequal. Find a 99% confidence interval for the difference in population means, $\mu_1 - \mu_2$.
b. Using the confidence interval in part (a), is there any evidence to suggest that the two population means are different? Justify your answer.

Applications

10.31 Fuel Consumption and Cars The durability and flatness of the front rotors on an automobile are important for braking and for a smooth ride. The flatness of a rotor can be determined by a special optical measuring device which measures the largest deviation from perfect flatness in micro inches. Suppose independent random samples of rotors from two different manufacturers were obtained. The largest deviation from perfect flatness of each rotor was measured, and the resulting summary statistics are given in the following table.

Manufacturer	Sample size	Sample mean	Sample standard deviation
Tire Rack	11	26.74	8.31
JC Whitney	14	29.53	6.85

Assume the underlying distributions are normal and the population variances are equal.

a. Is there any evidence to suggest that there is a difference in population mean deviations from perfect flatness for these two rotor brands? Use $\alpha = 0.05$.

b. Find bounds on the p value associated with this hypothesis test.

10.32 Manufacturing and Product Development The mean weight of an ordinary key is an important consideration, as most Americans carry a pocketful of keys. A manufacturer claims that a new process produces a lighter and more durable key. Independent random samples of both types of keys were obtained, and each key was carefully weighed and its weight (in ounces) was recorded. The resulting summary statistics are given in the following table, with the sample means and sample variances.

Key type	Sample size	Sample mean	Sample variance
Old process	10	0.321	0.0137
New process	10	0.199	0.0202

Assume the underlying distributions are normal and the population variances are equal.

a. Is there any evidence that the population mean weight of a new-process key is less than the population mean weight of an old-process key? Use $\alpha = 0.05$.

b. Find bounds on the p value associated with this hypothesis test.

10.33 Technology and Internet Record companies and artists/songwriters claim that sharing MP3 files via the Internet is a violation of copyright laws. A few computer users have been prosecuted, while some companies have started to sell these song files. In a recent study, the size of MP3 files was compared for two different genres—rap and jazz. Independent random samples of MP3 files from these two genres were obtained, and the file size (in kb) for each is given on the data CD and book's web site. Is there any evidence to suggest that the population mean file size for jazz MP3 files is greater than for rap MP3 files? Use $\alpha = 0.05$, and assume that the underlying distributions are normal and the population variances are equal.

10.34 Physical Sciences The *Great American Structures* web site claims that the Stratosphere Tower in Las Vegas, Nevada, has the world's fastest elevators. Other sources suggest that the Yokohama Landmark Tower in Yokohama, Japan, has faster elevators. In order to compare the two sets of elevators, independent random samples of elevator speeds (in feet per minute) were obtained. The summary statistics are given in the following table.

Building	Sample size	Sample mean	Sample standard deviation
Stratosphere Tower	23	2576.4	236.8
Yokohama Landmark Tower	25	2417.9	265.5

Is there any evidence to suggest that the population mean elevator speeds are different? Use $\alpha = 0.05$, and assume the underlying distributions are normal, with equal variances.

10.35 Medicine and Clinical Studies A study was conducted to determine standard reference values for musculoskeletal ultrasonography in healthy adults. Independent random samples of men and women were obtained and the sagittal diameter (in mm) of the biceps tendon was measured in each subject. The resulting summary statistics are given in the following table.

Group	Sample size	Sample mean	Sample standard deviation
Women	54	2.5	0.49
Men	48	2.8	0.49

(*Source*: W. A. Schmidt et al., Standard reference values for musculoskeletal ultrasonography, *Annals of Rheumatic Diseases*, Vol. 63, 2004, pp. 988–994.)

Assume the underlying populations are normal, with equal variances.

a. Is there any evidence to suggest that the population mean sagittal diameter of women's biceps tendons is different from that of men's biceps tendons? Use $\alpha = 0.01$.

b. Construct a 95% confidence interval for the difference in population mean sagittal diameters, $\mu_1 - \mu_2$.

10.36 Manufacturing and Product Development A company that produces hospital furniture has two assembly lines dedicated to cutting and drilling wood for medical cabinets. Each computer-controlled process is designed to drill holes in a certain cabinet part with depth 12.7 mm. Independent random samples of drilled holes were obtained from the two assembly lines. The resulting hole depths (in mm) are given on the data CD and book's web site. Assume the underlying populations are normal, with equal variances.

a. Is there any evidence to suggest that Line 2 is producing holes with a greater population mean depth than Line 1? Use $\alpha = 0.05$.

b. Find bounds on the p value associated with this hypothesis test.

10.37 Marketing and Consumer Behavior In order to help retailers plan ahead, each year a study is conducted to determine how much people intend to spend on holiday gifts. In a November 2008 survey, a sample of shoppers was obtained and they were asked to estimate their planned holiday spending (in dollars). The sample mean anticipated holiday spending was reported by gender, age group, and income level.[4] Consider the summary statistics given in the following table.

Group	Sample size	Sample mean	Sample standard deviation
Men	27	784.00	37.50
Women	25	652.00	29.90

Historically, men report that they will spend more than women on holiday gifts. Using the data from 2008, is there any evidence to suggest that the mean amount men intend to spend is greater than the mean amount women intend to spend? Use $\alpha = 0.001$, and assume the populations are normal, with equal variances.

10.38 Manufacturing and Product Development A recent study was conducted in order to determine the curing efficiency (time to harden) of dental composites (resins for the restoration of damaged teeth) using two different types of lights. Independent random samples of lights were obtained and a certain composite was cured for 40 seconds. The depth of each cure (in mm) was measured using a penetrometer. The summary statistics for the Halogen light were $n_1 = 10$, $\bar{x}_1 = 5.35$, and $s_1 = 0.07$. The summary statistics for the LuxOMax light were $n_2 = 10$, $\bar{x}_2 = 3.90$, and $s_2 = 0.08$.[5] Assume the underlying populations are normal, with equal variances.

a. The maker of the Halogen light claims that they produce a larger cure depth after 40 seconds than LuxOMax lights. Is there any evidence to support this claim? Use $\alpha = 0.01$.

b. Construct a 99% confidence interval for the difference in population mean cure depths.

10.39 Manufacturing and Product Development Certain masonry ties used in residential construction receive a hot-dipped galvanized finish for strength and protection against moisture. Independent random samples of masonry ties from two competing companies were obtained. The amount of coating on one side of each tie was measured in g/m^2. The resulting summary statistics are given in the following table.

Company	Sample size	Sample mean	Sample variance
Fero	10	331.4	201.64
Cintex	18	298.7	1190.25

Assume the underlying populations are normal, with unequal variances

a. Managers at Fero claim that their product has a larger mean coating than Cintex. Is there any evidence to support this claim? Use $\alpha = 0.01$.

b. Find a 95% confidence interval for the difference in population mean coatings.

10.40 Sports and Leisure The curve in a hockey stick is measured by first placing the face of the blade against a flat surface. The maximum distance from the surface to the bottom of the blade cannot exceed 1/2 inch.[6] Independent random samples of hockey sticks used by players on the Toronto Maple Leafs and Montreal Canadiens teams were obtained. The curve in each stick was measured (in inches), and the resulting data are summarized in the following table.

Team	Sample size	Sample mean	Sample standard deviation
Toronto	10	0.261	0.122
Montreal	20	0.325	0.051

Assume the underlying distributions are normal, with unequal variances. Is there any evidence to suggest that the mean curve in Toronto sticks is different from the mean curve in Montreal sticks? Use $\alpha = 0.001$.

10.41 Manufacturing and Product Development The tear strength, tensile strength, backing, and thickness all contribute to the durability of vinyl wallpaper. A new company (Aries Wallcoverings)

claims to sell the thickest vinyl wallpaper of any currently on the market. Independent random samples of Aries wallpaper and all others were obtained. The thickness of each wallpaper (in inches) is given on the data CD and book's web site. Assume the underlying distributions are normal, with unequal variances. Is there any evidence to support the claim made by Aries Wallcoverings? Use $\alpha = 0.01$.

10.42 Business and Management Contract negotiations between a labor union and management can be contentious and long affairs. The length of negotiations depends upon the labor/management relationship, contract issues, and economic conditions. Suppose independent random samples of contract negotiations involving airline carriers were obtained, and the duration of each (in months) was recorded. The summary statistics are given in the following table.

Carrier	Sample size	Sample mean	Sample standard deviation
American	10	15.7	13.1
Delta	4	15.2	2.9

(*Source*: Bureau of Labor Statistics.)

If the populations are assumed normal and the variances are assumed unequal, is there any evidence to suggest that the mean duration of negotiations for these two airlines is different? Use $\alpha = 0.05$.

10.43 Marketing and Consumer Behavior The mean amount spent on gift cards during the 2005 holiday season was projected to be $88.03 per consumer.[7] Suppose independent random samples of consumers on the East and West coasts were obtained, and the amount spent on gift cards (in dollars) during 2005 was recorded for each. The data are given on the data CD and book's web site. Assume the underlying populations are normal and the population variances are equal. Is there any evidence to suggest that the mean amount spent on gift cards per consumer is different on the East coast and the West coast? Use $\alpha = 0.01$.

10.44 Demographics and Population Statistics A random sample of workers in Santa Clara, CA and San Mateo, CA was obtained and the weekly wage of each person was recorded. The data are given on the data CD and book's web site.[8] Assume the underlying populations are normal and the population variances are equal.

a. Is there any evidence to suggest that the mean weekly wage is different in the two counties? Use $\alpha = 0.05$.

b. Find bounds on the p value for the hypothesis test in part (a).

10.45 Medicine and Clinical Studies It has been reported that human antitoxin is an effective treatment for babies with botulism. Babies with this disease are infected with one of two toxins, type A or B, that destroy nerve endings. These babies usually remain in intensive care for several weeks until their nerves regenerate. However, human antitoxin has been shown to decrease the length of hospital stay and the total cost of treatment. A random sample of babies with botulism was obtained, and each was treated with human antitoxin. The number of days in the hospital for each child, by toxin, is given on the data CD and book's web site.[9] Assume the underlying populations are normal and the population variances are equal. Is there any evidence to suggest that the mean number of days in the hospital is different for type A toxin and type B toxin? Use $\alpha = 0.05$ and justify your answer.

Extended Applications

10.46 Physical Sciences Tinted residential windows have become popular because they help a home absorb solar energy, keep out harmful ultraviolet rays, and add privacy. Two independent random samples of tinted windows were obtained, each produced by applying a thin film of a specified color and density. The shading coefficient of each tinted window (a unitless quantity) was measured, and the summary statistics are given in the following table.

Tinted window	Sample size	Sample mean	Sample standard deviation
Silver	8	0.601	0.113
Neutral	11	0.741	0.077

Assume the underlying populations are normal and the population variances are equal.
a. Is there any evidence to suggest that the population mean shading coefficients are different? Use $\alpha = 0.01$.
b. Construct a 99% confidence interval for the difference in population mean shading coefficients, $\mu_1 - \mu_2$.
c. Use the confidence interval in part (b) to determine whether there is any evidence to suggest that the shading coefficients are different. Does your answer agree with part (a)? If so, why? If not, why not?

10.47 Economics and Finance The Bureau of Engraving and Printing produces \$1, \$5, \$10, \$20, \$50, and \$100 dollar bills. The \$2 banknote is still legal tender, but is no longer in production. Each bill is designed to have the same width, but many people perceive larger denomination bills to be larger in size. Independent random samples of newly minted \$1 and \$20 dollar bills were obtained, and the width of each (in mm) was recorded. The summary statistics are given in the following table.

Bill	Sample size	Sample mean	Sample variance
\$1	23	66.5990	0.0132
\$20	24	66.6924	0.0057

Assume the underlying distributions are normal.
a. If the population variances are assumed equal, is there any evidence to suggest that the mean width of a \$20 bill is greater than the mean length of a \$1 bill? Use $\alpha = 0.01$.
b. If the population variances are assumed unequal, is there any evidence to suggest that the mean length of a \$20 bill is greater than the mean length of a \$1 bill? Use $\alpha = 0.01$.
c. Why do both tests lead to the same conclusion (with very similar p values)?

10.48 Medicine and Clinical Studies The cost for prescription drugs in the United States can vary dramatically depending on the pharmacy and the state. Thirty 10-mg tablets of Lipitor were obtained at pharmacies in New Jersey and Florida. The price for each purchase is given on the data CD and book's web site.[10] Assume the underlying populations are normal.

a. Compute the summary statistics for each group. Do you think the assumption of equal variances is reasonable? Why or why not?
b. Based on your answer to part (a), conduct the appropriate hypothesis test to determine whether there is any evidence that the mean price for Lipitor is higher in New Jersey than in Florida. Use $\alpha = 0.01$.
c. Find bounds on the p value associated with the hypothesis test in part (b).

10.49 Physical Sciences A pressure relief valve (PRV) is installed on a residential hot-water heater in order to protect against overheating and, of course, high pressure. Independent random samples of PRVs from different companies were obtained. Each valve was tested by recording the pressure (in psi) required to cause the valve to open. The summary statistics are given in the following table.

Company	Sample size	Sample mean	Sample variance
Delta	30	147.6	7.09
Gamma	35	147.8	13.70

Assume the underlying populations are normal, with unequal variances.
a. Is there any evidence to suggest that the mean pressure required to open each valve is different? Use $\alpha = 0.05$.
b. Find a 95% confidence interval for the difference in population mean pressure required to open each valve. Is this confidence interval consistent with the results in part (a)? Explain.

10.50 Physical Sciences The lifetime of a fuel rod in a commercial light water nuclear reactor is related to the internal pressure. Typically, a fuel rod lasts for 36 months, and one-third of all fuel rods are replaced each year during a plant shutdown. A new fuel rod includes a gas relief capsule and is designed to last longer. Independent random samples of the two types of fuel rods were obtained, and the lifetime of each (in months) was recorded. The summary statistics are given in the following table.

Fuel-rod design	Sample size	Sample mean	Sample standard deviation
Old	11	34.91	3.20
New	11	39.55	3.55

Assume the underlying populations are normal and the variances are equal.
a. Conduct the relevant hypothesis test to determine whether the new fuel rod does last longer. Use $\alpha = 0.01$.
b. Construct a 99% confidence interval for the difference in population mean lifetimes. Does this confidence interval support the hypothesis test conclusion in part (a)? Explain.

Challenge

10.51 Robust Statistics The two-sample t test for comparing population means when the variances are equal is a robust statistical procedure. If the population variances are different, as long as the underlying populations are normal and the sample sizes are equal, then the hypothesis test is still very reliable.

Generate a random sample of size 25 from a normal distribution with mean $\mu_1 = 100$ and standard deviation $\sigma_1 = 5$. Generate a second random sample of size 25 from a normal distribution with mean $\mu_2 = 100$ and standard deviation $\sigma_2 = 5$. Conduct a two-sided, two-sample t test for comparing population means assuming the population variances are equal and with $\alpha = 0.05$. Do this 100 times and record the number of times you reject the null hypothesis.

Repeat the same procedure but use $\sigma_2 = 7$. Record the number of times you reject the null hypothesis. Repeat the same procedure for $\sigma_2 = 10, 15, 20, 25, 30, 50$. Use your results to explain the robust nature of this hypothesis test.

10.3 PAIRED DATA

When we compared population means in the previous two sections, we made the necessary assumption that the samples were obtained independently. The n_1 observations from the first population and the n_2 observations from the second population were unrelated. There are, however, many experiments that involve only n individuals, or objects, in which two observations are made of each individual. A classic example involves a diet-and-exercise program designed to help people lose weight. A random sample of n individuals is selected and each is weighed. Each person follows the regimen for a specified time period and is weighed again at the end of the experiment. There are two observations of each individual, a *before* weight and an *after* weight. The data are used to determine whether the diet-and-exercise program is effective.

Trial-and-error learning or memory experiments in animals present another good example. In a typical psychology experiment, a random sample of rats is obtained and each is timed as it maneuvers through a maze. After several weeks of training, each rat is timed again. This produces two observations of each animal, a *before* time and an *after* time. This experiment might be designed to determine whether animals can *learn* the correct path through a maze and hence decrease the mean time needed to traverse the course.

The difference between the experiments described above and those in the previous two sections is that here, the *paired* observations are *dependent*. We are still interested in comparing population means (the *before* mean μ_1 and the *after* mean μ_2), and, therefore, are still interested in the difference $\mu_1 - \mu_2$. However, the sample means, $\overline{X}_1$ and $\overline{X}_2$, are *not* independent. The standardizations used previously are not applicable, because the variance of $\overline{X}_1 - \overline{X}_2$ is more complicated. Therefore, another method is necessary, one that addresses the dependence and yet still considers the difference $\mu_1 - \mu_2$.

The variance of $\overline{X}_1 - \overline{X}_2$ must account for the dependence.

Suppose that

For reference, these are the two-sample paired t test assumptions.

1. There are n individuals or objects, or n pairs of individuals or objects, that are related in an important way or share a common characteristic, and

2. There are two observations of each *individual*. The population of first observations is normal, and the population of second observations is also normal.

Even though the word *individual* is used, this means *individual or object*.

◀ Let X_1 represent a randomly selected first observation and let X_2 represent the corresponding second observation on the same individual. Consider the random variable $D = X_1 - X_2$, the difference in the observations and the n observed differences $d_i = (x_1)_i - (x_2)_i$, $i = 1, 2, \ldots, n$. Since X_1 and X_2 are both normal, D is also normal. And more importantly, the differences are independent. A hypothesis test concerning $\mu_1 - \mu_2$ is based on the sample mean of the differences $\overline{D}$. This random variable has the following properties.

PROPERTIES OF $\overline{D}$

1. $\mathrm{E}(\overline{D}) = \mu_1 - \mu_2$
   $\overline{D}$ is an unbiased estimator for the difference in means $\mu_1 - \mu_2$.
2. The variance of $\overline{D}$ is unknown, but can be estimated using the sample variance of the differences.
3. Since both underlying populations are normal, D is normal, and hence, $\overline{D}$ is also normal. ▶

Here's what all of these results mean for us. To compare population means, μ_1 and μ_2, when the data are paired, we focus on the difference $\mu_1 - \mu_2$. As in earlier two-sample tests, the null hypothesis H_0: $\mu_1 = \mu_2$ is equivalent to H_0: $\mu_1 - \mu_2 = 0$. A test to determine whether the underlying population means of two paired samples are equal is equivalent to a test to determine whether the population mean of the paired differences is zero. We compute the differences, $d_1, d_2, \ldots, d_n$, and conduct a one-sample t test (with $n - 1$ degrees of freedom) using the differences.

HYPOTHESIS TESTS CONCERNING TWO POPULATION MEANS WHEN DATA ARE PAIRED

Given the two-sample paired t test assumptions, a hypothesis test concerning the two population means in terms of the difference $\mu_D = \mu_1 - \mu_2$, with significance level α, has the following form.

H_0: $\mu_D = \mu_1 - \mu_2 = \Delta_0$

H_a: $\mu_D > \Delta_0$, $\mu_D < \Delta_0$,   or $\mu_D \neq \Delta_0$

TS: $T = \dfrac{\overline{D} - \Delta_0}{S_D/\sqrt{n}}$

where S_D is the sample standard deviation of the differences.

RR: $T \geq t_\alpha$, $T \leq -t_\alpha$,   or $|T| \geq t_{\alpha/2}$,   df $= n - 1$

This is the template for a hypothesis test concerning two population means when data are paired: a paired t test.

💡 ILLUMINATING THE CONCEPTS

1. Δ_0 is the hypothesized difference in the population means. Usually $\Delta_0 = 0$: the null hypothesis is that the two population means are equal. However, Δ_0 may be nonzero. For example, the null hypothesis H_0: $\mu_D = \mu_1 - \mu_2 = 5 = \Delta_0$ specifies that the difference in population means is 5.

2. A paired t test is valid even if the underlying population variances are unequal, that is, even if $\sigma_1^2 \neq \sigma_2^2$. The sample variance of the differences, S_D^2, is a good estimator of $\mathrm{Var}(\overline{X}_1 - \overline{X}_2)$ when the observations are paired.

3. If a paired t test is appropriate, the test statistic is based on $n - 1$ degrees of freedom. A two-sample t test (incorrect here) would be based on a test statistic with $n + n - 2 = 2n - 2$ degrees of freedom. Therefore, the correct analysis is based on a distribution with greater variability and is more conservative. ∎

The common characteristic is patient identity.

Example 10.8 Relaxing Music While there is no direct scientific measure of stress, some physical properties of the body that are believed to be related to stress include pulse rate, blood pressure, breathing rate, brain waves, muscle tension, skin resistance, and body temperature. Some researchers claim that music can be relaxing and, therefore, can reduce stress.[11] Twelve patients who claim to be suffering from job-related stress were selected at random. An initial resting pulse rate (in beats per minute, bpm) was obtained, and each person participated in a month-long, music-listening, relaxation-therapy program. A final resting pulse rate was taken at the end of the experiment. The data are given in the following tables.

Subject	1	2	3	4	5	6
Initial pulse rate	67	71	67	83	70	75
Final pulse rate	61	72	70	76	58	61
Difference	6	−1	−3	7	12	14

Subject	7	8	9	10	11	12
Initial pulse rate	71	68	72	88	78	70
Final pulse rate	74	59	61	64	71	77
Difference	−3	9	11	24	7	−7

Is there any evidence to suggest that the music-listening, relaxation-therapy program reduced the mean pulse rate and, therefore, the stress level? Assume the underlying distributions of initial and final pulse rate are normal, and use $\alpha = 0.05$.

SOLUTION

STEP 1 Traditionally, the *before* measurements are population 1 and the *after* measurements are population 2.

The null hypothesis is that the two population means are equal (i.e., the music-listening relaxation-therapy program has no effect): $\mu_1 = \mu_2 \Rightarrow \mu_1 - \mu_2 = \mu_D = 0\ (=\Delta_0)$.

Each population is assumed normal, and there are two observations on each individual. A paired t test is appropriate.

Since the therapy program is designed to reduce stress, the alternative hypothesis is $\mu_1 > \mu_2 \Rightarrow \mu_1 - \mu_2 = \mu_D > 0$. This is a one-sided, right-tailed test.

STEP 2 The four parts of the hypothesis test are

$H_0: \mu_D = 0$

$H_a: \mu_D > 0$

TS: $T = \dfrac{\overline{D} - \Delta_0}{S_D/\sqrt{n}}$

RR: $T \geq t_\alpha = t_{0.05} = 1.7959$ df = 11

STEP 3 In anticipation of a paired t test, the differences are given in the preceding table.

The sample mean of the differences is

$$\bar{d} = \frac{1}{12}[6 + (-1) + \cdots + (-7)] = 6.3333$$

Solution Trail

KEYWORDS

- Initial pulse rate; final pulse rate.
- Is there any evidence?
- Reduced the mean.

↓

TRANSLATION

- *Before* and *after* measurements on the same individual.
- Conduct a one-sided, right-tailed test to compare the *before* and *after* mean pulse rates.

↓

CONCEPTS

Hypothesis test concerning two population means when data are paired.

↓

VISION

The data are certainly paired—there is a *before* and an *after* measurement on each individual—and each population is assumed normal. Compute the differences, use a one-sided alternative hypothesis and the corresponding rejection region, find the value of the test statistic, and draw a conclusion.

The sample variance of the differences is

$$s_D^2 = \frac{1}{11}\left[1320 - \frac{1}{12}(76)^2\right] = 76.2424$$

The sample standard deviation of the differences is

$$s_D = \sqrt{76.2424} = 8.7317$$

The value of the test statistic is

$$t = \frac{\bar{d} - 0}{s_D/\sqrt{n}} = \frac{6.3333}{8.7317/\sqrt{12}} = 2.5126 \;\; (\geq 1.7959)$$

Try to find bounds for the *p* value associated with this hypothesis test.

STEP 4 The value of the test statistic, $t = 2.5126$, lies in the rejection region; thus, we reject the null hypothesis at the $\alpha = 0.05$ significance level. There is evidence to suggest that the music-listening, relaxation-therapy program does reduce a person's resting pulse rate (and therefore the stress level).

Figures 10.23 through 10.26 together show a technology solution.

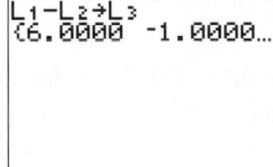

Figure 10.23 Compute the differences.

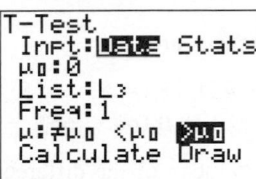

Figure 10.24 T-Test input screen.

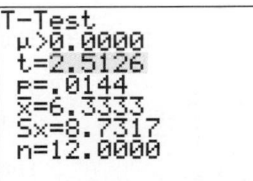

Figure 10.25 Hypothesis test results.

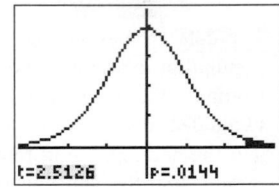

Figure 10.26 T-Test Draw results. ●

Example 10.9 Rapid-Fire Cash Registers A Stop-N-Go convenience store sells milk, noncarbonated drinks, breads, ice cream, and snack foods in high volume. Customers appreciate the convenience and discounted prices but often wait in long checkout lines. The manager of the store is considering new cash registers—rapid-fire point-of-sale registers—with keys designated for the most popular items. Cashiers can then tap a single key per item rather than two or three for the item price. This may improve accuracy and decrease checkout time or could cause more errors and actually increase checkout time. A random sample of seven typical purchases of store items was assembled. Each shopping bag of items was totaled by a cashier working with an old machine and then by a cashier working with a new machine. The times (in seconds) are given in the following table.

The common characteristic is the cashier.

Shopping bag	1	2	3	4	5	6	7
Old cash register	45	83	62	65	39	66	62
New cash register	42	55	45	44	17	66	69

Is there any evidence to suggest that the new cash register changes the mean checkout time? Assume the underlying populations are normal, use $\alpha = 0.01$, and find bounds on the *p* value associated with this hypothesis test.

Even though the new register was faster in five out of seven cases, there is insufficient evidence to reject the null hypothesis at the $\alpha = 0.01$ level. This is because the standard deviation of the differences is large and n is small.

p value illustration:

$p = 2P(T \geq 2.4023)$

$= \boxed{0.0531} > 0.01 = \alpha$

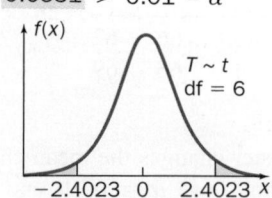

SOLUTION

STEP 1 Let population 1 be the checkout times associated with the old cash registers, and let population 2 be the checkout times using the new cash registers.

The null hypothesis is that the two population means are equal and the two different cash registers result in the same mean checkout time:

$$\mu_1 = \mu_2 \Rightarrow \mu_1 - \mu_2 = \mu_D = 0 \ (= \Delta_0).$$

Each population is assumed normal and there are two observations on each shopping bag of items. A paired t test is appropriate.

We are looking for any difference in the mean checkout times. Therefore, the alternative hypothesis is $\mu_1 \neq \mu_2 \Rightarrow \mu_1 - \mu_2 = \mu_D \neq 0$. This is a two-sided test.

STEP 2 The four parts of the hypothesis test are

$H_0: \mu_D = 0$

$H_a: \mu_D \neq 0$

TS: $T = \dfrac{\overline{D} - \Delta_0}{S_D/\sqrt{n}}$

RR: $|T| \geq t_{\alpha/2} = t_{0.005} = 3.7074$ df = 6

STEP 3 The differences are

$3 \ (= 45 - 42), 28 \ (= 83 - 55), 17 \ (= 62 - 45), 21 \ (= 65 - 44),$
$22 \ (= 39 - 17), 0 \ (66 - 66), -7 \ (= 62 - 69)$

The sample mean of the differences is

$$\overline{d} = \frac{1}{7}[3 + 28 + 17 + 21 + 22 + 0 + (-7)] = 12.00$$

The sample variance of the differences is

$$s_D^2 = \frac{1}{6}\left[2056 - \frac{1}{7}(84)^2\right] = 174.6667$$

The sample standard deviation is

$$s_D = \sqrt{174.6667} = 13.2162$$

The value of the test statistic is

$$t = \frac{\overline{d} - 0}{s_D/\sqrt{n}} = \frac{12.00}{13.2162/\sqrt{7}} = \boxed{2.4023}$$

STEP 4 The value of the test statistic, $t = 2.4023$, does not lie in the rejection region—thus, we do not reject the null hypothesis. There is no evidence to suggest that the new cash registers change the mean checkout time.

STEP 5 Using Table V in the Appendix, we can only bound the p value.

$|t| = |2.4023| = 2.4023$

In Table V, row $n - 1 = 7 - 1 = 6$, place 2.4023 in the ordered list of critical values.

$$1.9432 \leq 2.4023 \leq 2.4469$$
$$t_{0.05} \leq 2.4023 \leq t_{0.025}$$
$$0.025 \leq p/2 \leq 0.05$$

Therefore, $0.05 \leq \boxed{p} \leq 0.10$

Figure 10.27 shows a technology solution.

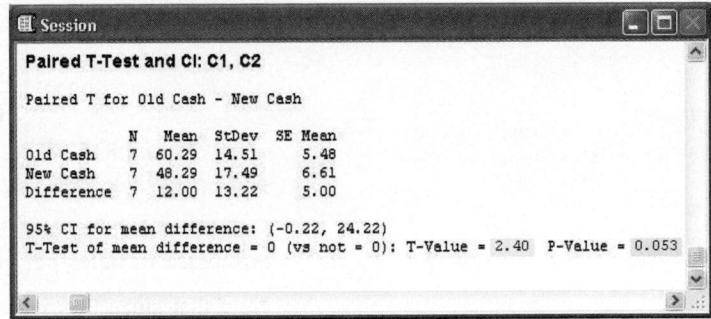

Figure 10.27 Hypothesis test (and confidence interval) results.

The random variable T here is

$$T = \frac{\overline{D} - \mu_D}{S_D/\sqrt{n}}$$

The usual technique can be used to construct a confidence interval for the difference in means, $\mu_D = \mu_1 - \mu_2$, when the observations are paired. Start with a symmetric interval about 0 such that the probability that T lies in this interval is $1 - \alpha$. Manipulate the inequality to sandwich μ_D.

HOW TO FIND A $100(1 - \alpha)\%$ CONFIDENCE INTERVAL FOR μ_D

Given the paired t test assumptions, a $100(1 - \alpha)\%$ confidence interval for μ_D has as endpoints the values

$$\overline{d} \pm t_{\alpha/2} \frac{s_D}{\sqrt{n}} \qquad (10.6)$$

with the critical value based on df $= n - 1$.

Example 10.10 Improved Mileage A local automotive repair shop advertises that their special maintenance package, including tire balancing, new spark plugs, engine oil additive, and a front-end alignment, will certainly improve gas mileage. In order to check this claim, 18 cars (and drivers) were randomly selected. Each car was driven on a specially designed route and the miles per gallon for each car was recorded. Then all the cars received the maintenance package. Each driver took the same route, and the miles per gallon were measured again. The summary statistics for the differences (before maintenance mpg − after maintenance mpg) were $\overline{d} = -1.28$, $s_D = 5.62$. Assuming normality, find a 99% confidence interval for the true difference in mean miles per gallon.

SOLUTION

STEP 1 The sample size and summary statistics are given, and the underlying distributions (before maintenance mpg and after maintenance mpg) are assumed normal.

Since the observations are paired, use Equation 10.6.

$1 - \alpha = 0.99 \Rightarrow \alpha = 0.01 \Rightarrow \alpha/2 = 0.005$ Find $\alpha/2$.

$t_{\alpha/2} = t_{0.005} = 2.8982$ Find the t critical value with df $= 17$.

STEP 2 Use Equation 10.6.

$$\bar{d} \pm t_{\alpha/2} \frac{s_D}{\sqrt{n}}$$

Equation 10.6.

$$= -1.28 \pm (2.8982) \frac{5.62}{\sqrt{18}}$$

Use summary statistics and critical value.

$$= -1.28 \pm 3.8391$$

Simplify.

$$= (-5.1191, 2.5591)$$

Compute endpoints.

$(-5.1191, 2.5591)$ is a 99% confidence interval for the true mean difference in miles per gallon, μ_D.

Note that since 0 is included in this confidence interval, there is no evidence to suggest that μ_D is different from 0, and there is no evidence to suggest that the maintenance program improves mileage.

Figures 10.28 and 10.29 together show a technology solution.

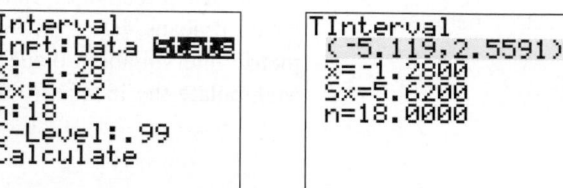

Figure 10.28 TInterval input screen.

Figure 10.29 Resulting confidence interval.

TECHNOLOGY CORNER

Procedure: Hypothesis tests and confidence intervals concerning paired data.
Reconsider: Example 10.9, page 475, solution, and interpretation.

TI-84 Plus

Compute the differences if necessary. Use the built-in functions T-Test and TInterval. Input is either summary statistics or data in lists.

1. Enter the Old cash register data into list L1 and the New cash register data into list L2.
2. Find the paired differences on the Home Screen, and store them in the list L3.
3. Select STAT; TESTS; T-Test. Highlight Data. Enter μ_0, the hypothesized difference in means and the list containing the paired differences. Set the frequency to 1 and highlight the alternative hypothesis. See Figure 10.30.
4. Highlight Calculate and press ENTER. The results are displayed on the Home Screen. See Figure 10.31.
5. Use STAT; TESTS; TInterval to construct a confidence interval for the paired differences. This procedure is described in the Technology Corner in Section 8.3. A 95% confidence interval for the paired differences in shown in Figure 10.32.

 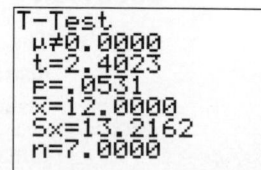 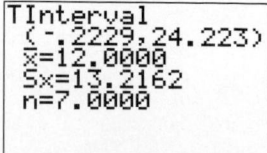

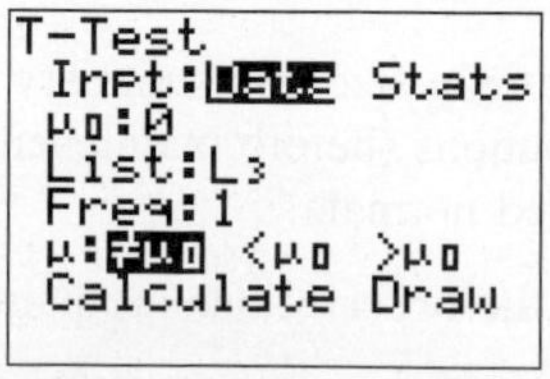

Figure 10.30 T-Test input screen.

Figure 10.31 Hypothesis test results.

Figure 10.32 A 95% confidence interval for the paired differences.

Minitab

Use the built-in function $\underline{P}$aired t to conduct a hypothesis test and to construct a confidence interval. Input is either paired data in two columns or summarized data for the paired differences.

1. Enter the Old cash register data into column C1 and the New cash register data into column C2.
2. Select $\underline{S}$tat; $\underline{B}$asic Statistics; $\underline{P}$aired t.
3. Choose Samples in columns and enter C1 in the First sample input window and C2 in the Second sample input window.
4. Choose the Options button. Enter a Confidence level, the Test Mean (Δ_0), and the Alternative hypothesis.
5. The hypothesis test results and the confidence interval are displayed in a session window. See Figure 10.27, page 477.

Excel

Use the Data Analysis toolkit function t-Test: Paired Two Sample for Means. If only summarized data for the paired differences is given, use the function TINV to find the critical value and compute the endpoints of the confidence interval (described in the Technology Corner in Section 8.3).

1. Enter the Old cash register data into column A and the New cash register data into column B.
2. Under the Data tab, select Data Analysis; t-Test: Paired Two Sample for Means.
3. Enter the Variable 1 Range (A1:A7), Variable 2 Range (B1:B7), and the Hypothesized Mean Difference (Δ_0).
4. Choose an Output option and click OK.
5. Summary statistics along with the value of the test statistic, critical values, and p values are displayed. See Figure 10.33.

D	E	F
t-Test: Paired Two Sample for Means		
	Variable 1	Variable 2
Mean	60.2857	48.2857
Variance	210.5714	305.9048
Observations	7	7
Pearson Correlation	0.6734	
Hypothesized Mean Difference	0	
df	6	
t Stat	2.4023	
P(T<=t) one-tail	0.0266	
t Critical one-tail	1.9432	
P(T<=t) two-tail	0.0531	
t Critical two-tail	2.4469	

Figure 10.33 Hypothesis test results.

SECTION 10.3 EXERCISES

Practice

10.52 In each experiment, determine whether the data are obtained independently or are paired. If the data are independent, indicate the two distinct populations. If the data are paired, indicate the common characteristic.

a. School Board members believe that adding a teacher's aide to each K–4 class will improve classroom management and increase instruction time. Twenty-six elementary classrooms were selected at random, and the daily instruction time for each was recorded. A teacher's aide was then added to each classroom, and the daily instruction time was recorded again. The data will be used to determine whether there is any evidence that adding a teacher's aide increases the mean daily instruction time.

b. A researcher investigating home insurance costs obtained a random sample of homes in the Northeast and another random sample of homes in the South. The yearly insurance cost for each home was recorded. The data will be used to determine whether there is any difference in the mean yearly home insurance costs between the Northeast and the South.

c. Officials at the transit authority of a large city would like to compare the route times during morning and evening rush hours. Eleven routes were selected at random. A morning and an evening route completion time were recorded for each. The data will be used to determine whether the mean evening route time is less than the mean morning route time.

d. A supplementary health insurance provider is investigating changes in claim patterns. A random sample of 32 policy holders was selected. The total amounts claimed in 2008 and in 2009 were recorded for each person. The data will be used to determine whether there is any evidence that the mean amount claimed has increased from 2008 to 2009.

e. Random samples of 45 new home sites in Kansas and 52 new home sites in upstate New York were selected. The *flatness coefficient* (a unitless quantity between 0 and 1) of each lot was measured. The data will be used to determine whether the mean flatness coefficient is less for new home sites in Kansas than in upstate New York.

10.53 In each experiment, determine whether the data are obtained independently or are paired. If the data are independent, indicate the two distinct populations. If the data are paired, indicate the common characteristic.

a. A surgeon is investigating the effect of physical therapy on patients who have had rotator cuff injuries. Twenty-eight patients were selected at random. The range of motion in the affected arm was measured in each patient prior to starting physical therapy. After three weeks of consistent therapy, the range of motion in each patient was measured again. The data will be used to determine whether physical therapy has increased the mean range of motion.

b. A woodwork manufacturer has several lathes used for shaving parts. The accuracy of each is determined by measuring the production width of a designed 1-mm wood strip. Nine lathes were selected at random. The accuracy of each was measured. Then the gib screws on each lathe were adjusted, and the width of another 1-mm wood strip was measured. The data will be used to determine if the mean width is different before and after adjustment.

c. A random sample of 60 twenty-year-old males and a random sample of 45 seventy-year-old males were obtained. The length of each person's right ear was measured (in inches). The data will be used to determine whether the mean ear length is greater in 70-years-old males than in 20-year-old males.

d. The manufacturer of a PDA allows the user to transfer information to a desktop computer via a USB port or a serial port. Ten different files on the PDA were selected at random. The times to transfer the file via a USB port and a serial port were recorded for each file. The data will be used to determine whether the mean time to transfer a file is different via USB and serial ports.

e. Random samples of 25 frequent flyers on United Airlines and on Delta Airlines were obtained. The total number of accumulated frequent-flyer miles for each person was recorded. The data will be used to determine whether the mean number of frequent-flyer miles is greater for United Airlines passengers than for Delta Airlines passengers.

10.54 The following summary statistics were obtained in a paired-data study: $\bar{d} = 15.68$, $s_D = 33.55$, and $n = 17$. Assume normality and conduct a test of H_0: $\mu_D = 0$ versus H_a: $\mu_D > 0$. Use $\alpha = 0.05$.

10.55 Consider the following paired data.

Subject	1	2	3	4	5
Before treatment	332.5	289.3	288.2	268.0	278.0
After treatment	317.3	302.5	312.9	325.4	267.3

a. Assume normality and conduct a test of H_0: $\mu_D = 0$ versus H_a: $\mu_D < 0$. Use $\alpha = 0.01$.

b. Find bounds on the p value associated with this hypothesis test.

10.56 Consider the paired data given on the data CD and book's web site.

a. Assume normality and conduct a test of H_0: $\mu_D = 0$ versus H_a: $\mu_D \neq 0$. Use $\alpha = 0.01$.

b. Find bounds on the p value associated with this hypothesis test.

Applications

10.57 Technology and Internet Twenty-one computer programmers from IT firms around the country were selected at random. Each was asked to write code in C++ and in Java for a specific application. The runtime (in seconds) for each program, by computer language, is given on the data CD and book's web site.

a. What is the common characteristic that makes these data paired?

b. Assume normality. Conduct the appropriate hypothesis test to determine whether there is any evidence that the mean runtime is greater for Java programs than for C++ programs. Use $\alpha = 0.001$.

c. Find bounds on the p value associated with this hypothesis test.

10.58 Physical Sciences A consultant working for a State Police barracks contends that service weapons will fire with a higher muzzle velocity if the barrel is properly cleaned. A random sample of Glock 9-mm handguns was obtained, and the muzzle velocity (in feet per second) of a single shot from each gun was measured. Each gun was professionally cleaned, and the muzzle velocity of a second shot (with the same bullet type) was measured. The data are given in the following table.

Gun	1	2	3	4	5	6
Before	1505	1419	1504	1494	1510	1506
After	1625	1511	1459	1441	1472	1521

a. What is the common characteristic that makes these data paired?

b. Assume normality. Conduct the appropriate hypothesis test to determine whether there is any evidence that a clean gun fires with a higher muzzle velocity. Use $\alpha = 0.01$.

c. Find bounds on the p value associated with this hypothesis test.

10.59 Sports and Leisure In 1969 major league pitching mounds were lowered from 15 to 10 inches. This decreased a pitcher's leverage and presumably the speed of a typical fastball. Fifteen major league pitchers were selected at random, and each threw his best fastball from a 15-inch mound and from a 10-inch mound. The speed of each pitch (in mph) is given on the data CD and book's web site.

a. Assume normality. Construct a 99% confidence interval for the true mean difference in fastball speeds from a 15-inch mound and a 10-inch mound.

b. Using the confidence interval in part (a), is there any evidence to suggest that pitching speed is, on average, faster from a higher mound? Justify your answer.

10.60 Public Health and Nutrition A filtration system made for small businesses is designed to remove particulate matter from the air. In order to test the new device, a random sample of businesses was obtained. The concentration of particulate matter in the air of their offices was measured (in $\mu g/m^3$). The filtration system was then allowed to run for 24 hours, and the concentration of particulate matter was measured again. The differences (before filtration − after filtration) are given on the data CD and book's web site. Assume normality. Is there any evidence to suggest that the new filtration system improves air quality by removing particulate matter? Use $\alpha = 0.05$.

10.61 Manufacturing and Product Development An autofocus feature is common on most digital cameras but can produce a very annoying shutter lag, the time between pressing the shutter release and the actual shutter opening. To reduce shutter lag, some manufacturers recommend using prefocus: partially depressing the shutter release to focus but without taking the picture. A random sample of 46 digital cameras was obtained, and a professional photographer agreed to help test the shutter lag. Each camera was used to take the same still photograph, once using full autofocus and once using prefocus. The shutter lag for each picture was measured in seconds. The summary statistics for the differences (autofocus − prefocus) were $\bar{d} = 0.1799$ and $s_D = 0.2096$.

a. Is there any evidence to suggest that using prefocus decreases the shutter lag? Assume normality and use $\alpha = 0.0001$.

b. Find bounds on the p value associated with this hypothesis test.

10.62 Physical Sciences It is important to maintain a low ammonia-ion concentration in freshwater aquariums to ensure healthy fish (and plants). An ammonia neutralizer is advertised to almost instantly detoxify ammonia (i.e., reduce the concentration of ammonia ions) in order to protect fish. Fourteen untreated 20-gallon aquariums were selected at random, and the ammonia-ion concentration (in ppm) in each was measured. One hour after the directed amount of the neutralizer was used, the ammonia-ion concentration was measured again. The resulting data are given on the data CD and book's web site. Assuming normality, is there any evidence to suggest that the neutralizer decreases the mean ammonia-ion concentration? Use $\alpha = 0.025$.

10.63 Public Health and Nutrition Beef boullion generally has a high salt content, which can cause health problems. A food columnist for a local newspaper suggested simmering boullion with slices of raw potato to remove salt. In order to check this claim, 10 different boullion brands were selected at random and the salt content in each was measured (in mg/cup of water). Five potato slices were then added to each broth and the mixtures were left to simmer for 15 minutes. Following this procedure, the salt content of the broths was measured again. The difference between the initial and the final salt content was computed for each boullion brand, and the data are given in the following table.

−169	−222	431	110	−168
353	−207	68	25	203

Assume normality. Is there evidence to suggest that simmering with raw potatoes decreases the mean salt content in beef boullion? Use $\alpha = 0.05$.

10.64 Technology and Internet Rendering three-dimensional graphics on a computer screen often takes a long time. High-performance graphics cards are designed specifically to speed up this process and to improve the quality of the image. An architectural firm decided to test two leading graphics cards before purchasing one. Two identical computers, except for the graphics cards, with the same software were used. The time (in seconds) to render 18 different, complex, three-dimensional objects was recorded for each graphics card. The data are given on the data CD and book's web site.

a. What is the common characteristic that makes this data paired?

b. Assume normality. Conduct the appropriate hypothesis test to determine whether there is any evidence to suggest a difference in the true mean times to render three-dimensional graphics. Use $\alpha = 0.01$.

10.65 Physical Sciences Every type of pager, cell phone, and personal security alarm emits electromagnetic energy. The amount of this radiation depends on the type of unit and the manufacturer. Eight pagers with transmitters were selected at random, and the electromagnetic energy emission from each was measured (in mW). The antenna in each pager was replaced with one designed to emit less radiation and the electromagnetic energy emission was measured again. The summary statistics for the differences in electromagnetic emission (old antenna − new antenna) were $\bar{d} = 0.213$ and $s_D = 1.271$.

a. Assume normality and conduct the appropriate hypothesis test to determine whether the new antenna reduces the true mean electromagnetic energy emission in pagers with transmitters. Use $\alpha = 0.05$.

b. Find bounds on the p value associated with this hypothesis test.

10.66 Manufacturing and Product Development The porosity of a concrete block is a measure (as a percentage) of the amount of empty space in the block. In residential homes with concrete-block foundations, a larger porosity leads to damper, colder basements. A contractor recommends pretreatment of concrete blocks with a product designed to decrease the porosity. A random sample of concrete blocks was obtained, and the porosity of each was measured. The clear, paint-like product was applied to each block, and the porosity was measured again. The differences (before treatment − after treatment) in porosities are given on the data CD and book's web site. Assume normality. Is there any evidence to suggest that the new product decreases the mean porosity of concrete blocks? Use $\alpha = 0.001$.

10.67 Psychology and Human Behavior Each employee hired at an electronics parts assembly line in Edmonton is given a general intelligence test. In order to determine which method of training is more effective, eight pairs of new hires were matched according to their exam scores. One set of employees were asked to read appropriate training manuals, while the other group watched interactive training videos. Each employee was then asked to assemble a part used in a locater beacon transmitter, and the time (in minutes) to completion was recorded. The data are given in the following tables.

Employee pair	1	2	3	4
Written manual	4.9	4.6	5.3	4.9
Interactive video	3.1	4.1	4.4	4.9

Employee pair	5	6	7	8
Written manual	4.9	5.4	5.5	5.0
Interactive video	3.6	3.9	6.5	5.3

a. What is the common characteristic that makes this data paired?

b. Is there any evidence to suggest that the true mean time difference, μ_D, is different from 0? Assume normality and use $\alpha = 0.05$.

10.68 Public Health and Nutrition Americans love hamburgers, but the high fat content in some cooked patties presents a severe health threat. Certain electric grills are designed to drain fat away from the patty, resulting in a healthier, although perhaps less tasty, meal. A random sample of ground beef packages with various fat contents was obtained. Two patties were made from each package. One was cooked in an electric grill, while the other was prepared in a frying pan on top of a stove. The fat content (as a percentage) in each cooked patty is given on the data CD and book's web site.

a. Conduct the appropriate hypothesis test to determine whether the true mean fat content in hamburgers cooked on an electric grill is less than the true mean fat content of hamburgers cooked in a frying pan. Assume normality and use $\alpha = 0.001$.

b. Find bounds on the p value associated with this hypothesis test.

Extended Applications

10.69 Medicine and Clinical Studies A new drug designed to reduce fever (and relieve aches and pains) is being tested for efficacy and side effects. Ten patients entering a hospital with a high fever were selected at random. The temperature (in °F) of each patient was measured, the drug was administered, and two hours later the temperature was measured again. The data are given in the following tables.

Patient	1	2	3	4	5
Before drug	102.6	99.2	102.3	101.1	102.7
After drug	99.8	98.8	97.5	100.3	99.6

Patient	6	7	8	9	10
Before drug	102.6	100.5	103.5	105.7	104.3
After drug	102.8	99.0	101.8	97.1	99.2

a. What is the common characteristic that makes these data paired?

b. Assume normality. Conduct the appropriate hypothesis test to determine whether there is any evidence that the new drug reduces the mean patient temperature after two hours. Use $\alpha = 0.05$.

c. Find bounds on the p value associated with this hypothesis test.

d. What characteristic of the differences suggests that a hypothesis test will be significant?

10.70 Fuel Consumption and Cars Biodiesel fuel has a cloud point, the temperature at which the fuel becomes cloudy, of approximately 13 °C. This clouding can lead to poor engine performance and can even cause an engine to stop completely. An industrial chemical company produces an additive designed to lower the cloud point of this type of fuel. A random sample of six different biodiesel fuels was obtained and the cloud point was measured for each. One ounce of the chemical additive was mixed in with every fuel sample and the cloud point was measured again. The resulting data are given in the following table (temperatures in °C).

Fuel	1	2	3	4	5	6
Before additive	11.7	12.9	14.2	12.7	11.3	12.4
After additive	10.3	10.7	14.1	10.0	11.2	12.1

a. Assume normality, and conduct the appropriate hypothesis test to determine whether the additive lowers the mean cloud point in biodiesel fuel. Use $\alpha = 0.05$.

b. Conduct an inappropriate two-sample t test to compare the population mean cloud point before treatment with the population mean cloud point after treatment. Assume the population variances are unequal and use $\alpha = 0.05$.

c. Compare the conclusions in parts (a) and (b). How are the test statistics the same, and how do they differ?

10.71 Manufacturing and Product Development A company makes packaging pellets from the starch in corn and potatoes and from biodegradable polymers. The resilience of a pellet is its ability to return to its original shape after a sustained force is applied for a fixed time. Random samples of old-formula and new-formula pellets were obtained. Various forces were applied to each pellet and the resilience was carefully measured (as a percentage). The data are given in the following table.

Force	1	2	3	4	5	6
Old formula	33.0	36.2	39.6	42.5	38.1	39.6
New formula	44.5	58.5	62.0	54.9	44.7	44.7

Force	7	8	9	10	11
Old formula	31.8	42.0	31.9	39.7	36.5
New formula	37.9	52.3	41.3	42.7	52.8

a. What is the common characteristic that makes this data paired?

b. Use a one-sided paired t test to determine whether new-formula pellets have a higher mean resilience. Assume normality and use $\alpha = 0.01$.

c. What characteristic of the differences suggests that the hypothesis test in part (b) will be significant?

10.4 COMPARING TWO POPULATION PROPORTIONS USING LARGE SAMPLES

The methods presented in this section can be used to compare two population proportions. For example, a social scientist may conduct an experiment to determine whether the true proportion of men who favor legalized gambling is the same as the true proportion of women. Or an advertising agency might be interested in comparing the true proportions of children who saw a certain television commercial in two different regions of the country.

Here is a quick review of the notation (presented at the beginning of this chapter) associated with populations 1 and 2 and samples 1 and 2.

Population proportion of successes:	$p_1, \ p_2$
Sample size:	$n_1, \ n_2$
Number of successes:	$x_1, \ x_2$
Corresponding random variables:	$X_1, \ X_2$
Sample proportion of successes:	$\hat{p}_1 = x_1/n_1, \ \hat{p}_2 = x_2/n_2$
Corresponding random variables:	$\hat{P}_1 = X_1/n_1, \ \hat{P}_2 = X_2/n_2$

The general null hypothesis is stated (as usual) in terms of a difference, H_0: $p_1 - p_2 = \Delta_0$. However, there are two cases to consider: (1) $\Delta_0 = 0$ and (2) $\Delta_0 \neq 0$. In both cases, a reasonable estimator for $p_1 - p_2$ is the difference in sample proportions, $\hat{P}_1 - \hat{P}_2$. The following properties are used to construct a hypothesis test (and confidence interval) concerning the difference of two population proportions.

PROPERTIES OF THE SAMPLING DISTRIBUTION OF $\hat{P}_1 - \hat{P}_2$

1. The mean of $\hat{P}_1 - \hat{P}_2$ is the true difference in population proportions $p_1 - p_2$. That is,

$$E(\hat{P}_1 - \hat{P}_2) = \mu_{\hat{P}_1 - \hat{P}_2} = p_1 - p_2.$$

2. The variance of $\hat{P}_1 - \hat{P}_2$ is

$$\mathrm{Var}(\hat{P}_1 - \hat{P}_2) = \sigma^2_{\hat{P}_1 - \hat{P}_2} = \frac{p_1(1 - p_1)}{n_1} + \frac{p_2(1 - p_2)}{n_2}.$$

The standard deviation of $\hat{P}_1 - \hat{P}_2$ is

$$\sigma_{\hat{P}_1 - \hat{P}_2} = \sqrt{\frac{p_1(1 - p_1)}{n_1} + \frac{p_2(1 - p_2)}{n_2}}.$$

3. If
  a. both n_1 and n_2 are large,
  b. $n_1 p_1 \geq 5$ and $n_1(1 - p_1) \geq 5$, and
  c. $n_2 p_2 \geq 5$ and $n_2(1 - p_2) \geq 5$, then the distribution of $\hat{P}_1 - \hat{P}_2$ is approximately normal.

In symbols, $\hat{P}_1 - \hat{P}_2 \stackrel{\cdot}{\sim} \mathrm{N}\left(p_1 - p_2, \dfrac{p_1(1 - p_1)}{n_1} + \dfrac{p_2(1 - p_2)}{n_2}\right).$

Items b and c are the nonskewness criteria.

The appropriate standardization will result in a Z distribution. The estimate of the standard deviation, $\sigma_{\hat{P}_1 - \hat{P}_2}$, is determined by the value of Δ_0.

Case 1: H_0: $p_1 - p_2 = 0$, or $p_1 = p_2$ ($\Delta_0 = 0$)

If this null hypothesis is true, there is one common value for the two population proportions, denoted p ($= p_1 = p_2$). The variance of $\hat{P}_1 - \hat{P}_2$ becomes

$$\sigma^2_{\hat{P}_1 - \hat{P}_2} = \frac{p(1-p)}{n_1} + \frac{p(1-p)}{n_2} = p(1-p)\left(\frac{1}{n_1} + \frac{1}{n_2}\right). \qquad (10.7)$$

Using the properties of $\hat{P}_1 - \hat{P}_2$, the random variable

$$Z = \frac{(\hat{P}_1 - \hat{P}_2) - 0}{\sqrt{p(1-p)\left(\dfrac{1}{n_1} + \dfrac{1}{n_2}\right)}} \qquad (10.8)$$

is approximately standard normal. As for the common variance in Section 10.2, an estimator for the *common* proportion, p, is obtained by using information from both samples.

The pooled or **combined estimate of the common population proportion** is

$$\hat{P}_c = \frac{X_1 + X_2}{n_1 + n_2} = \left(\frac{n_1}{n_1 + n_2}\right)\hat{P}_1 + \left(\frac{n_2}{n_1 + n_2}\right)\hat{P}_2. \qquad (10.9)$$

The general hypothesis test procedure is based on the standardization in Equation 10.8 with $\hat{p}_c$ as an estimate of p.

$\hat{P}_c$ is another weighted average.

This is the template for a large-sample hypothesis test concerning two population proportions when $\Delta_0 = 0$.

Again, use only one (appropriate) alternative hypothesis and the corresponding rejection region.

> ### HYPOTHESIS TESTS CONCERNING TWO POPULATION PROPORTIONS WHEN $\Delta_0 = 0$
>
> Given two random samples of sizes n_1 and n_2, a large-sample hypothesis test concerning two population proportions in terms of the difference $p_1 - p_2$ (with $\Delta_0 = 0$) with significance level α has the following form.
>
> H_0: $p_1 - p_2 = 0$
>
> H_a: $p_1 - p_2 > 0$, $p_1 - p_2 < 0$,   or $p_1 - p_2 \neq 0$
>
> TS: $Z = \dfrac{\hat{P}_1 - \hat{P}_2}{\sqrt{\hat{P}_c(1 - \hat{P}_c)\left(\dfrac{1}{n_1} + \dfrac{1}{n_2}\right)}}$
>
> RR: $Z \geq z_\alpha$, $Z \leq -z_\alpha$,   or $|Z| \geq z_{\alpha/2}$

There is no confidence interval for the difference in population proportions in this case. If we assume $p_1 = p_2$, then there is no reason to construct a confidence interval for the difference $p_1 - p_2 = 0$.

ILLUMINATING THE CONCEPTS

1. This test is valid as long as the nonskewness criteria hold for both samples. Use the estimates $\hat{p}_1$ and $\hat{p}_2$ to check the inequalities.

2. Just a reminder, the z critical values for this test are from the standard normal distribution.

3. Remember, we can also determine whether to reject or not to reject the null hypothesis by comparing the p value associated with the value of the test statistic to the significance level α. ∎

The following example illustrates this hypothesis test procedure.

Example 10.11 Daily Smokers Although smoking is banned in many public places in Europe, a high proportion of European adults (ages 15 and older) still smoke daily. In a random sample of 250 adults in Bulgaria, 92 were daily smokers, and in a random sample of 300 adults in Greece, 114 were daily smokers.[12] Is there any evidence to suggest that the true proportion of daily smokers is different in Bulgaria and in Greece? Use $\alpha = 0.05$.

SOLUTION

STEP 1 This is a two-sided test in which we are looking for *any* difference in population proportions. Therefore, $\Delta_0 = 0$, and case 1 is appropriate. Arbitrarily, let Bulgarian adults be population 1 and Greek adults be population 2. The given information is presented here:

	Bulgaria	Greece
Sample size	$n_1 = 250$	$n_2 = 300$
Number of successes	$x_1 = 92$	$x_2 = 114$
Sample proportion	$\hat{p}_1 = 92/250 = 0.3680$	$\hat{p}_2 = 114/300 = 0.3800$

STEP 2 Check the nonskewness criteria using estimates for p_1 and p_2.

$$n_1\hat{p}_1 = (250)(0.3680) = 92 \geq 5 \quad n_1(1 - \hat{p}_1) = (250)(0.6320) = 158 \geq 5$$
$$n_2\hat{p}_2 = (300)(0.3800) = 114 \geq 5 \quad n_2(1 - \hat{p}_2) = (300)(0.6200) = 186 \geq 5$$

Since all of the inequalities are satisfied, $\hat{P}_1 - \hat{P}_2$ is approximately normal, and the large-sample hypothesis test concerning population proportions can be used.

STEP 3 The four parts of the hypothesis test are

$$H_0: p_1 - p_2 = 0$$
$$H_a: p_1 - p_2 \neq 0$$

$$\text{TS}: Z = \frac{\hat{P}_1 - \hat{P}_2}{\sqrt{\hat{P}_c(1 - \hat{P}_c)\left(\frac{1}{n_1} + \frac{1}{n_2}\right)}}$$

$$\text{RR}: |Z| \geq z_{\alpha/2} = z_{0.025} = 1.9600$$

STEP 4 The estimate of the common population proportion is

$$\hat{p}_c = \frac{x_1 + x_2}{n_1 + n_2} = \frac{92 + 114}{250 + 300} = 0.3745$$

The value of the test statistic is

$$z = \frac{\hat{p}_1 - \hat{p}_2}{\sqrt{\hat{p}_c(1 - \hat{p}_c)\left(\frac{1}{n_1} + \frac{1}{n_2}\right)}} = \frac{0.3680 - 0.3800}{\sqrt{(0.3745)(0.6255)\left(\frac{1}{250} + \frac{1}{300}\right)}} = -0.2895$$

STEP 5 The value of the test statistic does *not* lie in the rejection region. We do not reject the null hypothesis. At the $\alpha = 0.05$ significance level, there is no evidence to suggest that the population proportion of adults who smoke daily in Bulgaria is different from the proportion in Greece.

p value illustration:
$$p = 2P(Z \leq -0.2895)$$
$$= 0.7722 > 0.05 = \alpha$$

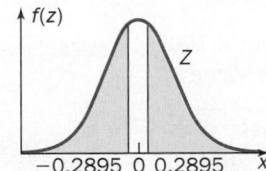

The technology solution indicates that the *p* value is 0.7722. Since $0.7722 > 0.05 = \alpha$, we cannot reject the null hypothesis.

Figures 10.34 through 10.36 together show a technology solution.

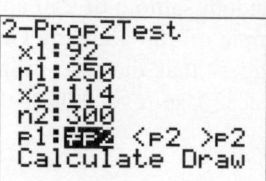

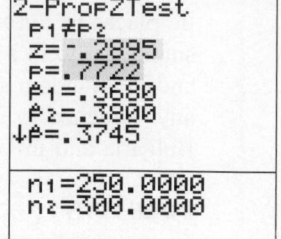

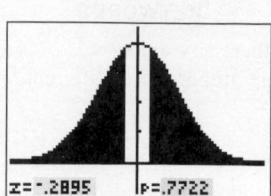

Figure 10.34
2-PropZTest input screen.

Figure 10.35
Hypothesis test results.

Figure 10.36
2-PropZTest Draw results.

Case 2: $H_0: p_1 - p_2 = \Delta_0 \neq 0$

This case, with $\Delta_0 \neq 0$, is less common. Since p_1 and p_2 are assumed unequal, there is no hypothesized, *common* population proportion. The hypothesis test follows routinely from the properties of $\hat{P}_1 - \hat{P}_2$.

HYPOTHESIS TESTS CONCERNING TWO POPULATION PROPORTIONS WHEN $\Delta_0 \neq 0$

This is the template for a large-sample hypothesis test concerning two population proportions when $\Delta_0 \neq 0$. The nonskewness criteria must also be met.

Given two random samples of sizes n_1 and n_2, a large-sample hypothesis test concerning two population proportions in terms of the difference $p_1 - p_2$ (with $\Delta_0 \neq 0$) with significance level α has the following form.

$H_0: p_1 - p_2 = \Delta_0$

$H_a: p_1 - p_2 > \Delta_0, \quad p_1 - p_2 < \Delta_0, \quad \text{or} \quad p_1 - p_2 \neq \Delta_0$

$$\text{TS: } Z = \frac{(\hat{P}_1 - \hat{P}_2) - \Delta_0}{\sqrt{\dfrac{\hat{P}_1(1 - \hat{P}_1)}{n_1} + \dfrac{\hat{P}_2(1 - \hat{P}_2)}{n_2}}}$$

$\text{RR: } Z \geq z_\alpha, \quad Z \leq -z_\alpha, \quad \text{or} \quad |Z| \geq z_{\alpha/2}$

Suppose two random samples of sizes n_1 and n_2 are obtained, and the nonskewness criteria are satisfied. Using the properties of $\hat{P}_1 - \hat{P}_2$, a confidence interval for $p_1 - p_2$ can be derived (p_1 and p_2 are assumed unequal). Start with a symmetric interval about 0 such that the probability that Z lies in this interval is $1 - \alpha$. As usual, manipulate the inequality to sandwich the parameter $p_1 - p_2$.

HOW TO FIND A $100(1 - \alpha)\%$ CONFIDENCE INTERVAL FOR $p_1 - p_2$

Given two (large) random samples of sizes n_1 and n_2, a $100(1 - \alpha)\%$ confidence interval for $p_1 - p_2$ has as endpoints the values

$$(\hat{p}_1 - \hat{p}_2) \pm z_{\alpha/2} \sqrt{\frac{\hat{p}_1(1 - \hat{p}_1)}{n_1} + \frac{\hat{p}_2(1 - \hat{p}_2)}{n_2}}. \tag{10.10}$$

Solution Trail

KEYWORDS

- Is there any evidence?
- True proportion is more than 0.03 greater than.

↓

TRANSLATION

- Conduct a one-sided hypothesis test about $p_1 - p_2$.
- $\Delta_0 = 0.03$.

↓

CONCEPTS

Large-sample hypothesis test concerning two population proportions when $\Delta_0 \neq 0$.

↓

VISION

Check the large-sample assumptions. Use the template for a one-sided test concerning $p_1 - p_2$ when $\Delta_0 \neq 0$. Use $\alpha = 0.05$ to find the critical value, compute the value of the test statistic, and draw a conclusion.

Example 10.12 Seniors in the Workforce The proportion of men and women aged 65 and older in the workforce declined from 1890 to 1990, with the proportion of elderly men traditionally higher than the proportion of elderly women. However, after 1999, more men and women elected not to retire at 65, and both proportions increased slightly. The U.S. Department of Labor projects that seniors 65 and older will make up 15.9% of the workforce by 2012, up from 11.5% in 1992. Suppose an employment study was conducted in 2009. In a random sample of 500 employed men, 90 were elderly, and in a random sample of 400 employed women, 40 were elderly.

a. Conduct the appropriate hypothesis test to determine whether there is evidence that the true proportion of elderly men in the workforce is more than 0.03 greater than the true proportion of elderly women in the workforce. Use $\alpha = 0.05$.

b. Find the p value associated with this hypothesis test.

SOLUTION

STEP 1 Let employed men be population 1 and employed women be population 2. We are looking for evidence that the difference $p_1 - p_2$ is greater than $0.03 = \Delta_0 \neq 0$. Therefore, case 2 is appropriate.

The given information is presented here.

	Men	**Women**
Sample size	$n_1 = 500$	$n_2 = 400$
Number of successes	$x_1 = 90$	$x_2 = 40$
Sample proportion	$\hat{p}_1 = 90/500 = 0.1800$	$\hat{p}_2 = 40/400 = 0.1000$

Check the nonskewness criteria using estimates for p_1 and p_2.

$$n_1\hat{p}_1 = (500)(0.1800) = 90 \geq 5 \quad n_1(1 - \hat{p}_1) = (500)(0.8200) = 410 \geq 5$$
$$n_2\hat{p}_2 = (400)(0.1000) = 40 \geq 5 \quad n_2(1 - \hat{p}_2) = (400)(0.9000) = 360 \geq 5$$

Since all of the inequalities are satisfied, $\hat{P}_1 - \hat{P}_2$ is approximately normal, and the large-sample hypothesis test concerning population proportions can be used.

STEP 2 The four parts of the hypothesis test are

$H_0: p_1 - p_2 = 0.03$
$H_a: p_1 - p_2 > 0.03$

$$\text{TS: } Z = \frac{(\hat{P}_1 - \hat{P}_2) - 0.03}{\sqrt{\dfrac{\hat{P}_1(1 - \hat{P}_1)}{n_1} + \dfrac{\hat{P}_2(1 - \hat{P}_2)}{n_2}}}$$

RR: $Z \geq z_\alpha = z_{0.05} = 1.6449$

STEP 3 The value of the test statistic is

$$z = \frac{(\hat{p}_1 - \hat{p}_2) - 0.03}{\sqrt{\dfrac{\hat{p}_1(1 - \hat{p}_1)}{n_1} + \dfrac{\hat{p}_2(1 - \hat{p}_2)}{n_2}}} = \frac{(0.18 - 0.10) - 0.03}{\sqrt{\dfrac{(0.18)(0.82)}{500} + \dfrac{(0.10)(0.90)}{400}}}$$

$$= 2.1922 \ (\geq 1.6449)$$

The value of the test statistic lies in the rejection region. Therefore, we reject the null hypothesis. At the $\alpha = 0.05$ significance level, there is evidence to suggest that the true proportion of elderly men in the workforce is more than 0.03 greater than the true proportion of elderly women in the workforce.

p value illustration:
$p = P(Z \geq 2.1922)$
$= 0.0142 \leq 0.05 = \alpha$

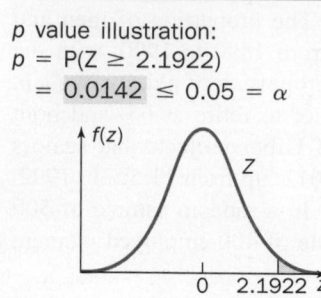

STEP 4 Since this is a one-sided right-tailed test, the p value is a right-tail probability.

$$p = P(Z \geq 2.1922)$$
$$= 1 - P(Z \leq 2.1922)$$
$$= 1 - 0.9858 = 0.0142 \, (\leq 0.05 = \alpha)$$

Definition of p value for H_a: $p_1 - p_2 > 0.03$.

The Complement Rule.

Use Table III in the Appendix.

Figure 10.37 shows a technology solution.

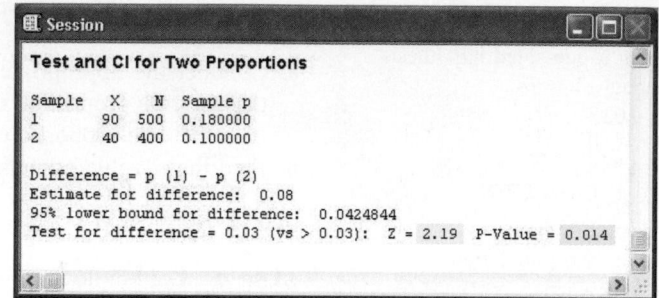

Figure 10.37 Hypothesis test results and (one-sided) confidence interval.

The computations for a confidence interval for $p_1 - p_2$ are illustrated in the next example.

Example 10.13 Storm Watch The Weather Channel (TWC) is one of the most popular cable TV networks. However, the number of viewers varies greatly according to geographic region and the current weather conditions. Random samples of cable TV viewers in the Northeast (population 1) and in the West (population 2) were obtained. The number of viewers who watched TWC in the past week was recorded. The data are given in the following table.[13]

	Northeast	West
Sample size	$n_1 = 1000$	$n_2 = 1500$
Number of successes	$x_1 = 446$	$x_2 = 303$
Sample proportion	$\hat{p}_1 = 446/1000 = 0.4460$	$\hat{p}_2 = 303/1500 = 0.2020$

Construct a 99% confidence interval for the true difference in proportions of cable TV viewers who watched TWC in the past week.

SOLUTION

STEP 1 The sample sizes, number of successes, and sample proportions are given. Check the nonskewness criteria using estimates for p_1 and p_2.

$n_1\hat{p}_1 = (1000)(0.4460) = 446 \geq 5$ $n_1(1 - \hat{p}_1) = (1000)(0.5540) = 540 \geq 5$
$n_2\hat{p}_2 = (1500)(0.2020) = 303 \geq 5$ $n_2(1 - \hat{p}_2) = (1500)(0.7980) = 1197 \geq 5$

Since all of the inequalities are satisfied, the distribution of the difference in sample proportions is approximately normal. A large-sample confidence interval is appropriate.

STEP 2 Find the critical value.

$$1 - \alpha = 0.99 \Rightarrow \alpha = 0.01 \Rightarrow \alpha/2 = 0.005$$

Find $\alpha/2$.

$$z_{\alpha/2} = z_{0.005} = 2.5758$$

Common critical value.

A technology solution:

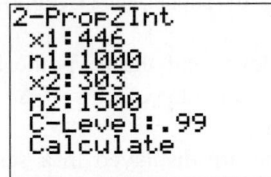

Figure 10.38
2-PropZInt input
screen.

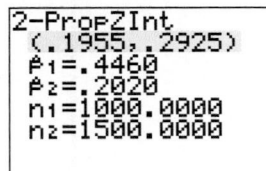

Figure 10.39 Resulting
confidence interval.

STEP 3   Use Equation 10.10.

$$(\hat{p}_1 - \hat{p}_2) \pm z_{\alpha/2}\sqrt{\frac{\hat{p}_1(1 - \hat{p}_1)}{n_1} + \frac{\hat{p}_2(1 - \hat{p}_2)}{n_2}}$$ Equation 10.10.

$$= (0.4460 - 0.2020) \pm (2.5758)\sqrt{\frac{(0.4460)(0.5540)}{1000} + \frac{(0.2020)(0.7980)}{1500}}$$

Use summary statistics and critical value.

$$= 0.2440 \pm 0.0485$$ Simplify.

$$= (0.1955, 0.2925)$$ Compute endpoints.

(0.1955, 0.2925) is a 99% confidence interval for the difference in the proportion of cable TV viewers who watched TWC in the past week in the Northeast and in the West, $p_1 - p_2$. Note that since 0 is not included in this interval, there is evidence to suggest that the two proportions are different.   ●

TECHNOLOGY CORNER

Procedure: Hypothesis tests and confidence intervals concerning two population proportions.
Reconsider: Example 10.11, page 485, solution, and interpretation.

TI-84 Plus

Use 2-PropZTest to conduct a hypothesis test concerning two population proportions, $\Delta_0 = 0$, and 2-PropZInt to construct a confidence interval for the difference in population proportions. There is no built-in function to conduct a hypothesis test if $\Delta_0 \neq 0$.

1. Select STAT; TESTS; 2-PropZTest.
2. Enter the number of successes and the number of trials for each sample: x_1, n_1, x_2, n_2. Highlight the appropriate alternative hypothesis. See Figure 10.34, page 486.
3. The Calculate and Draw results are shown in Figures 10.35 and 10.36, page 486.
4. To construct a confidence interval, select STAT; TESTS; 2-PropZInt.
5. Enter the number of successes and the number of trials for each sample, x_1, n_1, x_2, n_2, and the confidence level. See Figure 10.40.
6. Highlight Calculate and press ENTER. The confidence interval is displayed on the Home Screen. See Figure 10.41.

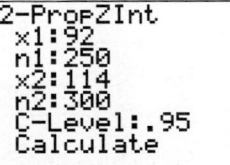

Figure 10.40
2-PropZInt input
screen.

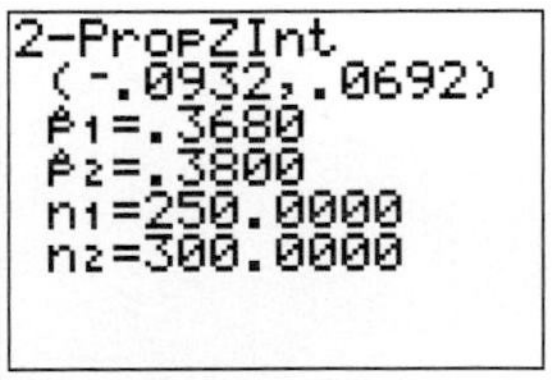

Figure 10.41
Resulting confidence
interval.

Minitab

Use the function 2 Proportions to conduct a hypothesis test and to construct a confidence interval. Input is samples in one column, samples in different columns, or summarized data.

1. Select <u>S</u>tat; <u>B</u>asic Statistics; 2 Pro<u>p</u>ortions.
2. Choose Summarized data and enter the number of successes (Events) and number of trials for each sample.
3. Choose the Options button. Enter a Confidence level and the (hypothesized) Test difference (Δ_0), and select the appropriate Alternative hypothesis. If $\Delta_0 = 0$, check the box for Use pooled estimate of p for test.
4. The hypothesis test results and confidence interval are displayed in a session window. See Figure 10.42.

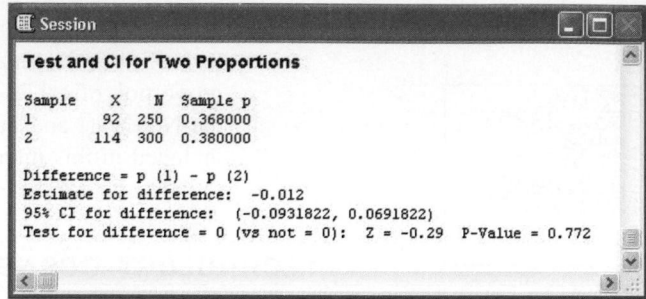

Figure 10.42 Hypothesis test results and confidence interval.

Excel

There are no built-in functions to conduct a hypothesis test concerning two population proportions or to construct a confidence interval for the difference in population proportions. However, functions associated with the standard normal distribution may be used to find critical values and p values. Use ordinary spreadsheet calculations where necessary.

1. Enter the number of successes and number of trials for each sample.
2. Compute $\hat{p}_1$, $\hat{p}_2$, and $\hat{p}_c$.
3. Compute the value of the test statistic, z, and use the function NORMSDIST to find the p value.
4. To construct the confidence interval, compute the difference, $\hat{p}_1 - \hat{p}_2$, use the function NORMSINV to find the critical value, and compute
$$\sqrt{\frac{\hat{p}_1(1 - \hat{p}_1)}{n_1} + \frac{\hat{p}_2(1 - \hat{p}_2)}{n_2}}.$$
5. Find the left endpoint and the right endpoint of the confidence interval. See Figure 10.43.

	A	B	C	D
1	92	= x_1	114	= x_2
2	250	= n_1	300	= n_2
3	0.3680	= A1/A2 = phat_1	0.3800	= C1/C2 = phat_2
4	0.3745	= (A1+C1)/(A2+c2) = phat_c		
5	-0.2895	= (A3-C3)/(SQRT(A4*(1-A4)*(1/A2+1/C2))) = z		
6	0.7722	= 2*NORMSDIST(A5) = p value		
7				
8	-0.0120	= A3 - C3 = phat_1 - phat_2		
9	1.9600	= NORMSINV(0.025) = critical value		
10	0.0414	= SQRT(A3*(1-A3)/A2+C3*(1-C3)/C2)		
11	-0.0932	= A8-A9*A10 = left endpoint		
12	-0.0125	= A8+A8*A10 = right endpoint		

Figure 10.43 Hypothesis test results and confidence interval.

SECTION 10.4 EXERCISES

Practice

10.72 In each of the following problems, the size and the number of individuals or objects with a certain characteristic are given for samples from two populations. Check the two nonskewness inequalities for both samples, and determine whether a large-sample test concerning two population proportions is appropriate.
a. $n_1 = 303$, $x_1 = 175$, $n_2 = 463$, $x_2 = 250$.
b. $n_1 = 560$, $x_1 = 140$, $n_2 = 530$, $x_2 = 125$.
c. $n_1 = 160$, $x_1 = 155$, $n_2 = 185$, $x_2 = 170$.
d. $n_1 = 1020$, $x_1 = 700$, $n_2 = 1277$, $x_2 = 950$.
e. $n_1 = 842$, $x_1 = 319$, $n_2 = 755$, $x_2 = 280$.
f. $n_1 = 4375$, $x_1 = 237$, $n_2 = 5005$, $x_2 = 245$.

10.73 In each of the following problems, the sample sizes and population proportions are given. Find the mean, variance, and standard deviation of the estimator $\hat{P}_1 - \hat{P}_2$, and compute each probability.
a. $n_1 = 645$, $p_1 = 0.24$, $n_2 = 650$, $p_2 = 0.26$,
$P(\hat{P}_1 - \hat{P}_2 \geq 0.045)$
b. $n_1 = 250$, $p_1 = 0.37$, $n_2 = 270$, $p_2 = 0.33$,
$P(\hat{P}_1 - \hat{P}_2 \leq -0.04)$
c. $n_1 = 144$, $p_1 = 0.87$, $n_2 = 156$, $p_2 = 0.86$,
$P(-0.05 < \hat{P}_1 - \hat{P}_2 < 0.05)$
d. $n_1 = 520$, $p_1 = 0.65$, $n_2 = 480$, $p_2 = 0.72$,
$P(\hat{P}_1 - \hat{P}_2 > -0.10)$
e. $n_1 = 1200$, $p_1 = 0.73$, $n_2 = 1150$, $p_2 = 0.85$,
$P(\hat{P}_1 - \hat{P}_2 < -0.06)$
f. $n_1 = 500$, $p_1 = 0.645$, $n_2 = 525$, $p_2 = 0.604$,
$P(-0.02 \leq \hat{P}_1 - \hat{P}_2 \leq 0.10)$

10.74 A hypothesis test concerning two population proportions is described in each of the following problems. Identify each population, and determine the appropriate null and alternative hypotheses in terms of p_1 and p_2.
a. A study was conducted to determine whether there is any difference in the proportion of people who have a satellite radio in California versus Tennessee.
b. Two random samples were obtained of men and women who tried to talk their way out of a traffic ticket. The number of each who said they missed a street sign was recorded. The data will be used to determine whether the proportion of women who say they missed a street sign is greater than the proportion of men.
c. Random samples of teens ages 13–19 from two different school districts were obtained. The number of teens who own a cell phone was recorded. The data will be used to determine whether there is any difference in the proportion of teens who own a cell phone in the two districts.
d. Random samples of Americans who received an income tax refund were obtained. The people were classified by income level (low versus high) and asked whether they intended to use their refund to pay outstanding bills. The data will be used to determine whether the proportion of high-income Americans who pay bills with tax refunds is 0.10 greater than the proportion of low-income Americans who do so.

10.75 In each of the following problems, use the given data to conduct the appropriate hypothesis test concerning two population proportions, find the p value, and state your conclusion.
a. $n_1 = 500$, $x_1 = 400$, $n_2 = 525$, $x_2 = 405$, H_0: $p_1 - p_2 = 0$, H_a: $p_1 - p_2 > 0$, $\alpha = 0.05$.
b. $n_1 = 646$, $x_1 = 280$, $n_2 = 680$, $x_2 = 330$, H_0: $p_1 - p_2 = 0$, H_a: $p_1 - p_2 < 0$, $\alpha = 0.01$.
c. $n_1 = 255$, $x_1 = 81$, $n_2 = 266$, $x_2 = 110$, H_0: $p_1 - p_2 = 0$, H_a: $p_1 - p_2 \neq 0$, $\alpha = 0.025$.
d. $n_1 = 1440$, $x_1 = 907$, $n_2 = 1562$, $x_2 = 970$, H_0: $p_1 - p_2 = 0$, H_a: $p_1 - p_2 \neq 0$, $\alpha = 0.001$.

10.76 In each of the following problems, use the given data to conduct the appropriate hypothesis test concerning two population proportions, find the p value, and state your conclusion.
a. $n_1 = 200$, $x_1 = 100$, $n_2 = 300$, $x_2 = 165$, H_0: $p_1 - p_2 = 0.05$, H_a: $p_1 - p_2 < 0.05$, $\alpha = 0.01$.
b. $n_1 = 480$, $x_1 = 384$, $n_2 = 490$, $x_2 = 367$, H_0: $p_1 - p_2 = 0.02$, H_a: $p_1 - p_2 > 0.02$, $\alpha = 0.05$.
c. $n_1 = 610$, $x_1 = 450$, $n_2 = 675$, $x_2 = 470$, H_0: $p_1 - p_2 = 0.10$, H_a: $p_1 - p_2 \neq 0.10$, $\alpha = 0.01$.
d. $n_1 = 2500$, $x_1 = 710$, $n_2 = 3100$, $x_2 = 770$, H_0: $p_1 - p_2 = 0.07$, H_a: $p_1 - p_2 \neq 0.07$, $\alpha = 0.001$.

10.77 In each of the following problems, use the given data and confidence level to construct a confidence interval for the difference of two population proportions, $p_1 - p_2$.
a. $n_1 = 388$, $x_1 = 230$, $n_2 = 402$, $x_2 = 250$, 95%.
b. $n_1 = 528$, $x_1 = 475$, $n_2 = 530$, $x_2 = 497$, 95%.
c. $n_1 = 180$, $x_1 = 92$, $n_2 = 194$, $x_2 = 100$, 99%.
d. $n_1 = 2300$, $x_1 = 1705$, $n_2 = 2404$, $x_2 = 1690$, 90%.

Applications

10.78 Fuel Consumption and Cars There were more than 1,095,769 automobile thefts in the United States in 2007, an increase of 1.1 percent from the previous year.[14] Despite this trend, many people still leave their cars unlocked. In a random sample of 200 men, 110 said they usually leave their car unlocked, and in a random sample of 250 women, 120 said they usually leave their car unlocked. Is there any evidence to suggest that the proportion of men who leave their car unlocked is different from the proportion of women who do so? Use $\alpha = 0.05$.

10.79 Marketing and Consumer Behavior Automobile dealers strive for repeat customers, that is, new-vehicle buyers and lessees who replace a vehicle with another of the same make. In 2008, Honda had the highest customer retention rate, 64.7%, followed by Toyota and Lexus.[15] A random sample of customers who purchased a new Ford or new Chevrolet was obtained and they were asked if they were repeat customers. The data are given in the following table.

Make	Sample size	Repeat customers
Ford	400	210
Chevrolet	550	286

Is there any evidence to suggest that the proportion of repeat customers is different for Ford and Chevrolet? Use $\alpha = 0.05$.

10.80 Psychology and Human Behavior Many young adults are choosing to live at home with their parents following college in order to start a career and build some savings. Independent random samples of young adults, 25–34 years old, were obtained and the residence of each was recorded. Of the 226 women, 18 were living at home, and of the 254 men, 36 were living at home. Is there any evidence to suggest that the proportion of young women living at home is less than the proportion of young men living at home? Use $\alpha = 0.01$.[16]

10.81 Marketing and Consumer Behavior Many critics have been known to say, "They sure don't make movies like they used to." In order to assess Americans' opinions of movies, a random sample of people was obtained and each was asked about the quality of movies. The data are given in the following table.

Age group	Sample size	Number who said movies are getting better
18–29	347	238
30–49	387	221

a. Conduct the appropriate hypothesis test to determine whether there is any evidence to suggest that the true proportion of 18- to 29-year-olds who believe movies are getting better is greater than the proportion of 30- to 49-year-olds. Use $\alpha = 0.001$.

b. Find the p value associated with the hypothesis test in part (a).

10.82 Public Health and Nutrition Several years ago, most doctors believed that it was not necessary to take any dietary supplement. Now, because many Americans do not eat a healthy, balanced diet, many physicians recommend a once-a-day multivitamin. A random sample of people was obtained, and they were asked whether they regularly take a multivitamin. The data are given in the following table.

Group	Sample size	Number who take a multivitamin
Men	490	181
Women	428	214

(*Source*: HighBeam Research, LLC; and *Mintel's Vitamin and Mineral Report*.)

Is there any evidence that the proportion of women who take a multivitamin is greater than the proportion of men? Use $\alpha = 0.005$.

10.83 Travel and Transportation Many people who commute to work by car in New York City every day use either the George Washington Bridge or the Lincoln Tunnel. A random sample of commuters who use one of these two routes was obtained, and they were asked whether they carpooled to work. The data are given in the following table.

Commuting route	Sample size	Number who carpool
Bridge	1055	530
Tunnel	1663	825

a. Verify that the nonskewness criteria inequalities are satisfied.

b. Is there any evidence to suggest that the proportion of carpoolers crossing the George Washington Bridge is greater than the proportion of carpoolers using the Lincoln Tunnel? Use $\alpha = 0.01$.

c. Find the p value associated with the hypothesis test in part (b).

10.84 Public Policy and Political Science Americans have become increasingly skeptical of news reports and, according to a recent Pew Research survey, they are watching and reading news from sources that most closely agree with their own ideologies. A random sample of American adults was obtained, and they were asked which television news program they regularly watch and their political affiliation. Of the 280 that watch Fox News Channel, 98 were Republicans, and 57 of the 300 who watch CNN were Republicans.[17] Conduct the appropriate hypothesis test to determine whether there is evidence that the true proportion of Fox News Channel viewers who are Republican is greater than the true proportion of CNN viewers who are Republicans. Use $\alpha = 0.05$.

10.85 Medicine and Clinical Studies According to the National Institute of Allergy and Infectious Diseases, approximately 54.6% of all U.S. citizens test positive to one or more allergens. Between 9 and 16 percent suffer from hay fever. A random sample of people who suffer from hay fever was obtained, and each was treated with either a conventional antihistamine or butterbur extract. The number of subjects who experienced relief from hayfever was recorded for each group. The resulting data are given in the following table.

Treatment	Sample size	Number who experienced relief
Antihistamine	255	71
Butterbur extract	237	55

a. Compute the sample proportion of people who experienced relief for each treatment.

b. Conduct the appropriate hypothesis test to determine whether the proportion of people who experience relief due to the antihistamine is different from the proportion of people who experience relief from butterbur extract. Use $\alpha = 0.01$.

10.86 Public Health and Nutrition A survey was conducted concerning physical activity of adults in two states. Random samples of adults were obtained from Arizona and from West Virginia, and participants were asked whether they considered themselves physically inactive. The data are given in the following table.

State	Sample size	Number who are physically inactive
Arizona	1122	163
West Virginia	1181	205

(*Source*: Centers for Disease Control and Prevention.)

Is there any evidence to suggest that the proportion of adults who consider themselves physically inactive is greater in West Virginia than in Arizona? Use $\alpha = 0.001$ and find the p value.

10.87 Manufacturing and Product Development Blenko Specs has two different processes for the manufacture of optical lenses supplied to the military. A random sample of finished lenses was obtained from each process, and each lens was carefully inspected for defects. Of the 106 lenses from Process A, 8 were defective, and 12 of the 121 lenses from Process B were defective.

a. Compute the sample proportion of defectives for each process.
b. Check the nonskewness criteria and verify that the inequalities are satisfied.
c. Conduct a hypothesis test to determine whether there is any evidence that the sample proportion of defective lenses is different for Process A and Process B. Use $\alpha = 0.05$.

10.88 Education and Child Development In October 2008, *Billboard* celebrated the 50th anniversary of the Hot 100 chart. It was determined that the most popular single in the past 50 years was "The Twist" sung by Chubby Checker. The methodology was based on chart position and length of time on the chart. (John Travolta and Uma Thurman performed The Twist in the movie *Pulp Fiction.*) Random samples of sixth-grade students in cities and in rural areas were obtained, and the students were asked whether they knew this number one song. The resulting data are given in the following table.

Area	Sample size	Number who knew the song
Rural	480	120
City	510	95

Do you believe that the proportion of rural schoolchildren who know "The Twist" is greater than the proportion of city schoolchildren? Use $\alpha = 0.01$.

10.89 Demographics and Population Statistics According to the U.S. Census Bureau (as reported by Stuckey and Gelles for *USA Today*), 23% of the residents in Virginia Beach are veterans. Other cities with large proportions of veterans include Colorado Springs, Jacksonville, Anchorage, and Corpus Christi. Random samples of residents in Colorado Springs and Jacksonville were obtained, and each was classified as a veteran or nonveteran. The resulting data are given in the following table.

City	Sample size	Veterans
Colorado Springs	625	113
Jacksonville	730	133

Is there any evidence to suggest that the true proportion of veterans living in Colorado Springs and Jacksonville is different? Use $\alpha = 0.05$.

10.90 Demographics and Population Statistics For people in the military, family life is often very stressful. According to Defense Department records, over 56,000 couples have divorced since the conflict in Afghanistan started in 2001. The Army has introduced special programs to help troops and their spouses to deal with long deployments. While the divorce rate for Army personnel is less than the divorce rate for Americans in general, there is still increased spending on quality-of-life services for troops and families. A random sample of Army Reserve and Army Guard couples was obtained, and the number of divorces during 2006 was recorded. The data are given in the following table.

Personnel	Number of couples	Number of divorces
Reserve	324	12
Guard	286	7

(*Source*: Jelinek, P., *Press Enterprise*, February 13, 2006.)

Is there any evidence to suggest that the true proportion of military couples who divorce is greater for those in the Reserve than in the Guard? Use $\alpha = 0.01$.

Extended Applications

10.91 Marketing and Consumer Behavior Americans have many sources for daily news, for example, local television shows, public radio, or national newspapers. A random sample of Americans was obtained and classified by age. Each person was asked whether he or she obtained news every day from three specific sources. The data are given in the following table.[18]

	Age group			
	18- to 29-year-olds		30- to 49-year-olds	
News source	Sample size	Number who obtained news every day	Sample size	Number who obtained news every day
Nightly network news	570	103	462	120
Cable news networks	450	108	520	182
Internet	546	197	568	239

a. Conduct the appropriate hypothesis test to determine whether there is evidence that the true proportion of 18- to 29-year-olds who obtain news every day from nightly network news shows is less than the true proportion of 30- to 49-year-olds who obtain news every day from nightly network news shows. Use $\alpha = 0.05$. Find the p value associated with this hypothesis test.
b. Repeat (a) for cable news networks. Use $\alpha = 0.01$. Find the p value associated with this hypothesis test.
c. Repeat (a) for the Internet. Use $\alpha = 0.005$. Find the p value associated with this hypothesis test.

10.92 Manufacturing and Product Development Two different machines in a manufacturing facility are designed to fill cans with 280 grams of chai tea latte powdered mix. A random sample of filled cans from each machine was obtained, and each can was carefully weighed. Of the 134 cans from machine A, 10 were underfilled, and 7 of 114 cans from machine B were underfilled.

a. Compute the sample proportion of underfilled cans for each machine.
b. Verify the nonskewness criteria.

c. Find a 95% confidence interval for the true difference in the proportion of underfilled cans for machine A and B.

d. Using the confidence interval in part (c), is there any evidence to suggest that the proportion of underfilled cans is different for the two machines? Justify your answer.

10.93 Psychology and Human Behavior Historically, the three most popular home-improvement projects (respectively) are interior decorating, landscaping, and expansion. A random sample of homeowners and condominium owners was obtained, and they were asked whether they planned any landscaping within the next year. The data are given in the following table.

Residence	Sample size	Number who are planning to landscape
Homeowner	261	90
Condominium owner	303	65

Is there any evidence to suggest that the proportion of homeowners planning a landscaping project is more than 0.10 greater than the proportion of condominium owners planning a landscaping project? Use $\alpha = 0.01$.

10.5 COMPARING TWO POPULATION VARIANCES OR STANDARD DEVIATIONS

Many practical business decisions are based on a comparison of variability, or manufacturing precision. For example, a company that produces a certain drug via fermentation would like to maintain a very small variability in yield. The company would like to choose a fermentation process that is more reliable and less variable than another. A hardware store wants very little variability in paint color from gallon to gallon. One paint mixer may be more precise (less variable) than another. Even food manufacturers strive to reduce the differences in product taste from one batch to the next. And we may even compare population variances in order to decide which two-sample t test is appropriate, the pooled test or the approximate test.

S_1^2 and S_2^2 are good (unbiased) estimators for the population variances σ_1^2 and σ_2^2, respectively. However, a hypothesis test for comparing σ_1^2 and σ_2^2 is based on a new standardization and an F distribution, introduced below.

An F distribution has positive probability only for nonnegative values. The probability density function for an F random variable is 0 for $x < 0$. Once again, it is important to focus on the properties of an F distribution and the method for finding critical values associated with this distribution.

PROPERTIES OF AN *F* DISTRIBUTION

The *numerator* and *denominator* designations will make more sense as you read on.

1. An F distribution is completely determined by two parameters, the number of degrees of freedom in the numerator and the number of degrees of freedom in the denominator, given in that order. Both values must be positive integers (1, 2, 3, . . .) and there is, of course, a different F distribution for every combination.

2. If X has an F distribution with ν_1 and ν_2, ($X \sim F$ with df ν_1 and ν_2), then

Why are these restrictions on ν_2 necessary? What do you suppose the mean is if $\nu_2 = 2$?

$$\mu_X = \frac{\nu_2}{\nu_2 - 2}, \nu_2 \geq 3; \quad \sigma_X^2 = \frac{2\nu_2^2(\nu_1 + \nu_2 - 2)}{\nu_1(\nu_2 - 2)^2(\nu_2 - 4)}, \nu_2 \geq 5. \quad (10.11)$$

3. Suppose $X \sim F$ with df ν_1 and ν_2. The density curve for X is positively skewed (*not* symmetric), and gets closer and closer to the x-axis but never touches it. As both degrees of freedom increase, the density curve becomes taller and more compact. See Figure 10.44.

The definition and notation for an F *critical value* are analogous to those for Z, t, and χ^2 critical values.

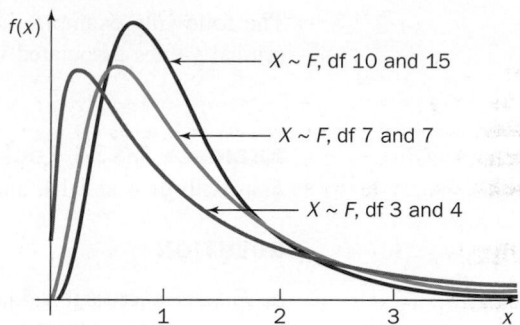

Figure 10.44 Density curves for several F distributions.

DEFINITION

F_α is a critical value related to an F distribution with df ν_1 and ν_2. If $X \sim F$ with df ν_1 and ν_2, then $P(X \geq F_\alpha) = \alpha$.

💡 **ILLUMINATING THE CONCEPTS**

1. F_α is a value on the measurement axis in an F world such that there is α of the area (probability) to the right of F_α. Remember, as for a chi-square distribution, there is no symmetry in F critical values.

2. Since critical values are defined in terms of right-tail probability, and the F distribution is not symmetric, the notation here is similar to that for chi-square critical values. It will be necessary to find critical values denoted $F_{1-\alpha}$, where $1 - \alpha$ is large. By definition, $P(X \geq F_{1-\alpha}) = 1 - \alpha$, and by the Complement Rule, $P(X \leq F_{1-\alpha}) = \alpha$. See Figure 10.45.

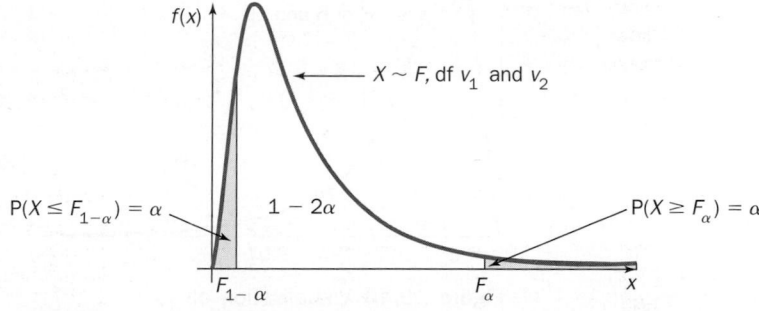

Figure 10.45 An illustration of F critical values.

3. F critical values are related according to the following equation.

$$F_{1-\alpha}\,(\text{df } \nu_1 \text{ and } \nu_2) = \frac{1}{F_\alpha(\text{df } \nu_2 \text{ and } \nu_1)}. \qquad (10.12)$$ ■

Notice how the degrees of freedom switch positions.

Table VII in the Appendix presents selected critical values associated with various F distributions and right-tail probabilities. The degrees of freedom in the numerator are given in the top row and the degrees of freedom in the denominator are given in the left column. In the body of the table, F_α is at the intersection of the appropriate row and column. Left-tail probabilities are found using Equation 10.12.

The following example illustrates the use of Table VII in the Appendix for finding critical values associated with an F distribution.

Example 10.14 Looking Up Critical Values Find each critical value: **a.** $F_{0.05}$ with df 8 and 10, and **b.** $F_{0.99}$ with df 9 and 15.

SOLUTION

a. $F_{0.05}$ is a critical value related to an F distribution with 8 and 10 degrees of freedom. By definition, if $X \sim F$ with df 8 and 10, then $P(X \geq F_{0.05}) = 0.05$. Using Table VII in the Appendix, for $\alpha = 0.05$, find the intersection of the $\nu_1 = 8$ column and the $\nu_2 = 10$ row.

$\alpha = 0.05$

		ν_1					
ν_2		6	7	8	9	10	
⋮	⋮	⋮	⋮	⋮	⋮	⋮	⋮
8	...	3.58	3.50	3.44	3.39	3.35	...
9	...	3.37	3.29	3.23	3.18	3.14	...
10	...	3.22	3.14	3.07	3.02	2.98	...
11	...	3.09	3.01	2.95	2.90	2.85	...
12	...	3.00	2.91	2.85	2.80	2.75	...
⋮	⋮	⋮	⋮	⋮	⋮	⋮	⋮

Therefore, $F_{0.05} = 3.07$ and if $X \sim F$ with df 8 and 10, then $P(X \geq 3.07) = 0.05$, as illustrated in Figure 10.46.

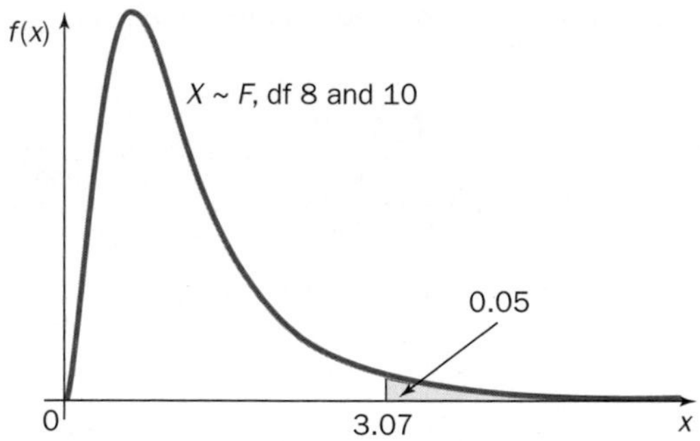

Figure 10.46 Visualization of $F_{0.05} = 3.07$.

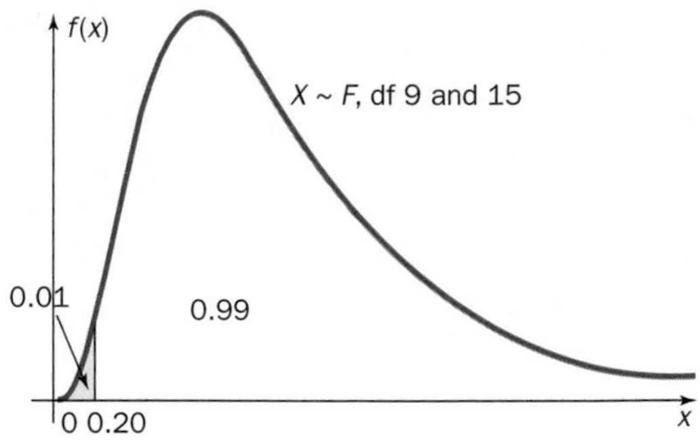

Figure 10.47 Visualization of $F_{0.99} = 0.202$.

b. $F_{0.99}$ is a critical value related to an F distribution with df 9 and 15. By definition, if $X \sim F$ with df 9 and 15, then $P(X \geq F_{0.99}) = 0.99$. Since $F_{0.99}$ is in the left tail of the distribution, use Equation 10.12.

$$F_{0.99}\text{ (df 9 and 15)} = F_{1-0.01} = \frac{1}{F_{0.01}\text{ (df 15 and 9)}}$$

Using Table VII in the Appendix, for $\alpha = 0.01$, find the intersection of the $\nu_1 = 15$ column and the $\nu_2 = 9$ row.

$\alpha = 0.01$

ν_2		9	10	15	20	30	
				ν_1			
$\vdots$	$\vdots$	$\vdots$	$\vdots$	$\vdots$	$\vdots$	$\vdots$	$\vdots$
7	...	6.72	6.62	6.31	6.16	5.99	...
8	...	5.91	5.81	5.52	5.36	5.20	...
9	...	5.35	5.26	4.96	4.81	4.65	...
10	...	4.94	4.85	4.56	4.41	4.25	...
11	...	4.63	4.54	4.25	4.10	3.94	...
$\vdots$	$\vdots$	$\vdots$	$\vdots$	$\vdots$	$\vdots$	$\vdots$	$\vdots$

A technology solution:

	A	B
3	0717	= FINV(0.05,8,10)
	0 2015	= FINV(0.99,9,15)

Figure 10.48 Use the Excel function `FINV` to find F critical values.

Therefore, $F_{0.99} = 1/4.96 = 0.202$. If the random variable $X \sim F$ with df 9 and 15 then $P(X \geq 0.202) = 0.99$ and $P(X \leq 0.202) = 0.01$, as illustrated in Figure 10.47. ●

 ILLUMINATING THE CONCEPTS

1. Table VII in the Appendix is very limited. There are only three values for α and a limited number of values for ν_1 and ν_2. The TI-84 Plus does *not* have a built-in function for finding F critical values. However, the `SOLVE` feature may be used to find a critical value related to any F distribution.

2. Minitab and Excel may also be used to find a critical value related to any F distribution. Minitab uses inverse cumulative probability and Excel uses right-tail probability. For example, to find $F_{0.99}$ with df 9 and 15 using Minitab, let the random variable $X \sim F$ with df 9 and 15.

   $P(X \geq F_{0.99}) = 0.99$ *Definition of F critical value.*

   $P(X \leq F_{0.99}) = 1 - 0.99 = 0.01$ *The Complement Rule.*

   $F_{0.99} = 0.2015$ *Use Minitab.* ◼

Hypothesis tests concerning two population variances and a confidence interval for the ratio of two population variances are based on the following results.

Let S_1^2 be the sample variance of a random sample of size n_1 from a normal distribution with variance σ_1^2, let S_2^2 be the sample variance of a random sample of size n_2 from a normal distribution with variance σ_2^2, and suppose the samples are independent.

For reference, we'll call these the two-sample F test assumptions.

1. The random variable

This is yet another kind of standardization, a transformation to an F distribution.

$$F = \frac{S_1^2/\sigma_1^2}{S_2^2/\sigma_2^2} \qquad (10.13)$$

has an F distribution with $n_1 - 1$ (from the numerator) and $n_2 - 1$ (from the denominator) degrees of freedom.

2. If the null hypothesis is $H_0: \sigma_1^2 = \sigma_2^2$, then the random variable simplifies to

If σ_1^2 and σ_2^2 are equal, they cancel out.

$F = \dfrac{S_1^2/\sigma_1^2}{S_2^2/\sigma_2^2} = S_1^2/S_2^2$. This simple ratio is the test statistic for comparing two population variances.

This is the template for a hypothesis test concerning two population variances, sometimes called a two-sample F test.

HYPOTHESIS TESTS CONCERNING TWO POPULATION VARIANCES

Given the two-sample F test assumptions, a hypothesis test concerning two population variances with significance level α has the following form.

$H_0: \sigma_1^2 = \sigma_2^2$

$H_a: \sigma_1^2 > \sigma_2^2, \quad \sigma_1^2 < \sigma_2^2, \quad \text{or} \quad \sigma_1^2 \neq \sigma_2^2$

TS: $F = S_1^2/S_2^2$

RR: $F \geq F_\alpha, \quad F \leq F_{1-\alpha}, \quad F \leq F_{1-\alpha/2}, \quad \text{or} \quad F \geq F_{\alpha/2}$
with df $n_1 - 1$ and $n_2 - 1$.

Using the same assumptions, we can derive a confidence interval for the ratio of two population variances. Let $X \sim F$ with df $n_1 - 1$ and $n_2 - 1$ and find an interval that captures $1 - \alpha$ in the *middle* of this F distribution. Manipulate the inequality to sandwich the ratio σ_1^2/σ_2^2.

HOW TO FIND A $100(1 - \alpha)\%$ CONFIDENCE INTERVAL FOR THE RATIO OF TWO POPULATION VARIANCES

Given the two-sample F test assumptions, a $100(1 - \alpha)\%$ confidence interval for σ_1^2/σ_2^2 is given by

$$\left(\frac{s_1^2}{s_2^2} \frac{1}{F_{\alpha/2}}, \frac{s_1^2}{s_2^2} \frac{1}{F_{1-\alpha/2}} \right) \qquad (10.14)$$

with the critical values based on df $n_1 - 1$ and $n_2 - 1$. Using Equation 10.12 the confidence interval can be written as

$$\left(\frac{s_1^2}{s_2^2} \frac{1}{F_{\alpha/2}}, \frac{s_1^2}{s_2^2} F_{\alpha/2} \right) \qquad (10.15)$$

The critical value in the right endpoint is based on df $n_2 - 1$ and $n_1 - 1$.

The hypothesis test procedure described above can also be used to compare two population standard deviations. And you can take the square root of each endpoint of Equation 10.14 in order to find a $100(1 - \alpha)\%$ confidence interval for the ratio of two population standard deviations. The following example illustrates the hypothesis test procedure.

Example 10.15 Long-Term Care Cost The cost of long-term care in a nursing home varies considerably by region and may be as much as $50,000 per year. Two independent samples of nursing homes in Connecticut and in Colorado were obtained, and the cost-per-day for each was recorded. The data are given in the following tables.

Connecticut (1)

270	294	174	180	314	274	160	210	255	187	271

Colorado (2)

161	150	164	109	168	172	133	148	120	157	138	94
166	116	98	168	153	118	138	116	120			

a. Conduct the appropriate hypothesis test to determine whether there is any evidence that the population variance in cost-per-day is different in Connecticut and Colorado. Assume the cost-per-day underlying populations are normal and use $\alpha = 0.02$.

b. Find bounds on the p value for the hypothesis test in part (a).

SOLUTION

STEP 1   The null hypothesis is that the two population variances are equal. Since we are looking for *any difference* in the variances, the alternative hypothesis is two-sided. The underlying populations are assumed normal and the samples were obtained independently. A two-sample F test is appropriate.

In this case, $n_1 = 11$ and $n_2 = 21$; $\alpha/2 = 0.01$ and $1 - \alpha/2 = 0.99$.

STEP 2   The four parts of the hypothesis test are

$$H_0: \sigma_1^2 = \sigma_2^2$$
$$H_a: \sigma_1^2 \neq \sigma_2^2$$
$$\text{TS: } F = S_1^2/S_2^2$$
$$\text{RR: } F \leq F_{1-\alpha/2} = F_{0.99} = 1/4.41 = 0.2268 \text{ or}$$
$$F \geq F_{\alpha/2} = F_{0.01} = 3.37$$

df 10 and 20.

STEP 3   The summary statistics are

$$s_1^2 = \frac{1}{10}\left[638819 - \frac{1}{11}(2589)^2\right] = 2946.25$$

$$s_2^2 = \frac{1}{20}\left[414601 - \frac{1}{21}(2907)^2\right] = 609.46$$

The value of the test statistic is

$$f = \frac{s_1^2}{s_2^2} = \frac{2946.25}{609.46} = 4.83 \; (\geq 3.37)$$

The value of the test statistic lies in the rejection region. Therefore, we reject the null hypothesis at the $\alpha = 0.02$ significance level. There is evidence to suggest that the two population variances are different.

STEP 4   Due to the limited tables of critical values for F distributions, we can only bound the p value. Place the value of the test statistic, $f = 4.83$, in an ordered list of critical values with df 10 and 20.

$$3.37 \leq 4.83 \leq 5.08$$
$$F_{0.01} \leq 4.83 \leq F_{0.001}$$

Therefore, $0.001 \leq p/2 \leq 0.01$
and $0.002 \leq p \leq 0.02$

Figures 10.49 through 10.51 together show a technology solution.

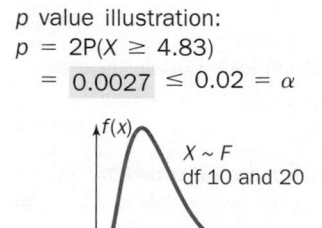

p value illustration:
$p = 2P(X \geq 4.83)$
   $= 0.0027 \leq 0.02 = \alpha$

$X \sim F$
df 10 and 20

4.83

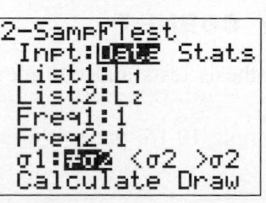

Figure 10.49
2-SampFTest input screen.

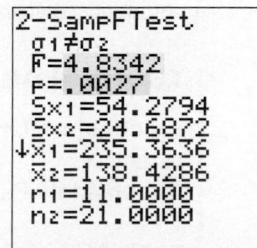

Figure 10.50
Hypothesis test results.

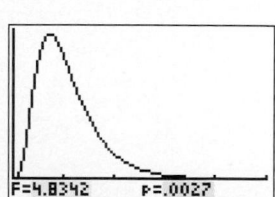

Figure 10.51
2-SampFTest Draw results.

(Louoates/Dreamstime.com)

The following example involves constructing a confidence interval for the ratio of two population variances.

Example 10.16 Tanker Tare Weight The state Department of Environmental Protection is concerned about the number of tanker trucks using major highways in Ohio and the amount of fuel carried by each truck. Independent random samples of tanker trucks on major highways during morning rush hours and evening rush hours were obtained. The amount of gasoline carried by each was recorded (in gallons). The data are summarized in the following table.

$$\text{Morning rush hour:} \quad n_1 = 16, \quad s_1^2 = 1250$$
$$\text{Evening rush hour:} \quad n_2 = 21, \quad s_2^2 = 1875$$

Construct a 90% confidence interval for the ratio of population variances in the amount of gasoline carried by tanker trucks during morning and evening rush hours.

SOLUTION

STEP 1 The samples are independent; the sample sizes and sample variances are given. A confidence interval for the ratio of two population variances is appropriate.

STEP 2 Find the critical values.

$$1 - \alpha = 0.90 \Rightarrow \alpha = 0.10 \Rightarrow \alpha/2 = 0.05 \qquad \text{Find } \alpha/2.$$
$$F_{\alpha/2} = F_{0.05} = 2.20 \qquad \text{Critical value, left endpoint. Use Table VII in the Appendix, df 15 and 20.}$$

$$F_{\alpha/2} = F_{0.05} = 2.33 \qquad \text{Critical value, right endpoint. Use Table VII in the Appendix, df 20 and 15.}$$

STEP 3 Use Equation 10.15.

$$\left(\frac{s_1^2}{s_2^2} \frac{1}{F_{\alpha/2}}, \ \frac{s_1^2}{s_2^2} F_{\alpha/2} \right) \qquad \text{Equation 10.15.}$$

$$= \left(\frac{1250}{1875} \left(\frac{1}{2.20} \right), \ \frac{1250}{1875} (2.33) \right) \qquad \text{Use sample variances and critical values.}$$

$$= (0.3030, 1.5533) \qquad \text{Simplify.}$$

(0.3030, 1.5533) is a 90% confidence interval for the ratio of the population variances. ●

Suppose that a two-sample *t* test will be used to compare two population means. The hypothesis test presented in this section is often used first to compare the population variances. The results and conclusion suggest the appropriate hypothesis test concerning population means from Section 10.2, according to whether or not there is evidence that the two population variances are unequal.

TECHNOLOGY CORNER

Procedure: Hypothesis tests and confidence intervals concerning two population variances.
Reconsider: Example 10.15, page 498, solutions, and interpretations.

TI-84 Plus

Use the built-in function 2-SampFTest to conduct a hypothesis test concerning two population variances. Input is either data in lists or summary statistics. There is no built-in function to construct a confidence interval for the ratio of population variances.

1. Enter the Connecticut data into list L1 and the Colorado data into list L2.
2. Select [STAT]; TESTS; 2-SampFTest.
3. Highlight Data and enter the two lists. Set each frequency to 1 and highlight the appropriate alternative hypothesis. See Figure 10.49, page 499.
4. Highlight Calculate and press [ENTER] to display the hypothesis test results. See Figure 10.50, page 499.

Minitab

Use the built-in function 2 Variances to conduct a hypothesis test concerning two population variances, two-sided only. Input is samples in one column (with subscripts), samples in different columns, or summarized data. This routine also returns confidence intervals for the population standard deviations, conducts an additional hypothesis test for equality of variances (Levene's test), displays graphs of the confidence intervals for the population standard deviations, and produces box plots for the data. There is no built-in function to construct a confidence interval for the ratio of population variances.

1. Enter the Connecticut data into column C1 and the Colorado data into column C2.
2. Select Stat; Basic Statistics; 2 Variances.
3. Choose Samples in different columns and enter C1 as the First column and C2 as the Second column.
4. Choose the Options button and enter a Confidence level (for the confidence intervals for the population standard deviations).
5. Click OK and the results are displayed in a graph window. See Figure 10.52.

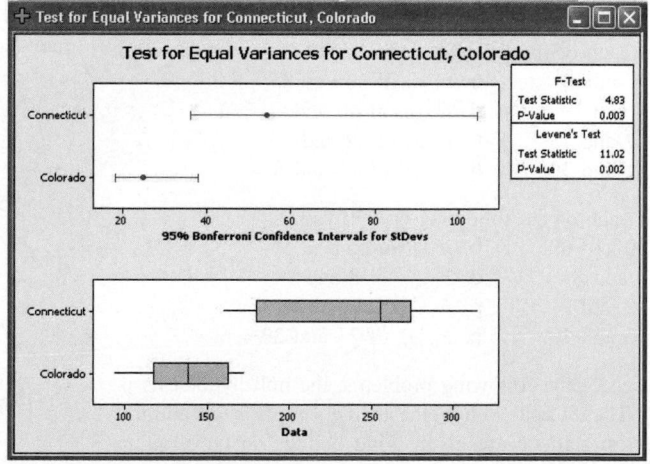

Figure 10.52 2 Variances output.

Excel

Use the built-in function F-Test Two-Sample for Variances to conduct a hypothesis test concerning two population variances. There is no built-in function to construct a confidence interval for the ratio of population variances.

1. Enter the Connecticut data into column A and the Colorado data into column B.
2. Under the Data tab, select Data Analysis; F-Test Two-Sample for Variances.

3. Enter the Variable 1 Range, Variable 2 Range, and a value for Alpha, and specify an Output option. There is no alternative hypothesis option.
4. Click OK to view the summary statistics and the hypothesis test results. See Figure 10.53. The *p* value displayed is for a one-sided hypothesis test. Double this value for a two-sided test.

F-Test Two-Sample for Variances		
	Variable 1	Variable 2
Mean	235.36	138.43
Variance	2946.25	609.46
Observations	11	21
df	10	20
F	4.83	
P(F<=f) one-tail	0.0013	
F Critical one-tail	2.35	

Figure 10.53 Hypothesis test results.

SECTION 10.5 EXERCISES

Practice

10.94 Find each of the following critical values.
a. $F_{0.05}$, df 7 and 19 **b.** $F_{0.05}$, df 30 and 25
c. $F_{0.01}$, df 6 and 19 **d.** $F_{0.001}$, df 40 and 40
e. $F_{0.95}$, df 17 and 15 **f.** $F_{0.95}$, df 12 and 10
g. $F_{0.99}$, df 21 and 30 **h.** $F_{0.999}$, df 11 and 8

10.95 Find each of the following critical values.
a. $F_{0.01}$, df 20 and 60 **b.** $F_{0.01}$, df 15 and 19
c. $F_{0.05}$, df 6 and 8 **d.** $F_{0.001}$, df 8 and 24
e. $F_{0.99}$, df 12 and 9 **f.** $F_{0.99}$, df 6 and 8
g. $F_{0.95}$, df 10 and 10 **h.** $F_{0.999}$, df 23 and 20

10.96 In each of the following problems, the null hypothesis is $H_0: \sigma_1^2 = \sigma_2^2$. The alternative hypothesis, the sample sizes, and the value of the test statistic are given. Find bounds on the *p* value associated with each hypothesis test.
a. $H_a: \sigma_1^2 > \sigma_2^2$, $n_1 = 16$, $n_2 = 17$, $f = 2.29$.
b. $H_a: \sigma_1^2 < \sigma_2^2$, $n_1 = 11$, $n_2 = 16$, $f = 0.34$.
c. $H_a: \sigma_1^2 \neq \sigma_2^2$, $n_1 = 7$, $n_2 = 10$, $f = 7.36$.
d. $H_a: \sigma_1^2 > \sigma_2^2$, $n_1 = 31$, $n_2 = 26$, $f = 4.26$.

10.97 Consider independent random samples of sizes $n_1 = 31$ and $n_2 = 25$ from normal populations.
a. Write the four parts for a one-sided, right-tailed hypothesis test concerning the population variances with $\alpha = 0.05$.
b. Suppose $s_1^2 = 44.89$ and $s_2^2 = 17.64$. Find the value of the test statistic, and draw a conclusion about the population variances.

c. Find bounds on the *p* value associated with this hypothesis test and carefully sketch a graph to illustrate this value.

10.98 Consider the two independent random samples from normal distributions given in the following tables.

Sample 1

89.6	61.1	83.7	74.2	60.6	50.4	82.4	79.0
56.5	72.2	72.4	77.1	58.2	72.3	71.6	70.3
76.1	73.2						

Sample 2

73.7	37.8	76.2	64.7	74.8	75.8	67.9	61.6
76.2	82.7	88.9	60.3	34.4	74.5	68.0	100.2
73.7	41.5	76.9	55.0	76.0			

a. Write the four parts for a one-sided, left-tailed hypothesis test concerning the two population variances with $\alpha = 0.05$.
b. Compute each sample variance, find the value of the test statistic, and draw a conclusion.
c. Find bounds on the *p* value associated with this hypothesis test.

10.99 Consider independent random samples of sizes $n_1 = 10$ and $n_2 = 16$ from normal populations.
a. Write the four parts for a two-sided hypothesis test concerning the population variances with $\alpha = 0.01$.

b. Suppose that $s_1^2 = 426.42$ and $s_2^2 = 88.36$. Find the value of the test statistic, and draw a conclusion about the population variances.

c. Find bounds on the p value associated with this hypothesis test.

10.100 In each of the following problems, the sample sizes and the confidence level are given. Find the appropriate F critical values for use in constructing a confidence interval for the ratio of the population variances.

a. $n_1 = 10$, $n_2 = 10$, 90%
b. $n_1 = 21$, $n_2 = 31$, 98%
c. $n_1 = 9$, $n_2 = 7$, 98%
d. $n_1 = 41$, $n_2 = 31$, 99.8%

10.101 In each of the following problems, the sample sizes, the sample variances, and the confidence level are given. Assume the underlying populations are normal and the samples were obtained independently. Find the associated confidence interval for the ratio of the population variances.

a. $n_1 = 10$, $s_1^2 = 17.2$, $n_2 = 9$, $s_2^2 = 15.6$, 90%.
b. $n_1 = 16$, $s_1^2 = 54.1$, $n_2 = 16$, $s_2^2 = 32.6$, 98%.
c. $n_1 = 16$, $s_1^2 = 3.35$, $n_2 = 31$, $s_2^2 = 4.59$, 98%.
d. $n_1 = 31$, $s_1^2 = 126.8$, $n_2 = 41$, $s_2^2 = 155.3$, 99.8%.

10.102 Use Table VII in the Appendix and linear interpolation to approximate each critical value. Verify each approximation using technology.

a. $F_{0.05}$, df 25 and 15
b. $F_{0.99}$, df 20 and 32
c. $F_{0.01}$, df 10 and 56
d. $F_{0.025}$, df 15 and 20
e. $F_{0.995}$, df 10 and 7
f. $F_{0.05}$, df 35 and 35

Applications

10.103 **Biology and Environmental Science** In a recent government study, the aerosol light absorption coefficient was measured (in Mm^{-1}) at randomly selected locations in Africa and in South America. The resulting data are summarized in the following table.

Country	Sample size	Sample variance
Africa	10	243.36
South America	21	51.84

(*Source*: National Oceanic and Atmospheric Administration, Global Monitoring Division.)

Is there any evidence that the population variance in aerosol light absorption coefficient is greater in Africa than in South America? Use $\alpha = 0.05$ and assume normality.

10.104 **Sports and Leisure** According to the Internet Movie Database, the top-grossing movie of 2008 was *The Dark Knight*. Independent random samples of films produced by 20th Century Fox and by Paramount Pictures were obtained, and the gross ticket sales (in millions of dollars) for each was recorded. For 20th Century Fox, $n_1 = 9$ and $s_1^2 = 985.96$; for Paramount Pictures, $n_2 = 9$ and $s_2^2 = 756.25$. Is there any evidence to suggest that the variability in gross ticket sales per film is different for these two studios? Use $\alpha = 0.05$ and assume normality.

10.105 **Manufacturing and Product Development** In 2008, Minnesota and North Carolina raised the most turkeys in the United States, approximately 49 million and 39 million birds each, respectively.[19] Independent random samples of frozen turkeys from each state were obtained, and each turkey was weighed. The resulting data (in pounds) are given in the following tables.

Minnesota								
10.1	11.5	17.1	13.4	15.9	17.9	14.7	9.5	11.4
14.5	12.5	14.2	16.8	13.7	16.0	19.4		

North Carolina								
19.9	14.0	19.9	12.3	17.0	25.2	23.9	7.8	15.8
21.2	13.8	7.4	15.1	10.1	3.2	17.4		

Is there any evidence to suggest that there is more variability in the weight of frozen turkeys from North Carolina than from Minnesota? Use $\alpha = 0.01$ and assume normality.

10.106 **Sports and Leisure** Many basketball purists believe that the three-point shot (a shot from behind the three-point line, 22 feet from the basket) has dramatically changed the game, for the worse. Independent random samples of attempted shots from National Basketball Association games played in 1975 (prior to the three-point shot) and in 2008 were obtained. The shot distance (in feet) was recorded for each attempt. The data are given in the following table.

Year	Sample size	Sample variance
1975	61	12.25
2008	61	26.01

Is there any evidence to suggest that the variability in shot distance is greater in the year 2008 than it was in 1975? Use $\alpha = 0.01$ and assume normality. (Why do you suppose there is greater variability in shot distance with a three-point shot?)

10.107 **Sports and Leisure** A Laurel Downs racetrack official believes that there is less variability in winning times for a race in which the purse is at least $10,000, called a stakes race. Independent random samples of ordinary races and stakes races were obtained, and the winning time (in seconds) for each race was recorded. The summary statistics for an ordinary race were $n_1 = 26$ and $s_1^2 = 110.25$, and for a stakes race were $n_2 = 26$ and $s_2^2 = 38.44$.

a. Write the four parts for a hypothesis test to check for evidence of the official's assertion. Use $\alpha = 0.01$, assume normality, and find the critical value using technology.

b. Construct a 98% confidence interval for the ratio of population variances.

10.108 **Sports and Leisure** A study was conducted to compare the variability in times for men and women involved in collegiate swimming events. Independent random samples of 800-meter freestyle competitors were obtained, and the time (in minutes) was recorded for each swimmer. The data are summarized in the following table.

Group	Sample size	Sample variance
Men	11	0.1025
Women	12	0.1241

(*Source*: Loyola and George Washington versus Navy swimming meet, October 15, 2005.)

a. Find the critical values necessary to construct a 95% confidence interval for the ratio of population variances.
b. Construct the confidence interval.

10.109 Travel and Transportation Additional airport security and heightened safety concerns have caused more and longer flight delays. Independent random samples of Delta and United flights into Chicago's O'Hare International Airport were obtained, and the arrival delay (in minutes) for each flight was recorded.[20] A delay of 0 means the flight arrived on time. The data are given on the data CD and book's web site.

a. Is there any evidence of a difference in variability of flight-delay times for Delta and United? Use $\alpha = 0.01$ and assume normality.
b. Do you think the normality assumption is reasonable? Why or why not?

10.110 Manufacturing and Product Development A sailboat manufacturer has two machines for constructing main mast poles with diameter designed to be 76.2 mm. Small variability in production is very important to ensure boat control and safety. Independent random samples of mast poles produced on each machine were obtained, and the diameter of each was carefully measured. The summary statistics for Machine A were $n_1 = 7$, $s_1^2 = 0.0231$, and for Machine B were $n_2 = 8$, $s_2^2 = 0.0096$. Conduct the appropriate hypothesis test to determine whether there is any evidence of a difference in variability of mast–pole diameter between the two machines. Assume normality, find the p value associated with this test, and use this value to draw a conclusion.

10.111 Business and Management Government employees traveling in the United States on business are reimbursed according to the domestic per diem rates. A random sample of counties in the northeast and west was obtained and the Fiscal Year 2009 lodging per diem rate was obtained for each.[21] The summary statistics are given in the following table.

Region	Sample size	Sample variance
Northeast	25	2284.827
West	16	254.867

Is there any evidence to suggest that the variability in lodging per diem is greater in the Northeast than in the West? Use $\alpha = 0.01$ and assume normality.

10.112 Education and Child Development Tuition at public and private U.S. colleges and universities has risen dramatically over the past three decades. Independent random samples of schools were obtained, and the current tuition and fees (in dollars) was recorded for each. The data are summarized in the following table.[22]

School type	Sample size	Sample variance
Tier 1	20	4.3477
Tier 3	20	37.5819

Is there any evidence to suggest that the variability in tuition at Tier 1 colleges is less than the variability in tuition at Tier 3 colleges? Use $\alpha = 0.001$ and assume normality.

10.113 Business and Management The *Chicago Sun-Times* and *The Boston Globe* have similar circulation rates. According to figures filed with the Audit Bureau of Circulations, they both have a circulation of approximately 480,000 copies per weekday. Independent weekdays were selected at random and each paper's circulation was recorded. For the *Sun-Times*, $n_1 = 35$, $s_1 = 9722.3$, and for *The Globe*, $n_2 = 25$, $s_2 = 13119.0$.

a. Conduct the appropriate test to determine whether there is any difference in the population variance of circulation rates. Use $\alpha = 0.02$.
b. Do you believe that the normality assumption is reasonable in this example? Why or why not?

Extended Applications

10.114 Physical Sciences Crude oil pumped from ocean wells contains salt that must be removed before the oil is refined. Otherwise, equipment would erode quickly. Independent random samples of unrefined crude oil from two ocean wells were obtained, and the percentage of salt in each sample was recorded. The data are given in the following table.

Oil well	Sample size	Sample variance
North Sea	21	56.40
Antarctica	31	82.42

a. Conduct the appropriate test to determine whether there is evidence of any difference in variability of salt content between these two wells. Use $\alpha = 0.10$ and assume normality.
b. Use Table VII in the Appendix to find bounds on the p value for this hypothesis test. Use technology to find an exact p value.

10.115 Public Health and Nutrition Saccharin is a low-calorie sweetener used in sugar-free foods and beverages. According to the FDA, the acceptable daily intake (ADI) of saccharin is 5 mg for a person with a body weight of 60 kg. If saccharin is used as an additive, it must be included on the food label and cannot exceed certain limits. Independent random samples of 12-ounce bottles of iced tea from two different manufacturers were obtained, and the amount of saccharin in each drink was measured (in mg). The summary statistics for Fishing Creek were $n_1 = 20$, $s_1^2 = 7.84$, and for Honest Tea were $n_2 = 15$, $s_2^2 = 2.89$.

a. Conduct the appropriate test to determine whether there is any difference in the population variance of saccharin amounts. Use $\alpha = 0.02$.

b. Find bounds on the p value associated with this hypothesis test.

Challenge

10.116 Physical Sciences When we are comparing two population variances, if both sample sizes, n_1 and n_2, are large and the null hypothesis, $H_0: \sigma_1^2 = \sigma_2^2$, is true, then the test statistic $F = S_1^2/S_2^2$ is approximately normal with

$$\mu_F = \frac{n_2 - 1}{n_2 - 3} \quad \text{and} \quad \sigma_F^2 = \frac{2(n_2 - 1)^2(n_1 + n_2 - 4)}{(n_1 - 1)(n_2 - 3)^2(n_2 - 5)}.$$

An approximate hypothesis test is based on standardizing F to a Z random variable.

The four parts of the hypothesis test are

$H_0: \sigma_1^2 = \sigma_2^2$

$H_a: \sigma_1^2 > \sigma_2^2, \quad \sigma_1^2 < \sigma_2^2, \quad \text{or} \quad \sigma_1^2 \neq \sigma_2^2$

$$\text{TS: } Z = \frac{(S_1^2/S_2^2) - [(n_2 - 1)/(n_2 - 3)]}{\sqrt{\dfrac{2(n_2 - 1)^2(n_1 + n_2 - 4)}{(n_1 - 1)(n_2 - 3)^2(n_2 - 5)}}}$$

RR: $Z \geq z_\alpha, \quad Z \leq -z_\alpha \quad \text{or} \quad |Z| \geq z_{\alpha/2}$

The National Wind Energy Assessment contains data from 975 stations and includes measurements of wind speed and wind power density. Suppose independent random samples of wind power density (in watts/m^2) during the winter were obtained from two stations. The data are summarized in the following table.

Location	Sample size	Sample mean	Sample variance
Chanute	31	207	95.35
Dodge City	31	283	53.68

(*Source:* National Renewable Energy Laboratory.)

a. Write the four parts of a large-sample, two-sided, approximate test based on the standard normal distribution to determine whether there is any evidence to suggest that the two population variances are different. Conduct the test using $\alpha = 0.05$.

b. Conduct an *exact* hypothesis test based on the F distribution. Compare your answer to your answer for part (a).

(Manchan/Getty Images)

Chapter 10 Challenge Wrap-Up

In the survey of participation in arts and culture, let adult New York City residents be population 1 and child residents be population 2. We are looking for any difference in population proportions of people who have visited an art museum within the past year. A two-sided hypothesis test concerning population proportions with $\Delta_0 = 0$ is appropriate. Use $\alpha = 0.05$.

The summary statistics are as follows.

	Adults (ages 18 and over)	Children (under 18)
Sample size	$n_1 = 125$	$n_2 = 150$
Number of successes	$x_1 = 54$	$x_2 = 89$
Sample proportion	$\hat{p}_1 = 54/125 = 0.4320$	$\hat{p}_2 = 89/150 = 0.5933$

Check the nonskewness criteria using estimates for p_1 and p_2.

$n_1\hat{p}_1 = (125)(0.4320) = 54 \geq 5 \qquad n_1(1 - \hat{p}_1) = (125)(0.5680) = 71 \geq 5$

$n_2\hat{p}_2 = (150)(0.5933) = 89 \geq 5 \qquad n_2(1 - \hat{p}_2) = (150)(0.4067) = 61 \geq 5$

Since all of the inequalities are satisfied, the difference in sample proportions is approximately normal, and the large-sample hypothesis test concerning population proportions can be used.

The four parts of the hypothesis test are

$H_0: p_1 - p_2 = 0$

$H_a: p_1 - p_2 \neq 0$

$$\text{TS: } Z = \frac{\hat{P}_1 - \hat{P}_2}{\sqrt{\hat{P}_c(1 - \hat{P}_c)\left(\dfrac{1}{n_1} + \dfrac{1}{n_2}\right)}}$$

RR: $|Z| \geq z_{\alpha/2} = z_{0.025} = 1.9600$

The estimate of the common population proportion is

$$\hat{p}_c = \frac{x_1 + x_2}{n_1 + n_2} = \frac{54 + 89}{125 + 150} = 0.5200$$

The value of the test statistic is

$$z = \frac{\hat{p}_1 - \hat{p}_2}{\sqrt{\hat{p}_c(1 - \hat{p}_c)\left(\dfrac{1}{n_1} + \dfrac{1}{n_2}\right)}} = \frac{0.4320 - 0.5933}{\sqrt{(0.5200)(0.4800)\left(\dfrac{1}{125} + \dfrac{1}{150}\right)}} = -2.6659$$

The value of the test statistic lies in the rejection region. We reject the null hypothesis at the $\alpha = 0.05$ significance level. There is evidence to suggest that the population proportion of adults is different from the population proportion of children who have visited an art museum within the past year.

Figure 10.54 shows a technology solution.

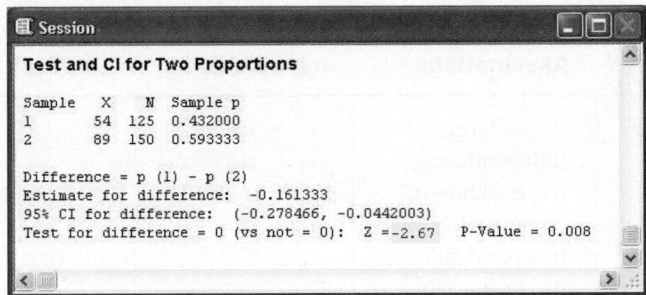

Figure 10.54
Hypothesis test
results and confidence
interval.

CHAPTER 10 SUMMARY

Concept	Page	Notation / Formula / Description
Independent samples	447	Two samples are independent if the process of selecting individuals or objects in sample 1 has no effect on the selection of individuals or objects in sample 2.
Paired data set	447	The result of matching each individual or object in sample 1 with a similar individual or object in sample 2.
Pooled estimator for the common variance	458	$S_p^2 = \dfrac{(n_1 - 1)S_1^2 + (n_2 - 1)S_2^2}{n_1 + n_2 - 2}$
Combined estimate of the common population proportion	484	$\hat{P}_c = \dfrac{X_1 + X_2}{n_1 + n_2}$

Summary of confidence intervals

Parameter	Assumptions	$100(1 - \alpha)\%$ Confidence Interval
$\mu_1 - \mu_2$	n_1, n_2 large, independence, σ_1^2, σ_2^2 known, or normality, independence, σ_1^2, σ_2^2 known.	$(\bar{x}_1 - \bar{x}_2) \pm z_{\alpha/2}\sqrt{\dfrac{\sigma_1^2}{n_1} + \dfrac{\sigma_2^2}{n_2}}$
$\mu_1 - \mu_2$	normality, independence, σ_1^2, σ_2^2 unknown but equal.	$(\bar{x}_1 - \bar{x}_2) \pm t_{\alpha/2}\sqrt{s_p^2\left(\dfrac{1}{n_1} + \dfrac{1}{n_2}\right)}$ $s_p^2 = \dfrac{(n_1 - 1)s_1^2 + (n_2 - 1)s_2^2}{n_1 + n_2 - 2}, \text{df} = n_1 + n_2 - 2$
$\mu_1 - \mu_2$	normality, independence, σ_1^2, σ_2^2 unknown and unequal.	$(\bar{x}_1 - \bar{x}_2) \pm t_{\alpha/2}\sqrt{\dfrac{s_1^2}{n_1} + \dfrac{s_2^2}{n_2}}$ $\text{df} = \dfrac{\left(\dfrac{s_1^2}{n_1} + \dfrac{s_2^2}{n_2}\right)^2}{\dfrac{(s_1^2/n_1)^2}{n_1 - 1} + \dfrac{(s_2^2/n_2)^2}{n_2 - 1}}$
$\mu_D = \mu_1 - \mu_2$	normality, n pairs, dependence.	$\bar{d} \pm t_{\alpha/2}\dfrac{s_D}{\sqrt{n}}, \text{df} = n - 1$
$p_1 - p_2$	n_1, n_2 large, nonskewness, independence.	$(\hat{p}_1 - \hat{p}_2) \pm z_{\alpha/2}\sqrt{\dfrac{\hat{p}_1(1 - \hat{p}_1)}{n_1} + \dfrac{\hat{p}_2(1 - \hat{p}_2)}{n_2}}$
$\dfrac{\sigma_1^2}{\sigma_2^2}$	normality, independence.	$\left(\dfrac{s_1^2}{s_2^2}\dfrac{1}{F_{\alpha/2}}, \dfrac{s_1^2}{s_2^2}\dfrac{1}{F_{1-\alpha/2}}\right), \text{df } n_1 - 1 \text{ and } n_2 - 1$

Summary of hypothesis tests

Null hypothesis	Assumptions	Alternative hypothesis	Test statistic	Rejection region
$\mu_1 - \mu_2 = \Delta_0$	n_1, n_2 large, independence, σ_1^2, σ_2^2 known, or normality, independence, σ_1^2, σ_2^2 known.	$\mu_1 - \mu_2 > \Delta_0$ $\mu_1 - \mu_2 < \Delta_0$ $\mu_1 - \mu_2 \neq \Delta_0$	$Z = \dfrac{(\overline{X}_1 - \overline{X}_2) - \Delta_0}{\sqrt{\dfrac{\sigma_1^2}{n_1} + \dfrac{\sigma_2^2}{n_2}}}$	$Z \geq z_\alpha$ $Z \leq -z_\alpha$ $|Z| \geq z_{\alpha/2}$
$\mu_1 - \mu_2 = \Delta_0$	normality, independence, σ_1^2, σ_2^2 unknown, $\sigma_1^2 = \sigma_2^2$.	$\mu_1 - \mu_2 > \Delta_0$ $\mu_1 - \mu_2 < \Delta_0$ $\mu_1 - \mu_2 \neq \Delta_0$	$T = \dfrac{(\overline{X}_1 - \overline{X}_2) - \Delta_0}{\sqrt{S_p^2\left(\dfrac{1}{n_1} + \dfrac{1}{n_2}\right)}}$ $S_p^2 = \dfrac{(n_1 - 1)S_1^2 + (n_2 - 1)S_2^2}{n_1 + n_2 - 2}$	$T \geq t_\alpha$ $T \leq -t_\alpha$ $|T| \geq t_{\alpha/2}$ $\mathrm{df} = n_1 + n_2 - 2$
$\mu_1 - \mu_2 = \Delta_0$	normality, independence, σ_1^2, σ_2^2 unknown, $\sigma_1^2 \neq \sigma_2^2$.	$\mu_1 - \mu_2 > \Delta_0$ $\mu_1 - \mu_2 < \Delta_0$ $\mu_1 - \mu_2 \neq \Delta_0$	$T' = \dfrac{(\overline{X}_1 - \overline{X}_2) - \Delta_0}{\sqrt{\dfrac{S_1^2}{n_1} + \dfrac{S_2^2}{n_2}}}$ $\nu = \dfrac{\left(\dfrac{s_1^2}{n_1} + \dfrac{s_2^2}{n_2}\right)^2}{\dfrac{(s_1^2/n_1)^2}{n_1 - 1} + \dfrac{(s_2^2/n_2)^2}{n_2 - 1}}$	$T' \geq t_\alpha$ $T' \leq -t_\alpha$ $|T'| \geq t_{\alpha/2}$ $\mathrm{df} = \nu$
$\mu_D = \Delta_0$	normality, n pairs, dependence.	$\mu_D > \Delta_0$ $\mu_D < \Delta_0$ $\mu_D \neq \Delta_0$	$T = \dfrac{\overline{D} - \Delta_0}{S_D/\sqrt{n}}$	$T \geq t_\alpha$ $T \leq -t_\alpha$ $|T| \geq t_{\alpha/2}$ $\mathrm{df} = n - 1$
$p_1 - p_2 = 0$	n_1, n_2 large, nonskewness, independence.	$p_1 - p_2 > 0$ $p_1 - p_2 < 0$ $p_1 - p_2 \neq 0$	$Z = \dfrac{\hat{P}_1 - \hat{P}_2}{\sqrt{\hat{P}_c(1 - \hat{P}_c)\left(\dfrac{1}{n_1} + \dfrac{1}{n_2}\right)}}$ $\hat{P}_c = \dfrac{X_1 + X_2}{n_1 + n_2}$	$Z \geq z_\alpha$ $Z \leq -z_\alpha$ $|Z| \geq z_{\alpha/2}$
$p_1 - p_2 = \Delta_0$	n_1, n_2 large, nonskewness, independence.	$p_1 - p_2 > \Delta_0$ $p_1 - p_2 < \Delta_0$ $p_1 - p_2 \neq \Delta_0$	$Z = \dfrac{(\hat{P}_1 - \hat{P}_2) - \Delta_0}{\sqrt{\dfrac{\hat{P}_1(1 - \hat{P}_1)}{n_1} + \dfrac{\hat{P}_2(1 - \hat{P}_2)}{n_2}}}$	$Z \geq z_\alpha$ $Z \leq -z_\alpha$ $|Z| \geq z_{\alpha/2}$
$\sigma_1^2 = \sigma_2^2$	normality, independence.	$\sigma_1^2 > \sigma_2^2$ $\sigma_1^2 < \sigma_2^2$ $\sigma_1^2 \neq \sigma_2^2$	$F = \dfrac{S_1^2}{S_2^2}$	$F \geq F_\alpha$ $F \leq F_{1-\alpha}$ $F \leq F_{1-\alpha/2}$ or $F \geq F_{\alpha/2}$ $\mathrm{df}\ n_1 - 1$ and $n_2 - 1$

CHAPTER 10 EXERCISES

Applications

10.117 Travel and Transportation The U.S. Department of Transportation requires vehicles transporting hazardous materials to use special placards indicating the type of cargo. There are many other regulations involving containers, separation of various materials, and gross weight. Independent random samples of trucks carrying corrosive materials were stopped on highways in North Carolina and in Virginia, and the weight (in kg) of the hazardous material was recorded. The summary statistics and known variances are given in the following table.

State	Sample size	Sample mean	Population variance
North Carolina	22	835.6	3192.25
Virginia	25	884.2	3956.41

Is there any evidence to suggest that the mean amount of corrosive material carried by trucks in North Carolina is different from the mean amount of corrosive material carried by trucks in Virginia? Use $\alpha = 0.01$ and assume each underlying distribution of weight is normal.

10.118 Psychology and Human Behavior According to a 2008 report from comScore, the most memorable advertisement during the 2007 Super Bowl was from Anheuser-Busch. A survey was conducted to determine viewers' favorite part of watching the Super Bowl. Independent random samples of men and women who watched the most recent Super Bowl were obtained, and the viewers were asked to identify their favorite part of watching the Super Bowl. Of the 902 men, 144 watched for the ads, and 217 of the 620 women watched for the ads.[23] Is there any evidence to suggest that the population proportions of men and women who watched for the ads are different? Use $\alpha = 0.001$.

10.119 Psychology and Human Behavior According to the National Funeral Directors Association, Nevada and Washington had two of the highest cremation rates in 2005. In a recent survey, independent random samples of deaths in these two states were obtained, and the number of cremations was recorded. The data are summarized in the following table.

State	Sample size	Number of cremations
Nevada	544	335
Washington	603	381

a. Find the sample proportion of cremation for each state. Verify the nonskewness criteria.
b. Conduct the appropriate hypothesis test to determine whether there is any evidence that the two population proportions are different. Use $\alpha = 0.01$.
c. Find the p value for the hypothesis test in part (b).

10.120 Biology and Environmental Science Maple syrup producers in New York and Vermont collect sweetwater sap from sugar maples and black maples in early spring. It takes approximately 30-50 gallons of sap to yield, through boiling and evaporation, 1 gallon of maple syrup. Independent random samples of maple trees in both states were obtained, and the amount of sap collected from each tree was recorded. The data are given on the data CD and book's web site. Assume the underlying populations are normal, with equal variances. Is there any evidence to suggest that the population mean amount of sap from trees in Vermont is different from the population mean amount of sap from trees in New York? Use $\alpha = 0.01$.

10.121 Physical Sciences Recycling of aluminum, glass, newspapers, and magazines is good for the environment and the economy. According to *Waste News*, Portland, Oregon, has the highest recycling rate of any state (recycling rate = tons collected for recycling/tons of all waste generated). Despite efforts to make the process easier, many people still do not recycle. Independent random samples of residents in Ohio and in Florida were obtained, and residents were asked whether they recycle newspapers. Of the 909 Ohio residents, 700 said they recycled newspapers, and of the 923 Florida residents, 691 said they recycled newspapers.
a. Is there any evidence to suggest that the population proportion of residents in Ohio who recycle newspapers is greater than the population proportion of residents in Florida? Use $\alpha = 0.01$.
b. Find the p value for this hypothesis test.

10.122 Sports and Leisure Archery target shooters use a variety of arrows made from wood, carbon, aluminum, or even platinum. One measure of the quality of an arrow (and bow) is the speed of the arrow when shot. A random sample of archers was obtained, and each was asked to shoot a carbon arrow and a similarly made aluminum arrow. The speed (in feet per second) of each arrow was measured. The data are given on the data CD and book's web site.
a. What is the common characteristic that makes this data paired?
b. Assume normality. Conduct the appropriate hypothesis test to determine whether there is any evidence that the aluminum arrow flies faster. Use $\alpha = 0.05$.
c. Find bounds on the p value associated with this hypothesis test.

10.123 Manufacturing and Product Development Raytheon Aircraft is now manufacturing business jets with a molded carbon fiber fuselage instead of aluminum. This reduces the overall weight of the plane, speeds production time, and increases cabin space. The total wall thickness of a carbon fiber fuselage is 0.81 inches versus 3 inches for aluminum, and the variability in thickness is theoretically much smaller also. Independent random samples of the two fuselage types were obtained, and the thickness of each was measured (in inches). The data are given in the following table.

Fuselage type	Sample size	Sample variance
Aluminum	9	0.0196
Carbon fiber	11	0.0025

Is there any evidence to suggest that the variability in fuselage thickness is less for carbon fiber fuselages? Use $\alpha = 0.01$.

10.124 Public Health and Nutrition Iron is an important dietary supplement that aids in oxygen transport and metabolism. There are many sources of iron. Some breads are a good source of protein, vitamins, calcium, and iron. Independent random samples of white breads and whole-wheat breads were obtained, and the iron content (in mg) in each slice was measured. The data are given on the data CD and book's web site. Assume the underlying populations are normal, with equal variances. Is there any evidence to suggest that the population mean amount of iron in whole wheat bread is greater than the population mean amount of iron in white bread? Use $\alpha = 0.005$.[24]

10.125 Biology and Environmental Science A study was conducted to evaluate the contaminants in fish caught in the Housatonic River. Independent random samples of Smallmouth Bass from two locations were obtained and each fish was tested for polychlorinated biphenyls (PCBs). Summary data for PCBs (in parts per million, ppm) are given in the following table.

Location	Sample size	Sample mean	Sample standard deviation
Bull's Bridge	10	1.08	0.32
Lake Zoar	10	0.58	0.52

(*Source*: Public Health Evaluation of Fish Contaminant Data in the Housatonic River, U.S. Department of Health and Human Services, July 8, 2008).

Assume the underlying populations are normal. However, it is likely that the population variances are unequal. Conduct the appropriate hypothesis test to determine whether the population mean PCB level in Smallmouth Bass is greater at Bull's Bridge than at Lake Zoar. Use $\alpha = 0.001$.

10.126 Biology and Environmental Science The moisture content in bulk grain is important because high levels can encourage the development of fungi. Potential buyers want to know how much water they are buying along with their grain. Two direct methods for measuring the moisture content are by means of a chemical reaction (with iodine in the presence of sulfur dioxide) and by distillation. A random sample of bulk grain from around the country was obtained, and the moisture content of each grain sample was measured using each method. The percentage of water measured with each method is given on the data CD and book's web site. Assuming normality, conduct the appropriate hypothesis test to determine whether there is any difference in the population mean moisture content of bulk grain measured by chemical reaction and by distillation. Use $\alpha = 0.05$.

10.127 Sports and Leisure During the last decade, concert ticket prices have risen sharply. In 2004, the mean ticket price for the 100 biggest tours was $52.39.[25] Suppose independent random samples of ticket prices for summer 2005 concerts by the Rolling Stones and Coldplay were obtained. The data are summarized in the following table.

Group	Sample size	Sample mean	Sample standard deviation
Rolling Stones	12	325.50	52.50
Coldplay	10	267.90	68.60

Assume normality and equal population variances. Is there any evidence to suggest that the population mean concert ticket price is greater for the Rolling Stones than for Coldplay? Use $\alpha = 0.001$.

10.128 Public Policy and Political Science California law requires fuel outlets to install special catch basins designed to contain gasoline leaks in underground storage tanks. Owners who do not comply can face stiff fines and other penalties. Independent random samples of gasoline stations around Los Angeles and around San Francisco were obtained, and each station was inspected for catch basins. Sixteen of 140 stations near Los Angeles had no catch basins, and 12 of 126 in San Francisco were not complying with the law.

a. Find the sample proportion of stations without catch basins near each city. Verify the nonskewness criteria.

b. Is there any evidence that the population proportion of stations in noncompliance with the law is different near Los Angeles and near San Francisco? Use $\alpha = 0.01$.

10.129 Biology and Environmental Science The ambient air quality is assessed by measuring the amount of carbon monoxide, lead, particulates, and other pollutants in the air. The data on the data CD and book's web site are a subset of measurements of the levels of fine particulates (in $\mu g/m^3$) in two areas in Texas on October 19, 2005.[26] Assume normality and equal variances. Is there any evidence to suggest that the population mean fine particulate measure is different in these two areas? Use $\alpha = 0.05$.

Extended Applications

10.130 Economics and Finance The Internal Revenue Service estimates that the average taxpayer takes approximately six hours to complete form 1040. A study was conducted to examine the amount of time it takes to complete this dreaded form, by income level. Independent random samples of federal filers in two income ranges were obtained, and the length of time (in hours) to complete form 1040 was recorded for each. The summary statistics are given in the following table.

Income level (in dollars)	Sample size	Sample mean	Sample variance
50,000- <100,000	17	4.56	1.5625
100,000- <200,000	14	6.58	15.0544

Assume the underlying populations are normal.

a. Conduct an F test to determine whether there is any evidence that the two population variances are different. Use $\alpha = 0.02$.

b. Using your conclusion from part (a), conduct the appropriate test for evidence that the mean time to complete form 1040 for the lower income level is less than the mean time for the higher income level. Use $\alpha = 0.05$. State your conclusion and find bounds on the p value.

10.131 Public Policy and Political Science In many states, lawyers are encouraged to do pro bono work by both their firms and judicial advisory councils. However, in recent years lawyers have been devoting more time to paying clients and less time to pro bono legal aid. Independent random samples of lawyers from two large firms were obtained, and the number of pro bono hours for the past year was recorded for each lawyer. The summary statistics are given in the following table.

Law firm	Sample size	Sample mean	Sample variance
Dewey, Cheatum, & Howe	26	75.1	5.92
Howard, Fine, & Howard	26	80.9	5.65

Assume the underlying populations are normal, with equal variances.

a. Is there any evidence to suggest that the mean number of yearly pro bono hours is different at these two law firms? Use $\alpha = 0.01$.

b. Construct a 99% confidence interval for the difference in mean pro bono hours.

c. Does the confidence interval in part (b) support your conclusion in part (a)? Explain.

10.132 Economics and Finance Online investing has grown with the Internet and with companies like E*TRADE and TD AMERITRADE. Independent random samples of investors were obtained, and each was asked whether they traded online within the past year. The data are given in the following table, by portfolio size.

Portfolio	Sample size	Number of online traders
Less than $100,000	348	132
At least $100,000	226	65

a. Compute the sample proportions and verify the non-skewness criteria.

b. Construct a 95% confidence interval for the difference in population proportions of online investors.

c. Using the interval in part (b), is there any evidence to suggest that the population proportion of online investors is different for these two portfolio classifications? Justify your answer.

10.133 Public Policy and Political Science Over the past two decades, the number of discrimination (race, age, sex, and disability) lawsuits has increased significantly. Independent random samples of jury awards in discrimination cases were obtained, and the summary statistics are given (in thousands of dollars) in the following table.

Discrimination lawsuit	Sample size	Sample mean	Sample standard deviation
Race	15	120.9	203.5
Age	15	268.0	185.6
Disability	16	175.0	195.6

(*Source*: M. McCogg, Employment lawsuit costs skyrocket, CMM Online, Jan. 24, 2002.)

Assume all three underlying populations are normal, with equal variances. Conduct the appropriate hypothesis tests to determine which pairs of population mean jury awards are different. Use $\alpha = 0.05$ in each test.

10.134 Physical Sciences Independent random samples of ore taken from two high-grade gold mines were obtained, and the gold value (in grams/tonne) was measured for each. The summary statistics were

El Aguila mine	$n_1 = 8$,	$\bar{x}_1 = 15.6$,	$s_1 = 5.2$
Dolaucothi mine	$n_2 = 11$,	$\bar{x}_2 = 26.8$,	$s_2 = 21.6$

Assume normality and unequal variances.

a. Conduct the appropriate hypothesis test to determine whether there is any evidence that the population mean gold value at the El Aguila mine is less than at the Dolaucothi mine. Use $\alpha = 0.05$.

b. Your conclusion in part (a) should be that there is no evidence to suggest a difference. Explain why this result is correct even though the sample means are *very far apart*.

The Analysis of Variance

Chapter 11 Challenge

Does one region grow the healthiest food?

As oxygen is used by your body, some molecules are converted to free radicals. These highly reactive groups of atoms can attack cell membranes, damage the genetic code in DNA, and trigger the development of cancer.

Fortunately, many of the foods we eat protect our bodies from free radicals. Antioxidants, which include vitamin C, vitamin E, beta-carotene, and selenium, are natural neutralizing agents found in fruits, vegetables, and grains. Empirical evidence suggests that people who include fruits and vegetables in their daily diet tend to have lower incidence rates of cancer.

(Juice Images/AgeFotostock)

Carrots contain a higher amount of beta-carotene than other vegetables. Researchers are investigating whether carrots grown in different parts of the country contain the same amount of beta-carotene. Random samples of 8-ounce carrot servings from four regions were obtained, and the amount of beta-carotene in each serving was measured (in mg). The summary statistics are given in the following table.

Region	Sample size	Sample mean	Sample standard deviation
North	10	11.93	0.623
South	10	11.53	0.821
Midwest	10	11.48	0.947
West	10	12.34	1.101

It looks like carrots grown in the West might be better in the fight against free radicals. The statistical techniques presented in this chapter can be used to determine whether *any* two population mean amounts of beta-carotene are different. The results are presented in the Chapter Challenge Wrap-Up (page 550). If at least two means are different, then we need to identify which pairs of means are contributing to an overall difference.

Review

- Recall the statistical tests we use to compare two population means.
- Remember the assumptions and the four parts to a two-sample *t* test.

CONTENTS

Preview

- Learn how to compare $k\ (> 2)$ population means.
- Understand the assumptions related to and conduct one-way, single-factor analysis of variance tests.
- Learn methods to isolate group differences if an analysis of variance test is significant.
- Understand the assumptions and hypothesis tests and learn how the total variation is decomposed in a two-way ANOVA.

11.1 ONE-WAY ANOVA

A one-way, or single-factor, ANOVA involves the analysis of data sampled from more than two populations. The only difference among the populations is a single factor. For example, consider a study in which random samples of the amount of carbon dioxide in underground train tunnels are obtained in four different cities. The data may be used to determine whether there is any difference in the mean amount of carbon dioxide (in train tunnels) among the four cities. The single factor that varies among the populations is the city. Or suppose a researcher is investigating techniques for controlling the amount of electricity lost during transmission over utility lines. Experimental results may be used to determine whether there is any difference in the mean amount of electricity lost for five differently designed lines. The single factor here is the design of the electricity line. In theory, everything else is the same among the five populations: the initial amount of electricity transmitted, the distance the electricity is transmitted, the weather conditions, etc.

Suppose three random samples are obtained, and a histogram is constructed using *all* of the data. The resulting graph is shown in Figure 11.1. Analysis of variance is used to determine whether the data came from a single population or whether at least two samples came from populations with different means.

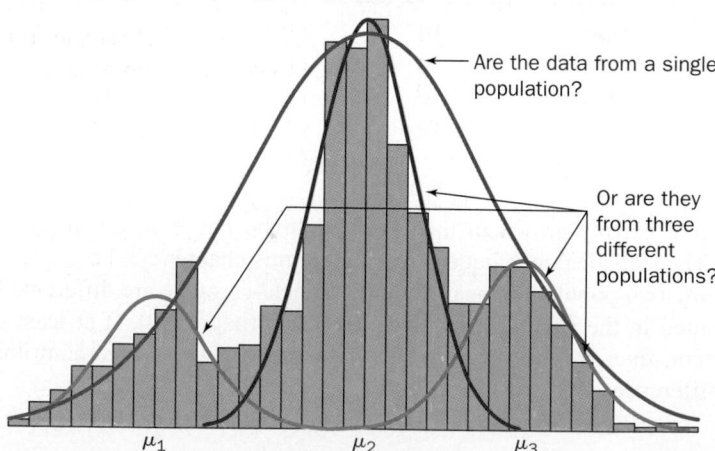

Figure 11.1 A typical ANOVA problem.

The notation used in a one-way ANOVA is similar to and is an extension of the notation used in the previous chapter.

ANOVA NOTATION

k = the number of populations under investigation.

Population	1	2	$\cdots$	i	$\cdots$	k
Population mean	μ_1	μ_2	$\cdots$	μ_i	$\cdots$	μ_k
Population variance	σ_1^2	σ_2^2	$\cdots$	σ_i^2	$\cdots$	σ_k^2
Sample size	n_1	n_2	$\cdots$	n_i	$\cdots$	n_k
Sample mean	$\bar{x}_{1.}$	$\bar{x}_{2.}$	$\cdots$	$\bar{x}_{i.}$	$\cdots$	$\bar{x}_{k.}$
Sample variance	s_1^2	s_2^2	$\cdots$	s_i^2	$\cdots$	s_k^2

$n = n_1 + n_2 + \cdots + n_k$
= the total number of observations in the *entire* data set.

The null and alternative hypotheses are stated in terms of the population means.

$H_0: \mu_1 = \mu_2 = \cdots = \mu_k$ (All k population means are equal.)

$H_a: \mu_i \neq \mu_j$ for some $i \neq j$ (At least two of the k population means differ.)

The assumptions for this test procedure are similar to those for a two-sample t test.

1. The k population distributions are normal.

2. The k population variances are equal; that is, $\sigma_1^2 = \sigma_2^2 = \cdots = \sigma_k^2$.

3. The samples are selected randomly and independently from the respective populations.

For reference, these are the one-way ANOVA assumptions.

To denote observations, we use a single letter with two subscripts. The first subscript indicates the sample number and the second subscript denotes the observation number within that sample. In general,

x_{ij} = the jth measurement taken from the ith population.

X_{ij} = the corresponding random variable.

A comma is placed between i and j if there is ambiguity. For example, $x_{1,23}$ is the 23rd observation in the first sample. But $x_{12,3}$ is the 3rd observation in the 12th sample.

The mean of the observations in the ith sample is

$$\bar{x}_{i.} = \frac{1}{n_i} \sum_{j=1}^{n_i} x_{ij} = \frac{1}{n_i}(x_{i1} + x_{i2} + \cdots + x_{in_i}) \tag{11.1}$$

The dot in the second subscript is used to indicate a sum over that subscript (j) while the first subscript (i) is held fixed.

The mean of *all* the observations, called the **grand mean**, is the sum of all the observations divided by n:

$$\bar{x}_{..} = \frac{1}{n} \sum_{i=1}^{k} \sum_{j=1}^{n_i} x_{ij.} \tag{11.2}$$

There is just a little more notation that will make some of the calculations easier.

A more theoretical development of a one-way ANOVA test includes the corresponding *random variables* $\bar{X}_{i.}$, $\bar{X}_{..}$, $T_{i.}$, and $T_{..}$.

$$t_{i.} = \sum_{j=1}^{n_i} x_{ij} = \text{sum of the observations in the } i\text{th sample.}$$

$$t_{..} = \sum_{i=1}^{k} \sum_{j=1}^{n_i} x_{ij} = \text{sum of } all \text{ the observations.}$$

Here's where the analysis of *variance* plays a role. The total variation in the data (the total sum of squares) is decomposed into a sum of *between-samples* variation (the sum of squares due to factor) and *within-samples* variation (the sum of squares due to error). The total variation is the variability of individual observations from the grand mean. The between-samples (or between-factor) variation is the variability in the sample means; this tells us how different the sample means are from each other. The within-samples variation is the variability of the observations from their sample mean; this is just like the sample variance we have already been using. The three sums of squares are defined in the following fundamental identity, which shows the decomposition of the total variation in the data.

ONE-WAY ANOVA IDENTITY

Let SST = total sum of squares, SSA = sum of squares due to factor, and SSE = sum of squares due to error.

$$\underbrace{\sum_{i=1}^{k}\sum_{j=1}^{n_i}(x_{ij}-\bar{x}_{..})^2}_{\text{SST}} = \underbrace{\sum_{i=1}^{k}n_i(\bar{x}_{i.}-\bar{x}_{..})^2}_{\text{SSA}} + \underbrace{\sum_{i=1}^{k}\sum_{j=1}^{n_i}(x_{ij}-\bar{x}_{i.})^2}_{\text{SSE}} \qquad (11.3)$$

 ILLUMINATING THE CONCEPTS

1. SSA is used to denote the sum of squares due to factor (instead of SSF) because in a two-way ANOVA there is a factor A and a factor B.

2. The sample size is used as a *weight* in the expression for SSA. ■

If the null hypothesis is true, then each observation comes from the same population with mean μ and variance σ^2. Therefore, the sample means, the $\bar{x}_{i.}$'s, should all be about the same and should all be close to the grand mean, $\bar{x}_{..}$, and SSA should be (relatively) *small*. If at least two population means are different, then at least two $\bar{x}_{i.}$'s should be very different, and these values will be far away from $\bar{x}_{..}$. In this case, SSA should be (relatively) *large*.

The one-way ANOVA test statistic is based on two separate estimates for the common variance, σ^2, computed using SSA and SSE.

DEFINITION

The **mean square due to factor**, **MSA**, is SSA divided by $k-1$:

$$\text{MSA} = \frac{\text{SSA}}{k-1}. \qquad (11.4)$$

The **mean square due to error**, **MSE**, is SSE divided by $n-k$:

$$\text{MSE} = \frac{\text{SSE}}{n-k}. \qquad (11.5)$$

If the null hypothesis is true, then the (random variable) **mean square due to factor**, **MSA**, is an unbiased estimator of σ^2. If H_a is true, then MSA tends to overestimate σ^2.

The **mean square due to error**, **MSE**, is an unbiased estimator of the common variance σ^2 regardless of whether H_0 or H_a is true.

Consider the ratio $F = \text{MSA/MSE}$. If the value of F is close to 1, then the two estimates of σ^2, or sources of variation, are approximately the same. There is no evidence to suggest that the population means are different. If the value of F is much greater than 1, then the variation *between* samples is greater than the variation *within* samples. This suggests that the alternative hypothesis is true.

If the one-way ANOVA assumptions are satisfied and H_0 is true, then the statistic $F = \text{MSA/MSE}$ has an F distribution with df $k - 1$ and $n - k$ degrees of freedom. Since *large* values of F suggest that H_a is true, the rejection region is *only* in the right tail of the distribution.

> One-way ANOVA does not mean a one-sided statistical test; it indicates that there is one factor.

ONE-WAY ANOVA TEST PROCEDURE

Given the one-way ANOVA assumptions, the test procedure with significance level α is

$H_0: \mu_1 = \mu_2 = \cdots = \mu_k$

$H_a: \mu_i \neq \mu_j$ for some $i \neq j$

$\text{TS}: F = \dfrac{\text{MSA}}{\text{MSE}}$

$\text{RR}: F \geq F_\alpha, \quad \text{df } k - 1 \quad \text{and} \quad n - k$

If the value of F is smaller than the critical value, then there is no evidence from the data to reject H_0. If the value of F is in the rejection region, we say that the F test is significant and that there is a statistically significant difference among the population means. Recall that we can also use the p value to conduct an equivalent test: if $p \leq \alpha$, then we reject H_0.

Remember that if any of the one-way ANOVA assumptions are violated, then the conclusion is unreliable. There are several methods to check for normality, presented in Section 6.3, and there are statistical procedures for testing equality of variances. The samples must also be selected randomly and independently from the appropriate populations.

The following *computational formulas* (rather than the definitions) are used to find the sums of squares, and then the mean squares.

> These equations are used for the same reasons the computational formula for the sample variance is used: they are easier, faster, and more accurate.

COMPUTATIONAL FORMULAS

$$\text{SST} = \underbrace{\sum_{i=1}^{k} \sum_{j=1}^{n_i} (x_{ij} - \bar{x}_{..})^2}_{\text{definition}} = \underbrace{\left(\sum_{i=1}^{k} \sum_{j=1}^{n_i} x_{ij}^2 \right) - \frac{t_{..}^2}{n}}_{\text{computational formula}}$$

$$\text{SSA} = \underbrace{\sum_{i=1}^{k} n_i (\bar{x}_{i.} - \bar{x}_{..})^2}_{\text{definition}} = \underbrace{\left(\sum_{i=1}^{k} \frac{t_{i.}^2}{n_i} \right) - \frac{t_{..}^2}{n}}_{\text{computational formula}}$$

$$\text{SSE} = \underbrace{\sum_{i=1}^{k} \sum_{j=1}^{n_i} (x_{ij} - \bar{x}_i)^2}_{\text{definition}} = \underbrace{\text{SST} - \text{SSA}}_{\text{computational formula}}$$

One last detail—one-way ANOVA calculations are often presented in an **analysis of variance table**, or ANOVA table. The values included in this table are associated with the three sources of variation and the calculation of the F statistic.

One-way ANOVA summary table

Source of variation	Sum of squares	Degrees of freedom	Mean square	F	p value
Factor	SSA	$k - 1$	$\text{MSA} = \dfrac{\text{SSA}}{k-1}$	$\dfrac{\text{MSA}}{\text{MSE}}$	p
Error	SSE	$n - k$	$\text{MSE} = \dfrac{\text{SSE}}{n-k}$		
Total	SST	$n - 1$			

The following example illustrates the computations involved in a one-way ANOVA and the process of making an inference based on the value of the test statistic.

Example 11.1 Maximum Heat   The Center for Fire Research is testing the maximum heat release rate of common U.S. Park Service lodging furniture during a fire. Independent random samples of four different night tables were examined. Each was set on fire, and the peak rate of heat release (in kW) was recorded. The data are given in the following table.

Sample	Observations				
Model 1	3101	3116	3110	3110	3084
Model 2	3101	3154	3162	3149	3150
Model 3	3093	3115	3078	3089	3136
Model 4	3099	3103	3097	3109	3134

(*Source*: NIST Center for Fire Research.)

Is there any evidence to suggest that at least two of the population mean heat release rates are different? Use $\alpha = 0.05$.

SOLUTION

STEP 1   Assume that the one-way ANOVA assumptions hold. There are $k = 4$ samples or groups and $n = 5 + 5 + 5 + 5 = 20$ total observations. Some summary statistics are given in the following table.

Sample	Sample size	Sample total	Sample mean	Sample variance
Model 1	$n_1 = 5$	$t_{1.} = 15{,}521$	$\bar{x}_{1.} = 3104.2$	$s_1^2 = 156.2$
Model 2	$n_2 = 5$	$t_{2.} = 15{,}716$	$\bar{x}_{2.} = 3143.2$	$s_2^2 = 582.7$
Model 3	$n_3 = 5$	$t_{3.} = 15{,}511$	$\bar{x}_{3.} = 3102.2$	$s_3^2 = 537.7$
Model 4	$n_4 = 5$	$t_{4.} = 15{,}542$	$\bar{x}_{4.} = 3108.4$	$s_4^2 = 225.8$

The sum of all the observations, or grand total, is

$$t_{..} = 15{,}521 + 15{,}716 + 15{,}511 + 15{,}542 = 62{,}290$$

Solution Trail

KEYWORDS

- Is there any evidence?
- At least two of the population means are different.
- Four types.

↓

TRANSLATION

- Statistical inference.
- Analysis of variance.

↓

CONCEPTS

One-way ANOVA test procedure.

↓

VISION

Compute the summary statistics necessary to complete the ANOVA summary table. Find the critical value, and draw the appropriate conclusion.

The summary-statistics table suggests a possible violation of the equal-variances assumption. There are formal statistical procedures for testing equality of (several) variances, but the ANOVA test statistic is robust. That is, even if the population variances are a little different, the test still provides reliable results. In addition, since the sample sizes are small, it is more difficult to detect a real difference in population variances.

STEP 2   The four parts of the hypothesis test are

$H_0: \mu_1 = \mu_2 = \mu_3 = \mu_4$ (all four population means are equal)
$H_a: \mu_i \neq \mu_j$ for some $i \neq j$ (at least two population means differ)

TS: $F = \dfrac{\text{MSA}}{\text{MSE}}$

RR: $F \geq F_\alpha = F_{0.05} = 3.24$ df 3 and 16.

STEP 3   Find the total sum of squares.

$$\text{SST} = \left(\sum_{i=1}^{k} \sum_{j=1}^{n_i} x_{ij}^2 \right) - \frac{t_{..}^2}{n} \qquad \text{Computational formula for SST.}$$

$$= (3101^2 + 3116^2 + \cdots + 3134^2) - \frac{62{,}290^2}{20} \qquad \text{Apply the formula.}$$

$$= 194{,}013{,}806.0 - 194{,}002{,}205 = 11{,}601 \qquad \text{Simplify.}$$

Find the sum of squares due to factor.

$$\text{SSA} = \left(\sum_{i=1}^{k} \frac{t_{i.}^2}{n_i} \right) - \frac{t_{..}^2}{n} \qquad \text{Computational formula for SSA.}$$

$$= \left(\frac{15{,}521^2}{5} + \frac{15{,}716^2}{5} + \frac{15{,}511^2}{5} + \frac{15{,}542^2}{5} \right) - \frac{62{,}290^2}{20}$$

                                             Use sample totals and grand total.

$$= 194{,}007{,}796.4 - 194{,}002{,}205 = 5591.4 \qquad \text{Simplify.}$$

Use these two values to find the sum of squares due to error.

$$\text{SSE} = \text{SST} - \text{SSA} = 11601 - 5591.4 = 6009.6$$

STEP 4   Compute MSA and MSE.

$\text{MSA} = \text{SSA}/(k - 1)$ Definition.
       $= 5591.4/(4 - 1) = 1863.8$ Use SSA and $k = 4$ groups.

$\text{MSE} = \text{SSE}/(n - k)$ Definition.
       $= 6009.6/(20 - 16) = 375.6$ Use SSE, $n = 20$ total observations, and $k = 4$.

STEP 5   The value of the test statistic is

$$f = \frac{\text{MSA}}{\text{MSE}} = \frac{1863.8}{375.6} = 4.96 \qquad (\geq 3.24)$$

The value of the test statistic lies in the rejection region. Reject the null hypothesis. There is evidence to suggest that at least two population means are different.

STEP 6   Recall, that due to the limited tables of critical values for F distributions, we can only bound the p value. Place the value of the test statistic, $f = 4.96$, in an ordered list of critical values with df 3 and 16.

$$3.24 \leq 4.96 \leq 5.29$$
$$F_{0.05} \leq 4.96 \leq F_{0.01}$$

Therefore, $0.01 \leq p \leq 0.05$

STEP 7   Here's how all of these calculations are presented in an ANOVA table.

The next reasonable question is, "Which pairs of means are contributing to this *overall* difference?" We'll address this issue in Section 11.2.

p value illustration:
$p = P(X \geq 4.96)$
   $= 0.0127 \leq 0.05 = \alpha$

$X \sim F$
df 3 and 16

The p value in this table is from a technology solution.

ANOVA summary table

Source of variation	Sum of squares	Degrees of freedom	Mean square	F	p value
Factor	5591.4	3	1863.8	4.96	0.0127
Error	6009.6	16	375.6		
Total	11,601.0	19			

Figure 11.2 shows a technology solution.

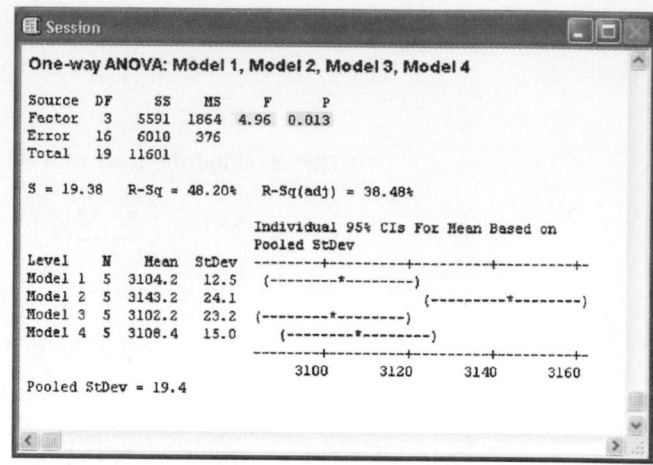

Figure 11.2 One-way ANOVA results.

In the next example, fewer detailed calculations are shown and the ANOVA table is presented, with the focus on the inference process.

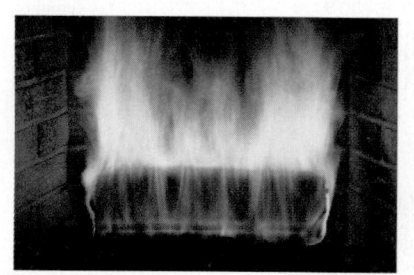

(Izzy Schwartz/Getty Images)

Solution Trail

KEYWORDS

- Is there any evidence?
- Difference in population means.
- Five brands.

↓

TRANSLATION

- Statistical inference.
- Analysis of variance.

↓

CONCEPTS

One-way ANOVA test procedure.

↓

VISION

Compute the summary statistics necessary to complete the ANOVA summary table. Find the critical value, and draw the appropriate conclusion.

Example 11.2 Fireplace Logs and Burn Times Fireplace logs are designed to ignite more easily and burn cleaner and longer than traditional firewood. Many are made from vegetable wax, wood shavings, old newspapers, and even spent coffee grounds and are available in hardware and grocery stores. Five different fireplace log brands were selected, and independent random samples of each type were obtained. The burn time (in hours) for each is given in the following table.

Brand (factor)	Observations					
Duraflame (1)	2.8	2.6	2.4	2.4	2.7	2.6
Hearthlog (2)	2.7	2.7	2.5	2.2	2.0	2.4
Tinderbox (3)	2.7	2.4	2.3	2.6	2.7	2.7
Hot Logs (4)	2.4	2.6	2.7	2.5	2.6	2.5
Wax Logs (5)	2.4	2.4	2.4	2.4	2.4	2.6

Is there any evidence to suggest that there is a difference in population mean burn times for the five brands of fireplace logs? Use $\alpha = 0.05$.[1]

SOLUTION

STEP 1 Assume that the one-way ANOVA assumptions hold. There are $k = 5$ groups and $n = 6 + 6 + 6 + 6 + 6 = 30$ total observations. We would like to know whether the data suggest that any two of the population mean burn times are different.

The four parts of the hypothesis test are

$H_0: \mu_1 = \mu_2 = \mu_3 = \mu_4 = \mu_5$

$H_a: \mu_i \neq \mu_j$ for some $i \neq j$

$$\text{TS: } F = \frac{\text{MSA}}{\text{MSE}}$$

$$\text{RR: } F \geq F_\alpha = F_{0.05} = 2.76 \qquad \text{df 4 and 25.}$$

STEP 2 Here are the sample totals and the grand total.

$$t_{1.} = 15.5 \qquad t_{2.} = 14.5 \qquad t_{3.} = 15.4$$
$$t_{4.} = 15.3 \qquad t_{5.} = 14.6 \qquad t_{..} = 75.3$$

Find the total sum of squares.

$$\text{SST} = \left(\sum_{i=1}^{k} \sum_{j=1}^{n_i} x_{ij}^2 \right) - \frac{t_{..}^2}{n} \qquad \text{Computational formula for SST.}$$

$$= (2.8^2 + 2.6^2 + \cdots + 2.6^2) - \frac{75.3^2}{30} \qquad \text{Apply the formula.}$$

$$= 189.9100 - 189.0030 = 0.9070 \qquad \text{Simplify.}$$

Find the sum of squares due to factor.

$$\text{SSA} = \left(\sum_{i=1}^{k} \frac{t_{i.}^2}{n_i} \right) - \frac{t_{..}^2}{n} \qquad \text{Computational formula for SSA.}$$

$$= \left(\frac{15.5^2}{6} + \frac{14.5^2}{6} + \frac{15.4^2}{6} + \frac{15.3^2}{6} + \frac{14.6^2}{6} \right) - \frac{75.3^2}{30}$$

Use sample totals and grand total.

$$= 189.1517 - 189.0030 = 0.1487 \qquad \text{Simplify.}$$

Use these two values to find the sum of squares due to error.

$$\text{SSE} = \text{SST} - \text{SSA} = 0.9070 - 0.1487 = 0.7583$$

STEP 3 Compute MSA and MSE.

$$\text{MSA} = \text{SSA}/(k-1) \qquad \text{Definition.}$$
$$= 0.1487/(5-1) = 0.0372 \qquad \text{Use SSA and } k = 5 \text{ groups.}$$

$$\text{MSE} = \text{SSE}/(n-k) \qquad \text{Definition.}$$
$$= 0.7583/(30-5) = 0.0303 \qquad \text{Use SSE, } n = 30 \text{ observations, and } k = 5.$$

STEP 4 The value of the test statistic is

$$f = \frac{\text{MSA}}{\text{MSE}} = \frac{0.0372}{0.0303} = 1.2277$$

The value of the test statistic does not lie in the rejection region. Do not reject the null hypothesis. There is no evidence to suggest that any two population mean burn times are different.

STEP 5 Here is the ANOVA summary table.

p value illustration:
$p = P(X \geq 1.2277)$
$= 0.3244 > 0.05 = \alpha$

The p value in this table is from a technology solution.

ANOVA summary table					
Source of variation	Sum of squares	Degrees of freedom	Mean square	F	p-value
Factor	0.1487	4	0.0372	1.23	0.3244
Error	0.7583	25	0.0303		
Total	0.9070	29			

Figure 11.3 shows a technology solution.

	AX	AY	AZ	BA	BB	BC	BD
Anova: Single Factor							
SUMMARY							
Groups		*Count*	*Sum*	*Average*	*Variance*		
Column 1		6	15.5	2.58	0.0257		
Column 2		6	14.5	2.42	0.0777		
Column 3		6	15.4	2.57	0.0307		
Column 4		6	15.3	2.55	0.0110		
Column 5		6	14.6	2.43	0.0067		
ANOVA							
Source of Variation	*SS*	*df*	*MS*	*F*	*P-value*	*F crit*	
Between Groups	0.1487	4	0.0372	1.2253	0.3253	2.76	
Within Groups	0.7583	25	0.0303				
Total	0.9070	29					

Figure 11.3 One-way ANOVA results.

TECHNOLOGY CORNER

Procedure: Conduct a one-way analysis of variance test.
Reconsider: Example 11.1, page 518, solution, and interpretations.

TI-84 Plus

Use the built-in function ANOVA.

1. Enter the data from sample 1 into list L1, from sample 2 into list L2, from sample 3 into list L3, and from sample 4 into list L4.
2. Select STAT; TESTS; ANOVA.
3. Enter the lists containing the data as function arguments. See Figure 11.4.
4. Press ENTER. The results are displayed on the Home Screen. See Figure 11.5.

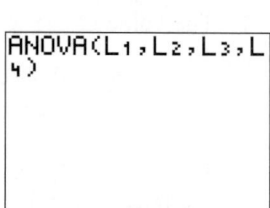

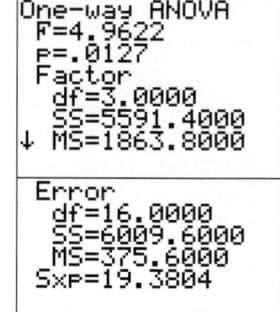

Figure 11.4 ANOVA function with the data lists as arguments.

Figure 11.5 One-way ANOVA results.

Minitab

Minitab has several built-in functions to conduct an analysis of variance test, depending upon the model. Use Qne-Way or One-Way (Unstacked) for the model presented in this section.

1. Enter the data from sample 1 into column C1, from sample 2 into column C2, from sample 3 into column C3, and from sample 4 into column C4. This is unstacked data. That is, each sample is in a separate column.
2. Select Stat; ANOVA; One-Way (Unstacked).
3. Enter the Responses—the columns containing the data.
4. Click OK. The output is displayed in a session window. See Figure 11.2, page 520.

Excel

Use the Data Analysis Toolkit function Anova: Single Factor.

1. Enter the data from sample 1 into column A, from sample 2 into column B, from sample 3 into column C, and from sample 4 into column D.
2. Under the Data tab, select Data Analysis; Anova: Single Factor.
3. Enter the Input Range, choose Grouped By Columns, and select an Output option. Click OK to display the results. See Figure 11.6.

	F	G	H	I	J	K	L
Anova: Single Factor							
SUMMARY							
Groups		Count	Sum	Average	Variance		
Column 1		5	15521	3104.2	156.2		
Column 2		5	15716	3143.2	582.7		
Column 3		5	15511	3102.2	537.7		
Column 4		5	15542	3108.4	225.8		
ANOVA							
Source of Variation	SS	df	MS	F	P-value	F crit	
Between Groups	5591.4	3	1863.8	4.96	0.0127	3.24	
Within Groups	6009.6	16	375.6				
Total	11601.0	19					

Figure 11.6 One-way ANOVA results.

SECTION 11.1 EXERCISES

Practice

11.1 Consider the following data in the context of an ANOVA test.

Group	Observations								
1	33	27	27	32	27	31	23	26	34
2	27	35	32	28	35	39	33		
3	30	36	33	35	33	28			

a. Find n_i, $i = 1, 2, 3$ (the number of observations in each sample), and n (the total number of observations).
b. Find $t_{i.}$, $i = 1, 2, 3$ (the total for each sample), and $t_{..}$ (the grand total).
c. Find $\sum_{i=1}^{3} \sum_{j=1}^{n_i} x_{ij}^2$ (the sum of all the squared observations).

11.2 Consider the following data in the context of an ANOVA test.

			Group		
1	2	3	4	5	6
2.8	4.8	6.3	6.8	6.4	6.3
5.6	4.2	3.9	3.3	5.8	4.7
3.7	4.4	5.4	5.7	6.8	5.9
4.4	2.8	4.4	4.0	5.7	3.4
5.8	5.1	3.0	6.1	5.7	5.7
6.1	3.2	6.3	7.5	4.3	2.3
5.0	4.6	6.9	2.9	2.5	4.5
5.9	2.3	2.2	5.1	6.8	4.9
5.2	6.3	6.8	3.8		3.6
6.8		3.9	5.5		5.6

a. Find n_i, $i = 1, 2, \ldots, 6$ (the number of observations in each sample), and n (the total number of observations).
b. Find $t_{i.}$, $i = 1, 2, \ldots, 6$ (the total for each sample), and $t_{..}$ (the grand total).
c. Find $\sum_{i=1}^{6} \sum_{j=1}^{n_i} x_{ij}^2$ (the sum of all the squared observations).

11.3 Consider the following data in the context of an ANOVA test.

Factor	Observations				
1	162	155	157	144	157
2	168	147	163	131	136
3	135	138	155	172	168
4	144	162	150	140	157

a. Find $t_{i.}$ for $i = 1, 2, 3, 4$. Find $t_{..}$.
b. Find $\sum_{i=1}^{4} \sum_{j=1}^{n_i} x_{ij}^2$.
c. Find SST (the total sum of squares), SSA (the sum of squares due to factor), and SSE (the sum of squares due to error).
d. Find MSA (the mean square due to factor), and MSE (the mean square due to error).
e. Compute the value of the test statistic.

11.4 Consider the data on the data CD and book's web site in the context of an ANOVA test.
a. Find SST, SSA, and SSE.
b. Find MSA and MSE.
c. Compute the value of the test statistic.
d. If $\alpha = 0.05$, would you reject the null hypothesis? Justify your answer.

11.5 Complete the following ANOVA table.

ANOVA summary table

Source of variation	Sum of squares	Degrees of freedom	Mean square	F	p value
Factor		4			
Error	12,062.1				
Total	12,646.2	51			

a. State the null and the alternative hypothesis in terms of the population means.
b. Find the rejection region for $\alpha = 0.05$.
c. Conduct the test. Is there any evidence to suggest that at least two population means are different? Justify your answer.

11.6 Complete the following ANOVA table.

ANOVA summary table

Source of variation	Sum of squares	Degrees of freedom	Mean square	F	p value
Factor			4.522		
Error		62	0.988		
Total		65			

a. State the null and the alternative hypothesis in terms of the population means.
b. Find the rejection region for $\alpha = 0.01$.
c. Conduct the test. Is there any evidence to suggest that at least two population means are different? Justify your answer.

Applications

11.7 Public Health and Nutrition A study was conducted to compare the amount of salt in potato chips. Random samples of four varieties were obtained and the amount of salt in each 1-ounce portion of potato chips was recorded (in mg of sodium). The data are given in the following table.

Variety	Observations					
BBQ	338	155	239	184	185	261
Cheese-Flavored	235	238	251	229	233	232
Olestra-Based	164	197	136	214	148	230
Baked	290	343	294	373	306	357

(*Source*: Nutrition-wise, how do your potato chips stack up? *Diet Bites*, 2009.)

Conduct an analysis of variance test to determine whether there is any evidence that the population mean amount of salt per serving is different for at least two varieties. Use $\alpha = 0.05$.
a. State the four parts of the hypothesis test.
b. Complete an ANOVA table.
c. Draw the appropriate conclusion.

11.8 Manufacturing and Product Development Breitling sells men's gold, silver, and titanium watch bands. A random sample of each type (in similar styles) was obtained, and the weight of each watchband (in grams) was recorded. The data are given in the following table.

Watchband	Observations							
Gold	7.9	7.2	7.8	8.1	7.9	8.3	9.9	
Silver	9.5	7.0	8.7	7.6	7.5	9.3	7.3	6.9
Titanium	6.7	7.1	6.5	7.1	5.5	6.7	4.9	3.9

a. Conduct an analysis of variance test to determine whether there is any evidence that the mean weights of any two watchband types are different. Include an ANOVA table. Use $\alpha = 0.01$.
b. Compute the sample mean weight for each sample. Given your conclusion in part (a), which pair(s) of population means do you think are different?

11.9 Travel and Transportation According to the National Traffic Scorecard, the city with the worst rush hour is Honolulu, HI.[2] Five other cities from the top 10 were selected. Random drivers were asked for the time (in minutes) it normally takes them to get through the traffic congestion during morning rush hour. The cities and summary statistics are given in the following table.

City	Sample size	Sample total	Sum of squared observations
Chicago	14	463	16,185
Los Angeles	18	620	22,458
New York	20	506	13,686
Seattle	15	356	8,964
Washington, D.C.	19	586	18,840

Conduct an analysis of variance test to determine whether there is any evidence that the population mean times through some pair of traffic bottlenecks are different. Use $\alpha = 0.05$.

11.10 Sports and Leisure A study was conducted to determine the tension needed by various types of guitar strings in order to produce the proper frequency. A high E was used for comparison, and tension was measured in newtons. The guitar string brands and summary statistics are given in the following table.

String brand	Sample size	Sample total	Sum of squared observations
Darco Acoustic	8	458.5	26,347.8
Ernie Ball	9	508.8	28,947.4
Martin	9	554.5	34,242.7
Gibson	8	523.9	34,373.9

Conduct the appropriate test to check the hypothesis of no difference in population mean tensions. Use $\alpha = 0.01$.

11.11 Marketing and Consumer Behavior The deli department in a Publix Supermarket conducted a survey to compare orders on certain items. A random sample of sliced ham, roast beef, and turkey orders was obtained, and the weight of each order (in pounds) is given on the data CD and book's web site. Conduct an analysis of variance test to determine whether there is any evidence to suggest that at least two of the population mean weights are different. Use $\alpha = 0.01$.

11.12 Public Health and Nutrition Nondairy creamers (for coffee and tea) contain vegetable fat, corn-syrup solids, casein, and other ingredients. Independent random samples of various nondairy creamers were obtained, and the percentage of fat in a serving size was measured. The data are given on the data CD and book's web site. Is there any evidence to suggest that at least two nondairy creamers have a different population mean percentage of fat per serving? Use $\alpha = 0.05$.

11.13 Travel and Transportation The wait times at airport security checkpoints vary according to day, time, and terminal, and the lines are often long and frustrating. Independent random samples of passengers on Fridays between 4:00 and 5:00 P.M. from four different terminals at Los Angeles International Airport were obtained, and the time required (in minutes) to pass through airport security was recorded. Summary statistics are given in the following table. Use this information with $\alpha = 0.01$ to test the null hypothesis of no difference in population mean security checkpoint time.

Terminal	Sample size	Sample total	Sum of squared observations
TBIT-SOUTH	16	90.2	565.30
1	19	160.1	1429.77
2	12	93.9	802.69
4	17	149.1	1388.95

(*Source*: Transportation Security Administration.)

11.14 Medicine and Clinical Studies The sweetener sorbitol has fewer calories than sucrose, adds texture to food, is noncarcinogenic, and is very useful for people with diabetes. This sweetener is also found in children's cough syrups. Independent random samples of four cough syrups were obtained, and the amount of sorbitol (in grams) was measured per teaspoon. The data are given on the data CD and book's web site. Conduct the appropriate test to determine whether there is any evidence that at least two population mean levels of sorbitol are different. Use $\alpha = 0.01$ and include an ANOVA table.

11.15 Public Policy and Political Science The Drug Enforcement Administration's Domestic Cannabis Eradication/ Suppression Program began in 1979 in Hawaii and California. It is the only nationwide law enforcement program that is specifically designed to halt the spread of marijuana cultivation. All 50 states and 102 state and local law enforcement agencies now participate in this program. Independent random samples of seized outdoor plots from 3 states were obtained, and the number of plants on each plot was recorded. The data are given on the data CD and book's web site.[3] Use these data to test the hypothesis of no difference in the true mean number of plants per seized plot in these 3 states with $\alpha = 0.05$.

11.16 Medicine and Clinical Studies One focus in the Canadian First Minister's report *10-Year Plan to Strengthen Health Care* was a reduction in wait times, especially in five areas: cancer, heart problems, diagnostic imaging, joint replacements, and sight restoration. In four provinces, patients waiting for chemotherapy were

selected at random. The time that each patient waited for treatment (in days) was recorded. The summary statistics are given in the following table. Conduct an analysis of variance test to determine whether there is evidence that some pair of population mean wait times for chemotherapy treatment are different. Use $\alpha = 0.01$ and include an ANOVA table.[4]

Province	Sample size	Sample total	Sum of squared observations
Ontario	15	382	15086
Saskatchewan	12	182	2778
Alberta	16	120	950
British Columbia	14	95	675

11.17 Biology and Environmental Science A study was conducted to examine the thaw depths in late August to early September in the floodplain sites of the Tanana River southwest of Fairbanks, Alaska. The data in the following table are a subset of the thaw depth measurements (in cm) from four locations.

Study site	Observations					
253	45	47	44	45	50	38
1133	47	58	46	50	55	46
254	69	34	37	39	40	62
1113	44	36	38	41	43	55

(*Source*: Bonanza Creek Long-Term Ecological Research.)

Is there any evidence to suggest that at least two population mean thaw depths are different? Use $\alpha = 0.05$.

11.18 Business and Management Robert Klaise, the president of the Bridgeville, Delaware, Chamber of Commerce, recently studied the hourly wages of workers in the area. Independent random samples of workers in private industry were obtained and classified, and the hourly wage of each person was recorded. The data are given on the data CD and book's web site.[5] Is there any evidence to suggest that at least two population mean hourly wages are different? Use $\alpha = 0.001$.

11.19 Public Health and Nutrition The Centers for Disease Control and Prevention has developed several measures in order to track the health status and the health-related quality of life of adults in the United States. An *unhealthy day* is a day in which an individual felt that his or her physical or mental health was not good. An experiment was conducted to determine whether the number of unhealthy days is related to the seasons. Over a one-year period, a random sample of Americans was obtained, and the number of unhealthy days during a 30-day period was recorded for each. The data are given on the data CD and book's web site.[6] Is there any evidence to suggest that at least two of the population mean number of unhealthy days per 30-day period are different? Use $\alpha = 0.05$.

11.20 Biology and Environmental Science Discharge from sewage treatment facilities along the Susquehanna River flows into the Chesapeake Bay and can greatly affect the water quality in the Bay. Nitrogen, which can cause more algae to grow, is the biggest pollutant affecting the Bay. In 2006, the commonwealth of Pennsylvania began limiting the amount of nitrogen each plant can discharge into the river. A random sample of days was obtained, and the nitrogen concentration of the plant discharge for each of three facilities was measured. The data are given on the data CD and book's web site.[7] Is there any evidence to suggest that at least two of the population mean nitrogen discharge concentrations are different? Use $\alpha = 0.01$.

Extended Applications

11.21 Manufacturing and Product Development Many cities, towns, and college campuses use a rotary broom for sweeping debris and snow from sidewalks. The pressure created by the broom is one measure of how effectively the machine removes debris. Random samples of rotary brooms from four manufacturers were obtained, and the pressure of each broom was measured (in psi). The data are given in the following table.

Company	Observations			
Ditch Witch	2081	1980	2210	2297
	2204	2765	2327	
Schwarze	2567	1799	2422	2437
	2367	2244	2245	
Elgin	2228	2581	2364	2375
	2066	2091	2543	
Holder	2905	2695	2503	2931
	2657	2591	2138	

a. Conduct the appropriate test to determine whether there is any evidence that at least two population mean pressures are different. Use $\alpha = 0.05$.
b. Which manufacturer would you recommend and why?

11.22 Fuel Consumption and Cars Automobile service clubs offer free maps, trip-interruption protection, help with some legal fees, and roadside assistance. Independent random samples of roadside service calls were selected for three different clubs. The time required (in minutes) for a tow truck to arrive was recorded for each call. The summary statistics are given in the following table.

Club	Sample size	Sample mean	Sample variance
AAA	15	36.8	10.18
Discover	17	43.8	9.97
Executive	20	34.8	12.15

a. Use the definitions to find SSA and SSE. Use these two values to find SST.
b. Complete an ANOVA table, and use this information to determine whether there is any evidence to suggest that at least two population mean waiting times are different. Use $\alpha = 0.01$.

11.23 Physical Sciences During the summer months, grocery stores, convenience stores, and gas stations sell bags of ice. Independent random sample of bags were obtained from various locations, and the weight (in pounds) of each was recorded. The summary statistics are given in the following table.

Location	Sample size	Sample total	Sum of squared observations
Giant	10	80.8	654.38
Sheetz	10	88.8	789.20
Star Market	14	119.9	1028.31
Unimart	12	104.1	903.57
Weis	15	139.7	1303.41

a. Do the data suggest that the population mean weight of bags of ice is the same at these five locations? Use $\alpha = 0.001$.
b. If the bags of ice cost approximately the same at all the stores, where would you make your purchase? Justify your answer.

Challenge

11.24 Public Policy and Political Science Every city has design specifications for streets, including minimum right of way, minimum vertical grade, and minimum centerline radii on curves. Independent random samples of city streets were obtained from Washington, D.C., and New York, and the width (in feet) of a randomly selected section was recorded. The data are given in the following table.

City	Observations					
New York	28.2	32.9	34.6	31.5	31.6	29.5
	30.3	29.2	25.6	28.6	28.8	31.5
Washington, D.C.	32.5	36.3	34.3	33.0	31.0	36.5
	29.8	30.0	30.2	34.7	31.9	30.0

a. Conduct a two-sample t test to determine whether there is any evidence to suggest that the population mean street widths are different. Assume the population variances are equal and use $\alpha = 0.05$. Find the exact p value associated with this test.
b. Conduct a one-way analysis of variance test to determine whether there is any difference among the $k = 2$ population mean street widths due to city. Use $\alpha = 0.05$. Find the exact p value associated with this test.
c. What is the relationship between the value of the test statistic in part (a) and the value of the test statistic in part (b)? How are the p values related? Why do these relationships make sense?

11.25 The Effect Size Generate a random sample of size $n_1 = 20$ from a normal distribution with mean $\mu_1 = 50$ and variance $\sigma_1^2 = 100$. Generate a second random sample of size $n_2 = 20$ from a normal distribution with mean $\mu_2 = 50$ and variance $\sigma_2^2 = 100$. Generate a third random sample of size $n_3 = 20$ from a normal distribution with mean $\mu_3 = 52$ and variance $\sigma_3^2 = 100$. Conduct a one-way ANOVA test with $\alpha = 0.05$ to determine whether there is any evidence to suggest that the population means differ.

Repeat this process 100 times and record the proportion of times you reject the null hypothesis, p_r.

Let $\mu_T = (\mu_1 + \mu_2 + \mu_3)/3$ and compute the *effect size*

$$e = \sqrt{\frac{\sum_{i=1}^{3}(\mu_i - \mu_T)^2/3}{100}}.$$

Plot the point (e, p_r).

Repeat this process for various values of μ_1, μ_2, and μ_3, and therefore, various effect sizes. Plot the points (e, p_r).

Explain the resulting graph. What concept related to an ANOVA test does this graph illustrate?

11.2 ISOLATING DIFFERENCES

If we reject H_0, then we would like to know which means are different.

Recall, a Type I error means rejecting H_0 when H_0 is true.

X = number of bills in which there is a mistake.

$X \sim B(3, 0.10)$.

$P(X \geq 1)$
$= 1 - P(X = 0)$
$= 1 - 0.7290 = 0.2710$.

When we fail to reject the null hypothesis in a one-way ANOVA, it means that there is no evidence to suggest any difference among population means. The statistical analysis stops here. However, if we do reject the null hypothesis, there is evidence to suggest an *overall* difference among means. The next logical step is to try and isolate the difference(s), to determine which pair(s) of means contributed to the overall (significant) difference. There are several **multiple comparison procedures** for isolating differences. Two will be presented in this section.

If two population means are being compared, then a t test (or Z test) is usually appropriate. However, if there are three or more population means to compare, the analysis requires a little more finesse. Here's why. Suppose in a one-way ANOVA with three groups, we reject H_0. It seems reasonable to conduct a test on every possible pair of means (μ_1 versus μ_2, μ_1 versus μ_3, and μ_2 versus μ_3). However, we cannot simply set the significance level in each individual hypothesis test. The probability α of making a type I error (a mistake) is set in each test under the assumption that *only one* test is conducted per experiment. Therefore, the more tests we do, the greater the chance of making an error.

Imagine a waiter is totaling a customer bill. Suppose the probability that the waiter makes a mistake on any single bill is 0.10. If the waiter must total three bills, the probability that he makes a mistake on *at least* one bill is 0.2710. The more bills he totals, the greater his chance of making at least one error. The same principle applies to hypothesis tests. Suppose three hypothesis tests (using data from the same experiment) are conducted, each with significance level 0.05. The probability of making at least one mistake is 0.1426. The more hypothesis tests (or the more comparisons) we do, the more likely we are to make at least one error. As we have seen, the probability of making at least one error is more than two times as big as α.

We would really like to control the (overall) probability of making at least one mistake. For example, we might want the probability of making at least one mistake in three hypothesis tests to be 0.10. We typically set this overall error probability and work backward to compute the individual error probabilities associated with each individual test. The procedures presented here provide methods for capping the probability of making at least one mistake in all of the comparisons.

While we could actually conduct hypothesis tests, usually we construct multiple confidence intervals for the difference between population means. Recall, if a confidence interval for $\mu_1 - \mu_2$ contains 0, there is no evidence to suggest that the population means are different. However, if 0 is not included in the confidence interval, there is evidence to suggest that the two population means are different. If we want to find a $100(1 - \alpha)\%$ confidence interval for all possible paired comparisons (so that the overall probability is α), we must make the intervals much wider that those for individual differences of means.

In a one-way ANOVA with three groups, suppose there is evidence to suggest that at least one pair of means is different and a multiple comparison procedure produces the following confidence intervals.

Difference	Confidence interval
$\mu_1 - \mu_2$	$(-1.21, \ 9.00)$
$\mu_1 - \mu_3$	$(\ 3.09, 13.31)$
$\mu_2 - \mu_3$	$(-0.81, \ 9.41)$

Zero is included in the confidence intervals for the differences $\mu_1 - \mu_2$ and $\mu_2 - \mu_3$. There is no evidence to suggest that these pairs of population means are different. The confidence interval for $\mu_1 - \mu_3$ does not include 0. There is evidence to suggest that μ_1 is different from μ_3.

The general form of each **Bonferroni confidence interval** is very similar to a confidence interval for the difference between two means based on a t distribution using a pooled estimate of the common variance (introduced in Section 10.2). Here, we use the mean square due to error (MSE) as an estimate of the common variance. And a t critical value is used in order to achieve a *simultaneous*, or *familywise*, *confidence level* of $100(1 - \alpha)\%$.

THE BONFERRONI MULTIPLE COMPARISON PROCEDURE

In a one-way analysis of variance, suppose there are k groups, $n = n_1 + n_2 + \cdots + n_k$ total observations, and H_0 is rejected.

1. There are $c = \dbinom{k}{2} = \dfrac{k(k-1)}{2}$ pairs of population means to compare.

2. The c simultaneous $100(1 - \alpha)\%$ **Bonferroni confidence intervals** have as endpoints the values

$$(\bar{x}_{i.} - \bar{x}_{j.}) \pm t_{\alpha/(2c)} \sqrt{\text{MSE}} \sqrt{\frac{1}{n_i} + \frac{1}{n_j}} \text{ for all } i \neq j \qquad (11.6)$$

with the critical value based on df $= n - k$.

Example 11.3 Activity Time Comparison   A survey was conducted to compare the amount of time (in minutes) that individuals age 15 and older spend in sports and exercise activities per day, according to body mass index (BMI). Independent random samples were obtained from four BMI groups, with 10 observations in each group. The resulting sample means and the ANOVA table are shown below.

Group number	1	2	3	4
Factor	Underweight	Normal weight	Overweight	Obese
Sample mean	7.77	16.19	15.50	10.07

ANOVA summary table

Source of variation	Sum of squares	Degrees of freedom	Mean square	F	p value
Factor	508.310	3	169.437	7.22	0.0007
Error	845.215	36	23.478		
Total	1353.525	39			

Source: Economic Research Service, U.S. Department of Agriculture, Eating and Health Module, 2006.)

The ANOVA test is significant at the $p = 0.0007$ level. There is evidence to suggest that at least one pair of population means is different (an overall difference). Construct the Bonferroni 95% confidence intervals and use them to isolate the pair(s) of means contributing to this overall experiment difference.

SOLUTION

STEP 1 The number of pairwise comparisons needed is

$$c = \frac{4 \cdot 3}{2} = 6$$

$$95\% = 100(1 - \alpha)\% \Rightarrow \alpha = 0.05 \Rightarrow \frac{\alpha}{2c} = \frac{0.05}{2 \cdot 6} = 0.0042$$

This (infrequent) right-tail probability is not specified in Table V in the Appendix; however, one can use linear interpolation or technology, with df $n - k = 40 - 4 = 36$, to find $t_{\alpha/(2c)} = t_{0.0042} = 2.7888$.

STEP 2 The Bonferroni confidence interval for the difference $\mu_1 - \mu_2$ is

$$(\bar{x}_{1.} - \bar{x}_{2.}) \pm t_{\alpha/(2c)} \sqrt{MSE} \sqrt{\frac{1}{n_1} + \frac{1}{n_2}}$$

$$= (\bar{x}_{1.} - \bar{x}_{2.}) \pm t_{0.0042} \sqrt{MSE} \sqrt{\frac{1}{n_1} + \frac{1}{n_2}}$$

$$= (7.77 - 16.19) \pm (2.7888) \sqrt{23.478} \sqrt{\frac{1}{10} + \frac{1}{10}}$$

$$= (-14.47, -2.37)$$

STEP 3 The other five confidence intervals are found in the same manner. Each Bonferroni confidence interval and its corresponding conclusion are shown in the following table.

Difference	Bonferroni confidence interval	Conclusion
$\mu_1 - \mu_2$	$(-14.47, -2.37)$	0 **not** in CI. Evidence to suggest that $\mu_1 \neq \mu_2$.
$\mu_1 - \mu_3$	$(-13.78, -1.68)$	0 **not** in CI. Evidence to suggest that $\mu_1 \neq \mu_3$.
$\mu_1 - \mu_4$	$(-8.36, 3.74)$	0 in CI. μ_1 and μ_4 are not significantly different.
$\mu_2 - \mu_3$	$(-5.36, 6.74)$	0 in CI. μ_2 and μ_3 are not significantly different.
$\mu_2 - \mu_4$	$(0.06, 12.17)$	0 **not** in CI. Evidence to suggest that $\mu_2 \neq \mu_4$.
$\mu_3 - \mu_4$	$(-0.62, 11.48)$	0 in CI. μ_3 and μ_4 are not significantly different.

Finally, the initial ANOVA test indicates that there is an overall difference among the four population means. The simultaneous 95% Bonferroni confidence intervals suggest that this overall difference is due to a difference between μ_1 and μ_2, μ_1 and μ_3, and μ_2 and μ_4. ●

There is another common, compact, graphical method to summarize the results of a multiple comparison procedure. Write the sample means in order from smallest to largest. Use the results from a multiple comparison procedure to draw a horizontal line under the groups of means that are *not* significantly different.

Here is the graphical summary for Example 11.3.

1. The sample means in order are

	$\bar{x}_{1.}$	$\bar{x}_{4.}$	$\bar{x}_{3.}$	$\bar{x}_{2.}$
Sample mean	7.77	10.07	15.50	16.19

2. There is no significant difference between μ_1 and μ_4. Draw a horizontal line under the sample means from population 1 and 4.

	$\bar{x}_{1.}$	$\bar{x}_{4.}$	$\bar{x}_{3.}$	$\bar{x}_{2.}$
Sample mean	7.77	10.07	15.50	16.19

3. There is no significant difference between μ_2 and μ_3. Draw a horizontal line under the sample means from population 2 and 3. There is no significant difference between μ_3 and μ_4. Draw a horizontal line under the sample means from population 3 and 4.

	$\bar{x}_{1.}$	$\bar{x}_{4.}$	$\bar{x}_{3.}$	$\bar{x}_{2.}$
Sample mean	7.77	10.07	15.50	16.19

Those pairs of means *not* connected by a horizontal line are significantly different.

Tukey's multiple comparison procedure also yields simultaneous confidence intervals for all pairwise differences. The form of the confidence intervals is similar to Equation 11.6, but it uses a Q critical value from the **Studentized range distribution**. This distribution is completely characterized by two parameters, the degrees of freedom in the numerator and denominator, m and ν, respectively.

Using the usual notation, let Q_α denote the right-tail critical value of the Studentized range distribution with df m and ν. Table VIII in the Appendix presents selected critical values associated with various Studentized range distributions. The degrees of freedom in the numerator are given in the top row and the degrees of freedom in the denominator are given in the left column. In the body of the table, Q_α is at the intersection of column m and row ν.

TUKEY'S MULTIPLE COMPARISON PROCEDURE

In a one-way analysis of variance, suppose there are k groups, $n = n_1 + n_2 + \cdots + n_k$ total observations, and H_0 is rejected. The set of $c = \binom{k}{2}$ simultaneous $100(1 - \alpha)\%$ confidence intervals have as endpoints the values

$$(\bar{x}_{i.} - \bar{x}_{j.}) \pm \frac{1}{\sqrt{2}} Q_\alpha \sqrt{\text{MSE}} \sqrt{\frac{1}{n_i} + \frac{1}{n_j}} \quad \text{for all } i \neq j \qquad (11.7)$$

with the critical value based on df k and $n - k$.

If all pairwise comparisons are considered, the Bonferroni procedure produces wider confidence intervals than the Tukey procedure. However, if only a subset of all pairwise comparisons is needed, then the Bonferroni method may be better. There are also other methods for comparing population means following an ANOVA test. No single comparison method is uniformly best.

In the following example, Tukey's procedure is used to isolate pairwise differences contributing to an overall significant ANOVA test.

(Paul Conrath/AgeFotostock)

Example 11.4 Armored Car Insurance Lloyd's of London plans to insure four armored car services in the United States during the next year. In order to help determine the premium, independent random samples of cash delivery trips were selected for each service, and the total amount of money transported was recorded (in thousands of dollars). The summary statistics and the ANOVA table are shown below.

Group number	1	2	3	4
Company	Brinks	Cambridge	Dunbar	Wells Fargo
Sample size	8	9	9	8
Sample mean	726.7	572.7	644.2	694.5

ANOVA summary table

Source of variation	Sum of squares	Degrees of freedom	Mean square	F	p value
Factor	115,519	3	38,506.33	4.88	0.007
Error	236,660	30	7,888.67		
Total	352,178	33			

The ANOVA test is significant at the $p = 0.007$ level. There is evidence to suggest that at least one pair of means is different (an overall difference). Construct the Tukey 95% simultaneous confidence intervals and use them to isolate the pair(s) of means contributing to this overall experiment difference.

SOLUTION

STEP 1   The number of pairwise comparisons needed is $\binom{4}{2} = 6$ and $\alpha = 0.05$. The Studentized range critical value (Table VIII in the Appendix, df 4 and 30) is

$$Q_\alpha = Q_{0.05} = 3.845.$$

The output representing simultaneous confidence intervals produced by computers and calculators varies greatly. For example, consider the Tukey CIs in Figure 11.8. The endpoints of each interval are denoted by Lower and Upper, and the midpoint of each interval is denoted by Center.

STEP 2   The first confidence interval, for the difference $\mu_1 - \mu_2$, based on Tukey's procedure is

$$(\bar{x}_{1.} - \bar{x}_{2.}) \pm \frac{1}{\sqrt{2}} Q_{0.05} \sqrt{\text{MSE}} \sqrt{\frac{1}{n_1} + \frac{1}{n_2}}$$

$$= (726.7 - 572.7) \pm \frac{1}{\sqrt{2}} (3.845) \sqrt{7888.67} \sqrt{\frac{1}{8} + \frac{1}{9}}$$

$$= 154.0 \pm 117.34 = (36.66, 271.34)$$

STEP 3   The remaining confidence intervals and conclusions are given in the following table.

Differences	Tukey confidence interval	Conclusion
$\mu_1 - \mu_2$	(36.66, 271.34)	0 **not** in CI. Evidence to suggest that $\mu_1 \neq \mu_2$.
$\mu_1 - \mu_3$	(−34.84, 199.84)	0 in CI. μ_1 and μ_3 are not significantly different.
$\mu_1 - \mu_4$	(−88.54, 152.94)	0 in CI. μ_1 and μ_4 are not significantly different.
$\mu_2 - \mu_3$	(−185.34, 42.34)	0 in CI. μ_2 and μ_3 are not significantly different.
$\mu_2 - \mu_4$	(−239.14, −4.46)	0 **not** in CI. Evidence to suggest that $\mu_2 \neq \mu_4$.
$\mu_3 - \mu_4$	(−167.64, 67.04)	0 in CI. μ_3 and μ_4 are not significantly different.

Here are the results presented in graphical form.

	$\bar{x}_{2.}$	$\bar{x}_{3.}$	$\bar{x}_{4.}$	$\bar{x}_{1.}$
Sample mean	572.7	644.2	694.5	726.7

The simultaneous 95% confidence intervals constructed using Tukey's procedure suggest that the overall significance is due to a difference between μ_1 and μ_2, and μ_2 and μ_4.

The Bonferroni multiple comparison procedure is conservative. This means that the *overall* resulting confidence level is probably higher than the *specified* value. To compute Bonferroni intervals, we simply divide the overall confidence level evenly among all pairwise comparisons. If we are only interested in some of the $\binom{k}{2} = c$ pairwise comparisons, we divide the overall confidence level accordingly. In this case, the resulting confidence level may be closer to the specified value. Tukey's procedure is based on a special distribution and tends to result in more accurate confidence levels. Therefore, if all pairwise comparisons are necessary, Tukey's multiple comparison procedure is usually better.

Consider the Bonferroni Confidence intervals for Example 11.4.

Difference	Bonferroni confidence interval	Conclusion
$\mu_1 - \mu_2$	(32.23, 275.77)	0 **not** in CI. Evidence to suggest that $\mu_1 \neq \mu_2$.
$\mu_1 - \mu_3$	(−39.27, 204.27)	0 in CI. μ_1 and μ_3 are not significantly different.
$\mu_1 - \mu_4$	(−93.10, 157.50)	0 in CI. μ_1 and μ_4 are not significantly different.
$\mu_2 - \mu_3$	(−189.63, 46.63)	0 in CI. μ_2 and μ_3 are not significantly different.
$\mu_2 - \mu_4$	(−243.57, −0.03)	0 **not** in CI. Evidence to suggest that $\mu_2 \neq \mu_4$.
$\mu_3 - \mu_4$	(−172.07, 71.47)	0 in CI. μ_3 and μ_4 are not significantly different.

Notice that each confidence interval is wider than the corresponding Tukey confidence interval. A wider, more conservative, Bonferroni confidence interval could result in a different conclusion concerning the difference between two population means. For example, compare the Tukey and Bonferroni confidence intervals for $\mu_2 - \mu_4$; the Bonferroni confidence interval barely excludes 0.

TECHNOLOGY CORNER

Procedure: Construct simultaneous confidence intervals to isolate the pair(s) of means contributing to an overall experiment difference detected using an analysis of variance test.

Reconsider: Example 11.3, page 528, solution, and interpretations.

The TI-84 and Excel have no built-in functions to construct Bonferroni nor Tukey confidence intervals.

Excel

There is no built-in function to construct simultaneous confidence intervals. However, use the function TINV and ordinary spreadsheet calculations to construct the Bonferroni confidence intervals.

1. Enter the data from sample 1 into column A, from sample 2 into column B, from sample 3 into column C, and from sample 4 into column D.
2. Under the Data tab, select Data Analysis; Anova: Single Factor.
3. Enter the Input Range, choose Grouped by Columns, and select an Output option. Click OK to display the results.
4. Use TINV to find the critical value and compute $\sqrt{\text{MSE}}$. Use these values, the sample means, and the sample sizes in Equation 11.6 to construct the confidence intervals. See Figure 11.7.

Minitab

There is an option under One-Way Analysis of Variance to construct the Tukey confidence intervals. There is no option for Bonferroni confidence intervals.

1. Enter the data from sample 1 into column C1, from sample 2 into column C2, from sample 3 into column C3, and from sample 4 into column C4.
2. Select Stat; ANOVA; One-Way (Unstacked).

F	G	H	I	J	K	L	M
Anova: Single Factor							
SUMMARY							
Groups	*Count*	*Sum*	*Average*	*Variance*			
Column 1	10	77.65	7.77	30.9091			
Column 2	10	161.86	16.19	23.2971			
Column 3	10	154.97	15.50	25.7130			
Column 4	10	100.71	10.07	13.9936			
ANOVA							
Source of Variation	*SS*	*df*	*MS*	*F*	*P-value*	*F crit*	
Between Groups	508.310	3	169.437	7.22	0.0007	2.87	
Within Groups	845.215	36	23.478				
Total	1353.525	39					
Multiple Comparisons	Left	Right		2.7888 = TINV(0.0084,36) = t			
$\mu_1 - \mu_2$	-14.46	-2.38		4.8454 = SQRT(I14) = SQRT(MSE)			
$\mu_1 - \mu_3$	-13.78	-1.69					
$\mu_1 - \mu_4$	-8.35	3.74					
$\mu_2 - \mu_3$	-5.35	6.73					
$\mu_2 - \mu_4$	0.07	12.16					
$\mu_3 - \mu_4$	-0.62	11.47					

Figure 11.7 Bonferroni confidence intervals.

3. Enter the Responses: the columns containing the data.
4. Choose the Comparisons options button. Check Tukey's, family error rate and enter 1 − confidence level.
5. Click OK. The output is displayed in a session window. See Figure 11.8. Note that the mean differences are reversed and each confidence interval is displayed graphically.

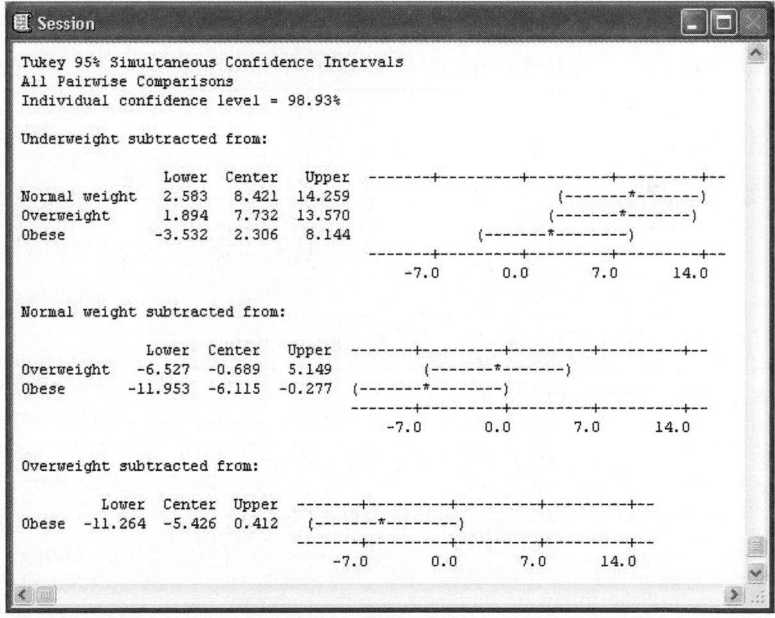

Figure 11.8 Tukey confidence intervals.

SECTION 11.2 EXERCISES

Practice

11.26 In each of the following problems, the number of groups, k, the total number of observations, n, and the overall confidence level for the Bonferroni multiple comparison procedure are given. Find the total number of comparisons, c, and the t critical value used in the calculation of each Bonferroni confidence interval.

a. $k = 3$, $n = 30$, 95%.　　**b.** $k = 3$, $n = 43$, 99%.
c. $k = 4$, $n = 32$, 99%.　　**d.** $k = 5$, $n = 55$, 90%.
e. $k = 6$, $n = 30$, 95%.

11.27 In each of the following problems, the number of observations in each group and the overall confidence level for Tukey's multiple comparison procedure are given. Find the critical value from the Studentized range distribution to be used in the calculation of each Tukey confidence interval.

a. $n_1 = 6$, $n_2 = 8$, $n_3 = 7$, 95%.
b. $n_1 = n_2 = 11$, $n_3 = 12$, $n_4 = 10$, 95%.
c. $n_1 = 14$, $n_2 = 12$, $n_3 = 10$, $n_4 = 18$, 99%.
d. $n_1 = n_2 = n_3 = n_4 = n_5 = 11$, 99%.
e. $n_1 = n_2 = n_3 = 5$, $n_4 = n_5 = 6$, $n_6 = 9$, 99.9%.

11.28 In each of the following problems, use the summary statistics and the Bonferroni confidence intervals to construct the corresponding graph indicating which pairs of means are and are not significantly different.

a.

Sample means	Difference	Confidence interval
$\bar{x}_{1.} = 52.8$	$\mu_1 - \mu_2$	$(-6.00, 17.85)$
$\bar{x}_{2.} = 46.9$	$\mu_1 - \mu_3$	$(-9.50, 14.35)$
$\bar{x}_{3.} = 50.4$	$\mu_2 - \mu_3$	$(-15.42, 8.42)$

b.

Sample means	Difference	Confidence interval
$\bar{x}_{1.} = 4.82$	$\mu_1 - \mu_2$	$(-5.69, -1.13)$
$\bar{x}_{2.} = 8.23$	$\mu_1 - \mu_3$	$(-4.49, 0.07)$
$\bar{x}_{3.} = 7.03$	$\mu_2 - \mu_3$	$(-1.08, 3.48)$

c.

Sample means	Difference	Confidence interval
$\bar{x}_{1.} = 16.09$	$\mu_1 - \mu_2$	$(-12.16, -1.07)$
$\bar{x}_{2.} = 22.71$	$\mu_1 - \mu_3$	$(-7.98, 3.11)$
$\bar{x}_{3.} = 18.53$	$\mu_1 - \mu_4$	$(-1.37, 9.72)$
$\bar{x}_{4.} = 16.33$	$\mu_2 - \mu_3$	$(-5.78, 5.31)$
	$\mu_2 - \mu_4$	$(0.83, 11.92)$
	$\mu_3 - \mu_4$	$(-3.34, 7.75)$

d.

Sample means	Difference	Confidence interval
$\bar{x}_{1.} = 201.7$	$\mu_1 - \mu_2$	$(9.98, 56.20)$
$\bar{x}_{2.} = 168.6$	$\mu_1 - \mu_3$	$(13.20, 59.42)$
$\bar{x}_{3.} = 165.4$	$\mu_1 - \mu_4$	$(-19.89, 26.33)$
$\bar{x}_{4.} = 219.7$	$\mu_2 - \mu_3$	$(-41.10, 5.12)$
	$\mu_2 - \mu_4$	$(-74.19, -27.98)$
	$\mu_3 - \mu_4$	$(-77.41, -31.20)$

11.29 In each of the following problems, use the summary statistics and the Tukey confidence intervals to construct the corresponding graph indicating which pairs of means are and are not significantly different.

a.

Sample means	Difference	Confidence interval
$\bar{x}_{1.} = -33.44$	$\mu_1 - \mu_2$	$(-37.57, 0.36)$
$\bar{x}_{2.} = -14.83$	$\mu_1 - \mu_3$	$(-52.88, -14.95)$
$\bar{x}_{3.} = 0.48$	$\mu_1 - \mu_4$	$(-56.71, -18.77)$
$\bar{x}_{4.} = 4.30$	$\mu_2 - \mu_3$	$(-34.28, 3.66)$
	$\mu_2 - \mu_4$	$(-38.10, -0.17)$
	$\mu_3 - \mu_4$	$(-22.79, 15.14)$

b.

Sample means	Difference	Confidence interval
$\bar{x}_{1.} = 1.62$	$\mu_1 - \mu_2$	$(0.086, 0.320)$
$\bar{x}_{2.} = 1.41$	$\mu_1 - \mu_3$	$(0.197, 0.432)$
$\bar{x}_{3.} = 1.30$	$\mu_1 - \mu_4$	$(-0.006, 0.229)$
$\bar{x}_{4.} = 1.50$	$\mu_2 - \mu_3$	$(0.001, 0.235)$
	$\mu_2 - \mu_4$	$(-0.203, 0.320)$
	$\mu_3 - \mu_4$	$(-0.314, -0.079)$

c.

Sample means	Difference	Confidence interval
$\bar{x}_{1.} = 64.35$	$\mu_1 - \mu_2$	$(0.91, 19.38)$
$\bar{x}_{2.} = 54.21$	$\mu_1 - \mu_3$	$(-5.68, 12.79)$
$\bar{x}_{3.} = 60.80$	$\mu_1 - \mu_4$	$(3.20, 21.66)$
$\bar{x}_{4.} = 51.92$	$\mu_1 - \mu_5$	$(-9.73, 8.73)$
$\bar{x}_{5.} = 64.85$	$\mu_2 - \mu_3$	$(-15.83, 2.64)$
	$\mu_2 - \mu_4$	$(-6.95, 11.52)$
	$\mu_2 - \mu_5$	$(-19.88, -1.41)$
	$\mu_3 - \mu_4$	$(-0.36, 18.11)$
	$\mu_3 - \mu_5$	$(-13.29, 5.18)$
	$\mu_4 - \mu_5$	$(-22.16, -3.70)$

11.30 In each of the following problems, the results from a multiple comparison procedure are shown graphically. Use each illustration to identify all pairs of population means that are significantly different.

a.

	$\bar{x}_{3.}$	$\bar{x}_{1.}$	$\bar{x}_{2.}$	$\bar{x}_{4.}$
Sample mean	12.69	14.64	15.94	16.21

b.

	$\bar{x}_{1.}$	$\bar{x}_{2.}$	$\bar{x}_{3.}$	$\bar{x}_{4.}$
Sample mean	29.98	32.59	32.62	37.25

c.

	$\bar{x}_{1.}$	$\bar{x}_{2.}$	$\bar{x}_{3.}$	$\bar{x}_{4.}$
Sample mean	39.01	41.40	50.83	51.62

d.

	$\bar{x}_{2.}$	$\bar{x}_{3.}$	$\bar{x}_{5.}$	$\bar{x}_{1.}$	$\bar{x}_{4.}$
Sample mean	4.67	6.08	6.23	6.30	7.20

11.31 Suppose data collected from an experiment resulted in a significant ANOVA test. Using the summary statistics and MSE = 29.83, construct the Bonferroni 99% confidence intervals for all pairwise comparisons. Use these intervals to identify the significantly different population means.

Group	1	2	3	4
Sample size	20	20	20	20
Sample mean	19.59	21.58	16.50	25.76

11.32 Suppose data collected from an experiment resulted in a significant ANOVA test. Using the summary statistics and MSE = 8111.8, construct the Tukey 95% confidence intervals for all pairwise comparisons. Use these intervals to identify the significantly different population means.

Group	1	2	3	4
Sample size	16	18	21	9
Sample mean	476.3	450.2	698.2	597.6

Applications

11.33 Marketing and Consumer Behavior A study was conducted to compare the cost of a Big Mac (in U.S. dollars) in three countries. Independent random samples were obtained, and the resulting ANOVA test was significant at the $p = 0.0003$ level. The summary statistics are given below. If MSE = 0.6849, construct the Bonferroni 95% confidence intervals to isolate the population mean prices that are significantly different.

Country	Sample size	Sample mean
Canada	25	3.14
Mexico	20	2.47
United States	22	3.57

(*Source*: FXBigMac, *The Economist*, Oanda Corporation, July 24, 2008.)

11.34 Demographics and Population Statistics A study was conducted to compare the population mean age of women at the time of their first marriage. Independent random samples of size 11 were obtained from weddings in three different years. The resulting ANOVA test was significant at the $p = 0.003$ level. The sample means are given below. If MSE = 17.68, construct the Tukey 95% confidence intervals to isolate the population mean ages that are significantly different.

Year	1985	1995	2003
Sample mean	23.3	24.5	25.3

(*Source*: U.S. Census Bureau.)

11.35 Manufacturing and Product Development An experiment was conducted to compare the true cost of microwave popcorn brands by examining the percentage of popped kernels. Independent random samples of six packages of buttered microwave popcorn (all weighing the same) were obtained for each of four different brands. Each bag was popped and the percentage of popped kernels was measured. An ANOVA test resulted in the test statistic $F = 7.69$ and MSE = 2.72. The sample means are given in the following table.

Microwave popcorn	Pop Secret	Best Choice	Act II	Orville Redenbacher's
Sample mean	90.8	86.9	90.9	89.2

(*Source*: popcorntest.com, 2008.)

a. Find bounds on the p value associated with this test statistic. State the conclusion of the ANOVA test.
b. Construct the Bonferroni 99% simultaneous confidence intervals to determine which pairs of population mean percentages of popped kernels are significantly different.

11.36 Psychology and Human Behavior The number of inmates in correctional facilities is rising due to longer sentences, mandatory terms, and aging baby boomers. Low salaries and tight budgets make it difficult to attract correctional officers. Consequently, the number of inmates per guard is increasing. In California, for example, there is one guard for every 5.28 inmates and in Texas, there is one guard for every 7.03 inmates.[8] Suppose independent random samples of state and federal correctional facilities were obtained, and the security guard to inmate ratio was computed for

each. An ANOVA test was conducted to determine whether there was any difference in population mean security guard to inmate ratio ($F = 4.70$ and MSE = 0.0488). The summary statistics by region are given in the following table.

Region	Sample size	Sample mean
Northeast	12	4.8133
Southeast	14	4.5771
Midwest	15	4.9440
West	17	4.7088

a. Show that there is evidence of an overall difference in population mean security guard to inmate ratio. Use $\alpha = 0.01$ and justify your answer.
b. Construct the Tukey 99% confidence intervals and indicate which pairs of means are contributing to the overall difference.

11.37 Biology and Environmental Science Soil permeability, the rate at which water can flow through the soil, is an important factor that construction companies must consider. Four different potential county development sites in Wisconsin were selected for testing. Independent random samples of locations within each development were selected, and the soil permeability was measured at each location (in inches per hour). An ANOVA test was conducted to determine whether there was any difference in population mean permeability rates ($F = 9.10$ and MSE = 0.02504). The summary statistics are given in the following table.

Development	Sample size	Sample mean
Grant	10	1.3164
Green	12	0.2567
Dane	11	0.5601
Rock	11	0.9206

(*Source*: Walker and Krug, *Soil Permeability in Wisconsin*, 2005.)

a. Find bounds on the p value associated with this (significant) F statistic.
b. Construct the Bonferroni 95% confidence intervals for all pairwise comparisons and draw a graph to represent the results.

11.38 Sports and Leisure There is a 35-second shot clock for men's NCAA basketball games. A team must shoot before the shot clock expires, or they turn the ball over to the opposing team. A team rarely uses the entire 35 seconds, and some teams shoot as quickly as possible. A study compared the mean time to take a shot in five different athletic conferences. Independent random samples of possessions were obtained, and the time (in seconds) to take a shot was recorded for each. An ANOVA test was conducted to determine whether there was any overall difference in population mean shot times ($F = 6.33$ and MSE = 65.201). The summary statistics are given in the following table.

Conference	Sample size	Sample mean
ACC	20	17.27
Big East	26	27.90
Pac-10	23	19.14
Sun Belt	25	22.97
WAC	26	20.05

Construct the Bonferroni 99% confidence intervals for all pairwise comparisons and draw a graph to represent the results. Identify the pairs of athletic conference population means that are significantly different.

11.39 Biology and Environmental Science Most honey comes from European honey bee colonies cultivated in the United States. Africanized killer bees tend to be aggressive and produce less honey. A study was conducted to compare the mean amount of honey produced by colonies in four parts of the country. Independent random samples of colonies were obtained in each area, and the amount of honey (in pounds) per year per colony was recorded. The data are given on the data CD and book's web site.

a. Conduct an analysis of variance test to show that there is evidence of an overall difference in population mean honey production. Use $\alpha = 0.05$.

b. Construct the Bonferroni 95% confidence intervals to isolate the pairs of means contributing to the overall difference.

c. Draw a graph to represent the results of the multiple comparison procedure in part (b).

11.40 Fuel Consumption and Cars Several new designs for an automobile gearbox were developed, and tests were conducted of the vibration inside the device when it was subject to a fundamental gear frequency of approximately 290 Hz. Independent random samples of gearboxes for each design were obtained, and the vibration was assessed by measuring acceleration in meters per second squared (m/s^2). The data are given in the following table.

Model	Observations				
A	26.3	14.1	17.4	38.6	31.4
B	44.6	61.2	54.6	35.5	35.1
C	43.9	53.0	57.8	62.5	60.0
D	18.4	6.9	51.1	14.4	47.0

a. Conduct an analysis of variance test to show that there is evidence of an overall difference in population mean vibration. Use $\alpha = 0.01$.

b. Construct the Tukey 99% confidence intervals to isolate the pairs of means contributing to the overall difference.

c. Draw a graph to represent the results of the multiple comparison procedure in part (b).

11.41 Demographics and Population Statistics A stun gun delivers a high-voltage, low-amperage electrical shock to an attacker. In certain states, cities, and countries the use of this self-protection device is restricted or even illegal. Independent random samples of stun gun owners in cities, suburban areas, and rural areas were obtained. Each stun gun was examined, and the voltage (in thousands of volts) on the device was carefully measured in a single test. The data are given on the data CD and book's web site.

a. Conduct an analysis of variance test to show that there is evidence of an overall difference in population mean stun gun voltage. Use $\alpha = 0.05$.

b. Construct the Tukey 95% confidence intervals to isolate the pairs of means contributing to the overall difference.

c. Draw a graph to represent the results of the multiple comparison procedure in part (b).

11.42 Public Health and Nutrition A simple gelatin is known today as "America's most famous dessert." Although it was patented

in 1845, Jell-O sales were minimal until the early 1900s. Gelatin is primarily made from processed collagen but also contains sugar, artificial flavors, and coloring. Independent random samples of three gelatin brands were obtained, and the amount of sugar (in grams) in one serving was measured. The data are given on the data CD and book's web site.

a. Conduct an analysis of variance test to show that there is evidence of an overall difference in population mean sugar content in gelatin servings. Use $\alpha = 0.01$.

b. Construct the Bonferroni 99% confidence intervals to isolate the pairs of means contributing to the overall difference.

c. Draw a graph to represent the results of the multiple comparison procedure in part (b).

11.43 Sports and Leisure The outside dimensions and playing lines of a tennis court are standard and well established. However, the distance between courts, or sidelines, varies. Many construction guidelines recommend 24 feet between courts, but to conserve space, builders may plan for as little as 12 feet. Independent random samples of public courts were obtained in five different cities, and the distance (in feet) between courts was recorded. The data are given in the following table.

City	Observations				
Atlanta	14.0	14.2	13.0	15.2	15.0
Boston	14.2	16.8	18.6	15.5	16.6
Los Angeles	14.3	14.9	16.5	15.1	14.4
Miami	14.3	17.3	17.3	14.9	16.4
New York	22.0	18.3	19.3	20.5	18.5

a. Conduct an analysis of variance test to show that there is evidence of an overall difference in population mean distances between tennis courts in public parks. Use $\alpha = 0.01$.

b. Construct the Bonferroni 95% confidence intervals to isolate the pairs of means contributing to the overall difference.

c. Draw a graph to represent the results of the multiple comparison procedure in part (b).

Extended Applications

11.44 Biology and Environmental Science Hatcheries around the country provide chicks to poultry farms where they are raised as roasters and broilers. Independent random samples of chicks from different hatcheries were obtained, and the shipping weight (in grams) of each chick was measured. The data are given in the following table.

Hatchery	Observations				
Bedwell Farms	47.1	48.8	52.6	49.1	53.3
Clinton Chicks	63.0	52.0	56.4	56.2	55.3
Sunny Creek	51.6	52.9	54.9	52.8	54.6
Wild Wings	57.4	57.4	55.2	56.0	56.0

a. Conduct an analysis of variance test to show that there is evidence of an overall difference in population mean chick weights. Use $\alpha = 0.05$.

b. Construct the Bonferroni 95% confidence intervals to isolate the pairs of means contributing to the overall difference.

c. Construct the Tukey 95% confidence intervals to isolate the pairs of means contributing to the overall difference.

d. Are your answers to parts (b) and (c) the same? If so, did you expect this to happen? If not, why not?

11.45 Public Health and Nutrition Butter and margarine contain saturated fat, which can increase the bad type of cholesterol in your blood. In order to advise clients, a dietitian examined four types of margarine. A random sample of each type was obtained, and the amount of saturated fat (in grams) per serving was recorded. The data are given on the data CD and book's web site.[9]

a. Conduct an analysis of variance test to show that there is evidence of an overall difference in population mean saturated fat per serving. Use $\alpha = 0.01$.

b. Construct the Tukey 95% confidence intervals to isolate the pairs of means contributing to the overall difference.

c. Are the differences found in part (b) consistent with the percentage of fat indicated on each product label? If not, can you explain any discrepancy?

Challenge

11.46 Manufacturing and Product Development An experiment was conducted to compare the acoustic properties of a control plastic (styrene) with four alternative treatment plastics. Independent random samples were selected, and the attenuation value of each piece was measured (in dB/mm at 5 MHz). The data are given on the data CD and book's web site.

a. Conduct a one-way analysis of variance to test for an overall difference among the five population means. Use $\alpha = 0.05$.

b. Construct 95% Bonferroni confidence intervals *only* for the differences between each *treatment* plastic population mean and the control, styrene, population mean. Distribute the confidence level among the four comparisons (1 versus 2, 1 versus 3, 1 versus 4, and 1 versus 5) so that the simultaneous confidence level is 95%. Which treatment plastic attenuation means are significantly different from the control?

11.3 TWO-WAY ANOVA

In the previous two sections, we considered the effect of a single factor on a response variable. In this section, we will consider experiments in which two factors may contribute to the overall variability in response. A two-way ANOVA is designed to compare the means of populations that can be classified in two different ways. For example, suppose we are interested in the weight of soup spoons. The weight may vary by the pattern (style), by the forge (where it is manufactured), or neither.

Without going into as much detail as Section 11.1, suppose there are a levels of factor A, b levels of factor B, and n observations for each combination of levels, for a total of abn observations. Using the soup spoon example, there could be $a = 5$ levels of factor A: 5 different patterns; and $b = 3$ levels of factor B: 3 different forges; and $n = 6$ observations for each combination of pattern and forge, for a total of $abn = (5)(3)(6) = 90$ observations.

In Table 11.1, there are $a = 3$ levels of factor A, $b = 4$ levels of factor B, and $n = 2$ observations per cell. Let x_{ijk} represent the kth observation for the ith level of factor A and the jth level of factor B. For example, in the table $x_{132} = 0.9$, the 2nd observation for the 1st level of factor A and the 3rd level of factor B.

Table 11.1 Presentation of data in a two-way ANOVA

		Factor B							
		1		2		3		4	
Factor A	1	2.7	1.6	0.8	1.5	0.5	0.9	1.3	1.2
	2	1.9	2.8	0.5	3.8	1.4	2.0	1.0	0.7
	3	3.0	1.9	2.3	2.0	2.3	1.5	1.3	2.7

An interaction effect is significant if one level of factor A and one level of factor B interact differently, or inconsistently, from other factor combinations.

The total variation in the data, sum of squares (**SST**), is composed of the sum of squares due to factor A (**SSA**), the sum of squares due to factor B (**SSB**), the sum of squares due to interaction [**SS(AB)**], and the sum of squares due to error (**SSE**).

Consistent with previous notation, dots in the subscript of $\bar{x}$ and t indicate the mean and the sum of x_{ijk}, respectively, over the appropriate subscript(s). For example,

$$\bar{x}_{.j.} = \frac{1}{an} \sum_{i=1}^{a} \sum_{k=1}^{n} x_{ijk} \quad \text{and} \quad t_{...} = \sum_{i=1}^{a} \sum_{j=1}^{b} \sum_{k=1}^{n} x_{ijk}.$$

Here are the fundamental identity, definitions, and computational formulas for a two-way ANOVA.

TWO-WAY ANOVA IDENTITY

Let SST = total sum of squares, SSA = sum of squares due to factor A, SSB = sum of squares due to factor B, SS(AB) = sum of squares due to interaction, and SSE = sum of squares due to error.

$$\text{SST} \quad = \underbrace{\sum_{i=1}^{a}\sum_{j=1}^{b}\sum_{k=1}^{n}(x_{ijk}-\bar{x}_{...})^2}_{\text{definition}} = \underbrace{\left(\sum_{i=1}^{a}\sum_{j=1}^{b}\sum_{k=1}^{n}x_{ijk}^2\right) - \frac{t_{...}^2}{abn}}_{\text{computational formula}}$$

$$\text{SSA} \quad = \underbrace{bn\sum_{i=1}^{a}(\bar{x}_{i..}-\bar{x}_{...})^2}_{\text{definition}} = \underbrace{\frac{\sum_{i=1}^{a}t_{i..}^2}{bn} - \frac{t_{...}^2}{abn}}_{\text{computational formula}}$$

$$\text{SSB} \quad = \underbrace{an\sum_{j=1}^{b}(\bar{x}_{.j.}-\bar{x}_{...})^2}_{\text{definition}} = \underbrace{\frac{\sum_{j=1}^{b}t_{.j.}^2}{an} - \frac{t_{...}^2}{abn}}_{\text{computational formula}}$$

$$\text{SS(AB)} = \underbrace{n\sum_{i=1}^{a}\sum_{j=1}^{b}(\bar{x}_{ij.}-\bar{x}_{i..}-\bar{x}_{.j.}+\bar{x}_{...})^2}_{\text{definition}}$$

$$= \underbrace{\frac{\sum_{i=1}^{a}\sum_{j=1}^{b}t_{ij.}^2}{n} - \frac{\sum_{i=1}^{a}t_{i..}^2}{bn} - \frac{\sum_{j=1}^{b}t_{.j.}^2}{an} + \frac{t_{...}^2}{abn}}_{\text{computational formula}}$$

$$\text{SSE} \quad = \underbrace{\sum_{i=1}^{a}\sum_{j=1}^{b}\sum_{k=1}^{n}(x_{ijk}-\bar{x}_{ij.})^2}_{\text{definition}} = \underbrace{\text{SST} - \text{SSA} - \text{SSB} - \text{SS(AB)}}_{\text{computational formula}}$$

$$\text{SST} \quad = \text{SSA} + \text{SSB} + \text{SS(AB)} + \text{SSE}$$

The assumptions for a two-way ANOVA are stated in terms of each *cell*, which is considered a *population*.

For reference, these are the two-way ANOVA assumptions.

1. The *ab* population distributions are normal.

2. The *ab* population variances are equal.

3. The samples are selected randomly and independently from the respective populations.

Two-way ANOVA calculations are also usually presented in a summary table. The values in this table are associated with the five sources of variation, and the F statistics are used to conduct appropriate hypothesis tests.

Two-way ANOVA summary table

Source of variation	Sum of squares	Degrees of freedom	Mean square	F	p value
Factor A	SSA	$a - 1$	$MSA = \dfrac{SSA}{a - 1}$	$F_A = \dfrac{MSA}{MSE}$	p_A
Factor B	SSB	$b - 1$	$MSB = \dfrac{SSB}{b - 1}$	$F_B = \dfrac{MSB}{MSE}$	p_B
Interaction	SS(AB)	$(a - 1)(b - 1)$	$MS(AB) = \dfrac{SS(AB)}{(a - 1)(b - 1)}$	$F_{AB} = \dfrac{MS(AB)}{MSE}$	p_{AB}
Error	SSE	$ab(n - 1)$	$MSE = \dfrac{SSE}{ab(n - 1)}$		
Total	SST	$abn - 1$			

Check to make sure that the expressions for the number of degrees of freedom do sum to $abn - 1$.

There are three hypothesis tests associated with a two-way ANOVA.

TWO-WAY ANOVA TESTS

Test 1. Test for an interaction effect.

H_0: There is no interaction effect.
H_a: There is an effect due to interaction.

TS: $F_{AB} = \dfrac{MS(AB)}{MSE}$

RR: $F_{AB} \geq F_{\alpha}$, df $(a - 1)(b - 1)$ and $ab(n - 1)$

Test 2. Test for an effect due to factor A.

H_0: There is no effect due to factor A.
H_a: There is an effect due to factor A.

TS: $F_A = \dfrac{MSA}{MSE}$

RR: $F_A \geq F_{\alpha}$, df $a - 1$ and $ab(n - 1)$

Test 3. Test for an effect due to factor B.

H_0: There is no effect due to factor B.
H_a: There is an effect due to factor B.

TS: $F_B = \dfrac{MSB}{MSE}$

RR: $F_B \geq F_{\alpha}$, df $b - 1$ and $ab(n - 1)$

The hypothesis test for an interaction effect is usually considered first. An interaction effect is present when the relationship between the two factors is not *linear*, or *additive*, for at least one combination of levels. For example, suppose the total production of corn depends on two factors, the amount of water and the amount of fertilizer. Intuitively, one might expect the total production of corn to increase as the amount of water increases and/or as the amount of fertilizer increases. Each change in water and/or fertilizer results in a predictable, additive change in the total amount of corn produced. However, consider a very high level of water and high level of fertilizer. The additive model predicts a huge total production in corn. But there could be an interaction or inconsistent effect associated with these two levels.

Too much water and too much fertilizer might combine to produce very little corn. An interaction plot is a scatter plot of each cell sample mean versus factor A level. Connect the points in the graph corresponding to the same factor B levels. Parallel lines suggest no evidence of interaction.

Case 1.

If the null hypothesis is *not* rejected, then the other two hypothesis tests can be conducted as usual, to see whether there are effects due to either (or both) factors.

Case 2.

If the null hypothesis is rejected, then there is evidence of a significant interaction. Interpretation of the other two hypothesis tests is tricky.

1. If we reject a null hypothesis of no effect due to factor A (and/or factor B), then the effect due to factor A (and/or factor B) is probably significant.

2. If we do not reject a null hypothesis of no effect due to factor A (and/or factor B), then the effect due to factor A (and/or factor B) is inconclusive.

Example 11.5 Satellite TV Quality The quality of the picture on a home satellite TV depends on the strength of the signal. Professional installation is often necessary to properly align a dish and tune in a satellite. A study was conducted to determine whether the signal strength was related to the satellite company and/or the geographic region. For each company and region, a random sample of satellite TV users was obtained, and the signal strength of each system was measured (in $dB\mu V$). The data are given in the following table.

		Geographic region							
		Northeast		**Southeast**		**Midwest**		**West**	
Company	**DIRECTV**	65.9	73.9	64.0	61.4	55.9	64.2	54.7	69.9
		70.9	74.9	55.0	52.1	67.2	74.3	62.2	65.5
	Dish Network	71.3	76.8	64.3	64.0	69.4	68.6	65.6	67.8
		71.4	65.2	61.3	68.2	61.2	64.9	55.2	66.0
	Echostar	64.0	64.2	60.8	62.9	58.3	64.4	62.4	62.9
		76.7	65.1	57.0	69.6	69.3	73.1	64.5	71.3

Conduct a two-way analysis of variance to determine whether signal strength is affected by company and/or geographic region. Use $\alpha = 0.05$.

SOLUTION

STEP 1 Assume the two-way ANOVA assumptions are true. There are $a = 3$ levels of factor A (satellite TV company), $b = 4$ levels of factor B (geographic region), and $n = 4$ observations in each cell. Sample totals are given in the following table.

		Geographic region				
		Northeast	**Southeast**	**Midwest**	**West**	
Company	**DIRECTTV**	$t_{11.} = 285.6$	$t_{12.} = 232.5$	$t_{13.} = 261.6$	$t_{14.} = 252.3$	$t_{1..} = 1032.0$
	Dish Network	$t_{21.} = 284.7$	$t_{22.} = 257.8$	$t_{23.} = 264.1$	$t_{24.} = 254.6$	$t_{2..} = 1061.2$
	Echostar	$t_{31.} = 270.0$	$t_{32.} = 250.3$	$t_{33.} = 265.1$	$t_{34.} = 261.1$	$t_{3..} = 1046.5$
		$t_{.1.} = 840.3$	$t_{.2.} = 740.6$	$t_{.3.} = 790.8$	$t_{.4.} = 768.0$	$t_{...} = 3139.7$

STEP 2 Find the sums of squares.

$$SST = \left(\sum_{i=1}^{3}\sum_{j=1}^{4}\sum_{k=1}^{4} x_{ijk}^2\right) - \frac{t_{...}^2}{(3)(4)(4)} \qquad \text{Computational formula.}$$

$$= (65.9^2 + 73.9^2 + \cdots + 71.3^2) - \frac{3139.7^2}{48} \qquad \text{Apply the formula.}$$

$$= 206{,}990.75 - 205{,}369.09 = 1621.66 \qquad \text{Simplify.}$$

$$SSA = \frac{\sum_{i=1}^{3} t_{i..}^2}{(4)(4)} - \frac{t_{...}^2}{(3)(4)(4)} \qquad \text{Computational formula.}$$

$$= \frac{1032.0^2 + 1061.2^2 + 1046.5^2}{16} - \frac{3139.7^2}{48} \qquad \text{Apply the formula.}$$

$$= 205{,}395.73 - 205{,}369.09 = 29.64 \qquad \text{Simplify.}$$

$$SSB = \frac{\sum_{j=1}^{4} t_{.j.}^2}{(3)(4)} - \frac{t_{...}^2}{(3)(4)(4)} \qquad \text{Computational formula.}$$

$$= \frac{840.3^2 + 740.6^2 + 790.8^2 + 768.0^2}{12} - \frac{3139.7^2}{48} \qquad \text{Apply the formula.}$$

$$= 205{,}815.09 - 205{,}369.09 = 446.00 \qquad \text{Simplify.}$$

$$SS(AB) = \frac{\sum_{i=1}^{3}\sum_{j=1}^{4} t_{ij.}^2}{4} - \frac{\sum_{i=1}^{3} t_{i..}^2}{(4)(4)} - \frac{\sum_{j=1}^{4} t_{.j.}^2}{(3)(4)} + \frac{t_{...}^2}{(3)(4)(4)} \qquad \text{Computational formula.}$$

$$= \frac{285.6^2 + 232.5^2 + \cdots + 261.1^2}{4}$$

$$- 205{,}395.73 - 205{,}815.09 + 205{,}369.09 \qquad \text{Apply and simplify.}$$

$$= 108.19 \qquad \text{Simplify.}$$

$$SSE = SST - SSA - SSB - SS(AB) \qquad \text{Computational formula.}$$

$$= 1621.66 - 26.64 - 446.00 - 108.19 = 1040.83 \qquad \text{Apply and simplify.}$$

STEP 3 Check for an interaction effect first.

H_0: There is no interaction effect.

H_a: There is an effect due to interaction.

$$TS: F_{AB} = \frac{MS(AB)}{MSE}$$

$$RR: F_{AB} \geq F_{\alpha} = F_{0.05} = 2.37 \qquad \text{df 6 and 36.}$$

$$MS(AB) = \frac{SS(AB)}{(a-1)(b-1)} = \frac{108.19}{6} = 18.0317$$

$$MSE = \frac{SSE}{ab(n-1)} = \frac{1040.83}{36} = 28.9119$$

The value of the test statistic is

$$f_{AB} = \frac{MS(AB)}{MSE} = \frac{18.0317}{28.9119} = 0.6237$$

The value of the test statistic does not lie in the rejection region. There is no evidence of an interaction effect. The tests for factor effects can be conducted as usual.

p value illustration:
$p = P(X \geq 0.6237)$
$\quad = 0.7101 > 0.05 = \alpha$

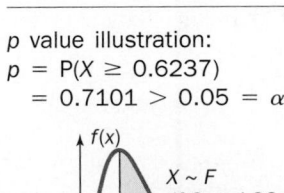

p value illustration:
$p = P(X \geq 0.4607)$
$= 0.6345 > 0.05 = \alpha$

$f(x)$

$X \sim F$
df 2 and 36

0   0.4607 x

---

p value illustration:
$p = P(X \geq 5.14)$
$= 0.0046 \leq 0.05 = \alpha$

$f(x)$

$X \sim F$
df 3 and 36

0   5.14 x

The *p* values in this table were found using technology but are slightly different from those given below due to round-off error in computing f_A, f_B, and f_{AB}.

STEP 4 Check for an effect due to satellite company (factor A).

H_0: There is no effect due to factor A.

H_a: There is an effect due to factor A.

TS: $F_A = \dfrac{\text{MSA}}{\text{MSE}}$

RR: $F_A \geq F_\alpha = F_{0.05} = 3.26$.   df 2 and 36

The value of the test statistic is

$$f_A = \frac{\text{MSA}}{\text{MSE}} = \frac{\text{SSA}/(a-1)}{\text{SSE}/ab(n-1)} = \frac{26.64/2}{1040.83/36} = 0.4607$$

The value of the test statistic does not lie in the rejection region. There is no evidence to suggest that satellite company has an effect on the signal strength.

STEP 5 Check for an effect due to geographic region (factor B).

H_0: There is no effect due to factor B.

H_a: There is an effect due to factor B.

TS: $F_B = \dfrac{\text{MSB}}{\text{MSE}}$

RR: $F_B \geq F_\alpha = F_{0.05} = 2.87$ df 3 and 36

The value of the test statistic is

$$f_B = \frac{\text{MSB}}{\text{MSE}} = \frac{\text{SSB}/(b-1)}{\text{SSE}/ab(n-1)} = \frac{446.00/3}{1040.83/36} = 5.14 \quad (\geq 2.87)$$

The value of the test statistic lies in the rejection region. There is evidence to suggest that signal strength is affected by geographic region.

STEP 6 Finally, here is the ANOVA summary table.

Source of variation	Sum of squares	Degrees of freedom	Mean square	F	p value
Factor A	26.64	2	13.32	0.46	0.6345
Factor B	446.00	3	148.67	5.14	0.0046
Interaction	108.19	6	18.03	0.62	0.7101
Error	1040.83	36	28.91		
Total	1621.66	47			

Figure 11.9 shows a technology solution.

The Minitab commands and options to produce this output are detailed in the Technology Corner at the end of this section.

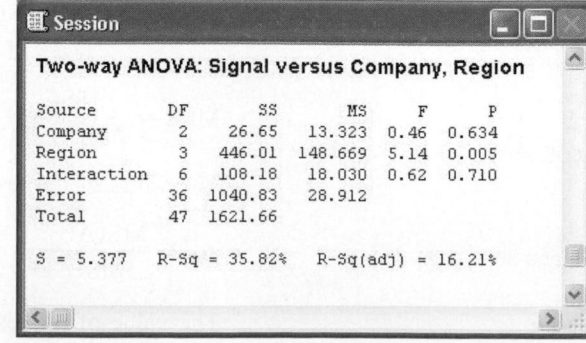

Figure 11.9 Two-way ANOVA results.

(Gen Nishino/Getty Images)

Example 11.6 Pinball Points   Beelan Enterprises, a manufacturer of arcade games, is designing a new pinball game. The company would like the game to be challenging but rewarding by allowing players to score a lot of points. A study was conducted to determine whether the total number of points scored in a single game is related to the design of game (factor A) and/or gender of the player (factor B). For each of five designs ($a = 5$) and gender ($b = 2$), independent random samples of size $n = 3$ were obtained, and the total score for one game was recorded (in millions of points). The following ANOVA summary table was obtained.

Source of variation	Sum of squares	Degrees of freedom	Mean square	F	p value
Factor A	752.71	4	188.18	1.75	0.1787
Factor B	777.24	1	777.24	7.23	0.0141
Interaction	1293.94	4	323.49	3.01	0.0428
Error	2150.59	20	107.53		
Total	4974.48	29			

a. Is there any evidence of interaction? Use $\alpha = 0.05$.

b. Is there any evidence that design affects the total score? Use $\alpha = 0.05$.

c. Is there any evidence that gender affects the total score? Use $\alpha = 0.05$.

SOLUTION

a. Check for an interaction effect.

H_0: There is no interaction effect.

H_a: There is an effect due to interaction.

TS: $F_{AB} = \dfrac{\text{MS(AB)}}{\text{MSE}}$

RR: $F_{AB} \geq F_\alpha = F_{0.05} = 2.87$ df 4 and 20.

Using the ANOVA summary table, the value of the test statistic is

$$f_{AB} = \frac{\text{MS(AB)}}{\text{MSE}} = \frac{323.49}{107.53} = 3.01 \quad (\geq 2.87)$$

The value of the test statistic lies in the rejection region; $p = 0.0428 \leq 0.05$. Therefore, we reject the null hypothesis at the $\alpha = 0.05$ significance level. There is evidence of an interaction effect.

Since there is evidence of interaction, interpretation of the factor effects is not as clear-cut and explicit.

b. Check for an effect due to design (factor A).

H_0: There is no effect due to factor A.

H_a: There is an effect due to factor A.

TS: $F_A = \dfrac{\text{MSA}}{\text{MSE}}$

RR: $F_A \geq F_\alpha = F_{0.05} = 2.87$ df 4 and 20.

Using the ANOVA summary table, the value of the test statistic is

$$f_A = \frac{\text{MSA}}{\text{MSE}} = \frac{188.18}{107.53} = 1.75$$

---

p value illustration:

$p = P(X \geq 3.01)$

$\quad = 0.0428 \leq 0.05 = \alpha$

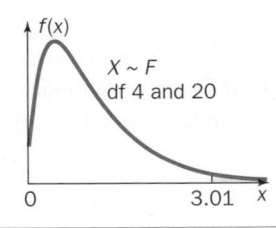

---

p value illustration:

$p = P(X \geq 1.75)$

$\quad = 0.1787 > 0.05 = \alpha$

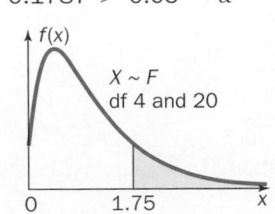

The value of the test statistic does not lie in the rejection region ($p = 0.1787 > 0.05$). Therefore, we do not reject the null hypothesis. However, since there is an interaction effect, the effect due to design is inconclusive.

c. Check for an effect due to gender (factor B).

H_0: There is no effect due to factor B.

H_a: There is an effect due to factor B.

$$\text{TS}: F_B = \frac{\text{MSB}}{\text{MSE}}$$

$$\text{RR}: F_B \geq F_\alpha = F_{0.05} = 4.35 \qquad \qquad \text{df 1 and 20}$$

Using the ANOVA summary table, the value of the test statistic is

$$f_B = \frac{\text{MSB}}{\text{MSE}} = \frac{777.24}{107.53} = 7.23 \quad (\geq 4.35)$$

The value of the test statistic lies in the rejection region ($p = 0.0141 \leq 0.05$). Therefore, we reject the null hypothesis. Since there is an interaction effect, the effect due to gender is probably significant.   ●

p value illustration:

$p = P(X \geq 7.23)$

$\quad = 0.0141 \leq 0.05 = \alpha$

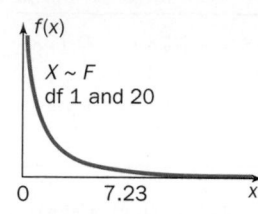

TECHNOLOGY CORNER

Procedure: Two-way analysis of variance.

Reconsider: Example 11.5, page 540, solution, and interpretations.

Minitab

Use the built-in function Two-Way.

There is no built-in TI-84 Plus function to conduct a two-way analysis of variance.

1. Enter the (response) data in a single worksheet column (C3). Enter the company (row factor) level in column C1 and the region (column factor) level in column C2.
2. Select Stat; ANOVA; Two-Way.
3. Enter the Response column, the Row factor column, and the Column factor column. To include an interaction term, do not check the box for Fit additive model.
4. Click OK. The results are displayed in a session window. See Figure 11.9, page 542.

Excel

Use the Data Analysis function Anova: Two-Factor With Replication.

1. Enter the data in an array as shown in Figure 11.10.

A	B	C	D	E
	Northeast	Southeast	Midwest	West
Direct TV	65.9	64.0	55.9	54.7
	73.9	61.4	64.2	69.9
	70.9	55.0	67.2	62.2
	74.9	52.1	74.3	65.5
Dish Network	71.3	64.3	69.4	65.6
	76.8	64.0	68.6	67.8
	71.4	61.3	61.2	55.2
	65.2	68.2	64.9	66.0
Echostar	64.0	60.8	58.3	62.4
	64.2	62.9	64.4	62.9
	76.7	57.0	69.3	64.5
	65.1	69.6	73.1	71.3

Figure 11.10 Data input array for Anova: Two-Factor With Replication.

2. Under the Data tab, select Data Analysis; Anova: Two-Factor With Replication.

3. Enter the Input Range, including the row and column headings, the Rows per sample, and the Alpha level.
4. Choose an Output option and click OK.
5. The analysis of variance table is displayed along with summary statistics by row and column. See Figure 11.11.

	G	H	I	J	K	L	M
	Anova: Two-Factor With Replication						
	SUMMARY	Northeast	Southeast	Midwest	West	Total	
	Direct TV						
	Count	4	4	4	4	16	
	Sum	285.60	232.50	261.60	252.30	1032.00	
	Average	71.40	58.13	65.40	63.08	64.50	
	Variance	16.33	30.44	58.05	41.12	53.48	
	Dish Network						
	Count	4	4	4	4	16	
	Sum	284.70	257.80	264.10	254.60	1061.20	
	Average	71.18	64.45	66.03	63.65	66.33	
	Variance	22.47	8.07	14.19	32.65	24.62	
	Echostar						
	Count	4	4	4	4	16	
	Sum	270.00	250.30	265.10	261.10	1046.50	
	Average	67.50	62.58	66.28	65.28	65.41	
	Variance	37.85	27.90	40.95	16.94	28.24	
	Total						
	Count	12	12	12	12		
	Sum	840.30	740.60	790.80	768.00		
	Average	70.03	61.72	65.90	64.00		
	Variance	24.39	25.79	31.02	25.69		
	ANOVA						
	Source of Variation	SS	df	MS	F	P-value	F crit
	Sample	26.65	2	13.32	0.46	0.6344	3.26
	Columns	446.01	3	148.67	5.14	0.0046	2.87
	Interaction	108.18	6	18.03	0.62	0.7101	2.36
	Within	1040.83	36	28.91			
	Total	1621.66	47				

Figure 11.11
Anova: Two-Factor With Replication output.

SPSS

Use the built-in function General Linear Model; Univariate.

1. Enter the (response) data in a single column (S). Enter the company (row factor) level in column R and the region (column factor) level in column C.
2. Select Analyze; General Linear Model; Univariate.
3. Enter the Dependent Variable, (S), and the two Fixed Factor(s), C and R.
4. Choose the Model option and select Full factorial. Click Continue and OK. The analysis of variance table is displayed in an output window. See Figure 11.12.

Univariate Analysis of Variance

Tests of Between-Subjects Effects

Dependent Variable: S

Source	Type III Sum of Squares	df	Mean Square	F	Sig.
Corrected Model	580.832[a]	11	52.803	1.826	.086
Intercept	205369.085	1	205369.085	7103.244	.000
C	26.645	2	13.323	.461	.634
R	446.006	3	148.669	5.142	.005
C * R	108.181	6	18.030	.624	.710
Error	1040.832	36	28.912		
Total	206990.750	48			
Corrected Total	1621.665	47			

a. R Squared = .358 (Adjusted R Squared = .162)

Figure 11.12 SPSS General Linear Model; Univariate results.

SECTION 11.3 EXERCISES

Remember to use technology wherever possible in order to obtain numerical results. Focus on interpretation of results.

Practice

11.47 Consider the following data in the context of a two-way ANOVA test.

Factor A	Factor B 1			Factor B 2			Factor B 3		
1	5	7	9	8	9	14	14	12	11
2	8	13	7	10	11	11	16	17	15

a. Find a, the number of levels for factor A; b, the number of levels for factor B; and n, the number of observations for each combination of levels.
b. Find $t_{ij.}$ for $i = 1, 2$ and $j = 1, 2, 3$.
c. Find $t_{i..}$ for $i = 1, 2$. Find $t_{.j.}$ for $j = 1, 2, 3$. Find $t_{...}$.

11.48 Consider the following data in the context of a two-way ANOVA test.

Factor A	Factor B 1				2			
1	1.5	2.5	0.7	1.8	3.1	4.4	4.1	3.9
2	3.6	2.8	2.8	3.4	5.7	4.7	7.0	6.0
3	1.0	2.1	1.5	1.6	4.6	3.2	2.0	4.5
4	3.8	2.6	3.3	1.3	4.4	4.1	3.6	4.8

a. Find $t_{i..}$ for $i = 1, 2, 3, 4$, the sample total for each level of factor A. Find $t_{.j.}$ for $j = 1, 2$, the sample total for each level of factor B. Find $t_{ij.}$, the sample total for each *cell*. Find $t_{...}$, the grand total.
b. Find $\sum_{i=1}^{4}\sum_{j=1}^{2}\sum_{k=1}^{4} x_{ijk}^2$.
c. Compute SST, SSA, SSB, SS(AB), and SSE.

11.49 Consider the following data in the context of a two-way ANOVA test.

	Factor B 1		2		3	
1	11.4	16.4	4.9	9.6	17.7	14.9
	9.6		12.6		16.8	
2	8.9	8.5	8.6	12.7	6.5	17.3
	11.7		8.5		9.0	
3	8.7	6.4	9.4	6.6	8.4	10.0
	11.2		9.2		11.7	
4	14.8	13.9	13.8	11.7	19.8	15.3
	9.3		15.6		16.2	

(Factor A rows: 1, 2, 3, 4)

a. Compute SST, SSA, SSB, SS(AB), and SSE.
b. Compute MSA, MSB, MS(AB), and MSE.
c. Compute f_A, f_B, and f_{AB}.
d. Conduct the hypothesis tests to check for effects due to interaction, factor A, and factor B. Use $\alpha = 0.05$ for each test. State your conclusions.

11.50 Consider the following data in the context of a two-way ANOVA test.

Factor A	Factor B 1		2		3	
1	34.2	33.3	36.3	25.9	29.7	37.6
	30.4	27.9	28.5	34.7	38.9	35.6
2	45.5	39.7	37.5	37.9	33.2	45.0
	38.3	35.9	42.4	37.5	40.1	44.5
3	31.0	35.1	38.3	35.7	41.4	39.8
	29.5	39.1	36.3	33.5	39.0	36.7

a. Complete an ANOVA summary table.
b. Conduct the hypothesis tests to check for effects due to interaction, factor A, and factor B. Use $\alpha = 0.01$ for each test. State your conclusions.

11.51 Complete the following ANOVA table.

Source of variation	Sum of squares	Degrees of freedom	Mean square	F	p value
Factor A		4	40.66		
Factor B		2			
Interaction	144.23				
Error	1206.87	75			
Total	1670.28	89			

Conduct the hypothesis tests to check for effects due to interaction, factor A, and factor B. Use $\alpha = 0.05$ for each test. State your conclusions.

11.52 Complete the following ANOVA table.

Source of variation	Sum of squares	Degrees of freedom	Mean square	F	p value
Factor A	121.25			1.57	
Factor B	91.12	4			
Interaction					
Error	1446.41	75			
Total	1833.33	99			

Conduct the hypothesis tests to check for effects due to interaction, factor A, and factor B. Use $\alpha = 0.001$ for each test. State your conclusions.

Applications

11.53 Public Health and Nutrition William Soeltz, the food editor for *The Boston Herald*, conducted a study to examine the quality of wine (measured by age only) at local restaurants. Independent random samples of three types of wines (red, white, and rosé) were obtained from four exclusive restaurants (The Federalist, Top of the Hub, Capital Grille, and Azure). The age (in years) of each wine was recorded. Consider the following partial ANOVA summary table.

Source of variation	Sum of squares	Degrees of freedom	Mean square	F	p value
Type	90.14	2			
Restaurant	266.49	3			
Interaction	56.59				
Error		48			
Total	1128.82	59			

a. Complete the ANOVA table.
b. What was the total number of observations?
c. Conduct the hypothesis test for any effect due to interaction. Use $\alpha = 0.05$. What does your conclusion imply about the other two hypothesis tests?
d. Conduct the hypothesis tests for factor effects. Use $\alpha = 0.05$. State your conclusions.

11.54 Business and Management A temp agency recently conducted a study to determine the effects of job type and gender on length of employment. Independent random samples of employees who worked in service, technology, sales, security, and labor were obtained. The length (in weeks) of each assignment was recorded. Consider the following partial ANOVA summary table.

Source of variation	Sum of squares	Degrees of freedom	Mean square	F	p value
Gender	16.33	1			
Job type	184.39	4			
Interaction		4			
Error	202.42	50			
Total	422.48				

a. Complete the ANOVA table.
b. Is there any evidence of interaction? Use $\alpha = 0.01$.
c. Is there any evidence that gender or job type affects the length of employment? Use $\alpha = 0.01$.

11.55 Manufacturing and Product Development The quality of a digital photograph is partially determined by the number of megapixels per photo. Higher megapixel cameras produce better quality photographs. However, high-quality photographs may take longer to print. A study was conducted to examine the time taken to print randomly selected 4×6 digital photographs of different quality on two printers. The times (in seconds) are given in the following table.

	Printer							
Megapixels	Epson				HP			
2	16.8	17.3	15.3	13.3	18.7	15.8	17.9	12.0
4	16.0	19.8	17.6	18.8	17.5	19.4	14.9	18.8
5	18.8	17.5	19.6	21.1	17.3	18.2	21.6	20.3
8	21.8	21.5	15.7	18.1	20.5	19.0	24.7	20.6

Construct a two-way ANOVA summary table. Use $\alpha = 0.05$ to test the following hypotheses: (a) there is no effect due to interaction; (b) there is no effect due to megapixels; and (c) there is no effect due to printer.

11.56 Sports and Leisure Tailgate parties prior to college football games have become long and lavish, and some present an added security risk. A study was conducted to determine whether the length of a tailgate party is affected by the outside temperature and/or by the college team. The data presented in the following table are the lengths (in hours) of randomly selected tail-gate parties (prior to a game) by school and temperature (Cold:C, Moderate:M, Hot:H).

		School							
Temperature		Michigan		Miami		Ohio State		UCLA	
C		5.1	5.0	4.5	7.0	3.4	4.6	2.9	2.5
		6.0	2.2	6.6	2.8	0.2	3.6	3.1	3.6
M		5.9	5.5	4.7	3.4	3.1	5.7	6.7	7.4
		4.4	4.9	5.4	7.9	5.0	7.2	5.3	5.8
H		4.8	2.8	6.5	5.1	6.3	5.8	4.4	5.8
		4.7	3.4	7.5	5.3	8.0	8.5	9.8	7.2

Use $\alpha = 0.05$ to test the following hypotheses: (a) there is no effect due to interaction; (b) there is no effect due to temperature; and (c) there is no effect due to school. Include the ANOVA summary table.

11.57 Public Policy and Political Science The city of Austin, Texas, recently introduced new procedures in order to streamline the review of commercial and residential building proposals. A random sample of proposals was obtained and classified by type of development and by the city office conducting the review. The length of the review process (in business days) for each proposal is given in the following table.

		Type of development							
Office		Commercial				Residential			
1		5	18	32	32	30	27	27	30
2		20	23	23	28	31	26	31	30
3		33	24	24	20	31	31	26	34

Use $\alpha = 0.05$ to test the following hypotheses: (a) there is no effect due to interaction; (b) there is no effect due to type of development; and (c) there is no effect due to school. Include the ANOVA summary table.

11.58 Medicine and Clinical Studies Alzheimer's disease affects approximately 4 million Americans. While there is no known cure, several drugs may slow the progression of the disease. According to *MH Today*, an Alzheimer's patient typically lives for 6–8 years after the onset of symptoms. However, the range is from less than 2 to 20 years. In a study conducted by a research hospital, a random sample of newly diagnosed Alzheimer's patients was obtained for each combination of type of care and medication (if any). The number of years each patient lived following diagnosis is given in the following table.

		Type of care					
Medication		Live at home with family care			Nursing home with paid care		
Aricept		13.6	11.1	16.9	5.6	10.4	10.9
		14.3	13.8		12.9	11.2	
Exelon		13.0	9.4	15.1	8.4	11.5	16.0
		13.3	16.9		9.4	9.2	
Tacrine		10.0	10.8	15.2	10.0	14.3	6.9
		16.7	14.8		13.6	10.2	
NGF		11.8	13.2	17.0	8.5	8.5	9.8
		12.4	13.4		9.7	8.3	
None		10.7	16.9	11.8	11.8	8.3	6.8
		14.8	9.8		14.9	10.4	

(*Source*: Alzheimer's Disease Education and Referral Center.)

Conduct an analysis of variance with $\alpha = 0.05$ to test for interaction and factor effects. State your conclusions.

11.59 Biology and Environmental Science Wetlands, sometimes called the *nurseries of life*, are home to thousands of species of plants and animals and are found on every continent except Antarctica. Canada has more than 127 million hectares of wetlands, approximately 25% of all the wetlands on our planet.[10] Suppose, in a recent U.S. Department of Agriculture study, a random sample of wetlands was obtained for each combination of state and cover type—for example, meadow or swamp. The cover type classifications are defined by the U.S. Fish and Wildlife service and are determined by the representative plant species. The area covered by water was measured (in hectares), and the data are given in the following table.

		State						
		Louisiana		Mississippi		California		Arkansas	

Cover type		Louisiana		Mississippi		California		Arkansas	
	AB3	0.8	1.1	1.2	0.7	1.2	1.8	3.3	0.2
	FO1	2.0	1.7	0.5	0.1	2.4	3.0	1.4	1.5
	OW	2.6	0.6	2.7	1.6	1.3	1.8	2.7	1.8
	SS1	2.7	2.3	1.1	0.8	1.3	2.4	1.8	1.3

a. Construct the ANOVA summary table. Is there any evidence of an effect due to interaction of state and cover type? Use $\alpha = 0.01$.
b. Is there any evidence of an effect due to state or cover type? Use $\alpha = 0.01$ for both hypothesis tests. State your conclusions.

11.60 Sports and Leisure The Genesis Diving Institute of Florida certifies scuba divers using a variety of different systems of education. This organization recently studied the time spent underwater by scuba divers exploring caves and in open water. Each diver was also classified by age group. Independent dives were selected at random for each combination of age group and dive type. The data (in minutes) are given in the following table.

		Dive type						
		Cave				Open water			
Age group	**20–30**	39	38	41	39	42	40	40	39
		41	42	40	40	37	45	36	40
	30–40	43	40	38	36	45	39	43	43
		41	46	40	37	47	50	46	44
	40–50	42	41	50	47	37	40	40	38
		43	38	46	47	34	45	34	41
	≥50	38	34	35	36	33	37	42	38
		40	38	42	38	30	39	41	39

Construct the two-way ANOVA summary table. Interpret the results. Use $\alpha = 0.001$ for each hypothesis test.

Extended Applications

11.61 Travel and Transportation The Swedish Highway Department conducted a study to examine the speed of cars on major highways. Three regions were selected and rated using four different quality scores (based on retroreflective properties) for road markings. (The four classes of road markings in Sweden are K1, certainly approved; K2, probably approved; K3, probably rejected; and K4, certainly rejected.) The speed of randomly

selected cars (in km/hr) was recorded, the sample totals for each cell are given in the following table ($n = 4$), and SST $= 1680.66$.

		Road marking quality			
		K1	K2	K3	K4
Region	**North**	402.4	393.4	386.6	353.3
	Mälardalen	398.6	418.3	381.6	374.2
	Mitt	419.4	425.3	385.2	386.3

Use $\alpha = 0.05$ to test the following hypotheses: (a) there is no effect due to interaction, (b) there is no effect due to region, and (c) there is no effect due to road-marking quality.

11.62 Psychology and Human Behavior The FBI studied the amount of money stolen (in thousands of dollars) during randomly selected bank robberies at different types of banks and in different locations. The sample totals for each combination of bank type and location are given in the following table ($n = 3$) and

$$\sum_{i=1}^{3}\sum_{j=1}^{4}\sum_{k=1}^{3} x_{ijk}^2 = 2514.28$$

	Bank type			
Location	Commercial	Savings	Savings & Loan	Credit Union
City	25.2	23.7	26.8	24.6
Suburban	32.2	20.2	24.4	31.6
Rural	20.5	22.5	16.0	22.0

(*Source*: Federal Bureau of Investigation).

Use $\alpha = 0.01$ to test the following hypotheses: (a) there is no effect due to interaction, (b) there is no effect due to location, and (c) there is no effect due to bank type. Construct the ANOVA summary table.

11.63 Public Health and Nutrition Jobs in the logging industry are tedious, tiring, and dangerous. There were approximately 2900 nonfatal on-the-job injuries in the logging industry in 2007.[11] Suppose a recent OSHA study examined the number of workdays lost due to on-the-job injuries in three different states. In a random sample of logging injury reports, SST $= 465.91$ and the sample total number of workdays lost for each combination of state and injury cause is given in the following table ($n = 6$).

		State		
		Alaska	Oregon	Washington
Injury cause	**Machines**	84.4	82.6	80.9
	Metal items	95.2	73.4	85.5
	Rubber products	74.3	96.0	90.6
	Vehicles	84.1	74.7	72.4
	Working surfaces	75.4	70.4	72.0

a. Construct the summary ANOVA table. Is there any evidence of an effect due to interaction of state and injury cause? Use $\alpha = 0.05$.
b. Is there any evidence of an effect due to state or injury cause? Use $\alpha = 0.05$ for both hypothesis tests.

11.64 Public Health and Nutrition Cerner Corporation recently presented their vision of future hospital rooms containing technology to improve medical efficiency and decrease the chance of human error.[12] The new rooms would have a system that scans staff IDs and a monitor that displays all relevant patient medical information, and might be a different size from current rooms. Suppose a random sample of existing hospital rooms was obtained, and the amount of square feet per patient in each room was measured. Each room was classified by location and type and the data are given in the following table.

Location		Room type: Private		Semiprivate two beds		Semiprivate four beds	
City		206	190	210	244	182	183
		263	212	233	201	142	240
		217	205	185	213	177	176
Suburban		206	203	228	192	143	200
		198	191	204	216	170	195
		197	179	178	200	124	187

a. Construct the ANOVA summary table. Is there any evidence of an effect due to interaction of location and hospital room type? Use $\alpha = 0.05$.

b. Is there any evidence of an effect due to location or hospital room type? Use $\alpha = 0.05$ for both hypothesis tests. State your conclusions.

c. Does this analysis suggest there is one combination of location and room type that is *on average* the smallest in square feet per patient? If so, what is the location and room type?

(Juice Images/AgeFotostock)

Chapter 11 Challenge Wrap-Up

Here are the beta-carotene data (in mg) for the carrot experiment described in the Chapter Challenge.

Region	Observations				
North	11.87	11.86	11.32	11.02	12.01
	12.03	12.92	11.96	11.39	12.93
South	12.04	13.29	11.00	12.05	11.49
	11.44	11.35	10.40	10.62	11.59
Midwest	11.96	10.43	11.33	10.90	12.54
	9.59	11.70	11.78	11.87	12.66
West	11.55	13.59	13.74	12.32	11.51
	11.39	10.96	13.43	13.48	11.46

A one-way ANOVA will be used to determine whether there is any difference among the regions in the population mean beta-carotene per carrot serving. Use the computational formulas associated with a one-way ANOVA and $\alpha = 0.05$. In this case, $k = 4$ and $n = 40$.

$$t_{..} = 11.87 + 11.86 + \cdots + 11.46 = 472.77$$

$$\text{SST} = \left(\sum_{i=1}^{k} \sum_{j=1}^{n_i} x_{ij}^2 \right) - \frac{t_{..}^2}{n}$$

$$= (11.87^2 + 11.86^2 + \cdots + 11.46^2) - \frac{472.77^2}{40} = 33.4327$$

$$\text{SSA} = \left(\frac{119.31^2}{10} + \frac{115.27^2}{10} + \frac{114.76^2}{10} + \frac{123.43^2}{10} \right) - \frac{472.77^2}{40} = 4.9003$$

$$\text{SSE} = \text{SST} - \text{SSA} = 33.4327 - 4.9003 = 28.5324$$

Compute the mean square due to factor and the mean square due to error.

$$\text{MSA} = \text{SSA}/(4 - 1) = 4.9003/3 = 1.6334$$

$$\text{MSE} = \text{SSE}/(40 - 4) = 28.5324/36 = 0.7926$$

The four parts of the hypothesis test are

$H_0: \mu_1 = \mu_2 = \mu_3 = \mu_4$ (all four population means are equal)

$H_a: \mu_i \neq \mu_j$ for some $i \neq j$ (at least two population means differ)

$$\text{TS: } F = \frac{\text{MSA}}{\text{MSE}}$$

RR: $F \geq F_\alpha = F_{0.05} = 2.87$ **df 3 and 36.**

The value of the test statistic is

$$f = \frac{\text{MSA}}{\text{MSE}} = \frac{1.6334}{0.7926} = 2.06$$

p value illustration:
$$p = P(X \geq 2.06)$$
$$= 0.1228 > 0.05 = \alpha$$

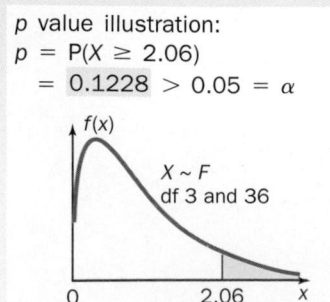

The value of the test statistic does *not* lie in the rejection region. At the $\alpha = 0.05$ significance level, there is no evidence to suggest that any two population mean beta-carotene levels are different. There is no need to construct simultaneous Bonferroni or Tukey confidence intervals or only eat carrots grown in the West. Here are the ANOVA summary table and the technology solution.

The *p* value in this table was computed using technology.

ANOVA summary table

Source of variation	Sum of squares	Degrees of freedom	Mean square	F	p value
Factor	4.9003	3	4.9003	2.06	0.1228
Error	28.5324	36	0.7926		
Total	33.4327	39			

Figure 11.13 shows a technology solution.

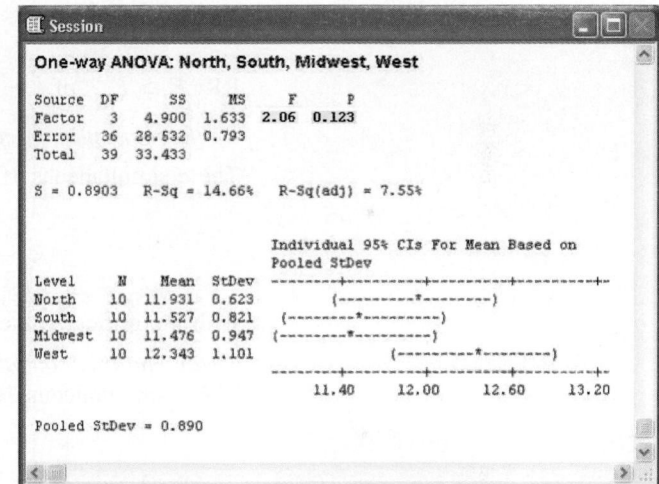

Figure 11.13 One-way ANOVA results.

CHAPTER 11 SUMMARY

One-Way ANOVA

Assumptions:

1. The k population distributions are normal.

2. The k population variances are equal; that is, $\sigma_1^2 = \sigma_2^2 = \cdots = \sigma_k^2$.

3. The samples are selected randomly and independently from the respective populations.

A one-way, or single-factor, analysis of variance is a statistical technique used to determine whether there is any difference among k population means.

The sums of squares:

The fundamental identity: SST = SSA + SSE

$$\text{SST} = \text{total sum of squares} = \sum_{i=1}^{k}\sum_{j=1}^{n_i}(x_{ij} - \bar{x}_{..})^2 = \left(\sum_{i=1}^{k}\sum_{j=1}^{n_i}x_{ij}^2\right) - \frac{t_{..}^2}{n}$$

$$\text{SSA} = \text{sum of squares due to factor} = \sum_{i=1}^{k}n_i(\bar{x}_{i.} - \bar{x}_{..})^2 = \left(\sum_{i=1}^{k}\frac{t_{i.}^2}{n_i}\right) - \frac{t_{..}^2}{n}$$

$$\text{SSE} = \text{sum of squares due to error} = \sum_{i=1}^{k}\sum_{j=1}^{n_i}(x_{ij} - \bar{x}_{i.})^2 = \text{SST} - \text{SSA}$$

ANOVA summary table

Source of variation	Sum of squares	Degrees of freedom	Mean square	F	p value
Factor	SSA	$k - 1$	$\text{MSA} = \dfrac{\text{SSA}}{k - 1}$	$\dfrac{\text{MSA}}{\text{MSE}}$	p
Error	SSE	$n - k$	$\text{MSE} = \dfrac{\text{SSE}}{n - k}$		
Total	SST	$n - 1$			

Hypothesis test:

$H_0: \mu_1 = \mu_2 = \cdots = \mu_k$ (all k population means are equal)

$H_a: \mu_i \neq \mu_j$ for some $i \neq j$ (at least two population means differ)

TS: $F = \dfrac{\text{MSA}}{\text{MSE}}$

RR: $F \geq F_\alpha$ df $k - 1$ and $n - k$

Bonferroni multiple comparison procedure:

The c simultaneous $100(1 - \alpha)\%$ Bonferroni confidence intervals have as endpoints

$$(\bar{x}_{i.} - \bar{x}_{j.}) \pm t_{\alpha/(2c)} \sqrt{\text{MSE}} \sqrt{\frac{1}{n_i} + \frac{1}{n_j}} \quad \text{for all } i \neq j$$

with the critical value based on df $= n - k$.

Tukey multiple comparison procedure:

The c simultaneous $100(1 - \alpha)\%$ confidence intervals have as endpoints

$$(\bar{x}_{i.} - \bar{x}_{j.}) \pm \frac{1}{\sqrt{2}} Q_\alpha \sqrt{\text{MSE}} \sqrt{\frac{1}{n_i} + \frac{1}{n_j}} \quad \text{for all } i \neq j$$

with the critical value based on k and $n - k$.

Two-Way ANOVA

A two-way analysis of variance is a statistical procedure used to determine the effect of two factors on a response variable.

The sums of squares:

The fundamental identity: SST = SSA + SSB + SS(AB) + SSE

SST = total sum of squares

$$= \sum_{i=1}^{a} \sum_{j=1}^{b} \sum_{k=1}^{n} (x_{ijk} - \bar{x}_{...})^2 = \left(\sum_{i=1}^{a} \sum_{j=1}^{b} \sum_{k=1}^{n} x_{ijk}^2 \right) - \frac{t_{...}^2}{abn}$$

SSA = sum of squares due to factor A

$$= bn \sum_{i=1}^{a} (\bar{x}_{i..} - \bar{x}_{i..})^2 = \frac{\displaystyle\sum_{i=1}^{a} t_{i..}^2}{bn} - \frac{t_{...}^2}{abn}$$

SSB = sum of squares due to factor B

$$= an \sum_{j=1}^{b} (\bar{x}_{.j.} - \bar{x}_{...})^2 = \frac{\displaystyle\sum_{j=1}^{b} t_{.j.}^2}{an} - \frac{t_{...}^2}{abn}$$

SS(AB) = sum of squares due to interaction

$$= n \sum_{i=1}^{a} \sum_{j=1}^{b} (\bar{x}_{ij.} - \bar{x}_{i..} - \bar{x}_{.j.} + \bar{x}_{...})^2 = \frac{\sum_{i=1}^{a} \sum_{j=1}^{b} t_{ij.}^2}{n} - \frac{\sum_{i=1}^{a} t_{i..}^2}{bn} - \frac{\sum_{j=1}^{b} t_{.j.}^2}{an} + \frac{t_{...}^2}{abn}$$

SSE = sum of squares due to error

$$= \sum_{i=1}^{a} \sum_{j=1}^{b} \sum_{k=1}^{n} (x_{ijk} - \bar{x}_{ij.})^2 = \text{SST} - \text{SSA} - \text{SSB} - \text{SS(AB)}$$

ANOVA summary table

Source of variation	Sum of squares	Degrees of freedom	Mean square	F	p value
Factor A	SSA	$a - 1$	$\text{MSA} = \dfrac{\text{SSA}}{a - 1}$	$F_A = \dfrac{\text{MSA}}{\text{MSE}}$	p_A
Factor B	SSB	$b - 1$	$\text{MSB} = \dfrac{\text{SSB}}{b - 1}$	$F_B = \dfrac{\text{MSB}}{\text{MSE}}$	p_B
Interaction	SS(AB)	$(a - 1)(b - 1)$	$\text{MS(AB)} = \dfrac{\text{SS(AB)}}{(a - 1)(b - 1)}$	$F_{AB} = \dfrac{\text{MS(AB)}}{\text{MSE}}$	p_{AB}
Error	SSE	$ab(n - 1)$	$\text{MSE} = \dfrac{\text{SSE}}{ab(n - 1)}$		
Total	SST	$abn - 1$			

Hypothesis tests:

1. Test for an interaction effect.

H_0: There is no interaction effect.

H_a: There is an effect due to interaction.

TS: $F_{AB} = \dfrac{\text{MS(AB)}}{\text{MSE}}$

RR: $F_{AB} \geq F_\alpha$ df $(a - 1)(b - 1)$ and $ab(n - 1)$.

2. Test for an effect due to factor A.

H_0: There is no effect due to factor A.

H_a: There is an effect due to factor A.

TS: $F_A = \dfrac{\text{MSA}}{\text{MSE}}$

RR: $F_A \geq F_\alpha$ df $a - 1$ and $ab(n - 1)$

3. Test for an effect due to factor B.

H_0: There is no effect due to factor B.

H_a: There is an effect due to factor B.

TS: $F_B = \dfrac{\text{MSB}}{\text{MSE}}$

RR: $F_B \geq F_\alpha$ df $b - 1$ and $ab(n - 1)$

CHAPTER 11 EXERCISES

Applications

11.65 Sports and Leisure Backing a trailer down a boat ramp in order to launch a boat can be awkward and tricky. A steep ramp angle often makes this task even more intimidating, especially for new boat owners. Independent random samples of boat ramps in various regions of Florida were obtained, and the angle of each was measured (in degrees). Consider the following partial ANOVA table.

Source of variation	Sum of squares	Degrees of freedom	Mean square	F	p value
Factor		4			
Error	470.355	45			
Total	502.650				

a. Complete the ANOVA summary table.
b. How many regions in Florida were considered? How many total observations were obtained?
c. Is there any evidence that the population mean boat-ramp angle differs among these populations? Use $\alpha = 0.05$.

11.66 Physical Sciences It can get very hot for actors on stage. Not only do they have to remember all of their lines, but theater spotlights are usually bright and intense. Independent random samples of spotlights were obtained from four different Broadway theaters (Cort Theater, Imperial Theater, Majestic Theater, and the New Amsterdam Theater). The wattage of each light was measured, and a partial ANOVA summary table is shown.

Source of variation	Sum of squares	Degrees of freedom	Mean square	F	p value
Factor	106,568	3			
Error		20			
Total	258,565				

Complete the table and use this information to determine whether there is any evidence that the population mean wattage of spotlights is different among the four theaters. Use $\alpha = 0.05$.

11.67 Manufacturing and Product Development Before construction of the Burlington Public Library in California began, stress tests were conducted on four types of anchor bolts. Independent random samples of bolts were obtained, and each was subjected to a motion simulating an earthquake. The horizontal displacement of the attached panel was measured (in inches), and the resulting data are given on the data CD and book's web site. Construct the one-way ANOVA table and test the hypothesis of no difference in population mean horizontal displacement. Use $\alpha = 0.01$.

11.68 Manufacturing and Product Development Microlithography is a process that includes baking a semiconductor wafer. An experiment was conducted to study the temperature of a 200-mm diameter wafer at different locations on the wafer during a new baking process. Independent random samples of wafers were selected, and a temperature sensor was placed on each wafer at one of four distances from the center of the wafer. The temperature was recorded (in °C) 80 seconds into the process. The summary statistics are given in the following table.

Location	T1	T2	T3	T4
Sample size	10	10	10	10
Sample mean	106.7	104.6	104.3	98.6

A one-way analysis of variance test was significant at the $p = 0.009$ level. Use MSE $= 26.92$ to find the Bonferroni 95% confidence intervals for all pairwise differences. Draw a graph to represent the results.

11.69 Technology and Internet An experiment was conducted to compare the accuracy of five different computer algorithms for translating Chinese into English. Each algorithm was tested 25 times on randomly selected passages, and an accuracy score (between 0 and 1) was recorded for each trial. A one-way analysis of variance test was significant at the $p = 0.003$ level. Use MSE $=$ 0.0276 and the summary statistics below to find the Tukey 95% confidence intervals for all pairwise differences. Draw a graph to represent the results.

Algorithm	OO	CP	SLCP	AC	T
Sample mean	0.5566	0.7020	0.6190	0.7023	0.7115

11.70 Manufacturing and Product Development Using a new atomic imaging technique, a study was conducted to measure randomly selected thin films of NaCl on a metal substrate. Four different growth modes were used, and the step height was measured (in nm) in each case. The data are given on the data CD and book's web site.

a. Conduct a one-way analysis of variance test to determine whether there is an overall difference in population mean step heights. Use $\alpha = 0.05$.
b. Construct the Bonferroni 99% confidence intervals to isolate any population means contributing to an overall difference, and draw a graph to represent the results.

11.71 Public Health and Nutrition A root canal is an endodontic procedure to remove the infected or damaged nerve tissue of a tooth. The American Dental Association conducted a study to determine whether the major canal diameter of a tooth is related to the tooth type and/or the patient's race. Independent random samples of adult male root canal patients were selected for each combination of tooth type and race. The canal diameters (in mm) are given in the following table.

		Tooth type			
		Maxillary		Mandibular	
Group	White	0.97	1.00	1.09	1.39
		0.75	1.06	1.17	1.19
	African American	1.17	1.00	1.21	1.48
		1.09	0.88	0.94	1.16
	Native American	0.86	1.03	1.10	1.45
		0.77	0.88	1.58	1.18
	Hispanic	1.13	0.81	0.90	1.01
		0.78	0.86	1.57	1.42

Construct the two-way ANOVA summary table. Interpret the results. Use $\alpha = 0.05$ for each hypothesis test.

Extended Applications

11.72 Physical Sciences There are approximately 440 nuclear power plants in operation around the world, with an estimated power production of 368,611 MWh per year. Independent random samples of nuclear power plants in three countries were obtained, and the net generating capacity per year (in MW) of each was obtained.[13] Summary statistics are given in the following table.

Company	Sample size	Sample total	Sum of squared observations
United States	5	5848	6,861,942
France	6	6470	7,623,250
Germany	5	5212	5,726,814

Conduct the appropriate test to check the hypothesis of no difference in population mean net generating capacity. Use $\alpha = 0.01$.

11.73 Sports and Leisure The biggest catfish caught in the U.S. Catfish Anglers Tournament Series, Kansas District, Mid-West Championship in 2008 was 74.60 pounds.[14] In preparation for the next catfish tournament, a fisherman (familiar with statistics) randomly selected anglers from five different locations and recorded the weight (in pounds) of the last catfish they caught. The data (in pounds) are given in the following table.

Location	Observations					
Hill's Landing	22.7	14.1	9.6	5.9	21.5	28.7
	45.7	17.4				
Eagle's Nest	38.7	44.6	34.6	19.9	59.3	23.2
	28.3	35.5	38.8	34.5	39.1	
Santee	37.5	53.2	47.0	31.8	19.9	35.9
	50.0	27.6	30.6	22.4		
Campground	23.7	32.1	38.9	31.1	34.2	37.2
	49.8	56.6	52.2	34.2	44.5	38.9
Rock Hill	42.0	41.6	63.2	41.6	25.5	28.4
	33.8					

a. Conduct a one-way analysis of variance test to determine whether there is an overall difference in population mean catfish weights. Use $\alpha = 0.05$.
b. Construct the Tukey 99% confidence intervals to isolate any population means contributing to an overall difference, and draw a graph to represent the results. Which site would you recommend in order for the fisherman to have a good chance at winning the tournament? Why?

11.74 Marketing and Consumer Behavior A real-estate agent, Sarah Harvey, conducted a study to compare the effect of location and season on weekly time-share costs (in dollars). The sample totals for each combination of island (Aruba: A, Martinique: M, St. Kitts: SK, St. Lucia: SL) and season are given in the following table ($n = 6$) and SST $= 487,902,980.64$.

		Season			
		Spring	Summer	Fall	Winter
Island	A	18,045	12,168	19,894	26,214
	M	14,495	1,925	13,890	17,538
	SK	18,075	32,887	24,398	18,834
	SL	25,457	20,757	27,505	36,428

Use $\alpha = 0.01$ to test the following hypotheses: (a) there is no effect due to interaction, (b) there is no effect due to island, and (c) there is no effect due to season.

11.75 Biology and Environmental Science A random sample of male baby California sea lions was captured, tagged, and monitored for several years. The weight of each sea lion was measured (in pounds) once, and each sea lion was classified by species ID and age (in years). The sample totals are given in the following table. Here, $n = 4$ and SST $= 103152.38$.

		Age		
		1	2	3
Species ID	50011	1768	1735	1752
	50023	1794	1563	1517
	50037	1750	1742	1328

Use $\alpha = 0.05$ to test the following hypotheses: (a) there is no effect due to interaction, (b) there is no effect due to species, and (c) there is no effect due to age.

11.76 Public Health and Nutrition A consumer group recently studied the amount of partially hydrogenated oils (trans fat) in various peanut butter brands. Four brands were selected in smooth and chunky varieties. The amount of trans fat was measured (in grams) per serving (2 tablespoons) in each randomly selected jar, and the data are given in the table below.
a. Construct the two-way ANOVA summary table. Interpret the results. Use $\alpha = 0.05$ for each hypothesis test.
b. If a consumer wanted to avoid trans fat as much as possible, which combination of brand and style would you recommend? Why?

		Peanut butter brand						
		Jif		Peter Pan		Skippy		Smucker's
Style	Smooth	0.51	0.72	0.89	0.69	0.75	0.85	0.73	0.71
	Chunky	0.46	0.63	0.88	0.81	0.76	0.69	0.73	0.76

Correlation and Linear Regression

12

Chapter 12 Challenge

Are windmills too noisy?

Modern windmills are quickly becoming an efficient, clean alternative to fossil fuels for producing energy in the United States and other countries. On a typical three-blade windmill, the length of a blade is approximately 50–60 feet with diameter 12 feet, and each blade is angled. While a cost-effective windmill can utilize low wind velocity, the ideal locations for wind farms are along ocean coastlines or in mountainous areas, where there is a consistent wind speed of at least 15 mph.

Windmills cannot be constructed too near a town due to noise-level regulations. A typical windmill produces approximately 56 dB 200 feet away. This is softer than the sound of human speech (which is about 70 dB). Suppose a small coastal community is considering constructing a windmill in order to generate electricity for the town hall. An experiment is conducted in order to measure the windmill's noise level (in dB) at various distances (in feet) from the proposed site. The data are summarized in the following table.

(Beverett/Dreamstime.com)

Distance	10	50	75	120	150	160	200	250	400	500
Noise level	75	110	73	52	58	77	56	57	28	4

(*Source*: American Wind Energy Association.)

The techniques presented in this chapter can be used to determine whether there is a significant linear relationship between distance and noise level. The Chapter Challenge Wrap-Up (page 623) shows how regression analysis can be used to predict a value of the noise level for a given distance from the windmill.

Review

- Recall the properties and the methods to compute probabilities associated with a normal distribution.
- Remember the procedures for conducting hypothesis tests and constructing confidence intervals based on a t distribution.
- Think about the procedures for conducting hypothesis tests based on an F distribution.

Preview

- Learn how to visualize and measure a linear relationship between two variables.
- Construct scatter plots, compute the correlation coefficient, and consider linear regression concepts and computations.
- Investigate methods for selecting the best mathematical model.

CONTENTS

12.1   SIMPLE LINEAR REGRESSION

Remember function notation from your algebra class? $f(x)$ is read as "f of x" and is the rule for determining y for a value of x.

A *deterministic* relationship between two variables x and y is one in which the value of y is completely determined by the value of x. In general, $y = f(x)$ is a deterministic relationship between x and y. The value of y *depends* on the value of x. The *independent* variable is x and the *dependent* variable is y. We are free to choose x, but y depends on the value selected. For example, $y = x^2 + 2x + 5$ is a deterministic relationship between x and y. If $x = 1$, then y is completely determined and $y = 1^2 + 2(1) + 5 = 8$.

Remember $y = mx + b$? This is the same idea, rewritten. Here, the Greek letter β is not related to a type II error, as discussed in Section 9.3.

One of the simplest deterministic relationships between x and y is a *linear* relationship: $y = \beta_0 + \beta_1 x$, where β_0 and β_1 are constants. This equation specifies the set of ordered pairs (x, y) such that $y = \beta_0 + \beta_1 x$ forms a straight line with slope β_1 and y-intercept β_0. For example, the graph of $y = 3 + 7x$ is a straight line with slope 7 and y-intercept 3. (Can you sketch this line?)

There is often a need to predict the value of the dependent variable for a given value of the independent variable. For example, if the price is $110, predict the number of tickets that will be sold. Simple linear regression allows us to find the best prediction equation.

An extension of a deterministic relationship is a *probabilistic model*. For a fixed value x, the value of the second variable is randomly distributed. For example, suppose we are investigating the relationship between one-way airfare (in dollars) from New York City to Los Angeles and the number of tickets sold on the 10:00 A.M. flight. If we select $x = 99$ (dollars), then the number of tickets sold (for $x = 99$) is a random variable Y. On one particular day the observed value of Y associated with $x = 99$ may be $y = 135$ tickets.

The *independent variable*, fixed by the experimenter, is usually denoted by x. For a fixed value x, the second variable is randomly distributed. This random variable is the *dependent variable* and is usually denoted by Y. Consistent with previous notation, an observed value of Y is denoted by y.

In a general *additive probabilistic model* there is a deterministic part and a random part. The value of Y differs from $f(x)$ (the deterministic part) by a random amount. The model can be written as

$$Y = (\text{deterministic function of } x) + (\text{random deviation})$$
$$= f(x) + E \tag{12.1}$$

where E is a random variable, called the *random error*.

A value of the random variable E is denoted by e. Here, e is not the base of the natural logarithm.

Suppose we choose a value of x, say $x = x_0$, and observe a value of Y for this value of x, denoted y. If the value of the random variable E is positive, then $y > f(x_0)$. Similarly, if $e < 0$, then $y < f(x_0)$. And if $e = 0$, then $y = f(x_0)$. Geometrically, the value y lies either above, below, or on the graph of $y = f(x)$. Figure 12.1 illustrates the case where $e > 0$.

Figure 12.1 An illustration of a probabilistic model. An observed value y is composed of a deterministic part, $f(x_0)$, and a random error, e.

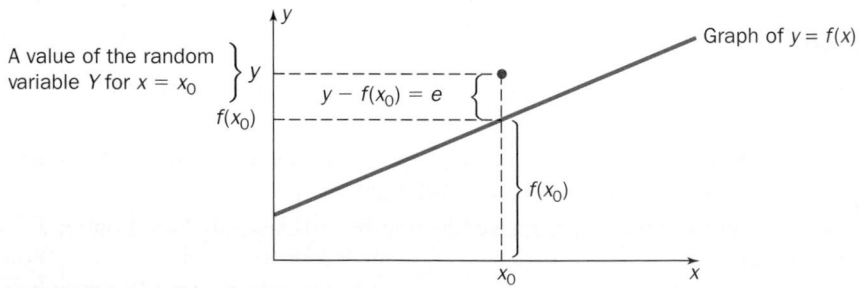

Suppose there are n observations on specific, fixed values of the independent variable. Here is the notation we use.

1. The observed values of the independent variable are denoted $x_1, x_2, \ldots, x_n$.

2. Y_i and y_i are the random variable and the observed value of the random variable associated with x_i, for $i = 1, 2, \ldots, n$. This is tricky. For each x_i, there is a corresponding random variable Y_i. So there are really n random variables in these problems.

3. The data set consists of n ordered pairs: $(x_1, y_1), (x_2, y_2), \ldots, (x_n, y_n)$.

Example 12.1 Temperature and Cattle Stress   Because heat and humidity can cause stress and lower milk production,[1] a dairy farmer is investigating the relationship between cattle respiration rate (breaths per minute) and ambient temperature (°F). The data for a random sample of days and Senepol cattle are given in the following table.

Temperature, x	85	89	81	86	91	92	87	84	80	82	83	85
Breaths per minute, y	58	61	42	56	67	65	62	44	46	52	45	53

a. Identify the independent and dependent variables.
b. List the ordered pairs in the data set.
c. Construct a scatter plot for these data. What does the plot suggest about the relationship between the variables?

SOLUTION

a. The independent variable is temperature, and the dependent variable is breaths per minute. The dairy farmer believes that there is a relationship between these two variables and hopes to be able to predict cattle breaths per minute as a function of temperature.
b. The data set consists of 12 ordered pairs: (85, 58), (89, 61), . . . , (85, 53). For each (fixed) value of temperature, there is a corresponding observation on a random variable for breaths per minute.
c. Figure 12.3 is a *scatter plot* of breaths per minute (y) versus temperature (x). This plot suggests that as temperature increases, breaths per minute also increase, and therefore the relationship might be positive linear.

A technology solution:

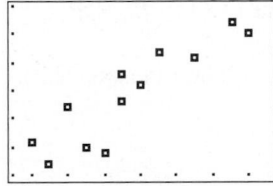

Figure 12.2 A TI-84 Plus scatter plot of breaths per minute versus temperature.

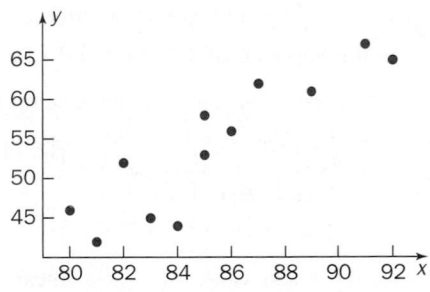
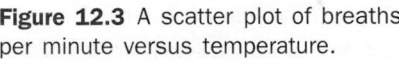

Figure 12.3 A scatter plot of breaths per minute versus temperature.

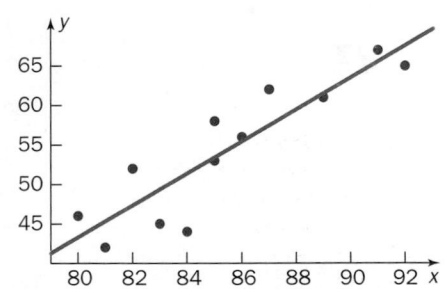

Figure 12.4 Cattle breaths scatter plot with a (deterministic) line added.

In a **simple linear regression model**, the deterministic function $f(x)$ in Equation 12.1 is assumed to be linear (i.e., $f(x) = \beta_0 + \beta_1 x$). The graph of this regression equation is a straight line that describes how a dependent variable y changes as an independent variable x changes. Figure 12.4 shows the graph of a possible deterministic straight line added to the scatter plot. Notice that the data points lie *close* to the line; the vertical distance between a point and the line depends on the value of the random error. And the line has positive slope, which conveys the relationship between the two variables: as the temperature rises, so do the breaths per minute.

This section presents a method for finding the *best* deterministic straight line for a given set of data. Hypothesis tests will be used to determine whether there is a *significant* linear relationship between two variables. As with all other hypothesis tests, certain assumptions must be true in order for the related statistical procedures to be valid.

In just a few pages, we'll define what is meant by *best*.

SIMPLE LINEAR REGRESSION MODEL

Let (x_1, y_1), (x_2, y_2), . . . , (x_n, y_n) be n pairs of observations such that y_i is an observed value of the random variable Y_i. We assume that there exist constants β_0 and β_1 such that

$$Y_i = \beta_0 + \beta_1 x_i + E_i$$

where $E_1, E_2, \ldots, E_n$ are independent, normal random variables with mean 0 and variance σ^2. That is,

1. The E_i's are normally distributed (which implies that the Y_i's are normally distributed).
2. The expected value of E_i is 0 (which implies that $E(Y_i) = \beta_0 + \beta_1 x_i$).
3. $\text{Var}(E_i) = \sigma^2$ (which implies that $\text{Var}(Y_i) = \sigma^2$).
4. The E_i's are independent (which implies that the Y_i's are independent).

ILLUMINATING THE CONCEPTS

1. The E_i's are the **random deviations** or **random error terms**.

2. $y = \beta_0 + \beta_1 x$ is the **true regression line**. Each point, (x_i, y_i), lies *near* the true regression line, depending on the value of the random error term, e_i.

3. The four assumptions in the simple linear regression model definition can be stated compactly in terms of the random error term: $E_i \overset{\text{ind}}{\sim} N(0, \sigma^2)$. ∎

Just a little more notation is necessary in order to understand the simple linear regression model and the resulting properties.

Consider each random variable $Y_i = Y|x_i$.

> Recall, $Y|x_i$ means "Y given x_i."

$\mu_{Y|x_i} = E(Y|x_i)$ is the expected value of Y for a fixed value x_i, and

$\sigma^2_{Y|x_i}$ is the variance of Y for a fixed value x_i.

The simple linear regression model assumptions imply that

$$\mu_{Y|x_i} = E(\beta_0 + \beta_1 x_i + E_i) = \beta_0 + \beta_1 x_i + E(E_i) = \beta_0 + \beta_1 x_i$$
$$\sigma^2_{Y|x_i} = \text{Var}(\beta_0 + \beta_1 x_i + E_i) = \sigma^2$$
$$Y|x_i \text{ is normal}$$

Therefore, the mean value of Y is a linear function of x. The true regression line passes through the *line of mean values*.

The variability in the distribution of Y is the *same* for every value of x (this is called **homogeneity of variance**).

Figure 12.5 illustrates the model assumptions and resulting properties. Each Y_i has a normal distribution, centered at $\beta_0 + \beta_1 x_i$. All the distributions have the same width, or variance.

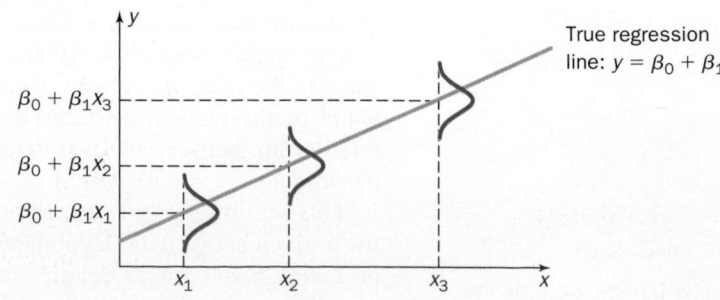

Figure 12.5 The true regression line connects the mean values $\beta_0 + \beta_1 x_i$.

Solid bone matrix | Weakened bone matrix

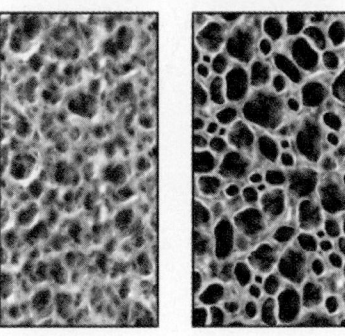

(Nucleus Medical Art, Inc/Phototake)

Example 12.2 Bone Mineral Density Many medical studies suggest that bone mineral density (y, measured in mg/cm^2) is (linearly) related to calcium intake (x, measured in mg/day).[2] For women between ages 40 and 50, suppose that the true regression line is $y = -1575.2 + 2.8x$.

a. Find the expected bone mineral density when the calcium intake is 750 mg/day.

b. How much change in bone mineral density is expected if the calcium intake increases by 10 mg/day? What if it decreases by 20 mg/day?

c. Suppose $\sigma = 12$ mg/cm^2. Find the probability that an observed value of bone mineral density exceeds 470 mg/cm^2 when the calcium intake is 725 mg/day.

SOLUTION

a. We need the expected value of Y for the value $x = 750$.

$$E(Y|x) = \beta_0 + \beta_1 x \qquad \text{Simple linear regression model implication.}$$
$$E(Y|750) = -1575.2 + 2.8(750) \qquad \text{Use the true regression line with } x = 750.$$
$$= 524.8 \qquad \text{Simplify.}$$

The expected bone mineral density for a woman aged 40–50 with a calcium intake of 750 mg/day is 524.8 mg/cm^2. See Figure 12.6.

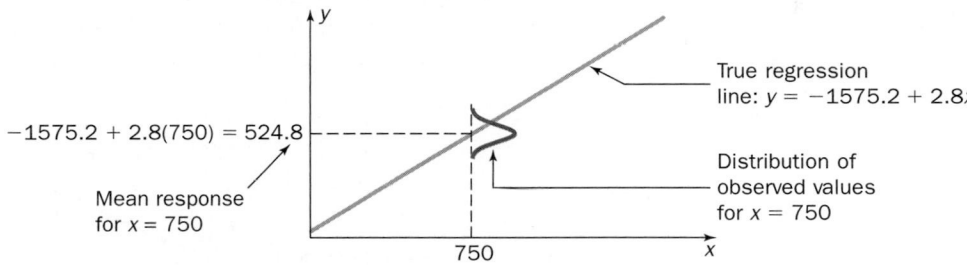

Figure 12.6 The true regression line, the mean response for $x = 750$, and the distribution of observed values around the mean response.

b. The slope of the true regression line, $\beta_1 = 2.8$, is the change in bone mineral density associated with a 1 mg/day change in calcium intake.

If calcium increases by 10 mg/day, the expected change in bone mineral density is (β_1)(change in x) = $(2.8)(10) = 28$ mg/cm^2.

If calcium decreases by 20 mg/day, the expected change in bone mineral density is (β_1)(change in x) = $(2.8)(-20) = -56$ mg/cm^2.

The change in x is negative because calcium intake is *decreasing*.

c. If $\sigma = 12$, then for $x = 725$ the random variable Y is normally distributed with mean $-1575.2 + 2.8(725) = 454.8$ and variance $\sigma^2 = 12^2 = 144$: $Y \sim N(454.8, 144)$.

Find the probability that Y exceeds 470.

$$P(Y > 470) = P\left(\frac{Y - 454.8}{12} > \frac{470 - 454.8}{12}\right) \qquad \text{Standardize.}$$
$$= P(Z > 1.2667) \qquad \text{Use Equation 6.8, simplify.}$$
$$= 1 - P(Z \le 1.2667) \qquad \text{Use cumulative probability.}$$
$$= 1 - 0.8974 = 0.1026 \qquad \text{Table III in the Appendix.}$$

A technology solution:

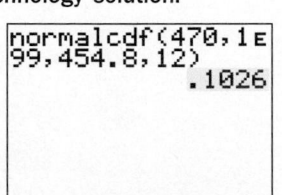

Figure 12.7 Probability calculation using `normalcdf`.

For a calcium intake of 725 mg/day, the probability of observing a bone mineral density greater than 470 mg/cm^2 is 0.1026. See Figure 12.8.

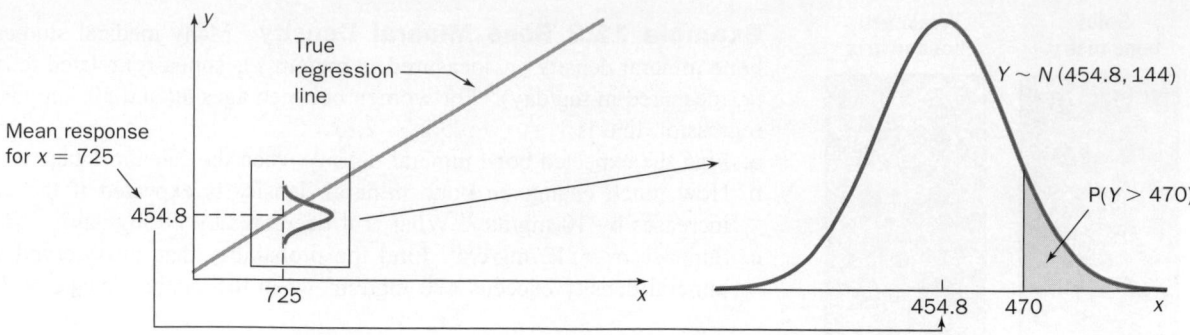

Figure 12.8 The distribution of Y for $x = 725$.

Suppose that two variables are related via a simple linear regression model. The parameters β_0 and β_1 are usually unknown. However, if we assume that the observations (x_1, y_1), (x_2, y_2), . . . , (x_n, y_n) are independent, then these sample data can be used to estimate the model parameters β_0 and β_1.

The **line of best fit**, or **estimated regression line**, is obtained using the **principle of least squares**. Figure 12.9 illustrates this concept: Minimize the sum of the squared deviations, or vertical distances from the observed points to the line. Consider the vertical distances from the points (x_1, y_1), (x_2, y_2), . . . , (x_n, y_n) to the line. The principle of least squares produces an estimated regression line such that the sum of all squared vertical distances is a minimum.

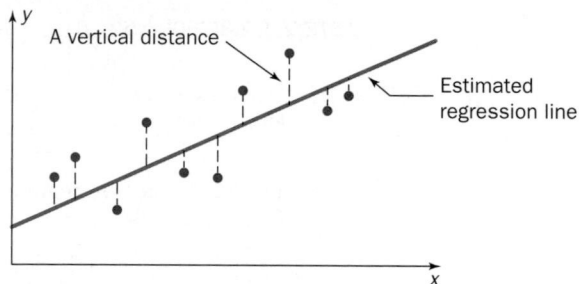

Figure 12.9 An illustration of the principle of least squares.

The focus in the remaining portion of this section is on the calculations associated with finding the line of best fit. In Section 12.2, several hypothesis tests utilizing these various statistics will be introduced.

LEAST-SQUARES ESTIMATES

$\hat{\beta}_0$ and $\hat{\beta}_1$ are estimates of β_0 and β_1, respectively, but $\hat{\beta}_1$ must be found first, because its value is used in the calculation of $\hat{\beta}_0$.

The least-squares estimates of the y-intercept (β_0) and the slope (β_1) of the true regression line are

$$\hat{\beta}_1 = \frac{n\sum x_i y_i - \left(\sum x_i\right)\left(\sum y_i\right)}{n\sum x_i^2 - \left(\sum x_i\right)^2} \tag{12.2}$$

and

$$\hat{\beta}_0 = \frac{\sum y_i - \hat{\beta}_1 \sum x_i}{n} = \bar{y} - \hat{\beta}_1 \bar{x} \tag{12.3}$$

The estimated regression line is $y = \hat{\beta}_0 + \hat{\beta}_1 x$.

💡 **ILLUMINATING THE CONCEPTS**

1. Before using these equations, always consider a scatter plot and compute the sample correlation coefficient (presented in Section 12.2) to make sure a linear model is reasonable. For example, in Figure 12.10 a linear model seems reasonable; the relationship between x and y appears to be negative linear. In Figure 12.11, a linear model is not reasonable. The relationship between the variables appears to be quadratic.

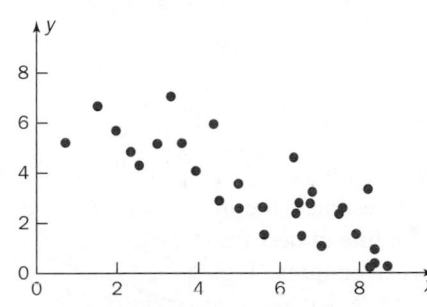

Figure 12.10 A scatter plot of data in which the relationship appears linear.

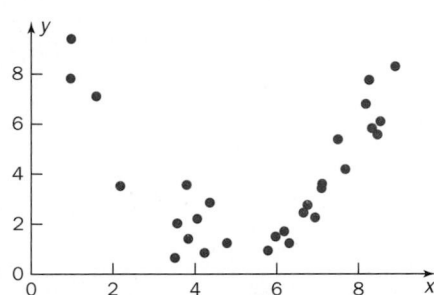

Figure 12.11 A scatter plot of data in which the relationship does not appear to be linear.

2. If x^* is a specific value of the independent, or *predictor*, variable x, let $y^* = \hat{\beta}_0 + \hat{\beta}_1 x^*$.
   a. y^* is an estimate of the *mean* value of Y for $x = x^*$, denoted $\mu_{Y|x^*}$.
   b. y^* is also an estimate of an *observed* value of Y for $x = x^*$. ■

The predictor variable is the independent variable, and the response variable is the dependent variable.

Example 12.3 Growth Charts Doctors often use growth charts in order to compare the heights and weights of children of similar ages. The clinical growth charts of the Centers for Disease Control and Prevention show that the weight of infant girls is approximately linear between the ages of 12 and 36 months. A random sample of infant girls was obtained, and the weight (in kilograms) and age (in months) for each is given in the following table.

Age	33	32	14	20	15	16	30	17	21	23
Weight	12.9	13.8	8.2	12.2	8.5	12.9	13.7	11.2	11.9	10.4

a. Find the estimated regression line.
b. Estimate the true mean weight of a 22-month-old girl.

SOLUTION

a. Age (x) is the independent, or predictor, variable, and weight (y) is the dependent, or response, variable. Find the necessary summary statistics for the $n = 10$ pairs of observations, and use Equations 12.2 and 12.3.

$$\sum x_i = 221.0 \qquad \sum y_i = 115.7 \qquad \sum x_i y_i = 2650.5$$
$$\sum x_i^2 = 5349.0 \qquad \sum y_i^2 = 1374.5$$

Solution Trail

KEYWORDS

- Estimated regression line.
- True mean weight of a 22-month-old girl.

↓

TRANSLATION

- $y = \hat{\beta}_0 + \hat{\beta}_1 x$.
- $E(Y \mid 22)$

↓

CONCEPTS

- Least-squares estimates.
- Estimate of the mean value of Y for $x = x^*$.

↓

VISION

Find the necessary summary statistics. Use Equation 12.2 to find $\hat{\beta}_1$ and Equation 12.3 to find $\hat{\beta}_0$. Compute $y^* = \hat{\beta}_0 + \hat{\beta}_1(22)$ as an estimate of the mean value of Y for $x = 22$.

$$\hat{\beta}_1 = \frac{n\sum x_i y_i - (\sum x_i)(\sum y_i)}{n\sum x_i^2 - (\sum x_i)^2} \qquad \text{Use Equation 12.2.}$$

$$= \frac{(10)(2650.5) - (221.0)(115.7)}{(10)(5349.0) - (221.0)^2} \qquad \text{Use summary statistics.}$$

$$= \frac{935.3}{4649} = 0.2012 \qquad \text{Simplify.}$$

$$\hat{\beta}_0 = \frac{\sum y_i - \hat{\beta}_1 \sum x_i}{n} \qquad \text{Use Equation 12.3.}$$

$$= \frac{115.7 - (0.2012)(221.0)}{10} \qquad \text{Use summary statistics and } \hat{\beta}_1.$$

$$= 7.123 \qquad \text{Simplify.}$$

The estimated regression line is $y = 7.123 + 0.2012x$. Figure 12.12 shows a scatter plot of the data and the graph of the estimated regression line.

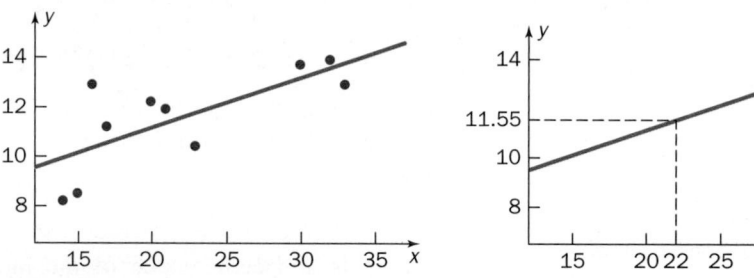

Figure 12.12 A scatter plot of the data and the graph of the estimated regression line.

Figure 12.13 The expected weight for a 22-month-old girl.

b. The estimated true mean weight of a 22-month-old girl is found by substituting $x^* = 22$ into the estimated regression line equation.

$$y = 7.123 + 0.2012x \qquad \text{Estimated regression line equation.}$$
$$= 7.123 + (0.2012)(22) = 11.55 \qquad \text{Substitute and simplify.}$$

An estimate for the expected weight of a 22-month-old girl is 11.55 kilograms. See Figure 12.13.

A technology solution:

Figure 12.14 The estimated regression line, an analysis of variance table, and the results from relevant hypothesis tests (explanations to follow in Section 12.2).

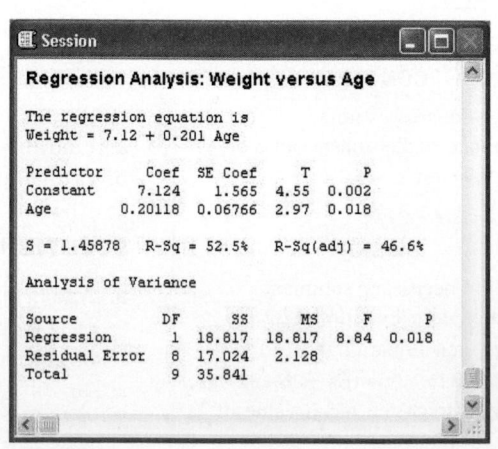

The variance σ^2 is a measure of the underlying variability in the simple linear regression model. A *large* σ^2 means that the data will vary widely from the true regression line. A *small* σ^2 implies that the observed values will lie close to the true regression line.

An estimate of σ^2 is used to conduct hypothesis tests and to construct confidence intervals related to simple linear regression. The estimate is based on the deviations of the observed values from the estimated regression line. There will be more on hypothesis tests and confidence intervals in Section 12.2.

One method to assess the accuracy of a simple linear regression model involves an analysis of variance table, similar to those constructed in Chapter 11. In order to explain the ANOVA table entries and to make computations easier, a little more notation is necessary.

For computational purposes, define

$$S_{xx} = \underbrace{\sum(x_i - \bar{x})^2}_{\text{definition}} = \underbrace{\sum x_i^2 - \frac{1}{n}\left(\sum x_i\right)^2}_{\text{computational formula}}$$

$$S_{yy} = \underbrace{\sum(y_i - \bar{y})^2}_{\text{definition}} = \underbrace{\sum y_i^2 - \frac{1}{n}\left(\sum y_i\right)^2}_{\text{computational formula}}$$

$$S_{xy} = \underbrace{\sum(x_i - \bar{x})(y_i - \bar{y})}_{\text{definition}} = \underbrace{\sum x_i y_i - \frac{1}{n}\left(\sum x_i\right)\left(\sum y_i\right)}_{\text{computational formula}}$$

The *ith predicted*, or *fitted*, *value*, denoted $\hat{y}_i$, is $\hat{y}_i = \hat{\beta}_0 + \hat{\beta}_1 x_i$. This is simply the *estimated* regression line evaluated at x_i.

The *ith residual* is $y_i - \hat{y}_i$. This quantity is a measure of how far away the observed value of Y is from the estimated value of Y.

The total variation in the data (the **total sum of squares**, denoted **SST**) is composed of a sum of the variation explained by the model (the **sum of squares due to regression**, denoted **SSR**) and the variation about the regression line (the **sum of squares due to error**, denoted **SSE**). These three sums are defined in the following identity, which shows the decomposition of the total variation in the data.

SUMS OF SQUARES

$$\underbrace{\sum(y_i - \bar{y})^2}_{\text{SST}} = \underbrace{\sum(\hat{y}_i - \bar{y})^2}_{\text{SSR}} + \underbrace{\sum(y_i - \hat{y}_i)^2}_{\text{SSE}} \tag{12.4}$$

Here are the *computational formulas* for these sums of squares.

$$\text{SST} = S_{yy}, \qquad \text{SSR} = \hat{\beta}_1 S_{xy}, \qquad \text{SSE} = \text{SST} - \text{SSR}$$

You can probably guess that a large F statistic suggests a significant regression. The formal hypothesis test is presented in Section 12.2.

Regression computations are often summarized in an analysis of variance table, as shown below. As in Chapter 11, **mean squares** are corresponding sums of squares divided by the corresponding degrees of freedom. The F test for a significant regression and the associated p value are discussed in Section 12.2.

ANOVA summary table for simple linear regression

Source of variation	Sum of squares	Degrees of freedom	Mean square	F	p value
Regression	SSR	1	$\text{MSR} = \dfrac{\text{SSR}}{1}$	$\dfrac{\text{MSR}}{\text{MSE}}$	p
Error	SSE	$n - 2$	$\text{MSE} = \dfrac{\text{SSE}}{n - 2}$		
Total	SST	$n - 1$			

COEFFICIENT OF DETERMINATION

The **coefficient of determination**, denoted r^2, is a measure of the proportion of the variation in the data that is explained by the regression model, and is defined by

$$r^2 = \text{SSR/SST}. \tag{12.5}$$

Since $0 \le \text{SSR} \le \text{SST}$, the **coefficient of determination**, r^2, is always a number between 0 and 1 (inclusive). The higher r^2 is, the better the model is. Many statistical software packages report $100r^2$, the percentage of variation explained by the regression model.

The following example illustrates the computations necessary to produce the estimated regression line and the ANOVA table.

$r^2 = 1$ implies a perfect fit. What does $r^2 = 0$ imply?

Solution Trail

KEYWORDS

- Estimated regression line.
- ANOVA table.
- Coefficient of determination.

↓

TRANSLATION

- $y = \hat{\beta}_0 + \hat{\beta}_1 x$.
- ANOVA summary table.
- r^2.

↓

CONCEPTS

- Least-squares estimates.
- ANOVA summary table for simple linear regression.
- The proportion of variation in the data explained by the regression model.

↓

VISION

Find the necessary summary statistics. Use Equation 12.2 to find $\hat{\beta}_1$ and Equation 12.3 to find $\hat{\beta}_0$. Complete the ANOVA summary table and compute r^2.

Example 12.4 Cooling Degree Days The developer of a residential community in the Midwest, in which each all-electric home is equipped with air conditioning, believes that the monthly electricity bill (in dollars) is linearly related to the cooling degree days (CDD, the cumulative number of degrees the mean temperature per day is greater than 75 °F). A random sample of months and homes was selected. The electricity bill and the cooling degree days were recorded for each home. The data are given in the following table.

CDD	18	18	328	211	261	94	261	110	110
Bill	41.99	49.53	70.31	69.79	65.61	37.03	100.22	56.05	70.33

CDD	261	328	94	328	328	94	94	94	211
Bill	103.64	134.99	82.27	89.06	91.38	112.40	107.82	94.23	75.31

(*Source*: C. McLaren and B. McLaren, Electric bill data, *Journal of Statistics Education*, Vol. 11, No. 1, 2003.)

a. Find the estimated regression line and explain the meaning of the estimated coefficient $\hat{\beta}_1$.

b. Complete the ANOVA table (without the p value), and find and interpret the coefficient of determination.

SOLUTION

a. CDD (x) is the independent variable and electricity bill (y) is the dependent variable. The summary statistics for the $n = 18$ pairs of observations are

$$\sum x_i = 3243 \qquad \sum y_i = 1451.96 \qquad \sum x_i y_i = 283{,}792.15$$
$$\sum x_i^2 = 792{,}769 \qquad \sum y_i^2 = 128{,}662.066$$

$$\hat{\beta}_1 = \frac{n\sum x_i y_i - (\sum x_i)(\sum y_i)}{n\sum x_i^2 - (\sum x_i)^2}$$ Use Equation 12.2.

$$= \frac{(18)(283,792.15) - (3243)(1451.96)}{(18)(792,769) - (3243)^2}$$ Use summary statistics.

$$= \frac{399,552.42}{3,752,793} = 0.1065$$ Simplify.

$$\hat{\beta}_0 = \frac{\sum y_i - \hat{\beta}_1 \sum x_i}{n}$$ Use Equation 12.3.

$$= \frac{1451.96 - (0.1065)(3243)}{18}$$ Use summary statistics and $\hat{\beta}_1$.

$$= 61.48$$ Simplify.

The estimated regression line is $y = 61.48 + 0.1065x$. The value $\hat{\beta}_1 = 0.1065$ suggests that an increase of 1 CDD costs approximately an additional 11 cents. Figure 12.15 shows a scatter plot of the data and the graph of the estimated regression line.

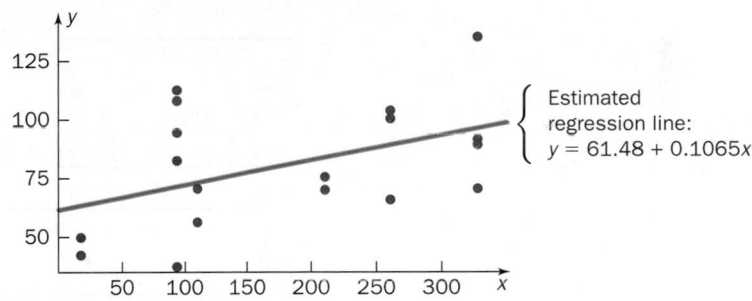

Figure 12.15 A scatter plot of the data and the graph of the estimated regression line.

b. Complete the ANOVA table, except for the p value ($n = 18$).

$$\text{SST} = S_{yy} = \sum y_i^2 - \frac{1}{18}(\sum y_i)^2$$ Computational formula.

$$= 128,662.066 - \frac{1}{18}(1451.96)^2 = 11,540.52$$ Use summary statistics.

$$\text{SSR} = \hat{\beta}_1 S_{xy} = \hat{\beta}_1 \left[\sum x_i y_i - \frac{1}{18}(\sum x_i)(\sum y_i) \right]$$ Computational formula.

$$= 0.1065 \left[283,796.15 - \frac{1}{18}(3243)(1451.96) \right]$$ Use summary statistics.

$$= 2364.44$$ Simplify.

$$\text{SSE} = \text{SST} - \text{SSR}$$ Computational formula.

$$= 11,540.52 - 2364.44 = 9176.08$$

Compute the mean squares, MSR and MSE.

$$\text{MSR} = \text{SSR}/1$$ Definition.

$$= 2364.44/1 = 2364.44$$ Use SSR.

$$\text{MSE} = \text{SSE}/(n-2)$$ Definition.

$$= 9176.08/16 = 573.51$$ Use SSE and $n = 18$.

The value of the test statistic is

$$f = \frac{\text{MSR}}{\text{MSE}} = \frac{2364.44}{573.51} = 4.12$$

You can probably guess that we will reject the null hypothesis if the value of the test statistic is large. The formal hypothesis test is presented in Section 12.2.

Here are all of these calculations presented in an ANOVA table.

ANOVA summary table for simple linear regression

Source of variation	Sum of squares	Degrees of freedom	Mean square	F
Regression	2364.44	1	2364.44	4.12
Error	9176.08	16	573.51	
Total	11,540.52	17		

The coefficient of determination is

$$r^2 = \text{SSR/SST} = 2364.44/11{,}540.52 = 0.2049$$

Approximately 0.2049, or 20%, of the variation in the data is explained by the regression model.

Figure 12.16 shows a technology solution.

D	E	F	G	H	I	J
SUMMARY OUTPUT						
Regression Statistics						
Multiple R	0.4525					
R Square	0.2048					
Adjusted R Square	0.1551					
Standard Error	23.9494					
Observations	18					
ANOVA						
	df	*SS*	*MS*	*F*	*Significance F*	
Regression	1	2363.31	2363.31	4.12	0.0593	
Residual	16	9177.21	573.58			
Total	17	11540.52				
	Coefficients	*Standard Error*	*t Stat*	*P-value*	*Lower 95%*	*Upper 95%*
Intercept	61.4825	11.0076	5.59	0.0000	38.1475	84.8175
X Variable 1	0.1065	0.0525	2.03	0.0593	-0.0047	0.2177

Figure 12.16 Excel output: regression coefficients and analysis of variance table.

TECHNOLOGY CORNER

Procedure: Construct a scatter plot, find the estimated regression line, and construct an ANOVA table.

Reconsider: Example 12.4, page 566, solutions, and interpretations.

TI-84 Plus

A scatter plot is one of the six built-in statistical plots. Use the function LinReg(a+bx) to compute the regression coefficients. There is no built-in function to produce the ANOVA table.

1. Enter the CCD data into list L1 and the Bill data into list L2.
2. Choose STATPLOT; STAT PLOTS; Plot1. Turn the plot On, select Type scatter plot (the first graph icon), enter the Xlist, L1, the Ylist, L2, and choose a Mark (for the points on the graph).
3. Enter appropriate window settings and press GRAPH to view the scatter plot. See Figure 12.17.
4. On the Home Screen, choose STAT; CALC; LinReg(a+bx), and enter two arguments: the independent variable list, L1, and the dependent variable list, L2. Press ENTER to execute the command. The regression coefficients are displayed on the Home Screen. See Figure 12.18.

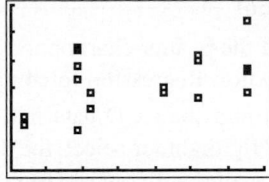

Figure 12.17 TI-84 Plus
scatter plot.

Figure 12.18 The
regression coefficients.

The function `LinReg(a+bx)` has a optional third argument, a function variable, for example `Y1`, for storing the regression equation. Other calculator functions will also produce the regression coefficients, for example `LinReg(ax+b)` and `LinRegTTest`.

Minitab

Minitab has a Scatterplot option in the Graph menu, and there are several built-in functions to compute the regression coefficients and the ANOVA table, in the Stat; Regression menu.

1. Enter the CCD data into column C1 and the Bill data into column C2.
2. Select Graph; Scatterplot. Choose a Simple scatter plot.
3. Enter C2 under Y variables and C1 under X variables. Click OK. The scatter plot is displayed in a graph window. Adjust the labels and axes as necessary. See Figure 12.19.

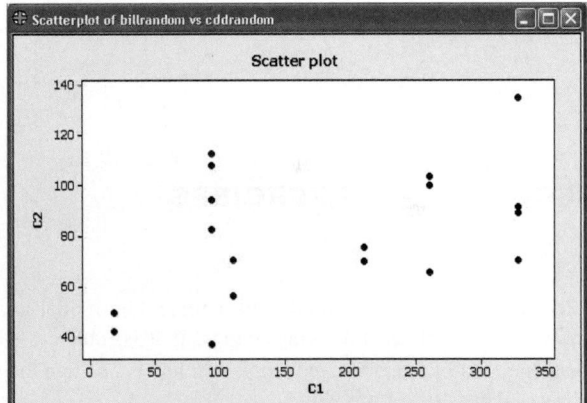

Figure 12.19 Minitab
scatter plot

4. Select Stat; Regression; Regression. Enter the Response variable, C2, and the Predictors, C1. Click OK. The regression coefficients and the ANOVA table are displayed in a session window. See Figure 12.20.

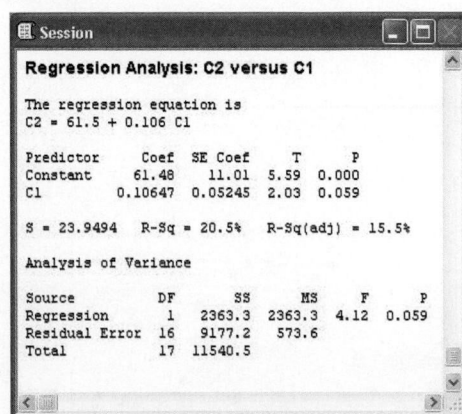

Figure 12.20 Regression
coefficients and ANOVA table.

Excel

Use the Scatter chart option to construct a scatter plot and the Data Analysis Toolkit function Regression to compute the regression coefficients and ANOVA table.

1. Enter the CCD data into column A and the Bill data into column B.
2. Highlight, or select, the data range, A1:B18. Under the Insert tab, select Scatter; Scatter with only Markers. The scatter plot is displayed in the current worksheet. Edit the plot options as necessary. See Figure 12.21.
3. Under the Data tab, select Data Analysis; Regression. Enter the Y Range (B), the X Range (A), and the Output Range. Click OK to display the regression coefficients and ANOVA table. See Figure 12.16, page 568.

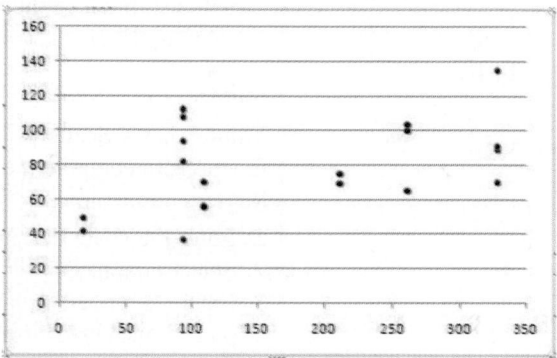

Figure 12.21 Excel scatter plot.

SECTION 12.1 EXERCISES

Practice

12.1 Decide whether a simple linear regression model is appropriate for each of the following graphs. If it is, indicate whether the slope ($\widehat{\beta}_1$) of the estimated regression line is positive, negative, or zero. If it is not, state why.

a.
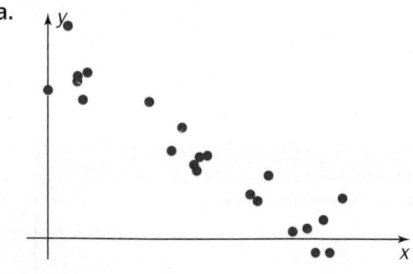

b.
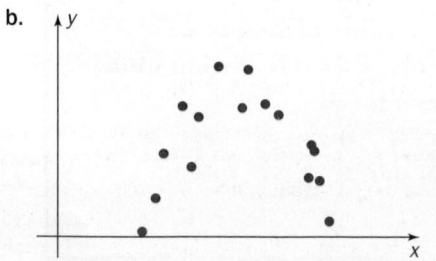

c.
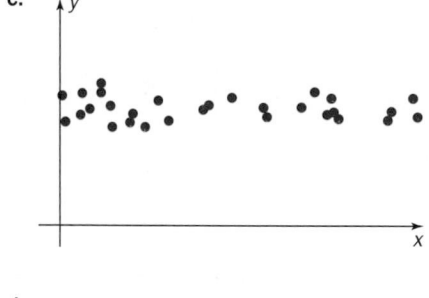

d.
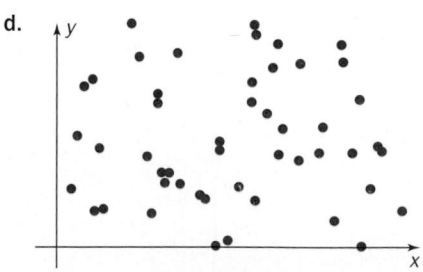

12.2 In each of the following problems, construct a scatter plot for the data. Without doing any calculations for ($\widehat{\beta}_0$) and ($\widehat{\beta}_1$), add a straight line to each graph that you believe would be *close* to the estimated regression line. That is, try to draw a line through the *middle* of the points, so that there are approximately the same number of points above and below the line.

a.

x	8.3	1.9	8.4	8.6	2.3	4.6	2.4
y	11.6	3.4	9.3	10.1	5.6	7.6	4.3
x	7.7	8.0	5.6	6.9	1.8	4.8	9.4
y	10.5	11.5	6.6	9.8	2.9	6.8	12.8
x	6.8	4.9	7.8	5.3	1.1	0.3	
y	10.2	4.9	9.4	8.2	3.6	3.2	

b.

x	98.2	56.4	52.9	99.3	73.5	82.7
y	88.7	167.0	185.5	72.5	149.5	113.0
x	54.1	68.2	70.3	71.2	53.4	90.2
y	163.0	135.4	135.8	130.1	168.4	82.0
x	53.8	99.4	85.5	72.6	92.7	89.6
y	184.2	90.1	130.3	132.9	93.2	103.6
x	80.9	76.4	54.5	97.3	58.7	92.6
y	121.0	115.7	174.1	79.1	159.3	86.4

12.3 Suppose the true regression line relating the variables x and y, for values of x between 100 and 200, is $y = 75.0 + 3.6x$.
a. Find the expected value of Y when $x = 150$, $E(Y|150)$.
b. How much change in the dependent variable is expected when x increases by one unit? Justify your answer.
c. Suppose $\sigma = 6$. Find the probability that an observed value of Y is less than 500 when $x = 120$.

12.4 Suppose the true regression line relating the variables x and y, for values of x between 25 and 35, is $y = -35.1 - 7.2x$.
a. Find the expected value of Y when $x = 27.4$.
b. How much change in the dependent variable is expected when x decreases by five units? Justify your answer.
c. Suppose $\sigma = 1.5$. Find the probability that an observed value of Y is between -250 and -252 when $x = 30$.

12.5 Suppose a simple linear regression model is used to explain the relationship between x and y. A random sample of $n = 12$ values for the independent variable was selected, and the corresponding values of the dependent variable were observed. The summary statistics were

$$\sum x_i = 460.53 \qquad \sum y_i = -6349.7$$
$$\sum x_i^2 = 17,875.1 \qquad \sum y_i^2 = 3,421,892$$
$$\sum x_i y_i = -246,677$$

a. Find the estimated regression line.
b. Estimate the true mean value of Y when $x = 41$.

12.6 Suppose a simple linear regression model is used to explain the relationship between x and y. A random sample of $n = 15$ values for the independent variable was selected, and the corresponding values of the dependent variable were observed. The data are given in the following table.

x	4.9	9.9	0.3	3.8	9.8	4.6	9.8	5.4
y	49.1	53.4	21.7	40.4	78.6	60.4	73.9	37.4
x	3.1	3.0	4.2	4.0	1.7	1.2	7.0	
y	41.6	29.6	46.7	28.5	23.8	15.2	72.3	

a. Construct a scatter plot for these data. Is a simple linear regression model reasonable? Justify your answer.
b. Find the estimated regression line. Add a graph of this line to your scatter plot.
c. What is the predicted value of Y when $x = 5.7$?

12.7 Suppose a simple linear regression model is used to explain the relationship between x and y. A random sample of $n = 21$ values for the independent variable was selected, and the corresponding values of the dependent variable were observed. The data are given on the data CD and book's web site.
a. Construct a scatter plot for these data. Is a simple linear regression model reasonable? Justify your answer.
b. Find the estimated regression line. Add a graph of this line to your scatter plot.
c. Estimate the true mean value of Y when $x = 62$.
d. Construct an ANOVA summary table, except for the p value.

Applications

12.8 Biology and Environmental Science Agricultural research suggests that the final corn yield in bushels per acre (y) is linearly related to the number of inches between rows (x). Suppose that the true regression line is $y = 197.5 - 6.1x$.
a. Find the expected yield when there are 15 inches between rows.
b. How much change in yield is expected if the distance between rows decreases by 2 inches?
c. Suppose $\sigma = 3.2$ bushels per acre. Find the probability that an observed value of yield is between 90 and 95 bushels per acre when rows are 17 inches apart.

12.9 Physical Sciences Once a PVC pipe joint is assembled using solvent cement, the new joint needs a certain amount of time to set. For 4- to 8-inch diameter pipes, suppose the set time (y, measured in hours) is linearly related to the ambient temperature (x, measured in °F). The true regression line is assumed to be $y = 8.3 - 0.09x$.
a. Find an estimate of an observed setting time for an ambient temperature of 65 °F.
b. How much less time does a joint take to set if the temperature rises by 10 °F?
c. If $\sigma = 15$ minutes, find the probability that an observed set time is less than one hour if the ambient temperature is 80 °F.

12.10 Biology and Environmental Science In a report prepared for the Vermont Independent Power Producers Association, it was determined that precipitation is the major factor influencing hydroelectric power generation. Let y be the summer seasonal hydroelectric power produced state-wide expressed as a percent of maximum and x be the total precipitation per month in inches. The estimated regression line reported was $y = 0.0146x - 0.0375$.[3]
a. Find the expected value of y if the precipitation is 7.5 inches.
b. Find an estimate of an observed y value for $x = 3.4$.
c. Suppose $\sigma = 0.2$. Find the probability that an observed y value is more than 0.13 when precipitation is 10.0.

12.11 Manufacturing and Product Development As soon as a bottle of soda is opened, it begins to lose its carbonation. Fourteen 12-ounce bottles of cola were obtained, and each was assigned a randomly selected period (in hours). Each bottle was opened and allowed to stand at room temperature. The carbonation (y) in each bottle was measured (by volume) after the prescribed time period (x). The summary statistics are

$$\Sigma x_i = 8.6 \qquad \Sigma y_i = 37.4 \qquad \Sigma x_i y_i = 19.61$$
$$\Sigma x_i^2 = 7.26 \qquad \Sigma y_i^2 = 110.24$$

a. Find the estimated regression line.

b. Estimate the true mean carbonation after 1 hour and 15 minutes.

12.12 Biology and Environmental Science Some research suggests that higher levels of carbon dioxide (CO_2) increase the rate at which plants and trees grow.[4] In fact, even without ideal conditions (for example, normal precipitation and temperature), some trees grow more in elevated levels of CO_2. Suppose that in a research study, data were collected on oak trees in northern U.S. forests. For randomly selected levels of CO_2 (x, measured in parts per million, ppm), the most recent tree-ring growth (y, in cm) was measured. The summary statistics are ($n = 10$)

$$\Sigma x_i = 5035.0 \qquad \Sigma y_i = 13.7 \qquad \Sigma x_i y_i = 7012.5$$
$$\Sigma x_i^2 = 2,607,575 \qquad \Sigma y_i^2 = 19.25$$

a. Find the estimated regression line.

b. Find an estimate of an observed value of tree-ring growth if the level of CO_2 is 600 ppm.

12.13 Fuel Consumption and Cars Although Americans are reluctant to change their driving habits, the recent high price of gasoline has slightly decreased the demand. For a random sample of prices (x, in dollars per gallon) the demand for gasoline in the United States was recorded (y, in millions of gallons per day). The summary statistics are ($n = 20$)

$$\Sigma x_i = 50.95 \qquad \Sigma y_i = 214.4 \qquad \Sigma x_i y_i = 545.653$$
$$\Sigma x_i^2 = 130.9151 \qquad \Sigma y_i^2 = 2298.96$$

a. Find the estimated regression line.

b. Find the expected demand when the price of gasoline is $2.65 per gallon.

c. Find an estimate of an observed demand when the price of gasoline is $2.89 per gallon.

12.14 Sports and Leisure Sailboat enthusiasts believe that the wind speed (x, in miles per hour) is linearly related to (downwind) boat speed (y, in knots). For wind speeds between 10 and 30 mph and Hobie catamarans, the following data were recorded.

x	26.2	11.8	27.3	24.3	20.4
y	16.3	7.7	17.3	18.2	16.9

x	24.6	20.7	14.0	15.5	20.3
y	10.3	12.7	13.3	10.0	6.8

a. Construct a scatter plot for these data.

b. Find the estimated regression line. Add a graph of the estimated regression line to the scatter plot in part (a).

c. Estimate the mean speed of a Hobie catamaran for a wind speed of 18 mph.

12.15 Psychology and Human Behavior A development officer at a large university believes that the amount of money donated to the general fund by alumni each year is linearly related to the football team's winning percentage. A random sample of Division I football teams was selected. The winning percentage (x) and subsequent alumni donation amount (y, in millions of dollars) are given in the following table.

x	92	67	92	25	83	83	58	8	75
y	15.8	11.1	13.5	7.0	17.4	10.8	11.2	3.9	23.7

x	100	92	58	92	17	25	75	92	33
y	27.8	23.3	11.3	6.6	7.3	10.3	15.8	12.0	6.8

a. Construct a scatter plot for these data.

b. Find the estimated regression line. Add a graph of the estimated regression line to the scatter plot in part (a).

c. Find an estimate of an observed value of the donation amount for a football team with a winning percentage of 75%.

Extended Applications

12.16 Medicine and Clinical Studies Many automobile accidents are caused by tired drivers. Several research studies have shown that changes in the pupils of the eyes are related to fatigue.[5] A random sample of 25 drivers was obtained, and the oscillations in pupil size (x, in millimeters per second) were measured using a pupillograph. Each person's tiredness (y) was also recorded using the pupil unrest index (PUI). The summary statistics are

$$\Sigma x_i = 7.1 \qquad \Sigma y_i = 192.0 \qquad \Sigma x_i y_i = 49.22$$
$$\Sigma x_i^2 = 2.1064 \qquad \Sigma y_i^2 = 2094.0$$

a. Find the estimated regression line.

b. Find the expected PUI for $x = 0.3$ millimeters per second.

c. Suppose a driver is considered to be too tired to drive if the PUI score is 15 (or higher). What value of x yields an expected PUI score of 15?

12.17 Fuel Consumption and Cars Each new car has a sticker on a side window with the Environmental Protection Agency (EPA) estimated miles per gallon (mpg). *Consumer Reports, Motor Trends Magazine,* and *Edmunds* publish actual mpg for certain cars for comparison. In addition, the U.S. Department of Energy publishes the EPA combined estimated mpg and the actual combined mpg as reported by consumers (*Source:* www.fueleconomy.gov). A random sample of 2008 automobiles from this site was obtained and the EPA combined mpg and the actual mpg were recorded for each. The data are given in the following table.

Vehicle	EPA mpg (x)	CR mpg (y)
Honda Accord	24	25.6
Toyota Camry	25	24.0
Volkswagen Passat	24	30.5
Subaru Forester AWD	22	25.0
Saturn Aura	20	22.3
Pontiac Grand Prix	21	28.9
Lexus EX 350	22	23.3
Infiniti G35	20	18.7
Ford Focus	28	33.9
Chevrolet Impala	22	26.9

a. Find the estimated regression line.

b. Complete the ANOVA table (without the p value).

c. Find the actual mpg for an automobile rated at 30 combined mpg.

12.18 Medicine and Clinical Studies Ultrasound measurements are often used to predict the birth weight of a newborn. However, if ultrasound is not available or is not used, the mother's prepregnancy weight is believed to be linearly related to birth weight.[6] A random sample of Caucasian pregnant women experiencing no complications was obtained. The weight prepregnancy (x, in pounds) of each was recorded and the newborn weight (y, in pounds) was also recorded. The data are given on the data CD and book's web site.

a. Construct a scatter plot for these data.
b. Find the estimated regression line.
c. Complete the ANOVA table (without the p value).
d. Estimate the expected weight of a newborn if the mother's prepregnancy weight is 112 pounds.

12.19 Physical Sciences Deep-water (>300 m) wave forecasts are important for large cargo ships. One method of prediction suggests that the wind speed (x, in knots) is linearly related to the wave height (y, in feet). A random sample of buoys was obtained, and the wind speed and wave height was measured at each. The data are given in the following table.

Wind speed	9	11	10	10	11	9	9	6
Wave height	2.9	1.4	1.7	0.9	1.2	1.0	1.5	0.7

Wind speed	9	5	8	9	12	9	12	9
Wave height	1.9	0.1	2.0	2.6	3.0	1.7	2.1	1.5

Wind speed	12	8	7	13	9	8	6	8
Wave height	3.1	2.7	0.4	2.5	1.7	0.6	0.7	1.4

(*Source*: National Data Buoy Center, National Oceanic and Atmospheric Administration.)

a. Find the estimated regression line.
b. Complete the ANOVA table (without the p value).
c. Find the coefficient of determination. Interpret this value.
d. Suppose a 10-foot wave is considered to be the storm threshold. What wind speed yields an expected storm threshold?

12.20 Manufacturing and Product Development There is good evidence to suggest that the depth of a bounce on a certain circular trampoline is linearly related to the stiffness of the springs around the edges, or spring constant k. A random sample of production trampolines was obtained and the spring constant of each was recorded (in lb/in). The bounce was measured (in feet) with a testing weight of 200 pounds released from a height of 2 feet. The data are given in the following table.

Spring constant	0.43	0.77	0.63	0.73	0.55	0.42
Bounce	1.1	2.6	2.3	3.0	3.0	1.9

Spring constant	0.38	0.39	0.20	0.73	0.56	0.57
Bounce	1.4	1.4	1.3	2.2	2.0	1.2

a. Find the estimated regression line.
b. Complete the ANOVA table (without the p value).
c. Find the coefficient of determination. Interpret this value.
d. Suppose the trampoline sits 4 feet above the ground. For what value of k can a 200-pound person expect to hit the ground?

12.21 Demographics and Population Statistics Infant mortality and life expectancy are often used to characterize the overall health of a country. A random sample of countries was obtained and the infant mortality rate, x, (infant deaths per 1000 live births) and the life expectancy, y, (in years) in 2008 was recorded for each. The data are given on the data CD and book's web site.[7]

a. Find the estimated regression line.
b. Complete the ANOVA table (without the p value).
c. Find the coefficient of determination.
d. Find the expected life expectancy in a country with infant mortality rate 15.5.

Challenge

12.22 Medicine and Clinical Studies Some physicians use the cholesterol ratio (CR = total cholesterol/HDL cholesterol) as a measure of a patient's risk of heart disease.[8] In addition, the triglyceride concentration (TG) is associated with coronary artery disease in many patients. In a study of the relationship between these two variables, a random sample of adults was obtained, and the triglyceride level (x_1, mg/dL) and cholesterol ratio (y) was obtained for each person. The data are given on the data CD and book's web site.

a. Carefully sketch a scatter plot of these data (y versus x_1). The relationship does not appear to be linear. Can you describe this relationship between y and x_1?
b. Compute the natural logarithm of each difference ($x_1 - 129$). That is, find the values of a new predictor variable, $x_2 = \ln(x_1 - 129)$.
c. Carefully sketch a scatter plot of y versus the new predictor variable x_2. Describe this relationship.
d. Find the estimated regression line, and complete the ANOVA table using x_2 as the predictor variable.

12.2 HYPOTHESIS TESTS AND CORRELATION

Suppose there is theoretical or empirical evidence that two variables are linearly related. The mean squares (in the regression ANOVA table) are used to determine whether the linear relationship is statistically significant. The null hypothesis states that the variation in Y is completely random and is independent of the value of x; knowing the value of x provides no additional information about the value of Y. In this case, a scatter plot would have a *shotgun* appearance, with points scattered randomly in the plane and no discernible linear pattern (Figure 12.22).

Even though a true regression line is shown in Figure 12.23, there is really no significant linear relationship. The true regression line is of absolutely no use for predicting or estimating values of Y.

Remember that β_1 is the slope of the true regression line. A horizontal line has slope, β_1, equal to 0.

For simple linear regression, a test of significance is equivalent to testing the hypothesis H_0: $\beta_1 = 0$. If H_0 is true, then the model assumptions imply that the mean value of Y for any value of x is the same. That is, $\mu_{Y|x_i} = \beta_0$. Therefore, the values of Y vary around the horizontal line $y = \beta_0$, and knowing the value of x adds no additional information. See figure 12.23.

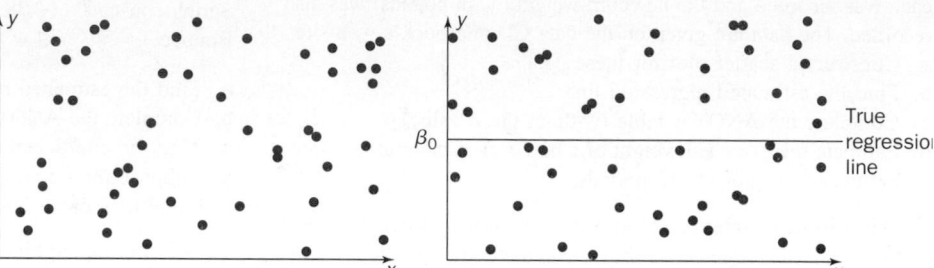

Figure 12.22 Scatter plot of data showing no linear pattern.

Figure 12.23 If $\beta_1 = 0$, the values of Y vary randomly around the true regression line $y = \beta_0$.

Here is a summary of an F test for a significant regression with significance level α.

HYPOTHESIS TEST FOR A SIGNIFICANT LINEAR REGRESSION

H_0: There is no significant linear relationship ($\beta_1 = 0$)

H_a: There is a significant linear relationship ($\beta_1 \neq 0$)

TS: $F = \dfrac{\text{MSR}}{\text{MSE}}$

RR: $F \geq F_\alpha$ df 1 and $n - 2$

💡 ILLUMINATING THE CONCEPTS

1. The null hypothesis is rejected only for large values of the test statistic. The associated p value is a right-tail probability. The F ratio will be larger when $\beta_1 \neq 0$ than when $\beta_1 = 0$.

2. This is often called a **model utility test**. In general, if H_0 is rejected, then the value of r^2 is usually large.

3. Alternatively, we can conduct this hypothesis test by using the p value. Recall, if H_0 is true, the p value is the probability of obtaining a value of the test statistic at least as large as the observed value. If $p \leq \alpha$, we reject the null hypothesis. If $p > \alpha$, we do not reject the null hypothesis. ∎

Recall, a specific value of B_1 is $\hat{\beta}_1$.

Consider the random variable B_1, an estimator for β_1, and $S^2 = \text{MSE} = \text{SSE}/(n - 2)$, an estimator for the underlying variance σ^2. If the simple linear regression assumptions are true, then S^2 is an unbiased estimator for σ^2, and the estimator B_1 has the following properties.

1. B_1 is an unbiased estimator for β_1: $E(B_1) = \mu_{B_1} = \beta_1$.

2. The variance of B_1 is $\text{Var}(B_1) = \sigma_{B_1}^2 = \sigma^2/S_{xx}$.
   If we use s^2 as an estimate of σ^2, then an estimate of the variance of B_1 is $s_{B_1}^2 = s^2/S_{xx}$.

3. The random variable B_1 has a normal distribution.

The hypothesis test procedure concerning β_1 is based on the following theorem.

THEOREM

If the simple linear regression assumptions are true, then the random variable

$$T = \frac{B_1 - \beta_1}{S/\sqrt{S_{xx}}} = \frac{B_1 - \beta_1}{S_{B_1}}$$

has a t distribution with $n - 2$ degrees of freedom.

No proof is given here that the two tests are equivalent, but there is some numerical evidence in Example 12.5.

The null and alternative hypotheses are stated in terms of the parameter β_1. The most common test has $H_0: \beta_1 = 0$. If $\beta_1 = 0$, there is no significant linear relationship between the two variables. The true regression line is $y = \beta_0$ (a horizontal line), and knowing the value of x is of no use in predicting the value of $Y|x$. For simple linear regression, the following hypothesis test (with $\beta_{10} = 0$) is equivalent to an F test for a significant regression with significance level α.

HYPOTHESIS TEST AND CONFIDENCE INTERVAL CONCERNING β_1

The hypothesized value of β_1, β_{10}, can be any constant. We can conduct a test for evidence that the regression coefficient β_1 is different from any value. However, usually $\beta_{10} = 0$. This means we are looking for any evidence to suggest that the value of x helps to predict the value of $Y|x$.

$H_0: \beta_1 = \beta_{10}$

$H_a: \beta_1 > \beta_{10}, \qquad \beta_1 < \beta_{10}, \qquad$ or $\qquad \beta_1 \neq \beta_{10}$

TS: $T = \dfrac{B_1 - \beta_{10}}{S_{B_1}}$

RR: $T \geq t_\alpha, \qquad T \leq -t_\alpha, \qquad$ or $\qquad |T| \geq t_{\alpha/2}, \qquad$ df $= n - 2$

A $100(1 - \alpha)\%$ confidence interval for β_1 has as endpoints the values

$$\hat{\beta}_1 \pm t_{\alpha/2} s_{B_1} \tag{12.6}$$

and the critical value is based on $n - 2$ degrees of freedom.

There are similar properties, a hypothesis test procedure, and a confidence interval concerning the simple linear regression parameter β_0. The estimator B_0 has the following properties.

1. B_0 is an unbiased estimator for β_0: $E(B_0) = \mu_{B_0} = \beta_0$.

2. The variance of B_0 is $\text{Var}(B_0) = \sigma_{B_0}^2 = \dfrac{\sigma^2 \sum x_i^2}{n S_{xx}}$.

If we use s^2 as an estimate of σ^2, then an estimate of the variance of B_0 is

$s_{B_0}^2 = \dfrac{s^2 \sum x_i^2}{n S_{xx}}$.

3. The random variable B_0 has a normal distribution.

THEOREM

If the simple linear regression assumptions are true, then the random variable

$$T = \frac{B_0 - \beta_0}{S\sqrt{\sum x_i^2/(n S_{xx})}} = \frac{B_0 - \beta_0}{S_{B_0}}$$

has a t distribution with $n - 2$ degrees of freedom.

β_{00} is the hypothesized value of β_0.

What does $\beta_0 = 0$ imply about the true regression line?

HYPOTHESIS TEST AND CONFIDENCE INTERVAL CONCERNING β_0

$H_0: \beta_0 = \beta_{00}$

$H_a: \beta_0 > \beta_{00}, \qquad \beta_0 < \beta_{00}, \qquad \text{or} \qquad \beta_0 \ne \beta_{00}$

TS: $T = \dfrac{B_0 - \beta_{00}}{S_{B_0}}$

RR: $T \ge t_\alpha, \qquad T \le -t_\alpha, \qquad \text{or} \qquad |T| \ge t_{\alpha/2}, \qquad \text{df} = n - 2$

A $100(1 - \alpha)\%$ confidence interval for β_0 has as endpoints the values

$$\widehat{\beta}_0 \pm t_{\alpha/2}\, s_{B_0} \tag{12.7}$$

with the critical value based on $n - 2$ degrees of freedom.

Solution Trail

KEYWORDS

- Estimated regression line.
- ANOVA table.
- t test (concerning β_1) for significant regression.

$\downarrow$

TRANSLATION

- $y = \widehat{\beta}_0 + \widehat{\beta}_1 x$.
- ANOVA summary table.
- Hypothesis test with $\beta_{10} = 0$.

$\downarrow$

CONCEPTS

- Least-squares estimates.
- ANOVA summary table for simple linear regression.
- Hypothesis test concerning β_1.

$\downarrow$

VISION

Find the necessary summary statistics. Use Equation 12.2 to find $\widehat{\beta}_1$ and Equation 12.3 to find $\widehat{\beta}_0$. Complete the ANOVA summary table. Use the template for a hypothesis test concerning β_1 with $\beta_{10} = 0$.

Example 12.5 Ready to Operate The Pulsar Corporation sells a large sterilizer with four extendable shelves for medical tools. Company engineers believe that the time to reach operating temperature from a cold start (y, measured in minutes) is linearly related to the thickness of insulation (x, in inches). A random sample of $n = 12$ thicknesses was selected, and the time to reach operating temperature was recorded for each. The data and the summary statistics are as follows.

x	1.3	1.8	0.9	1.6	2.6	1.5	2.1	3.0	0.8	2.4	2.5	2.6
y	8.0	6.9	8.1	7.0	6.3	6.5	6.4	5.8	8.3	8.3	6.6	6.6

$\sum x_i = 23.1 \qquad\qquad \sum y_i = 84.8 \qquad\qquad \sum x_i y_i = 158.5$

$\sum x_i^2 = 50.13 \qquad\qquad \sum y_i^2 = 607.66$

a. Find the estimated regression line.

b. Complete the ANOVA table and conduct an F test for a significant regression. Use a significance level of 0.05.

c. Conduct a t test (concerning β_1) for a significant regression. Use a significance level of 0.05.

SOLUTION

a. Use the summary statistics to find $\widehat{\beta}_0$ and $\widehat{\beta}_1$.

$$
\begin{aligned}
\widehat{\beta}_1 &= \frac{n \sum x_i y_i - \left(\sum x_i\right)\left(\sum y_i\right)}{n \sum x_i^2 - \left(\sum x_i\right)^2} && \text{Use Equation 12.2.} \\[2mm]
&= \frac{(12)(158.5) - (23.1)(84.8)}{(12)(50.13) - (23.1)^2} && \text{Use summary statistics.} \\[2mm]
&= \frac{-56.88}{67.95} = -0.8371 && \text{Simplify.} \\[2mm]
\widehat{\beta}_0 &= \frac{\sum y_i - \widehat{\beta}_1 \sum x_i}{n} && \text{Use Equation 12.3.} \\[2mm]
&= \frac{84.8 - (-0.8371)(23.1)}{12} && \text{Use summary statistics and } \widehat{\beta}_1. \\[2mm]
&= 8.6781 && \text{Simplify.}
\end{aligned}
$$

The estimated regression line is $y = 8.6781 - 0.8371x$.

b. Here are the calculations for the ANOVA table and the F test for a significant regression.

$$\text{SST} = S_{yy} = \Sigma y_i^2 - \frac{1}{n}(\Sigma y_i)^2 \qquad \text{Computational formula.}$$

$$= 607.66 - \frac{1}{12}(84.8)^2 = 8.4067 \qquad \text{Use summary statistics.}$$

$$\text{SSR} = \hat{\beta}_1 S_{xy} = \hat{\beta}_1 \left[\Sigma x_i y_i - \frac{1}{n}(\Sigma x_i)(\Sigma y_i) \right] \qquad \text{Computation formula.}$$

$$= -0.8371 \left[158.5 - \frac{1}{12}(23.1)(84.8) \right] \qquad \text{Use summary statistics.}$$

$$= 3.9679 \qquad \text{Simplify.}$$

$$\text{SSE} = \text{SST} - \text{SSR} \qquad \text{Computational formula.}$$

$$= 8.4067 - 3.9679 = 4.4388$$

Compute the mean squares.

$$\text{MSR} = \text{SSR}/1 = 3.9679/1 = 3.9679$$

$$\text{MSE} = \text{SSE}/(n-2) = 4.4388/10 = 0.4439$$

The F test for a significant regression with significance level 0.05 is

H_0: There is no significant linear relationship.

H_a: There is a significant linear relationship.

TS: $F = \text{MSR}/\text{MSE}$

RR: $F \geq F_\alpha = F_{0.05} = 4.96$ 　　　　　　　　　df 1 and 10

The value of the test statistic is

$$f = \text{MSR}/\text{MSE} = 3.9679/0.4439 = \boxed{8.9387} \quad (\geq 4.96)$$

Since f lies in the rejection region, there is evidence to suggest that insulation thickness is linearly related to time to reach operating temperature.

Recall, we can only use the tables of critical values for F distributions to bound the p value. Place the value of the test statistic in an ordered list of critical values with $\nu_1 = 1$ and $\nu_2 = 10$.

$$4.96 \leq 8.9387 \leq 10.04$$
$$F_{0.05} \leq 8.9387 \leq F_{0.01}$$

Therefore, 　　　　　$0.01 \leq \quad p \quad \leq 0.05.$

Here are all the calculations presented in the ANOVA table, with some help from technology to find the exact p value.

p value illustration:
$$p = P(X \geq 8.9387)$$
$$= 0.0136 \leq 0.05 = \alpha$$

ANOVA summary table for simple linear regression

Source of variation	Sum of squares	Degrees of freedom	Mean square	F	p value
Regression	3.9679	1	3.9679	8.9387	0.0136
Error	4.4388	10	0.4439		
Total	8.4067	11			

c. The t test for a significant regression (concerning β_1) is two-sided and has $\beta_{10} = 0$. Use a significance level of 0.05.

H_0: $\beta_1 = 0$

H_a: $\beta_1 \neq 0$

Remember, $s^2 = $ MSE.

TS: $T = \dfrac{B_1}{S_{B_1}}$

RR: $|T| \geq t_{\alpha/2} = t_{0.025} = 2.2281$

df = 10.

$$S_{xx} = \sum x_i^2 - \frac{1}{n}\left(\sum x_i\right)^2$$

Computational formula.

$$= 50.13 - \frac{1}{12}(23.1)^2 = 5.6625$$

Use summary statistics.

$$s_{B_1}^2 = s^2/S_{xx} = 0.4439/5.6625 = 0.0784$$

and the value of the test statistic is

$$t = \frac{\hat{\beta}_1}{s_{B_1}} = \frac{-0.8371}{\sqrt{0.0784}} = \boxed{-2.9896}; \quad |-2.9896| = 2.9896 \geq 2.2281$$

Since $|t|$ lies in the rejection region, there is evidence to suggest that $\beta_1 \neq 0$, and the regression is significant. The tables of critical values for t distributions can be used to bound the p value. Using technology,

$$p/2 = P(T \geq 2.9896) = 0.0068 \Rightarrow p = 0.0136$$

p value illustration:
$p = 2P(T \geq 2.9896)$
  $= 0.0136 \leq 0.05 = \alpha$

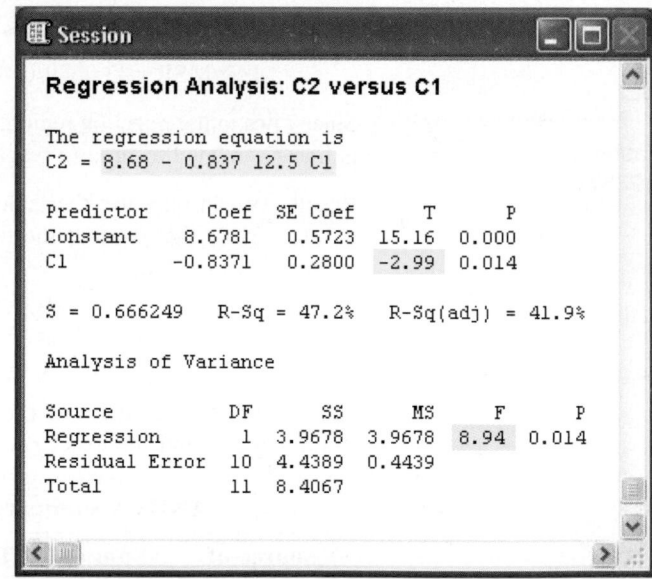

Note: Comparing the two tests for a significant regression, we notice that $t^2 \approx f$. In fact, for simple linear regression $t^2 = f$ in every case, subject to round-off error. The p values associated with these two tests are also always the same. It seems reasonable that the two tests should always lead to the same conclusion, since they are both overall tests of a significant regression.

Figure 12.24 shows a technology solution.

So, why have two hypothesis tests that are essentially the same? In multiple linear regression there may be many (more than one) hypothesized explanatory variables. An overall F test for a significant regression is conducted first. If the regression is significant, then a t test is conducted for each variable to see whether it contributes significantly to the variability in y.

```
Session

Regression Analysis: C2 versus C1

The regression equation is
C2 = 8.68 - 0.837 12.5 C1

Predictor      Coef   SE Coef       T       P
Constant     8.6781    0.5723   15.16   0.000
C1          -0.8371    0.2800   -2.99   0.014

S = 0.666249   R-Sq = 47.2%   R-Sq(adj) = 41.9%

Analysis of Variance

Source          DF      SS       MS       F       P
Regression       1   3.9678   3.9678    8.94   0.014
Residual Error  10   4.4389   0.4439
Total           11   8.4067
```

Figure 12.24 Minitab regression analysis output.

Remember that correlation, or an association between two variables, does not imply causation.

Correlation is a statistical term indicating a relationship between two variables. For example, the temperature is correlated with the number of cars that will not start in the morning, and childhood exposure to lead is correlated with IQ level. In each case, a change in one variable is associated with a steady, consistent change in the other variable. As the temperature decreases, for example, the number of cars that will not start in the morning increases.

The **sample correlation coefficient** is a measure of the strength of a linear relationship between two continuous variables, x and y. Suppose there are n pairs of observations (x_1, y_1), (x_2, y_2), . . . , (x_n, y_n). If large values of x are associated with large values of y (as x increases, the corresponding value of y tends to increase), then x and y are positively related. If small values of x are associated with large values of y (as x increases, the corresponding value of y tends to decrease), then x and y are negatively related.

To understand the formula for the sample correlation coefficient, consider the scatter plots in Figures 12.25 and 12.26 and the quantity $S_{xy} = \sum(x_i - \bar{x})(y_i - \bar{y})$.

The horizontal line $y = \bar{y}$ and the vertical line $x = \bar{x}$ divide the plane region into four parts. The signs, plus (+) or minus (−), on each graph indicate the sign of the product $(x_i - \bar{x})(y_i - \bar{y})$ in each part. For example, for any ordered pair in the top right corner of Figure 12.25, $x_i > \bar{x}$ and $y_i > \bar{y}$. Therefore, $x_i - \bar{x} > 0$ and $y_i - \bar{y} > 0$, and the product is positive. Similarly, for any ordered pair in the bottom right corner of Figure 12.26, $x_i > \bar{x}$ and $y_i < \bar{y}$. Therefore, $x_i - \bar{x} > 0$ and $y_i - \bar{y} < 0$, and the product is negative.

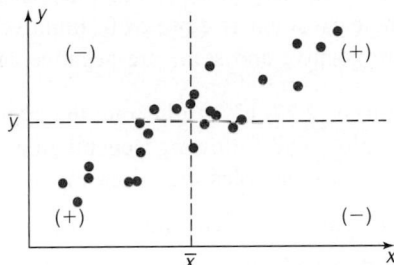

Figure 12.25 If x and y have a positive linear relationship, then most of the products $(x_i - \bar{x})(y_i - \bar{y})$ are positive.

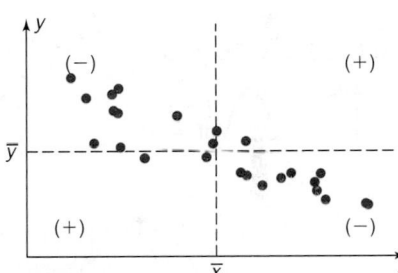

Figure 12.26 If x and y have a negative linear relationship, then most of the products $(x_i - \bar{x})(y_i - \bar{y})$ are negative.

If x and y are positively related, as in Figure 12.25, then most of the products $(x_i - \bar{x})(y_i - \bar{y})$ are positive. If x and y are negatively related, as in Figure 12.26, then most of the products $(x_i - \bar{x})(y_i - \bar{y})$ are negative. Therefore, it seems reasonable to find the sum of all of these products, S_{xy}, and use this as a measure of the linear relationship between the two variables. Large positive values of S_{xy} should indicate a positive linear relationship, and large negative values should indicate a negative linear relationship.

This approach is intuitive, but must be modified slightly. The magnitude of S_{xy} depends on the units of x and y. For example, if we multiply every value of x by 10, the inherent linear relationship between x and y should not change, but the value of S_{xy} increases. The sample correlation coefficient adjusts S_{xy} so that it is unit-independent.

The sample correlation coefficient, r, is an estimate of the population correlation coefficient, ρ.

SAMPLE CORRELATION COEFFICIENT

Suppose there are n pairs of observations (x_1, y_1), (x_2, y_2), . . . , (x_n, y_n). The **sample correlation coefficient** for these n pairs is

$$r = \frac{S_{xy}}{\sqrt{S_{xx}S_{yy}}} = \frac{\sum x_i y_i - \frac{1}{n}\left(\sum x_i\right)\left(\sum y_i\right)}{\sqrt{\left[\sum x_i^2 - \frac{1}{n}\left(\sum x_i\right)^2\right]\left[\sum y_i^2 - \frac{1}{n}\left(\sum y_i\right)^2\right]}} \qquad (12.8)$$

ILLUMINATING THE CONCEPTS

1. The value of r does not depend on the order of the variables and is independent of units, or is unitless.

2. The value of r is always between -1 and $+1$, that is, $-1 \leq r \leq +1$. r is exactly $+1$ if and only if all of the ordered pairs lie on a straight line with positive slope. r is exactly -1 if and only if all of the ordered pairs lie on a straight line with negative slope.

3. The square of the sample correlation coefficient is the coefficient of determination in a simple linear regression model. Since $-1 \leq r \leq +1$, $0 \leq r^2 \leq 1$.

4. r is a measure of the strength of a *linear* relationship. If r is near 0, there is no evidence of a linear relationship, but x and y may be related in another way.

5. Suppose there is a horizontal line ($y = \beta_0$) with zero slope, and all the data points lie very close to this line. There is no association between the variables; the correlation is close to 0. Intuitively, some of the products $(x_i - \bar{x})(y_i - \bar{y})$ are positive and some are negative, and they tend to cancel out one another. ■

Figures 12.27–12.30 illustrate the approximate value of r corresponding to each scatter plot. The following general rule is used to describe the linear relationship between two variables, based on the value of the sample correlation coefficient.

1. If $0 \leq |r| \leq 0.5$, then there is a *weak* linear relationship.

2. If $0.5 < |r| \leq 0.8$, then there is a *moderate* linear relationship.

3. If $|r| > 0.8$, then there is a *strong* linear relationship.

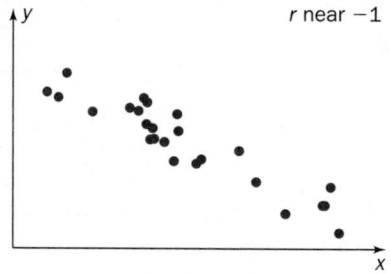

Figure 12.27 The variables x and y are (strongly) negatively related. The points all fall near a straight line with negative slope. The sample correlation coefficient is near -1.

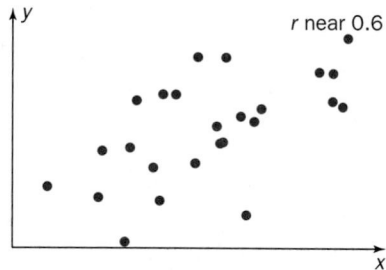

Figure 12.28 The variables x and y are (moderately) positively related. As the values of x increase, the values of y tend to increase. The sample correlation coefficient is near 0.6.

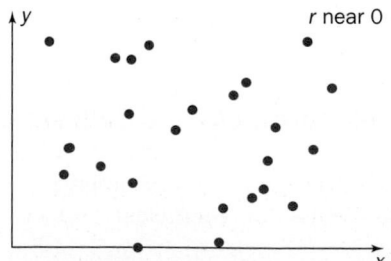

Figure 12.29 The scatter plot shows no linear relationship between the variables x and y. The sample correlation coefficient is near 0.

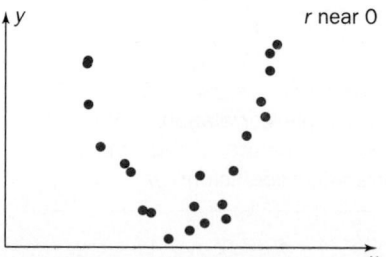

Figure 12.30 The variables x and y appear to be related but not linearly. There is a pattern in the scatter plot, but because the sample correlation coefficient measures linear association, r is near 0.

Try the Correlation and Regression statistical applet on the text web site.

(Science Photo Library/Alamy)

A technology solution:

```
LinReg
y=a+bx
a=.7710
b=2.7324
r²=.8653
r=.9302
```

Figure 12.31 Sample correlation coefficient.

Remember, *r* is a measure of *association*, not *causation*. Two variables may be strongly related, but that does not mean one variable causes the other. For example, shoe size and height are associated, but probably both are caused by a third variable, age.

Example 12.6 Fluoride in Drinking Water Most municipal water supplies include fluoride, a chemical believed to promote dental health. However, some researchers believe that high levels of fluoride in drinking water increase a child's absorption of lead into the bloodstream. In a recent study of $n = 10$ children two to nine years old in Bennington, Oklahoma, the amount of fluoride added to their drinking water was recorded (in mg/L) and the amount of lead in each child's blood was measured (in micrograms per deciliter, μg/dL). The data are given in the following table.

Fluoride	2.5	1.1	3.2	2.9	0.8	3.6	1.2	0.7	1.3	0.3
Lead	8.6	5.3	9.5	9.2	4.9	9.5	2.2	1.4	4.5	0.7

Find the sample correlation coefficient between fluoride level and blood lead level, and interpret this value.

SOLUTION

STEP 1 Compute the summary statistics needed to find the sample correlation coefficient.

$$S_{xx} = \sum x_i^2 - \frac{1}{n}\left(\sum x_i\right)^2 = 43.42 - \frac{1}{10}(17.6)^2 = 12.444$$

$$S_{yy} = \sum y_i^2 - \frac{1}{n}\left(\sum y_i\right)^2 = 418.74 - \frac{1}{10}(55.8)^2 = 107.376$$

$$S_{xy} = \sum x_i y_i - \frac{1}{n}\left(\sum x_i\right)\left(\sum y_i\right)$$

$$= 132.21 - \frac{1}{10}(17.6)(55.8) = 34.002$$

STEP 2 Using Equation 12.8, the sample correlation coefficient is

$$r = \frac{S_{xy}}{\sqrt{S_{xx}S_{yy}}} = \frac{34.002}{\sqrt{(12.444)(107.376)}} = \boxed{0.9302}$$

Since $r = 0.9302 > 0.8$, there is a strong positive linear relationship between fluoride levels and blood lead levels in children two to nine years old. High levels of fluoride are associated with high levels of blood lead. ●

TECHNOLOGY CORNER

Procedure: Find the estimated regression line, complete the ANOVA table, and conduct a *t* test (concerning β_1) for a significant regression.
Reconsider: Example 12.5, page 576, solution, and interpretations.

TI-84 Plus

Execute `DiagonisticsOn` on the Home Screen in order to display the coefficient of determination and sample correlation coefficient with the estimated regression line.
1. Enter the thickness of insulation values into list `L1` and the times into list `L2`.
2. Select `STAT`; CALC; `LinReg(a+bx)`. Enter the name of the list containing the observed values of the independent variable, `L1`, and the name of the list containing the observed values of the dependent variable, `L2`. Press `ENTER`. The estimated regression coefficients are displayed on the Home Screen. See Figure 12.32.
3. In order to test for a significant regression in terms of β_1, select `STAT`; TESTS; `LinRegTTest`. Enter the name of the list containing the observed values of the independent variable, `L1`, and the name of the list containing the

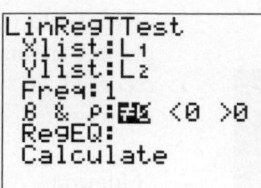

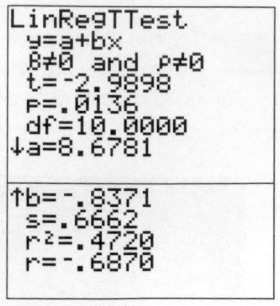

Figure 12.32 LinReg results.

Figure 12.33 LinRegTTest setup screen.

Figure 12.34 LinRegTTest results.

observed values of the dependent variable, L2, set Freq to 1, highlight ≠ 0, and optionally store the regression equation in a function variable. See Figure 12.33.

4. Highlight Calculate and press ENTER. The output is displayed on the Home Screen. See Figure 12.34.

Minitab

There are several ways to find the estimated regression line in Minitab, in a session window and in the <u>Stat</u>; <u>Regression</u> menu.

1. Enter the thickness of insulation values into column C1 and the times into column C2.

2. Select <u>Stat</u>; <u>Regression</u>; <u>Regression</u>. Enter the Response variable, C2, and the Predictor variable, C1. Click OK. The results are displayed in a session window. See Figure 12.24, page 578.

Excel

Regression analysis is part of the Data Analysis tool pack.

1. Enter the thickness of insulation values into column A and the times into column B.

2. Under the Data tab, select Data Analysis; Regression. Enter the Y Range, X Range, and Output Range. Click OK. The output is displayed in the specified range. See Figure 12.35.

C	D	E	F	G	H	I	J	K
SUMMARY OUTPUT								
Regression Statistics								
Multiple R	0.6870							
R Square	0.4720							
Adjusted R Square	0.4192							
Standard Error	0.6662							
Observations	12							
ANOVA								
	df	*SS*	*MS*	*F*	*Significance F*			
Regression	1	3.9678	3.9678	8.94	0.0136			
Residual	10	4.4369	0.4439					
Total	11	8.4067						
	Coefficients	*Standard Error*	*t Stat*	*P-value*	*Lower 95%*	*Upper 95%*	*Lower 95.0%*	*Upper 95.0%*
Intercept	8.6781	0.5723	15.1646	0.0000	7.4030	9.9531	7.4030	9.9531
X Variable 1	-0.8371	0.2800	-2.9898	0.0136	-1.4609	-0.2132	-1.4609	-0.2132

Figure 12.35 Regression analysis results.

Procedure: Compute the sample correlation coefficient.

Reconsider: Example 12.6, page 581, solution, and interpretations.

TI-84 Plus

Use the built-in function LinReg(ax+b) or LinReg(a+bx) to find the sample correlation coefficient. Execute DiagonisticsOn on the Home Screen in order to display the coefficient of determination and sample correlation coefficient with the estimated regression line.

1. Enter the values for fluoride into list L1 and the values for lead into list L2.
2. Select $\boxed{\text{STAT}}$; CALC; LinReg(a+bx). Enter the name of the list containing the observed values of the independent variable, L1, and the name of the list containing the observed values of the dependent variable, L2. Press $\boxed{\text{ENTER}}$.
3. The sample correlation coefficient, r, is displayed at the bottom of the screen. See Figure 12.31, page 581.

Minitab

Minitab has a built-in function to compute the sample correlation coefficient, in the Stat; Basic Statistics menu.

1. Enter the values for fluoride into column C1 and the values for lead into column C2.
2. Select Stat; Basic Statistics; Correlation.
3. Select the Variables (columns) containing the data. Click OK. The sample correlation coefficient is displayed in a session window. See Figure 12.36.

Figure 12.36 Minitab: sample correlation coefficient.

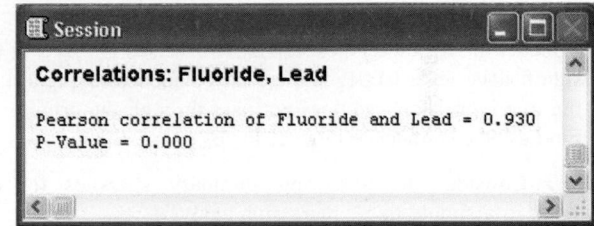

Excel

Use the built-in function CORREL to compute the sample correlation coefficient.

1. Enter the values for fluoride into column A and the values for lead into column B.
2. Use the function CORREL with the two arrays. See Figure 12.37.

Figure 12.37 Excel function CORREL.

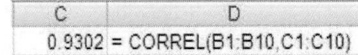

SECTION 12.2 EXERCISES

Practice

12.23 Consider the following (partial) ANOVA summary table from a simple linear regression analysis.

ANOVA summary table for simple linear regression

Source of variation	Sum of squares	Degrees of freedom	Mean square	F	p value
Regression	11,691.9				
Error		23			
Total	116,064.0				

a. Complete the ANOVA summary table.
b. Conduct an F test for a significant regression. Use $\alpha = 0.05$.
c. Find the coefficient of determination.
d. Use your answer in part (c) to find the sample correlation coefficient.

12.24 Consider the following (partial) ANOVA summary table from a simple linear regression analysis.

ANOVA summary table for simple linear regression

Source of variation	Sum of squares	Degrees of freedom	Mean square	F	p value
Regression	2772.93				
Error	12,988.70				
Total		35			

a. Complete the ANOVA summary table.
b. Conduct an F test for a significant regression. Use $\alpha = 0.01$.
c. Find the coefficient of determination.
d. Use your answer in part (c) to find the sample correlation coefficient. Is the relationship between these two variables positive or negative? Justify your answer.

12.25 Consider the following summary statistics for data obtained in a simple linear regression analysis.

$$n = 23 \qquad MSE = 561.088$$
$$S_{xx} = 151.086 \qquad \hat{\beta}_1 = 4.4285$$

a. Conduct a hypothesis test for a significant regression based on the value of β_1 ($H_0: \beta_1 = 0$ versus $H_a: \beta_1 \neq 0$). Use a significance level of 0.05.
b. Find a 95% confidence interval for the slope of the true regression line, β_1.
c. Using your confidence interval in part (b), is there any evidence to suggest that $\beta_1 \neq 0$? Justify your answer.

12.26 Consider the following data obtained in a simple linear regression study.

x	3.27	1.26	4.55	0.86	4.07	4.79	3.25
y	16.67	19.93	14.65	17.48	18.18	13.58	15.70

a. Find the estimated regression line, and complete the ANOVA table.
b. Conduct the hypothesis test $H_0: \beta_0 = 0$ versus $H_a: \beta_0 \neq 0$ with significance level 0.001. Is there any evidence to suggest that the true regression line does not pass through the origin?
c. Find a 99% confidence interval for β_0.

12.27 Consider the following summary statistics for data obtained in a simple linear regression analysis.

$$n = 16 \qquad SSR = 1155.9 \qquad SSE = 3912.82$$
$$S_{xx} = 1980.24 \qquad \hat{\beta}_1 = 0.7640$$

a. Conduct an F test for a significant regression. Use $\alpha = 0.05$. Bound the p value and use technology to find the exact p value.
b. Conduct a t test for a significant regression. Use $\alpha = 0.05$. Bound the p value and use technology to find the exact p value.
c. Square the value of the test statistic in part (a). Compare this value with the test statistic in part (b).
d. How do the exact p values compare in parts (a) and (b)? Explain this result.

12.28 Consider the following summary statistics for data obtained in a study of the linear relationship between two variables.

$$S_{xx} = 199.418 \qquad S_{yy} = 81.430 \qquad S_{xy} = 111.774$$

a. Find the sample correlation coefficient.
b. Use the value of r from part (a) to describe the relationship between the two variables.

12.29 Consider the following data obtained in a study of the relationship between two variables.

x	40.4	44.8	40.7	31.7	41.3	38.1
y	-320.3	-303.9	-264.1	-197.0	-311.8	-280.3

a. Find the estimated regression line and complete the ANOVA table.
b. Conduct an F test for a significant regression. Use $\alpha = 0.05$.
c. Find the coefficient of determination.
d. Find the sample correlation coefficient. Use this value to describe the relationship between the two variables. How does the estimate $\hat{\beta}_1$ support this answer?

e. Square the sample correlation coefficient. Verify that this value is the coefficient of determination.

Applications

12.30 **Sports and Leisure** Many factors affect the length of a professional football game. A study was conducted to determine the relationship between the total number of penalty yards (x) and the time required to complete a game (y, in hours). The following data were obtained.

x	196	164	167	35	111	78	150	121	40
y	4.2	4.1	3.5	3.2	3.2	3.6	4.0	3.1	1.9

a. Find the estimated regression line, and complete the ANOVA table.
b. Conduct a t test (concerning β_1) for a significant regression. Use a significance level of 0.05.
c. What proportion of the observed variation in game length can be explained by this simple linear regression model?

12.31 **Biology and Environmental Science** In a recent study, the weight of an orange (x, in pounds) was compared with the amount of fresh-squeezed juice from the orange (y, in ounces). A random sample of $n = 15$ oranges was obtained. Each was carefully weighed and squeezed. The following summary statistics were obtained.

$$\sum x_i = 9.09 \qquad \sum y_i = 50.44 \qquad \sum x_i y_i = 30.72$$
$$\sum x_i^2 = 5.73 \qquad \sum y_i^2 = 173.21$$

a. Find the estimated regression line, and complete the ANOVA table.
b. Conduct an F test for a significant regression. Use a significance level of 0.01.
c. Find a 95% confidence interval for the regression parameter β_0. Does this interval suggest that β_0 is different from 0?

12.32 **Physical Sciences** The temperature of the upper layer of ocean water is affected by sunlight and wind. There is often a very sharp difference in temperature between the surface zone and the more stationary deep zone. The thermocline layer marks the abrupt drop-off in temperature. The following data were obtained in a study of temperature (x, measured in °C) versus depth (y, measured in meters) above the thermocline layer in the Mediterranean Sea.

x	76	54	146	7	91	130	131	117
y	11.0	16.9	8.0	24.7	17.4	16.0	11.1	16.5

a. Complete the ANOVA summary table for simple linear regression.
b. Conduct an F test for a significant regression and find bounds on the p value associated with this test.
c. Find a 95% confidence interval for the true value of β_1.

12.33 **Physical Sciences** Permafrost is soil or rock that remains at or below 0 °C for at least two years. A recent Arctic study analyzed the observed changes in mean annual air temperature and the depth of the freezing layer.[9] Suppose the depth of the upper surface of the permafrost layer (y, in feet) was measured for $n = 26$ mean annual air temperatures (x, in °F). The summary statistics are

$$S_{xx} = 8.6786, \qquad S_{yy} = 2.2771, \qquad S_{xy} = -2.8600$$

a. Find the sample correlation coefficient.

b. Use the value of r from part (a) to describe the relationship between mean annual temperature and the depth of the permafrost layer.

12.34 Public Health and Nutrition Crimini mushrooms are more common than white mushrooms, and they contain a high amount of copper, which is an essential element according to the U.S. Food and Drug Administration. A study was conducted to determine whether the weight of a mushroom is linearly related to the amount of copper it contains. A random sample of crimini mushrooms was obtained, and the weight (in grams) and the total copper content (in mg) were measured for each. The data are given on the data CD and book's web site.

a. Construct a scatter plot of the data.

b. Find the sample correlation coefficient.

c. Use your results in parts (a) and (b) to describe the relationship between crimini mushroom weight and copper content.

d. If a simple linear regression analysis were conducted, identify the independent and dependent variables that would be used.

12.35 Demographics and Population Statistics Many studies suggest that parents' salaries, or socioeconomic status, are passed on to, or inherited by, their children.[10] Recently, economists have even increased their estimate of the association in income between generations. Suppose a random sample of parents and children was obtained. Using IRS records, the mean salary for each parent and each child was computed for specific years. For parents, the 16-year period 1980–1995 was used, and for children, the 5 years 1998–2002 were used. The salaries (in thousands of dollars) are given on the data CD and book's web site.

a. Construct a scatter plot of the data.

b. Find the sample correlation coefficient.

c. Do your results in parts (a) and (b) support or refute this economic association theory? Justify your answer.

12.36 Fuel Consumption and Cars An investigative reporter believes that certain automobile service stations that offer state vehicle inspections routinely charge for unnecessary repair work. Preliminary data suggest that the cost of the repair work may be related to the age of the car. A random sample of automobiles inspected at these stations was obtained, and the age (in years) along with the cost of the repairs (in dollars) were recorded for each vehicle. The data are given in the following table.

Age	Repair cost	Age	Repair cost
9.1	1882	3.2	17
3.8	193	3.3	1268
5.0	368	6.7	1126
1.7	1047	3.9	646
1.9	315	9.7	955
5.3	1631	2.0	801
5.9	652	5.4	973
6.4	475		

a. Find the sample correlation coefficient, and describe the linear relationship.

b. Find the estimated regression line for age (x) and repair cost (y). Conduct an F test for a significant regression, and find the bounds on the p value for this test.

c. Using your results in parts (a) and (b), do you believe the reporter's claim? Justify your answer.

12.37 Medicine and Clinical Studies Some research suggests that there is a relationship between psychological factors and coronary artery disease.[11] Suppose in a recent study at the Walter Reed Army Medical Center in Washington, D.C., physicians investigated this relationship. Measures of anxiety (x) and of coronary artery calcification (y) were obtained for each randomly selected patient (both are unitless measures). The data are given in the following table.

x	y	x	y	x	y	x	y
21	157	22	37	21	121	19	2
42	56	46	29	48	60	42	4
44	96	8	14	41	136	8	27
3	80	7	21	33	58	42	10
21	254	12	60	12	33	18	72

a. Construct a scatter plot of these data.

b. Find the sample correlation coefficient, and describe the linear relationship.

12.38 Sports and Leisure Luge competitions consist of a descent down an iced track in the prone position, feet pointing down the run. The best teams practice intensely, have dedicated coaches, and are helped by natural weather conditions. A random sample of men's and women's single competition runs, by country, from the 2008 FIL World Championships in Oberhof, Germany, was obtained. The finish time (in seconds) for each run is given in the following table.

Country	Men's time	Women's time
Russia	45.527	43.813
Germany	44.996	43.019
United States	45.529	43.633
Latvia	45.870	43.615
Austria	45.688	43.472
Canada	45.735	43.696
Slovakia	45.857	45.011
Japan	46.384	45.052
Switzerland	46.141	43.569
Poland	46.843	46.498
Italy	45.342	44.208

a. Construct a scatter plot of these data.

b. Compute the sample correlation coefficient, and describe the linear relationship. What does this suggest about men's and women's luge competitors?

Extended Applications

12.39 Business and Management The owner of a small ice cream stand believes that total weekly revenue $(y$, in dollars) during the summer months is related to money spent on advertising $(x$, dollars per week). A random sample of summer weeks was selected, and the resulting data are given in the following table.

x	30	300	380	275	350	190	85
y	957	1125	1202	1028	1134	1124	1062

a. Find the estimated regression line, and complete the ANOVA table.

b. What proportion of the observed variation in weekly revenue is explained by this regression model?

c. Find a 99% confidence interval for the regression parameter β_1.

d. Conduct a hypothesis test of $H_0: \beta_0 = 0$ versus $H_a: \beta_0 > 0$. Interpret the results. (What happens if the owner spends nothing on advertising in a week?)

12.40 Public Health and Nutrition Salt intake is related to fluid intake in adults and soft drink consumption in children and adolescents.[12] Suppose a random sample of 18 children and adolescents was obtained and the amount of salt consumed (x, g/week) and the amount of sugar-sweetened soft drinks consumed (y, g/week) was recorded for each. The following summary statistics were obtained.

$\sum x_i = 430$ $\sum y_i = 17{,}512$ $\sum x_i y_i = 441{,}608$
$\sum x_i^2 = 11{,}138$ $\sum y_i^2 = 18{,}197{,}768$

a. Find the estimated regression line, and complete the ANOVA table.

b. Conduct an F test for a significant regression. Find bounds on the p value for this test.

c. Conduct the hypothesis test $H_0: \beta_1 = 25$ versus $H_a: \beta_1 \neq 25$. Use a significance level of 0.05.

d. Find a 99% confidence interval for the true value of β_1.

12.41 Biology and Environmental Science The water quality in many lakes and rivers is routinely monitored by tracking the turbidity, a measure of the total amount of suspended solids in the water. High turbidity measured in nephelometric turbidity units, or NTUs, suggests murky water. Heavy rains tend to raise river levels and increase turbidity. A sample of weekly rainfall total (x, measured in inches) and turbidity (y, in NTUs) was obtained for the Wide Waters site at Owasco Lake in New York. The data are given in the following table.

x	0.67	0.08	0.64	0.98	0.34	0.37	0.85
y	5.29	3.21	5.68	1.77	3.66	3.38	3.19

x	0.75	0.07	0.68	0.64	0.26	0.09	1.44
y	2.45	2.22	2.44	2.26	2.55	1.66	8.41

(*Source:* Owasco Watershed Lake Association, Inc.)

a. Find the estimated regression line.

b. Conduct a test of $H_0: \beta_1 = 0$ versus $H_a: \beta_1 \neq 0$ with a significance level of 0.05.

c. Find an estimate of the expected turbidity if 0.55 inches of rain has fallen within the past week.

d. What proportion of the observed variation in turbidity is explained by this regression model? How do you think this model could be improved?

12.42 Economics and Finance Stock brokers and even casual investors are always searching for better methods to predict the movement in the price of stocks. The "skirt-length theory" suggests that if women's skirts are short, then the markets will rise. If skirts are long, then the markets will be headed down. In order to test this theory, a random sample of years was obtained, and the length of women's skirts was measured (x, in inches) for a typical fashion model. The change in the S&P 500 market index (y) was also recorded for that year. The data are given in the following table.

x	24	15	24	16	16	16	19	15
y	7	−7	3	36	−35	23	−35	79

x	19	25	15	18	17	22	16	19
y	−5	−55	−29	8	42	7	−9	−45

a. Find the estimated regression line, and complete the ANOVA table.

b. Conduct an F test for a significant regression. Use a significance level of 0.01.

c. Construct a scatter plot of the data, and find the sample correlation coefficient.

d. Using your answers to parts (b) and (c), do you think the skirt-length theory has merit? Justify your answer.

12.43 Physical Sciences The rate of evaporation at the surface of the water in a swimming pool (kg/h) is believed to be related to the air velocity (m/s) or the relative humidity (measured as a percentage). A pool-and-spa company obtained the data that appears on the data CD and book's web site.

a. Find the estimated regression line, and complete the ANOVA table for evaporation rate (y) and air velocity (x).

b. Find the estimated regression line and complete the ANOVA table for evaporation rate (y) and relative humidity (x).

c. Which of these two models do you think is better? Justify your answer.

Challenge

12.44 Sports and Leisure Let ρ be the population correlation coefficient between two variables. A hypothesis test for a correlation different from zero is based on the sample correlation coefficient, R (the random variable), and involves the t distribution.

$H_0: \rho = 0$

$H_a: \rho > 0, \quad \rho < 0, \quad$ or $\quad \rho \neq 0$

$\text{TS: } T = \dfrac{R\sqrt{n-2}}{\sqrt{1-R^2}}$

$\text{RR: } T \geq t_\alpha, \quad T \leq -t_\alpha, \quad$ or $\quad |T| \geq t_{\alpha/2} \quad \text{df} = n-2$

A new mountain climber believes that there is a linear relationship between the diameter of a single rope and the dry weight. A random sample of climbing ropes was obtained. The diameter of each was measured (x, in mm) and the weight of each (y, in grams per meter) was also recorded. The data are given in the following table.

x	10.07	9.54	10.34	10.85	9.85	9.45	9.95
y	71.2	64.5	67.2	73.5	64.5	65.2	68.2

x	10.63	9.73	10.34
y	70.9	67.8	72.0

a. Find the sample correlation coefficient, and conduct a two-sided hypothesis test with $H_0: \rho = 0$. Use a significance level of 0.05.

b. Find the estimated regression line, and conduct a test of $H_0: \beta_1 = 0$ versus $H_a: \beta_1 \neq 0$. Use a significance level of 0.05.

c. Explain the similarities between the values of the test statistics in parts (a) and (b). Explain why this relationship makes sense.

12.3 INFERENCES CONCERNING THE MEAN VALUE AND AN OBSERVED VALUE OF Y FOR $x = x^*$

Remember that $E(Y|x^*)$ is the mean of *all* values of Y for which $x = x^*$. And y^* is a single observation for $x = x^*$.

Recall from Section 12.1: Suppose x^* is a specific value of the independent, or predictor, variable x and $y = \hat{\beta}_0 + \hat{\beta}_1 x$ is the estimated regression line. The value $y^* = \hat{\beta}_0 + \hat{\beta}_1 x^*$ is

1. an estimate of the *mean* value of Y for $x = x^*$, and

2. an estimate of an *observed* value of Y for $x = x^*$.

The error in estimating the *mean* value of Y is less than the error in estimating an *observed* value of Y. In the first case, y^* is used to estimate a *single value*, the mean. However, an estimate of an *observed* value of Y is a guess at the next value selected from an entire distribution; the error in estimation must be greater. In this section, we will first consider a hypothesis test and confidence interval concerning the mean value of Y for $x = x^*$. Then, we consider a **prediction interval** for an observed value of Y if $x = x^*$.

Suppose the simple linear regression model assumptions are true. For $x = x^*$, the random variable $B_0 + B_1 x^*$ has a normal distribution with expected value

$$E(B_0 + B_1 x^*) = \beta_0 + \beta_1 x^* \qquad (12.9)$$

and variance

$$\text{Var}(B_0 + B_1 x^*) = \sigma^2 \left[\frac{1}{n} + \frac{(x^* - \bar{x})^2}{S_{xx}} \right] \qquad (12.10)$$

The standard deviation is the square root of the expression in Equation 12.10 and an estimate of the standard deviation is obtained by using s as an estimate for σ.

The numerator, $(x^* - \bar{x})^2$, is 0 when $x^* = \bar{x}$.

The variance of $B_0 + B_1 x^*$ is smallest when $x = \bar{x}$. The farther x^* is from $\bar{x}$, the greater the squared difference $(x^* - \bar{x})^2$ and the greater the variance. Therefore, the estimator, $B_0 + B_1 x^*$, for the mean value of Y is most precise near $\bar{x}$. And a confidence interval for the mean value of Y would be narrower for values of x^* near $\bar{x}$. Intuitively, think about predicting the weather. The farther into the future we try to predict the weather, the more inaccurate the forecast is.

The following theorem is used to conduct a hypothesis test and construct a confidence interval concerning the mean value of Y given $x = x^*$.

THEOREM

If the simple linear regression assumptions are true, then the random variable

$$T = \frac{(B_0 + B_1 x^*) - (\beta_0 + \beta_1 x^*)}{S\sqrt{(1/n) + [(x^* - \bar{x})^2/S_{xx}]}}$$

has a t distribution with $n - 2$ degrees of freedom.

The null and alternative hypotheses are stated in terms of the parameter $y^* = \beta_0 + \beta_1 x^*$.

HYPOTHESIS TEST AND CONFIDENCE INTERVAL CONCERNING THE MEAN VALUE OF Y FOR $x = x^*$

$H_0: y^* = y_0^*$

$H_a: y^* > y_0^*, \quad y^* < y_0^*, \quad \text{or} \quad y^* \neq y_0^*$

TS: $T = \dfrac{(B_0 + B_1 x^*) - y_0^*}{S\sqrt{(1/n) + [(x^* - \bar{x})^2/S_{xx}]}}$

RR: $T \geq t_\alpha, \quad T \leq -t_\alpha, \quad \text{or} \quad |T| \geq t_{\alpha/2} \quad df = n - 2$

A $100(1 - \alpha)\%$ confidence interval for $\mu_{Y|x^*}$, the mean value of Y for $x = x^*$, has as endpoints the values

$$(\hat{\beta}_0 + \hat{\beta}_1 x^*) \pm t_{\alpha/2} s \sqrt{\frac{1}{n} + \frac{(x^* - \bar{x})^2}{S_{xx}}} \tag{12.11}$$

with the critical value based on $n - 2$ degrees of freedom.

y_0^* can be any constant. We can conduct a test for evidence that the mean value of Y for $x = x^*$ is different from any value.

Solution Trail

KEYWORDS

- Hypothesis test.
- 150 feet.
- Mean uranium contamination greater than.

↓

TRANSLATION

Conduct a one-sided, right-tailed hypothesis test concerning the mean when $x = 150$.

↓

CONCEPTS

Hypothesis test concerning the mean value of Y for $x = x^*$.

↓

VISION

The hypothesized value of the mean uranium contamination for a 150-foot well is $y_0^* = 0.37$. Use this value and the summary statistics to conduct a one-sided, right-tailed hypothesis test concerning the mean value of Y for $x = x^* = 150$.

Example 12.7 Well Contamination   Many state, local, and residential wells are contaminated with uranium. Some research suggests that the amount of contamination is related to the depth of the well.[13] Suppose a random sample of wells in Pennsylvania was obtained, and the depth (x, in feet) and the uranium contamination (y, in parts per billion, ppb) were measured for each. The data are given in the following table.

x	168	185	130	40	85	170	85	90	50	100
y	0.231	0.434	0.159	0.042	0.076	0.729	0.100	0.074	0.070	0.284

x	50	75	65	190	80	87	100	90	125	230
y	0.044	0.037	0.026	0.721	0.068	0.015	0.145	0.052	0.028	0.066

The estimated regression line is $y = -0.107 + 0.0025244x$, and the summary ANOVA table is as follows.

ANOVA summary table for simple linear regression

Source of variation	Sum of squares	Degrees of freedom	Mean square	F	p value
Regression	0.33801	1	0.33801	11.10	0.0037
Error	0.54820	18	0.03046		
Total	0.88621	19			

In addition, $s = \sqrt{\text{MSE}} = \sqrt{0.03046} = 0.1745$, $\bar{x} = 109.8$, and $S_{xx} = 53041.8$.

a. For $x = 150$ feet, conduct a hypothesis test to determine whether there is any evidence that the mean uranium contamination is greater than 0.25 ppb.

b. Construct a 95% confidence interval for the true mean uranium contamination in a well that is 200 feet deep.

SOLUTION

a. ◀ The hypothesis test is one-sided and right-tailed with $y_0^* = 0.25$. Use a significance level of 0.05.

H_0: $y^* = 0.25$
H_a: $y^* > 0.25$

TS: $T = \dfrac{(B_0 + B_1 x^*) - y_0^*}{S\sqrt{(1/n) + [(x^* - \bar{x})^2/S_{xx}]}}$ df = 18

RR: $T \geq t_\alpha = t_{0.05} = 1.7341$

The value of the test statistic is

$$t = \dfrac{(\hat{\beta}_0 + \hat{\beta}_1 x^*) - y_0^*}{s\sqrt{(1/n) + [(x^* - \bar{x})^2/S_{xx}]}}$$

$$= \dfrac{[-0.107 + 0.0025244(150)] - 0.25}{0.1745\sqrt{(1/20) + [(150 - 109.8)^2/53041.8]}} = 0.4376$$

The value of the test statistic does not lie in the rejection region. At the $\alpha = 0.05$ significance level, there is no evidence to suggest that the mean uranium contamination is greater than 0.25 ppb for 150-foot wells. ▮

b. To construct the confidence interval, first find the appropriate critical value.

$1 - \alpha = 0.95 \rightarrow \alpha = 0.05 \rightarrow \alpha/2 = 0.025$ Find $\alpha/2$.

$t_{\alpha/2} = t_{0.025} = 2.1009$ Use Table V in the Appendix, df = 18, to find the critical value.

Use Equation 12.11.

$$(\hat{\beta}_0 + \hat{\beta}_1 x^*) \pm t_{\alpha/2}\, s\sqrt{\dfrac{1}{20} + \dfrac{(x^* - \bar{x})^2}{S_{xx}}}$$ Use Equation 12.11.

$$= (\hat{\beta}_0 + \hat{\beta}_1 x^*) \pm t_{0.025}\, s\sqrt{\dfrac{1}{20} + \dfrac{(x^* - \bar{x})^2}{S_{xx}}}$$ Use the values of α and n.

$$= [-0.107 + 0.0025244(200)] \pm (2.1009)(0.1745)\sqrt{\dfrac{1}{20} + \dfrac{(200 - 109.8)^2}{53041.8}}$$

    Use summary statistics, values for s and $t_{0.025}$.

$= 0.3979 \pm 0.1653$ Simplify.

$= (0.2326, 0.5632)$ Compute endpoints.

$(0.2326, 0.5632)$ is a 95% confidence interval for the true mean uranium contamination in a well that is 200 feet deep.

Note: Figure 12.38 shows a scatter plot of the data, the estimated regression line, and the 95% **confidence bands** for the true mean uranium contamination for each well depth. The confidence bands allow us to visualize 95% confidence intervals for the true mean value of *Y* for any value *x = x**. The top confidence band connects the right, or upper, endpoint of each 95% confidence interval, and the bottom confidence band connects the left, or lower, endpoint of each 95% con-

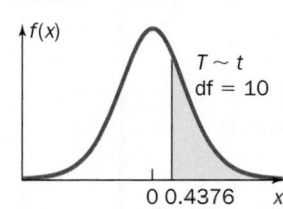

p value illustration:
$p = P(T \geq 0.4376)$
$\quad = 0.3334 > 0.05 = \alpha$

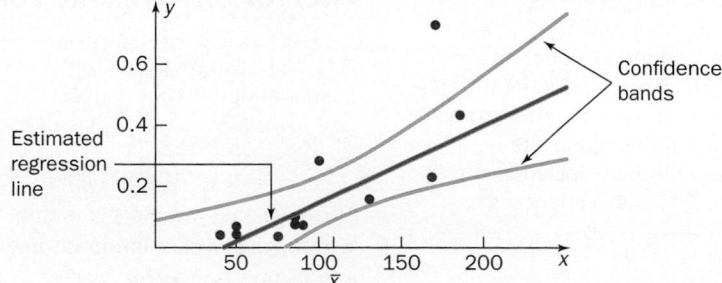

Figure 12.38 Scatter plot, estimated regression line, and confidence bands for the well depth–uranium contamination data.

fidence interval. To estimate the 95% confidence interval for the true mean value of Y for $x = 200$ from the graph, draw a vertical line at $x = 200$. The intersection of the line and the confidence bands yields the endpoints of the interval. As x^* moves away from $\bar{x}$, the width of the confidence interval increases.

Figure 12.39 shows a technology solution.

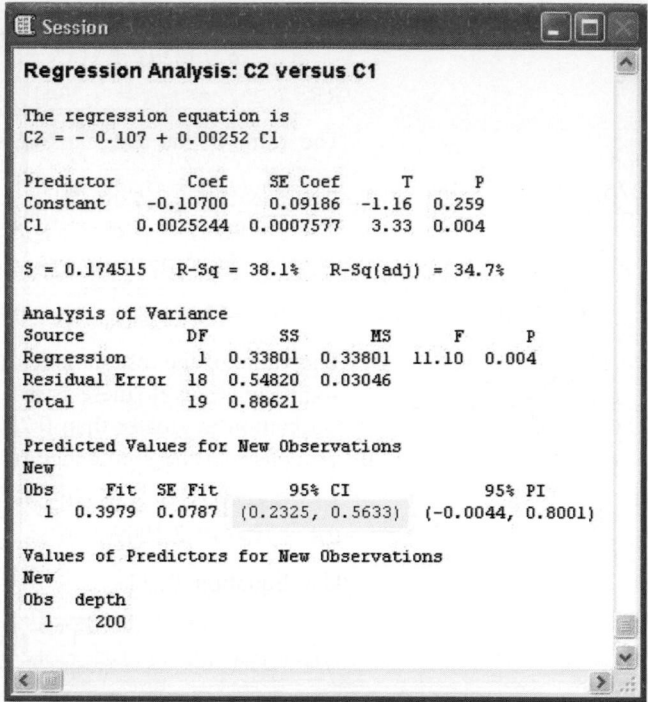

Figure 12.39 Minitab output: regression analysis and a 95% confidence interval for the true mean uranium contamination in a well that is 200 feet deep.

An investigator may be interested in constructing an interval of possible values for an *observed* value of Y if $x = x^*$. For example, suppose temperature is used to predict the amount of expansion (in inches) of a certain type of vinyl siding. We may need an interval of possible values of expansion when the temperature is 90 °F.

An observed value of Y (for $x = x^*$) is a value of a random variable, not a fixed parameter. This helps to explain why the error of estimation for an observed value is larger than the error of estimation for a single mean value of Y. The interval of possible values is called a **prediction interval**.

The random variable $(B_0 + B_1 x^*) - (\beta_0 + \beta_1 x^* + E^*)$ is used in order to derive a prediction interval for Y. Using the properties of this random variable, we can find a prediction interval.

> Note that we construct a *confidence interval* for the mean value of Y for $x = x^*$, and we construct a *prediction interval* for an observed value of Y when $x = x^*$.

> The only difference between this prediction interval and the confidence interval in Equation 12.10 is the extra 1 underneath the square root symbol. This extra 1 is reasonable because the random variable here includes an extra term, E^*, with variance s^2.

PREDICTION INTERVAL FOR AN OBSERVED VALUE OF Y

A $100(1 - \alpha)\%$ prediction interval for an observed value of Y when $x = x^*$ has as endpoints the values

$$(\hat{\beta}_0 + \hat{\beta}_1 x^*) \pm t_{\alpha/2}\, s \sqrt{1 + \frac{1}{n} + \frac{(x^* - \bar{x})^2}{S_{xx}}} \qquad (12.12)$$

with critical value based on $n - 2$ degrees of freedom.

Example 12.8 Well Contamination (Continued)   Use the well depth–uranium contamination data presented in the previous example. Suppose a single 175-foot well is selected at random. Find a 90% prediction interval for the amount of uranium contamination in this well.

SOLUTION

STEP 1   Find the appropriate critical value.

$$1 - \alpha = 0.90 \longrightarrow \alpha = 0.10 \longrightarrow \alpha/2 = 0.05 \qquad \text{Find } \alpha/2.$$
$$t_{\alpha/2} = t_{0.05} = 1.7341 \qquad \text{Use Table V in the Appendix, df = 18, to find the critical value.}$$

STEP 2   Use Equation 12.12.

$$(\hat{\beta}_0 + \hat{\beta}_1 x^*) \pm t_{\alpha/2}\, s \sqrt{1 + \frac{1}{n} + \frac{(x^* - \bar{x})^2}{S_{xx}}} \qquad \text{Use Equation 12.12.}$$

$$= (\hat{\beta}_0 + \hat{\beta}_1 x^*) \pm t_{0.02}\, s \sqrt{1 + \frac{1}{20} + \frac{(x^* - \bar{x})^2}{S_{xx}}} \qquad \text{Use the values of } \alpha \text{ and } n.$$

$$= [-0.107 + 0.0025244(175)]$$

$$\pm (1.7341)(0.1745)\sqrt{1 + \frac{1}{20} + \frac{(175 - 109.8)^2}{53041.8}} \qquad \begin{array}{l} \text{Use summary statistics,} \\ \text{values for } s \text{ and } t_{0.05}. \end{array}$$

$$= 0.3348 \pm 0.3217 \qquad \text{Simplify.}$$
$$= (0.0131, 0.6565) \qquad \text{Compute endpoints.}$$

(0.0131, 0.6565) is a 90% prediction interval for an observed value of uranium contamination in a 175-foot well.

Figure 12.40 shows a technology solution.

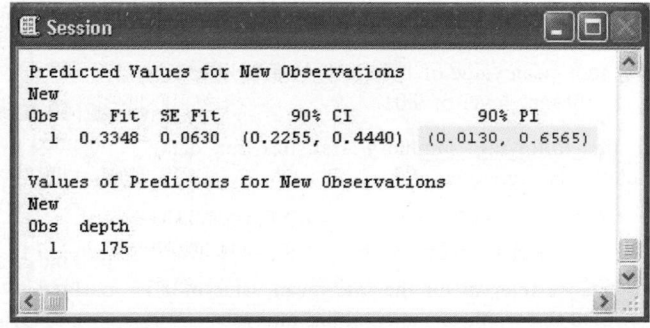

Figure 12.40 Minitab output: a 90% prediction interval (and a 90% confidence interval).

💡 **ILLUMINATING THE CONCEPTS**

Can you explain why the prediction interval is wider the farther *x** is from $\bar{x}$? Look carefully at Equation 12.12.

1. A confidence interval for the mean value of *Y* and a prediction interval for an observed value of *Y* (for *x* = *x**) are centered at the same value. Compare the endpoints for these two intervals in Equations 12.11 and 12.12. The only difference is a 1 underneath the radical in the prediction interval.

2. As *x** moves farther away from $\bar{x}$, the width of the corresponding prediction interval increases. ∎

TECHNOLOGY CORNER

Procedure: Construct a confidence interval for the mean value of Y and a prediction interval for an observed value of Y when $x = x^*$.
Reconsider: Examples 12.7 and 12.8, pages 588 and 591, solutions, and interpretations.

Minitab

Use regression options to construct a confidence interval and a prediction interval.

1. Enter the values for depth into column C1 and the values for uranium contamination into column C2.
2. Select Stat; Regression; Regression. Enter the Response column, C2, and the Predictor column, C1.
3. Select Options and enter 200 (175) under Prediction intervals for new observations.
4. Click OK. The regression analysis, confidence interval(s), and prediction interval(s) are displayed in a session window. See Figures 12.39, page 590, and 12.40, page 591.

Note: The TI-84 Plus and Excel do not have a built-in function for constructing a confidence interval for Y nor a prediction interval for y when $x = x^*$.

SECTION 12.3 EXERCISES

Practice

12.45 Consider the following summary statistics for data obtained in a simple linear regression analysis.

$\hat{\beta}_0 = -34.38$	$n = 18$	MSE $= 103.27$
$\hat{\beta}_1 = 3.38$	$\bar{x} = 15.367$	$S_{xx} = 138.14$

a. Conduct a hypothesis test to determine whether there is any evidence that the true mean value of Y for $x = 16.2$ is greater than 20. Use a significance level of 0.05.
b. Conduct a hypothesis test to determine whether there is any evidence that the true mean value of Y for $x = 11.5$ is different from 5. Use a significance level of 0.01.

12.46 Consider the following summary statistics for data obtained in a simple linear regression analysis.

$\hat{\beta}_0 = 38.86$	$n = 21$	MSE $= 21.23$
$\hat{\beta}_1 = -4.318$	$\bar{x} = 30.891$	$S_{xx} = 6.298$

a. Find a 95% confidence interval for the true mean value of Y when $x = 31.5$. Find the width of the resulting interval.
b. Find a 95% confidence interval for the true mean value of Y when $x = 31.9$. Find the width of the resulting interval.
c. Explain why the width of the confidence interval in part (b) is greater than the width of the confidence interval in part (a).

12.47 Consider the following summary statistics for data obtained in a simple linear regression analysis.

$\hat{\beta}_0 = 23.69$	$n = 7$	MSE $= 277.0$
$\hat{\beta}_1 = 1.452$	$\bar{x} = 19.1$	$S_{xx} = 15.131$

a. Find a 99% prediction interval for an observed value of Y when $x = 19.25$. Find the width of the resulting interval.
b. Find a 99% prediction interval for an observed value of Y when $x = 18.10$. Find the width of the resulting interval.
c. Explain why the width of the prediction interval in part (b) is greater than the width of the prediction interval in part (a).

12.48 Consider the following data obtained in a simple linear regression analysis.

x	5.6	5.1	7.6	3.9	6.5	6.7	5.1	2.5
y	5.7	2.8	10.5	1.6	7.7	8.8	3.4	6.1

a. Find the estimated regression line, and complete the summary ANOVA table. Conduct an F test for a significant regression.
b. Find s, an estimate of the standard deviation.
c. Conduct a hypothesis test to determine whether there is any evidence that the true mean value of Y for $x = 6$ is greater than 4. Use a significance level of 0.05.
d. Find a 99% confidence interval for the true mean value of Y when $x = 8.5$. Give an interpretation of this interval.

12.49 Consider the data on the data CD and book's web site obtained in a simple linear regression analysis.

a. Find the estimated regression line, and complete the summary ANOVA table. Conduct an F test for a significant regression. Use a significance level of 0.01.
b. Find s, an estimate of the standard deviation.
c. Find a 95% prediction interval for an observed value of Y when $x = 37$. On the basis of this interval, do you think it is likely that an observed value of Y (for $x = 37$) will be greater than 170? Justify your answer.

Applications

12.50 Psychology and Human Behavior The linguistics department at University of Massachusetts at Amherst recently studied the relationship between external facial movements and the acoustics of speech sounds. Facial motions were captured using infrared markers and summarized using a unitless measure (x). A special unidirectional microphone was used for acoustic recording and to measure the sound level of speech $(y,$ in dB). A random sample of individuals was selected, and each was asked to read a certain word. The facial expression and sound level were recorded for each person, and the summary statistics are as follows.

$\hat{\beta}_0 = 11.669$ $n = 16$ MSE = 65.42
$\hat{\beta}_1 = 28.009$ $\bar{x} = 0.4269$ $S_{xx} = 0.8615$

a. Find an estimate of the true mean decibel level for a facial expression value of $x = 0.40$.
b. Find a 95% confidence interval for the true mean decibel level for a facial expression value of $x = 0.40$.
c. Suppose the facial expression value is 0.55. Conduct a hypothesis test to determine whether there is any evidence that the true mean decibel level is greater than 30. Use a significance level of 0.01.

12.51 Physical Sciences A new solar collector is being tested for use in charging batteries that can provide electricity for an entire home. A random sample of days was selected and the amount of solar radiation was measured (x, in langleys) for each. The total battery charge was measured as a proportion (y, between 0 and 1). The summary statistics are as follows.

$\hat{\beta}_0 = 0.2007$ $n = 21$ MSE = 0.06135
$\hat{\beta}_1 = 0.00446$ $\bar{x} = 103.095$ $S_{xx} = 12335.8$

a. What proportion of a charge can you expect the batteries to take if the amount of solar radiation is 100 langleys?
b. Find a 95% confidence interval for the true mean battery charge proportion if the amount of solar radiation is 130 langleys.
c. A value of 80 langleys indicates a typical cloudy day. On such a day, is there any evidence to suggest that the true mean charge proportion is greater than 0.06 (the proportion needed to ensure that a home will have sufficient energy until the next day)? Use a significance level of 0.01.

12.52 Fuel Consumption and Cars An automobile mechanic believes that the weight of a tire is related to the overall diameter of the tire. A random sample of automobile tires was obtained and the diameter (x, inches) and weight (y, pounds) were measured for each.[14] The estimated regression line was $y = -37.8316 + 2.46657x$ with SSR = 87.1673, SSE = 258.8706, $\bar{x} = 25.3474$, and $S_{xx} = 14.3274$.

a. Complete the ANOVA summary table and conduct an F test for a significant regression. Explain the relationship between tire weight and tire diameter.
b. Find an estimate of the observed tire weight for a tire diameter of 25.2 inches.
c. Find a 99% prediction interval for an observed tire weight if the tire diameter is 24.8 inches.
d. Conduct the hypothesis test $H_0: \beta_0 = 0$ versus $H_a: \beta_0 \neq 0$ with significance level 0.01. Is there any evidence to suggest that the true regression line does not pass through the origin? Does this result make practical sense? Why or why not?

12.53 Biology and Environmental Science A study was conducted to investigate the relationship between asthma prevalence and annual rainfall in countries around the world. A random sample of 11 countries was obtained. The total annual rainfall (x, in mm) was recorded for each country, as well as the percentage of adults 22–44 years old who were treated for asthma (y). The following summary statistics were reported.

$S_{xx} = 178{,}661$ $S_{yy} = 49.2655$ $S_{xy} = 380.955$
$\bar{x} = 792.636$ $\bar{y} = 5.26$

a. Find the estimated regression line.
b. Complete the ANOVA summary table, and conduct an F test for a significant regression. Does annual rainfall help to explain the variation in asthma prevalence? Justify your answer.
c. Find a 95% prediction interval for an observed asthma prevalence if the total annual rainfall is 1000 mm. Comment on anything odd about this interval.

12.54 Economics and Finance Many factors contribute to a company's yearly financial statement. Some investors believe that yearly profit is strongly related to yearly revenue. A sample of worldwide companies was obtained, and the yearly revenue (x, in billions of dollars) and the yearly profit (y, in billions of dollars) in 2008 was recorded for each. The data are given in the following table.

Revenue	Profit	Revenue	Profit
378.799	12.731	201.516	12.649
164.877	5.467	159.229	3.617
122.644	8.861	106.444	5.063
96.758	5.234	89.630	8.874
88.364	1.931	87.879	3.209
84.740	4.395	82.096	2.916
79.322	5.565	77.682	3.235
76.330	2.637	67.205	9.511

(*Source:* CNNMoney.com, Global 500.)

a. Find the estimated regression line.
b. Find an estimate of the true mean profit for yearly revenue of 75 billion dollars.
c. Suppose the yearly revenue for a company is 90 billion dollars. Is there any evidence to suggest that the true mean profit is less than 5.5 billion dollars? Find the p value associated with this test.

12.55 Travel and Transportation Highway engineers have long argued that roads designed with high skid resistance help to prevent accidents, especially in wet conditions. A random sample of two-lane highways was selected from across the United States, and the skid resistance was measured (in skid numbers). The accident rate (per 10,000 vehicles) was computed for 25-mile sections of each highway, during wet conditions. The data are given on the data CD and book's web site.

a. Identify the independent variable and the dependent variable.
b. Construct a scatter plot for these data, and describe the relationship between the two variables.
c. Find the estimated regression line.
d. Suppose the skid resistance is 0.50. Conduct a hypothesis test to determine whether there is any evidence that the true mean accident rate is less than 0.60. Use a significance level of 0.05.

12.56 Business and Management A consumer group recently investigated the relationship between the cost for overseas telephone calls from New York and the distance between the cities. A random sample of foreign cities was obtained, and the distance from New York City (x, in miles) was computed for each. The cost per minute (y, in cents) using a prepaid phone card was also recorded. The data are given in the following table.

City	Distance	Cost
Bogota	2480	7
Warsaw	4261	5
Sao Paulo	4755	4
Moscow	4680	2
Beijing	6847	2
Copenhagen	3850	2
Tokyo	6760	6
New Delhi	7301	15
Bucharest	4770	10
Riyadh	6556	7
Caracas	2120	9
Cairo	5621	15

(*Source*: PennyTalk and indo.com.)

a. Find the estimated regression line. Explain the relevance of the sign on the estimate of β_1.

b. Conduct an F test for a significant regression. Find the p value associated with this test.

12.57 Psychology and Human Behavior Several studies suggest that the number of violent crimes is related to temperature.[15] Suppose the FBI investigated this relationship between temperature and the number of violent crimes in several large cities in the United States. A random sample of days was obtained, and the *average* temperature (x, in °F) and the number of violent crimes per 100,000 people (y) were recorded for each day. The data are given on the data CD and book's web site.

a. Construct a scatter plot for these data, and describe the relationship between the two variables.

b. Find the estimated regression line.

c. Find a 95% prediction interval for an observed number of violent crimes if the average temperature is 80 °F.

d. Find a 95% prediction interval for an observed number of violent crimes if the average temperature is 60 °F.

Extended Applications

12.58 Marketing and Consumer Behavior Lynne Kaminski, a marketing executive at Multidigital Record Company, believes that the number of electronic downloads of a newly released song can be used to predict the total album sales. A random sample of new songs and associated albums was selected. The number of electronic downloads during the first week of availability (x, in thousands) and the first month's album sales (y, in thousands) were recorded. Summary statistics are

$\hat{\beta}_0 = 22.95$	$n = 19$	MSE $= 1658.82$
$\hat{\beta}_1 = 2.9855$	$\bar{x} = 0.51$	$S_{xx} = 1131.07$

a. Find an estimate of the true mean number of album sales in the first month if there are 1000 downloads in the first week.

b. Suppose the number of downloads is 2500. Is there any evidence that the true mean number of album sales will exceed 35,000? Use a significance level of 0.05.

c. Construct a 99% confidence interval for the true mean number of album sales when $x = 1.5$ (1500 downloads). On the basis of this interval, do you think that the mean number of album sales is less than 25,000? Justify your answer.

12.59 Public Policy and Political Science A global study by Transparency International found that a country's environmental performance is related to political corruption. A random sample of 15 countries was selected. The Transparency International Corruption Perceptions Index (CPI, x), a measure of perceived corruption existing among public officials and politicians, was computed for each country. In addition, the Environmental Sustainability Index (ESI, y), a measure of overall progress toward environmental sustainability, was computed for each country. The estimated regression line was $y = 26.432 + 4.546x$, with SSR $= 853.50$, SSE $= 1234.23$, $\bar{x} = 5.4333$, and $S_{xx} = 41.2933$.

a. Complete the ANOVA summary table, and conduct an F test for a significant regression. Explain the relationship between CPI and ESI.

b. Find an estimate of the observed ESI for a CPI score of 6.7.

c. Find a 95% prediction interval for an observed ESI if the CPI is 8.2. On the basis of this interval, do you think a randomly selected country with CPI 8.2 will have an ESI of 90 or greater? Justify your answer.

12.60 Economics and Finance Most people shopping for a diamond ring believe that the price is related to the size of the stone. A sample of ladies' diamond rings was obtained. The size of the diamond (x, in carats) and the price (y, in dollars) was recorded for each. The data are given on the data CD and book's web site.[16]

a. Construct a scatter plot for these data, and describe the relationship between the two variables.

b. Find the estimated regression line. Does the estimate of β_0 seem reasonable? Why or why not?

c. Conduct an F test for a significant regression. Find the p value associated with this test.

d. Find a 99% confidence interval for the true mean price of a diamond ring if the size is 0.40 carats.

e. Using the confidence interval in part (d), is there any evidence that the mean price is different from $750 (for a 0.40-carat ring)?

12.61 Medicine and Clinical Studies Golden Rule, a medical insurance company, recently investigated the relationship between the number of patients per registered nurse in a hospital and the patient's length of stay. A random sample of hospitals was selected, and the number of patients per registered nurse was computed (x). A patient was randomly selected for each hospital, and the length of stay was recorded (y, in hours). The data are given on the data CD and book's web site.

a. Construct a scatter plot for these data, and describe the relationship between the two variables.

b. Find the estimated regression line.

c. Find a 99% prediction interval for an observed length of stay if the number of patients per nurse is 3.7.

d. Find a 99% prediction interval for an observed length of stay if the number of patients per nurse is 3.3.

e. Why is the prediction interval in part (c) wider than the prediction interval in part (d)?

12.4   REGRESSION DIAGNOSTICS

Recall, the assumptions in a simple linear regression model are stated in terms of the random deviations, $E_i, i = 1, 2, \ldots, n$. It is assumed that the E_i's are independent, normal random variables with mean 0 and (constant) variance σ^2. If any one of these assumptions is violated, then the results and subsequent inferences are in doubt.

If the true regression line were known, the set of actual random errors,

$$
\begin{aligned}
e_1 &= y_1 - (\beta_0 + \beta_1 x_1) \\
e_2 &= y_2 - (\beta_0 + \beta_1 x_2) \\
&\vdots \\
e_n &= y_n - (\beta_0 + \beta_1 x_n)
\end{aligned}
$$

could be computed and used to check the assumptions. However, since we usually do not know the values for β_0 and β_1, the **residuals**, or deviations from the estimated regression line,

$$
\begin{aligned}
\hat{e}_1 &= y_1 - (\hat{\beta}_0 + \hat{\beta}_1 x_1) \\
\hat{e}_2 &= y_2 - (\hat{\beta}_0 + \hat{\beta}_1 x_2) \\
&\vdots \\
\hat{e}_n &= y_n - (\hat{\beta}_0 + \hat{\beta}_1 x_n)
\end{aligned}
$$

are used to check for assumption violations. In practice, these estimates of the random errors are used in a variety of diagnostic checks. This section presents several preliminary graphical procedures used to reveal assumption violations.

Recall from Section 6.3, a normal probability plot is a scatter plot of each observation versus its corresponding expected value from a Z distribution, or normal score. For observations from a normal distribution, the points will fall along a straight line. The data axis can be horizontal or vertical. If the scatter plot is nonlinear, there is evidence to suggest that the data did not come from a normal distribution.

A normal probability plot of the residuals may be used to check the normality assumption. A simple histogram or stem-and-leaf plot may also be used to reveal departures from normality.

A scatter plot of the residuals versus the independent variable values is also used to check the simple linear regression assumptions. The ordered pairs in this plot are $(x_i, \hat{e}_i)$. Figure 12.41 illustrates the definition of a residual and a scatter plot of the residuals versus the independent variable. The first graph is a scatter plot of y versus x with the estimated regression line. Next to this is a scatter plot of the resulting residuals $\hat{e}_i$ versus x_i. If there are no violations in assumptions, this scatter plot should look like a horizontal band around zero with randomly distributed points and no discernible pattern. See Figure 12.42. There should be no relation between the residuals and the predictor variable.

Multiple linear regression is used if an additional predictor variable is necessary. This statistical technique is a natural extension of the simple linear regression model and is presented in Section 12.5.

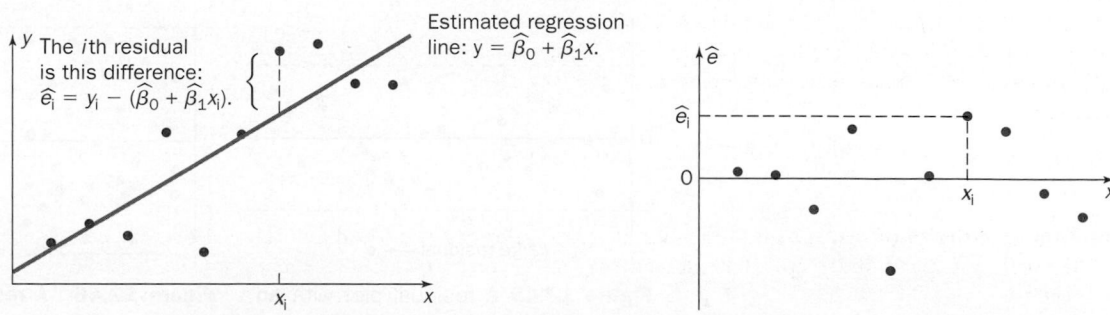

Figure 12.41 An illustration of the definition of a residual and a scatter plot of residuals versus the independent variable.

Figure 12.42 A desirable residual plot—no discernible pattern in a horizontal band.

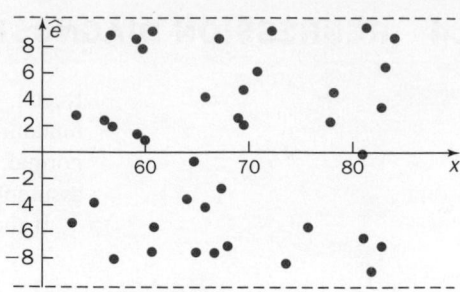

Here is a list of *patterns* in a residual plot that indicate a possible violation in assumptions.

1. Check for a distinct *curve* in the plot, either mound- or bowl-shaped (parabolic). See Figure 12.43. A curved plot suggests that an additional (or different) predictor variable may be necessary. A *linear* model for this graph is not appropriate.

2. Check for a *nonconstant* spread. See Figure 12.44. If there is not a uniform horizontal band, or if the spread of the residuals varies outside this band, this suggests that the variance is not constant.

3. Check for any *unusually large* (in magnitude) residual (Figure 12.45). This suggests that one observation is very far away from the rest. The data may have been recorded or entered incorrectly. Often, the offending point is omitted and a new estimated regression line is computed.

4. Check for any *outliers* (Figure 12.46). If the observation is correct, an outlying residual suggests that one observation has an unusually large influence on the estimated regression line. This point is also often omitted, and a new line is computed.

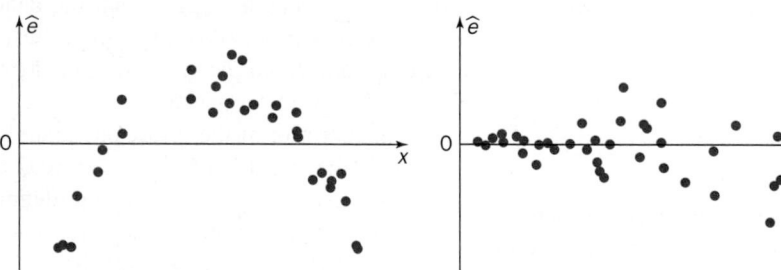

Figure 12.43 A *curved* residual plot. This suggests that a linear model is not appropriate.

Figure 12.44 A residual plot with *nonconstant* spread. This suggests that the variance is not the same for each value of *x*.

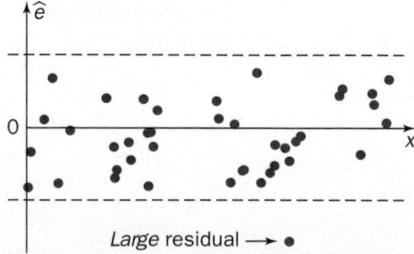

Figure 12.45 A residual plot with an *unusually large* (negative) residual. This suggests that an observation is far away from the rest.

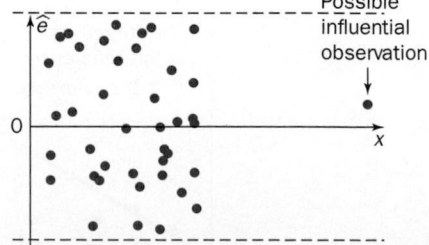

Figure 12.46 A residual plot with an *outlying* residual. This observation is very influential when the estimated regression line is computed.

(Ryan Pike/FeaturePics)

(Percho/Dreamstime)

Reminder: The residuals should be placed in order from smallest to largest when the normal scores are assigned.

Example 12.9 Pillbugs and Red Clover Isopods, or pillbugs, use moss, ferns, and even poison ivy for shelter. Some scientists believe that the preferred natural shelter is red clover, and that the density of pillbugs can be predicted from the density of red clover. A random sample of fields was obtained. The number of red clover plants per square meter (x) and the number of pillbugs per square meter (y) were measured for each field. The data are given below, and the estimated regression line is $y = -12.360 + 14.213x$. Compute the residuals, construct a normal probability plot of the residuals, and carefully sketch a graph of the residuals versus the predictor variable values.

SOLUTION

STEP 1   Use the estimated regression line to find the predicted value, $\hat{y}_i$, for each x value: $\hat{y}_i = -12.360 + 14.213x_i$. Then, compute each residual: $\hat{e}_i = y_i - \hat{y}_i$. The normal scores for $n = 20$ are given in Table IV in the Appendix.

x_i	y_i	$\hat{y}_i$	$\hat{e}_i$	Normal score
5.9	108	71.5	36.5	0.92
4.4	12	50.2	-38.2	-0.92
3.6	86	38.8	47.2	1.13
9.1	136	117.0	19.0	0.45
8.8	103	112.7	-9.7	-0.19
6.3	99	77.2	21.8	0.59
11.7	136	153.9	-17.9	-0.31
10.9	94	142.6	-48.6	-1.13
6.3	9	77.2	-68.2	-1.87
5.7	145	68.7	76.3	1.87
5.0	4	58.7	-54.7	-1.40
8.4	124	107.0	17.0	0.31
4.5	68	51.6	16.4	0.19
13.3	177	176.7	0.3	0.06
4.9	91	57.3	33.7	0.74
14.2	259	189.5	69.5	1.40
10.4	101	135.5	-34.5	-0.74
10.2	101	132.6	-31.6	-0.45
5.0	56	58.7	-2.7	-0.06
3.6	7	38.8	-31.8	-0.59

STEP 2   The normal probability plot of the residuals is shown in Figure 12.47. The points lie along an approximate straight line. There is no evidence to suggest that the normality assumption is violated.

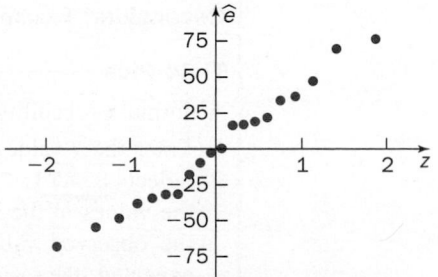

Figure 12.47 The normal probability plot of the residuals.

STEP 3 Figure 12.48 shows the residual plot ($\hat{e}_i$ versus x_i). There is no discernible pattern in this graph. One point in the upper right, corresponding to $x = 14.2$, may be an outlier, but the graphical evidence is not convincing enough. These two graphs suggest that there is no violation in the simple linear regression assumptions.

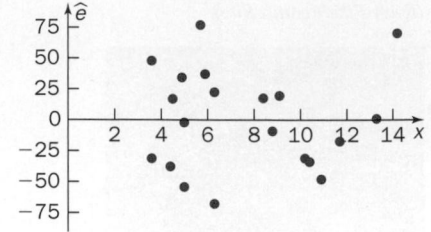

Figure 12.48 The residual plot of residuals versus predictor variable values.

Figures 12.49 and 12.50 together show a technology solution.

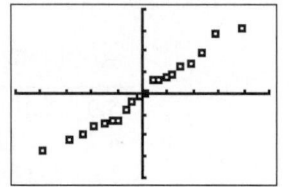

Figure 12.49 Normal probability plot of the residuals.

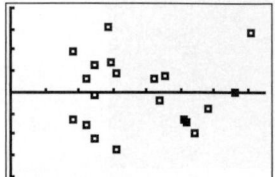

Figure 12.50 Scatter plot of the residuals versus the predictor variable values.

 ILLUMINATING THE CONCEPTS

1. The sum of the residuals is always 0, subject to round-off error. We'll check a few examples empirically.

2. Other diagnostic plots include a plot of the residuals versus the predicted values $\hat{y}_i$. Any distinct pattern suggests a violation in the simple linear regression assumptions.

3. Other types of residuals are also used to check for violations in assumptions. **Standardized residuals** utilize the standard deviation of each residual and are useful for identifying residuals with large magnitudes. **Studentized residuals** are also standardized but by using a model without the current observation. ■

TECHNOLOGY CORNER

Procedure: Construct a normal probability plot of the residuals and a scatter plot of the residuals versus the predictor variable values.
Reconsider: Example 12.9, page 597, solution, and interpretations.

TI-84 Plus

A normal probability plot and a scatter plot are built-in statistical plots.

1. Enter the x values into list L1 and the observed y values into list L2.
2. Select $\boxed{\text{STAT}}$; CALC; LinReg(a+bx). Enter the name of the list containing the values of the independent variable, L1, and the name of the list containing the observed values of the dependent variable, L2. When this command is executed, the residuals are automatically stored in the named list RESID.
3. Construct a normal probability plot for the residuals (in the list RESID as described in Section 6.3). See Figure 12.49.

4. Construct a scatter plot of the residuals versus the predictor variable values as described in Section 12.1. See Figure 12.50.

Minitab

Use the built-in function NSCORES to compute the normal scores. Construct a scatter plot of the residuals versus the normal scores. Construct a scatter plot of the residuals versus the predictor variable values.

1. Enter the x values into column C1 and the observed y values into column C2.
2. Select Stat; Regression; Regression. Enter the Response and Predictor columns.
3. Select Storage and check Diagnostic Measures; Residuals. The residuals will be stored in a column labeled RESIn.
4. Click OK to conduct the regression analysis and store the residuals.
5. Construct a normal probability plot for the residuals as described in Section 6.3. See Figure 12.51.

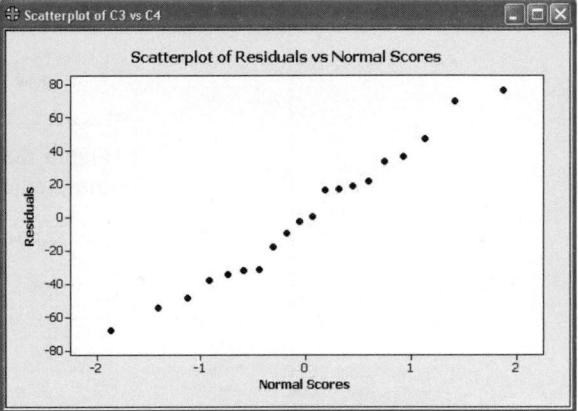

Figure 12.51 Normal probability plot of the residuals.

6. Construct a scatter plot of the residuals versus the predictor variable values as described in Section 12.1. See Figure 12.52.

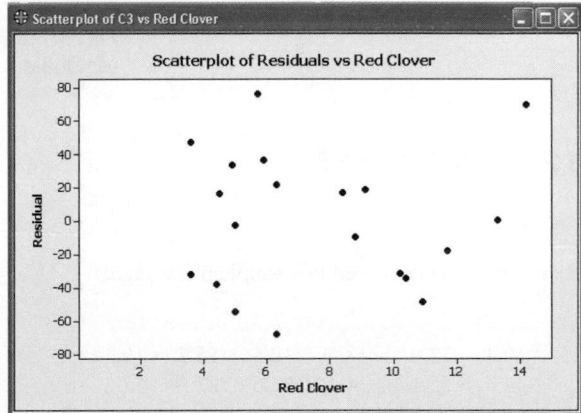

Figure 12.52 Scatter plot of the residuals versus the predictor variable values.

Excel

Compute the normal scores for the residuals as described in Section 6.3. Construct scatter plots as described in Section 12.1.

1. Enter the x values into column A and the observed y values into column B.
2. Under the Data tab, select Data Analysis; Regression. Enter the Y Range, the X Range, and an Output Range. Check Residuals to save the residuals and

In the Graphs tab, there are options to automatically save the residuals and construct a histogram of the residuals, a normal probability plot of the residuals, a scatter plot of the residuals versus the predicted values, and a scatter plot of the residuals versus any predictor variable.

Residual Plots to construct a plot of the residuals versus the predictor variable values.

3. Click OK. The scatter plot of the residuals versus predictor variable values is shown in Figure 12.53.

4. Compute the normal scores and construct the normal probability plot as described in Section 6.3. See Figure 12.54.

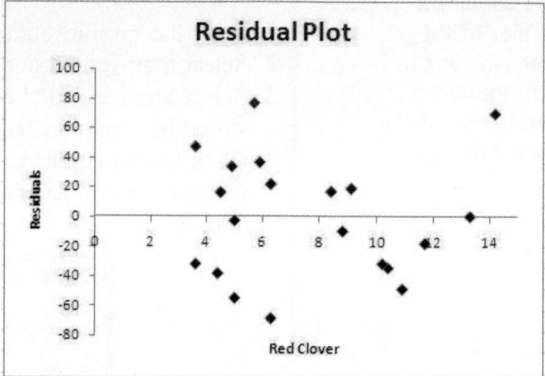

Figure 12.53 Scatter plot of the residuals versus the predictor variable values.

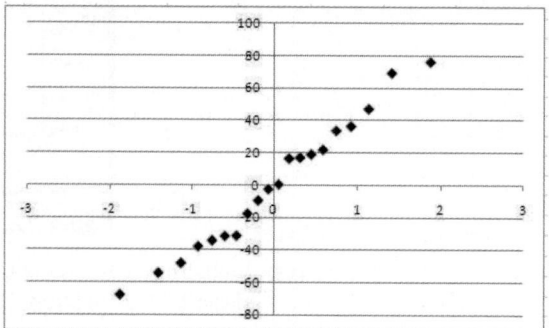

Figure 12.54 Normal probability plot of the residuals.

SECTION 12.4 EXERCISES

Practice

12.62 Consider the following data used in a simple linear regression analysis.

x	11.9	11.3	10.4	12.4	18.5	12.5	15.7	10.2
y	45.2	33.1	34.4	25.1	−4.4	37.6	27.5	52.6

The estimated regression line is $y = 97.63 - 5.15x$.
a. Find the residuals.
b. Find the sum of the residuals.

12.63 For the data on the data CD and book's web site, the estimated regression line is $y = 4.7245 + 3.3717x$.
a. Find the residuals.
b. Find the sum of the residuals.

12.64 Examine each normal probability plot of residuals. Is there any evidence to suggest that the random error terms are not normal?

a.

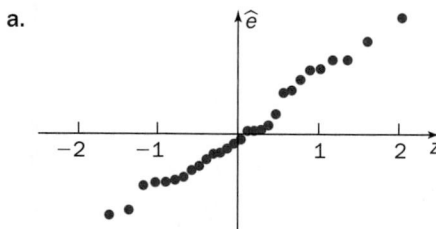

b.

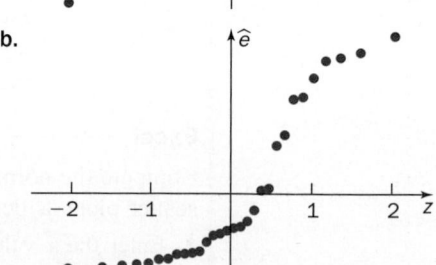

c.

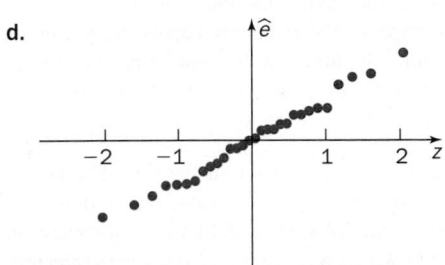

d.

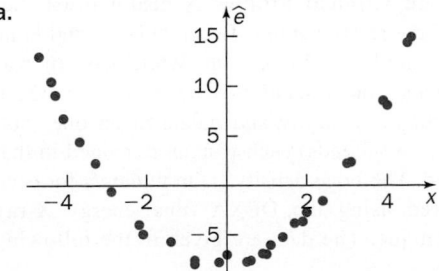

c.

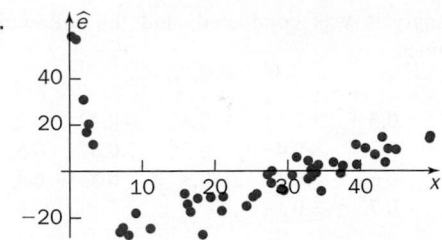

d.

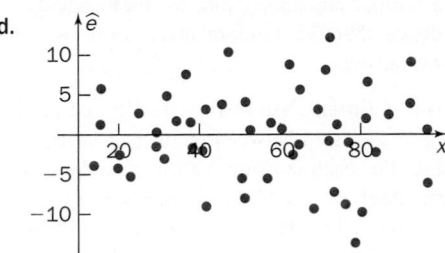

12.65 The following residuals were obtained in a simple linear regression analysis.

−22.2	−67.9	−1.1	−6.3	−12.4	−33.0	−24.2
35.0	−13.2	33.7	42.3	14.2	13.6	−10.7
−1.6	23.7	38.0	0.4	−20.3	12.1	

a. Construct a normal probability plot for these residuals.
b. Is there any evidence to suggest that the random errors are not normal? Justify your answer.

12.66 The residuals on the data CD and book's web site were obtained in a simple linear regression analysis.
a. Construct a normal probability plot for these residuals.
b. Is there any evidence to suggest that the random errors are not normal? Justify your answer.

12.67 Examine each residual plot (residuals versus predictor variable). Is there any evidence to suggest a violation in the simple linear regression model assumptions? If the graph suggests a violation, indicate which assumption is in doubt.

a.

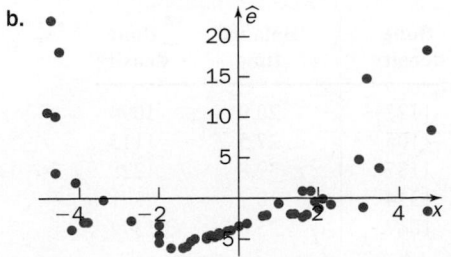

b.

12.68 The data on the data CD and book's web site were used in a simple linear regression analysis. The estimated regression line is $y = 91.74 − 3.7384x$.
a. Find the residuals.
b. Carefully sketch a graph of the residuals versus the predictor variable, x. Is there any evidence of a violation in the regression assumptions? Justify your answer.

12.69 The data on the data CD and book's web site were used in a simple linear regression analysis.
a. Find the estimated regression line.
b. Compute the residuals.
c. Carefully sketch a graph of the residuals versus the predictor variable, x. Is there any evidence of a violation in the regression assumptions? Justify your answer.

Applications

12.70 Physical Sciences The Atlantic Elevator Company conducted a study to examine the relationship between the total weight of passengers on an elevator (x, in pounds) and the total energy (y, in kW) required to lift the passengers one floor in an office building. A simple linear regression analysis was conducted, and the following residuals were obtained.

−1.2	−1.1	1.5	−8.5	−13.1	−3.3	−17.4
7.5	−6.0	8.2	15.6	6.9	−4.3	−19.2
18.9	−9.4	16.3	0.8	−3.0	10.9	

a. Carefully sketch a normal probability plot for the residuals.
b. Is there any evidence that the random error terms are not normal? Justify your answer.

12.71 Manufacturing and Product Development Suppose the research department for Century Tire Company is developing a snow tire based on a new formulation. A study was conducted to examine the relationship between the amount of the new additive per liter (x, in mL) added in the first stage and the amount of wear after 50,000 indoor test miles (y, in mm). A simple linear regression analysis was conducted, and the following residuals were obtained.

0.7	−0.2	0.5	−1.1	−0.8	−0.2	0.2
−0.1	0.7	0.3	−0.4	0.5	0.3	0.5
0.8	0.2	0.6	0.1	−0.8	−0.3	0.4
−0.7	−0.4	−0.7	−0.1			

a. Carefully sketch a normal probability plot for the residuals.
b. Is there any evidence that the random error terms are not normal? Justify your answer.

12.72 Medicine and Clinical Studies A family counselor believes that there is a relationship between number of years married and blood pressure. For each married man in a random sample, the number of years married (x) and the systolic blood pressure (y, in mmHg) were recorded. The data are given in the following table.

Years married	Blood pressure	Years married	Blood pressure
4.9	133	3.4	94
10.3	157	3.1	114
11.9	129	12.0	144
13.0	141	12.9	147
13.0	166	7.1	124

The estimated regression line is $y = 97.95 + 4.034x$.
a. Compute the residuals, and verify that they sum to 0.
b. Carefully sketch a normal probability plot for the residuals. Is there any evidence that the random error terms are not normal? Justify your answer.

12.73 Medicine and Clinical Studies Some research suggests that the size of a person's brain is related to the person's intelligence.[17] To investigate this theory, suppose a random sample of adults was obtained. Magnetic resonance imaging (MRI) was used to determine the brain size of each subject (x, in pixels) and a full-scale IQ score was derived from four Wechsler subtests (y). The data are given on the data CD and book's web site.
a. Find the estimated regression line and compute the residuals.
b. Carefully sketch a normal probability plot of the residuals and a plot of the residuals versus the predictor variable.
c. Do these graphs provide any evidence that the simple linear regression assumptions are invalid? Justify your answer.

12.74 Biology and Environmental Science Agriculture researchers recently investigated the relationship between wheat yield and wind speed. A random sample of wheat-field plots was obtained. The mean daily wind speed over the growing period (x, in m/s) was recorded for each field, along with the yield (y, in bushels per acre). The data are given on the data CD and book's web site. The estimated regression line is $y = 50.052 − 1.103x$.
a. Compute the residuals, and verify that they sum to 0.
b. Carefully sketch a graph of the residuals versus the predictor variable (wind speed). Is there any evidence that the simple linear regression assumptions are violated? Justify your answer.

12.75 Education and Child Development Some academic research involving predominantly undergraduate institutions suggests that grant activity is related to the number of students who go on to graduate school. A random sample of undergraduate colleges was obtained, and the total amount of grant awards (x, in thousands of dollars) for 2009 was recorded. In addition, the number of students per thousand (y) who continued on to graduate school was computed. The data are given on the data CD and book's web site. The estimated regression line is $y = 13.151 + 0.014839x$.
a. Compute the residuals, and verify that they sum to 0.
b. Carefully sketch a graph of the residuals versus the predictor variable (grant amount). Is there any evidence that the simple linear regression assumptions are violated? Justify your answer.

12.76 Physical Sciences The solar wind, or the flow of charged particles from the Sun, affects communications, navigation, and the Earth's static pressure (the force exerted by the atmosphere on the surface of the Earth). A random sample of days was selected, and the solar wind density was measured (x, in protons/cm^3) using a special NASA satellite. For each day selected, the static pressure (y, in MPa) was also recorded. The data are given in the following table.

Solar wind density	Static pressure	Solar wind density	Static pressure
5.6	0.099	9.9	0.117
7.4	0.080	9.4	0.110
8.0	0.115	11.0	0.138
8.7	0.110	6.7	0.104
10.1	0.110	6.9	0.094

a. Find the estimated regression line, and compute the residuals.
b. Carefully sketch a normal probability plot of the residuals and a plot of the residuals versus the predictor variable.
c. Do these graphs provide any evidence that the simple linear regression assumptions are invalid? Justify your answer.

12.77 Medicine and Clinical Studies A health researcher recently investigated the relationship between balance and bone density. A random sample of 25-year-old white women was obtained. Each woman took a standing balance test in which she held her arms out horizontally and balanced on one foot. The length of time (x, in seconds) each woman remained in that position was recorded. The bone density (y, in mg/cm^2) for each woman was measured using the DEXA (dual-energy X-ray absorptiometry) technique. The data are given in the following table.

Balance time	Bone density	Balance time	Bone density
29.8	1127	20.9	1094
26.3	1105	27.5	1115
26.0	1187	39.8	1228
37.2	1334	10.7	871
30.1	1067	35.6	1377

a. Find the estimated regression line, and compute the residuals.

b. Carefully sketch a normal probability plot of the residuals and a plot of the residuals versus the predictor variable.

c. Do these graphs provide any evidence that the simple linear regression assumptions are invalid? Justify your answer.

12.78 Medicine and Clinical Studies A coal miner's job is extremely hazardous and can cause serious health problems, for example, black lung disease and vibration-induced white finger. A study was conducted to examine the hearing loss of coal miners exposed to years of vibration and cool temperatures underground. A random sample of coal miners was selected, and the length of time on the job (x, in years) was recorded for each person. The hearing threshold level (HTL, y, in dB) of each miner was measured. The data are given on the data CD and book's web site.

a. Find the estimated regression line, and compute the residuals.

b. Carefully sketch a normal probability plot of the residuals and a plot of the residuals versus the predictor variable.

c. Do these graphs provide any evidence that the simple linear regression assumptions are invalid? Justify your answer. How do you think this regression model could be improved?

Extended Applications

12.79 Sports and Leisure There have been many studies to determine how golfers can produce long drives. A recent investigation considered grip strength. A random sample of golfers was obtained and a hand-grip dynamometer was used to measure the grip strength (x, in newtons) of each person. Each golfer then drove a ball, and the length of the drive was measured (y, in yards). The data are given on the data CD and book's web site.

a. Find the estimated regression line, and compute the residuals.

b. Conduct an F test for a significant regression. Use a significance level of 0.01.

c. Carefully sketch a normal probability plot of the residuals and a plot of the residuals versus the predictor variable.

d. Do these graphs provide any evidence that the simple linear regression assumptions are invalid? Justify your answer. How do you think this regression model could be improved?

12.80 Business and Management Suppose the Human Resources director at NVR, a large construction company, believes that the number of sick hours per year taken by an employee is related to the commuting distance. A random sample of employees was obtained, and each person's travel distance to work (x, in miles) was recorded. Personnel records were used to obtain the number of hours off for sickness in a year (y) for each employee involved in the study. The data are given on the data CD and book's web site.

a. Find the estimated regression line, and compute the residuals.

b. Conduct an F test for a significant regression. Use a significance level of 0.10. Do you believe that there is a relationship between commuting distance and sick hours? Why or why not?

c. Carefully sketch a normal probability plot of the residuals and a plot of the residuals versus the predictor variable.

d. Do these graphs provide any evidence that the simple linear regression assumptions are invalid? Justify your answer.

12.81 Physical Sciences Researchers at a marine institute investigated the relationship between the dissolved barium concentration and the silicate/nitrate ratio in the North Pacific Ocean. The silicate/nitrate ratio (x) and the dissolved barium concentration (y, in nmol/kg) were measured at randomly selected locations. The data are given on the data CD and book's web site.

a. Find the estimated regression line, and compute the residuals.

b. Conduct an F test for a significant regression. Use a significance level of 0.001.

c. Carefully sketch a normal probability plot of the residuals and a plot of the residuals versus the predictor variable.

d. Do these graphs provide any evidence that the simple linear regression assumptions are invalid? Justify your answer.

12.82 Public Health and Nutrition Although many people buy fast food because it is cheap, quick, and heavily advertised, these meals can contain a large amount of sodium, or salt. Some research suggests that the amount of sodium is linearly related to the total calories. A sample of Wendy's sandwiches was obtained, and the total calories (x) and sodium (y, in grams) for one serving was measured. The data are given in the following table.

x	y	x	y	x	y
220	490	260	690	300	730
310	670	360	810	220	490
260	700	430	870	700	1440
960	2010	830	1880	320	950
440	1300	430	1120	540	1360
330	680	310	630	260	760
320	860	320	960		

(*Source*: Wendy's U.S. Nutrition Information as of November 2008.)

a. Find the estimated regression line, and compute the residuals.

b. Conduct an F test for a significant regression. Find bounds on the p value associated with this test.

c. Carefully sketch a plot of the residuals versus the predictor variable. Does this graph provide any evidence that the simple linear regression assumptions are invalid? Justify your answer.

Challenge

12.83 Sum of the Residuals In a simple linear regression analysis, show that the sum of the residuals is always 0. *Hint*: In the definition of the ith residual, use the definition of $\hat{y}_i$, and then the formula for $\hat{\beta}_0$.

12.5   MULTIPLE LINEAR REGRESSION

Although simple linear regression analysis has practical uses, many real-world applications involve a model with a dependent variable Y and at least two independent variables, $x_1, x_2, \ldots, x_k$. For example, the number of days it takes to complete construction of a new home (Y) might be predicted by the square footage of the home (x_1) and the total linear feet of wiring in the home (x_2). The temperature of an automobile engine (Y) might be modeled (predicted) by the amount of coolant (x_1), the amount of engine oil (x_2), and the rate of air flow (x_3). The purpose of this section is to extend the simple linear regression model to those involving k (≥ 2) predictor variables. The formal model and assumptions are given below. The procedure is similar to that used for the simple linear regression model.

MULTIPLE LINEAR REGRESSION MODEL

Let $(x_{11}, x_{21}, \ldots, x_{k1}, y_1), (x_{12}, x_{22}, \ldots, x_{k2}, y_2), \ldots, (x_{1n}, x_{2n}, \ldots, x_{kn}, y_n)$ be n sets of observations such that y_i is an observed value of the random variable Y_i. We assume that there exist constants $\beta_0, \beta_1, \ldots, \beta_k$ such that

$$Y_i = \beta_0 + \beta_1 x_{1i} + \beta_2 x_{2i} + \cdots + \beta_k x_{ki} + E_i \qquad (12.13)$$

where $E_1, E_2, \ldots, E_n$ are independent, normal random variables with mean 0 and variance σ^2. The following assumptions apply:

1. The E_i's are normally distributed (which means that the Y_i's are normally distributed).
2. The expected value of E_i is 0 (which implies that $E(Y_i) = \beta_0 + \beta_1 x_{1i} + \beta_2 x_{2i} + \cdots + \beta_k x_{ki}$).
3. $\text{Var}(E_i) = \sigma^2$ (which implies that $\text{Var}(Y_i) = \sigma^2$).
4. The E_i's are independent (which implies that the Y_i's are independent).

ILLUMINATING THE CONCEPTS

1. The double subscript notation on x is necessary in order to indicate both the variable and the observation. For example, x_{21} is the value of the variable x_2 that corresponds to the observed value of y_1. Similarly, x_{12} is the value of the variable x_1 that corresponds to the observed value of y_2.

2. In the multiple linear regression model, the E_i's again represent the **random deviations** or **random error terms**.

3. The **true regression equation** is $y = \beta_0 + \beta_1 x_1 + \beta_2 x_2 + \cdots + \beta_k x_k$. Notice that this equation is a *linear* function of the unknown parameters $\beta_0, \beta_1, \ldots, \beta_k$. The graph of the true regression equation is, in general, a *surface*, not a line. However, we often refer to this equation as the true regression *line*.

4. The unknown constants $\beta_0, \beta_1, \ldots, \beta_k$ are called partial regression coefficients. Recall that in Section 12.1, β_1 represented the mean amount of change in y for every one-unit increase in x_1. In a multiple linear regression model, β_i represents the mean change in y for every increase of one unit in x_i if the values of all the other predictor variables are kept fixed. ∎

This section focuses on the method for finding the best deterministic linear model, that is, finding estimates of the unknown parameters $\beta_0, \beta_1, \ldots, \beta_k$. Hypothesis tests will be used to determine whether the overall model explains a significant amount of the variability in the dependent variable and to evaluate the contribution of each independent variable. The principle of least squares can be used again to minimize the sum of the squared deviations between the observations and the estimated values (the sum of squares due to error).

Although hand calculations are possible, it is much more efficient to use technology in order to compute the estimates $\hat{\beta}_0, \hat{\beta}_1, \ldots, \hat{\beta}_k$ for the true regression parameters, or coefficients, $\beta_0, \beta_1, \ldots, \beta_k$. For specific values of the independent variables, $(x_1^*, x_2^*, \ldots, x_k^*) = \boldsymbol{x}^*$, let $y^* = \hat{\beta}_0 + \hat{\beta}_1 x_1^* + \hat{\beta}_2 x_2^* + \cdots + \hat{\beta}_k x_k^*$. The value y^* is an estimate of the mean value of Y for $\boldsymbol{x} = \boldsymbol{x}^*$, denoted $\mu_{Y|\boldsymbol{x}^*}$, and also an estimate of an observed value of Y for $\boldsymbol{x} = \boldsymbol{x}^*$.

Solution Trail

KEYWORDS

- Estimated regression equation.
- True mean words per minute.

↓

TRANSLATION

- Multiple linear regression.
- Find the expected number of words per minute for 68 °F and 55%.

↓

CONCEPTS

- Least-squares estimates.
- Estimate of the mean value of Y for $x = x^*$.

↓

VISION

Use technology to find the estimates of the regression coefficients. Compute $y^* = \hat{\beta}_0 + \hat{\beta}_1(68) + \hat{\beta}_2(55)$ as an estimate of the mean value of Y for $x = (68, 55)$.

Example 12.10 Worker Productivity An efficiency expert claims that productivity in a closed office environment is directly affected by the temperature and the relative humidity. In order to test this theory, a random sample of office workers at an insurance company was obtained and each was asked to type a certain part of an automobile policy. The words per minute (WPM), the office temperature (in °F), and relative humidity (as a percentage) for each person are given in the following table.

WPM	52	56	69	58	61	65	51	63	63	67
Temperature	74	77	69	73	72	71	76	73	74	72
Relative humidity	51	45	42	45	45	47	41	48	44	45

a. Find the estimated regression equation.
b. Estimate the true mean words per minute for a temperature of 68 °F and a relative humidity of 55%.

SOLUTION

a. Words per minute (y) is the dependent, or response, variable. Temperature (x_1) and relative humidity (x_2) are the independent, or predictor, variables. Use technology to find the estimated regression coefficients. The results from Minitab are shown in Figure 12.55.

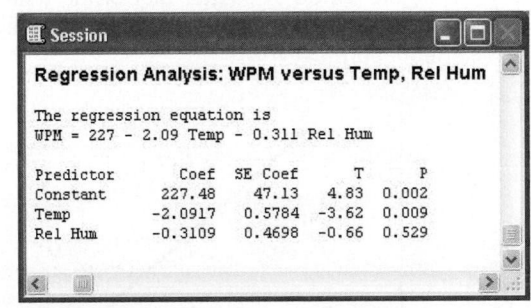

Figure 12.55 Estimated regression coefficients.

The sign of $\hat{\beta}_i$ represents the way in which the surface is tilted along the x_i axis.

The estimated regression equation is $y = 227.48 - 2.0917x_1 - 0.3109x_2$. That is, $\hat{\beta}_0 = 227.48, \hat{\beta}_1 = -2.0917$, and $\hat{\beta}_2 = -0.3109$. Note that the estimated regression coefficient on temperature is negative. This suggests that as temperature increases, words per minute, or productivity, decreases. The estimated regression coefficient on relative humidity is also negative. This suggests that as relative humidity increases, words per minute decreases. The signs on both estimated regression coefficients seem reasonable.

b. The estimated true mean words per minute is found by substituting $x_1 = x_1^* = 68$ and $x_2 = x_2^* = 55$ into the estimated regression equation.

$$y = 227.48 - 2.0917x_1 - 0.3109x_2 \quad \text{Estimated regression equation.}$$
$$= 227.48 - 2.0917(68) - 0.3109(55) \quad \text{Substitute.}$$
$$= 68.1449 \quad \text{Simplify.}$$

An estimate for the expected number of words per minute for a temperature of 68 °F and 55% relative humidity is approximately 68.

In this example, the graph of $y = 227.48 - 2.0917x_1 - 0.3109x_2$ is a plane in three dimensions. Figure 12.56 is a three-dimensional scatter plot of the data and the graph of the estimated regression equation. Note that the remaining points are behind the plane and are not visible in this particular view. Figures 12.57, 12.58, and 12.59 illustrate how y depends on each predictor variable separately. In addition to the graph of the estimated regression equation, Figure 12.57 includes the graphs of two lines (in three dimensions), one for $x_1 = 73$ held constant, and one for $x_2 = 47$ held constant. Figures 12.58 and 12.59 each represent a *slice* through the three-dimensional plane at a particular value of the other predictor variable. In Figure 12.58, $\hat{\beta}_1 = -2.0917$ is the slope of the line when $x_2 = 47$. And in Figure 12.59, $\hat{\beta}_2 = -0.3109$ is the slope of the line when $x_1 = 73$.

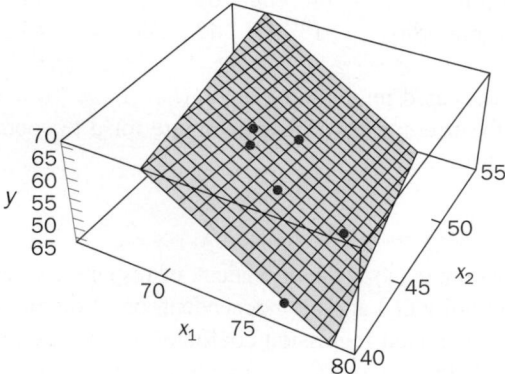

Figure 12.56 A scatter plot of the data and the graph of the estimated regression equation.

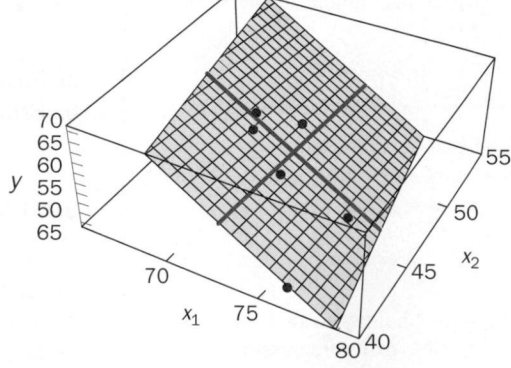

Figure 12.57 A scatter plot of the data, the graph of the estimated regression equation, and two lines.

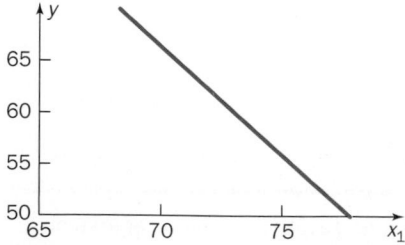

Figure 12.58 A graph of y versus x_1 when $x_2 = 47$.

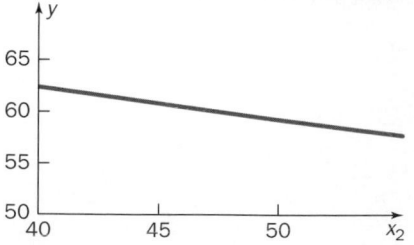

Figure 12.59 A graph of y versus x_2 when $x_1 = 73$.

The following concepts from simple linear regression also apply to multiple linear regression.

1. The variance σ^2 is a measure of the underlying variability in the model. An estimate of σ^2 is used to conduct hypothesis tests and to construct confidence intervals related to multiple linear regression.

2. The ith predicted value is $\hat{y}_i = \hat{\beta}_0 + \hat{\beta}_1 x_{1i} + \cdots + \hat{\beta}_k x_{ki}$.

3. The ith residual is $y_i - \hat{y}_i$.

4. The total sum of squares, the sum of squares due to regression, and the sum of squares due to error have the same definitions. The sum of squares identity, Equation 12.4, is also true. The total variation in the data (SST) is a sum of the variation explained by the model (SSR) and the variation about the regression equation (SSE).

An analysis of variance table is used to summarize the computations in multiple linear regression. The differences are in the degrees of freedom associated with the sum of squares due to regression and the sum of squares due to error. As usual, the mean squares are corresponding sums of squares divided by the corresponding degrees of freedom.

ANOVA summary table for multiple linear regression

Source of variation	Sum of squares	Degrees of freedom	Mean square	F	p value
Regression	SSR	k	$MSR = \dfrac{SSR}{k}$	$\dfrac{MSR}{MSE}$	p
Error	SSE	$n - k - 1$	$MSE = \dfrac{SSE}{n - k - 1}$		
Total	SST	$n - 1$			

The coefficient of determination, $r^2 = SSR/SST$, is a measure of the proportion of variation in the data that is explained by the regression model. For multiple linear regression, a test of significance is equivalent to testing the hypothesis that all regression coefficients (except the constant term) are zero. That is, the null hypothesis is $H_0: \beta_1 = \beta_2 = \cdots = \beta_k = 0$; none of the predictor variables helps to explain any variation in the dependent variable. Suppose $k = 2$ and H_0 is true. Graphically, this means that the values of Y vary around the plane (in three dimensions) $y = \beta_0$ with no discernible pattern.

A test for a significant multiple linear regression model is based on the ratio of the mean square due to regression and the mean square due to error. Here is a summary of an F test for a significant regression with significance level α.

This is the model utility test to determine whether the multiple linear regression model is appropriate/useful (similar to the simple linear regression case).

HYPOTHESIS TEST FOR A SIGNIFICANT MULTIPLE LINEAR REGRESSION

$H_0: \beta_1 = \beta_2 = \cdots = \beta_k = 0$
   (None of the predictor variables helps to predict y.)

$H_a: \beta_i \neq 0$ for at least one i
   (At least one predictor variable helps to predict y.)

TS: $F = \dfrac{MSR}{MSE}$

RR: $F \geq F_\alpha$ df k and $n - k - 1$

Solution Trail

KEYWORDS

- The effects of...
- *F* test.
- r^2

↓

TRANSLATION

- Three predictor variables.
- *F* test for a significant regression.
- Coefficient of determination.

↓

CONCEPTS

- Multiple linear regression; summary ANOVA table.
- Hypothesis test for a significant regression.
- Computation of r^2.

↓

VISION

Use the sums of squares, number of predictor variables, and number of observations to complete the ANOVA summary table for multiple linear regression. Use the template for a hypothesis test for a significant regression with $\alpha = 0.05$. The sum of squares due to regression and the total sum of squares are used to compute r^2.

Example 12.11 Price at the Pump A study was conducted to investigate the effects of the cost of crude oil (x_1, in dollars per barrel), the state tax on gasoline (x_2, in dollars per gallon), and the number of publicly owned automobiles in a state (x_3, in thousands) on the price of gasoline (in dollars per gallon). Data from 29 randomly selected days were used to produce the following multiple linear regression equation:

$$y = 2.59 + 0.008x_1 + 0.191x_2 + 0.0146x_3.$$

In addition, SSR = 0.3517 and SSE = 0.5085.

a. Complete the summary ANOVA table and conduct an *F* test for a significant regression. Use a significance level of 0.05.

b. Compute r^2 and interpret this value.

SOLUTION

a. Use the sum of squares identity to compute the total sum of squares.

SST = SSR + SSE = 0.3517 + 0.5085 = 0.8602   Use Equation 12.4.

There are $n = 29$ observations and $k = 3$ predictor variables. Use these values to compute the mean squares.

MSR = SSR/k = 0.3517/3 = 0.1172
MSE = SSE/$(n - k - 1)$ = SSE/$(29 - 3 - 1)$ = 0.5085/25 = 0.0203

The *F* test for a significant regression with $\alpha = 0.05$ is

$H_0: \beta_1 = \beta_2 = \beta_3 = 0$
$H_a: \beta_1 \neq 0$ for at least i
TS: $F = \dfrac{\text{MSR}}{\text{MSE}}$
RR: $F \geq F_\alpha = F_{0.05} = 2.99$ df 3 and 25.

The value of the test statistic is

$f = \text{MSR/MSE} = 0.1172/0.0203 = 5.7734 \; (\geq 2.99)$

Since *f* lies in the rejection region, there is evidence to suggest that at least one of the regression coefficients is different from 0. At least one of the predictor variables can be used to explain a significant amount of variation in the price per gallon of gasoline. The next reasonable step is to determine which of the three predictors is really contributing to this overall significant regression.

Recall, we can only use the tables of critical values for *F* distributions to bound the *p* value. Place the value of the test statistic in an ordered list of critical values with $\nu_1 = 3$ and $\nu_2 = 25$.

$$4.68 \; \leq 5.7734 \leq 7.45$$
$$F_{0.01} \; \leq 5.7734 \leq F_{0.001}$$
Therefore, $0.001 \leq \quad p \quad \leq 0.01$

Here are all the calculations presented in the ANOVA table, with some help from technology to find the exact *p* value.

p value illustration:
$p = P(X \geq 5.57734)$
$= 0.0038 \leq 0.05 = \alpha$

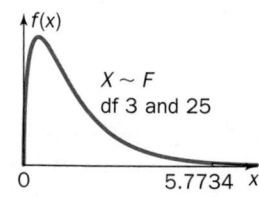

$X \sim F$
df 3 and 25

0   5.7734 *x*

ANOVA summary table for multiple linear regression

Source of variation	Sum of squares	Degrees of freedom	Mean square	F	p value
Regression	0.3517	3	0.1172	5.7734	0.0038
Error	0.5085	25	0.0203		
Total	0.8602	28			

b. The coefficient of determination is

$$r^2 = \text{SSR/SST} = 0.3517/0.8602 = 0.4089$$

Approximately 0.4089, or 41%, of the variation in the price per gallon of gasoline is explained by this regression model. ●

If the overall multiple linear regression F test is significant, then there is evidence to suggest that at least one of the independent variables can be used to predict the value of Y. Hypothesis tests (or confidence intervals) can be used in order to determine whether x_i helps to predict the value of $Y|\boldsymbol{x}$, or equivalently, whether β_i is different from 0.

Consider the random variable B_i, an estimator for β_i, and the statistic $S^2 = \text{MSE} = \text{SSE}/(n - k - 1)$, an estimator for the underlying variance σ^2. If the multiple linear regression assumptions are true, then S^2 is an unbiased estimator for σ^2 and is used in constructing a hypothesis test concerning β_i.

β_{i0} is the hypothesized value of β_i. β_{i0} can be any constant. We can conduct a test for evidence that the regression coefficient β_i is different from any value. However, usually $\beta_{i0} = 0$ and $H_a\colon \beta_i \neq 0$. This means we are looking for any evidence to suggest that the value of x_i helps to predict the value of $Y|x$.

HYPOTHESIS TEST AND CONFIDENCE INTERVAL CONCERNING β_i

$H_0\colon \beta_i = \beta_{i0}$

$H_a\colon \beta_i > \beta_{i0}, \quad \beta_i < \beta_{i0}, \quad \text{or} \quad \beta_i \neq \beta_{i0}$

$\text{TS: } T = \dfrac{B_i - \beta_{i0}}{S_{B_i}}$

$\text{RR: } T \geq -t_\alpha, \quad T \leq -t_\alpha, \quad \text{or} \quad |T| \geq t_{\alpha/2} \quad \text{df} = n - k - 1$

A $100(1 - \alpha)\%$ confidence interval for β_i has as endpoints the values

$$\widehat{\beta}_i \pm t_{\alpha/2}\, s_{B_i} \tag{12.14}$$

with the critical value based on $n - k - 1$ degrees of freedom.

The estimate for the standard deviation of each B_i, s_{B_i}, is reported in the multiple linear regression output of most statistical software packages. In fact, most regression routines automatically present calculations associated with $k + 1$ hypothesis tests concerning $\beta_0, \beta_1, \ldots, \beta_k$. The null hypothesis in each case is $H_0\colon \beta_i = 0$, and the associated test statistic and the p value are usually given.

Example 12.12 Diesel Engine Exhaust In a recent study of heavy-duty diesel engine exhaust emissions, a random sample of city buses in Atlanta was obtained and the composite nitrogen oxide and nonmethane hydrocarbon emission was measured for each (y, in grams per brake horsepower hour, g/bhp-hr). In addition, the curb weight (x_1, in thousands of pounds) and engine size (x_2, in liters) were also measured. Multiple linear regression was used to investigate the effects of curb weight and engine size on exhaust emissions. The resulting (Minitab) output is shown in Figure 12.60.

a. Verify that the multiple linear regression is significant at the $\alpha = 0.01$ level.

b. Conduct separate hypothesis tests to determine whether each predictor variable contributes to the overall significant regression. Use $\alpha = 0.05$ in each test.

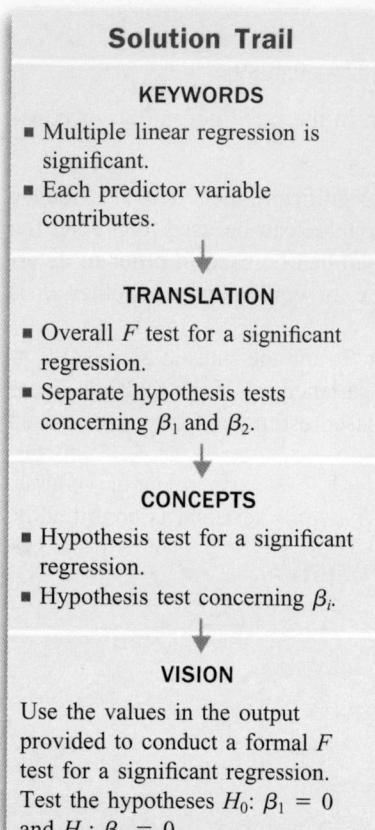

Solution Trail

KEYWORDS

- Multiple linear regression is significant.
- Each predictor variable contributes.

↓

TRANSLATION

- Overall F test for a significant regression.
- Separate hypothesis tests concerning β_1 and β_2.

↓

CONCEPTS

- Hypothesis test for a significant regression.
- Hypothesis test concerning β_i.

↓

VISION

Use the values in the output provided to conduct a formal F test for a significant regression. Test the hypotheses H_0: $\beta_1 = 0$ and H_0: $\beta_2 = 0$.

Notice that Minitab rounds the value of the test statistic, f, to two decimal places.

Notice again that Minitab rounds the value of the test statistic, t, to two decimal places.

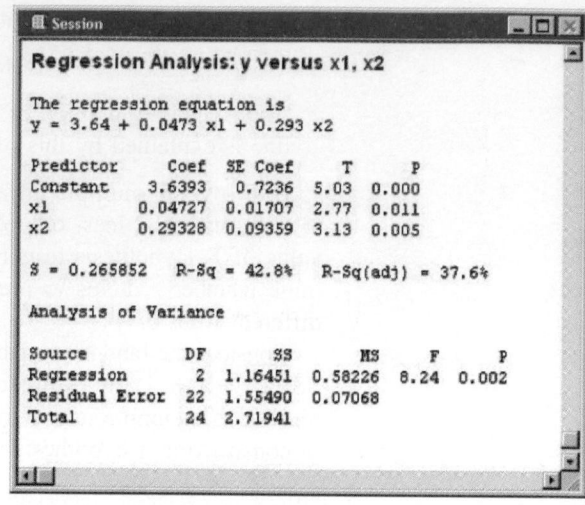

Figure 12.60 Minitab regression analysis.

SOLUTION

a. There are $k = 2$ predictor variables and, from the ANOVA table, there are $n = 25$ observations ($n - 1 = 24$).

The F test for a significant regression with $\alpha = 0.05$ is

H_0: $\beta_1 = \beta_2 = 0$.

H_a: $\beta_i \neq 0$ for at least one i

TS: $F = \text{MSR}/\text{MSE}$

RR: $F \geq F_\alpha = F_{0.01} = 5.72$ df 2 and 22.

As given in the ANOVA table, the value of the test statistic is

$f = \text{MSR}/\text{MSE} = 0.58226/0.07068 = 8.238 \ (\geq 5.72)$

Since f lies in the rejection region (or equivalently, $p = 0.002 \leq 0.01$), there is evidence to suggest that at least one of the regression coefficients is different from 0.

b. In order to test whether x_1, curb weight, is a significant predictor, conduct a hypothesis test concerning the regression coefficient β_1.

H_0: $\beta_1 = 0$

H_a: $\beta_1 \neq 0$

TS: $T = \dfrac{B_1 - 0}{S_{B_i}}$

RR: $|T| \geq t_{\alpha/2} = t_{0.025} = 2.0739$ df = 22.

Using the Minitab output, the value of the test statistic is

$t = \dfrac{\hat{\beta}_1 - 0}{s_{B_1}} = \dfrac{0.04727}{0.01707} = 2.7692$

The value of the test statistic lies in the rejection region, $|t| = |2.7692| = 2.7692 \geq 2.0739$ (equivalently, $p = 0.011 \leq 0.05$). There is evidence to suggest that the regression coefficient β_1 is different from 0. This suggests that the predictor variable curb weight is contributing to the overall regression effect and is contributing significantly to the variability in y.

Conduct a similar hypothesis test concerning the regression coefficient β_2.

$H_0: \beta_2 = 0$

$H_a: \beta_2 \neq 0$

TS: $T = \dfrac{B_2 - 0}{S_{B_2}}$

RR: $|T| \geq t_{\alpha/2} = t_{0.025} = 2.0739$ df = 22.

Using the Minitab output, the value of the test statistic is

$t = \dfrac{\hat{\beta}_2 - 0}{S_{B_2}} = \dfrac{0.29328}{0.09359} = 3.1337$

The value of the test statistic lies in the rejection region, $|t| = |3.1337| = 3.1337 \geq 2.0739$ (equivalently, $p = 0.005 \leq 0.05$). There is evidence to suggest that the regression coefficient β_2 is different from 0. This implies that predictor variable engine size is also contributing to the overall regression effect and is contributing significantly to the variability in y.

ILLUMINATING THE CONCEPTS

Recall, if a confidence interval for β_i contains 0, there is no evidence to suggest that the ith predictor variable is significant. However, if 0 is not included in the confidence interval, there is evidence to suggest that the ith regression coefficient is different from 0.

1. The Minitab output in Example 12.12 also includes the results of a hypothesis test concerning the constant term, β_0, with $H_0: \beta_0 = 0$. If we fail to reject this null hypothesis, a model without a constant term might be more appropriate.

2. If a model utility test is significant in a multiple linear regression model, then there are $k \geq 2$ hypothesis tests to consider in order to isolate those variables contributing to the overall effect. Recall from Chapter 11, we cannot set the significance level in each *individual* hypothesis test. The probability α of making a Type I error (a mistake) is set in each test under the assumption that *only one* test is conducted per experiment. Therefore, the more tests we do, the greater the chance is of making an error. In order to control the probability of making at least one mistake, we can use the Bonferroni technique (presented in Chapter 11) applied to simultaneous hypothesis tests or confidence intervals.
   a. If k hypothesis tests are conducted, then the significance level in each case is α/k.
   b. The k simultaneous $100(1 - \alpha)\%$ confidence intervals have as endpoints the values $\hat{\beta}_i \pm t_{\alpha/(2k)} s_{B_i}$, with the critical value based on $n - k - 1$ degrees of freedom.

Consider doing additional reading about a partial F test, stepwise, and best subsets regression. A good text is *Applied Linear Regression Models-Revised Edition with Student CD*, 4th Edition, by Kutner, M., Nachtsheim, C., and Neter, J.

3. There are many different statistical procedures to select the *best* regression model. For now, the most reasonable method is to simply keep only those variables in the model that have regression coefficients significantly different from 0. Eliminate the others, and calculate a new, *reduced* model for prediction. ∎

Two very useful inferences in multiple linear regression involve an estimate of the *mean* value of Y for $x = x^*$ and an estimate of an *observed* value of Y for $x = x^*$. Recall from the simple linear regression case, the error in estimating the mean value of Y is less than the error in estimating an observed value of Y. The difference is due to estimating a single value versus estimating the next value from an entire distribution.

If the multiple linear regression assumptions are true, then a hypothesis test and confidence interval concerning the mean value of Y for $x = x^*$ is based on the *t* distribution. Calculating the standard deviation by hand is taxing. An appropriate symbol is used below and we will rely on technology to provide the necessary calculations. The null and alternative hypotheses are stated in terms of the parameter $y^* = \beta_0 + \beta_1 x_1^* + \beta_2 x_2^* + \cdots + \beta_k x_k^*$. The random variable Y^* is used as an estimate of y^*.

HYPOTHESIS TEST AND CONFIDENCE INTERVAL CONCERNING THE MEAN VALUE OF Y FOR x = x*

Recall, we can conduct a test of $y^* = y_0^*$ using either a formal hypothesis test or a confidence interval. Most statistical software packages produce a confidence interval rather than conducting a hypothesis test.

H_0: $y^* = y_0^*$

H_a: $y^* > y_0^*$, $y^* < y_0^*$,   or $y^* \neq y_0^*$

TS: $T = \dfrac{(B_0 + B_1 x_1^* + \cdots + B_k x_k^*) - y_0^*}{S_{Y*}}$

RR: $T \geq t_\alpha$,   or $T \leq -t_\alpha$, $|T| \geq t_{\alpha/2}$ df $= n - k - 1$

A $100(1 - \alpha)\%$ confidence interval for $\mu_{Y|x*}$, the mean value of Y for $x = x^*$, has as endpoints the values

$$(\hat{\beta}_0 + \hat{\beta}_1 x_1^* + \cdots + \hat{\beta}_k x_k^*) \pm t_{\alpha/2}\, s_{Y*}. \tag{12.15}$$

with the critical value based on $n - k - 1$ degrees of freedom.

PREDICTION INTERVAL FOR AN OBSERVED VALUE OF Y

Recall, $s = \sqrt{\text{MSE}}$.

A $100(1 - \alpha)\%$ prediction interval for an observed value of Y when $x = x^*$ has as endpoints the values

$$(\hat{\beta}_0 + \hat{\beta}_1 x_1^* + \cdots + \hat{\beta}_k x_k^*) \pm t_{\alpha/2} \sqrt{s^2 + s_{Y*}^2}. \tag{12.16}$$

with the critical value based on $n - k - 1$ degrees of freedom.

Example 12.13 More Administrators The National Center for Education Statistics conducts an annual survey of public elementary and secondary schools in the United States. Data from the school year 2002–2003 (without Guam and the Bureau of Indian Affairs) were used to fit a multiple linear regression equation of the form $y = \beta_0 + \beta_1 x_1 + \beta_2 x_2 + \beta_3 x_3$ where

y = School administrators, in thousands, SCHADM.
x_1 = Total full-time-equivalent teachers, in thousands, TOTTCH.
x_2 = Administrative support staff, in thousands, LEASUP.
x_3 = Total students, in thousands, MEMBER.

a. Construct a 95% confidence interval for the mean number of school administrators when $x_1 = 70$, $x_2 = 3$, and $x_3 = 900$. Use this confidence interval to determine whether there any evidence that the mean number of school administrators for these values is different from 4.5.

b. Construct a 95% prediction interval for an observed value of school administrators when $x_1 = 70$, $x_2 = 3$, and $x_3 = 900$.

SOLUTION

a. Using Minitab, the Stat; Regression; Regression; Options menu is used to enter the specific values for x_1, x_2, and x_3, and the confidence level. The relevant output is shown in Figure 12.61.

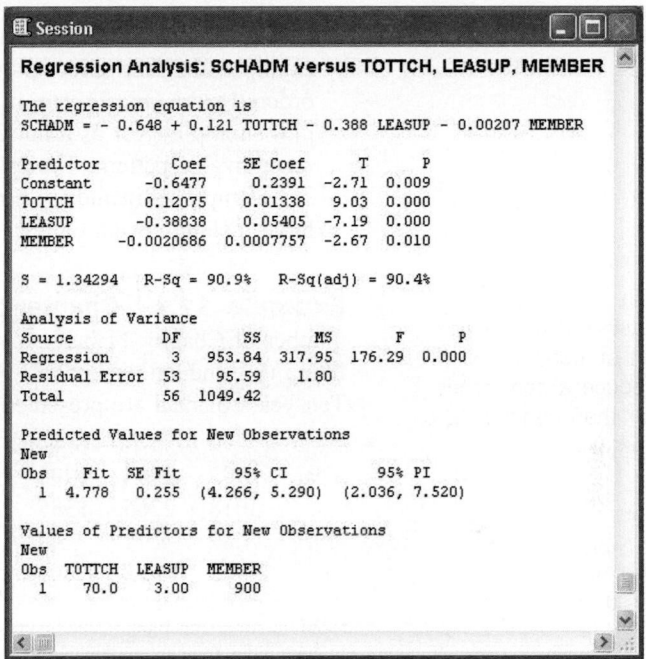

```
Session

Regression Analysis: SCHADM versus TOTTCH, LEASUP, MEMBER

The regression equation is
SCHADM = - 0.648 + 0.121 TOTTCH - 0.388 LEASUP - 0.00207 MEMBER

Predictor       Coef    SE Coef      T      P
Constant      -0.6477    0.2391   -2.71  0.009
TOTTCH        0.12075   0.01338    9.03  0.000
LEASUP       -0.38838   0.05405   -7.19  0.000
MEMBER     -0.0020686  0.0007757  -2.67  0.010

S = 1.34294   R-Sq = 90.9%   R-Sq(adj) = 90.4%

Analysis of Variance
Source            DF       SS      MS       F      P
Regression         3   953.84  317.95  176.29  0.000
Residual Error    53    95.59    1.80
Total             56  1049.42

Predicted Values for New Observations
New
Obs    Fit  SE Fit      95% CI           95% PI
  1  4.778   0.255  (4.266, 5.290)  (2.036, 7.520)

Values of Predictors for New Observations
New
Obs  TOTTCH  LEASUP  MEMBER
  1    70.0    3.00     900
```

Figure 12.61 Minitab regression analysis output including a confidence interval and a prediction interval.

The estimated regression equation is

$$y = -0.648 + 0.121x_1 - 0.388x_2 - 0.00207x_3$$

The model utility test (using the ANOVA table) is significant at the $p < 0.001$ level, and each predictor variable is also significant. These results together with $r^2 = 0.909$ suggest that the model is very effective in predicting the total number of school administrators.

Using the Minitab output, a 95% confidence interval for the true mean number of school administrators when $x_1 = 70$, $x_2 = 3$, and $x_3 = 900$ is (4.266, 5.290). Since 4.5 is included in this interval, there is no evidence to suggest that the mean number of school administrators is different from 4.5.

b. Using the Minitab output, a 95% prediction interval for a single observation of the number of school administrators when $x_1 = 70$, $x_2 = 3$, and $x_3 = 900$ is (2.036, 7.520). Notice that this 95% prediction interval is larger (wider) than the corresponding confidence interval. ●

A residual, $\hat{e}_i$, is the difference between the observed value of Y_i and the predicted value for Y_i.

The assumptions in a multiple linear regression model are also given in terms of the random deviations, E_i, $i = 1, 2, \ldots, n$. As in a simple linear regression model, it is assumed that the E_i's are independent, normal random variables with mean 0 and constant variance σ^2. If any of these assumptions are violated, then the resulting analysis is unreliable. The residuals, $\hat{e}_i = y_i - (\hat{\beta}_0 + \hat{\beta}_1 x_{1i} + \cdots + \hat{\beta}_k x_{ki})$, $i = 1$, $2, \ldots, n$, are estimates of the random errors and are used to check the regression assumptions.

The following graphical procedures, developed for a simple linear regression model, can be extended to a multiple linear regression model.

1. Construct a histogram, stem-and-leaf plot, scatter plot, and/or normal probability plot of the residuals. These graphs are all used to check the normality assumption.

2. Construct a scatter plot of the residuals versus *each* independent variable. For example, the first scatter plot has the ordered pairs $(x_{1i}, \hat{e}_i)$, the second has the ordered pairs $(x_{2i}, \hat{e}_i)$, etc. If there are no violations in assumptions, each scatter plot should appear as a horizontal band around 0. There should be no recognizable pattern. Typical patterns in a residual plot that suggest a violation in assumptions include a distinct curve, a nonconstant spread, an unusually large residual, or an outlier.

Example 12.14 Chapter 11 Bankruptcy   An economist believes that the number of Chapter 11 bankruptcies (y) in the United States is related to the Misery Index (x_1, unemployment rate + inflation rate) and the consumer price index (x_2). Ten years of data are presented in the following table.

Year	1993	1994	1995	1996	1997	1998	1999	2000	2001	2002
y_i	3018	2265	1385	1197	1093	888	722	700	789	997
x_{1i}	9.87	8.71	8.40	8.34	7.28	6.05	6.41	7.35	7.59	7.37
x_{2i}	144.5	148.2	152.4	156.9	160.5	163.0	166.6	172.2	177.1	179.9

(*Source*: American Bankruptcy Institute, NPA Services, Inc., and U.S. Bureau of Labor Statistics.)

The estimated regression equation is $y = 2668 + 387.1x_1 - 26.88x_2$ as shown in Figure 12.62. Compute the residuals and construct a normal probability plot of the residuals. Sketch a graph of the residuals versus each predictor variable and discuss any indication of violations in regression assumptions.

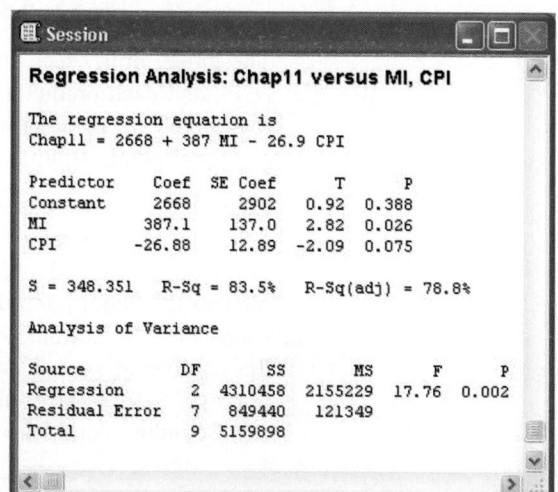

Figure 12.62 Minitab regression analysis.

SOLUTION

STEP 1   Use the estimated regression line to find the predicted value, $\hat{y}_i$, for each $x = (x_1, x_2)$ value. Compute each residual ($\hat{e}_i = y_i - \hat{y}_i$) and use the normal scores for $n = 10$ given in Table IV in the Appendix.

x_{1i}	x_{2i}	y_i	$\hat{y}_i$	$\hat{e}_i$	Normal score
9.87	144.5	3018	2605.0	413.0	1.55
8.71	148.2	2265	2056.5	208.5	0.38
8.40	152.4	1385	1823.6	−438.6	−1.00
8.34	156.9	1197	1679.4	−482.4	−1.55
7.28	160.5	1093	1172.3	−79.3	−0.38
6.05	163.0	888	628.9	259.1	0.66
6.41	166.6	722	671.5	50.5	0.12
7.35	172.2	700	884.9	−184.9	−0.66
7.59	177.1	789	846.1	−57.1	−0.12
7.37	179.9	997	685.7	311.3	1.00

STEP 2   The normal probability plot of the residuals is shown in Figure 12.63. The points lie along an approximate straight line. There is no overwhelming evidence to suggest that the normality assumption is violated.

STEP 3   Figures 12.64 and 12.65 show the residual plots: $\hat{e}_i$ versus x_{1i}, and $\hat{e}_i$ versus x_{2i}. In the first plot there is some indication that the spread of the residuals increases as x_1 increases. However, with only $n = 10$ observations, the graphical evidence is not convincing enough. These two graphs suggest that there is no violation in the multiple linear regression model assumptions.

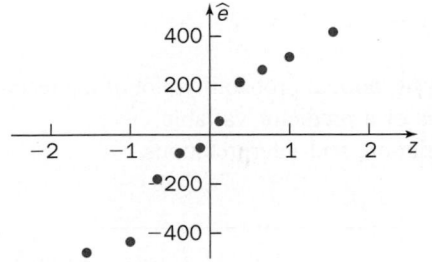

Figure 12.63 The normal probability plot of the residuals.

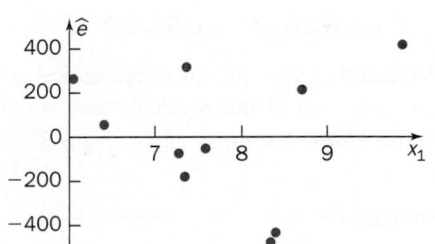

Figure 12.64 The plot of residuals versus the first predictor variable.

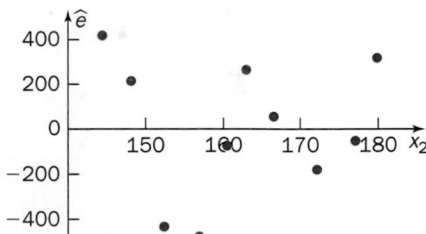

Figure 12.65 The plot of residuals versus the second predictor variable.

Figures 12.66 through 12.68 together show a technology solution.

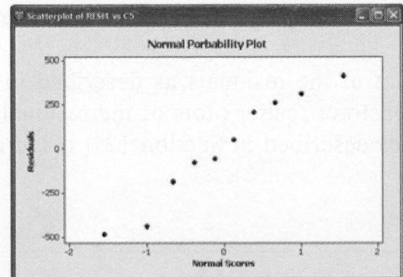

Figure 12.66 Normal probability plot.

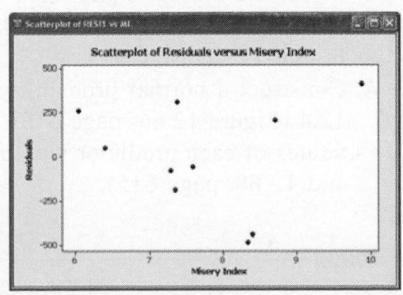

Figure 12.67 Plot of the residuals versus the first predictor.

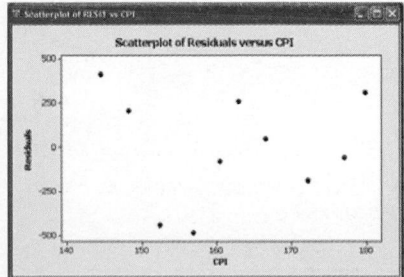

Figure 12.68 Plot of the residuals versus the second predictor.

Many other possible models could explain the variation in a dependent variable y. For example, a quadratic model with one predictor variable has the form

$$Y_i = \beta_0 + \beta_1 x_i + \beta_2 x_i^2 + E_i.$$

A more general kth degree polynomial model with one predictor has the form

$$Y_i = \beta_0 + \beta_1 x_i + \beta_2 x_i^2 + \cdots + \beta_k x_i^k + E_i.$$

And a model with two predictor variables and an interaction term has the form

$$Y_i = \beta_0 + \beta_1 x_{1i} + \beta_2 x_{2i} + \beta_3 x_{1i} x_{2i} + E_i.$$

These models are still considered *linear* regression models because each is a linear combination of the regression coefficients—that is, a sum of terms of the form $\beta_i v_i$, where v_i is a function of one or more predictor variables.

There are other models that do not appear to be linear, but are *intrinsically* linear. These models can be transformed into linear models and the techniques presented in this chapter can be used to estimate the regression coefficients. For example, the exponential model

$$Y_i = \beta_0 e^{\beta_1 x_i} E_i$$

is intrinsically linear. The general growth model

$$Y_i = \beta_0 + \beta_1 * e^{\beta_2 x_i} + E_i$$

cannot be made into a linear model. The power model and the general logistic model are other common models. No matter what model you decide is best, you should use technology to estimate the unknown parameters.

TECHNOLOGY CORNER

Procedure: Multiple linear regression analysis, normal probability plot of the residuals, and plot of the residuals versus values of a predictor variable.
Reconsider: Example 12.14, page 614, solution, and interpretations.

Minitab

There are several ways to conduct multiple linear regression analysis in Minitab, in a session window, and in the <u>Stat</u>; <u>Regression</u> menu.

1. Enter values of the dependent variable into column C1 and values for the predictor variables into columns C2 and C3.
2. Select <u>Stat</u>; <u>Regression</u>; <u>Regression</u>. Enter the Response column, C1, and the Predictor columns, C2 and C3.
3. Select the Storage button and check Residuals in order to store the residuals in a Worksheet column. Click OK to display the regression analysis. See Figure 12.62, page 614.
4. Construct a normal probability plot of the residuals as described in Section 12.4 (Figure 12.66, page 615). Construct scatter plots of the residuals versus values of each predictor variable as described in Section 12.4 (Figures 12.67 and 12.68, page 615).

Excel

Regression analysis is part of the Data Analysis tool pack.

1. Enter values of the dependent variable into column A and values for the predictor variables into columns B and C.

2. Under the Data tab, select Data Analysis; Regression. Enter the Y Range, X Range, and Output Range. Check Residuals to save the residuals and Residual Plots to construct a plot of the residuals versus the predictor variable values. The summary output is shown in Figure 12.69.

SUMMARY OUTPUT

Regression Statistics	
Multiple R	0.91399
R Square	0.83538
Adjusted R Square	0.78834
Standard Error	348.35129
Observations	10

ANOVA

	df	SS	MS	F	Significance F
Regression	2	4310458.1	2155229.0	17.76	0.0018
Residual	7	849440.3	121348.6		
Total	9	5159898.4			

	Coefficients	Standard Error	t Stat	P-value	Lower 95%	Upper 95%	Lower 95.0%	Upper 95.0%
Intercept	2667.9256	2901.9790	0.9193	0.3885	-4194.1642	9530.0154	-4194.1642	9530.0154
X Variable 1	387.1323	137.0403	2.8250	0.0256	63.0833	711.1812	63.0833	711.1812
X Variable 2	-26.8782	12.8870	-2.0857	0.0754	-57.3511	3.5946	-57.3511	3.5946

Figure 12.69 Excel regression analysis summary output.

3. Compute the normal scores and construct a normal probability plot of the residuals as described in Section 6.4. See Figure 12.70. The plots of the residuals versus the predictor variable values are shown in Figures 12.71 and 12.72.

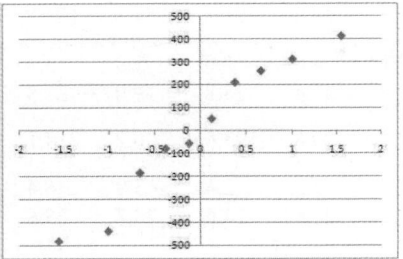

Figure 12.70 Normal probability plot.

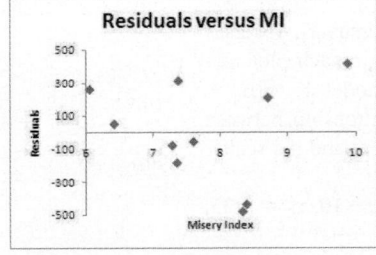

Figure 12.71 Plot of the residuals versus the first predictor.

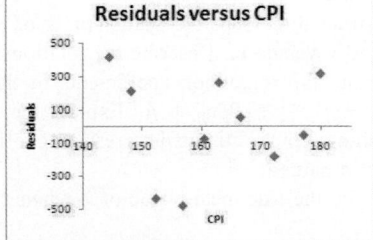

Figure 12.72 Plot of the residuals versus the second predictor.

Note: There is no built-in function on the TI-84 Plus to conduct multiple regression analysis.

SECTION 12.5 EXERCISES

Practice

12.84 Suppose the true regression equation relating the variables x_1, x_2, and y, for values of x_1 between 50 and 100 and for values of x_2 between 4.5 and 11, is $y = 32.0 + 7.5x_1 - 5.3x_2$.
a. Find the expected value of Y when $x_1 = 65$ and $x_2 = 7$.
b. Use the sign of the coefficient of x_1 to explain the relationship between the x_1 and y.
c. How much change in the dependent variable is expected when x_2 increases by 1 unit?
d. Suppose $\sigma = 3.7$. Find the probability that an observed value of Y is less than 554 when $x_1 = 77$ and $x_2 = 9.8$.

12.85 Suppose the true regression equation relating the variables x_1, x_2, x_3, x_4, and y is $y = -10.7 + 5x_1 - 14x_2 - 23x_3 + 6.7x_4$.
a. Find the expected value of Y when $x_1 = 1$, $x_2 = 2$, $x_3 = 2$, and $x_4 = 4.5$ [or $x = (1, 2, 2, 4.5)$].
b. How much change in the dependent variable is expected when x_2 increases by 2 units and x_4 decreases by 5 units?
c. Suppose $\sigma = 24$. Find the probability that an observed value of Y is between -460 and 400 when $x = (10, 12.5, 15, 7)$.

12.86 Suppose a multiple linear regression model is used to explain the relationship among y, x_1, and x_2. A random sample of $n = 12$ pairs for the independent variables were selected, and the corresponding values of the dependent variable were observed. The data are given in the following table.

y	196.7	203.4	227.0	221.3	154.8	185.1
x_1	7.8	8.7	4.0	2.6	1.9	3.1
x_2	22.5	24.0	27.4	27.8	19.0	21.8
y	198.7	200.9	220.8	193.6	206.2	214.1
x_1	8.3	7.8	2.2	2.4	3.9	6.4
x_2	22.1	23.4	26.6	24.6	24.3	25.6

a. Find the estimated regression equation.
b. What is the predicted value of Y when $x_1 = 3.3$ and $x_2 = 20$?

12.87 An experiment resulted in observations on a single dependent variable and three independent variables. The data are given on the data CD and book's web site.
a. Construct three separate scatter plots of y versus x_1, y versus x_2, and y versus x_3. Describe the relationship in each plot.
b. Estimate the regression coefficients in the model $Y_i = \beta_0 + \beta_1 x_{1i} + \beta_2 x_{2i} + \beta_3 x_{3i} + E_i$. Explain the relationship between the sign of each estimated regression coefficient and the scatter plots in part (a).
c. Estimate the true mean value of Y when $x_1 = -10$, $x_2 = 1.35$, and $x_3 = 52.6$.

12.88 Data from an observational study were used to fit a multiple linear regression model and the following summary statistics were obtained.

$n = 23$ SSE $= 80.75$ SST $= 152.5$ MSE $= 4.75$

a. Complete the ANOVA summary table.
b. How many predictor variables are in the multiple linear regression model?

c. Conduct a model utility test with $\alpha = 0.05$. Find bounds on the p value associated with this test.
d. Find the value of r^2 and explain the meaning of this value.

12.89 An experiment was conducted and the data obtained were used to fit the multiple linear regression model $Y_i = \beta_0 + \beta_1 x_{1i} + \beta_2 x_{2i} + \beta_3 x_{3i} + \beta_4 x_{4i} + E_i$. Minitab was used to estimate the regression coefficients and the output is as follows.

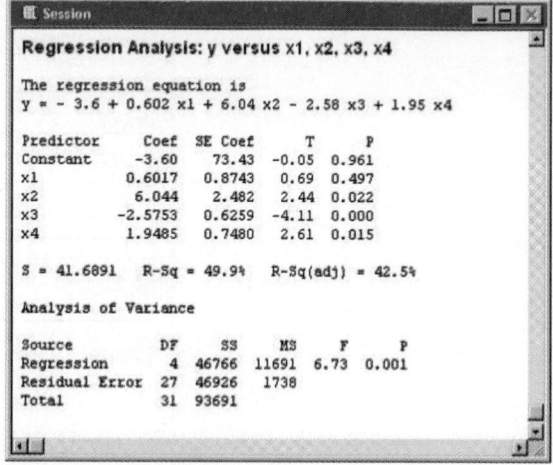

a. Is the overall regression significant? Conduct the appropriate hypothesis test and justify your answer.
b. Conduct four hypothesis tests with H_0: $\beta_i = 0$ for $i = 1, 2, 3, 4$ and $\alpha = 0.05$ in each test. Which regression coefficients are significantly different from 0, and, therefore, which predictor variables are significant?
c. Conduct four hypothesis tests with H_0: $\beta_i = 0$ for $i = 1, 2, 3, 4$ with an *overall* Type I error of $\alpha = 0.05$. Using these results, which predictor variables are significant? Compare your answers in parts (b) and (c).

12.90 An experiment resulted in the following observations on a single dependent variable and three independent variables.

y	x_1	x_2	x_3
-31.56	3.76	0.94	16.5
-44.48	4.47	0.23	17.2
-26.28	1.97	0.52	15.7
-27.54	1.16	0.45	15.8
-22.68	0.73	0.48	16.7
-20.34	0.82	0.63	16.5
-46.19	5.84	0.27	18.2
-29.00	2.69	0.61	16.4
-31.60	3.08	0.46	16.3
-34.31	5.38	0.73	17.5

a. Estimate the regression coefficients in the model $Y_i = \beta_0 + \beta_1 x_{1i} + \beta_2 x_{2i} + \beta_3 x_{3i} + E_i$.
b. Conduct an F test for a significant regression. Use $\alpha = 0.05$. Use technology to find the exact p value.
c. Find the coefficient of determination.
d. Which predictor variable(s) is(are) significant in explaining the variation in the dependent variable? Conduct the appropriate hypothesis tests.

e. Construct a 95% confidence interval for the *constant* regression coefficient, β_0. Use this interval to determine whether there is evidence to suggest that the constant regression coefficient is different from 0.

12.91 An experiment resulted in observations on a single dependent variable and four independent variables. The data are given on the data CD and book's web site.

a. Estimate the regression coefficients in the model $Y_i = \beta_0 + \beta_1 x_{1i} + \beta_2 x_{2i} + \beta_3 x_{3i} + \beta_4 x_{4i} + E_i$. Find the coefficient of determination.

b. Conduct the appropriate hypothesis tests to determine which predictor variables are significant (equivalently, which regression coefficients are significantly different from 0).

c. Using the results from part (b), write the *reduced* multiple linear regression model (including only the significant predictor variables). Estimate the regression coefficients and find the coefficient of determination in this *reduced* model.

d. Compare the value of r^2 in both models. Which model do you think is better? Why?

12.92 An experiment was conducted in order to determine whether the freezing point of a certain solution can be predicted from the concentrations of three chemicals. Twenty combinations of chemical concentrations were studied and the data are given on the data CD and book's web site.

a. Estimate the regression coefficients in the model $Y_i = \beta_0 + \beta_1 x_{1i} + \beta_2 x_{2i} + \beta_3 x_{3i} + E_i$.

b. Compute the residuals and construct a normal probability plot for the residuals. Is there any evidence to suggest that the random error terms are not normal? Justify your answer.

c. Construct a graph of the residuals versus each predictor variable. Is there any evidence of a violation in the regression assumptions? If so, what modification(s) could you make to improve the model?

12.93 The amount of sludge build-up in an automobile engine can severely affect performance. In an article in the Vanagon Maintenance Guide, a technical editor discussed the advantages of synthetic lubricants. Suppose an experiment was conducted to determine whether the sludge build-up in an automobile engine can be predicted by the oil viscosity, oxidation inhibitors, detergents, dispersants, and/or anti-wear additives. The data obtained were used to fit the multiple linear regression model $Y_i = \beta_0 + \beta_1 x_{1i} + \beta_2 x_{2i} + \beta_3 x_{3i} + \beta_4 x_{4i} + \beta_5 x_{5i} + E_i$. Minitab was used to estimate the regression coefficients and the output is as follows.

```
Session                                         _ □ X □ X

Regression Analysis: y versus x1, x2, x3, x4, x5

The regression equation is
y = - 1.70 + 1.01 x1 - 1.08 x2 + 3.44 x3 - 1.89 x4 + 1.32 x5

Predictor    Coef    SE Coef      T       P
Constant    -1.702    4.792    -0.36   0.726
x1           1.0138   0.4416    2.30   0.033
x2          -1.0846   0.5023   -2.16   0.043
x3           3.4354   0.4472    7.68   0.000
x4          -1.8904   0.4319   -4.38   0.000
x5           1.3154   0.3965    3.32   0.003

S = 2.38509   R-Sq = 82.2%   R-Sq(adj) = 77.7%

Analysis of Variance

Source          DF     SS      MS      F      P
Regression       5   525.10  105.02  18.46  0.000
Residual Error  20   113.77    5.69
Total           25   638.87
```

Consider the value $x^* = (1.5, 4.6, 4.7, 4.9, 4.7)$ and suppose an estimate of the standard deviation of Y^* is $s_{Y^*} = 1.049$.

a. Find a 95% confidence interval for the true mean value of Y when $x = x^*$. Give an interpretation of this interval.

b. Find a 95% prediction interval for an observed value of Y when $x = x^*$. Give an interpretation of this interval.

Applications

12.94 Economics and Finance A researcher believes that the six-month certificate of deposit interest rate can be predicted from the price of gold (x_1, in dollars per troy ounce) and the M1 money supply (x_2, in billions of dollars). A random sample of months was selected and the following data were obtained.

y	x_1	x_2
8.17	410.11	800.66
6.50	363.33	833.72
4.13	338.50	953.69
3.22	329.35	1021.90
5.17	380.35	1143.66
5.73	386.23	1145.18
5.69	352.06	1067.52
6.07	283.41	1147.99
6.68	273.68	1089.66
2.66	441.76	1393.11
1.92	310.25	1187.34

(*Source*: Financial Forecast Center.)

a. Find the estimated regression equation.

b. Find an estimate of the true mean six-month certificate of deposit interest rate for $x_1 = 400$ and $x_2 = 1200$.

c. Conduct an F test for a significant regression with $\alpha = 0.05$.

12.95 Business and Management In March 2008 the biopharmaceutical firm Shire decided to move its tax base out of the United Kingdom. This action raised concern about main indicators that affect the location for pharmaceutical companies in the United Kingdom compared with other countries, including tax rates and rules, and available skilled workforce.[18] Suppose the data on the data CD and book's web site resulted from observations associated with randomly selected countries for the year 2008. The variables are the number of new pharmaceutical businesses created minus the number of such businesses closed (y), the number of new graduates with degrees in sciences relevant to the pharmaceutical industry (x_1), and the corporation tax rate (x_2, as a percentage).

a. Estimate the regression coefficients in the model $Y_i = \beta_0 + \beta_1 x_{1i} + \beta_2 x_{2i} + E_i$.

b. Do these predictor variables explain a significant amount of variation in Y? Conduct the appropriate model utility test using $\alpha = 0.05$.

c. Find an estimate of an observed value of Y for $x_1 = 10,000$ and $x_2 = 25.0$.

12.96 Physical Sciences An experiment was conducted to test the effect of temperature (x_1, °F), contact area (x_2, cm^2), and wood density (x_3, g/cm^3) on the bonding strength (y) of a certain wood glue. Each piece of wood was glued to the same control block and the entire fixture was placed in an industrial oven. The bonding strength was measured by recording the time (in minutes) until the test piece of wood separated from the control block. The multiple linear regression model equation is $Y_i = \beta_0 + \beta_1 x_{1i} + \beta_2 x_{2i} + \beta_3 x_{3i} + E_i$. Minitab was used to analyze the data and a portion of the output follows.

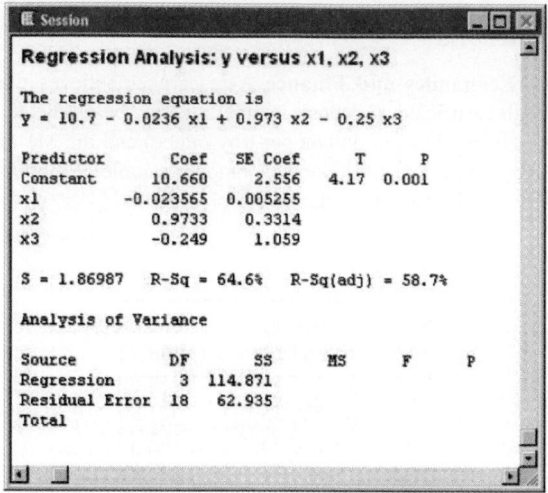

a. Complete the ANOVA table and conduct a model utility test.

b. Conduct three hypothesis tests with $H_0: \beta_i = 0$ and $\alpha = 0.05$ in each test. Find bounds on the p value associated with each test. Use these results to determine which predictor variables are the most important in determining the bonding strength of glue.

12.97 Physical Sciences A Daniel cell uses zinc and copper solutions to produce electricity. An experiment was conducted to determine whether voltage (y) is affected by the temperature of the solutions (x_1, °F), the concentration of the solutions (x_2, M), and/or the surface area of the electrodes (x_3, cm²). The data are given on the data CD and book's web site.

a. Estimate the regression coefficients in the model $Y_i = \beta_0 + \beta_1 x_{1i} + \beta_2 x_{2i} + \beta_3 x_{3i} + E_i$.

b. Conduct a model utility test and the other appropriate tests to determine which variables are the most important predictors of voltage.

c. Construct a normal probability plot of the residuals. Use this plot to determine whether there is any evidence of violation in multiple linear regression assumptions.

12.98 Physical Sciences In December 2008 the Navy tested the Combat SkySat communications balloon.[19] This huge balloon carries GPS systems, radios, and a hanging antenna, and floats up to 120,000 feet. At high altitudes, ozone and ultraviolet radiation destroy the balloon rather quickly. Suppose an experiment was conducted to determine whether altitude (x_1, in thousands of feet), ozone level (x_2, in Dobson units), and/or ultra-violet radiation level (x_3, W/m²) can be used to predict the time until a balloon fails (y, in days). The data are given on the data CD and book's web site.

a. Estimate the regression coefficients in the model $Y_i = \beta_0 + \beta_1 x_{1i} + \beta_2 x_{2i} + \beta_3 x_{3i} + E_i$.

b. Construct the ANOVA table and conduct the model utility test.

c. Conduct the necessary hypothesis tests to determine the most important variables in predicting the time to failure.

d. Check the model assumptions by constructing a normal probability plot of the residuals and the appropriate scatter plots.

12.99 Public Policy and Political Science During the Cold War, the United States increased its nuclear weapons stockpile to a maximum of approximately 32,000 in 1966. Other countries responded and postured by increasing their nuclear arms in this same period. The Moscow Treaty, signed in 2002 by Presidents Bush and Putin, is an agreement to limit strategic nuclear warheads. Suppose a senior political strategist believes that the number of nuclear weapons in the United States (y) can be predicted by the number of nuclear weapons in Russia (x_1), in the United Kingdom (x_2), in France (x_3), and in China (x_4). A random sample of years between 1964 and 2007 was obtained, and the data are given on the data CD and book's web site.[20]

a. Estimate the regression coefficients in the model $Y_i = \beta_0 + \beta_1 x_{1i} + \beta_2 x_{2i} + \beta_3 x_{3i} + \beta_4 x_{4i} + E_i$.

b. Conduct a model utility test using $\alpha = 0.05$ and find the value of the coefficient of determination.

c. Conduct the appropriate hypothesis tests to determine which predictor(s) is(are) are significant.

d. Conduct a hypothesis test to determine whether there is any evidence to suggest that β_1 is greater than 0.20. Use $\alpha = 0.05$.

Extended Applications

12.100 Biology and Environmental Science There is some evidence to suggest that the variation in the size of penguin colonies is related to the extent of sea ice.[21] Suppose a study was conducted to investigate the effect of sea ice extent and stormy weather on the size of penguin colonies along the west coast of the Antarctic Peninsula. Fifteen years were selected at random, and the data are given on the data CD and book's web site.

$$y = \text{Size of the penguin colony population}$$
$$x_1 = \text{Sea ice extent (as a percentage)}$$
$$x_2 = \text{Number of stormy days (yearly total)}$$

a. Estimate the regression coefficients in the model $Y_i = \beta_0 + \beta_1 x_{1i} + \beta_2 x_{2i} + E_i$.

b. Conduct the model utility test. Use $\alpha = 0.01$.

c. Find the value of r^2 and interpret this value.

d. Estimate the mean value of Y for a sea ice extent of 12.5% and 35 stormy days.

12.101 Physical Sciences Potassium ferricyanide crystals are used for etching and in electroplating applications. Paul Boyle of North Carolina State University recently wrote about some of the factors that affect crystal growth and therefore the size of the crystals.[22] These factors include solubility of the compound and the evaporation process. An experiment was conducted to determine whether the weight of the final crystal (y, in grams) is affected by the amount of initial solid (x_1, g), the amount of water (x_2, ml), and/or the temperature of the water (x_3, °F). Using a multiple linear regression model of the form $Y_i = \beta_0 + \beta_1 x_{1i} + \beta_2 x_{2i} + \beta_3 x_{3i} + E_i$. Minitab was used to analyze the data and a portion of the output follows.

Regression Analysis: y versus x1, x2, x3

The regression equation is
y = 39.9 + 0.344 x1 - 0.181 x2 + 0.199 x3

Predictor	Coef	SE Coef	T	P
Constant	39.91	22.55	1.77	0.091
x1	0.3437	0.1212	2.83	0.010
x2	-0.18058	0.08648	-2.09	0.049
x3	0.19874	0.06962	2.85	0.009

S = 5.34244 R-Sq = 47.0% R-Sq(adj) = 39.4%

Analysis of Variance

Source	DF	SS	MS	F	P
Regression					
Residual Error	21	599.38			
Total	24	1130.92			

Predicted Values for New Observations

New Obs	Fit	SE Fit	95% CI	95% PI
1	55.63	1.44		
2	62.37	2.96		

a. Complete the ANOVA table and conduct a model utility test.
b. Use the output to determine which of the three predictor variables contributes to the overall significant regression.
c. Suppose $x_1^* = (93, 200, 100)$ and $x_2^* = (100, 220, 140)$. The predicted value for each is given in the Minitab output, along with the standard error of each estimate (SE Fit). For example, $s_{\hat{Y}_1}^* = 1.44$. Find a 95% confidence interval for the true mean value of Y when $x = x_1^*$. Find a 95% confidence interval for the true mean value of Y when $x = x_2^*$.
d. Find a 95% prediction interval for an observed value of Y when $x = x_1^*$. Find a 95% prediction interval for an observed value of Y when $x = x_2^*$.
e. Why do you suppose $s_{Y_1^*} < s_{Y_2^*}$?

12.102 Demographics and Population Statistics In a recent report prepared for the Department of Health and Human Services Office of Child Support Enforcement, the economic, demographic, and programmatic factors affecting state child support enforcement programs were studied. Data from all 50 states were used. One multiple linear regression model was used to explain the variation in the percentage of cases with child support orders. The model consisted of seven predictor variables, one of which was a *dummy* variable (a variable with either a value of 0 or 1). The independent variables are given in the following table.

STPOVRT	State poverty rate.
PCTURB	Percentage of the state's population in an urban area.
CASEFTE	Number of cases per full-time-equivalent staff.
JUDADMIN	Taxonomy scale, a measure of the judicial process.
POPSTAB	Percentage of population stability.
TANFNOW	Percentage of TANF cases (Temporary Assistance for Needy Families).
CWODUM	Dummy variable, indicates whether the state passed the audit for cases with orders.

A portion of the analysis associated with this model follows.

Predictor	Coefficient estimate	Standard error	t	p
Constant	67.43889	49.03867	1.38	0.1773
STPOVRT	-2.12308	0.51548		
PCTURB	-0.23829	0.07220		
CASEFTE	-0.45600	0.01244		
JUDADMIN	-0.83898	0.43343		
POPSTAB	0.67485	0.59123		
TANFNOW	-0.28561	0.20797		
CWODUM	8.59857	9.53486		

ANOVA summary table

Source of variation	Sum of squares	Degrees of freedom	Mean square	F	p value
Regression	8128.67				
Error	2942.32	37			
Total		44			

a. Complete the ANOVA table and conduct a model utility test. Use technology to find the p value associated with this test. Find the value of r^2. Use these results to describe the overall significance of this model.
b. Find the value of the test statistic and the p value associated with each hypothesis test for a significant regression coefficient. Use these results to determine the most important predictor variables in this model.
c. Use the estimates of the regression coefficients to answer the following questions.
   (i) Explain the effect of an increase of 1 percentage point in a state's urban population on the percentage of cases with orders.
   (ii) Explain the effect of an increase of 1 percentage point in a state's population stability on the percentage of cases with orders.

12.103 Physical Sciences The Hormel Corporation claims that two of the variables that affect home freezer storage time are the rate of freezing, and the temperature of the freezer during storage.[23] An experiment was conducted in which 2-pound portions of ground beef with identical wrapping material were frozen, stored in separate freezers, and carefully monitored. Consider the following variables and the data given in the table.
y = Storage time (in days) in which the concentration of dissolved solute remained above a threshold value.
x_1 = The temperature of the initial freezer (in °F), which is directly related to the rate of freezing.
x_2 = The temperature of the storage freezer (in °F).

y	x_1	x_2
146	−20	10
188	−26	21
231	−18	9
67	−3	27
154	−6	8
89	−2	19
120	−25	22
77	−6	14
201	−27	16
276	−13	5

a. Estimate the regression coefficients in the model $Y_i = \beta_0 + \beta_1 x_{1i} + \beta_2 x_{2i} + E_i$.

b. Conduct the model utility test. Use $\alpha = 0.05$.

c. Find the value of r^2 and interpret this value.

d. Conduct the appropriate hypothesis tests to determine which regression coefficient(s) is(are) significantly different from 0. Use $\alpha = 0.05$ in each test.

e. Do you believe the claim made by the Hormel Corporation? Why or why not?

12.104 Public Health and Nutrition A recent experiment was designed to determine whether the percentage of lactic acid (y) in cultured buttermilk (produced from whole milk) can be predicted by the temperature to which the milk is heated (x_1, °C), the temperature to which the milk is cooled before the started culture is added (x_2, °C), the percentage of started culture added (x_3), and/or the fermentation time (x_4, in hours). The data are given on the data CD and book's web site.

a. Estimate the regression coefficients in the model $Y_i = \beta_0 + \beta_1 x_{1i} + \beta_2 x_{2i} + \beta_3 x_{3i} + \beta_4 x_{4i} + E_i$.

b. Conduct a model utility test using $\alpha = 0.05$ and find the value of the coefficient of determination.

c. Conduct the appropriate hypothesis tests to determine which predictor(s) is(are) significant.

d. Conduct a hypothesis test to determine whether there is any evidence to suggest that β_3 is less than 0.93. Use $\alpha = 0.05$.

e. Find a 95% confidence interval for the true mean amount of lactic acid in cultured buttermilk when $x_1 = 95$, $x_2 = 22$, $x_3 = 1.5$, and $x_4 = 20$.

Challenge

12.105 Geothermal Gradient The temperature naturally increases as one digs deeper into the continental crust. This phenomenon is called the geothermal gradient. A random sample of continental crust depths (x, in km) was obtained, and the temperature at each depth was recorded (y, in °C). The data are given in the following table.

Depth	192	306	75	375	96	120	399
Temperature	212	682	15	855	93	49	943

Depth	469	439	209	224	313	379	296
Temperature	1456	1156	271	336	604	911	606

Depth	37	316	41	242	21	230
Temperature	35	576	31	462	10	274

a. Carefully sketch a scatter plot of these data.

b. Find the estimated regression line, and complete the ANOVA summary table.

c. There is some evidence to suggest that the rate of change in temperature is different according to two depth zones. For example, at small depths, the temperature changes only slightly for each kilometer. However, at large depths, the temperature change is much greater for each kilometer. Use your scatter plot in part (1) to divide the data into two depth zones. Find the estimated regression line and the ANOVA summary table for each zone.

d. Some researchers believe that the relationship between depth and temperature is nonlinear and is better modeled by a quadratic curve. Square each depth to create a new list of observations (x^2). Find the estimated regression line using x^2 as the independent variable.

e. Which of these three models do you think is the best? Justify your answer.

Chapter 12 Challenge Wrap-Up

In the windmill experiment, residents would like to be able to predict the noise level at a certain distance from a windmill. The distance (x, in feet) is the predictor variable, and the noise level (y, in dB) is the response variable. A scatter plot of the data (Figure 12.73) suggests a linear relationship. Find the estimated regression line, and complete the ANOVA table.

(Beverett/Dreamstime.com)

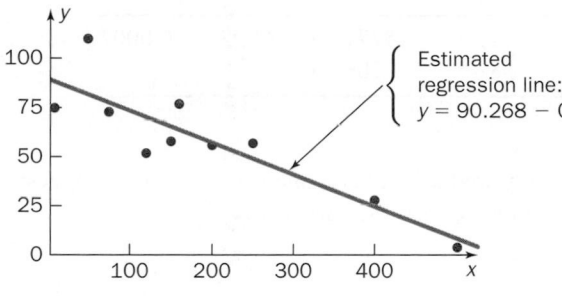

Figure 12.73 Scatter plot of the windmill data and the estimated regression line.

Here are the summary statistics.

$\sum x_i = 1915$ $\qquad$ $\sum y_i = 590$ $\qquad$ $\sum x_i y_i = 77{,}635$

$\sum x_i^2 = 583{,}225$ $\qquad$ $\sum y_i^2 = 42{,}236$

$$\hat{\beta}_1 = \frac{n \sum x_i y_i - \left(\sum x_i\right)\left(\sum y_i\right)}{n \sum x_i^2 - \left(\sum x_i\right)^2} \qquad \text{Use Equation 12.2.}$$

$$= \frac{(10)(77{,}635) - (1915)(590)}{(10)(583{,}225) - (1915)^2} \qquad \text{Use summary statistics.}$$

$$= \frac{-353{,}500}{2{,}165{,}025} = -0.16328 \qquad \text{Simplify.}$$

$$\hat{\beta}_0 = \frac{\sum y_i - \hat{\beta}_1 \sum x_i}{n} \qquad \text{Use Equation 12.3.}$$

$$= \frac{590 - (-0.16328)(1915)}{10} \qquad \text{Use summary statistics and } \hat{\beta}_1.$$

$$= 90.268 \qquad \text{Simplify.}$$

The estimated regression line is $y = 90.268 - 0.16328x$. Complete the calculations for the ANOVA table.

$$\text{SST} = S_{yy} = \sum y_i^2 - \frac{1}{n}\left(\sum y_i\right)^2 \qquad \text{Computational formula.}$$

$$= 42{,}236 - \frac{1}{10}(590)^2 = 7426.0 \qquad \text{Use summary statistics.}$$

$$\text{SSR} = \hat{\beta}_1 S_{xy} = \hat{\beta}_1\left[\sum x_i y_i - \frac{1}{10}\left(\sum x_i\right)\left(\sum y_i\right)\right] \qquad \text{Computational formula.}$$

$$= -0.16328\left[77{,}635 - \frac{1}{10}(1915)(590)\right] \qquad \text{Use summary statistics.}$$

$$= 5771.9 \qquad \text{Simplify.}$$

$$\text{SSE} = \text{SST} - \text{SSR} \qquad \text{Computation formula.}$$

$$= 7426.0 - 5771.9 = 1654.1$$

Compute the mean squares, MSR and MSE.

$$MSR = SSR/1$$
$$= 5771.9/1 = 5771.9$$
$$MSE = SSE/(n - 2)$$
$$= 1654.1/8 = 206.76$$

Definition.
Use SSR.
Definition.
Use SSE and $n = 10$.

The value of the test statistic is

$$f = \frac{MSR}{MSE} = \frac{5771.9}{206.76} = 27.92$$

Here are all of these calculations presented in an ANOVA table.

The p value in this table was found using technology.

ANOVA summary table for simple linear regression

Source of variation	Sum of squares	Degrees of freedom	Mean square	F	p value
Regression	5771.9	1	5771.90	27.92	0.0007
Error	1654.1	8	206.76		
Total	7426.0	9			

The F test for a significant regression produces a very small p value. This suggests that there is a significant linear relationship between distance and noise level.

The coefficient of determination is

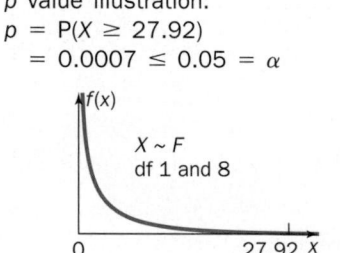

p value illustration:
$p = P(X \geq 27.92)$
$= 0.0007 \leq 0.05 = \alpha$

$X \sim F$
df 1 and 8

$$r^2 = SSR/SST = 5771.9/7426.0 = 0.777$$

Approximately 0.777, or 78%, of the variation in noise level is explained by the regression model.

Suppose a windmill is being built 350 feet from your home. An estimate of the mean noise level (at 350 feet) is

$$90.268 - 0.16328(350) = 33.12$$

The predicted noise level at your home is 33.12 dB (which is between a soft whisper and a quiet office).

Figure 12.74 shows a technology solution.

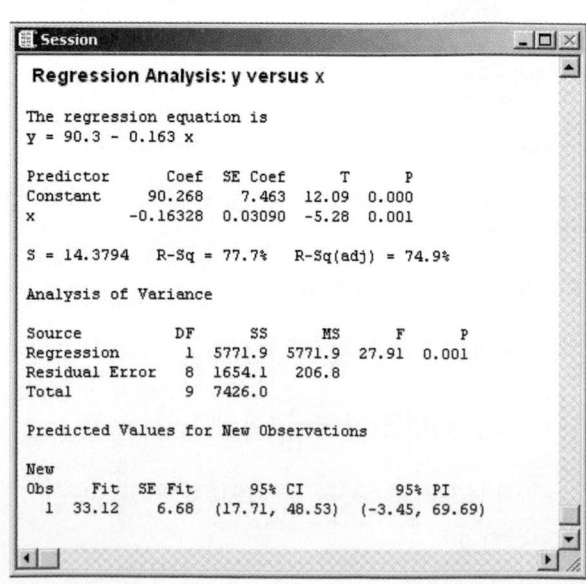

Figure 12.74 Minitab regression analysis.

Figures 12.75 and 12.76 are a normal probability plot of the residuals and a plot of the residuals versus the predictor variable. There is no evidence of non-normality. There is some evidence that the variance in the random error terms decreases as distance increases. However, for $n = 10$ observations, the pattern is not distinct enough to be evidence of a violation in the simple linear regression assumptions.

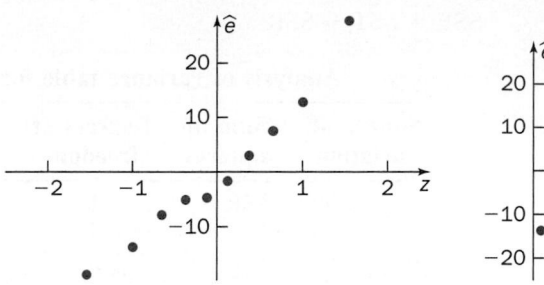

Figure 12.75 A normal probability plot of the residuals.

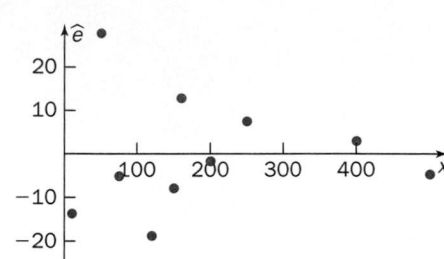

Figure 12.76 A scatter plot of the residuals versus distance.

CHAPTER 12 SUMMARY

Simple Linear Regression Model

Let $(x_1, y_1), (x_2, y_2), \ldots, (x_n, y_n)$ be n pairs of observations such that y_i is an observed value of the random variable Y_i. We assume that there exist constants β_0 and β_1 such that

$$Y_i = \beta_0 + \beta_1 x_i + E_i$$

where $E_1, E_2, \ldots, E_n$ are independent, normal random variables with mean 0 and variance σ^2. The following assumptions hold:

1. The E_i's are normally distributed (which implies that the Y_i's are normally distributed).

2. The expected value of E_i is 0 (which implies that $\mathrm{E}(Y_i) = \beta_0 + \beta_1 x_i$).

3. $\mathrm{Var}(\mathrm{E}_i) = \sigma^2$ (which implies that $\mathrm{Var}(Y_i) = \sigma^2$).

4. The E_i's are independent (which implies that the Y_i's are independent).

The E_i's are the random deviations or random error terms. $y = \beta_0 + \beta_1 x$ is the true regression line.

Principle of least squares:
The estimated regression line is obtained by minimizing the sum of the squared deviations, or vertical distances from the observed points to the line.

Least squares estimates:

$$\hat{\beta}_1 = \frac{n\sum x_i y_i - (\sum x_i)(\sum y_i)}{n\sum x_i^2 - (\sum x_i)^2} \qquad \hat{\beta}_0 = \frac{\sum y_i - \hat{\beta}_1 \sum x_i}{n} = \bar{y} - \hat{\beta}_1 \bar{x}$$

The estimated regression line is $y = \hat{\beta}_0 + \hat{\beta}_1 x$.

The ith predicted (fitted) value is $\hat{y}_i = \hat{\beta}_0 + \hat{\beta}_1 x_i$ ($i = 1, 2, \ldots, n$).

The ith residual is $\hat{e}_i = y_i - \hat{y}_i$.

Sum of squares:

$$\underbrace{\sum (y_i - \bar{y})^2}_{\text{SST}} = \underbrace{\sum (\hat{y}_i - \bar{y})^2}_{\text{SSR}} + \underbrace{\sum (y_i - \hat{y}_i)^2}_{\text{SSE}}$$

Computation formulas:

$$\text{SST} = S_{yy} = \sum(y_i - \bar{y})^2 = \sum y_i^2 - \frac{1}{n}\left(\sum y_i\right)^2$$

$$\text{SSR} = \hat{\beta}_1 S_{xy} = \hat{\beta}_1 \sum(x_i - \bar{x})(y_i - \bar{y}) = \hat{\beta}_1\left[\sum x_i y_i - \frac{1}{n}\left(\sum x_i\right)\left(\sum y_i\right)\right]$$

$$\text{SSE} = \text{SST} - \text{SSR}$$

Analysis of variance table for simple linear regression

Source of variation	Sum of squares	Degrees of freedom	Mean square	F	p value
Regression	SSR	1	$\text{MSR} = \dfrac{\text{SSR}}{1}$	$\dfrac{\text{MSR}}{\text{MSE}}$	p
Error	SSE	$n - 2$	$\text{MSE} = \dfrac{\text{SSE}}{n-2}$		
Total	SST	$n - 1$			

Coefficient of determination: $r^2 = \text{SSR}/\text{SST}$

Estimate of variance: $s^2 = \text{SSE}/(n - 2)$

F test for a significant linear regression:

H_0: There is no significant linear relationship ($\beta_1 = 0$)

H_a: There is a significant linear relationship ($\beta_1 \neq 0$)

TS: $F = \dfrac{\text{MSR}}{\text{MSE}}$

RR: $F \geq F_\alpha$ df 1 and $n - 2$

Hypothesis test and confidence interval concerning β_1:

H_0: $\beta_1 = \beta_{10}$

H_a: $\beta_1 > \beta_{10}$, $\beta_1 < \beta_{10}$,   or $\beta_1 \neq \beta_{10}$

TS: $T = \dfrac{B_1 - \beta_{10}}{S/\sqrt{S_{xx}}} = \dfrac{B_1 - \beta_{10}}{S_{B_1}}$

RR: $T \geq t_\alpha$, $T \leq -t_\alpha$,   or $|T| \geq t_{\alpha/2}$ df $= n - 2$

A $100(1 - \alpha)\%$ confidence interval for β_1 has as endpoints $\hat{\beta}_1 \pm t_{\alpha/2}s_{B_1}$ with the critical value based on $n - 2$ degrees of freedom.

Hypothesis test and confidence interval concerning β_0:

H_0: $\beta_0 = \beta_{00}$

H_a: $\beta_0 > \beta_{00}$, $\beta_0 < \beta_{00}$,   or $\beta_0 \neq \beta_{00}$

TS: $T = \dfrac{B_0 - \beta_{00}}{S\sqrt{\sum x_i^2/(nS_{xx})}} = \dfrac{B_0 - \beta_{00}}{S_{B_0}}$

RR: $T \geq t_\alpha$, $T \leq -t_\alpha$,   or $|T| \geq t_{\alpha/2}$ df $= n - 2$

A $100(1 - \alpha)\%$ confidence interval for β_0 has as endpoints $\hat{\beta}_0 \pm t_{\alpha/2}s_{B_0}$ with the critical value based on $n - 2$ degrees of freedom.

Sample correlation coefficient:

$$r = \frac{S_{xy}}{\sqrt{S_{xx}S_{yy}}} = \frac{\sum x_i y_i - \dfrac{1}{n}\left(\sum x_i\right)\left(\sum y_i\right)}{\sqrt{\left[\sum x_i^2 - \dfrac{1}{n}\left(\sum x_i\right)^2\right]\left[\sum y_i^2 - \dfrac{1}{n}\left(\sum y_i\right)^2\right]}}$$

Hypothesis test and confidence interval concerning the mean value of Y for x = x*:

$H_0: y^* = y_0^*$

$H_a: y^* > y_0^*, \quad y^* < y_0^*, \quad \text{or} \quad y^* \neq y_0^*$

TS: $T = \dfrac{(B_0 + B_1 x^*) - y_0^*}{S\sqrt{(1/n) + [(x^* - \bar{x})^2/S_{xx}]}}$

RR: $T \geq t_\alpha, \quad T \leq -t_\alpha, \quad \text{or} \quad |T| \geq t_{\alpha/2} \quad \text{df} = n - 2$

A $100(1 - \alpha)\%$ confidence interval for $\mu_{Y|x^*}$, the mean value of Y for $x = x^*$, has as endpoints the values

$$(\hat{\beta}_0 + \hat{\beta}_1 x^*) \pm t_{\alpha/2}\, s \sqrt{\frac{1}{n} + \frac{(x^* - \bar{x})^2}{S_{xx}}}$$

with the critical value based on $n - 2$ degrees of freedom.

Prediction interval for an observed value of Y:

A $100(1 - \alpha)\%$ prediction interval for an observed value of Y when $x = x^*$ has as endpoints the values

$$(\hat{\beta}_0 + \hat{\beta}_1 x^*) \pm t_{\alpha/2}\, s \sqrt{1 + \frac{1}{n} + \frac{(x^* - \bar{x})^2}{S_{xx}}}$$

with the critical value based on $n - 2$ degrees of freedom.

Regression diagnostics:

1. Normal probability plot of residuals. The points should lie along an approximately straight line.

2. Scatter plot of residuals versus predictor variable. There should be no distinct pattern.

Multiple Linear Regression Model

Let $(x_{11}, x_{21}, \ldots, x_{k1}, y_1), (x_{12}, x_{22}, \ldots, x_{k2}, y_2), \ldots, (x_{1n}, x_{2n}, \ldots, x_{kn}, y_n)$ be n sets of observations such that y_i is an observed value of the random variable Y_i. We assume that there exist constants $\beta_0, \beta_1, \ldots, \beta_k$ such that

$$Y_i = \beta_0 + \beta_1 x_{1i} + \beta_2 x_{2i} + \cdots + \beta_k x_{ki} + E_i$$

where $E_1, E_2, \ldots, E_n$ are independent, normal random variables with mean 0 and variance σ^2. The following assumptions hold:

1. The E_i's are normally distributed (which means that the Y_i's are normally distributed).

2. The expected value of E_i is 0 (which implies that $\mathrm{E}(Y_i) = \beta_0 + \beta_1 x_{1i} + \beta_2 x_{2i} + \cdots + \beta_k x_{ki}$).

3. $\mathrm{Var}(E_i) = \sigma^2$ (which implies that $\mathrm{Var}(Y_i) = \sigma^2$).

4. The E_i's are independent (which implies that the Y_i's are independent).

The E_i's are the random deviations or random error terms.
$y = \beta_0 + \beta_1 x_1 + \beta_2 x_2 + \cdots + \beta_k x_k$ is the true regression line.

Principle of least squares:
The estimated regression equation is obtained by minimizing the sum of the squared deviations between the observations and the estimated values.

The estimated regression equation is $y = \hat{\beta}_0 + \hat{\beta}_1 x_1 + \hat{\beta}_2 x_2 + \cdots + \hat{\beta}_k x_k$.
The ith predicted (fitted) value is $\hat{y}_i = \hat{\beta}_0 + \hat{\beta}_1 x_{1i} + \hat{\beta}_2 x_{2i} + \cdots + \hat{\beta}_k x_{ki}$ ($i = 1, 2, \ldots, n$).
The ith residual is $\hat{e}_i = y_i - \hat{y}_i$.

Sum of squares:

$$\underbrace{\sum (y_i - \bar{y})^2}_{\text{SST}} = \underbrace{\sum (\hat{y}_i - \bar{y})^2}_{\text{SSR}} + \underbrace{\sum (y_i - \hat{y}_i)^2}_{\text{SSE}}$$

Analysis of variance table for multiple linear regression

Source of variation	Sum of squares	Degrees of freedom	Mean square	F	p value
Regression	SSR	k	$\text{MSR} = \dfrac{\text{SSR}}{k}$	$\dfrac{\text{MSR}}{\text{MSE}}$	p
Error	SSE	$n - k - 1$	$\text{MSE} = \dfrac{\text{SSE}}{n - k - 1}$		
Total	SST	$n - 1$			

Coefficient of determination: $r^2 = \text{SSR}/\text{SST}$

Estimate of variance: $s^2 = \text{SSE}/(n - k - 1)$

F test for a significant multiple linear regression:

$H_0: \beta_1 = \beta_2 = \cdots = \beta_k = 0$
  (None of the predictor variables helps to predict y.)

$H_a: \beta_i \neq 0$ for at least one i
  (At least one predictor variable helps to predict y.)

TS: $F = \dfrac{\text{MSR}}{\text{MSE}}$

RR: $F \geq F_\alpha$ df k and $n - k - 1$

Hypothesis test and confidence interval concerning β_i:

$H_0: \beta_i = \beta_{i0}$
$H_a: \beta_i > \beta_{i0}, \quad \beta_i < \beta_{i0}, \quad$ or $\quad \beta_i \neq \beta_{i0}$

TS: $T = \dfrac{B_i - \beta_{i0}}{S_{B_i}}$

RR: $T \geq t_\alpha, \quad T \leq -t_\alpha, \quad$ or $\quad |T| \geq t_{\alpha/2}$ df $= n - k - 1$

A $100(1 - \alpha)\%$ confidence interval for β_i has as endpoints the values

$$\hat{\beta}_i \pm t_{\alpha/2} s_{B_i}$$

with the critical value based on $n - k - 1$ degrees of freedom.

Hypothesis test and confidence interval concerning the mean value of Y for $\boldsymbol{x} = \boldsymbol{x}^*$:

$H_0: y^* = y_0^*$
$H_a: y^* > y_0^*, \quad y^* < y_0^*, \quad$ or $\quad y^* \neq y_0^*$

TS: $T = \dfrac{(B_0 + B_1 x_1^* + \cdots + B_k x_k^*) - y_0^*}{S_{Y^*}}$

RR: $T \geq t_\alpha, \quad$ or $\quad T \leq -t_\alpha, \quad |T| \geq t_{\alpha/2}$ df $= n - k - 1$

A $100(1 - \alpha)\%$ confidence interval for $\mu_{Y|x^*}$, the mean value of Y for $\boldsymbol{x} = \boldsymbol{x}^*$, has as endpoints the values

$$(\hat{\beta}_0 + \hat{\beta}_1 x_1^* + \cdots + \hat{\beta}_k x_k^*) \pm t_{\alpha/2} s_{Y^*}$$

with the critical value based on $n - k - 1$ degrees of freedom.

Prediction interval for an observed value of Y:

A $100(1 - \alpha)\%$ prediction interval for an observed value of Y when $x = x^*$ has as endpoints the values

$$(\hat{\beta}_0 + \hat{\beta}_1 x_1^* + \cdots + \hat{\beta}_k x_k^*) \pm t_{\alpha/2}\sqrt{s^2 + s_{\hat{Y}^*}^2}$$

with the critical value based on $n - k - 1$ degrees of freedom.

Regression diagnostics:

1. Construct a histogram, stem-and-leaf plot, scatter plot and/or normal probability plot of the residuals. These graphs are all used to check the normality assumption.

2. Construct a scatter plot of the residuals versus *each* independent variable. If there are no violations in assumptions, each scatter plot should appear as a horizontal band around 0. There should be no recognizable pattern.

CHAPTER 12 EXERCISES

Applications

12.106 Medicine and Clinical Studies Officials at the U.S. Environmental Protection Agency worked with hospital physicians in Anaheim, California, to examine the relationship between environmental pollution and illness. A random sample of summer days was selected, and at 2:00 P.M. on each day the air quality was assessed by measuring the concentration of carbon monoxide (x, in ppm). Illness was measured by counting the number of new patients (y) seen at the hospital that day for respiratory problems. Suppose the true regression line is $y = -5.1 + 0.9x$.

a. Find the expected number of respiratory patients at the hospital on a day when the carbon monoxide concentration is 12 ppm.

b. What is the expected change in the number of patients per day if the carbon monoxide concentration decreases by 5 ppm?

c. Suppose $\sigma = 1.2$. Find the probability that between 5 and 7 patients are observed when the carbon monoxide concentration is 13 ppm.

12.107 Economics and Finance An economist believes that a credit card company sets the spending limits for customers on the basis of loyalty, that is, the number of years the customer has been a cardholder. A random sample of 22 cardholders was selected. The number of years since obtaining the card (x) was recorded as well as the spending limit (y, in thousands of dollars) on the card. The summary statistics are

$$\sum x_i = 155.2 \qquad \sum y_i = 392.0$$
$$\sum x_i^2 = 1508.8 \qquad \sum y_i^2 = 10{,}090.0$$
$$\sum x_i y_i = 3644.2$$

a. Find the estimated regression line.

b. Estimate the true mean spending limit for a customer who has had this credit card for 10 years.

c. Find an estimate of an observed value of the spending limit for a customer who has had this credit card for 2 years.

12.108 Biology and Environmental Science Farmers in northern Sweden sell one of the most expensive cheeses in the world, made from moose milk.[24] Suppose a random sample of female moose was obtained, and the weight of each was measured (x, in kilograms). The amount of milk produced by each moose in one day was also measured (y, in liters). The data are given on the data CD and book's web site.

a. Construct a scatter plot for these data.

b. Find the estimated regression line. Add a graph of the estimated regression line to the scatter plot in part (a).

c. Complete the ANOVA summary table, and conduct an F test for a significant regression. Use a significance level of 0.05.

d. Remove the first observation, a possible outlier, from the data set. Find the new estimated regression line. Is this new regression significant at the 0.05 level?

12.109 Physical Sciences Marine biologists recently studied the relationship between the density of seagrass and the quality of water in coastal waters. A random sample of 100-by-100-meter plots near California coastlines was selected. The density of seagrass (x, number of plants per square meter) was carefully measured for each coastal plot. The water quality was assessed by measuring the clarity: A Secchi disk and measuring tape were used to record the depth at which the disk disappeared from view (y, in meters). The data are given on the data CD and book's web site.

a. Find the estimated regression line, and complete the ANOVA table.

b. Find the coefficient of determination, r^2.

c. Find the sample correlation coefficient using the definition. Verify that the square of this value is your answer in part (b).

d. Do you think that seagrass density is a good measure of water quality? Why or why not?

12.110 Physical Sciences In a recent study of climate and precipitation, the North Atlantic oscillation (NAO, a climatic phenomenon) and the monthly precipitation in Dallas, Texas, were considered. A random sample of months between 1950 and 2008 was selected. The NAO index (x, a unitless quantity) and the rainfall (y, in inches) were recorded. The data are given in the following table.

NAO	Rainfall	NAO	Rainfall
1.500	4.33	−0.100	1.62
−1.800	2.17	−0.299	0.56
−1.252	3.35	0.200	0.22
0.600	3.07	0.300	0.85
−0.200	1.90	0.000	2.53
−0.100	5.17	0.100	0.08
−0.100	1.85	0.000	3.99
−1.300	0.78	0.900	3.15
0.600	0.96	1.600	3.17
0.900	6.91	−1.300	2.38
−0.300	3.93	−1.07	3.85
−1.39	0.84	−0.28	0.27

(*Source*: National Weather Service, Climate Prediction Center.)

a. Carefully sketch a scatter plot of these data.
b. Find the sample correlation coefficient, and describe the relationship between the NAO index and the precipitation in Dallas.

12.111 Biology and Environmental Science In a recent study by the U.S. Forestry Service, the relationship between certain chemical elements and oak tree defoliation was studied. A random sample of oak trees was obtained, and the concentration of phosphorus (x, in g/kg) in leaves was measured for each tree. The percentage of defoliation (y) for each tree was carefully estimated at the end of the growing season. The data are given on the data CD and book's web site.
a. Carefully sketch a scatter plot of these data. Describe the relationship between these two variables.
b. Find the estimated regression line. Is this regression significant? Conduct an appropriate test at the 0.05 level. Add a graph of this line to your scatter plot in part (a).
c. Find the residuals, and sketch a graph of the residuals versus the predictor variable.
d. How could this model be improved? Justify your answer.

12.112 Medicine and Clinical Studies In a study of hair regeneration, 30 men over 45 years old who had used minoxidil for six months were randomly selected. The daily dose of minoxidil (x, in mg) and the hair density (y, in hairs/mm^2) were recorded for each man. The following summary statistics were reported.

$S_{xx} = 21423.0$ $S_{yy} = 80.119$ $S_{xy} = 165.593$
$\bar{x} = 49.033$ $\bar{y} = 2.207$

a. Find the estimated regression line.
b. Complete the ANOVA table, and conduct an F test for a significant regression. Does the dosage of minoxidil help to explain variation in hair density? Justify your answer.
c. Find a 95% confidence interval for the mean hair density for a man taking a daily 50-mg dose of minoxidil.

12.113 Economics and Finance A statistician (and cigar smoker) recently studied the relationship between *Cigar Aficionado*'s blind ratings of cigars and price. The purpose of

this study was to determine whether premium cigars are really more expensive. A random sample of 50 cigars was selected, and the rating (x) and price per cigar (y, in dollars) were recorded for each. The following summary statistics were obtained.

$S_{xx} = 185.725$ $S_{yy} = 1481.83$ $S_{xy} = 111.877$
$\bar{x} = 5.6084$ $\bar{y} = 11.7654$

a. Find the estimated regression line.
b. Complete the ANOVA table.
c. Conduct a test of H_0: $\beta_1 = 0$ versus H_a:$\beta_1 \neq 0$ with a significance level of 0.05.
d. Do you believe that the saying, "You get what you pay for" applies to cigars? Justify your answer.

12.114 Manufacturing and Product Development The Consumer Federation of America recently studied the relationship between tensile strength and amount of nickel in household stainless steel products. The percentage of nickel by weight (x) and the tensile strength (y, in MPa) were measured for each product. The data are given in the following table.

x	5.9	2.8	5.3	4.6	4.6	3.5	3.8	5.0	5.3	4.2
y	948	859	921	909	915	876	828	964	964	900

a. Find the estimated regression line, and complete the ANOVA table.
b. Compute the residuals.
c. Carefully sketch a normal probability plot of the residuals, and interpret the graph.
d. Carefully sketch a scatter plot of the residuals versus the predictor variable. Is there any indication of a violation in the regression model assumptions?

12.115 Physical Sciences A publisher conducted a study to determine the relationship between the amount of sulfate ash in leather used for archival book-binding and the surface shrinkage after one year. A random sample of books ready for archiving was obtained, and each leather bookbinding was chemically tested for the percentage of sulfate ash (x). After one year, the bookbinding was measured for the percentage of shrinkage (y). The data are given on the data CD and book's web site.
a. Find the estimated regression line, and complete the ANOVA table.
b. Find a 95% prediction interval for an observed value of shrinkage if the sulfate ash content is 1.6%.
c. Suppose the sulfate ash content is 0.8%. Is there any evidence to suggest that the true mean shrinkage is greater than 1%? Use a significance level of 0.01.

12.116 Economics and Finance Scientists at the U.S. Mint believe that the number of years a quarter is in circulation is linearly related to the condition of the coin. In order to test this theory, a random sample of quarters was obtained. The number of years in circulation (x) was recorded for each coin, and each quarter was assessed for wear using the Official American Numismatic Association Grading Standards for U.S. Coins (y, a 70-point scale). A simple linear regression analysis was performed, and the following partial ANOVA table was obtained.

ANOVA summary table					
Source of variation	**Sum of squares**	**Degrees of freedom**	**Mean square**	**F**	**p value**
Regression					
Error	7671.0				
Total	11,148.4	54			

Cycles	Length	Cycles	Length	Cycles	Length
128.0	0.193	118.1	0.165	199.0	0.191
246.4	0.167	100.7	0.191	100.8	0.177
188.8	0.209	205.9	0.208	131.1	0.207
152.3	0.180	107.6	0.153	145.8	0.193
129.9	0.153	182.5	0.169	229.1	0.191
101.5	0.172	132.8	0.187		

a. Complete the ANOVA table.

b. Find an estimate of the variance of the random error terms.

12.117 Biology and Environmental Science Every 13 or 17 years a brood of cicadas, sometimes in biblical proportions, emerges from underground in the northeastern part of the United States. Although harmless to humans, cicadas can cause damage to small trees and shrubs, and they can make a piercing, irritating sound. Some research suggests that the density of cicadas is related to the density of moles, natural predators of cicadas. Plots were randomly selected, and the density of moles per acre (x) and the density of cicadas per acre (y, in millions) were carefully estimated. The data are given in the following table.

x	4	6	3	0	2	4	0	5	4
y	1.5	0.8	1.2	1.4	1.5	1.5	1.7	1.2	1.1

a. Carefully sketch a scatter plot of the data. Describe the relationship between the density of moles and the density of cicadas.

b. Find the estimated regression line. Conduct a hypothesis test of $H_0: \beta_1 = 0$ versus $H_a: \beta_1 \neq 0$. Use a significance level of 0.05.

c. Construct a normal probability plot of the residuals and a scatter plot of the residuals versus the predictor variable. Is there any evidence that the regression assumptions are not satisfied?

12.118 Medicine and Clinical Studies High levels of C-reactive protein (CRP) have been associated with cardiovascular diseases.[25] Suppose a group of physicians studied the relationship between CRP level and body mass index (BMI), a measure of body fat, on the basis of height and weight. A random sample of 50-year-old women was obtained, and the CRP level (y, in mg/L) was measured for each subject. The BMI (x) for each woman was computed using her weight and height. The data are given on the data CD and book's web site.

a. Find the estimated regression line, and complete the ANOVA table.

b. Conduct a hypothesis test of $H_0: \beta_1 = 0.07$ versus $H_a: \beta_1 < 0.07$. Find the p value for this test.

c. Find the sample correlation coefficient, r.

d. Explain why r and $\hat{\beta}_1$ have the same sign.

12.119 Manufacturing and Product Development In structural tests of material to be used in a plane's fuselage, rivets are cycled (tapped) until they crack. The number of cycles (x, in thousands) is recorded along with the detectable crack length (y, in mm). The data are given in the following table.

a. Carefully sketch a scatter plot of these data. Does there appear to be a linear relationship between the number of cycles and the length of the crack? Explain.

b. Find the estimated regression line, and complete the ANOVA table.

c. How does the F test for a significant regression support your answer to part (a)?

12.120 Medicine and Clinical Studies Twenty-five adults were selected at random for a study of the effect of castor oil on the immune system. Participants took a certain amount of castor oil (x, in milliliters) each day for 3 months. At the end of this regimen, the white blood cell count was measured (y, thousands of cells per microliter of blood) for each person. The data are given on the data CD and book's web site. Analyze these data using any appropriate methods to determine whether there is a significant linear relationship between the amount of castor oil consumed and the number of white blood cells. Check the relevant assumptions. Assume that the more white blood cells there are, the stronger the immune system is, and draw a conclusion about the effect of castor oil on the immune system.

Extended Applications

12.121 Manufacturing and Product Development *Consumer Reports* recently conducted a study to determine the effect of thread count (x_1) and the percentage of cotton (x_2) on the lifetime of standard twin bed sheets (y, in years). A machine was constructed to simulate continued usage and to allow a measurement of the lifetime. Suppose an experiment was conducted and the data are given on the data CD and book's web site.

a. Estimate the regression coefficients in the model $Y_i = \beta_0 + \beta_1 x_{1i} + \beta_2 x_{2i} + E_i$.

b. Construct the ANOVA table and conduct the model utility test. Use technology to find the p value associated with this test. Find the value of r^2. Explain the meaning of this value.

c. Conduct the necessary hypothesis tests to determine whether both predictor variables are significant. Use an *overall* significance level of $\alpha = 0.05$.

d. Find a 99% prediction interval for an observed value of Y for $x_1 = 320$ and $x_2 = 100$.

e. Suppose a certain brand of bed sheets is sold with 50% cotton and various thread counts. What thread count would guarantee an estimate of the mean lifetime of at least 15 years?

12.122 Travel and Transportation Many of the highways in Pennsylvania are paved with concrete rather than tar. Suppose an experiment was conducted to determine whether the compression strength in cured concrete pavement (y, psi) is affected by the water-to-cement ratio (x_1), the sand-to-cement ratio (x_2), the percentage of fly ash (x_3), and/or the ambient temperature during curing (x_4, °F). The data are given on the data CD and book's web site.

a. Estimate the regression coefficients in the model $Y_i = \beta_0 + \beta_1 x_{1i} + \beta_2 x_{2i} + \beta_3 x_{3i} + \beta_4 x_{4i} + E_i$. Use the sign of each estimated regression coefficient to explain the effect of each predictor variable on the compression strength of concrete.

b. Construct the ANOVA table and conduct the model utility test. Use technology to find the p value associated with this test.

c. Conduct the appropriate hypothesis tests to determine the significant predictor variables.

d. Construct a normal probability plot of the residuals and the plots of the residuals versus each predictor variable. Is there any evidence of a violation in the multiple linear regression assumptions?

12.123 Sports and Leisure There were 28 college football bowl games at the conclusion of the 2005–2006 season, starting with the Wyndham New Orleans Bowl on December 20 and ending with the Rose Bowl on January 4. While the Rose Bowl was considered the national championship game, money is the real issue for many teams. A sample of these bowl games was obtained. The total number of season wins (before the bowl game) for both teams (x) and the per-team payout (y, in millions of dollars) for each game is given in Table 12.1.

a. Carefully sketch a scatter plot of these data. Describe the relationship between these two variables.

b. Find the estimated regression line, and complete the ANOVA table. Is there evidence to suggest that the total number of wins can be used to predict the per-team payout? Justify your answer.

c. Find a 95% confidence interval for the mean per-team payout for 20 total wins.

Table 12.1

Bowl	Teams	Wins	Per-team payout
Motor City	Memphis, Akron	13	0.750
Champ Sports	Clemson, Colorado	14	0.862
MasterCard Alamo	Nebraska, Michigan	14	1.650
Pacific Life Holiday	Oklahoma, Oregon	17	2.000
Gaylord Hotels Music City	Virginia, Minnesota	13	0.780
Vitalis Sun	UCLA, Northwestern	16	1.575
Independence	Missouri, South Carolina	13	1.200
Chick-fil-A Peach	LSU, Miami	19	2.400
AutoZone Liberty	Tulsa, Fresno State	16	1.500
Outback	Florida, Iowa	15	2.850
AT&T Cotton	Alabama, Texas Tech	18	2.500
Capital One	Wisconsin, Auburn	18	5.312
Tostitos Fiesta	Ohio State, Notre Dame	18	14.900
Rose	USC, Texas	23	14.900

(*Source*: *USA Today*, January 3, 2006.)

Categorical Data and Frequency Tables

13

Chapter 13 Challenge

Are all pharmaceutical companies researching the same types of drugs?

(Alexraths/Dreamstime.com)

The pharmaceutical industry is regulated by government agencies that review research results, clinical trials, safety, and effectiveness. A pharmaceutical company may test 5,000–10,000 different substances before discovering a single marketable drug. Once a new drug is discovered, it takes, on average, approximately 15 years to deliver the drug to a patient. This lengthy process is one reason the cost of drugs is so high.

Although companies search for new drugs to treat all types of diseases, the focus is on treatments for cancer, diseases of aging, and heart disease and stroke. A random sample of these potential treatments in clinical trials or awaiting federal approval was obtained, and each treatment was classified by pharmaceutical company and type of drug. The resulting frequencies, or counts, are given in the following two-way, or contingency, table.

	Type of drug		
Company	**Cancer**	**Diseases of aging**	**Heart disease and stroke**
Abbott Laboratories	18	10	5
Boehringer Ingelheim	20	15	6
Bristol-Myers Squibb	25	12	8
Merck	16	8	7

The techniques presented in this chapter will be used to determine whether the type of drug being developed is independent of pharmaceutical company. In the context of this problem, we would like to know whether these companies all have the same development strategies, or whether the type of drug research is dependent on company. The results are presented in the Chapter Challenge Wrap-Up (page 658).

Review

- Remember the definition of univariate categorical data: nonnumerical observations that may be placed in categories.
- Recall how bivariate categorical data is obtained and characterized: two nonnumerical observations on each individual or object.
- Think about the natural summary measures for categorical data: frequency and relative frequency.

CONTENTS

> **Preview**
> • Learn the background, computations, and interpretations of a goodness-of-fit test concerning the true population proportions.
> • Conduct a test for homogeneity or independence.

13.1 UNIVARIATE CATEGORICAL DATA, GOODNESS-OF-FIT TESTS

Categorical data are often displayed in a frequency distribution. In this section, we will focus only on the number of observations in each category and will display these results in a one-way frequency table. This table lists each possible category and the number of times each category occurred (the observed count or frequency for each category). For example, consider a survey of 200 people interested in reality TV shows. Each person was asked to select their favorite type of reality TV. There were four possible categorical responses, and the total for each response is given in the following one-way frequency table.

Favorite type of reality TV	Escaping remote locations	Dares	Breaking world records	Racing across country
Frequency	70	55	45	30

The hypothesis test procedure presented in this section is designed to compare a set of *hypothesized* proportions with the set of *true* proportions, to check the **goodness of fit**. For example, a producer might use the reality TV show data to determine whether there is any evidence that the true proportions are different from the following hypothesized proportions: 0.4 (for escaping remote locations), 0.3 (for dares), 0.2 (for breaking world records), and 0.1 (for racing across country).

The notation used here is an extension of the notation used to represent sample and population proportions. Suppose each observation falls into one of k categories.

<table>
<tr><th></th><th>True proportion</th><th>Hypothesized proportion</th></tr>
<tr><td>Category 1</td><td>p_1</td><td>p_{10}</td></tr>
<tr><td>Category 2</td><td>p_2</td><td>p_{20}</td></tr>
<tr><td>⋮</td><td>⋮</td><td>⋮</td></tr>
<tr><td>Category i</td><td>p_i</td><td>p_{i0}</td></tr>
<tr><td>⋮</td><td>⋮</td><td>⋮</td></tr>
<tr><td>Category k</td><td>p_k</td><td>p_{k0}</td></tr>
</table>

The subscript in the hypothesized proportion p_{10} is read as "one zero" not "ten." The first number in the subscript (1) denotes the category, and the second number (0) indicates that this is a hypothesized value.

The true proportion and the hypothesized proportion are both population proportions. The sum of the hypothesized proportions (like the sum of the true proportions) must be 1. That is, $p_{10} + p_{20} + \cdots + p_{k0} = 1$.

The goodness-of-fit test is done to determine whether there is any evidence (from the observed, or sample, counts) that the true population proportions differ from the hypothesized population proportions. The null hypothesis and the alternative hypothesis are stated in terms of the true and hypothesized category proportions.

H_0 is a composite null hypothesis. It involves several parameters and is true only if all the equalities hold.

$H_0: p_1 = p_{10}, p_2 = p_{20}, \ldots, p_k = p_{k0}$
(Each true category proportion is equal to a specified hypothesized value.)

$H_a: p_i \neq p_{i0}$ for at least one i.
(There is at least one true category proportion that is not equal to the corresponding specified hypothesized value.)

In this context, a *cell* is simply a category.

Suppose a random sample of size n is selected. Let n_i $(i = 1, 2, \ldots, k)$ be the number of observations falling into each category. In order to decide whether the sample data *fit* the hypothesized proportions, *observed* cell counts (the n_i's) are compared with *expected* cell counts. If the null hypothesis is true, then the expected frequency, or count, for category 1, or in cell 1, is $e_1 = np_{10}$. For cell 2, the expected frequency is $e_2 = np_{20}$, etc. For example, if the sample size is $n = 100$ and $p_{10} = 0.25$, then we would expect the count for category 1 to be, on average, $np_{10} = (100)(0.25) = 25$.

The test statistic is a measure of how far away the observed cell counts are from the expected cell counts. If the null hypothesis is true, the random variable

The X in X^2 is the uppercase Greek letter chi (χ).

$$X^2 = \sum_{\text{All cells}} \frac{(\text{Observed cell count} - \text{Expected cell count})^2}{\text{Expected cell count}} = \sum \frac{(n_i - e_i)^2}{e_i}$$

has approximately a chi-square distribution with $k - 1$ degrees of freedom. This approximation is good if $e_i = np_{i0} \geq 5$ for all i—that is, if all expected cell counts are at least 5.

If the observed cell counts are *close* to the expected cell counts, then the value of X^2 will be small. If the observed cell counts are considerably different from the expected cell counts, then the value of X^2 will be large. Therefore, the null hypothesis is rejected only for large values of the test statistic. Now we have all of the pieces for the complete hypothesis test.

GOODNESS-OF-FIT TEST

Let n_i be the number of observations falling into the ith category ($i = 1, 2, \ldots, k$), and let $n = n_1 + n_2 + \cdots + n_k$. A hypothesis test about the true category population proportions with significance level α has the following form:

$H_0: p_1 = p_{10}, \ p_2 = p_{20}, \ldots, \ p_k = p_{k0}$

$H_a: p_i \neq p_{i0}$ for at least one i.

TS: $X^2 = \sum \dfrac{(n_i - e_i)^2}{e_i}$ where $e_i = np_{i0}$

RR: $X^2 \geq \chi_\alpha^2$ df $= k - 1$

A reminder: This test is appropriate if all expected cell counts are at least 5 ($e_i = np_{i0} \geq 5$ for all i). The chi-square distribution was introduced in Section 8.5 and a hypothesis test based on this distribution was discussed in Section 9.7.

Example 13.1 Hobby Shop Purchases The Discount Hobby Warehouse sells parts, accessories, and complete kits related to model rockets, planes, trains, and race cars. A random sample of purchases was obtained, and each was classified into one of these four categories. Use the following one-way frequency table to test the hypothesis that the four possible hobby-type purchases occur with equal frequency. Use $\alpha = 0.05$.

Hobby	Rockets	Planes	Trains	Race cars
Frequency	27	31	22	20

Solution Trail

KEYWORDS

- Test the hypothesis.
- Four possible hobby-type purchases occur with equal frequency.

↓

TRANSLATION

- Statistical inference.
- True category population proportions are equal.

↓

CONCEPTS

Goodness-of-fit test.

↓

VISION

Find the expected cell counts, and use the goodness-of-fit test procedure to determine whether there is any evidence that any one of the true population proportions differs from 0.25.

SOLUTION

STEP 1 If the four hobby types are equally likely, the proportion of purchases falling into each category is $1/k = 1/4 = 0.25$.

The four parts of the hypothesis test are

$$H_0: p_1 = 0.25, \quad p_2 = 0.25, \quad p_3 = 0.25, \quad p_4 = 0.25$$
$$H_a: p_i \neq p_{i0} \text{ for at least one } i$$

$$\text{TS}: X^2 = \sum \frac{(n_i - e_i)^2}{e_i}$$

$$\text{RR}: X^2 \geq \chi_\alpha^2 = \chi_{0.05}^2 = 7.8147 \qquad \text{From Table VI in the Appendix, df = 3.}$$

STEP 2 There are $n = 27 + 31 + 22 + 20 = 100$ total observations. The expected cell counts are given in the following table.

Cell	Category	Observed cell count	Expected cell count
1	Rockets	27	$e_1 = np_{10} = (100)(0.25) = 25$
2	Planes	31	$e_2 = np_{20} = (100)(0.25) = 25$
3	Trains	22	$e_3 = np_{30} = (100)(0.25) = 25$
4	Race cars	20	$e_4 = np_{40} = (100)(0.25) = 25$

All four expected cell counts are greater than 5. The chi-square goodness-of-fit test is appropriate.

STEP 3 The value of the test statistic is

$$\chi^2 = \sum \frac{(n_i - e_i)^2}{e_i}$$

$$= \frac{(27 - 25)^2}{25} + \frac{(31 - 25)^2}{25} + \frac{(22 - 25)^2}{25} + \frac{(20 - 25)^2}{25}$$

Use observed and expected cell counts.

$$= 0.16 + 1.44 + 0.36 + 1.00 = 2.96 \; (< 7.8147) \qquad \text{Simplify.}$$

STEP 4 The value of the test statistic does not lie in the rejection region. At the $\alpha = 0.05$ significance level, there is no evidence to suggest that any one of the true population proportions differs from 0.25.

Figure 13.1 shows a technology solution.

---

p value illustration:
$$p = P(X \geq 2.96)$$
$$= 0.3978 > 0.05 = \alpha$$

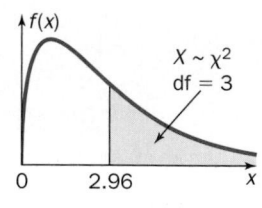

Recall that using Table VI in the Appendix, we can only *bound* the *p* value for this hypothesis test. Minitab or the TI-84 can be used to find the exact *p*-value for this test.

	A	B	C	D
1		Hypothesized	Observed	Expected
2	Category	proportion	cell count	cell count
3	Rockets	0.25	27	25
4	Planes	0.25	31	25
5	Trains	0.25	22	25
6	Race cars	0.25	20	25
7	Total	1.00	100	100
8				
9	0.3978	=CHITEST(C3:C6,D3:D6)		
10	2.96	=CHIINV(A9,3)		

Figure 13.1 Excel goodness-of-fit test.

Example 13.2 Favorite TV Sport A television executive is trying to decide which sport to bid on in order to obtain the broadcast rights for next season. A random sample of adults was obtained, and each was asked to name their favorite sport (from a list of nine) to watch on TV. The data and historical proportions are given in the following table.

TV sport	Frequency	Historical proportion
Auto racing	84	0.10
Baseball	119	0.22
Fishing	16	0.05
Football	213	0.33
Golf	36	0.06
Hunting	17	0.03
Soccer	79	0.13
Tennis	43	0.05
Track and field	13	0.03

(*Source*: strangesports.com.)

Is there evidence to suggest that any of the true cell proportions differs from the historical proportions? Use $\alpha = 0.01$.

SOLUTION

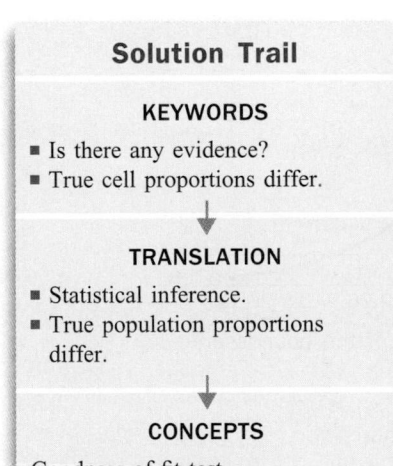

Solution Trail

KEYWORDS

- Is there any evidence?
- True cell proportions differ.

↓

TRANSLATION

- Statistical inference.
- True population proportions differ.

↓

CONCEPTS

Goodness-of-fit test.

↓

VISION

Find the expected cell counts, and use the goodness-of-fit test procedure to determine whether there is any evidence that any one of the true population proportions differs from the historical proportions.

STEP 1 A goodness-of-fit test is appropriate to determine whether there is evidence that any of the true category population proportions have changed. There are $k = 9$ categories.

The four parts of the hypothesis test are

$H_0: p_1 = 0.10, \quad p_2 = 0.22, \quad p_3 = 0.05, \quad p_4 = 0.33, \quad p_5 = 0.06,$
$\qquad p_6 = 0.03, \quad p_7 = 0.13, \quad p_8 = 0.05, \quad p_9 = 0.03$

$H_a: p_i \neq p_{i0}$ for at least one i

TS: $X^2 = \sum \dfrac{(n_i - e_i)^2}{e_i}$

RR: $X^2 \geq \chi^2_\alpha = \chi^2_{0.01} = 20.0902$ Table VI in the Appendix, df = 8.

STEP 2 There are $n = 84 + 119 + 16 + 213 + 36 + 17 + 79 + 43 + 13 = 620$ total observations. The expected cell counts are given in the following table.

Cell	TV sport	Observed cell count	Expected cell count
1	Auto racing	84	$e_1 = np_{10} = (620)(0.10) = 62.00$
2	Baseball	119	$e_2 = np_{20} = (620)(0.22) = 136.40$
3	Fishing	16	$e_3 = np_{30} = (620)(0.05) = 31.00$
4	Football	213	$e_4 = np_{40} = (620)(0.33) = 204.60$
5	Golf	36	$e_5 = np_{50} = (620)(0.06) = 37.20$
6	Hunting	17	$e_6 = np_{60} = (620)(0.03) = 18.60$
7	Soccer	79	$e_7 = np_{70} = (620)(0.13) = 80.60$
8	Tennis	43	$e_8 = np_{80} = (620)(0.05) = 31.00$
9	Track and field	13	$e_9 = np_{90} = (620)(0.03) = 18.60$

All expected cell counts are greater than 5. The chi-square goodness-of-fit test is appropriate.

STEP 3 The value of the test statistic is

$$\chi^2 = \sum \frac{(n_i - e_i)^2}{e_i}$$

$$= \frac{(84 - 62.00)^2}{62.00} + \frac{(119 - 136.40)^2}{136.40} + \frac{(16 - 31.00)^2}{31.00}$$

$$+ \frac{(213 - 204.60)^2}{204.60} + \frac{(36 - 37.20)^2}{37.20} + \frac{(17 - 18.60)^2}{18.60}$$

$$+ \frac{(79 - 80.60)^2}{80.60} + \frac{(43 - 31.00)^2}{31.00} + \frac{(13 - 18.60)^2}{18.60}$$

Use observed and expected cell counts.

$$= 7.81 + 2.22 + 7.26 + 0.34 + 0.04 + 0.14 + 0.03 + 4.65 + 1.69$$

$$= 24.18 \ (\geq 20.0902)$$ Simplify.

If you were bidding for broadcasting rights, which sport would you choose? Why?

STEP 4 The value of the test statistic ($\chi^2 = 24.18$) lies in the rejection region. At the $\alpha = 0.01$ significance level, there is evidence to suggest that at least one population proportion has changed from its historical value. *Note*: We can use Table VI in the Appendix to bound the p value for this hypothesis test. In row $k - 1 = 8$ (degrees of freedom), place 24.18 in the ordered list of critical values.

$$21.9550 \leq 24.18 \leq 26.1245$$
$$\chi^2_{0.005} \leq 24.18 \leq \chi^2_{0.001}$$

Therefore, $\quad 0.001 \leq \boxed{p} \leq 0.005$ See Figure 13.2.

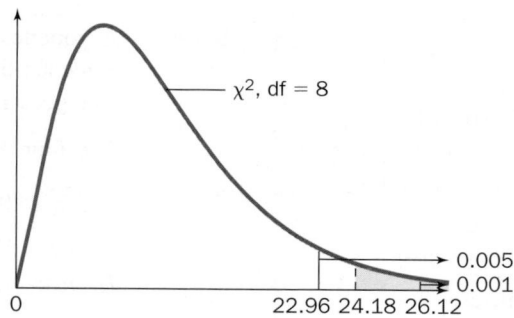

χ^2, df = 8

0.005
0.001

0   22.96 24.18 26.12

Figure 13.2 Visualization of the bounds on the p value.

Figure 13.3 shows a technology solution.

	A	B	C	D
1		Hypothesized	Observed	Expected
2	TV Sport	proportion	cell count	cell count
3	Auto racing	0.10	84	62.00
4	Baseball	0.22	119	136.40
5	Fishing	0.05	16	31.00
6	Football	0.33	213	204.60
7	Golf	0.06	36	37.20
8	Hunting	0.03	17	18.60
9	Soccer	0.13	79	80.60
10	Tennis	0.05	43	31.00
11	Track & Field	0.03	13	18.60
12	Total	1.00	620	620
13				
14	0.0021	=CHITEST(C3:C11,D3:D11)		
15	24.17	=CHIINV(A14,8)		

Figure 13.3 Excel goodness-of-fit test. ●

TECHNOLOGY CORNER

Procedure: Chi-square goodness-of-fit test.
Reconsider: Example 13.2, page 637, solution, and interpretations.

TI-84 Plus

Use the built-in function χ^2GOF-Test.

1. Enter the observed frequencies into list L1 and the expected frequencies into list L2.
2. Select [STAT]; TESTS; χ^2GOF-Test. Enter the list containing the observed frequencies, the list containing the expected frequencies, and the degrees of freedom associated with the test statistics.
3. Highlight Calculate and press [ENTER]. The hypothesis test results are displayed on the Home Screen. See Figure 13.4. The Draw results are shown in Figure 13.5.

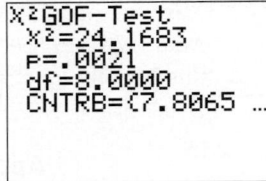

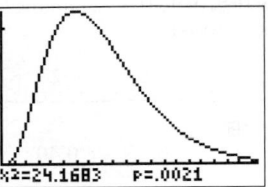

Figure 13.4
χ^2GOF-Test
hypothesis test results.

Figure 13.5
χ^2GOF-Test Draw
results.

Minitab

The chi-square goodness-of-fit test is in the Stat; Tables menu.

1. Enter the category names into column C1 (optional), the hypothesized proportions into column C2, and the observed counts into column C3.
2. Select Stat; Tables; Chi-Square Goodness-of-Fit Test (One variable). Enter the Observed counts column, the Category names column, and the Specific proportions column.
3. Click OK. The results are displayed in a session window. See Figure 13.6. This Minitab procedure also automatically produces a chart of observed and expected values by category, and a chart of the contribution to the value of the chi-square statistic by category.

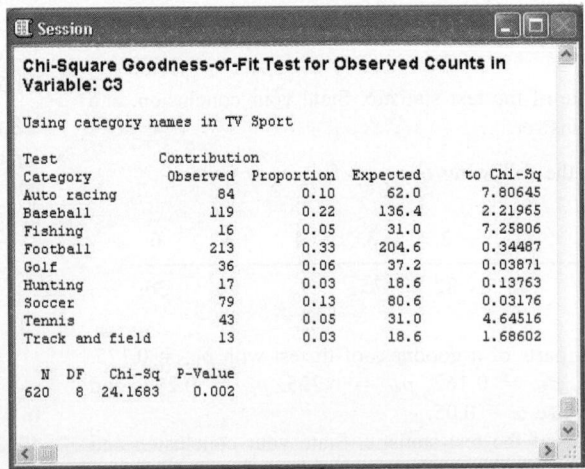

Figure 13.6 Minitab goodness-of-fit test.

> **Excel** ——————————————
>
> Use the built-in function CHITEST.
>
> **1.** Enter the observed cell counts into column B and compute or enter the expected cell counts into column C.
> **2.** Use the function CHITEST to find the *p* value associated with the goodness-of-fit test and the function CHIINV to find the value of the test statistic. See Figure 13.3, page 638.

SECTION 13.1 EXERCISES

Practice

13.1 Use the hypothesized proportions in the following frequency table to find the expected count for each category.

Category	1	2	3	4
Frequency	42	58	62	56
p_{i0}	0.25	0.40	0.20	0.15

13.2 Use the hypothesized proportions in the following frequency table to find the expected count for each category.

Category	1	2	3	4	5
Frequency	125	150	300	220	205
p_{i0}	0.15	0.14	0.31	0.23	0.17

13.3 Consider the following one-way frequency table.

Category	1	2	3	4
Frequency	115	85	70	30

a. Suppose the hypothesized proportions are $p_{10} = 0.4$, $p_{20} = 0.3$, $p_{30} = 0.2$, and $p_{40} = 0.1$. Find the four parts of the appropriate goodness-of-fit test. Use $\alpha = 0.01$.
b. Find the expected count for each cell. Verify that each is at least 5.
c. Find the value of the test statistic. State your conclusion, and justify your answer.

13.4 Consider the following one-way frequency table.

Category	1	2	3	4	5	6
Frequency	90	82	75	95	96	36

a. Find the four parts of a goodness-of-fit test with $p_{10} = 0.175$, $p_{20} = 0.171$, $p_{30} = 0.162$, $p_{40} = 0.225$, $p_{50} = 0.202$, and $p_{60} = 0.065$. Use $\alpha = 0.05$.
b. Find the value of the test statistic. State your conclusion and justify your answer.
c. Find bounds on the *p* value.

13.5 Consider the following one-way frequency table.

Category	1	2	3	4	5
Frequency	140	135	155	152	168

Conduct a goodness-of-fit test to check the hypothesis that all five categories are equally likely. Use $\alpha = 0.05$. Find bounds on the *p* value.

Applications

13.6 Psychology and Human Behavior A random sample of adult males was obtained and each was asked whether he belonged to a fraternal organization. The results are given in the following one-way frequency table.

Lodge	Elks	Moose	Raccoon	None
Frequency	150	60	60	230

Past research indicated that 25% of all men belonged to an Elks Lodge, 15% belonged to a Moose Lodge, and 10% belonged to a Raccoon Lodge. Conduct a goodness-of-fit test to determine whether there is evidence that at least one of the population proportions associated with the four categories above has changed. Use $\alpha = 0.05$.

13.7 Marketing and Consumer Behavior In a random sample of Caribou coffee-shop customers who purchased flavor shots, 185 bought almond, 180 purchased vanilla, 230 chose raspberry, 220 opted for caramel, and 185 went with hazelnut. Use a goodness-of-fit test to determine whether there is any evidence that customers who use flavor shots prefer one over the others. Use $\alpha = 0.05$.

13.8 Travel and Transportation A random sample of adults staying at hotels, motels, or bed-and-breakfast establishments was obtained, and each was asked to indicate his or her total trip duration. The data are summarized in the following one-way frequency table.

Trip duration	1–2 nights	3–6 nights	7 or more nights
Frequency	125	115	40

In a 2008 survey by the Travel Industry Association of America, 47% indicated one or two nights, 38% indicated three to six nights, and 15% said seven or more nights. Is there evidence to suggest that any of these percentages have changed? Use $\alpha = 0.01$.

13.9 Biology and Environmental Science In a recent study of adult eye color, the hypothesized distribution was given as blue 25%, green 10%, brown 50%, and black 15%. A random sample of adults was obtained, and the resulting eye colors are summarized in the following one-way frequency table.

Eye color	Blue	Green	Brown	Black
Frequency	121	65	242	72

Conduct a goodness-of-fit test to determine whether these data provide any evidence that the true proportions differ from the hypothesized proportions. Use $\alpha = 0.05$.

13.10 Marketing and Consumer Behavior Original Designs sells home plans in five different styles. A random sample of purchases was obtained, and the number falling into each category was recorded. The data are given in the following table.

Style	A-frame	Cape Cod	Colonial	Log home	Ranch
Frequency	75	180	268	58	385

Conduct a goodness-of-fit test to determine whether there is evidence that any of the true proportions differs from the hypothesized values 0.10, 0.20, 0.25, 0.05, and 0.40. Use $\alpha = 0.05$.

13.11 Sports and Leisure A random sample of 4-H variety show acts was obtained and classified. The resulting data along with the hypothesized proportions are given in the following table.

Act type	Frequency	p_{i0}
Singing	90	0.16
Instrumental	82	0.18
Juggling	21	0.03
Magic	33	0.07
Dance	80	0.21
Comedy	130	0.35

Conduct a goodness-of-fit test to determine whether there is evidence that any of the true proportions differs from the hypothesized value. Use $\alpha = 0.01$.

13.12 Marketing and Consumer Behavior The manager at an HEB grocery store is trying to determine which types of lettuce to order and sell to customers. A random sample of shoppers was obtained, and each was asked to select his or her favorite lettuce type. The data are given in the following table.

Lettuce	Frequency
Crisphead	45
Butterhead	17
Romaine	130
Leafy	90
Celtuce	28

Test the fit of these data to the hypothesized proportions 0.14, 0.06, 0.37, 0.30, and 0.13. Use $\alpha = 0.05$.

13.13 Marketing and Consumer Behavior A Girl Scout troop is preparing to order extra cookies for a Saturday morning sale in a store parking lot. A random sample of potential buyers was obtained, and each was asked to name his or her favorite Girl Scout cookie. The results and the percentages sold nationally last year are given in the following table.

Cookie	Frequency	Last year percentages
Thin Mints	120	25
Samoas/ Caramel deLites	95	19
Peanut Butter Patties/ Tagalongs	70	13
Peanut Butter Sandwich/ Do-Si-Do's	58	11
Shortbread/Trefoils	60	9
All other varieties	101	23

(*Source:* girlscouts.org.)

Conduct a goodness-of-fit test to determine whether there is evidence that any of this year's true population proportions differs from last year's proportion. Use $\alpha = 0.05$.

13.14 Business and Management In a recent study, a random sample of small-business owners was obtained, and each was asked to name the biggest problem facing his or her company. The results are given in the following one-way frequency table.

Problem	Frequency
Cost of health insurance	430
Liability insurance	145
Worker's compensation	135
Cost of natural gas, propane, gasoline, diesel, fuel oil	90

(*Source*: National Federation of Independent Business.)

In an economic report, the true proportions for each category were given as 0.50, 0.20, 0.20, and 0.10. Do these data provide any evidence to contradict the economic report? Find bounds on the *p* value associated with this test.

13.15 Demographics and Population Statistics The admissions offices at the University of California campuses maintain very careful historical records of applicants. In 2008, the proportions of students applying to colleges within the university system by California location were as follows: Los Angeles (LA), 29.2%; San Francisco Bay Area (SF), 26.1%; Orange County (OC), 9.9%; Riverside/San Bernadino (RS), 7.7%; all other locations (AO), 27.1%.[1] Suppose a random sample of applicants in 2009 was obtained and the frequency associated with each location is given in the following table.

Location	LA	SF	OC	RS	AO
Frequency	125	96	45	44	128

Is there any evidence of a shift in the proportion of applications by California location? Use $\alpha = 0.005$.

13.16 Sports and Leisure As soon as the gates at an amusement park open, there's mad rush to the most attractive rides. A random sample of people waiting in line was obtained, and each was asked which ride he or she was headed to first. The results and the hypothesized proportions are given in the following table.

Ride	Frequency	p_{i0}
Tower of Terror	63	0.40
Rockin' Roller Coaster	35	0.30
Star Tours	16	0.15
Studios Backlot Tour	14	0.15

Conduct a goodness-of-fit test to determine whether any of the true population proportions differs from the hypothesized proportion. Use $\alpha = 0.01$.

13.17 Education and Child Development According to a 2003 study by the Department of Education, approximately 30 million Americans have a below-basic prose literacy level. For example, they do not have the skills to read simple text to determine what a patient is allowed to drink before a medical test. Suppose a random sample of adults was obtained in 2009, and the prose literacy of each person was determined. The data are given in the following table.

Literacy level	Frequency
Below basic	131
Basic	295
Intermediate	456
Proficient	110

In 2003, the proportions for each category were as follows: below basic, 0.14; basic, 0.29; intermediate, 0.44; proficient, 0.13.[2] Conduct a goodness-of-fit test to determine whether there is any evidence to suggest that the true population proportions have changed. Use $\alpha = 0.01$.

Extended Applications

13.18 Psychology and Human Behavior Berkeley Breathed stopped writing his Pulitzer-Prize–winning comic strip *Bloom County* in 1989. Since then, he has written several books and animations. Recently, Mr. Breathed has become involved in writing a children's movie and two sequel comic strips, *Outland* and *Opus*, featuring some of the old *Bloom County* gang. A random sample of people who read the original *Bloom County* comic strip was obtained, and each was asked to name his or her favorite character. The results are given in the following table.

Character	Frequency
Opus	250
Michael Binkley	210
Oliver Wendell Jones	205
Milo Bloom	190
Bill the Cat	260
Cutter John	195
Steve Dallas	201
Portnoy	206
Hodge Podge	185
Rosebud	204

Is there any evidence to suggest that one or more characters are more popular than the others? Find bounds on the p value associated with this test.

13.19 Travel and Transportation One consequence of 9/11 has been an increase in the quality and quantity of airport shopping malls. The volume of travelers offers a great retail opportunity, especially when thousands of people are stuck in an airport terminal due to delayed flights.[3] Suppose a random sample of frequent business travelers was obtained, and each was asked to select the best airport mall from a list of 10. The data are given in the following table.

Airport	Frequency
Portland International (PDX)	38
Minneapolis–St. Paul International (MSP)	45
San Francisco International (SFO)	55
Charlotte Douglas International (CLT)	53
Denver International (DIA)	46
Detroit Metro (DTW)	62
Dallas/Fort Worth International (DFW)	35
Nashville International (BNA)	45
Logan International (BOS)	56
Orlando International (MCO)	65

Conduct a goodness-of-fit test to determine whether there is any evidence to suggest that one airport mall is preferred over the rest. Use $\alpha = 0.05$.

13.20 Business and Management In 2007, Airbus had 1458 orders and delivered 453 aircraft, the highest level ever for an aircraft manufacturer. The following table contains the historical total orders worldwide for all Airbus families through November 2008[4] and a random sample of Airbus orders from the last quarter of 2009.

Aircraft family	Total orders	Sample 2009 orders
A318/A319/A320/A321	6303	1270
A300/A310	816	208
A330/A340/A350	1877	375
A380	198	51

a. Compute the proportion of historical orders for each aircraft family.
b. Conduct a goodness-of-fit test to determine whether there is any evidence that the true historical proportions of orders by aircraft family have changed in 2009. Use $\alpha = 0.01$.

13.21 Fuel Consumption and Cars There were approximately 13 million automobiles sold in the United States in 2008, a decrease of 18% over 2007 sales. General Motors and Toyota reported the highest sales, and despite the overall decrease in sales, Asian automobiles continued to increase their share of the U.S. market. The following table contains the 2008 U.S. auto sales data[5] and a summary of a random sample of automobile sales in the United States in 2009.

Company	Total 2008 sales	Sample 2009 sales
General Motors	2,955,860	440
Ford Motor Company	1,942,041	279
DaimlerChrysler	1,697,458	255
Toyota Motors	2,217,660	320
American Honda	1,428,765	175
Nissan N. America	951,446	130
Hyundai Motor	401,742	53
Volkswagen	310,888	35
BMW	303,190	33
Kia Motors	273,397	30
Other	712,116	81

Compute the proportion of 2008 sales for each company. Use this information to conduct a goodness-of-fit test to determine whether there is any evidence to suggest that the true 2008 population proportions of sales by company have changed in 2009. Use $\alpha = 0.01$.

Challenge

13.22 Business and Management In Britain, a small baker does not have a fully automatic plant and sells the majority of his or her production on-site or from vehicles. According to the Birmingham City Council trading standards, the law states that the average weight of one loaf type must be 400 grams. Suppose the Small Baker's Association (SBA) claims that the weight of each loaf of bread of this type is approximately normal with mean 400 grams and standard deviation 15 grams. In order to check this claim, a random sample of small-baker loaves was obtained, and each was carefully weighed. The observed frequency of weights in each specified interval (in grams) is summarized in the following one-way table.

Interval	Frequency
<370	18
370–385	67
385–400	175
400–415	184
415–430	75
≥430	14

a. Assume the SBA claim is true. Find the probability that a randomly selected loaf of bread falls into each interval.

b. Use the probabilities computed in part (a) as the hypothesized population proportions associated with each interval. Conduct a goodness-of-fit test to determine whether the observed weights *fit* the hypothesized distribution. Use $\alpha = 0.05$. State your conclusion.

Note: The goodness-of-fit test provides a formal test for normality. It complements the methods used to check for normality in Section 6.3.

13.2 BIVARIATE CATEGORICAL DATA, TESTS FOR HOMOGENEITY AND INDEPENDENCE

If *any* two observations are made on an individual or object, the data set is bivariate. For example, one observation might be categorical and the other numerical, or both observations could be numerical.

If two categorical observations are made on the same individual or object, the data set is **bivariate**. This type of data arises in two common ways.

1. Random samples are obtained from two or more populations, and each individual is classified by values of a categorical variable.

2. Suppose there are two categorical variables of interest. In a (single) random sample, a value of each variable is recorded for each individual.

The test for homogeneity applies to the first kind of data (samples from two or more populations) and the test for independence applies to the second kind of data (data from a single sample, with two categorical variables).

Let's focus on the first type of bivariate categorical data. Suppose random samples of college-bound students are obtained from three western Texas school districts, and the type of college that each will attend is recorded. The districts are the *populations* and the college type is the *categorical variable*. The data is *bivariate* because there are two values (district and college type) for, or associated with, each student. The data may be recorded in the following manner.

Student	District	College type
1	Lubbock	Community college
2	Amarillo	Technical college
3	Midland	State school
4	Amarillo	Private school
⋮	⋮	⋮
250	Amarillo	State school

The natural summary for this type of bivariate data set (in which each observation is categorical) is the number of observations in each category *combination*. For example, compute the number of students (or frequency) in the Lubbock school district *and* attending a community college, the number of students in the Lubbock school district *and* attending a technical college, etc. This summary information can be displayed in a 3 × 4 **two-way frequency table**, or **contingency table**, shown in Table 13.1. Each cell in this table contains the number of observations (observed count or frequency) in a category combination, or pairing.

Two-way tables are described by the number of rows and the number of columns. For example, a 2 × 6 contingency table has two rows and six columns.

Table 13.1 A two-way frequency table for the data obtained from college-bound students in western Texas

	College type			
	State	**Private**	**Technical**	**Community**
Amarillo	14	20	12	18
Lubbock	22	16	10	24
Midland	26	30	28	30

The columns of the two-way table in Figure 13.1 correspond to college types, and the rows correspond to districts. Each *cell* in the body of the table contains a frequency, or observed cell count. For example, there were 10 students from the Lubbock district who will be attending a technical school, and there were 30 students from the Midland district who will be attending a community college.

If we consider a *single* district (population), say Amarillo, then a bar chart may be used to represent the distribution of college type (Figure 13.7). A bar chart may be constructed for each district. For example, Figure 13.8 is another bar chart associated with these data, a summary of college type for Lubbock.

Bar charts were introduced in Chapter 2.

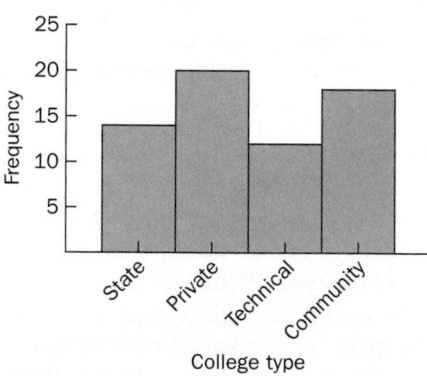

Figure 13.7 A bar chart showing the frequency of college types for students from the Amarillo district.

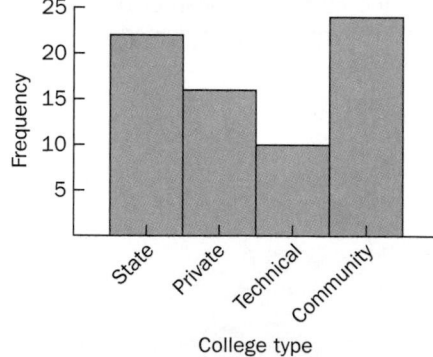

Figure 13.8 A bar chart showing the frequency of college types for students from the Lubbock district.

A side-by-side or stacked bar chart may be used to compare categorical data from two or more sources, or populations. Figure 13.9 shows a side-by-side bar chart of college-type frequencies grouped by district. Figure 13.10 shows a stacked bar chart of college-type frequencies grouped by district.

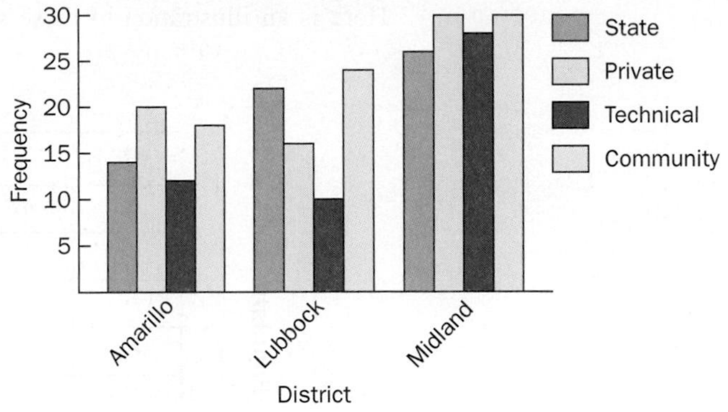

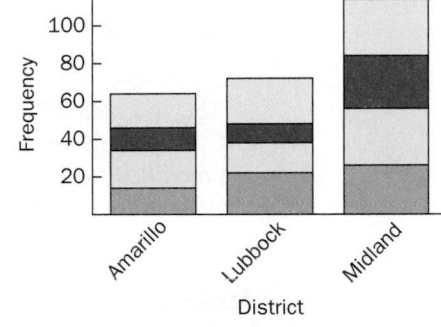

Figure 13.9 A side-by-side bar chart showing the frequency of occurrence of each college type, by district.

Figure 13.10 A stacked bar chart showing the frequency of occurrence of each college type, by district.

Although these graphs are useful for suggesting possible differences among populations, there is a more precise statistical test. The practical problem to consider is, are all of the true category proportions the same for each population? This is a test for **homogeneity** of populations. Homogeneity is the state of having identical properties of values; in this case, it refers to populations having identical true category proportions. In the district and college-type example, it seems reasonable to ask whether the proportion of students attending each type of college is the same for each district. The statistical procedure used to analyze this problem is based on observed and expected cell counts (as in Section 13.1). Under the null hypothesis that the populations have the same category proportions, the test statistic also has a chi-square distribution.

Suppose there are I rows and J columns in a two-way frequency table. The notation associated with this table includes the *dot* notation in a subscript (introduced in Chapter 11) to indicate a sum over that subscript while the other subscript is held fixed.

n_{ij} = observed cell count, or frequency, in the (ij) cell (the intersection of the ith row and the jth column)

$$n_{i.} = \sum_{j=1}^{J} n_{ij}$$

= ith row total, the sum of the cell counts, or observed frequencies, in the ith row

$$n_{.j} = \sum_{i=1}^{I} n_{ij}$$

= jth column total, the sum of the cell counts, or observed frequencies, in the jth column

$$n = \sum_{i=1}^{I} \sum_{j=1}^{J} n_{ij}$$

= grand total, the total of all cell counts, or observed frequencies

Here is an illustration of these symbols in an $I \times J$ two-way frequency table.

		Category						
		1	2	...	j	...	J	Row total
Population	1	n_{11}	n_{12}	...	n_{1j}	...	n_{1J}	$n_{1.}$
	2	n_{21}	n_{22}	...	n_{2j}	...	n_{2J}	$n_{2.}$
	⋮	⋮	⋮	⋮	⋮	⋮	⋮	⋮
	i	n_{i1}	n_{i2}	...	n_{ij}	...	n_{iJ}	$n_{i.}$
	⋮	⋮	⋮		⋮	⋮	⋮	⋮
	I	n_{I1}	n_{I2}	...	n_{Ij}	...	n_{IJ}	$n_{I.}$
Column total		$n_{.1}$	$n_{.2}$	...	$n_{.j}$	...	$n_{.J}$	n

The row and column totals are used to compute the expected cell counts. The district and college-type data will be used to illustrate these calculations. Table 13.2 is a modified two-way table containing the observed cell counts, the row and column totals, and the grand total.

Table 13.2 A modified two-way frequency table including the row and column totals and the grand total for the data obtained from college-bound students in western Texas

		College type				
		State	Private	Technical	Community	Row total
District	Amarillo	14	20	12	18	64
	Lubbock	22	16	10	24	72
	Midland	26	30	28	30	114
Column total		62	66	50	72	250

There were 250 individuals in this study, and 62 will be attending a state school. The proportion of all individuals in the data set attending a state school is $62/250 = 0.248$. Suppose there is no difference in the proportion of students attending a state school among districts. Then we would expect 24.8% of the students from the Amarillo district to attend a state school. This expected frequency in the (11) cell is denoted e_{11} and is computed by

Here, as in Section 13.1, the expected cell count may not be an integer.

$$e_{11} = 0.248 \times 64 = \frac{62}{250} \times 64 = \frac{(64)(62)}{250}$$

$$= \frac{(\text{1st row total})(\text{1st column total})}{\text{Grand total}} = \frac{n_{1.} \times n_{.1}}{n} = 15.87$$

Similarly, we expect 24.8% of the students from the Lubbock district to attend a state school. The expected cell count in the (21) cell is

The *e* here is used to denote expected frequency and is not connected in any way with the random errors or residuals in Chapter 12.

$$e_{21} = 0.248 \times 72 = \frac{62}{250} \times 72 = \frac{(72)(62)}{250}$$

$$= \frac{(\text{2nd row total})(\text{1st column total})}{\text{Grand total}} = \frac{n_{2.} \times n_{.1}}{n} = 17.86$$

The expected counts in column 2 are computed in a similar manner. There are 66 students who will be attending a private college. The proportion of all students attending a private college is $66/250 = 0.264$. If there is no difference in the proportion of students attending a private college among districts, then we expect 26.4% of the students, or

$$e_{12} = 0.264 \times 64 = \frac{66}{250} \times 64 = \frac{(64)(66)}{250}$$

$$= \frac{(1\text{st row total})(2\text{nd column total})}{\text{Grand total}} = \frac{n_{1.} \times n_{.2}}{n} = 16.90$$

students from the Amarillo district to attend a private school. We continue in the same manner to compute all the expected counts. Table 13.3 shows each expected count in parentheses beneath the corresponding observed count.

Table 13.3 A modified two-way table including the row and column totals, the grand total, and the expected cell counts for the data obtained from college-bound students in western Texas

		College type				
		State	Private	Technical	Community	Row total
District	**Amarillo**	14 (15.87)	20 (16.90)	12 (12.80)	18 (18.43)	64
	Lubbock	22 (17.86)	16 (19.01)	10 (14.40)	24 (20.74)	72
	Midland	26 (28.27)	30 (30.10)	28 (22.80)	30 (32.83)	114
	Column total	62	66	50	72	250

The computations in this example suggest an easy formula for finding the expected frequencies. In an $I \times J$ two-way table, the expected count, or frequency, in the (ij) cell can be written as

$$e_{ij} = \frac{(i\text{th row total})(j\text{th column total})}{\text{Grand total}} = \frac{n_{i.} \times n_{.j}}{n}$$

The test statistic is a measure of how far away the observed cell counts are from the expected cell counts. If there is no difference in category proportions among populations, the random variable

$$X^2 = \sum_{\text{All cells}} \frac{(\text{Observed cell count} - \text{Expected cell count})^2}{\text{Expected cell count}} = \sum_{i=1}^{I} \sum_{j=1}^{J} \frac{(n_{ij} - e_{ij})^2}{e_{ij}}$$

has approximately a chi-square distribution with $(I - 1)(J - 1)$ degrees of freedom. This approximation is good if $e_{ij} \geq 5$ for all i and j; that is, if all expected cell counts are at least 5.

If the observed cell counts are close to the expected cell counts, then the value of X^2 will be small. If the observed cell counts are considerably different from the expected cell counts, then the value of X^2 will be large. As in Section 13.1, the null hypothesis is rejected only for large values of the test statistic.

For the data obtained from the college-bound students in western Texas, the value of the test statistic is

$$\chi^2 = \sum_{i=1}^{3} \sum_{j=1}^{4} \frac{(n_{ij} - e_{ij})^2}{e_{ij}}$$

$$= \frac{(14 - 15.87)^2}{15.87} + \frac{(20 - 16.90)^2}{16.90} + \cdots + \frac{(30 - 32.83)^2}{32.83} = 5.760$$

At the $\alpha = 0.05$ level, this test is not significant, since for $\alpha = 0.05$ the critical value is

$$\chi_\alpha^2 = \chi_{0.05}^2 = 12.5916$$

$$\text{df} = (I - 1)(J - 1) = (2)(3) = 6.$$

TEST FOR HOMOGENEITY OF POPULATIONS

In an $I \times J$ two-way frequency table, let n_{ij} be the observed count in the (ij) cell and let e_{ij} be the expected count in the (ij) cell. A hypothesis test for homogeneity of populations with significance level α has the following form:

H_0: The true category proportions are the same for all populations (homogeneity of populations)

H_a: The true category proportions are not the same for all populations

TS: $X^2 = \sum_{i=1}^{I} \sum_{j=1}^{J} \frac{(n_{ij} - e_{ij})^2}{e_{ij}}$

where $e_{ij} = \dfrac{(i\text{th row total})(j\text{th column total})}{\text{Grand total}} = \dfrac{n_{i.} \times n_{.j}}{n}$

RR: $X^2 \geq \chi_\alpha^2$ df $= (I - 1)(J - 1)$

A reminder: This test is appropriate if all expected cell counts are at least 5 ($e_{ij} \geq 5$ for all i and j). Although this procedure is called a test of homogeneity, this property is stated in the null hypothesis, so that we are really testing for evidence of inhomogeneity. We cannot *prove* homogeneity; we can only *test* for evidence of inhomogeneity.

Example 13.3 Dining Out Random samples of men and women were obtained, and each person was asked to indicate the number of times a week he or she eats the evening meal out, or simply, dines out. The observed frequencies are given in the table below. Is there any evidence to suggest that the true category proportions of dining out are different for men and for women? Use $\alpha = 0.05$.

		Dine out				
		Three or more times	Twice	Once	Less than once	Never
Gender	**Men**	23	19	34	10	12
	Women	14	14	39	17	15

(*Source*: Food Marketing Institute, 2004.)

SOLUTION

STEP 1 This is a test for homogeneity of populations. There are two populations, men and women, and five categories for dining ($I = 2$, $J = 5$). We would like to know whether the true proportions of dining out categories are the same for each population.

(Banana Stock/AgeFotostock)

Solution Trail

KEYWORDS

- Is there any evidence?
- True category proportions . . . are different.

↓

TRANSLATION

- Statistical inference.
- Population proportions are different.

↓

CONCEPTS

Test for homogeneity of populations.

↓

VISION

We need to compare the proportion of the number of times a week two populations—men and women—dine out. Find the expected cell counts, and use the test for homogeneity of populations procedure to compute the value of the test statistic and draw the appropriate conclusion.

---

p value illustration:

$p = P(X \geq 5.43)$

$= 0.2458 > 0.05 = \alpha$

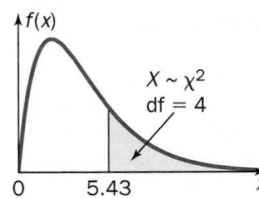

STEP 2 The four parts of the hypothesis test are

H_0: The true dining out category proportions are the same for men and for women

H_a: The true dining out category proportions are not the same

TS: $X^2 = \sum_{i=1}^{2} \sum_{j=1}^{5} \frac{(n_{ij} - e_{ij})^2}{e_{ij}}$

RR: $X^2 \geq \chi_\alpha^2 = \chi_{0.05}^2 = 9.4877$ df = 4.

STEP 3 Find each expected cell count. Here is one calculation.

$e_{11} = \frac{n_{1.} \times n_{.1}}{n} = \frac{(37)(98)}{197} = 18.41$

All of the expected cell counts are given in Table 13.4.

Table 13.4 The two-way table for the dining out example, including the row and column totals, the grand total, and the expected cell counts

		Dine out					
		Three or more times	Twice	Once	Less than Once	Never	Row total
Gender	Men	23 (18.41)	19 (16.42)	34 (36.31)	10 (13.43)	12 (13.43)	98
	Women	14 (18.59)	14 (16.58)	39 (36.69)	17 (13.57)	15 (13.57)	99
Column total		37	33	73	27	27	197

STEP 4 The value of the test statistic is

$\chi^2 = \sum_{i=1}^{2} \sum_{j=1}^{5} \frac{(n_{ij} - e_{ij})^2}{e_{ij}}$

$= \frac{(23 - 18.41)^2}{18.41} + \frac{(19 - 16.42)^2}{16.42} + \cdots + \frac{(15 - 13.57)^2}{13.57}$

$= 5.43 \ (< 9.4877)$

STEP 5 The value of the test statistic does not lie in the rejection region. At the $\alpha = 0.05$ significance level, there is no evidence to suggest that the true dining out category proportions are different for men and women.

Figure 13.11 shows a technology solution.

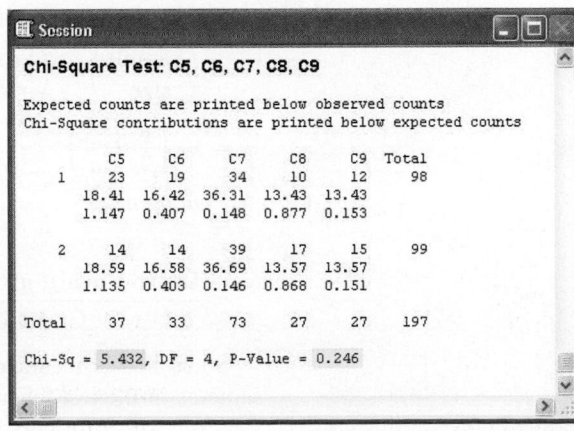

Figure 13.11 Minitab chi-square test results.

Suppose bivariate data arise from a single random sample in which the values of two categorical variables are recorded for each individual or object. For example, in a random sample of home burglaries, the type of item stolen and the method of entry might be recorded. Or suppose a random sample of employed people is obtained, and the occupation and any job-related injury is recorded for each person. In each instance, the data are bivariate, and there are observations on two categorical variables.

Given this type of bivariate data, it seems reasonable to ask whether the values of one variable affect the values of the other (are the variables dependent? or are the two variables independent?). Knowing the value of one variable may suggest nothing special about the value of the other. For example, suppose the proportion of all employed people who suffer from job-related respiratory illnesses is 0.15 for farm workers, 0.12 for industry and construction workers, 0.03 for service workers, and 0.04 for all other occupations. Now, suppose we know an employed person suffered a job-related broken bone. Does this change the proportions? If so, then the variables (occupation and job-related injury) are dependent. If knowing the occupation does not alter the job-related injury proportions, then the variables are independent.

As in the test for homogeneity, this is actually a test for dependence. We cannot prove independence, we can only test for evidence of dependence.

We use the same notation as in the test for homogeneity. In an $I \times J$ two-way table, there are I categories for the first variable (instead of I populations), and J categories for the second variable. The test for independence is based on observed and expected counts once again. The test statistic is exactly the same as in the test for homogeneity. Here's why.

Recall from Chapter 4: If two events A and B are independent, then the probability of A and B is the product of the corresponding probabilities. That is,

Remember, *and* means intersection.

$$P(A \cap B) = P(A) \cdot P(B) \qquad \text{if } A \text{ and } B \text{ are independent.}$$

Suppose the two categorical variables are independent and an individual or object falls into the (ij) cell. Consider the following probability.

P[an individual falls into the (ij) cell]

$$= P \left(\begin{matrix} \text{An individual responds with the } i\text{th value of the first variable} \\ \textit{and} \text{ the } j\text{th value of the second variable} \end{matrix} \right)$$

$$= P[(i\text{th value for first variable}) \cap (j\text{th value for second variable})].$$

And means intersection.

$$= P(i\text{th value for first variable}) \cdot P(j\text{th value for second variable})$$

Independent events; multiply corresponding probabilities.

$$= \frac{n_{i.}}{n} \cdot \frac{n_{.j}}{n}$$

Using the notation introduced in this section, the probability of falling into the ith row times the probability of falling into the jth column.

Since there is a total of n individuals, the expected count, or frequency, in the (ij) cell is

$$e_{ij} = \text{Expected count in the } (ij) \text{ cell}$$

$$= \left(\begin{matrix} \text{Sample} \\ \text{size} \end{matrix} \right) \cdot \left(\begin{matrix} \text{Probability that an individual falls} \\ \text{into the } (ij) \text{ cell} \end{matrix} \right)$$

$$= n \cdot \left(\frac{n_{i.}}{n} \cdot \frac{n_{.j}}{n} \right)$$

$$= \frac{n_{i.} \times n_{.j}}{n}$$

Simplify: cancel an n.

$$= \frac{(i\text{th row total})(j\text{th column total})}{\text{Grand total}}$$

Symbol translation.

Since this is identical to the expression we used before, the test statistic is the same as in the test for homogeneity. The test statistic is again a measure of how far away the observed cell counts are from the expected cell counts. The formal hypothesis test follows.

TEST FOR INDEPENDENCE OF TWO CATEGORICAL VARIABLES

In a random sample of n individuals, suppose the values of two categorical variables are recorded. In the resulting $I \times J$ two-way frequency table, let n_{ij} be the observed count in the (ij) cell and let e_{ij} be the expected count in the (ij) cell. A hypothesis test for independence of the two categorical variables with significance level α has the following form:

H_0: The two variables are independent

H_a: The two variables are dependent

$$\text{TS: } X^2 = \sum_{i=1}^{I} \sum_{j=1}^{J} \frac{(n_{ij} - e_{ij})^2}{e_{ij}}$$

$$\text{where } e_{ij} = \frac{(i\text{th row total})(j\text{th column total})}{\text{Grand total}} = \frac{n_{i.} \times n_{.j}}{n}$$

$$\text{RR: } X^2 \geq \chi_\alpha^2 \quad \text{df} = (I - 1)(J - 1)$$

This test is appropriate if all expected cell counts are at least 5 ($e_{ij} \geq 5$ for all i and j).

Example 13.4 Rest Stop Preferences The Pilot Travel Plaza in Pennsylvania is located at the intersection of Routes 487 and 80 so that travelers on each road have easy-on/easy-off access in both directions. Research is being conducted to summarize food preferences and to attract vendors. A random sample of people who purchased food at this plaza was obtained, and the traveling direction and the food vendor were recorded for each person. The observed frequencies are given in the table that follows. Is there any evidence to suggest that traveling direction and food vendor are dependent? Use $\alpha = 0.01$.

<table>
<tr><th></th><th></th><th colspan="5">Food vendor</th></tr>
<tr><th></th><th></th><th>Aunt
Annie's</th><th>Pizza
Hut</th><th>Taco
Bell</th><th>Mrs.
Fields</th><th>Hot Dog
Company</th></tr>
<tr><td rowspan="4">Traveling
direction</td><td>North</td><td>25</td><td>30</td><td>17</td><td>38</td><td>56</td></tr>
<tr><td>South</td><td>40</td><td>22</td><td>25</td><td>45</td><td>41</td></tr>
<tr><td>East</td><td>34</td><td>24</td><td>20</td><td>43</td><td>48</td></tr>
<tr><td>West</td><td>28</td><td>27</td><td>25</td><td>31</td><td>32</td></tr>
</table>

SOLUTION

STEP 1 This is a test for independence of two categorical variables: traveling direction and food vendor. There are four possible values for traveling direction and five possible responses for food vendor ($I = 4$ and $J = 5$).

STEP 2 The four parts of the hypothesis test are

H_0: Traveling direction and food vendor are independent

H_a: Traveling direction and food vendor are dependent

$$\text{TS: } X^2 = \sum_{i=1}^{4} \sum_{j=1}^{5} \frac{(n_{ij} - e_{ij})^2}{e_{ij}}$$

$$\text{RR: } X^2 \geq \chi_\alpha^2 = \chi_{0.01}^2 = 26.2170 \qquad \text{df = 12.}$$

Solution Trail

KEYWORDS

- Is there any evidence?
- Traveling direction and food vendor are dependent.

↓

TRANSLATION

- Statistical inference.
- Are these two categorical variables dependent?

↓

CONCEPTS

Test for independence of two categorical variables.

↓

VISION

We need to determine whether there is any evidence that traveling direction and food vendor are dependent. Compute the expected cell counts, and use the test for independence of two categorical variables procedure to find the value of the test statistic and draw the appropriate conclusion.

STEP 3 Find each expected cell count. Here is one calculation.

$$e_{11} = \frac{n_{1.} \times n_{.1}}{n} = \frac{(166)(127)}{651} = 32.38$$

All of the expected cell counts are given in Table 13.5.

Table 13.5 The two-way table for the travel-plaza example, including the row and column totals, the grand total, and the expected cell counts

		Food vendor					
		Aunt Annie's	Pizza Hut	Taco Bell	Mrs. Fields	Hot Dog Company	Row total
Traveling direction	**North**	25 (32.38)	30 (26.26)	17 (22.18)	38 (40.03)	56 (45.13)	166
	South	40 (33.75)	22 (27.37)	25 (23.12)	45 (41.72)	41 (47.04)	173
	East	34 (32.97)	24 (26.74)	20 (22.59)	43 (40.76)	48 (45.95)	169
	West	28 (27.90)	27 (22.63)	25 (19.11)	31 (34.49)	32 (38.88)	143
Column total		127	103	87	157	177	651

p value illustration:
$p = P(X \geq 14.598)$
$= 0.2642 > 0.01 = \alpha$

$X \sim \chi^2$
df = 12

0   14.598   X

STEP 4 The value of the test statistic is

$$\chi^2 = \sum_{i=1}^{4} \sum_{j=1}^{5} \frac{(n_{ij} - e_{ij})^2}{e_{ij}}$$
$$= \frac{(25 - 32.38)^2}{32.38} + \frac{(30 - 26.26)^2}{26.26} + \cdots + \frac{(32 - 38.88)^2}{38.88}$$
$$= 14.598 \ (<26.2170)$$

STEP 5 The value of the test statistic does not lie in the rejection region. At the $\alpha = 0.01$ significance level, there is no evidence to suggest that the two categorical variables are dependent.

Figure 13.12 shows a technology solution.

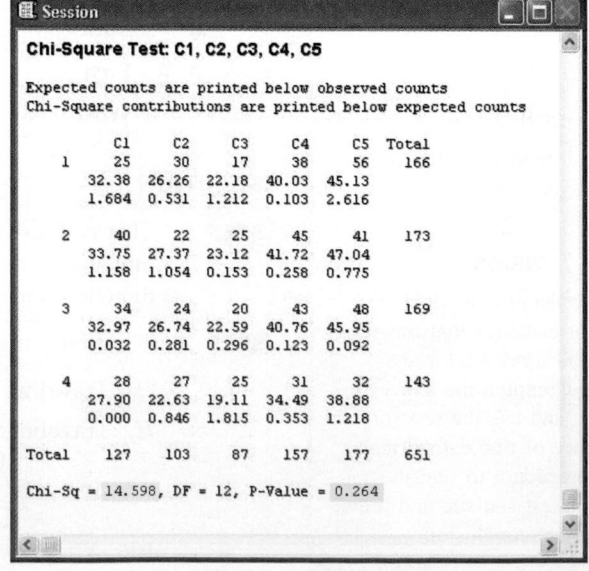

Figure 13.12 Minitab chi-square test results.

TECHNOLOGY CORNER

Procedure: Chi-square test for homogeneity or independence.
Reconsider: Example 13.4, page 651, solution, and interpretations.

TI-84 Plus

Use the built-in function χ^2-Test.

1. Enter the observed cell counts into the matrix [A]. See Figure 13.13.
2. Select $\boxed{\text{STAT}}$; TESTS; χ^2-Test. Enter the matrix containing the observed cell counts, [A], and specify a matrix for the expected frequency, [B]. See Figure 13.14.

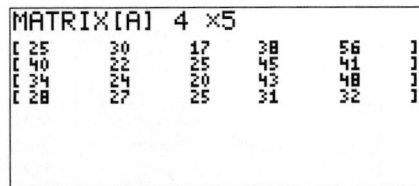

Figure 13.13 Observed cell counts.   **Figure 13.14** χ^2-Test input screen.

3. Highlight Calculate and press $\boxed{\text{ENTER}}$. The expected frequencies are shown in Figure 13.15 and the chi-square test results are shown in Figure 13.16. The Draw results are shown in Figure 13.17.

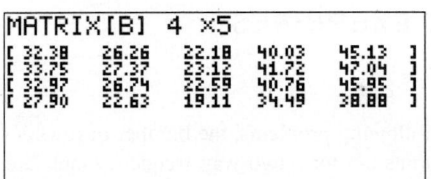

Figure 13.15 Expected frequencies.

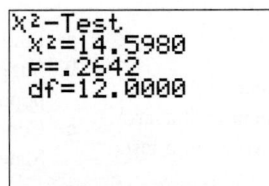
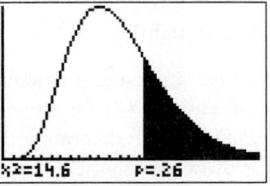

Figure 13.16 χ^2-Test results.   **Figure 13.17** χ^2-Test Draw results.

Minitab

Use the built-in function Chi-Square Test in the Stat; Tables menu or on a command line in a session window.

1. Enter the observed category frequencies into the columns C1-C5.
2. Select Stat; Tables; Chi-Square Test (Two-Way Table in Worksheet). Enter the columns containing the table.
3. Click OK. The chi-square hypothesis test results are displayed in a sessions window. See Figure 13.12, page 652.

Excel

Use the built-in function CHITEST. This function accepts two arguments (ranges): the observed counts and the expected counts.

1. Enter the observed cell counts in a rectangular array (matrix) and the expected cell counts in another rectangular array.
2. Use the function CHITEST to find the p value associated with the goodness-of-fit test and the function CHIINV to find the value of the test statistic. See Figure 13.18.

	A	B	C	D	E	F	G
1				Observed counts			
2		Aunt Annie	Pizza Hut	Taco Bell	Mrs. Fields	Hot Dog Co	Total
3	North	25	30	17	38	56	166
4	South	40	22	25	45	41	173
5	East	34	24	20	43	48	169
6	West	28	27	25	31	32	143
7	Total	127	103	87	157	177	651
8				Expected counts			
9	North	32.38	26.26	22.18	40.03	45.13	
10	South	33.75	27.37	23.12	41.72	47.04	
11	East	32.97	26.74	22.59	40.76	45.95	
12	West	27.90	22.63	19.11	34.49	38.88	
13							
14	0.2642	=CHITEST(B3:F6,B9:F12)					
15	14.60	=CHIINV(A14,12)					

Figure 13.18 Rectangular arrays and the goodness-of-fit hypothesis test results.

SECTION 13.2 EXERCISES

Practice

13.23 In each of the following problems, the number of rows (I) and the number of columns (J) for a two-way frequency table are given. Use the value of α to determine the critical value in a test for homogeneity of populations.
a. $I = 3, J = 4, \alpha = 0.05$.
b. $I = 2, J = 6, \alpha = 0.01$.
c. $I = 4, J = 3, \alpha = 0.025$.
d. $I = 5, J = 3, \alpha = 0.001$.

13.24 In each of the following problems, the number of rows (I) and the number of columns (J) for a two-way frequency table are given. Use the value of α to determine the critical value in a test for independence of two categorical variables.
a. $I = 6, J = 4, \alpha = 0.05$.
b. $I = 2, J = 8, \alpha = 0.005$.
c. $I = 3, J = 7, \alpha = 0.0005$.
d. $I = 4, J = 6, \alpha = 0.0001$.

13.25 Find the missing observed cell counts and row and column totals in the following two-way frequency table.

		Category			Row
	1	2	3	4	total
Population 1	18	14	18		65
Population 2	25			12	
Population 3		33	26	28	119
Column total		68	60		258

13.26 Consider the following two-way frequency table with three populations and three values of a categorical variable.

		Category		
		1	2	3
Population	1	58	62	33
	2	65	55	40
	3	70	60	25

a. Find the row and column totals and the grand total.
b. Suppose the true category proportions are the same for each population. Find the expected count for each cell.
c. Conduct a test for homogeneity of populations. Use $\alpha = 0.025$. State your conclusion.

13.27 Consider the following two-way frequency table in which the values of two categorical variables were recorded for each individual.

		Variable 2			
		1	2	3	4
Variable 1	1	235	267	245	386
	2	241	264	280	305
	3	228	254	270	394
	4	219	235	263	363

Conduct a test for independence of the two categorical variables. Use $\alpha = 0.05$. State your conclusion.

Applications

13.28 Sports and Leisure Random samples of gamblers at four Las Vegas casinos were obtained, and each gambler was asked which game he or she played most. The results are given in the following two-way frequency table.

	Game			
	Blackjack	**Poker**	**Roulette**	**Slots**
Bellagio	22	20	38	66
Caesar's	30	38	22	68
Golden Nugget	28	25	21	81
Harrah's	38	25	29	84

(Casino labels along left margin)

Conduct a test for homogeneity of populations. Is there any evidence to suggest that the true proportion of gamblers at each game is not the same for all casinos? Use $\alpha = 0.05$.

13.29 Marketing and Consumer Behavior A marketing manager obtained random samples of shoppers from three different grocery stores and asked each shopper to name his or her favorite Tastykake product. The results are summarized in the following two-way frequency table.

	Product			
	Krimpets	**Cupcakes**	**Kandy Kakes**	**Creamies**
Giant	90	80	95	92
Shaw	81	66	87	56
Weis	94	83	92	55

(Grocery store labels along left margin)

Is there any evidence to suggest that the true proportion of each favorite is not the same for all populations? Use $\alpha = 0.01$.

13.30 Marketing and Consumer Behavior Random samples of customers at two different office-supply stores were obtained, and each customer was asked which type of writing implement he or she prefers. The results are summarized in the following two-way frequency table.

	Writing implement		
	Traditional pencil	**Mechanical pencil**	**Pen**
Office Max	183	164	480
Staples	130	202	420

(Store labels along left margin)

Conduct a test for homogeneity of populations. Use $\alpha = 0.01$. State your conclusion, justify your answer, and find bounds on the p value associated with this test.

13.31 Public Policy and Political Science CBC/Radio Canada is Canada's public broadcaster and airs regional and cultural programs in several languages. A survey was conducted concerning the funding level for this broadcasting corporation. A random sample of Canadians was selected and each was asked whether funding should be increased, decreased, or maintained at the current levels. The data are summarized in the following two-way table.

	Funding for CBC			
	Decrease	**Maintain**	**Increase**	**Unsure**
Atlantic	17	58	36	9
Quebec	27	167	65	37
Ontario	59	191	79	26
West	68	212	106	44

(Region labels along left margin)

(*Source*: Nanos, survey conducted in September 2008.)

Conduct a test for homogeneity of populations. Use $\alpha = 0.005$. State your conclusion, and justify your answer.

13.32 Education and Child Development Random samples of boys and girls in elementary and secondary schools were obtained, and their parents were asked how often they helped with homework. The data are summarized in the following two-way frequency table.

	Number of times spent helping with homework per week			
	Less than once	**1 or 2 times**	**3 or 4 times**	**5 or more times**
Girls	259	369	254	118
Boys	274	335	262	129

(*Source*: Office of Educational Research and Improvements.)

Is there any evidence that the number of times parents spent helping with homework is different for boys and girls? Use $\alpha = 0.01$. Find bounds on the p value associated with this test.

13.33 Psychology and Human Behavior While food-and-wine pairing is subjective and an inexact science, traditionally red wine goes with red meat, and white wine goes with fish and poultry. A random sample of diners at four-star restaurants was obtained, and each diner was classified according to the food and wine that he or she ordered. Here is the resulting two-way frequency table.

	Wine	
	Red	**White**
Red meat	86	46
Fish or poultry	50	64

(Food labels along left margin)

Is there any evidence that food and wine are dependent? Test the relevant hypothesis with $\alpha = 0.005$. Does this data suggest that diners are still following the traditional food-and-wine pairings?

13.34 Sports and Leisure The Whitewater Ski Resort in British Columbia offers downhill skiing, cross-country skiing, snowboarding, and snow tubing. The manager is planning to develop a new, targeted advertising campaign. A random sample of customers was obtained, and each was classified by age group and activity. The data are summarized in the following two-way table.

Resort activity

Age group	Downhill skiing	Cross-country skiing	Snow-boarding	Snow tubing
< 16	24	35	23	48
16–20	50	25	40	32
20–30	21	27	27	32
30–40	29	39	29	29
≥ 40	26	37	35	30

Is there any evidence to suggest that resort activity is dependent upon age group? Conduct the appropriate hypothesis test with $\alpha = 0.01$.

13.35 Medicine and Clinical Studies A long-term medical study was conducted to examine the relationship between diet and the risk of colon cancer. A random sample of adults 40–50 years old was obtained, and each was monitored for 10 years. Each patient received a complete physical exam every year and at the end of the study was classified according to diet and whether or not he or she developed colon cancer. The results are given in the following two-way frequency table.

Diet

Cancer	Meat	Fish, poultry	Vegetarian	Mediterr-anean
Yes	56	23	21	20
No	700	780	180	220

(*Source*: Ann Chao et al., *Journal of the American Medical Association*, Vol. 293, January 12, 2005, pp. 172–182.)

Is there any evidence to suggest that the risk of colon cancer and choice of diet are dependent? Conduct the appropriate hypothesis test with $\alpha = 0.001$. Find bounds on the p value associated with this test.

13.36 Medicine and Clinical Studies Some research suggests that college student athletes who are subject to high stress levels (due to academic demands, personal problems, or other sources) are more likely to be injured during a game. A random sample of student athletes was obtained, and a questionnaire was used to determine the stress level of each prior to a game. Following the game, the injury status of each athlete was also recorded. The data are summarized in the following table.

Stress level

Injury	Low	Medium	High
Yes	12	17	25
No	335	292	288

Is there any evidence to suggest that stress level and injury are dependent? Conduct the appropriate hypothesis test with $\alpha = 0.05$.

Extended Applications

13.37 Public Policy and Political Science Public swimming pools are routinely inspected by city health officials. In a random sample of pool code violations, the type of pool and the violation class were recorded. The resulting two-way frequency table follows.

Violation

Pool type	Water chemistry	Filtration system	Policy / management
Hotel / motel	369	326	165
Condominium / apartments	1425	1207	650
School / university	34	45	11
Private club	195	185	84
Wading / children's	258	209	90
Water park	31	40	5
Medical / therapy	12	11	5
Municipal	29	45	8
Campground	39	35	18

(*Source*: Centers for Disease Control and Prevention.)

Is there any evidence to suggest that the type of violation and the type of pool are dependent? Conduct the appropriate hypothesis test with $\alpha = 0.01$ and find bounds on the p value associated with this test.

13.38 Psychology and Human Behavior Suppose the Oregon State Police are interested in a possible relationship between incident types and days of the week. A random sample of incident reports was obtained, and each incident was classified by type and day. The data are summarized in the two-way frequency table that follows. Is there any evidence that incident type and day are dependent? Use $\alpha = 0.001$.

Incident type

Day	Burglary	Robbery	Assault
Mon	3487	365	924
Tue	3217	361	849
Wed	3262	417	793
Thu	3283	400	900
Fri	3610	381	931
Sat	3208	453	1091
Sun	3108	359	1072

(*Source*: Oregon Uniform Crime Reporting.)

13.39 Public Policy and Political Science The United Nations is over 60 years old. However many people in countries all over the world are still unfamiliar with the UN and do not understand its function. A survey concerning familiarity with the United Nations was conducted in six nations. The data are summarized in the two-way frequency table that follows. Is there any evidence to suggest that the familiarity with the UN and the country are dependent? Conduct the appropriate hypothesis test with $\alpha = 0.01$.

		Response			
		Very familiar	Somewhat familiar	Not that familiar	Not at all familiar
Country	US	340	1020	553	213
	UK	33	402	457	196
	France	31	261	669	84
	Italy	32	547	400	74
	Spain	60	423	403	121
	Germany	21	238	610	165

(*Source*: HarrisInteractive, December 17, 2008.)

Challenge

13.40 Public Health and Nutrition During the first three months of 2008, the AFL-CIO and Working America sponsored a survey concerning health care in the United States. Participants responded to questions concerning insurance coverage, Medicare, and satisfaction with the current health care system. Suppose that in a random sample of 500 adults in Arizona, 135 said they were satisfied with the quality of health care in our country. In a random sample of 600 adults in New Jersey, 204 indicated they were satisfied with the quality of health care in our country.[6]

a. Conduct a hypothesis test concerning two population proportions to determine whether there is any evidence that the proportion of adults satisfied with the quality of health care is different in Arizona and in New Jersey. Find the p value associated with this test.

b. Consider "Arizona" and "New Jersey" as populations and "satisfied" and "not satisfied" as categories. Using the data given in this problem, construct a two-way frequency table and

conduct a test for homogeneity of populations. Use technology to find the p value for this test.

c. What is the relationship between the value of the test statistic in part (a) and the value of the test statistic in part (b)? How are the p values related? Why do these relationships make sense?

13.41 Public Health and Nutrition The American Dental Association recently conducted a survey regarding dental hygiene habits. A random sample of elderly people was obtained, and each person was classified according to brushing and flossing frequency. The following codes and categories were used for each variable.

Code	Floss/brush frequency
1	Never
2	Once per month
3	A few times per month
4	Once per week
5	A few times per week
6	Once per day
7	More than once per day

The survey results are summarized in the following two-way frequency table.

		Floss						
		1	2	3	4	5	6	7
Brush	1	35	37	48	52	55	80	101
	2	36	38	42	47	54	75	97
	3	38	43	46	51	58	77	94
	4	33	51	42	42	51	52	86
	5	76	79	85	87	93	102	115
	6	81	41	46	78	107	103	116
	7	98	94	126	136	142	198	252

Is there any evidence to suggest that flossing frequency and brushing frequency are dependent? Conduct the appropriate hypothesis test with $\alpha = 0.01$. Find the p value associated with this test.

(Alexraths/Dreamstime.com)

Chapter 13 Challenge Wrap-Up

The two-way frequency table summarizing type of drug and pharmaceutical company will be used to conduct a test for independence of categorical variables. We'll use $\alpha = 0.05$. There are $I = 4$ companies and $J = 3$ drug types.

The four parts of the hypothesis test are

H_0: Pharmaceutical company and drug type are independent.

H_a: Pharmaceutical company and drug type are not dependent.

TS: $X^2 = \displaystyle\sum_{i=1}^{4} \sum_{j=1}^{3} \frac{(n_{ij} - e_{ij})^2}{e_{ij}}$

RR: $X^2 \geq \chi_\alpha^2 = \chi_{0.05}^2 = 12.5916$ df = 6.

Find each expected cell count. Here is one calculation.

$$e_{11} = \frac{n_{1.} \times n_{.1}}{n} = \frac{(79)(33)}{150} = 17.8$$

All the expected cell counts are given in Table 13.6.

Table 13.6 The two-way table for the Chapter Challenge, including the row and column totals, the grand total, and the expected cell counts

		Type of drug			
		Cancer	**Diseases of aging**	**Heart disease and stroke**	**Row Total**
Company	**Abbott Laboratories**	18 (17.38)	10 (9.90)	5 (5.72)	33
	Boehringer Ingelheim	20 (21.59)	15 (12.30)	6 (7.11)	41
	Bristol-Myers Squibb	25 (23.70)	12 (13.50)	8 (7.80)	45
	Merck	16 (16.33)	8 (9.30)	7 (5.37)	31
	Column total	79	45	26	150

The value of the test statistic is

$$\chi^2 = \sum_{i=1}^{4} \sum_{j=1}^{3} \frac{(n_{ij} - e_{ij})^2}{e_{ij}}$$
$$= \frac{(18 - 17.38)^2}{17.38} + \frac{(10 - 9.90)^2}{9.90} + \cdots + \frac{(7 - 5.37)^2}{5.37} = 1.92 \, (<12.5916)$$

The value of the test statistic does not lie in the rejection region. At the $\alpha = 0.05$ significance level, there is no evidence to suggest that drug type and pharmaceutical company are dependent.

p value illustration:
$p = P(X \geq 1.92)$
$= \boxed{0.9269} > 0.05 = \alpha$

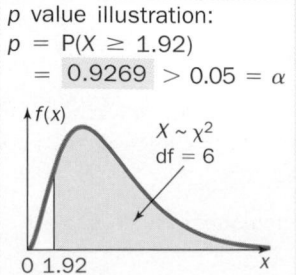

Figure 13.19 shows a technology solution.

	A	B	C	D	E
1		Observed counts			
2		Cancer	Aging	Stroke	
3	Abbott	18	10	5	33
4	Boehringer	20	15	6	41
5	Bristol-Myers	25	12	8	45
6	Merck	16	8	7	31
7	Total	79	45	26	150
8		Expected counts			
9	Abbott	17.38	9.90	5.72	
10	Boehringer	21.59	12.30	7.11	
11	Bristol-Myers	23.70	13.50	7.80	
12	Merck	16.33	9.30	5.37	
13					
14	0.9269	=CHITEST(B3:D6,B9:D12)			
15	1.92	=CHIINV(A14,6)			

Figure 13.19 Chi-square test for independence.

CHAPTER 13 SUMMARY

One-Way Frequency Table

A one-way frequency table is a method for summarizing a univariate categorical data set. The table lists each possible category and the number of times each category occurred (the observed count or frequency for each category).

Goodness-of-Fit Test

Let n_i be the number of observations falling into the ith category ($i = 1, 2, \ldots, k$), and let $n = n_1 + n_2 + \cdots + n_k$. A hypothesis test about the true category population proportions with significance level α has the following form:

$H_0: p_1 = p_{10}, \ p_2 = p_{20}, \ \ldots, p_k = p_{k0}$

$H_a: p_i \neq p_{i0}$ for at least one i

TS: $X^2 = \displaystyle\sum_{i=1}^{k} \frac{(n_i - np_{i0})^2}{np_{i0}}$

RR: $X^2 \geq \chi_\alpha^2 \quad$ df $= k - 1$

This test is appropriate if all expected cell counts are at least 5 ($np_{i0} \geq 5$ for all i).

Two-Way Frequency, or Contingency, Table

A two-way frequency table is a method for summarizing a bivariate categorical data set. Each *cell* contains the number of observations (observed count or frequency) in a category *combination*, or *pairing*.

Test for Homogeneity of Populations

In an $I \times J$ two-way frequency table, let n_{ij} be the observed count in the (ij) cell and let e_{ij} be the expected count in the (ij) cell. A hypothesis test for homogeneity of populations with significance level α has the following form:

H_0: The true category proportions are the same for all populations (homogeneity of populations)

H_a: The true category proportions are not the same for all populations

$$\text{TS: } X^2 = \sum_{i=1}^{I} \sum_{j=1}^{J} \frac{(n_{ij} - e_{ij})^2}{e_{ij}}$$

$$\text{where } e_{ij} = \frac{(i\text{th row total})(j\text{th column total})}{\text{Grand total}} = \frac{n_{i.} \times n_{.j}}{n}$$

$$\text{RR: } X^2 \geq \chi_\alpha^2 \quad \text{df} = (I-1)(J-1)$$

This test is appropriate if all expected cell counts are at least 5 ($e_{ij} \geq 5$ for all i and j).

Test for Independence of Two Categorical Variables

In a random sample of n individuals, suppose that the values of two categorical variables are recorded. In the resulting $I \times J$ two-way frequency table, let n_{ij} be the observed count in the (ij) cell and let e_{ij} be the expected count in the (ij) cell. A hypothesis test for independence of the two categorical variables with significance level α has the following form:

H_0: The two variables are independent

H_a: The two variables are dependent

$$\text{TS: } X^2 = \sum_{i=1}^{I} \sum_{j=1}^{J} \frac{(n_{ij} - e_{ij})^2}{e_{ij}}$$

$$\text{where } e_{ij} = \frac{(i\text{th row total})(j\text{th column total})}{\text{Grand total}} = \frac{n_{i.} \times n_{.j}}{n}$$

$$\text{RR: } X^2 \geq \chi_\alpha^2 \quad \text{df} = (I-1)(J-1)$$

This test is appropriate if all expected cell counts are at least 5 ($e_{ij} \geq 5$ for all i and j).

CHAPTER 13 EXERCISES

Applications

13.42 Medicine and Clinical Studies Unfortunately, foreign objects are sometimes left inside patients during surgery. In the United States, approximately 1500 people have foreign objects left inside them each year, and approximately 67% of these objects are sponges.[7] Suppose a random sample of patients who had foreign objects left inside them was obtained, and the type of object was recorded for each. The data are given in the following table.

Object	Sponge	Clamp	Retractor	Other
Frequency	55	12	16	12

Suppose that the proportion of sponges left inside a patient is 0.67 and that the other three proportions are all equal. Conduct a goodness-of-fit test to determine whether there is any evidence to suggest that the data are inconsistent with these proportions. Use $\alpha = 0.05$.

13.43 Psychology and Human Behavior The Rescue Pet Store sells five breeds of dogs, and the owner is trying to determine whether one breed is preferred over the others. A random sample of recent dog sales was obtained, and the number of dogs of each breed purchased is given in the following table.

Dog breed	Frequency
American Bulldog	46
Collie	54
Golden Retriever	32
German Shepherd	30
Yorkshire Terrier	46

Is there evidence to suggest that the true population proportions of sales for any breed is different from 0.20? Use $\alpha = 0.05$.

13.44 Marketing and Consumer Behavior The manager of a CVS in Madison, Wisconsin, Paul Coleman, obtained a random sample of customers who purchased adhesive bandages. The brands and frequencies are given in the following table.

Brand	Band-Aid	Curad	Nexcare	Generic
Frequency	220	215	95	510

Historical records indicate that the population proportions are Band-Aid 0.2, Curad 0.2, Nexcare 0.1, and Generic 0.5. Conduct a goodness-of-fit test to determine whether there is any evidence to suggest that the data are not consistent with the past proportions. Use $\alpha = 0.01$.

13.45 Public Policy and Political Science Kim Schwartz, a Lubbock County government official, discovered some extra money in the budget that must be spent by the end of the fiscal year. A random sample of county residents was obtained, and each was asked how the money should be spent. The data and the hypothesized population proportions (from past county referendum votes) are summarized in the following one-way frequency table.

Project	Frequency	p_{i0}
Road construction	103	0.3
Road resurfacing	119	0.4
Bicycle paths	40	0.1
New sidewalks	35	0.1
Park improvements	20	0.1

Is there evidence to suggest that any of the true population proportions is different from the hypothesized proportion? Use $\alpha = 0.05$.

13.46 Economics and Finance Many financial institutions offer customers with an Individual Retirement Account (IRA) four different plans for automatic transfer of funds from a checking or savings account to their IRA. Random samples of IRA customers from each of five different companies were obtained, and the transfer plan was recorded for each. The data are summarized in the following two-way table (ING; Commonfund, CF; Lincoln, LI; Prudential, PR; Ultimus: UL).

Transfer plan

	Monthly	Quarterly	Semi-annually	Annually
ING	71	70	67	59
CF	90	51	56	61
LI	75	82	70	60
PR	69	57	78	69
UL	93	92	77	91

Company (row label)

Conduct a test for homogeneity of populations with $\alpha = 0.05$. Is there any evidence to suggest that the true proportions associated with transfer plans are different for any of the populations? Justify your answer.

13.47 Psychology and Human Behavior Americans are accustomed to shopping at grocery stores any time of the day, and on any day of the week. However, the grocery shopping patterns vary widely in other countries. A random sample of adults in Latin American countries was obtained, and each individual was asked to indicate his or her grocery shopping frequency. The data are summarized in the following two-way frequency table (Columbia, Col; Costa Rica, CR; Ecuador, Ec).

Grocery shopping frequency

	Monthly	Weekly	Daily	Other
Col	102	112	100	90
CR	101	189	28	71
Ec	103	180	42	101

Country (row label)

(*Source*: Strategy Research Corporation.)

Is there any evidence to suggest that the true proportion of grocery shopping frequency is not the same for all countries? Use $\alpha = 0.005$.

13.48 Marketing and Consumer Behavior One of the most common home remodeling jobs involves the kitchen. In almost every house, this room is heavily used and often needs to be expanded to accommodate personal tastes. Three building-supply stores were selected, and a random sample of individuals purchasing kitchen countertops was obtained from each. The type of countertop was recorded for each person, and the data are summarized in the following two-way frequency table (Home Depot, HD; Lowe's, LO; TrueValue, TV).

Countertop

	Concrete	Corian	Marble	Granite
HD	52	24	37	90
LO	76	36	36	87
TV	53	43	31	78

Supply Store (row label)

Is there any evidence to suggest that the true proportion of each type of countertop purchased is not the same for all supply stores? Use $\alpha = 0.05$.

13.49 Education and Child Development Each year, schools in Pennsylvania administer the Pennsylvania System of School Assessment (PSSA) exams to 6th-, 9th-, and 11th-grade students. Although schools may implement their own curricula, the statewide exams are designed to impose standards in reading, writing, speaking, listening, and mathematics. There are four general performance-level descriptors. A sample of 11th-grade students from two districts was obtained, and each was classified according to his or her performance on the mathematics PSSA exam and district. The data are summarized in the following table (Bethel Park, BP; State College, SC).

Performance level

	Advanced	Proficient	Basic	Below basic
BP	174	130	60	60
SC	326	168	69	84

District (row label)

(*Source*: Pennsylvania Department of Education.)

Is there any evidence to suggest that the performance on the mathematics PSSA exam is associated with school district? Conduct the appropriate test with $\alpha = 0.01$.

13.50 Psychology and Human Behavior Some research suggests that music influences how much time people spend in a store. To investigate this theory, a random sample of customers at a Paramount retail store was obtained (over a long period of time), and each was classified by the amount of time spent shopping (in minutes) and the type of music played during the day. The results are given in the following table.

Type of Music

	Classical	Easy listening	Rock	Country
<15	49	35	17	14
15–30	51	27	19	41
30–60	67	41	41	22
≥60	47	26	26	33

Time (row label)

Is there any evidence of an association between music type and time spent shopping? Conduct the appropriate hypothesis test with $\alpha = 0.01$.

13.51 Physical Sciences Recent research suggests that the Titanic luxury liner broke into three sections, causing it to sink faster than previously believed.[8] The collision with an iceberg was a terrifying event, and some experts believe that the chance of survival was associated with location aboard the vessel. The following table presents the class and survival status of passengers aboard the Titanic.

	Survival status	
	Died	**Survived**
First	122	203
Second	167	118
Third	528	178
Crew	673	212

(Class is the row label)

(*Source*: British Board of Trade Inquiry.)

Is there any evidence of an association between class and survival status? Conduct the appropriate hypothesis test with $\alpha = 0.05$.

Extended Applications

13.52 Marketing and Consumer Behavior A new radio station in Geneva, Ohio, WKKY, plays a wide variety of music from 10 different genres. In order to assess customer preference and narrow its focus, the station obtained a random sample of listeners, and each was asked to select his or her favorite type of music. The data are summarized in the following one-way frequency table.

Genre	Frequency	Genre	Frequency
Country	50	Jazz	44
Hits	62	Dance	36
Christian	62	Latin	38
Rock	68	World	42
Urban	70	Classical	35

Is there any evidence to suggest that one music genre is preferred? Conduct the appropriate hypothesis test with $\alpha = 0.001$. Find bounds on the p value associated with this test.

13.53 Economics and Finance The online investment company Stock Tracker obtained a random sample of customers and asked each person to rate the chance of economic recovery within the next year and to identify the largest part of his or her portfolio. The data are summarized in the following table.

	Portfolio majority		
	Mutual funds	**Stocks**	**Bonds**
Optimistic	28	62	42
Neutral	35	45	75
Pessimistic	24	33	86

(Outlook is the row label)

a. Is there any evidence to suggest that portfolio majority and outlook for economic recovery are dependent? Use $\alpha = 0.001$.
b. Find bounds on the p value associated with the hypothesis test in part (a).
c. Suppose you are optimistic regarding economic recovery. Using the table and the hypothesis test results as guides, how would you allocate your portfolio? Justify your answer.

Nonparametric Statistics

14

Chapter 14 Challenge

Does the outside color of a silo affect the inside temperature?

The temperature inside a grain silo during the mid-summer months can reach as high as 60 °C (or about 140 °F). This intense heat can cause physical harm to anyone entering a silo and can also lead to explosions and fire. There are many different methods for reducing the temperature inside a silo, including unique structural designs and venting systems.

A very simple proposal to keep silo temperatures low involves the outside color of the structure. The claim is that older, galvanized silos tend to absorb sunlight and therefore warm up to extreme temperatures. White or off-white silos should reflect more sunlight and therefore remain cooler.

(Kgiszewski/Dreamstime.com)

Farms across the Midwest with at least one galvanized grain silo were selected. A new (similar) silo was erected on each farm and painted white. Each farmer randomly selected a midsummer day on which the amount of grain stored in both silos was nearly identical. The temperature (in °C) inside each silo was measured. The data are given in the following table.

Farm	1	2	3	4	5	6	7	8	9	10
Galvanized	39	58	53	50	38	49	53	48	46	41
White	40	48	55	42	39	37	43	46	41	46

It seems reasonable to conduct a paired t test on the temperature differences. However, it is known that silo temperatures are not normally distributed, and therefore the differences are not normal. Since the normality assumption is violated, a paired t test is not valid.

The methods presented in this chapter allow for comparison of continuous distributions, with few assumptions necessary. These nonparametric procedures are handy when very little is known about the underlying distributions. A statistical test based on ranks will be used to compare the temperatures in these two types of silos. The results are presented in the Chapter Challenge Wrap-Up (pg. 703).

Review

- Recall the parametric methods discussed in previous chapters: statistical techniques based on the normality assumption.
- Remember that if any assumptions are violated, the conclusions may be invalid.

CONTENTS

> **Preview**
> - Learn several nonparametric, or distribution-free, procedures that require very few assumptions about the underlying population(s).
> - Understand the intuitive reasons for using many nonparametric test statistics.
> - Compute and use ranks in several nonparametric test procedures.

14.1 THE SIGN TEST

The normality assumption is very reasonable, because almost all distributions are normal or approximately normal.

Each of the hypothesis tests presented in the previous chapters depends on a set of assumptions. If any of the assumptions are violated, the conclusions may be invalid. Most of the statistical procedures include a normality assumption, which states that the random sample(s) is(are) drawn from a normal distribution. Statistical techniques based on this assumption are called **parametric methods**. This chapter presents some alternative statistical techniques called **nonparametric**, or **distribution-free**, **procedures**. These techniques usually require very few assumptions about the underlying population(s).

There are usually very few assumptions necessary in order for a nonparametric test to be valid. In addition, many statisticians perceive the formula, or rule, for computing the test statistic in a nonparametric procedure to be more intuitive and easier to apply than a comparable parametric test. For example, the test statistic in a **sign test** is simply a count of the number of observations greater than the hypothesized median.

There are, however, some disadvantages to nonparametric tests. Because there are few assumptions, these procedures usually do not utilize all of the information captured in a sample. Nonparametric tests may ignore certain inherent facts, for simplicity or ease of use. This means that there is a greater chance of making an error when you use a nonparametric test. Therefore, if there is ever a case in which either a parametric or a nonparametric test can be used, it is usually better to use the parametric procedure. Nonparametric tests are most useful when we cannot assume normality, or for analyzing certain nonnumerical data sets.

Recall, the population median divides the distribution in half.

Suppose a random sample is obtained from a continuous (nonnormal) distribution. Consider a test concerning the population median with null hypothesis H_0: $\tilde{\mu} = \tilde{\mu}_0$. If the null hypothesis is true, then approximately half of the observations should lie above $\tilde{\mu}_0$, and the other half should fall below $\tilde{\mu}_0$.

To use the *sign test*, we replace each observation above $\tilde{\mu}_0$ with a plus sign and each observation below $\tilde{\mu}_0$ with a minus sign. If H_0 is true, then the number of plus signs and the number of minus signs should be about the same. A problem arises if an observation is equal to $\tilde{\mu}_0$. Since the underlying distribution is continuous, theoretically the probability of obtaining an observation exactly equal to $\tilde{\mu}_0$ is zero. However, in practice it is common to obtain such an observation. Any observations equal to $\tilde{\mu}_0$ are excluded from the analysis.

The binomial distribution was defined in Section 5.4.

The test statistic, X, is a count of the number of plus signs (or number of observations greater than $\tilde{\mu}_0$). If the null hypothesis is true, then the probability of a plus sign (an observation is greater than $\tilde{\mu}_0$) is 1/2. Therefore, if H_0 is true, the random variable X has a binomial distribution with number of trials equal to n, the number of observations included in the test, and $p = 1/2$: $X \sim \mathrm{B}(n, 0.5)$. We should reject the null hypothesis for very large or very small values of X.

14.1 The Sign Test **665**

THE SIGN TEST CONCERNING A POPULATION MEDIAN

Suppose a random sample of size n is obtained from a continuous distribution. A hypothesis test concerning a population median $\tilde{\mu}$ with significance level α has the following form:

$H_0: \tilde{\mu} = \tilde{\mu}_0$

$H_a: \tilde{\mu} > \tilde{\mu}_0, \quad \tilde{\mu} < \tilde{\mu}_0, \quad \text{or} \quad \tilde{\mu} \neq \tilde{\mu}_0$

TS: X = the number of observations greater than $\tilde{\mu}_0$

RR: $X \geq c_1, \quad X \leq c_2, \quad X \geq c \quad \text{or} \quad X \leq n - c$

The critical values c_1, c_2, and c are obtained from Table I in the Appendix: Binomial Distribution Cumulative Probabilities with parameters n and $p = 0.5$ to yield a significance level of approximately α. That is, the critical values are chosen so that $P(X \geq c_1) \leq \alpha$, $P(X \leq c_2) \leq \alpha$, and $P(X \geq c) \leq \alpha/2$.

Observations equal to $\tilde{\mu}_0$ are excluded from the analysis, and the sample size is reduced accordingly.

 ILLUMINATING THE CONCEPTS

1. If the underlying (continuous) distribution is symmetric, then $\mu = \tilde{\mu}$ and the sign test can be used to test a hypothesis about a population mean.

2. We really do not need to literally replace observations with plus or minus signs. Discard any observations equal to $\tilde{\mu}_0$, and simply count the number of observations greater than $\tilde{\mu}_0$.

3. Because the binomial distribution is discrete, we usually cannot find critical values that yield an exact level-α test. Use critical values such that the significance level is as close to α as possible but not greater than α. ∎

(Banana Stock)

Example 14.1 Time in Child Care A growing number of preschool children spend at least some time each week in a child-care center. Although the cost can be high, children are spending more time in child care because parents have work requirements or are concerned about their children's school readiness. Responses from the 2002 National Survey of America's Families (NSAF) indicated that 42% of children under the age of five with employed mothers spent at least 35 hours per week in child care.[1] A random sample of 15 children under five with employed mothers was obtained. The number of hours per week that the children spent in child care are given in the following table.

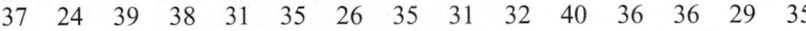

37	24	39	38	31	35	26	35	31	32	40	36	36	29	35

Is there any evidence to suggest that the median time per week spent in child care is greater than 30 hours? Use a significance level of $\alpha = 0.10$.

SOLUTION

STEP 1 The assumed median is $\tilde{\mu} = 30 \ (= \tilde{\mu}_0)$, the sample size is $n = 15$, and we will use $\alpha = 0.10$.

We are looking for any evidence that the median time the children spent in child care each week is greater than 30 hours. Therefore, the relevant alternative hypothesis is one-sided, right-tailed.

Since the population is not assumed to be normal and this is a test concerning the median, the sign test should be used.

STEP 2 The four parts of the hypothesis test are

$H_0: \tilde{\mu} = 30$

$H_a: \tilde{\mu} > 30$

TS: X = the number of observations greater than 30

RR: $X \geq c_1 = 11$

To find the critical value c_1, follow these steps:

a. There are no observations equal to 30, the hypothesized median. Therefore, no values are excluded from the analysis.

If the null hypothesis is true, X is a binomial random variable with $n = 15$ and $p = 0.5$: $X \sim B(15, 0.5)$. We need to find a value c_1 such that $P(X \geq c_1)$ is as close to $\alpha = 0.10$ as possible without going over. Find the smallest c_1 such that $P(X \geq c_1) \leq 0.10$.

b. Use the Complement Rule applied to a discrete random variable to convert this equation to cumulative probability.

$P(X \geq c_1) = 1 - P(X < c_1) = 1 - P(X \leq c_1 - 1) \leq 0.10$

Or find the smallest c_1 such that $P(X \leq c_1 - 1) \geq 0.90$.

c. Using Table I in the Appendix, with $n = 15$ and $p = 0.5$, $c_1 - 1 = 10$. Therefore, $c_1 = 11$. The actual significance level using this critical value is

$P(X \geq 11) = 1 - P(x < 11) = 1 - P(X \leq 10) = 1 - 0.9408 = 0.0592.$

Solution Trail

KEYWORDS

- Is there any evidence?
- Median time greater than.

↓

TRANSLATION

One-sided, right-tailed hypothesis test concerning a population median.

↓

CONCEPTS

Sign test concerning a population median.

↓

VISION

We cannot assume anything about the shape of the underlying distribution of time spent in child care each week. Since the normality assumption is not justified and this is a test concerning the median, the (nonparametric) sign test is appropriate.

Table I in the Appendix, Binomial Distribution Cumulative Probabilities, $n = 15$:

x	...	p 0.50	...
⋮		⋮	
8		0.6964	
9		0.8491	
10	...	0.9408	
11		0.9824	
12		0.9963	
⋮		⋮	

STEP 3 Using signs, classify each observation as either greater than or less than the hypothesized median.

Observation	37	24	39	38	31	35	26	35	31	32	40	36	36	29	35
Sign	+	−	+	+	+	+	−	+	+	+	+	+	+	−	+

The value of the test statistic is the number of plus signs, or the number of observations greater than $\tilde{\mu} = 30$. Therefore, $x = 12$.

STEP 4 The value of the test statistic lies in the rejection region. At the $\alpha = 0.10$ significance level, there is evidence to suggest that the population median hours per week spent in child care is greater than 30.

STEP 5 Since it is unlikely that a critical value will yield the exact desired significance level, it is often more appropriate to find a p value when using the sign test.

$p = P(X \geq 12)$ *Definition of a p value.*

$= 1 - P(X \leq 11)$ *The Complement Rule; discrete random variable.*

$= 1 - 0.9824 = 0.0176$ *Table I in the Appendix, $n = 15$, $p = 0.5$.*

Using the p value to draw a conclusion, since $p = 0.0176 \leq 0.10 = \alpha$, we reject the null hypothesis. There is evidence to suggest that the population median hours per week spent in child care is greater than 30.

Figure 14.1 shows a technology solution.

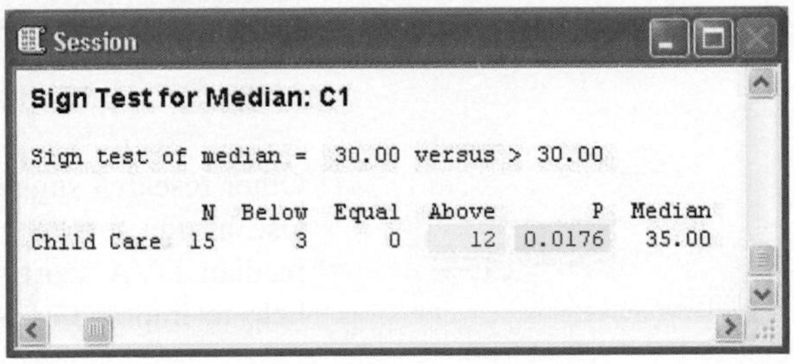

```
Session                                               □ ⊡ ✕

Sign Test for Median: C1

Sign test of median =  30.00 versus > 30.00

                    N  Below  Equal  Above      P  Median
Child Care  15      3      0     12  0.0176   35.00
```

Figure 14.1 Minitab sign test.

If the data are paired (two observations on each individual or object) and the underlying distributions are not normal, then the sign test can be used to compare population medians. Compute each pairwise difference, disregard the magnitude of the difference, and only consider the sign of the difference. Replace each positive difference with a plus sign and each negative difference with a minus sign. Use these signs and the test procedure described above. This analysis is appropriate for comparing two population medians, $\tilde{\mu}_1$ and $\tilde{\mu}_2$.

> This is the nonparametric counterpart to a paired t test.

THE SIGN TEST TO COMPARE TWO POPULATION MEDIANS

Suppose there are n independent pairs of observations such that the population of first observations is continuous and the population of second observations is also continuous. A hypothesis test concerning the two population medians in terms of the difference, $\tilde{\mu}_D = \tilde{\mu}_1 - \tilde{\mu}_2$, with significance level α has the following form:

H_0: $\tilde{\mu}_D = \Delta_0$

H_a: $\tilde{\mu}_D > \Delta_0$, $\tilde{\mu}_D < \Delta_0$, or $\tilde{\mu}_D \neq \Delta_0$

TS: X = the number of pairwise differences greater than Δ_0

RR: $X \geq c_1$, $X \leq c_2$, $X \geq c$ or $X \leq n - c$

The critical values c_1, c_2, and c are obtained from Table I in the Appendix: Binomial Distribution Cumulative Probabilities with parameters n and $p = 0.5$ to yield a significance level of approximately α; that is, so that $P(X \geq c_1) \leq \alpha$, $P(X \leq c_2) \leq \alpha$, and $P(X \geq c) \leq \alpha/2$.

Differences equal to Δ_0 are excluded from the analysis, and the sample size is reduced accordingly.

Solution Trail

KEYWORDS

- Sign test.
- Median score before, after.
- Eggs help to improve eyesight.

↓

TRANSLATION

One-sided, left-tailed hypothesis test concerning the difference in population medians.

↓

CONCEPTS

Sign test to compare two population medians.

↓

VISION

There is a before and after measurement on each individual; the data are paired. Compute each difference, and use the sign test to compare two medians procedure to find the value of the test statistic and draw the appropriate conclusion.

Example 14.2 Eggs: Good or Bad   There have been numerous studies concerning the benefits and possible harmful effects of eating eggs. Some researchers believe that the vitamins in eggs promote good eyesight.[2] A random sample of healthy male adults, ages 25–45, was obtained. The dynamic visual acuity (DVA, logMAR scale) was used to measure each subject's ability to detect objects while moving his head. A higher DVA score indicates better perception. Each person was put on a diet that included one egg per day. After four weeks, each subject was tested again for dynamic visual acuity. The data are given in the following table.

Subject	1	2	3	4	5	6	7	8	9	10
Before	0.58	0.78	0.76	0.54	0.60	0.73	0.56	0.45	0.59	0.45
After	0.60	0.49	0.41	0.74	0.80	0.77	0.52	0.58	0.69	0.71

Subject	11	12	13	14	15	16	17	18	19	20
Before	0.42	0.62	0.76	0.61	0.67	0.49	0.48	0.56	0.57	0.52
After	0.58	0.70	0.73	0.43	0.52	0.70	0.47	0.44	0.64	0.68

Other research suggests that the underlying DVA score populations are not normal. Use a sign test to compare the median DVA score before the egg diet with the median DVA score after the egg diet. Is there any evidence to suggest that eggs help to improve eyesight, as measured by DVA? Use a significance level of 0.01.

SOLUTION

STEP 1 The data are certainly paired; there are before and after measurements on each individual. Typically, the before measurements are considered population 1 and the after measurements are population 2. Since the underlying populations are not normal, a paired t test is not appropriate.

STEP 2 The null hypothesis is that the two population medians are equal; the egg-a-day diet has no effect on visual acuity: $\tilde{\mu}_1 = \tilde{\mu}_2 \Rightarrow \tilde{\mu}_1 - \tilde{\mu}_2 = \tilde{\mu}_D = 0 \, (= \Delta_0)$.

Since we are searching for evidence that the egg diet improves visual acuity, the alternative hypothesis is $\tilde{\mu}_1 < \tilde{\mu}_2 \Rightarrow \tilde{\mu}_1 - \tilde{\mu}_2 = \tilde{\mu}_D < 0$. This is a one-sided, left-tailed test.

Table I in the Appendix, Binomial Distribution Cumulative Probabilities, $n = 20$:

x	...	p 0.50	...
⋮		⋮	
2		0.0002	
3		0.0013	
4	...	0.0059	
5		0.0207	
6		0.0577	
⋮		⋮	

What would the significance level be if $c_2 = 5$?

STEP 3 The four parts of the hypothesis test are

H_0: $\tilde{\mu}_D = 0$

H_a: $\tilde{\mu}_D < 0$

TS: X = the number of differences greater than 0

RR: $X \le c_2 = 4$

To find the critical value c_2, follow these steps:

a. There are no differences equal to $\Delta_0 = 0$, the hypothesized difference. Therefore, no pairs are excluded from the analysis.

If the null hypothesis is true, X has a binomial distribution with $n = 20$ and $p = 0.5$: $X \sim B(20, 0.5)$. Find the largest value c_2 such that $P(X \le c_2) \le 0.01 = \alpha$.

b. Using Table I in the Appendix, with $n = 20$ and $p = 0.5$, the critical value is $c_2 = 4$. The actual significance level using this critical value is $P(X \le 4) = 0.0059 \, (\le 0.01)$.

STEP 4 Compute each pairwise difference (before–after) and determine the signs.

Subject	1	2	3	4	5	6	7	8	9	10
Before	0.58	0.78	0.76	0.54	0.60	0.73	0.56	0.45	0.59	0.45
After	0.60	0.49	0.41	0.74	0.80	0.77	0.52	0.58	0.69	0.71
Difference	−0.02	0.29	0.35	−0.20	−0.20	−0.04	0.04	−0.13	−0.10	−0.26
Sign	−	+	+	−	−	−	+	−	−	−

Subject	11	12	13	14	15	16	17	18	19	20
Before	0.42	0.62	0.76	0.61	0.67	0.49	0.48	0.56	0.57	0.52
After	0.58	0.70	0.73	0.43	0.52	0.70	0.47	0.44	0.64	0.68
Difference	−0.16	−0.08	0.03	0.18	0.15	−0.21	0.01	0.12	−0.07	−0.16
Sign	−	−	+	+	+	−	+	+	−	−

STEP 5 The value of the test statistic is the number of plus signs, or the number of differences greater than $\Delta_0 = 0$. Therefore, $x = \boxed{8}$.

STEP 6 The value of the test statistic does not lie in the rejection region. At the $\alpha = 0.01$ significance level, there is no evidence to suggest that the egg diet increased the median DVA score.

STEP 7 The p-value for this hypothesis test is

$p = P(X \leq 8)$ Definition of p value.

$\quad = \boxed{0.2517}$ Cumulative probability; Table I in the Appendix, $n = 20$, $p = 0.5$.

Using the p value to draw a conclusion, since $p = 0.2517 > 0.01 = \alpha$, we cannot reject the null hypothesis. There is no evidence to suggest that the egg diet increased the median DVA score.

Figure 14.2 shows a technology solution.

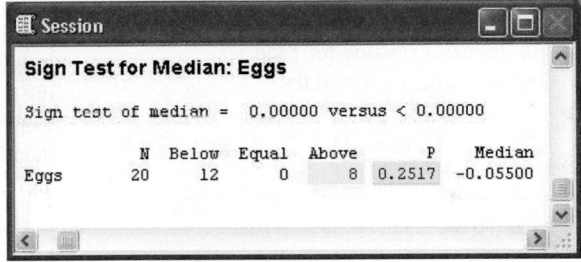

Figure 14.2 Minitab sign test.

TECHNOLOGY CORNER

Procedure: Conduct a sign test concerning a population median.
Reconsider: Example 14.1, page 665, solution, and interpretations.

Minitab

The built-in one-sample sign test is accessible in the <u>S</u>tat; <u>N</u>onparametrics menu and may also be executed using commands in a session window.

1. Enter the hours per week data into column C1.
2. Select <u>S</u>tat; <u>N</u>onparametrics; 1-Sample Sign. Enter the Variable (column containing the data, C1), select Test median and enter the hypothesized value, 30, and select the Alternative hypothesis (greater than).
3. Click OK. The test results are displayed in a session window. See Figure 14.1, page 667.

Excel

There is no built-in function to conduct a sign test. Use other functions to count the number of observations greater than the hypothesized median and to compute the p value associated with the test.

1. Enter the data into column A.
2. Count the number of observations greater than 30 using the function COUNTIF.
3. Use the function BINOMDIST to compute the p value. See Figure 14.3.

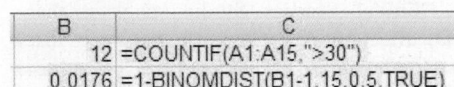

Figure 14.3 Excel functions to conduct a sign test.

SECTION 14.1 EXERCISES

Practice

14.1 In each of the following problems a data set and a null hypothesis (concerning the population median) are given. Assume that the sign test will be used, and find the value of the test statistic x.

a. $\{20, 20, 13, 20, 16, 19, 19, 11, 20, 14\}$ H_0: $\tilde{\mu} = 15$.

b. $\{66, 90, 77, 68, 70, 56, 56, 75, 57, 65, 65, 56, 66, 70, 54\}$ H_0: $\tilde{\mu} = 70$.

c. $\{46.4, 42.6, 50.9, 49.2, 46.4, 47.6, 50.7, 49.3, 59.3, $
$51.5, 45.3, 52.9, 54.5, 51.1, 50.7, 41.3, 57.8, 59.5, $
$52.3, 47.2\}$ H_0: $\tilde{\mu} = 51.5$.

d. $\{-4, -7, 8, -7, 3, 9, 1, 1, 3, 10, -9, -7, -5, 8, -1, $
$-9, -2, 6, 7, -6, -7, -9, 7, 9, 3\}$ H_0: $\tilde{\mu} = 0$.

14.2 Use the random sample in each problem to conduct a sign test with the indicated null hypothesis, alternative hypothesis, and significance level. Find the exact p value for each test.

a. H_0: $\tilde{\mu} = 16$, H_a: $\tilde{\mu} < 16$, $\alpha = 0.05$.

13.1	17.4	13.6	18.4	11.2	15.9	14.6
13.6	13.6	14.5				

b. H_0: $\tilde{\mu} = -25$, H_a: $\tilde{\mu} > -25$, $\alpha = 0.01$.

−19	−20	−25	−28	−17	−25	−16	−11
−12	−28	−23	−11	−11	−25	−28	

c. H_0: $\tilde{\mu} = 8$, H_a: $\tilde{\mu} \neq 8$, $\alpha = 0.05$.

8.90	7.68	9.41	9.47	8.98	8.41	9.51	7.40
9.66	8.17	5.32	9.06	7.67	7.87	8.16	8.77
7.01	9.22	7.41	8.57				

d. H_0: $\tilde{\mu} = 125$, H_a: $\tilde{\mu} \neq 125$, $\alpha = 0.001$.

196	126	187	168	111	196	164	181	110
113	187	136	159	154	177	132	103	151
187	191	192	139	187	190	112		

14.3 Using the paired data on the data CD and book's web site, conduct a sign test to compare population medians with H_0: $\tilde{\mu}_1 - \tilde{\mu}_2 = 0$, H_a: $\tilde{\mu}_1 - \tilde{\mu}_2 > 0$, and significance level $\alpha = 0.05$.

14.4 Using the paired data on the data CD and book's web site, conduct a sign test to compare population medians with H_0: $\tilde{\mu}_1 - \tilde{\mu}_2 = 3$, H_a: $\tilde{\mu}_1 - \tilde{\mu}_2 \neq 3$, and significance level $\alpha = 0.01$. Find the exact p value for this test.

Applications

14.5 Medicine and Clinical Studies Studies suggest that Parkinson's disease (PD) affects nerve endings in the heart that produce the chemical norepinephrine. Low levels of norepinephrine decrease the patient's ability to control muscle movement, producing the shaking symptoms associated with PD. A random

sample of women with PD was obtained, and the concentration of norepinephrine (in nmol/L) was measured in each. The data are given in the following table.

1.36	1.29	1.05	1.25	1.34	1.50	1.33	1.31
1.08	1.11	1.29	1.04	1.09	1.38	1.44	

Assume the underlying distribution is continuous. Use the sign test to determine whether there is any evidence that the median concentration of norepinephrine in women with PD is less than 1.38 nmol/L (the normal level). Use $\alpha = 0.05$.

14.6 Fuel Consumption and Cars Disc brake pads are critical safety features on an automobile and should be replaced when the thickness has worn to approximately 2 mm. A random sample of automobiles entering a New Jersey State Inspection Station was obtained, and the brake-pad thickness on the left front tire was measured on each. The data are given on the data CD and book's web site. Assume that the underlying distribution of brake-pad thicknesses is continuous. Use the sign test to determine whether there is any evidence that the median brake-pad thickness is less than 2 mm. Use $\alpha = 0.10$ for this test and find the exact p value.

14.7 Manufacturing and Product Development Modular homes are constructed in sections in a factory, with no delays due to weather conditions. Sections are transported to a home site on truck beds, attached together, and placed on a premade foundation. A random sample of modular-home deliveries was obtained, and the mileage from the factory to the home site was recorded for each. The data are given on the data CD and book's web site. Assume that the underlying distribution of mileage is continuous. Use the sign test to determine whether there is any evidence that the median mileage is greater than 100. Use $\alpha = 0.05$ for this test and find the exact p value.

14.8 Manufacturing and Product Development Grade 40s grease mohair must have a fiber diameter of approximately 24 microns.[3] A manufacturer of mohair jackets received a large shipment of 40s grade raw material and obtained 30 random measurements of fiber diameter. The data are given on the data CD and book's web site. The manufacturer can only assume that the underlying distribution of fiber diameters is continuous. Is there any evidence to suggest that the median diameter is different from 24 microns? Use a significance level of $\alpha = 0.05$.

14.9 Sports and Leisure In 2004, enthusiastic fans spent approximately $16 billion on licensed sports merchandise. However, there is some fear that this spending spree cannot continue since no major U.S. professional sports fan base grew between 2000 and 2004. In addition, the median fan age for the four major U.S. professional sports is 42.[4] This is seven years older than the median age for all Americans. Suppose a random sample of sports fans in 2009 was obtained. The age of each (in years) is given on the data CD and book's web site. Assume the underlying distribution of age is continuous. Is there any evidence to suggest that the median age of sports fans has increased (from 42)? Use $\alpha = 0.01$.

14.10 Economics and Finance According to an article about real estate[5] the median down payment by first-time home buyers was 4 percent of the total cost of the home. In 2009, suppose a random sample of first-time home buyers was obtained, and the down payment percentage of each home sale is given in the following table.

3.5	4.6	7.7	6.6	2.3	5.9	5.4	7.0
3.8	6.1	3.8	9.8	7.4	2.0	5.3	5.6
2.3	5.2	6.0	3.1	5.7	4.7	5.5	5.7
5.4	3.3	6.1	5.0	3.2	6.8		

Assume the underlying distribution of down payment percentage is continuous. Is there any evidence to suggest that the median down payment percentage has changed (from 4%)? Use $\alpha = 0.05$.

14.11 Biology and Environmental Science Chlorophyll plays an important part in photosynthesis in plants and is also present in microscopic algae and other phytoplankton. A random sample of 20 locations along the western coastline of the United States was selected. Using fluorometry, the quantity of chlorophyll in the surface water was measured (in mg/m^3) in early April and in late August. The pairwise differences (April measurement–August measurement) are given in the following table.

6.24	6.51	−10.96	9.52	5.39	5.88	−9.03
4.58	−8.16	6.99	4.92	−9.35	4.88	−1.99
4.96	3.90	−8.13	7.55	12.61	7.68	

The distribution of chlorophyll in surface water is assumed to be continuous but not normal. Is there any evidence to suggest that the median chlorophyll amount in surface water is different in April than in August? Use a significance level of $\alpha = 0.05$.

14.12 Physical Sciences There are two basic methods for dredging channels and moving sediments to open-water sites: (1) hydraulic cutterhead suction dredge and (2) clamshell bucket dredge. A study was conducted using the hydraulic cutterhead suction method to examine the dredged channel area. A random sample of dredged waterways was obtained, and the bulk density of the channel was measured before and after (in g/cm^3) dredging. The pairwise differences (before − after) are given on the data CD and book's web site.[6] There is no evidence to suggest that the distribution of bulk densities is normal. Use the sign test to determine whether there is any evidence that the median bulk density has decreased after dredging. Use a 0.01 significance level, and find the exact p value for this test.

Extended Applications

14.13 Biology and Environmental Science Jet engine emissions at high altitudes cause vapor trails that contribute to cloud cover and pollution, and may even affect weather patterns. An experiment was conducted to measure the concentration of volatile organic compounds (VOC) directly behind an operating engine. A random sample of jet engines was obtained. The VOC concentration was measured (in mg/m^3) before and after a special exhaust scrubber was installed. The data are given in the following table.

Before	13.0	14.9	14.0	13.4	9.8	14.9	12.0	11.2
After	11.0	13.6	12.3	9.9	10.3	10.7	9.6	9.5

Before	14.9	15.0	9.7	13.1	14.4	11.3	14.6
After	11.8	13.1	9.2	10.0	11.9	13.7	12.3

Use the sign test with $\alpha = 0.05$ to determine whether there is any evidence that the median VOC concentration is smaller when the scrubber is installed. Find the p value associated with this test.

14.14 Medicine and Clinical Studies Calcium blockers generally decrease your heart pumping strength and ease tension in blood vessels.[7] Suppose an experiment was conducted to determine whether calcium blockers affect heart rhythms in patients with arrhythmias. A random sample of patients was obtained, and their resting pulse rates were measured (in beats per minute). After a two-week regimen of a calcium blocker, each person's resting pulse rate was measured again. The data are given on the data CD and book's web site. Suppose no assumptions can be made about the shape of the continuous distributions of the before and after data for pulse rates.

a. Conduct a sign test to determine whether there is any evidence that the median pulse rate before the calcium blocker medication is different from the median pulse rate after the medication. Use a significance level of $\alpha = 0.01$.

b. Do you believe the assumptions for the sign test are valid in this case? Why or why not?

14.2 THE SIGNED-RANK TEST

This test was developed in 1945 by Frank Wilcoxon and is usually called the Wilcoxon signed-rank test.

The sign test concerning a population median or to compare two population medians uses only the signs (plus or minus) of the relevant differences. The signed-rank test also utilizes the magnitude of each difference and, of course, the ranks. The test statistic does not have a common distribution, but if n is sufficiently large, a normal approximation may be used.

Suppose a random sample of size n is obtained from a continuous, symmetric distribution, and the null hypothesis is H_0: $\tilde{\mu} = \tilde{\mu}_0$.

1. Subtract $\tilde{\mu}_0$ from each observation; that is, compute the differences
$x_1 - \tilde{\mu}_0, x_2 - \tilde{\mu}_0, \ldots, x_n - \tilde{\mu}_0.$

2. Consider the magnitude, or absolute value, of each difference; compute $|x_1 - \tilde{\mu}_0|, |x_2 - \tilde{\mu}_0|, \ldots, |x_n - \tilde{\mu}_0|.$

The rank of an observation is its position in the ordered list.

3. Place the absolute values in increasing order, and assign a rank to each from smallest (rank 1) to largest (rank n).

4. Equal absolute values are assigned the mean rank of their positions in the ordered list. For example, if the fifth, sixth, and seventh absolute values were all equal, then each would be assigned the rank $(5 + 6 + 7)/3 = 6$.

5. Add the ranks associated with the positive differences.

If the null hypothesis is true, then approximately half of the observations should be above the median and approximately half below the median. Since the distribution is assumed to be symmetric, for every positive difference there should be a corresponding negative difference of approximately the same magnitude. Therefore, the sum of the ranks associated with the positive differences should be approximately equal to the sum of the ranks associated with the negative differences. If the sum of the ranks associated with the positive differences is very large or very small, there is evidence to suggest that the population median is different from $\tilde{\mu}_0$. Consider the dot plots in Figures 14.4 and 14.5.

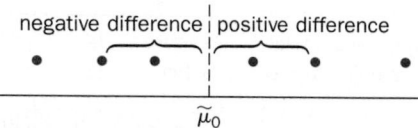

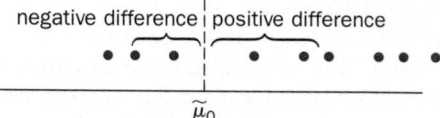

Figure 14.4 If H_0 is true, for every positive difference there should be a negative difference of about the same magnitude.

Figure 14.5 If $\tilde{\mu} > \tilde{\mu}_0$, then there should be more positive differences of larger magnitude. The sum of the ranks associated with the positive differences would be large.

THE WILCOXON SIGNED-RANK TEST

Suppose a random sample of size n is obtained from a continuous, symmetric distribution. A hypothesis test concerning a population median $\tilde{\mu}$ with significance level α has the following form:

H_0: $\tilde{\mu} = \tilde{\mu}_0$

H_a: $\tilde{\mu} > \tilde{\mu}_0$, $\tilde{\mu} < \tilde{\mu}_0$,   or $\tilde{\mu} \neq \tilde{\mu}_0$
Rank the absolute differences $|x_1 - \tilde{\mu}_0|, |x_2 - \tilde{\mu}_0|, \ldots, |x_n - \tilde{\mu}_0|$. Equal absolute values are assigned the mean rank for their positions.

TS: T_+ = the sum of the ranks corresponding to the positive differences $x_i - \tilde{\mu}_0$.

RR: $T_+ \geq c_1$, $T_+ \leq c_2$, $T_+ \geq c$ or $T_+ \leq n(n + 1) - c$
The critical values c_1, c_2, and c are obtained from Table IX in the Appendix such that $P(T_+ \geq c_1) \approx \alpha$, $P(T_+ \leq c_2) \approx \alpha$, and $P(T_+ \geq c) \approx \alpha/2$.

Differences equal to 0 $(x_i - \tilde{\mu}_0 = 0)$ are excluded from the analysis, and the sample size is reduced accordingly.

The Normal Approximation: As n increases $(n \geq 20)$, the statistic T_+ approaches a normal distribution with mean and variance

$$\mu_{T_+} = \frac{n(n + 1)}{4} \quad \text{and} \quad \sigma^2_{T_+} = \frac{n(n + 1)(2n + 1)}{24}.$$

Therefore, the random variable $Z = \dfrac{T_+ - \mu_{T_+}}{\sigma_{T_+}}$ has approximately a standard normal distribution. In this case $(n \geq 20)$, Z is the test statistic and the rejection region is

RR: $Z \geq z_\alpha$, $Z \leq -z_\alpha$,   or $|Z| \geq z_{\alpha/2}$

Since we assume that the underlying population is symmetric (when using the signed-rank test), the median is equal to the mean. Therefore, the test procedure just described can be used to test a hypothesis concerning a population mean (when the underlying population is not normal but is symmetric).

Solution Trail

KEYWORDS

- Signed-rank test.
- Median weight less than.

↓

TRANSLATION

One-sided, left-tailed hypothesis test concerning a population median.

↓

CONCEPTS

Wilcoxon signed-rank test.

↓

VISION

Rank the absolute differences, and find the value of the test statistic. Use this value to draw the appropriate conclusion.

Example 14.3 Packaging Materials Lighter packaging generally leads to savings in transportation costs, since the total load weighs less. A manufacturer of three-hole punches for business use has been packaging this product in cardboard for many years. One assembly line was reconfigured and the finished product was packaged using plastic. A random sample of 10 packaged products from this assembly line was obtained, and the weight of the plastic packaging on each three-hole punch was carefully measured (in grams). The data are given in the following table.

| 9.2 | 10.2 | 8.9 | 10.3 | 8.4 | 9.0 | 8.7 | 8.4 | 11.0 | 9.0 |

Assume that the underlying distribution of the weights of plastic packaging is continuous and symmetric. Use a signed-rank test to determine whether there is any evidence that the median weight of the packaging is less than 10 grams, with a significance level of approximately 0.05.

SOLUTION

STEP 1 The distribution of the weights of plastic packaging is assumed to be continuous and symmetric. Since this problem involves the population median, the signed-rank test is appropriate. We are looking for evidence that the population median is less than 10; the hypothesis test is one-sided, left-tailed. The number of observations is $n = 10$ (< 20); the test statistic is T_+.

STEP 2 The four parts of the hypothesis test are

$H_0: \tilde{\mu} = 10$

$H_a: \tilde{\mu} < 10$

TS: $T_+ =$ the sum of the ranks corresponding to the positive differences $x_i - \tilde{\mu}_0$.

RR: $T_+ \leq c_2 = 11$

We need a value for c_2 such that $P(T_+ \leq c_2) \approx 0.05$. Using Table IX in the Appendix, $P(T_+ \leq 11) = 0.053$.

STEP 3 The following table shows the data, each difference $x_i - 10$, the absolute value of each difference $|x_i - 10|$, the rank associated with each absolute difference, and the signed rank. The positive or negative sign from the pairwise difference has been attached to the rank to create the signed rank. Note that there are no zero differences. Therefore, no observations are excluded from the analysis.

Observation	9.2	10.2	8.9	10.3	8.4	9.0	8.7	8.4	11.0	9.0
Difference	−0.8	0.2	−1.1	0.3	−1.6	−1.0	−1.3	−1.6	1.0	−1.0
Absolute difference	0.8	0.2	1.1	0.3	1.6	1.0	1.3	1.6	1.0	1.0
Rank	3.0	1.0	7.0	2.0	9.5	5.0	8.0	9.5	5.0	5.0
Signed rank	−3.0	+1.0	−7.0	+2.0	−9.5	−5.0	−8.0	−9.5	+5.0	−5.0

To assign each rank, consider the ordered list of absolute differences and their position in the list.

Absolute difference	0.2	0.3	0.8	1.0	1.0	1.0	1.1	1.3	1.6	1.6
Position	1	2	3	4	5	6	7	8	9	10
Rank	1.0	2.0	3.0	5.0	5.0	5.0	7.0	8.0	9.5	9.5

The absolute difference 0.2 is in the first position. It is assigned the rank 1.0. Similarly, 0.3 and 0.8 are assigned the ranks 2.0 and 3.0, respectively.

The absolute differences in positions 4, 5, and 6 are equal. Each is assigned the mean rank of these three positions, $(4 + 5 + 6)/3 = 5.0$.

The absolute differences 1.1 and 1.3 are assigned ranks 7.0 and 8.0, respectively.

The absolute differences in positions 9 and 10 are equal. Each is assigned the mean rank, $(9 + 10)/2 = 9.5$.

STEP 4 The value of the test statistic is the sum of the positive signed ranks.

$$t_+ = 1.0 + 2.0 + 5.0 = \boxed{8.0} \ (\leq 11)$$

STEP 5 The value of the test statistic lies in the rejection region. At a significance level of approximately 0.05, there is evidence to suggest that the true median (mean) weight of plastic packaging is less than 10 grams.

Figure 14.6 shows a technology solution.

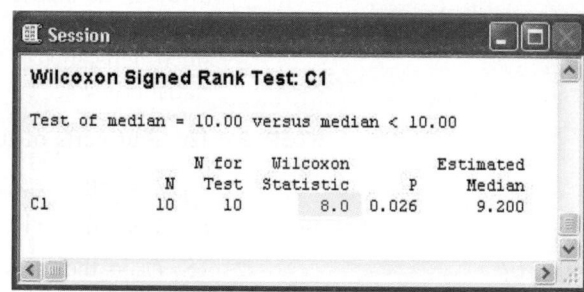

Figure 14.6 Minitab Wilcoxon signed-rank test.

The signed-rank test may also be used to compare population medians when the data are paired and the underlying distributions are not normal. We assume that the distribution of the pairwise differences is continuous and symmetric, and that each pair of values is independent of all the other pairs. If the null hypothesis is $H_0: \tilde{\mu}_1 - \tilde{\mu}_2 = \tilde{\mu}_D = \Delta_0$, we apply the one-sample test procedure to the pairwise differences.

1. Subtract Δ_0 from each pairwise difference. That is, compute $d_i - \Delta_0$ $(i = 1, 2, \ldots, n)$.

2. Rank the absolute differences, $|d_i - \Delta_0|$ $(i = 1, 2, \ldots, n)$.

3. Find the sum of the ranks associated with the positive differences, t_+.

There are two kinds of *differences* here: the pairwise differences are $d_i = x_i - y_i$, and the differences used in the calculation of the test statistic are $d_i - \Delta_0$.

(Courtesy Minnesota Soybean Organization)

Example 14.4 Soy-Based Foods Health-conscious Americans are consuming more soy-based foods because protein from soy contains less saturated fat than protein from meat and no cholesterol. A study was conducted to compare the amount of protein in traditional foods versus their soy-based imitations. For example, the amount of protein was measured in a traditional hamburger and a soy burger. Traditional and imitation (soy) products were selected at random, and the amount of protein in each serving was measured (in grams). The data are given in the following table.

Traditional food	Soy-based imitation	Traditional food	Soy-based imitation
6.8	5.7	3.4	7.7
5.6	11.6	14.6	5.6
12.8	6.7	3.1	4.3
5.4	6.6	12.2	17.5
3.7	9.8	10.8	12.4
5.3	10.6	11.9	5.2
14.3	8.2	4.0	4.8
7.3	16.8	4.2	13.8
4.7	17.2	7.4	11.9
9.3	12.4	10.9	5.0
14.9	11.4	7.8	12.9
4.9	3.4	9.3	13.6

Assume that the underlying distribution of the pairwise differences is continuous and symmetric. Use the Wilcoxon signed-rank test to determine whether there is any evidence that the median amount of protein in soy-based products is different from the median amount of protein in traditional foods. Use $\alpha = 0.05$.

SOLUTION

STEP 1 The underlying distributions are not assumed to be normal, but the distribution of the differences is assumed to be continuous and symmetric. The assumptions for the Wilcoxon signed-rank test are met. Note that the sign test to compare two medians could also be used, but the Wilcoxon signed-rank test is a more reliable test since it uses more of the information in the sample.

Since we are searching for evidence of any difference in the medians, the hypothesis test is two-sided with $\Delta_0 = 0$.

Let population 1 be the amount of protein in servings of traditional foods and population 2 be the amount of protein in servings of imitation, soy-based foods.

Since $\Delta_0 = 0$, we do not need to modify the pairwise differences. No pairwise differences are equal to zero; no pairs are excluded from the analysis.

Since there are $n = 24 \geq 20$ observations (pairwise differences), the normal approximation will be used. The four parts of the hypothesis test are

$H_0: \tilde{\mu}_1 - \tilde{\mu}_2 = \tilde{\mu}_D = \Delta_0 = 0$

$H_a: \tilde{\mu}_1 - \tilde{\mu}_2 = \tilde{\mu}_D \neq \Delta_0 = 0$

$\text{TS: } Z = \dfrac{T_+ - \mu_{T_+}}{\sigma_{T_+}}$

$\text{RR: } |Z| \geq z_{\alpha/2} = z_{0.025} = 1.9600$

Solution Trail

KEYWORDS

- Signed-rank test.
- Median amount different from.

↓

TRANSLATION

Two-sided hypothesis test concerning the difference in population medians.

↓

CONCEPTS

Wilcoxon signed-rank test.

↓

VISION

Compute each pairwise difference, and rank the absolute differences. Find the value of the test statistic, and draw the appropriate conclusion.

STEP 2 Compute the absolute value of each pairwise difference, and rank these numbers. Equal values are assigned the mean rank for their positions. The following table shows the original data, the pairwise differences, the absolute value of each pairwise difference, the rank associated with each pairwise difference, and the signed rank.

Traditional food	Soy-based imitation	Pairwise difference	Absolute difference	Rank	Signed rank
6.8	5.7	1.1	1.1	2.0	+2.0
5.6	11.6	−6.0	6.0	16.0	−16.0
12.8	6.7	6.1	6.1	18.0	+18.0
5.4	6.6	−1.2	1.2	3.5	−3.5
3.7	9.8	−6.1	6.1	18.0	−18.0
5.3	10.6	−5.3	5.3	13.5	−13.5
14.3	8.2	6.1	6.1	18.0	+18.0
7.3	16.8	−9.5	9.5	22.0	−22.0
4.7	17.2	−12.5	12.5	24.0	−24.0
9.3	12.4	−3.1	3.1	7.0	−7.0
14.9	11.4	3.5	3.5	8.0	+8.0
4.9	3.4	1.5	1.5	5.0	+5.0
3.4	7.7	−4.3	4.3	9.5	−9.5
14.6	5.6	9.0	9.0	21.0	+21.0
3.1	4.3	−1.2	1.2	3.5	−3.5
12.2	17.5	−5.3	5.3	13.5	−13.5
10.8	12.4	−1.6	1.6	6.0	−6.0
11.9	5.2	6.7	6.7	20.0	+20.0
4.0	4.8	−0.8	0.8	1.0	−1.0
4.2	13.8	−9.6	9.6	23.0	−23.0
7.4	11.9	−4.5	4.5	11.0	−11.0
10.9	5.0	5.9	5.9	15.0	+15.0
7.8	12.9	−5.1	5.1	12.0	−12.0
9.3	13.6	−4.3	4.3	9.5	−9.5

STEP 3 The sum of the positive signed ranks is

$$t_+ = 2.0 + 5.0 + 8.0 + 15.0 + 18.0 + 18.0 + 20.0 + 21.0 = \boxed{107.0}$$

The mean and variance of the random variable T_+ are

$$\mu_{T_+} = \frac{n(n+1)}{4} = \frac{(24)(25)}{4} = 150.0$$

$$\sigma_{T_+}^2 = \frac{n(n+1)(2n+1)}{24} = \frac{(24)(25)(49)}{24} = 1225.0$$

The value of the test statistic is

$$z = \frac{t_+ - \mu_{T_+}}{\sigma_{T_+}} = \frac{107.0 - 150.0}{\sqrt{1225.0}} = -1.2286$$

STEP 4 The value of the test statistic does not lie in the rejection region; we do not reject the null hypothesis. At the $\alpha = 0.05$ significance level, there is no evidence to suggest that the median amount of protein in traditional foods is different from the median amount of protein in soy imitations.

To find the p value associated with this test, find the left-tail probability and multiply by 2.

$$p/2 = P(Z \leq -1.2286)$$

Definition of p value.

$$= 0.1096$$

Cumulative probability; use Table III in the Appendix.

$$p = 2(0.1096) = \boxed{0.2192}$$

Solve for p.

Using the p value to draw a conclusion, since $p = 0.2192 > 0.05 = \alpha$, we do not reject the null hypothesis. There is no evidence to suggest that the median amount of protein in traditional foods is different from the median amount of protein in soy imitations.

Figure 14.7 shows a technology solution.

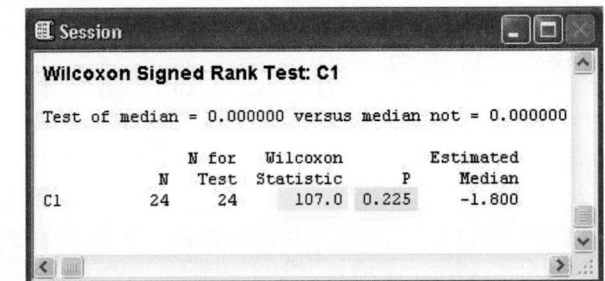

Figure 14.7
Minitab Wilcoxon signed-rank test.

TECHNOLOGY CORNER

Procedure: Conduct a Wilcoxon signed-rank test.
Reconsider: Example 14.3, page 673, solution, and interpretations.

Minitab

The built-in Wilcoxon signed-rank test is accessible in the Stat; Nonparametrics menu and may also be executed using commands in a session window. In the case of paired data, use the differences $d_i - \Delta_0$ ($i = 1, 2, \ldots, n$).

1. Enter the data into column C1.
2. Select Stat; Nonparametrics; 1-Sample Wilcoxon. Enter the Variable (column containing the data, C1), select Test median and enter the hypothesized value, 10, and select the Alternative hypothesis (less than).
3. Click OK. The test results are displayed in a session window. See Figure 14.6, page 674.

Excel

There is no built-in function to conduct a Wilcoxon signed-rank test. Use other arithmetic functions to compute the sum of the signed ranks.

1. Enter the data into column A.
2. Compute the differences and the absolute differences. Enter or compute the ranks. The built-in function RANK does not produce the appropriate ranks for this test. However, it can be used with the COUNT function to produce the correct ranks for this test. Use the built-in function IF to compute the signed ranks.
3. Sum the positive signed ranks. See Figure 14.8.

Figure 14.8 Excel calculations to compute the value of the test statistic in a Wilcoxon signed-rank test.

	A	B	C	D	E
1			Absolute		Signed
2	Weight	Difference	difference	Rank	rank
3	9.2	-0.8	0.8	3.0	-3.0
4	10.2	0.2	0.2	1.0	1.0
5	8.9	-1.1	1.1	7.0	-7.0
6	10.3	0.3	0.3	2.0	2.0
7	8.4	-1.6	1.6	9.5	-9.5
8	9.0	-1.0	1.0	5.0	-5.0
9	8.7	-1.3	1.3	8.0	-8.0
10	8.4	-1.6	1.6	9.5	-9.5
11	11.0	1.0	1.0	5.0	5.0
12	9.0	-1.0	1.0	5.0	-5.0
13					
14		8.0	= SUMIF(E3:E12,">0",E3:E12)		

SECTION 14.2 EXERCISES

Practice

14.15 A data set and null hypothesis are given in each of the following problems. Assume that a Wilcoxon signed-rank test will be used to test H_0. Find (i) the differences, (ii) the absolute differences, and (iii) the rank associated with each absolute difference.

a. H_0: $\tilde{\mu} = 60$.

41	66	36	72	33	22	24	36	47
53	28	77	31	34	38	59	60	45

b. H_0: $\tilde{\mu} = 20$.

21.4	20.3	18.5	18.8	21.5	21.6	20.8	21.9
19.8	19.6	19.4	20.8	19.6	18.7	20.4	20.2
21.6	19.2	20.1	18.5	18.5	19.7	19.7	

c. H_0: $\tilde{\mu} = -5$.

-2	-8	-9	-8	-7	-2	-2	-1	-5	-4
-9	-7	-9	-7	-5	-5	-8	-8	-3	-1
-4	-6	-7	-7	-2					

d. H_0: $\tilde{\mu} = 305.4$.

296	303	271	263	288	312	260	305
250	308	315	264	254	258	274	314
279	267	312	310	309	273	269	293
264	278	255	285	272	307		

14.16 Use the random sample in each problem to conduct a Wilcoxon signed-rank test with the indicated null hypothesis, alternative hypothesis, and significance level. Find the p value associated with each test.

a. H_0: $\tilde{\mu} = 70$, H_a: $\tilde{\mu} \neq 70$, $\alpha = 0.10$.

67.1	63.5	70.1	62.5	72.2	63.7	63.7	67.7
79.1	61.6	79.7	69.5	72.2	64.5	69.3	

b. H_0: $\tilde{\mu} = 0.7$, H_a: $\tilde{\mu} < 0.7$, $\alpha = 0.05$.

0.072	0.348	0.319	0.502	0.733	0.603	0.052
0.493	0.721	0.762	0.166	0.965	0.616	0.904
0.882	0.773	0.771	0.506	0.735		

c. H_0: $\tilde{\mu} = -45$, H_a: $\tilde{\mu} > -45$, $\alpha = 0.02$.

-50	-35	-35	-31	-41	-32	-32
-40	-46	-35	-38	-31	-36	-32

d. H_0: $\tilde{\mu} = 450$, H_a: $\tilde{\mu} \neq 450$, $\alpha = 0.01$.

461	424	436	485	476	457	463	409	424	450
406	402	435	457	420	410	402	438	428	450
410	418	423	466	497	492	411	426		

14.17 Using the paired data on the data CD and book's web site, conduct a Wilcoxon signed-rank test to compare population medians with H_0: $\tilde{\mu}_1 - \tilde{\mu}_2 = 0$, H_a: $\tilde{\mu}_1 - \tilde{\mu}_2 \neq 0$, and significance level $\alpha = 0.02$. Find the p value associated with this test.

Applications

14.18 Medicine and Clinical Studies A normal adult brain oxidizes approximately 120 grams of glucose per day.[8] Some researchers believe that the rate of metabolism is higher in adults whose jobs require them to make many decisions daily. A random sample of basketball referees was obtained, and the brain oxidation rate of each was measured. The data are given in the following table.

120	121	119	122	121	122	121	119	121
122	124	121	123	122	123	119	120	119

The underlying distribution of oxidation rates is assumed to be continuous and symmetric.

a. Use the Wilcoxon signed-rank test to determine whether there is any evidence that the mean oxidation rate for basketball referees is greater than 120 grams per day. Use a significance level of $\alpha = 0.05$. Find the p value for this test.

b. Why can the signed-rank test be used in this case to test a hypothesis concerning the population mean (rather than the population median)?

14.19 Medicine and Clinical Studies A random sample of patients with a certain kidney disorder was obtained, and the renal blood-flow rate was measured (in L/min) for each. The data are given on the data CD and book's web site. In healthy patients, the median renal blood-flow rate is known to be 3.00 L/min, and the distribution of flow rates is assumed to be continuous. Use the Wilcoxon signed-rank test to determine whether there is any evidence that the median renal blood-flow rate in patients with this kidney disorder is different from 3.00 L/min. Use $\alpha = 0.01$, and find the p value for this test.

14.20 Public Policy and Political Science Regulations for a new office building specify that the median strength for nonmetallic, nonshrink grout should be 6000 psi when it is supporting concrete. A random sample of grout from various locations in an office building project was obtained, and the strength of each batch (in psi) was determined using the grout-cube test at 28 days. The data are given on the data CD and book's web site. Assume that the underlying distribution of grout strength is continuous and symmetric. Use the Wilcoxon signed-rank test with $\alpha = 0.01$ to determine whether there is any evidence that the median grout strength is different from 6000 psi. Find the p value for this test.

14.21 Medicine and Clinical Studies An individual's A1C value is a percentage that indicates the average blood glucose level for the last 3 months. For people with diabetes, an A1C level of less than 7% is recommended. A new intervention program for diabetics, sponsored by ALR Technologies, has been shown to significantly change the A1C values for participants.[9] Suppose a random sample of diabetics was selected and the A1C value for each was measured. After three months of participating in the intervention program, the A1C value for each was measured again. The data are given on the data CD and book's web site. Assume that the distribution of pairwise differences is continuous and symmetric. Use the Wilcoxon signed-rank test with $\alpha = 0.025$ to determine whether there is any evidence that intervention program decreases the median A1C value.

14.22 Medicine and Clinical Studies One measure of collagen breakdown and bone degradation in older men is the Dpd, or deoxypyridinoline, value. If the Dpd value is more than 5.4, this may indicate excess bone degradation.[10] A random sample of adult men with Dpd values greater than 5.4 was obtained, and each was placed on a diet that included a synthetic vitamin D supplement. After six months, each patient's Dpd value was measured again (in nmol Dpd/mmol creatinine). The data are given on the data CD and book's web site. Assume that the distribution of pairwise differences is continuous and symmetric. Use the Wilcoxon signed-rank test with $\alpha = 0.01$ to determine whether there is any evidence that the vitamin D supplement decreases the median Dpd value.

14.23 Economics and Finance Two counties in Pennsylvania handle delinquent property-tax payments in different manners. The first county has very strict regulations that include the levying of interest on the unpaid balance and eventually property seizure. The second county uses a more personal touch. Someone calls or visits the property owner to discuss the overdue bill and, if necessary, a payment schedule is created. A random sample of delinquent payments in each county was obtained, and the data were paired according to the size of the original property-tax bill. The amount of time (in months) until the full payment was made was recorded for each property owner. The data are given on the data CD and book's web site. Assume that the distribution of pairwise differences is continuous and symmetric. Use the Wilcoxon signed-rank test to determine whether there is any evidence of a difference in median collection time due to the method of handling delinquent payments. Use $\alpha = 0.10$, and find the p value associated with this test.

14.24 Biology and Environmental Science The 154-day weights (in pounds) of 16 randomly selected Rambouillet lambs on a farm in Nebraska were recorded. The data are given in the following table.

118	119	120	115	114	113	119	119
118	117	120	116	113	115	118	119

Assume that the underlying distribution of weights is continuous and symmetric. Use the Wilcoxon signed-rank test to determine whether there is any evidence that the median 154-day weight is less than 118 pounds. Use a significance level of 0.05.

Extended Applications

14.25 Travel and Transportation Subway lines are often rated by the scheduled frequency of trains, the number of breakdowns, seat availability, and cleanliness. In each of two years, the trains on each subway line in New York City were rated for on-time performance. The percentage of trains arriving at the end of the line within five minutes of the scheduled time for each line is given in the following table.

Line	2006	2007	Line	2006	2007
1	92	88	2	80	81
3	86	88	4	80	70
5	86	82	6	92	89
7	94	90	A	92	93
B	96	96	C	96	96
D	96	96	E	96	95
F	94	92	G	98	98
J/Z	99	98	L	95	92
M	99	99	N	95	96
Q	98	97	R	95	95
V	94	95	W	95	97

(*Source*: Metropolitan Transportation Authority, as reported by R. Rivera, Subway delays rise, and the No. 4 line is the slowest, *The New York Times*, July 22, 2008.)

Assume that the distribution of pairwise differences is continuous and symmetric. Use the Wilcoxon signed-rank test with $\alpha = 0.05$ to determine whether there is any evidence that the difference in median on-time percentages is different from zero. What does your conclusion suggest about the change in on-time performance of the New York City Transit System from 2006 to 2007?

14.26 Physical Sciences In a study of several Chesapeake Bay tributaries, a random sample of nontidal freshwater locations on the Patuxent River was obtained. The total arsenic concentration was measured at each location (in μg/L), and the data are given in the following table.

0.67	0.22	0.56	0.27	0.17	0.57	0.55	0.55
0.53	0.09	0.61	0.45	0.49	0.04	0.44	0.19

Suppose the distribution of arsenic concentrations is continuous and symmetric and that the safe level of arsenic concentration is 0.30 μg/L.

a. Conduct a sign test to determine whether there is any evidence that the median arsenic concentration in freshwater locations is greater than 0.30 μg/L. Use $\alpha = 0.05$.

b. Conduct a signed-rank test to determine whether there is any evidence that the median arsenic concentration in freshwater locations is greater than 0.30 μg/L. Use $\alpha = 0.05$.

c. Compare your conclusions in parts (a) and (b). Which test do you think is more accurate? Why?

14.3 THE RANK-SUM TEST

The nonparametric Wilcoxon rank-sum test is used to compare two population medians. Suppose two independent random samples are obtained from continuous, nonnormal distributions. Assume that the first sample has size n_1, the second sample has size n_2, and $n_1 \leq n_2$.

Combine all of the data for a total of $n_1 + n_2$ observations. Place these combined data in increasing order, and assign a rank to each from smallest (rank 1) to largest (rank $n_1 + n_2$). Equal values are assigned the mean rank of their positions in the ordered list (just as in the signed-rank test). Let w be the sum of the ranks associated with observations from the first sample.

Suppose, for example, that the sample sizes are equal. If $\tilde{\mu}_1 < \tilde{\mu}_2$, then the observation in the first sample would tend to be smaller than the observations in the second sample, and w would be small. If $\tilde{\mu}_1 > \tilde{\mu}_2$, then the observations in the first sample would tend to be larger than the observations in the second sample, and w would be large. There is also a normal approximation that is valid when both n_1 and n_2 are large.

> Can you figure out the smallest and the largest possible value of w?

THE WILCOXON RANK-SUM TEST

Suppose that two independent random samples of sizes n_1 and n_2 ($n_1 \leq n_2$) are obtained from continuous distributions. A hypothesis test concerning the two population medians with significance level α has the following form:

H_0: $\tilde{\mu}_1 - \tilde{\mu}_2 = \Delta_0$

H_a: $\tilde{\mu}_1 - \tilde{\mu}_2 > \Delta_0$, $\tilde{\mu}_1 - \tilde{\mu}_2 < \Delta_0$,   or $\tilde{\mu}_1 - \tilde{\mu}_2 \neq \Delta_0$

Subtract Δ_0 from each observation in the first sample. Combine these differences and the observations in the second sample, and rank all of these values. Equal values are assigned the mean rank for their positions.

TS: W = the sum of the ranks corresponding to the differences from the first sample.

RR: $W \geq c_1$, $W \leq c_2$, $W \geq c$ or $W \leq n_1(n_1 + n_2 + 1) - c$

The critical values c_1, c_2, and c are obtained from Table X in the Appendix such that $P(W \geq c_1) \approx \alpha$, $P(W \leq c_2) \approx \alpha$, and $P(W \geq c) \approx \alpha/2$.

The Normal Approximation: As n_1 and n_2 increase, the statistic W approaches a normal distribution with

$$\mu_W = \frac{n_1(n_1 + n_2 + 1)}{2} \quad \text{and} \quad \sigma_W^2 = \frac{n_1 n_2 (n_1 + n_2 + 1)}{12}$$

Therefore, the random variable $Z = \dfrac{W - \mu_W}{\sigma_W}$ has approximately a standard normal distribution. The normal approximation is good when both n_1 and n_2 are greater than 8. In this case, Z is the test statistic and the rejection region is

RR: $Z \geq z_\alpha$, $Z \leq -z_\alpha$, $|Z| \geq z_{\alpha/2}$

 ILLUMINATING THE CONCEPTS

1. This general test procedure allows for any hypothesized difference between the two population medians, Δ_0. Usually, $\Delta_0 = 0$, and the null hypothesis is that the two population medians are equal.

2. If $\Delta_0 \neq 0$, we subtract this quantity from each observation in the first sample. Intuitively, if H_0 is true, this *shifts* sample 1 so that the two samples have the same median.

3. Like other nonparametric tests, this procedure is appropriate when the underlying distributions are nonnormal, or when the data are ranks.

4. Suppose both underlying populations are symmetric. The Wilcoxon rank-sum test can be used to compare population means because in each population, the mean is equal to the median. In this case, one might also consider the two-sample t test, discussed in Section 10.2. However, if the underlying populations are not normal then the t test results are not valid.

5. Sometimes, a slightly different test statistic is used in this test procedure. The Mann–Whitney U statistic is a function of W and also approaches a normal distribution as n_1 and n_2 increase. The two statistics lead to similar conclusions. ■

Many software packages call this the Mann–Whitney rank-sum test.

Example 14.5 Predicting Heart Disease Recent research suggests that folic acid supplements may reduce the amount of homocysteine in the blood, which is believed to be a good predictor of heart disease.[11] Flour is a natural source of folic acid, and grain could be supplemented to provide higher levels of this important vitamin. Independent random samples of white enriched unbleached flour from two manufacturers were obtained, and the amount of folic acid in each sample was carefully measured (in mg/kg). The data are given in the following table.

| King Arthur | 0.200 | 0.302 | 0.235 | 0.160 | 0.189 | |
| Gold Medal | 0.257 | 0.230 | 0.381 | 0.245 | 0.380 | 0.296 |

Suppose the underlying distributions of folic acid levels are continuous. Conduct a Wilcoxon rank-sum test to determine whether there is any evidence to suggest a difference in the population folic acid medians. Use $\alpha = 0.05$.

SOLUTION

STEP 1 The samples are independent, and the populations are assumed to be continuous. Since we are searching for *any* difference in the medians, $\Delta_0 = 0$.

The sample sizes are $n_1 = 5$ and $n_2 = 6$. Since n_1 and n_2 are both less than 8, the test statistic is W.

The four parts of the hypothesis test are

H_0: $\tilde{\mu}_1 - \tilde{\mu}_2 = \Delta_0 = 0$

H_a: $\tilde{\mu}_2 - \tilde{\mu}_2 \neq \Delta_0 = 0$

TS: W = the sum of the ranks corresponding to the differences from the first sample.

Since $\Delta_0 = 0$, W is the sum of the ranks corresponding to the first (smaller) sample.

RR: $W \geq 41$ or $W \leq 5(5 + 6 + 1) - 41 = 19$

We need a value for c such that $P(W \geq c) \approx 0.05/2 = 0.025$. Using Table X in the Appendix, $P(W \geq 41) = 0.026$. Therefore, the actual significance level for this test is $2(0.026) = 0.052 \approx 0.05$.

Solution Trail

KEYWORDS

- Rank-sum test.
- Difference in population medians.

↓

TRANSLATION

Two-sided hypothesis test concerning two population medians.

↓

CONCEPTS

Wilcoxon rank-sum test.

↓

VISION

Combine the samples, and rank the ordered observations. Find the value of the test statistic, W, and draw the appropriate conclusion.

STEP 2 The following table shows the combined samples in order and the rank associated with each value. The shaded columns correspond to observations from the first sample.

Observation	0.160	0.189	0.200	0.230	0.235	0.245	0.257	0.296	0.302	0.380	0.381
Rank	1	2	3	4	5	6	7	8	9	10	11

STEP 3 The value of the test statistic is the sum of the ranks corresponding to the observations from the first sample.

$$w = 1 + 2 + 3 + 5 + 9 = 20$$

The value of the test statistic does not lie in the rejection region. At the $\alpha = 0.05$ significance level, there is no evidence to suggest that the population median folic acid levels are different.

Figure 14.9 shows a technology solution.

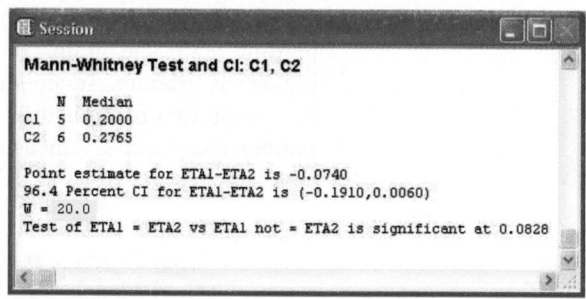

Figure 14.9 Minitab Mann-Whitney test.

Example 14.6 Parking Garage Construction Two companies were erecting several parking garages in major cities in the United States. As a measure of the quality of the concrete being used, the water–cement material (w/cm) ratio was computed for every slab. Independent random samples of garages being constructed by each company were obtained, and the w/cm ratio was computed on the first floor of each structure. The data are given in the following table.

Bechtel	0.33	0.27	0.33	0.37	0.35	0.38	0.26
	0.24	0.34	0.21	0.32	0.39		

Owens	0.32	0.49	0.46	0.49	0.41	0.43	0.44
	0.49	0.50	0.40	0.32	0.45	0.35	0.31

(*Source*: concretenetwork.com.)

Assume that the underlying distributions of w/cm ratios are continuous. Use a Wilcoxon rank-sum test to determine whether there is any evidence to suggest a difference in the population median w/cm ratios. Use $\alpha = 0.01$.

SOLUTION

STEP 1 The samples are independent, and the populations are assumed to be continuous. Since we are searching for *any* difference in the medians, $\Delta_0 = 0$.

The sample sizes are $n_1 = 12$ and $n_2 = 14$. Since both sample sizes are greater than eight, the normal approximation will be used.

The four parts of the hypothesis test are

$H_0: \tilde{\mu}_1 - \tilde{\mu}_2 = \Delta_0 = 0$

$H_a: \tilde{\mu}_1 - \tilde{\mu}_2 \neq \Delta_0 = 0$

$\text{TS: } Z = \dfrac{W - \mu_W}{\sigma_W}$

$\text{RR: } |Z| \geq z_{\alpha/2} = z_{0.005} = 2.5758$

STEP 2 The following table shows the combined samples in order and the rank associated with each value. The shaded columns correspond to observations from the first sample.

Observation	0.21	0.24	0.26	0.27	0.31	0.32	0.32	0.32	0.33
Rank	1.0	2.0	3.0	4.0	5.0	7.0	7.0	7.0	9.5

Observation	0.33	0.34	0.35	0.35	0.37	0.38	0.39	0.40	0.41
Rank	9.5	11.0	12.5	12.5	14.0	15.0	16.0	17.0	18.0

Observation	0.43	0.44	0.45	0.46	0.49	0.49	0.49	0.50
Rank	19.0	20.0	21.0	22.0	24.0	24.0	24.0	26.0

Note: There are several ties. Equal values are assigned the mean rank for their positions.

STEP 3 The value w is the sum of the ranks corresponding to the observations from the first sample.

$w = 1.0 + 2.0 + 3.0 + 4.0 + 7.0 + 9.5 + 9.5 + 11.0 + 12.5 + 14.0$
$\quad + 15.0 + 16.0 = 104.5$

The mean and variance of the random variable W are

$\mu_W = \dfrac{n_1(n_1 + n_2 + 1)}{2} = \dfrac{12(12 + 14 + 1)}{2} = 162.0$

$\sigma_W^2 = \dfrac{n_1 n_2 (n_1 + n_2 + 1)}{12} = \dfrac{(12)(14)(12 + 14 + 1)}{12} = 378.0$

The value of the test statistic is

$z = \dfrac{w - \mu_W}{\sigma_W} = \dfrac{104.5 - 162.0}{\sqrt{378.0}} = -2.9575$

The value of the test statistic lies in the rejection region. At the $\alpha = 0.01$ significance level, there is evidence to suggest that the population median w/cm ratios are different.

Figure 14.10 shows a technology solution.

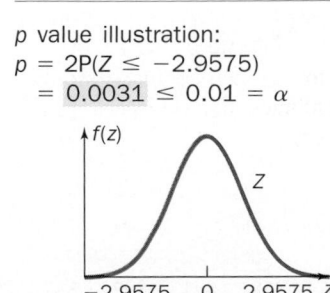

p value illustration:
$p = 2P(Z \leq -2.9575)$
$= 0.0031 \leq 0.01 = \alpha$

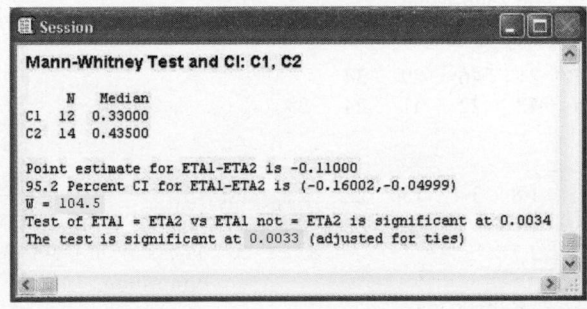

Figure 14.10 Minitab Mann-Whitney test.

TECHNOLOGY CORNER

Procedure: Conduct a rank-sum test.
Reconsider: Example 14.6, page 682, solution, and interpretations.

Minitab

Use the built-in function Mann-Whitney, accessible in the Stat; Nonparametric menu or by using the appropriate commands in a session window.

1. Enter the Bechtel data into column C1 and the Owens data into column C2.
2. Select Stat; Nonparametrics; Mann-Whitney. Enter the First Sample column, C1, the Second Sample column, C2, and select an Alternative hypothesis (not equal).
3. Click OK. The test results are displayed in a session window. See Figure 14.10, page 683.

Excel

There is no built-in function to conduct a rank-sum test. Use other arithmetic functions to compute the values of the test statistics w and z and the p value associated with this test.

1. Enter the data from both samples into Column A and the corresponding sample number into column B.
2. Compute the ranks and store them in column C.
3. Sum the ranks corresponding to the first (smaller) sample, w. Compute the value of the test statistic z and the p value. See Figure 14.11.

	A	B	C	D	E
1	0.33	1	9.5		
2	0.27	1	4.0		
3	0.33	1	9.5		
4	0.37	1	14.0		
5	0.35	1	12.5		
27					
28	104.5	=SUMIF(B1:B26,"=1",C1:C26) = w			
29					
30	162.0	=(12)*(12+14+1)/2 = mean of W			
31	378.0	=(12)*(14)*(12+14+1)/12 = variance of W			
32	-2.9575	=(A28-A30)/SQRT(A31) = z			
33					
34	0.0031	=NORMDIST(A32,0,1,TRUE) * 2 = p			

Figure 14.11 Excel calculations to compute the value of the test statistics in a rank-sum test and the p value.

SECTION 14.3 EXERCISES

Practice

14.27 In each of the following problems, two independent random samples are given. Combine all of the observations, and find the rank associated with each value.

a.
Sample 1	37	21	46	29	34	
Sample 2	45	42	22	41	24	39

b.
Sample 1	4.5	1.8	3.4	1.4	2.2	2.1	1.5	3.7
Sample 2	4.6	6.6	1.2	6.0	2.4	2.3	2.2	0.4
	2.5	2.7						

c. Sample 1

820	872	814	825	876	858	841	892	882
809	826	887	884	862	846	801	871	803

Sample 2

850	842	888	879	832	827	831	810	871
840	870	816	821	865	899	830	875	800
813	839	822	865	818	818			

14.28 In each of the following problems, assume a Wilcoxon rank-sum test will be used to compare population medians. Two independent random samples and the value of Δ_0 are given. Find w, the value of the test statistic.

a. $\Delta_0 = 0$

Sample 1	90	90	87	87	81
Sample 2	83	84	80	84	87

b. $\Delta_0 = 0$

Sample 1	58	69	52	53	57	58	60	56	59
Sample 2	53	52	60	55	70	66	65	69	51
	69	66							

c. $\Delta_0 = 5$

Sample 1

36.2	33.5	32.5	33.2	32.5	34.6	39.0	39.7
32.1	39.5	38.3	30.6	32.5	38.4	38.4	

Sample 2

31.7	34.4	32.3	25.6	27.3	32.7	33.0	28.4
32.6	28.6	27.8	30.5	30.6	33.1	25.4	32.4
31.5	33.5						

14.29 In each of the following problems, assume that a Wilcoxon rank-sum test will be used to compare population medians. Use the sample sizes and alternative hypothesis to find the best rejection region for the given significance level. Report the exact significance level for your rejection region.

a. $n_1 = 3$, $n_2 = 5$, $H_a: \tilde{\mu}_1 < \tilde{\mu}_2$, $\alpha = 0.05$.
b. $n_1 = 4$, $n_2 = 4$, $H_a: \tilde{\mu}_1 > \tilde{\mu}_2$, $\alpha = 0.05$.
c. $n_1 = 4$, $n_2 = 10$, $H_a: \tilde{\mu}_1 \neq \tilde{\mu}_2$, $\alpha = 0.05$.
d. $n_1 = 6$, $n_2 = 8$, $H_a: \tilde{\mu}_1 < \tilde{\mu}_2$, $\alpha = 0.01$.
e. $n_1 = 7$, $n_2 = 9$, $H_a: \tilde{\mu}_1 \neq \tilde{\mu}_2$, $\alpha = 0.10$.
f. $n_1 = 8$, $n_2 = 8$, $H_a: \tilde{\mu}_1 > \tilde{\mu}_2$, $\alpha = 0.01$.

14.30 In each of the following problems, assume that a Wilcoxon rank-sum test with the normal approximation will be used to compare population medians. Find the mean, the variance, and the standard deviation for the test statistic W.

a. $n_1 = 15$, $n_2 = 21$.
b. $n_1 = 18$, $n_2 = 18$.
c. $n_1 = 11$, $n_2 = 27$.
d. $n_1 = 12$, $n_2 = 16$.
e. $n_1 = 23$, $n_2 = 24$.
f. $n_1 = 25$, $n_2 = 30$.

14.31 Two independent random samples from continuous distributions are given in the following table. Conduct a Wilcoxon rank-sum test of $H_0: \tilde{\mu}_1 = \tilde{\mu}_2$ versus $H_a: \tilde{\mu}_1 > \tilde{\mu}_2$ using a significance level of 0.01.

Sample 1	51.8	55.6	57.6	58.7	56.2	53.1
Sample 2	57.2	58.2	57.1	63.3	59.7	60.6
	61.1	56.9				

14.32 Two independent random samples from continuous distributions are given on the data CD and book's web site. Conduct a Wilcoxon rank-sum test (using the normal approximation) of $H_0: \tilde{\mu}_1 = \tilde{\mu}_2$ versus $H_a: \tilde{\mu}_1 \neq \tilde{\mu}_2$ using a significance level of 0.05.

Applications

14.33 Biology and Environmental Science Planning managers in Michigan have been working to reduce the amount of salt used on county roads. The state is trying to reach a balance between traffic safety and environmental concerns. A new "Dream Machine" snow-plow has been designed to make winter driving safer and to reduce the need for deicing chemicals.[12] Two counties were selected, and independent random samples of the salt usage (in tons) per mile for last winter were obtained. The data are given in the following table.

Alcona County

12.5	12.9	12.4	11.5	12.7	12.8	14.2	11.0

Otsego County

12.5	13.3	13.8	13.2	13.5	13.9	13.6	13.5	12.9

Use the Wilcoxon rank-sum test with $\alpha = 0.05$ to determine whether there is any evidence that the two population median salt usages per mile are different. Assume that the underlying distributions are continuous.

14.34 Public Health and Nutrition The Providence College dining service has started to publish nutrition information tables. One information sheet indicates that an apple–pecan muffin and a slice of banana bread have approximately the same fat content (in grams). In order to check this claim, independent random samples of apple–pecan muffins and banana bread slices were obtained. The fat content of each item was measured (in grams), and the data are given in the following table.

Apple–pecan muffin

7.7	8.4	7.0	7.2	8.2	8.0

Banana bread

8.0	8.1	8.3	8.5	8.8	8.2	8.4

(*Source*: USDA Nutrient Database.)

Use the Wilcoxon rank-sum test with $\alpha = 0.05$ to determine whether there is any evidence to suggest that the population median fat contents are different. Assume that the underlying distributions are continuous.

14.35 Manufacturing and Product Development Better World Technologies claims to have produced a revolutionary new jack-hammer with less vibration, a reduction in noise, lower energy input, fewer moving parts, and greater impact strength. Independent random samples of conventional jackhammers and new jackhammers were obtained. The impact strength of each, using heavy-duty concrete breakers, was measured (in ft-lb) at 75 psi. The data are given in the following table.

Conventional jackhammer

95	94	102	100	93	93	93

New jackhammer

119	130	99	126	114	130	101

a. Use the Wilcoxon rank-sum test with $\alpha = 0.01$ to determine whether there is any evidence to suggest that the population median impact strength is higher for the new jackhammer than for the conventional jackhammer. Assume that the underlying distributions are continuous.

b. Find the p value for this hypothesis test.

14.36 Sports and Leisure Many hockey players use a slapshot—a short, quick, powerful swing with the stick—to try to score a goal. The speed of this shot makes it very difficult for a goalie to react and make a save. The fastest slapshot recorded was 106.6 mph in 2006 by Chad Kilger.[13] A National Hockey League (NHL) scout believes that defensemen have faster slapshots than forwards. In order to test this claim, independent random samples from each group of players were obtained. Each player took a slapshot from the blue line and the speed of the puck was recorded (in mph) using a radar gun. The data are given in the following table.

Defensemen

94	98	97	94	94	94	91	100	98	96

Forwards

92	88	86	94	90	85	86	95	89	85

Use the Wilcoxon rank-sum test with $\alpha = 0.01$ to determine whether there is any evidence to suggest that the population median slapshot speed of NHL defensemen is greater than the population median slapshot speed of NHL forwards. Assume that the underlying distributions are continuous.

14.37 Business and Management Kroll's South restaurant serves dairy products in two separate buffet lines. Independent random samples of the dairy products holding temperature (in °F) in each line were obtained. The data are given on the data CD and book's web site. Use the Wilcoxon rank-sum test to determine whether there is any evidence to suggest that the population median holding temperatures are different. Use $\alpha = 0.05$, and find the p value associated with this test.

Extended Applications

14.38 Public Health and Nutrition Independent random samples of almonds from two suppliers in the United Kingdom were obtained, and the amount of protein (in grams) per 100 grams of edible portion was measured for each sample. The data are given on the data CD and book's web site.[14] SBP claims that their almonds contain more protein than any other brand. Use the Wilcoxon rank-sum test with $\alpha = 0.01$ to compare the population median amounts of protein in SBP and WTL International almonds. Do you believe the SBP claim? Why or why not? Find the p value for this test.

Challenge

14.39 Biology and Environmental Science The indoor air quality (the concentration of particles less than 2.5 microns in diameter) in 15 bars, restaurants, and other public venues in Louisville was measured just before new smoking regulations took effect. Two months after all public venues were required to be smoke-free, the air quality was measured again. The data (in $\mu g/m^3$) are given in the following table.

Venue	1	2	3	4	5	6	7	8
Before	353	386	104	198	597	62	412	273
After	56	35	28	21	83	10	27	34

Venue	9	10	11	12	13	14	15
Before	38	156	35	87	105	101	324
After	27	31	13	26	26	18	25

(*Source*: *Morbidity and Mortality Weekly Report*, Vol. 53, 2004, pp. 1038–1041.)

Suppose that the underlying before and after distributions of air quality are continuous.

a. Conduct a sign test with $\alpha = 0.05$ to determine whether there is any evidence that the median amount of particulates decreased after the smoking regulations went into effect.

b. Conduct a Wilcoxon signed-rank test with $\alpha = 0.05$ to determine whether there is any evidence that the median amount of particulates was greater before the smoking regulations went into effect than after.

c. Use the Wilcoxon rank-sum test with $\alpha = 0.05$ to determine whether there is any evidence that the median amount of particulates decreased after the smoking regulations went into effect.

d. Compare the results of these three statistical tests. Which test(s) is(are) appropriate and why?

14.4 THE KRUSKAL–WALLIS TEST

The Kruskal–Wallis test is a nonparametric analysis of variance of ranks. This procedure is an alternative to the analysis of variance F test that does not require the assumptions concerning normality or equal population variances. Although there is a table of critical values associated with this procedure, even for small sample sizes the test statistic has approximately a chi-square distribution.

Suppose that $k > 2$ independent random samples are obtained from continuous distributions. Assume that the first sample has size n_1, the second sample has size n_2, and so on, such that the kth sample has size n_k.

Combine all of the data for a total of $n = n_1 + n_2 + \cdots + n_k$ observations. Place these combined data in increasing order and assign a rank to each from smallest (rank 1) to largest (rank n). Equal values are assigned the mean rank for their positions in the ordered list (just as in the signed-rank test). Let r_i be the sum of the ranks associated with observations from sample i.

If all of the populations have identical distributions, then the sum of the ranks associated with each sample should be approximately equal. We expect the sum of the ranks associated with a sample from a different population to be distinct and separate from the rest. The test statistic in the Kruskal–Wallis test assesses the differences in the sums of the ranks.

THE KRUSKAL–WALLIS TEST

Suppose $k > 2$ independent random samples of sizes $n_1, n_2, \ldots, n_k$ are obtained from continuous distributions. A hypothesis test concerning the general populations with significance level α has the following form:

H_0: The k samples are from identical populations

H_a: At least two of the populations are different

Combine all observations, and rank these values from smallest (1) to largest (n). Equal values are assigned the mean rank for their positions. Let R_i be the sum of the ranks associated with the ith sample.

TS: $H = \left[\dfrac{12}{n(n+1)} \sum \dfrac{R_i^2}{n_i} \right] - 3(n+1)$

Critical values for the Kruskal–Wallis test statistic are available. However, if H_0 is true and either

1. $k = 3$, $n_i \geq 6$, $(i = 1, 2, 3)$ or
2. $k > 3$, $n_i > 5$, $(i = 1, 2, \ldots, k)$

then H has an approximate chi-square distribution with $k - 1$ degrees of freedom.

RR: $H \geq \chi_\alpha^2$ df $= k - 1$

The chi-square distribution was defined in Section 8.5.

If the sample sizes are small, then the distribution of H may not be close enough to chi-square to draw a reliable conclusion.

💡 **ILLUMINATING THE CONCEPTS**

1. We reject the null hypothesis only for *large* values of the test statistic H. If the sample sizes are large enough, the rejection region is always in the right tail of the appropriate chi-square distribution.

2. If we reject the null hypothesis, there is evidence to suggest that at least two populations are different. Further analysis is needed to determine which pairs of populations are different and how they differ; the means, medians, variances, shapes of the distributions, or other characteristics could be dissimilar. ◼

(Courtesy UPS)

Example 14.7 Shipping Distances A study was conducted to compare the populations of package shipping distance within the United States for three major companies. A sample of packages from each company was randomly selected, and the shipping distance (in miles) from the point of origin was recorded for each. The data are given in the following table; the sample numbers are in parentheses.

Solution Trail

KEYWORDS

- Kruskal–Wallis test.
- Is there any evidence?
- Populations are different.

↓

TRANSLATION

Hypothesis test concerning the general populations.

↓

CONCEPTS

Kruskal–Wallis test.

↓

VISION

The underlying distributions are assumed to be continuous. Use the Kruskal–Wallis test procedure: Combine and rank all of the observations. Find the value of the test statistic, and draw the appropriate conclusion.

Airborne Express (1)	UPS (2)	Federal Express (3)
834	245	1617
2954	600	1538
2845	915	1298
1889	998	1580
2006	284	1968
1318	325	1526
1675	558	1002
1959	493	

The underlying populations are assumed to be continuous. Use the Kruskal–Wallis test to determine whether there is any evidence that the package shipping distance populations are different. Use a significance level of $\alpha = 0.05$.

SOLUTION

STEP 1 Since there are $k = 3$ independent random samples of sizes $n_1 = 8 \geq 6$, $n_2 = 8 \geq 6$, and $n_3 = 7 \geq 6$, the Kruskal–Wallis test statistic has an approximate chi-square distribution with $k - 1 = 3 - 1 = 2$ degrees of freedom.

STEP 2 The four parts of the hypothesis test are

H_0: The three samples are from identical populations
H_a: At least two of the populations are different

TS: $H = \left[\dfrac{12}{n(n+1)} \sum \dfrac{R_i^2}{n_i} \right] - 3(n+1)$

RR: $H \geq \chi_\alpha^2 = \chi_{0.05}^2 = 5.9915$ df = 2

STEP 3 There are $n = n_1 + n_2 + n_3 = 8 + 8 + 7 = 23$ total observations. The following table shows all 23 observations, sorted in ascending order, with the associated rank. Each sample is color-coded to provide a visual comparison of the ranks.

Observation	245	284	325	493	558	600	834	915
Rank	1	2	3	4	5	6	7	8

Observation	998	1002	1298	1318	1526	1538	1580	1617
Rank	9	10	11	12	13	14	15	16

Observation	1675	1889	1959	1968	2006	2845	2954
Rank	17	18	19	20	21	22	23

STEP 4 If the populations are identical, we expect the ranks to be evenly distributed among the three samples. The shaded cells suggest that there are more low ranks associated with sample 2, middle ranks associated with sample 3, and high ranks associated with sample 1; this implies that the populations are different.

All the observations, the associated ranks, and the rank sums are given in the following table.

Airborne Express (1)	Rank	UPS (2)	Rank	Federal Express (3)	Rank
834	7	245	1	1617	16
2954	23	600	6	1538	14
2845	22	915	8	1298	11
1889	18	998	9	1580	15
2006	21	284	2	1968	20
1318	12	325	3	1526	13
1675	17	558	5	1002	10
1959	19	493	4		
Rank sum	**139**		**38**		**99**

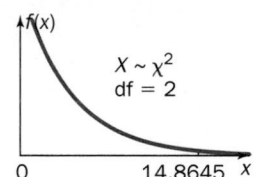

p value illustration:

$p = P(X \geq 14.8645)$
$= \boxed{0.0006} \leq 0.05 = \alpha$

$X \sim \chi^2$
df = 2

STEP 5 The value of the test statistic is

$$h = \left[\frac{12}{n(n+1)} \sum \frac{R_i^2}{n_i} \right] - 3(n+1)$$
$$= \left[\frac{12}{(23)(24)} \left(\frac{139^2}{8} + \frac{38^2}{8} + \frac{99^2}{7} \right) \right] - 3(24) = \boxed{14.8645} \ (\geq 5.9915)$$

The value of the test statistic ($h = 14.8645$) lies in the rejection region. At the $\alpha = 0.05$ significance level, there is evidence to suggest that the package shipping distance populations are different.

Figure 14.12 shows a technology solution.

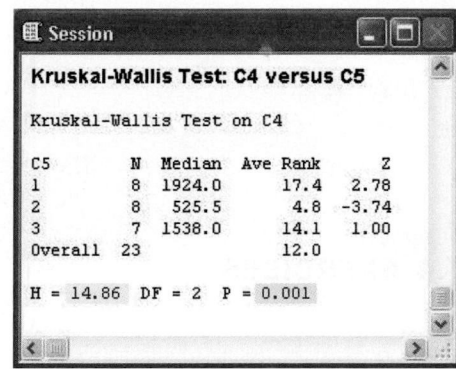

Figure 14.12 Minitab Kruskal–Wallis test results.

TECHNOLOGY CORNER

Procedure: Conduct a Kruskal–Wallis test.
Reconsider: Example 14.7, page 687, solution, and interpretations.

Minitab

The Kruskal–Wallis test is accessible in the <u>S</u>tat; <u>N</u>onparametrics menu and may also be executed using commands in a session window. The data from all samples must be in a single column and the corresponding group numbers must be in a second column.

1. Enter the data from each sample into a separate column (C1, C2, and C3).
2. Stack the three samples into column C4 and store the subscripts (group numbers) in column C5.

3. Select <u>S</u>tat; <u>N</u>onparametric; <u>K</u>ruskal-Wallis. Enter the Response column, C4, and the Factor column, C5.
4. Click OK. The test results are displayed in a session window. See Figure 14.12, page 689.

Excel

There is no built-in function to conduct a Kruskal–Wallis test. Use other arithmetic functions to compute the ranks, the value of the test statistic h, and the p value.

1. Enter all three samples into column A and the corresponding sample numbers into column B.
2. Compute the ranks and store them in column C.
3. Compute the value of the test statistic h and the p value. See Figure 14.13.

	A	B	C	D
1	834	1	7	
2	2954	1	23	
3	2845	1	22	
4	1889	1	18	
5	2006	1	21	
24				
25	Sample	1	2	3
26	R_i	139	38	99
27	R_i^2	19321	1444	9801
28	R_i^2 / n_i	2415.125	180.5	1400.143
29	h	14.8645		
30	p value	0.0006		

Figure 14.13 Excel calculations for a Kruskal–Wallis test.

SECTION 14.4 EXERCISES

Practice

14.40 Three independent random samples are given in the following table. Assume that a Kruskal–Wallis test will be used to compare populations. Find the rank sum associated with each sample.

Sample 1

88	94	94	79	91	76	77	79

Sample 2

19	72	77	79	89	72	87	74	82	90

Sample 3

85	96	95	96	93	85	95	93	87	95	93	93

14.41 Five independent random samples were obtained; assume that a Kruskal–Wallis test will be used to compare populations. Use the sample sizes and the rank sums in the following table to conduct this test at the $\alpha = 0.01$ level of significance.

Sample	Size	Rank sum
1	$n_1 = 12$	$r_1 = 395.5$
2	$n_2 = 14$	$r_2 = 428.5$
3	$n_3 = 12$	$r_3 = 287.0$
4	$n_4 = 16$	$r_4 = 620.0$
5	$n_5 = 20$	$r_5 = 1044.0$

14.42 Four independent random samples are given on the data CD and book's web site. Conduct a Kruskal–Wallis test to compare populations. Use $\alpha = 0.05$.

14.43 Three independent random samples were obtained, and the data are given on the data CD and book's web site. Use the Kruskal–Wallis test with $\alpha = 0.05$ to test the hypothesis that all three populations are identical.

Applications

14.44 Travel and Transportation A marketing manager at a rental-car agency believes that the number of miles a customer drives per week is related to the type of car rented. Independent random samples of week-long rental reservations were obtained, and the number of miles driven was recorded for each car. The summary statistics are given in the following table.

Car classification	Sample size	Rank sum
Compact car	$n_1 = 10$	$r_1 = 201.0$
Standard car	$n_2 = 10$	$r_2 = 227.0$
Luxury car	$n_3 = 12$	$r_3 = 100.0$

Use a Kruskal–Wallis test to compare these three populations. Is there any evidence to suggest that the populations are different? Use $\alpha = 0.005$, and assume that the underlying populations are continuous.

14.45 Physical Sciences Raytheon is testing different propellants for the Tomahawk missile used by the U.S. Navy. Independent random samples of missiles were obtained, and each was totally fueled with one of four types of solid propellant. Each missile was fired at the Naval test range near Point Mugu, California, and the total distance traveled was recorded (in km). The summary statistics are given in the following table.

Solid propellant	Sample size	Rank sum
1	$n_1 = 14$	$r_1 = 525.0$
2	$n_2 = 15$	$r_2 = 355.5$
3	$n_3 = 12$	$r_3 = 373.0$
4	$n_4 = 16$	$r_4 = 399.5$

Use the Kruskal–Wallis test to determine whether there is any evidence to suggest that at least two populations of distance traveled are different. Use $\alpha = 0.05$, and assume that the underlying populations are continuous.

14.46 Psychology and Human Behavior Many prisons have become overcrowded and officials have reduced the time served at correctional facilities for some offenders in order to ease the space problem. A random sample of first-time criminals was obtained. Each prisoner was classified by offense, and the length of time (in months) until his or her release from jail was recorded. The data are given on the data CD and book's web site.[15]
a. Use the Kruskal–Wallis test with $\alpha = 0.05$ to test the hypothesis that all three jail-time populations are identical. Assume that the underlying populations are continuous.
b. Find bounds on the p value associated with this test.

14.47 Psychology and Human Behavior Our emotional reaction to different colors affects advertising, product design, and even architecture. Three subway tunnel walkways of identical length were painted different colors. A random sample of adults was selected in each tunnel and secretly timed (in seconds) as they walked through the tunnel. The data are given in the following table.

Red		Orange		Yellow		Black	
12	19	21	14	29	23	22	23
16	23	27	12	18	24	11	27
15	22	26	24	25	30	10	24
18	25	22	30	25	12	24	14

Use the Kruskal–Wallis test with $\alpha = 0.01$ to determine whether there is any evidence that the tunnel-walking-time populations are different. Assume that the underlying populations are continuous.

14.48 Public Health and Nutrition Independent random samples of four types of 16-ounce steaks were obtained, and the amount of fat (in grams) in each was measured. The data are given on the data CD and book's web site. Assume that the underlying populations are continuous. Is there any evidence to suggest that the fat populations are different? Use $\alpha = 0.025$.

14.49 Public Policy and Political Science The U.S. Product Safety Commission issues playground specifications and guidelines to help communities build safe playgrounds. One concern is the uncompressed depth of wood chips and other loose-fill material (used as shock absorbers). Independent random samples of playgrounds in various cities were obtained, and the uncompressed depth of wood chips (in inches) was measured for each playground. The data are given in the following table.

Atlanta			Dallas			Denver		
10.5	12.8	12.9	11.8	11.6	11.3	10.9	11.8	11.2
10.2	11.5	10.5	11.6	10.0	10.2	12.4	12.0	11.6

Assume that the underlying distributions of uncompressed depths are continuous. Is there any evidence to suggest that the populations are different? Use $\alpha = 0.05$, and find bounds on the p value associated with this test.

Extended Applications

14.50 Business and Management One indication of morale in county government positions is the length of service in the current position. Independent random samples of county employees were obtained, and the length of service (in years) was recorded for each person. The data, by job classification, are given on the data CD and book's web site.
a. Assume that the underlying distributions are continuous. Use the Kruskal–Wallis test with $\alpha = 0.05$ to show that there is evidence to suggest that at least two populations are different.
b. Which pair(s) of populations do you think is(are) different? Why?

14.5   THE RUNS TEST

Some practical applications of the runs test involve the sequence of positive or negative gains of a certain stock, the strength of signals from an object in space, and projects that monitor stream pollution.

In all of the inference procedures presented in this text, it is very important for the sample, or samples, to be selected *randomly* from the underlying population(s). Otherwise, the results are not valid. Although it is not a test for verifying whether a sample has been randomly selected, the procedure described in this section can be used to examine the *order* in which observations were drawn from a population. The **runs test** is used to assess only whether there is evidence that the sequence of observations is not random.

Suppose an usher at a Broadway play records the seating section for the next 15 patrons, O for the orchestra section and B for the balcony. If all 15 people sat in the balcony or all sat in the orchestra section, we would certainly conclude that the order of patrons entering the theater was not random. Similarly, if the first 10 sat in the balcony, and the last 5 sat in the orchestra section, we would still question whether the order was random.

To determine whether the order of observations is random, we first separate the entire sequence into smaller subsequences in which the observations are the same. Consider the sequence of theater patrons in the following table, grouped by seating section.

O O	B B B B	O	B B	O O O O	B B

The grouped subsequences of similar observations, or symbols, are called **runs**. In the sequence of observations above, there are six runs. The test for randomness is based on the total number of runs.

DEFINITION

A **run** is a series, or subsequence, of one or more identical observations.

💡 **ILLUMINATING THE CONCEPTS**

1. The runs test is appropriate for testing whether a sequence of observations is not random. It can be used if the data can be divided into two mutually exclusive categories, for example, defective or satisfactory, citizen or foreigner, pass or fail. This test may also be used for quantitative data that can be classified into one of two categories, for example, above or below the median, or dangerous versus acceptable temperature.

2. The smallest possible number of runs in any sample is 1. This would occur if every observation in the sample fell into the same category or had the same attribute. The largest possible number of runs in a sample depends on the number of observations in each category.

   Note: A run can have length 1.

3. The statistical test is based on the total number of runs. It seems reasonable that if the order of observations is random, then the total number of runs should not be very large or very small. There is a table of critical values that uses the exact distribution for the number of runs. There is also a normal approximation that can be used if the number of observations in each category is large. ∎

The runs test is a nonparametric procedure because no assumptions are made about the underlying population.

THE RUNS TEST

Suppose a sample is obtained in which each observation is classified into one of two mutually exclusive categories. Assume that there are m observations in one category and n observations in the other.

H_0: The sequence of observations is random

H_a: The sequence of observations is not random

TS: V = the total number of runs

RR: $V \geq v_1$ or $V \leq v_2$

The critical values v_1 and v_2 are obtained from Table XI in the Appendix such that
$P(V \geq v_1) \approx \alpha/2$ and $P(V \leq v_2) \approx \alpha/2$.

The Normal Approximation: As m and n increase, the statistic V approaches a normal distribution with

$$\mu_V = \frac{2mn}{m+n} + 1 \quad \text{and} \quad \sigma_V^2 = \frac{2mn(2mn - m - n)}{(m+n)^2(m+n-1)}.$$

Therefore, the random variable $Z = \dfrac{V - \mu_V}{\sigma_V}$ has approximately a standard normal distribution. The normal approximation is good when both m and n are greater than 10. In this case, Z is the test statistic and the rejection region is

RR: $|Z| \geq z_{\alpha/2}$

Solution Trail

KEYWORDS

- Any evidence.
- Order in which the sample was selected was not random.

↓

TRANSLATION

Hypothesis test to determine whether there is evidence that the sequence of observations is not random.

↓

CONCEPTS

Runs test.

↓

VISION

Use the runs test procedure: compute the total number of runs, and draw the appropriate conclusion.

Example 14.8 Economic Confidence   A sample of delegates entering the 2004 Republican National Convention to hear President Bush's acceptance speech was obtained, and each was asked whether they had confidence in the economy. The sequence of responses (C for confidence and N for no confidence) is given in the following table.

C	C	N	N	C	N	C	C	N	N	N	N	N	C

Use the runs test with $\alpha = 0.05$ to determine whether there is any evidence that the order in which the sample was selected was not random.

SOLUTION

STEP 1   Each observation is classified into one of two mutually exclusive categories: confidence in the economy (C) or no confidence in the economy (N). We are looking for evidence to suggest that the sequence of observations was not random. The runs test is appropriate. There are $m = 6$ confidence observations and $n = 8$ no-confidence observations.

STEP 2   The four parts of the hypothesis test are

H_0: The sequence of observations is random

H_a: The sequence of observations is not random

For $v_1 = 11$, $P(V \geq 11) = 0.0629$, which leads to a much higher significance level.

TS: V = the total number of runs

RR: $V \geq v_1 = 12$ or $V \leq v_2 = 4$

Find the value v_1 such that $P(V \geq v_1) \approx 0.05/2 = 0.025$. Using Table XI in the Appendix, with $m = 6$ and $n = 8$, $P(V \geq 12) = 1 - P(V \leq 11) = 1 - 0.9837 = 0.0163 \approx 0.025$. Therefore, $v_1 = 12$.

STEP 3 The following table shows the original sequence of observations separated into runs.

C	C	N	N	C	N	C	C	N	N	N	N	N	C

There are seven runs. The value of the test statistic, $v = \boxed{7}$, does not lie in the rejection region. At the $\alpha = 0.05$ significance level, there is no evidence to suggest that the order of observations was not random.

Figure 14.14 shows a technology solution.

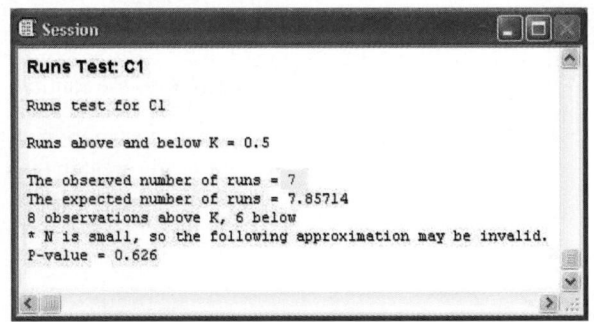

Figure 14.14 Minitab runs test.

The original data in the following example are quantitative but can be classified into two mutually exclusive categories. Since the number of observations in each category is large, the normal approximation will be used in the runs test.

Solution Trail

KEYWORDS

- Threshold value is 1.00 Sv.
- Is there any evidence?
- Order of observations is not random.
- $m = 13$, $n = 15$.

TRANSLATION

- Hypothesis test to determine whether there is evidence that the sequence of observations is not random with respect to the threshold value.
- m and n greater than 10.

CONCEPTS

- Runs test.
- Normal approximation.

VISION

Use the runs test procedure: Compute the total number of runs, find the value of the test statistic, and draw the appropriate conclusion.

Example 14.9 X-Ray Radiation The Panoramic Corporation is developing a new, powerful X-ray machine for use in hospitals and medical imaging centers. There is some concern that the machine emits an amount of radiation with each X-ray that could cause radiation sickness. According to the National Council on Radiation Protection and Measurements, acute radiation sickness follows exposure to a radiation dose generally greater than 1 Sv (sieverts) in a time generally less than one day. A sample of X-rays was taken using the new machine, and the amount of radiation emitted during each X-ray was measured (in Sv). The data are given in the following table, in order from left to right.

1.01	1.16	1.15	1.03	1.02	1.03	1.02	0.89	0.73	0.90
0.80	1.09	1.02	0.84	0.94	0.85	1.06	0.90	0.97	0.90
1.05	1.07	1.03	0.84	0.98	0.96	0.88	0.90		

Suppose that the threshold value for radiation sickness is 1.00 Sv. Is there any evidence to suggest that the order of the observations is not random with respect to this threshold value? Use $\alpha = 0.05$.

SOLUTION

STEP 1 Replace each observation above the threshold value with an A and each observation below the threshold value with a B. Any observation equal to 1.00 would be excluded from the analysis and the sample size would be reduced. Each observation now falls into one of two mutually exclusive categories: above or below the threshold.

STEP 2 The runs test will be used to determine whether there is evidence to suggest that the sequence of observations is not random. There are $m = 13$ observations above 1.00 and $n = 15$ observations below 1.00.

STEP 3 Since both m and n are greater than 10, the normal approximation will be used. The four parts of the hypothesis test are

H_0: The sequence of observations is random
H_a: The sequence of observations is not random

TS: $Z = \dfrac{V - \mu_V}{\sigma_V}$

RR: $|Z| \geq z_{\alpha/2} = z_{0.025} = 1.9600$

STEP 4 Compute the number of runs using the original sequence of observations.

1.01	1.16	1.15	1.03	1.02	1.03	1.02	0.89	0.73	0.90
A	A	A	A	A	A	A	B	B	B

0.80	1.09	1.02	0.84	0.94	0.85	1.06	0.90	0.97	0.90
B	A	A	B	B	B	A	B	B	B

1.05	1.07	1.03	0.84	0.98	0.96	0.88	0.90
A	A	A	B	B	B	B	B

There are $\nu = 8$ runs.

STEP 5 The mean and the variance of the random variable V are

$$\mu_V = \frac{2mn}{m + n} + 1 = \frac{2(13)(15)}{13 + 15} + 1 = 14.9286$$

$$\sigma_V^2 = \frac{2mn(2mn - m - n)}{(m + n)^2(m + n - 1)} = \frac{2(13)(15)[2(13)(15) - 13 - 15]}{(13 + 15)^2(13 + 15 - 1)}$$

$$= 6.6695$$

The value of the test statistic is

$$z = \frac{\nu - \mu_V}{\sigma_V} = \frac{8 - 14.9286}{\sqrt{6.6695}} = -2.6829$$

STEP 6 Since $|z| = |-2.6829| = 2.6829 \geq 1.9600$, the value of the test statistic lies in the rejection region. At the $\alpha = 0.05$ significance level, there is evidence to suggest that the order of the observations is not random.

To find the p value associated with this test, find the left-tail probability and multiply by 2.

$p/2 = P(Z \leq -2.6829)$ *Definition of p value.*

$= 0.0037$ *Cumulative probability; use Table III in the Appendix.*

$p = 2(0.0037) = \boxed{0.0074}$ *Solve for p.*

p value illustration:
$p = 2P(Z \leq -2.6829)$
$= 0.0074 \leq 0.05 = \alpha$

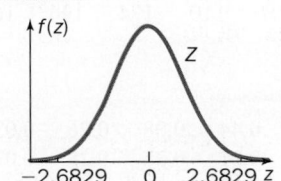

Figure 14.15 shows a technology solution.

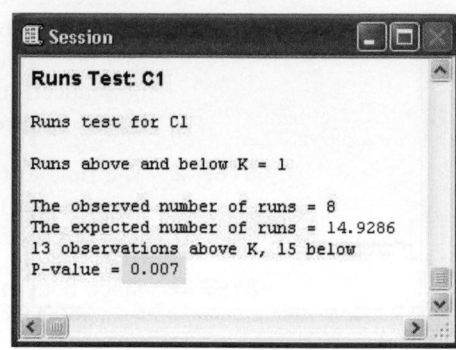

Figure 14.15 Minitab runs test.

TECHNOLOGY CORNER

Procedure: Conduct a runs test.
Reconsider: Example 14.9, page 694, solution, and interpretations.

Minitab

The runs test is accessible in the Stat; Nonparametrics menus and may also be executed using commands in a session window. Categorical data must be converted to numerical values.

1. Enter the data into column C1.
2. Select Stat; Nonparametrics; Runs Test. Enter the Variable, the column containing the data, C1.
3. Select Above and below, and enter 1.
4. Click OK. The results are displayed in a session window. See Figure 14.15.

SECTION 14.5 EXERCISES

Practice

14.51 Find the number of runs in each sequence of observations.
a. A A B B B A B A B B A B
b. G B B B B B G G B B G B G G B
c. F F S F F S F F S S F S F F S S S
   F F F
d. + − + − − + − − − + + − + −
   + + − − + − − +

14.52 In each of the following problems, use the values for m and n to find the critical values in a runs test with approximate significance level α. Find the exact significance level for your choice of critical values.
a. $m = 4$, $n = 7$, $\alpha = 0.05$.
b. $m = 5$, $n = 8$, $\alpha = 0.01$.
c. $m = 6$, $n = 6$, $\alpha = 0.05$.
d. $m = 8$, $n = 9$, $\alpha = 0.01$.

14.53 In each of the following problems, use the population median to classify each observation in the ordered sample as either above or below the median. Find the number of runs in each sequence of observations. *Note*: The order of observations is left to right, then down.

a. $\tilde{\mu} = 20.0$

22.4	15.1	22.0	25.6	19.0	25.1	18.7	21.2
11.5	19.6						

b. $\tilde{\mu} = 4.00$

4.14	3.26	3.23	3.12	4.80	5.52	5.81	5.71
3.77	3.68	3.79	3.03	4.36	5.29		

c. $\tilde{\mu} = 150$

115	169	139	101	102	138	123	195	107
175	178	151	181	107	110	174	196	167

d. $\tilde{\mu} = 0.44$

0.07	0.19	0.10	0.12	0.44	0.08	0.26	0.03
0.10	0.14	0.41	0.20	0.01	0.38	0.28	0.03
0.15	0.03	0.00	0.04	0.19	0.04		

14.54 In each of the following problems, use the values for m and n to find the expected number of runs, μ_V, and the variance of the number of runs, σ_V^2.

a. $m = 10$, $n = 15$.   **b.** $m = 15$, $n = 21$.
c. $m = 2$, $n = 23$.   **d.** $m = 26$, $n = 26$.

14.55 A sample was obtained, and each observation was classified into one of two mutually exclusive categories. The sequence of observations is given in the following table.

B	A	B	A	A	B	B	A	B	A
B	B	B	B	A	A	B	A	A	B

Conduct a runs test with $\alpha = 0.05$. Is there any evidence to suggest that the order of observations is not random?

14.56 A sample was obtained from a population with hypothesized median $\tilde{\mu} = 10.0$. The sequence of observations is given on the data CD and book's web site.

a. Conduct a runs test using the normal approximation with $\alpha = 0.01$. Is there any evidence to suggest that the order of observations is not random?
b. Find the p value associated with this test.

Applications

14.57 Public Policy and Political Science During freshman move-in day at the University of Michigan, the College Republicans staffed an information table in order to solicit new members. A sample of consecutive new members was obtained, and each was classified by gender (M or F). The ordered observations are given in the following table.

M	M	M	M	M	M	M	M
F	F	M	F	M	M		

Use the runs test with $\alpha \approx 0.05$ to determine whether there is any evidence to suggest that the order of observations is not random with respect to gender.

14.58 Public Health and Nutrition Health care costs continue to rise and national surveys suggest that one reason many people are uninsured is because insurance is simply too expensive. The National Coalition on Health Care reports that approximately half of all Americans are worried about the rising cost of health care.[16] Suppose a sample of Americans was obtained, and each was classified as to whether he or she was worried about rising health care costs (W) or not worried (N). The ordered observations are given in the following table.

W	W	W	N	W	N	N	W	N
W	W	W	N	N	N	N	W	W

Is there any evidence to suggest that the order of observations is not random? Use the runs test with $\alpha \approx 0.05$.

14.59 Marketing and Consumer Behavior A one-hour photo shop prints pictures in either a glossy (G) or a matte (M) finish. A sample of customers was obtained, and each job was classified

according to finish. The ordered observations are given in the following table.

M	M	G	M	G	M	G	M	G
M	G	M	M	G	M	M	G	M
G	G	G	M	M	M	G	M	M
G	G	G	M	M	M	G	G	

Is there any evidence to suggest that the order of observations is not random? Use the runs test with the normal approximation and a 0.02 level of significance.

14.60 Business and Management Many national surveys suggest that more and more employee emails are not work related. In fact, slightly more than half of all corporate email is either spam or personal.[17] A sample of employee email at Liberty Mutual was obtained, and each was classified as work related (W) or personal (P). The sequence of emails is given in the following table.

W	P	W	W	P	W	W	W	P	W	W	W	P

a. Use the runs test with $\alpha \approx 0.10$ to determine whether there is any evidence that the order of observations is not random.
b. Find the p value associated with this test.

14.61 Manufacturing and Product Development Composite decks are made of recycled wood and plastic, are low maintenance and sturdy, and are designed to last much longer than traditional wood decks. A manufacturer of composite decks routinely checks the surface wear as part of quality control. A sample of planks was obtained, and each was measured for surface wear using a Taber tester. The ordered observations (in Taber units) are given in the following table.

204	211	188	208	190	203	203	193	204
214	176	209						

The planks are manufactured to have a median wear index of 200 units. Use the runs test with $\alpha \approx 0.02$ to determine whether there is any evidence that the order of observations is not random with respect to the median.

14.62 Business and Management The lengths of MTV videos vary considerably, but the station tries not to show too many long or short videos in a row. A sample of MTV videos aired during prime time was obtained, and the length (in seconds) of each was recorded. The ordered observations are given on the data CD and book's web site. Suppose a video under three minutes is considered short, and over three minutes, long. Use the runs test with the normal approximation and $\alpha = 0.05$ to determine whether there is any evidence that the order of MTV videos is not random with respect to length.

Challenge

14.63 Random Number Generators The purpose of this exercise is to determine whether a random number generator really produces a random sequence of observations. Using your graphing calculator or favorite statistical software,

a. Generate 100 observations from a standard normal distribution.

b. Classify each observation as either above or below the mean (and median) 0.

c. Conduct a runs test using the normal approximation to determine whether there is any evidence to suggest that the sequence

of observations is not random with respect to the mean. Use $\alpha = 0.05$.

Do this 100 times. How many times did you reject the null hypothesis H_0: the sequence is random? Draw a conclusion about the sequence of observations produced by your random number generator.

14.6 SPEARMAN'S RANK CORRELATION COEFFICIENT

The sample correlation coefficient, r, was introduced in Chapter 12 as a measure of the strength of the linear relationship between two continuous variables. **Spearman's rank correlation coefficient** is a nonparametric alternative and is computed using ranks. Without any assumptions being made about the underlying populations, each observation is converted to a rank, and the sample correlation coefficient is computed using the ranks in place of the actual observations.

SPEARMAN'S RANK CORRELATION COEFFICIENT

Suppose that there are n pairs of observations (x_1, y_1), (x_2, y_2), . . . , (x_n, y_n). Rank the observations on each variable separately, from smallest to largest. Let u_i be the rank of the ith observation on the first variable and let v_i be the rank of the ith observation on the second variable. **Spearman's rank correlation cofficient**, r_S, is the sample correlation coefficient between the ranks and is computed using the equation

$$r_S = 1 - \frac{6\sum d_i^2}{n(n^2 - 1)} \tag{14.1}$$

where $d_i = u_i - v_i$.

 ILLUMINATING THE CONCEPTS

1. As usual, equal values within each variable are assigned the mean rank of their positions in the ordered list. Equation 14.1 is not exact when there are tied observations within either variable. In this case, one should compute r_S by finding the sample correlation coefficient between the ranks.

The sample correlation coefficient was defined in Section 12.2.

2. Since r_S is really a sample correlation coefficient, the value is always between -1 and $+1$. Values near -1 indicate a strong negative linear relationship, and values near $+1$ suggest a strong positive linear relationship between the ranks.

3. Remember, correlation does not imply causation; r_S is a measure of the linear association between the ranks. And a strong linear relationship between the ranks does not imply that the relationship between the original variables is also linear. ■

Example 14.10 Vitamin B12 and Memory Recall A 5-year study suggested that higher levels of vitamin B12 may actually prevent brain shrinkage and thus memory loss.[18] Suppose another study was conducted to determine whether levels of B12 are associated with performance on memory tests. A sample of healthy adults over 75 years old was randomly selected, and the B12 level (in pmol/L) was measured for each. Twenty-four words were read to each adult, and after a period of two minutes the subject was asked to recall as many words as possible within two minutes. The data are given in the following table.

Adult	1	2	3	4	5	6	7	8
B12 level	317	263	135	111	235	350	144	165
Word recall	23	22	6	11	20	21	9	13

Compute Spearman's rank correlation coefficient and interpret this value.

SOLUTION

STEP 1 Let x represent the B12 level and y represent the word-recall count. For each variable, order the observations from smallest to largest and assign a rank to each value. Compute the difference between each pair of ranks.

The following table shows each observation, its associated rank, and each difference.

Adult i	B12 level x_i	Rank u_i	Word recall y_i	Rank v_i	Difference d_i
1	317	7	23	8	−1
2	263	6	22	7	−1
3	135	2	6	1	1
4	111	1	11	3	−2
5	235	5	20	5	0
6	350	8	21	6	2
7	144	3	9	2	1
8	165	4	13	4	0

STEP 2 There are $n = 8$ pairs and no ties within either variable. Use Equation 14.1 to compute Spearman's rank correlation coefficient.

$$r_S = 1 - \frac{6\sum d_i^2}{n(n^2 - 1)}$$
Use Equation 14.1.

$$= 1 - \frac{6[(-1)^2 + (-1)^2 + (1)^2 + (-2)^2 + (0)^2 + (2)^2 + (1)^2 + (0)^2]}{8(8^2 - 1)}$$
Use the d_i's.

$$= 1 - \frac{72}{504} = \boxed{0.8571}$$

STEP 3 Since $r_S = 0.8571 > 0.8$, there is a strong positive linear relationship between the B12 ranks and the word-recall ranks. This suggests that there is a strong positive correlation between B12 levels and word recall; high levels of B12 are associated with better memory.

Figure 14.16 shows a technology solution.

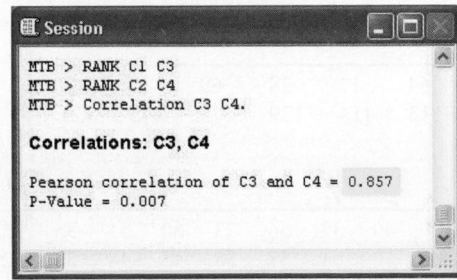

Figure 14.16 Minitab session window commands to compute Spearman's rank correlation coefficient.

TECHNOLOGY CORNER

Procedure: Compute Spearman's rank correlation cofficient.
Reconsider: Example 14.10, page 698, solution, and interpretations.

Minitab

There is no built-in function to compute r_S. Rank each sample separately, and find the sample correlation cofficient between the ranks.

1. Enter the B12 data into column C1 and the word recall data into column C2.
2. Rank the observations in C1 and store the results in column C3. Rank the observations in C2 and store the results in column C4.
3. Find the sample correlation coefficient between the ranks, columns C3 and C4. See Figure 14.16, page 699.

Excel

There is no built-in function to compute r_S. Rank each sample separately, and find the sample correlation cofficient between the ranks.

1. Enter the B12 data into column A and the word recall data into column B.
2. Rank the observations in A and store the results in column B. Rank the observations in B and store the results in column D.
3. Use the function CORREL to find the sample correlation coefficient between the ranks, columns C and D. See Figure 14.17.

	A	B	C	D
1	B12	Word	B12 ranks	Word ranks
2	317	23	7	8
3	263	22	6	7
4	135	6	2	1
5	111	11	1	3
6	235	20	5	5
7	350	21	8	6
8	144	9	3	2
9	165	13	4	4
10				
11	0.8571	= CORREL(C2:C9,D2:D9)		

Figure 14.17 Excel, Spearman's rank correlation coefficient.

SECTION 14.6 EXERCISES

Practice

14.64 Random samples of pairs of observations are given in each of the following problems. Rank the observations in each sample separately, from smallest to largest, and find each difference between the ranks, d_i.

a.

Observation	1	2	3	4	5	6
Sample 1	54	17	28	69	13	49
Sample 2	113	114	139	173	145	121

b.

Observation	1	2	3	4	5	6	7	8	9
Sample 1	57	40	32	56	33	60	51	35	53
Sample 2	35	50	51	57	38	45	44	52	32

c.

Observation	1	2	3	4	5	6
Sample 1	22.5	27.0	22.8	26.5	29.9	20.8
Sample 2	27.0	30.5	21.6	27.9	33.0	22.9

Observation	7	8	9	10	11	12
Sample 1	19.3	28.3	19.5	20.3	28.2	27.9
Sample 2	24.7	22.2	24.3	26.2	25.1	29.4

Observation	13	14	15
Sample 1	27.8	31.6	25.3
Sample 2	22.6	24.8	20.3

d.

Observation	1	2	3	4	5	6
Sample 1	49.0	51.2	46.7	54.9	53.6	48.9
Sample 2	78.1	60.8	70.4	72.2	64.9	70.8

Observation	7	8	9	10	11	12
Sample 1	46.8	46.2	55.8	40.8	46.8	55.7
Sample 2	74.3	68.4	66.5	75.9	65.9	70.3

Observation	13	14	15	16	17	18
Sample 1	54.9	45.0	55.1	43.2	43.8	53.9
Sample 2	71.4	75.6	72.0	69.4	71.7	78.3

14.65 In each of the following problems, consider the data obtained in a study of the relationship between two variables, find Spearman's rank correlation coefficient, and use it to describe the relationship between the two variables.

a.

x	156	268	262	206	162	166	148
y	131	258	235	189	180	262	207

b.

x	5.39	5.60	5.12	4.53	6.01	5.00
y	-15.3	-15.4	-16.5	-16.3	-14.2	-13.6

x	5.62	5.04	4.65	6.19
y	-14.9	-14.4	-15.6	-15.2

c.

x	25.51	25.80	24.10	24.19	25.52	26.16	24.79
y	24.45	26.98	25.72	27.20	28.56	25.69	27.38

x	23.95	25.14	25.09	25.35	26.64	25.63	23.84
y	25.84	24.04	25.87	25.51	26.94	26.64	26.53

d.

x	y	x	y	x	y
418	754	411	755	414	776
484	725	378	759	453	781
448	712	425	749	478	745
458	718	436	733	465	785
476	756	443	727	438	775
421	772	442	716	459	735
480	728	435	763	490	751
461	753	440	752	407	744

14.66 Consider the following data obtained in a study of the relationship between two variables.

x	y	x	y
1.5	7.3	1.4	7.4
1.8	6.8	1.3	7.2
1.5	7.4	1.2	7.2
1.6	7.5	1.4	7.3
1.7	6.7	1.2	6.6

a. Rank the observations in each sample separately, and find each difference between the ranks, d_i.

b. Find the sample correlation coefficient between the ranks.

c. Find Spearman's rank correlation coefficient using the differences, d_i.

d. Explain why the values in parts (b) and (c) are different.

Applications

14.67 Travel and Transportation Dynasty Travel offers cruises to the Caribbean and often discounts unsold tickets as the sailing date approaches. These last-minute deals were studied by comparing the price of an inside stateroom (x, in dollars) with the number of days before sailing (y). A random sample of last-minute cruise deals was obtained, and the data are given in the following table.

x	998	956	763	1313	1071	1091
y	9	3	4	13	10	6

Find Spearman's rank correlation coefficient, and interpret this value. Do these data suggest that there is an association between the number of days before the cruise and cost?

14.68 Biology and Environmental Science Farmers consider heavy molasses to be a high-energy food for cattle. In a recent study, the percentage of digestible energy (x) was compared with the percentage of protein (y) in heavy molasses. Random samples of various brands of molasses were obtained. The percentage of digestible energy and of protein were carefully measured, and the data are given on the data CD and book's web site. Compute Spearman's rank correlation coefficient. Use this value to describe the relationship between the percentage of digestible energy and the percentage of protein in heavy molasses.

14.69 Biology and Environmental Science The bulk density of soil (x, in g/cm^3) affects plant growth, specifically, how easily roots can penetrate the soil. A group of carrot growers conducted a study to compare soil bulk density with soil texture (y), measured on a scale from 1 to 10, with 1 being very sandy soil and 10 corresponding to clay. A random sample of plots was obtained, and the data are given on the data CD and book's web site. Compute Spearman's rank correlation coefficient, and use this value to describe the relationship between these two variables.

14.70 Public Health and Nutrition There is some evidence to suggest that central body fat is related to overall lifestyle.[19] A random sample of adults over age 35 was obtained. The central body fat of each person (x) was computed using the waist/hip ratio (WHR). In addition, each person completed a questionnaire concerning nutrition, occupation, medical condition, and physical activity. A lifestyle score (y, between 0 and 100) was computed from the responses, with 100 representing a very healthy lifestyle. The data are given on the data CD and book's web site. Compute Spearman's rank correlation coefficient. What does this value suggest about the relationship between lifestyle and central body fat?

14.71 Marketing and Consumer Behavior The *Publishers Weekly* Bestseller List is produced using data from large-city bookstores, bookstore chains, and local bestseller lists. A sample of books from the January 12, 2009, list was obtained, and the price (x, in dollars) and the number of weeks on the bestseller list (y)

for each book was recorded. The data are given in the following table.

Book	Cost	Weeks on list
Black Ops	26.95	1
Scarpetta	27.95	4
Running Hot	25.95	1
The Host	25.99	32
Fire and Ice	26.00	1
The Story of Edgar Sawtelle	25.95	28
Cross Country	27.99	6
The Christmas Sweater	19.99	7
The Hour I First Believed	29.95	7
Arctic Drift	27.95	5
Just After Sunset: Stories	28.00	7
A Mercy	23.95	7
The Lucky One	24.99	13
Your Heart Belongs to Me	27.00	5
Charlemagne Pursuit	26.00	4

Compute Spearman's rank correlation coefficient. What does this value suggest about the relationship between the cost of a book and the number of weeks it spends on the bestseller list?

Extended Applications

14.72 Sports and Leisure The owner of a snow tubing resort in the Pocono Mountains believes that the total snowfall (x, in feet) during a winter is related to the number of customers (y, in thousands). A random sample of winters was obtained, and the data are given on the data CD and book's web site.
a. Construct a scatter plot for these data, and describe the relationship between the two variables.
b. Compute Spearman's rank correlation coefficient, and use this value to describe the relationship between the two variables.
c. Explain any differences in your answers to parts (a) and (b).

14.73 Physical Sciences A certain solder joint on the undercarriage of a city bus is being tested for shear strength. A sample of solder joints was randomly selected, and the shear strength for

each was measured (in N) before (x) and after (y) temperature cycle stress. The data are given in the following table.

x	71	69	66	69	62	67	69	64	73	69
y	64	56	65	62	65	64	57	63	57	63

(*Source*: Micrel Semiconductor Lead-Free Solder Joint Reliability.)

a. Rank the observations in each sample separately, and find each difference between the ranks, d_i.
b. Find the sample correlation coefficient between the ranks.
c. Find Spearman's rank correlation coefficient using the d_i's.
d. Explain why the values in parts (b) and (c) are different, and use these results to explain the relationship between the two variables.

Challenge

14.74 Fuel Consumption and Cars Suppose there are n pairs of observations and ρ_S is the population correlation between ranks. The four parts of a hypothesis test concerning ρ_S based on a normal approximation are

H_0: $\rho_S = 0$ (no population correlation between ranks)

H_a: $\rho_S > 0$, $\quad \rho_S < 0$, $\quad$ or $\quad \rho_S \neq 0$

TS: $Z = R_S \sqrt{n-1}$

$\quad R_S$ is Spearman's rank correlation coefficient.

RR: $Z \geq z_\alpha$, $\quad Z \leq -z_\alpha$, $\quad$ or $\quad |Z| \geq z_{\alpha/2}$

A turbocharger is a compact way to add more power to an automobile by increasing the amount of air going into the cylinders. A study was conducted to examine the relationship between the added pressure (x, in psi) and power (y, percentage change in horsepower) provided by a certain turbocharger. A variety of automobiles was selected, and a randomly selected turbocharger was installed in each. The additional air pressure and power measurements are given on the data CD and book's web site.
a. Compute Spearman's rank correlation coefficient, and use this value to describe the relationship between added air pressure and change in engine power.
b. Conduct a hypothesis test to determine whether there is any evidence that the true population correlation between ranks is greater than 0. Use $\alpha = 0.01$.

Chapter 14 Challenge Wrap-Up

(Kgiszewski/Dreamstime.com)

The underlying silo temperature distributions are not assumed to be normal, but the differences are assumed to be continuous and symmetric. Use the Wilcoxon signed-rank test to determine whether there is any evidence that the galvanized-silo median temperature is greater than the white-silo median temperature. The hypothesis test is one-sided, with $\Delta_0 = 0$.

Let population 1 be the galvanized-silo temperatures and population 2 be the white-silo temperatures. Since $\Delta_0 = 0$, we do not need to modify the pairwise differences. Each pairwise difference, absolute difference, absolute difference rank, and signed rank is given in the following table.

Galvanized silo	White silo	Pairwise difference	Absolute difference	Rank	Signed rank
39	40	−1	1	1.5	−1.5
58	48	10	10	8.5	8.5
53	55	−2	2	3.5	−3.5
50	42	8	8	7.0	7.0
38	39	−1	1	1.5	−1.5
49	37	12	12	10.0	10.0
53	43	10	10	8.5	8.5
48	46	2	2	3.5	3.5
46	41	5	5	5.5	5.5
41	46	−5	5	5.5	−5.5

The number of observations is $n = 10$ and the test statistic is T_+. The four parts of the hypothesis test are

H_0: $\tilde{\mu}_1 - \tilde{\mu}_2 = \Delta_0 = 0$
H_a: $\tilde{\mu}_1 - \tilde{\mu}_2 > \Delta_0 = 0$
TS: T_+ = the sum of the ranks corresponding to the positive differences.
RR: $T_+ \geq c_1 = 44$

Using Table IX in the Appendix, $P(T_+ \geq 44) = 0.053 \approx 0.05$. The value of the test statistic is

$$t_+ = 8.5 + 7.0 + 10.0 + 8.5 + 3.5 + 5.5 = \boxed{43}$$

The value of the test statistic does not lie in the rejection region. There is no evidence to suggest that the true median temperature of the galvanized silos is greater than the true median temperature of the white silos.

Figure 14.18 shows a technology solution.

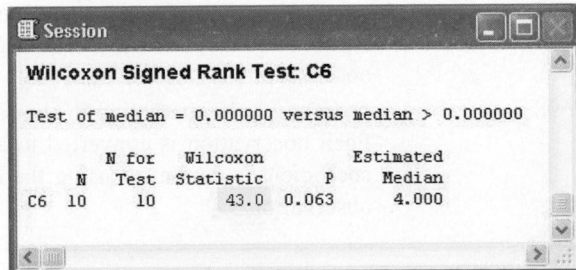

Figure 14.18 Minitab Wilcoxon rank sum test.

Note: Remember, if an inappropriate test is used, then we may reach the wrong conclusion. The technology solution in Figure 14.19 presents the results of the inappropriate paired *t* test. The *p* value in this case is 0.015. This test indicates that there is evidence to suggest that the true median temperature of the galvanized silos is greater than the true median temperature of the white silos, contrary to the results from the Wilcoxon signed-rank test.

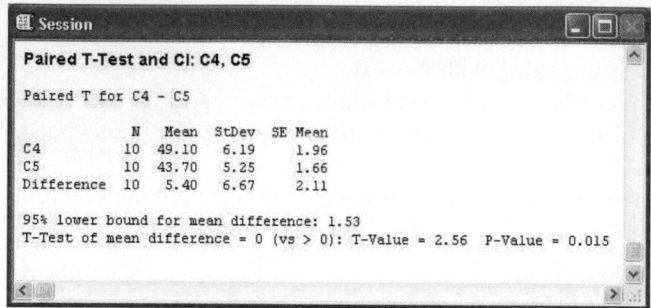

Figure 14.19 Inappropriate paired *t* test results.

CHAPTER 14 SUMMARY

Concept	Page	Notation / Formula / Description
Sign test	665	A nonparametric test concerning a population median and based on the number of observations greater than $\tilde{\mu}_0$. It can also be used to compare two population medians on the basis of the number of pairwise differences greater than Δ_0.
Wilcoxon signed-rank test	672	A nonparametric test concerning a population median and based on the sum of the ranks corresponding to the positive differences $x_i - \tilde{\mu}_0$. It can also be used to compare two population medians on the basis of the sum of the ranks corresponding to the positive differences $d_i - \Delta_0$.
Wilcoxon rank-sum test	680	A nonparametric test concerning the difference of two population medians and based on the sum of the ranks corresponding to the positive differences $x_i - \Delta_0$.
Kruskal–Wallis test	687	A nonparametric test to determine, on the basis of the ranks of all observations combined, whether at least two of k samples are from different populations.
Run	692	A series, or subsequence, or one or more identical observations.
Runs test	693	A nonparametric test to determine, on the basis of the total number of runs, whether there is evidence that a sequence of observations is not random.
Spearman's rank correlation coefficient	698	A nonparametric alternative to the sample correlation coefficient. Each observation is converted to a rank, and the correlation coefficient is computed using the ranks in place of the actual observations.

A comparison of parametric and nonparametric procedures:

Case	Nonparametric procedure		Parametric procedure	
	Null hypothesis	**Statistical test**	**Null hypothesis**	**Statistical test**
One sample	$\tilde{\mu} = \tilde{\mu}_0$	Sign test	$\mu = \mu_0$	One-sample t test
	$\tilde{\mu} = \tilde{\mu}_0$	Wilcoxon signed-rank test	$\mu = \mu_0$	One-sample t test
	Sequence of observations is random	Runs test	No comparable parametric test	
Two independent samples	$\tilde{\mu}_1 - \tilde{\mu}_2 = \Delta_0$	Wilcoxon rank-sum test	$\mu_1 - \mu_2 = \Delta_0$	Two-sample t test
Paired data	$\tilde{\mu}_D = \Delta_0$	Sign test	$\mu_D = \Delta_0$	Paired t test
	$\tilde{\mu}_D = \Delta_0$	Wilcoxon signed-rank test	$\mu_D = \Delta_0$	Paired t test
		Spearman's rank correlation coefficient		Sample correlation coefficient
k independent samples	k samples are from independent populations	Kruskal–Wallis test	$\mu_1 = \cdots = \mu_k$	One-way ANOVA

Note: Nonparametric procedures are applicable in a very broad range of situations. However, if the underlying population is normal, then the corresponding parametric procedure is more efficient.

CHAPTER 14 EXERCISES

Applications

14.75 Biology and Environmental Science Many heavy-duty trucks are equipped with large diesel engines. These engines tend to last longer than conventional gasoline engines but also emit higher quantities of pollutants. A random sample of 1994–1997 diesel engines was obtained, and the amount of nitrogen oxide emitted from each was recorded. The data (in grams/brake horsepower-hour; g/bhp-hr) are given in the following table.

5.68	5.79	3.41	4.59	5.81	5.77	4.67	3.93
5.10	5.12	6.61	5.39	6.13	6.64	4.09	

According to the Bureau of Transportation Statistics, federal exhaust emissions certification standards require the nitrogen oxide emissions to be at most 5 g/bhp-hr in 1994–1997 engines. Assume that the underlying distribution of nitrogen oxide emissions is continuous. Use the sign test with $\alpha = 0.05$ to determine whether there is any evidence that the median nitrogen oxide emissions amount is greater than 5 g/bhp-hr.

14.76 Biology and Environmental Science Farms across the country have old agricultural chemicals stored in deteriorating containment vessels that pose a threat to humans. The U.S. Agriculture Department has started a program to clean up and properly dispose of these unusable chemicals. A sample of farms was randomly selected, and the amount of DDT still stored on each farm was recorded. The data (in kg) are given in the following table.

3.5	4.3	7.0	3.0	6.0	2.1	2.5	4.9	5.9	2.4
3.5	8.6	0.8	1.3	2.9	2.3	2.1	1.0	5.5	1.7

Assume that the underlying distribution of stored DDT is continuous.
a. Use the sign test with $\alpha = 0.01$ to determine whether there is any evidence that the median amount of stored DDT on farms is greater than 2.0 kg.
b. Find the p value associated with this test.

14.77 Economics and Finance Insurance companies recommend a personal umbrella policy for most customers, even if they already have a home and an automobile policy. Basic liability coverage may not be adequate as lawsuits become more common and jury awards escalate. A random sample of adults with umbrella policies was obtained, and the amount of coverage was recorded for each (in millions of dollars). The data are given on the data CD and book's web site. Assume that the underlying distribution of umbrella coverage amounts is continuous. Use the sign test with $\alpha = 0.05$ to determine whether there is any evidence that the median coverage amount is different from 2 million dollars.

14.78 **Biology and Environmental Science** The national bird of New Zealand is the kiwi, a flightless, secretive bird that lives mostly in native forests. There is some concern that these birds are not growing normally because of environmental hazards. A random sample of male great spotted kiwis was obtained, and each was carefully weighed (in grams). The data are given in the following table.

1988	1759	1838	2500	2545	2938	1919	3000
2456	2418	2288	2315	2399	2316	2633	2480

(*Source*: New Zealand Department of Conservation.)

The underlying distribution of kiwi weights is assumed to be continuous and symmetric. Use the Wilcoxon signed-rank test with $\alpha = 0.05$ to determine whether there is any evidence that the mean kiwi weight is less than 2370 grams. Find the p value for this test.

14.79 **Physical Sciences** The fountain beneath the Gateway Arch in St. Louis is designed to spray water 630 feet straight up, the same height as the Arch. A random sample of days was selected, and on each day the height of the spray was measured (in feet) during the fountain show. The data are given in the following table.

603	633	612	619	624	627	622	619	630	609
630	626	630	615	630	643	639	620	606	

The underlying distribution is assumed to be continuous.
a. Use the Wilcoxon signed-rank test with $\alpha = 0.01$ to determine whether there is any evidence that the median spray height is less than 630 feet.
b. Find the p value associated with this test.
c. Why can't this procedure be used in a test concerning the mean spray height?

14.80 **Biology and Environmental Science** Pelicans have a long flat bill and expandable pouch and use a spectacular dive-bomb type plunge into the water to capture fish. A random sample of pelicans at Tigertail Beach on Marco Island was obtained. The height of a plunge was measured (in feet) for each bird, and the data are given in the following table.

41	43	48	46	44	40	43	47	40

The underlying distribution is assumed to be continuous. Use the Wilcoxon signed-rank test with $\alpha = 0.04$ to determine whether the median plunge height is different from 42 feet. Find the p value associated with this test.

14.81 **Economics and Finance** Some economists believe that a family's financial situation is associated with the total hours worked each week. A random sample of couple families aged 20–50 years with a child under age six from several countries was obtained. Each family was classified by perceived financial situation and the amount of hours worked weekly for each family was recorded. The data are given in the following table.

	Financial situation	
Country	Well off	Just managing
Denmark	73	60
Finland	72	60
Norway	68	58
Sweden	70	59
Greece	65	64
Italy	62	55
Portugal	78	68
Spain	61	46
Ireland	66	55
United Kingdom	66	60
Austria	67	59
Germany	62	55
Netherlands	58	47
Belgium	67	58
France	61	60
Luxembourg	56	58

(*Source*: Organisation for Economic Co-operation and Development.)

Assume that the distribution of pairwise difference is continuous and symmetric. Use the Wilcoxon signed-rank test with $\alpha = 0.05$ to determine whether there is any evidence that the median time spent working per week is greater for families that are well off than for families that just manage.

14.82 **Travel and Transportation** A random sample of Georgia drivers renewing their vehicle registrations was obtained. Using motor vehicle records, the number of miles driven in the past year was recorded for each. The data (in thousands of miles) were classified by whether the driver had an organ donor card and are given in the following table.

Organ donor card

13.5	13.8	14.1	13.3	14.5

No organ donor card

13.1	12.7	12.9	13.4	14.1	12.3	13.0

Assume that the underlying distributions are continuous. Is there any evidence to suggest that the median number of miles driven is different for people who carry an organ donor card and for those who do not? Use the Wilcoxon rank-sum test with $\alpha = 0.05$.

14.83 **Public Health and Nutrition** Irradiation is a method of food preservation and is used to reduce bacteria and microorganisms in meat, poultry, and spices. The dose of an ionizing energy source varies according to the food type and is measured in gray (Gy), the amount of radiation absorbed. In a recent study, two types of radiation sources were compared. A random sample of spices was selected, and each machine was set for the maximum allowed dose of ionizing energy. The absorbed radiation was measured for each food sample (in kGy), and the data are given in the following table.

X-ray generator

22.6	25.9	26.3	22.2	23.5	24.9	27.2	22.9

Electron accelerator

27.5	26.3	22.1	19.5	26.7	21.6	21.2	21.4

Assume that the underlying distributions are continuous.
a. Is there any evidence to suggest that there is a difference in the absorbed radiation by machine? Use the Wilcoxon rank-sum test with $\alpha = 0.10$.
b. Find the p value associated with this test.

14.84 Manufacturing and Product Development The pressure for brewing espresso is necessary to help form cream and to distinguish this drink from strong drip coffee. Random samples of espressos from each of two commercial machines were obtained, and the pressure (in atmospheres) while brewing each cup was recorded. The data are given on the data CD and book's web site. Assume that the underlying distributions are continuous. Is there any evidence to suggest that the espresso setting on each machine produces a different median pressure? Use the Wilcoxon rank-sum test with $\alpha = 0.05$, and find the p value associated with this test.

14.85 Economics and Finance According to the 2005 Iowa Land Value Survey by Iowa State University, the value of Iowa farmland increased by approximately $285 an acre over the previous year.[20] Suppose a study was conducted to compare the farmland value per acre in two counties. A random sample of farmland was obtained from each county and the value (in dollars) per acre was recorded for each. The data are given in the following table.

County					
Winneshiek			**Howard**		
2607	2548	2584	2502	2349	2481
2468	2535	2543	2401	2494	2296
2491	2450	2583	2412	2483	2387
2641	2468	2457	2300	2409	2415
2380	2474	2347	2278	2363	2317
2331	2679	2450	2351	2438	2556
2541	2218		2381	2433	2504
			2387		

Is there any evidence to suggest that the land value for farmland is different in these two (adjacent) counties? Assume that the underlying distributions are continuous, and use a Wilcoxon rank-sum test with $\alpha = 0.001$.

14.86 Manufacturing and Product Development The slate in a pool table may be one piece, but because it is prone to fracturing during transport, it is often split into three slabs. Slate slabs from three pool-table manufacturers were randomly selected. The weight of each slab (in kg) was recorded, and the summary statistics are given in the following table.

Company	Sample size	Rank sum
Brunswick	$n_1 = 12$	$r_1 = 263.0$
AMF	$n_2 = 14$	$r_2 = 185.0$
Olhausen	$n_3 = 18$	$r_3 = 542.0$

Assume that the underlying weight populations are continuous. Use the Kruskal–Wallis test with $\alpha = 0.05$ to determine whether there is any evidence to suggest that at least two slate weight populations are different.

14.87 Physical Sciences Amateur radio operators (hams) communicate with one another all over the world, and many help coordinate relief efforts during a natural disaster. Whereas there is a 5-watt power limit on a CB radio, ham radios can have as much as 1500 watts of power. A random sample of amateur radio operators was obtained from four states. The power on each transmitter was measured (in watts), and the summary statistics are given in the following table.

State	Sample size	Rank sum
New Hampshire	$n_1 = 15$	$r_1 = 611.0$
Alabama	$n_2 = 16$	$r_2 = 605.5$
Texas	$n_3 = 18$	$r_3 = 609.0$
California	$n_4 = 20$	$r_4 = 589.5$

Assume that the underlying transmitter power distributions are continuous. Use the Kruskal–Wallis test with $\alpha = 0.025$ to determine whether there is any evidence to suggest that at least two transmitter power populations are different. Find the p value associated with this test.

14.88 Sports and Leisure Paintball has become a very popular sport that includes teams, leagues, and organized tournaments. A paintball is a glob of paint enclosed in a gelatin capsule that bursts on impact. A random sample of paintballs from three manufacturers was obtained, and each was carefully weighed. The data (in grams) are given in the following table.

Brass Eagle		**RP Scherer Premium**		**Zap Pro Series**	
3.228	3.163	3.194	3.185	3.217	3.194
3.192	3.220	3.184	3.187	3.195	3.196
3.234	3.247	3.194	3.182	3.208	3.202
3.238	3.168	3.183	3.194	3.196	3.199
3.291		3.186		3.195	

(*Source*: Gary and Neal Dyrkacz, *Some Statistics on Paintballs*, http://home.comcast.net/~dyrgcmn/pball/paintstats.html.)

Is there any evidence to suggest that at least two of the paintball weight population distributions are different? Assume that the underlying populations are continuous, and use the Kruskal–Wallis test with $\alpha = 0.05$.

14.89 Technology and Internet All students who enroll at Temple College are given an email account with the default password abc123. Students are instructed to change their passwords when they log in for the first time. Two weeks after classes began, a sample of students entering a public computer lab in the library was obtained, and each was classified by email password: default (D) or changed (C). The ordered observations are given in the following table.

D	C	D	C	D	C	C	D	C	D	C

a. Use the runs test with $\alpha \approx 0.05$ to determine whether there is any evidence to suggest that the order of observations is not random with respect to email password.
b. Find the p value associated with this test.

14.90 Fuel Consumption and Cars A sample of automobiles entering a parking garage in Monterey before 6:00 A.M. on a Monday morning was obtained, and each was classified as either foreign (F) or domestic (D). The ordered observations are given in the following table.

D	F	D	F	F	D	F	F	D	F	D	F	D	D	F
D	F	D	F	F	F	F	D	F	D	F	F	D	D	F

Is there any evidence to suggest that the order of automobiles entering the parking garage is not random? Use the runs test with the normal approximation and a 0.05 level of significance.

14.91 Public Health and Nutrition The Mayflower Health Insurance Company is conducting a survey of customers to estimate the number who have received a routine physical exam during the past year. A sample of customers was contacted by phone during the evening, and each was classified by exam (E) or no exam (N). The sequence of observations is given in the following table.

E	E	N	E	E	E	N	E	E	N	E	N	E	E	N
N	E	N	N	N	E	E	E	N	E	E	E	N	E	E
N	N	E	N	E	N	E	N	E	N					

Is there any evidence to suggest that the order of observations is not random? Use the runs test with the normal approximation and $\alpha = 0.01$. Find the p value associated with this test.

14.92 Medicine and Clinical Studies A study was conducted to examine the relationship between depressed mood and blood pressure.[21] A random sample of healthy women was obtained. Each patient's mood (x), measured using the Beck Depression Inventory (BDI), and systolic blood pressure (y, in mmHg) were measured. The data are given in the following table.

x	32.6	27.9	25.2	18.7	16.7	21.8	18.3	24.7
y	101	106	112	117	109	116	119	94

Compute Spearman's rank correlation coefficient, and use this value to describe the relationship between these two variables.

14.93 Public Health and Nutrition A study was conducted to examine the relationship between the volume and the quality of health care. A random sample of patients admitted to various hospitals in Indianapolis was obtained. After being discharged, each patient was asked to complete a questionnaire regarding the quality of care, and the results were evaluated to yield a quality score (x) between 1 (bad) and 50 (good). The total number of people in the hospital when the patient was admitted (y) was also recorded. The data are given on the data CD and book's web site. Compute Spearman's rank correlation coefficient. What does this value suggest about the relationship between the volume and the quality of health care?

14.94 Physical Sciences Researchers analyzed the optical night sky brightness at the Indian Astronomical Observatory during the years 2000–2008. A typical night sky spectrum was presented and the identified lines/bands wavelength (in Angstroms) and strengths (E. widths, in Angstroms) are given on the data CD and book's web site.[22] Compute Spearman's rank correlation coefficient. Use this value to explain the relationship between the wavelength and strength.

Extended Applications

14.95 Manufacturing and Product Development Most air conditioners use freon gas to provide cooling in a typical evaporation cycle. Although an air conditioner is theoretically a closed system, units usually lose freon and must be recharged every few years. A random sample of 10,000 BTU air conditioners was obtained, and the amount of freon (in pounds) in each was carefully measured. Each unit was recharged by a technician, and the amount of freon in each was measured after the service. The data are given on the data CD and book's web site. Suppose no assumptions can be made about the shape of the continuous distributions of before and after freon weights.
a. Conduct a sign test to determine whether there is any evidence that the median freon weight before service is less than the median freon weight after service. Use a significance level of $\alpha = 0.01$.
b. On the basis of your results, do you believe that recharging an air conditioner really increases the amount of freon in the system? Why or why not?

14.96 Manufacturing and Product Development A nail gun is a handy tool, especially if you need to install a wood floor or replace a roof, or if you hit your thumb a lot with a hammer. These devices propel a nail at incredible speeds and can save time and energy. A random sample of nail guns from four companies was obtained, and the speed of the nail was measured for each gun (in feet per second, fps). The data are given on the data CD and book's web site. Assume that the underlying speed distributions are continuous.
a. Use the Kruskal–Wallis test with $\alpha = 0.05$ to determine whether there is any evidence that at least two of the nail-gun speed population distributions are different.
b. Find bounds on the p value associated with this test.
c. Suppose you would like to purchase the brand of nail gun that fires a nail at the highest speed. Which brand would you choose, and why?

14.97 Marketing and Consumer Behavior A home-building company offers a variety of styles in only two exterior finishes:

vinyl siding (V) or brick (B). Immediately following an advertising campaign explaining the advantages of vinyl siding, a sample of consecutive customers was obtained, and each was classified according to the exterior finish that he or she had chosen. The ordered observations are given in the following table.

B   V   V   V   V   B   V   V   V   V   B   V   B   B

a. Conduct a runs test with $\alpha \approx 0.05$ to determine whether there is any evidence to suggest that the order of observations is not random with respect to exterior finish.
b. Using this sample, do you believe that the advertising campaign was successful? Why or why not?

14.98 Psychology and Human Behavior The manager of concessions at Fenway Park believes that the number of runs scored by the Red Sox (x) is related to the number of Fenway Franks consumed by fans (y). A random sample of nine-inning games was obtained. The hot dog and run totals are given on the data CD and book's web site.
a. Construct a scatter plot for these data, and describe the relationship between the two variables.
b. Rank the observations in each sample separately, and find each difference between the ranks, d_i.
c. Find the sample correlation coefficient between the ranks.
d. Find Spearman's rank correlation coefficient using the d_i's.
e. Explain why the values in parts (c) and (d) are different.
f. Suppose that the manager of concessions would like to sell as many hot dogs as possible. How many runs would he or she like the Red Sox to score? Why?

14.99 Psychology and Human Behavior A full moon has been linked to mental illness, increased incidence of disasters and crimes, and even reports of werewolves. In a study by the Sussex Police, a link was found between full moons and violent crimes and aggressive behavior among drinkers.[23] Suppose a random sample of days was obtained. Each day was classified by moon cycle and the number of drunk drivers arrested in Sussex was recorded for each. The data are given in the following table.

Moon cycle					
Two days before a full moon		Full moon cycle		Other times	
195	146	168	202	122	113
178	200	182	157	129	121
175	187	166	165	116	111
160	190	145	176	115	125
168	183	167	180	121	117
170	196	184	133	103	120
160	186	157	151	102	124
164	190	150	176	122	136

Assume that the underlying number of drunk drivers distributions are continuous.
a. Use the Kruskal–Wallis test with $\alpha = 0.01$ to determine whether there is any evidence that at least two of the moon cycle population distributions are different.
b. Find bounds on the p value associated with this test.
c. What do these data suggest about the safest time to drive?

Notes and Data Sources

CHAPTER 1

1. National Resources Defense Council.
2. E. Flossmann and R. Rothwell, Effect of aspirin on long term risk of colorectal cancer: consistent evidence from randomized controlled trials and observational studies, *The Lancet*, Vol. 369, No. 9573, 2007, pp. 1603–1613.
3. National Association of Colleges and Employers Salary Survey, Summer 2007.
4. P. Hanley and D. Forkenbrock, Safety of passing longer combination vehicles on two-lane highways, *Transportation Research Part A*: *Policy and Practice*, Vol. 39, No. 1, 2005, pp. 1–15.
5. M. Dolan and J. Felch, DNA matches aren't always a lock, *Los Angeles Times*, May 4, 2008.
6. National Weather Service.
7. World Natural Health Organization.
8. Nielsen Media Research.
9. American Petroleum Institute.
10. National Aquarium, Baltimore.
11. baneclene.com.
12. Liver Specialists of Texas.
13. E. Frauenheim, *ZDNet News*, December 11, 2002.
14. U.S. Department of Energy, 2005.
15. S. Lindberg et al., Exhaustion measured by the SF-36 vitality scale is associated with flattened diurnal cortisol profile, *Psychoneuroendocrinology*, Vol. 33, No. 4, May 2008, pp. 471–477.

CHAPTER 2

1. C. Bonder Embse, *Eightysomething*, Spring 1994, Vol. 3, No. 2.
2. Current Population Survey, Bureau of Labor Statistics, 2007.
3. National Cholesterol Education Program and the American Heart Association.
4. Canadian Centre for Occupational Health and Safety.
5. *National Health Statistics Reports*, No. 7, August 6, 2008.
6. Maine State Climate Office.
7. L.G. Thompson, et al., Kilimanjaro Ice Core Records: Evidence of Holocene Climate Change in Tropical Africa, *Science*, Vol. 298, October 2002, pp. 589–593.
8. American Road Builders Association.
9. *Federal Register*, Vol. 66, No. 118, June 2001, p. 32914.

CHAPTER 3

1. J. Surowiecki, No Work and No Play, *The New Yorker,* November 28, 2005, p. 68.
2. United Nations Food and Agriculture Organization Statistical Databases.
3. www.fueleconomy.gov.
4. United Nations Food and Agriculture Organization Fisheries Department.
5. *Detroit News* Sports Insider Statistics.
6. G. Giraud, Cooperative Learning and Statistics Instruction, *The Journal of Statistics Education*, Vol. 5, No. 3, November, 1997.
7. DiveMeets.com.
8. Seattle Post-Intelligencer, June 14, 2005.
9. Motorcycle Info and Accessories, motorcycleinfo. calsci.com, by Mark Lawrence.
10. World Cube Association.
11. InflationData.com.

CHAPTER 4

1. Society for Human Resources Management, September 24, 2007.
2. Teens, Video Games, and Civics, Pew Internet and American Life Project, September 16, 2008.
3. McDonald's and U.S. Consumer Product Safety Division.
4. National Center for Statistics and Analysis, National Highway Safety Administration.
5. Amtrak.
6. Zagat Company overview.
7. Eileen Alt Powell, Pensions got a future? A rickety one, *Seattle Times*, May 12, 2005.
8. Energy Information Administration, Department of Energy, *Characteristics of Residential Housing Units by Ceiling Fans*.
9. National Conference of State Legislators, *Gambling Developments in the States, 2008*.
10. Data Transmission Services—1st quarter 2005, Anacom.
11. G. Gutierrez, Medication nation: half of Americans now taking daily pills, NaturalNews.com, September 16, 2008.
12. Centers for Disease Control and Prevention.
13. Bayer HealthCare, *Cold and Flu Facts*.
14. A. Manning, Immunization rate exceeds federal goals; 8 out of 10 toddlers in USA vaccinated; adults lay behind, *USA Today,* July 27, 2005 p. D-5.

15. Centers for Disease Control and Prevention, U.S. National Immunization Survey, 2004.
16. Julian Barbassa, Herbicide-resistant weed plagues California, *Associated Press*, August 9, 2005.
17. H.S. Bawaskar and P.H. Bawaskar, Envenoming by the Common Krait (Bungarus caeruleus) and Asian Cobra (Naja naja): Clinical Manifestations and Their Management in a Rural Setting, Wilderness and Environmental Medicine, Vol. 15, 2004, pp. 257–266.
18. U.S. Census Bureau, International Database.
19. T. Higgins, Debt Valley, The *Business Gazette*, March 18, 2005.
20. Mosaic Oil.
21. *USA Today*, November 7, 2005.
22. www.celticsstats.com.
23. American Bone Marrow Donor Registry.
24. *Tuberculosis Facts, Health Promotion and Education.*
25. Bureau of Transportation Statistics.
26. GlobalSecurity.org.
27. National Partnership for Immunization.
28. Fedco Seeds, Waterville, Maine.
29. BondsOnline and Federal Citizen Information Center.
30. T. McNeeney, Harris Interactive, Results from a Harris Poll. The religious and other beliefs of Americans, *Business Wire,* November 29, 2007.
31. CNNMoney.com and U.S. Census Bureau, June 2007.
32. *SMU News,* December 6, 2005, and L. King, et al., Health Insurance and Cardiac Transplantation: A Call for Reform, *Journal of the American College of Cardiology,* Vol. 45, No. 9, 2005, pp.1388–1391.
33. Norman Herr, *Sourcebook of Teaching Science,* John Wiley/Jossey-Boss publication, 2007.
34. American College of Asthma, Allergy, and Immunology.
35. Based on information from Boeing.
36. American Forest and Paper Association, 2005 Community Survey.

CHAPTER 5

1. National Center for Infectious Diseases, Centers for Disease Control and Prevention.
2. NASA John F. Kennedy Space Center.
3. Auto Insurance Compendium.
4. Uri Simonshon and Don Ariely, eBay's Happy Hour: Non-Rational Herding in Online Auctions, 2005 draft.
5. Bureau of Transportation Statistics, U.S. Department of Transportation.
6. Rick Mattoon, Sun Wellness, *Getting the Most Out of Your Tanning Session*, Phoenix, AZ : Virgo Publishing, 2007.
7. Hoopsvibe.com, November 21, 2008.
8. U.S. Mint, 2008.
9. RxList, Inc.
10. Lifehacker, October 12, 2008.

11. Long Island Rail Road.
12. Consumeraffairs.com, 2006.
13. Center for Women's Business Research.
14. PNC News Release, December 13, 2007.
15. Marketwire, June 24, 2008.
16. Idaho Transportation Department, Economics Research Section.
17. Federal Bureau of Investigation.
18. Associated Content, January 10, 2007.
19. BIGEAST.org.
20. Canadian Pacific Railway.
21. Dell LCD Display Pixel Policy, 2008.
22. Viruslist.com.
23. V. Fletcher, How 4 cups of tea a day cut risk of heart attack, *UK News*, June 11, 2008.
24. TexasGasPrices.com.
25. *Colorado Department of Public Health News*, 2008.
26. Gallup, Inc., November 17, 2008.
27. Club and District Chair Workshops, Surf Life Saving, May 2008.
28. HawaiiCam, Beaches, April 25, 2008.
29. U.S. Census Bureau.
30. Linked in on CNBC, September 5, 2008.
31. HarrisInteractive, December 1, 2005.
32. National Ski Areas Association, September 12, 2008.

CHAPTER 6

1. U.S. Fish and Wildlife Service.
2. Illinois Natural History Survey.
3. M. Molinari, QOL evaluation of patients undergoing TACE for HCC, *Canadian Journal of Surgery*, Vol. 51, August 2008 (supplement), pp. 539–543.
4. Defense Technical Information Center.
5. National Data Buoy Center.
6. The *Physics Factbook*, Edited by Glenn Elert, An educational, fair use web site.
7. U.S. Department of Agriculture, Agriculture Research Service, 2008.
8. American Time Use Survey, U.S. Bureau of Labor Statistics.
9. Leigh Jones, Galveston residents frustrated with long building permit lines, *Galveston County Daily News*, October 30, 2008.
10. U.S. Forest Service.
11. BatteryCountry.com.
12. Richard Ward, Best bow, *Strings*, No. 117, March 2004.
13. TransUnion.
14. Student Monitor annual financial services study, 2008.
15. Near Earth Objects Program at the Jet Propulsion Laboratory.
16. K.M. Stadler and J.S. Essa, *The ABC's of Eating Out*, Department of Human Nutrition, Foods and Exercise, Virginia Tech.

17. National Highway Safety Administration.
18. Center for Science in the Public Interest.
19. Niagara Fall Chamber of Commerce, September 19, 2005.

CHAPTER 7

1. *Whittmanhart Benefits Summary.*
2. MayoClinic.com.
3. Ray Lilley, Fishermen catch huge squid, Associated Press, February 22, 2009.
4. Landels-Hill Big Creek Reserve.
5. Bartleby.com.
6. nascar.com.
7. M. Yusouf Rathor, Typhoid fever, *Indian Journal for the Practising Doctor*, Vol. 1, No. 3, 2004, pp. 11–12.
8. topendsports.com.
9. NOAA Climate Monitoring and Diagnostics Laboratory.
10. U.S. Department of Commerce.
11. FOXNews.com, April 1, 2008.
12. PGATour.com, November 14, 2008.
13. *Harris Poll*, No. 37, April 7, 2008.
14. *Milwaukee and Southeastern Wisconsin Business News*, November 21, 2008.
15. Reuters, April 16, 2008.
16. USNews.com.
17. Harris Interactive, 2007.
18. The Ocean Conservancy, *International Coastal Cleanup Report*, 2007.
19. National Health and Nutrition Examination Survey, National Center for Health Statistics.
20. blueplanetbiomes.org.
21. National Institutes of Health.
22. Canadian Union of Postal Workers, September 2, 2008.

CHAPTER 8

1. U.S. Census Bureau.
2. USA Weekend.com.
3. D. Butler, "vampire" devices sucking power; Federal energy initiative could save enough juice to light every home in N.B., *Montreal Gazette*, April 24, 2008, p. A12.
4. The Geyser Observation and Study Organization.
5. Students for the Exploration and Development of Space.
6. U.S. Department of Agriculture.
7. J.L. Bamber and J.A. Dowdeswell, Keel depths of modern Antarctic icebergs and implications for sea-floor scouring in the geologic record, *Marine Geology*, April 2007, pp. 120–131.
8. koalaexpress.co.au.
9. Alison Damast, College tuition just keeps rising, BusinessWeek.com, October 29, 2008.
10. Center for Educational and Training Technology, Mississippi State University.

11. Avantifix (Fixing Solutions) Limited.
12. Save the Manatee Club.
13. CNN.com.
14. K. Terry, 2008 exclusive survey—malpractice premiums: dropping, but still high, *Medical Economics*, August 1, 2008.
15. Prevent Blindness America.
16. U.S. Census Bureau, General Mobility, 2006 to 2007.
17. *U.S. Mint Annual Report*, 2004.
18. J. McReady et al., Correlates of gambling-related problems among older adults in Ontario, *Journal of Gambling Issues*, December 2008, pp. 174–194.
19. Kan Haidong et al., Dietary fiber, lung function, and chronic obstructive pulmonary disease in the Atherosclerosis Risk in Communities study, *American Journal of Epidemiology*, March 1, 2008, pp. 570–578.
20. Georgia Geoscience Online.
21. cyclingnews.com.
22. Canadian TV, November 11, 2008.
23. Bankrate.com.
24. Texas Commission on Environmental Quality.
25. D. Guttierez, Long duration flights triple risk of blood clots, Natural News.com, April 19, 2008.
26. U.S. Department of Agriculture, Economic Research Service.
27. Theft: Retail's Real Grinch, 2005 national study shows marked increase in retail employee theft, *MSN Business* News, reported in www.onsitesecurity.com.

CHAPTER 9

1. lovemagazine.com.
2. The College Board.
3. Kicking Tires, blogs.cars.com.
4. Entertainment Software Association.
5. Canwest News Service, May 27, 2008.
6. California Franchise Tax Board.
7. New Hampshire Department of Transportation.
8. The Waikiki Roughwater Swim Committee, Inc.
9. Mauro Serafini et al., Plasma antioxidants from chocolate, *Nature*, Vol. 424, 2003, p. 1013.
10. Société Générale, Cross Asset Research, *Global Economic Research*, December 5, 2008.
11. www.bloodbook.com.
12. Seeds-by-Size, Watermelon Seeds, 2008/2009 list, www.seeds-by-size.co.uk.
13. Lee Siebert and Tom Simkin, *Volcanoes of the World*, Smithsonian Institution, Global Volcanism Program, Digital Information Series.
14. Australian Bureau of Meteorology.
15. NHTSA Vehicle Database.
16. ConAgra Foods, Inc.
17. National Renewable Energy Laboratory, U.S. Department of Energy.

18. Alaska Ocean Observing System, Airborne Ice Thickness Profiles.

19. Agency of Toxic Substances and Disease Registry.

20. NFL.com and *NFL News*.

21. Nitterhouse Masonry Products, LLC, product descriptions.

22. www.boeing.com.

23. National Fitness Survey, U.K.

24. Moose hunting permit winners are drawn, *The Barre (Vermont) World*.

25. Bureau of Transportation Statistics.

26. www.actsinc.org.

27. *Forbes*, May 1, 2008, reported by the Council on International and Public Affairs.

28. U.S. Army Corps of Engineers, *FY 2008 Corps Dredge Report*.

29. Iowa Farm Bureau.

30. National Weather Service, Climate Prediction Center.

31. American Dental Association.

32. Jane Black, Americans' Food stamp use nears all-time high, *Washington Post*, November 26, 2008.

33. U.S. Postal Service.

34. Amber Wallace, Canadian executives unaware of gender gap in business schools, Queen's University School of Business, Kingston, Ontario, April 7, 2008.

35. CNNMoney.com, May 18, 2008.

36. Reuters, August 5, 2008.

37. Focalyst, September 22, 2008.

38. www.americanheart.org.

39. Florida Department of Education.

40. Institute of International Education, *Open Doors Report 2004*.

41. *USA Today*, Kaiser Family Foundation, and Harvard School of Public Health poll, reported by Harvard School of Public Health, March 4, 2008.

42. Josephson Institute, Ethics of American Youth Survey—2008 Summary.

43. Rachel Easley et al., Urban landscaping and the heat island effect in Mérida, Mexico, paper presented at the conference on Globalization and Sustainable Development in Latin America, REU Program funded by NSF.

44. EOS Archive, NASA Goddard Space Flight Center.

45. World Data Center for Aerosols.

46. *Estes Product Catalog*.

47. BBC, *Science and Nature: Animals*, December 17, 2008.

48. Office of Attorney General Bill McCollum, Florida Gang Reduction Strategy, 2008–2012.

49. *The Indian News*, March 13, 2008.

50. U.S. National Library of Medicine, National Institutes of Health.

51. *The Seattle Times*, December 7, 2005, and *Tax & Business Law Commentary*, November 10, 2005.

52. Transene Company, Inc.

CHAPTER 10

1. Fordham Institute for Innovation in Social Policy.

2. Media Life.

3. Conceptualized Reference Database for Building Envelope Research.

4. ABC News Poll: Holiday Shopping, November 19, 2008.

5. R. W. Mills et. al., Optical power outputs, spectra and dental composite depths of cure, obtained with blue light emitting diode (LED) and halogen light curing units (LCUs), *British Dental Journal,* vol. 193, 2002, pp. 459–463.

6. usahockey.com.

7. *USA Today,* December 11, 2005.

8. County Employment and Wages Summary, Bureau of Labor Statistics, October 17, 2008.

9. *USA Today,* February 1, 2006.

10. New Jersey Prescription Drug Registry and Florida Agency for Health Care Administration, 2008.

11. L. Bernardi et. al., Cardiovascular, cerebrovascular, and respiratory changes induced by different types of music in musicians and non-musicians: the importance of silence, *Heart*, Vol. 92, No. 4, 2006, pp. 445–452.

12. World Health Organization and NationMaster.com.

13. The Media Audit, November 2008.

14. Rocky Mountain Insurance Information Assoc.

15. J. D. Power and Associates reports, December 10, 2008.

16. U.S. Census Bureau and Information Please Database.

17. Brookings Institution.

18. Gallup Poll, December 15, 2008.

19. Agricultural Statistics Board, NASS, USDA, as reported by the National Turkey Federation.

20. Bureau of Transportation Statistics, U.S. Department of Transportation.

21. U.S. General Services Administration.

22. *U.S. News and World Report*, Best colleges 2009.

23. comScore, Inc., January 31, 2008.

24. USDA National Nutritional Database.

25. Randolph Heaster, Concert tickets are still pricey, *Kansas City Star*, June 26, 2005.

26. Texas Commission on Environmental Quality.

CHAPTER 11

1. Duraflame, Inc., 2008; Nancy Yoshihara, Better, cleaner burn, *Los Angeles Times*, November 15, 2007; and U.S. Recycled Wood Products, Inc.

2. Inrix, National Traffic Scorecard, 2007.

3. U.S. Department of Justice, *Sourcebook of Criminal Justice Statistics*.

4. Canadian Institute for Health Information, *Wait Time Tables: A Comparison by Province*.

5. Bureau of Labor Statistics, National Compensation Survey, Occupational Earnings Tables: United States, December 2006–January 2008.
6. U.S. Department of Health and Human Services.
7. Chesapeake Bay Foundation; Pennsylvania Department of Environmental Protection, and *Press Enterprise*, December 22, 2005.
8. Beaumont Enterprise, as reported in Grits for Breakfast, April 7, 2008.
9. U.S. Department of Agriculture.
10. Ducks Unlimited Canada, 2008.
11. National Institute for Occupational Safety and Health, Bureau of Labor Statistics.
12. Robert Wood Johnson Foundation, Cerner unveils hospital room of the future, *Quality/Equality*, October 14, 2008.
13. Inernational Atomic Energy Agency.
14. U.S.C.A.T.S., fish caught by Brandon Shumway.

CHAPTER 12

1. Combating heat stress in dairy cows, *Dairy Connection*, June 2008.
2. See, for example, L. Barclay and D. Lee, Calcium may improve bone mineral density in men, *Medscape Medical News*, November 11, 2008; and National Dairy Council, Dairy's role in adolescent bone health, 2009.
3. Lyndon State College Institute of Applied Meteorology, *The Effect of Temperature and Precipitation on Hydro-electric Power Generation in Vermont*, May 2008.
4. V.C. LaMarche et al., Increasing atmospheric carbon dioxide: tree ring evidence for growth enhancement in natural vegetation, *Science*, Vol. 225, September 7, 1984, pp. 1019 to 1021.
5. See B. Wilhelm et al., Pupillographic assessment of sleepiness in sleep-deprived healthy subjects, *Sleep*, Vol. 21, 1998, pp. 158–165; and several sources cited in J. Andreassi, *Psychophysiology, Human Behavior and Physiological Response*, 5th ed., Philadelphia: Psychology Press, 2006.
6. E.T.C. Lam et al., The effects of excrcise on birth weight: a meta-analysis, *American Journal of Health Studies*, Winter 2002.
7. U.S. Census Bureau, International Database, as reported in Information Please Database, Pearson Education, Inc.
8. American Heart Association, *What Are Healthy Levels of Cholesterol?*
9. O.W. Frauenfeld et al., Interdecadal changes in seasonal freeze and thaw depths in Russia, *Journal of Geophysical Research*, Vol. 109, 2004.
10. See, for example, K.K. Charles and E. Hurst, the correlation of wealth across generations, *Journal of Political Economy*, Vol. 111, 2003, pp. 1155–1192; and *The Undercover Economist: Wealth Generations*, March 3, 2008.
11. See, for example, H. Nabi et al., Do psychological factors affect inflammation and incident coronary heart disease, *Arteriosclerosis, Thrombosis, and Vascular Biology*, Vol. 28, 2008, pp. 1398–1406; and W. Kop, *Behavioral and Immunological Factors in Coronary Disease*, National Heart, Lung, and Blood Institute, U.S. National Institutes of Health, May 2008.
12. F. He et al., Salt intake is related to soft drink consumption in children and adolescents, *Hypertension*, Vol. 51, 2008, pp. 629–634; American Heart Association.
13. See, for example, K. Roberts, *A Survey of Naturally Occurring Uranium in Groundwater in Southwestern North Dakota*, North Dakota Department of Health, 2008; and M. Gleason, *Uranium Enriched Ground Water, Knox Mountain Pluton, Vermont; Occurrence and Geological Controls*, Department of Geology, Middlebury College, May 2007.
14. Kumho Tire USA, Specification Data, 2009.
15. R. Doolittle, Baby, it's safe outside, The Star.com, N. Read, Crime goes up with temperature, psychologist finds, Can-West News Service, July 21, 2006; and I. Simister and G. Cooper, Thermal stress in the U.S.A.: effects on violence and on employee behavior, *Stress and Health*, Vol. 21, 2005, pp. 3–15.
16. Diamonds-USA.com, January 7, 2009.
17. See, for example, Canadian Press, Study links brain size, intelligence, December 23, 2005, and S. Novella, Brain size and intelligence, *Neurolgicabloy*, Feb 2, 2007.
18. K. Allen, Drug maker is latest firm to flee UK's corporate tax regime, *The Guardian*, April 16, 2008.
19. The Pentagon Brief, December 14, 2008.
20. National Resources Defense Council; Nuclear Weapons Stockpile Chart, Carnegie Endowment for Internal Peace, July 2005; Nuclear Weapons Stockpiles, Nuclear Threat Initiative, 2007.
21. L. Emmerson and C. Southwell, Sea ice cover and its influence on Adélie Penguin reproductive Performance, *Ecology*, Vol. 89, No. 8, 2008, pp. 2006–2012.
22. Digital Library for Earth System Education.
23. Hormel Always Tender Products, Cooking and Storage FAQs.
24. www.discoverychannel.co.uk, January 2009; The most expensive cheese in the world: cheers to Sweden, *Elite Choice*, October 2007.
25. R. Arena et al., The relationship between C-reactive protein and other cardiovascular risk factors in men and women, *Journal of Cardiopulmonary Rehabilitation and Prevention*, Vol. 26, No. 5, September/October 2006, pp. 323–327; WebMD, January 7, 2009.

CHAPTER 13

1. University of California, Office of the President, Student Affairs, January 4, 2008.
2. Department of Education, 2003 National Assessment of Adult Literacy.
3. C. Walters, Airports are being transformed into shopping malls, The Consumerist, November 5, 2007.
4. Airbus for Analysts.
5. Ward's AutoInfoBank, 2009.
6. 2008 Health Care for America Survey, Summary of Findings.
7. Surgical objects accidentally left inside about 1500 patients in US each year, *Science Daily*, December 9, 2007.
8. *The Boston Globe*, December 6, 2005.

CHAPTER 14

1. The Urban Institute.
2. About, Inc., and Canadian Egg Marketing Agency.
3. U.S. Department of Agriculture.
4. Scarborough USA, as reported in *Press Enterprise*, November 13, 2005.
5. More first-time buyers entering the market, *Daily Real Estate News*, November 10, 2008.
6. Coastal and Estuarine Geology Program, Maryland Geological Survey.
7. American Heart Association.
8. K. Colabray, Metabolism: powering the human body, *Muhlenberg Weekly*, September 29, 2005.
9. See Average A1C reduced from 7.6 to 6.7 after three months, Diabetes Intervention Program, Alabama, *Medical News Today*, December 12, 2008.
10. Hormone-Refractory Prostate Cancer Association, Inc.

11. F. V. van Oort et al., Folic acid and reduction of plasma homocysteine concentrations in older adults: a dose-response study, *American Journal of Clinical Nutrition*, Vol. 77, pp. 1318–1323.
12. Office of Environment and Planning, Federal Highway Administration.
13. B. Lankoff, Long, winding road, but Kilger has arrived, *Canoe, Inc.*, September 29, 2007; and P. Hickey, Canadians unlikely to win Forsberg sweepstakes, Canada.com, February 10, 2007.
14. USDA Nutrient Database.
15. Federal Justice Statistics Resource Center, 2007.
16. National Coalition on Health Care, Health Insurance Cost, www.nchc.org/facts/cost.shtml.
17. MarketingHire.com.
18. A. Vogiatzoglou et al., Vitamin B_{12} status and rate of brain volume loss in community-dwelling elderly, *Neurology*, Vol. 71, No. 9, September 9, 2008, pp. 826–832.
19. K. Matthews, *Central Body Fat*, John D. and Catherine T. MacArthur Research Network on Socioeconomic Status and Health, 1996.
20. G. Lucht, Land values soar to record levels, *Iowa Farmer Today,* December 14, 2005.
21. K.M. Grewen et al., Depressive symptoms are related to higher ambulatory blood pressure in people with a family history of hypertension, *Psychosomatic Medicine*, Vol. 66, 2004, pp. 9–16.
22. C.S. Stalin et al., Night sky at the Indian Astronomical Observatory during 2000–2008, *Bulletin of the Astronomical Society of India,* Vol. 36, 2008, pp. 111–127.
23. L. Acford, Cops to hit streets to tackle lunar-tic drunks, *The Argus*, June 5, 2007.

Tables Appendix

Table I Binomial Distribution Cumulative Probabilities

Let X be a binomial random variable with parameters n and p: $X \sim B(n, p)$. This table contains cumulative probabilities:

$$P(X \le x) = \sum_{k=0}^{x} P(X = k) = P(X = 0) + P(X = 1) + P(X = 2) + \cdots + P(X = x).$$

n = 5　　　　　　　　　　　　　　　　　　　　　　p

x	0.01	0.05	0.10	0.20	0.25	0.30	0.40	0.50	0.60	0.70	0.75	0.80	0.90	0.95	0.99
0	0.9510	0.7738	0.5905	0.3277	0.2373	0.1681	0.0778	0.0313	0.0102	0.0024	0.0010	0.0003	0.0000		
1	0.9990	0.9774	0.9185	0.7373	0.6328	0.5282	0.3370	0.1875	0.0870	0.0308	0.0156	0.0067	0.0005	0.0000	
2	1.0000	0.9988	0.9914	0.9421	0.8965	0.8369	0.6826	0.5000	0.3174	0.1631	0.1035	0.0579	0.0086	0.0012	0.0000
3		1.0000	0.9995	0.9933	0.9844	0.9692	0.9130	0.8125	0.6630	0.4718	0.3672	0.2627	0.0815	0.0226	0.0010
4			1.0000	0.9997	0.9990	0.9976	0.9898	0.9688	0.9222	0.8319	0.7627	0.6723	0.4095	0.2262	0.0490

n = 10　　　　　　　　　　　　　　　　　　　　　　p

x	0.01	0.05	0.10	0.20	0.25	0.30	0.40	0.50	0.60	0.70	0.75	0.80	0.90	0.95	0.99
0	0.9044	0.5987	0.3487	0.1074	0.0563	0.0282	0.0060	0.0010	0.0001	0.0000					
1	0.9957	0.9139	0.7361	0.3758	0.2440	0.1493	0.0464	0.0107	0.0017	0.0001	0.0000	0.0000			
2	0.9999	0.9885	0.9298	0.6778	0.5256	0.3828	0.1673	0.0547	0.0123	0.0016	0.0004	0.0001	0.0000		
3	1.0000	0.9990	0.9872	0.8791	0.7759	0.6496	0.3823	0.1719	0.0548	0.0106	0.0035	0.0009	0.0000		
4		0.9999	0.9984	0.9672	0.9219	0.8497	0.6331	0.3770	0.1662	0.0473	0.0197	0.0064	0.0001	0.0000	
5		1.0000	0.9999	0.9936	0.9803	0.9527	0.8338	0.6230	0.3669	0.1503	0.0781	0.0328	0.0016	0.0001	
6			1.0000	0.9991	0.9965	0.9894	0.9452	0.8281	0.6177	0.3504	0.2241	0.1209	0.0128	0.0010	0.0000
7				0.9999	0.9996	0.9984	0.9877	0.9453	0.8327	0.6172	0.4744	0.3222	0.0702	0.0115	0.0001
8				1.0000	1.0000	0.9999	0.9983	0.9893	0.9536	0.8507	0.7560	0.6242	0.2639	0.0861	0.0043
9						1.0000	0.9999	0.9990	0.9940	0.9718	0.9437	0.8926	0.6513	0.4013	0.0956

n = 15　　　　　　　　　　　　　　　　　　　　　　p

x	0.01	0.05	0.10	0.20	0.25	0.30	0.40	0.50	0.60	0.70	0.75	0.80	0.90	0.95	0.99
0	0.8601	0.4633	0.2059	0.0352	0.0134	0.0047	0.0005	0.0000							
1	0.9904	0.8290	0.5490	0.1671	0.0802	0.0353	0.0052	0.0005	0.0000						
2	0.9996	0.9638	0.8159	0.3980	0.2361	0.1268	0.0271	0.0037	0.0003	0.0000					
3	1.0000	0.9945	0.9444	0.6482	0.4613	0.2969	0.0905	0.0176	0.0019	0.0001	0.0000				
4		0.9994	0.9873	0.8358	0.6865	0.5155	0.2173	0.0592	0.0093	0.0007	0.0001	0.0000			
5		0.9999	0.9978	0.9389	0.8516	0.7216	0.4032	0.1509	0.0338	0.0037	0.0008	0.0001			
6		1.0000	0.9997	0.9819	0.9434	0.8689	0.6098	0.3036	0.0950	0.0152	0.0042	0.0008			
7			1.0000	0.9958	0.9827	0.9500	0.7869	0.5000	0.2131	0.0500	0.0173	0.0042	0.0000		
8				0.9992	0.9958	0.9848	0.9050	0.6964	0.3902	0.1311	0.0566	0.0181	0.0003	0.0000	
9				0.9999	0.9992	0.9963	0.9662	0.8491	0.5968	0.2784	0.1484	0.0611	0.0022	0.0001	
10				1.0000	0.9999	0.9993	0.9907	0.9408	0.7827	0.4845	0.3135	0.1642	0.0127	0.0006	
11					1.0000	0.9999	0.9981	0.9824	0.9095	0.7031	0.5387	0.3518	0.0556	0.0055	0.0000
12						1.0000	0.9997	0.9963	0.9729	0.8732	0.7639	0.6020	0.1841	0.0362	0.0004
13							1.0000	0.9995	0.9948	0.9647	0.9198	0.8329	0.4510	0.1710	0.0096
14								1.0000	0.9995	0.9953	0.9866	0.9648	0.7941	0.5367	0.1399

Table I Binomial Distribution Cumulative Probabilities (*Continued*)

n = 20

x	0.01	0.05	0.10	0.20	0.25	0.30	0.40	0.50	0.60	0.70	0.75	0.80	0.90	0.95	0.99
0	0.8179	0.3585	0.1216	0.0115	0.0032	0.0008	0.0000								
1	0.9831	0.7358	0.3917	0.0692	0.0243	0.0076	0.0005	0.0000							
2	0.9990	0.9245	0.6769	0.2061	0.0913	0.0355	0.0036	0.0002							
3	1.0000	0.9841	0.8670	0.4114	0.2252	0.1071	0.0160	0.0013	0.0000						
4		0.9974	0.9568	0.6296	0.4148	0.2375	0.0510	0.0059	0.0003						
5		0.9997	0.9887	0.8042	0.6172	0.4164	0.1256	0.0207	0.0016	0.0000					
6		1.0000	0.9976	0.9133	0.7858	0.6080	0.2500	0.0577	0.0065	0.0003	0.0000				
7			0.9996	0.9679	0.8982	0.7723	0.4159	0.1316	0.0210	0.0013	0.0002	0.0000			
8			0.9999	0.9900	0.9591	0.8867	0.5956	0.2517	0.0565	0.0051	0.0009	0.0001			
9			1.0000	0.9974	0.9861	0.9520	0.7553	0.4119	0.1275	0.0171	0.0039	0.0006			
10				0.9994	0.9961	0.9829	0.8725	0.5881	0.2447	0.0480	0.0139	0.0026	0.0000		
11				0.9999	0.9991	0.9949	0.9435	0.7483	0.4044	0.1133	0.0409	0.0100	0.0001		
12				1.0000	0.9998	0.9987	0.9790	0.8684	0.5841	0.2277	0.1018	0.0321	0.0004		
13					1.0000	0.9997	0.9935	0.9423	0.7500	0.3920	0.2142	0.0867	0.0024	0.0000	
14						1.0000	0.9984	0.9793	0.8744	0.5836	0.3828	0.1958	0.0113	0.0003	
15							0.9997	0.9941	0.9490	0.7625	0.5852	0.3704	0.0432	0.0026	
16							1.0000	0.9987	0.9840	0.8929	0.7748	0.5886	0.1330	0.0159	0.0000
17								0.9998	0.9964	0.9645	0.9087	0.7939	0.3231	0.0755	0.0010
18								1.0000	0.9995	0.9924	0.9757	0.9308	0.6083	0.2642	0.0169
19									1.0000	0.9992	0.9968	0.9885	0.8784	0.6415	0.1821

n = 25

x	0.01	0.05	0.10	0.20	0.25	0.30	0.40	0.50	0.60	0.70	0.75	0.80	0.90	0.95	0.99
0	0.7778	0.2774	0.0718	0.0038	0.0008	0.0001	0.0000								
1	0.9742	0.6424	0.2712	0.0274	0.0070	0.0016	0.0001								
2	0.9980	0.8729	0.5371	0.0982	0.0321	0.0090	0.0004	0.0000							
3	0.9999	0.9659	0.7636	0.2340	0.0962	0.0332	0.0024	0.0001							
4	1.0000	0.9928	0.9020	0.4207	0.2137	0.0905	0.0095	0.0005	0.0000						
5		0.9988	0.9666	0.6167	0.3783	0.1935	0.0294	0.0020	0.0001						
6		0.9998	0.9905	0.7800	0.5611	0.3407	0.0736	0.0073	0.0003						
7		1.0000	0.9977	0.8909	0.7265	0.5118	0.1536	0.0216	0.0012	0.0000					
8			0.9995	0.9532	0.8506	0.6769	0.2735	0.0539	0.0043	0.0001					
9			0.9999	0.9827	0.9287	0.8106	0.4246	0.1148	0.0132	0.0005	0.0000				
10			1.0000	0.9944	0.9703	0.9022	0.5858	0.2122	0.0344	0.0018	0.0002	0.0000			
11				0.9985	0.9893	0.9558	0.7323	0.3450	0.0778	0.0060	0.0009	0.0001			
12				0.9996	0.9966	0.9825	0.8462	0.5000	0.1538	0.0175	0.0034	0.0004			
13				0.9999	0.9991	0.9940	0.9222	0.6550	0.2677	0.0442	0.0107	0.0015			
14				1.0000	0.9998	0.9982	0.9656	0.7878	0.4142	0.0978	0.0297	0.0056	0.0000		
15					1.0000	0.9995	0.9868	0.8852	0.5754	0.1894	0.0713	0.0173	0.0001		
16						0.9999	0.9957	0.9461	0.7265	0.3231	0.1494	0.0468	0.0005		
17						1.0000	0.9988	0.9784	0.8464	0.4882	0.2735	0.1091	0.0023	0.0000	
18							0.9997	0.9927	0.9264	0.6593	0.4389	0.2200	0.0095	0.0002	
19							0.9999	0.9980	0.9706	0.8065	0.6217	0.3833	0.0334	0.0012	
20							1.0000	0.9995	0.9905	0.9095	0.7863	0.5793	0.0980	0.0072	0.0000
21								0.9999	0.9976	0.9668	0.9038	0.7660	0.2364	0.0341	0.0001
22								1.0000	0.9996	0.9910	0.9679	0.9018	0.4629	0.1271	0.0020
23									0.9999	0.9984	0.9930	0.9726	0.7288	0.3576	0.0258
24									1.0000	0.9999	0.9992	0.9962	0.9282	0.7226	0.2222

Table II Poisson Distribution Cumulative Probabilities

Let X be a Poisson random variable with parameter λ. This table contains cumulative probabilities:

$$P(X \leq x) = \sum_{k=0}^{x} P(X = k) = P(X = 0) + P(X = 1) + P(X = 2) + \cdots + P(X = x).$$

					λ					
x	0.05	0.10	0.15	0.20	0.25	0.30	0.35	0.40	0.45	0.50
0	0.9512	0.9048	0.8607	0.8187	0.7788	0.7408	0.7047	0.6703	0.6376	0.6065
1	0.9988	0.9953	0.9898	0.9825	0.9735	0.9631	0.9513	0.9384	0.9246	0.9098
2	1.0000	0.9998	0.9995	0.9989	0.9978	0.9964	0.9945	0.9921	0.9891	0.9856
3		1.0000	1.0000	0.9999	0.9999	0.9997	0.9995	0.9992	0.9988	0.9982
4				1.0000	1.0000	1.0000	1.0000	0.9999	0.9999	0.9998
5								1.0000	1.0000	1.0000

					λ					
x	0.55	0.60	0.65	0.70	0.75	0.80	0.85	0.90	0.95	1.00
0	0.5769	0.5488	0.5220	0.4966	0.4724	0.4493	0.4274	0.4066	0.3867	0.3679
1	0.8943	0.8781	0.8614	0.8442	0.8266	0.8088	0.7907	0.7725	0.7541	0.7358
2	0.9815	0.9769	0.9717	0.9659	0.9595	0.9526	0.9451	0.9371	0.9287	0.9197
3	0.9975	0.9966	0.9956	0.9942	0.9927	0.9909	0.9889	0.9865	0.9839	0.9810
4	0.9997	0.9996	0.9994	0.9992	0.9989	0.9986	0.9982	0.9977	0.9971	0.9963
5	1.0000	1.0000	0.9999	0.9999	0.9999	0.9998	0.9997	0.9997	0.9995	0.9994
6		1.0000	1.0000	1.0000	1.0000	1.0000	1.0000	1.0000	0.9999	0.9999
7									1.0000	1.0000

					λ					
x	1.1	1.2	1.3	1.4	1.5	1.6	1.7	1.8	1.9	2.0
0	0.3329	0.3012	0.2725	0.2466	0.2231	0.2019	0.1827	0.1653	0.1496	0.1353
1	0.6990	0.6626	0.6268	0.5918	0.5578	0.5249	0.4932	0.4628	0.4337	0.4060
2	0.9004	0.8795	0.8571	0.8335	0.8088	0.7834	0.7572	0.7306	0.7037	0.6767
3	0.9743	0.9662	0.9569	0.9463	0.9344	0.9212	0.9068	0.8913	0.8747	0.8571
4	0.9946	0.9923	0.9893	0.9857	0.9814	0.9763	0.9704	0.9636	0.9559	0.9473
5	0.9990	0.9985	0.9978	0.9968	0.9955	0.9940	0.9920	0.9896	0.9868	0.9834
6	0.9999	0.9997	0.9996	0.9994	0.9991	0.9987	0.9981	0.9974	0.9966	0.9955
7	1.0000	1.0000	0.9999	0.9999	0.9998	0.9997	0.9996	0.9994	0.9992	0.9989
8			1.0000	1.0000	1.0000	1.0000	0.9999	0.9999	0.9998	0.9998
9						1.0000	1.0000	1.0000	1.0000	1.0000

Table II Poisson Distribution Cumulative Probabilities (*Continued*)

					λ					
x	2.1	2.2	2.3	2.4	2.5	2.6	2.7	2.8	2.9	3.0
0	0.1225	0.1108	0.1003	0.0907	0.0821	0.0743	0.0672	0.0608	0.0550	0.0498
1	0.3796	0.3546	0.3309	0.3084	0.2873	0.2674	0.2487	0.2311	0.2146	0.1991
2	0.6496	0.6227	0.5960	0.5697	0.5438	0.5184	0.4936	0.4695	0.4460	0.4232
3	0.8386	0.8194	0.7993	0.7787	0.7576	0.7360	0.7141	0.6919	0.6696	0.6472
4	0.9379	0.9275	0.9162	0.9041	0.8912	0.8774	0.8629	0.8477	0.8318	0.8153
5	0.9796	0.9751	0.9700	0.9643	0.9580	0.9510	0.9433	0.9349	0.9258	0.9161
6	0.9941	0.9925	0.9906	0.9884	0.9858	0.9828	0.9794	0.9756	0.9713	0.9665
7	0.9985	0.9980	0.9974	0.9967	0.9958	0.9947	0.9934	0.9919	0.9901	0.9881
8	0.9997	0.9995	0.9994	0.9991	0.9989	0.9985	0.9981	0.9976	0.9969	0.9962
9	0.9999	0.9999	0.9999	0.9998	0.9997	0.9996	0.9995	0.9993	0.9991	0.9989
10	1.0000	1.0000	1.0000	1.0000	0.9999	0.9999	0.9999	0.9998	0.9998	0.9997
11				1.0000	1.0000	1.0000	1.0000	1.0000	0.9999	0.9999
12									1.0000	1.0000

					λ					
x	3.1	3.2	3.3	3.4	3.5	3.6	3.7	3.8	3.9	4.0
0	0.0450	0.0408	0.0369	0.0334	0.0302	0.0273	0.0247	0.0224	0.0202	0.0183
1	0.1847	0.1712	0.1586	0.1468	0.1359	0.1257	0.1162	0.1074	0.0992	0.0916
2	0.4012	0.3799	0.3594	0.3397	0.3208	0.3027	0.2854	0.2689	0.2531	0.2381
3	0.6248	0.6025	0.5803	0.5584	0.5366	0.5152	0.4942	0.4735	0.4532	0.4335
4	0.7982	0.7806	0.7626	0.7442	0.7254	0.7064	0.6872	0.6678	0.6484	0.6288
5	0.9057	0.8946	0.8829	0.8705	0.8576	0.8441	0.8301	0.8156	0.8006	0.7851
6	0.9612	0.9554	0.9490	0.9421	0.9347	0.9267	0.9182	0.9091	0.8995	0.8893
7	0.9858	0.9832	0.9802	0.9769	0.9733	0.9692	0.9648	0.9599	0.9546	0.9489
8	0.9953	0.9943	0.9931	0.9917	0.9901	0.9883	0.9863	0.9840	0.9815	0.9786
9	0.9986	0.9982	0.9978	0.9973	0.9967	0.9960	0.9952	0.9942	0.9931	0.9919
10	0.9996	0.9995	0.9994	0.9992	0.9990	0.9987	0.9984	0.9981	0.9977	0.9972
11	0.9999	0.9999	0.9998	0.9998	0.9997	0.9996	0.9995	0.9994	0.9993	0.9991
12	1.0000	1.0000	1.0000	0.9999	0.9999	0.9999	0.9999	0.9998	0.9998	0.9997
13				1.0000	1.0000	1.0000	1.0000	1.0000	0.9999	0.9999
14									1.0000	1.0000

Table II Poisson Distribution Cumulative Probabilities (*Continued*)

λ

x	4.1	4.2	4.3	4.4	4.5	4.6	4.7	4.8	4.9	5.0
0	0.0166	0.0150	0.0136	0.0123	0.0111	0.0101	0.0091	0.0082	0.0074	0.0067
1	0.0845	0.0780	0.0719	0.0663	0.0611	0.0563	0.0518	0.0477	0.0439	0.0404
2	0.2238	0.2102	0.1974	0.1851	0.1736	0.1626	0.1523	0.1425	0.1333	0.1247
3	0.4142	0.3954	0.3772	0.3594	0.3423	0.3257	0.3097	0.2942	0.2793	0.2650
4	0.6093	0.5898	0.5704	0.5512	0.5321	0.5132	0.4946	0.4763	0.4582	0.4405
5	0.7693	0.7531	0.7367	0.7199	0.7029	0.6858	0.6684	0.6510	0.6335	0.6160
6	0.8786	0.8675	0.8558	0.8436	0.8311	0.8180	0.8046	0.7908	0.7767	0.7622
7	0.9427	0.9361	0.9290	0.9214	0.9134	0.9049	0.8960	0.8867	0.8769	0.8666
8	0.9755	0.9721	0.9683	0.9642	0.9597	0.9549	0.9497	0.9442	0.9382	0.9319
9	0.9905	0.9889	0.9871	0.9851	0.9829	0.9805	0.9778	0.9749	0.9717	0.9682
10	0.9966	0.9959	0.9952	0.9943	0.9933	0.9922	0.9910	0.9896	0.9880	0.9863
11	0.9989	0.9986	0.9983	0.9980	0.9976	0.9971	0.9966	0.9960	0.9953	0.9945
12	0.9997	0.9996	0.9995	0.9993	0.9992	0.9990	0.9988	0.9986	0.9983	0.9980
14	0.9999	0.9999	0.9998	0.9998	0.9997	0.9997	0.9996	0.9995	0.9994	0.9993
15	1.0000	1.0000	1.0000	0.9999	0.9999	0.9999	0.9999	0.9999	0.9998	0.9998
16				1.0000	1.0000	1.0000	1.0000	1.0000	0.9999	0.9999
17									1.0000	1.0000

λ

x	5.5	6.0	6.5	7.0	7.5	8.0	8.5	9.0	9.5	10.0
0	0.0041	0.0025	0.0015	0.0009	0.0006	0.0003	0.0002	0.0001	0.0001	0.0000
1	0.0266	0.0174	0.0113	0.0073	0.0047	0.0030	0.0019	0.0012	0.0008	0.0005
2	0.0884	0.0620	0.0430	0.0296	0.0203	0.0138	0.0093	0.0062	0.0042	0.0028
3	0.2017	0.1512	0.1118	0.0818	0.0591	0.0424	0.0301	0.0212	0.0149	0.0103
4	0.3575	0.2851	0.2237	0.1730	0.1321	0.0996	0.0744	0.0550	0.0403	0.0293
5	0.5289	0.4457	0.3690	0.3007	0.2414	0.1912	0.1496	0.1157	0.0885	0.0671
6	0.6860	0.6063	0.5265	0.4497	0.3782	0.3134	0.2562	0.2068	0.1649	0.1301
7	0.8095	0.7440	0.6728	0.5987	0.5246	0.4530	0.3856	0.3239	0.2687	0.2202
8	0.8944	0.8472	0.7916	0.7291	0.6620	0.5925	0.5231	0.4557	0.3918	0.3328
9	0.9462	0.9161	0.8774	0.8305	0.7764	0.7166	0.6530	0.5874	0.5218	0.4579
10	0.9747	0.9574	0.9332	0.9015	0.8622	0.8159	0.7634	0.7060	0.6453	0.5830
11	0.9890	0.9799	0.9661	0.9467	0.9208	0.8881	0.8487	0.8030	0.7520	0.6968
12	0.9955	0.9912	0.9840	0.9730	0.9573	0.9362	0.9091	0.8758	0.8364	0.7916
13	0.9983	0.9964	0.9929	0.9872	0.9784	0.9658	0.9486	0.9261	0.8981	0.8645
14	0.9994	0.9986	0.9970	0.9943	0.9897	0.9827	0.9726	0.9585	0.9400	0.9165
15	0.9998	0.9995	0.9988	0.9976	0.9954	0.9918	0.9862	0.9780	0.9665	0.9513
16	0.9999	0.9998	0.9996	0.9990	0.9980	0.9963	0.9934	0.9889	0.9823	0.9730
17	1.0000	0.9999	0.9998	0.9996	0.9992	0.9984	0.9970	0.9947	0.9911	0.9857
18		1.0000	0.9999	0.9999	0.9997	0.9993	0.9987	0.9976	0.9957	0.9928
19			1.0000	1.0000	0.9999	0.9997	0.9995	0.9989	0.9980	0.9965
20					1.0000	0.9999	0.9998	0.9996	0.9991	0.9984
21						1.0000	0.9999	0.9998	0.9996	0.9993
22							1.0000	0.9999	0.9999	0.9997
23								1.0000	0.9999	0.9999
24									1.0000	1.0000

Table III Standard Normal Distribution Cumulative Probabilities

Let Z be a standard normal random variable: $\mu = 0$ and $\sigma = 1$.
This table contains cumulative probabilities: $P(Z \leq z)$.

$P(Z \leq z)$

z	.00	.01	.02	.03	.04	.05	.06	.07	.08	.09
−3.4	0.0003	0.0003	0.0003	0.0003	0.0003	0.0003	0.0003	0.0003	0.0003	0.0002
−3.3	0.0005	0.0005	0.0005	0.0004	0.0004	0.0004	0.0004	0.0004	0.0004	0.0003
−3.2	0.0007	0.0007	0.0006	0.0006	0.0006	0.0006	0.0006	0.0005	0.0005	0.0005
−3.1	0.0010	0.0009	0.0009	0.0009	0.0008	0.0008	0.0008	0.0008	0.0007	0.0007
−3.0	0.0013	0.0013	0.0013	0.0012	0.0012	0.0011	0.0011	0.0011	0.0010	0.0010
−2.9	0.0019	0.0018	0.0018	0.0017	0.0016	0.0016	0.0015	0.0015	0.0014	0.0014
−2.8	0.0026	0.0025	0.0024	0.0023	0.0023	0.0022	0.0021	0.0021	0.0020	0.0019
−2.7	0.0035	0.0034	0.0033	0.0032	0.0031	0.0030	0.0029	0.0028	0.0027	0.0026
−2.6	0.0047	0.0045	0.0044	0.0043	0.0041	0.0040	0.0039	0.0038	0.0037	0.0036
−2.5	0.0062	0.0060	0.0059	0.0057	0.0055	0.0054	0.0052	0.0051	0.0049	0.0048
−2.4	0.0082	0.0080	0.0078	0.0075	0.0073	0.0071	0.0069	0.0068	0.0066	0.0064
−2.3	0.0107	0.0104	0.0102	0.0099	0.0096	0.0094	0.0091	0.0089	0.0087	0.0084
−2.2	0.0139	0.0136	0.0132	0.0129	0.0125	0.0122	0.0119	0.0116	0.0113	0.0110
−2.1	0.0179	0.0174	0.0170	0.0166	0.0162	0.0158	0.0154	0.0150	0.0146	0.0143
−2.0	0.0228	0.0222	0.0217	0.0212	0.0207	0.0202	0.0197	0.0192	0.0188	0.0183
−1.9	0.0287	0.0281	0.0274	0.0268	0.0262	0.0256	0.0250	0.0244	0.0239	0.0233
−1.8	0.0359	0.0351	0.0344	0.0336	0.0329	0.0322	0.0314	0.0307	0.0301	0.0294
−1.7	0.0446	0.0436	0.0427	0.0418	0.0409	0.0401	0.0392	0.0384	0.0375	0.0367
−1.6	0.0548	0.0537	0.0526	0.0516	0.0505	0.0495	0.0485	0.0475	0.0465	0.0455
−1.5	0.0668	0.0655	0.0643	0.0630	0.0618	0.0606	0.0594	0.0582	0.0571	0.0559
−1.4	0.0808	0.0793	0.0778	0.0764	0.0749	0.0735	0.0721	0.0708	0.0694	0.0681
−1.3	0.0968	0.0951	0.0934	0.0918	0.0901	0.0885	0.0869	0.0853	0.0838	0.0823
−1.2	0.1151	0.1131	0.1112	0.1093	0.1075	0.1056	0.1038	0.1020	0.1003	0.0985
−1.1	0.1357	0.1335	0.1314	0.1292	0.1271	0.1251	0.1230	0.1210	0.1190	0.1170
−1.0	0.1587	0.1562	0.1539	0.1515	0.1492	0.1469	0.1446	0.1423	0.1401	0.1379
−0.9	0.1841	0.1814	0.1788	0.1762	0.1736	0.1711	0.1685	0.1660	0.1635	0.1611
−0.8	0.2119	0.2090	0.2061	0.2033	0.2005	0.1977	0.1949	0.1922	0.1894	0.1867
−0.7	0.2420	0.2389	0.2358	0.2327	0.2296	0.2266	0.2236	0.2206	0.2177	0.2148
−0.6	0.2743	0.2709	0.2676	0.2643	0.2611	0.2578	0.2546	0.2514	0.2483	0.2451
−0.5	0.3085	0.3050	0.3015	0.2981	0.2946	0.2912	0.2877	0.2843	0.2810	0.2776
−0.4	0.3446	0.3409	0.3372	0.3336	0.3300	0.3264	0.3228	0.3192	0.3156	0.3121
−0.3	0.3821	0.3783	0.3745	0.3707	0.3669	0.3632	0.3594	0.3557	0.3520	0.3483
−0.2	0.4207	0.4168	0.4129	0.4090	0.4052	0.4013	0.3974	0.3936	0.3897	0.3859
−0.1	0.4602	0.4562	0.4522	0.4483	0.4443	0.4404	0.4364	0.4325	0.4286	0.4247
−0.0	0.5000	0.4960	0.4920	0.4880	0.4840	0.4801	0.4761	0.4721	0.4681	0.4641

Table III Standard Normal Distribution Cumulative Probabilities (*Continued*)

z	.00	.01	.02	.03	.04	.05	.06	.07	.08	.09
0.0	0.5000	0.5040	0.5080	0.5120	0.5160	0.5199	0.5239	0.5279	0.5319	0.5359
0.1	0.5398	0.5438	0.5478	0.5517	0.5557	0.5596	0.5636	0.5675	0.5714	0.5753
0.2	0.5793	0.5832	0.5871	0.5910	0.5948	0.5987	0.6026	0.6064	0.6103	0.6141
0.3	0.6179	0.6217	0.6255	0.6293	0.6331	0.6368	0.6406	0.6443	0.6480	0.6517
0.4	0.6554	0.6591	0.6628	0.6664	0.6700	0.6736	0.6772	0.6808	0.6844	0.6879
0.5	0.6915	0.6950	0.6985	0.7019	0.7054	0.7088	0.7123	0.7157	0.7190	0.7224
0.6	0.7257	0.7291	0.7324	0.7357	0.7389	0.7422	0.7454	0.7486	0.7517	0.7549
0.7	0.7580	0.7611	0.7642	0.7673	0.7704	0.7734	0.7764	0.7794	0.7823	0.7852
0.8	0.7881	0.7910	0.7939	0.7967	0.7995	0.8023	0.8051	0.8078	0.8106	0.8133
0.9	0.8159	0.8186	0.8212	0.8238	0.8264	0.8289	0.8315	0.8340	0.8365	0.8389
1.0	0.8413	0.8438	0.8461	0.8485	0.8508	0.8531	0.8554	0.8577	0.8599	0.8621
1.1	0.8643	0.8665	0.8686	0.8708	0.8729	0.8749	0.8770	0.8790	0.8810	0.8830
1.2	0.8849	0.8869	0.8888	0.8907	0.8925	0.8944	0.8962	0.8980	0.8997	0.9015
1.3	0.9032	0.9049	0.9066	0.9082	0.9099	0.9115	0.9131	0.9147	0.9162	0.9177
1.4	0.9192	0.9207	0.9222	0.9236	0.9251	0.9265	0.9279	0.9292	0.9306	0.9319
1.5	0.9332	0.9345	0.9357	0.9370	0.9382	0.9394	0.9406	0.9418	0.9429	0.9441
1.6	0.9452	0.9463	0.9474	0.9484	0.9495	0.9505	0.9515	0.9525	0.9535	0.9545
1.7	0.9554	0.9564	0.9573	0.9582	0.9591	0.9599	0.9608	0.9616	0.9625	0.9633
1.8	0.9641	0.9649	0.9656	0.9664	0.9671	0.9678	0.9686	0.9693	0.9699	0.9706
1.9	0.9713	0.9719	0.9726	0.9732	0.9738	0.9744	0.9750	0.9756	0.9761	0.9767
2.0	0.9772	0.9778	0.9783	0.9788	0.9793	0.9798	0.9803	0.9808	0.9812	0.9817
2.1	0.9821	0.9826	0.9830	0.9834	0.9838	0.9842	0.9846	0.9850	0.9854	0.9857
2.2	0.9861	0.9864	0.9868	0.9871	0.9875	0.9878	0.9881	0.9884	0.9887	0.9890
2.3	0.9893	0.9896	0.9898	0.9901	0.9904	0.9906	0.9909	0.9911	0.9913	0.9916
2.4	0.9918	0.9920	0.9922	0.9925	0.9927	0.9929	0.9931	0.9932	0.9934	0.9936
2.5	0.9938	0.9940	0.9941	0.9943	0.9945	0.9946	0.9948	0.9949	0.9951	0.9952
2.6	0.9953	0.9955	0.9956	0.9957	0.9959	0.9960	0.9961	0.9962	0.9963	0.9964
2.7	0.9965	0.9966	0.9967	0.9968	0.9969	0.9970	0.9971	0.9972	0.9973	0.9974
2.8	0.9974	0.9975	0.9976	0.9977	0.9977	0.9978	0.9979	0.9979	0.9980	0.9981
2.9	0.9981	0.9982	0.9982	0.9983	0.9984	0.9984	0.9985	0.9985	0.9986	0.9986
3.0	0.9987	0.9987	0.9987	0.9988	0.9988	0.9989	0.9989	0.9989	0.9990	0.9990
3.1	0.9990	0.9991	0.9991	0.9991	0.9992	0.9992	0.9992	0.9992	0.9993	0.9993
3.2	0.9993	0.9993	0.9994	0.9994	0.9994	0.9994	0.9994	0.9995	0.9995	0.9995
3.3	0.9995	0.9995	0.9995	0.9996	0.9996	0.9996	0.9996	0.9996	0.9996	0.9997
3.4	0.9997	0.9997	0.9997	0.9997	0.9997	0.9997	0.9997	0.9997	0.9997	0.9998

Special critical values: $P(Z \geq z_\alpha) = \alpha$

α	0.10	0.05	0.025	0.01	0.005	0.001	0.0005	0.0001	
z_α	1.2816	1.6449	1.9600	2.3263	2 .5758	3.0902	3.2905	3.7190	

α	0.00009	0.00008	0.00007	0.00006	0.00005	0.00004	0.00003	0.00002	0.00001
z_α	3.7455	3.7750	3.8082	3.8461	3.8906	3.9444	4.0128	4.1075	4.2649

Table IV Standardized Normal Scores

This table contains the standardized normal scores, z_i, for selected values of n.

i	10	20	25	30	40	50
1	−1.55	−1.87	−1.96	−2.04	−2.16	−2.24
2	−1.00	−1.40	−1.52	−1.61	−1.75	−1.85
3	−0.66	−1.13	−1.26	−1.36	−1.51	−1.62
4	−0.38	−0.92	−1.06	−1.18	−1.34	−1.46
5	−0.12	−0.74	−0.90	−1.02	−1.20	−1.33
6	0.12	−0.59	−0.76	−0.89	−1.08	−1.22
7	0.38	−0.45	−0.64	−0.78	−0.98	−1.12
8	0.66	−0.31	−0.52	−0.67	−0.88	−1.03
9	1.00	−0.19	−0.41	−0.57	−0.79	−0.95
10	1.55	−0.06	−0.30	−0.47	−0.71	−0.87
11		0.06	−0.20	−0.38	−0.63	−0.80
12		0.19	−0.10	−0.29	−0.56	−0.73
13		0.31	0.00	−0.21	−0.49	−0.67
14		0.45	0.10	−0.12	−0.42	−0.61
15		0.59	0.20	−0.04	−0.35	−0.55
16		0.74	0.30	0.04	−0.28	−0.49
17		0.92	0.41	0.12	−0.22	−0.44
18		1.13	0.52	0.21	−0.16	−0.38
19		1.40	0.64	0.29	−0.09	−0.33
20		1.87	0.76	0.38	−0.03	−0.28
21			0.90	0.47	0.03	−0.23
22			1.06	0.57	0.09	−0.18
23			1.26	0.67	0.16	−0.13
24			1.52	0.78	0.22	−0.07
25			1.96	0.89	0.28	−0.02
26				1.02	0.35	0.02
27				1.18	0.42	0.07
28				1.36	0.49	0.13
29				1.61	0.56	0.18
30				2.04	0.63	0.23
31					0.71	0.28
32					0.79	0.33
33					0.88	0.38
34					0.98	0.44
35					1.08	0.49
36					1.20	0.55
37					1.34	0.61
38					1.51	0.67
39					1.75	0.73
40					2.16	0.80
41						0.87
42						0.95
43						1.03
44						1.12
45						1.22
46						1.33
47						1.46
48						1.62
49						1.85
50						2.24

Table V Critical Values for the *t* Distribution

This table contains critical values associated with the *t* distribution, t_α, defined by the degrees of freedom and α.

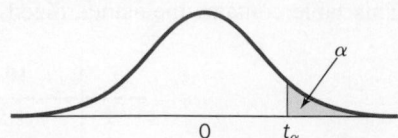

df	α								
	0.20	0.10	0.05	0.025	0.01	0.005	0.001	0.0005	0.0001
1	1.3764	3.0777	6.3138	12.7062	31.8205	63.6567	318.3088	636.6192	3183.0988
2	1.0607	1.8856	2.9200	4.3027	6.9646	9.9248	22.3271	31.5991	70.7001
3	0.9785	1.6377	2.3534	3.1824	4.5407	5.8409	10.2145	12.9240	22.2037
4	0.9410	1.5332	2.1318	2.7764	3.7469	4.6041	7.1732	8.6103	13.0337
5	0.9195	1.4759	2.0150	2.5706	3.3649	4.0321	5.8934	6.8688	9.6776
6	0.9057	1.4398	1.9432	2.4469	3.1427	3.7074	5.2076	5.9588	8.0248
7	0.8960	1.4149	1.8946	2.3646	2.9980	3.4995	4.7853	5.4079	7.0634
8	0.8889	1.3968	1.8595	2.3060	2.8965	3.3554	4.5008	5.0413	6.4420
9	0.8834	1.3830	1.8331	2.2622	2.8214	3.2498	4.2968	4.7809	6.0101
10	0.8791	1.3722	1.8125	2.2281	2.7638	3.1693	4.1437	4.5869	5.6938
11	0.8755	1.3634	1.7959	2.2010	2.7181	3.1058	4.0247	4.4370	5.4528
12	0.8726	1.3562	1.7823	2.1788	2.6810	3.0545	3.9296	4.3178	5.2633
13	0.8702	1.3502	1.7709	2.1604	2.6503	3.0123	3.8520	4.2208	5.1106
14	0.8681	1.3450	1.7613	2.1448	2.6245	2.9768	3.7874	4.1405	4.9850
15	0.8662	1.3406	1.7531	2.1314	2.6025	2.9467	3.7328	4.0728	4.8800
16	0.8647	1.3368	1.7459	2.1199	2.5835	2.9208	3.6862	4.0150	4.7909
17	0.8633	1.3334	1.7396	2.1098	2.5669	2.8982	3.6458	3.9651	4.7144
18	0.8620	1.3304	1.7341	2.1009	2.5524	2.8784	3.6105	3.9216	4.6480
19	0.8610	1.3277	1.7291	2.0930	2.5395	2.8609	3.5794	3.8834	4.5899
20	0.8600	1.3253	1.7247	2.0860	2.5280	2.8453	3.5518	3.8495	4.5385
21	0.8591	1.3232	1.7207	2.0796	2.5176	2.8314	3.5272	3.8193	4.4929
22	0.8583	1.3212	1.7171	2.0739	2.5083	2.8188	3.5050	3.7921	4.4520
23	0.8575	1.3195	1.7139	2.0687	2.4999	2.8073	3.4850	3.7676	4.4152
24	0.8569	1.3178	1.7109	2.0639	2.4922	2.7969	3.4668	3.7454	4.3819
25	0.8562	1.3163	1.7081	2.0595	2.4851	2.7874	3.4502	3.7251	4.3517
26	0.8557	1.3150	1.7056	2.0555	2.4786	2.7787	3.4350	3.7066	4.3240
27	0.8551	1.3137	1.7033	2.0518	2.4727	2.7707	3.4210	3.6896	4.2987
28	0.8546	1.3125	1.7011	2.0484	2.4671	2.7633	3.4082	3.6739	4.2754
29	0.8542	1.3114	1.6991	2.0452	2.4620	2.7564	3.3962	3.6594	4.2539
30	0.8538	1.3104	1.6973	2.0423	2.4573	2.7500	3.3852	3.6460	4.2340
40	0.8507	1.3031	1.6839	2.0211	2.4233	2.7045	3.3069	3.5510	4.0942
50	0.8489	1.2987	1.6759	2.0086	2.4033	2.6778	3.2614	3.4960	4.0140
60	0.8477	1.2958	1.6706	2.0003	2.3901	2.6603	3.2317	3.4602	3.9621
70	0.8468	1.2938	1.6669	1.9944	2.3808	2.6479	3.2108	3.4350	3.9257
80	0.8461	1.2922	1.6641	1.9901	2.3739	2.6387	3.1953	3.4163	3.8988
90	0.8456	1.2910	1.6620	1.9867	2.3685	2.6316	3.1833	3.4019	3.8780
100	0.8452	1.2901	1.6602	1.9840	2.3642	2.6259	3.1737	3.3905	3.8616
200	0.8434	1.2858	1.6525	1.9719	2.3451	2.6006	3.1315	3.3398	3.7891
500	0.8423	1.2832	1.6479	1.9647	2.3338	2.5857	3.1066	3.3101	3.7468
∞	0.8416	1.2816	1.6449	1.9600	2.3263	2.5758	3.0902	3.2905	3.7190

Table VI Critical Values for the Chi-Square Distribution

This table contains critical values associated with the chi-square distribution, χ^2_α, defined by the degrees of freedom and α.

df	0.9999	0.9995	0.999	0.995	0.99	0.975	0.95	0.90
1	0.0^7157	0.0^6393	0.0^5157	0.0^4393	0.0002	0.0010	0.0039	0.0158
2	0.0002	0.0010	0.0020	0.0100	0.0201	0.0506	0.1026	0.2107
3	0.0052	0.0153	0.0243	0.0717	0.1148	0.2158	0.3518	0.5844
4	0.0284	0.0639	0.0908	0.2070	0.2971	0.4844	0.7107	1.0636
5	0.0822	0.1581	0.2102	0.4117	0.5543	0.8312	1.1455	1.6103
6	0.1724	0.2994	0.3811	0.6757	0.8721	1.2373	1.6354	2.2041
7	0.3000	0.4849	0.5985	0.9893	1.2390	1.6899	2.1673	2.8331
8	0.4636	0.7104	0.8571	1.3444	1.6465	2.1797	2.7326	3.4895
9	0.6608	0.9717	1.1519	1.7349	2.0879	2.7004	3.3251	4.1682
10	0.8889	1.2650	1.4787	2.1559	2.5582	3.2470	3.9403	4.8652
11	1.1453	1.5868	1.8339	2.6032	3.0535	3.8157	4.5748	5.5778
12	1.4275	1.9344	2.2142	3.0738	3.5706	4.4038	5.2260	6.3038
13	1.7333	2.3051	2.6172	3.5650	4.1069	5.0088	5.8919	7.0415
14	2.0608	2.6967	3.0407	4.0747	4.6604	5.6287	6.5706	7.7895
15	2.4082	3.1075	3.4827	4.6009	5.2293	6.2621	7.2609	8.5468
16	2.7739	3.5358	3.9416	5.1422	5.8122	6.9077	7.9616	9.3122
17	3.1567	3.9802	4.4161	5.6972	6.4078	7.5642	8.6718	10.0852
18	3.5552	4.4394	4.9048	6.2648	7.0149	8.2307	9.3905	10.8649
19	3.9683	4.9123	5.4068	6.8440	7.6327	8.9065	10.1170	11.6509
20	4.3952	5.3981	5.9210	7.4338	8.2604	9.5908	10.8508	12.4426
21	4.8348	5.8957	6.4467	8.0337	8.8972	10.2829	11.5913	13.2396
22	5.2865	6.4045	6.9830	8.6427	9.5425	10.9823	12.3380	14.0415
23	5.7494	6.9237	7.5292	9.2604	10.1957	11.6886	13.0905	14.8480
24	6.2230	7.4527	8.0849	9.8862	10.8564	12.4012	13.8484	15.6587
25	6.7066	7.9910	8.6493	10.5197	11.5240	13.1197	14.6114	16.4734
26	7.1998	8.5379	9.2221	11.1602	12.1981	13.8439	15.3792	17.2919
27	7.7019	9.0932	9.8028	11.8076	12.8785	14.5734	16.1514	18.1139
28	8.2126	9.6563	10.3909	12.4613	13.5647	15.3079	16.9279	18.9392
29	8.7315	10.2268	10.9861	13.1211	14.2565	16.0471	17.7084	19.7677
30	9.2581	10.8044	11.5880	13.7867	14.9535	16.7908	18.4927	20.5992
31	9.7921	11.3887	12.1963	14.4578	15.6555	17.5387	19.2806	21.4336
32	10.3331	11.9794	12.8107	15.1340	16.3622	18.2908	20.0719	22.2706
33	10.8810	12.5763	13.4309	15.8153	17.0735	19.0467	20.8665	23.1102
34	11.4352	13.1791	14.0567	16.5013	17.7891	19.8063	21.6643	23.9523
35	11.9957	13.7875	14.6878	17.1918	18.5089	20.5694	22.4650	24.7967
36	12.5622	14.4012	15.3241	17.8867	19.2327	21.3359	23.2686	25.6433
37	13.1343	15.0202	15.9653	18.5858	19.9602	22.1056	24.0749	26.4921
38	13.7120	15.6441	16.6112	19.2889	20.6914	22.8785	24.8839	27.3430
39	14.2950	16.2729	17.2616	19.9959	21.4262	23.6543	25.6954	28.1958
40	14.8831	16.9062	17.9164	20.7065	22.1643	24.4330	26.5093	29.0505
50	21.0093	23.4610	24.6739	27.9907	29.7067	32.3574	34.7643	37.6886
60	27.4969	30.3405	31.7383	35.5345	37.4849	40.4817	43.1880	46.4589
70	34.2607	37.4674	39.0364	43.2752	45.4417	48.7576	51.7393	55.3289
80	41.2445	44.7910	46.5199	51.1719	53.5401	57.1532	60.3915	64.2778
90	48.4087	52.2758	54.1552	59.1963	61.7541	65.6466	69.1260	73.2911
100	55.7246	59.8957	61.9179	67.3276	70.0649	74.2219	77.9295	82.3581

Table VI Critical Values for the Chi-Square Distribution (*Continued*)

df	0.10	0.05	0.025	0.01	0.005	0.001	0.0005	0.0001
1	2.7055	3.8415	5.0239	6.6349	7.8794	10.8276	12.1157	15.1367
2	4.6052	5.9915	7.3778	9.2103	10.5966	13.8155	15.2018	18.4207
3	6.2514	7.8147	9.3484	11.3449	12.8382	16.2662	17.7300	21.1075
4	7.7794	9.4877	11.1433	13.2767	14.8603	18.4668	19.9974	23.5127
5	9.2364	11.0705	12.8325	15.0863	16.7496	20.5150	22.1053	25.7448
6	10.6446	12.5916	14.4494	16.8119	18.5476	22.4577	24.1028	27.8563
7	12.0170	14.0671	16.0128	18.4753	20.2777	24.3219	26.0178	29.8775
8	13.3616	15.5073	17.5345	20.0902	21.9550	26.1245	27.8680	31.8276
9	14.6837	16.9190	19.0228	21.6660	23.5894	27.8772	29.6658	33.7199
10	15.9872	18.3070	20.4832	23.2093	25.1882	29.5883	31.4198	35.5640
11	17.2750	19.6751	21.9200	24.7250	26.7568	31.2641	33.1366	37.3670
12	18.5493	21.0261	23.3367	26.2170	28.2995	32.9095	34.8213	39.1344
13	19.8119	22.3620	24.7356	27.6882	29.8195	34.5282	36.4778	40.8707
14	21.0641	23.6848	26.1189	29.1412	31.3193	36.1233	38.1094	42.5793
15	22.3071	24.9958	27.4884	30.5779	32.8013	37.6973	39.7188	44.2632
16	23.5418	26.2962	28.8454	31.9999	34.2672	39.2524	41.3081	45.9249
17	24.7690	27.5871	30.1910	33.4087	35.7185	40.7902	42.8792	47.5664
18	25.9894	28.8693	31.5264	34.8053	37.1565	42.3124	44.4338	49.1894
19	27.2036	30.1435	32.8523	36.1909	38.5823	43.8202	45.9731	50.7955
20	28.4120	31.4104	34.1696	37.5662	39.9968	45.3147	47.4985	52.3860
21	29.6151	32.6706	35.4789	38.9322	41.4011	46.7970	49.0108	53.9620
22	30.8133	33.9244	36.7807	40.2894	42.7957	48.2679	50.5111	55.5246
23	32.0069	35.1725	38.0756	41.6384	44.1813	49.7282	52.0002	57.0746
24	33.1962	36.4150	39.3641	42.9798	45.5585	51.1786	53.4788	58.6130
25	34.3816	37.6525	40.6465	44.3141	46.9279	52.6197	54.9475	60.1403
26	35.5632	38.8851	41.9232	45.6417	48.2899	54.0520	56.4069	61.6573
27	36.7412	40.1133	43.1945	46.9629	49.6449	55.4760	57.8576	63.1645
28	37.9159	41.3371	44.4608	48.2782	50.9934	56.8923	59.3000	64.6624
29	39.0875	42.5570	45.7223	49.5879	52.3356	58.3012	60.7346	66.1517
30	40.2560	43.7730	46.9792	50.8922	53.6720	59.7031	62.1619	67.6326
31	41.4217	44.9853	48.2319	52.1914	55.0027	61.0983	63.5820	69.1057
32	42.5847	46.1943	49.4804	53.4858	56.3281	62.4872	64.9955	70.5712
33	43.7452	47.3999	50.7251	54.7755	57.6484	63.8701	66.4025	72.0296
34	44.9032	48.6024	51.9660	56.0609	58.9639	65.2472	67.8035	73.4812
35	46.0588	49.8018	53.2033	57.3421	60.2748	66.6188	69.1986	74.9262
36	47.2122	50.9985	54.4373	58.6192	61.5812	67.9852	70.5881	76.3650
37	48.3634	52.1923	55.6680	59.8925	62.8833	69.3465	71.9722	77.7977
38	49.5126	53.3835	56.8955	61.1621	64.1814	70.7029	73.3512	79.2247
39	50.6598	54.5722	58.1201	62.4281	65.4756	72.0547	74.7253	80.6462
40	51.8051	55.7585	59.3417	63.6907	66.7660	73.4020	76.0946	82.0623
50	63.1671	67.5048	71.4202	76.1539	79.4900	86.6608	89.5605	95.9687
60	74.3970	79.0819	83.2977	88.3794	91.9517	99.6072	102.6948	109.5029
70	85.5270	90.5312	95.0232	100.4252	104.2149	112.3169	115.5776	122.7547
80	96.5782	101.8795	106.6286	112.3288	116.3211	124.8392	128.2613	135.7825
90	107.5650	113.1453	118.1359	124.1163	128.2989	137.2084	140.7823	148.6273
100	118.4980	124.3421	129.5612	135.8067	140.1695	149.4493	153.1670	161.3187

Table VII Critical Values for the F Distribution

This table contains critical values associated with the F distribution, F_α, defined by α and degrees of freedom v_1 and v_2.

$\alpha = 0.05$

v_2 \ v_1	1	2	3	4	5	6	7	8	9	10	15	20	30	40	50	60	100
1	161.45	199.50	215.71	224.58	230.16	233.99	236.77	238.88	240.54	241.88	245.95	248.01	250.10	251.14	251.77	252.20	253.04
2	18.51	19.00	19.16	19.25	19.30	19.33	19.35	19.37	19.38	19.40	19.43	19.45	19.46	19.47	19.48	19.48	19.49
3	10.13	9.55	9.28	9.12	9.01	8.94	8.89	8.85	8.81	8.79	8.70	8.66	8.62	8.59	8.58	8.57	8.55
4	7.71	6.94	6.59	6.39	6.26	6.16	6.09	6.04	6.00	5.96	5.86	5.80	5.75	5.72	5.70	5.69	5.66
5	6.61	5.79	5.41	5.19	5.05	4.95	4.88	4.82	4.77	4.74	4.62	4.56	4.50	4.46	4.44	4.43	4.41
6	5.99	5.14	4.76	4.53	4.39	4.28	4.21	4.15	4.10	4.06	3.94	3.87	3.81	3.77	3.75	3.74	3.71
7	5.59	4.74	4.35	4.12	3.97	3.87	3.79	3.73	3.68	3.64	3.51	3.44	3.38	3.34	3.32	3.30	3.27
8	5.32	4.46	4.07	3.84	3.69	3.58	3.50	3.44	3.39	3.35	3.22	3.15	3.08	3.04	3.02	3.01	2.97
9	5.12	4.26	3.86	3.63	3.48	3.37	3.29	3.23	3.18	3.14	3.01	2.94	2.86	2.83	2.80	2.79	2.76
10	4.96	4.10	3.71	3.48	3.33	3.22	3.14	3.07	3.02	2.98	2.85	2.77	2.70	2.66	2.64	2.62	2.59
11	4.84	3.98	3.59	3.36	3.20	3.09	3.01	2.95	2.90	2.85	2.72	2.65	2.57	2.53	2.51	2.49	2.46
12	4.75	3.89	3.49	3.26	3.11	3.00	2.91	2.85	2.80	2.75	2.62	2.54	2.47	2.43	2.40	2.38	2.35
13	4.67	3.81	3.41	3.18	3.03	2.92	2.83	2.77	2.71	2.67	2.53	2.46	2.38	2.34	2.31	2.30	2.26
14	4.60	3.74	3.34	3.11	2.96	2.85	2.76	2.70	2.65	2.60	2.46	2.39	2.31	2.27	2.24	2.22	2.19
15	4.54	3.68	3.29	3.06	2.90	2.79	2.71	2.64	2.59	2.54	2.40	2.33	2.25	2.20	2.18	2.16	2.12
16	4.49	3.63	3.24	3.01	2.85	2.74	2.66	2.59	2.54	2.49	2.35	2.28	2.19	2.15	2.12	2.11	2.07
17	4.45	3.59	3.20	2.96	2.81	2.70	2.61	2.55	2.49	2.45	2.31	2.23	2.15	2.10	2.08	2.06	2.02
18	4.41	3.55	3.16	2.93	2.77	2.66	2.58	2.51	2.46	2.41	2.27	2.19	2.11	2.06	2.04	2.02	1.98
19	4.38	3.52	3.13	2.90	2.74	2.63	2.54	2.48	2.42	2.38	2.23	2.16	2.07	2.03	2.00	1.98	1.94
20	4.35	3.49	3.10	2.87	2.71	2.60	2.51	2.45	2.39	2.35	2.20	2.12	2.04	1.99	1.97	1.95	1.91
21	4.32	3.47	3.07	2.84	2.68	2.57	2.49	2.42	2.37	2.32	2.18	2.10	2.01	1.96	1.94	1.92	1.88
22	4.30	3.44	3.05	2.82	2.66	2.55	2.46	2.40	2.34	2.30	2.15	2.07	1.98	1.94	1.91	1.89	1.85
23	4.28	3.42	3.03	2.80	2.64	2.53	2.44	2.37	2.32	2.27	2.13	2.05	1.96	1.91	1.88	1.86	1.82
24	4.26	3.40	3.01	2.78	2.62	2.51	2.42	2.36	2.30	2.25	2.11	2.03	1.94	1.89	1.86	1.84	1.80
25	4.24	3.39	2.99	2.76	2.60	2.49	2.40	2.34	2.28	2.24	2.09	2.01	1.92	1.87	1.84	1.82	1.78
30	4.17	3.32	2.92	2.69	2.53	2.42	2.33	2.27	2.21	2.16	2.01	1.93	1.84	1.79	1.76	1.74	1.70
40	4.08	3.23	2.84	2.61	2.45	2.34	2.25	2.18	2.12	2.08	1.92	1.84	1.74	1.69	1.66	1.64	1.59
50	4.03	3.18	2.79	2.56	2.40	2.29	2.20	2.13	2.07	2.03	1.87	1.78	1.69	1.63	1.60	1.58	1.52
60	4.00	3.15	2.76	2.53	2.37	2.25	2.17	2.10	2.04	1.99	1.84	1.75	1.65	1.59	1.56	1.53	1.48
100	3.94	3.09	2.70	2.46	2.31	2.19	2.10	2.03	1.97	1.93	1.77	1.68	1.57	1.52	1.48	1.45	1.39

Table VII Critical Values for the F Distribution (Continued)

α = 0.01

v_2	v_1 1	2	3	4	5	6	7	8	9	10	15	20	30	40	50	60	100
2	98.50	99.00	99.17	99.25	99.30	99.33	99.36	99.37	99.39	99.40	99.43	99.45	99.47	99.47	99.48	99.48	99.49
3	34.12	30.82	29.46	28.71	28.24	27.91	27.67	27.49	27.35	27.23	26.87	26.69	26.50	26.41	26.35	26.32	26.24
4	21.20	18.00	16.69	15.98	15.52	15.21	14.98	14.80	14.66	14.55	14.20	14.02	13.84	13.75	13.69	13.65	13.58
5	16.26	13.27	12.06	11.39	10.97	10.67	10.46	10.29	10.16	10.05	9.72	9.55	9.38	9.29	9.24	9.20	9.13
6	13.75	10.92	9.78	9.15	8.75	8.47	8.26	8.10	7.98	7.87	7.56	7.40	7.23	7.14	7.09	7.06	6.99
7	12.25	9.55	8.45	7.85	7.46	7.19	6.99	6.84	6.72	6.62	6.31	6.16	5.99	5.91	5.86	5.82	5.75
8	11.26	8.65	7.59	7.01	6.63	6.37	6.18	6.03	5.91	5.81	5.52	5.36	5.20	5.12	5.07	5.03	4.96
9	10.56	8.02	6.99	6.42	6.06	5.80	5.61	5.47	5.35	5.26	4.96	4.81	4.65	4.57	4.52	4.48	4.41
10	10.04	7.56	6.55	5.99	5.64	5.39	5.20	5.06	4.94	4.85	4.56	4.41	4.25	4.17	4.12	4.08	4.01
11	9.65	7.21	6.22	5.67	5.32	5.07	4.89	4.74	4.63	4.54	4.25	4.10	3.94	3.86	3.81	3.78	3.71
12	9.33	6.93	5.95	5.41	5.06	4.82	4.64	4.50	4.39	4.30	4.01	3.86	3.70	3.62	3.57	3.54	3.47
13	9.07	6.70	5.74	5.21	4.86	4.62	4.44	4.30	4.19	4.10	3.82	3.66	3.51	3.43	3.38	3.34	3.27
14	8.86	6.51	5.56	5.04	4.69	4.46	4.28	4.14	4.03	3.94	3.66	3.51	3.35	3.27	3.22	3.18	3.11
15	8.68	6.36	5.42	4.89	4.56	4.32	4.14	4.00	3.89	3.80	3.52	3.37	3.21	3.13	3.08	3.05	2.98
16	8.53	6.23	5.29	4.77	4.44	4.20	4.03	3.89	3.78	3.69	3.41	3.26	3.10	3.02	2.97	2.93	2.86
17	8.40	6.11	5.18	4.67	4.34	4.10	3.93	3.79	3.68	3.59	3.31	3.16	3.00	2.92	2.87	2.83	2.76
18	8.29	6.01	5.09	4.58	4.25	4.01	3.84	3.71	3.60	3.51	3.23	3.08	2.92	2.84	2.78	2.75	2.68
19	8.18	5.93	5.01	4.50	4.17	3.94	3.77	3.63	3.52	3.43	3.15	3.00	2.84	2.76	2.71	2.67	2.60
20	8.10	5.85	4.94	4.43	4.10	3.87	3.70	3.56	3.46	3.37	3.09	2.94	2.78	2.69	2.64	2.61	2.54
21	8.02	5.78	4.87	4.37	4.04	3.81	3.64	3.51	3.40	3.31	3.03	2.88	2.72	2.64	2.58	2.55	2.48
22	7.95	5.72	4.82	4.31	3.99	3.76	3.59	3.45	3.35	3.26	2.98	2.83	2.67	2.58	2.53	2.50	2.42
23	7.88	5.66	4.76	4.26	3.94	3.71	3.54	3.41	3.30	3.21	2.93	2.78	2.62	2.54	2.48	2.45	2.37
24	7.82	5.61	4.72	4.22	3.90	3.67	3.50	3.36	3.26	3.17	2.89	2.74	2.58	2.49	2.44	2.40	2.33
25	7.77	5.57	4.68	4.18	3.85	3.63	3.46	3.32	3.22	3.13	2.85	2.70	2.54	2.45	2.40	2.36	2.29
30	7.56	5.39	4.51	4.02	3.70	3.47	3.30	3.17	3.07	2.98	2.70	2.55	2.39	2.30	2.25	2.21	2.13
40	7.31	5.18	4.31	3.83	3.51	3.29	3.12	2.99	2.89	2.80	2.52	2.37	2.20	2.11	2.06	2.02	1.94
50	7.17	5.06	4.20	3.72	3.41	3.19	3.02	2.89	2.78	2.70	2.42	2.27	2.10	2.01	1.95	1.91	1.82
60	7.08	4.98	4.13	3.65	3.34	3.12	2.95	2.82	2.72	2.63	2.35	2.20	2.03	1.94	1.88	1.84	1.75
100	6.90	4.82	3.98	3.51	3.21	2.99	2.82	2.69	2.59	2.50	2.22	2.07	1.89	1.80	1.74	1.69	1.60

Table VII Critical Values for the F Distribution (Continued)

α = 0.001

v_2	\multicolumn{17}{c}{v_1}
	1	2	3	4	5	6	7	8	9	10	15	20	30	40	50	60	100
2	998.50	999.00	999.17	999.25	999.30	999.33	999.36	999.37	999.39	999.40	999.43	999.45	999.47	999.47	999.48	999.48	999.49
3	167.03	148.50	141.11	137.10	134.58	132.85	131.58	130.62	129.86	129.25	127.37	126.42	125.45	124.96	124.66	124.47	124.07
4	74.14	61.25	56.18	53.44	51.71	50.53	49.66	49.00	48.47	48.05	46.76	46.10	45.43	45.09	44.88	44.75	44.47
5	47.18	37.12	33.20	31.09	29.75	28.83	28.16	27.65	27.24	26.92	25.91	25.39	24.87	24.60	24.44	24.33	24.12
6	35.51	27.00	23.70	21.92	20.80	20.03	19.46	19.03	18.69	18.41	17.56	17.12	16.67	16.44	16.31	16.21	16.03
7	29.25	21.69	18.77	17.20	16.21	15.52	15.02	14.63	14.33	14.08	13.32	12.93	12.53	12.33	12.20	12.12	11.95
8	25.41	18.49	15.83	14.39	13.48	12.86	12.40	12.05	11.77	11.54	10.84	10.48	10.11	9.92	9.80	9.73	9.57
9	22.86	16.39	13.90	12.56	11.71	11.13	10.70	10.37	10.11	9.89	9.24	8.90	8.55	8.37	8.26	8.19	8.04
10	21.04	14.91	12.55	11.28	10.48	9.93	9.52	9.20	8.96	8.75	8.13	7.80	7.47	7.30	7.19	7.12	6.98
11	19.69	13.81	11.56	10.35	9.58	9.05	8.66	8.35	8.12	7.92	7.32	7.01	6.68	6.52	6.42	6.35	6.21
12	18.64	12.97	10.80	9.63	8.89	8.38	8.00	7.71	7.48	7.29	6.71	6.40	6.09	5.93	5.83	5.76	5.63
13	17.82	12.31	10.21	9.07	8.35	7.86	7.49	7.21	6.98	6.80	6.23	5.93	5.63	5.47	5.37	5.30	5.17
14	17.14	11.78	9.73	8.62	7.92	7.44	7.08	6.80	6.58	6.40	5.85	5.56	5.25	5.10	5.00	4.94	4.81
15	16.59	11.34	9.34	8.25	7.57	7.09	6.74	6.47	6.26	6.08	5.54	5.25	4.95	4.80	4.70	4.64	4.51
16	16.12	10.97	9.01	7.94	7.27	6.80	6.46	6.19	5.98	5.81	5.27	4.99	4.70	4.54	4.45	4.39	4.26
17	15.72	10.66	8.73	7.68	7.02	6.56	6.22	5.96	5.75	5.58	5.05	4.78	4.48	4.33	4.24	4.18	4.05
18	15.38	10.39	8.49	7.46	6.81	6.35	6.02	5.76	5.56	5.39	4.87	4.59	4.30	4.15	4.06	4.00	3.87
19	15.08	10.16	8.28	7.27	6.62	6.18	5.85	5.59	5.39	5.22	4.70	4.43	4.14	3.99	3.90	3.84	3.71
20	14.82	9.95	8.10	7.10	6.46	6.02	5.69	5.44	5.24	5.08	4.56	4.29	4.00	3.86	3.77	3.70	3.58
21	14.59	9.77	7.94	6.95	6.32	5.88	5.56	5.31	5.11	4.95	4.44	4.17	3.88	3.74	3.64	3.58	3.46
22	14.38	9.61	7.80	6.81	6.19	5.76	5.44	5.19	4.99	4.83	4.33	4.06	3.78	3.63	3.54	3.48	3.35
23	14.20	9.47	7.67	6.70	6.08	5.65	5.33	5.09	4.89	4.73	4.23	3.96	3.68	3.53	3.44	3.38	3.25
24	14.03	9.34	7.55	6.59	5.98	5.55	5.23	4.99	4.80	4.64	4.14	3.87	3.59	3.45	3.36	3.29	3.17
25	13.88	9.22	7.45	6.49	5.89	5.46	5.15	4.91	4.71	4.56	4.06	3.79	3.52	3.37	3.28	3.22	3.09
30	13.29	8.77	7.05	6.12	5.53	5.12	4.82	4.58	4.39	4.24	3.75	3.49	3.22	3.07	2.98	2.92	2.79
40	12.61	8.25	6.59	5.70	5.13	4.73	4.44	4.21	4.02	3.87	3.40	3.14	2.87	2.73	2.64	2.57	2.44
50	12.22	7.96	6.34	5.46	4.90	4.51	4.22	4.00	3.82	3.67	3.20	2.95	2.68	2.53	2.44	2.38	2.25
60	11.97	7.77	6.17	5.31	4.76	4.37	4.09	3.86	3.69	3.54	3.08	2.83	2.55	2.41	2.32	2.25	2.12
100	11.50	7.41	5.86	5.02	4.48	4.11	3.83	3.61	3.44	3.30	2.84	2.59	2.32	2.17	2.08	2.01	1.87

Table VIII Critical Values for the Studentized Range Distribution

This table contains critical values associated with the Studentized Range Distribution, Q_α, defined by α, and degrees of freedom m and v where m is the number of degrees of freedom in the numerator (the number of treatment groups) and v is the number of degrees of freedom in the denominator.

$\alpha = 0.05$

v \ m	2	3	4	5	6	7	8	9	10	11	12	13	14	15	16	17	18	19	20
1	17.97	26.98	32.82	37.08	40.41	43.12	45.40	47.36	49.07	50.59	51.96	53.20	54.33	55.36	56.32	57.22	58.04	58.83	59.56
2	6.085	8.331	9.798	10.88	11.74	12.44	13.03	13.54	13.99	14.39	14.75	15.08	15.38	15.65	15.91	16.14	16.37	16.57	16.77
3	4.501	5.910	6.825	7.502	8.037	8.478	8.853	9.177	9.462	9.717	9.946	10.15	10.35	10.53	10.69	10.84	10.98	11.11	11.24
4	3.927	5.040	5.757	6.287	6.707	7.053	7.347	7.602	7.826	8.027	8.208	8.373	8.525	8.664	8.794	8.914	9.028	9.134	9.233
5	3.635	4.602	5.218	5.673	6.033	6.330	6.582	6.802	6.995	7.168	7.324	7.466	7.596	7.717	7.828	7.932	8.030	8.122	8.208
6	3.461	4.339	4.896	5.305	5.628	5.895	6.122	6.319	6.493	6.649	6.789	6.917	7.034	7.143	7.244	7.338	7.426	7.508	7.587
7	3.344	4.165	4.681	5.060	5.359	5.606	5.815	5.998	6.158	6.302	6.431	6.550	6.658	6.759	6.852	6.939	7.020	7.097	7.170
8	3.261	4.041	4.529	4.886	5.167	5.399	5.597	5.767	5.918	6.054	6.175	6.287	6.389	6.483	6.571	6.653	6.729	6.802	6.870
9	3.199	3.949	4.415	4.756	5.024	5.244	5.432	5.595	5.739	5.867	5.983	6.089	6.186	6.276	6.359	6.437	6.510	6.579	6.644
10	3.151	3.877	4.327	4.654	4.912	5.124	5.305	5.461	5.599	5.722	5.833	5.935	6.028	6.114	6.194	6.269	6.339	6.405	6.467
11	3.113	3.820	4.256	4.574	4.823	5.028	5.202	5.353	5.487	5.605	5.713	5.811	5.901	5.984	6.062	6.134	6.202	6.265	6.326
12	3.082	3.773	4.199	4.508	4.751	4.950	5.119	5.265	5.395	5.511	5.615	5.710	5.798	5.878	5.953	6.023	6.089	6.151	6.209
13	3.055	3.735	4.151	4.453	4.690	4.885	5.049	5.192	5.318	5.431	5.533	5.625	5.711	5.789	5.862	5.931	5.995	6.055	6.112
14	3.033	3.702	4.111	4.407	4.639	4.829	4.990	5.131	5.254	5.364	5.463	5.554	5.637	5.714	5.786	5.852	5.915	5.974	6.029
15	3.014	3.674	4.076	4.367	4.595	4.782	4.940	5.077	5.198	5.306	5.404	5.493	5.574	5.649	5.720	5.785	5.846	5.904	5.958
16	2.998	3.649	4.046	4.333	4.557	4.741	4.897	5.031	5.150	5.256	5.352	5.439	5.520	5.593	5.662	5.727	5.786	5.843	5.897
17	2.984	3.628	4.020	4.303	4.524	4.705	4.858	4.991	5.108	5.212	5.307	5.392	5.471	5.544	5.612	5.675	5.734	5.790	5.842
18	2.971	3.609	3.997	4.277	4.495	4.673	4.824	4.956	5.071	5.174	5.267	5.352	5.429	5.501	5.568	5.630	5.688	5.743	5.794
19	2.960	3.593	3.977	4.253	4.469	4.645	4.794	4.924	5.038	5.140	5.231	5.315	5.391	5.462	5.528	5.589	5.647	5.701	5.752
20	2.950	3.578	3.958	4.232	4.445	4.620	4.768	4.896	5.008	5.108	5.199	5.282	5.357	5.427	5.493	5.553	5.610	5.663	5.714
24	2.919	3.532	3.901	4.166	4.373	4.541	4.684	4.807	4.915	5.012	5.099	5.179	5.251	5.319	5.381	5.439	5.494	5.545	5.594
30	2.888	3.486	3.845	4.102	4.302	4.464	4.602	4.720	4.824	4.917	5.001	5.077	5.147	5.211	5.271	5.327	5.379	5.429	5.475
40	2.858	3.442	3.791	4.039	4.232	4.389	4.521	4.635	4.735	4.824	4.904	4.977	5.044	5.106	5.163	5.216	5.266	5.313	5.358
60	2.829	3.399	3.737	3.977	4.163	4.314	4.441	4.550	4.646	4.732	4.808	4.878	4.942	5.001	5.056	5.107	5.154	5.199	5.241
120	2.800	3.356	3.685	3.917	4.096	4.241	4.363	4.468	4.560	4.641	4.714	4.781	4.842	4.898	4.950	4.998	5.044	5.086	5.126
∞	2.772	3.314	3.633	3.858	4.030	4.170	4.286	4.387	4.474	4.552	4.622	4.685	4.743	4.796	4.845	4.891	4.934	4.974	5.012

Table VIII Critical Values for the Studentized Range Distribution (Continued)

$\alpha = 0.01$

v										m									
	2	3	4	5	6	7	8	9	10	11	12	13	14	15	16	17	18	19	20
1	90.03	135.0	164.3	185.6	202.2	215.8	227.2	237.0	245.6	253.2	260.0	266.2	271.8	277.0	281.8	286.3	290.4	294.3	298.0
2	14.04	19.02	22.29	24.72	26.63	28.20	29.53	30.68	31.69	32.59	33.40	34.13	34.81	35.43	36.00	36.53	37.03	37.50	37.95
3	8.261	10.62	12.17	13.33	14.24	15.00	15.64	16.20	16.69	17.13	17.53	17.89	18.22	18.52	18.81	19.07	19.32	19.55	19.77
4	6.512	8.120	9.173	9.958	10.58	11.10	11.55	11.93	12.27	12.57	12.84	13.09	13.32	13.53	13.73	13.91	14.08	14.24	14.40
5	5.702	6.976	7.804	8.421	8.913	9.321	9.669	9.972	10.24	10.48	10.70	10.89	11.08	11.24	11.40	11.55	11.68	11.81	11.93
6	5.243	6.331	7.033	7.556	7.973	8.318	8.613	8.869	9.097	9.301	9.485	9.653	9.808	9.951	10.08	10.21	10.32	10.43	10.54
7	4.949	5.919	6.543	7.005	7.373	7.679	7.939	8.166	8.368	8.548	8.711	8.860	8.997	9.124	9.242	9.353	9.456	9.554	9.646
8	4.746	5.635	6.204	6.625	6.960	7.237	7.474	7.681	7.863	8.027	8.176	8.312	8.436	8.552	8.659	8.760	8.854	8.943	9.027
9	4.596	5.428	5.957	6.348	6.658	6.915	7.134	7.325	7.495	7.647	7.784	7.910	8.025	8.132	8.232	8.325	8.412	8.495	8.573
10	4.482	5.270	5.769	6.136	6.428	6.669	6.875	7.055	7.213	7.356	7.485	7.603	7.712	7.812	7.906	7.993	8.076	8.153	8.226
11	4.392	5.146	5.621	5.970	6.247	6.476	6.672	6.842	6.992	7.128	7.250	7.362	7.465	7.560	7.649	7.732	7.809	7.883	7.952
12	4.320	5.046	5.502	5.836	6.101	6.321	6.507	6.670	6.814	6.943	7.060	7.167	7.265	7.356	7.441	7.520	7.594	7.665	7.731
13	4.260	4.964	5.404	5.727	5.981	6.192	6.372	6.528	6.667	6.791	6.903	7.006	7.101	7.188	7.269	7.345	7.417	7.485	7.548
14	4.210	4.895	5.322	5.634	5.881	6.085	6.258	6.409	6.543	6.664	6.772	6.871	6.962	7.047	7.126	7.199	7.268	7.333	7.395
15	4.168	4.836	5.252	5.556	5.796	5.994	6.162	6.309	6.439	6.555	6.660	6.757	6.845	6.927	7.003	7.074	7.142	7.204	7.264
16	4.131	4.786	5.192	5.489	5.722	5.915	6.079	6.222	6.349	6.462	6.564	6.658	6.744	6.823	6.898	6.967	7.032	7.093	7.152
17	4.099	4.742	5.140	5.430	5.659	5.847	6.007	6.147	6.270	6.381	6.480	6.572	6.656	6.734	6.806	6.873	6.937	6.997	7.053
18	4.071	4.703	5.094	5.379	5.603	5.788	5.944	6.081	6.201	6.310	6.407	6.497	6.579	6.655	6.725	6.792	6.854	6.912	6.968
19	4.046	4.670	5.054	5.334	5.554	5.735	5.889	6.022	6.141	6.247	6.342	6.430	6.510	6.585	6.654	6.719	6.780	6.837	6.891
20	4.024	4.639	5.018	5.294	5.510	5.688	5.839	5.970	6.087	6.191	6.285	6.371	6.450	6.523	6.591	6.654	6.714	6.771	6.823
24	3.956	4.546	4.907	5.168	5.374	5.542	5.685	5.809	5.919	6.017	6.106	6.186	6.261	6.330	6.394	6.453	6.510	6.563	6.612
30	3.889	4.455	4.799	5.048	5.242	5.401	5.536	5.653	5.756	5.849	5.932	6.008	6.078	6.143	6.203	6.259	6.311	6.361	6.407
40	3.825	4.367	4.696	4.931	5.114	5.265	5.392	5.502	5.599	5.686	5.764	5.835	5.900	5.961	6.017	6.069	6.119	6.165	6.209
60	3.762	4.282	4.595	4.818	4.991	5.133	5.253	5.356	5.447	5.528	5.601	5.667	5.728	5.785	5.837	5.886	5.931	5.974	6.015
120	3.702	4.200	4.497	4.709	4.872	5.005	5.118	5.214	5.299	5.375	5.443	5.505	5.562	5.614	5.662	5.708	5.750	5.790	5.827
∞	3.643	4.120	4.403	4.603	4.757	4.882	4.987	5.078	5.157	5.227	5.290	5.348	5.400	5.448	5.493	5.535	5.574	5.611	5.645

Table VIII Critical Values for the Studentized Range Distribution (*Continued*)

$\alpha = 0.001$

v	2	3	4	5	6	7	8	9	10	11	12	13	14	15	16	17	18	19	20
1	900.3	1351.	1643.	1856.	2022.	2158.	2272.	2370.	2455.	2532.	2600.	2662.	2718.	2770.	2818.	2863.	2904.	2943.	2980.
2	44.69	60.42	70.77	78.43	84.49	89.46	93.67	97.30	100.5	103.3	105.9	108.2	110.4	112.3	114.2	115.9	117.4	118.9	120.3
3	18.28	23.32	26.65	29.13	31.11	32.74	34.12	35.33	36.39	37.34	38.20	38.98	39.69	40.35	40.97	41.54	42.07	42.58	43.05
4	12.18	14.99	16.84	18.23	19.34	20.26	21.04	21.73	22.33	22.87	23.36	23.81	24.21	24.59	24.94	25.27	25.58	25.87	26.14
5	9.714	11.67	12.96	13.93	14.71	15.35	15.90	16.38	16.81	17.18	17.53	17.85	18.13	18.41	18.66	18.89	19.10	19.31	19.51
6	8.427	9.960	10.97	11.72	12.32	12.83	13.26	13.63	13.97	14.27	14.54	14.79	15.01	15.22	15.42	15.60	15.78	15.94	16.09
7	7.648	8.930	9.768	10.40	10.90	11.32	11.68	11.99	12.27	12.52	12.74	12.95	13.14	13.32	13.48	13.64	13.78	13.92	14.04
8	7.130	8.250	8.978	9.522	9.958	10.32	10.64	10.91	11.15	11.36	11.56	11.74	11.91	12.06	12.21	12.34	12.47	12.59	12.70
9	6.762	7.768	8.419	8.906	9.295	9.619	9.897	10.14	10.36	10.55	10.73	10.89	11.03	11.18	11.30	11.42	11.54	11.64	11.75
10	6.487	7.411	8.006	8.450	8.804	9.099	9.352	9.573	9.769	9.946	10.11	10.25	10.39	10.52	10.64	10.75	10.85	10.95	11.03
11	6.275	7.136	7.687	8.098	8.426	8.699	8.933	9.138	9.319	9.482	9.630	9.766	9.892	10.01	10.12	10.22	10.31	10.41	10.49
12	6.106	6.917	7.436	7.821	8.127	8.383	8.601	8.793	8.962	9.115	9.254	9.381	9.498	9.606	9.707	9.802	9.891	9.975	10.06
13	5.970	6.740	7.231	7.595	7.885	8.126	8.333	8.513	8.673	8.817	8.948	9.068	9.178	9.281	9.376	9.466	9.550	9.629	9.704
14	5.856	6.594	7.062	7.409	7.685	7.915	8.110	8.282	8.434	8.571	8.696	8.809	8.914	9.012	9.103	9.188	9.267	9.343	9.414
15	5.760	6.470	6.920	7.252	7.517	7.736	7.925	8.088	8.234	8.365	8.483	8.592	8.693	8.786	8.872	8.954	9.030	9.102	9.170
16	5.678	6.365	6.799	7.119	7.374	7.585	7.766	7.923	8.063	8.189	8.303	8.407	8.504	8.593	8.676	8.755	8.828	8.897	8.963
17	5.608	6.275	6.695	7.005	7.250	7.454	7.629	7.781	7.916	8.037	8.148	8.248	8.342	8.427	8.508	8.583	8.654	8.720	8.784
18	5.546	6.196	6.604	6.905	7.143	7.341	7.510	7.657	7.788	7.906	8.012	8.110	8.199	8.283	8.361	8.434	8.502	8.567	8.628
19	5.492	6.127	6.525	6.817	7.049	7.242	7.405	7.549	7.676	7.790	7.893	7.988	8.075	8.156	8.232	8.303	8.369	8.432	8.491
20	5.444	6.065	6.454	6.740	6.966	7.154	7.313	7.453	7.577	7.688	7.788	7.880	7.966	8.044	8.118	8.186	8.251	8.312	8.370
24	5.297	5.877	6.238	6.503	6.712	6.884	7.031	7.159	7.272	7.374	7.467	7.551	7.629	7.701	7.768	7.831	7.890	7.946	7.999
30	5.156	5.698	6.033	6.278	6.470	6.628	6.763	6.880	6.984	7.077	7.162	7.239	7.310	7.375	7.437	7.494	7.548	7.599	7.647
40	5.022	5.528	5.838	6.063	6.240	6.386	6.509	6.616	6.711	6.796	6.872	6.942	7.007	7.067	7.122	7.174	7.223	7.269	7.312
60	4.894	5.365	5.653	5.860	6.022	6.155	6.268	6.366	6.451	6.528	6.598	6.661	6.720	6.774	6.824	6.871	6.914	6.956	6.995
120	4.771	5.211	5.476	5.667	5.815	5.937	6.039	6.128	6.206	6.276	6.339	6.396	6.448	6.496	6.542	6.583	6.623	6.660	6.695
∞	4.654	5.063	5.309	5.484	5.619	5.730	5.823	5.903	5.973	6.036	6.092	6.144	6.191	6.234	6.274	6.312	6.347	6.380	6.411

Table IX Critical Values for the Wilcoxon Signed-Rank Statistic

This table contains critical values and probabilities for the Wilcoxon signed-rank statistic T_+: n is the sample size, c_1 and c_2 are defined by $P(T_+ \leq c_1) = \alpha$ and $P(T_+ \geq c_2) = \alpha$.

n	c_1	c_2	α
1	0	1	0.5000
2	0	3	0.2500
3	0	6	0.1250
4	0	10	0.0625
	1	9	0.1250
5	0	15	0.0313
	1	14	0.0625
	2	13	0.0938
	3	12	0.1563
6	0	21	0.0156
	1	20	0.0313
	2	19	0.0469
	3	18	0.0781
	4	17	0.1094
	5	16	0.1563
7	0	28	0.0078
	1	27	0.0156
	2	26	0.0234
	3	25	0.0391
	4	24	0.0547
	5	23	0.0781
	6	22	0.1094
	7	21	0.1484
8	0	36	0.0039
	1	35	0.0078
	2	34	0.0117
	3	33	0.0195
	4	32	0.0273
	5	31	0.0391
	6	30	0.0547
	7	29	0.0742
	8	28	0.0977
	9	27	0.1250
9	0	45	0.0020
	1	44	0.0039
	2	43	0.0059
	3	42	0.0098
	4	41	0.0137
	5	40	0.0195
	6	39	0.0273
	7	38	0.0371
	8	37	0.0488
	9	36	0.0645
	10	35	0.0820
	11	34	0.1016
	12	33	0.1250

n	c_1	c_2	α
10	0	55	0.0010
	1	54	0.0020
	2	53	0.0037
	3	52	0.0049
	4	51	0.0068
	5	50	0.0098
	6	49	0.0137
	7	48	0.0186
	8	47	0.0244
	9	46	0.0322
	10	45	0.0420
	11	44	0.0527
	12	43	0.0654
	13	42	0.0801
	14	41	0.0967
	15	40	0.1162
	16	39	0.1377
11	0	66	0.0005
	1	65	0.0010
	2	64	0.0015
	3	63	0.0024
	4	62	0.0034
	5	61	0.0049
	6	60	0.0068
	7	59	0.0093
	8	58	0.0122
	9	57	0.0161
	10	56	0.0210
	11	55	0.0269
	12	54	0.0337
	13	53	0.0415
	14	52	0.0508
	15	51	0.0615
	16	50	0.0737
	17	49	0.0874
	18	48	0.1030
	19	47	0.1201
	20	46	0.1392

n	c_1	c_2	α
12	0	78	0.0002
	1	77	0.0005
	2	76	0.0007
	3	75	0.0012
	4	74	0.0017
	5	73	0.0024
	6	72	0.0034
	7	71	0.0046
	8	70	0.0061
	9	69	0.0081
	10	68	0.0105
	11	67	0.0134
	12	66	0.0171
	13	65	0.0212
	14	64	0.0261
	15	63	0.0320
	16	62	0.0386
	17	61	0.0461
	18	60	0.0549
	19	59	0.0647
	20	58	0.0757
	21	57	0.0881
	22	56	0.1018
	23	55	0.1167
	24	54	0.1331

n	c_1	c_2	α
13	0	91	0.0001
	1	90	0.0002
	2	89	0.0004
	3	88	0.0006
	4	87	0.0009
	5	86	0.0012
	6	85	0.0017
	7	84	0.0023
	8	83	0.0031
	9	82	0.0040
	10	81	0.0052
	11	80	0.0067
	12	79	0.0085
	13	78	0.0107
	14	77	0.0133
	15	76	0.0164
	16	75	0.0199
	17	74	0.0239
	18	73	0.0287
	19	72	0.0341
	20	71	0.0402
	21	70	0.0471
	22	69	0.0549
	23	68	0.0636
	24	67	0.0732
	25	66	0.0839
	26	65	0.0955
	27	64	0.1082
	28	63	0.1219
	29	62	0.1367

n	c_1	c_2	α
14	0	105	0.0001
	1	104	0.0001
	2	103	0.0002
	3	102	0.0003
	4	101	0.0004
	5	100	0.0006
	6	99	0.0009
	7	98	0.0012
	8	97	0.0015
	9	96	0.0020
	10	95	0.0026
	11	94	0.0034
	12	93	0.0043
	13	92	0.0054
	14	91	0.0067
	15	90	0.0083
	16	89	0.0101
	17	88	0.0123
	18	87	0.0148
	19	86	0.0176
	20	85	0.0209
	21	84	0.0247
	22	83	0.0290
	23	82	0.0338
	24	81	0.0392
	25	80	0.0453
	26	79	0.0520
	27	78	0.0594
	28	77	0.0676
	29	76	0.0765
	30	75	0.0863
	31	74	0.0969
	32	73	0.1083
	33	72	0.1206
	34	71	0.1338
	35	70	0.1479

Table IX Critical Values for the Wilcoxon Signed-Rank Statistic (*Continued*)

n	c_1	c_2	α	n	c_1	c_2	α	n	c_1	c_2	α	n	c_1	c_2	α
15	0	120	0.0000	16	0	136	0.0000	17	0	153	0.0000	18	0	171	0.0000
	1	119	0.0001		1	135	0.0000		1	152	0.0000		1	170	0.0000
	2	118	0.0001		2	134	0.0000		2	151	0.0000		2	169	0.0000
	3	117	0.0002		3	133	0.0001		3	150	0.0000		3	168	0.0000
	4	116	0.0002		4	132	0.0001		4	149	0.0001		4	167	0.0000
	5	115	0.0003		5	131	0.0002		5	148	0.0001		5	166	0.0000
	6	114	0.0004		6	130	0.0002		6	147	0.0001		6	165	0.0001
	7	113	0.0006		7	129	0.0003		7	146	0.0001		7	164	0.0001
	8	112	0.0008		8	128	0.0004		8	145	0.0002		8	163	0.0001
	9	111	0.0010		9	127	0.0005		9	144	0.0003		9	162	0.0001
	10	110	0.0013		10	126	0.0007		10	143	0.0003		10	161	0.0002
	11	109	0.0017		11	125	0.0008		11	142	0.0004		11	160	0.0002
	12	108	0.0021		12	124	0.0011		12	141	0.0005		12	159	0.0003
	13	107	0.0027		13	123	0.0013		13	140	0.0007		13	158	0.0003
	14	106	0.0034		14	122	0.0017		14	139	0.0008		14	157	0.0004
	15	105	0.0042		15	121	0.0021		15	138	0.0010		15	156	0.0005
	16	104	0.0051		16	120	0.0026		16	137	0.0013		16	155	0.0006
	17	103	0.0062		17	119	0.0031		17	136	0.0016		17	154	0.0008
	18	102	0.0075		18	118	0.0038		18	135	0.0019		18	153	0.0010
	19	101	0.0090		19	117	0.0046		19	134	0.0023		19	152	0.0012
	20	100	0.0108		20	116	0.0055		20	133	0.0028		20	151	0.0014
	21	99	0.0128		21	115	0.0065		21	132	0.0033		21	150	0.0017
	22	98	0.0151		22	114	0.0078		22	131	0.0040		22	149	0.0020
	23	97	0.0177		23	113	0.0091		23	130	0.0047		23	148	0.0024
	24	96	0.0206		24	112	0.0107		24	129	0.0055		24	147	0.0028
	25	95	0.0240		25	111	0.0125		25	128	0.0064		25	146	0.0033
	26	94	0.0277		26	110	0.0145		26	127	0.0075		26	145	0.0038
	27	93	0.0319		27	109	0.0168		27	126	0.0087		27	144	0.0045
	28	92	0.0365		28	108	0.0193		28	125	0.0101		28	143	0.0052
	29	91	0.0416		29	107	0.0222		29	124	0.0116		29	142	0.0060
	30	90	0.0473		30	106	0.0253		30	123	0.0133		30	141	0.0069
	31	89	0.0535		31	105	0.0288		31	122	0.0153		31	140	0.0080
	32	88	0.0603		32	104	0.0327		32	121	0.0174		32	139	0.0091
	33	87	0.0677		33	103	0.0370		33	120	0.0198		33	138	0.0104
	34	86	0.0757		34	102	0.0416		34	119	0.0224		34	137	0.0118
	35	85	0.0844		35	101	0.0467		35	118	0.0253		35	136	0.0134
	36	84	0.0938		36	100	0.0523		36	117	0.0284		36	135	0.0152
	37	83	0.1039		37	99	0.0583		37	116	0.0319		37	134	0.0171
					38	98	0.0649		38	115	0.0357		38	133	0.0192
					39	97	0.0719		39	114	0.0398		39	132	0.0216
					40	96	0.0795		40	113	0.0443		40	131	0.0241
					41	95	0.0877		41	112	0.0492		41	130	0.0269
					42	94	0.0964		42	111	0.0544		42	129	0.0300
					43	93	0.1057		43	110	0.0601		43	128	0.0333
									44	109	0.0662		44	127	0.0368
									45	108	0.0727		45	126	0.0407
									46	107	0.0797		46	125	0.0449
									47	106	0.0871		47	124	0.0494
									48	105	0.0950		48	123	0.0542
									49	104	0.1034		49	122	0.0594
													50	121	0.0649
													51	120	0.0708
													52	119	0.0770
													53	118	0.0837
													54	117	0.0907
													55	116	0.0982
													56	115	0.1061

Table IX Critical Values for the Wilcoxon Signed-Rank Statistic (*Continued*)

n	c_1	c_2	α	n	c_1	c_2	α	n	c_1	c_2	α	n	c_1	c_2	α
19	0	190	0.0000	19	41	149	0.0145	20	0	210	0.0000	20	41	169	0.0077
	1	189	0.0000		42	148	0.0162		1	209	0.0000		42	168	0.0086
	2	188	0.0000		43	147	0.0180		2	208	0.0000		43	167	0.0096
	3	187	0.0000		44	146	0.0201		3	207	0.0000		44	166	0.0107
	4	186	0.0000		45	145	0.0223		4	206	0.0000		45	165	0.0120
	5	185	0.0000		46	144	0.0247		5	205	0.0000		46	164	0.0133
	6	184	0.0000		47	143	0.0273		6	204	0.0000		47	163	0.0148
	7	183	0.0000		48	142	0.0301		7	203	0.0000		48	162	0.0164
	8	182	0.0000		49	141	0.0331		8	202	0.0000		49	161	0.0181
	9	181	0.0001		50	140	0.0364		9	201	0.0000		50	160	0.0200
	10	180	0.0001		51	139	0.0399		10	200	0.0000		51	159	0.0220
	11	179	0.0001		52	138	0.0437		11	199	0.0001		52	158	0.0242
	12	178	0.0001		53	137	0.0478		12	198	0.0001		53	157	0.0266
	13	177	0.0002		54	136	0.0521		13	197	0.0001		54	156	0.0291
	14	176	0.0002		55	135	0.0567		14	196	0.0001		55	155	0.0319
	15	175	0.0003		56	134	0.0616		15	195	0.0001		56	154	0.0348
	16	174	0.0003		57	133	0.0668		16	194	0.0002		57	153	0.0379
	17	173	0.0004		58	132	0.0723		17	193	0.0002		58	152	0.0413
	18	172	0.0005		59	131	0.0782		18	192	0.0002		59	151	0.0448
	19	171	0.0006		60	130	0.0844		19	191	0.0003		60	150	0.0487
	20	170	0.0007		61	129	0.0909		20	190	0.0004		61	149	0.0527
	21	169	0.0008		62	128	0.0978		21	189	0.0004		62	148	0.0570
	22	168	0.0010		63	127	0.1051		22	188	0.0005		63	147	0.0615
	23	167	0.0012						23	187	0.0006		64	146	0.0664
	24	166	0.0014						24	186	0.0007		65	145	0.0715
	25	165	0.0017						25	185	0.0008		66	144	0.0768
	26	164	0.0020						26	184	0.0010		67	143	0.0825
	27	163	0.0023						27	183	0.0012		68	142	0.0884
	28	162	0.0027						28	182	0.0014		69	141	0.0947
	29	161	0.0031						29	181	0.0016		70	140	0.1012
	30	160	0.0036						30	180	0.0018				
	31	159	0.0041						31	179	0.0021				
	32	158	0.0047						32	178	0.0024				
	33	157	0.0054						33	177	0.0028				
	34	156	0.0062						34	176	0.0032				
	35	155	0.0070						35	175	0.0036				
	36	154	0.0080						36	174	0.0042				
	37	153	0.0090						37	173	0.0047				
	38	152	0.0102						38	172	0.0053				
	39	151	0.0115						39	171	0.0060				
	40	150	0.0129						40	170	0.0068				

Table X Critical Values for the Wilcoxon Rank-Sum Statistic

This table contains critical values and probabilities for the Wilcoxon rank-sum statistic W, the sum of the ranks of the m observations in the smaller sample: m and n are the sample sizes, c_1 and c_2 are defined by $P(W \leq c_1) = \alpha$ and $P(W \geq c_2) = \alpha$.

m	n	c_1	c_2	α
2	3	3	9	0.1000
2	4	3	11	0.0667
		4	10	0.1333
2	5	3	13	0.0476
		4	12	0.0952
2	6	3	15	0.0357
		4	14	0.0714
		5	13	0.1429
2	7	3	17	0.0278
		4	16	0.0556
		5	15	0.1111
2	8	3	19	0.0222
		4	18	0.0444
		5	17	0.0889
		6	16	0.1333
2	9	3	21	0.0182
		4	20	0.0364
		5	19	0.0727
		6	18	0.1091
2	10	3	23	0.0152
		4	22	0.0303
		5	21	0.0606
		6	20	0.0909
		7	19	0.1364
3	3	6	15	0.0500
		7	14	0.1000
3	4	6	18	0.0286
		7	17	0.0571
		8	16	0.1143
3	5	6	21	0.0179
		7	20	0.0357
		8	19	0.0714
		9	18	0.1250
3	6	6	24	0.0119
		7	23	0.0238
		8	22	0.0476
		9	21	0.0833
		10	20	0.1310
3	7	6	27	0.0083
		7	26	0.0167
		8	25	0.0333
		9	24	0.0583
		10	23	0.0917
		11	22	0.1333

m	n	c_1	c_2	α
3	8	6	30	0.0061
		7	29	0.0121
		8	28	0.0242
		9	27	0.0424
		10	26	0.0667
		11	25	0.0970
		12	24	0.1394
3	9	6	33	0.0045
		7	32	0.0091
		8	31	0.0182
		9	30	0.0318
		10	29	0.0500
		11	28	0.0727
		12	27	0.1045
		13	26	0.1409
3	10	6	36	0.0035
		7	35	0.0070
		8	34	0.0140
		9	33	0.0245
		10	32	0.0385
		11	31	0.0559
		12	30	0.0804
		13	29	0.1084
		14	28	0.1434
4	4	10	26	0.0143
		11	25	0.0286
		12	24	0.0571
		13	23	0.1000
4	5	10	30	0.0079
		11	29	0.0159
		12	28	0.0317
		13	27	0.0556
		14	26	0.0952
		15	25	0.1429
4	6	10	34	0.0048
		11	33	0.0095
		12	32	0.0190
		13	31	0.0333
		14	30	0.0571
		15	29	0.0857
		16	28	0.1286

m	n	c_1	c_2	α
4	7	10	38	0.0030
		11	37	0.0061
		12	36	0.0121
		13	35	0.0212
		14	34	0.0364
		15	33	0.0545
		16	32	0.0818
		17	31	0.1152
4	8	10	42	0.0020
		11	41	0.0040
		12	40	0.0081
		13	39	0.0141
		14	38	0.0242
		15	37	0.0364
		16	36	0.0545
		17	35	0.0768
		18	34	0.1071
		19	33	0.1414
4	9	10	46	0.0014
		11	45	0.0028
		12	44	0.0056
		13	43	0.0098
		14	42	0.0168
		15	41	0.0252
		16	40	0.0378
		17	39	0.0531
		18	38	0.0741
		19	37	0.0993
		20	36	0.1301
4	10	10	50	0.0010
		11	49	0.0020
		12	48	0.0040
		13	47	0.0070
		14	46	0.0120
		15	45	0.0180
		16	44	0.0270
		17	43	0.0380
		18	42	0.0529
		19	41	0.0709
		20	40	0.0939
		21	39	0.0939

m	n	c_1	c_2	α
5	5	15	40	0.0040
		16	39	0.0079
		17	38	0.0159
		18	37	0.0278
		19	36	0.0476
		20	35	0.0754
		21	34	0.1111
5	6	15	45	0.0022
		16	44	0.0043
		17	43	0.0087
		18	42	0.0152
		19	41	0.0260
		20	40	0.0411
		21	39	0.0628
		22	38	0.0887
		23	37	0.1234
5	7	15	50	0.0013
		16	49	0.0025
		17	48	0.0051
		18	47	0.0088
		19	46	0.0152
		20	45	0.0240
		21	44	0.0366
		22	43	0.0530
		23	42	0.0745
		24	41	0.1010
		25	40	0.1338
5	8	15	55	0.0008
		16	54	0.0016
		17	53	0.0031
		18	52	0.0054
		19	51	0.0093
		20	50	0.0148
		21	49	0.0225
		22	48	0.0326
		23	47	0.0466
		24	46	0.0637
		25	45	0.0855
		26	44	0.1111
		27	43	0.1422

Table X Critical Values for the Wilcoxon Rank-Sum Statistic (Continued)

m	n	c_1	c_2	α
5	9	15	60	0.0005
		16	59	0.0010
		17	58	0.0020
		18	57	0.0035
		19	56	0.0060
		20	55	0.0095
		21	54	0.0145
		22	53	0.0210
		23	52	0.0300
		24	51	0.0415
		25	50	0.0559
		26	49	0.0734
		27	48	0.0949
		28	47	0.1199
		29	46	0.1489
5	10	15	65	0.0003
		16	64	0.0007
		17	63	0.0013
		18	62	0.0023
		19	61	0.0040
		20	60	0.0063
		21	59	0.0097
		22	58	0.0140
		23	57	0.0200
		24	56	0.0276
		25	55	0.0376
		26	54	0.0496
		27	53	0.0646
		28	52	0.0823
		29	51	0.1032
		30	50	0.1272
6	6	21	57	0.0011
		22	56	0.0022
		23	55	0.0043
		24	54	0.0076
		25	53	0.0130
		26	52	0.0206
		27	51	0.0325
		28	50	0.0465
		29	49	0.0660
		30	48	0.0898
		31	47	0.1201

m	n	c_1	c_2	α
6	7	21	63	0.0006
		22	62	0.0012
		23	61	0.0023
		24	60	0.0041
		25	59	0.0070
		26	58	0.0111
		27	57	0.0175
		28	56	0.0256
		29	55	0.0367
		30	54	0.0507
		31	53	0.0688
		32	52	0.0903
		33	51	0.1171
		34	50	0.1474
6	8	21	69	0.0003
		22	68	0.0007
		23	67	0.0013
		24	66	0.0023
		25	65	0.0040
		26	64	0.0063
		27	63	0.0100
		28	62	0.0147
		29	61	0.0213
		30	60	0.0296
		31	59	0.0406
		32	58	0.0539
		33	57	0.0709
		34	56	0.0906
		35	55	0.1142
		36	54	0.1412
6	9	21	75	0.0002
		22	74	0.0004
		23	73	0.0008
		24	72	0.0014
		25	71	0.0024
		26	70	0.0038
		27	69	0.0060
		28	68	0.0088
		29	67	0.0128
		30	66	0.0180
		31	65	0.0248
		32	64	0.0332
		33	63	0.0440
		34	62	0.0567
		35	61	0.0723
		36	60	0.0905
		37	59	0.1119
		38	58	0.1361

m	n	c_1	c_2	α
6	10	21	81	0.0001
		22	80	0.0002
		23	79	0.0005
		24	78	0.0009
		25	77	0.0015
		26	76	0.0024
		27	75	0.0037
		28	74	0.0055
		29	73	0.0080
		30	72	0.0112
		31	71	0.0156
		32	70	0.0210
		33	69	0.0280
		34	68	0.0363
		35	67	0.0467
		36	66	0.0589
		37	65	0.0736
		38	64	0.0903
		39	63	0.1099
		40	62	0.1317
7	7	28	77	0.0003
		29	76	0.0006
		30	75	0.0012
		31	74	0.0020
		32	73	0.0035
		33	72	0.0055
		34	71	0.0087
		35	70	0.0131
		36	69	0.0189
		37	68	0.0265
		38	67	0.0364
		39	66	0.0487
		40	65	0.0641
		41	64	0.0825
		42	63	0.1043
		43	62	0.1297

m	n	c_1	c_2	α
7	8	28	84	0.0002
		29	83	0.0003
		30	82	0.0006
		31	81	0.0011
		32	80	0.0019
		33	79	0.0030
		34	78	0.0047
		35	77	0.0070
		36	76	0.0103
		37	75	0.0145
		38	74	0.0200
		39	73	0.0270
		40	72	0.0361
		41	71	0.0469
		42	70	0.0603
		43	69	0.0760
		44	68	0.0946
		45	67	0.1159
		46	66	0.1405
7	9	28	91	0.0001
		29	90	0.0002
		30	89	0.0003
		31	88	0.0006
		32	87	0.0010
		33	86	0.0017
		34	85	0.0026
		35	84	0.0039
		36	83	0.0058
		37	82	0.0082
		38	81	0.0115
		39	80	0.0156
		40	79	0.0209
		41	78	0.0274
		42	77	0.0356
		43	76	0.0454
		44	75	0.0571
		45	74	0.0708
		46	73	0.0869
		47	72	0.1052
		48	71	0.1261
		49	70	0.1496

Table X Critical Values for the Wilcoxon Rank-Sum Statistic (Continued)

m	n	c_1	c_2	α
7	10	28	98	0.0001
		29	97	0.0001
		30	96	0.0002
		31	95	0.0004
		32	94	0.0006
		33	93	0.0010
		34	92	0.0015
		35	91	0.0023
		36	90	0.0034
		37	89	0.0048
		38	88	0.0068
		39	87	0.0093
		40	86	0.0125
		41	85	0.0165
		42	84	0.0215
		43	83	0.0277
		44	82	0.0351
		45	81	0.0439
		46	80	0.0544
		47	79	0.0665
		48	78	0.0806
		49	77	0.0966
		50	76	0.1148
		51	75	0.1349
8	8	36	100	0.0001
		37	99	0.0002
		38	98	0.0003
		39	97	0.0005
		40	96	0.0009
		41	95	0.0015
		42	94	0.0023
		43	93	0.0035
		44	92	0.0052
		45	91	0.0074
		46	90	0.0103
		47	89	0.0141
		48	88	0.0190
		49	87	0.0249
		50	86	0.0325
		51	85	0.0415
		52	84	0.0524
		53	83	0.0652
		54	82	0.0803
		55	81	0.0974
		56	80	0.1172
		57	79	0.1393

m	n	c_1	c_2	α
8	9	37	107	0.0001
		38	106	0.0002
		39	105	0.0003
		40	104	0.0005
		41	103	0.0008
		42	102	0.0012
		43	101	0.0019
		44	100	0.0028
		45	99	0.0039
		46	98	0.0056
		47	97	0.0076
		48	96	0.0103
		49	95	0.0137
		50	94	0.0180
		51	93	0.0232
		52	92	0.0296
		53	91	0.0372
		54	90	0.0464
		55	89	0.0570
		56	88	0.0694
		57	87	0.0836
		58	86	0.0998
		59	85	0.1179
8	10	38	114	0.0001
		39	113	0.0002
		40	112	0.0003
		41	111	0.0004
		42	110	0.0007
		43	109	0.0010
		44	108	0.0015
		45	107	0.0022
		46	106	0.0031
		47	105	0.0043
		48	104	0.0058
		49	103	0.0078
		50	102	0.0103
		51	101	0.0133
		52	100	0.0171
		53	99	0.0217
		54	98	0.0273
		55	97	0.0338
		56	96	0.0416
		57	95	0.0506
		58	94	0.0610
		59	93	0.0729
		60	92	0.0864
		61	91	0.1015
		62	90	0.1185

m	n	c_1	c_2	α
9	9	47	124	0.0001
		48	123	0.0001
		49	122	0.0002
		50	121	0.0004
		51	120	0.0006
		52	119	0.0009
		53	118	0.0014
		54	117	0.0020
		55	116	0.0028
		56	115	0.0039
		57	114	0.0053
		58	113	0.0071
		59	112	0.0094
		60	111	0.0122
		61	110	0.0157
		62	109	0.0200
		63	108	0.0252
		64	107	0.0313
		65	106	0.0385
		66	105	0.0470
		67	104	0.0567
		68	103	0.0680
		69	102	0.0807
		70	101	0.0951
		71	100	0.1112
9	10	48	132	0.0001
		49	131	0.0001
		50	130	0.0002
		51	129	0.0003
		52	128	0.0005
		53	127	0.0007
		54	126	0.0011
		55	125	0.0015
		56	124	0.0021
		57	123	0.0028
		58	122	0.0038
		59	121	0.0051
		60	120	0.0066
		61	119	0.0086
		62	118	0.0110
		63	117	0.0140
		64	116	0.0175
		65	115	0.0217
		66	114	0.0267
		67	113	0.0326
		68	112	0.0394
		69	111	0.0474
		70	110	0.0564
		71	109	0.0667
		72	108	0.0782
		73	107	0.0912
		74	106	0.1055

m	n	c_1	c_2	α
10	10	59	151	0.0001
		60	150	0.0001
		61	149	0.0002
		62	148	0.0002
		63	147	0.0004
		64	146	0.0005
		65	145	0.0008
		66	144	0.0010
		67	143	0.0014
		68	142	0.0019
		69	141	0.0026
		70	140	0.0034
		71	139	0.0045
		72	138	0.0057
		73	137	0.0073
		74	136	0.0093
		75	135	0.0116
		76	134	0.0144
		77	133	0.0177
		78	132	0.0216
		79	131	0.0262
		80	130	0.0315
		81	129	0.0376
		82	128	0.0446
		83	127	0.0526
		84	126	0.0615
		85	125	0.0716
		86	124	0.0827
		87	123	0.0952
		88	122	0.1088
		89	121	0.1237

Table XI Critical Values for the Runs Test

This table contains cumulative probabilities associated with the runs test. Let m be the number of observations in one category and n the number of observations in the other category ($m \leq n$), and V be the number of runs. The values in this table are the probabilities $P(V \leq v)$ if the order of observations is random.

m	n	2	3	4	5	6	7	8	9
2	2	0.3333	0.6667	1.0000					
2	3	0.2000	0.5000	0.9000	1.0000				
2	4	0.1333	0.4000	0.8000	1.0000				
2	5	0.0952	0.3333	0.7143	1.0000				
2	6	0.0714	0.2857	0.6429	1.0000				
2	7	0.0556	0.2500	0.5833	1.0000				
2	8	0.0444	0.2222	0.5333	1.0000				
2	9	0.0364	0.2000	0.4909	1.0000				
2	10	0.0303	0.1818	0.4545	1.0000				
3	3	0.1000	0.3000	0.7000	0.9000	1.0000			
3	4	0.0571	0.2000	0.5429	0.8000	0.9714	1.0000		
3	5	0.0357	0.1429	0.4286	0.7143	0.9286	1.0000		
3	6	0.0238	0.1071	0.3452	0.6429	0.8810	1.0000		
3	7	0.0167	0.0833	0.2833	0.5833	0.8333	1.0000		
3	8	0.0121	0.0667	0.2364	0.5333	0.7879	1.0000		
3	9	0.0091	0.0545	0.2000	0.4909	0.7455	1.0000		
3	10	0.0070	0.0455	0.1713	0.4545	0.7063	1.0000		
4	4	0.0286	0.1143	0.3714	0.6286	0.8857	0.9714	1.0000	
4	5	0.0159	0.0714	0.2619	0.5000	0.7857	0.9286	0.9921	1.0000
4	6	0.0095	0.0476	0.1905	0.4048	0.6905	0.8810	0.9762	1.0000
4	7	0.0061	0.0333	0.1424	0.3333	0.6061	0.8333	0.9545	1.0000
4	8	0.0040	0.0242	0.1091	0.2788	0.5333	0.7879	0.9293	1.0000
4	9	0.0028	0.0182	0.0853	0.2364	0.4713	0.7455	0.9021	1.0000
4	10	0.0020	0.0140	0.0679	0.2028	0.4186	0.7063	0.8741	1.0000

Table XI Critical Values for the Runs Test (*Continued*)

		v																	
m	*n*	2	3	4	5	6	7	8	9	10	11	12	13	14	15	16	17	18	19	20
5	5	0.0079	0.0397	0.1667	0.3571	0.6429	0.8333	0.9603	0.9921	1.0000										
5	6	0.0043	0.0238	0.1104	0.2619	0.5216	0.7381	0.9113	0.9762	0.9978	1.0000									
5	7	0.0025	0.0152	0.0758	0.1970	0.4242	0.6515	0.8535	0.9545	0.9924	1.0000									
5	8	0.0016	0.0101	0.0536	0.1515	0.3473	0.5758	0.7933	0.9293	0.9837	1.0000									
5	9	0.0010	0.0070	0.0390	0.1189	0.2867	0.5105	0.7343	0.9021	0.9720	1.0000									
5	10	0.0007	0.0050	0.0290	0.0949	0.2388	0.4545	0.6783	0.8741	0.9580	1.0000									
6	6	0.0022	0.0130	0.0671	0.1753	0.3918	0.6082	0.8247	0.9329	0.9870	0.9978	1.0000								
6	7	0.0012	0.0076	0.0425	0.1212	0.2960	0.5000	0.7331	0.8788	0.9662	0.9924	0.9994	1.0000							
6	8	0.0007	0.0047	0.0280	0.0862	0.2261	0.4126	0.6457	0.8205	0.9371	0.9837	0.9977	1.0000							
6	9	0.0004	0.0030	0.0190	0.0629	0.1748	0.3427	0.5664	0.7622	0.9021	0.9720	0.9944	1.0000							
6	10	0.0002	0.0020	0.0132	0.0470	0.1369	0.2867	0.4965	0.7063	0.8636	0.9580	0.9895	1.0000							
7	7	0.0006	0.0041	0.0251	0.0775	0.2086	0.3834	0.6166	0.7914	0.9225	0.9749	0.9959	0.9994	1.0000						
7	8	0.0003	0.0023	0.0154	0.0513	0.1492	0.2960	0.5136	0.7040	0.8671	0.9487	0.9879	0.9977	0.9998	1.0000					
7	9	0.0002	0.0014	0.0098	0.0350	0.1084	0.2308	0.4266	0.6224	0.8059	0.9161	0.9748	0.9944	0.9993	1.0000					
7	10	0.0001	0.0009	0.0064	0.0245	0.0800	0.1818	0.3546	0.5490	0.7433	0.8794	0.9571	0.9895	0.9981	1.0000					
8	8	0.0002	0.0012	0.0089	0.0317	0.1002	0.2145	0.4048	0.5952	0.7855	0.8998	0.9683	0.9911	0.9988	0.9998	1.0000				
8	9	0.0001	0.0007	0.0053	0.0203	0.0687	0.1573	0.3186	0.5000	0.7016	0.8427	0.9394	0.9797	0.9958	0.9993	1.0000				
8	10	0.0000	0.0004	0.0033	0.0134	0.0479	0.1170	0.2514	0.4194	0.6209	0.7822	0.9031	0.9636	0.9905	0.9981	0.9998	1.0000			
9	9	0.0000	0.0004	0.0030	0.0122	0.0445	0.1090	0.2380	0.3992	0.6008	0.7620	0.8910	0.9555	0.9878	0.9970	0.9996	1.0000			
9	10	0.0000	0.0002	0.0018	0.0076	0.0294	0.0767	0.1786	0.3186	0.5095	0.6814	0.8342	0.9233	0.9742	0.9924	0.9986	0.9998	1.0000		
10	10	0.0000	0.0001	0.0010	0.0045	0.0185	0.0513	0.1276	0.2422	0.4141	0.5859	0.7578	0.8724	0.9487	0.9815	0.9955	0.9990	0.9999	1.0000	

Table XII Greek Alphabet

This table contains the Greek alphabet: the letter name, the lowercase letter, the variant of the lowercase letter where applicable, and the uppercase letter.

Name	Lowercase letter	Lowercase variant	Uppercase letter
Alpha	α		A
Beta	β		B
Gamma	γ		Γ
Delta	δ		Δ
Epsilon	ϵ	ε	E
Zeta	ζ		Z
Eta	η		H
Theta	θ	ϑ	Θ
Iota	ι		I
Kappa	κ		K
Lambda	λ		Λ
Mu	μ		M
Nu	ν		N
Xi	ξ		Ξ
Omicron	o		O
Pi	π	ϖ	Π
Rho	ρ	ϱ	R
Sigma	σ	ς	Σ
Tau	τ		T
Upsilon	υ		Υ
Phi	ϕ	φ	Φ
Chi	χ		X
Psi	ψ		Ψ
Omega	ω		Ω

Answers to Odd-Numbered Exercises

CHAPTER 1

Sections 1.1–1.2

1.1 (a) Descriptive. (b) Inferential. (c) Inferential. (d) Inferential. (e) Descriptive. (f) Descriptive.

1.3 (a) Open-heart patients operated on in the last year. (b) 30 patients selected. (c) Length of stay.

1.5 Population: employees at Citigroup, Inc. Sample: 35 employees selected.

1.7 Population: 10,000 families affected by the flood. Sample: 75 affected families selected.

1.9 (a) Population: All Americans. (b) Sample: 1000 Americans selected. (c) Variable: Whether or not each believes sharks are dangerous.

1.11 Population: People diagnosed with hepatitis C. Sample: 50 patients selected. Variable: Liver enzyme levels.

1.13 (a) Population: all cheddar cheeses. Sample: 20 cheddar cheeses selected. (b) Probability question: What is the probability that at least 10 of the cheddar cheeses selected are aged less than two years? Statistics questions: Suppose 12 of the cheddar cheeses selected are aged less than two years. Does this suggest that the true proportion of all cheddars aged less than two years has decreased?

1.15 (a) Population: All American companies. (b) Sample: 75 companies selected. (c) Variable: Whether each company has overseas IT workers. (d) Probability question: What is the probability that exactly 30 of the 75 companies selected have overseas IT workers? Statistics question: Use the resulting data to determine whether there is evidence that the proportion of companies with overseas IT workers has changed.

Section 1.3

1.17 (a) Observational study. (b) Sample: 25 volunteer fire companies selected. (c) Not a random sample, largest companies selected.

1.19 Assign a number to each shipped weather station. Select numbers by using a random number generator, and examine each weather station corresponding to the numbers selected.

1.21 (a) Population: All men who use a disposable razor. Sample: 100 men selected. (b) Not a random sample. Just selected men observed buying a razor.

1.23 Assign a number to each challenge. Randomly select numbers from a random number table or random number generator.

1.25 (a) Experimental study. (b) Variable: Lifetime of each blossom. (c) Flip a coin: heads is treated, tails is untreated.

1.27 (a) Population: All ceramic tile from this manufacturer. Sample: 25 tiles selected. (b) Not a random sample, all tiles from the same box.

Chapter Exercises

1.29 (a) Descriptive. (b) Descriptive. (c) Inferential. (d) Inferential.

1.31 Population: All teenagers. Sample: Teenagers obtained. Variable: Whether each teenager can cook.

1.33 Population: All individuals. Sample: 78 individuals selected. Variable: Cortisol level 30 minutes after awakening.

1.35 (a) Observational. (b) People who rode the train and were asked if they used the Sightseer Lounge. (c) No. Assign a number to each passenger. Select numbers by using a random number generator, and ask each passenger corresponding to the numbers selected.

1.37 (a) Every installed software title. (b) 1000 software titles selected. (c) Whether the software title is pirated. (d) Probability question: What is the probability that more than 250 of the 1000 software titles selected are pirated? Statistics question: Use the resulting data to determine whether the software piracy rate in the United States has decreased.

CHAPTER 2

Section 2.1

2.1 (a) Numerical, continuous. (b) Numerical, discrete. (c) Categorical. (d) Numerical, discrete. (e) Numerical, continuous. (f) Categorical.

2.3 (a) Numerical, discrete. (b) Numerical, discrete. (c) Categorical. (d) Numerical, continuous. (e) Numerical, continuous. (f) Categorical.

2.5 (a) Continuous. (b) Continuous. (c) Discrete. (d) Continuous. (e) Continuous. (f) Discrete.

2.7 (a) Continuous. (b) Discrete. (c) Discrete. (d) Continuous. (e) Discrete. (f) Discrete.

2.9 (a) Discrete. (b) Categorical. (c) Continuous. (d) Continuous. (e) Categorical. (f) Continuous.

Section 2.2

2.11

Art	Frequency	Relative frequency
Abstract	15	0.3571
Expressionist	6	0.1429
Realist	12	0.2857
Surrealist	9	0.2143
Total	42	1.0000

2.13

(a)

County	Frequency	Relative frequency
Adair	915	0.0946
Carroll	1081	0.1118
Chariton	1095	0.1132
Grundy	735	0.0760
Linn	969	0.1002
Livingston	903	0.0934
Macon	1351	0.1397
Mercer	569	0.0588
Putnam	723	0.0748
Schuyler	480	0.0496
Sullivan	850	0.0879
Total	9671	1.0000

(b)

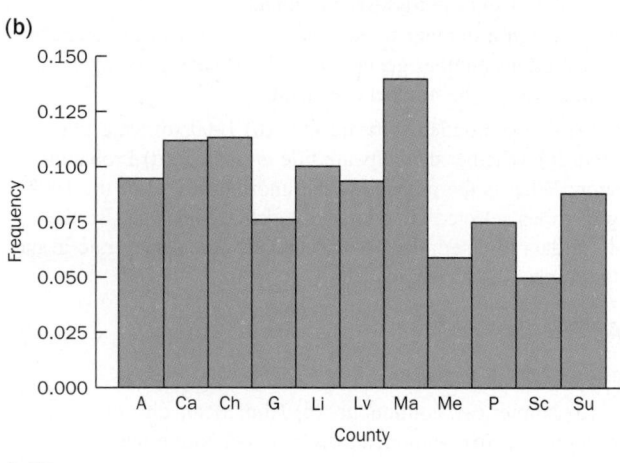

2.15

(a)

Political affiliation	Frequency	Relative frequency
D	23	0.3833
I	19	0.3167
R	18	0.3000
Total	60	1.0000

(b)

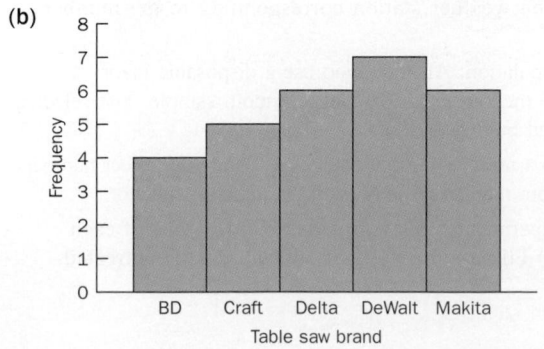

2.17

(a)

Grade	Frequency	Relative frequency
A	10	0.0676
B	43	0.2905
C	54	0.3649
D	26	0.1757
F	15	0.1014
Total	148	1.0000

(b)

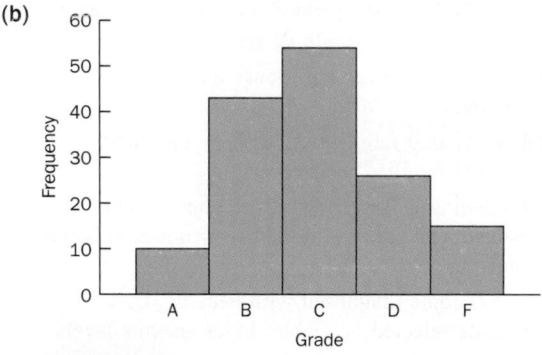

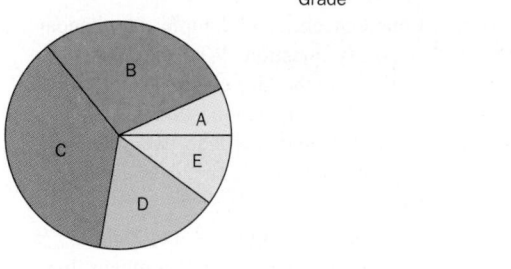

(c) 148; 0.8986

2.19

(a)

Table saw brand	Frequency	Relative frequency
B & D	4	0.1429
Craftsman	5	0.1786
Delta	6	0.2143
DeWalt	7	0.2500
Makita	6	0.2143
Total	28	1.0000

(b)

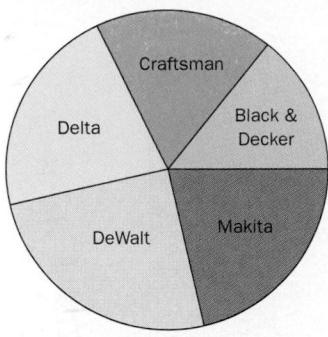

(c) 0.3215 (d) 0.7857

2.21

(a)

Book type	Frequency	Relative frequency
Education	5	0.1667
Law	3	0.1000
Literature	4	0.1333
Medicine	7	0.2333
Science	5	0.1667
Technology	6	0.2000

(b)

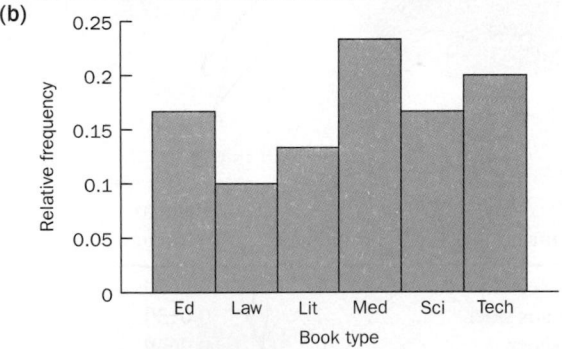

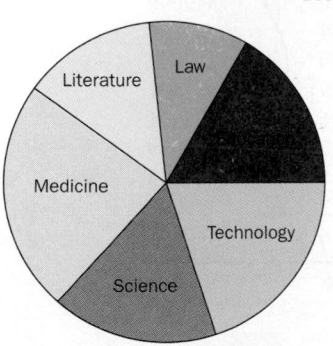

(c) Perhaps medicine. However, no type of book is overwhelmingly borrowed.

2.23

(a)

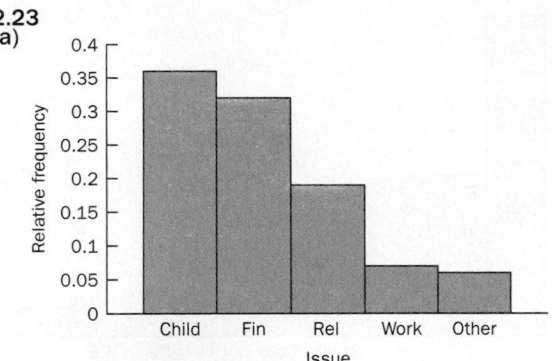

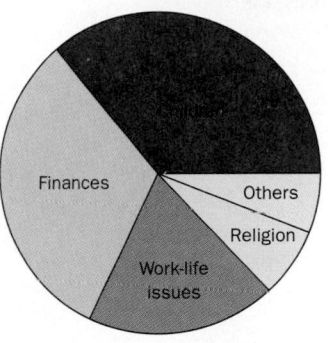

(b)

Issue	Frequency
Children	361
Finances	321
Religion	190
Work-life issues	70
Others	60

2.25

(a)

Think tank	Frequency	Relative frequency
Brookings Institution	2380	0.3167
Council on Foreign Relations	1191	0.1585
Heritage Foundation	1168	0.1554
RAND Corporation	740	0.0985
Cato Institute	640	0.0852
Urban Institute	558	0.0743
Carter Center	341	0.0454
Carnegie Endowment	287	0.0382
Aspen Institute	209	0.0278
Total	7514	1.0000

(b)

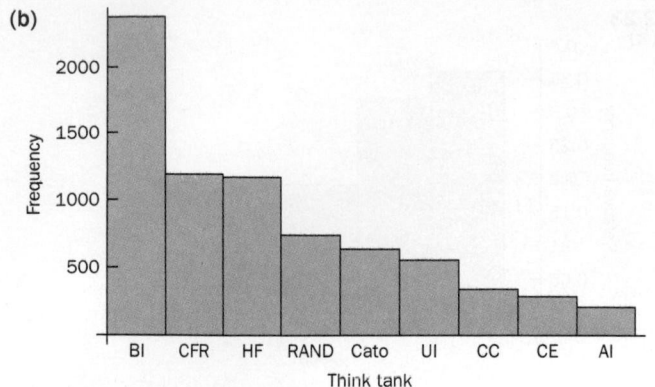

(b)

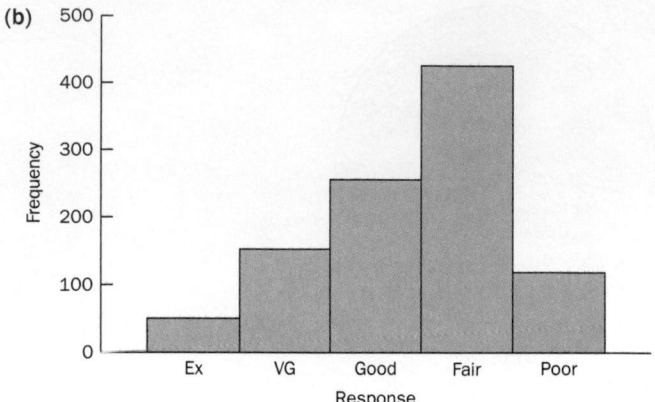

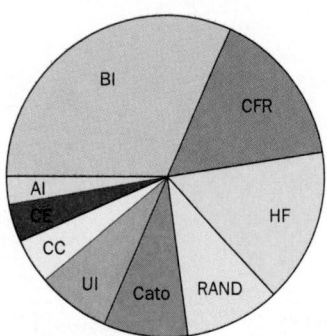

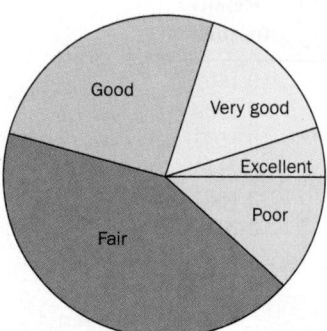

(c) 0.7980

2.27

Class	Frequency	Relative frequency
Bally's	40	0.200
Caesars	25	0.125
Harrah's	32	0.160
Resorts	22	0.110
Sands	25	0.125
Trump Plaza	56	0.280
Total	200	1.000

(a) 200 **(b)** Trump Plaza: largest (relative) frequency.

2.29

(a)

Response	Frequency	Relative frequency
Excellent	50	0.0500
Very good	152	0.1520
Good	255	0.2250
Fair	425	0.4250
Poor	118	0.1180
Total	1000	1.0000

2.31

(a)

Response	Frequency	Relative frequency
Misc. Recreation	4	0.0130
In bounds skier	4	0.0130
Snowshoer	20	0.0649
Snowboarder	34	0.1104
Out of bounds skier	30	0.0974
Backcountry skier	49	0.1591
Snowmobiler	126	0.4091
Climber/hiker	32	0.1039
Highway personnel	1	0.0032
Others at work	2	0.0065
Resident	6	0.0195
Total	308	1.0000

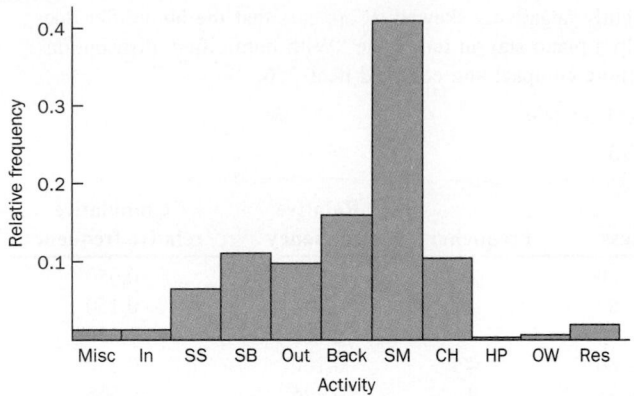

The two graphs are identical, except for the label on the vertical axis. Relative frequency might be more useful for avalanche rescuers.

2.33

(a)

Rating	Men relative frequency	Women relative frequency
Excellent	0.1840	0.2333
Very good	0.2750	0.2500
Good	0.2130	0.1100
Fair	0.2250	0.2400
Poor	0.1030	0.1667

(b)

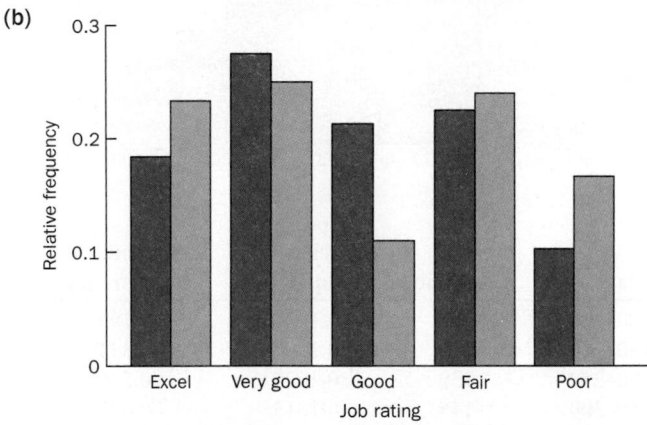

(c) The total number of men who responded is different from the total number of women who responded.

Section 2.3

2.35

```
2 | 79
3 |
3 | 56669
4 | 112
4 | 779
5 | 01124
5 | 57789
6 | 1444
6 | 68
7 | 1
```

Stem: ones; leaf: tenths.
The center of the data is between 5.0 and 5.5. A typical value is 5.1.

2.37

```
53 | 0344
53 | 799
54 | 111344
54 | 566677777889
55 | 112334
55 | 67777
56 | 002
56 | 9
```

Stem: hundreds and tens; leaf: ones.
The center of the data is between 545 and 550. A typical value is 547.

2.39 (a) 543, 543, 549 **(b)** 574 **(c)** The data tail off slowly on the low end. **(d)** There do not appear to be any outliers.

2.41

(a)

```
0 | 4
1 | 16
2 | 12779
3 | 3449
4 | 0111125555799
5 | 699
6 | 023444
7 | 0123679
8 | 258
9 | 3
```

Stem: ones; leaf: tenths.

(b)

```
0 | 4
1 | 16
2 | 02378
3 | 0355
4 | 001222256668
5 | 00799
6 | 133445
7 | 1134779
8 | 269
9 | 3
```

Stem: ones; leaf: tenths.
(c) These two graphs are slightly different. However, they both suggest the same general shape, center, and variability. Typical value is 4.5.

2.43

(a)

Lower floors		Upper floors
	10	14
	10	
4300	11	1
99887777777666665555	11	55578
33333222211111110000000	12	1244
665	12	55556777899
0	13	122234
	13	567888999
	14	02234
	14	5888
	15	244

(b) The lower floors distribution is more compact and has, on average, smaller values. The upper floors distribution has more variability and has, on average, larger values.

2.45 (a)

1	00699
2	244557
3	11245678899
4	000011555669
5	2259
6	8
7	1

Stem: tens; leaf: ones.
(b) Typical value: 38.5. No outliers.

2.47 (a)

2	135667
3	13456777788999
4	0001122355
5	002245566
6	
7	
8	0

Stem: tens; leaf: ones.
(b) Typical value: 40. One outlier: 80.

2.49 (a)

0	899
1	0
1	333
1	455
1	67777
1	8
2	001
2	233
2	45
2	
2	
3	
3	3

Stem: tens; leaf: ones.
(b) Typical weight: 17. One outlier: 33.

2.51 (a)

With		Without
	250	24
	251	
84	252	145
3	253	788
63	254	49
8742221	255	023
9866653110	256	149
873	257	1248
6553	258	23367
4	259	3
	260	26
	261	2
	262	3

Stem: hundreds, tens, and ones; leaf: tenths.
(b) With distribution: unimodal, compact, approximately symmetric. Without distribution: unimodal, lots of variability,

slightly negatively skewed. It appears that the humidifier does help a piano stay in tune. The "With humidifier" distribution is more compact and centered near 256.

Section 2.4

2.53

Class	Frequency	Relative frequency	Cumulative relative frequency
78–80	2	0.050	0.050
80–82	4	0.100	0.150
82–84	4	0.100	0.250
84–86	4	0.100	0.350
86–88	9	0.225	0.575
88–90	6	0.150	0.725
90–92	9	0.225	0.950
92–94	2	0.050	1.000
Total	40	1.000	

2.55

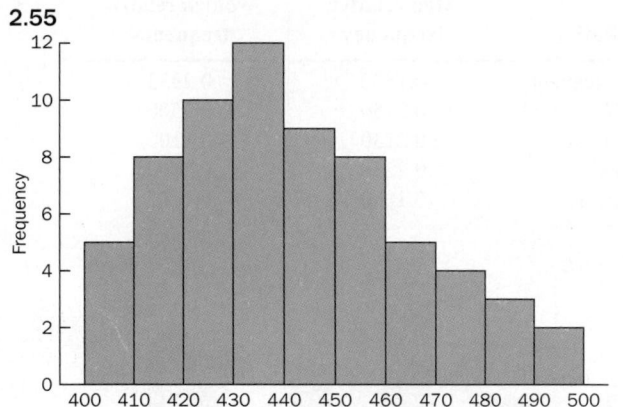

2.57

Class	Frequency	Relative frequency	Cumulative relative frequency
100–150	155	0.1938	0.1938
150–200	120	0.1500	0.3438
200–250	130	0.1625	0.5063
250–300	145	0.1813	0.6875
300–350	150	0.1875	0.8750
350–400	100	0.1250	1.0000
Total	800	1.0000	

2.59

Class	Frequency	Relative frequency	Cumulative relative frequency
0–25	150	0.150	0.150
25–50	200	0.200	0.350
50–75	175	0.175	0.525
75–100	150	0.150	0.675
100–125	125	0.125	0.800
125–150	100	0.100	0.900
150–175	75	0.075	0.975
175–200	25	0.025	1.000
Total	1000	1.000	

2.61 (a)

Class	Frequency	Relative frequency	Cumulative relative frequency
0–10	1	0.0167	0.0167
10–20	3	0.0500	0.0667
20–30	9	0.1500	0.2167
30–40	11	0.1833	0.4000
40–50	12	0.2000	0.6000
50–60	10	0.1667	0.7667
60–70	7	0.1167	0.8833
70–80	5	0.0833	0.9667
80–90	2	0.0333	1.0000
Total	60	1.0000	

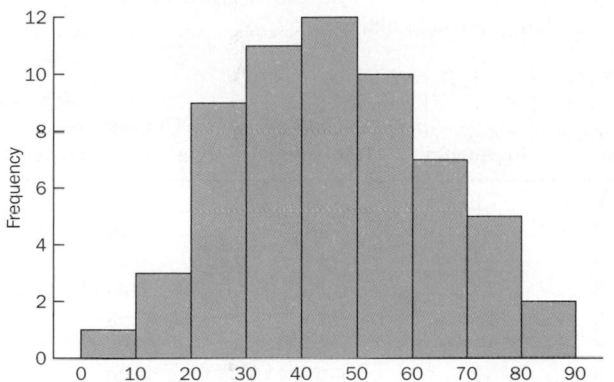

(b) Unimodal, symmetric, bell-shaped. **(c)** $M \approx 43$ **(d)** $Q_1 \approx 31.5$
(e) $Q_3 \approx 58.5$

2.63 (a)

Class	Frequency	Relative frequency	Cumulative relative frequency
0–50	5	0.1667	0.1667
50–100	9	0.3000	0.4667
100–150	5	0.1667	0.6333
150–200	3	0.1000	0.7333
200–250	2	0.0667	0.8000
250–300	1	0.0333	0.8333
300–350	4	0.1333	0.9667
350–400	0	0.0000	0.9667
400–450	0	0.0000	0.9667
450–500	0	0.0000	0.9667
500–550	1	0.0333	1.0000
Totals	30	1.0000	

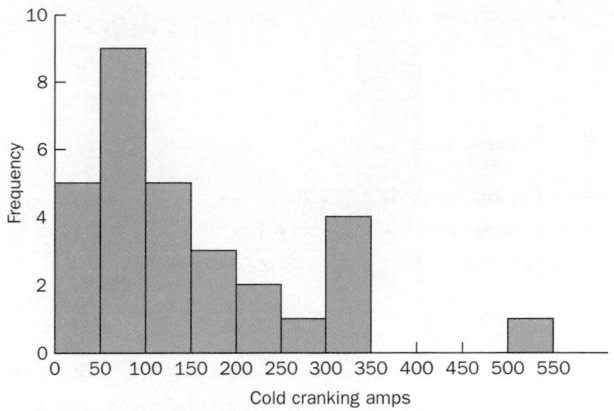

(b) Positively skewed. **(c)** $M \approx 110$

2.65

(a)

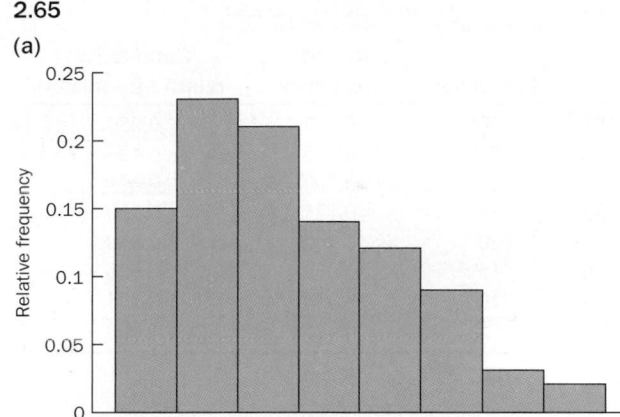

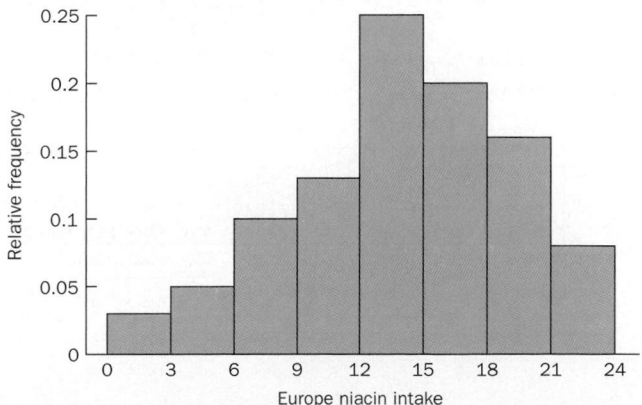

(b) United States: unimodal, positively skewed.
Europe: unimodal, negatively skewed. On average,
it appears that Europeans have a greater daily niacin intake.

2.67

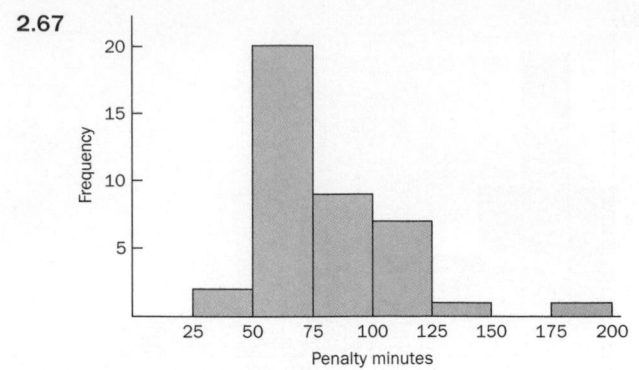

(b) Positively skewed. Center around 73. Lots of variability.
(c) 117.5

2.69 (a)

Class	Frequency	Relative frequency	Cumulative relative frequency
100–105	10	0.050	0.050
105–110	75	0.375	0.425
110–115	40	0.200	0.625
115–120	25	0.125	0.750
120–125	20	0.100	0.850
125–130	15	0.075	0.925
130–135	10	0.050	0.975
135–140	5	0.025	1.000
Total	200	1.000	

(b)

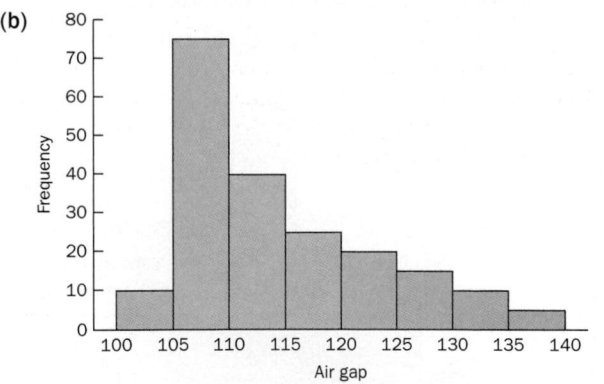

(c) 0.425

2.71
(a)

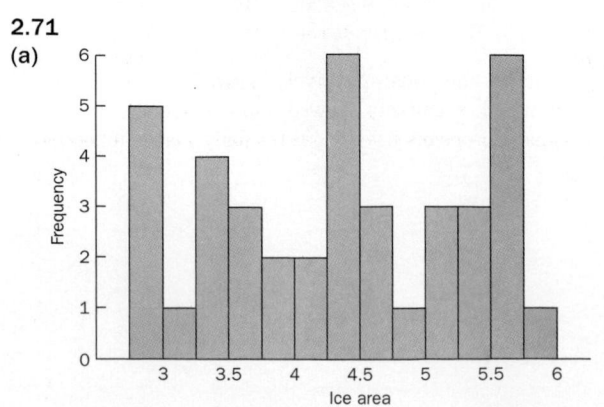

(b) No discernible shape. Center around 4.5. Lots of variability.
(c) $Q_1 \approx 3.45$, $Q_3 \approx 5.25$ **(d)** Should be $0.5 \times 40 = 20$ values between Q_1 and Q_3. There are 20 values between Q_1 and Q_3.

Chapter Exercises

2.73 (a)

1	5789
2	01134
2	5567889999
3	0011223344
3	55566679
4	01234
4	89
5	2

Stem: tenths; leaf: hundredths.
(b) Unimodal, approximately symmetric, center around 0.31, little variability, possible outlier: 0.52.

2.75 (a)

Class	Frequency	Relative frequency	Cumulative relative frequency
65–70	1	0.02	0.02
70–75	4	0.08	0.10
75–80	7	0.14	0.24
80–85	10	0.20	0.44
85–90	12	0.24	0.68
90–95	15	0.30	0.98
95–100	1	0.02	1.00
Total	50	1.0000	

(b)

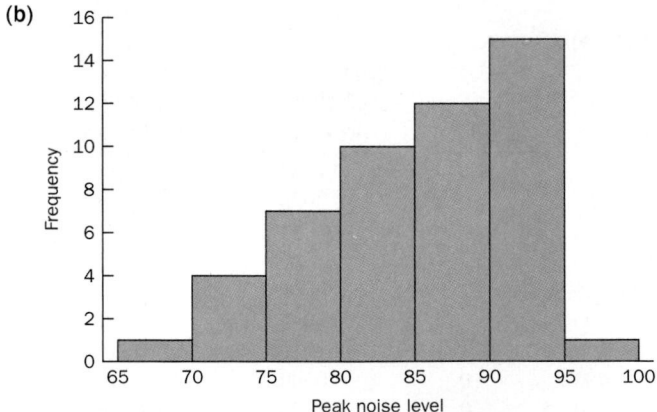

(c) 0.24 **(d)** 0.32

2.77

(a)

New		Traditional
	0	89
	1	5
87660	2	6
765443200	3	34
99774444322	4	02568
110	5	579
43	6	122888
	7	33567
	8	033589
	9	178
	10	034
	11	3
	12	033
	13	03
	14	16

Stem: tens and ones, leaf: tenths.

(b) New equipment times tend to be smaller, and the distribution is more compact. Traditional equipment times are more spread out and tend to be larger.

(c) The new equipment times tend to be better, shorter response times. The majority of the times are less than the traditional equipment times.

2.79 (a)

Class	Frequency	Relative frequency	Cumulative relative frequency
000–100	3303	0.3820	0.3820
100–200	3613	0.4179	0.7999
200–300	1117	0.1292	0.9291
300–400	375	0.0434	0.9725
400–500	128	0.0148	0.9873
500–600	61	0.0071	0.9944
600–700	29	0.0034	0.9978
700–800	10	0.0012	0.9990
800–900	9	0.0010	1.0000
Total	8646		

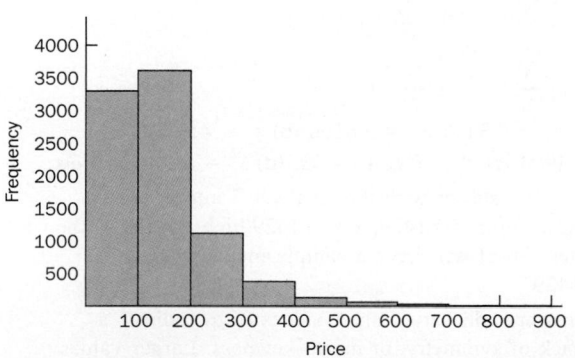

(b) Positively skewed. Center around 150. Lots of variability.

(c) Typical selling price: 150 thousand dollars. A few outliers in the 800–900 class. **(d)** 0.0709

2.81

(a)

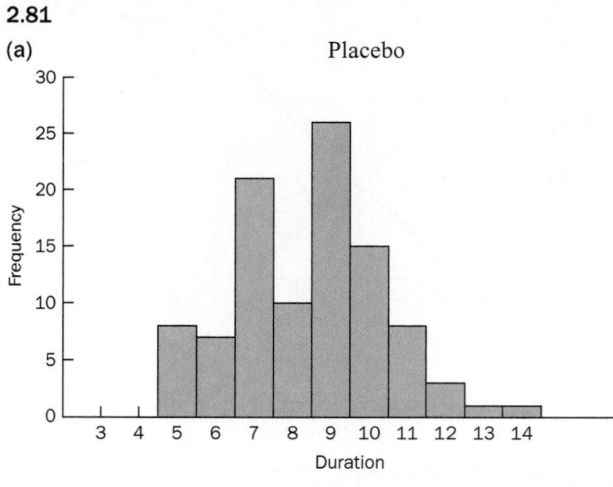

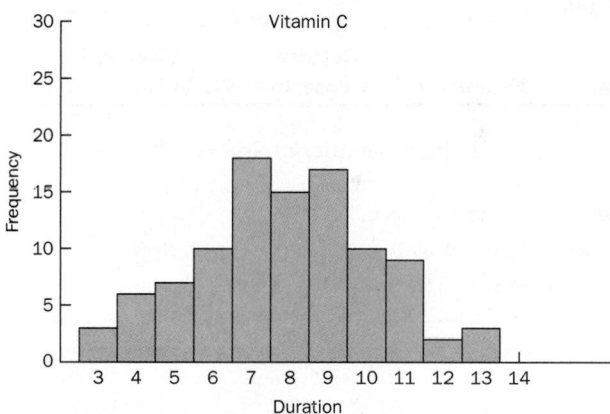

(b) Both graphs appear to be centered at about the same duration. Both appear to be symmetric and bell-shaped. The placebo durations are slightly more compact than the vitamin C durations. **(c)** There is no graphical evidence to suggest that vitamin C reduced the duration.

2.83 (a)

Class	Frequency	Relative frequency	Cumulative relative frequency
100–110	1	0.02	0.02
110–120	2	0.04	0.06
120–130	9	0.18	0.24
130–140	7	0.14	0.38
140–150	9	0.18	0.56
150–160	9	0.18	0.74
160–170	11	0.22	0.96
170–180	1	0.02	0.98
180–190	1	0.02	1.00
Total	50	1.00	

(b)

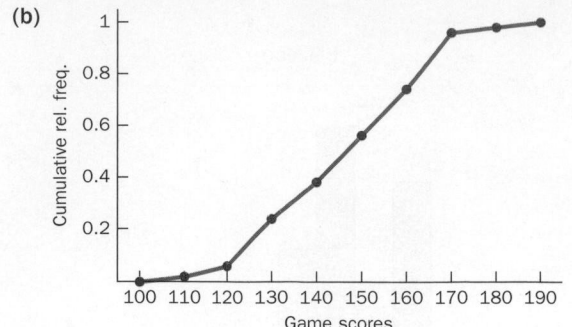

CHAPTER 3

Section 3.1

3.1 (a) 82 (b) 3474 (c) 32 (d) 2779 (e) 164 (f) 164

3.3 (a) 105.7 (b) 13.1852 (c) 6.9583 (d) 0.1232 (e) -2.4933
(f) 17.7432

3.5 (a) 6.6667, 7 (b) 6.6364, 9 (c) 10.6889, 7.7
(d) -107.69, -109.1

3.7 (a) Skewed left. (b) Symmetric. (c) Skewed left.
(d) Skewed left.

3.9 (a) 6 (b) 0 (c) No mode.

3.11 (a) 68.5238 (b) 67.0 (c) Slightly skewed right.

3.13 (a) 6.5083, 6.5150 (b) 6.7306, 6.5150 The mean is pulled
in the direction of the outlier. The median remains the same.

3.15 (a) $\bar{x} = 619.5$, $\tilde{x} = 620.0$ (b) 619.1667 (c) Approximately
symmetric.

3.17 (a) $\bar{x} = 108960.2$, $\tilde{x} = 9673$ (b) The median is a better
measure of central tendency, since there are extreme outliers.

3.19 (a) $\bar{x} = 59.8667$, $\tilde{x} = 60.5000$ (b) $\bar{x}_{\text{tr}(0.10)} = 60.1667$
(c) Two modes: 60, 61

3.21 (a) $\bar{x} = 6.4565$, $\tilde{x} = 6.5000$ (b) Mode: 6.5 (c) 17.4326;
this is 2.7 times the original mean.

3.23 (a) $\bar{x} = 1480.2692$ (b) 17763.2308; this is 12 times the
June mean.

3.25 (a) $\bar{x} = 16.5588$, $\tilde{x} = 15.1$ (b) Skewed right. (c) There is
no value that will make the sample mean equal to the sample
median.

3.27 (a) $x_5 = 9414$ (b) $x_5 = 6250.75$

3.29 (a) $\bar{x}_F = 57.1667$, $\tilde{x}_F = 57.5$ (b) $\bar{x}_C = 13.9815$
(c) $\bar{x}_C = (\bar{x}_F - 32)/1.8$

Section 3.2

3.31 (a) $R = 3.9$, $s^2 = 2.1690$, $s = 1.4728$ (b) $R = 20.6$,
$s^2 = 27.2812$, $s = 5.2231$ (c) $R = 98.32$, $s^2 = 1096.2665$,
$= 5$ 33.1099 (d) $R = 5.7$, $s^2 = 2.5843$, $s = 1.6076$

3.33 (a) 15.5, 45.5 (b) 10, 28 (c) 25.5, 75.5 (d) 12.5, 36.5

3.35 (a) $s^2 = 430.4$, $s = 20.7461$ (b) $s^2 = 430.4$, $s = 20.7461$
Same. (c) $s^2 = 172160.0$, $s = 414.9217$. The sample variance is
multiplied by $20^2 = 400$, and the sample standard deviation is
multiplied by 20.

3.37 (a) $R = 1.54$ (b) $s^2 = 0.2562$, $s = 0.5062$ (c) $Q_1 = 29.98$,
$Q_3 = 30.8$, $IQR = 0.82$

3.39 (a) $s^2 = 773.2292$, $s = 27.8070$ (b) $Q_1 = 667.5$, $Q_3 = 700$
(c) $IQR = 32.5$, $QD = 16.25$

3.41 (a) 158.1429 (b) 158.1429 (c) Same.

3.43 (a) $Q_1 = 291$, $Q_3 = 313$, $IQR = 22$ (b) $s^2 = 536.4889$,
$s = 23.1622$ (c) $IQR = 22$, $s^2 = 1160.3222$ (d) IQR is the
same, and s^2 is larger. s^2 is more sensitive to outliers.

3.45 (a) $s^2 = 11975.7018$, $s = 109.4335$ (b) $Q_1 = 265$,
$Q_3 = 352$, $IQR = 87$ (c) $s^2 = 3580.9853$, $s = 59.8413$,
$Q_1 = 270$, $Q_3 = 352$, $IQR = 82$ Both values are smaller in
the modified data set. By eliminating the two smallest values
(outliers), we find that the variability in the modified data set
is smaller.

3.47 (a) $Q_1 = 1.125$, $Q_3 = 2.005$, $IQR = 0.88$ (b) 1.12
(c) 28.1150

3.49 (a) $Q_1 = 300$, $Q_3 = 401$, $IQR = 101$ (b) $Q_1 = 300$,
$Q_3 = 401$, $IQR = 101$ (c) 401 (d) 300

3.51 (a) -4.6667, 9.3333, -13.6667, 5.3333, -11.6667,
15.3333 (b) Sum is 0.

(c)
$$\sum_{i=1}^{n}(x_i - \bar{x}) = \sum_{i=1}^{n}x_i - \sum_{i=1}^{n}\bar{x}$$
$$= \sum_{i=1}^{n}x_i - n\bar{x}$$
$$= \sum_{i=1}^{n}x_i - n\frac{1}{n}\sum_{i=1}^{n}x_i$$
$$= \sum_{i=1}^{n}x_i - \sum_{i=1}^{n}x_i = 0$$

3.53
$$\frac{1}{n-1}\sum_{i=1}^{n}(x_i - \bar{x})^2$$
$$= \frac{1}{n-1}\sum_{i=1}^{n}(x_i^2 - 2x_i\bar{x} + \bar{x}^2)$$
$$= \frac{1}{n-1}\left[\sum_{i=1}^{n}x_i^2 - 2\bar{x}\sum_{i=1}^{n}x_i + n\bar{x}^2\right]$$
$$= \frac{1}{n-1}\left[\sum_{i=1}^{n}x_i^2 - 2\bar{x}n\frac{1}{n}\sum_{i=1}^{n}x_i + n\bar{x}^2\right]$$
$$= \frac{1}{n-1}\left[\sum_{i=1}^{n}x_i^2 - 2n\bar{x}^2 + n\bar{x}^2\right]$$
$$= \frac{1}{n-1}\left[\sum_{i=1}^{n}x_i^2 - n\bar{x}^2\right]$$
$$= \frac{1}{n-1}\left[\sum_{i=1}^{n}x_i^2 - n\left(\frac{1}{n}\sum_{i=1}^{n}x_i\right)^2\right]$$
$$= \frac{1}{n-1}\left[\sum_{i=1}^{n}x_i^2 - \frac{1}{n}\left(\sum_{i=1}^{n}x_i\right)^2\right]$$

3.55 (a) $s_x^2 = 9.3420$, $s_x = 3.0566$ (b) $s_y^2 = 457.7902$,
$s_y = 21.3960$ (c) $s_y^2 = 7^2 s_x^2$, $s_y = 7 s_x$ (d) $s_y^2 = a^2 s_x^2$, $s_y = |a| s_x$

3.57 No. The subset with the smallest 7 numbers has
a sample mean $\bar{x} = 5.1429$, $\tilde{x} = 5.1429$,hich is greater than 5.
Any other subset will have a sample mean greater
than 5.1429.

3.59 Answers will vary. Larger values of g_1 indicate a
greater lack of symmetry, or more skewness. Larger values
(in magnitude) of g_2 suggest a more defined peak in the
distribution. Smaller values of g_2 suggest a flatter, more
uniform distribution.

Section 3.3

3.61 (a) (15, 25), (10, 30), (5, 35)

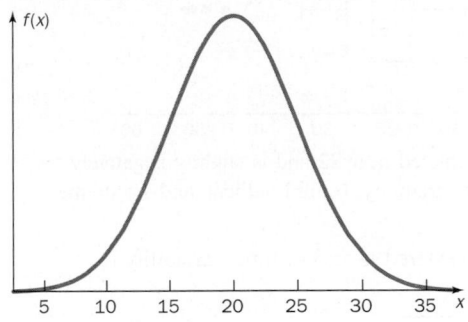

(b) (36.8, 37.2), (36.6, 37.4), (36.4, 37.6)

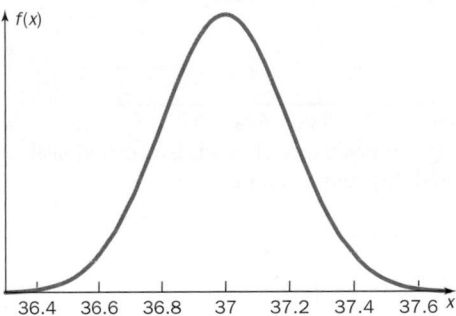

(c) (425, 925), (175, 1175), (−75, 1425)

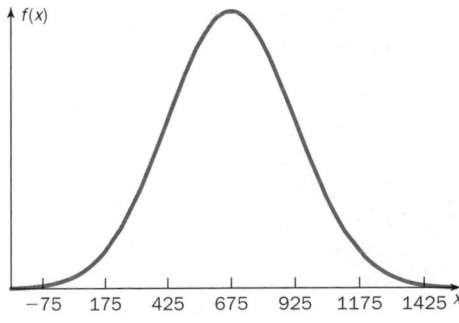

(d) (−17.5, 6.5), (−29.5, 18.5), (−41.5, 30.5)

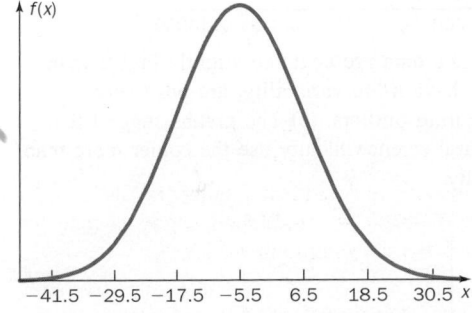

(e) (96.9, 100.3), (95.2, 102.0), (93.5, 103.7)

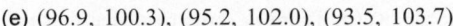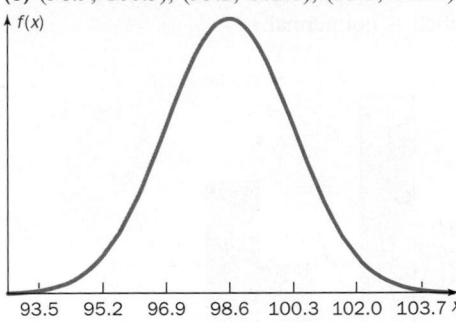

(f) (5130, 5430), (4980, 5580), (4830, 5730)

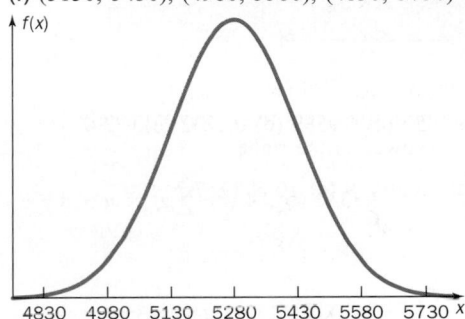

3.63 (a) 36.5 **(b)** 8.96 **(c)** −409.75 **(d)** 26.036 **(e)** 55.175
(f) 3.78 **(g)** 0.0 **(h)** 1.574

3.65 (a) (22.2, 30.8), (17.9, 35.1) **(b)** At least 0.75

3.67 85% of all fish caught in the tournament weighed less than the one caught by Ruskey, and 15% weighed more.

3.69 (a) At least 0.75. **(b)** At least 0.8889. **(c)** At most 0.1111. **(d)** At least 0.5556

3.71 (a) 0.95 **(b)** 0.8385 **(c)** 0.975

3.73 $z_1 = -0.8$, $z_2 = -1.0$. The second service actually performed *better*. The second service had a time that was farther away from the mean to the left in standard deviations.

3.75 (a)

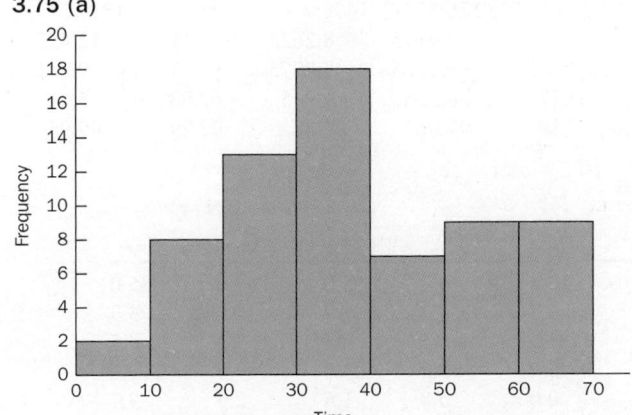

(b) 33, 52, 15 **(c)** $p_{45} = 32$, $p_{80} = 53$, $p_{10} = 14$

3.77 (a) 0.6333, 0.9667 1.0 **(b)** There is some evidence that the shape of the distribution is not normal.

(c)

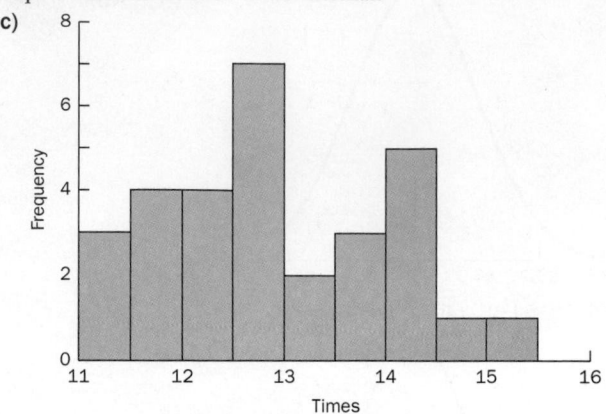

The shape of the distribution appears to be approximately bimodal. It is slightly skewed to the right.

3.79 (a) $\bar{x} = 40.3867$, $s = 3.3532$ **(b)** 14 **(c)** $\sum_{i=1}^{n} z_i^2 = n - 1$

Section 3.4

3.81 (a) $x_{min} = 28.0$, $Q_1 = 32.0$, $\tilde{x} = 34.5$, $Q_3 = 35.0$, $x_{max} = 40.0$ **(b)** $x_{min} = 52.0$, $Q_1 = 57.0$, $\tilde{x} = 66.5$, $Q_3 = 70.5$, $x_{max} = 78.0$ **(c)** $x_{min} = 80.0$, $Q_1 = 83.0$, $\tilde{x} = 91.5$, $Q_3 = 94.0$, $x_{max} = 98.0$ **(d)** $x_{min} = 0.4$, $Q_1 = 1.0$, $\tilde{x} = 1.95$, $Q_3 = 2.4$, $x_{max} = 10.9$ **(e)** $x_{min} = 103.1$, $Q_1 = 119.9$, $\tilde{x} = 141.9$, $Q_3 = 159.7$, $x_{max} = 196.9$ **(f)** $x_{min} = -40.1$, $Q_1 = -33.8$, $\tilde{x} = -28.0$, $Q_3 = -18.5$, $x_{max} = -9.8$

3.83

	IQR	IF$_L$	IF$_H$	OF$_L$	OF$_H$
(a)	24.0	−14.0	82.0	−50.0	118.0
(b)	51.0	1178.5	1382.5	1102.0	1459.0
(c)	9.46	51.56	89.4	37.37	103.59
(d)	225.6	576.5	1478.9	238.1	1817.3
(e)	2.795	−2.9175	8.2625	−7.11	12.455
(f)	2.245	−3.1025	5.8775	−6.47	9.245
(g)	9.77	−48.325	−9.245	−62.98	5.41
(h)	0.38	97.86	99.38	97.29	99.95

3.85

	x_{min}	Q_1	$\tilde{x}$	Q_3	x_{max}
(a)	20.0	40.0	55.0	60.0	85.0
(b)	−1.5	1.4	1.9	2.8	3.9
(c)	4.5	5.5	6.3	7.2	9.5
(d)	75.0	93.0	103.0	109.0	119.0
(e)	0.0	0.8	1.6	2.9	9.2
(f)	0.0	2.5	5.5	9.0	34
(g)	−80.0	−58.0	−51.0	−45.0	−22.0

3.87

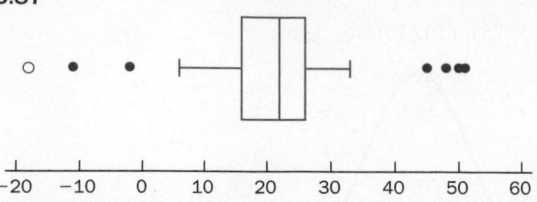

The distribution is centered near 22 and is slightly negatively skewed, with lots of variability, 6 mild outliers, and 1 extreme outlier.

3.89 Skewed left, centered near 3.4, little variability.

3.91

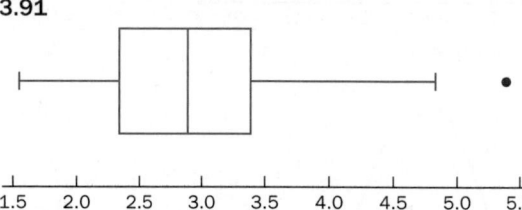

The distribution is slightly positively skewed, is centered near 2.8, and has little variability and 1 mild outlier.

3.93

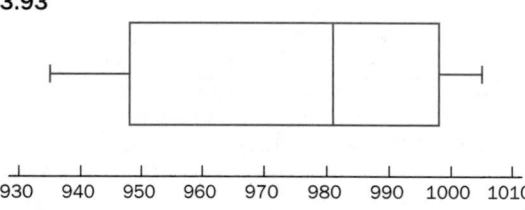

Slightly skewed left, centered near 981, little variability. There are no outliers.

3.95 (a)

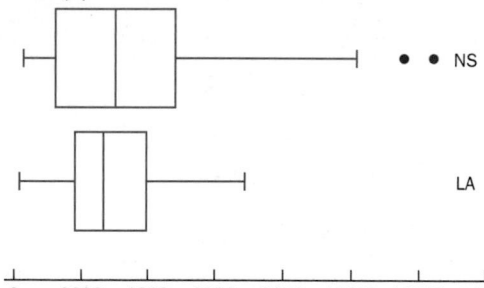

(b) The natural science data are centered slightly higher than the liberal arts data, have more variability, are positively skewed, and have 3 mild outliers. **(c)** The graphs suggest that, on average, the natural science faculty use the copier more than the liberal arts faculty.

3.97 (a)

(b) Centered near 14, lots of variability, positively skewed, no outliers.

(c)

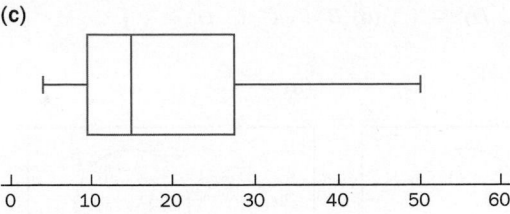

The new box plot looks nearly the same.

Chapter Exercises

3.99 (a) $\tilde{x} = 177.3500$, $s^2 = 207.5026$, $s = 14.4050$
(b) Within 1: 0.70; within 2: 1.0; within 3: 1.0.
(c) Since these proportions are close to the Empirical Rule proportions, there is no evidence to suggest that the distribution is nonnormal.

3.101 (a) $\tilde{x} = 243.5$, $Q_1 = 200.0$, $Q_3 = 361.0$, $IQR = 161.0$
(b) $p_{30} = 201$, $p_{95} = 588$ (c) $p_{93} = 438$, 438 lies in the 93rd percentile.

3.103 (a)

	R	s^2	IQR	CV	CQV
Over	5.4000	1.8023	1.7000	2.0707	1.3127
Mid	10.1000	7.9009	4.3000	4.6919	0.0361

(b) The summary statistics in part (a) suggest that the mid–over racket tensions have more variability.

(c)

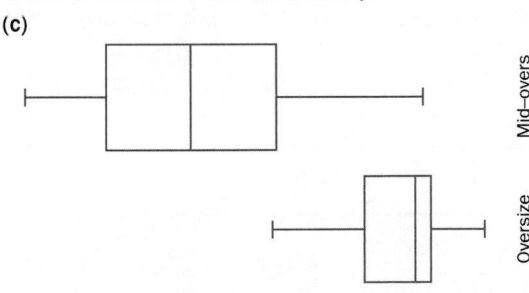

The box plots also suggest that the mid–over racket tensions have more variability.

3.105 (a) $\tilde{x} = 97.1111$, $\tilde{x} = 96.5$, $s^2 = 156.5752$, $s = 12.5130$

(b)

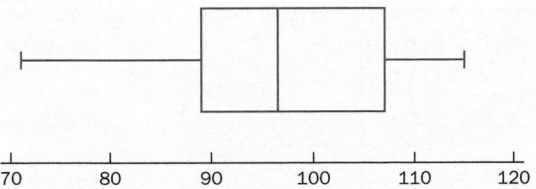

(c) The summary statistics and the box plot suggest that the distribution is approximately symmetric. The distribution is centered around 97, with lots of variability and no outliers.
(d) A person who drinks three cups of coffee has, on average,

around 291 mg ($= 3 \times 97$) of caffeine. This is under the moderate amount of 300 mg.

3.107 (a)

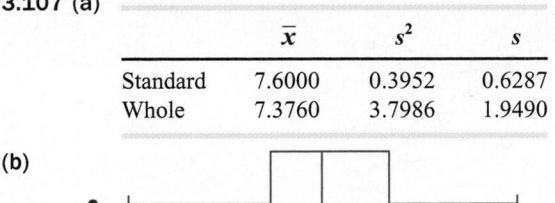

	$\bar{x}$	s^2	s
Standard	7.6000	0.3952	0.6287
Whole	7.3760	3.7986	1.9490

(b)

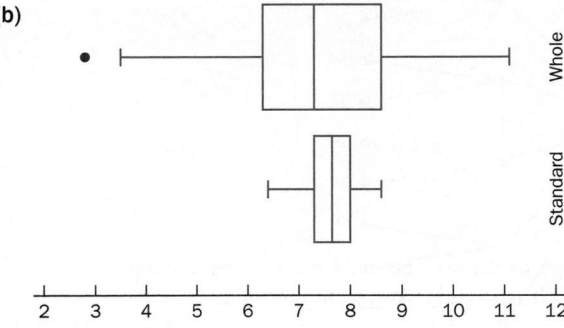

(c) The whole language reading speeds have much more variability and the center of the distribution is slightly smaller.

3.109 (a)

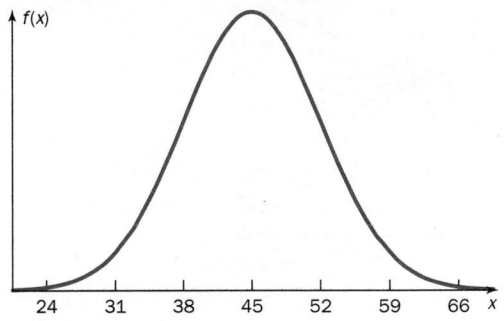

(b) Yes. 30 is over 2 standard deviations below the mean.
(c) 52

3.111 (a) 0.68 (b) 0.0015 (c) 0.4985 (d) No. This observation is within 2 standard deviations of the mean, a reasonable observation.

3.113 (a) At least 0.75. At most 0.1111. (b) There is evidence to suggest the manufacturer's claim is false. This is a very unusual observation.

3.115 (a) It is unlikely that a fisherman will catch a smallmouth bass with mercury level greater than 1, because this is 3 standard deviations from the mean. (b) It is even more unlikely that a fisherman will catch a smallmouth bass with mercury level greater than 1, because this is 6 standard deviations from the mean.

(c)

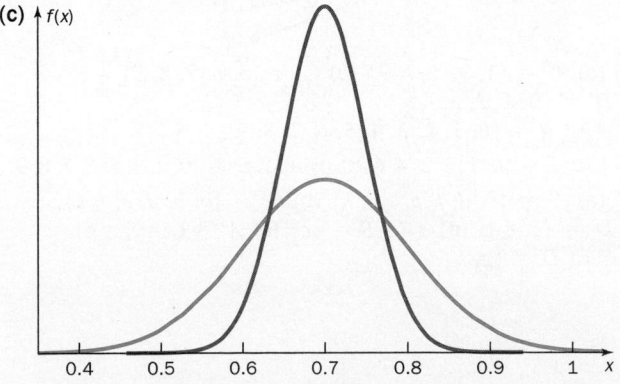

CHAPTER 4

Section 4.1

4.1

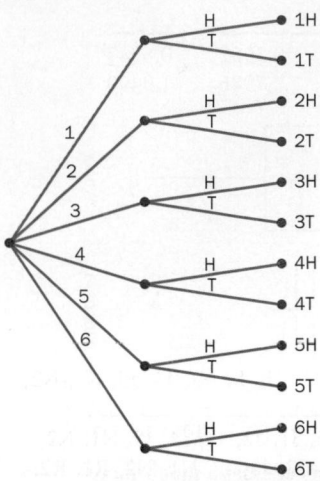

$S = \{1H, 2H, 3H, 4H, 5H, 6H, 1T, 2T, 3T, 4T, 5T, 6T\}$

4.3 25 outcomes.

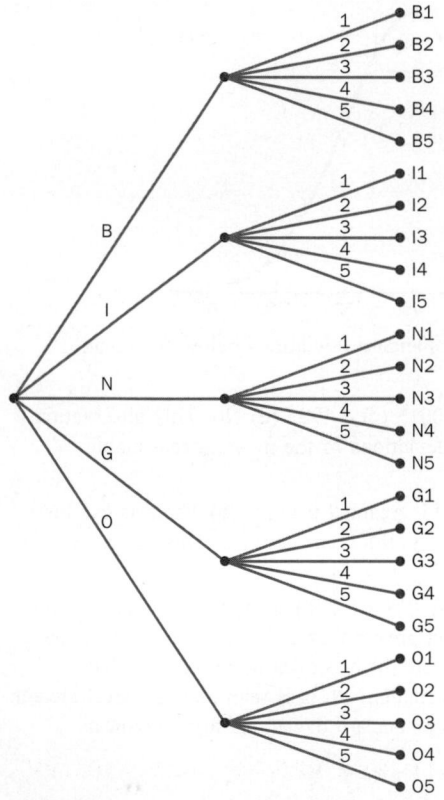

4.5 (a) $A' = \{1, 3, 5, 7, 9\}$ **(b)** $C' = \{5, 6, 7, 8, 9\}$
(c) $D' = \{0, 1, 2, 3, 4\}$
(d) $A \cup B = \{0, 1, 2, 3, 4, 5, 6, 7, 8, 9\} = S$
(e) $A \cup C = \{0, 1, 2, 3, 4, 6, 8\}$ **(f)** $A \cup D = \{0, 2, 4, 5, 6, 7, 8, 9\}$
4.7 (a) $A' = \{b, d, f, h, i, j, k\}$ **(b)** $C' = \{a, b, d, e, j, k\}$
(c) $D' = \{c, f, i\}$ **(d)** $A \cap B = \{c\}$ **(e)** $A \cap C = \{c, g\}$
(f) $C \cap D = \{g, h\}$

4.9 (a) $(A \cap B \cap C)' = \{a, b, d, e, f, g, h, i, j, k\}$
(b) $A \cup B \cup C \cup D = \{a, b, c, d, e, f, g, h, i, j, k\}$
(c) $(B \cup C \cup D)' = \{\ \}$ **(d)** $B' \cap C' \cap D' = \{\ \}$

4.11

(a)

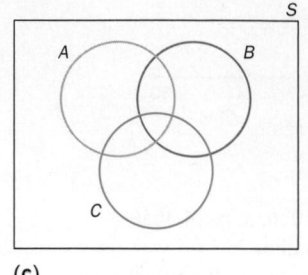

(b)

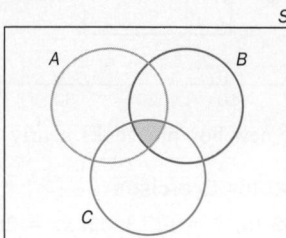

(c)

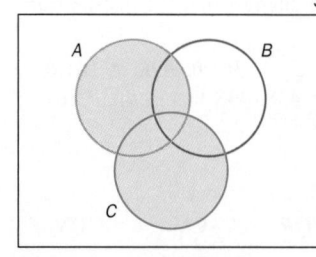

(d)

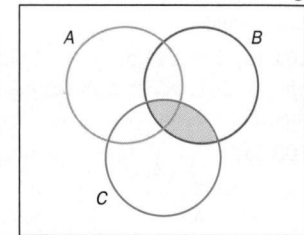

(e)

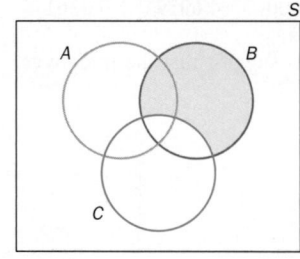

(f)

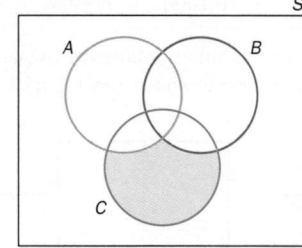

(g)

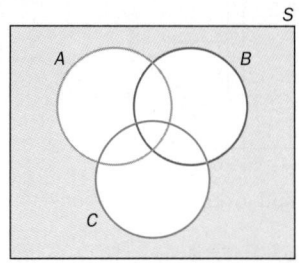

(h)

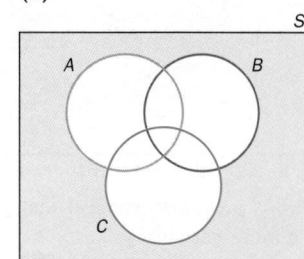

(i)

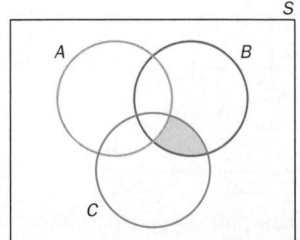

4.13

(a)   **(b)**

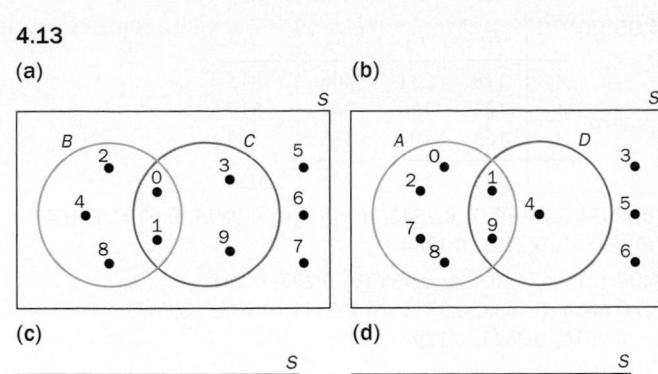

(c)   **(d)**

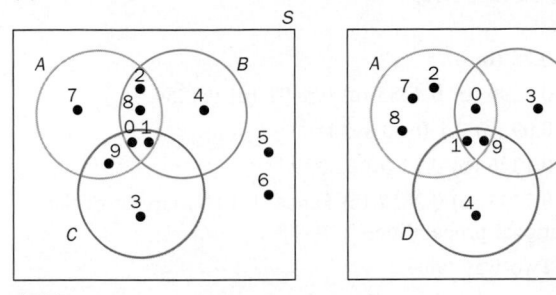

4.15

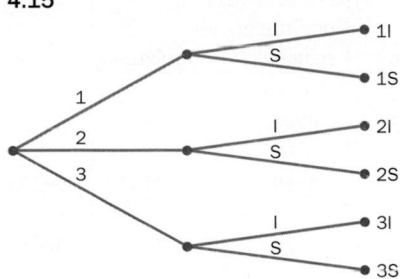

$S = \{1I, 1S, 2I, 2S, 3I, 3S\}$

4.17

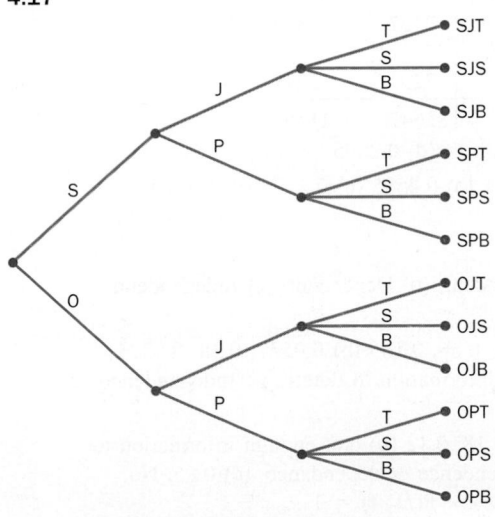

$S = \{$SJT, SJS, SJB, SPT, SPS, SPB,
OJT, OJS, OJB, OPT, OPS, OPB$\}$

4.19 (a) 4 **(b)** No. The experiment is over as soon as the bad battery is found.

4.21 398

4.23 (a) $S = \{$LS, LU, LV, LP, RS, RU, RV, RP, SS, SU, SV, SP$\}$ **(b)** $A = \{$LV, RV, SV$\}$, $B = \{$LS, LP, RS, RP, SS, SP$\}$, $C = \{$LS, LU, LV, LP$\}$, $D = \{$RS, RU, RV, RP, SS, SU, SV, SP$\}$ **(c)** $C \cup D = S$, $C \cap D = \{\ \}$

4.25 (a) $S = \{$A0, A1, A2, A3, A4, A5, F0, F1, F2, F3, F4, F5$\}$ **(b)** A = The passenger has 0 bags. B = The passenger is foreign. C = The passenger has 1 or 2 bags. D = The passenger is foreign and has 0 or 5 bags. E = The passenger has an odd number of bags.

4.27 (a)

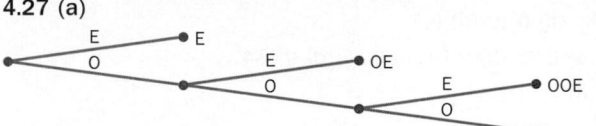

(b) $S = \{$E, OE, OOE, OOOE, . . .$\}$

4.29 (a) $S = \{$R1, R2, R3, R4, R5, J1, J2, J3, J4, J5, N1, N2, N3, N4, N5, C1, C2, C3, C4, C5$\}$
(b) (i) $A' = \{$C1, C2, C3, C4, C5, J1, J2, J3, J4, J5, N1, N2, N3, N4, N5$\}$ **(ii)** $A \cup C = \{$C1, C2, J1, J2, N1, N2, R1, R2, R3, R4, R5$\}$ **(iii)** $A \cap D = \{\ \}$ **(iv)** $C \cap D = \{$C1$\}$
(v) $A \cap C \cap D = \{\ \}$ **(vi)** $A \cap B = \{\ \}$

4.31

(a) $S = \{$B0, B1, B2, B3, B4, P0, P1, P2, P3, P4$\}$
(b) (i) $A \cup B = \{$B0, B1, B2, B3, B4, P0$\}$
(ii) $A \cap B = \{$B0$\}$ **(iii)** $B \cup C = \{$B0, B1, P0, P1$\}$
(iv) $B \cap C = \{$B0, P0$\}$ **(v)** $A \cap D = \{$B3$\}$
(vi) $A \cap B \cap C \cap D = \{\ \}$

Section 4.2

4.33 (a) 0.29 **(b)** 0.45 **(c)** 0.84 **(d)** 0.78 **(e)** 0.07 **(f)** 0.49 **(g)** 0.71 **(h)** 0.22 **(i)** 0.71 **(j)** 0.55 **(k)** 0.16 **(l)** 0.13 Probabilities sum to 1.

4.35 (a) 0.2727 **(b)** 0.5 **(c)** 0.5 **(d)** 0.1364

4.37 (a) 0.14 **(b)** 0.74 **(c)** 0.86 **(d)** 0.20

4.39

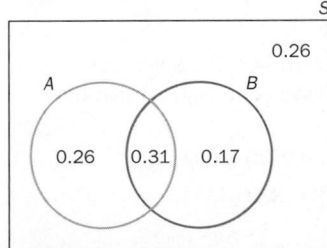

4.41 (a) 10, $\{$HLP, HLD, HLV, HPD, HPV, HDV, LPD, LPV, LDV, PDV$\}$ **(b)** 0.6 **(c)** 0.3

4.43 (a) 0.591 **(b)** 0.033 **(c)** 0.999

4.45 (a) 0.09 **(b)** 0.897 **(c)** 0.297

4.47 (a) 0.23, 0.52, 0.40 **(b)** 0.75, 0, 0.12
(c) 0.77, 0.17, 0 **(d)** 0.20, 0.20

4.49 (a) 0.509, 0.655, 0.133 **(b)** 0.509, 0.642, 0 **(c)** 0.491, 1

4.51 (a)

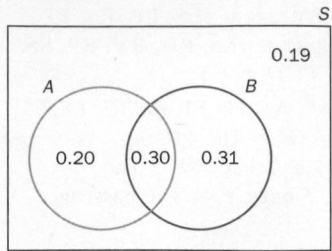

(b) 0.81 **(c)** 0.19 **(d)** 0.2

4.53 (a) 0.85 **(b)** 0.10 **(c)** 0.56 **(d)** 0.184

4.55 (a)

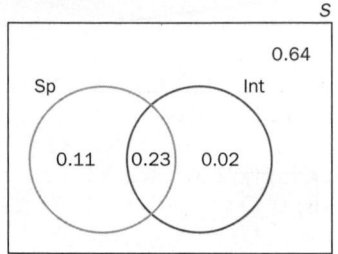

(b) 0.36 **(c)** 0.64 **(d)** 0.13

Section 4.3

4.57 (a) 1,680 **(b)** 1,663,200 **(c)** 11,880 **(d)** 3,628,800 **(e)** 10 **(f)** 1 **(g)** 72 **(h)** 380 **(i)** 9,900

4.59 362,880

4.61 390,700,800

4.63 6840

4.65 (a) 64,000 **(b)** 0.0156 **(c)** 59,280, 0.0121

4.67 (a) 216 **(b)** 144 **(c)** 36

4.69 (a) 0.3626 **(b)** 0.0088 **(c)** 0.6374

4.71 (a) 19,958,400 **(b)** 0.0152 **(c)** 0.0909

4.73 (a) 455 **(b)** 0.022 **(c)** 0.2637 **(d)** 0.8

4.75 (a) 40,320 **(b)** 0.125

4.77 (a) 2550 **(b)** 0.84 **(c)** 0.4706

4.79 (a) 20 **(b)** P (two girls selected) = 0.10. Since this probability is so small, there is evidence to suggest that the process was not random.

4.81 (a) 3,628,800 **(b)** 0.0222 **(c)** 0.2 **(d)** 0.004

4.83 (a) 1326 **(b)** 0.0045 **(c)** 0.0588 **(d)** 0.2353

4.85 (a) 4 **(b)** 8 **(c)** 16 **(d)** 2^n

4.87 $(n - 1)!$

4.89 (a) 7.3321×10^{21} **(b)** 5.6147×10^{-10} (pretty close to 0) **(c)** 0.2185

Section 4.4

4.91 (a) Conditional. **(b)** Unconditional. **(c)** Unconditional. **(d)** Conditional. **(e)** Unconditional.

4.93 (a) 0.118, 0.396, 0.486 **(b)** 0.455, 0.442, 0.103 **(c)** 0.095, 0.188, 0.093 **(d)** 0.2088, 0.8051, 0.5220 **(e)** 0.4747, 0.1914

4.95 (a)

	B_1	B_2	B_3	
A_1	178	231	406	815
A_2	123	150	244	517
A_3	165	202	335	702
	466	583	985	2034

(b) 2034 **(c)** 0.4007, 0.2542, 0.3451 **(d)** 0.0875, 0.0737, 0.1647 **(e)** 0.3541, 0.2901, 0.1804

4.97 (a) 0.574, 0.488, 0.465 **(b)** 0.297, 0.218 **(c)** 0.6086, 0.4688, 0.3333 **(d)** 0.2523, 0.1355, 0.7477 **(e)** 0.2910, 0.2623, 0.1291

4.99 0.6

4.101 (a) 0.75 **(b)** 0.75

4.103 (a) 0.5299 **(b)** 0.4085 **(c)** 0.5491 **(d)** 0.3254

4.105 (a) 0.09 **(b)** 0.1 **(c)** 0.8934

4.107 (a) 0.1326 **(b)** 0.22 **(c)** 0.5339

4.109 (a) 0.7234 **(b)** 0.2857 **(c)** Fence. $P(F|N)$ largest of the three conditional probabilities.

4.111 (a) Two-way table:

| | | Type of sentence | | |
| | | | Conditional | |
Criminal code		Prison	sentence	Probation	
Crimes of violence		15302	2791	36466	54559
Property crimes		24443	3619	33193	61255
Administration of justice		21412	1026	13635	36073
Other criminal code offenses		6871	671	9940	17482
Criminal code offenses (traffic)		7327	833	6659	14819
Other federal statute		7292	2214	8584	18090
		82647	11154	108477	202278

(b) 0.0179 **(c)** 0.0614 **(d)** 0.2805

4.113 (a) 0.0071 **(b)** 0.4595 **(c)** 0.2194 **(d)** 0.1285

Section 4.5

4.115 (a) Dependent. **(b)** Dependent. **(c)** Independent. **(d)** Dependent.

4.117 (a) 0.085, 0.66, 0.165 **(b)** 0.0527, 0.38, 0.0323 **(c)** Not enough information to determine independence or dependence.

4.119 (a) 0.1, 0.18, 0.12 **(b)** Not enough information to determine independence or dependence. **(c)** 0.75. No, $P(B|A) + P(C|A) + P(D|A) = 1$

4.121 (a) $P(A') = 0.65$, $P(C|A) = 0.18$, $P(B|A') = 0.36$ **(b)** 0.0630, 0.234 **(c)** 0.451

4.123 (a) 0.0000000615 **(b)** 0.9995 **(c)** 0.00049588

4.125 (a) 0.25 **(b)** 0.5 **(c)** 0.75

4.127 (a) 0.2549 **(b)** 0.0032 **(c)** 0.9968 **(d)** All four volcanoes erupted in 2008: 0.2549.

4.129 (a) 0.0016 **(b)** 0.4096 **(c)** 0.1808

4.131 (a) 0.2646 (b) 0.1554 (c) 0.3402

4.133 (a) 0.7531 (b) 0.0057 (c) 0.2412

4.135 (a) 0.2255 (b) 0.0583 (c) 0.2788

4.137 (a) 0.24 (b) 0.695 (c) 0.0072

4.139 (a) 0.0475 (b) 0.0665 (c) 0.7143

4.141 (a) 0.992 (b) 0.008 (c) 6

4.143 (a) $P(L) = 0.366$, $P(T'|D) = 0.774$, $P(T|L) = 0.156$, $P(B'|D \cap T) = 0.545$, $P(B'|D \cap T') = 0.622$, $P(B'|L \cap T) = 0.105$, $P(B'|L \cap T') = 0.005$ (b) 0.0652 (c) 0.3909 (d) 0.5803

4.145 Hotel C

Chapter Exercises

4.147 (a) $S = \{$BL, BM, BH, GL, GM, GH, EL, EM, EH$\}$ (b) $A = \{$EL, EM, EH$\}$, $B = \{$BH, GH, EH$\}$, $C = \{$BL, BM, BH, GL, EL$\}$, $D = \{$GM$\}$ (c) $A \cup B = \{$BH, EH, EL, EM, GH$\}$, $B \cup C = \{$BH, BL, BM, EH, EL, GH, GL$\}$, $D' = \{$BH, BL, BM, EH, EL, EM, GH, GL$\}$ (d) $A \cap B = \{$EH$\}$, $C \cap D = \{\ \}$, $(B \cup D)' = \{$BL, BM, EL, EM, GL$\}$

4.149 (a) 0.234, 0.92, 0.193 (b) 0.234, 0.427, 0.92 (c) 0.807, 1.0, 0.08

4.151 (a) 0.0016 (b) 0.4096 (c) 0.1536

4.153 (a) 0.2857 (b) 0.2449 (c) 0.5974 (d) 0.4998

4.155 (a) 0.0673 (b) 0.1904 (c) 0.1152 (d) 0.3057

4.157 (a) 0.0039 (b) 0.3164 (c) 0.2109

4.159 (a) 0.16 (b) 0.265 (c) 0.2264

4.161 (a) 0.0630 (b) 0.4475 (c) 0.5520

4.163 (a)

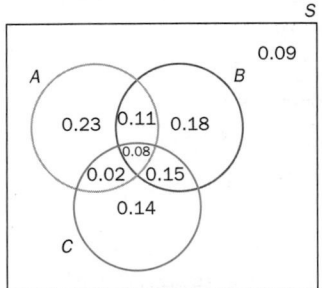

(b) 0.23 (c) 0.09 (d) 0.2564, 0.8, 0.1818

4.165 (a) 0.0509 (b) 0.5586 (c) 0.1593 (d) 0.1608 (e) 0.9491 (f) The events White and All others TR are not independent. (g) 0.0358

CHAPTER 5

Section 5.1

5.1 (a) Discrete. (b) Continuous. (c) Continuous. (d) Discrete. (e) Discrete. (f) Continuous. (g) Discrete. (h) Discrete.

5.3 (a) Discrete. (b) Continuous. (c) Continuous. (d) Discrete. (e) Continuous. (f) Discrete.

5.5 Continuous. Measuring a length of time.

5.7 Continuous. Measuring a distance.

Section 5.2

5.9 (a) 0.07 (b) 0.62, 0.42 (c) 0.7 (d) 0.38

5.11 (a) 0.65, 0.35 (b) 0.6 (c) 0.4615

(d)

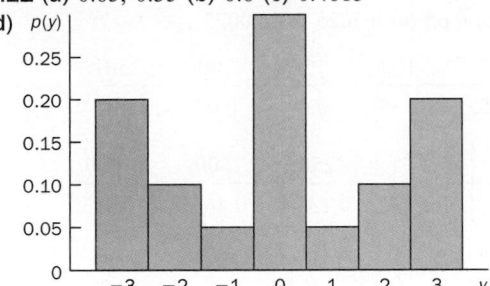

5.13

x	1	2	3	4
$p(x)$	0.01	0.08	0.27	0.64

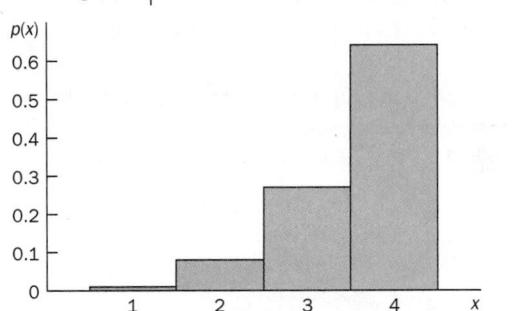

5.15 (a) $p(x) \geq 0$ for all x and $\sum_{x=1}^{6} p(x) = 1$ (b) 0.1786 (c) 0.9286 (d) 0.2857

(e)

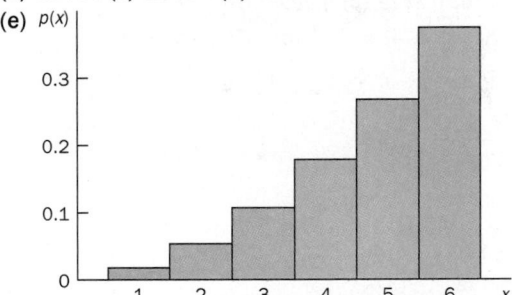

5.17 (a) 0.35 (b) 0.95 (c) 0.04 (d) 0.01 (e) 0.5775

5.19 (a)

x	0	1	2	3
$p(x)$	0.343	0.441	0.189	0.027

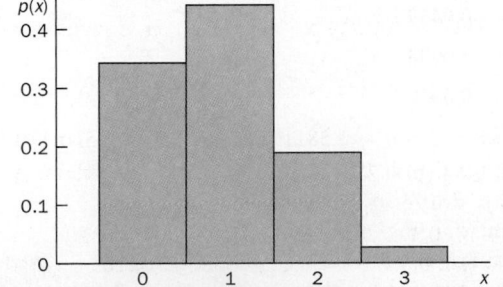

(b) 0.027 (c) 0.657

5.21

x	0	1	2
$p(x)$	0.4000	0.5333	0.0667

5.23 (a) 0.1 (b) 0.65 (c) 0.3025 (d) 0.0023
(e)

y	50	100	150	200	250
$p(y)$	0.55	0.35	0.07	0.02	0.01

5.25 (a)

m	100	250	500	1000
$p(m)$	0.0667	0.1333	0.4667	0.3333

(b) 0.1111

Section 5.3

5.27 $\mu = 7.2$, $\sigma^2 = 8.96$, $\sigma = 2.9933$
5.29 (a) $\mu = 0$, $\sigma^2 = 270$, $\sigma = 16.4317$ (b) 1 (c) 0.55, 0.45
5.31 (a) $\mu = 7.35$, $\sigma^2 = 24.6275$, $\sigma = 4.9626$ (b) 0.4 (c) 0.95
5.33 (a) $\mu = 1.085$ (b) $\sigma^2 = 0.1503$, $\sigma = 0.3877$ (c) 0.7 (d) 0.84
5.35 (a) $\mu = 15.55$, $\sigma^2 = 40.4475$, $\sigma = 6.3598$ (b) 0.82
(c) 0.92 (d) 386
5.37 (a) $\mu = 7.998$, $\sigma^2 = 6.706$, $\sigma = 2.5896$ (b) 0.6 (c) 0.000144
5.39 (a) $\mu = 12.725$, $\sigma^2 = 2.1594$, $\sigma = 1.4695$ (b) 0.81
(c) $\mu = 16.45$, $\sigma^2 = 8.6375$, $\sigma = 2.939$
5.41 (a) $p(x) \geq 0$ for all x and $\sum_{\text{all } x} p(x) = 1$
(b) $\mu = 0$, $\sigma^2 = 5.2632$, $\sigma = 2.2942$ (c) 0.0028
5.43 $\mu_Y = a\mu_X + b$, $\sigma_Y^2 = a^2\sigma_X^2$
5.45 $\mu_Y = 0$, $\sigma_Y^2 = 1$, $\sigma_Y = 1$

Section 5.4

5.47 (a) 0.0565 (b) 0.8829 (c) 0.9997 (d) 0.5920
5.49 (a) $\mu = 20$, $\sigma^2 = 4$, $\sigma = 2$ (b) 0.7927 (c) 0.0211
5.51 (a)

x	$p(x)$
0	0.0010
1	0.0098
2	0.0439
3	0.1172
4	0.2051
5	0.2461
6	0.2051
7	0.1172
8	0.0439
9	0.0098
10	0.0010

(b) $\mu = 5$, $\sigma^2 = 2.5$, $\sigma = 1.5811$ (c) $\mu = 5$, $\sigma^2 = 2.5$, $\sigma = 1.5811$
5.53 (a) 0.1484 (b) 0.2361
(c) Claim: $p = 0.75 \Rightarrow X \sim B(15, 0.75)$
Experiment: $x = 9$
Likelihood: $P(X \leq 9) = 0.1484$
Conclusion: There is no evidence to suggest that the claim is false.
5.55 (a) 0.1171 (b) 0.1256 (c) 0.5841
(d) Claim: $p = 0.60 \Rightarrow X \sim B(20, 0.60)$

Experiment: $x = 19$
Likelihood: $P(X \geq 19) = 0.0005$
Conclusion: There is evidence to suggest that the claim is false.
5.57 (a) $\mu = 7.5$, $\sigma^2 = 5.625$, $\sigma = 2.3717$ (b) 0.6008 (c) 0.0322
(d) Claim: $p = 0.25 \Rightarrow X \sim B(30, 0.25)$
Experiment: $x = 10$
Likelihood: $P(X \geq 10) = 0.1966$
Conclusion: There is no evidence to suggest that the claim is false.
5.59 (a) 0.1326 (b) 2 (c) 0.3233
(d) Claim: $p = 0.02 \Rightarrow X \sim B(100, 0.02)$
Experiment: $x = 6$
Likelihood: $P(X \geq 6) = 0.0155$
Conclusion: There is evidence to suggest that the claim is false.
5.61 (a) $\mu = 6.837$, $\sigma^2 = 5.2788$, $\sigma = 2.2976$ (b) 0.0607 (c) 0.0024
(d) Claim: $p = 0.2279 \Rightarrow X \sim B (30, 0.2279)$
Experiment: $x = 5$
Likelihood: $P(X \leq 5) = 0.29$
Conclusion: There is no evidence to suggest that the claim is
false.
5.63 (a) 0.0087 (b) 0.3331
(c) Claim: $p = 0.44 \Rightarrow X \sim B(50, 0.44)$
Experiment: $x = 17$
Likelihood: $P(X \leq 17) = 0.0990$
Conclusion: There is no evidence to suggest that the claim is false.
5.65 (a) 0.0432 (b) $\mu = 2$, 0.2852 (c) 0.1271
5.67 (a) 0.2880, 0.0640 (b) 0.9744 (c) 0.9898 (d) $n \geq 8$
5.69 (a) 0.0014 (b) 0.9659 (c) 0.9064

Section 5.5

5.71 (a) 0.0961 (b) 0.4225 (c) 0.5775 (d) 0.4225
5.73 (a) 0.1353 (b) 0.5938 (c) 0.0166 (d) 0.9955
5.75 (a) 0.3788 (b) 0.0076 (c) $\mu = 2.5$, $\sigma^2 = 0.7955$, $\sigma = 0.8919$
5.77 (a) 0.0630 (b) 0.3 (c) 1.4286 (d) 71.4286
5.79 (a) 0.09 (b) 0.0000009 (c) 1.1111 (d) 0.0001
5.81 (a) 0.9857 (b) 0.0537 (c) 0.0009119
5.83 (a) 0.0821 (b) 0.0042 (c) 0.9580
5.85 (a) 0.3297 (b) 0.0037 (c) 0.8462
5.87 (a) 0.4789 (b) 0.0789 (c) 0.6 (d) 10
5.89 (a) 0.1280 (b) 0.5379 (c) 0.2621 (d) 0.5120 (e) 0.5120
5.91 (a) 0.0183, 0.0733, 0.1465 (b) 0.0003, 0.0027, 0.0107
(c) 0.0003, 0.0027, 0.0107 (d) Same. Poisson random variable
is additive.

Chapter Exercises

5.93 (a) 0.2794 (b) 0.9838
(c) Claim: $p = 0.93 \Rightarrow X \sim B(30, 0.93)$
Experiment: $x = 20$
Likelihood: $P(X \leq 20) = 0.00002296$
Conclusion: There is evidence to suggest that the claim is false.
5.95 (a) $\mu = 5.93$, $\sigma^2 = 4.4651$, $\sigma = 2.1131$ (b) 0.24
(c) 0.16 (d) 0.75
5.97 (a) 0.4966 (b) 0.00009 (c) 0.6016
5.99 (a) 0.1653 (b) 0.5346
(c) Claim: $\mu = 1.8 \Rightarrow X$ is a Poisson random variable with
$\lambda = 1.8$.
Experiment: $x = 9$
Likelihood: $P(X \geq 9) = 0.0.0001$

Conclusion: There is evidence to suggest that the claim is false.
(d) 0.0001167

5.101 (a) 0.2721 (b) 0.0004 (c) 84

5.103 (a) $\mu = 42$, $\sigma^2 = 6.72$, $\sigma = 2.5923$ (b) 0.8339 (c) 0.9687

5.105 (a) 0.2194 (b) 0.4512
(c) Claim: $p = 0.85 \Rightarrow X \sim B(50, 0.85)$
Experiment: $x = 35$
Likelihood: $P(X \le 35) = 0.0053$
Conclusion: There is evidence to suggest that the claim is false—that the poll results are wrong.

5.107 (a) $\mu = 3$, $\sigma^2 = 2$, $\sigma = 1.4142$ (b) $\mu = 3.5$,
$\sigma^2 = 2.9167$, $\sigma = 1.7078$ (c) $\mu = (n + 1)/2$, $\sigma^2 = (n^2 - 1)/12$,
$\sigma = \sqrt{(n^2 - 1)/12}$

5.109 (a)

x	10	20	30	40
$p(x)$	0.2500	0.4725	0.1800	0.0975

(b) $\mu = 21.25$, $\sigma^2 = 80.4375$, $\sigma = 8.9687$ (c) 0.75

5.111 Answers will vary.
(a)

Complete shipments	Relative frequency	$p(y)$
0	0.0000	0.0000
1	0.0000	0.0007
2	0.0060	0.0052
3	0.1600	0.0234
4	0.0790	0.0701
5	0.1300	0.1471
6	0.2020	0.2207
7	0.2640	0.2365
8	0.1850	0.1774
9	0.0830	0.0887
10	0.0320	0.0266
11	0.0030	0.0036

(b)

Complete shipments	Relative frequency	$p(y)$
0	0.0000	0.0000
1	0.0000	0.0000
2	0.0000	0.0001
3	0.0000	0.0008
4	0.0050	0.0040
5	0.0140	0.0142
6	0.0390	0.0392
7	0.0960	0.0840
8	0.1210	0.1417
9	0.1980	0.1889
10	0.1910	0.1983
11	0.1590	0.1623
12	0.1050	0.1014
13	0.0580	0.0468
14	0.0120	0.0150
15	0.0020	0.0030
16	0.0000	0.0003

(c) 0.1612, 0.1538, 0.2735, 15

CHAPTER 6

Section 6.1

6.1 (a)

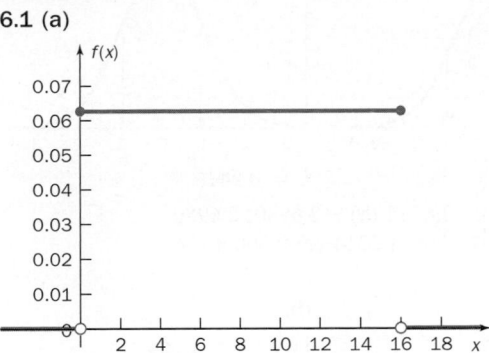

(b) $\mu = 8$, $\sigma^2 = 21.3333$, $\sigma = 4.6188$ (c) 0.75
(d) 0.625 (e) 0.4375

6.3 (a) $\mu = 75$, $\sigma^2 = 208.3333$, $\sigma = 14.4338$ (b) 0.5774
(c) 0 (d) $c = 60$

6.5 (a) 0.0625 (b) 0.4375 (c) 0 (d) 0.3125 (e) 0.4444
(f) 2.8284. The distribution is not symmetric.

6.7 (a) 0.2143 (b) 0.3 (c) 0.0571

6.9 (a) 0.5 (b) 0.3333 (c) 37.5 (d) 0.1667

6.11 (a) $f(x) \ge 0$ and total area is 1. (b) 0.4375
(c) 0.3125 (d) $t = 5.8579$ (e) 0.0056, 0.8741, 0.2751

6.13 (a) $f(x) \ge 0$ and total area is 1. (b) 0.75
(c) 0.75 (d) 0.25

6.15 (a) 0.6321 (b) 0.2231 (c) 0.2492

6.17 (a) 0.8647 (b) 0.6065 (c) 0.5578 (d) 0.4551

Section 6.2

6.19 (a) 0.9846 (b) 0.9846 (c) 0.3192 (d) 0.7673
(e) 0.0401 (f) 0.3790 (g) 1 (h) 0 (i) 1

6.21 (a) 0.6827 (b) 0.9545 (c) 0.9973. These are the Empirical Rule probabilities.

6.23

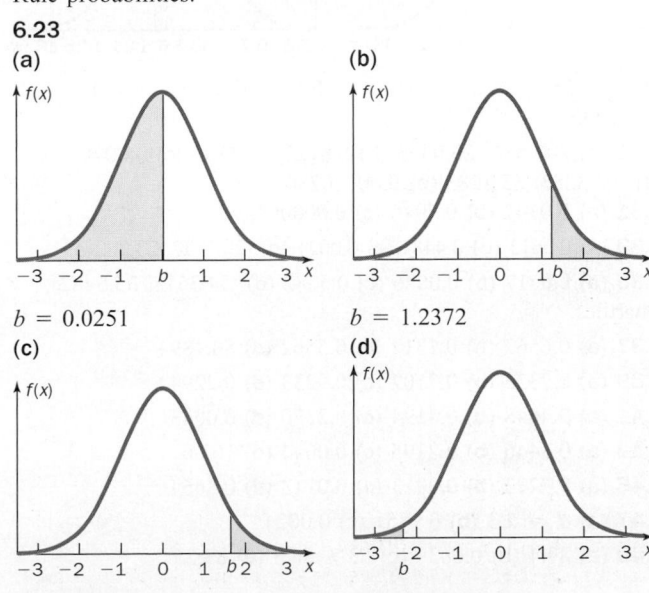

(a)

$b = 0.0251$

(b)

$b = 1.2372$

(c)

$b = 1.6449$

(d)

$b = -2.3264$

(e)

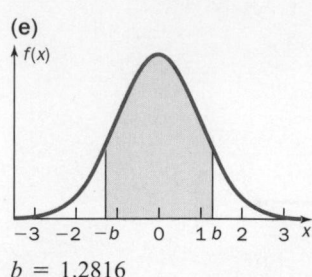

$b = 1.2816$

(f)

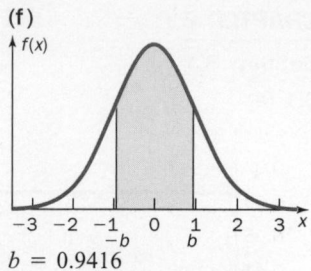

$b = 0.9416$

6.25 **(a)** -0.6745, 0.6745 **(b)** -2.6980, 2.6980
(c) 0.0070 **(d)** -4.7214, 4.7214 **(e)** 0.00000234

6.27

(a)

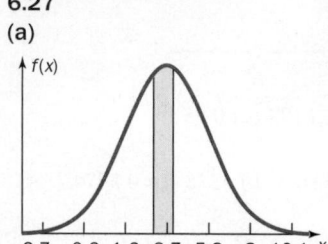

$P(3 \leq X \leq 4) = 0.1845$

(b)

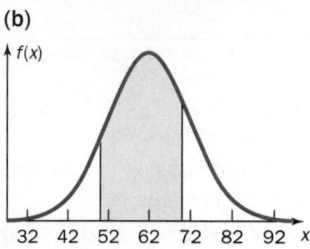

$P(50 < X < 70) = 0.6731$

(c)

$P(X \geq 45) = 0.0088$

(d)

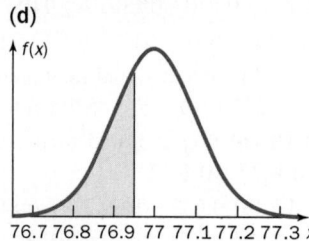

$P(X < 76.95) = 0.3085$

(e)

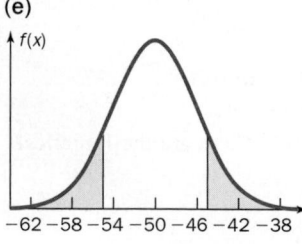

$P(X < -55 \cup X > -45)$
$= 0.2113$

(f)

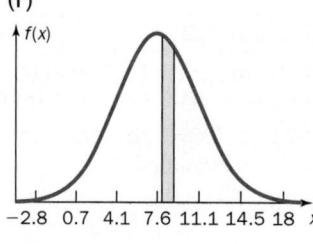

$P(8 \leq X \leq 9) = 0.1110$

6.29 **(a)** 20.9531, 29.0469 **(b)** 8.8123, 41.1876 **(c)** 0.0070
(d) -3.3283, 53.3283 **(e)** 0.00000234
6.31 **(a)** 0.0147 **(b)** 0.7946 **(c)** 0.000003598
6.33 **(a)** 0.0613 **(b)** 0.4481 **(c)** 0.0021 **(d)** $(20.7172, 23.3828)$
6.35 **(a)** 0.0217 **(b)** 0.0026 **(c)** 0.5382 **(d)** $(34.3518, 35.5482)$,
Quartiles.
6.37 **(a)** 0.6562 **(b)** 0.1312 **(c)** 0.5562 **(d)** 80.1894
6.39 **(a)** 0.7335 **(b)** 0.7107 **(c)** 0.9233 **(d)** 0.2294
6.41 **(a)** 0.4648 **(b)** 0.4194 **(c)** 0.2370 **(d)** 0.0043
6.43 **(a)** 0.3446 **(b)** 0.2195 **(c)** 0.0044 **(d)** 103.6
6.45 **(a)** 0.7273 **(b)** 0.0013 **(c)** 0.0118 **(d)** 0.3664
6.47 **(a)** $\sigma = 2.3$ **(b)** 0.9851 **(c)** 0.0024
6.49 **(a)** 89.4 **(b)** 0.0039 **(c)** 83.7, 86.3 **(d)** 83.25

Section 6.3

6.51

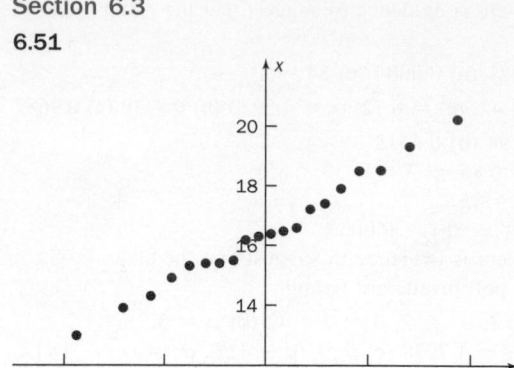

There is no evidence to suggest a non-normal population.
6.53 **(a)** There is no evidence to suggest that the data are from
a nonnormal population. **(b)** There is evidence to suggest that
the data are from a nonnormal population. The points do not
appear to fall on a straight line. **(c)** There is evidence to suggest
that the data are from a nonnormal population. The points do
not appear to fall on a straight line. **(d)** There is no evidence to
suggest that the data are from a nonnormal population.

6.55

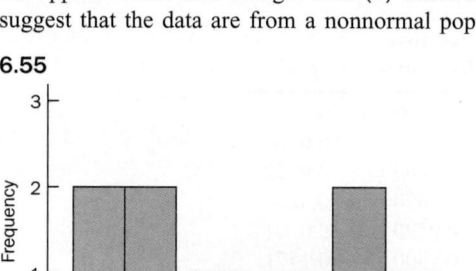

Backward Empirical Rule

Interval	Proportion
$(5.83, 10.51)$	0.5
$(3.48, 12.85)$	1.0
$(1.14, 15.20)$	1.0

$IQR/s = 1.93$
Normal probability plot:

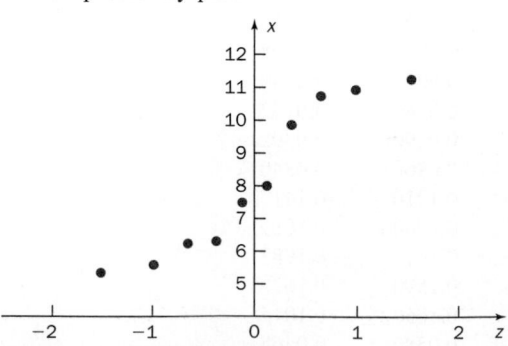

There is evidence to suggest that the data are from a
nonnormal population.

6.57

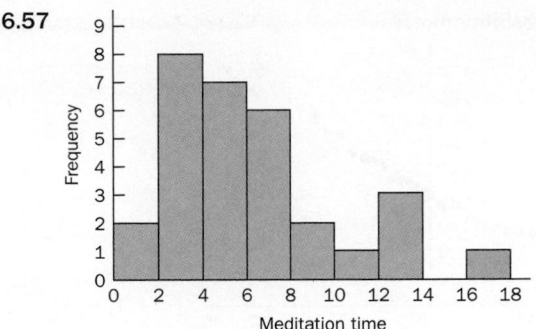

Backward Empirical Rule

Interval	Proportion
(2.37, 10.14)	0.73
(−1.51, 14.02)	0.97
(−5.39, 17.90)	1.00

$IQR/s = 1.083$

Normal probability plot:

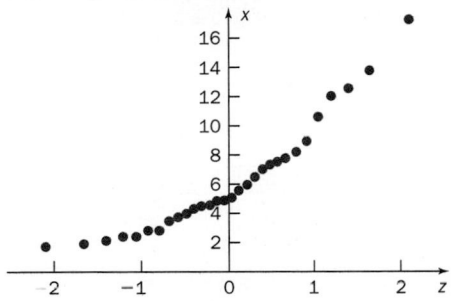

There is evidence to suggest that the data are from a nonnormal population.

6.59 (a) $\bar{x} = 10.2894$, $s = 1.7593$ **(b)** (8.53, 12.05), (6.77, 13.81), (5.01, 15.57) **(c)** 0.72, 0.96, 1.00. There is no evidence to suggest that the data are from a nonnormal population.

6.61 (a)

37	3
38	36
39	
40	9
41	5
42	348
43	03446
44	025
45	6
46	36
47	
48	
49	2

Stem: tenths; leaf: hundredths.

(b) $IQR/s = 0.8551$

(c)

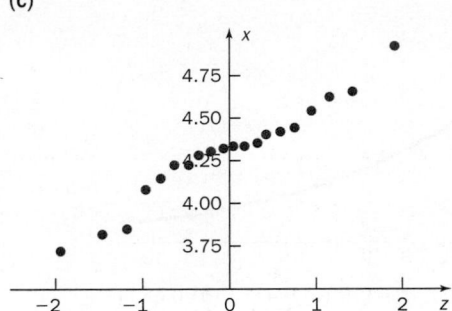

(d) There is some evidence to suggest that these data are from a nonnormal distribution. The stem-and-leaf plot has some outliers, the ratio IQR/s is far from 1.3, and the normal probability plot exhibits nonlinearity.

6.63

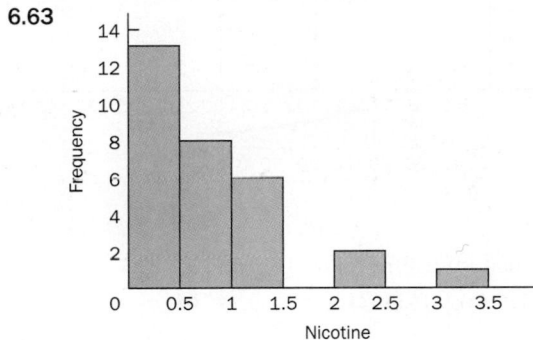

Backward Empirical Rule

Interval	Proportion
(0.0078, 1.5349)	0.90
(−0.7558, 2.2984)	0.93
(−1.5193, 3.0620)	0.97

$IQR/s = 1.0864$

Normal probability plot:

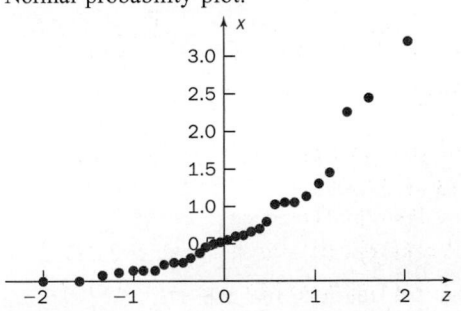

There is evidence to suggest that the data are from a nonnormal population.

Section 6.4

6.65 (a)

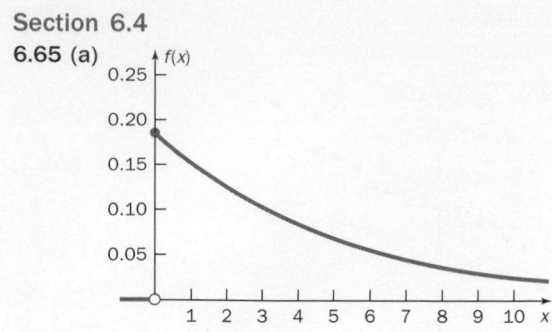

(b) $\mu = 10$, $\sigma^2 = 100$, $\sigma = 10$ **(c)** 0.4512 **(d)** 0.6534

6.67 (a) 0.4724

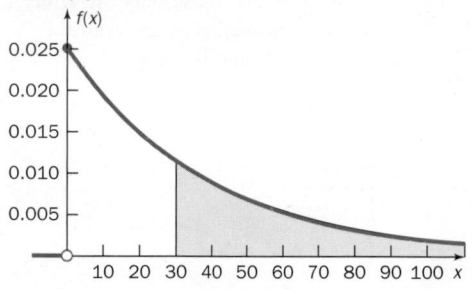

(b) 0.6065 **(c)** 0.6065

6.69 0.0015

6.71 (a) 0.00002 **(b)** 0.3012 **(c)** 0.2474 **(d)** 0.3297

6.73 (a) 0.9436 **(b)** 0.1534 **(c)** $\mu = 8$, $\sigma^2 = 64$, $\sigma = 8$
(d) 2.3015

6.75 (a) 0.6065 **(b)** 46.0517 **(c)** 0.6065

6.77 (a) 0.3935 **(b)** 0.4109 **(c)** 0.3679 **(d)** 0.0067

6.79 (a) $\mu = 30$, $\sigma^2 = 900$, $\sigma = 30$ **(b)** 0.2636
(c) 0.1790 **(d)** 69.0776 **(e)** 0.3893

Chapter Exercises

6.81 (a) 2.5 **(b)** 0.1813 **(c)** 0.1353 **(d)** 9.7801

6.83 (a) 0.0401 **(b)** 0.2417 **(c)** 0.9796 **(d)** 0.000149

6.85 Histogram:

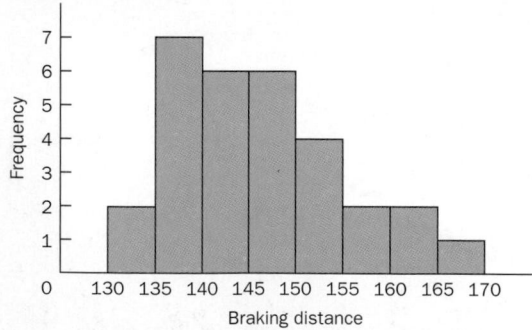

Backward Empirical Rule

Interval	Proportion
(138.13, 155.40)	0.70
(129.49, 164.04)	0.97
(120.85, 172.68)	1.00

$IQR/s = 1.52$

Normal probability plot:

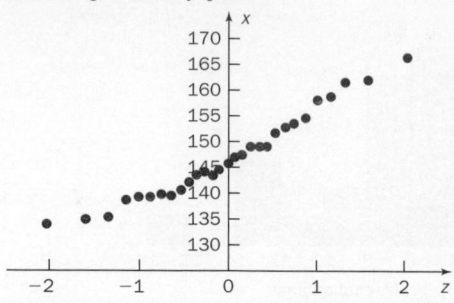

There is some evidence to suggest that the data are from a nonnormal population. The histogram is positively skewed, IQR/s is not close to 1.3, and the normal probability plot has a slight arc.

6.87 (a) 0.8054 **(b)** 0.0013 **(c)** 0.000337 **(d)** 0.0040

6.89 Histogram:

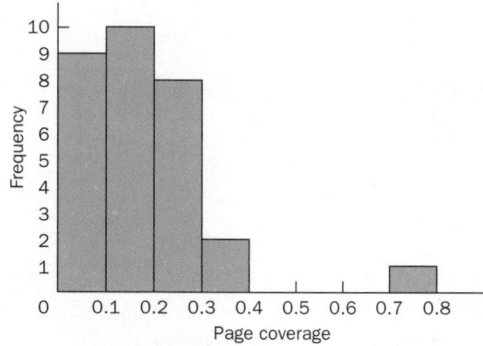

Backward Empirical Rule

Interval	Proportion
(0.0441, 0.3019)	0.87
(−0.0847, 0.4307)	0.97
(−0.2136, 0.5596)	0.97

$IQR/s = 1.241$

Normal probability plot:

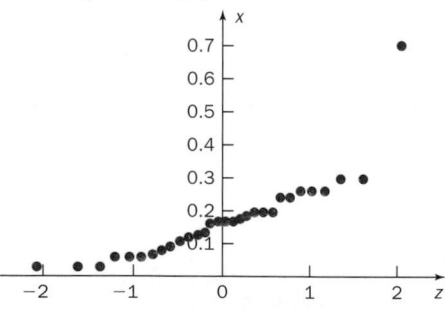

There is some evidence to suggest that the data are from a nonnormal population. The histogram and the normal probability plot indicate an outlier, and the backward Empirical Rule proportions are inconsistent.

6.91 (a) 0.0985 **(b)** 0.6421 **(c)** 0.0049

6.93 (a) 0.0599 **(b)** 0.4191 **(c)** (8180.16, 25819.84) **(d)** 0.0025

6.95 (a) 0.5 **(b)** 0.0923 **(c)** 0.000687

6.97 (a)

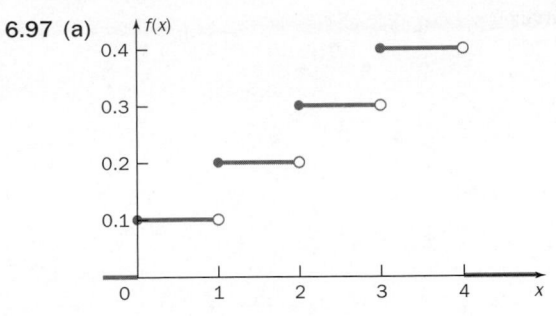

(b) 0.2 **(c)** 0.4 **(d)** 0.7 **(e)** 0.1667

6.99 (a)

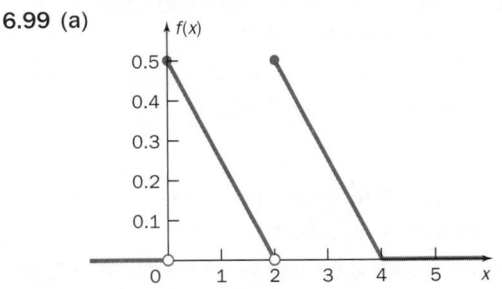

(b) 0.375 **(c)** 0.125 **(d)** 0.5

CHAPTER 7

Section 7.1

7.1 (a) Statistic. **(b)** Parameter. **(c)** Statistic. **(d)** Parameter.
(e) Statistic.

7.3 (a) $\mu = 16$, $\tilde{u} = 15$ **(b)** Sampling distribution:

$\bar{x}$	12.33	13.33	14.33	15.00	15.67
$p(\bar{x})$	0.1	0.1	0.1	0.1	0.1

$\bar{x}$	16.67	17.33	17.67	18.33	19.33
$p(\bar{x})$	0.1	0.1	0.1	0.1	0.1

$u_{\bar{X}} = 16$, $\sigma^2_{\bar{X}} = 4.6$, $\sigma_{\bar{X}} = 2.1448$

(c) Sampling distribution:

$\tilde{x}$	12	15	18
$p(\tilde{x})$	0.3	0.4	0.3

$\mu_{\tilde{X}} = 15$, $\sigma^2_{\tilde{X}} = 5.4$, $\sigma_{\tilde{X}} = 2.3238$ **(d)** The mean of the sample mean is the population mean. The mean of the sample median is the population median.

7.5 (a) Random samples will vary. **(b)** Histogram:

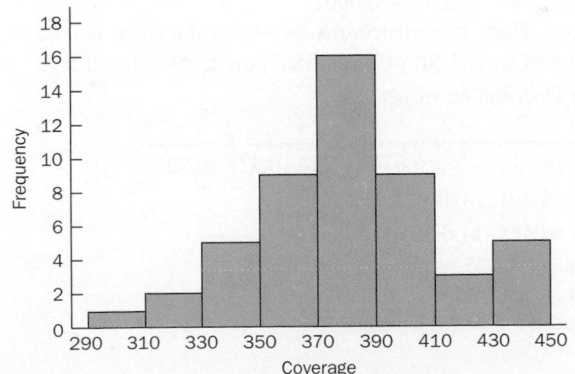

(c) Approximately normal. Approximate mean: 380
(d) $\mu = 379.7$, almost the same.

7.7 (a) Distribution of $\tilde{X}$:

$\tilde{x}$	0.0	0.5	1.0	1.5	2.0
$p(\tilde{x})$	0.2500	0.2500	0.1625	0.1500	0.1100

$\tilde{x}$	2.5	3.0	3.5	4.0
$p(\tilde{x})$	0.0450	0.0200	0.0100	0.0025

(b) $\mu_{\tilde{X}} = 0.95$, $\sigma^2_{\tilde{X}} = 0.7238$, $\sigma_{\tilde{X}} = 0.8507$

7.9 (a) Distribution of $\bar{X}$:

$\bar{x}$	6.90	6.95	7.05	8.25	8.35
$p(\bar{x})$	0.1	0.1	0.1	0.1	0.2

$\bar{x}$	8.40	8.45	8.50	9.80
$p(\bar{x})$	0.1	0.1	0.1	0.1

(b) Distribution of the total:

t	13.8	13.9	14.1	16.5	16.7
$p(t)$	0.1	0.1	0.1	0.1	0.2

t	16.8	16.9	17.0	19.6
$p(t)$	0.1	0.1	0.1	0.1

7.11 (a) 0.7875 **(b)** Distribution of S^2:

s^2	0.0	0.5	2.0	4.5
$p(s^2)$	0.365	0.405	0.180	0.050

(c) 0.7875, same.

7.13 (a) Distribution of the maximum weight:

m	83	95	100
$p(m)$	0.1667	0.3333	0.5000

(b) Distribution of the total weight:

t	153	165	170	178	183	195
$p(t)$	0.1667	0.1667	0.1667	0.1667	0.1667	0.1667

7.15 (a) 25.5 **(b)** The sample means should vary around 25.5.
(c) The histogram should be centered near 25.5.

Section 7.2

7.17 (a) $\bar{X} \sim N(10, 6.25/7)$, 0.1450
(b) $\bar{X} \sim N(10, 6.25/12)$, 0.0188
(c) $\bar{X} \sim N(10, 6.25/15)$, 0.5614
(d) $\bar{X} \sim N(10, 6.25/25)$, 0.3085
(e) $\bar{X} \sim N(10, 6.25/100)$, 0.4237

7.19 (a) $\bar{X} \sim N(50, 49/38)$. The shape of the underlying distribution is not known. **(b)** 0.1893 **(c)** 0.0391 **(d)** 0.5769 **(e)** 51.1769

7.21 (a) $\bar{X} \sim N(30, 62.5)$

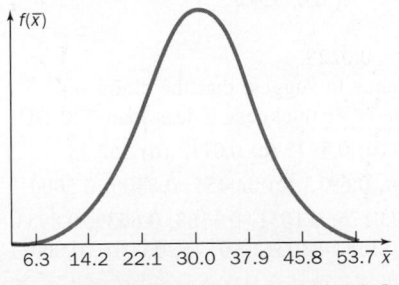

(b) 0.1558 **(c)** 0.7941 **(d)** 0.0289 **(e)** 5.57

7.23 Blue curve: X. Green curve: $\overline{X}$, $n = 5$. Red curve: $\overline{X}$, $n = 15$.

7.25 (a) $\overline{X} \stackrel{.}{\sim} N(8.25, 0.0002857)$

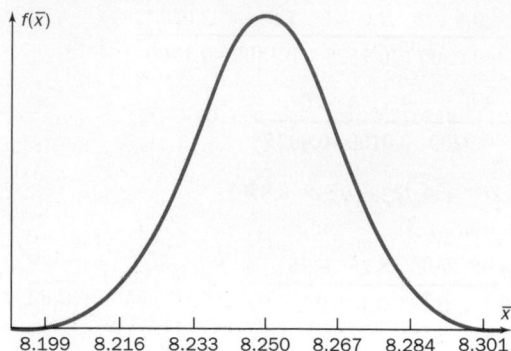

(b) Approximately 1. **(c)** 0.9985
(d) $\overline{X} \stackrel{.}{\sim} N(8.25, 0.0006429)$

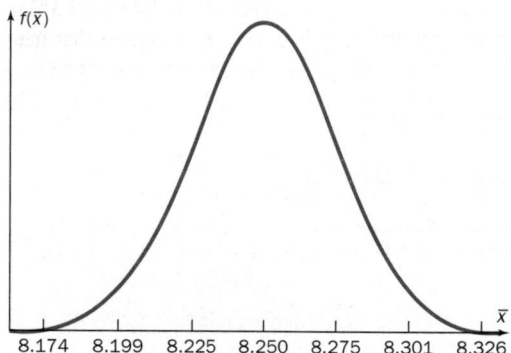

Approximately 1, 0.9757.

7.27 (a) 0.1459 **(b)** 0.2422
(c) Claim: $\mu = 100 \Rightarrow \overline{X} \stackrel{.}{\sim} N(100, 144/40)$
Experiment: $\bar{x} = 98.5$
Likelihood: $P(\overline{X} \le 98.5) = 0.2146$
Conclusion: There is no evidence to suggest that the claim is false—that μ is less than 100.

7.29 (a) $\overline{X} \stackrel{.}{\sim} N(6.5, 0.4571)$ **(b)** 0.2298
(c) Claim: $\mu = 6.5 \Rightarrow \overline{X} \stackrel{.}{\sim} N(6.5, 0.4571)$
Experiment: $\bar{x} = 5.1$
Likelihood: $P(\overline{X} \le 5.1) = 0.0192$
Conclusion: There is evidence to suggest that the claim is false—that the mean police standoff time is lower.

7.31 (a) 0.1168 **(b)** 0.3832 **(c)** 0.0095

7.33 (a) $\overline{X} \sim N(4.125, 0.0667)$ **(b)** 0.0732 **(c)** 0.4982 **(d)** 0.0077

7.35 (a) 0.0228 **(b)** 0.6827
(c) Claim: $\mu = 320 \Rightarrow \overline{X} \stackrel{.}{\sim} N(320, 25)$
Experiment: $\bar{x} = 310$
Likelihood: $P(\overline{X} \le 310) = 0.0228$
Conclusion: There is evidence to suggest that the claim is false—that the mean ozone-layer thickness is less than 320 DU.

7.37 (a) $T \stackrel{.}{\sim} N(525, 140)$ **(b)** 0.8975 **(c)** 0.0112 **(d)** 552.53

7.39 (a) 0.0124 **(b)** 0.9330, 0.6915 **(c)** 0.0455, 0.8400, 0.5000

7.41 (a) 0.0001, 0.0016, 0.0176, 0.1031, 0.3368, 0.6632, 0.8953, 0.9648, 0.8953, 0.6632, 0.3368, 0.1031, 0.0176, 0.0016, 0.0001

(b) OC curve:

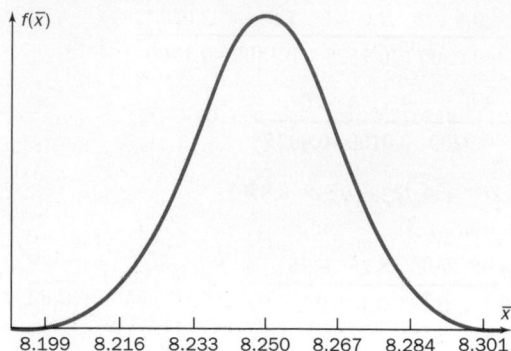

7.43 (a) 30.85 **(b)** Answers will vary. **(c)** Histogram should be approximately normal, centered near 30.85.

Section 7.3
7.45 (a) 0.1932 **(b)** 0.0745 **(c)** 0.4363 **(d)** 0.0433
7.47 (a) 0.2817 **(b)** 0.4741 **(c)** 0.1045
7.49 (a) $\hat{P} \stackrel{.}{\sim} N(0.7, 0.00084)$ **(b)** 0.0838 **(c)** 0.3650 **(d)** 0.7520
7.51 (a) $\hat{P} \stackrel{.}{\sim} N(0.48, 0.0021)$ **(b)** 0.3305 **(c)** 0.0397
(d) (0.4050, 0.5550)
7.53 (a) 0.3841 **(b)** 0.1193 **(c)** 0.4158

7.55 (a) 0.6585 **(b)** 0.1331
(c) Claim: $p = 0.40 \Rightarrow \hat{P} \stackrel{.}{\sim} N(0.40, 0.0024)$
Experiment: $\hat{p} = 0.47$
Likelihood: $P(\hat{P} \ge 0.47) = 0.0765$
Conclusion: There is no evidence to suggest that the claim is false—that the acceptance rate has increased.
7.57 (a) $\hat{P} \stackrel{.}{\sim} N(0.005, 0.000004975)$ **(b)** 0.0893
(c) 0.0125 **(d)** 0.0031

7.59 (a) 0.0383 **(b)** 0.7514
(c) Claim: $p = 0.272 \Rightarrow \hat{P} \stackrel{.}{\sim} N(0.272, 0.0004605)$
Experiment: $\hat{p} = 0.2674$
Likelihood: $P(\hat{P} \le 0.2674) = 0.4151$
Conclusion: There is no evidence to suggest that the claim is false—that the true proportion of cigarette debris has changed.

7.61 (a) 0.0047 **(b)** 0.000000328 **(c)** 0.3187
7.63 0.5

Chapter Exercises
7.65 Claim: $\mu = 0.5 \Rightarrow \overline{X} \stackrel{.}{\sim} N(0.5, 0.0008)$
Experiment: $\bar{x} = 0.6$
Likelihood: $P(\overline{X} \ge 0.6) = 0.0002$
Conclusion: There is evidence to suggest that the claim is false—that the mean coefficient of static friction is greater than 0.5.
7.67 (a) Distribution of M:

m	1	2	3	4	5	6
$p(m)$	0.0004	0.102	0.074	0.342	0.328	0.1536

(b) 4.356, 1.3013, 1.1407

7.69 (a) 0.0455 **(b)** 0.3010
(c) Claim: $\mu = 7.5 \Rightarrow \overline{X} \stackrel{.}{\sim} N(7.5, 0.0875)$
Experiment: $\bar{x} = 8.1$
Likelihood: $P(\overline{X} \ge 8.1) = 0.0213$

Conclusion: There is evidence to suggest that the claim is false—that the mean oxygen produced is greater than 7.5.
(d) 0.2151, 0.1534
Claim: $\mu = 7.5 \Rightarrow \overline{X} \sim N(7.5, 0.4018)$
Experiment: $\bar{x} = 8.1$
Likelihood: $P(\overline{X} \geq 8.1) = 0.1719$
Conclusion: There is no evidence to suggest that the claim is false.

7.71 (a) 0.0095 **(b)** 0.2075 **(c)** 135.25

7.73 (a) Statistic. **(b)** Parameter. **(c)** Statistic. **(d)** Parameter. **(e)** Statistic.

7.75 (a) f_1: underlying distribution; f_2: distribution of the sample mean. **(b)** f_1: underlying distribution; f_2: distribution of the sample mean. **(c)** f_1: distribution of the sample mean; f_1: underlying distribution. **(d)** f_1: underlying distribution; f_2: distribution of the sample mean.

7.77 (a) $\hat{P} \sim N(0.12, 0.000422)$

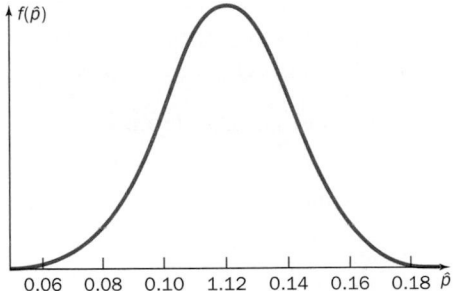

(b) 0.0722 **(c)** 0.0906
(d) Claim: $p = 0.12 \Rightarrow \hat{P} \sim N(0.12, 0.000422)$
Experiment: $\hat{p} = 0.09$
Likelihood: $P(\hat{P} \leq 0.09) = 0.0722$
Conclusion: There is no evidence to suggest the claim is false.

7.79 (a) $\hat{P} \sim N(0.65, 0.0002275)$, $np = 650 \geq 5$, $n(1-p) = 350 \geq 5$. **(b)** 0.2537 **(c)** 0.6539 **(d)** 0.6149

7.81 (a) 0.3661 **(b)** 0.0857 **(c)** 644

7.83 (a) 0.0124 **(b)** 0.0062 **(c)** 0.0000002871

CHAPTER 8

Section 8.1

8.1 $\hat{\theta}_2$; unbiased and small variance.

8.3 The value of the unbiased estimator is, on average, θ.

8.5 0.8125

8.7 0.18

8.9 (a) 95 **(b)** 104 **(c)** (95, 104)

8.11 (a) 0.7535 **(b)** 0.2052 **(c)** 0.30, 1.17

Section 8.2

8.13 (a) (13.507, 17.693) **(b)** (6232.2, 6411.8)
(c) (−51.06, −40.50) **(d)** (0.0763, 0.0827) **(e)** (36.287, 39.073)

8.15 (a) 9.7 **(b)** 95% CI: (8.55, 10.85); 99.9% CI: (8.40, 11.0). For a higher confidence level (all else being equal), the CI has to be larger.

8.17 (95.914, 116.37)

8.19 (a) (1.377, 1.588) **(b)** (1.344, 1.621) **(c)** Larger confidence level.

8.21 (a) (250.8, 269.2) **(b)** Yes. 200 is not in the CI found in part **(a)**.

8.23 (a) (19.422, 22.078) **(b)** Yes. The CI does not include 23.

8.25 (a) (44205, 50795) **(b)** 22 **(c)** 196

8.27 (a) (0.1248, 0.1372) **(b)** It's close, but $1/8 = 0.125$ is captured by the CI in part **(a)**. Therefore, there is no evidence to suggest that the true mean is greater than 0.125. The town should not embark on the safety program.

8.29 (a) (6.4539, 6.7461) **(b)** (6.5270, 6.6730) **(c)** (6.4539, 6.7461) is an interval in which we are 95% confident that the true mean wingspan lies ($\sigma = 0.5$). (6.5270, 6.6730) is an interval in which we are 95% confident that the true mean wingspan lies ($\sigma = 0.25$).

8.31 (a) (122496, 127904) **(b)** (145000, 166800) **(c)** Yes. The CIs do not overlap.

8.33 (a) Football: (61.109, 70.431); basketball: (49.427, 58.373); hockey: (64.899, 72.001) **(b)** There is evidence to suggest that the mean coping skills level is different for football and basketball players. The CIs do not overlap. **(c)** Football: 191; basketball: 151; hockey: 101

8.35 (a) Cashew: (5.0591, 5.2809); filbert: (4.0737, 4.4063); pecan: (2.3367, 2.8633). Cashews and pecans: yes. The CIs do not overlap. Filberts and pecans: yes. The CIs do not overlap. **(b)** Cashew: (4.9852, 5.3548); filbert: (3.9628, 4.5172); pecan: (2.1611, 3.0389). Cashews and pecans: yes. The CIs do not overlap. Filberts and pecans: yes. The CIs do not overlap.

8.37 It is the *shortest* $100(1 - \alpha)\%$ CI for μ.

Section 8.3

8.39 (a) 1.4759 **(b)** 0.8569 **(c)** 2.8609 **(d)** 2.3646 **(e)** 2.9467 **(f)** 5.2076 **(g)** 3.7676 **(h)** 22.2037

8.41 (a) 1.0931 **(b)** 1.5286 **(c)** 3.1534 **(d)** 2.4377 **(e)** 2.6981 **(f)** 1.9921 **(g)** 2.1150 **(h)** 2.8652

8.43 (a) (0.1908, 0.2772) **(b)** (217.5618, 301.6382) **(c)** (19.005, 26.695) **(d)** (367.9725, 393.8275) **(e)** (72.4005, 103.7995)

8.45 (a) (0.2345, 0.2963) **(b)** No. The CI in part (a) includes 0.25. **(c)** Check for evidence of nonnormality:

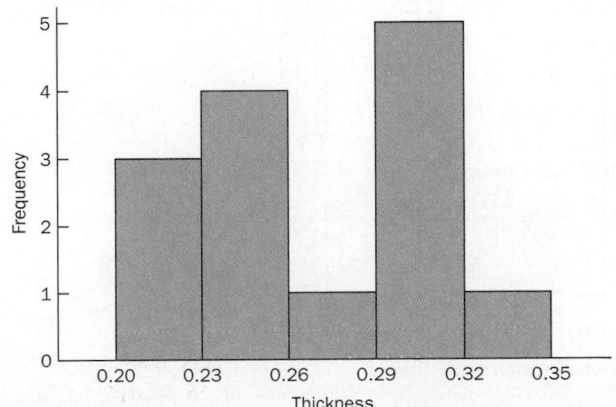

Backward Empirical Rule

Interval	Proportion
(0.2271, 0.3038)	0.64
(0.1887, 0.3421)	1.00
(0.1503, 0.3805)	1.00

$IQR/s = 1.615$

Normal probability plot:

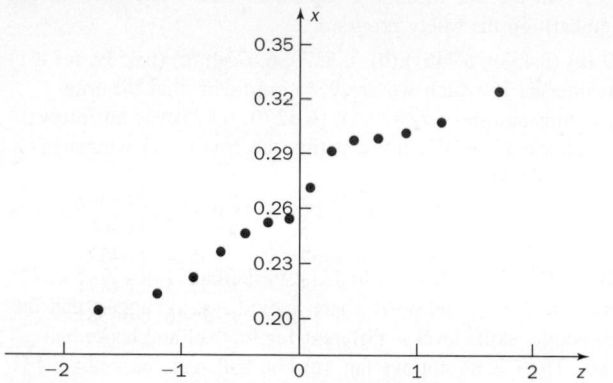

There is evidence to suggest that the data are from a nonnormal population.

8.47 (a) (964.0170, 1043.8164) (b) No. The CI in part (a) includes 1000.

8.49 (a) (15.3816, 18.0184) (b) No. The CI in part (a) suggests that the true mean length is under 20 miles.

8.51 (a) (29.2575, 29.7908) (b) Yes. The CI does not contain 30.

8.53 (a) (664.68, 1056.82) (b) Yes. The lower bound on the CI in part (a) is greater than 500.

8.55 (11.0687, 15.8695)

8.57 (a) (68.6122, 71.7878) (b) (71.0587, 73.1413) (c) The underlying populations are normal. (d) No. The two CIs overlap.

8.59 (a) (281.7269, 288.9397) (b) (249.4027, 266.0973) (c) Yes. The CIs do not overlap. (d) Yes. Atrial-flutter is a measurement, probably not a skewed distribution.

8.61 (a) (11.6981, 14.2989) (b) (7.8028, 9.7044) (c) Check for evidence of nonnormality: rural areas.

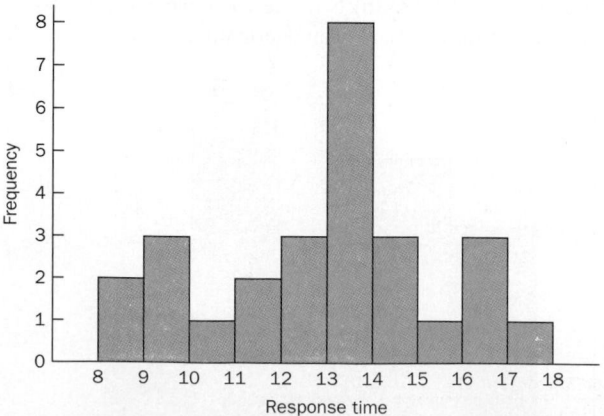

Backward Empirical Rule

Interval	Proportion
(10.5668, 15.4302)	0.67
(8.1351, 17.8619)	1.00
(5.7035, 20.2936)	1.00

$IQR/s = 1.213$

Normal probability plot:

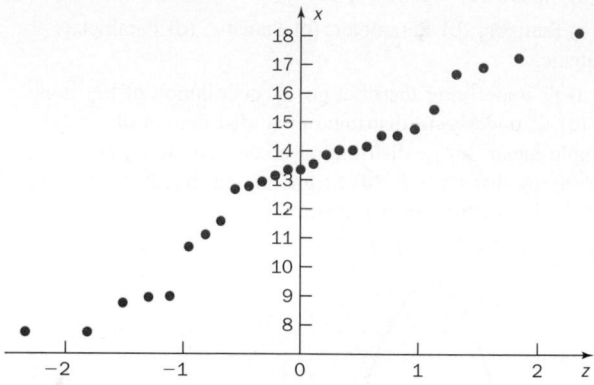

There is no evidence to suggest that the data are from a nonnormal population.

Check for evidence of nonnormality: cities.

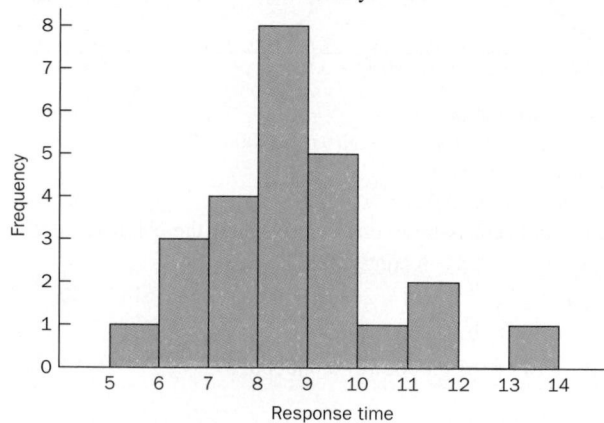

Backward Empirical Rule

Interval	Proportion
(7.0539, 10.4533)	0.72
(5.3543, 12.1529)	0.96
(3.6546, 13.8526)	1.00

$IQR/s = 1.153$

Normal probability plot:

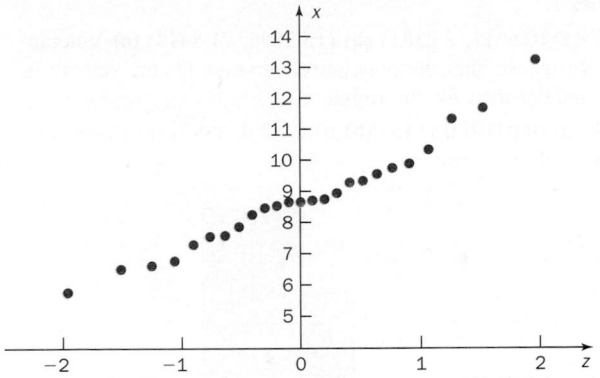

There is no evidence to suggest that the data are from a nonnormal population.
(d) Yes. The two CIs do not overlap.

8.63 (a) Men: (36.8592, 40.9408); women: (34.1608, 37.0392).
(b) No. The CIs overlap. **(c)** (240.1594, 274.8406) is an interval in which we are 99% confident that the true mean distance traveled lies.

Section 8.4

8.65

	np	$n(1 - n\hat{p})$	Approx. normal Yes	Approx. normal No
(a)	85	20	X	
(b)	1645	105	X	
(c)	220	5	X	
(d)	3	180		X
(e)	350	27	X	
(f)	478	2		X

8.67 (a) (0.7187, 0.7798) **(b)** (0.8543, 0.9005) **(c)** (0.3761, 0.5886)
(d) (0.8223, 0.9090) **(e)** (0.0449, 0.1212)

8.69 (a) 461 **(b)** 97 **(c)** 338244 **(d)** 271 **(e)** 68

8.71 (a) Increases. **(b)** Increases. **(c)** Decreases. **(d)** Decreases.

8.73 (a) (0.3750, 0.4424) **(b)** 1025

8.75 (a) (0.0524, 0.1876) **(b)** 542

8.77 (a) (0.4192, 0.4614) **(b)** (0.2903, 0.3296) **(c)** (0.2217, 0.2580)

8.79 (a) (0.2575, 0.3914) **(b)** 1153

8.81 (a) (0.2901, 0.3509) **(b)** No. 0.30 is included in the CI (just barely).

8.83 (a) Northeast: (0.7313, 0.8687); Midwest: (0.7510, 0.8722); South Central: (0.7084, 0.8332); South Atlantic: (0.7850, 0.8758); West: (0.7801, 0.8811). **(b)** Northeast. $\hat{p}$ is farthest from 0.5 and the sample size is the smallest.

8.85 (a) Secondary school: (0.0348, 0.1209); Other post-secondary: (0.0371, 0.1050); Post-secondary: (0.0086, 0.0636)
(b) No. The CIs overlap.

8.87 (a) (0.3171, 0.4029) **(b)** (0.3701, 0.4299) **(c)** No. The CIs overlap.

8.89 (a) Treatment: (0.1005, 0.1619); placebo: (0.0506, 0.1442).
(b) No. The two CIs overlap. **(c)** Treatment: (0.0398, 0.0936); placebo: (0.0145, 0.1024). **(d)** No. The two CIs overlap.

8.91 (a) (0.5040, 0.6960), (0.5020, 0.6906). The Wilson interval is shorter. It is more precise.

(b)

n	Traditional CI	Wilson CI
120	(0.5123, 0.6877)	(0.5106, 0.6832)
140	(0.5188, 0.6812)	(0.5172, 0.6774)
160	(0.5241, 0.6759)	(0.5226, 0.6727)
180	(0.5284, 0.6716)	(0.5271, 0.6688)
200	(0.5321, 0.6679)	(0.5308, 0.6654)
220	(0.5353, 0.6647)	(0.5341, 0.6625)
240	(0.5380, 0.6620)	(0.5369, 0.6599)
260	(0.5405, 0.6595)	(0.5394, 0.6577)
280	(0.5426, 0.6574)	(0.5416, 0.6557)
300	(0.5446, 0.6554)	(0.5436, 0.6538)
320	(0.5463, 0.6537)	(0.5454, 0.6522)
340	(0.5479, 0.6521)	(0.5471, 0.6507)
360	(0.5494, 0.6506)	(0.5486, 0.6493)
380	(0.5507, 0.6493)	(0.5500, 0.6480)
400	(0.5520, 0.6480)	(0.5513, 0.6468)
420	(0.5531, 0.6469)	(0.5524, 0.6457)
440	(0.5542, 0.6458)	(0.5535, 0.6447)
460	(0.5552, 0.6448)	(0.5546, 0.6438)
480	(0.5562, 0.6438)	(0.5555, 0.6429)
500	(0.5571, 0.6429)	(0.5565, 0.6420)

As n increases in the Wilson CI, $\hat{p}$ is closer to the center of the interval.

Section 8.5

8.93 (a) 9.2364 **(b)** 61.0983 **(c)** 26.2962 **(d)** 35.4789 **(e)** 3.0535
(f) 7.2609 **(g)** 11.6886 **(h)** 1.7349

8.95 (a) 10.2829, 35.4789 **(b)** 17.8867, 61.5812 **(c)** 2.5582, 23.2093
(d) 18.4927, 43.7730 **(e)** 0.4844, 11.1433 **(f)** 14.4012, 70.5881

8.97 (a) (1.3427, 11.2313), (1.1587, 3.3513) **(b)** (31.2940, 197.4147), (5.5941, 14.0504) **(c)** (31.9769, 183.4121), (5.6548, 13.5430)
(d) (3.0994, 26.5425), (1.7605, 5.1519) **(e)** (18.5739, 64.0850), (4.3097, 8.0053) **(f)** (5.8969, 36.9707), (2.4284, 6.0804)

8.99 (a) (2.5912, 8.2250) **(b)** (1.6097, 2.8679)

8.101 (a) (1.7229, 8.9829) **(b)** Check for evidence of nonnormality:

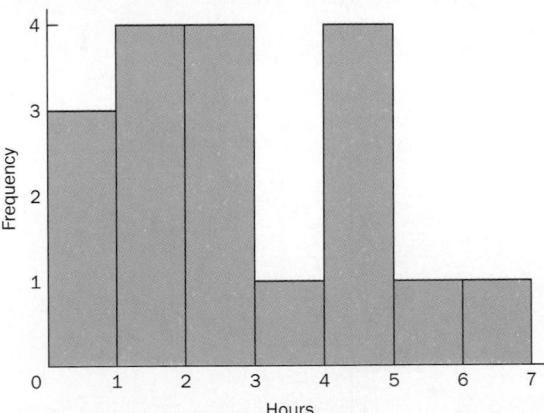

Backward Empirical Rule

Interval	Proportion
(0.9894, 4.6695)	0.67
(−0.8507, 6.5096)	0.94
(−2.6908, 8.3497)	1.00

$IQR/s = 1.315$

Normal probability plot:

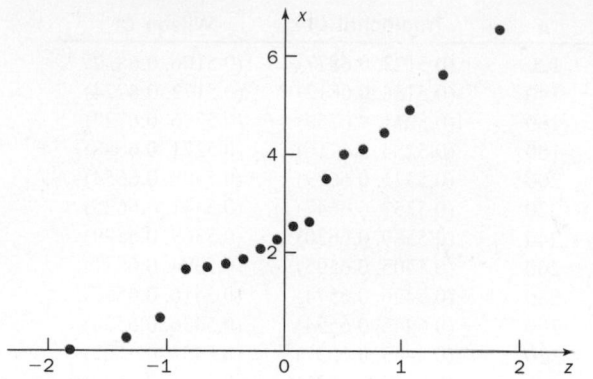

There is no overwhelming evidence to suggest that the data are from a nonnormal population.

8.103 (a) (3.4877, 9.9374) **(b)** Check for evidence of nonnormality:

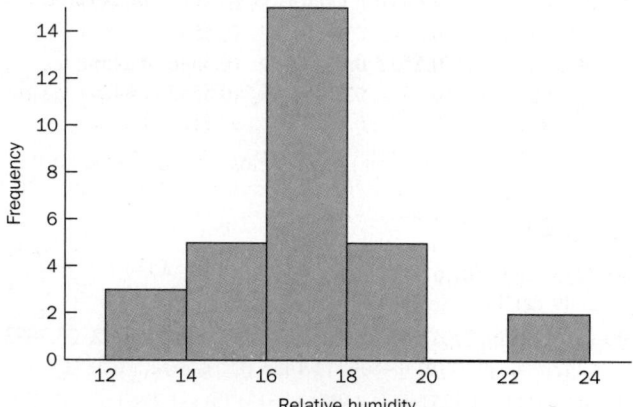

Backward Empirical Rule

Interval	Proportion
(14.1884, 18.8783)	0.63
(11.8434, 21.2233)	0.93
(9.4984, 23.5682)	1.00

$IQR/s = 0.853$

Normal probability plot:

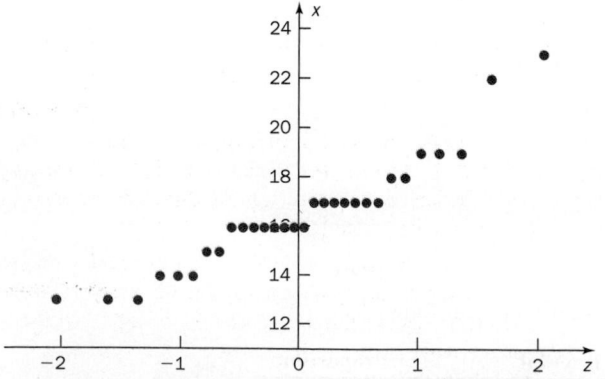

There is some evidence to suggest that the data are from a nonnormal population. IQR/s is far away from 1.3, and the normal probability plot is not very linear.

8.105 (a) (0.3346, 1.1053) **(b)** (0.5785, 1.0513) **(c)** No. The CI includes 1.

8.107 (a) (0.5611, 2.2381) **(b)** (15.5896, 71.8413) **(c)** Veteran. The CI suggests that the population variance for the veteran is much smaller than for the rookie.

8.109 (a) (0.0319, 0.1515) **(b)** (0.1787, 0.3892) **(c)** Check for evidence of nonnormality:

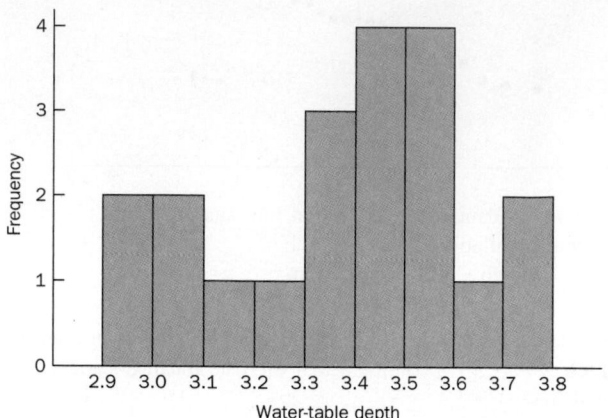

Backward Empirical Rule

Interval	Proportion
(3.1243, 3.6177)	0.65
(2.8777, 3.8643)	1.00
(2.6310, 4.1110)	1.00

$IQR/s = 1.257$

Normal probability plot:

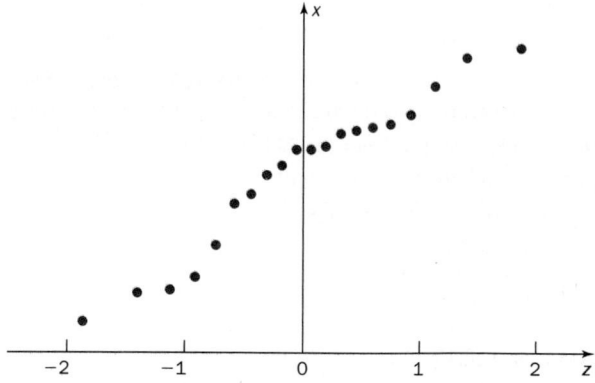

There is some evidence to suggest that the data are from a nonnormal population. The histogram does not appear normal, and the normal probability plot has distinct curves.

8.111 (a) $(1.3481 \times 10^{16}, 6.2555 \times 10^{16})$ **(b)** $(571509.18, 2.6520 \times 10^6)$ **(c)** Absolutely. The CI does not include 1 and is nowhere near 1.

8.113 (a) (6.5503, 23.3373) **(b)** No. The CI includes 12.

8.115 (a) (0.9514, 2.7108) **(b)** (0.5268, 1.8176)
(c) No. The CIs overlap.

8.117 (a) (0.0238, 0.2783), (94.1776, 627.5906) **(b)** (0.0400, 0.2152), (122.6953, 515.8717) **(c)** Column water vapor: no. The CIs overlap. IB: no. The CIs overlap.

8.119 (a) $0 < \sigma^2 < \dfrac{(n-1)s^2}{\chi^2_{1-\alpha}}$, $\sigma^2 > \dfrac{(n-1)s^2}{\chi^2_\alpha}$.

(b) $0 < \sigma^2 < 880,691,497$

Chapter Exercises

8.121 (a) (32.3143, 40.1274) (b) (51.7008, 168.4160)
(c) Normal probability plot:

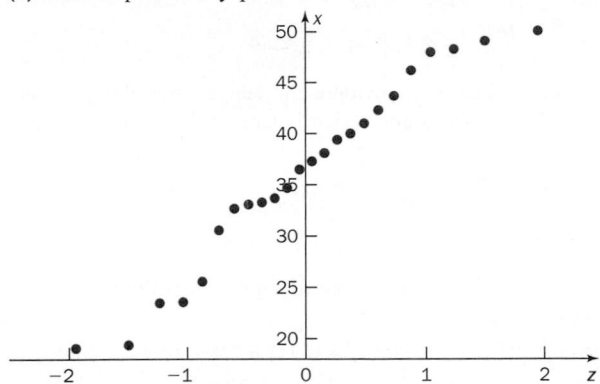

There is some evidence to suggest that the data are from a nonnormal population.

8.123 (a) $\hat{p} = 0.70$. $n\hat{p} = 189 \geq 5$, $n(1 - \hat{p}) = 81 \geq 5$.
(b) (0.6282, 0.7718) (c) 664

8.125 (a) $\hat{p} = 0.56$. $n\hat{p} = 280 \geq 5$, $n(1 - \hat{p}) = 220 \geq 5$.
(b) (0.5028, 0.6172) (c) No. The CI for p includes 0.60.

8.127 (a) (0.0016, 0.0034) (b) No. The CI in part (a) is less than 0.005.

8.129 (a) (14.8793, 17.3779) (b) (4.2313, 15.1195) (c) The underlying population is normal. (d) No. The CI in part (a) includes 15.

8.131 (a) (74.52, 181.48) (b) No. The CI includes 100.
(c) (2661.55, 17736.36) (d) Yes. The CI does not include 2500.

8.133 (a) White: (0.1168, 0.1553); African American: (0.1664, 0.2132); Hispanic: (0.3050, 0.3593). (b) Yes. African American and Hispanic. These two CIs do not include 0.146.

8.135 (a) (0.2832, 0.3168) (b) (0.3292, 0.3708) (c) Yes. The CIs do not overlap.

8.137 (a) (4.5138, 4.8102) (b) (1.9568, 2.2012) (c) No. Both CIs lie entirely below 5. (d) There is evidence to suggest that the mean mercury concentration is different for the two groups. The CIs do not overlap.

8.139 (a) (0.2740, 0.4294) (b) (0.3567, 0.4403)
(c) No. The CIs overlap.

CHAPTER 9

Section 9.1

9.1 (a) Valid, null hypothesis. (b) Invalid. (c) Invalid.
(d) Invalid. (e) Valid, alternative hypothesis. (f) Valid, alternative hypothesis. (g) Invalid. (h) Valid, null hypothesis.

9.3 (a) Valid. (b) Invalid. Should be H_a: $\mu > 9.7$. (c) Invalid. Should be H_a: $\sigma^2 \neq 98.6$. (d) Valid.

9.5 (a) Valid. (b) Valid. (c) Invalid. The null hypothesis should be stated so that μ (a parameter) equals a single value. (d) Valid.

9.7 H_0: $\mu = 1511$, H_a: $\mu > 1511$.

9.9 H_0: $\mu = 17060$, H_a: $\mu < 17060$.

9.11 (a) is appropriate. The software company is looking for evidence that the mean age is greater than 25.

9.13 H_0: $p = 0.75$, H_a: $p > 0.75$.

9.15 (c) is appropriate. The bus company is looking for evidence that the true proportion of parents who favor seat-belt installation is greater than 0.50.

9.17 H_0: $p = 0.35$, H_a: $p < 0.35$.

9.19 H_0: $\sigma = 7$, H_a: $\sigma < 7$.

9.21 H_0: $p = 0.60$, H_a: $p > 0.60$.

9.23 H_0: $p = 0.80$, H_a: $p > 0.80$.

9.25 H_0: $\tilde{u} = 125.50$, H_a: $\tilde{u} < 125.50$.

Section 9.2

9.27 (a) Correct decision. (b) Type II error. (c) Type II error. (d) Correct decision.

9.29 (a) $\beta(11) > \beta(15)$. As the alternative value of μ moves farther from the hypothesized value, the probability of a type II error decreases. There is a better chance of detecting the difference. (b) The probability of a type II error decreases.

9.31 α and β are inversely related. A very small α means that β, the probability of a type II error, is very large.

9.33 (a) Type I error: decide that $\mu > 10$ when the true mean is really 10 (or less). Type II error: decide that $\mu = 10$ (or less) when the true mean is really greater than 10. (b) Type II error. The files are really very old and need to be archived. (c) Type I error. The files are really not that old and the money does not need to be spent to archive them.

9.35 (a) H_0: $p = 0.08$, H_a: $p < 0.08$. (b) Type II error. The new academic policy is really working, but there is no evidence for that. (c) Type I error. The new academic policy is not working, but fewer students are showing up late for exams. This probably means that the new policy would remain in effect, but it really isn't necessary.

9.37 (a) H_0: $p = 0.60$, H_a: $p > 0.60$. (b) Type I error: decide that $p > 0.60$ when the true proportion is really 0.60 (or less). Type II error: decide that $p = 0.60$ (or less) when the true proportion is really greater than 0.60. (c) Type II error. Residents are in favor of the extended structure, but the evidence suggests that they are not. (d) Type I error. The city council believes that residents are in favor of the extended structure, but they really aren't.

9.39 (a) H_0: $\mu = 1,367$, H_a: $\mu > 1,367$. (b) Type I error: decide that $\mu > 1,367$ when the true mean is really 1,367 (or less). Type II error: decide that $\mu = 1,367$ (or less) when the true mean is really greater than 1,367.

9.41 (a) H_0: $\sigma^2 = 15$, H_a: $\sigma^2 < 15$. (b) Type I error: decide that $\sigma^2 < 15$ when the true population variance is really 15 (or more). Type II error: decide that $\sigma^2 = 15$ (or more) when the true population variance is really less than 15. (c) Type I error. NSF would commit more money, but there is really no evidence that TM decreases brain activity. Type II error. There is no evidence of decreased brain activity, but TM really works!

9.43 (a) H_0: $p = 0.15$, H_a: $p > 0.15$. **(b)** Type I error: decide that $p > 0.15$ when the true proportion is really 0.15 (or less). Type II error: decide that $p = 0.15$ (or less) when the true proportion is really greater than 0.15. **(c)** The probability of a type I error becomes smaller.

Section 9.3

9.45 (a) $Z = (\overline{X} - 170)/(15/\sqrt{38})$
(b) (i) $Z \le -2.3263$ **(ii)** $Z \le -1.96$ **(iii)** $Z \le -1.6449$
(iv) $Z \le -1.2816$ **(v)** $Z \le -3.0902$ **(vi)** $Z \le -3.7190$
9.47 (a) $Z = (\overline{X} + 11)/(4.5/\sqrt{21})$
(b) (i) $|Z| \ge 2.5758$ **(ii)** $|Z| \ge 1.2816$ **(iii)** $|Z| \ge 1.96$
(iv) $|Z| \ge 1.6449$ **(v)** $|Z| \ge 3.2905$ **(vi)** $|Z| \ge 3.7190$
9.49 (a) 0.05 **(b)** 0.10 **(c)** 0.005 **(d)** 0.001 **(e)** 0.20 **(f)** 0.02
9.51 (a) H_0: $\mu = 212$; H_a: $\mu > 212$; TS: $Z = (\overline{X} - \mu_0)/(\sigma/\sqrt{n})$;
RR: $Z \ge 2.3263$ **(b)** The underlying population is normal and the population standard deviation is known. **(c)** $z = 2.6042$ (≥ 2.3263). There is evidence to suggest that the population mean is greater than 212.
9.53 (a) H_0: $\mu = 365.25$; H_a: $\mu^\circ \ne 365.25$;
TS: $Z = (\overline{X} - \mu_0)/(\sigma/\sqrt{n})$; RR: $|Z| \ge 1.96$ **(b)** The sample size is large and the population standard deviation is known.
(c) $z = -1.6311$. There is no evidence to suggest that the population mean is different from 365.25.
9.55 H_0: $\mu = 51500$; H_a: $\mu < 51500$;

TS: $Z = \dfrac{\overline{X} - \mu_0}{\sigma/\sqrt{n}}$;   RR: $Z \le -2.3263$

$z = -2.8570 \le -2.3263$. There is evidence to suggest that the mean income per year of corporate communications workers has decreased.

9.57 H_0: $\mu = 295$; H_a: $\mu > 295$;

TS: $Z = \dfrac{\overline{X} - \mu_0}{\sigma/\sqrt{n}}$;   RR: $Z \ge 2.3263$

$z = 1.5056$. There is no evidence to suggest that the mean length of international calls has increased. Therefore, there is no evidence to suggest that the advertising campaign was successful.

9.59 (a) H_0: $\mu = 35$; H_a: $\mu > 35$;

TS: $Z = \dfrac{\overline{X} - \mu_0}{\sigma/\sqrt{n}}$;   RR: $Z \ge 2.3263$

$z = 3.2720 \ge 2.3263$. There is evidence to suggest that the mean LOA is greater than 35 feet. **(b)** No. For $\alpha = 0.1$, the critical value is even smaller.

9.61 (a) H_0: $\mu = 2200$; H_a: $\mu < 2200$;

TS: $Z = \dfrac{\overline{X} - \mu_0}{\sigma/\sqrt{n}}$;   RR: $Z \le -1.6449$

$z = -1.8860 \le -1.6449$. There is evidence to suggest that the mean caloric intake is less than 2200. **(b)** RR: $Z \le -2.3263$. $z = -1.8860$ does not lie in the rejection region. There is no evidence to suggest that the mean caloric intake is less than 2200.

9.63 (a) H_0: $\mu = 3.9$; H_a: $\mu > 3.9$;

TS: $Z = \dfrac{\overline{X} - \mu_0}{\sigma/\sqrt{n}}$;   RR: $Z \ge 2.3263$

$z = 1.0312$. There is no evidence to suggest that the true mean daily temperature in 2008 was higher than the long-term mean. **(b)** $z = 1.8244$. Same conclusion.

9.65 (a) H_0: $\mu = 12$; H_a: $\mu < 12$;

TS: $Z = \dfrac{\overline{X} - \mu_0}{\sigma/\sqrt{n}}$;   RR: $Z \le -1.6449$

$z = -1.9985 \le -1.6449$. There is evidence to suggest that the mean impact velocity is less than 12. **(b)** If $\alpha = 0.01$, then RR: $Z \le -2.3263$. There is no evidence to suggest that the mean impact velocity is less than 12.

9.67 (a) H_0: $\mu = 225$; H_a: $\mu < 225$;

TS: $Z = \dfrac{\overline{X} - \mu_0}{\sigma/\sqrt{n}}$;   RR: $Z \le -1.6449$

$z = -1.6025$. There is no evidence to suggest that the mean is less than 225. **(b)** The sample size is large and the population standard deviation is known.

9.69 H_0: $\mu = 42$; H_a: $\mu^\circ \ne 42$;

TS: $Z = \dfrac{\overline{X} - \mu_0}{\sigma/\sqrt{n}}$;   RR: $|Z| \ge 1.96$

$z = 1.3902$. There is no evidence to suggest that the mean is different from 42.

9.71 (a) 0.3668 **(b)** 0.1570, 0.0471 **(c)** 0.2399, 0.0848, 0.0207
9.73 (a) H_0: $\mu = 23.625$; H_a: $\mu^\circ \ne 23.625$;

TS: $Z = \dfrac{\overline{X} - \mu_0}{\sigma/\sqrt{n}}$;   RR: $|Z| \ge 1.96$

$z = 1.5811$. There is no evidence to suggest that the mean is different from 23.625. The assembly line should not be shut down. **(b)** 23.532, 23.718

9.75 (a) H_0: $\mu = 3.32$; H_a: $\mu^\circ \ne 3.32$;

TS: $Z = \dfrac{\overline{X} - \mu_0}{\sigma/\sqrt{n}}$;   RR: $|Z| \ge 2.5758$

$z = 1.3530$. There is no evidence to suggest that the mean ice thickness has changed. **(b)** 0.9749
(c)

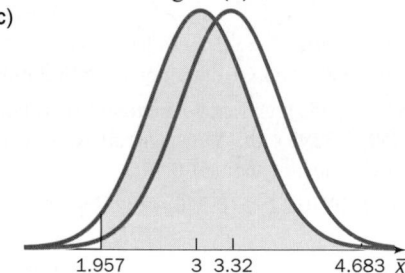

(d) 0.9442
9.77 (a) H_0: $\mu = 1250$; H_a: $\mu > 1250$;

TS: $Z = \dfrac{\overline{X} - \mu_0}{\sigma/\sqrt{n}}$;   RR: $Z \ge 2.3263$

$z = 2.3803 \ge 2.3263$. There is evidence to suggest that the mean is greater than 1250. **(b)** 0.1280
(c)

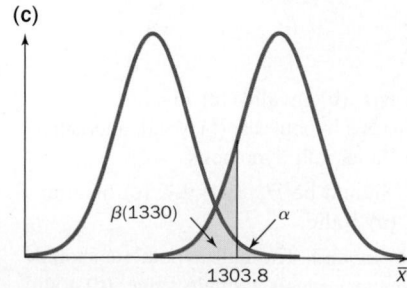

9.79 (a) 1047.7536, 1052.2464
(b) H_0: $\mu = 1050$; H_a: $\mu^\circ \neq 1050$;

TS: $Z = \dfrac{\overline{X} - \mu_0}{\sigma/\sqrt{n}}$; RR: $|Z| \geq 2.5758$

$z = -1.1467$ does not lie in the rejection region.
$-2.5758 < -1.1467 < 2.5758$. There is no evidence to
suggest that the mean is different from 1050. Similarly,
$\overline{x} = 1049$ does not lie in the rejection region, using the
distribution of $\overline{X}$. $1047.7536 < 1049 < 1052.2464$.

Section 9.4

9.81 (a) Do not reject. **(b)** Reject. **(c)** Do not reject.
(d) Do not reject. **(e)** Reject. **(f)** Do not reject.

9.83 (a) 0.0202 **(b)** 0.0764 **(c)** 0.0006 **(d)** 0.2514 **(e)** 0.000002325
(f) 0.5987

9.85 (a) 0.0764. Do not reject. **(b)** 0.0202. Reject. **(c)** 0.0801.
Reject. **(d)** 0.0009. Reject. **(e)** 0.1230. Do not reject. **(f)** 0.0188.
Do not reject.

9.87 (a) 0.2000. Do not reject. **(b)** 0.1671. Do not reject.
(c) 0.0021. Reject. **(d)** 0.0068. Do not reject. **(e)** 0.7288. Do not
reject. **(f)** 0.0094. Reject.

9.89 H_0: $\mu = 87.6$; H_a: $\mu > 87.6$; TS: $Z = \dfrac{\overline{X} - \mu_0}{\sigma/\sqrt{n}}$

$z = 2.1719$, $p = 0.0149$. There is evidence to suggest that the
mean is greater than 87.6.

9.91 (a) H_0: $\mu = 30$; H_a: $\mu > 30$;

TS: $Z = \dfrac{\overline{X} - \mu_0}{\sigma/\sqrt{n}}$; RR: $Z \geq 2.3263$

$z = 2.9394$. There is evidence to suggest that the mean is
greater than 30. **(b)** 0.0016

9.93 H_0: $\mu = 40$; H_a: $\mu \neq 40$; TS: $Z = \dfrac{\overline{X} - \mu_0}{\sigma/\sqrt{n}}$

$z = 1.2905$, $p = 0.1969$. There is no evidence to suggest that
the mean is different from 40.

9.95 H_0: $\mu = 185$; H_a: $\mu \neq 185$; TS: $Z = \dfrac{\overline{X} - \mu_0}{\sigma/\sqrt{n}}$

$z = 0.9363$, $p = 0.3491$. There is no evidence to suggest that
the mean is different from 185.

9.97 (a) H_0: $\mu = 60$; H_a: $\mu < 60$; TS: $Z = \dfrac{\overline{X} - \mu_0}{\sigma/\sqrt{n}}$

$z = -4.2693$, $p = 0.00098$. There is evidence to suggest that
the mean is less than 60. **(b)** Yes. The p value is very small.
(c) p value illustration:

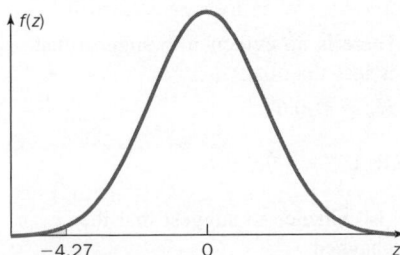

9.99 (a) H_0: $\mu = 85$; H_a: $\mu < 85$; TS: $Z = \dfrac{\overline{X} - \mu_0}{\sigma/\sqrt{n}}$

$z = -0.1660$, $p = 0.4341$. There is no evidence to suggest that
the mean is less than 85. **(b)** 0.4341.

Section 9.5

9.101 (a) $T = (\overline{X} - \mu_0)/(S/\sqrt{n})$ **(b) (i)** $T \leq -2.6245$
(ii) $T \leq -4.5869$ **(iii)** $T \leq -1.7247$ **(iv)** $T \leq -1.3195$
(v) $T \leq -7.1732$ **(vi)** $T \leq -4.2340$

9.103 (a) 0.025 **(b)** 0.001 **(c)** 0.01 **(d)** 0.001

9.105 (a) 0.10 **(b)** 0.01 **(c)** 0.002 **(d)** 0.0002

9.107 (a) $0.25 \leq p \leq 0.05$ **(b)** $p > 0.20$
(c) $0.01 \leq p \leq 0.025$ **(d)** $0.0005 \leq p \leq 0.001$

9.109 (a) H_0: $\mu = 1.618$; H_a: $\mu < 1.618$;
TS: $T = (\overline{X} - \mu_0)/(S/\sqrt{n})$; RR: $T \leq -1.7291$
(b) $t = -1.1727$. There is no evidence to suggest that the mean
is less than 1.618. **(c)** 0.1277

9.111 (a) H_0: $\mu = 9.96$; H_a: $\mu \neq 9.96$;
TS: $T = (\overline{X} - \mu_0)/(S/\sqrt{n})$; RR: $|T| \geq 3.4210$
(b) $t = -4.0568 \leq -3.4210$. There is evidence to suggest that
the mean is different from 9.96. **(c)** 0.0004

9.113 H_0: $\mu = 871$; H_a: $\mu > 871$;
TS: $T = (\overline{X} - \mu_0)/(S/\sqrt{n})$; RR: $T \geq 1.7959$
$t = 0.9793$. There is no evidence to suggest that the mean
weight is greater than 871.

9.115 H_0: $\mu = 31.9$; H_a: $\mu < 31.9$;
TS: $T = \dfrac{\overline{X} - \mu_0}{S/\sqrt{n}}$; RR: $T \leq -2.4851$
$t = -2.0842$. There is no evidence to suggest that the mean is
less than 31.9.

9.117 H_0: $\mu = 40$; H_a: $\mu > 40$;
TS: $T = \dfrac{\overline{X} - \mu_0}{S/\sqrt{n}}$; RR: $T \geq 2.6503$
$t = 4.8568$. There is no evidence to suggest that the population
mean is greater than 40. Assumption: the underlying population
is normal.

9.119 H_0: $\mu = 350$; H_a: $\mu \neq 350$;
TS: $T = \dfrac{\overline{X} - \mu_0}{S/\sqrt{n}}$; RR: $|T| \geq 2.4469$
$t = 1.7953$. There is no evidence to suggest that the population
mean is different from 350.

9.121 (a) H_0: $\mu = 25$; H_a: $\mu > 25$;
TS: $T = \dfrac{\overline{X} - \mu_0}{S/\sqrt{n}}$; RR: $T \geq 2.4377$
$t = 2.8889 \geq 2.4377$. There is evidence to suggest that the
population mean is greater than 25. **(b)** $0.001 \leq p \leq 0.005$

9.123 H_0: $\mu = 55.5$; H_a: $\mu > 55.5$;
TS: $T = \dfrac{\overline{X} - \mu_0}{S/\sqrt{n}}$; RR: $T \geq 1.7959$
$t = 0.9727$. There is no evidence to suggest that the population
mean is greater than 55.5.

9.125 (a) H_0: $\mu = 2$; H_a: $\mu \neq 2$;
TS: $T = \dfrac{\overline{X} - \mu_0}{S/\sqrt{n}}$; RR: $|T| \geq 2.8453$

$t = 1.0819$. There is no evidence to suggest that the population mean is different from 2. (b) $0.2 \leq p \leq 0.4$

9.127 (a) $H_0: \mu = 15$; $H_a: \mu < 15$;

TS: $T = \dfrac{\overline{X} - \mu_0}{S/\sqrt{n}}$;   RR: $T \leq -2.2281$

$t = -0.5749$. There is no evidence to suggest that the population mean is less than 15. (b) 14.35 is a reasonable observation, subject to natural variability. (c) $p > 0.20$

9.129 (a) $H_0: \mu = 1659$; $H_a: \mu < 1659$;

TS: $T = \dfrac{\overline{X} - \mu_0}{S/\sqrt{n}}$;   RR: $T \leq -1.7139$

$t = -1.7733 \leq -1.7139$. There is evidence to suggest that the population mean is less than 1659. We cannot conclude that the heat wave caused this decrease. (b) $0.025 \leq p \leq 0.05$

9.131 (a) $H_0: \mu = 0$; $H_a: \mu \neq 0$;

TS: $T = \dfrac{\overline{X} - \mu_0}{S/\sqrt{n}}$;   RR: $|T| \geq 2.8609$

$t = -0.6902$. There is no evidence to suggest that the mean is different from 0. $p > 0.40$. (b) There is no evidence to suggest prevailing drought or wet conditions.

9.133 (a) $H_0: \mu = 1000$; $H_a: \mu > 1000$;

TS: $T = \dfrac{\overline{X} - \mu_0}{S/\sqrt{n}}$;   RR: $T \geq 1.7033$

$t = 0.9399$. There is no evidence to suggest that the population mean is greater than 1000. (b) $0.10 \leq p \leq 0.20$

9.135 (a) $H_0: \mu = 2748$; $H_a: \mu > 2748$;

TS: $T = \dfrac{\overline{X} - \mu_0}{S/\sqrt{n}}$;   RR: $T \geq 2.7638$

$t = 1.9885$. There is no evidence to suggest that the population mean is greater than 2748. (b) $0.025 \leq p \leq 0.05$ (c) 22

9.137 (a) $H_0: \lambda = 11$; $H_a: \lambda > 11$;

TS: $Z = (X - \lambda_0)/\sqrt{\lambda_0}$,   RR: $Z \geq 2.3263$

(b) $z = 2.1106$. There is no evidence to suggest that the population mean number of dog bites per day is greater than 11. (c) 0.0174

Section 9.6

9.139

	RR	Value of the TS	Conclusion
(a)	$Z \geq 1.6449$	0.3033	Do not reject.
(b)	$Z \geq 1.2816$	1.4882	Reject.
(c)	$Z \geq 2.3263$	2.3327	Reject.
(d)	$Z \geq 1.9600$	0.6872	Do not reject.
(e)	$Z \geq 2.3263$	0.3307	Do not reject.

9.141

	RR	Value of the TS	Conclusion
(a)	$|Z| \geq 2.2414$	-2.0142	Do not reject.
(b)	$|Z| \geq 2.3263$	2.3451	Reject.
(c)	$|Z| \geq 1.9600$	-2.0294	Reject.
(d)	$|Z| \geq 2.8070$	-1.4025	Do not reject.
(e)	$|Z| \geq 2.5758$	2.6455	Reject.

9.143

	Value of the TS	p value	Conclusion
(a)	-2.0285	0.0213	Reject.
(b)	0.0574	0.5229	Do no reject.
(c)	-0.7611	0.2233	Do no reject.
(d)	-2.2166	0.0133	Reject.
(e)	-2.6187	0.0044	Reject.

9.145 (a) 500, 16, 0.02. (b) $np_0 = 10 \geq 5$, $n(1 - p_0) = 490 \geq 5$. The large-sample test is appropriate. (c) $H_0: p = 0.02$; $H_a: p > 0.02$;

TS: $Z = \dfrac{\hat{P} - p_0}{\sqrt{\frac{p_0(1 - p_0)}{n}}}$;   RR: $Z \geq 1.6449$

$z = 1.9166 \geq 1.6449$. There is evidence to suggest that the population proportion is greater than 0.02. (d) 0.0276.

9.147 (a) 225, 189, 0.90. (b) $np_0 = 202.5 \geq 5$, $n(1 - p_0) = 22.5 \geq 5$. The large-sample test is appropriate. (c) $H_0: p = 0.90$; $H_a: p \neq 0.90$;

TS: $Z = \dfrac{\hat{P} - p_0}{\sqrt{\frac{p_0(1 - p_0)}{n}}}$;   RR: $|Z| \geq 1.96$

$z = -3.0 \leq -1.96$. There is evidence to suggest that the population proportion is different from 0.90. (d) 0.0027

9.149 $H_0: p = 0.30$; $H_a: p > 0.30$;

TS: $Z = \dfrac{\hat{P} - p_0}{\sqrt{\frac{p_0(1 - p_0)}{n}}}$;   RR: $Z \geq 3.0902$

$z = 3.7243 \geq 3.0902$. There is evidence to suggest that the population proportion is greater than 0.30.

9.151 $H_0: p = 0.79$; $H_a: p < 0.79$;

TS: $Z = \dfrac{\hat{P} - p_0}{\sqrt{\frac{p_0(1 - p_0)}{n}}}$;   RR: $Z \leq -2.3263$

$z = -1.1771$. There is no evidence to suggest that the population proportion is less than 0.79.

9.153 $H_0: p = 0.40$; $H_a: p < 0.40$;

TS: $Z = \dfrac{\hat{P} - p_0}{\sqrt{\frac{p_0(1 - p_0)}{n}}}$;   RR: $Z \leq -2.3263$

$z = -2.4244 \leq -2.3263$. There is evidence to suggest that the population proportion is less than 0.40. The politician should enter the race for mayor.

9.155 $H_0: p = 0.10$; $H_a: p < 0.10$;

TS: $Z = \dfrac{\hat{P} - p_0}{\sqrt{\frac{p_0(1 - p_0)}{n}}}$;   RR: $Z \leq -1.6449$

$z = 0.6667$, $p = 0.7475$. There is no evidence to suggest that the population proportion is less than 0.10.

9.157 $H_0: p = 0.64$; $H_a: p \neq 0.64$;

TS: $Z = \dfrac{\hat{P} - p_0}{\sqrt{\frac{p_0(1 - p_0)}{n}}}$;   RR: $|Z| \geq 1.96$

$z = 2.4249 \geq 1.96$. There is evidence to suggest that the population proportion has changed.

9.159 (a) $H_0: p = 0.95$; $H_a: p < 0.95$;

TS: $Z = \dfrac{\hat{P} - p_0}{\sqrt{\frac{p_0(1 - p_0)}{n}}}$;   RR: $Z \leq -1.6449$

$z = -2.2711 \leq -1.6449$. There is evidence to suggest that the population proportion is less than 0.95. **(b)** 0.0116 **(c)** No. There is evidence to suggest that less than 95% of all batteries last at least three years.

9.161 (a) $H_0: p = 0.29$; $\quad H_a: p < 0.29$;

TS: $Z = \dfrac{\hat{P} - p_0}{\sqrt{\frac{p_0(1 - p_0)}{n}}}$; $\quad$ RR: $Z \leq -2.3263$

$z = -2.0907$. There is no evidence to suggest that the population proportion of adults who have not filled a prescription is less than 0.29. **(b)** 0.5987 **(c)** 0.2393 **(d)** 2543

Section 9.7

9.163 (a) $X^2 = \dfrac{(n - 1)S^2}{\sigma_0^2}$

(b) (i) 19.6751 **(ii)** 31.5264 **(iii)** 40.2894 **(iv)** 37.9159 **(v)** 20.5150 **(vi)** 42.5793

9.165 (a) $X^2 = \dfrac{(n - 1)S^2}{\sigma_0^2}$

(b)

	Rejection region		
(i)	$X^2 \leq 23.6543$	or	$X2 \$ 58.1201$
(ii)	$X^2 \leq 13.7867$	or	$X2 \$ 53.6720$
(iii)	$X^2 \leq 5.8957$	or	$X^2 \geq 49.0108$
(iv)	$X^2 \leq 5.2293$	or	$X^2 \geq 30.5779$
(v)	$X^2 \leq 10.3909$	or	$X^2 \geq 56.8923$
(vi)	$X^2 \leq 1.0636$	or	$X^2 \geq 7.7794$

9.167 (a) 0.0005 **(b)** 0.01 **(c)** 0.025 **(d)** 0.005

9.169 (a) $0.01 \leq p \leq 0.025$ **(b)** $0.05 \leq p \leq 0.10$ **(c)** $0.001 \leq p \leq 0.005$ **(d)** $p \leq 0.0001$

9.171 (a) $0.02 \leq p \leq 0.05$ **(b)** $0.0002 \leq p \leq 0.001$ **(c)** $0.05 \leq p \leq 0.10$ **(d)** $0.001 \leq p \leq 0.002$

9.173 (a) $H_0: \sigma^2 = 36.8$; $\quad H_a: \sigma^2 \neq 36.8$; TS: $X^2 = (n - 1)S^2/\sigma_0^2$;
RR: $X^2 \leq 6.2621$ or $X^2 \geq 27.4884$ **(b)** $s^2 = 105.863$, $\chi^2 = 43.1508 \geq 27.4884$. There is evidence to suggest that the population variance is different from 36.8. **(c)** $0.0002 \leq p \leq 0.01$

9.175 $H_0: \sigma^2 = 0.25$; $\quad H_a: \sigma^2 > 0.25$;
TS: $X^2 = (n - 1)S^2/\sigma_0^2$; $\quad$ RR: $X^2 \geq 48.6024$
$\chi^2 = 42.3455$. There is no evidence to suggest that the population variance is greater than 0.25.

9.177 $H_0: \sigma^2 = 324$; $\quad H_a: \sigma^2 > 324$;
TS: $X^2 = (n - 1)S^2/\sigma_0^2$; $\quad$ RR: $X^2 \geq 24.7250$
$\chi^2 = 15.7814$. There is no evidence to suggest that the population variance is greater than 324. There is no evidence to refute the manufacturer's claim.

9.179 $H_0: \sigma^2 = 62.5$; $\quad H_a: \sigma^2 > 62.5$;
TS: $X^2 = (n - 1)S^2/\sigma_0^2$; $\quad$ RR: $X^2 \geq 21.6660$
$\chi^2 = 10.0944$. There is no evidence to suggest that the population variance is greater than 62.5.

9.181 $H_0: \sigma^2 = 40000$; $\quad H_a: \sigma^2 < 40000$;
TS: $X^2 = (n - 1)S^2/\sigma_0^2$; $\quad$ RR: $X^2 \leq 44.0379$
$\chi^2 = 55.7408$. There is no evidence to suggest that the population variance is less than 40000; there is no evidence to suggest that the population standard deviation is less than 200.

9.183 $H_0: \sigma^2 = 22.5$; $\quad H_a: \sigma^2 < 22.5$;
TS: $X^2 = (n - 1)S^2/\sigma_0^2$; $\quad$ RR: $X^2 \leq 23.2686$
$\chi^2 = 24.9600$. There is no evidence to suggest that the population variance in ride times is less than 22.5; there is no evidence to suggest that the bull riding has become less exciting.

9.185 (a) $H_0: \sigma^2 = 0.36$; $\quad H_a: \sigma^2 > 0.36$;
TS: $X^2 = (n - 1)S^2/\sigma_0^2$; $\quad$ RR: $X^2 \geq 30.1435$
$\chi^2 = 22.1667$. There is no evidence to suggest that the population variance is greater than 0.36; **(b)** $p > 0.10$

9.187 (a) $H_0: \sigma^2 = 230$; $\quad H_a: \sigma^2 > 230$;
TS: $X^2 = (n - 1)S^2/\sigma_0^2$; $\quad$ RR: $X^2 \geq 38.8851$
$\chi^2 = 21.9352$. There is no evidence to suggest that the population variance is greater than 230. There is no evidence to suggest an inconsistent signal. **(b)** $p > 0.10$

9.189 (a) $H_0: \sigma^2 = 7.5625$; $\quad H_a: \sigma^2 > 7.5625$;
TS: $X^2 = (n - 1)S^2/\sigma_0^2$; $\quad$ RR: $X^2 \geq 32.6706$
$\chi^2 = 40.3239 \geq 32.6706$. There is evidence to suggest that the population variance in the wingspan is greater than 7.5625. **(b)** $0.005 \leq p \leq 0.01$

9.191 (a) $H_0: \sigma^2 = 1.56$; $\quad H_a: \sigma^2 \neq 1.56$;
TS: $Z = \frac{S^2 - \sigma_0^2}{\sqrt{2}\sigma_0^2/\sqrt{n - 1}}$; $\quad$ RR: $|Z| \geq 1.96$
(b) $z = 1.9888 \geq 1.96$. There is evidence to suggest that the population variance in the exchange rate is different from 1.56. **(c)** 0.0467
(d) $H_0: \sigma^2 = 1.56$; $\quad H_a: \sigma^2 \neq 1.56$;
TS: $X^2 = (n - 1)S^2/\sigma_0^2$;
RR: $X^2 \leq 29.9562$ or $X^2 \geq 67.8206$
$\chi^2 = 66.2821$. There is no evidence to suggest that the population variance in the exchange rate is different from 1.56. $p = 0.0666$. Note: The conclusion and p value are different.

Chapter Exercises

9.193 $H_0: \mu = 4$; $\quad H_a: \mu \neq 4$;
TS: $Z = (\bar{x} - \mu_0)/(\sigma/\sqrt{n})$; $\quad$ RR: $|Z| \geq 2.5758$
(a) $z = 1.7709$. There is no evidence to suggest that the population mean thickness is different from 4. The process should not be stopped. **(b)** $z = -2.6563 \leq -2.5758$. There is evidence to suggest that the population mean thickness is different from 4. The process should be stopped.

9.195 (a) $H_0: \mu = 23$; $\quad H_a: \mu > 23$;
TS: $Z = (\bar{x} - \mu_0)/(\sigma/\sqrt{n})$; $\quad$ RR: $Z \geq 2.3263$
$z = 2.9773 \geq 2.3263$. There is evidence to suggest that the population mean width is greater than 23. **(b)** 0.0015

9.197 $H_0: \mu = 3$; $\quad H_a: \mu > 3$;
TS: $T = \dfrac{\bar{X} - \mu_0}{S/\sqrt{n}}$; $\quad$ RR: $T \geq 1.7613$

$t = 1.4660$. There is no evidence to suggest that the population mean FEF is greater than 3.

9.199 (a) $H_0: p = 0.75$; $\quad H_a: p < 0.75$;
TS: $Z = \dfrac{\hat{P} - p_0}{\sqrt{\frac{p_0(1 - p_0)}{n}}}$; $\quad$ RR: $Z \leq -2.3263$

$z = -2.7325 \leq -2.3263$. There is evidence to suggest that the population proportion of residents who favor additional power to tap phones is less than 0.75. **(b)** 0.0031

9.201 H_0: $p = 0.80$; H_a: $p < 0.80$;

TS: $Z = \dfrac{\hat{P} - p_0}{\sqrt{\frac{p_0(1 - p_0)}{n}}}$;   RR: $Z \leq -3.0902$

$z = -3.8025 \leq -3.0902$. There is evidence to suggest that the population proportion of counterfeit goods from China has decreased.

9.203 H_0: $\sigma^2 = 0.0015$; H_a: $\sigma^2 > 0.0015$;
TS: $X^2 = (n - 1)S^2/\sigma_0^2$;   RR: $X^2 \geq 23.6848$
$\chi^2 = 24.2667 \geq 23.6848$. There is evidence to suggest that the population variance in the diameter of viruses has increased.

9.205 H_0: $\sigma^2 = 62500^2$; H_a: $\sigma^2 > 62500^2$;
TS: $X^2 = (n - 1)S^2/\sigma_0^2$;   RR: $X^2 \geq 67.9852$
$\chi^2 = 39.2593$. There is no evidence to suggest that the population variance in the blood platelet count has increased.

9.207 (a) H_0: $\mu = 1800$; H_a: $\mu > 1800$;

TS: $Z = (\bar{x} - \mu_0)/(\sigma/\sqrt{n})$;   RR: $Z \geq 1.6449$
$z = 1.5446$. There is no evidence to suggest that the population mean amount of ore extracted each day is greater than 1800. There is no evidence to suggest that the new machinery has improved production. (b) 0.2800, 0.0193

9.209 (a) H_0: $\mu = 40$; H_a: $\mu < 40$;

TS: $T = \dfrac{\bar{X} - \mu_0}{S/\sqrt{n}}$;   RR: $T \leq -2.5083$

$t = -1.1733$. There is no evidence to suggest that the population mean brightness is less than 40. (b) $0.10 \leq p \leq 0.20$

9.211 (a) H_0: $p = 0.60$; H_a: $p < 0.60$;

TS: $Z = \dfrac{\hat{P} - p_0}{\sqrt{\frac{p_0(1 - p_0)}{n}}}$;   RR: $Z \leq -2.3263$

$z = -2.7951 \leq -2.3263$. There is evidence to suggest that the population proportion of cast-iron pans with harmful bacteria is less than 0.60. (b) 0.0026 (c) No. The people who brought their pans for testing self-selected.

9.213 (a) H_0: $\sigma^2 = 3.1^2$; H_a: $\sigma^2 < 3.1^2$;
TS: $X^2 = (n - 1)S^2/\sigma_0^2$;   RR: $X^2 \leq 6.5706$
$\chi^2 = 9.3511$. There is no evidence to suggest that the population variance in b value has decreased. (b) $p > 0.10$

CHAPTER 10

Section 10.1

10.1 (a) $\mu_1 - \mu_2 = 0$ (b) $\mu_1 - \mu_2 < 0$ (c) $\mu_1 - \mu_1 \neq 7$
(d) $\mu_1 - \mu_2 > -4$ (e) $\mu_1 - \mu_2 \neq 0$ (f) $\mu_1 - \mu_2 = 10$
10.3 (a) H_0: $\mu_1 - \mu_2 = 0$; H_a: $\mu_1 - \mu_2 > 0$;

TS: $Z = \dfrac{(\bar{X}_1 - \bar{X}_2) - 0}{\sqrt{\frac{\sigma_1^2}{n_1} + \frac{\sigma_2^2}{n_2}}}$;   RR: $z \geq 1.6449$

(b) $z = 2.0951 \geq 1.6449$. There is evidence to suggest that population mean 1 is greater than population mean 2. (c) 0.0181
10.5 (a) H_0: $\mu_1 - \mu_2 = 0$; H_a: $\mu_1 - \mu_2 \neq 0$;

TS: $Z = \dfrac{(\bar{X}_1 - \bar{X}_2) - 0}{\sqrt{\frac{\sigma_1^2}{n_1} + \frac{\sigma_2^2}{n_2}}}$;   RR: $|Z| \geq 3.2905$

(b) $z = -1.9379$. There is no evidence to suggest that population mean 1 is different from population mean 2. (c) No. Both sample sizes are large.

10.7 (a) $\bar{X}_1 - \bar{X}_2$ is normal with mean 15, variance 2.4286, and standard deviation 1.5584. (b) Probability distribution:

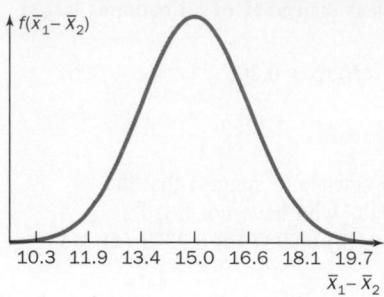

(c) 0.0997 (d) 0.2063 (e) 0.2605
10.9 H_0: $\mu_1 - \mu_2 = 0$; H_a: $\mu_1 - \mu_2 \neq 0$;

TS: $Z = \dfrac{(\bar{X}_1 - \bar{X}_2) - 0}{\sqrt{\frac{\sigma_1^2}{n_1} + \frac{\sigma_2^2}{n_2}}}$;   RR: $|Z| \geq 2.5758$

$z = -3.0814 \leq -2.5758$. There is evidence to suggest that the mean Nordstrom gift-certificate value is different from the mean Macy's gift-certificate value.

10.11 (a) $(-1.6135, 0.4735)$ (b) There is no evidence to suggest that the mean power-output ratings for the two brands differ. 0 is in the CI.

10.13 (a) H_0: $\mu_1 - \mu_2 = 5$; H_a: $\mu_1 - \mu_2 < 5$;

TS: $Z = \dfrac{(\bar{X}_1 - \bar{X}_2) - 5}{\sqrt{\frac{\sigma_1^2}{n_1} + \frac{\sigma_2^2}{n_2}}}$;   RR: $Z \leq -2.3263$

$z = -0.7546$. There is no evidence to refute the claim; there is no evidence to suggest that the differences in mean weights is less than 5 pounds. (b) 0.2252 (c) No. Both sample sizes are large.

10.15 (a) $(3.4032, 7.1968)$ (b) Yes. 0 is not in the CI, and the CI is entirely above 0.

10.17 (a) H_0: $\mu_1 - \mu_2 = 0$; H_a: $\mu_1 - \mu_2 > 0$;

TS: $Z = \dfrac{(\bar{X}_1 - \bar{X}_2) - 0}{\sqrt{\frac{\sigma_1^2}{n_1} + \frac{\sigma_2^2}{n_2}}}$;   RR: $Z \geq 2.3263$

$z = 3.1283 \geq 2.3263$. There is evidence to suggest that the population mean standby time for the Motorola phone is greater than the population mean standby time for the Uniden phone. (b) 0.000879

10.19 (a) H_0: $\mu_1 - \mu_2 = 0$; H_a: $\mu_1 - \mu_2 \neq 0$;

TS: $Z = \dfrac{(\bar{X}_1 - \bar{X}_2) - 0}{\sqrt{\frac{\sigma_1^2}{n_1} + \frac{\sigma_2^2}{n_2}}}$;   RR: $|Z| \geq 2.5758$

$z = -1.2536$. There is no evidence to suggest that the population mean magnesium in each serving of baked beans and potatoes is different. (b) $z = -1.8214$. Still no evidence to refute the claim. $p = 0.0685$. (c) $n_1 = n_2 = 76$

10.21 (a) H_0: $\mu_1 - \mu_2 = 0$; H_a: $\mu_1 - \mu_2 < 0$;

TS: $Z = \dfrac{(\bar{X}_1 - \bar{X}_2) - 0}{\sqrt{\frac{\sigma_1^2}{n_1} + \frac{\sigma_2^2}{n_2}}}$;   RR: $Z \leq -2.3263$

$z = -1.4998$. There is no evidence to suggest that the mean hourly wage for a plumber in Utica is less than the mean hourly wage for a plumber in Atlanta. (b) 0.0668 (c) Large population variances.

Section 10.2

10.23 (a) RR: $T \leq -1.7011$, $t = -0.1181$. There is no evidence to suggest that μ_1 is less than μ_2. (b) $p > 0.20$

10.25 (a) RR: $|T| \geq 2.0484$, $t = 3.0096 \geq 2.0484$. There is evidence to suggest that the two population means are different. (b) $0.001 \leq p \leq 0.01$

10.27 (a) 24 (b) 15 (c) 33 (d) 62

10.29 (a) RR: $|T| \geq 2.0796$, $t = 1.8502$. There is no evidence to suggest that the two population means are different. (b) RR: $|T'| \geq 2.1009$, $t' = 2.6994 \geq 2.1009$. There is evidence to suggest that the two population means are different. (c) Population variances are assumed unequal. The sample standard deviations suggest unequal variances.

10.31 (a) H_0: $\mu_1 - \mu_2 = 0$; H_a: $\mu_1 - \mu_2 \neq 0$;
TS: $T = \dfrac{(\bar{X}_1 - \bar{X}_2) - 0}{\sqrt{S_p^2\left(\frac{1}{n_1} + \frac{1}{n_2}\right)}}$; RR: $|T| \geq 2.0687$

$t = -0.9209$. There is no evidence to suggest that the population mean deviations from perfect flatness are different. (b) $0.20 \leq p \leq 0.40$

10.33 H_0: $\mu_1 - \mu_2 = 0$; H_a: $\mu_1 - \mu_2 < 0$;
TS: $T = \dfrac{(\bar{X}_1 - \bar{X}_2) - 0}{\sqrt{S_p^2\left(\frac{1}{n_1} + \frac{1}{n_2}\right)}}$; RR: $T \leq -1.7011$

$t = -2.4051. \leq -1.7011$. There is evidence to suggest that the mean file size for rap is less than the mean file size for jazz.

10.35 (a) H_0: $\mu_1 - \mu_2 = 0$; H_a: $\mu_1 - \mu_2 \neq 0$;
TS: $T = \dfrac{(\bar{X}_1 - \bar{X}_2) - 0}{\sqrt{S_p^2\left(\frac{1}{n_1} + \frac{1}{n_2}\right)}}$; RR: $|T| \geq 2.6259$

$t = -3.0863 \leq -2.6259$. There is evidence to suggest that the mean sagittal diameter of women's biceps tendons is different from that of men's biceps tendons. (b) $(-0.4928, -0.1072)$.

10.37 H_0: $\mu_1 - \mu_2 = 0$; H_a: $\mu_1 - \mu_2 > 0$;
TS: $T = \dfrac{(\bar{X}_1 - \bar{X}_2) - 0}{\sqrt{S_p^2\left(\frac{1}{n_1} + \frac{1}{n_2}\right)}}$; RR: $T \geq 3.2614$

$t = 13.9613 \geq 3.2614$. There is evidence to suggest that the mean amount men intend to spend is greater than the mean amount women intend to spend.

10.39 (a) H_0: $\mu_1 - \mu_2 = 0$; H_a: $\mu_1 - \mu_2 > 0$;
TS: $T' = \dfrac{(\bar{X}_1 - \bar{X}_2) - 0}{\sqrt{\left(\frac{S_1^2}{n_1} + \frac{S_2^2}{n_2}\right)}}$; RR: $T' \geq 2.4922$

$t' = 3.5202 \geq 2.4922$. There is evidence to suggest that the population mean amount of coating for Fero is greater than the population mean amount of coating for Cintex. (b) $(13.5281, 51.8719)$

10.41 H_0: $\mu_1 - \mu_2 = 0$; H_a: $\mu_1 - \mu_2 > 0$;
TS: $T' = \dfrac{(\bar{X}_1 - \bar{X}_2) - 0}{\sqrt{\left(\frac{S_1^2}{n_1} + \frac{S_2^2}{n_2}\right)}}$; RR: $T' \geq 2.4851$

$t' = 2.9891 \geq 2.4851$. There is evidence to suggest that the mean thickness of Aries wallpaper is greater than the mean thickness of all other wallpapers.

10.43 H_0: $\mu_1 - \mu_2 = 0$; H_a: $\mu_1 - \mu_2 \neq 0$;

TS: $T = \dfrac{(\bar{X}_1 - \bar{X}_2) - 0}{\sqrt{S_p^2\left(\frac{1}{n_1} + \frac{1}{n_2}\right)}}$; RR: $|T| \geq 2.7500$

$t = 0.3467$. There is no evidence to suggest that the population mean amount spent on gift cards per consumer is different on the East Coast and the West Coast.

10.45 H_0: $\mu_1 - \mu_2 = 0$; H_a: $\mu_1 - \mu_2 \neq 0$;
TS: $T = \dfrac{(\bar{X}_1 - \bar{X}_2) - 0}{\sqrt{S_p^2\left(\frac{1}{n_1} + \frac{1}{n_2}\right)}}$; RR: $|T| \geq 2.0244$

$t = 1.0183$. There is no evidence to suggest that the population mean number of days in the hospital is different for Type A toxin and Type B toxin.

10.47 (a) H_0: $\mu_1 - \mu_2 = 0$; H_a: $\mu_1 - \mu_2 < 0$;
TS: $T = \dfrac{(\bar{X}_1 - \bar{X}_2) - 0}{\sqrt{S_p^2\left(\frac{1}{n_1} + \frac{1}{n_2}\right)}}$; RR: $T \leq -2.4121$

$t = -3.3073 \leq -2.4121$. There is evidence to suggest that the mean width of a \$20 bill is greater than the mean width of a \$1 bill. (b) H_0: $\mu_1 - \mu_2 = 0$; H_a: $\mu_1 - \mu_2 < 0$;
TS: $T' = \dfrac{(\bar{X}_1 - \bar{X}_2) - 0}{\sqrt{\left(\frac{S_1^2}{n_1} + \frac{S_2^2}{n_2}\right)}}$; RR: $T' \leq -2.4314$

$t' = -3.28 \leq -2.4314$. There is evidence to suggest that the mean width of a \$20 bill is greater than the mean width of a \$1 bill. (c) The sample means are relatively close, the sample variances are relatively close, and the sample sizes are relatively large.

10.49 (a) H_0: $\mu_1 - \mu_2 = 0$; H_a: $\mu_1 - \mu_2 \neq 0$;
TS: $T' = \dfrac{(\bar{X}_1 - \bar{X}_2) - 0}{\sqrt{\left(\frac{S_1^2}{n_1} + \frac{S_2^2}{n_2}\right)}}$; RR: $|T'| \geq 1.9996$

$t' = -0.2524$. There is no evidence to suggest that the mean pressure required to open each valve is different. (b) $(-1.784, 1.3842)$. This CI is consistent with part (a); 0 is in the CI.

10.51 Should reject the null hypothesis approximately 5 times. As σ_2 increases, the null hypothesis will be rejected a greater number of times, but still approximately 5. This hypothesis test is robust. That is, we still reach the correct conclusion even though the underlying assumptions are false.

Section 10.3

10.53 (a) Paired; patients (arms). (b) Paired; lathes. (c) Independent; 20-year-old males, 70-year-old males. (d) Paired; files. (e) Independent; frequent flyers on United, frequent flyers on Delta.

10.55 (a) H_0: $\mu_D = 0$; H_a: $\mu_D < 0$;
TS: $T = \dfrac{\bar{D} - \Delta_0}{S_D/\sqrt{n}}$; RR: $T \leq -3.7469$

$t = -1.0551$. There is no evidence to suggest that the population mean before treatment is less than the population mean after treatment. (b) $0.10 \leq p \leq 0.20$

10.57 (a) Programmer. (b) H_0: $\mu_D = 0$; H_a: $\mu_D < 0$;
TS: $T = \dfrac{\bar{D} - \Delta_0}{S_D/\sqrt{n}}$; RR: $T \leq -3.5518$

$t = -3.8636 \leq -3.5518$. There is evidence to suggest that the population mean runtime for Java programs is greater than the population mean runtime for C++ programs. **(c)** $0.0001 \leq p \leq 0.0005$

10.59 (a) $(-0.3584, 5.9584)$ **(b)** No, 0 is included in the CI.

10.61 (a) $H_0: \mu_D = 0;$ $\quad H_a: \mu_D > 0;$

TS: $T = \dfrac{\overline{D} - \Delta_0}{S_D / \sqrt{n}};$ $\quad$ RR: $T \geq 4.0493$

$t = 5.8213 \geq 4.0493$. There is evidence to suggest that the population mean autofocus shutter lag is greater than the population mean prefocus shutter lag. **(b)** $p < 0.0001$

10.63 $H_0: \mu_D = 0;$ $\quad H_a: \mu_D > 0;$

TS: $T = \dfrac{\overline{D} - \Delta_0}{S_D / \sqrt{n}};$ $\quad$ RR: $T \geq 1.8331$

$t = 0.5689$. There is no evidence to suggest that the population mean salt content before potatoes is greater than the population mean salt content after potatoes.

10.65 (a) $H_0: \mu_D = 0;$ $\quad H_a: \mu_D > 0;$

TS: $T = \dfrac{\overline{D} - \Delta_0}{S_D / \sqrt{n}};$ $\quad$ RR: $T \geq 1.8946$

$t = 0.4740$. There is no evidence to suggest that the population mean electromagnetic emission from the old antenna is greater than the population mean electromagnetic emission from the new antenna. **(b)** $p > 0.20$

10.67 (a) Exam score. **(b)** $H_0: \mu_D = 0;$ $\quad H_a: \mu_D \neq 0;$

TS: $T = \dfrac{\overline{D} - \Delta_0}{S_D / \sqrt{n}};$ $\quad$ RR: $|T| \geq 2.3646$

$t = 1.7154$. There is no evidence to suggest that the written-manual population mean assembly time is different from the interactive-video population mean assembly time.

10.69 (a) Patient. **(b)** $H_0: \mu_D = 0;$ $\quad H_a: \mu_D > 0;$

TS: $T = \dfrac{\overline{D} - \Delta_0}{S_D / \sqrt{n}};$ $\quad$ RR: $T \geq 1.8331$

$t = 3.3746 \geq 1.8331$. There is evidence to suggest that the population mean temperature before the drug is greater than the population mean temperature after the drug. **(c)** $0.001 \leq p \leq 0.005$ **(d)** Nine of the 10 differences are positive.

10.71 (a) Force. **(b)** $H_0: \mu_D = 0;$ $\quad H_a: \mu_D < 0;$

TS: $T = \dfrac{\overline{D} - \Delta_0}{S_D / \sqrt{n}};$ RR: $T \leq -5.6938$

$t = -5.7564 \leq -5.6938$. There is evidence to suggest that the old-formula population mean resilience is less than the new-formula population mean resilience. **(c)** All differences are negative.

Section 10.4

10.73

	Mean	Variance	Standard deviation	Probability
(a)	-0.020	0.000579	0.0241	0.0034
(b)	0.040	0.001751	0.0418	0.0280
(c)	0.010	0.001557	0.0395	0.7804
(d)	-0.070	0.000858	0.0293	0.8472
(e)	-0.120	0.000275	0.0166	0.9999
(f)	0.041	0.000914	0.0302	0.9527

10.75 (a) $z = 1.1137, p = 0.1327$. Do not reject H_0.
(b) $z = -1.8938, p = 0.0291$. Do not reject H_0. **(c)** $z = -2.2705,$ $p = 0.0232$. Reject H_0. **(d)** $z = 0.5012, p = 0.6163$. Do not reject H_0.

10.77 (a) $(-0.0972, 0.0390)$ **(b)** $(-0.0710, -0.0052)$
(c) $(-0.1376, 0.1289)$ **(d)** $(0.0168, 0.0598)$

10.79 $H_0: p_1 - p_2 = 0;$ $\quad H_a: p_1 - p_2 \neq 0;$

TS: $Z = \dfrac{\widehat{P}_1 - \widehat{P}_2}{\sqrt{\widehat{P}_c(1 - \widehat{P}_c)\left(\frac{1}{n_1} + \frac{1}{n_2}\right)}};$ $\quad$ RR: $|Z| \geq 1.9600$

$z = 0.1523$. There is no evidence to suggest that the population proportion of repeat customers is different for Ford and Chevrolet.

10.81 (a) $H_0: p_1 - p_2 = 0;$ $\quad H_a: p_1 - p_2 > 0;$

TS: $Z = \dfrac{\widehat{P}_1 - \widehat{P}_2}{\sqrt{\widehat{P}_c(1 - \widehat{P}_c)\left(\frac{1}{n_1} + \frac{1}{n_2}\right)}};$ $\quad$ RR: $Z \geq 3.0902$

$z = 3.2086 \geq 3.0902$. There is evidence to suggest that the population proportion of 18–29-year-olds who believe that movies are getting better is greater than the population proportion of 30–49-year-olds who believe that movies are getting better. **(b)** 0.000667

10.83 (a) $n_1\widehat{p}_1 = 530 \geq 5, n_1(1 - \widehat{p}_1) = 525 \geq 5,$ $n_2\widehat{p}_2 = 825 \geq 5, n_2(1 - \widehat{p}_2) = 838 \geq 5$ **(b)** $H_0: p_1 - p_2 = 0;$ $H_a: p_1 - p_2 > 0;$

TS: $Z = \dfrac{\widehat{P}_1 - \widehat{P}_2}{\sqrt{\widehat{P}_c(1 - \widehat{P}_c)\left(\frac{1}{n_1} + \frac{1}{n_2}\right)}};$ $\quad$ RR: $Z \geq 2.3263$

$z = 0.3190$. There is no evidence to suggest that the population proportion of carpoolers crossing the George Washington Bridge is greater than the population proportion of carpoolers using the Lincoln Tunnel. **(c)** 0.3749

10.85 (a) $\widehat{p}_1 = 0.2784, \widehat{p}_2 = 0.2321$ **(b)** $H_0: p_1 - p_2 = 0;$ $H_a: p_1 - p_2 \neq 0;$

TS: $Z = \dfrac{\widehat{P}_1 - \widehat{P}_2}{\sqrt{\widehat{P}_c(1 - \widehat{P}_c)\left(\frac{1}{n_1} + \frac{1}{n_2}\right)}};$ RR: $|Z| \geq 2.5758$

$z = 1.1773$. There is no evidence to suggest that the population proportion of people who experience relief due to the antihistamine is different from the population proportion of people who experience relief from butterbur extract.

10.87 (a) $\widehat{p}_1 = 0.0755, \widehat{p}_2 = 0.0992$ **(b)** $n_1\widehat{p}_1 = 8 \geq 5,$ $n_1(1 - \widehat{p}_1) = 98 \geq 5, n_2\widehat{p}_2 = 12 \geq 5, n_2(1 - \widehat{p}_2) = 109 \geq 5$ **(c)** $H_0: p_1 - p_2 = 0;$ $\quad H_a: p_1 - p_2 \neq 0;$

TS: $Z = \dfrac{\widehat{P}_1 - \widehat{P}_2}{\sqrt{\widehat{P}_c(1 - \widehat{P}_c)\left(\frac{1}{n_1} + \frac{1}{n_2}\right)}};$ $\quad$ RR: $|Z| \geq 1.9600$

$z = -0.6286$. There is no evidence to suggest that the population proportion of defective lenses is different for Process A than Process B.

10.89 $H_0: p_1 - p_2 = 0;$ $\quad H_a: p_1 - p_2 \neq 0;$

TS: $Z = \dfrac{\widehat{P}_1 - \widehat{P}_2}{\sqrt{\widehat{P}_c(1 - \widehat{P}_c)\left(\frac{1}{n_1} + \frac{1}{n_2}\right)}};$ $\quad$ RR: $|Z| \geq 1.9600$

$z = -0.0663$. There is no evidence to suggest that the population proportion of veterans living in Colorado Springs and Jacksonville is different.

10.91 (a) $H_0: p_1 - p_2 = 0;$ $\quad H_a: p_1 - p_2 < 0;$

TS: $Z = \dfrac{\hat{P}_1 - \hat{P}_2}{\sqrt{\hat{P}_c(1 - \hat{P}_c)\left(\frac{1}{n_1} + \frac{1}{n_2}\right)}}$; RR: $Z \leq -1.6449$

$z = -3.0677 \leq -1.6449$. There is evidence to suggest that the population proportion of 18- to 29-year-olds who obtain news every day from nightly network news shows is less than the population proportion of 30- to 49-year-olds who obtain news every day from nightly network news shows. $p = 0.0011$

(b) H_0: $p_1 - p_2 = 0$; H_a: $p_1 - p_2 < 0$;

TS: $Z = \dfrac{\hat{P}_1 - \hat{P}_2}{\sqrt{\hat{P}_c(1 - \hat{P}_c)\left(\frac{1}{n_1} + \frac{1}{n_2}\right)}}$; RR: $Z \leq -2.3263$

$z = -3.7319 \leq -2.3263$. There is evidence to suggest that the population proportion of 18- to 29-year-olds who obtain news every day from nightly cable news shows is less than the population proportion of 30- to 49-year-olds who obtain news every day from nightly cable news shows. $p = 0.000095$

(c) H_0: $p_1 - p_2 = 0$; H_a: $p_1 - p_2 \neq 0$;

TS: $Z = \dfrac{\hat{P}_1 - \hat{P}_2}{\sqrt{\hat{P}_c(1 - \hat{P}_c)\left(\frac{1}{n_1} + \frac{1}{n_2}\right)}}$; RR: $Z \geq -2.5758$

$z = -2.0501$. There is no evidence to suggest that the population proportion of 18- to 29-year-olds who obtain news every day from the Internet is less than the population proportion of 30- to 49-year-olds who obtain news every day from the Internet. $p = 0.0202$

10.93 H_0: $p_1 - p_2 = 0.10$; H_a: $p_1 - p_2 > 0.10$;

TS: $Z = \dfrac{(\hat{P}_1 - \hat{P}_2) - \Delta_0}{\sqrt{\frac{\hat{P}_1(1 - \hat{P}_1)}{n_1} + \frac{\hat{P}_2(1 - \hat{P}_2)}{n_2}}}$; RR: $Z \geq 2.3263$

$z = 0.8038$. There is no evidence to suggest that the population proportion of homeowners planning a landscaping project is more than 0.10 greater than the population proportion of condominium owners planning a landscaping project.

Section 10.5

10.95 (a) 2.20 **(b)** 3.15 **(c)** 3.58 **(d)** 4.99 **(e)** 0.23 **(f)** 0.12 **(g)** 0.34 **(h)** 0.25

10.97 (a) H_0: $\sigma_1^2 = \sigma_2^2$; H_a: $\sigma_1^2 > \sigma_2^2$;

TS: $F = S_1^2 / S_2^2$; RR: $F \geq 1.94$
(b) $f = 2.5448 \geq 1.94$. There is evidence to suggest that population variance 1 is greater than population variance 2.
(c) $0.01 \leq p \leq 0.05$. p value illustration:

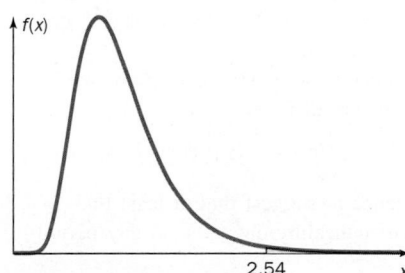

10.99 (a) H_0: $\sigma_1^2 = \sigma_2^2$; H_a: $\sigma_1^2 \neq \sigma_2^2$;

TS: $F = S_1^2 / S_2^2$; RR: $F \leq 0.17$ or $F \geq 4.54$
(b) $f = 4.8259 \geq 4.54$. There is evidence to suggest that population variance 1 is different from population variance 2.
(c) $0.002 \leq p \leq 0.02$

10.101 (a) (0.3254, 3.5608) **(b)** (0.4712, 5.8451) **(c)** (0.2703, 2.3458) **(d)** (0.2843, 2.5079)

10.103 H_0: $\sigma_1^2 = \sigma_2^2$; H_a: $\sigma_1^2 > \sigma_2^2$;

TS: $F = S_1^2 / S_2^2$; RR: $F \geq 2.39$

$f = 4.6944 \geq 2.39$. There is evidence to suggest that the population variance in the aerosol light absorption coefficient is greater in Africa than in South America.

10.105 H_0: $\sigma_1^2 = \sigma_2^2$; H_a: $\sigma_1^2 > \sigma_2^2$;

TS: $F = S_1^2 / S_2^2$; RR: $F \geq 3.52$

$f = 4.6786 \geq 3.52$. There is evidence to suggest that the population variance in the weight of frozen turkeys from North Carolina is greater than the population variance in the weight of frozen turkeys from Minnesota.

10.107 (a) H_0: $\sigma_1^2 = \sigma_2^2$; H_a: $\sigma_1^2 > \sigma_2^2$;

TS: $F = S_1^2 / S_2^2$; RR: $F \geq 2.6041$

$f = 2.8681 \geq 2.6041$. There is evidence to suggest that the population variance in winning times for an ordinary race is greater than the population variance in winning times for a stakes race. **(b)** (1.1014, 7.4689)

10.109 H_0: $\sigma_1^2 = \sigma_2^2$; H_a: $\sigma_1^2 \neq \sigma_2^2$;

TS: $F = S_1^2 / S_2^2$; RR: $F \leq 0.37$ or $F \geq 2.67$

$f = 0.5552$. There is no evidence to suggest that the population variance in flight-delay times for Delta is different from the population variance in flight-delay times for United. **(b)** No. The distributions are probably skewed right.

10.111 H_0: $\sigma_1^2 = \sigma_2^2$; H_a: $\sigma_1^2 > \sigma_2^2$;

TS: $F = S_1^2 / S_2^2$;   RR: $F \geq 3.2940$

$f = 8.9648 \geq 3.2940$. There is evidence to suggest that the population variance in lodging per diem is greater in the Northeast than in the West.

10.113 (a) H_0: $\sigma_1^2 = \sigma_2^2$; H_a: $\sigma_1^2 \neq \sigma_2^2$;

TS: $F = S_1^2 / S_2^2$;   RR: $F \leq 0.42$ or $F \geq 2.54$

$f = 0.5492$. There is no evidence to suggest that the population variance in circulation for the *Sun-Times* is different from the population variance in circulation for the *Globe*. **(b)** No. The circulation distribution could be skewed right.

10.115 (a) H_0: $\sigma_1^2 = \sigma_2^2$; H_a: $\sigma_1^2 \neq \sigma_2^2$;

TS: $F = S_1^2 / S_2^2$;   RR: $F \leq 0.31$ or $F \geq 3.53$

$f = 2.7128$. There is no evidence to suggest that the population variance in the saccharin amount for Fishing Creek is different from the population variance in the saccharin amount for Honest Tea. **(b)** $0.02 \leq p \leq 0.10$

Chapter Exercises

10.117 H_0: $\mu_1 - \mu_2 = 0$; H_a: $\mu_1 - \mu_2 \neq 0$;

TS: $Z = \dfrac{(\bar{X}_1 - \bar{X}_2) - 0}{\sqrt{\frac{\sigma_1^2}{n_1} + \frac{\sigma_2^2}{n_2}}}$; RR: $|Z| \geq 2.5758$

$z = -2.7903 \leq -2.5758$. There is evidence to suggest that the population mean amount of corrosive material carried by trucks in North Carolina is different from the population meant amount of corrosive material carried by trucks in Virginia.

10.119 (a) $\hat{p}_1 = 0.6158$, $\hat{p}_2 = 0.6318$

$n_1\hat{p}_1 = 335 \geq 5$, $n_1(1 - \hat{p}_1) = 209 \geq 5$, $n_2\hat{p}_2 = 381 \geq 5$,

$n_2(1 - \hat{p}_2) = 222 \geq 5$ **(b)** $H_0: p_1 - p_2 = 0$; $H_a: p_1 - p_2 \neq 0$;

TS: $Z = \dfrac{\hat{P}_1 - \hat{P}_2}{\sqrt{\hat{P}_c(1 - \hat{P}_c)\left(\frac{1}{n_1} + \frac{1}{n_2}\right)}}$; RR: $|Z| \geq 2.5758$

$z = -0.5598$. There is no evidence to suggest that the two population proportions are different. **(c)** 0.5756

10.121 (a) $H_0: p_1 - p_2 = 0$; $H_a: p_1 - p_2 > 0$;

TS: $Z = \dfrac{\hat{P}_1 - \hat{P}_2}{\sqrt{\hat{P}_c(1 - \hat{P}_c)\left(\frac{1}{n_1} + \frac{1}{n_2}\right)}}$; RR: $Z \geq 2.3263$

$z = 1.0728$. There is no evidence to suggest that the population proportion of residents in Ohio who recycle newspapers is greater than the population proportion of residents in Florida who recycle newspapers. **(b)** 0.1417

10.123 $H_0: \sigma_1^2 = \sigma_2^2$; $H_a: \sigma_1^2 > \sigma_2^2$;

TS: $F = S_1^2 / S_2^2$; RR: $F \geq 5.06$

$f = 7.84 \geq 5.06$. There is evidence to suggest that the population variance in aluminum fuselage thickness is greater than the population variance in carbon-fiber fuselage thickness.

10.125 $H_0: \mu_1 - \mu_2 = 0$; $H_a: \mu_1 - \mu_2 > 0$;

TS: $T' = \dfrac{(\bar{X}_1 - \bar{X}_2) - 0}{\sqrt{\frac{S_1^2}{n_1} + \frac{S_2^2}{n_2}}}$; RR: $T' \geq 3.7874$

$t' = 2.5896$. There is no evidence to suggest that the population mean PCB level in smallmouth bass at Bull's Bridge is greater than the population mean PCB level in smallmouth bass at Lake Zoar.

10.127 $H_0: \mu_1 - \mu_2 = 0$; $H_a: \mu_1 - \mu_2 > 0$;

TS: $T = \dfrac{(\bar{X}_1 - \bar{X}_2) - 0}{\sqrt{S_p^2\left(\frac{1}{n_1} + \frac{1}{n_2}\right)}}$; RR: $T \geq 3.5518$

$t = 2.2317$. There is no evidence to suggest that the population mean Rolling Stones concert ticket price is greater than the population mean Coldplay concert ticket price.

10.129 $H_0: \mu_1 - \mu_2 = 0$; $H_a: \mu_1 - \mu_2 \neq 0$;

TS: $T = \dfrac{(\bar{X}_1 - \bar{X}_2) - 0}{\sqrt{S_p^2\left(\frac{1}{n_1} + \frac{1}{n_2}\right)}}$; RR: $|T| \geq 2.0010$

$t = -2.6898 \leq -2.0010$. There is evidence to suggest that the population mean fine particulate measure is different in these two areas.

10.131 (a) $H_0: \mu_1 - \mu_2 = 0$; $H_a: \mu_1 - \mu_2 \neq 0$;

TS: $T = \dfrac{(\bar{X}_1 - \bar{X}_2) - 0}{\sqrt{S_p^2\left(\frac{1}{n_1} + \frac{1}{n_2}\right)}}$; RR: $|T| \geq 2.6778$

$t = -8.6946 \leq -2.6778$. There is evidence to suggest that the population mean number of yearly pro bono hours is different at these two law firms. **(b)** $(-7.5863, -4.0137)$ **(c)** Yes, 0 is not in the CI.

10.133 Race versus Age: $t = -2.0685$, $p = 0.0479$. Reject H_0. Race versus Disability: $t = -0.7547$, $p = 0.4565$. Do not reject H_0. Age versus Disability: $t = 1.3559$, $p = 0.1856$. Do not reject H_0.

CHAPTER 11

Section 11.1

11.1 (a) 9, 7, 6 **(b)** 260, 229, 195 **(c)** 21602

11.3 (a) 775, 745, 768, 753 **(b)** 465193 **(c)** 2808.95, 112.55, 2696.40 **(d)** 37.5167, 168.525 **(e)** 0.2226

11.5

ANOVA summary table

Source of variation	Sum of squares	Degrees of freedom	Mean square	F	p value
Factor	584.1	4	146.0250	0.57	0.6864
Error	12,062.1	47	256.6404		
Total	12,646.2	51			

(a) $H_0: \mu_1 = \mu_2 = \mu_3 = \mu_4 = \mu_5$; $H_a: \mu_1 \neq \mu_j$ for some $i \neq j$ **(b)** RR: $F \geq 2.57$ **(c)** $f = 0.57$

There is no evidence to suggest at least two of the population means are different.

11.7 (a) $H_0: \mu_1 = \mu_2 = \mu_3 = \mu_4$; $H_a: \mu_i \neq \mu_j$ for some $i \neq j$; TS: $F =$ MSA/MSE; RR: $F \geq 3.10$

(b)

ANOVA summary table

Source of variation	Sum of squares	Degrees of freedom	Mean square	F	p value
Factor	67000.33	3	22333.44	12.39	0.0001
Error	36039.67	20	1801.98		
Total	103040.00	23			

(c) $f = 12.39 \geq 3.10$. There is evidence to suggest that at least two of the population means are different.

11.9 $H_0: \mu_1 = \mu_2 = \mu_3 = \mu_4 = \mu_5$; $H_a: \mu_i \neq \mu_j$ for some $i \neq j$; TS: $F =$ MSA/MSE; RR: $F \geq 2.48$

$f = 7.35 \geq 2.48$. There is evidence to suggest that at least two population mean times are different.

11.11 $H_0: \mu_1 = \mu_2 = \mu_3$; $H_a: \mu_i \neq \mu_j$ for some $i \neq j$; TS: $F =$ MSA/MSE; RR: $F \geq 5.19$

$f = 6.35 \geq 5.19$. There is evidence to suggest that at least two of the population mean weights are different.

11.13 $H_0: \mu_1 = \mu_2 = \mu_3 = \mu_4$; $H_a: \mu_i \neq \mu_j$ for some $i \neq j$; TS: $F =$ MSA/MSE; RR: $F \geq 4.13$

$f = 6.82 \geq 4.13$. There is evidence to suggest that at least two of the population means are different.

11.15 $H_0: \mu_1 = \mu_2 = \mu_3$; $H_a: \mu_i \neq \mu_j$ for some $i \neq j$; TS: $F =$ MSA/MSE; RR: $F \geq 3.16$

$f = 6.24 \geq 3.16$. There is evidence to suggest that at least two population mean numbers of plants per seized plot are different.

11.17 $H_0: \mu_1 = \mu_2 = \mu_3 = \mu_4$; $H_a: \mu_i \neq \mu_j$ for some $i \neq j$; TS: $F =$ MSA/MSE; RR: $F \geq 3.10$

$f = 0.81$. There is no evidence to suggest that at least two population mean thaw depths are different.

11.19 $H_0: \mu_1 = \mu_2 = \mu_3 = \mu_4$; $H_a: \mu_i \neq \mu_j$ for some $i \neq j$; TS: $F =$ MSA/MSE; RR: $F \geq 2.72$

$f = 1.00$. There is no evidence to suggest that at least two population mean numbers of unhealthy days per 30-day period are different.

11.21 (a) $H_0: \mu_1 = \mu_2 = \mu_3 = \mu_4$; $H_a: \mu_i \neq \mu_j$ for some $i \neq j$; TS: $F =$ MSA/MSE; RR: $F \geq 3.01$

$f = 3.40 \geq 3.01$. There is evidence to suggest that at least two of the population mean pressures are different. **(b)** Holder. This broom has the highest mean pressure.

11.23 (a) H_0: $\mu_1 = \mu_2 = \mu_3 = \mu_4 = \mu_5$; H_a: $\mu_i \neq \mu_j$ for some $i \neq j$; TS: $F = $ MSA/MSE; RR: $F \geq 5.36$
$f = 21.58 \geq 5.36$. There is evidence to suggest that at least two population mean weights are different. (b) Weis. These bags have the largest sample mean.

11.25

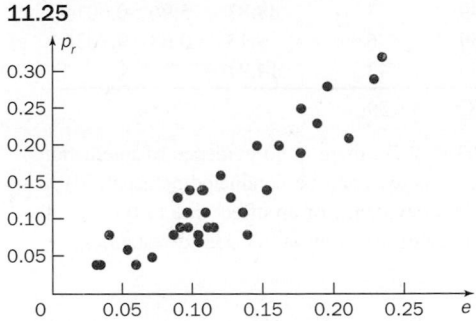

As effect size increases, the probability of rejecting H_0 also increases. This graph illustrates the power of the test—the probability of rejecting the null hypothesis for a specific alternative.

Section 11.2

11.27 (a) 3.609 (b) 3.791 (c) 4.634 (d) 4.863 (e) 6.469

11.29

(a)

$\bar{x}_1.$	$\bar{x}_2.$	$\bar{x}_3.$	$\bar{x}_4.$
-33.44	-14.83	0.48	4.30

(b)

$\bar{x}_3.$	$\bar{x}_2.$	$\bar{x}_4.$	$\bar{x}_1.$
1.30	1.41	1.50	1.62

(c)

$\bar{x}_4.$	$\bar{x}_2.$	$\bar{x}_3.$	$\bar{x}_1.$	$\bar{x}_5.$
51.92	54.21	60.80	64.35	64.85

11.31

Difference	Bonferroni CI	Significantly different
$\mu_1 - \mu_2$	(-7.62, 3.64)	No
$\mu_1 - \mu_3$	(-2.54, 8.72)	No
$\mu_1 - \mu_4$	(-11.80, -0.54)	Yes
$\mu_2 - \mu_3$	(-0.55, 10.71)	No
$\mu_2 - \mu_4$	(-9.81, 1.45)	No
$\mu_3 - \mu_4$	(-14.89, -3.63)	Yes

11.33

Difference	Bonferroni CI	Significantly different
$\mu_1 - \mu_2$	(0.06, 1.28)	Yes
$\mu_1 - \mu_3$	(-1.02, 0.16)	No
$\mu_2 - \mu_3$	(-1.73, -0.47)	Yes

11.35 (a) $0.001 \leq p \leq 0.01$. There is evidence to suggest that at least two population means are different.

(b)

Difference	Bonferroni CI	Significantly different
$\mu_1 - \mu_2$	(0.44, 7.35)	Yes
$\mu_1 - \mu_3$	(-3.56, 3.36)	No
$\mu_1 - \mu_4$	(-1.86, 5.06)	No
$\mu_2 - \mu_3$	(-7.45, -0.54)	Yes
$\mu_2 - \mu_4$	(-5.76, 1.16)	No
$\mu_3 - \mu_4$	(-1.76, 5.16)	No

11.37 (a) $p < 0.0001$

(b)

Difference	Bonferroni CI	Significantly different
$\mu_1 - \mu_2$	(0.87, 1.25)	Yes
$\mu_1 - \mu_3$	(0.56, 0.95)	Yes
$\mu_1 - \mu_4$	(0.20, 0.59)	Yes
$\mu_2 - \mu_3$	(-0.49, -0.12)	Yes
$\mu_2 - \mu_4$	(-0.85, -0.48)	Yes
$\mu_3 - \mu_4$	(-0.55, -0.17)	Yes

$\bar{x}_2.$	$\bar{x}_3.$	$\bar{x}_4.$	$\bar{x}_1.$
0.2567	0.5601	0.9206	1.3164

11.39 (a) H_0: $\mu_1 = \mu_2 = \mu_3 = \mu_4$; H_a: $\mu_i \neq \mu_j$ for some $i \neq j$; TS: $F = $ MSA/MSE; RR: $F \geq 3.10$
$f = 10.03 \geq 3.10$. There is evidence to suggest that at least two population means are different.

(b)

Difference	Bonferroni CI	Significantly different
$\mu_1 - \mu_2$	(-15.05, 0.31)	No
$\mu_1 - \mu_3$	(-6.53, 8.83)	No
$\mu_1 - \mu_4$	(-18.76, -3.40)	Yes
$\mu_2 - \mu_3$	(0.84, 16.20)	Yes
$\mu_2 - \mu_4$	(-11.40, 3.96)	No
$\mu_3 - \mu_4$	(-19.91, -4.55)	Yes

(c)

$\bar{x}_3.$	$\bar{x}_1.$	$\bar{x}_2.$	$\bar{x}_4.$
65.55	66.70	74.07	77.78

11.41 (a) H_0: $\mu_1 = \mu_2 = \mu_3$; H_a: $\mu_i \neq \mu_j$ for some $i \neq j$; TS: $F = $ MSA/MSE; RR: $F \geq 3.32$
$f = 30.85 \geq 3.32$. There is evidence to suggest that at least two population means are different.

(b)

Difference	Tukey CI	Significantly different
$\mu_1 - \mu_2$	(32.86, 63.56)	Yes
$\mu_1 - \mu_3$	(17.75, 49.08)	Yes
$\mu_2 - \mu_3$	(-29.76, 0.17)	No

(c)

$\bar{x}_2.$	$\bar{x}_3.$	$\bar{x}_1.$
67.80	82.59	116.01

11.43 (a) H_0: $\mu_1 = \mu_2 = \mu_3 = \mu_4 = \mu_5$; H_a: $\mu_i \neq \mu_j$ for some $i \neq j$; TS: $F = $ MSA/MSE; RR: $F \geq 4.43$
$f = 12.82 \geq 4.43$. There is evidence to suggest that at least two population means are different.

(b)

Difference	Bonferroni CI	Significantly different
$\mu_1 - \mu_2$	(−4.66, 0.54)	No
$\mu_1 - \mu_3$	(−3.36, 1.84)	No
$\mu_1 - \mu_4$	(−4.36, 0.84)	No
$\mu_1 - \mu_5$	(−8.04, −2.84)	Yes
$\mu_2 - \mu_3$	(−1.30, 3.90)	No
$\mu_2 - \mu_4$	(−2.30, 2.90)	No
$\mu_2 - \mu_5$	(−5.98, −0.78)	Yes
$\mu_3 - \mu_4$	(−3.60, 1.60)	No
$\mu_3 - \mu_5$	(−7.28, −2.08)	Yes
$\mu_4 - \mu_5$	(−6.28, −1.08)	Yes

(c)

$\bar{x}_{1.}$	$\bar{x}_{3.}$	$\bar{x}_{4.}$	$\bar{x}_{2.}$	$\bar{x}_{5.}$
14.28	15.04	16.04	16.34	19.72

11.45 (a) $H_0: \mu_1 = \mu_2 = \mu_3 = \mu_4$; $H_a: \mu_i \neq \mu_j$ for some $i \neq j$;
TS: $F = MSA/MSE$; RR: $F \geq 4.26$

$f = 19.51 \geq 4.26$. There is evidence to suggest that at least two population means are different.

(b)

Difference	Tukey CI	Significantly different
$\mu_1 - \mu_2$	(0.39, 1.34)	Yes
$\mu_1 - \mu_3$	(−0.18, 0.76)	No
$\mu_1 - \mu_4$	(0.75, 1.70)	Yes
$\mu_2 - \mu_3$	(−1.05, −0.10)	Yes
$\mu_2 - \mu_4$	(−0.11, 0.83)	No
$\mu_3 - \mu_4$	(0.46, 1.41)	Yes

(c) All differences are consistent, except between 68% fat and 40% fat. We would expect evidence to suggest that these two population means are different. The Tukey CI just barely includes 0.

Section 11.3

11.47 (a) 2, 3, 3 **(b)** $t_{11.} = 21$, $t_{12.} = 31$, $t_{13.} = 37$, $t_{21.} = 28$, $t_{22.} = 32$, $t_{23.} = 48$ **(c)** $t_{1..} = 89$, $t_{2..} = 108$, $t_{.1.} = 49$, $t_{.2.} = 63$, $t_{.3.} = 85$, $t_{...} = 197$
11.49 (a) SST = 481.880, SSA = 160.820, SSB = 76.827, SS(AB) = 45.480, SSE = 198.753 **(b)** MSA = 53.6067, MSB = 38.4133, MS(AB) = 7.5800, MSE = 8.2814
(c) $f_A = 6.47$, $f_B = 4.64$, and $f_{AB} = 0.92$ **(d)** $f_{AB} = 0.92 < 2.51$; There is no evidence of interaction. $f_A = 6.47 \geq 3.01$; There is evidence of an effect due to factor A. $f_B = 4.64 \geq 3.40$. There is evidence of an effect due to factor B.

11.51

Source of variation	Sum of squares	Degrees of freedom	Mean square	F	p value
Factor A	162.64	4	40.66	2.53	0.0476
Factor B	156.54	2	78.27	4.86	0.0103
Interaction	144.23	8	18.03	1.12	0.3596
Error	1206.87	75	16.09		
Total	1670.28	89			

$f_{AB} = 1.12 < 2.06$; there is no evidence of interaction. $f_A = 2.53 \geq 2.49$; there is evidence of an effect due to factor A. $f_B = 4.86 \geq 3.12$; there is evidence of an effect due to factor B.

11.53 (a)

Source of variation	Sum of squares	Degrees of freedom	Mean square	F	p value
Type	90.14	2	45.07	3.02	0.0580
Restaurant	266.49	3	88.83	5.96	0.0015
Interaction	56.59	6	9.43	0.63	0.7034
Error	715.60	48	14.91		
Total	1128.82	59			

(b) 60 **(c)** $f_{AB} = 0.63 < 2.29$; there is no evidence of interaction. The other two hypothesis tests can be conducted as usual. **(d)** $f_A = 3.02 < 3.19$; there is no evidence of an effect due to type. $f_B = 5.96 \geq 2.80$; there is evidence of an effect due to restaurant.

11.55

Source of variation	Sum of squares	Degrees of freedom	Mean square	F	p value
Megapixels	86.201	3	28.73	5.83	0.0039
Printer	2.101	1	2.10	0.43	0.5182
Interaction	6.011	3	2.00	0.41	0.7473
Error	118.315	24	4.93		
Total	212.629	31			

$f_{AB} = 0.41 < 3.01$; there is no evidence of interaction.
$f_A = 5.83 \geq 3.01$; there is evidence of an effect due to megapixels.
$f_B = 0.43 < 4.26$; there is no evidence of an effect due to printer.

11.57

Source of variation	Sum of squares	Degrees of freedom	Mean square	F	p value
Office	30.25	2	15.13	0.40	0.6761
Type	216.00	1	216.00	5.66	0.0286
Interaction	2.25	2	1.13	0.03	0.9705
Error	687.50	18	38.19		
Total	936.00	23			

$f_{AB} = 0.03$; there is no evidence of interaction. $f_A = 0.40$; there is no evidence of an effect due to office.
$f_B = 5.66 \geq 4.41$; there is evidence of an effect due to type of development.

11.59 (a)

Source of variation	Sum of squares	Degrees of freedom	Mean square	F	p value
Cover	1.5409	3	0.5137	0.87	0.4770
State	3.1184	3	1.0395	1.76	0.1953
Interaction	7.1078	9	0.7898	1.34	0.2919
Error	9.4550	16	0.5909		
Total	21.2222	31			

$f_{AB} = 1.34$; there is no evidence of interaction. **(b)** $f_A = 0.87$; there is no evidence of an effect due to cover type. $f_B = 1.76$; there is no evidence of an effect due to state.

11.61 (a) $f_{AB} = 1.68$; there is no evidence of interaction.
(b) $f_A = 7.23 \geq 3.26$; there is evidence of an effect due to region. **(c)** $f_B = 19.71 \geq 2.87$; there is evidence of an effect due to road marking quality.

11.63 (a)

Source of variation	Sum of squares	Degrees of freedom	Mean square	F	p value
Injury	68.89	4	17.22	4.324	0.0033
State	4.76	2	2.38	0.598	0.5525
Interaction	93.55	8	11.69	2.936	0.0066
Error	298.72	75	8.30		
Total	465.91	89			

$f_{AB} = 2.936$; there is no evidence of interaction. **(b)** $f_A = 4.76 \geq$ 2.49; there is no evidence of an effect due to state. $f_B = 4.324$; there is evidence of an effect due to injury.

Chapter Exercises

11.65 (a)

Source of variation	Sum of squares	Degrees of freedom	Mean square	F	p value
Factor	32.295	4	8.0737	0.77	0.5489
Error	470.355	45	10.4523		
Total	502.650	49			

(b) 5, 50 **(c)** $f = 0.77$; there is no evidence to suggest that at least two of the population mean boat-ramp angles are different.

11.67

Source of variation	Sum of squares	Degrees of freedom	Mean square	F	p value
Factor	3.7075	3	1.2358	1.67	0.1951
Error	20.6675	28	0.7381		
Total	24.3750	31			

$H_0: \mu_1 = \mu_2 = \mu_3 = \mu_4$; $H_a: \mu_i \neq \mu_j$ for some $i \neq j$; TS: $F = MSA/MSE$; RR: $F \geq 4.57$
$f = 1.67$. There is no evidence to suggest that at least two of the population mean displacements are different.

11.69

Difference	Tukey CI	Significantly different
$\mu_1 - \mu_2$	$(-0.28, -0.02)$	Yes
$\mu_1 - \mu_3$	$(-0.19, 0.07)$	No
$\mu_1 - \mu_4$	$(-0.28, -0.02)$	Yes
$\mu_1 - \mu_5$	$(-0.29, -0.02)$	Yes
$\mu_2 - \mu_3$	$(-0.05, 0.21)$	No
$\mu_2 - \mu_4$	$(-0.13, 0.13)$	No
$\mu_2 - \mu_5$	$(-0.14, 0.12)$	No
$\mu_3 - \mu_4$	$(-0.21, 0.05)$	No
$\mu_3 - \mu_5$	$(-0.22, 0.04)$	No
$\mu_4 - \mu_5$	$(-0.14, 0.12)$	No

$\bar{x}_{1.}$	$\bar{x}_{3.}$	$\bar{x}_{2.}$	$\bar{x}_{4.}$	$\bar{x}_{5.}$
0.5566	0.6190	0.7020	0.7023	0.7115

11.71

Source of variation	Sum of squares	Degrees of freedom	Mean square	F	p value
Group	0.0161	3	0.0054	0.15	0.9287
Tooth type	0.7200	1	0.7200	19.84	0.0002
Interaction	0.0827	3	0.0276	0.76	0.5276
Error	0.8709	24	0.0363		
Total	1.6896	31			

There is no evidence of interaction. There is no evidence of an effect due to group. There is evidence of an effect due to tooth type.

11.73 (a) $H_0: \mu_1 = \mu_2 = \mu_3 = \mu_4 = \mu_5$; $H_a: \mu_i \neq \mu_j$ for some $i \neq j$; TS: $F = MSA/MSE$; RR: $F \geq 2.57$
$f = 4.07 \geq 2.57$. There is evidence to suggest that at least two of the population mean catfish weights are different.
(b)

Difference	Tukey CI	Significantly different
$\mu_1 - \mu_2$	$(-33.32, 2.63)$	No
$\mu_1 - \mu_3$	$(-33.24, 3.46)$	No
$\mu_1 - \mu_4$	$(-36.41, -1.09)$	Yes
$\mu_1 - \mu_5$	$(-38.76, 1.28)$	No
$\mu_2 - \mu_3$	$(-16.45, 17.36)$	No
$\mu_2 - \mu_4$	$(-19.55, 12.74)$	No
$\mu_2 - \mu_5$	$(-22.10, 15.30)$	No
$\mu_3 - \mu_4$	$(-20.42, 12.70)$	No
$\mu_3 - \mu_5$	$(-22.91, 15.21)$	No
$\mu_4 - \mu_5$	$(-18.39, 18.40)$	No

$\bar{x}_{1.}$	$\bar{x}_{3.}$	$\bar{x}_{2.}$	$\bar{x}_{5.}$	$\bar{x}_{4.}$
20.70	35.59	36.05	39.44	39.45

Recommend Campground (OK, anywhere but Hill's Landing). On average, the largest catfish are caught at this location.

11.75 $f_{AB} = 2.43$; there is evidence of interaction. $f_A = 2.45$; there is no evidence of an effect due to species ID. $f_B = 5.68 \geq 3.35$; there is evidence of an effect due to age.

CHAPTER 12

Section 12.1

12.1 (a) Appropriate, slope negative. **(b)** Not appropriate, relationship is not linear. **(c)** Appropriate, slope zero. **(d)** Not appropriate, no linear relationship.

12.3 (a) 615 **(b)** 3.6; the coefficient on the independent variable. **(c)** 0.1217

12.5 (a) $y = 41.7004 - 14.8744x$ **(b)** -568.15

12.7 (a)

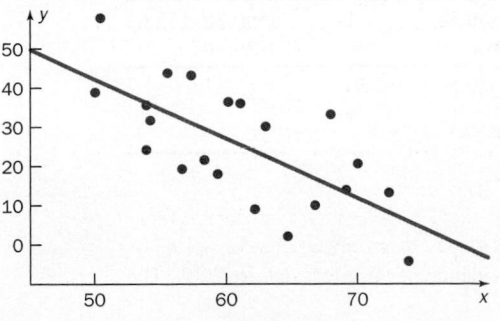

A simple linear regression model seems reasonable. The points appear to fall near a straight line. (b) $y = 117.91 - 1.5169x$ (c) 23.8622 (d)

Source of variation	Sum of squares	Degrees of freedom	Mean square	F
Regression	2290.75	1	2290.75	18.18
Error	2393.78	19	125.99	
Total	4684.53	20		

12.9 (a) 2.45 (b) 0.90 (c) 0.3446
12.11 (a) $y = 3.7167 - 1.7016$ (b) 1.5897
12.13 (a) $y = 11.9278 - 0.4741x$ (b) 10.6714 (c) 10.5577
12.15 (a)

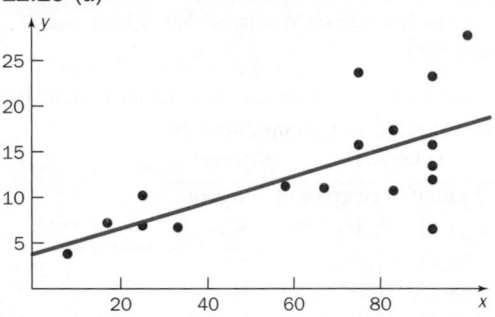

(b) $y = 3.8619 + 0.1423x$ (c) 14.5344
12.17 (a) $y = -2.5572 + 1.2486x$ (b) ANOVA table:

Source of variation	Sum of squares	Degrees of freedom	Mean square	F
Regression	86.68	1	86.68	8.20
Error	84.55	8	10.57	
Total	171.23	9		

(c) 34.90
12.19 (a) $y = -0.8107 + 0.2683x$ (b)

Source of variation	Sum of squares	Degrees of freedom	Mean square	F
Regression	6.81	1	6.81	14.95
Error	10.02	22	0.46	
Total	16.83	23		

(c) $r^2 = 0.405$. Approximately 40% of the variation in the data is explained by the regression model. (d) 40.2933
12.21 (a) $y = 79.1355 - 0.2950x$ (b)

Source of variation	Sum of squares	Degrees of freedom	Mean square	F
Regression	4560.38	1	4560.38	153.03
Error	1460.20	49	29.80	
Total	6020.59	50		

(c) 0.7575 (d) 74.563

Section 12.2

12.23 (a) ANOVA table:

Source of variation	Sum of squares	Degrees of freedom	Mean square	F	p value
Regression	11691.9	1	11691.90	2.58	0.1219
Error	104372.1	23	4537.92		
Total	116064.0	24			

(b) H_0: There is no significant linear relationship. H_a: There is a significant linear relationship.
TS: $F = MSR/MSE$; RR: $F \geq 4.28$
$f = 2.58$. There is no evidence of a significant linear relationship.
(c) 0.1007 (d) 0.3174
12.25 (a) H_0: $\beta_1 = 0$; H_a: $\beta_1 \neq 0$; TS: $T = B_1/S_{B_1}$;
RR: $|T| \geq 2.0796$. $t = 2.2980 \geq 2.0796$. There is evidence to suggest that $\beta_1 \neq 0$—the regression line is significant.
(b) (0.4209, 8.4361) (c) Yes. The CI does not include 0.
12.27 (a) H_0: There is no significant linear relationship.
H_a: There is a significant linear relationship.
TS: $F = MSR/MSE$; RR: $F \geq 4.60$
$f = 4.14$. There is no evidence of a significant linear relationship. $p > 0.05$, $p = 0.0614$ (b) H_0: $\beta_1 = 0$; $\beta_1 \neq 0$;
TS: $T = B_1/S_{B_1}$; RR: $|T| \geq 2.1448$. $t = 2.0337$. There is no evidence to suggest that β_1 is different from 0. $0.05 \leq p \leq 0.10$. 0.0614 (c) $t^2 = f$ (d) Same. These two hypothesis tests are testing the same null hypothesis.
12.29 (a) $y = 68.0071 - 8.7993x$

Source of variation	Sum of squares	Degrees of freedom	Mean square	F	p value
Regression	7462.54	1	7462.54	10.35	0.0324
Error	2884.77	4	721.19		
Total	10347.31	5			

(b) H_0: There is no significant linear relationship. H_a: There is a significant linear relationship.
TS: $F = MSR/MSE$; RR: $F \geq 7.71$
$f = 10.35 \geq 7.71$. There is evidence of a significant linear relationship. (c) 0.7212 (d) -0.8492. Negative relationship. β_1 is negative. (e) $r^2 = 0.7212$
12.31 (a) $y = 2.9430 + 0.6925x$

Source of variation	Sum of squares	Degrees of freedom	Mean square	F	p value
Regression	0.1062	1	0.1062	0.40	0.5403
Error	3.4909	13	0.2685		
Total	3.5971	14			

(b) H_0: There is no significant linear relationship. H_a: There is a significant linear relationship.
TS: $F = MSR/MSE$; RR: $F \geq 9.07$
$f = 0.40$. There is no evidence of a significant linear relationship.
(c) (1.4728, 4.4132). Yes. The CI does not include 0.
12.33 (a) -0.6434 (b) Negative relationship. As mean annual temperature increases, the depth of the permafrost layer decreases.

12.35 (a)

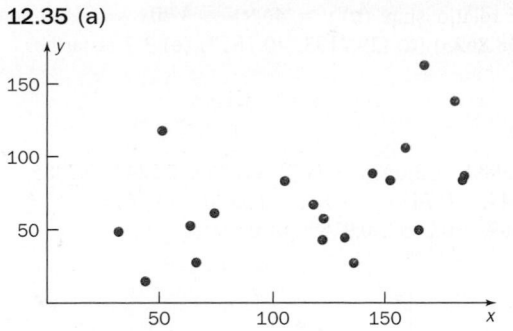

(b) 0.4794 **(c)** Support. There is a slight positive relationship.

12.37 (a)

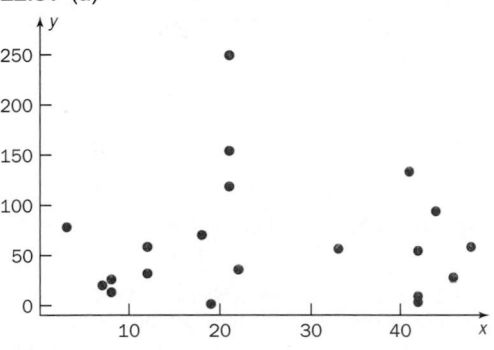

(b) 0.0101. There is no clear relationship.

12.39 (a) $y = 980.0067 + 0.4795x$

Source of variation	Sum of squares	Degrees of freedom	Mean square	F	p value
Regression	24472.35	1	24472.35	8.32	0.0344
Error	14705.08	5	2941.02		
Total	39177.43	6			

(b) 0.6247 **(c)** $(-0.1907, 1.1497)$ **(d)** $H_0: \beta_0 = 0$; $H_a: \beta_0 > 0$; TS: B_0/S_{B_0}; RR: $T \geq 2.0150$ $(\alpha = 0.05)$. $t = 22.5921 \geq 2.0150$. There is evidence to suggest that $\beta_0 > 0$. This suggests that even if the owner spends nothing on advertising in a week, he or she will still have a total weekly revenue greater than 0, close to 980.

12.41 (a) $y = 1.9195 + 2.7096x$ **(b)** $H_0: \beta_1 = 0$; $\beta_1 \neq 0$; TS: $T = B_1/S_{B_1}$; RR: $|T| \geq 2.1788$. $t = 2.3928 \geq 2.1788$. There is evidence to suggest that β_1 is different from 0. **(c)** 3.4097 **(d)** 0.3230. Obtain more data.

12.43 (a) $y = 0.1719 + 0.1157x$

Source of variation	Sum of squares	Degrees of freedom	Mean square	F	p value
Regression	0.3660	1	0.3660	11.67	0.0051
Error	0.3762	12	0.0313		
Total	0.7421	13			

(b) $y = 0.0895 + 0.0087x$

Source of variation	Sum of squares	Degrees of freedom	Mean square	F	p value
Regression	0.0246	1	0.0246	0.41	0.5334
Error	0.7176	12	0.0598		
Total	0.7421	13			

(c) Evaporation rate and air velocity. There is a significant linear relationship.

Section 12.3

12.45 (a) $H_0: y^* = 20$; $H_a: y^* > 20$;

TS: $T = \dfrac{(B_0 + B_1 x^*) - y_0^*}{S\sqrt{(1/n) + [(x^* - \bar{x})^2/S_{xx}]}}$; RR: $T \geq 1.7459$

$t = 0.1503$. There is no evidence to suggest that the mean value of Y for $x = 16.2$ is greater than 20.

(b) $H_0: y^* = 5$; $H_a: y^* \neq 5$;

TS: $T = \dfrac{(B_0 + B_1 x^*) - y_0^*}{S\sqrt{(1/n) + [(x^* - \bar{x})^2/S_{xx}]}}$; RR: $|T| \geq 2.9208$

$t = -0.1240$. There is no evidence to suggest that the mean value of Y for $x = 11.5$ is greater than 5.

12.47 (a) $(-20.1474, 123.4294)$, 143.5768 **(b)** $(-23.8157, 123.7581)$, 147.5738 **(c)** 18.1 is farther from the mean than 19.25.

12.49 (a) $y = 398.6420 - 8.3856x$

Source of variation	Sum of squares	Degrees of freedom	Mean square	F	p value
Regression	11954.82	1	11954.82	13.51	0.0028
Error	11502.28	13	884.79		
Total	23457.09	14			

H_0: There is no significant linear relationship.
H_a: There is a significant linear relationship.
TS: $F = MSR/MSE$; RR: $F \geq 9.07$
$f = 13.51$. There is evidence of a significant linear relationship.
(b) 29.7454 **(c)** (16.0985, 160.6499). No. The PI is completely below 170.

12.51 (a) 0.6467 **(b)** (0.6115, 0.9495) **(c)** $H_0: y^* = 0.06$; $H_a: y^* > 0.06$;

TS: $T = \dfrac{(B_0 + B_1 x^*) - y_0^*}{S\sqrt{(1/n) + [(x^* - \bar{x})^2/S_{xx}]}}$; RR: $T \geq 2.5395$

$t = 6.6636 \geq 2.5395$. There is evidence to suggest that the mean value of Y for $x = 80$ is greater than 0.06.

12.53 (a) $3.5699 + 0.0021x$ **(b)** ANOVA table:

Source of variation	Sum of squares	Degrees of freedom	Mean square	F	p value
Regression	0.8000	1	0.8000	0.15	0.7089
Error	48.4655	9	5.3851		
Total	49.2655	10			

H_0: There is no significant linear relationship.
H_a: There is a significant linear relationship.
TS: $F = MSR/MSE$; RR: $F \geq 5.12$ $(\alpha = 0.05)$
$f = 0.15$. There is no evidence of a significant linear relationship. No. **(c)** $(-0.3878, 11.7276)$. This PI includes some negative numbers.

12.55 (a) Independent: skid resistance. Dependent: accident rate.

(b) Scatter plot:

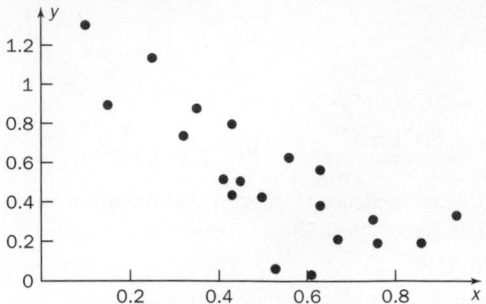

Relationship: linear, negative.

(c) $y = 1.1570 - 1.2285x$ (d) $H_0: y^* = 0.60$; $H_a: y^* < 0.60$;

$$\text{TS: } T = \frac{(B_0 + B_1 x^*) - y_0^*}{S\sqrt{(1/n) + [(x^* - \bar{x})^2/S_{xx}]}}; \quad \text{RR: } T \leq -1.7341$$

$(\alpha = 0.05)$
$t = -1.1942$. There is no evidence to suggest that the mean value of Y for $x = 0.50$ is less than 0.60.

12.57 (a)

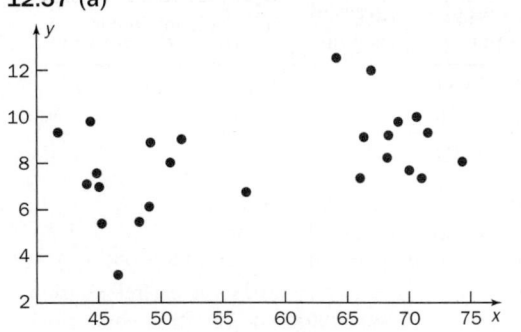

Slight positive relationship. (b) $y = 3.8228 + 0.0751x$ (c) (5.6068, 14.0548) (d) (4.3899, 12.2677)

12.59 (a) ANOVA summary table:

Source of variation	Sum of squares	Degrees of freedom	Mean square	F	p value
Regression	853.50	1	853.50	8.99	0.0103
Error	1234.23	13	94.94		
Total	2087.73	14			

H_0: There is no significant linear relationship.
H_a: There is a significant linear relationship.
TS: $F = $ MSR/MSE; RR: $F \geq 4.67$ ($\alpha = 0.05$)
$f = 8.99 \geq 4.67$. There is evidence of a significant linear relationship. As CPI increases, so does ESI. (b) 56.8902
(c) (40.1554, 87.2630). No. The PI does not include 90.

12.61 (a)

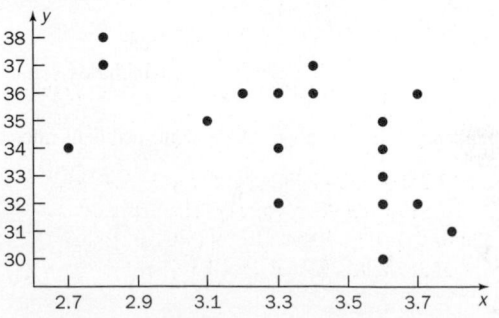

Slight negative relationship. (b) $y = 46.91 - 3.70x$
(c) (27.5775, 38.8625) (d) (29.2183, 40.1817) (e) 3.7 is farther from the mean than 3.3.

Section 12.4

12.63 (a) -0.9898, -3.0005, 3.1773, 0.1760, 2.5243, 0.7085, -0.0552, 2.0344, -0.7423, 1.0185, -0.3370, -3.5615, -2.0149, 0.4089, -0.2852, 0.9385 (b) 0

12.65 (a)

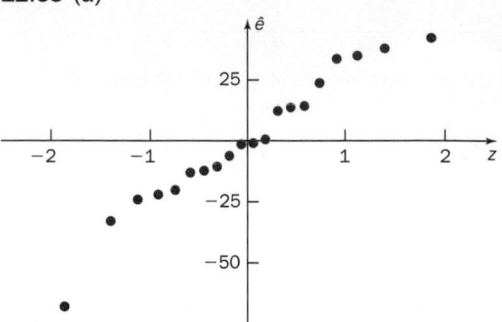

(b) Some. There appears to be an outlier, and the points are slightly wavy.

12.67 (a) Yes. The graph suggests that the relationship is not linear. (b) Yes. The graph suggests that the variance is not constant. (c) Yes. The graph suggests that the relationship is not linear. (d) Maybe. There is some evidence that the variance is not constant and increases as x increases.

12.69 (a) $y = 463.3508 - 3.5333x$ (b) -159.0214, 14.7795, 213.0710, 191.2377, -132.5613, -164.1746, -166.8346, 30.4313, 276.9243, -133.1347, 107.2645, -2.4806, -147.3346, -118.1416, -47.9406, 68.8129, -153.7613, 122.7197, 113.5130, -44.5406, -171.5279, 143.3864, 9.3647, 131.6464, 14.8262, 66.6262, 117.0730, -107.1617, 82.6596, -155.7210

(c)

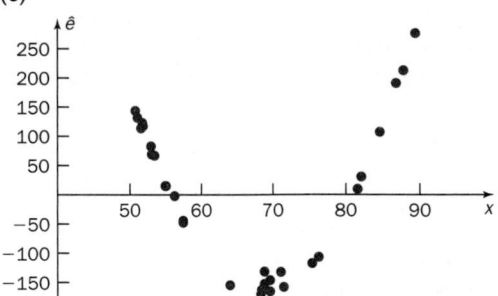

There is evidence to suggest a violation in the simple linear regression assumptions. There is a distinct curve in the plot.

12.71 (a)

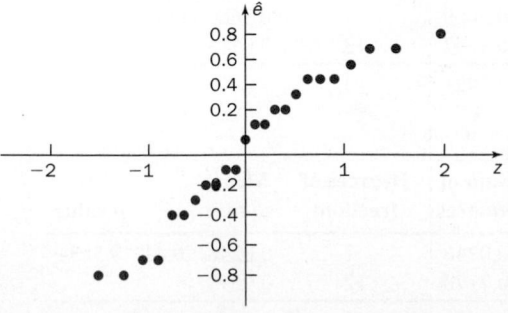

(b) There is some evidence of nonnormality. Each end of the plot flattens out slightly.

12.73 (a) $y = 5.167703 + 0.000119x$ **(b)** Normal probability plot:

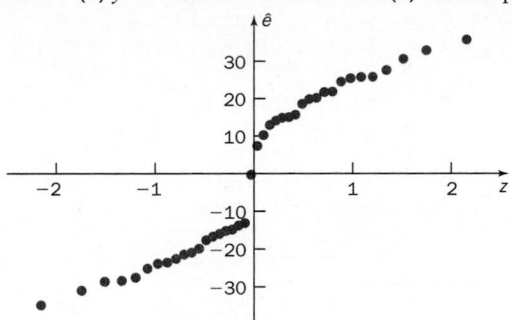

Residuals versus the predictor variable:

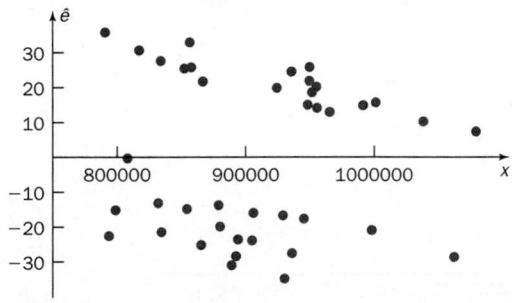

(c) There is evidence to suggest that the simple linear regression assumptions are invalid. There is a distinct pattern in the normal probability plot and in the plot of the residuals versus the predictor variable.

12.75 (a) -0.5180, 1.7744, -1.0140, 2.2575, 0.9086, -0.8920, 0.6604, -1.3632, -4.8381, 1.9684, 1.0153, -3.2186, 0.7614, -0.3538, -0.1420, 2.5750, 0.3094, -1.8818, 0.0585, 1.9326. Sum = 0 **(b)** Residuals versus the predictor variable:

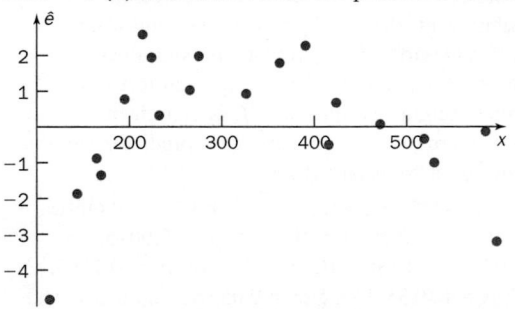

There is evidence that the simple linear regression assumptions are violated. There is a pattern in the graph.

12.77 (a) $y = 738.0426 + 14.5283x$. Residuals: -43.9849, -15.1359, 71.2226, 55.5060, -108.3433, 52.3167, -22.5698, -88.2675, -22.4950, 121.7512 **(b)** Normal probability plot:

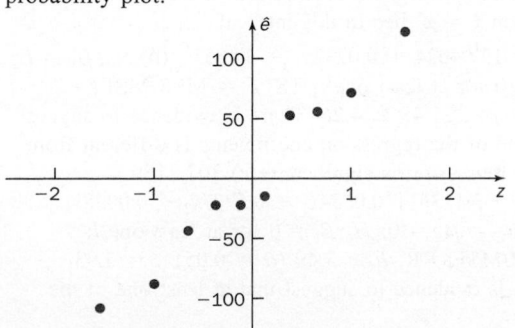

Residuals versus the predictor variable:

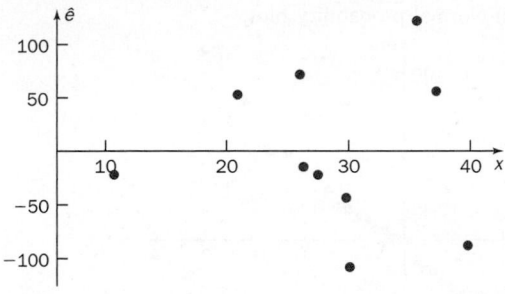

(c) There is no overwhelming evidence that the simple linear regression assumptions are invalid. There is a possible outlier, but the number of observations is small.

12.79 (a) $y = 187.2849 + 0.2440x$. Residuals: -1.4383, -4.0986, 0.0498, -2.3426, 4.2221, 0.4900, -3.9024, 0.5329, 1.9781, -3.3187, -5.0747, 0.0259, -1.9263, 14.8008, 3.4900, -3.2709, 1.0498, -1.1703, -2.8307, 2.7340 **(b)** H_0: There is no significant linear relationship. H_a: There is a significant linear relationship. TS: $F = MSR/MSE$; RR: $F \geq 8.29$ $f = 16.19 \geq 8.29$. There is evidence of a significant linear relationship. **(c)** Normal probability plot:

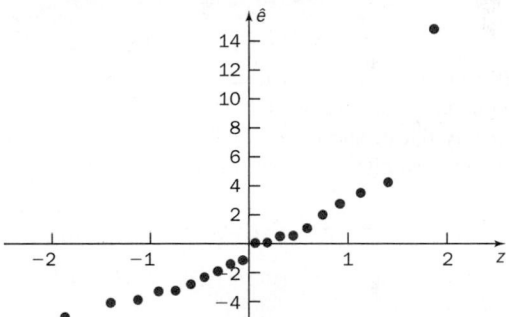

Residuals versus the predictor variable:

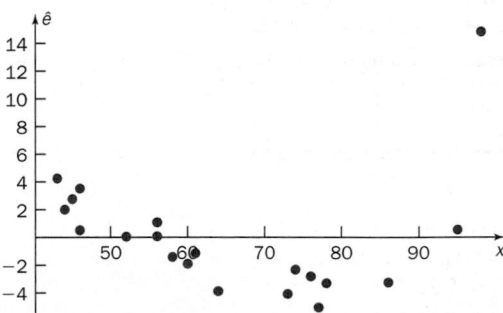

(d) There is evidence to suggest that the simple linear regression assumptions are invalid. Both plots suggest that there is an outlier, and the plot of the residuals versus the predictor variable suggests a parabolic pattern. We might try excluding the outlier from the data set or adding a quadratic term to the model.

12.81 (a) $y = 19.3238 + 24.7881x$. Residuals: 18.3040, 41.5141, 112.7136, -32.5726, -18.8476, -32.4605, -37.3222, 13.4905, 12.0668, 52.4850, 1.7290, 7.1883, 29.7337, -49.4007, -12.7228, -16.9850, -30.7366, -0.3410, -54.0453, 51.2467, 20.3345, -0.3261, -42.9850, -24.2732, 42.2475, -62.4481, 6.1188, -49.7961, -27.8173, 83.9075 **(b)** H_0: There is no significant linear relationship. H_a: There is a significant linear relationship. TS: $F = MSR/MSE$; RR: $F \geq 13.50$;

$f = 15.61 \geq 13.50$. There is evidence of a significant linear relationship. **(c)** Normal probability plot:

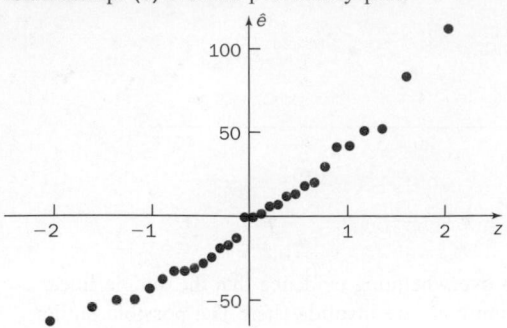

Residuals versus the predictor variable:

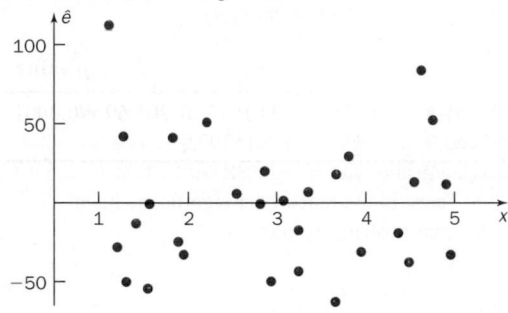

(d) The graphs do not provide any evidence that the simple linear regression assumptions are invalid. The normal probability plot is approximately linear, and the plot of the residuals versus the predictor variable exhibits no discernible pattern.

12.83

$$\sum_{i=1}^{n}(y_i - \hat{y}_i) = \sum_{i=1}^{n}(y_i - (\hat{\beta}_0 + \hat{\beta}_1 x_i))$$

$$= \sum_{i=1}^{n}(y_i - (\bar{y} - \hat{\beta}_1\bar{x} + \hat{\beta}_1 x_i))$$

$$= \sum_{i=1}^{n}(y_i - \bar{y}) - \hat{\beta}_1\sum_{i=1}^{n}(x_i - \bar{x})$$

$$= n\bar{y} - n\bar{y} - \hat{\beta}_1(n\bar{x} - n\bar{x}) = 0$$

Section 12.5

12.85 (a) -49.55 **(b)** -61.5 **(c)** 0.8625

12.87 (a) Scatter plots:

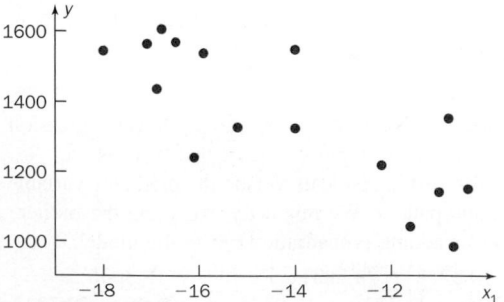

Negative linear relationship.

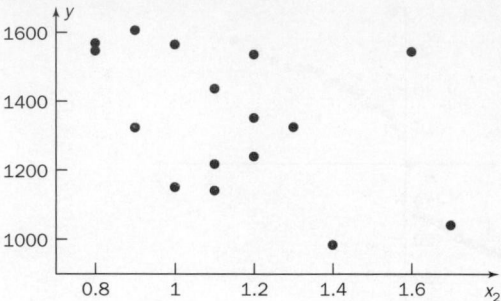

Slight indication of a negative linear relationship.

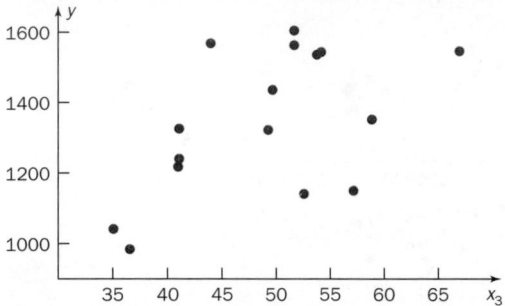

Slight indication of a positive linear relationship.
(b) $y = 221.9231 - 56.3497x_1 - 124.2215x_2 + 9.5798$. The sign of each estimated regression coefficient reflects the relationship in each scatter plot. **(c)** 1121.6179

12.89 (a) $H_0: \beta_1 = \cdots = \beta_4 = 0$; $H_a: \beta_i \neq 0$ for at least one i; TS: $F = \text{MSR/MSE}$; RR: $F \geq 2.73$ ($\alpha = 0.05$). $f = 6.73 \geq 2.73$. There is evidence to suggest that at least one of the regression coefficients is different from 0. The overall regression is significant. **(b)** β_2, β_3, and β_4 are significantly different from 0. Therefore, x_2, x_3, and x_4 are significant predictor variables. **(c)** The critical value in each test is 2.6763. Using the Minitab output, we find that β_3 is significantly different from 0; therefore, x_3 is a significant predictor variable. This result is different from part (b).

12.91 (a) $y = 114.4895 + 6.4722x_1 - 12.8017x_2 + 4.6091x_3 + 0.6409x_4$. $r^2 = 0.7791$ **(b)** $\beta_1: t = 4.3402$, $p = 0.0015$. $\beta_2: t = -0.5592$, $p = 0.5883$. $\beta_3: t = 1.4814$, $p = 0.1693$. $\beta_4: t = 2.9202$, $p = 0.0153$. Using $\alpha = 0.05$, we find that x_1 and x_4 are significant predictors. **(c)** $Y_i = \beta_0 + \beta_1 x_{1i} + \beta_4 x_{4i} + E_i$. $y = 105.4656 + 5.1074x_1 + 0.6863x_4$. $r^2 = 0.7306$ **(d)** The second model is better: fewer variables, and r^2 is only slightly lower.

12.93 (a) $(5.7072, 10.0835)$. We are 95% confident that the true mean value of Y when $x = x^*$ lies in this interval. **(b)** $(2.4597, 13.3310)$. We are 95% confident that an observed value of Y when $x = x^*$ lies in this interval.

12.95 (a) $y = 137.4024 + 0.0282x_1 - 4.4853x_2$ **(b)** $H_0: \beta_1 = \beta_2 = 0$; $H_a: \beta_i \neq 0$ for at least one i; TS: $F = \text{MSR/MSE}$; RR: $F \geq 4.26$. $f = 21.45 \geq 4.26$. There is evidence to suggest that at least one of the regression coefficients is different from 0. The overall regression is significant. **(c)** 307.2159

12.97 (a) $y = -3.1136 + 0.0554x_1 + 0.5777x_2 + 0.0028x_3$ **(b)** $H_0: \beta_1 = \beta_2 = \beta_3 = 0$; $H_a: \beta_i \neq 0$ for at least one i; TS: $F = \text{MSR/MSE}$; RR: $F \geq 3.49$ ($\alpha = 0.05$). $f = 3.93 \geq 3.49$. There is evidence to suggest that at least one of the regression coefficients is different from 0. The overall regression is significant. β_1: $t = 2.4743$, $p = 0.0293$. β_2: $t = 2.8634$, $p = 0.0143$. β_3: $t = 0.4409$, $p = 0.6671$. The temperature of the solutions and the concentration of the solutions are the most important (significant) variables. **(c)** Normal probability plot:

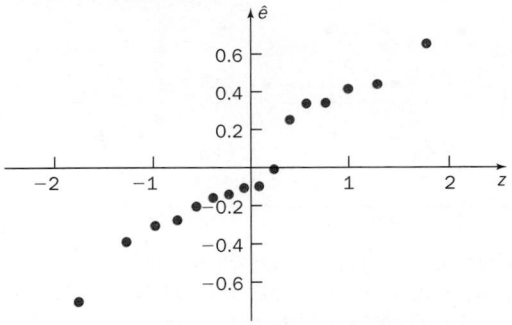

There is some evidence to suggest a violation in the multiple linear regression assumptions. The points in this plot are slightly nonlinear.

12.99 (a) $y = 12297.6651 + 0.2596x_1 + 51.1428x_2 - 22.6309x_3 - 17.8406x_4$ **(b)** H_0: $\beta_1 = \cdots = \beta_4 = 0$; H_a: $\beta_i \neq 0$ for at least one i; TS: $F = MSR/MSE$; RR: $F \geq 3.26$. $f = 15.17 \geq 3.26$. There is evidence to suggest that at least one of the regression coefficients is different from 0. The overall regression is significant. $r^2 = 0.8465$ **(c)** β_1: $t = 1.6341$, $p = 0.1305$. β_2: $t = 2.1100$, $p = 0.0586$. β_3: $t = -1.0963$, $p = 0.2964$. β_4: $t = -0.7300$, $p = 0.4806$. The variable x_2 is close to significant. **(d)** H_0: $\beta_1 = 0.20$; H_a: $\beta_1 > 0.20$; TS: $T = (B_1 - 0.20)/S_{B_1}$; RR: $T \geq 1.7823$. $t = 0.3753$. There is no evidence to suggest that $\beta_1 > 0.20$.

12.101 (a) ANOVA table:

Source of variation	Sum of squares	Degrees of freedom	Mean square	F	p value
Regression	531.54	3	177.18	6.21	0.0035
Error	599.38	21	28.5419		
Total	1130.92	24			

H_0: $\beta_1 = \beta_2 = \beta_3 = 0$; H_a: $\beta_i \neq 0$ for at least one i; TS: $F = MSR/MSE$; RR: $F \geq 3.07$ ($\alpha = 0.05$). $f = 6.21 \geq 3.07$. There is evidence to suggest that at least one of the regression coefficients is different from 0. The overall regression is significant. **(b)** All three variables contribute to the overall significant regression. **(c)** (52.64, 58.62), (56.21, 68.54) **(d)** (44.12, 67.14), (49.67, 75.08) **(e)** x_1^* is closer to the *mean* than x_2^*.

12.103 (a) $y = 197.1839 - 3.5821x_1 - 6.2638x_2$.
(b) H_0: $\beta_1 = \beta_2 = 0$; H_a: $\beta_i \neq 0$ for at least one i; TS: $F = MSR/MSE$; RR: $F \geq 4.74$. $f = 7.28 \geq 4.74$. There is evidence to suggest that at least one of the regression coefficients is different from 0. The overall regression is significant.
(c) 0.6752. Approximately 68% of the variation in y is explained by this regression model. **(d)** β_1: $t = -2.3709$, $p = 0.0495$. β_2: $t = -2.9934$, $p = 0.0201$. Both regression coefficients are significantly different from 0. **(e)** Yes. The overall regression is significant, and both variables contribute to the overall significance.

12.105 (a) Scatter plot:

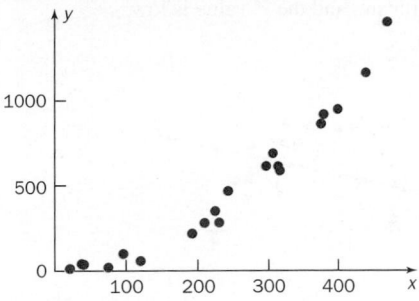

(b) $y = -215.1226 + 2.9043x$
ANOVA table:

Source of variation	Sum of squares	Degrees of freedom	Mean square	F	p value
Regression	3105671.6	1	3105671.6	201.69	<0.0001
Error	277162.9	18	15397.9		
Total	3382834.5	19			

(c) Small depths: $y = 7.4516 + 0.4828x$
ANOVA table:

Source of variation	Sum of squares	Degrees of freedom	Mean square	F	p value
Regression	1720.68	1	1720.68	2.47	0.1915
Error	2792.15	4	698.04		
Total	4512.83	5			

Large depths: $y = -598.5726 + 4.0383x$
ANOVA table:

Source of variation	Sum of squares	Degrees of freedom	Mean square	F	p value
Regression	1654572.7	1	1654572.7	309.29	<0.0001
Error	64194.7	12	5349.6		
Total	1718767.4	13			

(d) $y = 7.3195 + 0.0062x$ **(e)** The model using x^2 seems to be the best. A model with x^2 appears to fit the scatter plot, and the value of r^2 for this model is 0.987, which is very high.

Chapter Exercises

12.107 (a) $y = 2.8408 + 2.1231x$ **(b)** 24.0716 **(c)** 7.0870
12.109 (a) $y = 0.7218 + 0.0059x$. ANOVA table:

Source of variation	Sum of squares	Degrees of freedom	Mean square	F	p value
Regression	10.8922	1	10.8922	5.79	0.0285
Error	30.0878	16	1.8805		
Total	40.9800	17			

(b) 0.2658. **(c)** 0.5156 **(d)** No. The correlation is only moderate, the regression is barely significant, and the r^2 value is low.

12.111 (a) Scatter plot:

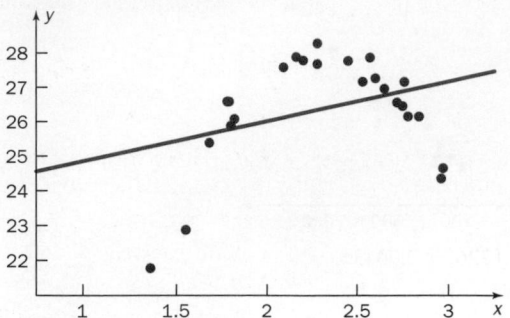

The relationship appears to be quadratic.
(b) $y = 23.6853 + 1.1767x$. H_0: There is no significant linear relationship; H_a: There is a significant linear relationship; TS: $F = MSR/MSE$; RR: $F \geq 4.30$ $f = 3.08$. There is no evidence of a significant linear relationship. **(c)** Residuals versus the predictor variable:

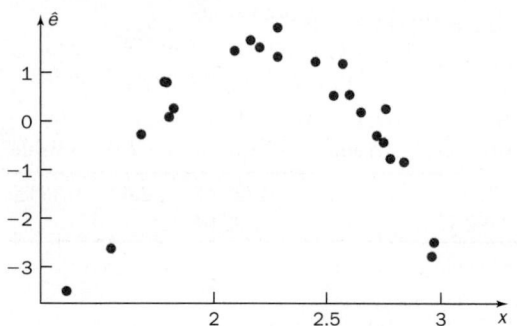

(d) Add a quadratic term: x^2.

12.113 (a) $y = 8.3870 + 0.6024x$ **(b)** ANOVA table:

Source of variation	Sum of squares	Degrees of freedom	Mean square	F	p value
Regression	67.39	1	67.39	2.29	0.1370
Error	1414.44	48	29.47		
Total	1481.83	49			

(c) H_0: $\beta_1 = 0$; H_a: $\beta_1 \neq 0$; TS: $T = B_1/S_{B_1}$; RR: $|T| \geq 2.0106$. $t = 1.5123$. There is no evidence to suggest that $\beta_1 \neq 0$—the regression line is not significant. **(d)** No. There is no significant relationship between rating and price.

12.115 (a) $y = 0.3168 + 0.9059x$.
ANOVA table:

Source of variation	Sum of squares	Degrees of freedom	Mean square	F	p value
Regression	2.8027	1	2.8027	17.23	0.0032
Error	1.3013	8	0.1627		
Total	4.1040	9			

(b) (0.776, 2.756) **(c)** H_0: $y^* = 1$; H_a: $y^* > 1$;
TS: $T = \dfrac{(B_0 + B_1 x^*) - y_0^*}{s\sqrt{(1/n) + [(x^* - \bar{x})^2/S_{xx}]}}$; RR: $T \geq 2.8965$

$t = 0.2554$. There is no evidence to suggest that the mean value of Y for $x = 0.8$ is greater than 1.

12.117 (a) Scatter plot:

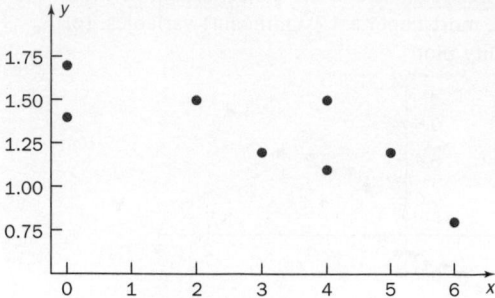

Negative linear relationship. **(b)** $y = 1.6096 - 0.0924x$.
H_0: $\beta_1 = 0$; H_a: $\beta_1 \neq 0$; TS: $T = B_1/S_{B_1}$; RR: $|T| \geq 2.3646$. $t = -2.6441 \leq -2.3646$. There is evidence to suggest that $\beta_1 \neq 0$ —the regression line is significant. **(c)** Normal probability plot:

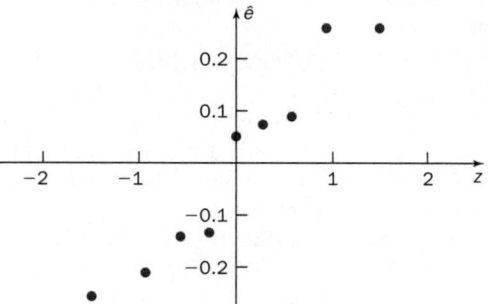

Residuals versus the predictor variable:

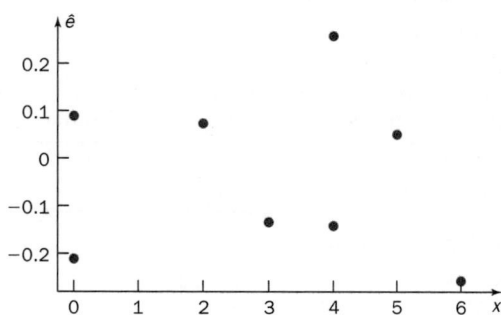

The normal probability plot suggests a violation in the normality assumption.

12.119 (a) Scatter plot:

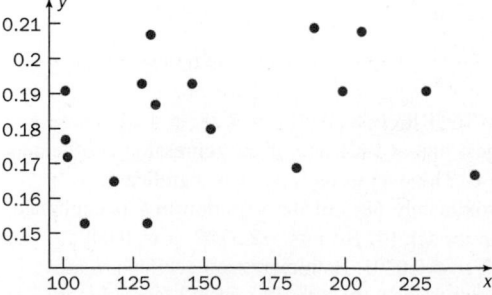

There does not appear to be a linear relationship. The scatter plot appears random.

(b) $y = 0.1673 + 0.0001x$. ANOVA table:

Source of variation	Sum of squares	Degrees of freedom	Mean square	F	p value
Regression	0.0004	1	0.0004	1.15	0.3008
Error	0.0047	15	0.0003		
Total	0.0050	16			

(c) The F test is not significant. There is no evidence to suggest a significant linear relationship.

12.121 (a) $y = 13.0865 + 0.0220x_1 - 0.0563x_2$
(b) ANOVA table:

Source of variation	Sum of squares	Degrees of freedom	Mean square	F	p value
Regression	34.3151	2	17.1576	16.73	0.0001
Error	17.4304	17	0.0253		
Total	51.7455	19			

$H_0: \beta_1 = \beta_2 = 0$; $H_a: \beta_i \neq 0$ for at least one i; TS: $F = $ MSR/MSE; RR: $F \geq 3.59$ ($\alpha = 0.05$). $f = 16.73 \geq 3.59$. There is evidence to suggest that at least one of the regression coefficients is different from 0. The overall regression is significant. $p = 0.0001$. $r^2 = 0.6632$. Approximately 66% of the variation in the data is explained by the regression model. **(c)** $\beta_1: t = 3.4520 \geq 2.4581$. $\beta_2: t = -3.7559 \geq -2.4581$. Both predictor variables are significant. **(d)** (10.944, 17.980) **(e)** 214.932

12.123 (a) Scatter plot:

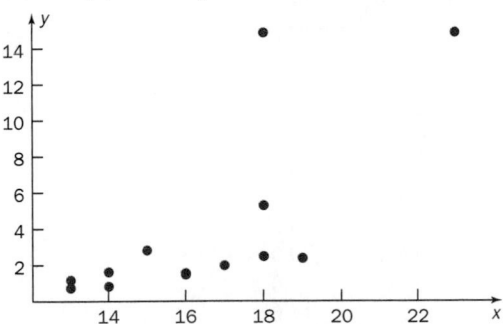

The relationship appears (positive) linear, with the exception of two outliers. **(b)** $y = -16.2434 + 1.2361x$. ANOVA table:

Source of variation	Sum of squares	Degrees of freedom	Mean square	F	p value
Regression	162.50	1	162.50	13.67	0.0031
Error	142.68	12	11.89		
Total	305.18	13			

There is evidence to suggest that the total number of wins can be used to predict the per-team payout. The overall test is significant, $p = 0.0031$. **(c)** (5.067, 11.889)

CHAPTER 13

Section 13.1

13.1 54.5, 87.2, 43.6, 32.7
13.3 (a) $H_0: p_1 = 0.4, p_2 = 0.3, p_3 = 0.2, p_4 = 0.1$; $H_a: p_i \neq p_{i0}$ for at least one i;

TS: $X^2 = \sum_{i=1}^{4}(n_i - e_i)^2/e_i$; RR: $X^2 \geq 11.3449$.
(b) 120, 90, 60, 30 **(c)** $\chi^2 = 2.1528$. There is no evidence to suggest that any one of the population proportions differs from its hypothesized value.

13.5 $H_0: p_i = 0.20$; $H_a: p_1 \neq p_{i0}$ for at least one i; TS:
$X^2 = \sum_{i=1}^{5}(n_i - e_i)^2/e_i$; RR: $X^2 \geq 9.4877$. $\chi^2 = 4.5200$.
There is no evidence to suggest that any one of the population proportions differs from its hypothesized value. $p > 0.10$

13.7 RR: $X^2 \geq 9.4877$. $\chi^2 = 10.75 \geq 9.4877$. There is evidence to suggest that at least one of the population proportions differs from its hypothesized value. There is evidence to suggest that customers who use flavor shots prefer one over the others.

13.9 RR: $X^2 \geq 7.8147$. $\chi^2 = 5.0040$. There is no evidence to suggest that any one of the population proportion eye colors differs from its hypothesized value.

13.11 RR: $X^2 \geq 15.0863$. $\chi^2 = 15.8340 \geq 15.0863$. There is evidence to suggest that at least one of the population proportion act types differs from its hypothesized value.

13.13 RR: $X^2 \geq 11.0705$. $\chi^2 = 7.3617$. There is no evidence to suggest that any one of this year's population proportions differs from last year's.

13.15 RR: $X^2 \geq 14.8603$. $\chi^2 = 6.9214$. There is no evidence to suggest a shift in the proportion of applications by California location.

13.17 RR: $X^2 \geq 11.3449$. $\chi^2 = 4.2939$. There is no evidence to suggest that any one of the literacy level population proportions has changed.

13.19 RR: $X^2 \geq 16.9190$. $\chi^2 = 17.48 \geq 16.9190$. There is evidence to suggest that at least one of the population proportions differs from its hypothesized value; evidence to suggest that one airport mall is preferred over the rest.

13.21 RR: $X^2 \geq 23.2093$. $\chi^2 = 15.8617$. There is no evidence to suggest that the proportion of sales by company has changed in 2009.

Section 13.2

13.23 (a) 12.5916 **(b)** 15.0863 **(c)** 14.4494 **(d)** 26.1245
13.25

		Category				Row total
		1	2	3	4	
Population	1	18	14	18	15	65
	2	25	21	16	12	74
	3	32	33	26	28	119
Col. total		75	68	60	55	258

13.27 RR: $X^2 \geq 16.9190$. $\chi^2 = 16.8226$. There is no evidence to suggest that the two categorical variables are dependent.

13.29 RR: $X^2 \geq 16.8119$. $\chi^2 = 9.2674$. There is no evidence to suggest that the true proportion of each favorite differs by grocery store.

13.31 RR: $X^2 \geq 23.5894$. $\chi^2 = 17.8692$. There is no evidence to suggest that funding opinion differs by region.

13.33 RR: $X^2 = 7.8794$. $\chi^2 = 11.2179 \geq 7.8794$. There is evidence to suggest that food and wine are dependent. This suggests that diners are still following the traditional food-and-wine pairings.

13.35 RR: $X^2 \geq 16.2662$. $\chi^2 = 26.0127 \geq 16.2662$. There is evidence to suggest that the risk of colon cancer and diet are dependent. $p < 0.0001$.

13.37 RR: $X^2 \geq 31.9999$. $\chi^2 = 34.7235 \geq 31.9999$. There is evidence to suggest that the type of violation and the type of pool are dependent. $0.001 \leq p \leq 0.005$.

13.39 RR: $X^2 \geq 30.5779$. $\chi^2 = 981.9745 \geq 30.5779$. There is evidence to suggest that familiarity with the UN and the country are dependent.

13.41 RR: $X^2 \geq 58.6192$. $\chi^2 = 62.4388 \geq 58.6192$. There is evidence to suggest that flossing frequency and brushing frequency are dependent. $p = 0.0041$.

Chapter Exercises

13.43 RR: $X^2 \geq 9.4877$. $\chi^2 = 10.0769 \geq 9.4877$. There is evidence to suggest at least one of the population proportion of sales of dog breed differs from 0.20.

13.45 RR: $X^2 \geq 9.4877$. $\chi^2 = 7.9711$. There is no evidence to suggest that any of the true population proportions differs from its hypothesized value.

13.47 RR: $X^2 \geq 18.5476$. $\chi^2 = 78.0454 \geq 18.5476$. There is overwhelming evidence to suggest that the true proportion of grocery shopping frequency is not the same for all countries.

13.49 RR: $X^2 \geq 11.3449$. $\chi^2 = 9.6684$. There is no evidence to suggest that the performance on the mathematics PSSA exam is associated with school district.

13.51 RR: $X^2 \geq 7.8147$. $\chi^2 = 190.4011 \geq 7.8147$. There is overwhelming evidence to suggest an association between class and survival status.

13.53 (a) RR: $X^2 \geq 18.4668$. $\chi^2 = 26.0201 \geq 18.4668$. There is evidence to suggest that portfolio majority and outlook for economic recovery are dependent. **(b)** $p \leq 0.0001$ **(c)** About half in stocks, 30% in bonds, and 20% in mutual funds.

CHAPTER 14

Section 14.1

14.1 (a) 7 (b) 3 (c) 6 (d) 13

14.3 $x = 16$, $p = 0.0022 \leq 0.05$. There is evidence to suggest that $\tilde{\mu}_1 > \tilde{\mu}_2$.

14.5 $x = 2$, $p = 0.0352 \leq 0.05$. There is evidence to suggest that the median is less than 1.38.

14.7 $x = 17$, $p = 0.0320 \leq 0.05$. There is evidence to suggest that the median mileage is greater than 100.

14.9 $x = 23$, $p = 0.1279$. There is no evidence to suggest that the median age of sports fans has increased.

14.11 $x = 14$, $p = 0.1153$. There is no evidence to suggest that the median chlorophyll amount in surface water is different in April and August.

14.13 $x = 13$, $p = 0.0037$. There is evidence to suggest that the median VOC concentration is smaller when the scrubber is installed.

Section 14.2

14.15

(a)

Difference	Absolute difference	Rank
−19	19	9.0
−7	7	4.0
6	6	3.0
−32	32	16.0
−24	24	11.5
17	17	8.0
12	12	5.0
−29	29	15.0
−27	27	14.0
−26	26	13.0
−38	38	18.0
−22	22	10.0
−36	36	17.0
−1	1	2.0
−24	24	11.5
0	0	1.0
−13	13	6.0
−15	15	7.0

(b)

Difference	Absolute difference	Rank
1.4	1.4	16.0
−0.2	0.2	2.5
1.6	1.6	21.5
0.3	0.3	5.0
−0.4	0.4	8.0
−0.8	0.8	12.0
−1.5	1.5	18.5
−0.6	0.6	10.0
0.1	0.1	1.0
−1.2	1.2	14.0
0.8	0.8	12.0
−1.5	1.5	18.5
1.5	1.5	18.5
−0.4	0.4	8.0
−1.5	1.5	18.5
1.6	1.6	21.5
−1.3	1.3	15.0
−0.3	0.3	5.0
0.8	0.8	12.0
0.4	0.4	8.0
−0.3	0.3	5.0
1.9	1.9	23.0
0.2	0.2	2.5

(c)

Difference	Absolute difference	Rank
3.0	3.0	16.5
−4.0	4.0	23.0
1.0	1.0	5.0
−3.0	3.0	16.5
−2.0	2.0	9.5
−1.0	1.0	5.0
−4.0	4.0	23.0
−4.0	4.0	23.0
−2.0	2.0	9.5
−3.0	3.0	16.5
−2.0	2.0	9.5
−2.0	2.0	9.5
−2.0	2.0	9.5
0.0	0.0	2.0
3.0	3.0	16.5
3.0	3.0	16.5
0.0	0.0	2.0
3.0	3.0	16.5
−3.0	3.0	16.5
4.0	4.0	23.0
−3.0	3.0	16.5
0.0	0.0	2.0
2.0	2.0	9.5
1.0	1.0	5.0
4.0	4.0	23.0

(d)

Difference	Absolute difference	Rank
−9.4	9.4	10.0
−55.4	55.4	30.0
−26.4	26.4	15.0
−41.4	41.4	23.5
−2.4	2.4	3.0
2.6	2.6	4.0
−38.4	38.4	22.0
−27.4	27.4	16.0
−34.4	34.4	20.0
9.6	9.6	11.0
6.6	6.6	7.5
−50.4	50.4	28.0
−42.4	42.4	25.0
−41.4	41.4	23.5
4.6	4.6	6.0
−20.4	20.4	14.0
−17.4	17.4	13.0
−51.4	51.4	29.0
3.6	3.6	5.0
−33.4	33.4	19.0
6.6	6.6	7.5
−47.4	47.4	27.0
−32.4	32.4	18.0
1.6	1.6	2.0
−45.4	45.4	26.0
−31.4	31.4	17.0
−36.4	36.4	21.0
−0.4	0.4	1.0
8.6	8.6	9.0
−12.4	12.4	12.0

14.17 $t_+ = 150.5, p = 0.0896$. There is no evidence to suggest that $\tilde{\mu}_1$ is different from $\tilde{\mu}_2$.

14.19 $t_+ = 30.5, p = 0.006$. There is evidence to suggest that the median renal blood-flow rate is different from 3.

14.21 $t_+ = 149.5, p = 0.0017$. There is evidence to suggest that the intervention program decreased the median A1C value.

14.23 $t_+ = 214.5, p = 0.3547$. There is no evidence to suggest a difference in median collection times.

14.25 $t_+ = 102.0, p = 0.0832$. There is no evidence to suggest a difference in median on-time percentages. This suggests that the on-time percentages have not changed from 2006 to 2007.

Section 14.3

14.27

(a)

Sample 1		Sample 2	
Obs	Rank	Obs	Rank
37	6	45	10
21	1	42	9
46	11	22	2
29	4	41	8
34	5	24	3
		39	7

(b)

Sample 1		Sample 2	
Obs	Rank	Obs	Rank
4.5	15.0	4.6	16.0
1.8	5.0	6.6	18.0
3.4	13.0	1.2	2.0
1.4	3.0	6.0	17.0
2.2	7.5	2.4	10.0
2.1	6.0	2.3	9.0
1.5	4.0	2.2	7.5
3.7	14.0	0.4	1.0
		2.5	11.0
		2.7	12.0

(c)

Sample 1		Sample 2	
Obs	Rank	Obs	Rank
820	11.0	850	25.0
809	4.0	840	21.0
872	33.0	813	6.0
826	15.0	842	23.0
814	7.0	870	30.0
887	39.0	839	20.0
825	14.0	888	40.0
884	38.0	816	8.0
876	35.0	822	13.0
862	27.0	879	36.0
858	26.0	821	12.0
846	24.0	865	28.5
841	22.0	832	19.0
801	2.0	865	28.5
892	41.0	818	9.5
871	31.5	827	16.0
882	37.0	899	42.0
803	3.0	818	9.5
		831	18.0
		830	17.0
		810	5.0
		875	34.0
		871	31.5
		800	1.0

14.29 (a) $W \le 7$, $\alpha = 0.0357$. (b) $W \ge 24$, $\alpha = 0.0571$. (c) $W \le 16$ or $W \le 44$, $\alpha = 0.0540$. (d) $W \le 27$, $\alpha = 0.0100$. (e) $W \le 44$ or $W \ge 75$, $\alpha = 0.1142$. (f) $W \ge 90$, $\alpha = 0.0103$.

14.31 $w = 28$, $p > 0.1412$. There is no evidence to suggest that $\tilde{\mu}_1 > \tilde{\mu}_2$.

14.33 $w = 49$, $p = 0.0274 \le 0.05$. There is evidence to suggest that the population medians are different.

14.35 (a) RR: $W \le 35$, ($\alpha = 0.0131$). $w = 31 \le 35$. There is evidence to suggest that the median impact strength is higher for the new jackhammer. (b) 0.0020

14.37 $w = 307.5$, $z = 1.8981$, $p = 0.0577$. There is no evidence to suggest that the population median holding temperatures are different.

14.39 (a) RR: $X \ge 11$ ($\alpha = 0.0592$). $x = 15$. There is overwhelming evidence to suggest that the median amount of particulates before the smoking regulations is less than the median amount after the regulations. (b) RR: $T_+ \ge 89$ ($\alpha = 0.0535$). $t_+ = 120 \ge 89$. There is evidence to suggest that the median amount of particulates before the smoking regulations is less than the median amount after the regulations. (c) RR: $Z \ge 1.6449$, $z = 4.4382 \ge 1.6449$. There is evidence to suggest that the median amount of particulates before the smoking regulations is less than the median amount after the regulations. (d) All three tests lead to the same conclusion. The rank-sum test, however, should not be used, since the samples are dependent.

Section 14.4

14.41 RR: $H \ge 13.2767$, $h = 16.1591 \ge 13.2767$. There is evidence to suggest that at least two of the populations are different.

14.43 RR: $H \ge 5.9915$, $h = 6.5184 \ge 5.9915$. There is evidence to suggest that at least two of the populations are different.

14.45 RR: $H \ge 7.8147$, $h = 6.3338$. There is no evidence to suggest that the populations are different.

14.47 RR: $H \ge 11.3449$, $h = 3.6953$. There is no evidence to suggest that the tunnel-walking-time populations are different.

14.49 RR: $H \ge 5.9915$, $h = 1.3713$. There is no evidence to suggest that the uncompressed depth populations are different. $p > 0.20$.

Section 14.5

14.51 (a) 8 (b) 8 (c) 11 (d) 1

14.53 (a) 8 (b) 5 (c) 8 (d) 15

14.55 RR: $V \le 7$ or $V \ge 17$. $v = 13$. There is no evidence to suggest that the order of observations is not random.

14.57 RR: $V \le 3$ or $V \ge 8$ ($\alpha = 0.0385$). $v = 5$. There is no evidence to suggest that the order of observations is not random with respect to gender.

14.59 RR: $|Z| \ge 2.3263$. $z = 1.2546$. There is no evidence to suggest that the order of observations is not random.

14.61 RR: $V \le 3$ or $V \ge 10$ ($\alpha = 0.0242$). $v = 9$. There is no evidence to suggest that the order of observations is not random.

14.63 Answers will vary. If the random number generator is good, then the number of times the null hypothesis is rejected should be close to 5.

Section 14.6

14.65 (a) 0.5357. Moderate positive relationship. (b) 0.3333. Weak positive relationship. (c) 0.0637. No definitive relationship. (d) -0.2922. Weak negative relationship.

14.67 0.7714. There is a positive relationship between x and y. As the price of a stateroom increases, so does the number of days before sailing. Therefore, this suggests that cruise prices are reduced at the last minute, although we must be careful about causation here.

14.69 -0.4429. This suggests a weak-to-moderate negative relationship. As the bulk density increases, the soil texture decreases.

14.71 -0.2207. This suggests a weak negative relationship. As the price increases, the number of weeks on the list decreases.

14.73 (a)

Sample 1		Sample 2		
Obs	Rank	Obs	Rank	d_i
71	9.0	64	7.5	1.5
69	6.5	56	1.0	5.5
66	3.0	65	9.5	-6.5
69	6.5	62	4.0	2.5
62	1.0	65	9.5	-8.5
67	4.0	64	7.5	-3.5
69	6.5	57	2.5	4.0
64	2.0	63	5.5	-3.5
73	10.0	57	2.5	7.5
69	6.5	63	5.5	1.0

(b) -0.5887 (c) -0.5212. (d) There are tied observations. There is a moderate negative relationship. As psi increases, the percentage change in horsepower decreases.

Chapter Exercises

14.75 $x = 10$, $p = 0.1509$. There is no evidence to suggest that the median nitrogen emissions amount is greater than 5.

14.77 $x = 15$, $p = 0.0414 \leq 0.05$. There is evidence to suggest that the median coverage amount is different from 2.

14.79 (a) RR: $T_+ \leq 20$ ($\alpha = 0.0108$). $t_+ = 17.5 \leq 20$. There is evidence to suggest that the median spray height is less than 630. (b) 0.0062 (c) The distribution is not assumed to be symmetric.

14.81 RR: $T_+ \geq 100$ ($\alpha = 0.0523$). $t_+ = 133 \geq 100$. There is evidence to suggest that the median time spent working per week for those well off is greater than for those who just manage.

14.83 (a) $W \leq 52$ or $W \geq 84$ ($\alpha = 0.1048$). $w = 78.5$. There is no evidence to suggest that there is a difference in the absorbed radiation by machine. (b) 0.2786

14.85 RR: $|Z| \geq 3.2905$, $w = 533$, $z = 2.5940$. There is no evidence to suggest that the land value for farmland is different in these two counties.

14.87 RR: $H \geq 9.3484$, $h = 3.1241$. There is no evidence to suggest that the transmitter power populations are different. $p = 0.3729$.

14.89 (a) RR: $V \leq 3$ or $V \geq 10$ ($\alpha = 0.0476$). $v = 10 \geq 10$. There is evidence to suggest that the order of observations is not random with respect to email password. (b) 0.0476.

14.91 RR: $|Z| \geq 2.5758$. $z = 1.7872$. There is no evidence to suggest that the order of observations is not random. $p = 0.0739$.

14.93 -0.4126. Weak to moderate negative relationship. This suggests that as the quality score increases, the total number of people in the hospital decreases.

14.95 (a) RR: $X \leq 5$ ($\alpha = 0.0207$). $x = 5 \leq 5$. There is evidence to suggest that the median freon weight before service is less than the median freon weight after service. (b) Yes. The sign test suggests that the median freon weight after service is larger.

14.97 (a) RR: $V \leq 4$ or $V \geq 11$ ($\alpha = 0.0709$). $v = 7$. There is no evidence to suggest that the order of observations is not random with respect to exterior finish. (b) Cannot tell. We don't know the historical proportion of home builders who use vinyl. Therefore, we cannot tell if this proportion has increased.

14.99 (a) RR: $H \geq 9.2103$, $h = 32.8940 \geq 9.2103$. There is excellent evidence to suggest that at least two of the populations are different. (b) $p < 0.0001$ (c) The safest time to drive is at *other times*.

Index

Table 6: Critical Values for the Chi-Square Distribution (*Continued*)

ν	0.10	0.05	0.025	0.01	0.005	0.001	0.0005	0.0001
1	2.7055	3.8415	5.0239	6.6349	7.8794	10.8276	12.1157	15.1367
2	4.6052	5.9915	7.3778	9.2103	10.5966	13.8155	15.2018	18.4207
3	6.2514	7.8147	9.3484	11.3449	12.8382	16.2662	17.7300	21.1075
4	7.7794	9.4877	11.1433	13.2767	14.8603	18.4668	19.9974	23.5127
5	9.2364	11.0705	12.8325	15.0863	16.7496	20.5150	22.1053	25.7448
6	10.6446	12.5916	14.4494	16.8119	18.5476	22.4577	24.1028	27.8563
7	12.0170	14.0671	16.0128	18.4753	20.2777	24.3219	26.0178	29.8775
8	13.3616	15.5073	17.5345	20.0902	21.9550	26.1245	27.8680	31.8276
9	14.6837	16.9190	19.0228	21.6660	23.5894	27.8772	29.6658	33.7199
10	15.9872	18.3070	20.4832	23.2093	25.1882	29.5883	31.4198	35.5640
11	17.2750	19.6751	21.9200	24.7250	26.7568	31.2641	33.1366	37.3670
12	18.5493	21.0261	23.3367	26.2170	28.2995	32.9095	34.8213	39.1344
13	19.8119	22.3620	24.7356	27.6882	29.8195	34.5282	36.4778	40.8707
14	21.0641	23.6848	26.1189	29.1412	31.3193	36.1233	38.1094	42.5793
15	22.3071	24.9958	27.4884	30.5779	32.8013	37.6973	39.7188	44.2632
16	23.5418	26.2962	28.8454	31.9999	34.2672	39.2524	41.3081	45.9249
17	24.7690	27.5871	30.1910	33.4087	35.7185	40.7902	42.8792	47.5664
18	25.9894	28.8693	31.5264	34.8053	37.1565	42.3124	44.4338	49.1894
19	27.2036	30.1435	32.8523	36.1909	38.5823	43.8202	45.9731	50.7955
20	28.4120	31.4104	34.1696	37.5662	39.9968	45.3147	47.4985	52.3860
21	29.6151	32.6706	35.4789	38.9322	41.4011	46.7970	49.0108	53.9620
22	30.8133	33.9244	36.7807	40.2894	42.7957	48.2679	50.5111	55.5246
23	32.0069	35.1725	38.0756	41.6384	44.1813	49.7282	52.0002	57.0746
24	33.1962	36.4150	39.3641	42.9798	45.5585	51.1786	53.4788	58.6130
25	34.3816	37.6525	40.6465	44.3141	46.9279	52.6197	54.9475	60.1403
26	35.5632	38.8851	41.9232	45.6417	48.2899	54.0520	56.4069	61.6573
27	36.7412	40.1133	43.1945	46.9629	49.6449	55.4760	57.8576	63.1645
28	37.9159	41.3371	44.4608	48.2782	50.9934	56.8923	59.3000	64.6624
29	39.0875	42.5570	45.7223	49.5879	52.3356	58.3012	60.7346	66.1517
30	40.2560	43.7730	46.9792	50.8922	53.6720	59.7031	62.1619	67.6326
31	41.4217	44.9853	48.2319	52.1914	55.0027	61.0983	63.5820	69.1057
32	42.5847	46.1943	49.4804	53.4858	56.3281	62.4872	64.9955	70.5712
33	43.7452	47.3999	50.7251	54.7755	57.6484	63.8701	66.4025	72.0296
34	44.9032	48.6024	51.9660	56.0609	58.9639	65.2472	67.8035	73.4812
35	46.0588	49.8018	53.2033	57.3421	60.2748	66.6188	69.1986	74.9262
36	47.2122	50.9985	54.4373	58.6192	61.5812	67.9852	70.5881	76.3650
37	48.3634	52.1923	55.6680	59.8925	62.8833	69.3465	71.9722	77.7977
38	49.5126	53.3835	56.8955	61.1621	64.1814	70.7029	73.3512	79.2247
39	50.6598	54.5722	58.1201	62.4281	65.4756	72.0547	74.7253	80.6462
40	51.8051	55.7585	59.3417	63.6907	66.7660	73.4020	76.0946	82.0623
50	63.1671	67.5048	71.4202	76.1539	79.4900	86.6608	89.5605	95.9687
60	74.3970	79.0819	83.2977	88.3794	91.9517	99.6072	102.6948	109.5029
70	85.5270	90.5312	95.0232	100.4252	104.2149	112.3169	115.5776	122.7547
80	96.5782	101.8795	106.6286	112.3288	116.3211	124.8392	128.2613	135.7825
90	107.5650	113.1453	118.1359	124.1163	128.2989	137.2084	140.7823	148.6273
100	118.4980	124.3421	129.5612	135.8067	140.1695	149.4493	153.1670	161.3187

Table 6: Critical Values for the Chi-Square Distribution

This table contains critical values associated with the chi-square distribution, χ_α^2, defined by the degrees of freedom and α.

df	α 0.9999	0.9995	0.999	0.995	0.99	0.975	0.95	0.90
1	0.0^7157	0.0^6393	0.0^5157	0.0^4393	0.0002	0.0010	0.0039	0.0158
2	0.0002	0.0010	0.0020	0.0100	0.0201	0.0506	0.1026	0.2107
3	0.0052	0.0153	0.0243	0.0717	0.1148	0.2158	0.3518	0.5844
4	0.0284	0.0639	0.0908	0.2070	0.2971	0.4844	0.7107	1.0636
5	0.0822	0.1581	0.2102	0.4117	0.5543	0.8312	1.1455	1.6103
6	0.1724	0.2994	0.3811	0.6757	0.8721	1.2373	1.6354	2.2041
7	0.3000	0.4849	0.5985	0.9893	1.2390	1.6899	2.1673	2.8331
8	0.4636	0.7104	0.8571	1.3444	1.6465	2.1797	2.7326	3.4895
9	0.6608	0.9717	1.1519	1.7349	2.0879	2.7004	3.3251	4.1682
10	0.8889	1.2650	1.4787	2.1559	2.5582	3.2470	3.9403	4.8652
11	1.1453	1.5868	1.8339	2.6032	3.0535	3.8157	4.5748	5.5778
12	1.4275	1.9344	2.2142	3.0738	3.5706	4.4038	5.2260	6.3038
13	1.7333	2.3051	2.6172	3.5650	4.1069	5.0088	5.8919	7.0415
14	2.0608	2.6967	3.0407	4.0747	4.6604	5.6287	6.5706	7.7895
15	2.4082	3.1075	3.4827	4.6009	5.2293	6.2621	7.2609	8.5468
16	2.7739	3.5358	3.9416	5.1422	5.8122	6.9077	7.9616	9.3122
17	3.1567	3.9802	4.4161	5.6972	6.4078	7.5642	8.6718	10.0852
18	3.5552	4.4394	4.9048	6.2648	7.0149	8.2307	9.3905	10.8649
19	3.9683	4.9123	5.4068	6.8440	7.6327	8.9065	10.1170	11.6509
20	4.3952	5.3981	5.9210	7.4338	8.2604	9.5908	10.8508	12.4426
21	4.8348	5.8957	6.4467	8.0337	8.8972	10.2829	11.5913	13.2396
22	5.2865	6.4045	6.9830	8.6427	9.5425	10.9823	12.3380	14.0415
23	5.7494	6.9237	7.5292	9.2604	10.1957	11.6886	13.0905	14.8480
24	6.2230	7.4527	8.0849	9.8862	10.8564	12.4012	13.8484	15.6587
25	6.7066	7.9910	8.6493	10.5197	11.5240	13.1197	14.6114	16.4734
26	7.1998	8.5379	9.2221	11.1602	12.1981	13.8439	15.3792	17.2919
27	7.7019	9.0932	9.8028	11.8076	12.8785	14.5734	16.1514	18.1139
28	8.2126	9.6563	10.3909	12.4613	13.5647	15.3079	16.9279	18.9392
29	8.7315	10.2268	10.9861	13.1211	14.2565	16.0471	17.7084	19.7677
30	9.2581	10.8044	11.5880	13.7867	14.9535	16.7908	18.4927	20.5992
31	9.7921	11.3887	12.1963	14.4578	15.6555	17.5387	19.2806	21.4336
32	10.3331	11.9794	12.8107	15.1340	16.3622	18.2908	20.0719	22.2706
33	10.8810	12.5763	13.4309	15.8153	17.0735	19.0467	20.8665	23.1102
34	11.4352	13.1791	14.0567	16.5013	17.7891	19.8063	21.6643	23.9523
35	11.9957	13.7875	14.6878	17.1918	18.5089	20.5694	22.4650	24.7967
36	12.5622	14.4012	15.3241	17.8867	19.2327	21.3359	23.2686	25.6433
37	13.1343	15.0202	15.9653	18.5858	19.9602	22.1056	24.0749	26.4921
38	13.7120	15.6441	16.6112	19.2889	20.6914	22.8785	24.8839	27.3430
39	14.2950	16.2729	17.2616	19.9959	21.4262	23.6543	25.6954	28.1958
40	14.8831	16.9062	17.9164	20.7065	22.1643	24.4330	26.5093	29.0505
50	21.0093	23.4610	24.6739	27.9907	29.7067	32.3574	34.7643	37.6886
60	27.4969	30.3405	31.7383	35.5345	37.4849	40.4817	43.1880	46.4589
70	34.2607	37.4674	39.0364	43.2752	45.4417	48.7576	51.7393	55.3289
80	41.2445	44.7910	46.5199	51.1719	53.5401	57.1532	60.3915	64.2778
90	48.4087	52.2758	54.1552	59.1963	61.7541	65.6466	69.1260	73.2911
100	55.7246	59.8957	61.9179	67.3276	70.0649	74.2219	77.9295	82.3581

(*Continued*)